Lecture Notes in Control and Information Sciences

Edited by M. Thoma and A. Wyner

Vol. 29: Optimization Techniques
Proceedings of the 9th IFIP Conference on

Vol. 32: Tsuyoshi Matsuo
Realization Theory of
Continuous-Time Dynamical Systems
VI, 329 pages. 1981.

Vol. 33: Peter Dransfield
Hydraulic Control Systems

Lecture Notes in Control and Information Sciences

Edited by M. Thoma and A. Wyner

Vol. 22: Optimization Techniques
Proceedings of the 9th IFIP Conference on Optimization Techniques,
Warsaw, September 4–8, 1979
Part 1
Edited by K. Iracki, K. Malanowski, S. Walukiewicz
XVI, 569 pages. 1980

Vol. 23: Optimization Techniques
Proceedings of the 9th IFIP Conference on Optimization Techniques,
Warsaw, September 4-8, 1979
Part 2
Edited by K. Iracki, K. Malanowski, S. Walukiewicz
XV, 621 pages. 1980

Vol. 24: Methods and Applications in Adaptive Control
Proceedings of an International Symposium
Bochum, 1980
Edited by H. Unbehauen
VI, 309 pages. 1980

Vol. 25: Stochastic Differential Systems – Filtering and Control
Proceedings of the IFIP-WG7/1 Working Conference
Vilnius, Lithuania, USSR, Aug. 28 – Sept. 2, 1978
Edited by B. Grigelionis
X, 362 pages. 1980

Vol. 26: D. L. Iglehart, G. S. Shedler
Regenerative Simulation of Response Times in Networks of Queues
XII, 204 pages. 1980

Vol. 27: D. H. Jacobson, D. H. Martin, M. Pachter, T. Geveci
Extensions of Linear-Quadratic Control Theory
XI, 288 pages. 1980

Vol. 28: Analysis and Optimization of Systems
Proceedings of the Fourth International Conference on Analysis and Optimization of Systems
Versailles, December 16–19, 1980
Edited by A. Bensoussan and J. L. Lions
XIV, 999 pages. 1980

Vol. 29: M. Vidyasagar,
Input-Output Analysis of Large-Scale Interconnected Systems – Decomposition, Well-Posedness and Stability
VI, 221 pages. 1981

Vol. 30: Optimization and Optimal Control
Proceedings of a Conference Held at Oberwolfach, March 16–22, 1980
Edited by A. Auslender, W. Oettli, and J. Stoer
VIII, 254 pages. 1981

Vol. 31: Berc Rustem
Projection Methods in Constrained Optimisation and Applications to Optimal Policy Decisions
XV, 315 pages. 1981

Vol. 32: Tsuyoshi Matsuo,
Realization Theory of Continuous-Time Dynamical Systems
VI, 329 pages, 1981

Vol. 33: Peter Dransfield
Hydraulic Control Systems – Design and Analysis of Their Dynamics
VII, 227 pages, 1981

Vol. 34: H.W. Knobloch
Higher Order Necessary Conditions in Optimal Control Theory
V, 173 pages, 1981

Vol. 35: Global Modelling
Proceedings of the IFIP-WG 7/1 Working Conference Dubrovnik, Yugoslavia, Sept. 1–5, 1980
Edited by S. Krčevinac
VIII, 232 pages, 1981

Vol. 36: Stochastic Differential Systems
Proceedings of the 3rd IFIP-WG 7/1 Working Conference
Visegrád, Hungary, Sept. 15–20, 1980
Edited by M. Arató, D. Vermes, A.V. Balakrishnan
VI, 238 pages, 1981

Vol. 37: Rüdiger Schmidt
Advances in Nonlinear Parameter Optimization
VI, 159 pages, 1982

Vol. 38: System Modeling and Optimization
Proceedings of the 10th IFIP Conference
New York City, USA, Aug. 31 – Sept. 4, 1981
Edited by R.F. Drenick and F. Kozin
XI, 894 pages. 1982

Vol. 39: Feedback Control of Linear and Nonlinear Systems
Proceedings of the Joint Workshop on Feedback and Synthesis of Linear and Nonlinear Systems
Bielefeld/Rom
XIII, 284 pages. 1982

Vol. 40: Y.S. Hung, A.G.J. MacFarlane
Multivariable Feedback: A Quasi-Classical Approach
X, 182 pages. 1982

Vol. 41: M. Gössel
Nonlinear Time-Discrete Systems – A General Approach by Nonlinear Superposition
VIII, 112 pages. 1982

Vol. 42: Advances in Filtering and Optimal Stochastic Control
Proceedings of the IFIP-WG 7/1 Working Conference
Cocoyoc, Mexico, February 1–6, 1982
VIII, 391 pages. 1982

For information about Vols. 1–21 please contact your bookseller or Springer-Verlag.

Lecture Notes in Control and Information Sciences

Edited by M. Thoma and A. Wyner

IFIP

84

System Modelling and Optimization

Proceedings of 12th IFIP Conference, Budapest, Hungary, September 2-6, 1985

Edited by
A. Prékopa, J. Szelezsán, and B. Strazicky

Springer-Verlag Berlin Heidelberg GmbH

Editors
A. Prékopa
Computer and Automation Institute
Hungarian Academy of Sciences
H-1502 Budapest P.O.B. 63

Dep. of Operation Research
L. Eötvös University
Budapest, 1088 Muzeum krt. 6-8

J. Szelezsán
Dep. of General Computer Sciences
L. Eötvös University
Budapest, 1088 Muzeum krt. 6-8

B. Strazicky
Computer and Automation Institute
Hungarian Academy of Sciences
H-1502 Budapest P.O.B. 63

ISBN 978-3-662-13587-7 ISBN 978-3-540-47138-7 (eBook)
DOI 10.1007/978-3-540-47138-7

Originally published by Springer-Verlag Berlin Heidelberg New York in 1986.

2161/3020-543210

P R E F A C E

The Proceedings contains papers presented at he 12th IFIP Conference on System Modelling and Optimization held in Budapest /Hungary/ from 2 to 6 September, 1985.

This biannual IFIP TC-7 general conference was organized by the John von Neumann Society for Computer Sciences.

The aim of the conference was to discuss recent advances in the mathematical reprezentation of engineering, sociotechnical, socio-economic systems as well as in the optimization of their performances.

Besides the theoretical lectures in accordance with the main objectives of the conference, most of the presentetions covered a variety of application areas.

The following topics were covered in separate sessions:

- Control of dynamic systems
- Control of distributed systems
- Stochastic systems
- Linear programming
- Nonlinear programming
- Discrete programming, networks
- Performance optimization in computer systems
- Simulation methods
- Applications:
    1. Transport, traffic systems
    2. Industrial, engineering systems
    3. Structural systems
    4. Economic systems
    5. Energy systems
    6. Biological systems.

Parallel with the conference a small exhibition and a poster section was organized.

A.Prékopa
J.Szelezsán
B.Strazicky

Contents

J. ABRAHAM, V. PEROV, K. NYIRI, L. IKLODI and I. KAPOSZTA: Mathematical Modelling and Optimization of PVC Powder Blending Process for Development of Multilevel, Optimized Process Control System 1

M.F. ALLAM and M.M. ALI: An Economic Approach for Optimum Long-Term Plant Mix Choice 10

I. ARANY: An Algorithm for Getting a Minimum Cut-Set of a Graph . . 17

F. ARCHETTI, A. SCIOMACHEN and C. VERCELLIS: Optimal Design of a Remote Heating Network 25

M.D. AŠIĆ and V.V. KOVAČEVIĆ-VUJČIĆ: Computational Complexity of Some Semiinfinite Programming Methods 34

O. BABIC: The Calculation of Menpower for Aircraft Ground Service . 43

G. BARTOLINI and T. ZOLEZZI: Discontinuous Adaptive Control of Non Minimum Phase Linear Plants 54

W. BAUER, E. LINDNER and Hj.WACKER: Optimal Control of the Storage Power Plant System Gosau 63

B. BETRÓ, F.SCHOEN and M.G. SPERANZA: Stochastic Control in Urban Traffic 75

R. BLAŠKO: Throughput Optimization of Packet Communication Networks.84

L.B. BOGUSLAVSKY, A.L. STOLYAR: Performance Analysis of Data Link and Communication Device Control Procedures in Distributed Micro/Mini Computer Systems 91

R. BREHMER and P. NEMETZ: Optimal Control Methods for Power System Operation 102

M. BRENNINGER-GÖTHE and K. JÖRNSTEN: Models and Methods for Estimating an Origin-Destination Trip Matrix from Network Data . 112

M. BROKATE: Optimal Control of Age-Structured Populations 125

C.G. BROYDEN: Skew-Symmetric Matrices, Staircase Functions and Theorems of the Alternative. 133

O.P. BURDAKOV: Stabilization of the Secant Method Via Quasi-Newton Approach 141

R.E. BURKARD: Assignment Problems: Recent Solution Methods and Applications*. 153

T. CHACON and O. PIRONNEAU: Optimal Control as a Tool for Solving the Stationnary Euler Equation with Periodic Boundary Conditions 170

A.P. CHERENKOV: Optimization Problems for Two-Stage Process of Resource Allocation 177

R.F. CURTAIN: Balanced Realisations for Infinite-Dimensional Discrete-Time Systems 181

T. CSENDES, B. DARÓCZY and Z. HANTOS: Nonlinear Parameter Estimation by Global Optimization: Comparison of Local Search Methods in Respiratory System Modelling 188

J.K. DELSON, R.T. JENKINS, W.L. STILLINGER, R. BEUNE and N.HALBERG: The "BENCHMARK" Chronological Simulation Model: A New Tool for Better Understanding the Economic and Physical Behaviour of Electrical Generating Systems 193

N.I. DJURANOVIĆ-MILIČIĆ: An Algorithm in Constrained Optimization . 203

A. DOURADO and J.L. CALVET: An Efficient Algorithm for the Optimal Operation of a Cascade Reactor. 209

E.DUDEK-DYDUCH: Evolutionary Method in Discrete Programming . . . 220

A. FARKAS: Environmental Systems Modelling with Respect to the Future 226

J.A. SOÉIRO FERREIRA and R.V.VALQUI VIDAL: On the Connections Between Mathematical Programming and Discrete Optimal Control. 234

S. FUKS: Exploration of Non-Renewable Resources: A Dynamic Approach 244

A. GARLIAUSKAS, A. BURKAUSKAS, G. LESINSKAS and R. VYGOVSKAYA: Dynamic and Stochastic Optimization of Transport Networks. . . 254

A. GARLIAUSKAS and A. GARLIAUSKIENE: Simulation of Biomedical Systems* 264

P. GRABOWSKI: Stability of Semilinear Systems in Hilbert Spaces. 282

W. GREGA: A Decomposition Algorithm for the Determination of Optimal Bus Frequencies 292

K.M. HANGOS and T. VIRÁG: The Effect of Point-Like Noise Sources on Chemical Distributed Parameter Systems 298

P.G. HARHAMMER and A. SCHADLER: Optimal Energy Management 304

CH. HENDRICH and B. DRAXLER: Effects of Power-Converters, Compensation Equipment and Filter Circuits on the Voltage Trajectories in Electrical Distribution Systems 312

A. HERTZ: ORELIA a Probabilistic Model for the Expansion Planning of the Ehv Electric Power System 321

P. HOLNICKI, J. SOKOŁOWSKI and A. ZOCHOWSKI: Sensitivity Analysis of an Optimal Control Problem Arising from Air Quality Control in Urban Area 331

T. HRABA and J. DOLEŽAL: On Mathematical Model of Immunological Tolerance 340

V.V. IVANOV and Ju.P. JACENKO: Problem of Optimal Control for Integral Dynamic Models with Controllable Prehistory 350

A. JANIAK: One-Machine Scheduling Problems with Resource Constraints 358

S. JENDO and W. MARKS: Multiobjective Structural Optimization . . 365

J.I. KALEV and L.AT. GUNCHEV: Linear Approximation of Surfaces with a Criterion "Minimum of Average Quadratic Time" Treating Optimal Control Problems 375

N.A. KARATUEVA and V.M. MATROSOV: Vector Lyapunov Functions Method for Difference-Differential Systems and its Applications to Immunology 384

I. KARPOUZAS and Y. CHERRUAULT: A New Global Optimization Technique for Solving Partial Differential Equations 394

P. KELLE: Modelling and Optimization of Buffer Stocks in a Production Line 399

A.M. KHAN: Simulating Expert Judgment: Actuarial Models and Applications. 406

H. KIVIJÄRVI and M. TUOMINEN: Corporate Planning and the Value of Information 412

É. KOMÁROMI: Duality in Probabilistic Constrained Linear Programming 423

F. KONECNY: Nonlinear Estimation Problems of Poisson Cluster Processes 430

A. KÓSA: On the Manifold of Control Processes in Lagrange Problems 437

W. KRABS: On the Numerical Solution of Certain Time-Minimal Problems of Second Order 442

M. KRAWCZAK: Nash Game with Regular Polynomial Perturbation . . 452

I. KUN: A Worst Case Analysis for the Optimal Gradient Method . 461

S. KYPREOS: Energy Modelling Approach at the Swiss Federal Institute for Reactor Research 468

F. LAMPARIELLO and S. LUCIDI: Analysis of DNA Distributions from Flow Cytometry by Means of an Optimization Procedure 478

A. LASCIAK: Planning Computation of a Long-Range Planning on the Basis of Macroeconomic Models 488

I. LASIECKA: Boundary stabilization Problems for Hyperbolic Dynamics - a review* 499

G. LEITMANN, E.P. RYAN and A. STEINBERG: Uncertain Systems: Robustness of Ultimate Boundedness Control with Respect to Neglected Dynamics 522

R. LEPP: Discrete Stability of Stochastic Programming Problems with Recourse 529

G. LEUGERING: Time Optimal Boundary Controllability of a Viscoelastic Beam 535

N. LEVAN: Stabilization by the Feedbacks $-B^{*}$ and $-B^{*}P$ 542

Y.V. LIVANOV: The Stochastic Models for Estimation of Functioning Quality of Communications and Their Application . 551

F.De LORENZO, V. GERVASIO and P. VERRECCHIA: MDS: An Instrument for the Construction of Models for Company Planning Using the Concepts of System Dynamics 557

A. MAGDZIARZ and Z. ZAGAN: Mathematical Simulation Model of Power Transformer for Electrical Power System Protective Schemes 567

P. MARINOV and P. KIRIAZOV: Time-Optimal Control of Acceleration Constrained Mechanical Systems 577

U.MASCHTERA and S. WALTER: A Relational View of a Linear Optimization Model and its Consequence for the Modelling Process 583

M. MICHNIEWICZ: An Adaptive Coordination Algorithm for the Interaction Balance Method with Feedback 591

S. MOLNÁR and F. SZIDAROVSZKY: A Stochastic Multiobjective Dynamic Programming Method with Applicaiton to Energy Modelling . . . 601

M. MUSCHICK and A. SCHADLER: Operation Planning of Power Systems .610

W. MÜLLER: Modelling and Numerical Simulation of Wind-Generated Circulation and Matter Transport in Shallow Lakes 620

H. MÜLLER, G. THELL and W. WALDMANN: Results of Evolutional Random Search Procedure for Load Flow Optimization in Electric Networks 628

P. NEITTAANMÄKI: On the Optimal Cooling of the Steel During Continuous Casting 637

H. NIEDERREITER: Good Lattice Points for Quasirandom Search Methods 647

R.C. OLIVEIRA: A Visual Interactive Simulation Model for the Design of a Railway Station 655

A. OLSBU, P.A. LOEKEN and I.E. GROSSMANN: Simulataneous Synthesis and Economic Optimization of the Power System on an Oil/Gas Production Platform 665

M.M.OSTOJIĆ: Dc Motor Drive Parameters Optimization 675

G.CH. PFLUG: Optimizing Simulated Markov Processes 684

G.DI PILLO, L. GRIPPO and S. LUCIDI: Globally Convergent Exact Penalty Algorithms for Constrained Optimization 694

J. PINTÉR and J.SZABÓ: Global Optimization Algorithms: Theory and Some Applications 704

G.DA PRATO: Periodic Solutions of an Infinite Dimensional Riccati Equation 714

G. RABENSTEINER: Computer Aided Expansion Planning of Combined Electric Power and District Heating Systems - a Mixed Integer Programming Model - 723

T. RAPCSÁK: Convex Programming on Riemannian Manifolds 733

N. RIJAVEC: Interactive Multicriteria Goal Programming by Reference Objectives 741

C. RICHTER and K.SCHÖNEFELD: Hybrid Methods in Nonlinear Programming 745

S.M. ROBINSON: Bundle-Based Decomposition: Description and Preliminary Results*. 751

R. GONZALEZ and E. ROFMAN: On the Optimization of a Short-Run Model of Energy Production Systems 757

J. RYCKBOSCH: Power System Corrective Switching: A New Approach Using Non-Linear Programming 768

C. SAGUEZ: Optimal Control of Free Boundary Problems* 776

P. SANDRIN: Optimal Control Methods for Large Power Systems Planning and Operation* 783

K. SCHITTKOWSKI: Mathematical Programming Software*. 804

C. SCHIEMANGK: Design, Analysis and Implementation of Thermodynamically Motivated Simulation for Optimization of Subgraphs 811

J. SIKORSKI: A Quasi-Subgradient Scheme for Calculating Surrogate Constraints 821

A.H. SILLABER: Linear Programming for Electric Power Distribution System Expansion Planning: A Dynamic Expansion Model for Loop Structured Networks 828

L. SŁOMINSKI: On Existence of Assignments in Zero - One Matrices . 838

J. SOKOŁOWSKI and J.P. ZOLESIO: Sensitivity Analysis of Elastic - -Plastic Torsion Problem 845

J. SOKOŁOWSKI: Differential Stability of Solutions to Boundary Optimal Control Problems for Parabolic Systems 854

GY. SONNEVEND: An "Analytical Centre" for Polyhedrons and New Classes of Global Algorithms for Linear (Smooth, Convex) Programming 866

J.D. SØRENSEN and P. THOFT-CHRISTENSEN: Structural Optimization with Reliability Constraints 876

J.A. MARQUES DE SOUZA, P. MOLINA, J.P. CHARPENTIER and K. SCHENK: IDEA's Activities in Electric System Expansion Planning 886

A. STACHURSKI: Monotone Sequences of Feasible Solutions for Quadratic Programming Problems with M-Matrices and Box Constraints 896

L.E. STANFEL: Clustering Theory in the Design of Information Systems 903

Ł. STETTNER: On Continuous Time Adaptive Impulsive Control . . . 913

N. STOICA and G. ROTH: Neuronal Networks Modelled by Petri Type Nets with Controllers 923

E.I. STOILOV and M.S. STRUGAROVA: The Use of Flying Horizon Method to Traffic Optimization in Process Control Communication Systems. 933

J. SZELEZSÁN: On the Numerical Solution of an Optimal Control Problem 943

F. SZIDAROVSZKY and K. OKUGUCHI: Linear and Nonlinear Oligopoly Models 949

J. SZILÁGYI: Solving Scheduling Problems by Computer 957

C. TÁNCZOS: Multiple Criteria Analysis and Evaluation in Decision Making of Transport Planning 961

P. THOFT-CHRISTENSEN: Reliability Modelling of Structural Systems* .970

Ph. TORRION, J. LEVEUGLE and C. HERY: Comparison of Different Methods for Solving a Large Scale, Stochastic Dynamic Problem . 982

J. TOMKÓ: Semi-Markov Analysis of the Inhomogeneous Machine Interference Model 992

A.D. TSVIRKUN and YU. YAKOVENKO: Lyapunov Functions and Turnpike Theory 1002

A. VÁSÁRHELYI and M. GRÓSZ: Elasto-Plastic Analysis of Continuum Structures by Nonlinear Programming 1008

G.E. VERESS: Modelling of Metrological Systems 1020

M. WIECEK and A. LIGEZA: Optimal Identification Problems of Regionalized Variables Systems 1028

R.J. WIJBRANDS: The Analysis of I/O Configurations: On the Robustness of the Central Server Model 1037

Invited papers are marked by "x".

MATHEMATICAL MODELLING AND OPTIMIZATION OF PVC POWDER BLENDING PROCESS FOR DEVELOPMENT OF MULTILEVEL, OPTIMIZED PROCESS CONTROL SYSTEM

Jozsef Abraham*, Vladimir Perov**, Karoly Nyiri*, Lajos Iklodi*, Istvan Kaposzta*

* Borsod Chemical Works, Kazincbarcika, Hungary

** UNIDO, Department of Industrial Operations, Wien, Austria.

Abstract

For optimized, multi-level process control system mathematical model of PVC powder blending process in high speed mixers was developed. For mathematical modelling of PVC powder blending process basic parameters of elementary processes were determined. On base of process analysis and data collection from PVC powder blending process in industrial high speed mixers with help of microcomputer determined most important factors from point of wiew of process control.
On base of energy balance of blending process and semi-empirical formulas of power consumption of mixers here suggested an mathematical model, which describes temperature changes and power consumption of high speed mixers during mixing process.
Formulated the optimization task: minimizing of consumed energy for production of PVC compound and maximizing of production capacity of mixers. Developed an optimization algorithm, which makes possible to control the blending process according to selected objectiv function. Suggested a new, multi-level structure of process control system for PVC powder blending, which uses developed mathematical model and optimization algorithm. Developed an supervison control program, which realises the upper level of optimized, multilevel on-line, real-time microprocessor based control system.

INTRODUCTION

In the literature reported about up to date methodes of description and mechanics of mixing process, but there are not enough information for modelling of blending process of powder like materials, which are typical for production of PVC compounds. Mainly studied diffusion blending [3] and mixing in horisontal drum mixers [3,4,5] , but not sufficiently studied for modelling blending process of PVC in high speed mixers, which are videly used in modern PVC processing plants [7-11]. Target of modelling reported in this paper was obtaining of such mathematical description of powder blending process, which makes possible optimal control of process. For mathematical modelling of PVC powder blending process basic parameters of elementary processes of PVC compound production were determined. Powder blending process was studied with laboratory and industrial high speed mixers. There were determined correlations between initial PVC powder quality characteristics and technological parameters of blending process, defined important and unimportant PVC quality characteristics.

DESCRIPTION OF BLENDING PROCESS OF PVC IN HIGH SPEED MIXERS

Scheme of industrial high speed mixer is shown on fig.1. Mixers operate as follow:

In the 1st step at minimum angular velocity of rotation of heating mixer rotor /$\omega_1 = \omega_{h\ min}$/ charged components of compound with mass M and initial temperature t_o, and mixed to temperature t_1 in so called compacting condition.

In the 2nd step ergular velocity of rotation of heating mixer rotor increased to maximum $\omega_2 = \omega_{h\ max}$. Mixing is continued up to temperature t_2 in fluidizing condition.

In the 3rd step at angular velocity of rotation of cooling mixer rotor $\omega_3 = \omega_c$ into cooling mixer charged hot compound with mass M and temperature t_2 from heating mixer and cooled to t_3 temperature.

In the 4th step into cooling mixer charged second batch of hot compound from the heating mixer. Compound cooled to end temperature t_4 discharged from cooling mixer and a new blending cycle begins from 1st step.

DEVELOPMENT OF MATHEMATICAL MODEL OF BLENDING PROCESS IN HIGH SPEED MIXERS

On basis of energy balance of blending process and semiempirical formulas of power consumption of mixers an mathematical model was developed, which describes temperature changes in high speed mixers during mixing process of PVC blends.

Detail description of development of this mathematical model was reported in papers [1,2] .

We would like summarize here basic results of mathematical modelling, which were used in development of multi-level optimized process control system.

Equation of mathematical model of heating mixer in final form is given as follow:

$$\frac{dT}{d\tau} = \begin{cases} K_{11} \cdot K_2 \cdot \omega_1^{\alpha_1} \cdot K_1 \cdot [1 + q_1 \cdot \exp/Z_1 \cdot \tau/] & \text{at } \omega_1 = \omega_{h\ min} \\ K_{12} \cdot K_2 \cdot \omega_2^{\alpha_2} \cdot K_2 \cdot \quad 1 - \exp/-Z_2/T/ & \text{at } \omega_2 = \omega_{h\ max} \end{cases} \qquad /1/$$

Where: $K_{11}, K_{12} = \dfrac{c \cdot \rho_p \cdot H}{C_p \cdot M}$ - technological constans /2/

$K_2 = L^{\beta} \cdot /b \cdot \sin d/^{\gamma} \cdot S^{\varepsilon}$ - constructional constant /3/

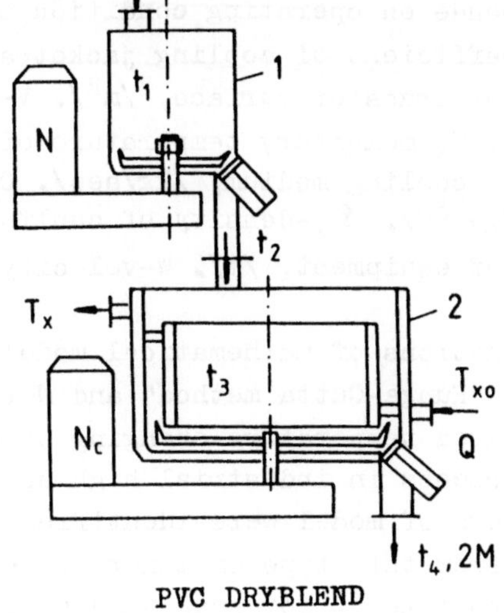

Fig.1. Scheme of investigated high speed mixer
1 - heating mixer, 2 - cooling mixer

Equations of mathematical model of cooling mixer are:

$$\frac{dT}{d\tau} = K_{1x} \cdot K_{2x} \cdot \omega_x^{\alpha 1} \cdot X_1 \left[1+q_1 \exp/Z_1 \cdot \tau/ - K_{4x}/T-T_x/\right] \quad /4/$$

$$\frac{\partial T_x/Z,\tau/}{\partial \tau} + W \frac{\partial T_x/Z,\tau/}{\partial Z} = \frac{K \cdot F}{Q \cdot \rho_x \cdot c_x}\left[T/\tau/ - T_x/Z,\tau/\right] \quad /5/$$

Where:

$$K_{1x} = \frac{c \cdot \rho_c \cdot H^{\delta}}{C_p \cdot M} \text{ - technological constant} \quad /6/$$

$$K_{2x} = L^{\beta} /B.\sin d/^{\gamma} \cdot S^{\varepsilon} \text{ -constructional constant} \quad /7/$$

$$K_{4x} = \frac{K \cdot F}{M \cdot C_p} \quad /8/$$

Where: . c-coefficient of resistance, ρ_c-density of mixed compound, /kg/m^3/. L-length of mixer blade, /m/. b-hight of mixer blade, /m/, H-hight of powder layer above blade, /m/. S-gap between edge of blade and wall of mixer tank, /m/. M-mass of one batch of compound,

/kg/, C_p-specific heat of compound, /kJ/kg.°C/. T-temperature of compound, /°C/. τ -mixing time, /min/. α, β, γ, δ, ε -semi-empirical exponents, which depende on operating condition of mixers.

K-heat transfer coefficient of cooling jacket and cooling ring, /kJ/ /m^2.sec. °C/. F-heat transfer surface, /m^2/. V-volume of cooling jacket and ring, /m^3/, T_x-momentary temperature of cooling medium, /°C/. Q-mass flow rate of cooling medium, /kg/sec/. C_x-specific heat of cooling medium, /kJ/kg.°C/. ϱ_x-density of cooling medium, /kg/m^3/. Z--linear dimension of equipment, /m/. W-velocity of cooling medium, m/sec/.

For solution of equations of mathematical model an algorithm was developed on base of "Runge-Cutta method" and Net method. Using this algorithm was calculated temperature changing of compound from 1. to 4. step of blending process in industrial high speed mixer type "MSHK 750/2000". Parameters of model were identified on base of experimental data obtained from this type of mixer, with help of microcomputer connected on line to mixers, for real time process control and data acquisition. Experimental data and results of calculation obtained with mathematical model are shown in fig.2. As it can be seen, good agreement of experimental data with calculated on base of developed mathematical model data was obtained. Adequacy of model was checked by Fischer criterion.

FORMULATION OF OPTIMIZATION TASK

1. Minimizing of consumed energy for production of PVC compound:/Exp./
Objective function is formulated as follow:

$$\text{MIN } E_{sp} = \text{MIN } \frac{/N_1 . \tau_1 + N_2 . \tau_2 + N_3 . \tau_3}{M} + \frac{N_4 . \tau_4/}{M} + $$

$$+ K. \frac{/Q_3 . \tau_3}{M} + \frac{Q_4 . \tau_4/}{2M} \text{ , /KJ/kg/} \qquad /9/$$

Constraints:

$$\tau_1 + \tau_2 \geqq \tau_3 \qquad /10/$$

$$\tau_1 + \tau_2 \geqq \tau_4 \qquad /11/$$

$$\tau_1 \geqq 0 \qquad /12/$$

$$\tau_q \geqq 0 \qquad /13/$$

$$M_{min} = M = M_{max} \qquad /14/$$

$$Q_{min} = Q_3 = Q_{max} \qquad /15/$$

$$Q_{min} = Q_4 = Q_{max} \qquad /16/$$

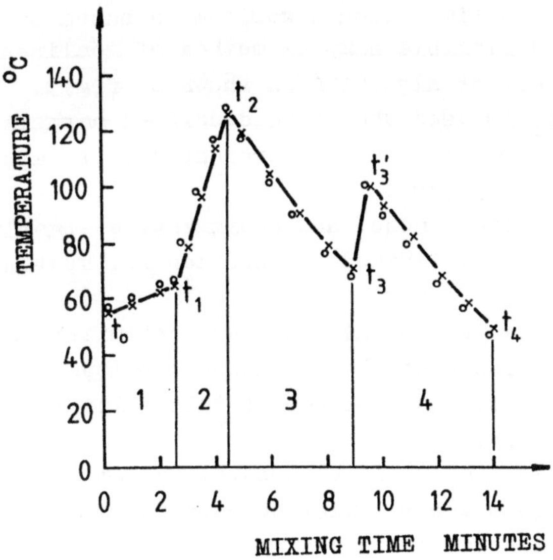

Fig.2. Identification of developed mathematical model
x-x- experimental data
o-o- calculated data

2. Maximizing of production capacity of mixers: G_{sp}

$$\text{MAX } G_{sp} = \frac{2M}{\tau_1 + \tau_2 + \tau_3 + \tau_4} \cdot 3600, \ /\text{kg/h}/ \qquad /17/$$

Constrains: Same as above /10-16/.

Where: N_1, N_2, N_3, N_4 - consumed by mixer engines power in 1., 2., 3., 4 - th step respectively, /kW/,

τ_1, τ_2, τ_3, τ_4 - mixing time in 1., 2., 3., 4 - th step respectively /sec/.

Q_3,Q_4 - mass flow rate of cooling medium, /kg/sec/.

K - coefficient.

M - mass of one mixed batch, /kg/.

For solution of aboveformulatedoptimization tasks developed an optimizing algorithm for calculation of optimal technological parameters of blending process, where as main control variable assumed change over temperature t_1 or change over time τ_1 of rotation speed of hea-

ring mixer rotor. The optimization algorithm is based on Nelder-Mead simplex method and flexible simplex method of nonlinear programming, general flowchart of algorithm is shown o fig.3.
For calculation of t_1 is used the above described mathematical model. This optimization algorithm is at the present time in approbation on process control microcomputers.
Application of mathematical model and optimization algorithm for development of multi-level optimized process control system of PVC powder blending process.
On basis of developed mathematical model and optimization algorithm a multi-level supervision control system was developed for PVC powder blending process. Structure scheme of this multi-level control system is shown on fig.4.
Lower level of optimized multi-level control system consists of on--line- real-time process control microcomputer programs /PCPROG, SEQCONT, DDCCONT/ which realise operative control of PVC blending process in high speed mixers.
Middle control level of optimized multi-level control system consists of MONITOR and MAINMIX programs, which coordinate operation of lower level process control microcomputers. Upper control level of optimized multi-level control system realises supervision control of full PVC blending process, and optimization /SUPERVISION, FLEXOPT, MIXMOD/. This upper control level has 4 operation modes.

1. Operation mode data acquisition.
   Programme MIXMOD collects process data of PVC blending, such as temperature of compound in heating and cooling mixers, consumed by mixers energy, etc, makes preliminary data processing and stores collected data.
2. Operation mode identification.
   Programme MIXMOD on base of collected data identifies parameters of model in strict controlled mixing conditions. Identified parameters of model are stored for supervisory control.
3. Operation mode supervision control.
   In supervision control mode programme SUPERVISION with help of mathematical model calculates theoretical standard mixing curve, and compares this theoretical mixing curve, to real changing of mixing parameters /temperature change, of compound and consumed power/during mixing time /in 1-4 step/. When an deviation is detected, which more then allowed, alarm signal is output, or the operation of mixer is stopped, if necessary.

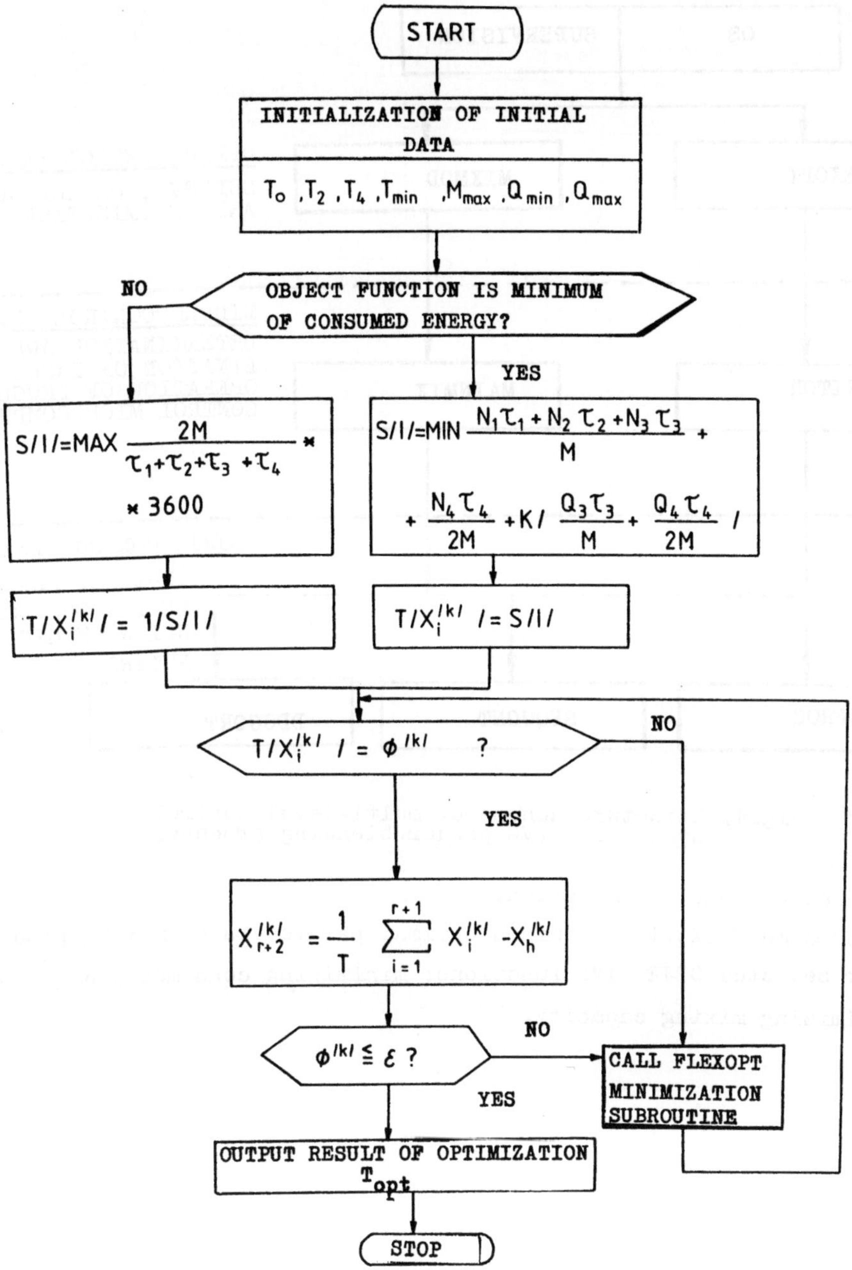

Fig.3. Algorithm of optimization program"FLEXOPT".

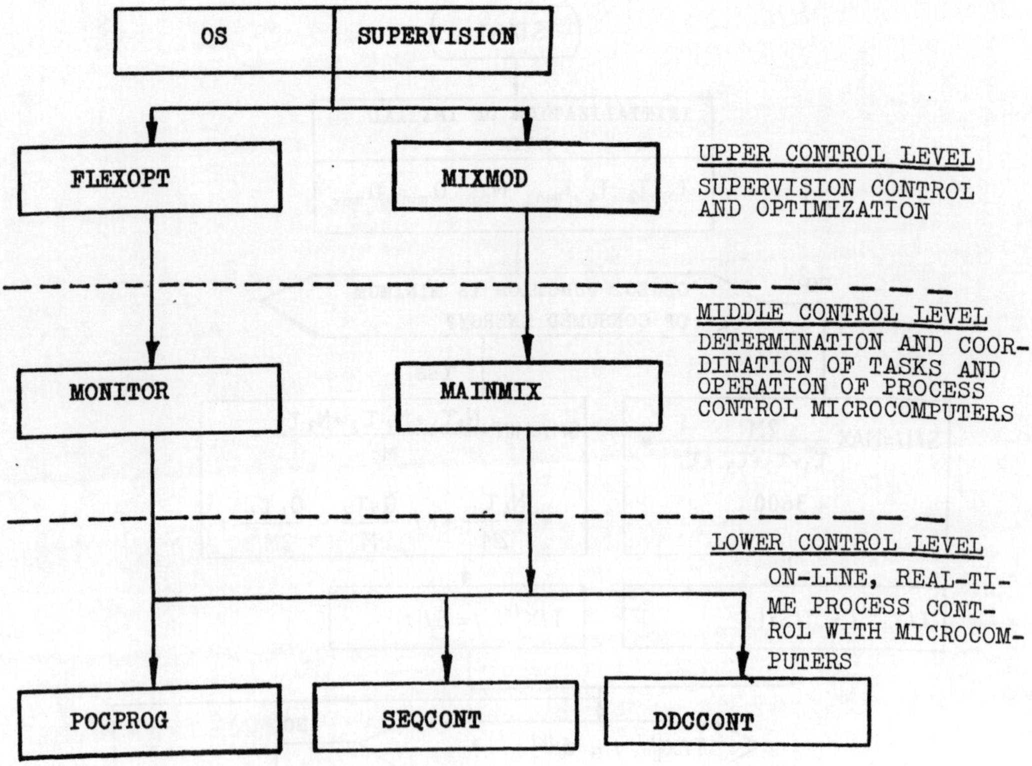

Fig.4. Structure scheme of multi-level control system for PVC powder blending process.

4. Operation mode optimization.

Programme FLEXOPT determines optimal parameters of mixing process, for selected OBJECTIVE functions: minimizing consumed energy or maximizing mixing capacity.

DISCUSSION OF RESULTS

1. Described above mathematical model describes temperature changes in high speed mixers during blending process with sufficient accuracy. Coefficients and exponents of model must be identifed for every recipe of PVC blend and for every mixer construction under consideration. This identification easy to make with help of process control microcomputer connected to mixer.

2. Developed optimization algorithm makes possible to control the blending process according to selected objectiv function: minimizing of production capacity of mixers.

3. Suggested multi-level, optimized process control system realises developed mathematical model algorithm and optimization algorithm on process control microcomputers.

References

1. Abrahám J., Perov V.L., Kiss L., Iklodi L., Nyiri K., Kaposzta I. Müanyag és Gumi, 21/7. 189-192., /1984/
2. J.Abraham, V.L. Perov., L.Kiss, L.Iklodi, K.Nyiri, I.Kaposzta Plaste und Kautschuk, /1985/ 2. p62-65.
3. V.V. Bogdanov, R.V. Torner, V.N. Kraszovszkij, E.O. Reger: Szemisenije polimerov, Leningrád, Himija, /1979/ pp.lo-14. 179-192.
4. K.Sommer: Chem. Ing. Techn. 49, /1977/ 4, pp. 3o5-311.
5. D.D. Rjabinin, Ju. E. Lukacs: Szemisitelnüe masinü dlja plasztmassz i rezinovüh szmeszej, Moszkva, Mashinostroenije /1972/ pp. 238-244.
6. V.A. Szilin, Himicseszkoje Masinosztroenije USSR, Kijev,Technika, N$^{\underline{o}}$4, /1966/ pp. 31-49.
7. Pukánszky B., Józsa S., Pukánszky B. Jn.: Müanyag és Gumi,15/11, 331-339., /1978/
8. Pukánszky B., Juhász E., Nyitrai Zs.: Müanyag és Gumi, 15/7, 2o2-2o6., /1978/
9. Dimitrov M., Pazonyi T., Pukánszky B.: Müanyag és Gumi, 13/7, 2o1-2o4., /1976/
1o. J. Schmiedke: Forschungsbericht,/1979/ jun. FIMUR. "Abhängigkeit des Temperaturverlaufs von der Zeit, der Drehzal und den Mischwerkzeug-Abständen bei der Aufbereitung von PVC."
11. J. Schmiedke: Plastverarbeiter /1982/ Vol.33, N$^{\underline{o}}$2, pp.157-162.

AN ECONOMIC APPROACH FOR OPTIMUM LONG-TERM PLANT MIX CHOICE

M.F. Allam
Elect. Eng. Dept. Faculty of Engineering, Alexandria University, El-Hadara, Alexandria, Egypt.

M.M. Ali,
Faculty of Engineering, Monofia University, Shibin El-Com, Egypt.

This paper introduces an application of linear programming to power system long-term generation expansion planning. A mathematical deterministic LP model is proposed and implemented to determine the optimum mix generation for such systems for a long-term plan. The model objective is to justify the construction of group of plants of particular types and sizes at certain locations which are committed by a long term strategy, to supply the system estimated demand to the end of a planning period.

I. Introduction

The general aim of power system planning is to insure the provision of reliable power supply at lowest possible cost with expected maximum percentage generating units capacities according, to fuel types and system uncertainties. The factors which should be considered in any plan are, future load forecast, availability of technology, real cost of money (cash flow), cost and availability of fuel, energy supply adequancy and enviromental effects. Thus the problem of long-term plant mix means how to analyse the economic alternatives of power system development according to such factors at maximum security and revenue.

One of the methods chosen for such process is the linear programming (LP) methods, as it becomes a very popular tool for use in optimization and can routinely handle very large problems with a thouthands of variables and several hundreds of constraints. The initial implementation of LP for power system was for production cost models[1] Then it was extended[2], to include the system generation planning problem as a simulation process where various expansion plans were postulated and computer would determine the associated costs and reliability. A direct optimization[3] for alternate conditions and a plan was determined for each. As the number of variables plays an important role in computer time to obtain a solution, a model[4] was proposed implementing only the variables representing the installation in the first few years of the plan. It was satisfactory approach for the planner, but it would ignore the significant long-term effects.

This paper introduces an economic approach for power system long term optimum mix generation using a deterministic LP model. As the system composes multiple fossile, nuclear and hydro plants lacing with its uncertainties, the proposed model incorporates:

i. a mathematical description of existing power system capacities.

ii. estimates of variables costs and capacity requirements of existing system,

iii. estimates of future load demand from the system during the whole planning period,

iv. future investement opportunities in term of the system additional capacities by building new plants of different types.

The model objective is to get the optimum decision, at less capital investements, less operating costs and maximum revenue to justify the construction of these future investements at the end of a corresponding planning period (say 5 years) from a long term plan.

The proposed model is discussed and implemented for the Egyptian power system to give a complete plan for it to the end of this century (year 2000).

II. Notation

P number of power plants in the system
C capital charge allowance for each plant during operating period.
C^p_{fk} cost of plant fuel/unit energy produced in the same period
C_T fixed charge for existing transmission network/unit power
C_{NT} fixed charge for new lines to be built/unit power
C_{NP} fixed charge for new plants to be built/unit power
C_{NK} cost of new plant fuel/unit generated
O^p existing plant operating cost/unit generated (ex. fuel)
O_{NP} operating cost of new plant/unit generated (ex. fuel)
X electric energy product by each power plant
X^p_B energy bought by the system from any neighbouring systems interconnected to it.
X_D energy fraction for domestic consumers
X_I energy fraction for industrial consumers
X^s energy send out to another interconnected systems
S_B buying cost/unit energy from neibouring systems
S_D selling cost/unit energy for domestic consumers
S_I selling cost/unit energy for industrial consumers
S^s selling cost/unit energy send out to interconnected systems
S_{ND} selling cost/unit energy of estimated domestic demand
S_{NI} selling cost/unit energy of estimated industrial demand

III. Power System LP models

III.1. Single Power plant LP model

The power plant can be regarded as a factory for the production of a given commodity, energy, from fuel as a raw material with a rate of manufacturing being the power capacity (in KW). Like the factory, the power plant is expected to return fair profit and to produce a satisfactory saleable product. Therefore, the plant must act as a unit, not as a collection of individual pieces of equipments. Breifly, it is the art of selecting and placing the necessary plant, so that a maximum revenue derived from domestic and industrial consumers it serves, will result from a minimum expenditure over its working life, and operating in a manner, to provide cheap, reliable and continueous service.

According to these activities defined for a power plant, a simplified flow diagram could be designed to illustrate them as shown in figure(1). Thus, it is a simple matter to set up a LP model by means of which the plant could operate of satisfy its activities .

Table (1), shows such proposed model. Numbers (other than 1 and -1) which appear in LP matrix are elucidated as follows:
(2) Fixed/cost/unit power for the plant installed capacity.
(3) Cost/unit fuel consumed.
(4) Operating unit energy generated (ex. fuel).
(5) Cost/unit energy received from interconnected power systems.
(6) Cost/unit energy retransmitted to the system.
(7) Selling cost/unit energy for domestic consumer.
(8) Selling cost/unit energy for industrial consumer.

(9) Fraction of generated power for the plant's auxiliaries.
(10) Local distribution system efficeincy.
(11) Amount of fuel consumed/unit energy generated by the plant.

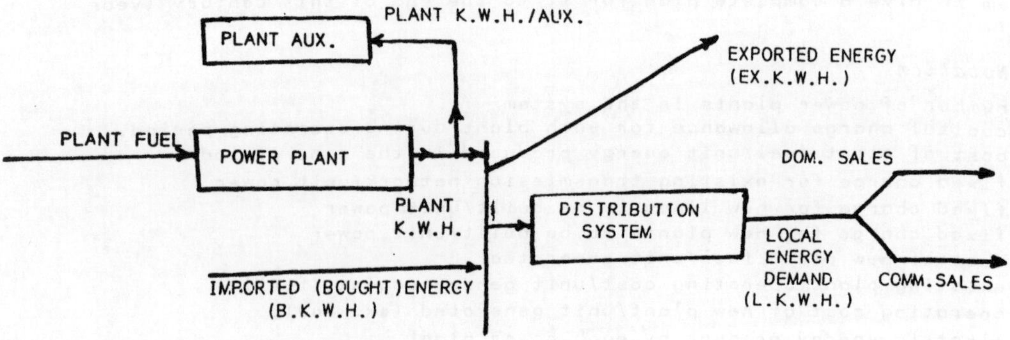

Fig. (1) Power Plant Static Model

Table 1. Power Plant Linear Programing model

Activities / Constraints	St.Installed Capacity	St.Fuel Consumption	St.Generated energy	St. Bought Energy	St. Gen. Energy for Aux.	St. Gen. Enrg. for Local dem.	St. Gen. En. Sod out Tr. System	St. Bought En. Local Demand	St. Bgught En. resend out	Total En. Sond out Tr. System	St. Local Domestic Demand	St. Local Ind. Domand	R.H.M.
Profit	2	3	4	5						6	7	8	
St.Energy Product Salance			1			1	1						≤ St. Aus
St.Bought Energy Balance				1				1	1				= 0
St. Output Energy Balance			9		1	10	1	10	1	1	1	1	= 0
St. Auxiliaries			9		1								= 0
St. Fuel Consumption Balance		-1	11										= 0
St. Basic capacity					1	1	1						≤ st. Capacity
St. Local Demand					1		1						> St. Local Demand
St. Surplus Energy	1		1										≥ 0

III.2 Power System short-term operational model

Following the model[5], developed by the Operational Research Division of BP for multi-refinary grouping in oil industries using LP technique, a power system model could be proposed on the same basis. The power plants models introduced in secion III, 1, are linked sequentially by the fact that:

i. They are using different sources of raw materials,

ii. They are supplying electric energy product from one plant to another to serve the system load centres.

The objective of this model, during a certain operating period (say) one year) is to minimize the system total operating costs. This is achieved by identifying the optimum generation level for each power plant in the system satisfying, in that same time, the system load demand with its expected growth during the same period. This optimum levels, sometimes do not agree with the optimum solution of each power plant operating individually, but they represent the system operational strategy to run it on an optimum economical basis.

The operational model objective function has the following form:

$$\sum_{P=1}^{P} C_p + \sum_{B=1}^{B} S_B X_B + \sum_{P=1}^{P} \sum_{K=1}^{K} (O_p + C_{fk}) X_p - \sum_{D=r}^{D} S_D X_D - \sum_{I=1}^{I} S_I X_I - \sum_{S=1}^{S} S_s X_s \qquad (1)$$

Subject to the following constraints

a) For each power plant in the system:

i) plant input/output balance = 0
ii) plant power product capacity ≤ Plant installed capacity
iii) plant fuel availability ≥ Plant fuel consumption (regarding fuel transport and storage).
iv) plant local demcnd = total (domestic industrial energy consumption in the plant region.

b) For the power system:
i. System input/output balance =0
ii. system generating and buying ≤ system total installed capacity and power units bough from neighbouring interconnected systems.
iii. total energy demand from the ≥ total energy demand at the system load centres and energy send out to other interconnected systems.

III.3 Power System Long-term LP Model

One of the purposes which is of direct relevance of present paper, is to build a long-term planning model for a power system. For this purpose, the short-term model, of section III.2, is modified by the inclusion of investments and estimated load demand activities[5,6,7] and constraints during a considered planning period.

Thus, the model objective function has the following form:

$$\sum_{T=1}^{T}\sum_{P=1}^{P}(C_T + C_P) + \sum_{P=1}^{P}\sum_{K=1}^{K}(O_p + C_{fk})X_p + \sum_{NT=1}^{NT} C_{NT}X_{NT} + \sum_{NP=1}^{NP} C_{NP}X_{NP} + \sum_{NP=1}^{NP}\sum_{NK=1}^{NK}(C_{NK}+O_{NP})X_{NP} - \sum_{D=1}^{D} S_D X_D - \sum_{I=1}^{I} S_I X_I - \sum_{ND=1}^{ND} S_{ND}X_{ND} - \sum_{NI=1}^{NI} S_{NI}X_{NI} \qquad (2)$$

For experimental evaluation of the long-term planning model, the data of the Egyptian power system is prepared to determine a complete long-term plan for it up to year-2000.

IV. Present & future Egyptian Power System. [8,9]

From the limited 33 KV systems in the Delta and the 60 KV in Cairo area were operated in 1930's the interconnected power system of Egypt stepped up in voltage and length to the present 500/220/132 KV system. This system is now supplied by a set of hydroelectric, steam and gas power plants with total generated power 4722 MW (table 2). As the demand of Egypt is over-growing up to year 2000, Table 2 summarises electric power generation, tossil fuel (oil and gas) utilizations, an itroduction of coal in some new steam plants, an introduction of nuclear energy production.

Table 2: Estimated electric power generation for Egypt up to year 2000.

Year	1952	1962	1972	1982	1990/1	2000/1
Population of Egypt (millions)	21.4	27.3	34.9	44	54	68
Generated electric energy (Twh)	0.423	3.163	7.384	23.35	52	101
Peak load (Mw)	125	548	1176	3900	9260	18220
Installed capacities of Hydro-electric plants (MW)						
Aswan & Aswan	--	345	345	345	345	345
High-Dam	--	--	2100	2100	2100	2100
Aswan II	--	--	--	--	270	270
Esna/Assut!Nag-Hammadi barrages	--	--	--	--	207	207
Qattara Project	--	--	--	--	--	1800
Pump storage at Galala et-Bahraiya (Gulf of Sues)	--	--	--	--	600	1200
Mini-Hydro.	--	--	--	--	15	28
Total Hydro-power	--	345	2445	2445	3267	5950
Oil/gas-steam Plants MW	154	593	1302	1815	5800	4968
Coal-steam plants (MW)	---	---	---	---	1800	4800
Oil/gas turbine Plants (MW)	--	--	28	462	1360	757
Nuclear Plants (MW)	--	--	--	--	2000	8000
Total generated power (MW)	154	938	3775	4722	12427	24475

V. The Model Computational Results

The proposed long-term planning model is computed four times for four 5 years planning periods. The model is provided by the data of the existing system, new power plants and transmission lines. The model results, for new investments, from each computational run, is recommended for for the next one. The model final output results are the decisions to build particular group of power plants of different types, at particular locations to supply the estimated load demand. Also, the results give the necessary decisions to extend the carring capacities of existing transmission network by addition of new lines.

For the model runs for two sucessive periods (1981-1990), if it is assumed that investment is to start now on huge Qattara hydroscheme, its civil work will take nearly ten years. Therefore, its expected output energy during these two periods will be zero. According to these assumption, the model during this ten years period looks for investments in other types of plants to supply the estimated load growth.

The model output computational results, for the four planning periods are summarized in table 3.

Table 3: % value of plants capacities recommended by the model to be installed at the end of each planning period.

End of period	1980-85	1986-90	1991-95	1996-2000
Existing system	90 %	100%	90%	90%
Aswan II	100%	100%	100%	95%
Nile Bar&Pump St.	100%	100%	100%	95.5 %
Qattara	---	---	74.4 %	98.8 %
Coal plants	37.5 %	100%	100%	100%
Oil-plants	---	100%	50 %	50 %
Gas Turbine	---	60.6%	60.6%	60.6%
Nuclear 1	---	53%	53%	98%
Nuclear II (1000 MW)	---	---	70%	98%
Nuclear III-VII (1000 MW each)	---	---	---	98%

The model capital investments decisions are to invest, first for, large hydraulic, nuclear, coal fired, oil fired and finally gas turbine plants. It also, recommended that there should be enough balance to develop the existing network to withstand the estimated load demand increase.

VI. Conclusion

The application of LP in making investment decisions for power system produces an important addition to the tools available to the system planner instead of the old cash flow net present value or rate of return. The electric energy production industry, is integrated, and at each level a range of choices exists. In this situation, investment projects just cannot be evaluated independently. The interdependence is, in fact, technical, goegraphical financial and temporal.

The model, introduced in this paper, gives the system generation, distribution and load demand for each specific planning period with its implicit investment decisions. Interrelation between operation and system plants capacities, and between investment and operation are examined in an integrating manner which permits these relations to be solved automatically.
Therefore, the proposed model has adequantly covered the geographical

and technical planning problems for power system.

VII. References

1. Day, J.T.: " Forcasting minimum production cost in linear programming". IEEE Trans. Power App. & Systems Vol. PAS-90, No.2, pp 814-823, March/April 1971.

2. Knight, U.G. : "Power system engineering and mathematics". Oxford Pergamon Press, England, PP. 114-117, 1972.

3. Day, J.T. and Menge, E.E.: "The role of direct optimization in generation planning". Proc. of American Power Conf. Vol. 36, PP. 884-890, 1974.

4. Day, J.T., Federowicz, A.J. and Menge, E.E. "Optimizing generation planning". Power Engineering, PP. 32-35, July 1973.

5. Deam, R.J., Bennet, J.W., Leather, J. and Carsberg, B.V. : "Firm: a computer model for financial planning". A report prepared for Research Committee, Institute of Chartered Accountant, England & Wales, February 1975.

6. Allam, M.F. R.J., Leather, J. and Laukhton, M.A., : " Power system long term financial planning by using FIrm model". Special report submitted to Energy Research Unit (ERU), Queen Mary College, London University, September 1975.

7. Porterfield, J.S.: " Investment decisions and capital costs". Prentice Hall, New Jersey 1965.

8. Abaza, M. : Egypt's energy strategy". Middle East Electricity Journal Vol. 7, No. 10, PP. 57-59, November 1983.

9. Abaza, M. : " Egypt's energy strategy". Middle East Electricity Journal Vol. 7, No. 11, PP. 57-58, December 1983.

AN ALGORITHM FOR GETTING A MINIMUM CUT-SET OF A GRAPH

ILONA ARANY
COMPUTER AND AUTOMATION INSTITUTE, HUNGARIAN ACADEMY OF SCIENCES
H-1052. BUDAPEST, KENDE U. 13-17., HUNGARY.

Abstract

Let x and y be two nodes being not connected by an edge in an undirected connected non-complete graph. We present here an algorithm which finds a minimum cut-set of the graph by which x and y are separated.

1. Introduction

In a number of problems derived from undirected graphs, it is often advisable to find a small cut-set of the graph. In this paper, we present an algorithm which (starting with a point pair x and y being not connected by an edge) finds a minimum cut-set $D(x,y)$ of the graph such that $D(x,y)$ separates x and y.

2. Some notations and terms

We use the notation $G=(X,E)$ for undirected connected non-complete graphs without loops and multiple edges. X is the node set; E is the set of the edges. For arbitrary $x,y \in X$, the *distance* $d(x,y)$ between them is the number of edges in a shortest path connecting them. For an arbitrary $x \in X$, its *eccentricity* is $e(x) = \max_{y \in X} d(x,y)$.

For a subset $Y \subset X$, its *neighbour set* is defined as

$$N(Y) = \{z \in X \setminus Y\colon d(Y,z) = 1\},$$

where $d(Y,z) = \min_{y \in Y} d(y,z)$.

For an arbitrary $x \in X$, the *level structure rooted at x* [1] is the partitioning $RLS(x) = \{L_0(x), L_1(x), \ldots, L_{e(x)}(x)\}$ of X such that $L_0(x) = \{x\}$; $L_1(x) = N(\{x\})$;

$$L_i(x) = N(L_{i-1}(x)) \setminus L_{i-2}(x), \quad (i=2,3,\ldots,e(x)).$$

For arbitrary $x,y \in X$, their *middle set* [2] is defined as

$$M(x,y) = \left\{z \in R(x,y)\colon d(x,z) = \left\lfloor \frac{d(x,y)}{2} \right\rfloor \right\},$$

where

$$R(x,y) = \{z \in X: d(x,z) + d(z,y) = d(x,y)\}.$$

3. Foundation for the algorithm

For a subset $Y \subset X$, the *section graph* [4] generated by Y is the subgraph $G(Y) = (Y,E(Y))$, where $E(Y) = \{(x,y) \in E: x \in Y, y \in Y\}$. A subset $Y \subset X$ is a *cut-set* (or separator [4]) of the graph, if $G(X\setminus Y)$ is disconnected. The we have $G(X\setminus Y) = \bigcup_{i=1}^{k} G_i(X_i)$, where $G_i(X_i)$ $(i=1,2,\ldots,k)$ are disjoint subgraphs of $G(X\setminus Y)$. For arbitrary $x,y \in X$, if they belong to different components of $G(X\setminus Y)$, then they are said to be *separated by Y*.

A cut-set $D \subset X$ is a *minimum cut-set* of the graph, if none of its subsets is a cut-set.

Our aim is to determine a minimum cut-set of the graph by which an arbitrarily chosen starting point pair $x,y \in X$ with $(x,y) \notin E$ are separated.

Theorem

Let $x,y \in X$ be with $(x,y) \notin E$. For every $z \in X$, let us define the following function

$$h_{xy}(z) = |d(z,x) - d(z,y)|.$$

Then we have

$$0 \le h_{xy}(z) \le d(x,y). \qquad (1)$$

Proof

Clearly, if $d(x,y)$ is even, then for every $s \in M(x,y)$, we have $h_{xy}(s) = 0$ by which the left hand side relation in (1) is proved.

Following from the triangular inequality, for every $z \in X$, we have

$$d(z,x) \le d(z,y) + d(x,y),$$
$$d(z,y) \le d(z,x) + d(x,y),$$

from where it follows

$$d(z,x) - d(z,y) \le d(x,y),$$
$$d(z,y) - d(z,x) \le d(x,y).$$

That is, we have

$$h_{xy}(z) \le d(x,y)$$

by which the right hand side relation in (1) is assured, as required.

Consequence

Obviously, $h_{xy}(x) = d(x,y)$ and $h_{xy}(y) = d(x,y)$ and for a node $p \in X\backslash(\{x\} \cup \{y\})$ satisfying $h_{xy}(p) = d(x,y)$, we have

$$\max(d(p,x),\ d(p,y)) > d(x,y).$$

In other words, nodes at which $h_{xy}(z)$ takes its maximum value are much farther from x (or y) than from y (or x). On the other hand, for a node $q \in X$ satisfying

$$h_{xy}(q) = \min_{z \in X} h_{xy}(z),$$

its distances from x and y are equal or nearly equal to each other.

Conclusion

$$D = \{z \in X: h_{xy}(z) < d(x,y)\} \subset X$$

is a cut-set of the graph by which x and y are separated. In addition,

$$D' = \{z \in X: h_{xy}(z) \le 1\} \subset D \qquad (2)$$

is also a cut-set of the graph by which x and y are separated.

Note, usually $|D'| \ll |D|$, but D' is not necessarily a minimum cut-set of the graph separating x and y, as the example in Figure 1. shows it. For every node in the graph in Figure 1., we indicate its distances (as a pair of numbers in brackets) measured from x and y, respectively. Here $d(x,y) = 4$; we have $|D'| \ll |D|$, and D' is not a minimum cut-set of the graph by which x and y are separated.

Note that in the example in Figure 1, the section graph generated by either D or D' consists of two disjoint subgraphs; i.e.

$$G(X\backslash D) = G_1(X_1) \cup G_2(X_2) \quad \text{and} \quad G(X\backslash D') = G_1(Y_1) \cup G_2(Y_2)$$

such that $X_1 \subset Y_1$ and $X_2 \subset Y_2$.

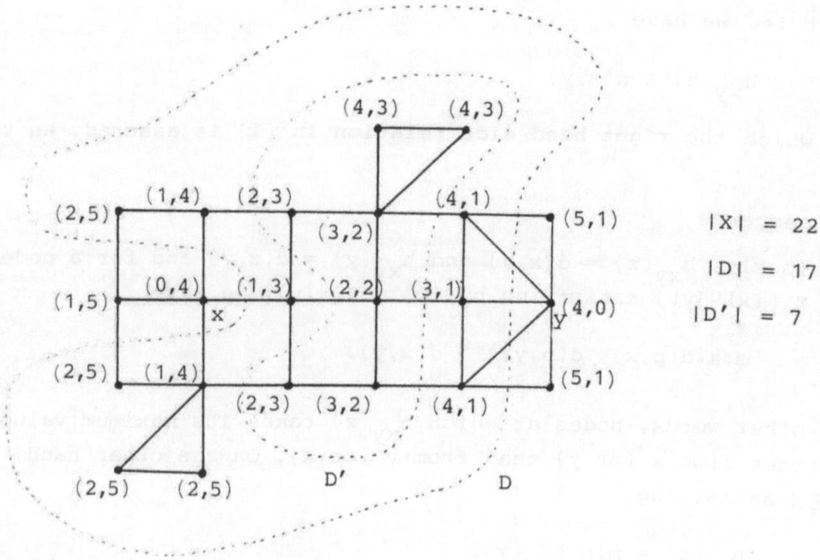

Fig. 1.

4. Present algorithm

Based on the discussion in the previous section, our present aim is to select some nodes from D' which form together a minimum cut-set D(x,y) of the graph separating x and y.

Let us describe D' in (2) by using RLS(x) and RLS(y), we get

$$D' = \bigcup_{i=1}^{e(x)} L_i(x) \cap (L_{i-1}(y) \cup L_i(y) \cup L_{i+1}(y)).$$

In other words, for getting D', we have to put together the nodes in $L_i(x) \cap L_{i-1}(y)$, $L_i(x) \cap L_i(y)$ and $L_i(x) \cap L_{i+1}(y)$, for all $1 \le i \le e(x)$. In order to select a minimum cut-set from D', in increasing order of indices, let us produce the levels of the same index in RLS(x) and RLS(y) after each other, and let us generate them only on X\D(x,y). (Note, at the beginning, D(x,y) is empty.) If a node is found in $L_i(x) \cap L_{i-1}(y)$ (or in $L_i(x) \cap L_i(y)$), then put it into D(x,y), and remove it from the two relevant levels. Turn to generate the next level

($L_i(y)$ or $L_{i+1}(x)$) on $X\backslash D(x,y)$ and by checking the common part of the corresponding two levels, continue the above procedure. Obviously, this process is terminated, if both the considered levels are empty.

Since the nodes having been put last into $D(x,y)$ are removed from the relevant two levels, and all the other levels to be generated are produced only on $X\backslash D(x,y)$, it can be seen easily that

- for certain index, both the levels generated last become empty;
- $D(x,y)$ obtained in such a way is a minimum cut-set of the graph by which x and y are separated.

The algorithm can be formed as follows:

Step 1 (Initialization)

Set $L_{-1}(x)\leftarrow\{\emptyset\}$; $L_{-1}(y)\leftarrow\{\emptyset\}$;

$L_0(x)\leftarrow\{x\}$; $L_0(y)\leftarrow\{y\}$;

$X_1\leftarrow\{x\}$; $X_2\leftarrow\{y\}$;

$D(x,y)\leftarrow\{\emptyset\}$;

$i\leftarrow 0$.

Step 2 (Generate the next level in RLS(x) and check it)

$i\leftarrow i+1$;

$L_i(x)\leftarrow N(L_{i-1}(x))\backslash(L_{i-2}(x)\cup D(x,y))$;

If $L_i(x)=\{\emptyset\}$ and $L_{i-1}(y)=\{\emptyset\}$, then go to Step 4;

If $L_i(x)=\{\emptyset\}$, and $L_{i-1}(y)\neq\{\emptyset\}$, then go to Step 3;

If $L_i(x)\neq\{\emptyset\}$, then set

$D(x,y)\leftarrow D(x,y)\cup(L_i(x)\cap L_{i-1}(y))$;

$L_i(x)\leftarrow L_i(x)\backslash D(x,y)$;

$L_{i-1}(y)\leftarrow L_{i-1}(y)\backslash D(x,y)$;

$X_1\leftarrow X_1\cup L_i(x)$;

$X_2\leftarrow X_2\backslash D(x,y)$;

Turn to Step 3.

Step 3 (Generate the next level in RLS(y) and check it)

$L_i(y) \leftarrow N(L_{i-1}(y)) \backslash (L_{i-2}(y) \cup D(x,y))$;

If $L_i(y)=\{\emptyset\}$ and $L_i(x)=\{\emptyset\}$, then go to Step 4;

If $L_i(y)=\{\emptyset\}$ and $L_i(x)\neq\{\emptyset\}$, then go to Step 2;

If $L_i(y)\neq\{\emptyset\}$, then set

$D(x,y) \leftarrow D(x,y) \cup (L_i(x) \cap L_i(y))$;

$L_i(y) \leftarrow L_i(y) \backslash D(x,y)$;

$L_i(x) \leftarrow L_i(x) \backslash D(x,y)$;

$X_2 \leftarrow X_2 \cup L_i(y)$;

$X_1 \leftarrow X_1 \backslash D(x,y)$;

Go to Step 2.

Step 4 (Exit)

D(x,y) is the resulting set determined, while $G_1(X_1)$ and $G_2(X_2)$ are connected subgraphs with $x \in X_1$ $y \in X_2$.

The algorithm presented here (which we refer to as algorithm CS) constructs a subset D(x,y) of D' which is a cut-set of the graph by which the starting point pair x and y are separated. On the other hand, following from the construction, D(x,y) is a minimum cut-set separating x and y, in fact. We should not forget that the resulting D(x,y) may contain a node which is a minimum cut-set of the graph itself, however, x and y are not separated by it. (See Figure 2.) We use the example in Figure 2 to illustrate how the algorithm works. Let the starting point pair be x and y. For every node in the graph, we indicate only the distances having been determined during the process.
The resulting D(x,y) is a minimum cut-set of the graph by which x and y are separated. Note that $\{z\} \subset D(x,y)$ is a minimum cut-set of the graph, however, x and y are not separated by {z}.

Note, the resulting D(x,y) depends on the order of the starting point pair; namely, D(y,x) may define another minimum cut-set separating x and y. (See Figure 3.)

In determining D(x,y) by using CS, $G(X \backslash D(x,y)) = \bigcup_{i=1}^{3} G_i(X_i)$ is generated, where $G_1(X_1)$ and $G_2(X_2)$ are connected subgraphs such that $x \in X_1$; $y \in X_2$. Note, $G_3(X_3)$ is not necessarily a connected subgraph. In

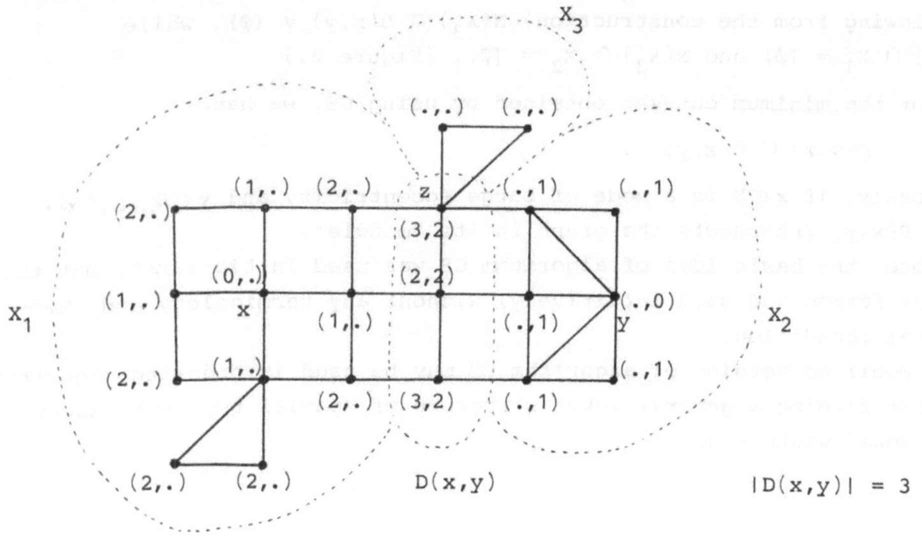
X3
(.,.) (.,.)
(1,.) (2,.) z (.,1) (.,1)
(2,.)
(3,2)
(0,.)
(2,2)
(1,.) (.,0)
X1 x (1,.) (.,1) y X2
(1,.) (.,1)
(2,.)
(2,.) (3,2) (.,1)
(2,.) (2,.) D(x,y)
|D(x,y)| = 3

Fig. 2.

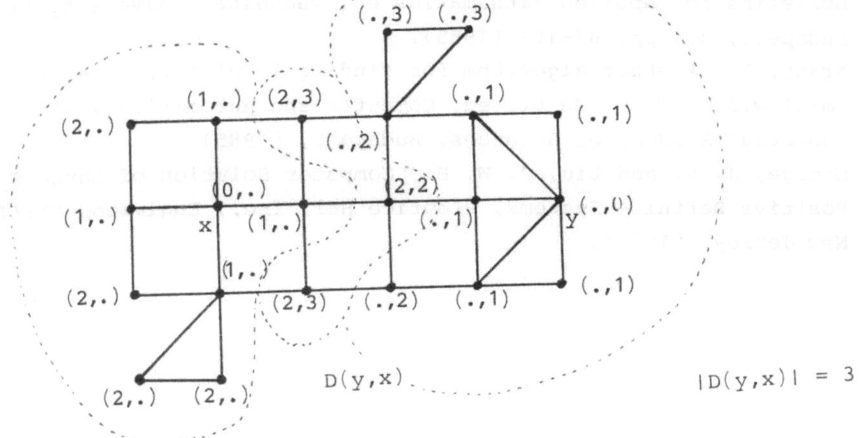
(.,3) (.,3)
(1,.) (2,3) (.,1)
(2,.) (.,1)
(.,2)
(2,2)
(0,.)
(1,.) (.,0)
x (1,.) (.,1) y
(1,.)
(2,.) (.,1)
(2,3) (.,2) (.,1)
D(y,x)
(2,.) (2,.)
|D(y,x)| = 3

Fig. 3.

addition, in many cases $X_3 = \{\emptyset\}$ (see Figure 3.). If $X_3 \neq \{\emptyset\}$, then (following from the construction) $N(X_3) \cap D(x,y) \neq \{\emptyset\}$, while $N(X_3) \cap X_1 = \{\emptyset\}$ and $N(X_3) \cap X_2 = \{\emptyset\}$. (Figure 2.)

For the minimum cut-set obtained by using CS, we have

$$M(y,x) \subsetneq D(x,y) \quad .$$

Obviously, if $x \in X$ is a node of large eccentricity and $y \in L_{e(x)}(x)$, then $D(x,y)$ intersects the graph in its "middle".

Note, the basic idea of algorithm CS was used in [1] first, but then it was formed and used intuitively, without any terminology and theoretical foundation.

A modified version of algorithm CS may be used in ordering algorithms [4] for finding a general level structure of special type with acceptably small width [3].

References

[1] Arany, I., Smyth, W. F. and Szóda, L., An improved method for reducing the bandwidth for sparse symmetric matrices, Proc. IFIP Congr. '71, North-Holland Publ. Co., Amsterdam, pp. 1246-1250 (1972).

[2] Arany, I., An algorithm for finding all peripheral nodes, Bulletins for Applied Mathematics ed. Technical University of Budapest, 38, pp. 89-101 (1985).

[3] Arany, I., Another algorithm for finding level structures with small width, W. P. MG/9., ed. Computer and Automation Institute, Hungarian Academy of Sciences, Budapest, (1985).

[4] George, J. A. and Liu, J. W. H., Computer Solution of Large Sparse Positive Definite Systems, Prentice Hall Inc., Englewood Cliffs, New Jersey, (1981).

OPTIMAL DESIGN OF A REMOTE HEATING NETWORK

F. Archetti, A. Sciomachen, C. Vercellis
Dipartimento di Matematica
Università di Milano
Milano, Italy

Abstract

In this paper the authors are concerned with a network design problem arising in the layout of a remote heating network. The problem can be modeled by a graph whose vertices represent the energy source, the demand points and the intermediate nodes and whose edges represent the possible pipe connections. Among the different arborescences rooted at the source and spanning all demand nodes and among the pipe diameters commercially available one must find a design of minimum cost which satisfies a number of physical constraints. In this paper, we propose an approximation algorithm in which a tentative design generated by a randomized algorithm is subsequently improved by a local search scheme. The computational results, obtained for a large graph of 166 vertices, 64 demand points and 213 edges representing a district of Milano, are reported.

1. Introduction and formulation of the problem

In this paper the authors are concerned with a network design problem arising in the layout of the remote heating network of an urban district.

The actual planning process is very complex because several factors are to be taken into account: a wide range of possible pipeline components, the plant control policy, the thermal specifications of the buildings and the topological design of the network.

The topology of the heating network can be represented by a graph $G=(V,E)$, whose n vertices $\{v_1,v_2\ldots v_n\}$ are grouped into a source s, a set U of demand points and a set I of intermediate vertices and whose m edges $\{e_1,e_2,\ldots,e_m\}$ represent the possible pipe connections between vertices. To each demand node $v_i \in U$ a value $r(v_i)>0$ is associated representing the energy requirement. The length of an edge $e_t = (v_i,v_j) \in E$ will be denoted as l_{e_t} or l_{ij}.

It is assumed in what follows that the amount of heating energy which is supplied by the source s is equal to the total requirement $R = \sum_{v_i \in V} r(v_i)$.

It is also assigned a set of commercially available pipe diameters, denoted by $D_k (k=1,2,\ldots q)$, together with the overall cost $c_{ij}(D_k)$ of a unit length of a pipe of diameter D_k placed along edge $(v_i,v_j) \in E$. Let d_{ij} be the decision variable representing the diameter of a pipe placed along edge $(v_i,v_j) \in E$.

Let T be the set of those arborescences in G which are rooted at the source s and do span all the demand nodes belonging to U. To each arborescence $T \in \mathcal{T}$ a (unique) flow $f(T) = \{f_{ij}, (v_i,v_j) \in T\}$ is associated through the balance equations at each node:

$$\sum_{(v_i,v_j) \in T} f_{ij} - \sum_{(v_k,v_i) \in T} f_{ki} = \begin{cases} 0 & \text{if } v_i \in I; \\ R & \text{if } v_i = s; \\ -r(v_i) & \text{if } v_i \in U. \end{cases}$$

An arborescence $T \in \mathcal{T}$ and an assignment $d(T) = \{d_{ij}, (v_i,v_j) \in T\}$ of pipe diameters to T, are said to be a feasible heating network (T, $d(T)$) if the following sets of constraints are satisfied:

(i) for each edge $(v_i,v_j) \in E$ the maximum capacity $q_{ij}(d_{ij})$ of the pipe is not exceeded: $f_{ij} \leq q_{ij}(d_{ij})$;

(ii) a global pressure constraint of the form $p(d(T), f(T)) \leq p_N$ holds, where p_N is a given value for the maximum pressure allowed, and the function $p(d,f)$ expresses the maximum pressure in the network, as described in Section 2.

A preliminary network model in which the constraint (ii) is not included is described in /1/.

Among the different feasible heating networks, one is required to find one of minimum cost, i.e. to solve the following combinatorial optimization problem (referred to in the sequel as Minimum Cost Heating Network Problem) (MCHNP):

$$\min_{T \in \mathcal{T}} \min_{d(T)} \left\{ c(d(T)) = \sum_{(v_i,v_j) \in T} l_{ij} c_{ij}(d_{ij}) \right\}$$

subject to the constraints (i) and (ii).

MCHNP is computationally difficult, i.e. NP-hard /4/ even if q=1, that is if there is only one available capacity and if the constraints (i) and (ii) are trivially satisfied. Indeed, with these assumptions, the problem becomes equivalent to the Steiner problem in graphs, known to be NP-hard /5/ , /2/.

It is therefore natural to develop approximation algorithms for MCHNP. One such algorithm has been proposed in /6/ for the design of a minimum cost drainage network, which can be thought of as a heating network with the edges reversed, but substantially differs due to the lack of the global constraint (ii): this algorithm first generates a starting feasible arborescence, which is next improved using articulation vertices (called Weber points) until no further improvement can be achieved. A modified version of this algorithm, taking into account constraint (ii) as well,has been considered for obtaining an approximate solution to MCHNP in order to assess the computational effectiveness of the algorithm proposed in this paper.

A different approximation algorithm has been proposed in /8/ for the design of a gas pipeline system, which is a problem similar to MCHNP in the particular case in which G is a complete graph and $I = \emptyset$, i.e. the use of intermediate nodes is disallowed. Hence, T turns out to be a spanning arborescence of G. This latter algorithm is based upon the use of local search techniques, a widely used scheme for developing heuristics in combinatorial optimization (see /7/ for a survey). This approach usually requires a structure of neighborhoods to be appropriately defined within the discrete space of feasible solutions, and consists of two main steps: first a starting feasible solution is generated by a randomized algorithm, then it is improved until a local minimum in its neighborhood is reached. Thus, the best of the minima obtained in a number of independent executions of the two basic steps is retained as an approximation of the optimum.

In Section 2 of this paper a model for the computation of the cost $c_{ij}(d_{ij})$ along with a procedure for solving the inner optimization problem in the definition of MCHNP is given. In the same Section it is also outlined how to evaluate the global pressure constraint.

In Section 3, we propose an approximation algorithm for solving MCHNP: it is based on local search schemes and extends to the case $I \neq \emptyset$ the structure of neighborhoods considered in /8/.

Two different randomized procedures are also proposed for generating the starting solutions: the first of them can be regarded as a randomized version of Kruskal's algorithm, while the second is obtained by incorporating random steps in the first stage of the algorithm proposed in /6/.

Finally, the computational results obtained for a large graph of 166 vertices, 63 demand points and 213 edges, representing a district of Milano, are reported in Section 4. The solution obtained by the proposed algorithm yields a better approximation than the one obtained using a modification of the algorithm proposed in /6/.

2. Selection of the pipe diameter

In this Section we shall briefly outline how to solve the optimization problem in the variables $\tilde{d}(T)$ in the formulation of MCHNP, i.e. how a value d_{ij} can be associated to each arc (v_i, v_j) of T in order to allow the flow f_{ij} at a minimum cost.

We shall also explain how the constraint (ii) is formulated and how its evaluation can be performed.

As far as the first point is concerned to each edge (v_i,v_j) of the network we associate the following costs, expressed yearly per unit length of the pipe:

$c^1_{ij}(D_k)$: investment required for placing a pipe of diameter D_k;

$c^2_{ij}(D_k)$: operating costs of the pumps;

$c^3_{ij}(D_k)$: cost of the heat loss.

For a given flow rate f_{ij} and for each available diameter D_k $(k=1,2,\ldots q)$ the total cost $c_{ij}(D_k) = c^1_{ij}(D_k)+c^2_{ij}(D_k)+c^3_{ij}(D_k)$ can be evaluated. The value of D_k, which minimizes $c_{ij}(D_k)$, is assigned to the diameter of the pipe along (v_i,v_j), taking into account a lower bound on D_k given implicitly by the requirement that the flow velocity along (v_i,v_j) must not exceed an assigned value empirically determined. In this way, for any given arborescence $T \in \mathcal{T}$ and for each $(v_i,v_j) \in T$, a diameter d_{ij} is determined which is economically optimal, i.e. minimizes $c_{ij}(D_k)$.

The constraint (ii) $p(\underset{\sim}{d}(T),f(T)) \leq p_N$ can be expressed as: $\max_{(v_i,v_j)\in T} p_{ij} \leq p_N$ where p_{ij} is the pressure in (v_i,v_j). Notice that p_{ij} is given by the formula

$$\sum_{(v_1,v_p)\in A_{ij}} \Delta H_{1,p} + K,$$

where A_{ij} denotes the set of arcs in the path from the source s to v_j and K is a given value depending on the maximum altitude of the network. The value ΔH is computed relating the flow velocity v and the pipe diameter D to the flow f by the relation $f = \frac{v \cdot \pi \cdot D^2}{4}$, in the so-called Darcy-Weisbach expression /3/:

$$\Delta H = \frac{L}{D}\,\lambda\,\frac{v^2}{2g} \quad ,$$

in which L is the arc length, λ is a friction factor and g is the gravity constant.

Incidentally, we remark that if the network spans areas of different altitude, a corresponding constraint expressing a lower bound on the global pressure is to be considered.

3. An approximation algorithm

As already mentioned in the Introduction, the approximation algorithm proposed here for solving MCHNP is based on a local search scheme. It consists therefore of two basic phases: first, a feasible design is generated using a randomized procedure, then, it is improved through a search among the feasible heating networks lying in a properly defined neighborhood of the current solution.

We first describe two different procedures - start1 and start2 - to be used for generating a random feasible solution in the first phase. In the sequel, procedure best-diam $(v_i, v_j, \underset{\sim}{f}(T))$ is intended to return an optimal diameter assignment d_{ij} to the edge (v_i, v_j) on the arborescence T, according to the method discussed in Section 2. Analogously, procedure feas-diam $(T, \underset{\sim}{d}(T))$ is intended to return success or failure, respectively, if the diameter assignment $\underset{\sim}{d}(T)$ on an arborescence T satisfies or not constraint (ii).

In the first procedure an arborescence T is generated using a randomized version of Kruskàl's algorithm; subsequently, a feasible diameter assignment $\underset{\sim}{d}(T)$ is obtained through a repeated increase of the diameters of those edges whose pressure exceeds the maximum allowed pressure. The basic structure of start1 can be summarized as follows: it takes as input the network G with the requirements $\underset{\sim}{r}$ and returns a feasible pair $(T, \underset{\sim}{d}(T))$.

<u>procedure</u> start1 $(G, s, U, \underset{\sim}{r})$

sort the edges in nondecreasing order of their lengths and set $T = \emptyset$:

<u>repeat</u>

the r-th edge e_{i_r} in E is chosen according to the realization r of a discrete r.v. whose probability mass function is a truncated geometric distribution over the integers $\{1,2,\ldots,|E|\}$ with real parameter belonging to the interval (0,1):

$$p = \frac{\sum_{e_i \in E} l_{e_i}}{|E| \cdot l_{e_{|E|}}}$$

Therefore, at each iteration, the probability of choosing the r-th edge is expressed by $\frac{p(1-p)^{(i-1)}}{Z}$ $(i=1,2,\ldots|E|)$, where $Z = 1-(1-p)^{|E|}$; the chosen edge is removed from E. Moreover, if $T \cup e_{i_r}$ does not contain a circuit the new edge is added to T;

<u>until</u> T spans the source s and the demand points U;

remove the redundant edges from T;

compute the flow $\underset{\sim}{f}(T)$;

<u>for</u> each edge $(v_i, v_j) \in T$ <u>do</u> d_{ij} = best-diam $(v_i, v_j, \underset{\sim}{f}(T))$;

<u>while</u> (<u>not</u> feas-diam$(T, \underset{\sim}{d}(T))$) <u>do</u>

increase the diameter of the edges on which $p_{ij} > p_N$;

<u>return</u> ($\underset{\sim}{d}$(T));

<u>end</u>.

In the second randomizad starting procedure, the edges of the network are chosen according to an approximation of their final cost. This means that the flow assignment on the part of the arborescence T already considered is computed before each edge selection, so that the new edge is added according to the cost of placing along it a pipe whose diameter is evaluated via best-diam ($v_i, v_j, \underset{\sim}{f}(T)$). Of course, the corresponding diameter assignments, and the associated costs, are subject to changes at the end of the procedure in order to ensure feasibility with respect to the pressure constraint,again by means of feas-diam (T,$\underset{\sim}{d}$(T)).

<u>procedure</u> start2 (G,s,U,$\underset{\sim}{r}$)

set T = Ø;

<u>repeat</u>

the vertex $v_i \in V$ farthest from the source s is chosen and removed from V;

a flow is assignad to each edge (v_i, v_j) incident to the vertex v_i, i.e. for $v_j \in \Gamma(v_i)$: $f_{ij} = r(v_i)$;

for each edge (v_i, v_j), $v_j \in \Gamma(v_i)$ the pipe diameter is computed via best-diam ($v_i, v_j, \underset{\sim}{f}(T)$) and the corresponding cost is evaluated: $c_{ij} = c_{ij}(d_{ij}) l_{ij}$;

the edges incident to the vertex v_i are ranked in non decreasing order according to their costs;

an edge e_t linking v_i to a node in $\Gamma(v_i)$ is chosen according to the realization t of a discrete r.v. whose p.m.f. is a truncated geometric distribution over the integers $\{1,2,\ldots|\Gamma(v_i)|\}$ with parameter

$$p = \frac{\sum_{e_j \in \Gamma(v_i)} c_{e_j}}{|\Gamma(v_i)| \cdot c_{e_{|\Gamma(v_i)|}}}$$

At each iteration the edge e_t has probability $\frac{p(1-p)^{(t-1)}}{Z}$ to be chosen ($t=1,2,\ldots |\Gamma(v_i)|$), where $Z=1-(1-p)^{|\Gamma|(v_i)}$;

the edge e_t is removed from E and added to T;

the requirement of the vertex v_j, for which $e_t = (v_i, v_j)$, is updated: $r(v_j) = r(v_j) + r(v_i)$;

the vertex v_i is removed from V;

<u>until</u> T spans the source s and all demand points U;

<u>while</u> (<u>not</u> feas-diam (T,$\underset{\sim}{d}$(T))) <u>do</u>

increase the diameter of the edges on which $p_{ij} > p_N$;

return ($\underset{\sim}{d}$,(T));

end.

We finally turn to the structure of neighborhoods for the local search scheme. Specifically, given an arborescence T, and an integer b, let γ(b) denote the elementary transformation which consists of the following steps:

(i) select any vertex v_j in T;

(ii) append to T one of the edges connecting v_j to one of its b nearest vertices, not already adjacent to v_j in T;

(iii) delete any of the previous edges which form a cycle in T with the new one.

Given a feasible heating network (T,$\underset{\sim}{d}$(T)), we define its neighborhood N(T,$\underset{\sim}{d}$(T)) as the set of all heating networks (T', $\underset{\sim}{d}$(T')) such that T' can be obtained from T by repeatedly applying the transformation γ(b); the network of minimum cost in N(T,$\underset{\sim}{d}$(T)) is said to be locally optimal.

The resulting approximation algorithm can be summarized as follows:

procedure approx-MCHNP (G,s,U,$\underset{\sim}{r}$)
  compute a starting feasible network (T,$\underset{\sim}{d}$(T)) using either start1 or start2;
  repeat
    search for a feasible network of least cost in N(T,$\underset{\sim}{d}$(T));
    if found update (T,$\underset{\sim}{d}$(T))
  until (T,$\underset{\sim}{d}$(T)) is locally optimal.
end.

The computational effort of each step within the "repeat" loop of the local search procedure can be bounded by observing that, if the cycle induced in the network by step (ii) via the transformation γ(b) has w edges, the cost of at most b•w•n different heating networks has to be computed. For small values of b, such as b=3 or b=4 used in the computational experience described in the next Section, this leads to neighborhoods of acceptable size, which can be searched in a reasonable amount of computing time.

4. Computational results

This Section is devoted to describe the computational results obtained for a graph representing a district of Milano. This graph has 166 vertices, 63 demand nodes, 1 source and 213 arcs. The number of available diameters is 20.

The algorithm has been implemented using the C-language. This choice ensures a wide portability of the program. Moreover C allows easy access to some Unix programming facilities and provides dynamic storage allocation which is particularly useful for the algorithms proposed in this paper. The representation of the graph G by means of its adjacency list has been preferred since the networks modelling urban districts are generally sparse.

As far as the computational times are concerned it takes nearly 5 minutes on a Gould 32/37 to generate a feasible distribution network and to improve it by local search. The solution obtained by the proposed algorithm is depicted in Fig. 1. The performance of the algorithm has been compared with that provided by the algorithm proposed in /6/, adapted to the case in which the articulation points are restricted to belong to the set I of intermediate nodes, showing a reduction in cost of about 20%.

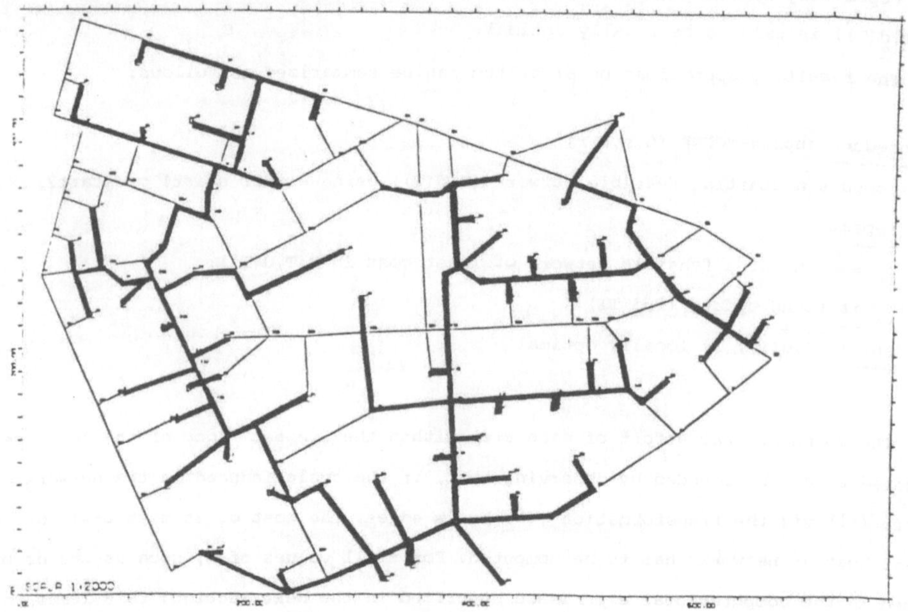

Fig. 1

References

/1/ F. Archetti, A. Sciomachen, C. Vercellis, "A Randomized Approximation Algorithm for a Network Design Problem, IAMI-CNR 84.11, 1984.

/2/ N. Christofides, "Graph theory. An algorithmic approach". Academic Press, New York, 1975.

/3/ "Flow of Fluids". Technical Paper n. 410, Crane Company, Chicago, 1976.

/4/ M.R.Garey, D.S.Johnson, "Computers and Intractability", Freeman,San Francisco,1979.

/5/ R.M. Karp, "Reducibility among combinatorial problems", in R.E. Miller and J.W. Tatcher (eds.), Complexity of Computer Computations, Plenum Press, New York,1972.

/6/ D.H. Lee, "Low cost drainage network", Networks, 6, 1976.

/7/ C.H. Papadimitriou , N. Steiglitz, "Combinatorial optimization: algorithms and complexity", Prentice Hall, Englewood Cliff, 1983.

/8/ B. Rothfarb, H. Frank, D.M. Rosenbamm, K. Steiglitz, D.J. Kleitman, "Optimal design of offshore natural-gas pipeline systems", Operations Research 18, 1970.

COMPUTATIONAL COMPLEXITY OF SOME SEMIINFINITE PROGRAMMING METHODS

M.D. AŠIĆ
Faculty of Mathematics, Belgrade University
Studentski Trg 16, Belgrade, Yugoslavia
V.V. KOVAČEVIĆ-VUJČIĆ
Faculty of Organizational Sciences, Belgrade University
Jove Ilića 154, Belgrade, Yugoslavia

1. Introduction

The aim of this paper is to analyze computational complexity of a class of methods for solving semiinfinite programming problems:

$$\min_{x\in X} f(x) \quad , \quad X=\{x\in R^n \mid c(x,t)\leq 0, \ t\in C\} \tag{1}$$

where $f:R^n\to R$, $c:R^n\times C\to R$ and $C\subseteq R^r$ is a compact index set. Problem (1) is a hard problem and the majority of the existing methods are either based on discretization (which involves consideration of exponentially growing number of grid points) or require global maximization problems (over the index set C) to be solved in order to perform the feasibility check. There are also methods which combine these two approaches; see e.g. [3]-[9]. In [1],[2] we have developed a class of methods based on 'selective' discretization. Considerable numerical evidence indicates that these methods effectively use only a linearly growing number of grid points. In this paper we give theoretical grounds for such a behaviour.

2. Description of the methods

Throughout the paper we will assume that Problem (1) satisfies the following

A s s u m p t i o n 2.1. (i) The function f is continuous

(ii) The feasible set X is bounded

(iii) The function c satisfies the (uniform) Lipschitz condition $|c(x,t)-c(x,s)|\leq L\|t-s\|$ for some $L>0$ and for all $t,s\in C$, $x\in X$

(iv) The function c is convex with respect to x for each $t\in C$

(v) The Slater's condition is fulfilled, i.e. there exists $\bar{x}\in R^n$ such that $c(\bar{x},t)<0$ for all $t\in C$

(vi) The index set C is compact

For the description of Algorithm 2.1 we need the following definitions

D e f i n i t i o n 2.1. Let $y\in R^n$, $t\in C$ and $\beta\geq 0$ be such that $c(y,t)\leq -\beta$. A nonnegative number α is said to be a reach of the inequality $c(y,t)\leq -\beta$ if $c(y,s)\leq -\beta$ whenever $\|t-s\|\leq\alpha$ and $s\in C$.

It is clear that for a fixed function c the reach α depends on y, t and β; we shall thus use the notation $\alpha_c(y,t,\beta)$. There are many ways to choose a reach function α_c (see [1],[2]).

Definition 2.2. Let $\gamma>0$. A finite set $M\subseteq C$ is said to be a γ-net for C if for each $s\in C$ there exists a $t\in M$ such that $\|s-t\|<\gamma$.

The definition of Algorithm 2.1 involves a sequence of γ_j-nets M_j, where $\gamma_j=\gamma_0/2^j$ and $M_j\subseteq M_{j+1}$, $j=0,1,\ldots$. The cardinality of these nets grows exponentially, but the algorithm effectively uses only M_0 and a small portion of the remaining nets.

Algorithm 2.1. Input parameters: β_0, where $0<\beta_0\leq\min\{-c(\bar{x},t)\mid t\in C\}$ for some Slater's point $\bar{x}\in X$; $L>0$ satisfying Assumption 2.1 (iii); reach function α_c; rule for generating γ-nets.

Step 0. Set $\gamma_0=\beta_0/L$. Find a γ_0-net $M_0\subseteq C$. Set $C_0=M_0$ and $k=0$.

Step 1. Solve mathematical programming problem

$$\min_{x\in Y_k} f(x), \quad Y_k=\{x\in R^n \mid c(x,t)\leq-\beta_k,\ t\in C_k\}$$

and let y be a solution. Set $j=0$, $E_0=C_0$.

Step 2. If $j=k$ go to Step 4. Otherwise let E'_{j+1} be the set of points t in E_j such that $\alpha_c(y,t,\beta_k)<\gamma_j$ and let E_{j+1} be the set of points in γ_{j+1}-net M_{j+1} whose distance from the set E'_{j+1} does not exceed $3\gamma_j/2$. Replace j by j+1.

Step 3. If $c(y,t)\leq-\beta_k$ for $t\in E_j$ go to Step 2. Otherwise, let T be the set defined by $T=\{t\in E_j \mid c(y,t)>-\beta_k\}$, replace C_k by $C_k\cup T$ and go to Step 1.

Step 4. Set $x^k=y$, $C_{k+1}=C_k$, $\beta_{k+1}=\beta_k/2$, $\gamma_{k+1}=\gamma_k/2$, replace k by k+1 and go to Step 1.

Algorithm 2.1 describes a whole class of methods depending on the choice of α_c and $M_0,M_1,\ldots$. In Section 3 we shall analyze the application of the algorithm to Problem (1) in the case when the index set C is the Cartesian product of closed intervals $I_i=[p_i,q_i]$, $i=1,\ldots,r$. A simple rule for generating the sequence (M_j) is then the following : Let $h_i^0=(q_i-p_i)/m_i$, $h_i^j=h_i^0/2^j$, $i=1,\ldots,r$, where m_i are appropriately chosen positive integers; then for $j=0,1,\ldots$ we can take

$$M_j=\{(p_1+k_1h_1^j,\ldots,p_r+k_rh_r^j) \mid k_i\in\{0,\ldots,2^jm_i\}\ ,\ i=1,\ldots,r\}\ .$$

In this case it is convenient to define the reach function in the following way. Let y,t and β be such that $c(y,t)\leq-\beta$ and let $h>0$. Let q(s) be a function of the form

$$q(s)=c(y,t)+\beta+\bar{A}_1(s_1-t_1)+\ldots+\bar{A}_r(s_r-t_r)+\bar{B}_1|s_1-t_1|+\ldots+\bar{B}_r|s_r-t_r| \qquad (2)$$

where $\bar{B}_i$'s are nonnegative and such that $q(s)\geq c(y,s)$ for $\|s-t\|\leq h$. Let δ be the distance from t to the set $S=\{s\in C\mid q(s)=0\}$ ($\delta=+\infty$ if $S=\emptyset$). Then we may take $\alpha_c(y,t,\beta)=\min\{h,\delta\}$. It can be shown that the actual computation of the reach in the algorithm can be avoided. Namely, the test $\alpha_c(y,t,\beta_k)<\gamma_j$ in Step 2 can be replaced by checking whether the function q of the type (2) satisfying

$$q(s)\geq c(y,s)+\beta_k \quad \text{for} \quad |s_i-t_i|\leq h_i^j, \; i=1,\dots,r$$

is nonnegative at the extreme points of the set F defined by

$$F=([t_1-h_1^j,t_1+h_1^j]\times\dots\times[t_r-h_r^j,t_r+h_r^j])\cap C .$$

In [2] we have proved the following

Theorem 2.1. Suppose that Assumption 2.1 is satisfied. Let (x^k) be the sequence generated by Algorithm 2.1. Then $x^k\in X$, k=0,1,... and each cluster point of (x^k) is a solution to Problem (1).

Remark : The proof of Theorem 2.1 also implies that the test in Step 3 of the algorithm cannot be satisfied unless $c(y,t)\leq-\beta_k$ for all $t\in M_j$. This fact will be used in the proofs of Lemmas 3.1, 3.2, 3.4 and 3.5.

Under the additional assumption that x* is the unique solution of Problem (1) it can be easily shown that the sequence of auxiliary points y defined in Step 1 of the algorithm converges to x* as well.

3. Computational complexity

In this section we shall analyze the cardinality of the sets E_j generated in Steps 1 and 2 of Algorithm 2.1. Namely, large cardinality of E_j's would inevitably mean a large number of tests to be performed. The aim of this section is to prove that under some reasonably general assumptions the cardinality of E_j is uniformly bounded by a constant (independent on j and k). Throughout the section we shall assume that the following assumption holds:

Assumption 3.1. (i) $C=I_1\times I_2 \dots\times I_r$, where I_i, i=1,...,r are closed intervals

(ii) x* is the unique solution of Problem (1)

(iii) $c(x^*,t)=0$ for finitely many t's; let T^* be a complete list

(iv) c(x,t) is twice continuously differentiable with respect to t for $(x,t)\in U_0\times C$, where U_0 is a neighbourhood of x*.

(v) For each $t^*\in T^*$ the following property is satisfied: if t_i^* is an endpoint of the interval I_i then $\partial c/\partial t_i\neq 0$ at (x^*,t^*). Moreover, the Hessian matrix with respect to the remaining variables is negative definite at (x^*,t^*).

Let G be any open subset of C which contains T^*. It is then clear that for each $t\in C\setminus G$ and x sufficiently close to x* we have $c(x,t)\leq w$, where w is a negative constant. Without loss of generality assume that this holds for $x\in U_0$. Using the Lipschitz condition we conclude without difficulty that then the reach of the points in $M_j\setminus G$ is satisfactory for j sufficiently large. This enables us to 'localize' the reasoning to small neighbourhoods of points in T^*.

For the sake of simplicity we shall consider the two-dimensional index set C. It will become clear from what follows that the proofs for the general case are quite similar. To simplify the formulae we shall use the following notation :

$$c_1' = \partial c/\partial t_1,\ c_2' = \partial c/\partial t_2,\ c_{11}'' = \partial^2 c/\partial t_1^2,\ c_{12}'' = \partial^2 c/\partial t_1\partial t_2,\ c_{22}'' = \partial^2 c/\partial t_2^2.$$

Since C is two-dimensional, the points in T^* can be classified in the following three classes :

$T_1 = \{(t_1^*, t_2^*) \in T^* \mid t_1^*$ and t_2^* are endpoints of I_1 and $I_2\}$
$T_2 = \{(t_1^*, t_2^*) \in T^* \mid$ one of t_1^*, t_2^* is an endpoint and the other isn't$\}$
$T_3 = \{(t_1^*, t_2^*) \in T^* \mid t_1^*$ and t_2^* are interior points of I_1 and $I_2\}$

Assumption 3.1 then implies that there exists a neighbourhood $U \subseteq U_0$ of x^* such that for each $t^* \in T^*$ there is a neighbourhood V so that the following holds:

if $t^* \in T_1$ then $0 < A_1 \le |c_i'(y,t)| \le B_1$, $i=1,2$, $(y,t) \in U \times V$;
if $t^* \in T_2$, t_1^* is interior point of I_1, t_2^* is endpoint of I_2 then

$$0 < A_1 \le |c_2'(y,t)| \le B_1,\quad (y,t) \in U \times V$$
$$0 > -B_2 \ge c_{11}''(y,t) \ge -A_2,\quad (y,t) \in U \times V;$$

if $t^* \in T_3$ then

$$0 > -B_3(s_1^2+s_2^2) \ge c_{11}''(y,t)s_1^2 + 2c_{12}''(y,t)s_1s_2 + c_{22}''(y,t)s_2^2 \ge -A_3(s_1^2+s_2^2),$$

for each $s_1, s_2 \in R$, $(y,t) \in U \times V$.

Remark: Note that for $t^* \in T_2$ it follows that $c_1'(y,t^*)=0$, so we may without loss of generality assume that the neighbourhood V is so small that $|c_1'(y,t)| \le A_1 h_2^0/(3h_1^0)$. Also, if $t^* \in T_3$ it is clear from the above inequalities that for all $(y,t) \in U \times V$ we have $|c_{11}''(y,t)| \le A_3$, $|c_{22}''(y,t)| \le A_3$, $|2c_{12}''(y,t)| \le A_3$.

Let $y \in U$, $t^* \in T^*$ and take the corresponding neighbourhood V of t^*. We shall now analyze the reach of $c(y,\bar t) \le -\beta_k$, $\bar t \in V \cap M_j$, assuming that the reach is being computed using the function q defined by

$$q(t) = c(y,\bar t) + \beta_k + c_1'(y,\bar t)(t_1 - \bar t_1) + c_2'(y,\bar t)(t_2 - \bar t_2) + A|t_1 - \bar t_1| + B|t_2 - \bar t_2| \quad (3)$$

Such a function can be obtained from the second order Taylor expansion at $\bar t$. In that case $A \to 0$, $B \to 0$ as $j \to \infty$. Recall that the actual computation of the reach can be replaced by the test whether $q(t) \le 0$ on the corresponding rectangle. The next lemma shows that the reach of $c(y,\bar t) \le -\beta_k$ is satisfactory provided that $\bar t$ is sufficiently close to $t^* \in T_1$. Note that then $t^* \in M_0 \subseteq M_j$, $j=1,2,\ldots$.

Lemma 3.1. Let $t^* \in T_1$. Then for j sufficiently large and for each $\bar t \in V \cap M_j$ we have $q(t) \le 0$ for all t satisfying $|t_1 - \bar t_1| \le h_1^j$, $|t_2 - \bar t_2| \le h_2^j$, $t \in C$.

Proof: Let $\bar t \in V \cap M_j$ and pick a t in the corresponding rectangle. Using

Taylor expansion we have

$$c(y,\bar{t})\leq c(y,t^*)-A_1|\bar{t}_1-t_1^*|-A_1|\bar{t}_2-t_2^*| \qquad (4)$$

If $\bar{t}_1=t_1^*$, $\bar{t}_2=t_2^*$ from (3) we obtain

$$q(t)\leq(A-A_1)|t_1-t_1^*|+(B-A_1)|t_2-t_2^*|\leq 0,$$

since A and B tend to zero as $j\to\infty$. If $\bar{t}_1=t_1^*$, $\bar{t}_2\neq t_2^*$, from (3) and (4) it follows

$$q(t)\leq c(y,t^*)+\beta_k-A_1|\bar{t}_2-t_2^*|+(A-A_1)|t_1-t_1^*|+c_2'(y,\bar{t})(t_2-\bar{t}_2)+B|t_2-\bar{t}_2|.$$

Since $|\bar{t}_2-t_2^*|\geq 2h_2^j$ we have

$$q(t)\leq(A-A_1)|t_1-t_1^*|+(|c_2'(y,\bar{t})|-2A_1+B)h_2^j\leq(A-A_1)|t_1-t_1^*|+(B-A_1)h_2^j\leq 0$$

for j sufficiently large. The case $\bar{t}_1\neq t_1^*$, $\bar{t}_2=t_2^*$ can be treated analogously.

Finally, let $\bar{t}_1\neq t_1^*$, $\bar{t}_2\neq t_2^*$. From (3) and (4) we conclude that

$$q(t)\leq(|c_1'(y,\bar{t})|-2A_1+A)h_1^j+(|c_2'(y,\bar{t})|-2A_1+B)h_2^j\leq(A-A_1)h_1^j+(B-A_1)h_2^j\leq 0$$

for j sufficiently large.

We now turn to the case $t^*\in T_2$.

Lemma 3.2. Let $t^*\in T_2$, let t_2^* be an endpoint of I_2 and let $\bar{t}\in V\cap M_j$ be a point in the interior of $I_1\times I_2$. Then for j sufficiently large we have $q(t)\leq 0$ whenever t satisfies $|t_1-\bar{t}_1|\leq h_1^j$, $|t_2-\bar{t}_2|\leq h_2^j$.

Proof: Let $\hat{t}=(\bar{t}_1,t_2^*)$; clearly $\hat{t}\in M_j$ and thus $c(y,\hat{t})+\beta_k\leq 0$. Using similar arguments as in the proof of Lemma 3.1 we have

$$q(t)\leq c(y,\hat{t})+\beta_k-|\bar{t}_2-t_2^*|A_1+(|c_1'(y,\bar{t})|+A)|t_1-\bar{t}_1|+(A_1+B)|t_2-\bar{t}_2|\leq$$

$$\leq(B-A_1)h_2^j+(|c_1'(y,\bar{t})|+A)h_1^j\leq 0$$

for j sufficiently large, since $|c_1'(y,\bar{t})|h_1^j\leq A_1h_2^j/3$ and A and B tend to zero.

For the analysis of the remaining cases we need the following simple

Lemma 3.3. (i) Let $g:[a,b]\to R$ be such that $g(a)\leq 0, g(b)\leq 0$ and $g''(u)\geq -A_2$ for $a\leq u\leq b$. Then $g(u)\leq A_2(b-a)^2/8$ for $u\in[a,b]$.

(ii) Let $w:[a_1,b_1]\times[a_2,b_2]\to R$ be such that $w(a_1,a_2)\leq 0$, $w(a_1,b_2)\leq 0$, $w(b_1,a_2)\leq 0$, $w(b_1,b_2)\leq 0$ and

$$w_{11}''(u_1,u_2)s_1^2+2w_{12}''(u_1,u_2)s_1s_2+w_{22}''(u_1,u_2)s_2^2\geq -A_3(s_1^2+s_2^2)$$

for $s_1,s_2\in R$ and $(u_1,u_2)\in[a_1,b_1]\times[a_2,b_2]$. Then for $(u_1,u_2)\in[a_1,b_1]\times[a_2,b_2]$

$$w(u_1,u_2)\leq A_3((b_1-a_1)^2+(b_2-a_2)^2)/8.$$

The proof is a simple exercise in Analysis and is thus omitted.

Lemma 3.4. Let, as in Lemma 3.2, $t^*\in T_2$, where t_2^* is an endpoint of I_2 and let now $\bar{t}\in V\cap M_j$ be such that $\bar{t}=t_2^*$. Furthermore, let $\tilde{t}=(\tilde{t}_1,t_2^*)$ be the (unique) local maximum of $c(y,t_1,t_2^*)$ on V. If $|\bar{t}_1-\tilde{t}_1|$ satisfies

$$|\bar{t}_1-\tilde{t}_1|\geq (A_2+(A_2^2+A_2B_2)^{1/2})h_1^j/B_2$$

then for j sufficiently large $q(t)\leq 0$.

P r o o f : By assumption 3.1 we have $c''_{11}(y,t)\leq -B_2<0$ for $y\in U$, $t\in V$ and thus for j sufficiently large in (3) we will have $A=0$. Hence

$$q(t)\leq c(y,\bar{t})+B_k+c'_1(y,\bar{t})(t_1-\bar{t}_1)+c'_2(y,\bar{t})(t_2-\bar{t}_2)+B|t_2-\bar{t}_2|\leq$$

$$\leq c(y,\bar{t})+B_k+c'_1(y,\bar{t})(t_1-\bar{t}_1)+(B-A_1)|t_2-\bar{t}_2|\leq c(y,\bar{t})+B_k+c'_1(y,\bar{t})(t_1-\bar{t}_1)$$

if j is large enough. Using Taylor expansions of c and c'_1 at $\tilde{t}$ and estimate (i) of Lemma 3.3 we have

$$c(y,\bar{t})+B_k+c'_1(y,\bar{t})(t_1-\bar{t}_1)=$$

$$=c(y,\tilde{t})+B_k+(1/2)c''_{11}(y,\tau)(\bar{t}_1-\tilde{t}_1)^2+c''_{11}(y,\tilde{\tau})(t_1-\bar{t}_1)(\bar{t}_1-\tilde{t}_1)\leq$$

$$\leq A_2(h_1^j)^2/2-B_2(\bar{t}_1-\tilde{t}_1)^2/2+A_2|\bar{t}_1-\tilde{t}_1|h_1^j .$$

The last expression is nonpositive if

$$|\bar{t}_1-\tilde{t}_1|\geq(A_2+(A_2^2+A_2B_2)^{1/2})h_1^j/B_2 .$$

This Lemma implies that at most $1+(A_2+(A_2^2+A_2B_2)^{1/2})/B_2$ points can fail the test in Step 2 if j is large enough.

L e m m a 3.5. Let $t^*\in T_3$ and $\bar{t}\in V\cap M_j$. Furthermore, let $\tilde{t}$ be the unique local maximum of $c(y,t)$ on V. If $\|\bar{t}-\tilde{t}\|$ satisfies

$$\|\bar{t}-\tilde{t}\|\geq(3A_3+(9A_3^2+A_3B_3)^{1/2})((h_1^j)^2+(h_2^j)^2)^{1/2}/B_3 .$$

then for j sufficiently large $q(t)\leq 0$.

P r o o f: Using Assumption 3.1 we conclude, as in the proof of Lemma 3.4 that

$$q(t)\leq c(y,\bar{t})+B_k+c'_1(y,\bar{t})(t_1-\bar{t}_1)+c'_2(y,\bar{t})(t_2-\bar{t}_2)$$

Using Taylor expansions of c, c'_1 and c'_2 around $\tilde{t}$ and inequalities $h_i^j\leq((h_1^j)^2+(h_2^j)^2)^{1/2}$, $|\bar{t}_i-\tilde{t}_i|\leq\|\bar{t}-\tilde{t}\|$, i=1,2 and applying (ii) of Lemma 3.3 to the function $c(y,t)$ we obtain

$$q(t)\leq A_3((h_1^j)^2+(h_2^j)^2)/2-B_3\|\bar{t}-\tilde{t}\|^2/2+3A_3((h_1^j)^2+(h_2^j)^2)^{1/2}\|\bar{t}-\tilde{t}\| .$$

The last expression is nonpositive if $\|\bar{t}-\tilde{t}\|$ satisfies

$$\|\bar{t}-\tilde{t}\|\geq(3A_3+(9A_3^2+A_3B_3)^{1/2})((h_1^j)^2+(h_2^j)^2)^{1/2}/B_3 .$$

From Lemmas 3.1, 3.2, 3.4, 3.5 it follows that the number of points in E_j is bounded above by a constant for j sufficiently large, say $j\geq j_0$. Since for $j<j_0$ this number is also bounded, we have thus proved:

T h e o r e m 3.1. Let the Problem (1) have the unique solution and let Assumptions 2.1 and 3.1 be satisfied. If (E_j) is the sequence of sets generated in Steps 1 and 2 of Algorithm 2.1 then card(E_j) is bounded by a constant independent of j and k.

Finally, let us point out that this theorem alone does not imply that the number of points in C_k grows linearly. To obtain such a conclusion one also needs a good estimate for the number of cycles of the type Step 1 →Step 2 →Step 3 →Step 1. Under some mild assumptions we are able to prove the uniform (with respect to k) boundedness of that number as well. Our computational experience shows that such cycles seldom occur more than twice in any iteration.

4. Numerical Examples

In order to test the method described in Section 2 we have solved a number of test problems, mostly taken from [4] and [11]. The obtained results agree very well with the data in the literature; the details are given in [2]. Therefore, only the average cardinality of sets E_j and the number of cycles of the type Step 1 →Step 2 →Step 3 →Step 1 at the k-th iteration will be given here.

Example 4.1. [5]. Solve

$$\min x_1+\frac{1}{2}x_2+\frac{1}{2}x_3+\frac{1}{3}x_4+\frac{1}{4}x_5+\frac{1}{3}x_6$$

subject to

$$\exp((t_1)^2+(t_2)^2)-(x_1+x_2t_1+x_3t_2+x_4(t_1)^2+x_5t_1t_2+x_6(t_2)^2)\le 0,$$
$$(t_1,t_2)\in[0,1]\times[0,1].$$

Input parameters: $L=55$, $\beta_o=3.3$, $h_1^o=h_2^o=1/12$.

The results are summarized in the following table, where $N(C_k)$ and $N(E_j)$ stand for the cardinality of C_k and the average cardinality of E_j in the k-th iteration, respectively. The row headed 'cycle' gives the number of cycles Step 1 →Step 2 →Step 3 →Step 1 in the k-th iteration.

k	0	1	2	3	4	5	6	7	8	9	10	11	12	13
$N(C_k)$	169	179	184	189	194	199	204	209	214	219	224	229	234	239
$N(E_j)$	0	48	40	39	31	29	34	30	23	26	25	22	25	27
cycle	0	2	1	1	1	1	1	1	1	1	1	1	1	1

Obtained solution: (2.5805,-4.1093,-4.1093,4.2474,4.5327,4.2474),
Exact solution: (2.5801,-4.1093,-4.1093,4.2474,4.5327,4.2474).

Example 4.2 [4]. Solve

$$\min x_1+\frac{1}{2}x_2+\frac{1}{2}x_3+\frac{1}{2}x_4+\frac{1}{4}x_5+\frac{1}{3}x_6$$

subject to the same constraints as in the previous example. Input parameters are the same as in Example 4.1.

k	0	1	2	3	4	5	6	7	8	9	10	11	12	13
$N(C_k)$	169	173	175	177	179	181	183	185	187	189	191	193	195	197
$N(E_j)$	0	22	19	19	18	17	15	15	16	14	14	15	14	14
cycle	0	2	1	1	1	1	1	1	1	1	1	1	1	1

Obtained solution: (2.0166,0.7758,-4.2137,0.0000,3.3127,5.4977),
Exact solution: (2.0164,0.7758,-4.2137,0.0000,3.3127,5.4977).

The next three examples arise from Chebyshev approximation problems. In this case the corresponding semiinfinite programming problem has a special structure:

$$\min x_{m+1} \quad \text{subject to} \quad x_{m+1} \geq | x_1 v_1(t) + \ldots + x_m v_m(t) - g(t) | \quad , \; t \in C,$$

where g is the function to be approximated on the set C by a linear combination of the functions $v_1, \ldots, v_m$.

Example 4.3 [11]. Approximate $(t_1)^2 t_2$ by $v_1=1$, $v_2=t_1$, $v_3=(t_1)^2$, $v_4=t_2$, $v_5=(t_2)^2$, $v_6=t_1t_2$ on $(t_1)^2+(t_2)^2 \leq 1$.

Following the authors the problem is reduced to approximation problem on $[0,1] \times [0,2\pi]$. Input parameters are: $L=10.5$, $\beta_0=4.4$, $h_1^0=1/2$, $h_2^0=2\pi/11$.

k	0	1	2	3	4	5	6	7	8	9	10	11	12	13
$N(C_k)$	36	42	46	50	54	58	62	66	70	74	78	82	86	90
$N(E_j)$	86	64	61	60	53	47	47	47	41	45	46	41	42	41
cycle	1	1	1	1	1	1	1	1	1	1	1	1	1	1

Obtained solution: (0.0000,0.0000,0.0000,0.2500,0.0000,0.0000),
Exact solution: (0,0,0,1/4,0,0).

Example 4.4 [10]. Approximate $\exp(-(t_1)^2-t_2)$ by functions $v_1=1$, $v_2=t_1$, $v_3=t_2$, $v_4=2(t_1)^2-1$, $v_5=t_1t_2$, $v_6=2(t_2)^2-1$ on the set $[0,1] \times [0,1]$.
Input parameters: $L=24.2$, $\beta_0=2.5$, $h_1^0=h_2^0=1/7$.

k	0	1	2	3	4	5	6	7	8	9	10	11	12	13
$N(C_k)$	64	70	76	79	83	85	88	93	97	100	103	106	108	111
$N(E_j)$	0	46	44	40	40	40	42	39	36	37	35	36	34	36
cycle	0	2	2	1	1	2	1	1	1	2	1	2	1	2

Obtained solution: (0.9858,-0.3480,-0.9027,-0.1446,0.4246,0.1129),
Solution in [10] : (0.9858,-0.3480,-0.9027,-0.1446,0.4246,0.1129).

Example 4.5 [11]. Approximate t^2 by functions $v_1=t$, $v_2=\exp(t)$ on the interval $[0,2]$.
Input parameters: $L=13$, $\beta_0=2$, $h_1=1/8$.

k	0	1	2	3	4	5	6	7	8	9	10	11	12	13
$N(C_k)$	9	11	12	13	14	15	16	17	18	19	20	21	22	23
$N(E_j)$	0	7	7	6	6	6	6	6	6	6	6	6	5	5
cycle	0	2	1	1	1	1	1	1	1	1	1	1	1	1

Obtained solution: (0.1842,0.4186),
Exact solution: (0.1842,0.4186).

The last example is a nonlinear semiinfinite programming problem:

Example 4.6 [4]. Solve

$$\min (x_1)^2+(x_2)^2+(x_3)^2$$

subject to

$$x_1(t_1+(t_2)^2+1)+x_2(t_1t_2-(t_2)^2)+x_3(t_1t_2+(t_2)^2+t_2)+1\leq 0 ,$$

$$(t_1,t_2)\in[0,1]\times[0,1] .$$

Input parameters: L=19, β_o=1, $h_1^o=h_2^o$=1/14.

k	0	1	2	3	4	5	6	7	8	9	10	11	12	13
$N(C_k)$	225	225	225	225	225	225	225	225	225	225	225	225	225	225
$N(E_j)$	0	0	0	0	0	0	0	0	0	0	0	0	0	0
cycle	0	0	0	0	0	0	0	0	0	0	0	0	0	0

In this example T^* contains a single point (0,0) which is contained in C_o. Due to that fact the cardinality of E_j is always 0.

Obtained solution: (-1.0001,0.0000,0.0000),

Exact solution :(-1,0,0).

References

[1] Ašić,M.D.;Kovačević-Vujčić,V.V.,An application of semiinfinite programming to approximation theory, Proc. Symp. SYM-OP-IS'84, Herceg Novi, 1984.

[2] Ašić,M.D.;Kovačević-Vujčić,V.V.,An interior semiinfinite programming method, submitted for publication.

[3] Watson,G.A.,Globally convergent methods for semiinfinite programming, BIT 21, 1981.

[4] Watson,G.A.,Numerical experiments with globally convergent methods for semiinfinite programming problems, Semi-Infinite Programming and Applications, Edited by A.V.Fiacco and K.O.Kortanek, Springer-Verlag, 1983.

[5] Coope,I.D.;Watson,G.A.,A projected Lagrangian algorithm for semi-infinite programming, Mathematical Programming 32, 1985.

[6] Polak,E.;Tits,A.L.,A recursive quadratic programming algorithm for semi-infinite optimization problems, Applied Mathematics and Optimization 8, 1982.

[7] Gustafson,S.A.,A three-phase algorithm for semi-infinite programs, Semi-Infinite Programming and Applications, Edited by A.V.Fiacco and K.O.Kortanek, Springer-Verlag, 1983.

[8] Hettich,R.;Zencke,P.,Numerische Methoden der Approximation und semi-infiniten Optimierung, B.G.Teubner, Stuttgart, 1982.

[9] Hettich,R.,A review of numerical methods for semi-infinite optimization, Semi-Infinite Programming and Applications, Edited by A.V. Fiacco and K.O.Kortanek, Springer-Verlag, 1983.

[10] Glashoff,K.;Gustafson,S.A.,Linear optimization and approximation, Applied Mathematical Sciences 45, Springer-Verlag, 1983.

[11] Andreassen,D.D.;Watson,G.A.,Linear Chebyshev Approximation without Chebyshev sets, BIT 16, 1976.

THE CALCULATION OF MENPOWER FOR AIRCRAFT ·GROUND SERVICE

BABIĆ OBRAD, University of Belgrade, Yugoslavia

ABSTRACT - This paper considers the activity of loading and unloading cargo and passenger luggage to/from aircraft in order to determine the minimum necessary number of workers and the number and beginning time of shifts needed to serve a set of aircraft which request loading and unloading within a specific time period. Constraints which appear include that loading/unloading has to start at a predetermined time, has to be finished at a specific time, etc. The paper offers one possible solution to the above-mentioned problem. The proposed problem solution is based on dynamic programming and monitoring the work of each worker during the shift. The proposed problem solution is illustrated by a numerical example.

1. INTRODUCTION

An airport apron is the location where ground handling activities take place. These activities include passenger, luggage, cargo and aircraft handling. This paper considers luggage and cargo handling, namely the activity of unloading and loading aircraft. This activity is performed by a group of workers trained for this job. The number of workers in a group depends on the quantity of cargo and luggage to be unloaded and/or loaded, as well as on the time interval available for execution of this activity under the constraint that there is no delay induced by this activity nor unnecessary prolonged occupation of given a parking position. The workers can usually have flexible working hours. The beginning of shifts for a particular group of workers, as well as the number of workers in the group can be decided based on the planned scheduled demand for servicing (airline schedule) for a given day. The total number of workers engaged and the number of shifts influence the economics of airport operations. This influence can be figured through the direct labor cost (wages) as well as the cost of transporting of workers to and from the airport (transport being provided by the airport). Minimizing these costs,i.e. the total number of workers engaged and number of shifts under the condition that no schedule delay is induced, is the subject of this paper. The problem of calculating the minimum necessary number of workers and assigning these workers to particular shifts when aircraft ground handling is considered has been studied by many authors such as Stern and Hersh (2) and Babić and Janković (5). In the first paper,Stern and Hersh (2) consider the determination of the minimum number of workers to clean an aircraft while the Babić and Janković (5) paper considers the problem of calculating the necessary number of workers to load and unload an aircraft. The differences between the model proposed in this paper for determining the minimum necessary number of workers and their assignmemt to particular shifts when loading and unloading aircraft and models proposed in the above-mentioned papers are as follows:

1. In the Stern and Hersh (2) paper the number of workers needed to clean an aircraft was obtained under the assumption that the number of shifts and their starting times were knowh in advance i.e. were given as inputs while in the model constructed in this paper the number of the shifts and their starting times were obtained during calculation of the optimum solution,

2. In the Babić and Janković (5) paper, the number of workers in the group which service a particular aircraft are given as input data while in the model constructed in this paper the number of workers in the group is considered as a variable i.e. is obtained during calculation of the optimum solution.

2. PROBLEM DEFINITION

Let $X = \{ x_1, x_2, \ldots, x_N \}$ be the set of N aircraft requesting servicing during time period T (e.g. during one day). The requests appear at times $t_i \in T$. Each request (aircraft) is serviced by a group of workers. The number of workers in a group is variable, i.e. ranging between s_{min} and s_{max} (minimum and maximum number of workers). Let k be the number of different worker groups which can be formed, each consisting of s_j, $j=1,2,\ldots,k$ workers, where $s_{min} \leq s_j \leq s_{max}$. Each aircraft $x_i \in X$ can be serviced by one of the k different groups with s_j, $j=1,2,\ldots,k$ workers. This results in the fact that the number of possible assignments of k worker groups to N aircraft is k^N. Figure 1 schematically presents the possible assignment of worker groups to aircraft. Each assignment q_1 can be associated with cost TR_1 which consists of two parts. The first part represents the total number of workers engaged which is necessary to serve the set of aircraft $x_i \in X$. The second part represents the number of shifts in which the total necessary number of workers engaged can be assigned. Thus, the optimum solution to the above-mentioned problem will be an assignment with an associated minimum total necessary number of workers engaged which also has the minimum number of shifts to which this minimum number of workers can be assigned. This optimum assignment is found under the condition that there is no delay inducted by this activity to any of the serviced aircraft.

The objective function of the above-described problem can be defined in the following way:

$$TR = \min_{1} \left\{ \overbrace{\sum_{i=1}^{N} g_i \, (s_{i,q_1} - n_i)}^{I} + \overbrace{\sum_{i=1}^{N} p_i \, \bar{s}_{i,q_1}}^{II} + \overbrace{\sum_{i=1}^{N} r_i}^{III} \right\} \qquad \ldots\ldots\ (1)$$

under the condition that $n_1=0$ and that all workers are trained to perform the same duties.

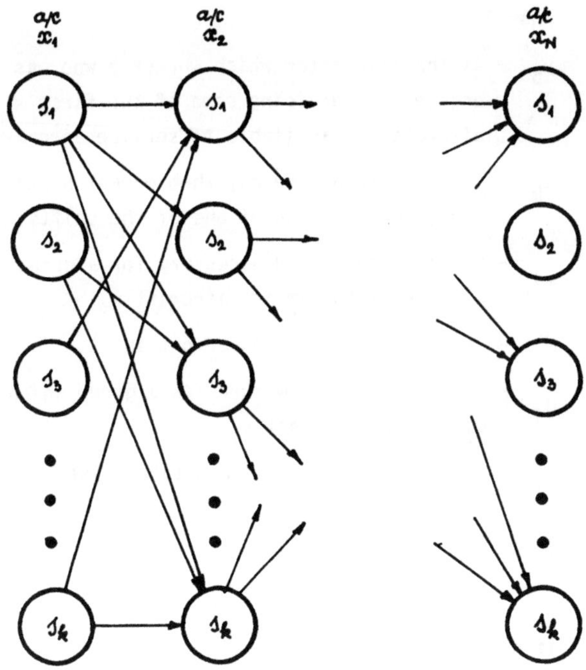

Figure 1. Schematic presentation of possible assignment

Expression (1) contains three parts. Part I represents the cost of engaging $(s_{i,q_1}-n_i)$ workers where:

- n_i is the number of workers which have already been engaged in servicing the first x_m, m= 1,2,....,i-1 aircraft, that can be assigned in the group to service aircraft x_i, i.e. workers which satisfy realtions (2) and (3) :

$$y_b \leqslant t_i \quad \ldots\ldots \quad (2)$$

$$t_i - z_b + a_{i,j} \leqslant M \quad \ldots\ldots \quad (3)$$

where:

y_b - is the time after which worker b who was previously engaged in servicing some of the first x_m,m=1,2,...,i-1 aircraft, is available to service aircraft x_i,

z_b - is the time of the day when worker b starts his working day (the beginning of one of the shifts),

$a_{i,j}$ - is the time period necessary for a group consisting of s_j workers to service aircraft x_i,

M - is the daily duration of one shift,

s_{i,q_1} - is the number of workers in a group which in assignment q_1 service aircraft x_i,

g_i - binary variable which can take values:

$$g_i = \begin{cases} 1 & \text{if } n_i < s_{i,q_1} \\ 0 & \text{if } n_i \geqslant s_{i,q_1} \text{ or } n_i = 0 \end{cases} \qquad \text{(4)}$$

Expression (2) represents the fact that worker b is available, i.e. no longer engaged in sevicing some of the first x_m,m=1,2,....,i-1 aircraft at time t_i when aircraft x_i is requesting servicing.

Expression (3) means that if worker b is engaged in servicing aircraft x_i at time t_i he will be able to finish servicing aircraft x_i before his working day elapses.

Part II of expression (1) represent the cost of engaging additional workers to service aircraft x_i when n_i = 0. Having this in mind, the binary variable p_i is defined in the following way:

$$p_i = \begin{cases} 1 & \text{if } n_i = 0 \\ 0 & \text{if } n_i \neq 0 \end{cases} \qquad \text{(5)}$$

Part III of expression (1) represents the cost of transporting workers to and from the airport, i.e. the number of shifts. The binary variable r_i is defined as follows:

$$r_i = \begin{cases} 1 & \text{if } n_i < s_{i,q_1} \text{ or } n_i = 0 \\ 0 & \text{if } n_i \geqslant s_{i,q1} \end{cases} \qquad \text{(6)}$$

The above - defined problem has a bicriterial objective function where the two criteria (the number of workers and the number of shifts) are not of the same importance. In multiattribute decision making and multiobjective programming problems, the relative importance of criteria can be defined using exact weights or descriptively using order of criteria importance (Hwang and Yoon, 3). This paper considers the problem such that the relative importance of the criteria can be defined using order of criteria importance. Namely, it is assumed here that the primary criterion is to minimize the total number of workers necessary to service a given set of aircraft, while the assignment of that minimum number of workers to a minimum number of shifts is considered as a secondary criterion. This criteria ranking is assumed due to the significant difference in the cost associated with the labor engaged and the cost of worker transportation to and from the airport.

In the above defined problem the constraints to be satisfied are the following:

- the total daily working hours for any worker can not be longer than the regulated maximum (M hours),
- servicing of aircraft $x_i \in X$ has to start at time t_i and has to be completed in time, i.e. without causing delay of that aircraft.

3. THE PROPOSED SOLUTION

The described problem of workers' assignment to service a set of aircraft is combinatorial by nature. The set of feasible solutions consists of k^N assignment of worker groups to the set of N aircraft these groups consisting of $s_j, j = 1,2,\ldots,k$ workers each. Assignments are represented schematically in Figure 2 using a tree.
Each of the nodes of the tree $y_{i,j}$ represent the fact that aircraft $x_i \in X$ is served by a group consisting of $s_j, j = 1,2,\ldots,k$ workers. Node $y_{o,o}$ represents the initial state of the system, i.e. at the time before the first aircraft request for servicing is delivered. Branches $(y_{i-1,g}\,;\,y_{i,j})$, $i = 2,3,\ldots, N$; $(g,j) = 1,2,\ldots,k$ which connect nodes at the adjacent level represent possible transitions from level i-1 to level i. The lengths of these branches represent the cost of additional labor necessary for a transition to the adjecent level (from i-1 to i). Let us denote the length of this branch with $R_i\,(y_{i-1,g};\,y_{i,j})$. Using the notation introduced above the length of the branch can be determined as:

$$R_i\,(y_{i-1,g};y_{i,j}) = \begin{cases} (s_j - n_i) + r_i & \text{if } n_i < s_j \\ s_j + r_i & \text{if } n_i = 0 \\ 0 & \text{if } n_i \geqslant s_j \end{cases} \qquad \ldots\ldots\ldots\ldots\ldots\ldots\ldots(7)$$

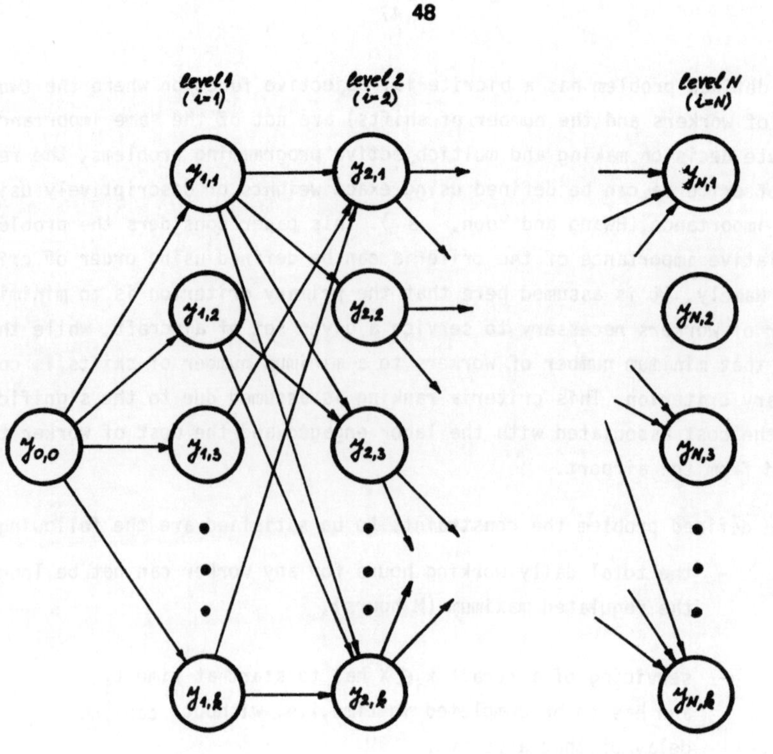

Figure 2. Schematic presentation of feasible solutions

Each path π_l, $l = 1,2, \ldots, k^N$ starting at node $y_{o,o}$ and ending at some node $y_{N,j}$, $j = 1,2,\ldots,k$ at level N represent a feasible solution. Each path π_l can be associated with cost TR^{π_l} which represents the number of necessary workers to service aircraft $x_i \in X$ as well as the number of necessary shifts. This cost is determined in the following way:

$$TR_{\pi_l} = \sum_{i=1}^{N} p_i \, s_{i,q_l} + \sum_{i=1}^{N} g_i \, (s_{i,q_l} - n_i) + \sum_{i=1}^{N} r_i \quad \ldots\ldots\ldots\ldots\ldots\ldots (8)$$

$$l = 1,2,\ldots,k^N$$

Among all feasible paths π_l, one (or more) paths π_l^* can be found along which cost

TR $\pi_1{}^*$ is at a minimum, i.e.

$$TR\ \pi_1{}^* = \min_1 TR\ \pi_1 \qquad 1 = 1,2,\ldots\ldots, k^N \quad \ldots\ldots(9)$$

The path satisfying relation (9) is the optimum solution for the problem defined by expressions (1) and (8). One of the possible ways to determine such a path is using dynamic programming. Dynamic programming equations corresponding to the problem considered here can be defined in the following way:

$$TR\ (0) = 0$$

$$TR_i\ (y_{i,j}) = \min\ (TR_{i-1}\ (y_{i-1,g}) + R_i\ (y_{i-1,g};\ y_{i,j}) \quad \ldots\ldots(10)$$

$$i = 1,2,\ \ldots\ldots, N \qquad (g,j) = 1,2,\ \ldots\ldots, K$$

where $TR_i\ (y_{i,j})$ is the cost of engaged labour along the shortest path betweens nodes $y_{o,o}$ and $y_{i,j}$. Expression (10) allows the path to be found which starts at the root of the tree and ends at level N,which satisfies expression (9). A computer program was developed to find such path using HPL language and an HP 9825A microcomputer.

4. NUMERICAL EXAMPLE

The numerical example presented here is based on actual data from the Belgrade Airport. One day during the summer period of 1984 is considered. The set of aircraft requesting servicing (unloading and loading) consists of N= 59 aircraft. Each of the requests can be serviced by a group consisting of a minimum of 3 and a maximum of 7 workers ($s_{min} = 3$, $s_{max} = 7$). It follows that the number of worker group types (each type having defferent number of workers, 3 to 7) is k=5 i.e. $s_1 = 3$, $s_2 = 4$, $s_3 = 5$, $s_4 = 6$, $s_5 = 7$. The data concerning times t_i when the servicing of aircraft x_i should start and time period $a_{i,j}$ which is necessary for a group consisting of s_j workers to service aircraft x_1, is given in table 1.

The optimum solution when daily working hours are M=8 for above presented input data is given in Table 2.

From Table 2 it can be seen that value of the objective function for the optimum solution based on the first criterion (necessary number of workers) is 35 workers, and based on the second criterion (number of shifts) is 11 shifts. The characteristics of the optimum solution obtained as an output of this model can also be seen in table 2. These characteristics consist of the following:

- The starting time for each shift
- Number of workers in each shift
- Number of workers in the group servicing each x_i aircraft
- Number of workers on duty at time t_i
- Cumulative number of workers and shifts until time t_i
- Cumulative number of workers whose working hours elapsed at time t_i.

5. CONCLUSIONS

This paper proposed a model which can be used to determine the minimum number of workers necessary to service given set of aircraft, and assignment of those workers to a minimum number of shifts. The model is very effecient which is illustrated by its short computing time (for the given numerical example this time was less than one minute) even using a microcomputer.

Slight modifications of the model presented in this paper can be used for modelling and optimizing other activities from the aircraft ground handling process (e.g. cabin cleaning etc).

The use of a microcomputer contributes to the fact that this model is easily applicable and thus attractive for practical purposes in airport operations and management. This means that the empirical manner of determining the necessary number of workers and the assignment of these workers to aircraft can be replaced with calculations of the optimum solution in an ecaxt manner using a microcomputer, entailing rather small funding.

Aircraft number	time t_i (h,min)	$a_{i,j}$: aircraft servicing time- min				
		number of workers in a group: s_j				
		3	4	5	6	7
1	05,00	20	15			
2	05,00	20	15			
3	05,15	20	15	10		
4	05,30	20	15	10		
5	05,50	20	15	10		
6	06,00	20	15			
7	06,50	25	20	15		
8	06,50	20	15	10		
9	07,10	25	20	15	10	
10	07,20	40	35	30	20	15
11	07,35			15	10	
12	07,50			20	15	
13	08,00		20	15	10	
14	08,05	20	15			
15	08,05	30	25	20	15	
16	08,20	35	30	25	20	
17	08,25		25	20	15	
18	08,30	25	20	15	10	
19	08,30	20	15	10		
20	08,50	30	25	20		
21	08,55		25	20	15	
22	09,20		30	25	20	
23	10,30			100	80	70
24	10,45	30	25	20	15	
25	11,00			80	65	55
26	14,15			25	20	15
27	14,25			20	15	
28	14,30			20	15	
29	14,40			25	20	15
30	14,45		25	20	15	
31	14,50	30	25	20	15	
32	14,50	30	25	20	15	
33	17,20	30	25	20	15	
34	17,30		20	15		
35	17,50			20	15	
36	18,10			75	65	55
37	18,30			20	15	
38	18,40			15	10	
39	18,50			20	15	
40	18,55		15	10		
41	19,20	40	35	30	20	15
42	19,20		30	25	20	15
43	19,50	30	25	20	15	
44	19,00				15	
45	20,05		15	10		
46	20,20		20	15		
47	20,45	35	30	25	20	15
48	20,50	20	15	10		
49	21,00		15	10		
50	22,05	30	25	20	15	
51	22,15		15	10		
52	22,30	35	30	25	15	
53	22,40		20	15	10	
54	22,45	25	20	15		
55	23,00	20	15	10		
56	23,15		20	15	10	
57	23,20	30	25	20	15	
58	23,35	20	15			
59	23,45	35	30	25	20	15

Ta bele 1. Input data

Aircraft number	Time t_1 (h/min)	Number of workers servicing aircraft x_1	Cumulative number of workers engaged until time t_1	Number of workers in a shift	Cumulative number of workers with elapsed working hours until time t_1	Number of workers on duty at time t_1	Cumulative number of shifts until time t_1
1	05,00	3	3	3	0	3	1
2	05,00	4	7	4	0	7	2
3	05,15	4	7	0	0	7	2
4	05,30	5	7	0	0	7	2
5	05,50	5	7	0	0	7	2
6	06,00	4	7	0	0	7	2
7	o6,50	3	7	0	0	7	2
8	06,50	4	7	0	0	7	2
9	07,10	6	7	0	0	7	2
10	07,20	7	7	0	0	7	2
11	07,35	6	7	0	0	7	2
12	07,50	5	7	0	0	7	2
13	08,00	4	9	2	0	9	3
14	08,05	3	10	1	0	10	4
15	08,05	3	11	1	0	11	5
16	08,20	5	11	0	0	11	5
17	08,25	6	11	0	0	11	5
18	08,30	5	11	0	0	11	5
19	08,30	5	11	0	0	11	5
20	08,50	5	11	0	0	11	5
21	08,55	6	11	0	0	11	5
22	09,20	7	11	0	0	11	5
23	10,30	5	11	0	0	11	5
24	10,45	6	11	0	0	11	5
25	11,00	7	11	0	0	11	5
26	14,15	5	12	1	7	5	6
27	14,25	5	17	5	7	10	7
28	14,30	5	22	5	7	15	8
29	14,40	5	22	0	7	15	8
30	14,45	6	22	0	7	15	8
31	14,50	6	22	0	7	15	8
32	14,50	6	22	0	7	15	8
33	17,20	6	22	0	11	11	8
34	17,30	4	26	4	11	15	9
35	17,50	6	26	0	11	15	9
36	18,10	7	26	0	11	15	9
37	18,30	6	26	0	11	15	9
38	18,40	6	26	0	11	15	9
39	18,50	6	26	0	11	15	9
40	18,55	5	26	0	11	15	9
41	19,20	7	26	0	11	15	9
42	19,20	7	26	0	11	15	9
43	19,50	5	26	0	11	15	9
44	19,50	5	26	0	11	15	9
45	20,05	4	26	0	11	15	9
46	20,20	4	26	0	11	15	9
47	20,45	7	26	0	11	15	9
48	20,50	5	26	0	11	15	9
49	21,00	5	26	0	11	15	9
50	22,05	6	26	0	11	15	9
51	22,15	5	26	0	11	15	9
52	22,30	6	26	0	11	15	9
53	22,40	6	26	0	11	15	9
54	22,45	5	26	0	11	15	9
55	23,00	5	26	0	11	15	9
56	23,15	6	26	0	11	15	9
57	23,20	6	32	6	12	10	10
58	23,35	4	32	0	22	10	10
59	23,45	3	35	3	22	13	11

Tabele 2. The optimum solution

R e f e r e n c e s:

1. Bellman R.E., and S.E. Dreyfus, Applied Dynamic Programming, Princeton University Press, 1962

2. Stern H.I., and M.Hersh,"Scheduling Aircraft Cleaning Crews", Transportation Science, Vol.14, pp 277-291, 1980

3. Hwang C.L., and K.Yoon, Multiple Attribute Decision Making, Springer - Verlag, Berlin, 1981

4. Taha H.A., Operations Research, Macmillan Publishing Co, INC, New York, 1982

5. Babić O., i S.Janković, "Odredjivanje potrebnog broja pretovarnih radnika za opslugu aviona, Zbornik radova sa XI Simpozijuma o operacionim istraživanjima SYM-OP-IS' 84, Herceg Novi, pp. 73-81, 1984.

   Babić O., and S.Janković, "The Calculation of Menpower for Aircraft Ground Service", Proceeding SYM-OP-IS' 84 (Symposium on Operations Research), Herceg Novi, Yugoslavia, pp. 73-81, 1984.

DISCONTINUOUS ADAPTIVE CONTROL OF NON MINIMUM PHASE LINEAR PLANTS

G. BARTOLINI - Dipartimento di Informatica, Sistemistica e Telematica - Università di Genova, Via Opera Pia, 11/A
16145 Genova - Italy
T. ZOLEZZI - Istituto Matematico - Università di Genova
Via L.B. Alberti, 4
16132 Genova - Italy

Abstract

An adaptive control scheme, effective when the controlled plant, supposed linear, has a transfer function with non hurwitzian numerator is presented. The solution to the problems involved is foreseen by means of the variable structure system theory and of the equivalence between Filippov's solution concept with a.e. solution of ordinary differential equation which, as it is well known, has been proved even for a class of systems wider than that of linear plants dealt with in this work.

0. Introduction

The model reference approach to the adaptive control of an unknown linear plant has given rise to control schemes whose global stability has been proved either when the system states are accessible /1/ or when only outputs are measurable /2,3,4/.

In the last work a great reduction in the complexity of the control scheme has been obtained, mainly avoiding the so called Augmented Error signal. If some a priori knowledge of bounds for system's parameters is available, the introduction of control laws that are discontinuous on suitably chosen manifold in the error state space /5,6/ further simplifies the control scheme, especially because it eliminates the integral parameter adaptation mechanism and the necessity of satisfying any real positivity condition, which was peculiarity of every previous scheme.

Nevertheless all the theory so far developed suffers from the limitation of requiring minimum phase plants. In this work a modification of the scheme in /4/ and /6/ is presented which is effective even when the plant zeros have positive real parts. If the plant parameters are known it is easy to build up an explicit model following control scheme which does not require non minimum phase zeros cancellation. The first section is devoted to the presentation of such a scheme. When plants are partially unknown it is impossible to satisfy the algebric equations required for the synthesis of the control law and an error equation arise in which only accessible states or bounded signals are involved. The second section deals with the deduction of the state error equation. Section three is devoted to find the conditions which assure the equivalence of the proposed adaptive control scheme with the case dealt with in the first section. The theoretical framework in which such research is carried out is Filippov's solution concept /7/ and the "so called" equivalent control method introduced in /8/ and recently extended by the authors in /9/ and /10/.

1. A model following control scheme for non minimum phase linear plant

It is well known, since the origin of control theory /11/ that models for plants with non minimum phase zeros must contain the same singularity in the numerator of their transfer function. On the other hand such a limitation may alter the behaviour of the model if transient phenomena are taken into account (overshoot, risetime, ...).

A model is chosen of the form

$$W_m(s) = \frac{B^+(s)}{B^+(-s)}\,\frac{P(s)}{Q(s)} = \frac{Y_m(s)}{R(s)} \tag{1}$$

in which $B(s)$ is the factor of the plant numerator whose roots are all in the positive halfplane, $P(s)/Q(s)$ is the transfer function of a "good model of reduced order and $R(s)$ is the Laplace transform of an input signal $r(t)$ L . If the Parseval's rule is applied to the integral square value of the output

$$\int_0^{\infty} y_m^2(\tau)d\tau = \frac{1}{2\pi j}\int_{c-j\infty}^{c+j\infty} Y_m(s)Y_m(-s)ds = \frac{1}{2\pi j}\int_{c-j\infty}^{c+j\infty} W_m(s)W_m(-s)R(s)R(-s)ds =$$

$$= \frac{1}{2\pi j}\int_{c-j\infty}^{c+j\infty} \frac{P(s)}{Q(s)}\,\frac{P(-s)}{Q(-s)}\,R(s)\,R(-s)ds$$

which coincides with $\int_0^{\infty} y^{*2}(\tau)d\tau$ being $y^*(t) = \mathcal{L}^{-1}\{[P(s)/Q(s)]R(s)\}$

The output of the model (1) and of the reduced order one are characterized by the same integral square value (if $r(t) \in L^{\infty}$ the previous conditions hold for the mean quadratic integral value.

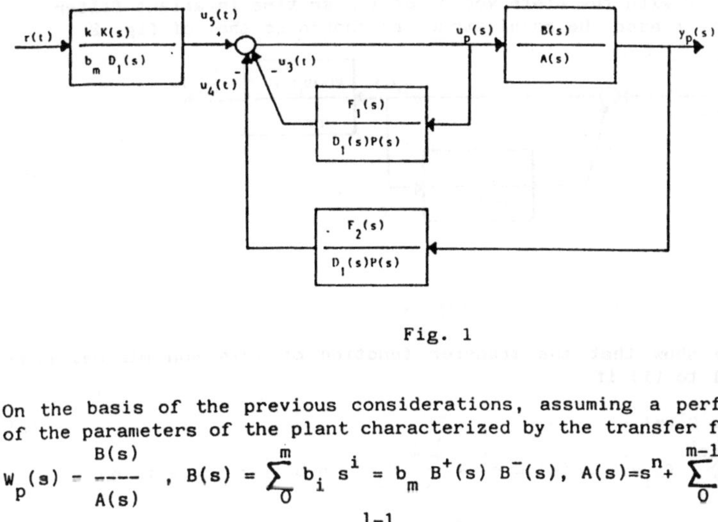

Fig. 1

On the basis of the previous considerations, assuming a perfect knowledge of the parameters of the plant characterized by the transfer function

$$W_p(s) = \frac{B(s)}{A(s)}, \quad B(s) = \sum_0^m b_i s^i = b_m B^+(s)\,B^-(s), \quad A(s)=s^n+\sum_0^{m-1} a_i s^i \tag{2}$$

l is the degree of $B^+(s) = s^l + \sum_0^{l-1} b_i^+ s^i$, the model parameters are

$$P(s) = K\left(s^l + \sum_0^{l-1} p_i s^i\right) \qquad l \le m, \quad Q(s) = s^n + \sum_0^{n-1} q_i s^i$$

It is easy to show that the control scheme shown in fig. 1 may satisfy

the control objective, i.e. y_p is related to r(t) by a transfer function identical to (1), if the regulators parameters are chosen in accordance to the following assumption:
k(s) is an arbitrary monic Hurwitzian polynomial $k(s) = s^{n-1-m} + \sum^{n-2-m} k_i s^i$
$D_1(s)$ is an arbitrary monic Hurwitzian polynomial

$$D_1(s)=s^r + \sum_0^{r-1} d_1 i\, s^i \quad r=n-1-1$$

and

$$F_1(s) = \sum_0^{n-2} f_{1i} s^i, \quad F_2(s) = \sum_0^{n-1} f_{2i} s^i$$

are the solutions to the Bezoutian equation

$$A(s)(F\ (s)+D_1(s)P(s))+B(s)F_2(s) = k(s)B^+(-s)B^-(s)Q(s) \qquad (3)$$

Remark 1.
Equation (3) does not admit an easy solution in terms of the coefficients a_i and b_i of the plant transfer function as in the case of minimum phase plant presented in /4/ and /6/. In this case Bezoutian equation AX+BY = BQ has the solution Y = Q-A X=B.

Remark 2.
The scheme of fig. 1 is characterized by a number of filters non minimal with respect to the number needed to satisfy the transfer function equation involved. This fact is due to the need, in the adaptive case, of representing the ideal control laws as the scalar product of a constant parameter vector with the state vector of linear time invariant filter.
For the same reason the model structure chosen is that of fig. 2

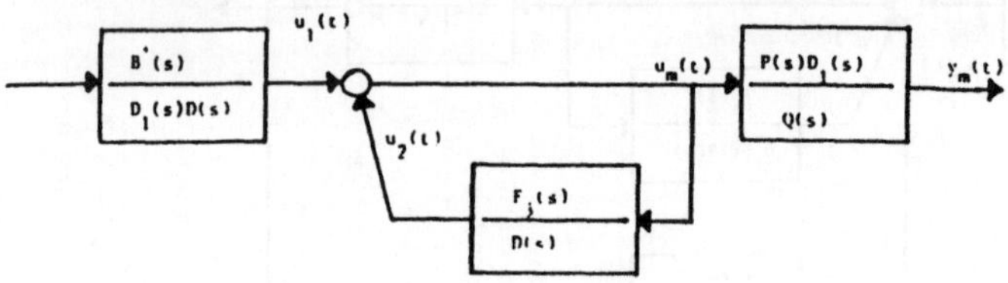

Fig. 2

it is easy to show that the transfer function of this non minimal order model is equal to (1) if

$$F_3(s) + D(s) = B^+(s) \qquad (4)$$

with $F_3(s) = \sum_0^{l-1} f_{3i} s^i$; $D(s) = s^l + \sum_0^{l+1} d_i s^i$, arbitrary hurwitzian.

2. Error Equation in the Adaptive case

When the plant parameters are unknown but bounded by known numbers i.e.

$$a_{im} \leq a_i \leq a_{iM} \quad i=0,\ldots,n-1; \ \leq b_{im} \leq b_i \leq b_{iM} \quad i=0,\ldots,m \qquad (5)$$

The equation involved in the determination of $u_m(t)$ and $u_p(t)$ cannot be solved.

If we indicate with $u_m^*(t)$ and $u_p^*(t)$ the ideal control laws, the problem arises of describing the evolution of the output error $y_m(t)-y_p(t)$ when the plant and the model are subjected to control law $u_m(t)$ and $u_p(t)$ different from the ideal ones in the following way:
From fig. 1 and 2

$$u_m(t) = u_1(t) - u_2(t), \qquad u_p(t) = u_5(t) - u_3(t) - u_4(t)$$

u_m and u_p may be written

$$u_m(t) = \theta_m x_m(t), \qquad \theta_m = |\theta_1 \; \theta_2| \; x_m^T = |x_1^T \; x_2^T| \tag{6}$$

θ_1 and θ_2 are respectively the vectors of coefficient of polynomial $B^+(s)$ and $F_3(s)$, x_1, x_2 are the state vectors of filters in companion controllable form with characteristic polynomials respectively equal to $D_1(s)D(s)$ and $D(s)$ and subjected respectively to $r(t)$ and $u_m(t)$ inputs

$$u_p(t) = \theta_p \; x_p(t) \tag{7}$$

$$\theta_p = |\theta_5 \; \theta_3 \; \theta_4| \quad x_p(t)^T = |x_5^T \; x_3^T \; x_4^T|$$

θ_5, θ_3, θ_4 are vectors of the coefficients of polynomials $K(s)$, $F_1(s)$, $F_2(s)$ respectively, x_5, x_3, x_4 are the state vectors of filters in companion controllable form with characteristic polynomials equal to $D_1(s)$, $D_1(s)\;P(s)$, $D_1(s)P(s)$ and subjected to $r(t)$, $u_p(t)$ and $y_p(t)$ inputs respectively.

If θ_m^* and θ_p^* are the vectors of the parameters in the case of known plant

$$u_m^*(t) = \theta_m^* x_m(t); \qquad u_p^*(t) = \theta_p^* x_p(t) \tag{8}$$

any control law in the adaptive case may be written as

$$u_m(t) = \theta_m x_m(t); \qquad u_p(t) = \theta_p x_p(t) \tag{9}$$

therefore

$$\begin{aligned} u_m(t) &= \theta_m^* \; x_m(t) + (\theta_m-\theta_m^*)x_m(t) = u_m^* + \mu_1(t) \\ u_p(t) &= \theta_p^* \; x_p(t) + (\theta_p-\theta_p^*)x_p(t) = u_p^* + \mu_2(t) \end{aligned} \tag{10}$$

from (10) it can be concluded that the control structure in the adaptive case is equivalent to the control in the known plant case if disturbances $\mu_1(t)$, $\mu_2(t)$ act respectively at the input of the reduced order model and at the plant input.

It is easy to show that the following equation holds for $\varepsilon(t) = y_m(t)-y_p(t)$

$$E(s)=k \frac{P(s)D_1(s)D(s)}{Q(s)B^+(-s)} M_1(s)+b_m \frac{D_1(s)P(s)B^+(s)}{K(s)B^+(-s)Q(s)} M_2(s) \tag{11}$$

E, M_1, M_2 are the Laplace transform of $\varepsilon, \mu_1, \mu_2$.

$$\frac{Q(s)B^+(-s)}{P(s)D_1(s)D(s)} E(s) = KM_1(s) + \frac{b_m B^+(s)}{D(s)K(s)} M_2(s)$$

If

$$E_F(s) = \frac{1}{P(s)D_1(s)D(s)} E(s) = \mathcal{L}\{\varepsilon_F(t)\}$$

$E_F(s)$ is the output of a state variable filter with accessible states whose eigenvalues are the roots of polynomial PD_1D having $\varepsilon(t)$ as input.

We obtain

$$Q(s)B^+(-s)E_F(s) = kM_1(s) + b_m \frac{B^+(s)}{D(s)K(s)} M_2(s) \tag{12}$$

moreover if

$$p = \deg B^+(s) + \deg Q(s)$$

let

$$B^+(-s)Q(s) = s^p + \sum_0^{p-1} a_i s^i$$

and

$$\underline{e}^T = \varepsilon_F, s\varepsilon_F, \ldots, s^{p-1}\varepsilon_F = e_1, e_2, \ldots, e_p \qquad s = d/dt$$

than (12) may be written:

$$\dot{\underline{e}} = \begin{vmatrix} 0 & 1 & 0 & . & . & . & 0 \\ 0 & 0 & 1 & . & . & . & 0 \\ & & & \vdots & & & \\ -a_0 & -a_1 & . & . & . & & -a_{p-1} \end{vmatrix} \underline{e} + K \begin{vmatrix} 0 \\ \\ \\ \mu_1(t) \end{vmatrix} + b_m \begin{vmatrix} 0 \\ . \\ \\ \eta \end{vmatrix} \tag{13}$$

being η the output of a filter of transfer function $B(s)/D(s)K(s)$ and input $\mu_2(t)$.

Equation (13) describe a system in canonical controllable form with unknown but bounded coefficients a_i with input $\mu_1(t)$ and subjected to the disturbance $\eta(t)$.

In this case it is easy to apply the V.S.S. theory /7,8/.

Chosen

$$s(e) = e_p + \sum_1^{p-1} \gamma_i e_i \tag{14}$$

on the manifold $s(e) = 0$ the system (13) behaves like a linear autonomous system whose eigenvalues are the roots of the polynomial

$(\lambda^n + \sum_1^{p-1} \gamma_i \lambda^i)$ which is stable if γ_i are suitably chosen.

If the condition

$s(e)\,\dot{s}(e) < 0$ (15)

globally holds everywhere in the error state space except on $s(e) = 0$ then the state trajectories are directed toward $s(e) = 0$ and a sliding mode may appear.

Condition (15) may be written

$$s(\dot{e}_p + \sum_{1}^{p-1} \gamma_i \dot{e}_i) < 0 \quad \text{from (13)}$$

$$s(-\sum_{0}^{p-1} \hat{a}_1 e_{i+1} + K\mu_1(t) + b_m\eta(t) + \sum_{0}^{p-1}\gamma_i e_{i+1}) < 0$$

therefore

$$\text{sign}\,(-\sum_{0}^{p-1} \hat{a}_i e_{i+1} + K\mu_1(t) + b_m\eta(t) + \sum_{0}^{p-1}\gamma_i e_{i+1}) = -\text{sign } s \tag{16}$$

Condition (16) can be satisfied if the following assumptions are true

a) a time function $\delta(t)$ is known which constitutes an upper bound of $|\eta(t, \mu_2, .)|$

b) bounds on $\hat{a}_i$ are known $|\hat{a}_i| < \alpha_i$ $i = 1, .., p-1$ and it is possible to generate a $\mu_1'(t)$ discontinuous on $s(e) = 0$

$$\mu_1'(t) = -\frac{1}{K}\left[b_m\delta(t) + \sum_{0}^{p-1} \alpha_i |e_i| + \sum_{0}^{p-1} \gamma_i |e_{i+1}| \right] \text{ sign } s \tag{17}$$

$\mu_1(t)$ is not directly available because it is the difference between the true control signal $u_m(t)$ and the ideal one $u_m^*(t)$. The following inequality must be satisfied

$$|u_m(t)| > |\mu_1'(t)| + |u_m^*(t)| \tag{18}$$

an upper bound on $u_m^*(t)$ is therefore needed.

From (6)

$$u_m^*(t) = \theta_1^* x_1 - \theta_2^* x_2$$

being the output of a butterworth filter, the following equation holds for the parameter vector θ_1^*, θ_2^*

$$\theta_{1j}^* = (-1)^j(\theta_{2j}^* + d_j) \tag{19}$$

$$u_m^*(t) = \sum_{1}^{l-1} \theta_{1j}^*(x_{2j} + (-1)^j x_{1j}) + \sum_{1}^{l-1} d_j \; x_{2j}$$

if numbers $|\theta_{1j}^M| < |\theta_{ij}|$ $j=1,...,l-1$ are known a bound on $|u_m^*(t)|$ is given by:

$$|w_1(t)| = \sum_{1}^{l-1} |\theta_{ij}^M (x_{2j} + (-1)^j x_{1j}| \sum_{1}^{l-1} (d_j \; x_{2j}) \tag{20}$$

upper bounds θ_{1j}^M of $\theta_{1j}^* = b_j$ are obtained if an upper bound on maximum modulus of the roots of $B^+(s)$ is known by applying the following lemma

Lemma 1

Given $y(t) = x_n(t) + \sum_1^{n-1} c_i x_i(t)$, if a number λ is known such that:

$\lambda > \lambda_i(C(s))$ i=1;..n

being $\lambda_i(.)$ a root of the polynomial $C(\dot{s}) = s^n + \sum_1^{n-1} c_i s^i$ then

$$|y(t)| \leq |x_n(t)| + \sum_1^{n-1} \binom{n}{n-1} \lambda^{n-i} |x_i(t)|$$

The proof is trivial and is omitted.

Now let us choose a control law

$$u_m(t) = \{\mu_1(t) + w_1(t)\} \text{ sign } s \tag{21}$$

$w_1(t)$ sign s, if Filippov's solution concept /7/is taken into account, belongs to a convex set of output of butterworth filters.

The above considerations may be synthetized in the following statement:

Theorem 1

Given the error state equation (13) if

i) a time function $\delta(t) > |\eta(t, u_p, ..)|$ is given

ii) bounds on coefficient $\hat{a}_i$ i=1..p and on maximum modulus of the roots of $B^+(s)$ are known.

It is possible to generate a control law $u_m(t)$ discontinuous on s(e)=0 satisfying the following conditions simultaneously:

a) $\forall t$ in $(0,\infty)$ $s(e)\,\dot{s}(e) < 0$

b) it belongs a.e. to the convex set of the output of Butterworth filters to which the desired control law $u_m^*(t)$ also belongs.

3. The determination of the plant control law

If we write the plant equation in the general form

$\dot{x}_p = f_p(t, x_p, u_p)$

where $f_p(.)$ is linear in x_p, u_p, and we choose a control

$$u_p(t) \triangleq |u_p(t)| \text{sign } s \tag{22}$$

discontinuous on s(e)=0 then from lemma /3/ of (7):

$$\dot{x}_p \in f_p(t, x_p\ (2\alpha - 1)u_p) \quad 0 \leq \alpha \leq 1 \tag{23}$$

if

$$|u_p| > |u_p^*| \tag{24}$$

the desired control belongs to set (22).

If a bound on modulus of u is known (24) is a necessary condition for the equivalence, in the sense of Filippov, of discontinuous control with the desired one.

From (7) we have

$u_p^*(t) = \theta_p^* x_p(t)$, where θ_p^* is a solutions of Bezoutian equation (3): upper bounds of the modulus of the components of such vector are therefore necessary. To this purpose we can prove the following.

Theorem 2

If assumption (5) holds and an upper bound of the modulus of the roots of B(s) is known the further assumption of knowing a number

$$\Delta \leq \min_{i,j} \left| \lambda_i(A) - \lambda_j(B) \right| \quad i=1,\ldots,n \quad , \quad j=1,\ldots,m$$

with $\lambda_k(P)$ denoting the kth root of a polynomial P(s), is a sufficient condition for the determination of a discontinuous control law (22) satisfying (24).

Proof:

Let us consider the polynomial $F_1^*(s) + D_1(s)P(s)$ in equation (3) which is the sum of the unknown polynomials

F_1^* with the known one D_1P if $F_1^*(s) + D_1(s)P(s) = G(s) = s^{n-1} + \sum_{i=0}^{n-2} g_i s^i$

If bounds on $|g_i|$ are known, bounds of coefficients of $F_1^*(s)$ are also known.

If the right hand side polynomial of (3) is written:

$$L(s) = k(s)\, B^+(-s)\, B^-(s)Q(s) = s^{2n-1} + \sum_{i=0}^{2n-2} l_i\, s^i$$

an upper bound of $|l_i|$ is known if the second assumption of the theorem holds.

Now, applying the identity of polynomials to (3) it is easy to show that the following algebraic system holds:

$$\Big|\underbrace{C}_{n-1}\ \ \underbrace{D}_{n}\Big| \begin{vmatrix} \underline{g} \\ \cdots \\ \underline{f} \end{vmatrix} \begin{matrix} \}\, n-1 \\ \\ \}\, n \end{matrix} + \begin{vmatrix} \underline{a} \\ \cdots \\ \underline{0} \end{vmatrix} \begin{matrix} \}\, n \\ \\ \}\, n-1 \end{matrix} = \Big| \underline{l} \Big| \}\, 2n-1 \qquad (26)$$

being g f a l vectors whose components are the coefficients of the polynomials G(s), $F_2(s)$, A(s), l(s) and

$$C = \left|\begin{array}{ccccc} 1 & 0 & . & . & 0 \\ a_{n-1} & 1 & 0 & . & 0 \\ a_{n-2} & a_{n-1} & 1 & & \\ & & & & \\ & & & & 1 \\ a_1 & & & & a_{n-1} \\ a_0 & & & & a_{n-2} \\ & . & & & \\ & . & & & \\ & . & & & \\ 0 & 0 & . & . & 0 \;\; 0 \end{array}\right| \begin{array}{l} \left.\begin{array}{c} \\ \\ \\ \\ \\ \end{array}\right\} n \\ \left.\begin{array}{c} \\ \\ \\ \\ \\ \end{array}\right\} n-1 \end{array} \qquad D = \left|\begin{array}{ccccc} 0 & & & & 0 \\ . & & & & . \\ . & & & & . \\ 0 & & & & 0 \\ 1 & C & & & \\ b_{m-1} & 1 & 0 & . & 0 \\ b_{m-2} & b_{m-1} & 1 & . & 0 \\ . & . & & & \\ . & . & & & 1 \\ . & . & & & b_{m-1} \\ . & & & & . \\ b_0 & & & & . \\ . & b_0 & & & b_1 \\ . & & & & \\ 0 & . & 0 & & b_0 \end{array}\right| \begin{array}{l} \left.\begin{array}{c} \\ \\ \\ \end{array}\right\} n-m-1 \\ \left.\begin{array}{c} \\ \\ \\ \\ \\ \end{array}\right\} n \\ \left.\begin{array}{c} \\ \\ \\ \\ \\ \end{array}\right\} m \end{array}$$

in simpler notation, with obvious meaning of symbols, equation (26) is written as:

$$Tx = l-r,\ x = T^{-1}[l-r]\ ;\ T^{-1} = (1/Det(T))\,(ADJ(T))$$

an upper bound of the minors of T appearing as elements of ADJ(T) is easy to find if (5) holds, a lower bound of Det(T) is found if the third assumption holds because as it is well known /12/ Det(T) is the so called <u>resultant</u> of the Bezoutian equation and is written as

$$Det(T) = \prod_{ij} (\lambda_i(B(s)) - \lambda_j(A(s))) \qquad i=1,\dots,m\ ;j=1,\dots,n$$

$$Det(T) \geqslant \Delta^{nm}$$

This means that an upper bound of induced norm of T is known the following disequations hold

$$|x_i| = |T^{-1}(l-r)|_i \leqslant \{\|T^{-1}\|\,|l-r|\}_i \leqslant \{\|\hat{T}\|\,|l'| + |r'|\}_i$$

$$\|\hat{T}\| > \|T\|^{-1},\ |l_i'| > |l_i|\ \ |r_i'| > |r_i|$$

a plant control law

$$u_p(t) = \sum |\theta_i\ x_{pi}|\ \text{sign } s$$

with $\theta_i = \{\|\hat{T}\|\,|l| + |r|\}_i$ satisfies disequation (24) and the theorem is proved. The knowledge of bounds of control law's parameters is not sufficient to prove the equivalence of the discontinuous adaptive control scheme with the model following one. On the manifold s(e)=0 many solutions are possible some of which may be unstable.

Further theoretical work must be done to eliminate, from the set of possible solutions, all the parasitic solutions except the desired one.

References

/1/ I.D. Landau, Adaptive Control. The Model Reference Approach. M. Dekker, 1979.

/2/ R.V. Monopoli, Model Reference Adaptive Control with an Augmented Error Signal. IEEE Trans. on A.C. Oct. 1974.

/3/ K.S. Narendra, L. Valavani, Stable Adaptive Controller Design. IEEE Trans. on A.C. Aug. 1978.

/4/ C.C. Hang, and H.S. Tang, Model Reference Adaptive Control Using Only Input and Output Measurements. INT - J. Control. Vol. 34. n. 5, 1981.

/5/ K.K.D. Young, Design of variable structure model following control systems. IEEE Trans. on AC. 1978.

/6/ G. Ambrosino, G. Celentano, and F. Garofalo, Variable Structure Model Reference Adaptive Control Systems. INT. J. Control. Vol. 39, n. 6, 1984.

/7/ A.F. Filippov, Differential Equations with Discontinuous Right Hand Side. Ann. Math. Soc. Transl. 42, 1964.

/8/ U.J. Utkin, Sliding Modes and their Application in Variable Structure Systems. M.I.R. Published, Moscow, 1978.

/9/ G. Bartolini, T. Zolezzi, Control of non linear variable structure systems. To appear on J. Math. Anal. Appl., 1986.

/10/ G. Bartolini, T. Zolezzi, Variable structure systems non linear in the control law, IEEE Trans. on AC. July 1985.

/11/ J. Truxal, Automatic feedback control synthesis. J. Wiley, 1955.

/12/ Van der Werden, Modern Algebra, 1953.

OPTIMAL CONTROL OF THE STORAGE POWER PLANT SYSTEM GOSAU

W.Bauer, E.Lindner, Hj.Wacker

I. Introduction

This work has been done in contract with the Upper Austrian Power Plant Company OKA to which the storage plant system belongs.
The system consists of three power plants - Gosau, Gosauschmied and Steeg - and its reservoirs - Lake Gosau, reservoir Gosauschmied and reservoir Klaushof.
Figure 1 gives a schematical description of the system:

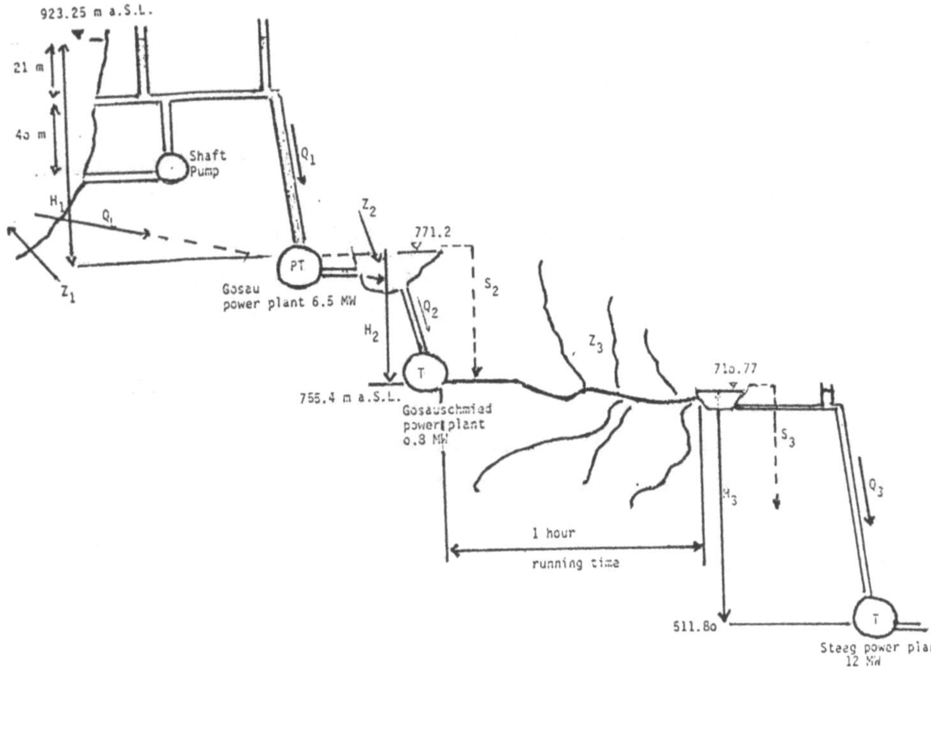

Notations:

Concerning to index i we have

i=1: storage power plant Gosau, i =2: storage power plant Gosauschmied, i=3: storage power plant Steeg.

$H_i(t)$ actual head of reservoir i: height difference between reservoir surface and the below water surface [m]

$h_i(t)$ level of reservoir i above sea level [m, a.S.L.]

$Q_i(t)$ discharges through the turbine(s) of plant i [m^3/s]

$Z_i(t)$ natural influx to reservoir i [m^3/s]

$S_i(t)$ spilling water at reservoir i [m^3/s] i = 2,3

$Q_L(h_1)$ seepage losses of Lake Gosau [m^3/s].

Hydrological description of the system

The highest point of the system is Lake Gosau which has a maximum water level of 923.25 m a.S.L.
Only a part of its content may be used directly while the remaining water (see level below 9o2.5 m) can only be used by activating a shuft pump. Having passed the pumping turbine of power station Gosau the water $Q_1(t)$ runs into the Gosauschmied reservoir just beside the Gosau plant. Alternatively water can be pumped from the Gosauschmied reservoir back into Lake Gosau. This implies that the actual level of reservoir Gosauschmied must be higher than the below water surface of the Gosau pumping turbine. The reservoir Lake Gosau is not thight, because it is situated in a karst region of the Austrian Alps. There are heavy unknown seepage losses $Q_L(h_1)$ depending on the level of the lake. We know, that these seepage losses are part of the influx to reservoir Gosauschmied. The water of the Gosauschmied reservoir runs through the turbine of power plant Gosauschmied $Q_2(t)$. Spilling water $S_2(t)$ has to be taken into account because of the small capacity of reservoir Gosauschmied. Spilling water and turbine-water runs as a natural brook down the Gosau valley and is collècted in the Klaushof reservoir which has a capacity much more smaller than that of reservoir Gosauschmied. Also for Klaushof we have to allow spilling water $S_3(t)$.

Finally the water, which flows through the turbines of power plant Steeg - $Q_3(t)$ runs into Lake Hallstatt.

Some operation conditions have to be observed for running the plants.

- by a supply treaty with the ÖBB (Austrian Railway Company) 1.5 MW are to be produced at Steeg from 6 a.m. to 1o p.m.
- to protect the landscape Lake Gosau must not fall below 915 m from June 15th till October 15th. In addition there is an agreement with the local tourist office to keep the level of the lake at about 921 m during July and August.

For a detailled description see [1].

II. Optimization Model

We want to maximize the "rated energy production" of the system Gosau in the time intervall [O,T] with respect to all restrictions, technical and other ones.

The "rated energy production" E(T) is a monetary value of dimension [Austrian Shillings]. E is a weighted energy production and gives the profit of producing energy.

We have

$$E(T) = c \int_0^T a(t)\, P(t)$$

with a(t) ... tariff rate (see Figure 2) [Austrian Shillings/KWh]
P(t) ... electrical power [W]
c ... constant (we will neglect this constant)

Figure 2: tariff rate

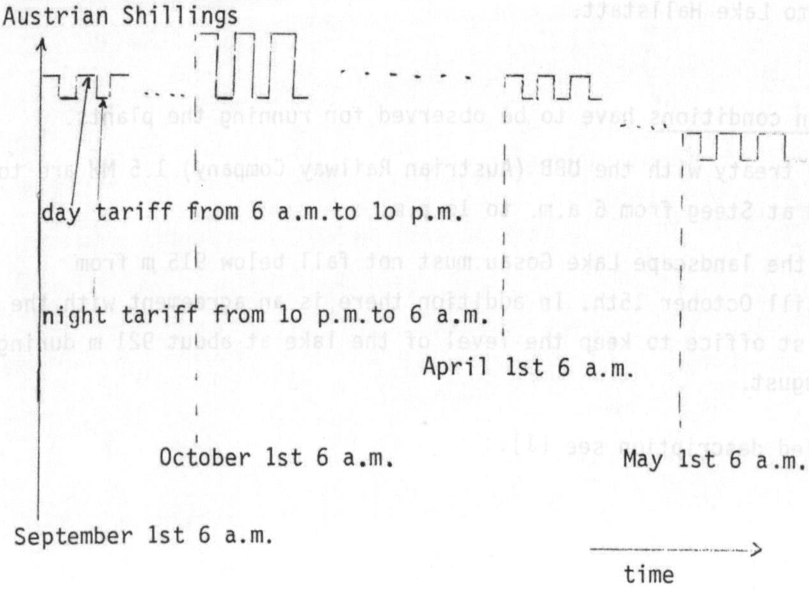

<u>Note:</u> The tariff rate changes four times a year and one time per day.
The optimal strategy will depend on the tariff rate.

2.1 <u>The Objective</u>

Neglecting the constant c we get the objective

$$E(T) = \int_0^T a(t) \sum_{i=1}^{3} P_i(t)\, dt$$

<u>Explanation:</u> (Index i omitted)

Assuming a discharge Q(t) and a head of H(t) we get the theoretical power $P_{Th}(t)$:

$$P_{Th}(t) = g.\rho.H(t).Q(t) \qquad [W]$$

with g: gravity constant
ρ: density of water

Note: In the special case of power plant Gosau P_{Th} depends on $h_1(t)-h_2(t)$ instead of $H_1(t)$ because of the pumping turbine (see Chapter I)

In reality $P_{Th}(t)$ is reduced by losses in the tunnel conduction system, losses in the turbine(s) and losses in the generator(s) and transformators. By multiplication with the efficiency η depending in general on H and Q we get the real output power P(t):

$$P(t) = P_{Th}(t) \cdot \eta(H(t),Q(t)) \quad [W]$$

Finally, to get the rated energy produced in the time interval [0,T] one has to integrate a(t).P(t).

We get a nonlinear nonconvex objective depending on $Q_i(t)$ and $H_i(t)$ resp. $h_i(t)$.

In reality the controlling time interval of storage power plant Gosau is one year, beginning at September 1st. That implies that the time intervall [0,T] is one year starting at September 1st. The controlling of storage plant Gosauschmied and Steeg is done within one day.

2.2 Continuity equations:

Discussing storage power plants the water balance in the reservoir has to be taken into account. This can be described by the continuity equation:

$$V(t) = V(o) + \int_0^t [Z(\tau)-Q(\tau)-S(\tau)]\, d\tau$$

where V(t) represents the water volume of the reservoir.

In detail we get:

Lake Gosau: $V_1(t) = V_1(o)+ \int_0^t [Z_1(\tau)-Q_1(\tau)-Q_L(h_1(\tau))]\, d\tau$

reservoir Gosauschmied: $V_2(t)=V_2(o)+\int_0^t [Z_2(\tau)+Q_1(\tau)+Q_L(h_1(\tau))-Q_2(\tau)-S_2(\tau)]d\tau$

reservoir Klaushof$^{(*)}$: $V_3(t)= V_3(o)+\int_0^t [Z_3(\tau)+Q_2(\tau)+S_2(\tau)-Q_3(\tau)-S_3(\tau)]\, d\tau$

(*) we neglect the running time from power plant Gosauschmied to reservoir Klaushof.

By help of the reservoir capacity function f_i we may express the height h_i resp. H_i in dependence on V_i.

Remark:

(i) Using V instead of H in the model the constraints become mostly linear.
(ii) h_i can be easily get from H_i and vice versa: $h_i = H_i + c_i$ (c_i = const.)

2.3 Restrictions

(a) Technical restrictions

(1) $V_{imin} \le V_i(t) \le V_{imax}$

$i \in \{1,2,3\}$

(2) $Q_{imin} \le Q_i(t) \le Q_{imax}$

(b) Further restrictions

(1) Supply treaty with the ÖBB:

$c.P_3(t) \ge 1.5$ [MW] $t \in$ [6 a.m, 10 p.m.]

(2) Tourist traffic
Between June 15th and October 15th the water level of Lake Gosau has to be above a prescribed level. Expressing this restriction by help of the volume V_1 we get

$V_1(t) \ge \tilde{V}_1(t)$ $t \in$ [June 15th, October 15th]

2.4 Boundary conditions

The water volume of V_{1end} is given by

$V_{1end} = \tilde{V}_1(t)$ t = Sept. 1st (~ 921 m)

The reservoirs of Gosauschmied and Steeg have a capacity of one day at most. We assume therefore

$$V_{iend} = V_{imax} \qquad i = 2,3.$$

Summing up we get a control problem with

- nonlinear nonconvex objective
- nonlinear state-equations
- linear and nonlinear restrictions of the control
- linear restrictions of the state
- control variable $Q_i(t)$
- state variable $V_i(t)$.

Looking at 2.1 and 2.2 we see that we need some pre-information.

3. Input Parameters

The model involves a lot of (partly) unknown input data:

- seepage losses of Lake Gosau
- efficiencies of the power plants
- influx $Z_i(t)$

(a) To determine the efficiency of a plant one has to take into account losses in the pipe system, losses in the turbine(s), losses in the generator(s), losses in the transformer(s) and, in our case, even losses by conduction. Based on measurements taken by the engineers of the OKA the efficiencies were determined using a newly develloped nonlinear least squares technique, see [5] .

(b) To get the seepage losses $Q_L(h_1)$ some thirty (and more) years had to be analysed. Two problems were to face. First it is not possible to measure the influx to Lake Gosau because it is subterraneous. Second only uncertain information was available on the way production was performed. Based on information concerning those time intervals when the plant had been inactive, an acceptable approximation of the function $Q_L(h_1)$ could be determined.

Additionally, information about some brooks which are situated down the valley was used to fix the three bends more exactly. Each time the level of Lake Gosau rises above those characteristic heights, the brooks start flowing. We will see that the optimization results depend on the shape of the curve of the seepage losses.

Figure 3: seepage losses of Lake Gosau

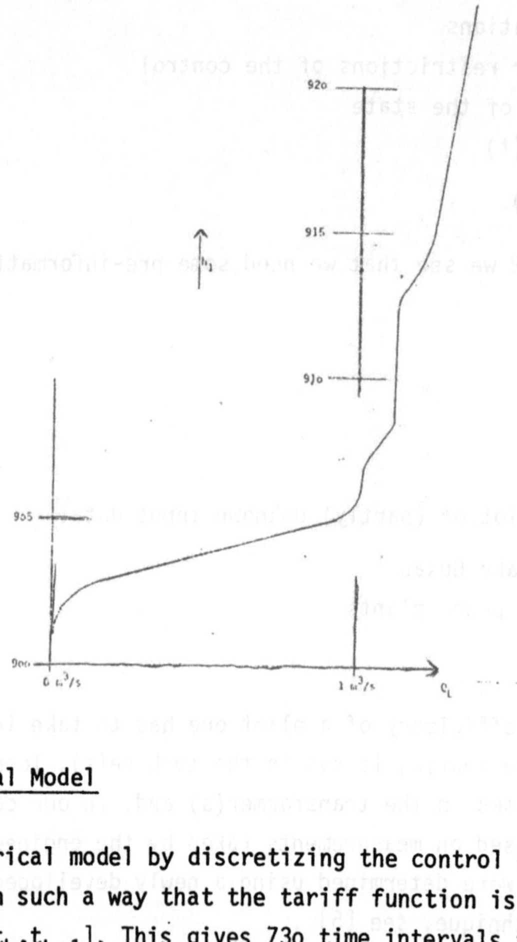

4. The Numerical Model

We get the numerical model by discretizing the control time interval [0,T] which is done in such a way that the tariff function is constant during each time interval $(t_k, t_{k+1}]$. This gives 73o time intervals for one year.

For each time interval $(t_k, t_{k+1}]$ we assume:

i) $Z_i(t)$ and $Q_i(t)$ are constant

ii) All functions depending on h_i (efficiencies, seepage losses) are determined at an average water level

$\bar{h}_i := \frac{1}{2}(h_i(t_k)+h_i(t_{k+1}))$

Constant discharges for a half day are welcomed by the engineer (no switching of the turbines!). To avoid low efficiencies resulting from small discharges it is allowed in those cases to work at the best point Q_i^* (for given h_i) for a time less than $\Delta t_k := t_{k+1}-t_k$.

We arrive at a nonlinear constrained optimization problem with 219o variables $Q_i(t_k) =: Q_{i_k}$ and 876o constraints. To reduce the numerical amount for this optimization model we meet a first simplification:

- we neglect the extremely small capacity of the reservoir Klaushof

The continuity equation for Klaushof is now substituted by

$$Q_3(t_k)+S_3(t_k) = Z_3(t_k)+Q_2(t_k)+S_2(t_k) \qquad k = o,1,\dots,N$$

and $Q_3(t_k)$ is given if we know $Q_2(t_k)$. There remain only two variable $Q_i(t)$ $i = 1,2$.

Because the resulting number of variables is still too high (146o) we tried to get a further reduction of the size of our problem. To this end we looked for the optimal control of storage plant Gosauschmied, $Q_2^*(t)$. This problem is solved in [3] [4].

Under some assumptions about the controlling of power plant Gosauschmied (avoiding spilling water in Gosauschmied and Klaushof) we can dispose of the variable $Q_2(t)$ setting the optimal $Q_2^*(t)$.

Now we get an one dimensional optimization problem with variable $Q_1(t)$ and $Q_2(t) \doteq Q_2^*(t)$ and an additional minimum problem

$$\int_o^T (S_2(t)+S_3(t))dt = \text{Min!} \qquad (*)$$

(This problem depends on $Q_1(t)$ and $Q_2(t)$!)

(*) means no spilling water if possible.

The resulting numerical optimization problem with 73o variables Q_{1_k} and 3286 restrictions is solved by classical Dynamic Programming.

The total amount of CPU-time was about 45 min (IBM 3o31)

For a detailed description of the numerical model see [2].

5. Results

We confine to present results only for storage power plant Gosau. Figure 4 gives schematically the optimal water level of Lake Gosau

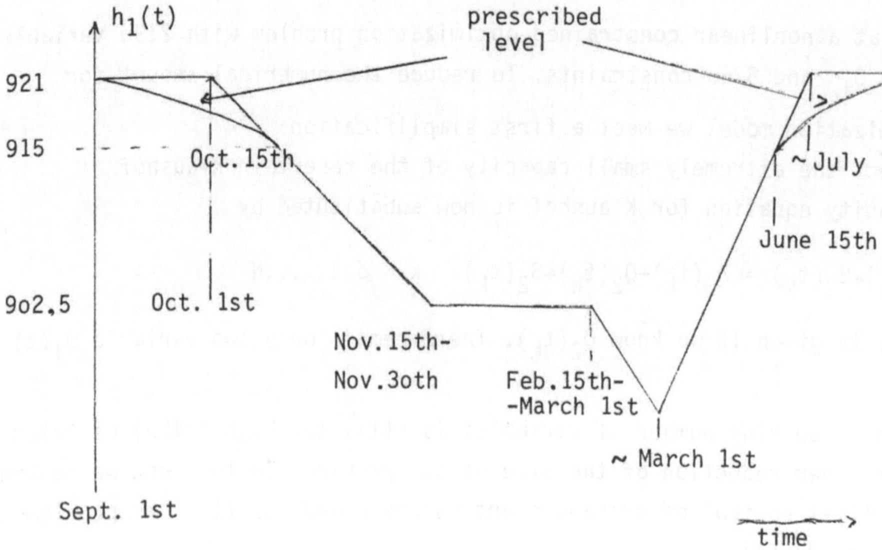

Discussion of the results:

The principles of optimal running storage power plant Gosau are the following:

(a) Main production is during October 1st and March 31st, because of the tariff rate. This period split into two parts:

Part (1): Sinking down Lake Gosau until the level 9o2.5 is reached (in the last half of November).
Production is mostly done during 6 a.m. and 1o p.m. when the high tariff is valid.

Part (2): For about 2 and a half months the level of Lake Gosau is held constant at 9o2.5 m. That means only the influx is used for production. There are two reasons for stopping the sinking down:

1. In general the level of 915 m is reached only by help of pumping activities on June 15th. Further sinking down of Lake Gosau during December and January would increase the amount of pumping water or even would make it impossible to reach 915 m on June 15th.
2. At the level of 9o2.5 m there is no need to activate the shaft pump for production and the seepage losses are almost zero.

Then the shaft pump is activated producing further winter energy until March 31st.

(b) After March 31st Lake Gosau is filled again partly by natural influx (snow melting) and pumping water. But there is no pumping activity during April, because of the tariff.
The level of 915 m is reached exactly on June 15th.

(c) The storaging of water goes on until the prescribed level of 921 m is reached.

(d) During summer (half of July, August and September) the level of Lake Gosau is held constant.

In reality the strategy of running the Gosau power plant was almost the same, but there are some exciting differences between the optimal and the usual strategy:

- sinking down was often started to early (at September 1st) or to late. The seelevel was above the prescribed level of 915 m on October 15th. The sinking down was stopping at a seelevel of about 9o5 m (where the seepage losses are still very high). Therefore, the optimization profit during this period is very high, about 25 %.
- usual the level reached on March 31st was above the optimal level. Therefore, the profit during the period-production by help of the shaft pump and storage of water until June 15th - is about 5-8 %, although the amount of pumping water is greater than it was in reality.
During July and August the strategy of running power plant Gosau is trivial, there are no differences between the optimal and the usual one.

We got this results comparing the data of the last 12 years with our optimization.

Acknowledgements:
The data were made available to us by the OKA Power Company. Thanks are given to the management of the OKA, Dir.Dipl.-Ing.W.Barwig and both to the engineers, Ing.Feichtinger and Ing.Peßl, for their substantial help which also includes laborious experiments. Our practical work has been supported by the Austrian Ministry of Science and Research and the OKA. Our theoretical work has been supported partly by the Austrian Fonds of Research (P4786).

References:

[1] W.Barwig/R.Peßl, Optimization of the Gosau System, in "Applied Optimization Techniques in Energy Problems (Proceedings), Teubner, pp. 94 - 118, 1985

[2] W.Bauer, H.Gfrerer, E.Lindner, A.Schwarz, Hj.Wacker, Optimization of the storage plant system Gosau-Gosauschmied-Steeg, in "Applied Optimization Techniques in Energy Problems (Proceedings), Teubner, 119-137, 1985

[3] W.Bauer, H.Reisinger, Hj.Wacker: Höhensteuerung eines Tagesspeicherkraftwerkes, ZOR/B, 26 (1982) pp.145-167

[4] H.Gfrerer, Optimization of hydro energy storage plant problems by variational methods, ZOR/B, Bd. 28, 87-1o1 (1984)

[5] E.Lindner/Hj.Wacker, Input Parameter for the Optimization of the System Partenstein: Forecasting of the Influx, Determination of the Efficiency Function, in: "Applied Optimization Techniques in Energy Problems (Proceedings), Teubner, 286-322, 1985

STOCHASTIC CONTROL IN URBAN TRAFFIC (+)

B. Betrò (*) F. Schoen (*) M.G. Speranza (**)

1. Introduction

A peculiarity of traffic flows in urban networks is the contemporaneous presence of variability due to large period events (e.g. rush hours) and variability due to short period events (e.g. platoons released by semaphorized intersections). For this reason, traffic flow models should have stochastic and dynamic nature.

In recent years, the availability of high-speed low-cost microprocessors has made it sensible the design of on-line control strategies of traffic lights, based upon the mentioned models, and making use of data collected by detectors counting vehicle passages.

Although the standard stochastic dynamic programming formulation gives, in principle, the optimal control strategy, its actual implementation in an on-line context is at present out of question. In this framework some suboptimal control strategies have been proposed in the literature (Baras and Dorsey, 1981; Camera et al., 1983; Gartner, 1983) and computer codes made available (Hunt et al., 1981; Hunt et al., 1982).

The aim of this paper is to propose a set of control strategies which possess the common feature of being simple, suitable for on-line implementation and of having a structure similar to that of the "μc-rule", which is the optimal policy in various situations of queueing theory. These strategies simply require a model which allows the estimation of some relevant quantities related to the departure and queue length processes. Stochastic dynamic models for a urban intersection as well as estimation formulas of the relevant quantities can be found, for instance, in Betrò et al. (1985b).

The effectiveness of the proposed control strategies is tested through simulation whose input data are taken from real observation of traffic flows at an intersection.

2. On-line control strategy definition

We consider the situation of a general single intersection regulated by a traffic light, such that on each street one or more detectors collect information about flows. We assume that time is discretized according to a certain time unit and indicate by y_t the set of information collected at time t. Moreover, we indicate by U_t the state of the traffic light at time t.

An on-line control strategy has the aim of determining the traffic light state in dependence of traffic flow conditions on the basis of the observations collected by the detectors, that is an on-line control strategy is a law which specifies, for each time instant t, the process $\{U_t\}$ so that:

$$U_t=U(t,y^t,U^{t-1})=U(t,y_1,\dots,y_t,U_1,\ .\ .\ .\ ,U_{t-1}). \qquad (2.1)$$

(+) This work has been partially supported by CNR Progetto Finalizzato Trasporti.

(*) CNR-IAMI, Milano, Italy

(**) Dip. di Matematica, Università di Milano, Milano, Italy.

Dependence of U_t upon previous traffic light states allows to take into account constraints on minimum green and maximum red duration, as well as the need for an amber phase.

In order to introduce the problem of selecting a traffic light control policy, it is necessary to define the geometry of the intersection under examination and the sets of flows which are simultaneously allowed to cross without conflicts the intersection.

The intersection can be described by an oriented graph $G=(V,E)$, where V is the set of vertices, with cardinality n, and E is the set of edges with cardinality m, in which the vertices represent the streets joining the intersection and the edges the allowed crossing flows. A certain number f of partial graphs $G_j=(V_j,E_j)$ can be associated to G so that each edge of the graph G belongs to at least one of the E_j's; each of these latter define a set of flows which may cross simultaneously the intersection and therefore a traffic light state, which will be referred to as phase, enabling the passage of these flows only. For example, in order to describe the intersection of figure 1,

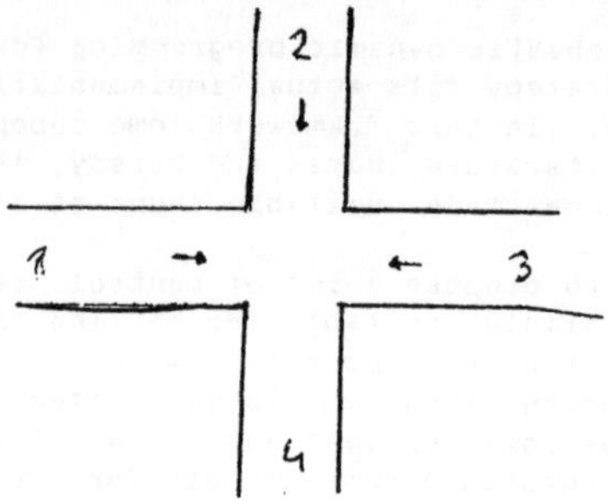

figure 1

the graph of the intersection geometry will present the following adjacency matrix:

	1	2	3	4
1	0	0	1	0
2	0	0	0	1
3	1	0	0	0
4	0	0	0	0

while the traffic light phases can be described through the partial graphs G_1, G_2 and G_3 with the following corresponding adjacency matrices:

$$A_1 = \begin{matrix} 0&0&1&0 \\ 0&0&0&0 \\ 1&0&0&0 \\ 0&0&0&0 \end{matrix} \quad A_2 = \begin{matrix} 0&0&0&0 \\ 0&0&0&1 \\ 0&0&0&0 \\ 0&0&0&0 \end{matrix} \quad A_3 = \begin{matrix} 0&0&0&0 \\ 0&0&0&0 \\ 0&0&0&0 \\ 0&0&0&0 \end{matrix} ,$$

where A_3 corresponds to the amber phase.

Let us indicate by F the set of adjacency matrices A_i corresponding to the different phases. Moreover, in order to deal with the control problem it is necessary to distinguish between the model parameters for the different streets joining the intersection. Therefore we introduce an index i, i=1,...,m, identifying the traffic flows, or equivalently, the edges of G according to a certain ordering.

3. Optimality of on-line control policies.

The aim of a traffic light control policy is to minimize the inconveniences suffered by the intersection "users". Different criteria have been introduced in the literature in order to provide measures of the degrees of effectiveness for a certain control policy (see e.g. Robertson, 1969; Hunt et al., 1981; Hunt et al., 1982; Gartner, 1983); among the others we recall the total fuel consumption due to stops and restarts, and the number of stops. However the most widely used index is the so called total aggregate delay, that is the total time spent by all vehicles in the queue during a certain time interval. Then the optimal traffic light control policy is defined as the policy which minimizes the expected aggregate delay, that is the solution of the following functional optimization problem:

$$\min_{U_1, U_2, \dots, U_T} E\{\sum_{i \in E} \sum_{t=0}^{T} \beta^t c^{(i)} Q_t^{(i)}\}, \qquad U_t(.) \in F, \tag{3.1}$$

where $c^{(i)}$ represents the cost due to the presence of a vehicle in the queue i and $\beta \in (0,1]$ is a discount factor.

It is clear from the problem statement itself that the solution of (3.1) is in general very cumbersome from a computational point of view (while theoretically the solution is given by the well known dynamic programming equations). Various attempts have therefore been made aimed at reducing such computational cost, generally through approximation techniques. It is nevertheless possible in some cases to obtain the optimal control policy in an explicit way; in particular, in the case of queues competing for a single server, Baras et al. (1985) and Buyukkok et al. (1985) recently proved that under wide assumptions the optimal server control policy has an extremely simple structure based on the (off-line) calculation of a parameter, called Gittins index (see e.g. Gittins, 1979). Defining as $Q_t^{(i)}$ the process describing the queue evolution on the street i and assuming that the queue is completely observed through the processes $A_t^{(i)}$ and $D_t^{(i)}$, describing respectively the arrivals at the queue i and the departures from the queue i, the result can be stated for the semaphorized intersection control problem as follows:

Theorem: Let the observation process be given by $Y_t = \langle A_t^{(i)}, D_t^{(i)}, \; i=1,\dots,m \rangle$. Under the following assumptions

- the departure process $D_t^{(i)}$ is Bernoullian with success probability $\mu^{(i)}$, that is satisfies the condition

$$p(d_t^{(i)} | q_{t-1}^{(i)}) = \begin{cases} 0 & \text{if } d_t^{(i)} > 1 \text{ or } q_{t-1}^{(i)} = 0 \\ {\mu^{(i)}}^{d_t^{(i)}} (1-\mu^{(i)})^{1-d_t^{(i)}} & \text{otherwise} \end{cases}$$

- at any instant the green phase is assigned to exactly one of the traffic flows;

the optimal control rule (i.e. the one which satisfies (3.1)), for any value of T and β, is the following (μc-rule):

at any instant t the control has to be assigned to flow i^* such that:

$$i^* = \arg \max_{i=1,\dots,m : Q_t^{(i)} > 0} \mu^{(i)} c^{(i)}. \tag{3.2}$$

The control policy (3.2) is interesting for several reasons. First of all, in queueing theory the μc-rule is the optimal policy for a broad class of problems (no assumption is made on the arrival process, the time horizon might be infinite, the hypothesis that departures are Bernoullian is acceptable under a fine enough time discretization); moreover the quantity which determines the choice of the flow to which the green has to be given is rather simple and can be computed off-line: the only on-line operation is to check which is the non empty queue to whom the largest of the indices "μc" corresponds. On the other hand, the applicability of the above theorem to concrete traffic light control situations is restricted by the hypotheses themselves, which rule out the case of partially observed queues, the case of multiple departures in a time unit, and the case in which the green can be simultaneously assigned to more than one flow. Furthermore no constraints on phase switching and duration are assumed, so that the control policy may alter the traffic light state instantaneously, switching from phase to phase without the necessary amber period and without considering that drivers do not like very short green as well as very long red phases.

It is therefore interesting to develop suboptimal policies able to keep more closely into account the traffic light control peculiarities, while keeping the computational cost at an acceptable level.

4. Suboptimal on-line control policies

In this section some policies are presented which are based upon the principle of assigning the control to the traffic light phase j^* which gives the maximum value to an index; this index, which is characteristic of each policy, can be computed independently for each flow.

Policy n. 1

$$j^* = \arg \max_{j=1,\dots,f} \sum_{i \in E_j} \mu^{(i)} c^{(i)} P\{Q_t^{(i)} > 0 \mid \text{observations}\} \tag{4.1}$$

where $\mu^{(i)} = E\{D_t^{(i)} \mid Q_{t-1} = \infty\}$ is the "virtual average departure number", i.e. the expected number of departures in the case that the traffic light is green on lane i and the queue is such that any number of departures is possible. This policy reduces to the μc-rule if the departures are 0-1 and the queue is completely observed and the cardinality of each E_j is 1.

Policy n. 2

$$j^* = \arg \max_{j=1,\dots,f} \sum_{i \in E_j} \mu^{(i)} c^{(i)} P\{Q_t^{(i)} > k \mid \text{observations}\} \tag{4.2}$$

where k>0 is a constant. In this policy the probability that a non negligible number of vehicles is queueing is taken into account.

Policy n. 3

$$j^* = \arg \max_{j=1,\dots,f} \sum_{i \in E_j} \mu^{(i)} c^{(i)} E(Q_t^{(i)} \mid \text{observations}) \tag{4.3}$$

This policy, like the preceding one, tries to take into account queue lengths, while avoiding the necessity of giving a threshold value k.

Policy n. 4

$$j^* = \arg \max_{j=1,\dots,f} \sum_{i \in E_j} c^{(i)} E(D_t^{(i)} \mid \text{observations}) \tag{4.4}$$

This is the so-called "one-step look-ahead" policy and is obtained from (3.1) assuming that T=1 and that the initial time is t.

We remark that, in order to implement policies 1-4, the computation of quantities like expected queue length and expected number of departures is required. This can be easily accomplished, for example, through the filtering formulas given in Betrò et al. (1985b).

5. Experimental results for on-line control policies

In order to check the feasibility of the proposed control policies, a computer code has been developed in FORTRAN 77 on a GOULD S.E.L. 32/27 minicomputer. We will now present the results related to the application of policies 1-4 to a set of data collected at a semaphorized intersection in Milan, whose structure is depicted in figure 2.

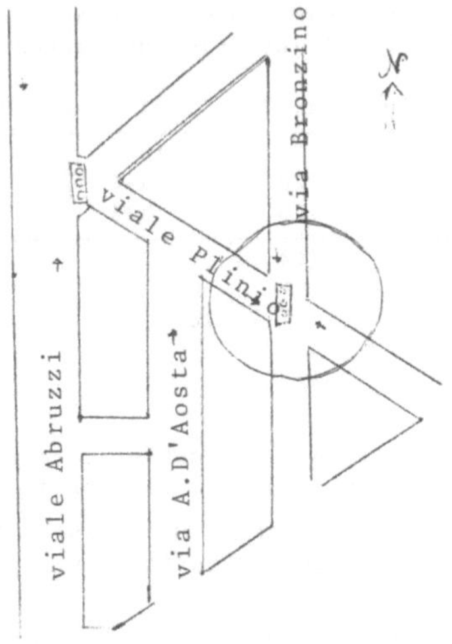

figure 2.

Arrival times have been collected on each street by means of a portable microcomputer which recorded the time of each keystroke. Data were collected at a distance of 60 meters from the traffic light. The unit time interval was chosen equal to 4 seconds.

In order to compare different policies in identical conditions only arrivals were collected from real observations, while departures had to be simulated, because of their dependency on the control policy. An analysis of departure processes during green phases has led to the conclusion that such processes are strongly regular and constant over time; in the experiments here reported departure processes during green phases were supposed to be Poisson with rate $\mu=2$.

For what concerns arrival processes, flow from street 1 to street 3 has been modelled as a Poisson process with stochastic rate switching between two values α_H and α_L, with $\alpha_H=1$, $\alpha_L=0.1$, probability of switching from $\alpha=0.1$ to $\alpha=1$ equal to 0.8 and probability of switching in the opposite way equal to 0.9. These values were inferred from the observation of the average platoon length and the average interplatoon gap (for details on arrival rate models, see Betrò et al., 1985a). For what concerns flows 2-4 and 3-1, which were characterized by rather low traffic volumes, they were modelled as Poisson processes with rates 0.3 and 0.2 respectively.

The first three plots in figures 3-6 show the evolution of queue lengths when policies 1-4 are used to control the traffic light. The horizontal axis represents the number of unit time intervals from the beginning of the experiment. The total duration of the experiment is 15 minutes, corresponding to 225 time units.

The last display in each figure represents the state of the traffic light in the direction 1-3 (or, equivalently, 3-1), where a vertical bar is used to represent a time interval during which the traffic light was green. In figure 7 is reported the effect of the presently operating control policy, which consists of 24 seconds of green, 4 of amber and 40 of red in directions 1-3 and 3-1.

Table I reports the total delay (in seconds) for each of the proposed policies and for the present one. It is evident how substantially each of the on-line control policies introduced reduces the queue lengths in each direction.

As a concluding remark, we observe that further experiments would be necessary in order to validate the efficiency of the proposed strategies. Policies are presently under investigation which are able to "look further" than policies 1-4: this could possibly lead to policies in which phase switching is less frequent than in the proposed ones and the total lost time decreased.

Direct.	Pol. 1	Pol. 2	Pol. 3	Pol. 4	Present
1-3	1056	1420	1096	1440	2320
2-4	1060	928	844	844	1060
3-1	364	420	376	376	888
Tot.	2480	2768	2316	2660	4268

Table I

References

[1] Baras J.S. and Dorsey A.J. (1981), "Stochastic control of two partially observed competing queues", IEEE Trans. on Automatic Control AC-26, 5, 1106-1117.

[2] Baras J.S., Dorsey A.J. and Makowski A.M. (1985), "Two competing queues with linear cost and geometric service requirements: the c-rule is often optimal", Advances in Applied Probability 17, 1, 186-209.

[3] Betrò B., Schoen F. and Speranza M.G. (1985a), "Markovian models for traffic evolution at an intersection", submitted to Methods of Operations Research.

[4] Betrò B., Schoen F. and Speranza M.G. (1985b), "Dynamic estimation of queue behaviour in urban traffic", submitted to EJOR.

[5] Buyukkoc, C., Varaiya, P. and Walrand, J. (1985), "The c rule revisited", Advances in Applied Probability 17, 1, 237-238.

[6] Camera, E., Peisino, M. and Piglione, M.C. (1983), "Full-adaptive traffic flow feedback control system design in a isolated intersection", Proceedings of the Operations Research Italian Society 693-707.

[7] Gartner, N.H. (1983), "OPAC: a demand-responsive strategy for traffic signal control", TRR 906, Transp. Res. Board, Washington.

[8] Gittins J.C. (1979), " Bandit processes and dynamic allocation indices (with discussion)", J. of Roy. Statist. Soc., Ser. B, 41, 148-177.

[9] Hunt, P.B., Robertson, D.I., Bretherton, R.D. and Royle, M.C. (1982), "The SCOOT on-line traffic optimisation technique", Traffic Engineering and Control, 23, 190-192.

[10] Hunt, P.B., Robertson, D.I., Bretherton, R.D. and Winton, R.I. (1981), "SCOOT-A traffic responsive method of coordinating signals", TRRL 1014, Dept. of Transport, Crowthorne, United Kingdom.

[11] Robertson D.I. (1969), "TRANSYT method for area traffic control", Traffic Engineering and Control, 11,276-281.

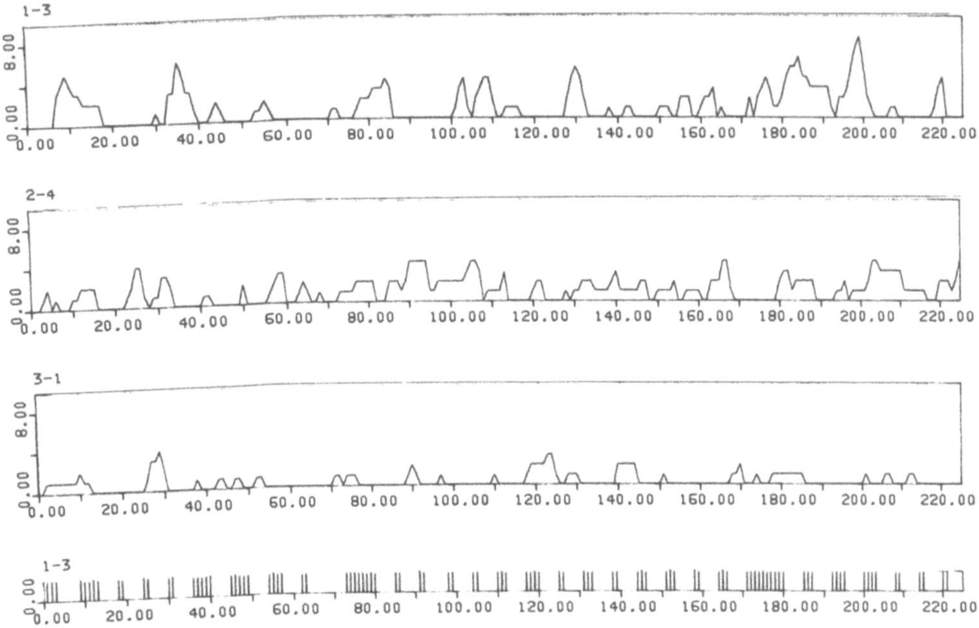

figure 3

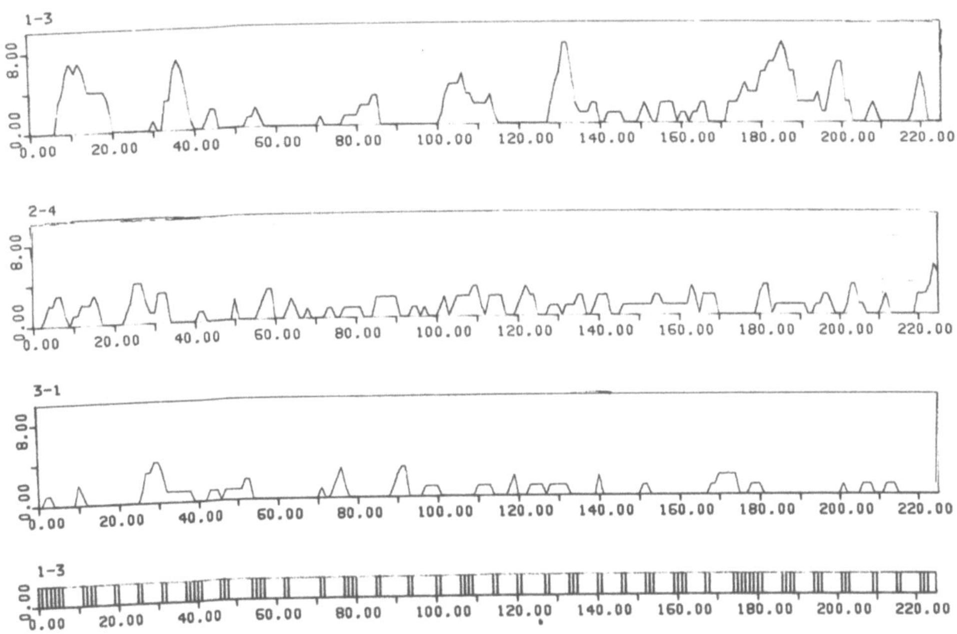

figure 4

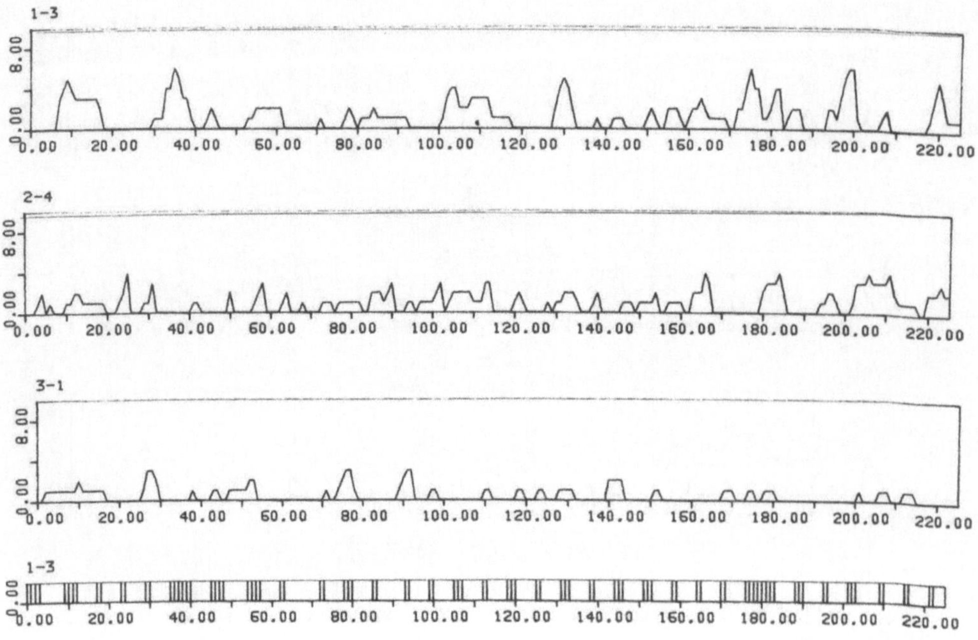

figure 5

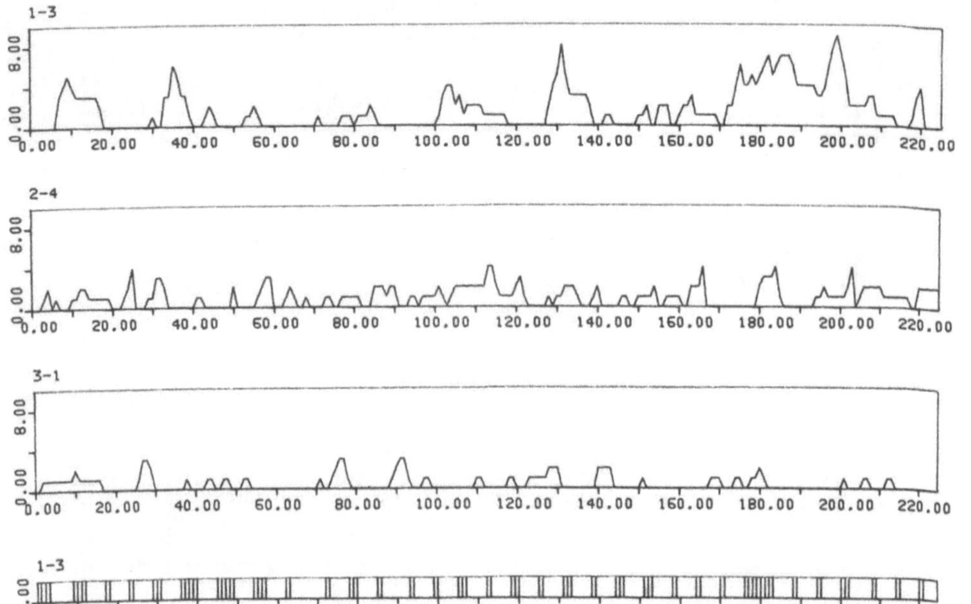

figure 6

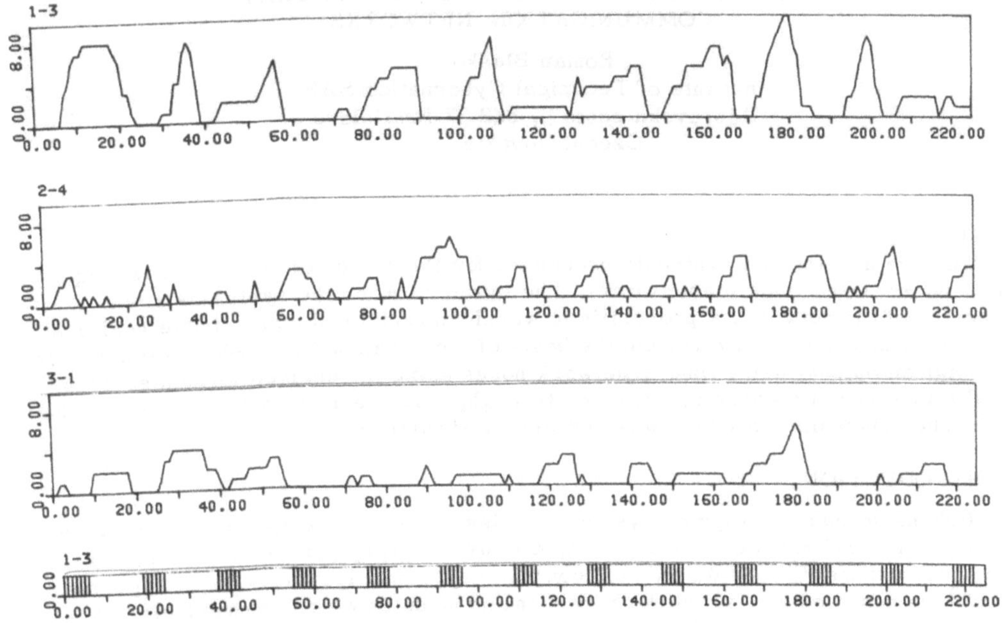
1-3
8.00
0.00
0.00 20.00 40.00 60.00 80.00 100.00 120.00 140.00 160.00 180.00 200.00 220.00
2-4
8.00
0.00
0.00 20.00 40.00 60.00 80.00 100.00 120.00 140.00 160.00 180.00 200.00 220.00
3-1
8.00
0.00
0.00 20.00 40.00 60.00 80.00 100.00 120.00 140.00 160.00 180.00 200.00 220.00
1-3
0.00
0.00 20.00 40.00 60.00 80.00 100.00 120.00 140.00 160.00 180.00 200.00 220.00

figure 7

THROUGHPUT OPTIMIZATION OF PACKET COMMUNICATION NETWORKS

Roman Blaško
Institute of Technical Cybernetics SAS
Dúbravská cesta 9, 842 37 Bratislava
Czechoslovakia

ABSTRACT

We have developed a synthesis procedure for the design of an optimal throughput in packet communication networks. The initial data for the synthesis are a structure of the network, a probabilistic workload model and approximate real delay time values in the network. On the basis of the developed procedure we are able to design an optimal delay time in network nodes using the derived formulae. This results in a network whose dynamics or throughput is specialized for a given workload. The synthesis procedure was verified by simulation.

1. INTRODUCTION

High-performance computer systems are developed as a composition or network of many functional modules. This principle is used in very perspective data-flow computer systems [5] and distributed computer systems [4]. Data-flow computers [2],[3] are designed as a network of independent modules which communicate asynchronously via unidirectional channels. A number of, or even all, modules can be active simultaneously. The information to be processed in the machine is transmitted throughout the hardware structure in the information packet form. The systems of this kind are called packet communication systems [3]. During processing the information packets flow throughout the structure. Their routes are determined by the attributes of the packets or by the situation arising in the network. The task to be processed in such network is the workload of the network.

The considered problem lies in the design of right delay times in nodes or modules of the network structure. We say that the network is dynamic balanced when all modules or nodes may be uninterruptedly active without blocking one another and their utilization approximates to the ideal, i.e. 100 %. In such situation we can say that the delay time in nodes is optimal. The possibility of an approximation to ideal behaviour of asynchronous networks depends on buffer capacity too [1]. In this case we suppose the proper and limited capacity of buffers between each pair of connected modules.

2. PARALLEL ASYNCHRONOUS NETWORKS

Now we concentrate upon basic parallel structures such as pipeline and parallel array which may be connected to hybrid clusters or networks. Modules communicate by a self-time ready/acknowledge protocol [3].

2.1. Pipeline Structure

The pipeline structure is a linear arrangement of modules. All modules can be active simultaneously by overlapping of active phases. The throughput of the pipeline structure with overlapping of active phasis is $h = 1/\Delta t$, where Δt is the delay time in each stage of the structure. By serial processing in "k" stages without overlapping the processing time of one item is $(k \cdot \Delta t)$ and the throughput of the structure is $h_s = 1/(k \cdot \Delta t)$. It implies that the throughput of the pipeline structure with overlapping i.e. by parallel processing is k times greater than by serial processing.

The speed-up coefficient S_r is defined as the ratio (time of serial processing) / (time of parallel processing). For mentioned pipeline structure [1] is $S_r \doteq k$, when $n \gg k$, where n is a number of processed items. Hence the throughput and the speed-up coefficient by parallel processing is rising proportionally to the number of stages in the pipeline structure, if $(k' \cdot \Delta t)$= constant. In a real asynchronous pipeline structure the delay times in stages are variable and thus the throughput and speed-up coefficient are computed as statistical values.

In the mentioned systems a stage can absorb and generate one or more items from an input and on an output respectively by one activation. The following conditions should be satisfied for a dynamic balanced structure. If the stage absorbs r_1 items and generates r_2 independent items by one activation, then for a dynamic balanced structure with the input period Δt_1 and the output period Δt_2 of items, the delay time "d" in the stage should satisfy these equations

$$d = r_1 \cdot \Delta t_1 \qquad (1) \qquad\qquad \Delta t_2 = d / r_2 \qquad (2)$$

2.2. Parallel Array

A parallel array of "k" modules represents "k" parallel routes. The throughput of the i-th module, with delay time Δt_i is $h_i = 1/\Delta t_i$, $i = 1,2,\ldots,k$. The throughput of parallel array with "k" modules is

$$h = \sum_{i=1}^{k} h_i = \sum_{i=1}^{k} \frac{1}{\Delta t_i} \qquad (3)$$

If $\Delta t_i = \Delta t_j = \Delta t'$, $i \neq j$, $i,j = 1,2,\ldots,k$, then $h_i = h_j = h'$ and

$$h = k \cdot h' = k \cdot \frac{1}{\Delta t'} \qquad (4)$$

The output period of the item generated by the structure is derived from (3)

$$\Delta t = \frac{1}{h} = \frac{1}{\sum_{i=1}^{k} \frac{1}{\Delta t_i}} \qquad (5)$$

If $\Delta t_i = \Delta t_j = \Delta t'$, $i \neq j$, $i,j = 1,2,\ldots,k$, then the speed-up coefficient is $S_r = k$. The parallel array, like the pipeline structure, has the speed-up coefficient equal to the number of modules in the structure in ideal dynamic balancing. In the case of variable delay times in the modules and an asynchronous communication mode, the throughput and speed-up coefficient are statistical values.

2.3. Distributor and Arbiter

Functional modules called distributor and arbiter are needed for a design of hybrid clusters or networks of basic structure introduced above.

D i s t r i b u t o r (see Fig. 1, modul D) is a functional module for a choise of one output to which an item should be transmitted from the input. The choice is due to the specified strategy. We suppose that distributor is incorporated in a network (see Fig. 1). Then the following equation has to be satisfied in dynamic balanced network

$$\Delta T = p_i \cdot \Delta t_i \qquad (6)$$

where: ΔT - delay time in the distributor, p_i - choice probability of the i-th output, Δt_i - delay time in the module connected to the i-th output, $i \in \{1,2,\ldots,k\}$ - index of the output, k - number of outputs.

If the delay time in the distributor is known, then the next formulae for the delay time in the output modules of dynamic balanced network are derived from (6) :

a / if $p_i \neq p_j$, $i \neq j$, $i,j = 1,2,\dots,k$, then $\Delta t_i = \Delta T / p_i$ (7)

b / if $p_i = p_j = 1/k$, $i \neq j$, $i,j = 1,2,\dots,k$, then $\Delta t_i = \Delta t = \Delta T \cdot k$ (8)

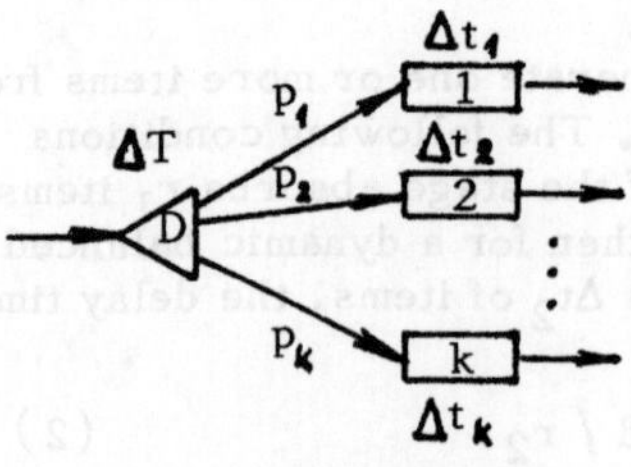

Fig.1. Distributor

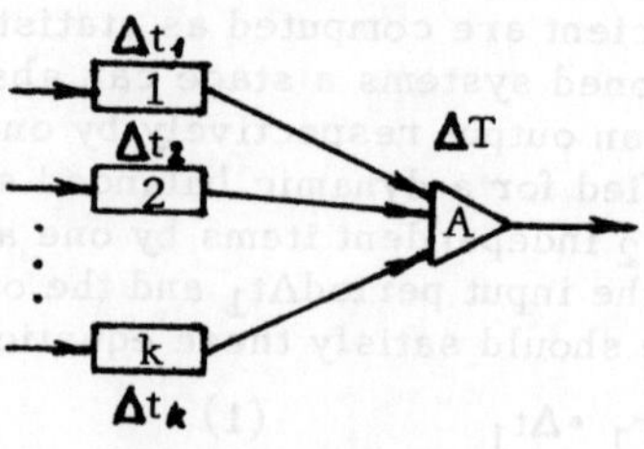

Fig.2. Arbiter

Let the items be distributed by a distributor according to their significant features. According to these we distinguish some types of items. Let a probability of each item type in the input stream be known. Then the probability of requirements for each output is given as a relative probability of the significant feature.

A r b i t e r (see Fig.2, modul A) is a functional unit for a choice of one input from which an item should be transmitted to the output. The choice is due the specified strategy, e.g. the round-robin strategy. We suppose that the arbiter is incorporated in the network (see Fig.2). Then this equation has to be satisfied in dynamic balanced network

$$\Delta T = \frac{1}{\sum_{i=1}^{k} \frac{1}{\Delta t_i}} \qquad (9)$$

where: ΔT - delay time in the arbiter, Δt_i - delay time in the module connected to the i-th input, $i \in \{1,2,\dots,k\}$ - index of the input, k - number of inputs.

If the delay times in modules are known, then the delay time in the arbiter is:

a / if $\Delta t_i \neq \Delta t_j$, $i \neq j$, $i,j = 1,2,\dots,k$ then ΔT is given by the formula (9),

b / if $\Delta t_i = \Delta t_j = \Delta t$, $i \neq j$, $i,j = 1,2,\dots,k$, then from (9) ΔT is

$$\Delta T = \frac{\Delta t}{k} \qquad (10)$$

2.4. Parallel Structure

We assume a parallel structure composed as pipeline structure with five stages, where the 3rd stage is a parallel array with "k" modules (see Fig.3). We suppose

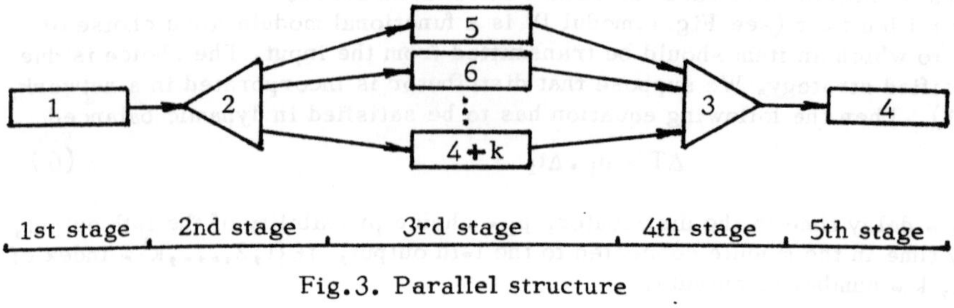

Fig.3. Parallel structure

that each independent item which enters the module "2" leaves the module "3". Then the time period of entering items has to be equal to the time period of leaving items. It is expressed by an equation derived from the formulae (6) and (9)

$$p_i \cdot \Delta t_i = \frac{1}{\sum_{i=1}^{k} \frac{1}{\Delta t_i}} \qquad (11)$$

For parallel routes the relationships $p_i \cdot \Delta t_i = p_2 \cdot \Delta t_2 = \dots = p_i \cdot \Delta t_i$ are derived from (6).

2.5. Network

Many types of networks [3] or network structures [4] can be composed of basic functional modules and parallel structures introduced above. The simple network

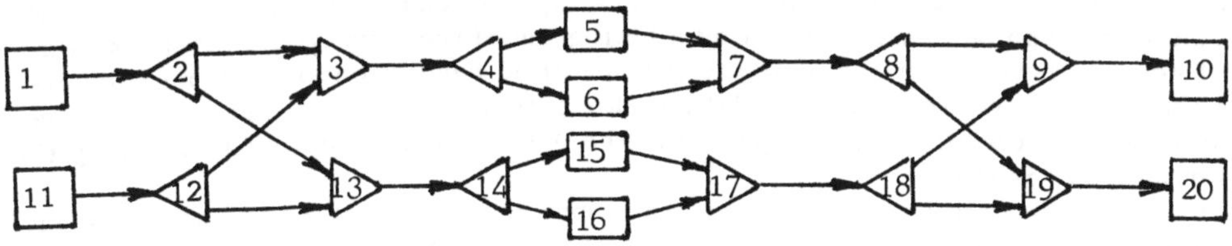

Fig.4. Network structure

structure of data-flow computer [2] is introduced as an example (see Fig.4.). The network can be homogeneous or heterogeneous from the point of view of structure or dynamics.

3. WORKLOAD OF A NETWORK

Requirements to performance of all modules and to transmission capability of channels in a network are determined by the workload. We assume the heterogeneous network from the functionality point of view. The routes of items depend on item attributes or a situation arising in a network. The stream of the items in each channel is characterized by a probabilistic characteristic such as a histogram of flowing items. The histogram distinguishes the items according to the attribute which is significant for a mentioned part of the network.

For an example, the requirement for processing in one of the four central modules {5,6,15,16} is a significant feature for routing in the first half of the network structure from Fig.4. The requirement for absorbtion in module "10" or "20" is a significant feature for transmission requirements in the second half of this network.

4. SYNTHESIS PROCEDURE

The synthesis procedure consists of ten rules. In the procedure we use the following concepts. An item representing an information packet is element flowing throughout the network and is processed by nodes of the network. The set of absorbed or generated items is called an entry. The procedure rules are the following.

1. Initial data for the synthesis are: structure of the network, histogram (-s) of the requirements to significant modules determining the item routes and the assumed real values of delay times in the nodes of the network.
2. The design of delay times begins at input modules of the given structure. The delay times of these modules have to be chosen with mutual ratio referring to their

real values. The absolute value of the delay times is not important in this step of the synthesis, but for the sake of clarity it may refer to the reality. The routes of the items in the next part of the structure are determined by histograms of entering items for each input of the structure.

3. The computation of delay times proceeds from the input modules in the transmission direction of items to the output modules of the structure.
4. If the given module is connected to a module whose output period of items is Δt_1, then the delay time d in the given module, being the input period of entries, is given by the formula (1). If $r_1>1$, then an entry contains more items, if $r_1=1$, then an entry is one item and $d=\Delta t_1$.
5. If the output period of entries d of a given module is known, then the output period of items Δt_2 is given by the formula (2). If $r_2=1$, then $\Delta t_2 = d$, similarly as in rule No.4.
6. If the given module is a distributor, then its delay time is equal to the input period of entries according to the rule No.4. The input stream of entries or items to the distributor can be characterized by a histogram of the item types. Distribution of items is carried out according to the specified resolution algorithm of the distributor. The requirement probabilities for outputs of the distributor are derived as relative probabilities. The period of the entries to each output is given by the formula (7) or (8).
7. If the given module is an arbiter and the input period of entries is known for each input (by rule No.4), then the delay time of the arbiter is given by the formula (9) or (10).
8. If the given module is a "simple" functional module (different from the arbiter or the distributor), then the delay time of the given module is given by the formula (1) and the output period of the items is given by the formula (2).
9. We obtain the rational values of all delay times in the given structure by computing required delay time in each module. The real absolute values of delay times (see rule No.2) in the network being designed are obtained by multiplication of all delay times by a proper constant. This constant may be determined e.g. by the ratio of the real delay time and the required delay time (according to the synthesis) in the fastest module of the network.
10. If the ratio of the delay times, obtained by the synthesis, is impossible for a real network, then we have to modify the part or the whole structure of the network. The synthesis of dynamics has to be carried out again for the modified part or for the whole structure.

Often the delay times in a real system are random with a mean value. In that case the procedure is valid for their mean values.

5. APPLICATION

We applied this procedure to the synthesis of dynamics for the network from Fig.4. The module "5" generates two items at the output as a response to the absorbtion of one item from the input. The distributor "14" inserts items in the buffer between modules "14" and "16" only when the buffer between modules "14" and "15" is full. Other distributors are working according to the significant attributes of the items. Arbiters choose an input according to the round-robin strategy. Items are generated by the modules "1" and "11" to the network and are absorbed by the modules "10" and "20" from the network. The histogram distinguishing the item types according to the significant feature for the first half of the network is given by probabilities $p_5= 0.5035$, $p_6= 0.0965$, $p_{15}= p_{16}= 0.4$. The histogram for the second half is given by probabilities $p_{10}= p_{20}= 0.5$. In this case the histograms are the same for both generating modules.

Table 1. Delay times in the network

i	1	2	3	4	5	6	7	8	9	10
d_i[nsec]	200	200	166.5	166.5	198.2	1034.2	90.45	90.45	132.8	132.8
i	11	12	13	14	15	16	17	18	19	20
d_i[nsec]	200	200	250	250	500	500	250	250	132.8	132.8

By the synthesis we obtained the delay times in the nodes given in Table 1. If the real delay time values are in accordance with the results of the synthesis, then the throughput in the entrance of the network would be

$$2 \cdot 1 / (200 \cdot 10^{-9})\ s^{-1} = 2 \cdot 5 \cdot 10^{6}\ s^{-1} = 10 \text{ MIPPS}^{*}.$$

The throughput of the central modules together is 10 MIPPS too. At the exit of the network the throughput would be

$$2 \cdot 1 / (132{,}8 \cdot 10^{-9})\ s^{-1} = 15 \text{ MIPPS}.$$

Due to the function of the modul "5", this throughput should be greater than in the central modules.

For the evaluation of simulation experiments we have used the following parameters. Total utilization of the network u_t is defined [1] as the sum of the utilization of each module u_i, which is expressed

$$u_t = \sum_{i=1}^{n} u_i \qquad (12)$$

where $u_t \in \langle 0, n \rangle$, $u_i \in \langle 0, 1 \rangle$ and n is the number of modules in the network. An ideal behaviour of the network is by $u_t = n$. In this case all modules of the network are uninterruptedly active. Throughput h of central modules $\{5,6,15,16\}$ is the sum of the throughput in four parallel routes.

For the given network the maximum value of total utilization $u_{t\,max}$ is 20 and the maximum value of the throughput in central modules h_{max} is 10 MIPPS. The simulation results are in Fig.5 , by the capacity of each buffer in the network for 10 items.

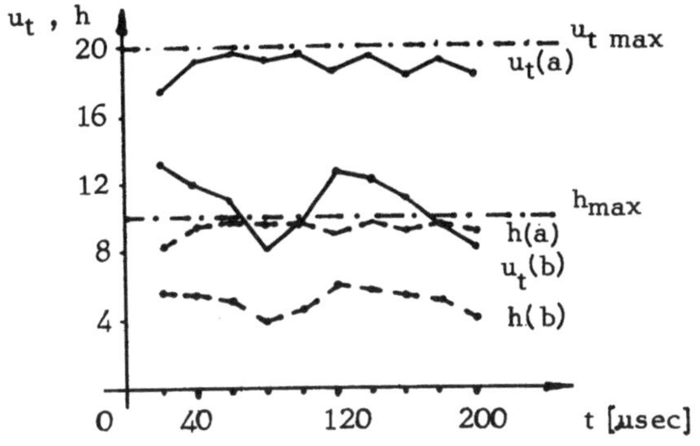

Fig.5. Performance parameters

According to these results the mean value of total utilization of the network $u_t(a)$ is 19.14 , that is 95 % of the maximum value and the mean value of the throughput for central modules h(a) is 9.55 MIPPS, that is 95.5 % of the maximum value.

With regards to the simulation results we can say that the synthesis is right, because values of the performance parameters approximate to the ideal value. The throughput of the network with the given structure is specialized for the given work-

---

* MIPPS denotes million information packets per second.

load. By the partial changing of the workload, according to the histogram $p_5 = p_6 = 0.2$, $p_{15} = p_{16} = 0.6$, the performance is degraded considerably. The dependence of the evaluated parameters for this workload is illustrated by curves $u_t(b)$ and $h(b)$ in Fig.5.

6. CONCLUSION

The synthesis procedure introduced above is relatively simple for the mentioned class of networks. These types of networks are used in many high-performance computers being designed at present. This procedure makes it possible to design a network specialized for a given workload, whereby performance measures approximate to ideal values.

An interesting research topic would be the automatic synthesis of a given network specialized for a given workload which would consist in connecting the above-described procedure with a powerful simulation and evaluation tool. Another objective would be not only the synthesis of optimal throughput but also the optimal structure with this throughput.

REFERENCES

[1] BLAŠKO,R.:"B-nets and Analysis of Data-Flow Systems",PhD Thesis, Institute of Technical Cybernetics,Slovak Academy of Sciences,Bratislava,December 1983, 140 pp. (in Slovak).

[2] CHUDÍK,J.-BLAŠKO,R.:"One Way of Achieving the Higher Performance of a Data-Flow Computer",The 5-th International Conference on Control Systems and Computer Science,Bucharest,June 1983,Preprints Vol.III.,pp.7-14.

[3] DENNIS,J.B.-BOUGHTON,G.A.-LEUNG,C.K.C.:"Building Blocks for Data-Flow Prototypes",Proc. of the 7-th Annual Symposium on Computer Architecture,1980,pp.1-8.

[4] DIAS,D.M.-JUMP,J.R.:"Packet Switching Interconnection Network for Modular Systems",Computer,December 1981,pp.43-53.

[5] TRELEAVEN,Ph.C.-BROWNBRIDGE,D.R.-HOPKINS,R.P.:"Data-Driven and Demand-Driven Computer Architecture",Computing Surveys,Vol.14., No.1,March 1982,pp.93-143.

to support arbitrary user distributed micro/minicomputer systems

2. MODELS OF A NODE WITH COMMUNICATION DEVICES

Consider a processor (the "main" node) with N lines connected to it through communication devices. The same kind of devices and processors are on the opposite ends of the lines. We investigate hard traffic conditions with transmitters always having packets to be transmitted. Packets are transmitted in both directions. Interrupt requests of receivers and transmitters share processor time. Only the receiving process in the node is considered. The time needed for processing of transmitter interrupt requests is recognized in the model by increasing the mean time for serving receiver interrupt requests. Information is transmitted in frames. The frames are of constant length ℓ byte. Each data character appears in receiver data buffer register, and can be stripped of start, stop, and parity bits.

2.1. The DL-11 interfaces

The DL-11 is an asynchronous interface (11 bits per character) which can support line speeds of up to 9.6 Kbaud. Each character received generates an interrupt request to the processor which is serviced by the software. There is a common queue of such requests from all communication devices of the node. The receiver register of the DL-11 is emptied as soon as an appropriate interrupt request is served. If the next character arrives when the register is busy, then the DATA OVERRUN error occurs and the device is switched into resynchronisation state. It receives and rejects all the ramaining characters of the frame up to the first character of the next frame. The number of instructions performed by the processor to serve an interrupt request depends on the type of the character received. We assume that the service time is exponentially distributed with mean $1/\mu$. This parameter depends on the processor type.

The performance measure of the system is the normalized throughput

$$V_o = n_c/\mu \tag{1}$$

where n_c is a number of characters of error free frames received by the node per unit time.

The model of the node is a single-server queue with the service rate μ and the N sources of customers. We assume that customers from ith source arrive according to a Poisson process with intensity λ_i , $i=\overline{1,N}$. The customers from ith source will be called "i-customers". If there is at least one i-customer in the system at the arrival of another i-customer the DATA OVERRUN error occurs. We may consider, that DATA OVERRUN occurs at the departure of an i-cus-

PERFORMANCE ANALYSIS OF DATA LINK AND COMMUNICATION DEVICE CONTROL PROCEDURES IN DISTRIBUTED MICRO/ MINI COMPUTER SYSTEMS

Leonid B. BOGUSLAVSKY, Aleksander L. STOLYAR
Institute of Control Sciences
Moscow, USSR

1. INTRODUCTION

Performance of distributed computer systems and networks depends on system configuration, control procedures and their parameters, and communication devices. We present a number of analytical results in performance evaluation of communication control procedures of DECnet which refers to a family of packet-switched network products developed by Digital Equipment Corporation [1]. Attention is paid to the communication devices (hardware interface between the processor and the communication line) and to the DDCMP data link protocol since computer network performance tend to be relatively sensitive to the choice of this components [2].

Three types of communication devices are considered: DL-11, asynchronous adapter, DZ-11asynchronous multiplixer and CAMAC module 500A. A slight modification in DECnet's DL-11 driver programm allows the use of CAMAC hardware interfaces in DECnet configurations [3]. Our results can be used also for performance evaluation of computer networks with MININET and SM NETWORK software [4, 5].

Performance analysis of data link control procedures is aimed to compare two different approaches to frame structure organization. In the first one, the transmission error within the head of a frame is detected with the help of a special 16-bit head checking sequence. It implies an additional overhead of control information. But on the other hand, the control information in the head of the frame can be used by a receive node even if an error occure in the information field of the frame. The DDCMP protocol [6] is an example of this approach. For the second approach, the head of a frame and the information field are controlled by one 16-bit frame checking sequence. The ISO link control procedure HDLC [7] is an example of this approach.

The object of our studies was to provide information which would help determine what resources (eg. processors, communication devices, communication lines) and protocol parameters are necessary tomer, if there is another i-customer in the system. Therefore

$$n_c = \sum_{i=1}^{N} \lambda_i P_i^{cor}$$

where P_i^{cor} is the steady-state probability of absence of DATA OVERRUN for the ℓ consecutive i-customers, which form one frame. Thus, M/M/1 queue with the intensity $\lambda=\sum_{i=1}^{N}\lambda_i$ of input stream is considered, and besides, every entering customer is regarded an i-customer, with probability $P_i=\lambda_i/\lambda$.

Assume that ℓ consecutive i-customers are numbered in order of entering the system. Denote as C_j, $j=\overline{1,N}$ the random number of customers in the system at the departure of jth i-customer (himself excluded), and the event {there is no other i-customer in the system at the departure of jth i-customer}, $j=\overline{1,\ell}$, as D_j.

$$\widetilde{D}_j = D_1\cdot\ldots\cdot D_j\,, \quad j=\overline{1,\ell}\,;$$

$$r_{jk}=\begin{cases} P\{C_j=k/\widetilde{D}_{j-1}\}, & j\geq 2\\ P\{C_1=k\}, & j=1\end{cases}, \quad k=\overline{0,\infty}\,;$$

$$r'_{jk}=P\{C_j=k/\widetilde{D}_j\}\,, \quad j=\overline{1,\infty}\,, \quad k=\overline{0,\infty}\,.$$

Then we can write

$$P_i^{cor}=P(\widetilde{D}_\ell)=\begin{cases} P(D_1)\cdot\prod_{j=2}^{\ell} P(D_j/\widetilde{D}_{j-1}), & \ell\geq 2\\ P(D_1), & \ell=1\,.\end{cases} \tag{2}$$

Since our process is Markovian,

$$P\{D_j/C_j=k,\widetilde{D}_{j-1}\}=P\{D_j/C_j=k\}=(1-P_i)^k, \quad k=\overline{0,\infty}, j\geq 2\,; \tag{3}$$

$$P\{C_{j+1}=k/C_j=m,\widetilde{D}_j\}=P\{C_{j+1}=k/C_j=m,D_j\}=f_{mk}, \quad m,k=\overline{0,\infty}, j\geq 1. \tag{4}$$

From (2) and (3) we have

$$P(\widetilde{D}_\ell)=\prod_{j=1}^{\ell}\sum_{k=0}^{\infty} r_{jk}(1-P_i)^k\,.$$

Thus, we are interested in the following sequence of distributions

$$\overline{r}_j=\{r_{jk}\}_{k=\overline{0,\infty}}\,, \quad j=\overline{1,\ell}\,.$$

Using (3) and (4) it is easy to show that $\overline{r}_{j+1}$ depends only on $\overline{r}_j$. The mapping $\overline{r}_j \longrightarrow \overline{r}_{j+1}$ can be represented as

$$\overline{r}_j \xrightarrow{A_1} \overline{r}'_j \xrightarrow{A_2} \overline{r}_{j+1}\,,$$

with $\overline{\pi}_j' = \{\pi_{jk}'\}_{k=\overline{0,\infty}}$ . .

The mapping A_1 is determined as

$$\pi_{jk}' = \frac{\pi_{jk}(1-P_i)^k}{\sum_{k=0}^{\infty} \pi_{jk}(1-P_i)^k}, \; k=\overline{0,\infty}, \; j \geqslant 1; \qquad (5)$$

and A_2, as

$$\overline{\pi}_{j+1} = \overline{\pi}_j' F, \qquad (6)$$

where $\overline{\pi}_{j+1}$ and $\overline{\pi}_j'$ are considered as row matrices. The elements of matrix $F=(f_{mk})_{m,k=\overline{0,\infty}}$ may be calculated using the following equations:

$$[-y^2+(1+\rho)y-(1-P_i)\rho]F(z,y)=[y-(1-P_i)\rho]F_0(z)+P_i\rho\frac{\Phi(z)y}{1-\Phi(z)y}, \qquad (7)$$

$$|z|\leqslant 1, \; |y|\leqslant 1; \qquad \rho=\lambda/\mu;$$

$$F(z,y)=\sum_{m=0}^{\infty}\sum_{k=0}^{\infty} f_{mk} z^k y^m; \quad \Phi(z)=\frac{1}{1+\rho-\rho z};$$

$$F_0(z)=\frac{1-\tilde{y}}{1-\tilde{y}z}, \quad \tilde{y}=\frac{1+\rho-\sqrt{(1-\rho)^2+4P_i\rho}}{2(1-P_i)}.$$

The expression (7) can be obtained from the following obvious set of equations:

$$f_{m,k}=a_3 f_{m-1,k}+a_1 f_{m+1,k}+a_2 u_{m+1,k}, \quad m=\overline{1,\infty}, \; k=\overline{0,\infty};$$

$$f_{0,k}=(1-P_i)f_{1,k}+P_i u_{1,k}, \qquad k=\overline{0,\infty};$$

$$a_1=\frac{(1-P_i)\rho}{1+\rho}, \quad a_2=\frac{P_i\rho}{1+\rho}, \quad a_3=\frac{1}{1+\rho},$$

$u_{mk}=P\{$In the course of serving m customers exactly k new customers have entered the system$\}$, $m=\overline{1,\infty}$, $k=\overline{0,\infty}$.

The expressions (5) and (6) are not applicable to straightforward calculation of P_i^{cor}. But probability P_i^{cor} can be estimated in the following way. Consider the system at the departure of jth i-customer. Let the DATA OVERRUN error be fixed if either there is another i-customer in the system or the total number of customers in the system exceeds some number M. In such a case, the probability of absence of DATA OVERRUN (in the frame) is equal to

$$\tilde{P}_i^{cor}=P\{D_1\cdot\ldots\cdot D_\ell\cdot\{c_1\leqslant M\}\cdot\ldots\cdot\{c_\ell\leqslant M\}\}.$$

Hence

$$P_i^{cor} = \tilde{P}_i^{cor} + \Delta P_i ,$$

where

$$\Delta P_i = \{\{D_1, C_1 > M\} \cup \ldots \cup \{D_\ell, C_\ell > M\}\} \leq \ell \rho^M (1-P_i)^{M+1} .$$

Thus, $\tilde{P}_i^{cor}$ can be calculated from (5) and (6) and $\tilde{P}_i^{cor} - P_i^{cor} \to 0$ as $M \to \infty$.

2.2. CAMAC Communication Module

This communication device operates in the following manner: an acknowledgement (ACK) character is sent to the transmitting device at the time when the receiving processor starts to serve the appropriate interrupt request. The next character is sent by the transmitting device only after receiving the ACK on the previous character. ACK character is 11 bits long. For these modules, the concept of a DATA OVERRUN error does not exist.

The model of the node with N lines and CAMAC communication devices is a two-station "closed" queueing network with N requests (customers). This network consists of an infinite server (IS) station (transmitters) and a one server FCFS station (processor and the queue of interrrupt requests).

The mean service time in IS station is 22/C, where C is line speed. The service time in the FCFS station is exponentially distributed with mean $1/\mu$.

Note that this queueing network is not closed in the classic form. A customer that leaves the queue of the FCFS station is divided into two customers. The first one occupies the server of this station, and the second enters the IS station. The normalized throughput V_0 is equal to the long run probability of a busy server in FCFS station.

We compare this queueing network (network $\bar{A}$) with the classic closed queueing network with IS and FCFS stations (network A). Let F(x) and Q(x) be distribution functions of the service times of customers in IS and FCFS stations respectively, for the both networks. Let P(N) and $\bar{P}(N)$ be utilizations (i.e. steady-state probabilities of being busy) of the server of FCFS station for the A and $\bar{A}$ networks, respectively. Then the following proposition is true. If the distribution functions F(x) and Q(x) are determined and the studied probabilities exist, then

$$P(N) \leq \bar{P}(N) \leq P(N+1). \tag{8}$$

The proof is carried out in following way. The control sequences $\{x_k\}_{k=\overline{1,\infty}}$ and $\{y_k\}_{k=\overline{1,\infty}}$ are considered, where elements of sequences are deterministic and nonnegative. The k th customer, entering the IS station, is served in this station during the time x_k and j th customer, entering the FCFS station, receives service time y_j. The initial state of the system is fixed. Then it can be shown by induction that a departure of a customer from the FCFS station in $\bar{A}$ network takes place earlier than in A network with N customers, but later than in A network with N+1 customers.

Analogously it can be shown that

$$\sup_{\mathcal{E}} \frac{\overline{P}(N)}{P(N)} = \sup_{\mathcal{E}} \frac{P(N+1)}{P(N)} = \frac{N+1}{N}, \tag{9}$$

where $\mathcal{E}$ is a set of networks for which these probabilities exist and are not equal to 0.

If service time in FCFS station is exponentially distributed and there is a given finite value of mathematical expectation of service time in the IS station, then the probability P(N) for A network is independent on the form of distribution function F(x) [9].

If we assume that service time in IS station of our network is also exponentially distributed, then

$$V_0 = \sum_{i=0}^{N} (i!\rho_e^i)^{-1} \Big/ \Big[\sum_{i=0}^{N} (i!\rho_e^i)^{-1} + (N!N\rho_e^{N+1})^{-1}\Big],$$

$$\rho_e = C/(22\mu).$$

By virtue of (8) and (9) this estimation for $\overline{P}(N)$ is acceptable.

2.3. The DZ-11 asynchronous multiplexer

This device can support up to 16 serial lines. Character size is 11 buts including 1 start and 2 stop bits. The receiver buffer has a 16-character capacity. At the begining of operation when buffer is empty the time-out is set. After the time-out expires an interrupt request to the processor is generated. The processor interruption causes the service of all characteres in the buffer. After that buffer is empty and the time-out again is set.

If the time-out is not yet expired but the buffer is filled up then the interrupt request is generated immedeately. The arrival of the additional character before the place for at least one character becomes available in the buffer causes the DATA OVERRUN error.

We consider two phases of buffer queue behavior: from the moment of time-out set up to the begining of the processor service, and from the begining of service to the moment of emptying the buffer. Let M be the buffer size and m_k be the average number of characters (customers) which entered the buffer during the second phase under the condition that there are exactly k customers at the begining of second phase $(k=\overline{0,M})$. Let m'_k be the average number of customers, damaged due to DATA OVERRUN during the second phase under the condition described above $(k=\overline{0,M})$. Then the following expressions take place: (It is assumed that the input stream of customers is Poisson with total intensity λ.)

$$\begin{cases} m_k = r\,(1+m_{k+1}) + q\, m_{k-1}, & k=\overline{1,M-1} \\ m_M = r\,(1+m_M) + q\, m_{M-1} \\ m_0 = 0 \end{cases}$$

$$\begin{cases} m'_k = r\, m'_{k+1} + q\, m'_{k-1}, & k=\overline{1,M-1} \\ m'_M = r\,(1+m'_M) + q\, m'_{M-1} \\ m'_0 = 0 \end{cases}$$

where $r=\rho/(1+\rho)$, $q=1/(1+\rho)$, $\rho=\lambda/\mu$.
Hence it can be shown that

$$m_k=\frac{\rho}{1-\rho}\left[k+\frac{\rho^{M+1}-\rho^{M-k+1}}{1-\rho}\right],$$

$$m'_k=\frac{\rho^{M-k+1}-\rho^{M+1}}{1-\rho}$$

Then the portion of damaged customers can be found as

$$P=\frac{\sum_{j=1}^{M-1} s_j m'_j + \left(\sum_{j=M}^{\infty} s_j\right) m'_M}{\sum_{j=1}^{M-1} s_j (j+m_j) + \left(\sum_{j=M}^{\infty} s_j\right)(M+m_M)},$$

where

$$s_j=\frac{(\lambda t_{out})^j}{j!}e^{-\lambda t_{out}},$$ t_{out} is a duration of time-out,

and the relative throughput V_0 can be estimated.

3. DDCMP-CONTROLLED DATA LINK MODEL

The head of the DDCMP protocol frame has the length of 8 bytes and includes a data frame identifier, the byte count field (it specifies the number of bytes in the DATA field), link control flags, the number of receiving frame (ACK), the number of given frame, node to which the frame is assigned and the block check on the frame header. The transparency is attained due to a simple count of the number of bytes in the data field. For this purpose, the head of the frame is first received and checked up. It provides the data field length counter subsequently used to receive the remaining part of the frame. The data field is terminated with a special checking sequence. Besides each frame is preceded by at least two synchrobytes. In a HDLC frame the control information takes up 6 bytes.

The asynchronous balance mode (ABM) of the HDLC protocol is analytically studied in [8]. This paper makes use of the approach suggested in [8] to compare the efficiency of the DDCMP protocol against of HDLC.

Two nodes A and B connected by a duplex data link are considered. Each I-frame contains one packet. The packets are of the fixed length ℓ bit. Data link transmission rate is C bit/s in both directions. The bit-error probability of channels is P_0. The rate of arrivals at node A(B) of packets to be transmitted to node B(A) is $\lambda_A (\lambda_B)$.

The protocol performance analysis shows, that the probability of error within an I-frame head or S-frame (i.e. within 8 bytes) can be neglected [8].

The characteristics to be estimated are:

$\overline{T}_f$: mean transfer time of I-frame. (Time intervals between the arrival of a packet at the sending node and end of its succesful transmission).

V: channel throughput in saturated case.

$\overline{T}_b$: mean buffer holding time. (The time a packet remains in the buffer of the sending node). Let $T_v(i)$ be virtual transmission time of I-frame no. i, i.e. the time interval between the start of transmission of the first I-frame no. i after the succesful transmission of I-frame no. (i-1) and the end of its succesful transmission. It is possible to assume [8], that λ_A is the parameter of exponential distribution, and $T_v(i)$ are random independent identically distributed variables.

Under the assumptions described above, each node is M/G/1 queue, for which T_v is random service time of customers and T_f is random sojourn time.

Therefore, the average value of T_f can be determined from the Pollaczek-Khintchine formula:

$$\overline{T}_f = \overline{T}_v + \frac{\lambda_A \overline{T_v^2}}{2(1-\lambda_A \overline{T}_v)} .$$

Hence, our problem is to determine the first two moments of the T_V. The probability of error within information field of I-frame is

$$P = 1-(1-P_0)^{\ell} .$$

Introduce the following denotation of random variables:

$T_c(i)$: the time interval between the end of ith transmission of disturbed I-frame and the start of its (i+1)st transmission;

T_1: the component of T_c, the time interval between the end of transmission of disturbed I-frame and the start of NAK-frame (negative receipt) transmission from nodeB;

T_2: the component of T_c, the time interval between the end of NAK-frame receiving at node A and the start of the next retransmission of disturbed I-frame.

Then we can write:

$$T_V = \sum_{i=1}^{N+1} t_I(i) + \sum_{i=1}^{N} T_c(i) , \tag{10}$$

where $t_I(i) = t_I$ is constant transmission time of an I-frame; N is random number of recovery cycles which is distributed according to

$$P[N=k] = (1-P)P^k, \qquad k=\overline{0,\infty} . \tag{11}$$

From this we have

$$\overline{T_V} = \frac{t_I}{1-P} + \frac{P\,\overline{T_c}}{1-P} ,$$

$$\overline{T_V^2} = \frac{t_I^2(1+P)}{(1-P)^2} + \frac{4P t_I \overline{T_c}}{(1-P)^2} + \frac{P\,\overline{T_c^2}}{1-P} + \frac{2P^2(\overline{T_c})^2}{(1-P)^2} , \tag{12}$$

$$T_c = T_1 + t_S + T_2 ,$$

where t_S is a constant transmission time of an S-frame. If we assume independence of variables T_1 and T_2, then

$$\overline{T_c} = \overline{T_1} + t_S + \overline{T_2} ,$$

$$\overline{T_c^2} = \overline{T_1^2} + \overline{T_2^2} + t_S^2 + 2\,\overline{T_1}\,\overline{T_2} + 2\,t_S\,\overline{T_1} + 2\,t_S\,\overline{T_2} ,$$

$$\overline{T_1} = Q(2)\,t_I/2 , \qquad \overline{T_2} = Q(1)\,t_I/2 , \tag{13}$$

$$\overline{T_1^2} = Q(2)\,t_I^2/3 , \qquad \overline{T_2^2} = Q(1)\,t_I^2/3 ,$$

with Q(1) and Q(2) are utilizations of the channels A $\longrightarrow$ B and B $\longrightarrow$ A, respectively. It is shown [8], that

$$Q(1) = \lambda_A t_I \left[1 + P(1 + Z_2(1))\right] ,$$

with

$$Z_2(1) = \lambda_A t_I + (1-\lambda_A t_I)\lambda_B t_I \left(1 - \frac{1-e^{-\lambda_A t_I}}{\lambda_A t_I}\right) . \tag{14}$$

Utilization Q(2) is determined in a similar way. Thus, first two moments of the T_v are obtained. Mean buffer holding time is

$$\overline{T}_b = \overline{T}_f + \overline{T}_{ACK}$$

where $\overline{T}_{ACK}$ is mean acknowledgement time (i.e. the time interval between the of succesful transmission of I-frame and the end of ACK-frame receiving at the sending node). The estimate of T_{ACK}, obtained in [8], is also valid for the protocol DDCMP:

$$\overline{T}_{ACK} = [1-Q(2)]e^{Q(2)}t_S + [1+Q(2)/2+Q(2)e^{Q(2)} - e^{Q(2)}]t_I. \quad (15)$$

For the computation of $\overline{T}_v$ in saturated case formulas (10) - (15) can be used with Q(1) = Q(2) = 1. The channel throughput in such a case is $V = \ell/\overline{T}_v$, and the relative throughput is $V_0 = V/C$.

4. NUMERICAL RESULTS

The performance of different types of processors with the devices described in Section 2 was considered under the condition that data transmission rate of channels is 9600 bit/s. If the DL-11 devices are used then with the increase of the number of links, the normalized throughput V_0 grows linearly first, then achieves some maximum value and just after that falls sharply. Fig. 1 shows a typical form of curves V_0 vs N. For example in the case of processor PDP-11/34, maximum 7 channels may be connected to the node. If the CAMAC modules are used, then with the number of links growing, the normalized throughput grows a lower speed than for the DL-11, but this growth is monotonic.

Fig. 2 presents a dependance of normalized throughput V_0 on length of frame's data field in a saturated case for the protocols DDCMP and HDLC. The numerical results for HDLC are obtained from [8]. The duration of time-out is $t_{out}=3t_I$. The protocol HDLC is better than DDCMP for small values of ℓ in a sence of a normalized throughput, because of a shorter head of I-frames and shorter S-frames. But if ℓ increases then the probability of error within information field of I-frames increases too; and as a result DDCMP becomes better than HDLC. The advantage of DDCMP increases with the

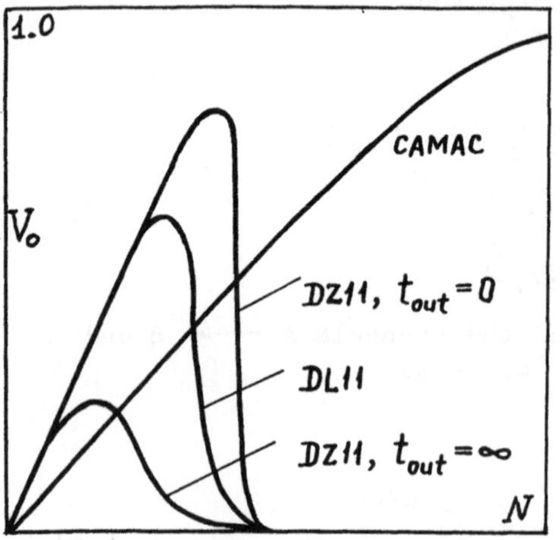

Fig. 1

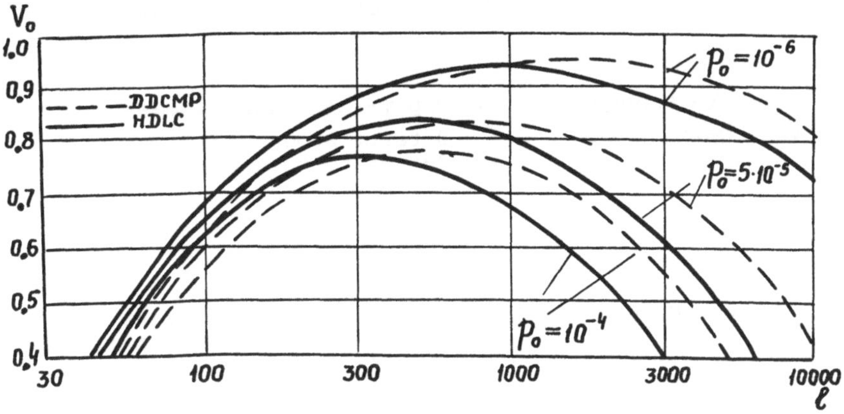

Fig. 2

increase of bit error probability. The protocol DDCMP quickly transmits the frames to a neighbouring node and withdraws the copies of frames in the fuffer because the losses of receipts practically take no place.

REFERENCES

[1] DECnet DIGITAL Network Architecture. General Description.Order No.AA-K 179 A-TK.

[2] Seligman D.R. On the performance evaluation of DECnet. 2-nd Intern. Conf. on Distrib. Comput. Syst., IEEE N.Y., 1981, 484-496.

[3] Boguslavsky L.B. et al. Performance evaluation and measurements of micro/minicomputer networks with DECnet software. 10th All-Union Workshop on computer networks, 1985, Scientific Council of Cybernetic, Moscow, pt.3, 254-259 [in Russian].

[4] Frank F. et al. MININET. GMD Gesellschaft Fuer Electron. Datenverarb., Tuebingen, West-Germany, 1981.

[5] Vasilyev G.P. et al. The software of SM computer networks. "Finans i Statistica", Moscow, 1983 [in Russian].

[6] DNA Digital Data Communications Message Protocol (DDCMP) Functional Specification, ver. 4.1.0, Ord. No. AA-K175-TK.

[7] HDLC-Proposed Balanced Class of Procedures. Document ISO/TC 97/SC6 - No.1444.

[8] Bux W. et al. Balanced HDLC procedures, a performance analysis. IEEE Trans. Commun. v.28, No.11, 1980, 1032-1041.

[9] König D. et al. Verallgemeinerunden der Engsetscher und Erlangsche Formeln.- Berlin: Springer-Verlag, 1967.

OPTIMAL CONTROL METHODS FOR POWER SYSTEM OPERATION

Dipl.-Ing. Reinhard Brehmer
Wr.Stadtwerke E-Werke (WSE)
Mariannengasse 4-6
A - 1090 Wien

Dipl.-Ing. Peter Nemetz
Österr. Brown-Boveri Werke
Pernerstorfergasse 94
A - 1101 Wien

Summary

Aspects of optimal power system control (structures, tasks etc.) are discussed in general and in the application for the new network control system of the Viennese Electricity Board (WSE), with the main emphasis on the district control centres.
The problems which arise from information overload due to enormous data collection and sophisticated on-line applications are shown and the steps undertaken within the new WSE control system to overcome them are presented.

1.0 Typical Structures and Tasks of Control Systems

Large electrical power systems are commonly structured in several hierarchical levels. This mainly reflects the structure of the power system itself - high-, medium- and low-voltage distribution network - and is found again in the structures of the control systems, which operate the process of power generation and distribution.

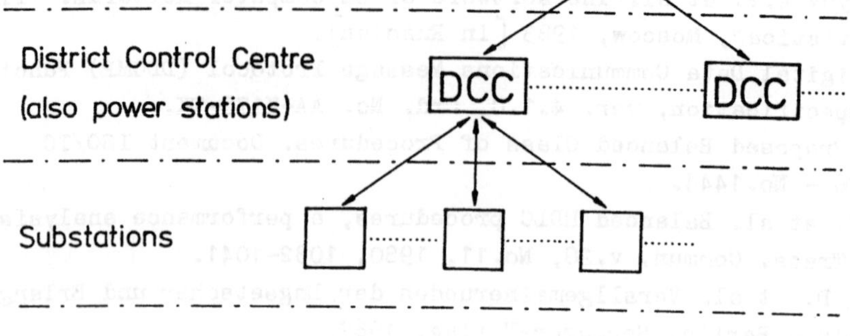

Fig.1. Structure of an electric energy network control system

Fig. 1 shows the model of an overall control structure for power system networks, used by IEC TC 57 (1). Typical tasks of system control can be found on each hierarchical level:

- Main and regional control centres are mainly concerned with load management, economic generation and distribution of energy, security analysis, short and medium term planning etc.(2,3). System control is restricted to the high-voltage distribution network (380 kV in Vienna).
- Network control and data acquisition are the principal tasks of district control centres (4). All data from the substations of the distribution network are transmitted via telecontrol equipment to the DCC and a reduced set of information - mainly high level-voltage network data - is passed to the regional control centre via computer links.

On all hierarchical levels a comfortable man-machine-interface (MMI), using colour display units, clearly arranged keyboards etc. is necessary to present the system state to the operators. This is a central part of every control system and must help to avoid problems, which arise from the handling of the great amount of data, which is transmitted to the control centres in case of blackouts.

With the growing size and complexity of power-systems, computer aid is necessary. PAS (Power Application Software) helps the operator to control the network in normal and emergency situations (5). Both actual network data and the results of PAS calculations are available to the operator, but it is of great importance that he is not overloaded by too much information (6).

Power systems have grown continuously in size and complexity over years, while the technological development of control equipment and methods has proceeded much faster during the same period. So every few years new solutions and concepts on process computer hardware and software, as well as on telecontrol systems become available. Today, control equipment used in most large power systems consists at least of two mainly different generations and/or different manufacturers. The introduction of new concepts of power system control at every hierarchical level has to take into account following main goals:

- system control must not be interrupted,
- optimization of control equipment availability and reliability,

- acceptance by the operators,
- data reduction to specific needs,
- cooperation with existing control equipment,
- minimization of implementation costs.

In these last years with rapidly increasing computer performance accompanied by decreasing costs, it becomes possible to include more and more highly sophisticated and computing time consuming programes and applications like database management, on-line PAS, comfortable MMI etc. While this significantly raises the probability of the results of e.g. state estimation or network security analysis and supplies a lot of reliable network data for the operator, it also slows down other, often more important tasks. In some cases to such an extent, that response times become almost unacceptable.

2.0 Network of the Viennese Electricity Board (fig.2)

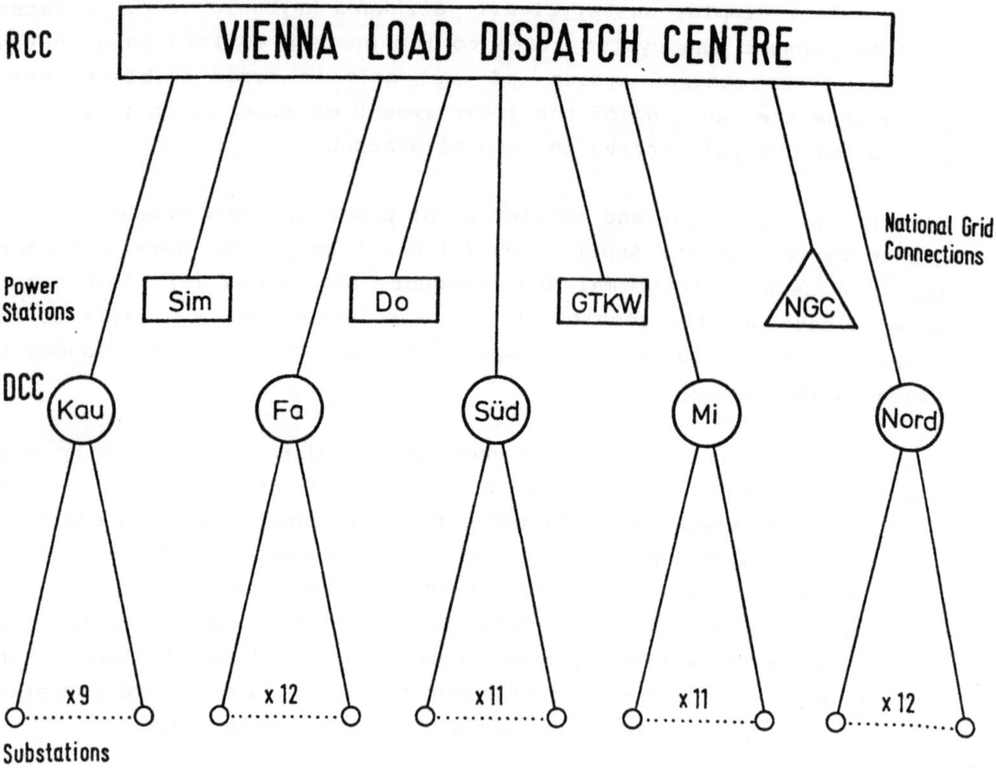

Fig.2. Vienna Electricity Board (WSE) network configuration.

In 1984 the WSE supplied about 6.6 TWh, while the peak load of 1985 was 1325 MW. Up to 60 % of the electrical energy is generated in own power stations - mainly in the two large thermal power plants Simmering and Donaustadt. The rest is supplied by the Austrian national grid via two main connections at Bisamberg and Sued Ost.

The high voltage network in Vienna consists of 380 kV and 110 kV lines, which transport the energy to 48 substations, where it is transformed to 20 kV or 10 kV. A bigger substation typically consists of 2 or 3 voltage levels with around 100 high voltage lines and 2000 alarms. The structure of the transportation network is meshed. These substations are controlled by five DCCs.

The medium voltage network consists of 10 kV cables in the town and of 20 kV lines in the surroundings of Vienna. The structure of the distribution network is of star configuration.

3.0 New Control System of WSE (7)

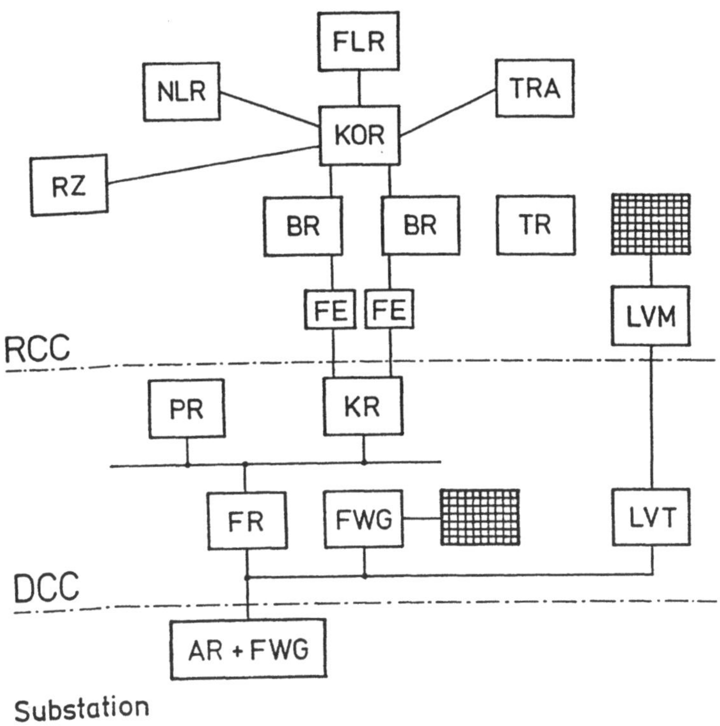

Fig.3. New WSE network control computer system

3.1 Control Equipment(fig.3)

The WSE network automation concept has a decentralized structure. Connected on-line to the operation computers (BR) via computer links and a communication computer (KOR) are seperate computers for main tasks. One is used for automatic generation control (NLR), one for ripple control (TRA) and one for energy interchange scheduling and monitoring (FLR). There also is a connection to WSE's commercial computer centre (RZ), which consists of IBM mainframes.

The operation system includes double computers Siemens PR 3267, working in a hot stand by mode, plus an additional one for testing purposes (TR). Coupling to the DCCs is achieved by front ends (FE). The DCC is explained in detail in 4.1.

All information from the substations is transmitted from the substation computer (AR) via telecontrol equipment (FWG) to the telecontrol computer (FR) of the DCC-system and to the mimic board of the DCC.

A condensed set of alarms and data is also brought to the mimic board of the RCC control room via the LVT- and LVM-processors. This serves as a reduced back-up system in case of computer breakdowns.

3.2 Control Methods

As shown in figure 3 and explained above, the control of the power system operation takes place on three distinct levels with their different tasks and needs.

3.2.1 Substation

A condensed alarm indication is available in case of failure of the telecontrol system. This alarm indication consists of a selection of the most important ones (approx. 1/10 of all) and their handling is very comfortable, via CRT, alarm memory and alarm descriptions (where, what to do, plan numbers etc).

3.2.2 District Control Centre

Network operation is the main task - besides data transmission to the RCC-, which means that fast and save data aquisition from the substations is needed. Important requirements are: safety, speed, selection.

The difficulty lies in the different selections needed for different operational tasks (details in 4.2).

From the viewpoint of system control the two thermal power stations Simmering and Donaustadt are of the same type as the district control centres. Similar control methods are used to present the operating state of the power station with its several thermal units.

3.2.3 Regional Control Centre

The three main areas of applications are:

- Control of the 380 und 110 kV network:
  Besides a comfortable MMI system, PAS functions, like state estimation, loadflow calculation, security analysis, single and three phase short circuit calculation are available to assist the operators.
- Energy management:
  Although an off-line package for optimal energy production is already available and in operation, on-line functions like energy optimization, load forecast, unit commitment, load shedding using ripple control etc. are planned as future implementations.
- Historical data:
  All alarms and a condensed set of measurements are stored on a long term basis. Sophisticated search and display routines allow fast access and comfortable handling of this information.

4.0 District Control Centre

4.1 Control Equipment (Partnerbus, fig.4)

The reliability of big centralised information systems is another main point of interest. One either has to double systems completly or install a reserve system, with often severly reduced information. The later means that detailed data are not any more available in case of single computer system failures.

So when the opportunity arose to redesign the district control centres, steps were undertaken to overcome the problems mentioned above. Instead of a single computer, distributed system technology will be used. A local area network (LAN) with three computers will be installed.

This becomes the main feature of the new DCC. Three minicomputers are connected via the PARTNERBUS (8) to a local area network. The Partnerbus is designed for high-speed, efficient and secure data communication between up to 32 computers. It supports broadcast communication in which all data are delivered simultaneously to all connected partners. Each Partnerbus interface is able to select types of messages, transmitted on the bus, and pass only that types to the partner, which it has specified. This specification of data types can be modified online. Important features are data transfer from the interface to each partner in DMA mode, keeping computer load for data communication low, and configuration control with automatic adaption of communication in case of changes in the number of connected partners.

The DCC functions can be seperated into three groups, which correspond with the chosen hardware configuration (fig 4.):

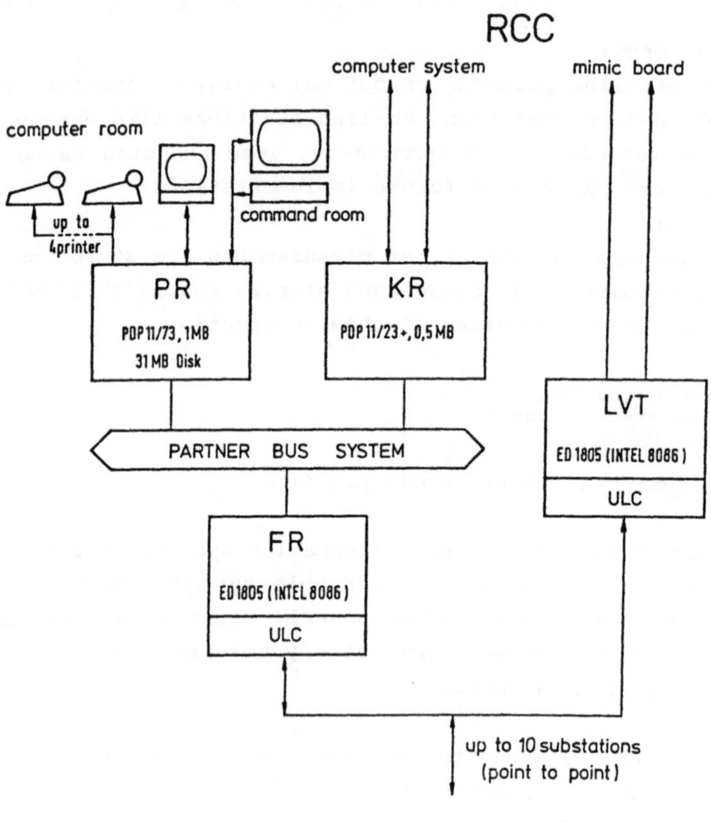

Fig.4. New WSE district control centre computer system

- FR - Telecontrol computer
  The FR is a multiprocessor system built up of several microcomputers ED 1805, using Intel 8086 CPU's. Two substations are connected to one ED 1805, which acts as a universal line coupler (ULC) and holds a small substation oriented data base. Network data is converted to a format which allows comfortable und efficient processing by all and/or transmission to all other computers. A master ED 1805 coordinates the information transfer from and to the Partnerbus.
- PR - Protocol Computer
  It performs all MMI functions and holds the central data base for the whole control system in the DCC. Therefore software maintainance, modifications and expansions are performed centrally on the PR. Subsets of the data base are generated on the PR and then downline loaded via the Partnerbus to the other computers.
- KR - Link Computer
  Its task is the communication with the RCC. Transmission takes place parallel on two lines, with each line connected to a frontend computer in the RCC. In the KR a data buffer of approx. 2000 messages is available, so no information is lost in case of transmission interruptions.

Further independent telecontrol lines from each DCC to the RCC are installed (LVT, LVM), as explained in 3.1. The equipment used is of the same type as the FR.

4.2. Operation

As this is the second computerized telecontrol system for WSE district control centres, experiences of the first one (9) have been used to improve network operation.

The first computerized remote control system by WSE was installed in 1977, already with computers in each of the substations too. Since then real time data from the substations are available in the control rooms both on CRT and data loggers. Back-up facilities are realized by a reserve system with mimic board and around 100 alarms per substation.

Besides the well known advantages of having exact data fast at hand, new problems did arise. Computers can handle much more data and do that much faster then man. With more and more substations connected to the system (48 in 1985 with up to twelve per control room) the amount of data presented to the operator in case of bigger break-downs reached such proportions, that it could be of no help at all. The effect was even to the contrary, with the operator using the reserve system only, instead of the data presented by the computerized system.

Instead of leaving out less important information altogether, all available data are still brought to the control room, but can now be sorted in several ways, regarding to the specific needs of the operator. Following features are available and easily used with a comfortable MMI based on menue technique:

- Coloured alarm list
  Each alarm group (e.g. busbar voltage, protection etc.) is easily recognized by its own colour.
- Data selection per function keys and menues
  It is possible to select information after time, substation, voltage level and type of information (like protection alarms, transformer faults, busbar voltage alarms etc.). Such a selection may for instance show all 20 kV protection and busbar voltage alarms of a specific substation between 8 a.m. and 9 a.m..
- Logic ties between certain alarms generate new alarms
  E.g. when a transformer trips, the system tells, whether the cause was a transformer fault or back up protection.
- Check and help lists for important breakdowns
  They help to find the cause and give advice what to do, e.g. tell what to do when a transformer was tripped by a differential current relay.
- Protection overview
  Protection information is shown in a condensed sketch of a substation - just busbars and transformers - at the point it occurred (voltage level, transformer).
- Training facilities with actual breakdown data
  To obtain that, actual break-down data (alarm list) are stored. After a description has been added, all is presented to the operator together with a dialog. He is asked what measures he would take and the answers he chooses are commented. In this training mode he can use all the normal selection functions on the alarm list of the break-down to help him in his decisions. In this way he not only learns what to do in typical break-down situations, but also to use the tools the computer provides.

The all too often neglected data presentation, which is at least equal important to data aquisition, becomes a main topic.

5.0 Conclusion

So after the first step of computerized control, that is including in the system as much information as is possibly needed, the next step is the reduction to few but highly relevant data presented by a well designed MMI. This second step is not only the more important one, but also the much more difficult one.

Accepting, transmitting and presenting of information is not any more enough. Concentrating and priority structuring to create more user friendliness becomes the main point in the use of computerized control methods in energy distribution systems. This helps to prevent serious problems, which operators otherwise would have to face, due to system information overload from increasingly sophisticated data collection sources.

References

(1) IEC Technical Committee No. 57: Telecontrol, Teleprotection and Associated Telecommunications for Electric Power Systems. Draft: Telecontrol Equipment and Systems. Part 1: General Principles

(2) G. Schaffer, M. Davis and H. Sandmayr: "Supervisory Network Control Systems Reflect User Requirements". Brown Boveri Rev. 71 1984 (9/10) 393-399

(3) Dy Liacco, E. Handschin, H. Glavitsch et al.: Report of Symposium: "Control Application for Power System Security." Florence 1983, Electra Nr.92, Jan. 1984

(4) P. Nemetz: "Design Considerations for Regional Control Centres". IEE Conference Publication 161, 1978, 67-71

(5) G. Schaffer: "Power Application Software for the Operation of Power Supply Systems". Brown Boveri Rev. 70 1983 (1/2) 28-35

(6) D.B. Goudie, M.Davis and A. Spatz: "The BECONTROL Family of Supervisory Network Control Systems". Brown Boveri Rev. 71 1984 (9/10) 406-415

(7) G. Schimkowitsch, J. Schubert, L. Hanzl: "Problemanalyse für Lastverteileraufgaben und das Lösungskonzept der WSE", E. + M. 1985

(8) J. Muheim: "BBC-Partnerbus - A Communication Bus Linking Computers in Local Area Networks". Brown Boveri Rev. 72 1985 (5) 241-248

(9) J. Schubert: "Das Netzautomatisierungskonzept der WSE" ÖZE 30. Jg. (1977), H4, Seite 170 - 1984

MODELS AND METHODS FOR ESTIMATING AN ORIGIN-DESTINATION TRIP MATRIX FROM NETWORK DATA

Maud Brenninger-Göthe and Kurt Jörnsten
Linköping Institute of Technology
Department of Mathematics
S-581 83 Linköping, Sweden

ABSTRACT

In this paper we present mathematical programming models for the problem of estimating a trip matrix from network data. The models presented are based on the results of Nguyen (1977) who has shown that trip matrices that reproduces observed linkflows in a congested network can be obtained by solving an elastic demand traffic assignment problem with a specific linear demand function.

It can be shown that this elastic traffic assignment problem has a unique optimal solution with respect to the linkflows but that the resulting trip matrix not necessarily is unique.

The mathematical programming models presented will deal with the problem of deriving one of the optimal trip matrices from Nguyens elastic demand traffic assignment problem.

Models with different choice criteria will be discussed and solution methods based on decomposition techniques will be presented for the various models.

The presented models will all be examples of implicit optimization models, i.e. optimization models in which the constraints involve another optimization model.

1. INTRODUCTION

In this paper we present some models and solution methods for the problem of estimating a trip matrix in a congested urban area. Nguyen (1977) has formulated an elastic demand traffic assignment model which solution reproduces the observed link flows. The assumption behind the model is that the traffic in the area which is beeing studied satisfies the equilibrium property.

The solution, which is obtained when Nguyen's model is used, is unique in the arc flows but not necessarily unique in terms of the trip matrix.

Another approach for estimation of a trip matrix is entropy maximization (Wilson (1970)) where the most probable trip matrix is obtained by the maximization of an entropy objective under some basic constraints.

The entropy function can also be used as a distance measure when a target trip table (i.e. a base year observation) is known. Other models for estimation of trip matrices based on different distance measure have been presented by Turnqvist, Gur (1977) and LeBlanc, Farhangian (1982).

In a paper by Jörnsten, Nguyen (1979) an entropy maximizing model is presented, which also includes equilibrium assignment conditions. The entropy function is here used to choose among all the optimal solutions to Nguyen's model the trip matrix which is the most probable. A cutting plane method for calculating this trip matrix is also presented. This solution method has been slightly extended by Brenninger-Göthe, Jörnsten (1983).

When using the entropy approach suggested by Jörnsten and Nguyen an overestimation of the short trips is likely to occur. This problem can be handled by using a target trip matrix. For the case when this information is not at hand Brenninger-Göthe, Jörnsten (1985) have proposed a model that can be used in order to obtain a more realistic trip distribution than the one obtained using a pure entropy objective.

2. THE ELASTIC TRAFFIC EQUILIBRIUM MODEL FOR TRIP MATRIX ESTIMATION

The problem of assigning traffic consistently with Wardrop's principles can be formulated as an optimizing problem, see Beckmann et al. (1956). Wardrop's first principle can be viewed as the case where the interzonal travel costs along all paths which are used between two zones are equal and where no unused path has an interzonal travel cost which is less than the one for the used paths. Due to the assumption that each user of the network seeks to minimize his own travel cost, no single traveller can reduce his own cost by choosing another path in the network. This is the so called equilibrium assignment or the user-optimized assignment principle.

The second principle formulated by Wardrop (1952) corresponds to a situation were the total network cost is minimized, i.e. a system-optimized assignment of traffic.

We consider an urban network described by arcs and nodes respectively corresponding to streets and street intersections/zones. Some of the nodes are origin and/or destination nodes. For each arc there exists an associated travel cost function $S_a(f_a)$, describing the travel time on arc a as a function of the total flow, f_a, on arc a. The travel cost function is assumed to be continuous and strictly increasing.

It should be noted that if the arc travel cost function S_a also depends on other arc flows than f_a, i.e. there is an interaction between arcs in the network, the Jacobian of the travel costs is usually asymmetric. In this case it is not possible to formulate the equilibrium problem as an optimization problem analogously to the today known optimization formulation for the symmetric case, see Brenninger-Göthe, Larsson (1985).

Let g_{pq} denote the number of trips between origin p and destination q. We assume that the monoton decreasing demand-function $D_{pq}(u_{pq})$, where u_{pq} denotes the interzonal travel time in relation p,q, for g_{pq} is known. We also assume that the inverse demand function $G_{pq}(g_{pq})$ exists and is integrable.

The problem of estimating a trip matrix can be formulated as the elastic traffic equilibrium problem

$$\text{P1:} \qquad \min\ F(f_a, g_{pq}) = \sum_a \int_0^{f_a} s_a(x)dx - \sum_p \sum_q \int_0^{g_{pq}} G_{pq}(x)dx \tag{2.1}$$

$$\text{s.t.} \quad \sum_\ell h_{\ell pq} - g_{pq} = 0 \qquad , \forall p,q \tag{2.2}$$

$$h_{\ell pq}\ ,\ g_{pq} \geq 0 \qquad , \forall \ell,p,q \tag{2.3}$$

$$f_a = \sum_p \sum_q \sum_\ell \delta_{a\ell pq}\ h_{\ell pq} \qquad , \forall a \tag{2.4}$$

where $h_{\ell pq}$ is the flow on path ℓ in relation p,q, and $\delta_{a\ell pq}$ is equal to 1 if arc a belongs to the path ℓ in relation p,q and 0 otherwise. By formulating the Kuhn-Tucker conditions of problem P1 it can be shown that the solution satisfies the equilibrium property. The constraints of problem P1 constitute the arc-chain formulation of the network flow equations. In the case of strict convexity the solution is unique in the arc flows f_a and in the demands g_{pq}. However, the path flows $h_{\ell pq}$ are not necessarily unique.

2.1. Nguyen's elastic traffic equilibrium model for trip matrix estimation

Nguyens model for trip matrix estimation can be viewed as a special case of problem P1 where the inverse of the demand function is constant and equal to $\hat{u}_{pq}$. We consider a situation where observed arc flows $\hat{f}_a$ for each arc in the network is known and where the arc flow pattern $\{\hat{f}_a\}$ is assumed to be at an equilibrium. By computing the shortest paths between each origin p and each destination q in the network, with arc lengths fixed to $s_a(\hat{f}_a)$, we obtain the equilibrium interzonal travel costs u_{pq}.

Nguyen's elastic traffic equilibrium problem that reproduces observed flows can be formulated

$$\text{P2:} \qquad \min\ F'(f_a, g_{pq}) = \sum_a \int_0^{f_a} s_a(x)dx - \sum_p \sum_q \hat{u}_{pq}\, g_{pq} \tag{2.5}$$

s.t. (2.2), (2.3), (2.4).

The solution of this problem yields a trip matrix $[g_{pq}]$, which when assigned onto the network according to the user-optimized routing closely reproduces the observed flows $\hat{f}_a$. Nguyen (1977) showed that the solution in the arc flows to problem P2 is unique and thus equal to $\hat{f}_a$. However, due to the fact that (2.5) is not strictly convex with respect to the trip matrix variables g_{pq}, there can exist more than one optimal trip matrix.

An elastic traffic equilibrium problem of the type P2 can efficiently be solved by the use of feasible-direction methods. One frequently used solution method is the Frank-Wolfe algorithm. The efficiency of these methods is due to the fact that the linearized subproblem can be interpreted as a mathematical formulation of the all or nothing traffic assignment and the subproblem decomposes into one shortest path problem for each origin.

For a more detailed description of solution methods see Nguyen (1976), Florian et al. (1977), Gartner (1980).

3. ENTROPY MAXIMIZING MODELS FOR TRIP MATRIX ESTIMATION

In transportation planning the entropy function has been used to estimate the most probable trip matrix (Wilson (1970), Erlander (1980)).

We consider a situation where the marginal totals, i.e. the total number of trips leaving and entering a zone is known. This information is given for all zones in the area we study together with the interzonal travel costs $\hat{u}_{pq}$.

Wilon's entropy maximization model is as follows

$$max \quad H(g_{pq}) = - \sum_p \sum_q g_{pq} \log g_{pq} \tag{3.1}$$

$$s.t. \quad \sum_q g_{pq} = S_p \qquad , \forall p \tag{3.2}$$

$$\sum_p g_{pq} = T_q \qquad , \forall q \tag{3.3}$$

$$\sum_p \sum_q \hat{u}_{pq} g_{pq} \leq \hat{C} \tag{3.4}$$

$$g_{pq} \geq 0 \qquad , \forall p,q \tag{3.5}$$

where S_p and T_q are the total number of trips leaving zone p and entering zone q respectively. $\hat{C}$ is an estimate of the cost of the total number of trips in the network. The constraints on the marginal totals can, in cases when this information is not at hand, be replaced by a more aggregate constraint on the total number of travellers in the network, i.e. $\sum_p \sum_q g_{pq} = \hat{G}$.

Wilson has shown that the solution to this problem is of the gravity model type. By stating the Kuhn-Tucker conditions we obtain the optimal solution

$$g^*_{pq} = exp\ (-\alpha^*_p - \beta^*_q - \gamma^* \hat{u}_{pq} - 1)$$

where the α^*_p's, β^*_q's and the γ^* are the dual variables to (3.2), (3.3) and (3.4), respectively.

Another approach is the minimum information principle (Snickars, Weibull (1977)), where the entropy function is defined as $H(g_{pq}) = - \sum_p \sum_q g_{pq} \log (g_{pq}/g^0_{pq})$. Here the element g^0_{pq} is a base year observation and the solution to the problem is a trip matrix $[g^*_{pq}]$ which is as "close" as possible to the information g^0_{pq}.

Erlander's models is based on the classical Hitchcock-Koopmans transportation model. Here the entropy-constraint $- \sum_p \sum_q g_{pq} \log g_{pq} \geq H_0$ is added, where H_0 is an estimate of the entropy demand from a base year.

3.1. A maximum entropy model including equilibrium assignment conditions

The entropy maximization model can be extended to include equilibrium assignment conditions if counts are available for the arcs in the network. Jörnsten, Nguyen (1979) have presented such a model, where the entropy is used to choose among all the optimal solutions to Nguyen's model (P2) the matrix which is the most probable. If a base year observation $[g^0_{pq}]$ is known the minimum information principle can also be used.

The most probable trip matrix $[g_{pq}] \in X$ = {optimal solutions to P2} is obtained by solving the problem

$$\text{P3:} \quad \left.\begin{aligned} & max \; - \sum_p \sum_q g_{pq} \log g_{pq} \\ & s.t. \; \sum_p \sum_q g_{pq} = \hat{G} \\ & \qquad g \in X \end{aligned}\right\} \tag{3.6}$$

The set X can be represented in a number of ways. Jörnsten-Nguyen used the formulation

$$\sum_p \sum_q \hat{u}_{pq} \, g_{pq} = \hat{C} \tag{3.7}$$

$$\left.\begin{aligned} & \hat{f}_a = \sum_\ell \sum_p \sum_q \delta_{a\ell pq} \, h_{\ell pq} & , \forall a \\ & \sum_\ell h_{\ell pq} - g_{pq} = 0 & , \forall p,q \\ & h_{\ell pq} \geq 0 & , \forall \ell,p,q \end{aligned}\right\} \tag{3.8}$$

$$g_{pq} \geq 0 \quad , \forall p,q \tag{3.9}$$

The problem (3.6) - (3.9) is a convex cost multicommodity network circulation flow problem with coupling constraints. In practical applications it has a very large number of constraints.

The solution method suggested by Jörnsten-Nguyen is based on a cutting plane (decomposition) technique. This is done by relaxing the network constraints (3.8) and then solving the entropy maximization problem (3.6), (3.7), (3.9). The solution $[g_{pq}^{(1)}]$ is then tested for feasibility by solving a fixed demand traffic assignment problem.

If $[g_{pq}^{(1)}] \in X$ then the trip matrix is optimal, otherwise a new constraint of the form

$$\sum_{p}\sum_{q} u_{pq}^{(k)} g_{pq} - \sum_{a} s_a(f_a^{(k)}) \hat{f}_a \leq 0$$

is added to the master entropy maximization problem which is then resolved for a new trip matrix suggestion $[g_{pq}^{(k+1)}]$. The interzonal travel costs $u_{pq}^{(k)}$ and the arc flows $f_a^{(k)}$ are obtained from the optimal solution to the fixed demand traffic assignment problem. Since no feasible trip matrix is obtained before the method has converged, the method can be viewed as a pure dual method. In Jörnsten-Nguyen (1979) the method is described more in detail.

A modification of the Jörnsten and Nguyen solution method has been presented in Brenninger-Göthe, Jörnsten (1983). Here an extra elastic demand problem is solved before the new cutting plane inequality is appended to the master problem. This extra step consists in solving the original Nguyen elastic traffic equlibrium model using $(f_a^{(k)}, [g_{pq}^{(k)}])$ as a starting point. By using this extension a feasible trip matrix that reproduces observed link flows $\hat{f}_a$ is obtained in every iteration and thus information giving both lower and upper bounds on the objective is at hand. The solution method has been tested on the Sioux-falls network and compared with the Nguyen-Jörnsten method. The test showed a reduction of the number of iterations from 6 to 2. The difference between the upper and lower bound of the entropy is within 5%.

Another class of methods based on Lagrangian relaxation techniques and another formulation of the set X are the dual methods (Jörnsten, Nguyen (1983), LeBlanc, Farhangian (1982)). Here one first solves the original

Nguyen model in order to obtain the optimal objective function value

$$\hat{F} = \sum_a \int_0^{f_a^*} S_a(x)dx - \sum_p \sum_q \hat{u}_{pq}\, g^*_{pq}$$

and a lower bound of the entropy

$$\hat{H} = - \sum_p \sum_q g^*_{pq} \log g^*_{pq}$$

Once this has been done either problem P4 or P5 is solved.

P4: $$\max \; - \sum_p \sum_q g_{pq} \log g_{pq} \tag{3.10}$$

$$\text{s.t.} \quad \sum_p \sum_q g_{pq} = \hat{G} \tag{3.11}$$

$$\sum_\ell h_{\ell pq} - g_{pq} = 0 \qquad , \forall p,q \tag{3.12}$$

$$\sum_a \int_0^{f_a} S_a(x)dx - \sum_p \sum_q \hat{u}_{pq}\, g_{pq} \leq \hat{F} \tag{3.13}$$

$$h_{\ell pq},\, g_{pq} \geq 0 \qquad , \forall \ell,p,q \tag{3.14}$$

where $$f_a = \sum_\ell \sum_p \sum_q \delta_{a\ell pq}\, h_{\ell pq} \tag{3.15}$$

P5: $$\min \; \sum_a \int_0^{f_a} S_a(x)dx - \sum_p \sum_q \hat{u}_{pq}\, g_{pq} \tag{3.16}$$

s.t. (3.11), (3.12), (3.14), (3.15)

$$- \sum_p \sum_q g_{pq} \log g_{pq} \geq H_0 \tag{3.17}$$

In the case when P4 is solved constraint (3.13) is relaxed using a multiplier θ. This creates a sequence of elastic demand subproblems with exponential demand functions. These problems can be solved efficiently using for instance Evans' partial linearization method (Evans (1976)) or a method based on the Frank-Wolfe algorithm (Florian et al. (1975)). In the second case when P5 is solved we dualize constraint (3.17) and iterate in H_0 starting with the lower bound $\hat{H}$ until the constraint (3.17) becomes binding. Theoretically we know that the optimal multiplier to (3.17) is 0 for all $[g_{pq}] \in X$.

3.2. A Bicriterion estimation model

The optimal trip matrix for the entropy problem belongs to the logit distribution class since

$$g_{pq}^{opt} = \frac{\hat{G}\ exp\ [-\beta \hat{u}_{pq} - \sum_k \alpha_k\, u_{pq}^{(k)}]}{\sum_p \sum_q exp\ [-\beta \hat{u}_{pq} - \sum_k \alpha_k\, u_{pq}^{(k)}]}$$

where β and α_k are dual multipliers corresponding to the master problem constraints. In computational tests using data from the city of Stockholm we have noted that the trip distribution $[g_{pq}^{opt}]$, compared with the observed trip distribution, has a higher frequency of short trips. To get a more reasonable trip distribution we suggest adding a constraint to the master problem, an expression following from the geometric mean for the cost of the total number of trips in the network.

Our model is the following

$$min \sum_p \sum_q g_{pq}\ log\ g_{pq} \tag{3.18}$$

$$s.t. \sum_p \sum_q g_{pq} = \hat{G} \tag{3.19}$$

$$\sum_p \sum_q \hat{u}_{pq}\, g_{pq} = \hat{C} \tag{3.20}$$

$$\prod_{p,q} (\hat{u}_{pq})^{g_{pq}} \geq \hat{C}' \tag{3.21}$$

$$g_{pq} \geq 0 \quad , \forall p,q \tag{3.22}$$

$$[g_{pq}] \in X \tag{3.23}$$

where $\hat{C}'$ is an estimate on the geometric mean travel cost.

By taking the logarithm of constraint (3.21) and then dualize it using a multiplier λ we obtain the Lagrangian problem for fixed λ:

$$\min \sum_p \sum_q g_{pq} \log(g_{pq}/\hat{u}^{\lambda}_{pq})$$

$$s.t. \quad (3.19),\ (3.20),\ (3.22),\ (3.23).$$

At the suggestion that the user of the model provides the parameter λ, this is nothing but the entropy problem where the trip distribution is weighted with the factor $\hat{u}^{\lambda}_{pq}$. In the case when $\lambda = 0$ we obtain the pure entropy problem presented in chapter 3.1. Hence, the above described solution methods can be used to solve this problem.

The optimal solution to the bicriterion model has the form

$$g_{pq}\ biopt = \frac{\hat{G}(\hat{u}^{\lambda}_{pq})\ exp\left[-\beta\hat{u}_{pq} - \sum_k \alpha_k\ u^{(k)}_{pq}\right]}{\sum_p \sum_q \hat{G}(\hat{u}_{pq})^{\lambda}\ exp\left[-\beta\hat{u}_{pq} - \sum_k \alpha_k\ u^{(k)}_{pq}\right]}$$

giving us a deterrence function of the type suggested by Tanner (1961).

Computational results obtained by testing this model on the Sioux-Falls network are presented in Brenninger-Göthe, Jörnsten (1985).

REFERENCES

Beckman M, McGuire, C B, Winsten, C (1956), Studies in the economics of transportation, Yale Univ. Press, New Haven, Conn.

Brenninger-Göthe, M, Jörnsten, K O (1983), A note on a cutting plane algorithm for estimating a trip matrix from network data, Linköping Institute of Technology, Report LiTH-MAT-R-83-14.

Brenninger-Göthe, M, Jörnsten, K O, (1985), A bicriterion model for estimation of a trip matrix from network data, Linköping Institute of Technology, Report (forthcoming).

Brenninger-Göthe, M, Larsson, T (1985), A note on the traffic equilibrium problem, Linköping Institute of Technology, Report (forthcoming).

Erlander, S (1980), Optimal spatial interaction and the gravity model, Lecture Notes in Economics and Mathematical Systems, 173, Springer-Verlag, Berlin, Heidelberg, New York.

Evans, S P (1976), Derivation and analysis of some models for combining trip distribution and assignment, Transportation Research, Vol 10, 37-57.

Florian, M, Nguyen, S, Ferland, J (1975), On the combined distribution-assignment of traffic, Transportation Science, Vol. 9, 43-53.

Florian, M, Nguyen, S (1977), A method for computing network equilibrium with elastic demands, Transportation Science 11, 166-179.

Gartner, G H (1980), Optimal traffic assignment with elastic demands: A Review, Part I. Analysis Framework, Part II. Algorithmic Approaches, Transportation Science Vol. 14, 174-208.

Jörnsten, K O, Nguyen, S (1979), On the estimation of a trip matrix from network data, Linköping Institute of Technology, Report LiTH-MAT-R-79-36.

Jörnsten, K O , Nguyen, S (1983), Estimation of an trip matrix from network data; dual approaches, Linköping Institute of Technology, Report LiTH-MAT-R-83-10.

LeBlanc, L J, Farhangian, K (1982), Selection of a trip table which reproduces observed link flows. Transportation Research, Vol. 16B, 83-88.

Nguyen, S (1976), Procedures for equilibrium traffic assignment with elastic demand, Centre de Recerche sur les Transport, Université de Montréal, Publication No. 39.

Nguyen, S (1977), Estimating an OD-matrix from network data. A network equilibrium approach, Centre de Recherche sur les Transport, Université de Montréal, Publication No. 60.

Snickars, F, Weibull, J W (1977), A minimum information principle, theory and practice, Regional Science and Urban Economics 7, 137-168.

Tanner, J C (1961), Factors affecting the amount of travel, Road Research Technical Paper 51, HMSO, London.

Turnquist, M Gur, Y (1979), Estimation of trip tables from observed volumes, Proceeding 58th annual meeting of the Transportation Research Board.

Wardrop, J G (1952), Some theoretical aspects of road traffic research, Proc. Inst. of Civil Eng. Part II, 325-378.

Wilson, A G (1970), Entropy in urban and regional modelling, Pion, London.

Optimal control of age-structured populations

MARTIN BROKATE
Institut für Mathematik
Universität Augsburg
Augsburg, F. R. G.

Abstract: We consider optimal control of a population with continuous age and time structure. We prove that an optimal control exists, give the necessary optimality conditions, and derive some consequences.

1. Problem formulation

We consider the following optimal harvesting problem.

$$\text{Maximize} \int_0^T \int_0^\infty u(a,t)\rho(a,t) \; da \; dt \tag{1}$$

subject to

$$\left(\frac{\partial}{\partial a} + \frac{\partial}{\partial t}\right)\rho(a,t) + \mu(a,p(t))\rho(a,t) + u(a,t)\rho(a,t) = 0$$
$$\rho(a,0) = \rho_0(a) \quad , \quad \rho(0,t) = b(t) \tag{2}$$

$$b(t) = \int_0^\infty \beta(a,p(t))\rho(a,t) \; da \tag{3}$$

$$p(t) = \int_0^\infty \rho(a,t) \; da \tag{4}$$

$$0 \leq u(a,t) \leq u_{max} \quad \text{for all} \quad (a,t) \tag{5}$$

where $0 < T < \infty$ is fixed, $u_{max} > 0$ and μ, β, ρ_0 are given functions. We obtain another version of the problem if we restrict the controls to functions of the form

$$u = u(t) \tag{6}$$

instead of $u = u(a,t)$. (The optimal solutions are quite different in this case.) For the latter version (1) - (4) and (6) together with the (more interesting) situation $T = \infty$, several results for special choices of μ and β are given in [1,2,3]. The system equations (2) - (4) without the harvesting term have been studied in [4].

We formulate our basic assumptions concerning μ, β and ρ_o. These are similar to [4].

Assumption 1

(i) $\mu, \beta: \mathbb{R}_+ \times \mathbb{R}_+ \to \mathbb{R}_+$ and $\rho_o: \mathbb{R}_+ \to \mathbb{R}_+$, i.e. they are nonnegative functions.

(ii) $\rho_o \in L_1(0,\infty) \cap L_\infty(0,\infty)$.

(iii) β is bounded on $\mathbb{R}_+ \times \mathbb{R}_+$.

(iv) $\mu(\cdot,p)$ and $\beta(\cdot,p)$ are measurable, $\mu(a,\cdot)$ and $\beta(a,\cdot)$ are C^2 and, together with their derivatives, bounded for bounded p uniformly with respect to $a \geq 0$.

■

We rewrite equation (2) as an integral equation. For later convenience we introduce the characteristic (x,t) - coordinates instead of (a,t) - coordinates:

$$x = t - a \ , \ a = t - x$$

If we set

$$y(x,t) = \rho(t-a,t)$$

and use the same letter u for the transformed control, the optimal harvesting problem (1) - (5) then takes the form

$$\text{Maximize} \int_0^T \int_{-\infty}^t u(x,t)y(x,t)\, dx\, dt \tag{7}$$

subject to

$$y(x,t) = y_o(x) - \int_0^t (\mu(s-x,p(s)) + u(x,s))y(x,s)\,ds \tag{8}$$

$$y_o(x) = \begin{cases} b(x) , & 0 < x \leq T \\ \rho_o(-x) , & -\infty < x \leq 0 \end{cases}$$

$$b(t) = \int_{-\infty}^{t} \beta(t-x,p(t))\,y(x,t)\,dx \tag{9}$$

$$p(t) = \int_{-\infty}^{t} y(x,t)\,dx \tag{10}$$

$$\begin{aligned} 0 \leq u(x,t) &\leq u_{max} , \quad t \geq x \\ u(x,t) &= 0 \qquad , \quad t < x \end{aligned} \tag{11}$$

where we use the convention $\mu(a,t) = \beta(a,t) = 0$ for $a < 0$.
We look for solutions

$$u \in L_\infty((-\infty,T] \times [0,T]) \tag{12}$$

$$\begin{aligned} &y \in C(0,T;L_\infty(-\infty,T)) \\ &b \in L_\infty(0,T) , \ p \in C[0,T] \end{aligned} \tag{13}$$

2. Existence

As a basic step in the solution of the optimal harvesting problem one needs the unique solvability of the system equations (8) - (10) for any given admissible control. This is provided by the following theorem which corresponds to theorems 3 and 4 in [4,p. 287].

Theorem 1

If assumption 1 holds, then for every u satisfying (11), (12) there exists a unique (y,b,p) which satisfies (8) - (10) and (13). Moreover, there exist constants C_1 and C_2 , which do not depend on u , such that for all x,t

$$0 \le b(t) \le C_1 e^{C_2 t}$$

$$0 \le p(t) \le C_1\, e^{C_2 t}$$

$$0 \le y(x,t) \le \rho_o(-x) \quad \text{if } x \le 0$$

$$0 \le y(x,t) \le C_1\, e^{C_2 t} \quad \text{if } x > 0$$

Proof: By Lemma 5.3 in [5], there exists a unique solution locally in t , i.e. in $(-\infty,T] \times [0,\bar{t})$ for some $\bar{t}$, which is nonnegative. Since μ and u are nonnegative, it follows from (8) and (9) that

$$y(x,t) \le \rho_o(-x) \quad \text{if} \quad x \le 0$$

$$b(t) \le \|\beta\|_\infty \int_{-\infty}^{t} y(x,t)\, dx \le$$

$$\le \|\beta\|_\infty \left(\|\rho_o\|_1 + \int_0^t |y_o(x)|\, dx\right)$$

$$\le \|\beta\|_\infty \left(\|\rho_o\|_1 + \int_0^t b(x)\, dx\right)$$

and the remaining inequalities (and therefore the global existence) are a consequence of Gronwall's lemma.

∎

We now can prove that the transformed version of the optimal harvesting problem has a solution.

Theorem 2

Let assumption 1 hold. Then the optimal harvesting problem (7) - (11) has a solution (y_*,b_*,p_*,u_*) in the space defined by (12) and (13).

Proof: Because of (11) and Theorem 1, the cost function (7) is bounded. Select a maximal sequence $\{u_n\}$ and denote by (y_n,b_n,p_n) the corresponding nonnegative solutions of (8) - (10) and define

$z_n \in L_\infty((-\infty,T) \times [0,T])$ by $z_n(x,t) = y_n(x,t)u_n(x,t)$.

We have $z_n \to z_*$, $y_n \to y_*$, $b_n \to b_*$ weak star in L_∞ for some subsequence. From Theorem 1 and (8) ,

$$\sup_{t\in[0,T]} \int_{-\infty}^{T} y_n(x,t)\, dx \le C$$

$$\sup_{t\in[0,T]} \int_{-\infty}^{T} |y_n(x,t+h) - y_n(x,t)|\, dx \le Ch$$

with a constant C independent from n and h . Hence, by the Arzelà-Ascoli theorem, $p_n \to p_*$ uniformly after passage to a subsequence for some p_* . Now consider

$$y_n(x,t) = y_{on}(x) - \int_0^t \mu(s-x,p_n(s))y_n(x,s) + z_n(x,s)\, ds$$

As a function of (x,t) , the integral converges in the sense of distributions to the function defined by

$$-\int_0^t \mu(s-x,p_*(s))y_*(x,s) + z_*(x,s)\, ds \quad .$$

If we now define

$$u_*(x,t) = \begin{cases} \dfrac{z_*(x,t)}{y_*(x,t)} , & y_*(x,t) \neq 0 \\ 0 , & y_*(x,t) = 0 \end{cases}$$

we see that (y_*,b_*,p_*,u_*) satisfies (8). In the same way, (9) and (10) are proved. (13) follows from (8), (9); (11) and (12) follow from

$$0 \le z_n(x,t) \le y_n(x,t) \cdot u_{max}$$

$$z_n(x,t) = 0 \quad , \quad t < x$$

Also, since

$$\lim_{x\to-\infty} \int_{-\infty}^{x}\int_0^{T} z_n(x,t)\, dt\, dx = 0$$

uniformly with respect to n , $J(z_n)$ converges to $J(z_*)$. ∎

3. Necessary optimality conditions

For the optimal harvesting problem (1) - (5), the necessary optimality conditions of first order are stated and proved in [5]. We formulate them in the special case

$$\beta(a,p) = \beta(a)$$
$$\mu(a,p) = \mu_n(a) + \mu_e(p) \tag{14}$$

which has been treated in [1,3] for controls of form $u = u(t)$.

Theorem 3

Let assumption 1 hold, let β and μ have the form (14). If (ρ_*, b_*, p_*, u_*) is a solution of the optimal harvesting problem (1) - (5), then there exist bounded measurable functions q, ζ, θ such that q is absolutely continuous along straight lines of slope 1, and the adjoint equations

$$\left(\frac{\partial}{\partial a} + \frac{\partial}{\partial t}\right) q(a,t) = (u_*(a,t) + \mu(a,p_*(t)))(1 + q(a,t)) - $$
$$- (\mu(a,p_*(t)) + \beta(a)\zeta(t) + \theta(t)) \tag{15}$$

$$q(a,T) = 0 \tag{16}$$
$$\zeta(t) = q(0,t) \tag{17}$$
$$\theta(t) = - \mu_e'(p_*(t)) \int_0^\infty \rho_*(a,t) q(a,t) \, da \tag{18}$$

hold. The optimal control is given by

$$u_*(a,t) = \begin{cases} u_{max} & , \text{ if } S(a,t) \le 0 \\ 0 & , \text{ if } S(a,t) > 0 \end{cases} \tag{19}$$

with the switching function

$$S(a,t) = -\rho_*(a,t)(1 + q(a,t)) \tag{20}$$

Proof: See Theorem 4 in [5].

■

For the following discussion we assume that μ and β are sufficiently smooth. We first pose the question: Are there singular regions (i.e. open sets, where $S(a,t) = 0$, $\rho_*(a,t) > 0$)? If G is a singular region, then by (20) we have $q(a,t) = -1$ in G and by (15)

$$\mu_n(a) + \mu_e(p_*(t)) + \beta(a)\zeta(t) + \theta(t) = 0 \tag{21}$$

If we differentiate twice with respect to a and combine both equations we get

$$\mu_n'(a)\beta''(a) - \mu_n''(a)\beta'(a) = 0$$

for all $a \in (a_1,a_2)$ as necessary condition for a singular region $G = (a_1,a_2) \times (t_1,t_2)$. This does not help if $\mu_n \equiv 0$. But then (21) implies $\beta'(a)\zeta(t) = 0$ in G which in turn implies $\zeta(t) = 0$ in (t_1,t_2) if $\beta'(a) \neq 0$ somewhere in (a_1,a_2). Therefore, (21) becomes

$$\theta(t) = -\mu_e(p_*(t)) , \quad t \in (t_1,t_2) \tag{22}$$

Differentiating (18), we see after some calculation that

$$\theta'(t) = \gamma(t)\theta(t) - \mu_e'(p_*(t)) \int_0^\infty \rho_*(a,t)u_*(a,t)\, da \quad .$$

for some function $\gamma(t)$. If μ_e is nondecreasing then because of $\theta(T) = 0$ it follows that

$$\theta(t) \geq 0 \quad , \quad t \in [0,T] \quad .$$

From (22) we obtain the result: If

$\mu_n(a) = 0$, $\mu_e(p) > 0$, $\mu_e'(p) \geq 0$ for all a,p
$\{a:\beta'(a) \neq 0\}$ is dense in $(0,A)$,

then there is no singular region $G \subset (0,A) \times (0,T)$.

A second question now is: If there is no singular region, can the singular set $\{(a,t): S(a,t) = 0\}$ be described as a union of finitely many switching curves? In which cases there are only one or two switching curves? It seems clear that some assumptions on β and ρ_o are necessary. Up to now, the author only has partial results available.

[1] GURTIN, M., MURPHY, L.: On the optimal harvesting of age-structured populations. In:
Differential equations and applications in ecology, epidermics, and population problems (eds. S. Busenberg, K. Cooke), pp. 115-129. New York: Academic Press 1981.

[2] GURTIN, M., MURPHY, L.: On the optimal harvesting of age-structured populations: Some simple models. Math. Biosci. 55 (1981), 115-136.

[3] GURTIN, M., MURPHY, L.: On the optimal harvesting of persistent age-structured populations. J. Math. Biol. 13 (1981), 131-148.

[4] GURTIN, M., MAC CAMY, R.: Non-linear age-dependent population dynamics. Arch. Rational Mech. Anal. 54 (1974), 281-300.

[5] BROKATE, M.: Pontryagin's principle for control problems in age-dependent population dynamics. J. Math. Biol. (in press)

Skew-symmetric Matrices, Staircase Functions and Theorems of the Alternative

C. G. BROYDEN
Computer Science Department,
University of Essex,
Wivenhoe Park,
Colchester,
Essex CO4 3SQ,
ENGLAND.

1. Introduction

One of the key ideas in the study of linear inequalities is embodied in a class of theorems known collectively as "theorems of the alternative". These theorems have much in common. They are quite difficult to prove from first principles, they are either equivalent (in the sense that once one has been established the others may be derived from it comparatively simply) or are very closely related and most may be expressed in the form "either A or B". One of the earliest is the Farkas Lemma (Ref. 1) which states that if A is some real m x n matrix and a is some n-th order vector then $\exists$ either a vector x such that $Ax \geq 0$ and $a^T x < 0$, or a vector $y \geq 0$ such that $a^T = y^T A$. Others are the Rockafellar-Fulkerson theorem (Ref. 2) and the so-called "Key Theorem" of linear programming (Ref. 3).

A more recent member of this class is the theorem (Ref. 4) that states that "for any skew-symmetric matrix A, $\exists$ a vector $x \geq 0$ such that $Ax \geq 0$ and $x + Ax > 0$". This theorem is equivalent (as defined above) to the Farkas Lemma but may also be derived from first principles. One such derivation introduces the concept of a "staircase function", which is not only integral to this proof but is an interesting function in its own right. It is the aim of this paper to provide an informal proof of this theorem in which mathematical rigour is not permitted to obscure the ideas underlying the proof. A more rigorous, but much longer, discussion may be found in Ref. 4.

2. Staircase Functions

If A is any skew-symmetric matrix, a vector $x \geq 0$ that satisfies $Ax \geq 0$ is referred to as a *feasible vector* of A, and any feasible vector of A that satisfies in addition the inequality $x + Ax > 0$ is termed a *strictly feasible vector* (sfv) of A. The theorem referred to in the introduction thus asserts the existence of an sfv for any skew-symmetric matrix A. The proof is by induction on the order of A. We assume, therefore, its validity for all r-th order matrices and investigate the sfvs x of the matrix M, where

$$M = A + \theta B, \tag{1a}$$

$$B = ab^T - ba^T, \tag{1b}$$

A is an r-th order skew-symmetric matrix, a and b are r-th order vectors and θ is a scalar parameter. In Ref. 4, a and b were required to be non-negative and to satisfy $a^Tb = 0$. It is simpler, however, to require that b is strictly positive and to impose no restrictions upon a. One result of this is that if x is an sfv of M then $b^Tx > 0$ for all θ. It follows that x may always be normalised to satisfy $b^Tx = 1$, and we implicitly assume such normalisation to have been carried out in what follows.

We shall further denote a^Tx by ρ. Clearly ρ depends on θ since x depends on θ, and we shall show that ρ is in fact a "staircase function" of θ.

Let θ_1 and θ_2 be distinct values of θ, and let x_1 and x_1 be the sfvs of $A + \theta_1 B$ and $A + \theta_2 B$ respectively. If y_1 and y_2 are defined by

$$y_1 = (A + \theta_1 B)x_1 \tag{2a}$$

and

$$y_2 = (A + \theta_2 B)x_2 \tag{2b}$$

it follows from the skew-symmetry of A, and equation (1), that

$$(\theta_1 - \theta_2)(\rho_2 - \rho_1) = x_2^T y_1 + x_1^T y_2 \geq 0 \tag{3}$$

Thus ρ decreases monotonically with θ. Now if we use the term *configuration* to denote the dispositions of the zero and non-zero elements of a non-negative vector, and define the configurations of x and y to be *complementary* if $x + y > 0$ and $x^T y = 0$, then it is trivial to show that if x is an sfv of $A = -A^T$ and $y = Ax$ then the configurations of x and y are complementary. Thus if, in equations (2a) and (2b) above, the configurations of x_1 and x_2 are identical it follows that the configurations of y_1 and y_2 are identical, and complementary to those of x_1 and x_2. Thus, since $\theta_1 \neq \theta_2$ and the configurations of x_1 and x_2 are identical, it follows from inequality (3) that $\rho_1 = \rho_2$. Now since the largest possible number of different configurations of an r-th order sfv is 2^r-1 it follows that if an sfv of M is evaluated for 2^r distinct values of θ then at least two of these will have the same configuration. Let θ_1 and θ_2, $\theta_1 < \theta_2$, be the corresponding values of θ. Then if

$$x = x_1 \left(\frac{\theta_2 - \theta}{\theta_2 - \theta_1}\right) + x_2 \left(\frac{\theta - \theta_1}{\theta_2 - \theta_1}\right) \tag{4}$$

and $y = Mx$, it follows from equations (1) and (2) that

$$y = y_1 \left(\frac{\theta_2 - \theta}{\theta_2 - \theta_1}\right) + y_2 \left(\frac{\theta - \theta_1}{\theta_2 - \theta_1}\right) \tag{5}$$

Thus if ε is sufficiently small, simple algebra indicates that x is an sfv of $A + B\theta$ for $\theta_1 - \varepsilon \leq \theta \leq \theta_2 + \varepsilon$. If we denote these intervals of constant ρ *step intervals* it follows from the above that step intervals are open intervals of θ, and from monotonicity that each interval corresponds to a particular value of ρ. Since each step interval also corresponds to a particular configuration of x, since the number of configurations is finite and since, by hypothesis, an sfv of A exists for all θ it follows that the real line is covered by the closure of a finite number of step intervals. Since, by hypotheses, sfvs exist at the limit points also it follows from the above that the configuration of an sfv at a limit point is unique to that point. It follows from inequality (3) that the corresponding value of ρ, ρ_C say, satisfies $\rho_L > \rho_C > \rho_R$, where ρ_L and ρ_R are the values of ρ on the left and right step interval respectively.

It may be inferred from equations (4) and (5) that the limit of the sfv x as θ approaches a limit point is a vector which is feasible but not strictly so. Denoting this vector by z, it follows that $z + Az \geq 0$ but where at least one element is zero. There are in fact two such vectors z, z_L and z_R, depending on whether the limit is approached from below or above and it is readily seen from (4) that $a^T z_L = \rho_L$ and $a^T z_R = \rho_R$, where $b^T z_L = b^T z_R = 1$. Thus, at a limit point, M has an sfv x_C say, and two feasible vectors, z_L and z_R. It is readily shown that if x is given by

$$x = z_L \sigma + x_C(1 - \sigma) \tag{6}$$

then x is a normalised sfv of M for $0 \leq \sigma < 1$. The corresponding value of ρ is

$$\rho = \rho_L \sigma + \rho_C(1 - \sigma).$$

A similar result holds for z_R and it follows from this that ρ can assume a n y value in the range $\rho_L > \rho > \rho_R$. Note that the inequalities are strict since ρ could only equal either ρ_L or ρ_R if the x as defined in equation (6) lost its strictly feasible status.

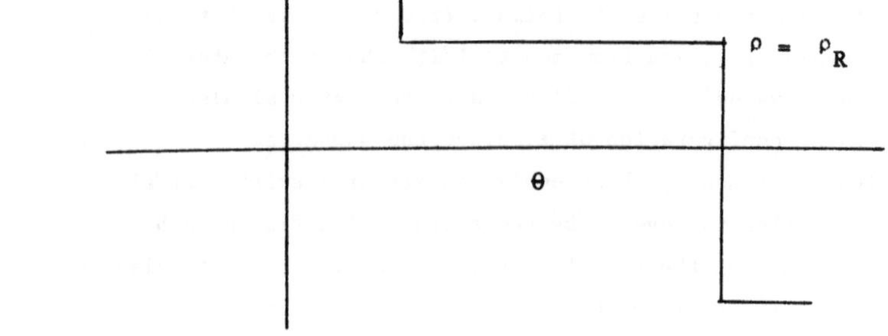

The essential properties of $\rho(\theta)$ required for the proof of the theorem are:

(A) It is monotonically decreasing, and

(B) If $\theta_0 < \theta_2$ and ρ assumes the values ρ_0 and ρ_2 at θ_0 and θ_2 respectively, then $\exists$ a value of θ such that $\rho(\theta) = \rho_1$ for any ρ_1 in the range $\rho_0 \leq \rho_1 \leq \rho_2$.

3. Proof of the Theorem

We now use the properties of the staircase function to prove our main theorem, but first we prove the following

Lemma

Let $e^T = [1, 1, \ldots, 1]$, x be the sfv of $M = A + \theta(ae^T - ea^T)$ where $e^Tx = 1$, and let $\rho = a^Tx$. Then if a is not strictly positive, $\rho \leq 0$ for θ sufficiently large.

Proof

We show that if a is not strictly positive, the assumption that $\rho > 0$ as θ increases yields a contradiction. Since x is an sfv of M,

$$Mx = Ax + \theta(a - e\rho) \geq 0 \tag{7}$$

Now $e^Tx = 1$ implies that $||x||_1 = 1$ so that $||Ax||_1 \leq ||A||_1$ and, a fortiori, $||Ax||_\infty \leq ||A||_1$. Let e_j denote the j-th column of the unit matrix, and assume that $e_j^Ta \leq 0$ and $\rho > 0$. Then $e_j^T(a - e\rho) < 0$ so that, if θ is sufficiently large, $\theta e_j^T(a - e\rho) < - ||A||_1 \leq - ||Ax||_\infty$, contradicting inequality (7) and proving the lemma. Note that if $e_j^Ta = 0$ the possibility that $\rho \to 0$ from above cannot arise as ρ is constant on each of a finite number of step intervals.

We now prove our main

Theorem

For any skew-symmetric matrix A, $\exists$ a vector $x \geq 0$ such that $Ax \geq 0$ and $x + Ax > 0$.

Proof

Assume the theorem to be true for all r-th order skew-symmetric matrices, let A be one such matrix and let

$$A_1 = \begin{bmatrix} A & a \\ -a^T & 0 \end{bmatrix}$$

We consider four cases, where x_0 is an sfv of A:

(1) $-a^T x_0 > 0$. Then $[x_0^T \; 0]^T$ is an sfv of A_1.

(2) $a > 0$. Then $[0^T \; 1]^T$, where 0^T is the r-th order null row vector, is an sfv of A_1.

(3) $-a^T x_0 < 0$, $a \not> 0$

If x is the sfv of $M = A + \theta(ae^T - ea^T)$ then x_0 is the sfv of M when $\theta = 0$, and the corresponding value of $\rho = a^T x_0$ is positive. Thus from the lemma, and the properties of the staircase function, $\exists$ a positive value of θ, θ_1 say, for which $\rho(\theta_1) = 0$. Denote the corresponding sfv of M by x_1. Then, since $\rho = 0$ and $Ax_1 + \theta_1(ae^T - ea^T)x_1 \geq 0$, $Ax_1 + \theta_1 a \geq 0$. But $a^T x_1 = 0$ so that $[x_1^T \; \theta_1]^T$ is an sfv of A_1.

(4) $-a^T x_0 = 0$.

Let $A_1 = \begin{bmatrix} 0 & -\bar{a}^T \\ \bar{a} & \bar{A} \end{bmatrix}$

and let $\bar{x}_0$ be an sfv of $\bar{A}$. If $\bar{a}^T\bar{x}_0 \neq 0$ the proof is as (1) to (3) above. If $\bar{a}^T\bar{x}_0 = 0$ and x_1 is given by $x_1^T = [x_0^T\ 0] + [0\ \bar{x}_0^T]$ then x_1 is an sfv of A_1.

Since these cases are collectively exhaustive, it follows that if an sfv exists for any skew-symmetric matrix of order r then one exists for any such matrix of order r + 1. Since, trivially, the theorem is true for r = 1 it is true for all $r \geq 1$.

Corollary The Farkas Lemma (for statement see introduction).

Apply the theorem to the system

$$\begin{bmatrix} 0 & 0 & A & -A \\ 0 & 0 & -a^T & a^T \\ -A^T & a & 0 & 0 \\ A^T & -a & 0 & 0 \end{bmatrix} \begin{bmatrix} y \\ t \\ z_1 \\ z_2 \end{bmatrix}$$

put $x = z_1 - z_2$ and consider the two cases t = 0 and t = 1.

Other "theorems of the alternative" may be similarly derived.

Acknowledgement

The author thanks W. E. Hart for his helpful comments on an earlier draft of the paper.

References

1. J. Farkas, "Uber die Theorie der Einfachen Ungleichungen", Jnl. fur die Reine und Angewandte Mathematik, 124 (1902), pp. 1 - 24.

2. George J. Minty, "A ´From Scratch´ Proof of a Theorem of Rockafellar and Fulkerson", Math. Prog. 7 (1974), pp. 368 - 375.

3. R. A. Good, "Systems of Linear Relations", S.I.A.M. Review 1 (1959), pp. 1 - 31.

4. C. G. Broyden, "A Novel Formulation and Proof of a Fundamental Duality Theorem", submitted to the Journal of Optimization Theory and Applications.

STABILIZATION OF THE SECANT METHOD VIA QUASI-NEWTON APPROACH

O. P. Burdakov

Computing Centre, USSR Academy of Sciences, Moscow, USSR

Abstract. Methods of the secant type for solving systems of nonlinear equations are considered. They are stable in contrast to the traditional secant method with respect to linear dependence of the search directions. A short survey of some variants of the secant method, that use quasi-Newton formulas to provide stability, is given. Some parallel algorithms are constructed on the basis of the stable secant approximations.

1. Introduction

Consider the system of nonlinear equations

$$f(x) = 0 \qquad (1.1)$$

where the mapping $f : D \subset R^n \longrightarrow R^n$ is continuously differentiable on the open convex set D. We suppose, that the system (1.1) has a solution $x_* \in D$, where the Jacobian matrix $f'(x_*)$ is nonsingular and the Lipschitz condition

$$\|f'(x) - f'_*\| \leqslant L \|x - x_*\| \qquad \forall x \in D$$

takes place. Here and below, we use notations: $f'_* = f'(x_*)$, $f_k = f(x_k)$, etc.; $\|\cdot\|$ - the Euclidean vector norm and induced matrix norm; $(a,b) = a^T b$ - scalar product.

The secant method was initially proposed for one equation with one unknown. It was generalized independently by L. Bittner [1], H. Heinrich (see [2]) and P. Wolfe [3] at the multivariate case. The detailed discussion of this method and its convergence theory are given in [2] , [4] .

The main idea of the secant method is as follows. Having the values of the mapping f in n+1 points we can construct a linear approximation of this mapping and then find the solution of the obtained linearized system of equations. This solution can be used as a new estimate to x_*. Thus, the secant method need not compute derivatives $f'(x)$.

The sequential (n+1)-point secant method is an important particular case of the secant method. At the k-th iteration of this method the value f_{k+1} is computed and used together with the values $f_k, \dots, f_{k-n+1}$ already computed at previous iterations for construction of a linearized system. It is seen, that only one value of f is computed at each iteration, nevertheless the rate of convergence $x_k \to x_*$ is superlinear.

But the sequential (n+1)-point secant method has a serious disadvantage which is unstability resulting from the fact, that in the multivariate case ($n > 1$) the secant plane may not converge to the tangent plane. A dange of unstability appears when the directions $\{\Delta x_i\}_{i=k-n+1}^{k}$, where $\Delta x_i = x_{i+1} - x_i$, tend to be linearly dependent as $k \to \infty$. Moreover, if the directions $\{\Delta x_i\}_{i=k-n+1}^{k}$ are linearly dependent, it usually results in complete failure of the method, because the linearization having the values $\{f_i\}_{i=k-n+1}^{k+1}$ at the points $\{x_i\}_{i=k-n+1}^{k+1}$ usually does not exist and if it even exists, the linearized system of equations may be inconsistent. For these reasons, the superlinear convergence of this method is proved generally under an assumption stronger than the linear independence of the directions $\{\Delta x_i\}_{i=k-n+1}^{k}$, namelly, under the assumption, that there is $\delta > 0$ such that

$$\left| \det \left(\frac{\Delta x_k}{\|\Delta x_k\|}, \ldots, \frac{\Delta x_{k-n+1}}{\|\Delta x_{k-n+1}\|} \right) \right| \geq \delta \quad \forall k \geq 0, \tag{1.2}$$

but as a role this inequality is violated in practice.

Some variants of the sequential (n+1)-point secant method are proposed in [5-12] . They have superlinear rate of convergence and above all they are stable with respect to linear dependence of the directions $\{\Delta x_i\}_{i=k-n+1}^{k}$. These variants can be devided in two groups.

The characteristic feature of one group is as follows. If the directions $\{\Delta x_i\}_{i=k-n+1}^{k}$ become nearly linearly dependent and as a result a danger of unstability arises, then the direction Δx_k is replaced by such one (i.g., orthogonal to the directions $\{\Delta x_i\}_{k-n+1}^{k-1}$) that the obtained directions would not be nearly linearly dependent. In this case the linearization needs one additional (with respect to the sequential (n+1)-point secant method) evaluation of the mapping f.

Another group consists of stable variants [10-12] which do not need additional evaluations of the mapping f. They use quasi-Newton formulas to construct the linearization.

The aim of the present paper is to give a short survey of stable variants of the secant method that use quasi-Newton approach and to propose some parallel algorithms based on these stable variants.

In Section 2 we give a quasi-Newton description of sequential (n+1)-point secant methods. Some of them were developed especially for systems of equations with symmetric Jacobian matrix.

In Sections 3 and 4 stable variants of the secant method developed in [10-12] on the base of quasi-Newton approach are considered. These variants use at the k-th iteration for construction of the linearization the previous linearization obtained at the (k-1)-th iteration and use also the directions $\{\Delta x_i\}_{i\in T(k)}$. The set T(k) consists of the index k and some other indices $i<k$. It should be noted that the set T(k) is such that the vectors $\{\Delta x_i\}_{i\in T(k)}$ are not nearly linearly dependent. In Section 3 the set T(k) results from excluding some indices from T(k-1) and including the index k. In Section 4 the set $T(k)=\{k, k-1,\dots,k-l(k)+1\}$. The value l(k) defines the number of vectors Δx_i used for linearization at the k-th iteration. If the vectors $\{\Delta x_i\}_{i\in T(k-1)\cup\{k\}}$ are nearly linearly dependent, then $l(k):=1$ and $T(k):=\{k\}$, i.e. so-called restart takes place.

In Section 5 we discuss a possibility of using algorithms described in Sections 3 and 4 for constructing parallel algorithms.

2. Quasi-Newton representation of the secant methods

The sequential (n+1)-point secant method may be written as

$$x_{k+1} = x_k - \alpha_k J_k^{-1} f_k \tag{2.1}$$

Here the matrix $J_k \in R^{n\times n}$ is an approximation to the Jacobian matrix f'_k and is given by the formula

$$J_k = F_k X_k^{-1}, \tag{2.2}$$

where $X_k = [\Delta x_{k-1}, \dots, \Delta x_{k-n}] \in R^{n\times n}$, $F_k = [\Delta f_{k-1},\dots,\Delta f_{k-n}] \in R^{n\times n}$, $\Delta f_i = f_{i+1} - f_i$. Value α_k is a step-length along the direction $-J_k^{-1} f_k$. Let it be chosen in a reasonable way to enlarge the region of convergence of the method (e.g., as in step-length rules proposed in [13] for Newton method). In further statements concerning local convergence properties of this and some other methods we suppose that $\alpha_k = 1$ for all sufficiently large k.

It should be noted that the approximate equality

$$f'_{k+1}\,\Delta x_i \approx \Delta f_i \tag{2.3}$$

takes place. It means that the less Δx_i is and the closer x_i is to x_{k+1}, then the better J_{k+1} approximates f'_{k+1} along the direction Δx_i, because according to (2.2) we have the exact equality

$$J_{k+1}\,\Delta x_i = \Delta f_i \tag{2.4}$$

for all i, $k-n+1 \le i \le k$.

It is known (see, e.g., [4]) that if condition (1.2) is fulfilled then the sequential (n+1)-point secant method (2.1), (2.2) converges superlinearly with R-order at least equal to the positive root of $t^{n+1} - t^n - 1 = 0$.

Systems of nonlinear equations (1.1) with symmetric Jacobian matrix $f'(x)$ are widely used, e.g., in optimization. Two symmetric secant methods taking into account this specific are proposed in [14].

One of them computes in contrast to (2.2) the symmetric matrix J_k by the formula

$$J_k = (X^T{}_k)^{-1} \text{ sym } (X^T{}_k F_k) X_k^{-1}, \qquad (2.5)$$

where the operation of symmetrization sym: $R^{n\times n} \to R^{n\times n}$ is such that

$$(\text{sym }(A))_{ij} = \begin{cases} A_{ij} & , i \geq j \\ A_{ij} & , i < j \end{cases}$$

In another method iterates are defined by

$$x_{k+1} = x_k - \alpha_k H_k f_k, \qquad (2.6)$$

where the symmetric matrix $H_k \in R^{n\times n}$ given by the formula

$$H_k = (F_k{}^T)^{-1} \text{ sym } (F_k{}^T X_k) F_k{}^{-1} \qquad (2.7)$$

approximates the inverse Jacobian matrix $(f'_k)^{-1}$.

Note that the method (2.1), (2.2) may be represented in the equivalent form (2.5), where $H_k = X_k F_k{}^{-1}$. (2.8)

In [14] a superlinear convergence of the symmetric secant methods given above is proved under the assumption (1.2). The estimate of the R-order is the same as in the method (2.1), (2.2).

Suppose the matrix $f'(x)$ is symmetric and compare the approximations (2.2) and (2.5). It can be easily shown that relation (2.4) is valid for the approximation (2.5) only in the subspace spanned on the vectors $\{\Delta x_j\}_{k-n+1}^{i}$, i.e the relation

$$(J_{k+1}\,\Delta x_i,\ p) = (\,\Delta f_i,\ p) \qquad (2.9)$$

takes place for all i, $k-n+1 \leq i \leq k$, and for all vectors p belonging to the subspace mentioned above. Note, that the vector Δf_i contains some information about the Jacobian matrix f'_{k+1} along the direction Δx_i (see (2.3)). The closer i to k is the more "fresh" (more reliable) this information is. We will compare approximations J_{k+1} from the viewpoint of reliability of information (about the matrix f'_{k+1}) used in construction of J_{k+1}. Reliability can be evaluated by the closeness of the number of iteration i, at which value Δf_i involved in approximation are obtained, to the number k. Suppose the matrix X_{k+1} is nonsingular. Consider a linear transformation of the variables $x = X_{k+1}\tilde{x}$. We transform also system of equations (1.1) to equivalent one $\tilde{f}(x) \equiv X_{k+1}{}^T f(X_{k+1}\tilde{x})=0$. As a result, Δf_i

and J_{k+1} will be transformed to $\Delta\widetilde{f}_i = X^T_{k+1}\,\Delta f_i$ and $\widetilde{J}_{k+1} = X^T_{k+1} J_{k+1} X_{k+1}$, respectively. Compare element-wise the matrices $\widetilde{J}_{k+1}$ given by formulas (2.2) and (2.5). As to (2.2), using (2.4) we have $\widetilde{J}_{k+1} e_j = \Delta\widetilde{f}_{k-j+1}$ for all j, $1 \le j \le n$, where e_j denotes j-th coordinate axis in R^n. Considering (2.5), from (2.9) we obtain

$$(\widetilde{J}_{k+1})_{ij} = \begin{cases} (e_i,\ \Delta f_{k-j+1}), & 1 \le j \le i \le n \\ (\Delta f_{k-i+1},\ e_j), & 1 \le i < j \le n \end{cases}$$

Thus the structure of the matrix $\widetilde{J}_{k+1}$ may be represented for (2.2) and (2.5), respectively, as follows

$$\left[\begin{array}{c|c|c|c} k & k-1 & & k-n+1 \\ \vdots & \vdots & \cdots & \vdots \\ k & k-1 & & k-n+1 \end{array}\right], \qquad \left[\begin{array}{cccc} k & \cdots & \cdots & k \\ \vdots & k-1 & \cdots & k-1 \\ \vdots & \vdots & \ddots & \\ k & k-1 & & k-n+1 \end{array}\right]$$

where it is shown for each element $(\widetilde{J}_{k+1})_{ij}$ at what iteration the information contained in Δf and used in construction of this element was obtained. Comparing the structures given above row by row we see that the approximation (2.2) uses additionally (with respect to the same row of the approximation (2.5)) more "old" and less reliable information.

Now, consider rank-one quasi-Newton methods (see [15] and [16]). They may be represented in two closely connected forms (2.1) and (2.6). The approximation J in (2.1) is updated by the formula

$$J_{k+1} = J_k + \frac{(\Delta f_k - J_k\,\Delta x_k)\,c_k^T}{(\Delta x_k,\ c_k)}\,. \tag{2.10}$$

In (2.6) the approximation H is updated as follows

$$H_{k+1} = H_k + \frac{(\Delta x_k - H_k\,\Delta f_k)\,d^T_k}{(\Delta f_k,\ d_k)} \tag{2.11}$$

The matrices J_o and H_o are supposed to be given. The vectors c_k, $d_k \in R^n$ are free parameters. Some concrete values of them correspond to known quasi-Newton methods.

Formula (2.10) has the following properties. At first, we have

$$J_{k+1} p = J_k p \qquad \forall\, p \in R^n,\ p \perp c_k, \tag{2.12}$$

i.e. there is a possibility to preserve some information about the Jacobian matrix accumulated in the matrix J_k. At second, we have

$$J_{k+1}\,\Delta x_k = \Delta f_k, \tag{2.13}$$

that unables us to insert in J_{k+1} the most "fresh" information about f'_{k+1} .

Let the matrices X_k are nonsingular for all $k \ge n$ and the

vector c_k is chosen in (2.10) such that $(c_k, \Delta x_i)=0$ (2.14)
for all $k-n+1 \leq i \leq k-1$. Then relation (2.4) follows from (2.12) and (2.13). In this case the quasi-Newton method (2.1),(2.9) coincides (see [5]) with the sequential (n+1)-point secant method (2.1),(2.2).

By analogy, the approximation (2.8) can be updated by the quasi-Newton formula (2.11), where the nonzero vector d_k is such that

$$(d_k, \Delta f_i) = 0 \tag{2.15}$$

for all $k-n+1 \leq i \leq k-1$.

If the Jacobian matrix is symmetric, then rank-two quasi-Newton methods can take into account this specific. In these methods the approximations J and H are updated by the formulas

$$J_{k+1}=J_k+\frac{(\Delta f_k - J_k \Delta x_k)c_k^T + c_k(\Delta f_k - J_k \Delta x_k)^T}{(\Delta x_k, c_k)} - \tag{2.16}$$

$$-\frac{(\Delta f_k - J_k \Delta x_k, \Delta x_k)\, c_k c_k^T}{(\Delta x_k, c_k)^2},$$

$$H_{k+1} = H_k - \frac{(\Delta x_k - H_k \Delta f_k)d_k^T + d_k(\Delta x_k - H_k \Delta f_k)^T}{(\Delta f_k, d_k)} - \tag{2.17}$$

$$-\frac{(\Delta x_k - H_k \Delta f_k, \Delta f_k)\, d_k d_k^T}{(\Delta f_k, d_k)^2}.$$

Formula (2.16) satisfies equality (2.13) and the relation

$$(J_{k+1}p_1, p_2) = (J_k p_1, p_2) \qquad \forall p_i \in R^n,\ p_i \perp c_k,\ i = 1,2. \tag{2.18}$$

If the matrix X_k is nonsingular and the vector c_k satisfies (2.14), then the matrix J_{k+1} satisfies relations (2.9). Hence, the symmetric sequential (n+1)-point secant method (2.1), (2.5) may be represented in the quasi-Newton form (2.16).

The choice of d_k according to (2.15) leads similarly to the quasi-Newton representation (2.6), (2.17) of the symmetric sequential (n+1)-point secant method (2.6),(2.7).

3. Stable variants of the secant method

Let us determine a number $\sigma \in (0,1)$ which will define the set of nearly linearly dependent vectors. We call vectors $a_1,\ldots,a_k \in R^n$ nearly linearly dependent if either $\|a_i\| = 0$ for some $i, 1 \leq i \leq k$, or

$$\det\left(\Gamma\left(\frac{a_1}{\|a_1\|},\ldots,\frac{a_k}{\|a_k\|}\right)\right) < \sigma^2,$$

where $\Gamma(\cdot)$ is the Gram matrix.

As it was mentioned, the secant method is unstable, because condition (1.2) is usually violated. If the vectors $\{\Delta x_i\}_{k-n+1}^{k}$ are linearly dependent, then, in general, there is no matrix J_{k+1}, satisfying (2.4) for all i, $k-n+1 \leq i \leq k$. If these vectors are not linearly dependent but they are nearly linearly dependent, then a danger of unstability appears. In [11] a method is proposed where the matrix J_{k+1} satisfies equality (2.4) not for all i, $k-n+1 \leq i \leq k$, but only for a part of these indices. In order to describe the approach developed in [11] we denote the set of indices with descending ordering, which satisfy (2.4), by $T(k)=\{t(k,1),t(k,2),\dots,t(k,l(k))\}$. The set T(k) is constructed so that the vectors $\{\Delta x_i\}_{i\in T(k)}$ are not nearly linearly dependent and $t(k,1)=k$.

Suppose we have a matrix J_k. According to (2.4) it containes some information about the Jacobian matrix along the directions $\{\Delta x_i\}_{i\in T(k-1)}$. After the k-th iteration (2.1) we obtain an additional information along the direction Δx_k. The problem is to compose a set T(k), using the index k and some of the indices T(k-1), so that the vectors $\{\Delta x_i\}_{i\in T(k)}$ would not be nearly linearly dependent. In other words, it is necessary to construct an approximation J_{k+1}, which would preserve some information about the Jacobian matrix accumulated in J_{k-1} and contain the "fresh" information obtained from the direction Δx_k. The properties (2.12) and (2.13) of the quasi--Newton methods give us such possibility. Since the set T(k) satisfying the requirements mentioned above can be constructed in various ways, we will take into account the fact that for $j < i \leq k$ the pair Δx_i, Δf_i containes more "fresh" (reliable) information about f'_{k+1} than the pair $\Delta x_j, \Delta f_j$. This natural preference ordering in indices leads to uniqueness in construction of the set T(k). Thus, first of all the index k is included in T(k), then subsequently for $i = 1,2,\dots,l(k-1)$ (i.e., in accordance with the preference ordering) the index t(k-1,i) is included in T(k) unless the vector $\Delta x_{t(k-1,i)}$ is nearly linearly dependent on the vectors Δx whose indices already included in T(k). A strict description of this algorithm looks like as follows.

<u>Algorithm 1</u>. Let an initial point $x_0 \in R^n$, a nonsingular matrix $J_0 \in R^{n\times n}$, a natural number $m \geq 0$ and real numbers $0 < \gamma \leq \sigma < 1$ be given. Compute f_0. Set $T(-1): = \emptyset$.

For k=0.1,2,... execute the following steps.

1. If $\|f_k\| = 0$ or J_k is singular, then stop.
2. Compute x_{k+1} by formula (2.1) and compute f_{k+1}.
3. Set $T(k): = \{k\}$. Then subsequently for $i=1,2,\dots,l(k-1)$,while $t(k-1,i) \geq k-m$, do the following step:

   3a) if the vectors $\{\Delta x_t\}_{t\in T(k)\cup\{t(k-1,i)\}}$ are not nearly linearly dependent, then set $T(k):=T(k)\cup\{t(k-1,i)\}$.

4. Find any nonzero vector $c_k\in R^n$ satisfying the inequality

$$|(\Delta x_k, c_k)| \geqslant \gamma \|\Delta x_k\| \|c_k\| \qquad (3.1)$$

and condition (2.14) for all $i\in T(k)\setminus\{k\}$

5. Compute the matrix J_{k+1} by formula (2.10).

<u>Algorithm 2</u>. Let an initial point $x_o\in R^n$, a nonsingular matrix $H_o\in R^{n\times n}$, a natural number $m\geqslant 0$ and real numbers $0<\gamma\leqslant\sigma<1$ be given. Compute f_o. Set $T(-1):=\emptyset$.

For k=0.1,2,... execute the following steps.

1. If $\|f_k\| = 0$, then stop.
2. Compute x_{k+1} by formula (2.6) and compute f_{k+1}.
3. Set $T(k):=\{k\}$. Then subsequently for $i=1,2,\ldots,l(k-1)$, while $t(k-1,i)\geqslant k-m$, do the following step:

   3a) if the vectors $\{\Delta x_t\}_{t\in T(k)\cup\{t(k-1,i)\}}$ are not nearly linearly dependent, then set $T(k):=T(k)\cup\{t(k-1,i)\}$.

4. Find any nonzero vector $d_k\in R^n$ satisfying the inequality

$$|(\Delta f_k, d_k)| \geqslant \gamma \|\Delta f_k\| \|d_k\| \qquad (3.2)$$

and condition (2.15) for all $i\in T(k)\setminus\{k\}$.

If the Jacobian matrix is symmetric then this specific can be taken into account by Algorithms 3 and 4. Let they differ from Algorithms 1 and 2 only in updating formulas, namely, (2.16) and (2.17) are used in Algorithms 3 and 4 instead of (2.10) and (2.11), respectively. In this case, the matrices J_o and H_o are required to be symmetric. Algorithms 3 and 4 have the same advantage over Algorithms 1 and 2 as (2.5) has over (2.2).

Note, that the check on nearly linear dependence in Step 3a) of Algorithms 1 and 3 can be performed by means of sequential orthogonalization of the vectors $\{\Delta x_t\}_{t\in T(k-1)}$. At the i-th step of the orthogonalization, where $i=1,2,\ldots,l(k-1)$, we find the orthogonal projection $\Delta x_k^{\perp}$ of the vector Δx_k on the subspace spanned on those vectors Δx, whose indices have been already included in $T(k)\setminus\{k\}$. The index $t(k-1,i)$ is included in $T(k)$ only if the inequality

$$\|\Delta x_k - \Delta x_k^{\perp}\| \geqslant \sigma^{1/l(k-1)} \|\Delta x_k\|$$

takes place. It guarantees that the vectors $\{\Delta x_t\}_{t\in T(k)}$ are not nearly linearly dependent and that the vector

$$c_k = \Delta x_k - P_k\,\Delta x_k, \qquad (3.3)$$

where $P_k\in R^{n\times n}$ denotes the matrix of orthogonal projection on the subspace spanned on $\{\Delta x_t\}_{t\in T(k)\setminus\{k\}}$, satisfies (3.1). Step 3a) of Algorithms 2 and 4 can be performed by analogy.

If (1.2) is satisfied then Algorithms 1 and 2 coincide with the secant method (2.1), and Algorithms 3 and 4 coincide with the symmetric secant methods (2.5) and (2.7), respectively. Superlinear convergence of some particular cases of Algorithms 1-4, corresponding to the choice of c_k by (3.3) (d_k in a similar way), is proved in [17] regardless of whether or not condition (1.2) is satisfied. The following more general result appears to take place. Namely, the quasi-Newton methods (2.1), (2.10) (or (2.16)), where c_k is any nonzero vector satisfying inequality (3.1), converge to x_* superlinearly. It means that superlinear convergence takes place also if (2.14) is violated, but the rate of convergence is seem to be higher when relation (2.14) and as a consequence the secant relation (2.4) are satisfied. A similar assertion concerning the quasi-Newton methods (2.6), (2.11) (or (2.17)), where d_k satisfies inequality (3.2), appears to take place. Using [18] we can simply prove these assertions for any $\gamma \in [17/18, 1]$. The problem of proving of superlinear convergence is still open for $\gamma \in (0, 17/18)$.

4. Stable variants of the secant method with restart

The distinctive feature of Algorithm 5 given below is that the secant relation (2.4) is valid for all i such that $k-l(k)+1 \leqslant i \leqslant k$ where the vectors $\{\Delta x_i\}^k_{k-l(k)+1}$ are not nearly linearly dependent. It is differed from Algorithm 1 by so-called restart, which is the result of setting $l(k)=1$ whenever the vector Δx_k is nearly linearly dependent on the vectors $\{\Delta x_i\}^{k-1}_{i=k-l(k-1)}$. In the case of restart, a part of information about the Jacobian matrix contained in J_k is lost, because having the relation $J_k \Delta x_i = \Delta f_i$ satisfied for all i, $k-l(k-1) \leqslant i \leqslant k-1$, we have no guarantee that the relation

$$J_{k+1}\Delta x_i = J_k \Delta x_i = \Delta f_i \tag{4.1}$$

will be valid for any i after restart. If restart does not happen at the k-th iteration then we set $l(k) := l(k-1)+1 > 1$ and, as a result, relation (4.1) is valid for all i such that $k-l(k-1) \leqslant i \leqslant k-1$.

Algorithm 5. Let an initial point $x_0 \in R^n$, a nonsingular matrix $J_0 \in R^{n \times n}$ and a real number $\sigma \in (0,1)$ be given. Compute f_0.

For k=0,1,2,... execute the following steps.

1. If $\|f_k\| = 0$ or J_k is singular, then stop.
2. Compute x_{k+1} by formula (2.1) and compute f_{k+1}.
3. If k=0, then set $P_k := 0$ where $P_k \in R^{n \times n}$, else compute $P_k \Delta x_k$, where
$$P_k = \sum_{i=k-l(k-1)}^{k-1} \frac{c_i c_i^T}{(c_i, c_i)} .$$
4. If $\|\Delta x_k - P_k \Delta x_k\| < \sigma^{1/l(k-1)} \|\Delta x_k\|$, then perform restart setting $l(k) := 1$ and $c_k := \Delta x_k$, else set $l(k) := l(k-1)+1$ and compute c_k in accordance with (3.3).
5. Compute the matrix J_{k+1} by formula (2.10).

Algorithm 6. Let an initial point $x_o \in R^n$, a nonsingular matrix $H_o \in R^{n\times n}$ and a real number $\sigma \in (0,1)$ be given. Compute f_o.

For k=0,1,2,..., execute the following steps.

1. If $\|f_k\| = 0$, then stop.
2. Compute x_{k+1} by formula (2.6) and compute f_{k+1}.
3. If k=0, then set $Q_k := 0$, where $Q_k \in R^{n\times n}$, else compute $Q_k \Delta f_k$ where

$$Q_k = \sum_{i=k-l(k-1)}^{k-1} \frac{d_i d_i^T}{(d_i, d_i)} .$$

4. If $\|\Delta f_k - Q_k \Delta f_k\| < \sigma^{1/l(k-1)} \|\Delta f_k\|$, then perform restart setting $l(k):=1$ and $d_k := \Delta f_k$, else set $l(k):=l(k-1)+1$ and $d_k := \Delta f_k - Q_k \Delta f_k$.
5. Compute the matrix H_{k+1} by formula (2.11).

Suppose the Jacobian matrix is symmetric. It can be taken into account by Algorithms 7 and 8. Let they differ from Algorithms 5 and 6 only by the updating formulas (2.16) and (2.17) that are used instead of (2.10) and (2.11), respectively. The matrices J_o and H_o are required in this case to be symmetric. The advantages of Algorithms 7 and 8 over Algorithms 5 and 6 are similar to those of the symmetric secant methods over the traditional one.

Algorithms 5 and 6 have been proposed in [10] and Algorithms 7 and 8 have been proposed in [12] . Superlinear convergence of these algorithms have been proved there. Although because of restart the size of information about the Jacobian matrix used in Algorithms 5-8 is less than one used in Algorithms 1-4, Algorithms 5-8 are simpler in realization and require lesser amount of arithmetic operations per iteration than Algorithms 1-4.

5. Parallel algorithms of secant type

Using the main ideas of Algorithms 1-8 we will propose some general scheme of parallel algorithms of secant type. This scheme admits various concrete realizations.

Possibilities of parallel computations seem to realize more completely when at each iteration we have not one approach x_k to the solution x_* but a set of approaches $\bar{x}_k = \{x_{ki}\}_{i=1}^m$, where $x_{ki} \in R^n$. This idea to iterate a set of approaches was used widely for solving one equation with one unknown (see survey [19]).

The general scheme of secant type parallel algorithms is as follows. Let we have a set of approaches $\bar{x}_k$, the respective set of values $\bar{f}_k = \{f_{ki}\}_{i=1}^m$ and approximations $\bar{J}_k = \{J_{ki}\}_{i=1}^m$ (or $\bar{H}_k$) to the Jacobian matrix (or its inverse). Then compute $\bar{x}_{k+1}$ and $\bar{J}_{k+1}$ (or $\bar{H}_{k+1}$) by some of Algorithms 1-8, where the quantities $x_{ki}-x_{kj}$ and $f_{ki}-f_{kj}$ are used as the required increments Δx and Δf, respectively. It is easily seen that the computation of each element of the sets $\bar{x}_{k+1}$, $\bar{f}_{k+1}$ and $\bar{J}_{k+1}$ can be performed parallely.

Consider a particular case corresponding to m=n+2. Let for j=1, ..., n+2 the points $x_{k+1,j}$ be find by the traditional secant method on the base of the n+1 points x_{ki} where $1 \leq i \leq n+2$ and $i \neq j$. We assume that a condition similar to (1.2) is satisfied. Then it can be simply shown that for each i=1,...,n+2 the sequence x_{ki} converges to x_* as $k \to \infty$ quadratically. If $x_{kj}=x_*$ for some j, then $x_{k+1,i} = x_*$ for all $i \neq j$, $1 \leq i \leq n+2$. Moreover, the inequality

$$\|x_{k+1,i} - x_*\| \leq c \, \| x_{kj} - x_*\| \sum_{s=1, s\neq i}^{n+2} \|x_{ks} - x_*\|$$

holds for each $i \neq j$, $1 \leq i \leq n+2$, where c is a constant determined by n and δ only (see [4] for instance). It means that, if we choose any j such that $1 \leq j \leq n+2$, fix all x_{ki} where $i \neq j$, $1 \leq i \leq n+2$, and suppose that $x_{kj} \to x_*$, then we see that the approach x_{kj} nearest to x_* influences essentially at the convergence of all other approaches to x_*. This fact gives us a possibility to compare this quadratically convergent parallel algorithm with various parallel realizations of the Newton method generating only one approach to x_* at each iteration. The first one seems to converge faster.

Note that the required above condition similar to (1.2) will be violated as a role in practice. For this reason it would be helpful to use Algorithms 1-8, that are stable with respect to location of the points x_{ki}, instead of the traditional secant method.

References

1. Bittner, L. Eine verallgemeinerung des Secantenverfahrens zur näherungsweisen Bereichnung der Nullstellen eines nichtlinearer Gleichungssystems. Wiss.Z. Techn.Univ.Dresden (1959).Bd.9, S. 325-329.
2. Schwetlick,H. Numerische Lösung nichtlinearer Gleichungen. Dtsch.Verl.Wiss., Berlin (1979).
3. Wolfe,P. The secant method for simultaneous nonlinear equations, Communs ACM (1959), vol.2, p.12-13.
4. Ortega, J.M.; Rheinboldt, W.C. Iterative Solution of Nonlinear Equations in Several Variables. Academic Press, New York (1970).
5. Barnes, J. An algorithm for solving nonlinear equations based on the secant method. Computer J. (1965), vol.8, p.66-72.
6. Schwetlick,H. Uber die Realisierung und Konvergenz von Mechrschrittverfahren zur iterativen Lösung nichtlinearer Gleichungen. Z. Angew.Math. und Mech. (1974), vol.54,p.479-493.
7. Gragg, W.B.;Stewart, G.W. A stable variant of the secant method

for solving nonlinear equations. SIAM J. Numer. Analys.(1976), vol.13, p.889-903.

8. Martinez, J.M. Three new algorithms based on the sequential secant method, BIT (1979), vol. 19, p. 236-243.
9. Martinez, J.M.; Lopes, T.L. Combination of the sequential secant method and Broyden's method with projected updates. Computing (1980), vol.25, p.379-386.
10. Gay,D.M.; Schnabel,R.B. Solving systems of nonlinear equations by Broyden's method with projected updates. In: Nonlinear Programming 3 (O.L.Mangasarian, S.M.Robinson, R.R.Meyer, eds.), Academic Press, New York (1978), p.245-281.
11. Burdakov,O.P. Stable versions of the secant method for solving systems of equations. U.S.S.R. Comput.Maths. Math.Phys.(1983), vol.23, N5, p.1-10.
12. Burdakov,O.P. On symmetric secant methods with restart.In:Methods of Mathematical Programming and their Software.Sverdlovsk (1984) p.33-34, in Russian (to appear in U.S.S.R. Comput.Maths.Math. Phys.).
13. Burdakov,O.P. Some globally convergent modifications of Newton's method for solving systems of nonlinear equations. Soviet Math. Dokl.(1980), vol.22,p.376-378.
14. Burdakov,O.P. Methods of the secant type for systems of equations with symmetric Jacobian matrix. Numer.Funct.Anal. and Optimiz (1983), vol.6,p.183-195.
15. Dennis,J.E.; More, J.J. Quasi-Newton methods, motivation and theory. SIAM Review (1977), vol.19,p.46-89.
16. Dennis,J.E.;Schnabel,R.B. Numerical Methods for Unconstrained Optimization and Nonlinear Equations. Prentice - Hell.Engbewood Cliffs, NJ (1983).
17. Burdakov, O.P. On superlinear convergence of some stable variants of the secant method. Z.Angew.Math. und Mech., to appear.
18. Broyden,C.G.; Dennis,J.E.; More,J.J. On the local and superlinear convergence of quasi-Newton methods. J.Inst.Math. and Applic. (1973), vol.12,p.223-245.
19. Miranker,W.L. A survey of parallelism in numerical analysis. SIAM Review (1971), vol.13,p.524-547.

ASSIGNMENT PROBLEMS: RECENT SOLUTION METHODS AND APPLICATIONS

Rainer E. BURKARD
Technische Universität Graz
Institut für Mathematik
Kopernikusgasse 24
A-8010 Graz, Austria

Abstract

Recent developments for linear and quadratic assignment problems are surveyed. In particular some new efficient solution techniques are outlined and a recent application concerning the assignment of time-slots in a time division multiple access system is described. Finally assignment problems are used to solve a problem in channel routing.

1. Linear assignment problems (LAP) and their underlying combinatorial structure.

Linear assignment problems have been studied since the early beginnings of mathematical programming (KUHN, 1955). Before we treat them as optimization problems, let us consider their underlying combinatorial structure.

Assignments are one-to-one mappings of finite sets. They can be described in different ways, for example by *permutations* or by *permutation matrices*. One of the basic problems consists in determining whether in a given situation an assignment exists. To be more concrete let us consider the following situation:

There are n persons $\{P_1,P_2,...,P_n\}$ and n jobs $\{J_1,J_2,...,J_n\}$. Every person can handle a subclass $J(P_i)$ of the jobs (cf. Fig. 1)

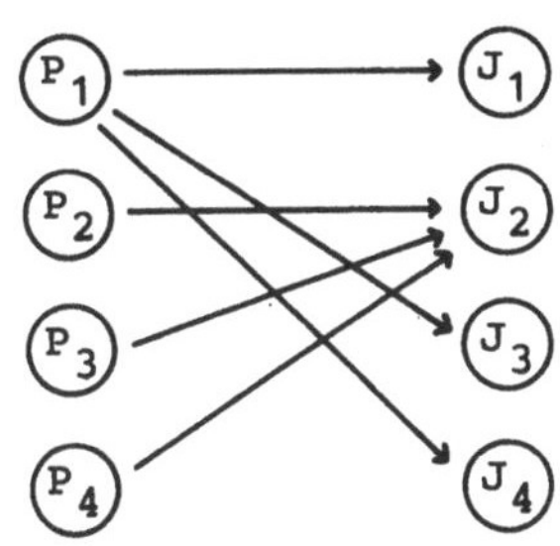

Fig. 1: Possible situation with four persons and jobs.

We ask, if we can select exactly one job for every person, such that all jobs are performed.

For any set P of the persons let us define

$$J(P) = \bigcup_{P_i \in P} J(P_i)$$

and let $|J(P)|$ the number of elements of the (finite) set $J(P)$. In the example of Fig. 1 we have $J(P_1) = \{J_1, J_3, J_4\}$ and $|J(P_1)| = 3$. Moreover, for $P = \{P_2, P_3, P_3\}$ we get $|J(P)| = 1$.

An answer to our question is given by the famous "marriage theorem":

Marriage theorem (Ph.HALL, 1935)

There is an assignment of persons to jobs, if and only if

$$|P| \leq |J(P)| \quad \text{for all} \quad P \subseteq \{P_1, P_2, \ldots, P_n\} \quad \text{"Hall condition"}.$$

H. WEYL gave an interpretation of this theorem where the persons correspond to boys who are in love with girls. The theorem answers the question if there is a marriage (=assignment) such that every boy marries one of his girlfriends. Since in our example shown by Fig. 1 we get $|J(P)| = 1$ for $P = \{P_1, P_2, P_3\}$, no assignemnt exists in this case.

Note that the Hall condition is a condition on *all* subsets of a given set $\{P_1, \ldots, P_n\}$. However it is possible to check this condition within $O(n^{5/2})$ steps by using a perfect matching algorithm in bipartite graphs (HOPCROFT and KARP, 1973). *Perfect matchings* in bipartite graphs is just another name for assignments.

Let us describe now our situation by matrices. We define an $(n \times n)$-matrix $X = (x_{ij})$ by

$$x_{ij} := \begin{cases} 1, & \text{if person } P_i \text{ handles job } J_j \\ 0, & \text{otherwise}. \end{cases}$$

If X corresponds to an assignment, we get

$$\sum_{i=1}^{n} x_{ij} = 1 \quad \text{for } j=1,2,\ldots,n \quad \text{namely every job is performed by one person} \tag{1.1}$$

$$\sum_{j=1}^{n} x_{ij} = 1 \quad \text{for } i=1,2,\ldots,n \quad \text{describing that every person handles a job} \tag{1.2}$$

and

$$x_{ij} \in \{0,1\} \quad \text{for all i and j} \tag{1.3}$$

Matrices with the properties (1.1) - (1.3) are called *permutation matrices*. They correspond uniquely to permutations as well as to assignments.

Example.

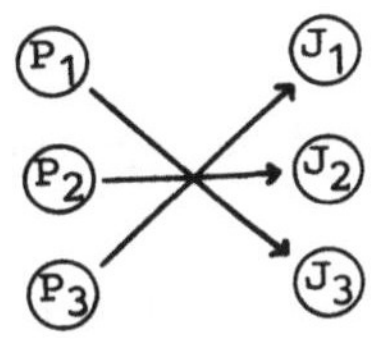

$$\Longleftrightarrow \quad X = \begin{pmatrix} 0 & 0 & 1 \\ 0 & 1 & 0 \\ 1 & 0 & 0 \end{pmatrix} \quad \Longleftrightarrow \quad \varphi = \begin{pmatrix} 1 & 2 & 3 \\ 3 & 2 & 1 \end{pmatrix}$$

perfect matching in a bipartite graph — permutation matrix — permutation

All these expressions are reformulations of the *same* assignment. □

Geometrically, assignments can be characterized as vertices of a polytope P_A, called assignment polytope (BIRKHOFF, 1944):

$$P_A = \{x \in \mathbb{R}^{n^2} \mid \sum_{j=1}^{n} x_{ij} = 1 \text{ for all } i, \sum_{i=1}^{n} x_{ij} = 1 \text{ for all } j, 0 \le x_{ij} \le 1\} .$$

Birkhoff's result that the vertices of the assignment polytope correspond uniquely to permutation matrices will be the basis for solving linear assignment problems via linear programs.

Now let us turn to the problem to find *best* assignments in a sense which has still to be specified. For this reason we introduce a measure c_{ij} indicating the cost which occur if P_i handles J_j. Thus we get an $(n \times n)$-cost matrix $C = (c_{ij})$. If we want to find an assignment (=permutation φ) which minimizes the cost, we get a *linear assignment problem*

$$\min_{\varphi} \sum_{i=1}^{n} c_{i\varphi(i)} \quad . \tag{1.4}$$

If the permutation φ is represented by a permutation matrix X, problem (1.4) can be written in the form

$$\begin{aligned} \min \quad & \sum_{i=1}^{n} \sum_{j=1}^{n} c_{ij}\, x_{ij} \\ \text{subject to} \quad & \sum_{i=1}^{n} x_{ij} = 1 \quad \text{for } j = 1,2,\ldots,n \\ & \sum_{j=1}^{n} x_{ij} = 1 \quad \text{for } i = 1,2,\ldots,n \\ & x_{ij} \in \{0,1\} \quad \text{for } i,j = 1,2,\ldots,n . \end{aligned}$$

This is n o t a linear program because of the integrality condition (1.3). Birkhoff's theorem together with the fact that a basic solution of a linear program corresponds to a vertex of the underlying polyhedron enable us to replace (1.3) by

$$0 \le x_{ij} \le 1 \,. \tag{1.5}$$

Therefore linear assignment problems can be considered as special linear programs.

In the next section we survey some efficient solution methods for linear assignment problems.

2. Solution methods for linear assignment problems.

Since linear assignment problems can be interpreted as linear programs, all solution techniques for LPs, in particular primal, primal-dual and dual simplex methods are applicable. Historically Kuhn's *Hungarian method* (KUHN, 1955) was the first solution method for LAPs. This algorithm, which can be conceived as a *dual* method consists of the following two steps:

Step 1. *Reduction step*: By adding constants to the rows and columns of the cost matrix C is tried to generate as many independent zero-elements in C as possible.

Step 2. The *Hall condition* is checked. If it is fulfilled the negative sum of all constants added yields the optimal objective function value. Otherwise a new reduction step is performed.

Example. Let the cost matrix C be given as below and add to its rows the constants -2, -1, 0:

$$C = \begin{pmatrix} 2 & 4 & 4 \\ 3 & 1 & 3 \\ 0 & 1 & 4 \end{pmatrix} \xrightarrow{\text{Row reduction}} \begin{pmatrix} 0 & 2 & 2 \\ 2 & 0 & 2 \\ 0 & 1 & 4 \end{pmatrix} \qquad \begin{matrix} u_1 = -2 \\ u_2 = -1 \\ u_3 = 0 \end{matrix}$$

Now add to the third column $v_3 = -2$ and we get an assignment with cost coefficients 0, which are marked by a star:

$$\xrightarrow{\text{Column reduction}} \begin{pmatrix} 0 & 2 & 0^* \\ 2 & 0^* & 0 \\ 0^* & 1 & 2 \end{pmatrix}, \quad v_3 = -2$$

Therefore the optimal assignment $\varphi = \begin{pmatrix} 1 & 2 & 3 \\ 3 & 2 & 1 \end{pmatrix}$ has the value

$z = -(u_1 + u_2 + u_3 + v_3) = 5$. □

In general it is necessary to perform beside row and column reductions a third kind of reduction, whose description is, however, beyond the scope of this short outline.

A few years ago BERTSEKAS (1981) has proposed some modifications and improvements which make this Hungarian method faster (see also the recent paper of GOLDFARB (1985)).

Shortest augmenting path methods can be conceived as primal-dual methods. As an example let us give an outline of Tomizawa's method (TOMIZAWA, 1971):

Let $C = \begin{pmatrix} 2 & 4 & 4 \\ 3 & 1 & 3 \\ 0 & 1 & 4 \end{pmatrix}$. We start with the first row and construct a network leading from a source s to the node R_1 (row 1) and from there we have arcs to the nodes C_1, C_2, C_3 which have costs $c(R_1,C_j)$. From every node C_j we have an arc to the sink t.

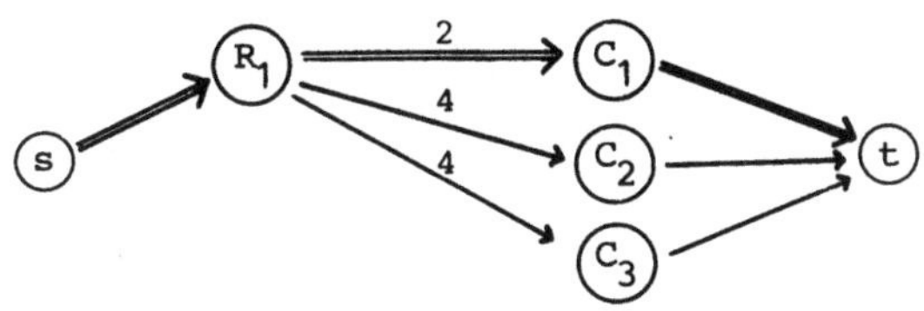

Fig. 2: Shortest path ▬▬

The shortest path in this network (which can be determined by Dijkstra's algorithm in $O(n^2)$ steps) leads from s via R_1, C_1 to t. Now we subtract $c_{11} = 2$ from all costs c_{ij} in our network, delete all arcs "⟹" but introduce instead a reverse arc (C_1,R_1) . Moreover, we introduce the node R_2, corresponding to the second row and the new arcs (R_2,C_j), j=1,2,3 with their corresponding cost. Again a shortest path from s to t in this new network has to be determined (see Fig. 3):

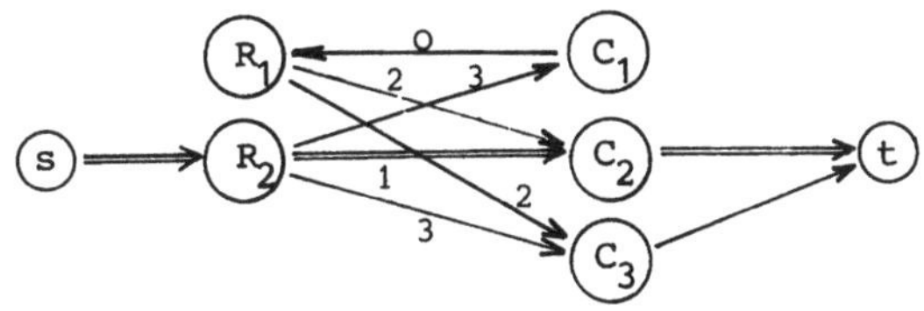

Fig. 3

In the third step we get the following network

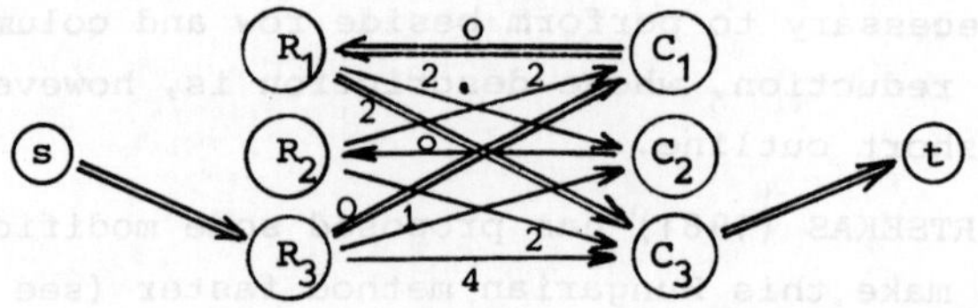

Fig. 4

After n shortest path computations the optimal assignment is found. Therefore we get an algorithm of complexity $O(n^3)$. Recently, DERIGS and METZ (1985) improved shortest augmenting path algorithms considerably by exploiting the fact that in the case of sparse problems shortest augmenting path algorithms are very fast. Let us give an outline of their technique:

For every row R_i (i=1,...,n) only the k smallest entries c_{ij} are considered (for example k=5) and this sparse LAP is solved by a shortest augmenting path technique. The optimal solution of this thinned-out problem is checked whether or not it is also optimal for the original problem. If it is not the case, some of the arcs for which the optimality condition is violated are added and the previous solution is modified to take also these newly introduced arcs into consideration. This algorithm yields astonishing short running times (see Table 1).

Computer codes based on shortest augmenting path techniques have been published in BURKARD and DERIGS (198o) (program "LSAP") and by CARPANETO and TOTH (198o) ("Algorithm 548") and belong to the fastest solution techniques for linear assignment problems. A special code for sparse problems ("Algorithm 50") has been published by CARPANETO and TOTH (1983).

Size of problems	CPU-time for LSAP	CPU-time for LAPIO
3oo	1.33	0.356
4oo	2.96	0.6oo
5oo	4.55	0.855

Table 1: Running times for the code LSAP and the new code LAPIO of Derigs-Metz in CPU seconds. The cost range of these randomly generated problems is [0,1000]. They were solved with FORTRAN HEXT-version on an IBM 3o81.

It has been known for a long while that *primal methods* are also computationally efficient for linear assignment problems (see for example GAVISH, SCHWEITZER and SHLIFER, 1977), but only in 1978 a polynomial primal method has been constructed (CUNNINGHAM and MARSH, 1978). [A remark in passing: The polynomial primal method runs on randomly generated test examples more slowly than a non-polynomial primal code.] Meanwhile also non-simplex methods for assignment problems have been developed. For such a method see THOMPSON (1981).

For some large-scale applications $O(n^3)$-algorithms for finding an optimal solution are too slow. Therefore there is a certain need for heuristics which can be applied to large scale assignment problems. A linear time heuristic has been described by IRI, MUROTA and MATZUI (1983) and has successfully been tested on finding optimal plotter movements.

3. An application: Time-slot assignments for TDMA-Systems.

Time-division-multiple-access (TDMA)-systems are used for instance in communication systems via satellites. Here the information is remitted from the ground stations via a satellite to a receiving station in very short data bursts. On board the satellite these data bursts are received by transponders and transmitted back to earth. A transponder can connect two different ground stations. The problem arises to determine optimal lengths for the data bursts such that all phone calls etc. can be handled in the shortest time possible.

Let $T = (t_{ij})$ be the *traffic matrix*, where t_{ij} is a measure for the amount of information which has to be remitted from ground station R_i to receiving station C_j $(i,j=1,2,\dots,n)$. The ground stations can be connected with the receiving station by a switch mode which is nothing else than an $(n\times n)$-permutation matrix P. We assume that on-board the satellite $2n$ fixed switch modes can be used, namely $P_1,\dots,P_n$ and $Q_1,\dots,Q_n$ which have the following two properties:

$$\sum_{k=1}^{n} P_k = \mathbb{1} , \quad \sum_{k=1}^{n} Q_k = \mathbb{1} , \qquad (3.1)$$

where $\mathbb{1}$ denotes a matrix with elements "1" only.

Every P_k has written every Q_l exactly on "1"-element in common. (3.2)

For $n = 3$ we can use for example

$$P_1 = \begin{pmatrix} 1 & 0 & 0 \\ 0 & 1 & 0 \\ 0 & 0 & 1 \end{pmatrix}, \quad P_2 = \begin{pmatrix} 0 & 1 & 0 \\ 0 & 0 & 1 \\ 1 & 0 & 0 \end{pmatrix}, \quad P_3 = \begin{pmatrix} 0 & 0 & 1 \\ 1 & 0 & 0 \\ 0 & 1 & 0 \end{pmatrix};$$

$$Q_1 = \begin{pmatrix} 0 & 0 & 1 \\ 0 & 1 & 0 \\ 1 & 0 & 0 \end{pmatrix}, \quad Q_2 = \begin{pmatrix} 0 & 1 & 0 \\ 1 & 0 & 0 \\ 0 & 0 & 1 \end{pmatrix}, \quad Q_3 = \begin{pmatrix} 1 & 0 & 0 \\ 0 & 0 & 1 \\ 0 & 1 & 0 \end{pmatrix}.$$

The problem arises to determine lengths $\lambda_1,\ldots,\lambda_n$, $\mu_1,\ldots,\mu_n$ for the time-slots (=length of the data bursts) such that their sum is minimal, but all messages can be transmitted by using switch mode P_1 during λ_1 time units, switch mode P_2 during λ_2 time units, ... , switch mode Q_n during μ_n time units.

In BURKARD (1985) it has been shown that this problem can be transformed to a linear assignment problem and thus be solved in $O(n^3)$-time.

Formally, the problem can be stated as follows:

Determine time lengths $\lambda_1,\lambda_2,\ldots,\lambda_n$, $\mu_1,\mu_2,\ldots,\mu_n$, all ≥ 0 such that

$$\begin{aligned} &\sum_{k=1}^{n} (\lambda_k + \mu_k) \to \min \\ \text{subject to} \quad &\sum_{k=1}^{n} (\lambda_k P_k + \mu_k Q_k) \geq T \quad \text{(elementwise)} \end{aligned} \tag{3.3}$$

(3.3) is a linear program the dual of which is (3.4), if s_{ij} denote the dual variables:

$$\begin{aligned} \min &- \sum_{i=1}^{n} \sum_{j=1}^{n} t_{ij}\, s_{ij} \\ \text{subject to} \quad &\sum_{i=1}^{n} \sum_{j=1}^{n} p_{ij}^{(k)} s_{ij} \leq 1 \\ &\sum_{i=1}^{n} \sum_{j=1}^{n} q_{ij}^{(l)} s_{ij} \leq 1, \quad s_{ij} \geq 0 \end{aligned} \tag{3.4}$$

Here, $p_{ij}^{(k)}$ are the elements of the permutation matrix P_k and $Q_l = (q_{ij}^{(l)})$. We introduce variables v_{kl} by

$$v_{kl} := \begin{cases} 1, & \text{if } \sum\limits_{ij}\sum p_{ij}^{(k)} q_{ij}^{(l)} s_{ij} \neq 0 \\ 0, & \text{otherwise} \end{cases} \quad (k,l=1,2,\ldots,n) \tag{3.5}$$

and new cost coefficients c_{kl} by

$$c_{kl} := a - t_{ij}, \quad \text{if} \quad p_{ij}^{(k)} = q_{ij}^{(l)} = 1 \tag{3.6}$$

where $a = \max_{i,j} t_{ij}$.

Thus, the dual program (3.4) is equivalent to the assignment problem

$$\begin{aligned} &\min \sum_{k=1}^{n} \sum_{l=1}^{n} c_{kl} v_{kl} \\ \text{subject to } &\sum_{k=1}^{n} v_{kl} = 1 \\ &\sum_{l=1}^{n} v_{kl} = 1, \quad v_{kl} \geq 0 \end{aligned} \tag{3.7}$$

Let a_k, b_l be the optimal values of the dual variables of this linear assignment problem (3.7). We determine δ by

$$\max_l b_l \leq \delta \leq \min_k (a-a_k) .$$

Then $\lambda_k := a - a_k - \delta \qquad (k=1,2,\ldots,n)$

$\mu_l := \delta - b_l \qquad (l=1,2,\ldots,n)$

are optimal lengths for the time slots using switch modes P_k and Q_l respectively.

Example. Let the traffic matrix $T = (t_{ij})$ be given as

$$T = \begin{pmatrix} 1 & 4 & 5 \\ 4 & 3 & 2 \\ 2 & 1 & 1 \end{pmatrix} .$$

We get $a = 5$ and the transformation (3.6) yields

$$C = \begin{pmatrix} 2 & 4 & 4 \\ 3 & 1 & 3 \\ 0 & 1 & 4 \end{pmatrix} .$$

The linear assignment problem with cost matrix C has already been solved in section 2. We get as dual variables $a_1=2$, $a_2=1$, $a_3=0$, $b_1=0$, $b_2=0$, $b_3=2$. Thus for $\delta=2$ we get an optimal solution

$$\lambda_1=1, \quad \lambda_2=2, \quad \lambda_3=3, \quad \mu_1=2, \quad \mu_2=2, \quad \mu_3=0 .$$

There is also a second optimal solution for $\delta=3$ with

$$\lambda_1=0, \quad \lambda_2=1, \quad \lambda_3=2, \quad \mu_1=3, \quad \mu_2=3, \quad \mu_3=1 .$$

4. Further results on linear assignment problems.

In this section we shall comment special cases of linear assignment problems which can be solved in less than $O(n^3)$-time and we shall consider assignment problems with generalized "linear" objective functions. A cost matrix $C = (c_{ij})$ has the Monge-property, if for $i=1,2,\ldots,n$ and all $j,k > i$

$$c_{ji} + c_{jk} \le c_{ik} + c_{ji} \tag{4.1}$$

holds. We get

Theorem (HOFFMAN, 1963; DERIGS, GOECKE and SCHRADER, 1984).
If C has the Monge-property then the identical permutation is optimal.

Let M denote the class of all those matrices C for which there is a permutation σ of their rows and a permutation τ of their columns such that

$$C_{\sigma,\tau} = (c_{\sigma(i),\tau(j)})$$

has the Monge-property.
It seems still to be an open problem to recognize matrices $C \in M$!

Now, let us define a stronger property. A matrix $C = (c_{ij})$ has the *strong Monge property*, if for all $1 \le i < j \le n$ and $1 \le k < l \le n$ the following relation (4.2) holds:

$$c_{ik} + c_{jl} \le c_{il} + c_{jk} \,. \tag{4.2}$$

Matrices with the strong Monge property have obviously also the Monge property (4.1). We define the class $\mathfrak{C}$ of matrices having the strong Monge property analogous to the class M. Within this class $\mathfrak{C}$ there are the following matrices:

Example 1. Let $u = (u_1,\ldots,u_n)$ and $v = (v_1,\ldots,v_n)$ be any real vectors. A matrix $C = (c_{ij})$ with $c_{ij} = u_i \cdot v_j$ has the strong Monge property.

Example 2 (GILMORE and GOMORY, 1964). Let again two vectors u and v be given and consider two integrable functions h_1 and h_2 with $h_1(t) + h_2(t) > 0$. If we define

$$c_{ij} := \begin{cases} \int_{u_i}^{v_j} h_1(t)dt & \text{for } u_i \le v_j \\ \int_{v_j}^{u_i} h_2(t)dt & \text{for } u_i > v_j \end{cases}$$

then $C = (c_{ij})$ has the strong Monge property.

Example 3 (BURDJUK and TROFIMOV, 1976). Let $f : \mathbb{R}^2 \to \mathbb{R}$ be a function with the property

$$f(x+\Delta x,\ y+\Delta y) + f(x,y) \ge f(x+\Delta x,\ y) + f(x,y+\Delta y) \quad \text{for all } \Delta x \ge 0,\ \Delta y \ge 0 .$$

Then any matrix $C = (c_{ij})$ defined by

$$c_{ij} = f(u_i, v_j) \qquad (u_i, v_j \ge 0)$$

has again the strong Monge property. In particular matrices defined by

$$c_{ij} := \min(u_i, v_j)$$
$$c_{ij} := \max(u_i, v_j)$$
$$c_{ij} := |u_i - v_j|^p, \quad p \ge 1$$

have the strong Monge property.

DEINEKO and FILONENKO (1979) have shown that matrices within the class $\mathfrak{C}$ can be recognized in $O(n^2)$ steps. Therefore the corresponding linear assignment problems are solvable in $O(n^2)$ time.

For another special linear assignment problem, which can be solved in $O(n)$ steps see KARP and LI (1975). In this problem the cost coefficients c_{ij} are defined by

$$c_{ij} := |u_i - v_j|$$

where u_i and v_j are arbitrary points on a line segment or on a circle.

Let us turn now to other objective functions than sums. If we replace the addition in the objective function of a linear assignment problem by the operation "max", we get a *Bottleneck Assignment Problem*

$$\min_{\varphi} \max_{1 \le i \le n} c_{i\varphi(i)} .$$

Using threshold methods paired with binary search bottleneck assignment problems can be solved in $O(n^{5/2} \log n)$ time. Efficient computer codes for this class of problems can be found in BURKARD and DERIGS (1980) as well as in CARPANETO and TOTH (1981).

Both, addition and maximization of cost elements are only special cases of a very general formulation for the operation between cost coefficients in objective function. We can consider the cost elements drawn from an ordered semigroup $(H,*,\leq)$ with semigroup operation "*". It has been shown that an assignment problem with such an *"algebraic" objective function* can also be solved by polynomial algorithms (BURKARD, HAHN and ZIMMERMANN, 1977; BURKARD and ZIMMERMANN, 1980). Thus for instance lexicographical problems or time-cost problems can be solved by this approach. Recently, MARTELLO, PULLEYBLANK, TOTH and DE WERRA (1984) considered an objective function which minimizes the maximum difference between two cost coefficients in the objective function

$$\min_{\varphi} \max_{i,j} |c_{i\varphi(i)} - c_{j\varphi(j)}|.$$

They showed that this *Balanced Assignment Problem* can be solved in $O(n^4)$ steps.

In the next section we shall describe assignment problems with a quadratic objective function. These "quadratic assignment problem" play an important rôle in applications.

5. Quadratic assignment problems (QAP).

The problem to minimize a quadratic objective function

$$\sum_{i=1}^{n} \sum_{j=1}^{n} \sum_{p=1}^{n} \sum_{q=1}^{n} d_{ijpq}\, x_{ij} \cdot x_{pq}$$

subject to assignment constraints

$$\sum_{i=1}^{n} x_{ij} = 1 \qquad \text{for} \quad j=1,2,\dots,n$$

$$\sum_{j=1}^{n} x_{ij} = 1 \qquad \text{for} \quad i=1,2,\dots,n$$

$$x_{ij} \in \{0,1\}$$

is called *Quadratic Assignment Problem (QAP)* and plays an important rôle in placement and plant location problems as well as in many other situations where the data are interrelated. There are several recent surveys on this important class of problems. The reader is referred to BURKARD (1983, 1984) and FINKE, BURKARD and RENDL (1986).

Since QAPs are NP-hard, only implicit enumeration methods are known for solving them optimally. A complete code for a branch and bound code using Gilmore-Lawler bounds can be found in BURKARD and DERIGS (198o). Since only problems with a small size (say up to $n \sim 15$ or 20) can be solved optimally, there is a special need for good heuristics. Recently good results were obtained with a simple simulated annealing algorithm, see BURKARD and RENDL (1984).

6. Asymptotic results for assignment problems.

Many papers deal recently with the probabilistic and asymptotic behaviour of assignment problems. Concerning linear (sum) assignment problems the following results are known:
WALKUP (1979) showed that for assignment problems with randomly generated cost coefficients

$$\max_{\varphi} \sum_{i=1}^{n} c_{i\varphi(i)} - \min_{\varphi} \sum_{i=1}^{n} c_{i\varphi(i)} \to \infty \quad \text{as} \quad n \to \infty \quad .$$

Stronger results were obtained for Euclidean linear assignment problems, where given n points in the plane the cost coefficient c_{ij} is the Euclidean distance between the points i and j.

Let $L_{min} = \min_{\varphi} \sum_{i=1} c_{i\varphi(i)}$. It can be shown that $\lim_{n\to\infty} \frac{1}{\sqrt{n}} L_{min}$ is a constant β almost surely. IRI et al. (1983) computed β to lie between 0.32 and 0.33. WEBER and LIEBLING (1985) determined β empirically to be 0.3189 ± 0.0052. Compare also PAPADIMITRIOU (1977).

Concerning asymptotic results of $L_{max} = \max_{\varphi} \sum_{k=1}^{n} c_{i\varphi(i)}$ it is known that $\lim_{n\to\infty} \frac{1}{n} L_{max}$ is a constant γ almost surely. DYER, FRIEZE and McDIARMID (1984) determined γ to be ≈ 0.3826. See also WEBER and LIEBLING (1985).

Quadratic assignment problems show a completely different asymptotic behaviour. In BURKARD and FINCKE (1983) it has been shown that

$$\text{Prob.} \left\{ \frac{L_{max}}{L_{min}} \le 1 + \varepsilon(n) \right\} \ge 1 - \rho(n) \tag{6.1}$$

with $\varepsilon(n)$, $\rho(n) \to 0$ as $n \to \infty$.

(Here L_{max} and L_{min} are defined in analogy to the linear case.)

This behaviour, which is also true for quadratic bottleneck assignment problems (BURKARD and FINCKE, 1982) is shown to hold for a larger class of combinatorial optimization problems, which can be described in the following way:
We consider a family (P_n), $n \in N$ of combinatorial optimization problems defined on finite ground sets E_n. The feasible solutions of a problem P_n are defined by a nonempty class T_n of subsets of E_n. Therefore a feasible solution S can be conceived as a subset of E_n. If there is a $\lambda_o > 0$ such that

$$\lambda_o \ |S| - \log |T_n| \to \infty \quad \text{as } n \to \infty \tag{6.2}$$

then the corresponding class of combinatorial optimization problems shows the behaviour (6.1) (cf. BURKARD and FINCKE, 1985).

7. Terminal assignments.

In the design of layouts for integrated circuits the problem arises to assign (route) entry terminals to exit terminals such that the *density* of the routing is minimized. The density is the maximum number of nets (connections) that cross any vertical line (cf. Fig. 5).

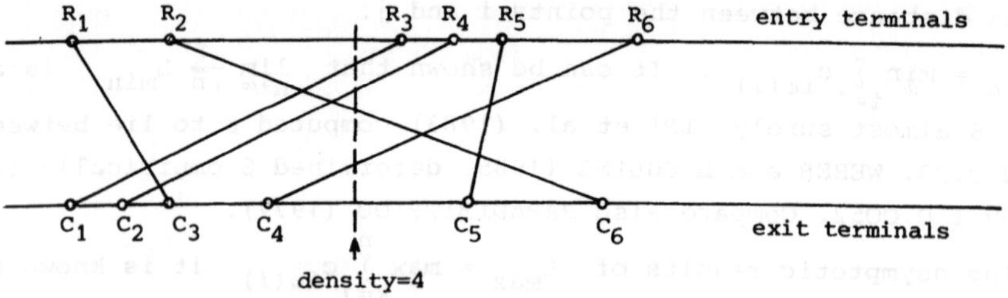

Fig. 5: Example of a 2-layer terminal assignment with density 4.

ATALLAH and HAMBRUSH (1984) characterized the best density which can be obtained and gave an O(n) algorithm for finding the corresponding optimal assignment. If more than two layers are involved, this can be modeled by colouring the entry terminals with different colours (corresponding to their respective layer) and taking the density as maximum density with respect to a single colour. ATALLAH and HAMBRUSH (1985) extended their approach to the case of 2 layers (colours) and showed that the optimal density can still computed in linear time and an optimal assignment can be found in O(n log n) steps. ROTHE and RENDL (1985) investigated terminal assignments with arbitrary many k ($k \geq 1$) colours. They could show that for a given density d it can be decided in linear time whether an assignment of density d exists or not. If it exists, it can be constructed in linear time, too. By using a recently developed recursive procedure they showed further

Theorem 7.1 (ROTHE and RENDL, 1985).

Given the entry and exit nodes sorted in increasing order an optimal terminal assignment minimizing the (k-colour) density can be determined in linear time.

References.

M.J.ATALLAH and S.E. HAMBRUSH (1984), On terminal assignments that minimize the density. Techn.Report CSD-TR-468, Purdue Univ., West Lafayette (Indiana), USA.

M.J.ATALLAH and S.E. HAMBRUSH (1985), An assignment algorithm with applications to layout design. Techn.Report, Purdue Univ., West Lafayette (Indiana), USA.

D.P.BERTSEKAS (1981), A new algorithm for the assignment problem. Mathematical Programming 21, 152-171.

G.BIRKHOFF (1944), Tres observaciones sobre el algebra lineal. Rev.univ.nac.Tucumán (Ser.A) 5, 147-151.

V.Y.BURDYUK and V.N.TROFIMOV (1976), Generalization of the results of Gilmore and Gomory on the solution of the traveling salesman problem. Izv.Akad.Nauk SSSR, Techn.Kibernet.3, 16-22 (Russian); translated as Eng.Cybernetics 14 (1976), 12-18.

R.E.BURKARD (1983), Locations with spatial interactions - the quadratic assignment problem. Report 83-31, Math.Institut, Techn.Universität Graz. To appear in R.L.Francis and P.B.Mirchandani (Eds.): *Discrete Location Theory*, Academic Press, 1986.

R.E.BURKARD (1984), Quadratic assignment problems. European J.of Operational Research 15, 283-289.

R.E.BURKARD (1985), Time-slot assignments for TDMA-systems. Computing 35, 99-112.

R.E.BURKARD and U.DERIGS (198o), *Assignment and Matching Problems: Solution Methods with FORTRAN-programs*. Springer (LN Econ.and Math.Systems, Vol.184), Berlin-Heidelberg-New York.

R.E.BURKARD and U.FINCKE (1982), On random quadratic bottleneck problems. Mathematical Programming 23, 227-232.

R.E.BURKARD and U.FINCKE (1983), The asymptotic probabilistic behaviour of quadratic sum assignment problems. Zeitschrift für Operations Research, Ser.A 27, 73-81.

R.E.BURKARD and U.FINCKE (1985), Probabilistic asymptotic properties of some combinatorial optimization problems. Discrete Applied Maths.12, 21-29.

R.E.BURKARD, W.HAHN and U.ZIMMERMANN (1977), An algebraic approach to assignment problems. Mathematical Programming 12, 318-327.

R.E.BURKARD and F.RENDL (1984), A thermodynamically motivated simulation procedure for combinatorial optimization problems. European J. of Operational Research 17, 169-174.

R.E.BURKARD and U.ZIMMERMANN (198o), Weakly admissible transformations for solving algebraic assignment and transportation problems. Math.Programming Study 12, 1-18.

G.CARPANETO and P.TOTH (198o), Algorithm 548 (Solution of the assignment problem). ACM Trans. on Math.Software 6, 1o4-111

G.CARPANETO and P.TOTH (1981), Algorithm 44: Algorithm for the solution of the bottleneck assignment problem. Computing 27, 179-187.

G.CARPANETO and P.TOTH (1983), Algorithm 50: Algorithm for the solution of the assignment problem for sparse matrices. Computing 31, 83-94.

W.M.CUNNINGHAM and A.B.MARSH, III (1978), A primal algorithm for optimum matching. Mathematical Programming Study 8, 5o-72.

V.G.DEINEKO and V.L.FILONENKO (1979), On the reconstruction of specially structured matrices. In: Aktualnye problemy EVM; programmirovanie. Dnepropetrovsk: DGU, 43-45 (Russian).

U.DERIGS and A.METZ (1985), An in-core/out of core method for solving large scale assignment problems. Report No. 85376-OR, Institut für Operations Research, Universität Bonn, F.R.G.

M.E.DYER, A.M.FRIEZE and C.J.H.McDIARMID (1984), Partitioning heuristics for two geometric maximization problems. Operations Research Letters 3, 267-27o.

G.FINKE, R.E.BURKARD and F.RENDL (1986), Quadratic assignment problems. To appear in: Annals of Discrete Maths., 1986.

B.GAVISH, P.SCHWEITZER and E.SHLIFER (1977), The zero pivot phenomenon in transportation and assignment problems and its computational implications. Mathematical Programming 12, 226-24o.

P.C.GILMORE and R.E.GOMORY (1964), Sequencing a one-state variable machine: a solvable case of the traveling salesman problem. Operations Research 12, 655-679.

D.GOLDFARB (1985), Efficient dual simplex algorithms for the assignment problem. Mathematical Programming 33, 187-2o3.

Ph.HALL (1935), On representations of subsets. J.London Math.Soc. 1o, 26-3o.

A.J.HOFFMAN (1963), On simple linear programming problems. Proc. of Symp. on Pure Maths., VII, Convexity, AMS, 317-327.

J.E.HOPCROFT and R.M.KARP (1973), An $n^{5/2}$ algorithm for maximum matchings in bipartite graphs. SIAM J.Comput. 2, 225-231.

M.IRI, K.MUROTA and Sh.MATSUI (1983), Heuristics for planar minimum-weight perfect matchings. Networks 13, 67-92.

R.M.KARP and S.R.LI (1975), Two special cases of the assignment problem. Discrete Maths. 13, 129-142.

H.W.KUHN (1955), The Hungarian method for assignment problems. Nav.Res.Log.Quart. 2, 83-97.

S.MARTELLO, W.R.PULLEYBLANK, P.TOTH and D.deWERRA (1984), Balanced optimization problems. Operations Research Letters 3, 275-278.

C.H.PAPADIMITRIOU (1977), The probabilistic analysis of matching algorithms. Proc. 15th Ann. Allerton Conf. on Communication, Control and Computing, 368-378.

G.L.THOMPSON (1981), A recursive method for solving assignment problems. In: P.Hansen (Ed.): *Studies on Graphs and Discrete Programming*, 319-343 (Annals of Discrete Mathematics, Vol.11), North Holland: Amsterdam-New York-Oxford.

N.TOMIZAWA (1971), On some techniques useful for solution of transportation network problems. Networks 1, 173-194.

D.W.WALKUP (1979), On the expected value of random assignment problems. SIAM J.Comput. 8, 440-442.

M.WEBER and Th.M.LIEBLING (1985), Euclidean matching problems and the Metropolis algorithm. Report, Département de Mathématiques, EPF-Lausanne, Switzerland.

OPTIMAL CONTROL AS A TOOL FOR SOLVING THE STATIONNARY EULER EQUATION WITH PERIODIC BOUNDARY CONDITIONS

T. CHACON
INRIA, B.P. 105, Rocquencourt, 78153 LE CHESNAY
France
&
Facultad de Matematicas, Universidad de Sevilla
41012 SEVILLA, Spain

O. PIRONNEAU
Université Paris 6 and INRIA, France

Abstract :
We study the numerical resolution of the 3D stationnary Euler equation, in the unit cube with periodic boundary conditions.

A formulation of the problem as distributed parameters optimal control problem is first given. This problem is discretized by the Finite Elements Method and solved by conjugate gradient algorithms. In order to set rid of a quadratic state construit several different formulations of the optimal control problem are studied. A quasi-Newton method is also tested.

1. Introduction :

In [1] McLaughlin et al. a model of turbulence is developed by homogeneization techniques ; there appears a subsidiary problem which involves a stationnary Euler equation with periodic boundary conditions :

$$w.\nabla w + \nabla \Pi = 0,\ \nabla . w = 0 \text{ in } Y =]0,1[^3 \tag{1a}$$

$$\langle w \rangle = 0 \quad \frac{1}{2}\langle |w| \rangle^2 = q \ ,\ \langle w.\nabla \times w \rangle = h \ ;\ q \geq 0,\ h \in \mathbb{R} \tag{1b}$$

$$w, \Pi \quad \text{periodic in } Y \tag{1c}$$

where w is the velocity field, Π is the associated pressure and the symbol $\langle . \rangle$ represents the mean in Y. Equation (1b) means that the invariants of (1a) (mean, kinetic energy q and global helicity h) are given.

The theoretical study and the numerical resolution of problem (1) is a hard task due essentially to the pure hyperbolic character of the Euler operator appearing in (1a). In the two dimensional case, an infinity of invariants have been found and for each set of values for them there is a unique solution(see Olver[6]).In the 3D case,we know only the three invariants appearing in (1b) and these are not sufficient to secure uniqueness of solutions : It is possible to obtain analitically solutions of (1) with only one Fourier mode (see Hemon [2] , Begue [3] , Chacon and Pironneau [7]). These solutions are also not unique, but they are degenerated in a suitable sense (see Chacon [4]). However, it is not unreasonable to believe that the solutions of (1) with full spectrum are isolated.

In this paper, we present a numerical resolution of the problem (1) based on the conjugate gradient algorithms. Classical methods which have been extensively used in transport problems (for instance, the method of characteristics) are not appropriated here, because we are looking for periodic solutions (see Begue and Pironneau [8]).

Solving the whole problem is very difficult (there are two nonlinear constraints, one of them envolving the derivatives of w). We will consider a reduced problem, with only one constraint (the kinetic energy), which is formulated in Section 2 as an optimal control problem. This formulation is discretized by the Finite Element Method in Section 3. Finally, in Section 4, we present the results obtained by using many algorithms of conjugate gradient type.

2. Optimal control formulation of a reduced problem

Problem (1) remains invariant under a set of orthogonal transformations which can be endowed with a group structure. In fact, this problem is invariant under changes of coordinates which conserve the orientation and transform the remit cube Y onto itself (see Spencer [10]).

In order to "eliminate" the 3rd constraint in (1b) (helicity), we shall consider the group G of all plane symmetries which leave invariant the unit cube. It is easily proved that if w is a solution of problem (1) which remains invariant under all the transformations of G, then w is an odd function and thus $\langle w\rangle = 0$, $\langle w.\nabla \times w\rangle = 0$. (see Chacon [5]).

This allow us to consider the following reduced problem :

$$\begin{aligned} & w.\nabla w + \nabla\Pi = 0, \quad \nabla.w = 0 \quad \text{in} \quad Y \\ & \tfrac{1}{2} \langle |w|^{2}\rangle = 1 \,, \\ & w,\ \Pi \quad \text{periodic in } Y \ ; \ \text{in} \quad G\text{-invariant} \end{aligned} \tag{2}$$

In problem (2), the constraint on helicity has been eliminated by including the G-invariance of w. Also, the dependence of the problem about the kinetic energy q has been eliminated because if w is solution of (2), then so is $\sqrt{q}w$.

Let us now give a formulation of problem (2) as an optimal control problems. Consider the functional spaces

$$\begin{aligned} \mathcal{H} &= \{v \in H^1(Y)/v \quad Y\text{-periodic}\} \\ \mathcal{V} &= \{w \in \mathcal{H}^3 \ / \ w \quad G\text{-invariant}\} \end{aligned} \tag{3}$$

For any given $v \in \mathcal{V}$, let us define the "error" ε as the solution of the problem

$$-\Delta\varepsilon = \nabla.(v \otimes v)+\nabla p$$
$$\varepsilon \in \mathcal{H}^3 , \langle\varepsilon\rangle = 0 \tag{4}$$

where $v \otimes v$ denotes the 3×3 tensor of elements $v_i v_j$, and where the associated pressure p is defined by

$$-\Delta p = \nabla.(v.\nabla v)$$
$$p \in \mathcal{H}, \langle p\rangle = 0 \tag{5}$$

Let us remark finally that problems (4) and (5) have unique solutions. Moreover, if v is G-invariant, it is easy to prove that ε is G-invariant too.
Under these definitions, it has been proved that problem (2) is equivalent to the following optimal control problem (see Begue [3]) :

$$\min_{v \in \mathcal{V}} \mathcal{C}(v) = \int_Y |\nabla\varepsilon|^2 \, dy. \tag{6}$$

3. Discretization by the Finite Element Method

First, from the analogy of the operators appearing in (1a) with those of the Navier-Stokes equations, it is reasonable to discretize velocities and pressures so as to verify the Brezzi-Babuska inf-sup condition (see Glowinski et al. [4]). Second, to obtain G-invariant discret solutions, the grid has to be symmetric about the three coordinate planes. The discretization obtained in this way is the following :
Let $\mathcal{T}_h$ be a regular triangulation of Y which is symmetric about the three coordinate planes. Let $\mathcal{T}_{h/2}$ be a triangulation obtained from $\mathcal{T}_h$ by subdividing each element into 8 tetraedra in such a way that $\mathcal{T}_{h/2}$ is also symmetric about the three coordinate planes ; and let $\mathcal{V}_h$, Q_h be the following discrete spaces :

$$\mathcal{V}_h = \{w_h \in \mathcal{C}^0(Y)^3 \mid v_{h|T} \in (\mathbb{P}^1)^3, \forall T \in \mathcal{T}_{h/2} \; ; v_h \text{ periodic and G-invariant}\}$$
$$Q_h = \{q_h \in \mathcal{C}^0(Y)/q_{h|T} \in (\mathbb{P}^1)^3, \forall T \in \mathcal{T}_h ; q_h \text{ periodic}, \langle q_h\rangle = 0\} \tag{7}$$

where $\mathcal{C}^0(Y)$ is the space of continuous functions on Y and $\mathbb{P}^1$ the space of affine functions on $\mathbb{R}^3$. It is easily seen that $v_h \subset \mathcal{V}$, $Q_h \subset \mathcal{H}$.
Under these definitions, problem (4)-(6) can be discretized by

$$\min_{v_h \in V_h} \mathcal{C}(v_h) = \int |\nabla\varepsilon_h|^2 \, dy \tag{8}$$

where $\mathcal{E}_h$ is defined by :

$$(\nabla \mathcal{E}_h, \nabla u_h) = -(v_h \otimes v_h, \nabla u_h) + (\nabla p_h, v_h), \; \forall u_h^{\cdot} \in \mathcal{V}_h$$
$$\mathcal{E}_h \in \mathcal{V}_h \tag{9}$$

$$(\nabla p_h, \nabla q_h) = -(v_h \cdot \nabla v_h, \nabla q_h), \ \forall q_h \in Q_h \qquad (10)$$
$$p_h \in Q_h$$

Problems (9) and (10) give well-posed linear systems where the unknowns are the values of ξ_h, p_h at the nodes of the respective grids. These linear systems have been solved by a fixed point algorithm which uses the fast Fourier transform to inverse the discret Laplacian, according to the periodic character of the solutions (see [5]).

4. Resolution of the discrete problem by Conjugate Gradient Algorithms

We have not yet considered the constraint appearing in problem (2). It is easily verified if one has a non zero solution w, by multiplying it by an appropriate constant $\sqrt{q}$.

To obtain non zero solutions of (2), let us first remark that the cost function $\mathcal{E}$ defined in (6)) is homogeneous of degree 4. Then, it will be more convenient to define a new cost function E as follows :

$$E(v) = \mathcal{E}\left(\frac{v}{\|v\|_{L^2}}\right), \ \forall v \in \mathcal{V} \qquad (11)$$

The problem of minimizing E has been solved in the functional frame defined in (7) by a standard Fletcher-Reeves algorithm which includes L^2-normalization of the approximation after each iteration. Thus normalization does not change the minimizing sequence because the corresponding gradient is proportional to the one obtained without normalization. The main advantage obtained by defining this new cost function E is that the minimizing sequence is included in the unit sphere of $L^2(Y)$ and thus (if the process converges, we obtain a non-zero solution (see Figure 1).

It can also be verified that the conjugate gradient method generates G-invariant functions if the initial one is G-variant.

Another way to obtain non-zero solutions of problem (2) which seems more appropriated is to define the following minimizing problem

$$\min_{v \in \mathcal{V}} \mathcal{J}(v) = E(v) + f(\|v\|_{L^2}) \qquad (12)$$

where f(x) is a differentiable real function with only one minimum X in R^+ and such that $\lim_{x \to 0} f(x) = \lim_{x \to +\infty} f(x) = +\infty$. Then, if $J(w) = \min_{x \geq 0} f(x) = E(w) = 0$ and $\|w\|_{L^2} = X \neq 0$. In fact, we have solved problem (12) by taking

$$f(x) = c(x^{\alpha} + x^{-\alpha}) \ , \quad \alpha > 0, \ c > 0 \qquad (13)$$

This numerical solution has been made in two ways : with a standard Fletcher-Reeves algorithm (Figure 2) and with a Quasi-Newton variable storage conjugate gradient algorithm (see Buckley et al. [11] (Figure 3). Computations have been made on a CRAY1 computer with a $(16)^3$ grid for velocities and a 9^3 grid for pressures, i.e. 13.000 unknowns !

As it can be appreciated in the figures, these two last algorithms have approximately the same speed of convergence. The first algorithm described in this section is much faster.

References

[1] D. MCLAUGHLIN G. PAPANICOLAOU, O.PIRONNEAU , Convection of microstructures and related problems, SIAM S. Appl. Math., Vol.45, No. 5, Oct. 1985, pp.780-797.

[2] M. HENON, Numerical exploration of Hamiltonian system (to appear).

[3] C. BEGUE, Simulation numérique de la turbulence par méthode d'homogénéisation. Thèse 3ème Cycle, Univ. Paris VI, Déc. 1983.

[4] T. CHACON, Etude d'un modèle pour la convection des microstructures. Thèse 3ème Cycle, Univ. Paris VI, April 1985.

[5] T. CHACON, Contribucion al estudio del modelo M.P.P. de turbulencia. Tesis Doctoral Univ. Sevilla (Spain), Sept. 1984.

[6] J.P. OLVER, A nonlinear Hamilton System structure for the Euler equations. Journal Math. Anal. Appl. 89, pp. 233-250 (1982).

[7] T. CHACON, O. PIRONNEAU, On the mathematical foundations of the k-ε turbulent model (to appear).

[8] C. BEGUE, O. PIRONNEAU, Hyperbolic systems with periodic boundary conditions, Comp. & Maths. with Appls., Vol.11, Nos. 1-3, pp. 113-128 (1985).

[9] R. GLOWINSKI, B. MANTEL, J. PERIAUX, O. PIRONNEAU, A Finite Element Approximation of Navier-Stokes equations for incompressible viscous fluids. Computer Methods in Fluids. Pentech Press, New-York (1980).

[10] A.J. SPENCER, Continuum Physics, Volume I. Academic Press, New-York and London, (1971).

[11] A. BUCKLEY, A. LENIR, ON-Like variable storage conjugate gradients. Mathematical Programming 27, 2, pp. 155-175 (1983).

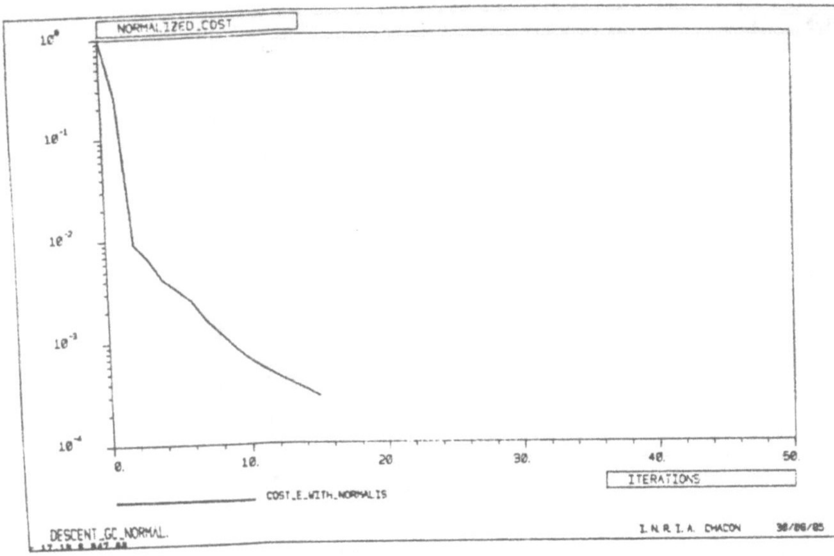

Figure 1

Cost E defined in (6) (normalized). The represented values are $E(v_o)/E(v_o)$. Corresponds to a conjugate gradient algorithm with normalization.

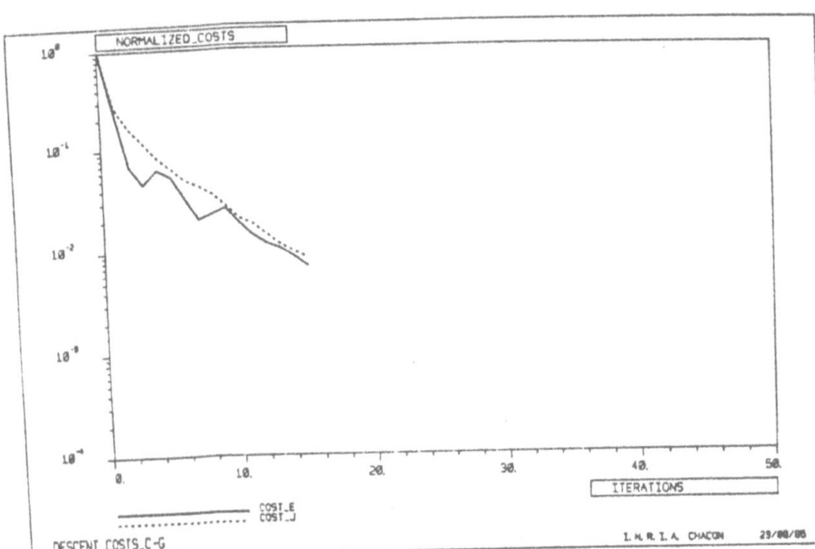

Figure 2

Costs E and J defined in (12) (normalized). The represented values for E are $E(v_n)/E(v_o)$ and those for J are $(J(v_n)-\min J)/(J(v_o)-\min J)$. Corresponds to a standard conjugate gradient algorithm.

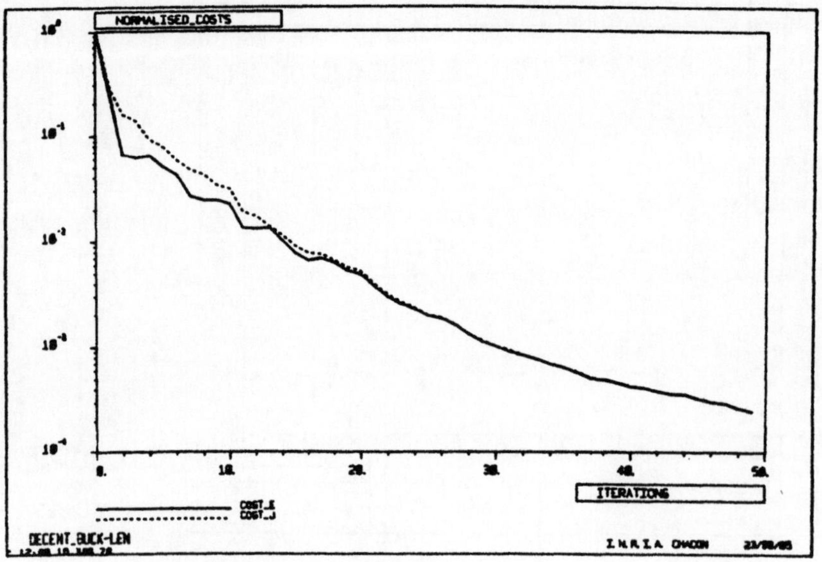

Figure 3

The same as figure 2, corresponding to a variable storage conjugate gradient algorithm.

OPTIMIZATION PROBLEMS FOR TWO-STAGE PROCESS OF RESOURCE ALLOCATION

A.P. CHERENKOV

Computing Center of the USSR Academy of Sciences
Moscow, USSR

1. INTRODUCTION

Consider the process of resource allocation consisting of two successively realized stages. Situation, that is the condition of the nature, is determined by an unknown parameter t. The strategy x of the first stage makes information about t more precise. After the first stage the estimate s of the parameter t is found. The main thing is the choice of the strategy of the second stage. The efficiency F of the process depends on x, y, t and s. Values x, y, t, s are vectorial. Quantities of the resources intended to accomplish the first and the second stages, C and D respectively, depend on corresponding strategies, and their sum does not exceed the known quantity K: $C(x) + D(y) \leqslant K$. It is necessary to choose a manner of the resource allocation between and within the stages.

The exact formulation of the problem depends on the character of information about t, which we have before every of two stagies of our process. The parameter t may be random or uncertain. It gives us four versions of the problem.

We shall consider more thoroughly the separable case, when the process takes place at several objects which are independent and functions F and D are linear concerning y.

As an example of practical realization of the described model we consider a model of search for oil or gas. In this case the parameter t characterizes location, shape and size of plots under consideration, where the mineral deposits can be found. The first stage is geophysical prospecting, its strategy x determines the intensity of use of different geophysical methods on different plots. The second stage is boring, its strategy y determines quantity of bore holes for each plot. The value F is prospective quantity of minerals.

2. THE VERSIONS OF INFORMATION.

It may be accepted one of two versions of information for each of two stages. Before the beginning of the stage the parameter t is random or uncertain.

Before the first stage either a priory frequency distribution function p(t) of the parameter t or the domain T of the permissible values of t is known: $t \in T$.

Before the second stage in case of the first version of information a posteriory frequency distribution function $p'(t;\ s,\ R(x,\ s))$ becomes known, here s and R are the parameters of the distribution; s is the mathematical expectation, R is the correlation matrix. In case of the second version of information before the second stage the domain $T'(s,\ R(x,\ s))$ of the permissible values of t becomes known: $t \in T'$, here s determines location of the centre and R does size of the domain T'.

In both cases s is the estimate of the parameter t. This estimate is found on the basis of results received after the first stage, but we shall not consider here this question. In both cases the function R(x, s) characterizing the efficiency of the strategy x is known beforehand. Information about s which is available before the first stage is restricted by a priory data concerning t.

3. FORMULATION OF PROBLEMS.

We set into accordance to each version of information the operator, which we shall name the information operator. This operator shows, what value needs to be maximized with accepted version of information. For probabilistic version it is the mathematical expectation in according with the corresponding frequency distribution function. For uncertain version it is the minimum over the domain of permissible values of t. The two-stage problems may be written symbolically so:

$$\max_x J_1 \max_y J_2 F,$$

where J_1, J_2 are the information operators for two stages. Depending on their form that is depending on the character of information about t before the stagies the following four versions of the problem are possible:

$$\max_{x\in X} \int \max_{y\in Y_x} (\int F(x, y, t, s)p'(t; s, R(x, s))dt)p(s)ds,$$

$$\max_{x\in X} \int \max_{y\in Y_x} \min_{t\in T'(s, R(x, s))} F(x, y, t, s)\ p(s)\ ds,$$

$$\max_{x\in X} \min_{s\in T} \max_{y\in Y_x} \int F(x, y, t, s)p'(t; s, R(x, s))dt,$$

$$\max_{x\in X} \min_{s\in T} \max_{y\in Y_x} \min_{t\in T'(s,R(x, s))} F(x, y, t, s).$$

Here $X = \{x: C(x) \leqslant K\}$, $Y_x = \{y: C(x) + D(y) \leqslant K\}$, and the variables in all the integrals vary over their natural domains.

The problems of the second stage are auxiliary for two-stage ones. For them the values x, s, R are given. When the first stage is over and the statistical work with its results is carried out, they have the independent signification.

It is necessarily to decide conceming the resource allocation between the stages and within the first state before accomplishment of the first stage, when the data of work with its results are absent. Therefore we consider the result of the solution of the problem of the second stage to be subjected to influence of information operator corresponding to the version of information for the first stage, the yet unknown estimate s of the parameter t playing a part of t. This imitates and replaces the procedure of estimation taking place under the practical realization of the process.

4. THE SPECIAL SEPARABLE CASE.

Consider the separable linear conceming y case in detail. Let us have n objects A_i, $i = 1, \dots, n$,

$$F(x,y,t,s) = \sum_{i=1}^{n} f_i(x_i, t_i, s_i)y_i, \quad D(y) = \sum_{i=1}^{n} d_i y_i,$$

where $x = (x_1, \ldots, x_n)$, $y = (y_1, \ldots, y_n)$, $t = (t_1, \ldots, t_n)$, $s = (s_1, \ldots, s_n)$, the values x_i, t_i, s_i are vectorial.

Let the objects be independent. Then

$$p(t) = \prod_{i=1}^{n} p_i(t_i), \quad T = T_1 \times \ldots \times T_n, \quad t_i \in T_n,$$

$$p'(t; s, R(x,s)) = \prod_{i=1}^{n} p_i'(t_i; s_i, R_i(x_i, s_i)),$$

$$T' = T_1' \times \ldots \times T_n', \quad t_i \in T_n';$$

here s_i is the estimate of t_i, p_i and p_i' – a priory and a posteriory frequency distribution functions, T_i and T_i' – the domains of localization of t_i.

Consider the influence of the operator J_2 on F. The vectors x and s are believed fixed.

In case of the first version of information

$$J_2 F = \int F p' dt = \int (\sum_i f_i y_i) \prod_i p_i' dt_i = \sum_i (\int f_i p_i' dt_i) y_i .$$

In case of the second version of information

$$J_2 F = \min_{t \in T'} F = \min_{t \in T'} \sum_i f_i y_i = \sum_i (\min_{t_i \in T_i'} f_i) y_i .$$

As we see in both cases

$$J_2 F = \sum_{i=1}^{n} q_i y_i ,$$

but the coefficients q_i depending on x_i and s_i are determined differently.

Under finding of the interior maximum, that is under solution of the problems of the second stage the Neyman–Pearson lemma and methods of discrete optimization can be used. It gives implicit analytical (sometimes approximate) solution, and it makes easier the search of the exterior extremum.

On solving the two-stage problems having in mind the multidimensionality of the vectors and integral form of the criteria of optimality it is worth-while to use the Monte-Carlo method of integral computation and the methods of stochastic optimization.

5. APPLICATION TO GEOLOGY: SEARCH OF OIL OR GAS.

The model of search of oil or gas is an example of practical applying of model under consideration. There are n plots on which deposits are possible. Projecting on horizontal plane the form of the deposits is elliptical. The parameter t characterizes their centre location, size, orientation. The first stage is geophysical prospecting, its strategy x determines the intensity of use of different geophysical methods (for example, the seismic prospecting) on different plots for two mutually perpendicular directions. The second stage is boring, its strategy y determines quantity of the bore holes for each plot. The problems under consideration are maximization of prospective quantity of minerals under restriction on the cost of work. The notion "prospective quantity" is specified depending on version of information.

For case when the probabilistic version of information takes place for both the stages the calculations have been carried out. The program for calculations have been composed by N.D. Katayeva. In this case it was believed that the only geophysical method was used, orientation of ellipses was given, the vector y was Boolean and in case $y_i = 1$ a bore hole was bored at supposed centre of ellipse. In this case n = 20, dim t = 80, dim x = 40, dim x = 20, For electronic computer EC-1055 the time of calculation was about 4 hours. It needed 15 iterations of the vector x.

BALANCED REALISATIONS FOR INFINITE-DIMENSIONAL DISCRETE-TIME SYSTEMS

Ruth F. Curtain

Abstract

For finite-dimensional systems the class of balanced realisations is defined as that whose controllability and observability gramians are both equal to the same positive diagonal matrix. The diagonal entries are in fact the singular values of the Hankel operator of the system and contain essential information about its behaviour. Balanced realisations have special structural properties and their truncations usually provide a good reduced order model of the original system. An account of balanced realisations for infinite-dimensional systems is given and its implications for the approximation of infinite-dimensional systems by finite-dimensional ones are discussed.

1. Introduction

Balanced realisations for finite dimensional systems are those in which the controllability and observability gramians are both equal to the same positive diagonal matrix. In [10] Moore studied further properties of balanced realisations and in [11], Pernebo and Silverman use truncations of balanced realisations as a means of obtaining a reduced-order approximating model for their original system. The error bounds were found by Glover in [8], but the main discussion concerned Hankel norm approximations which had also previously been studied by Kung and Lin in [9]. Of course system approximation for infinite-dimensional systems is a very important problem and in the recent results by Curtain and Glover in [4] and [6] on approximation of continuous-time infinite-dimensional systems with finite-dimensional inputs and output, the concept of balanced realisations played an central role. In finite dimensions the transition between continuous and discrete time cases is straight forward, but this is not the case in infinite-dimensions as is clear from Curtain [3]. So the theory of balanced realisations and approximation for discrete-time infinite-dimensional systems was developed directly in [5].

This conference presentation is a summary of the main results of Curtain [5]; for proofs and further details we refer to [5]. For infinite-dimensional discrete-time systems with an ℓ_1 impulse response the balanced realisation is explicitly given in terms of the singular values and Schmidt pairs of the Hankel operator and these can be calculated numerically. If the Hankel operator is Hilbert Schmidt, truncations of the balanced realisation generate a sequence of finite-dimensional impulse responses which converge in ℓ_2 to the original one and the transfer functions converge pointwise. Under the extra assumption that Γ is nuclear, the transfer functions converge in the L_∞ norm. So in this case the truncation of the balanced realisation provide good approximating reduced order models to the original infinite dimensional system.

Not surprisingly the discrete-time results are analagous to the continuous-time results obtained by Curtain and Glover in [4] and [6]. One significant difference lies in the conditions for nuclearity of the Hankel operator which were also studied by Curtain in [3]. For discrete-time systems it is sufficient for the impulse response h(k) be of order k^{-a} where $a > 3/2$ and this is a very weak form of stability for a discrete-time system. For continuous-time systems even exponential stability need not be sufficient.

These results were first announced together with the continuous-time results in [4], where at the same time related results were presented by N. Young [14]. Starting with the restricted shift realisation (Fuhrmann [7]), he obtains a balanced realisation for a very large class of systems: those for which G(1/z) is analytic and has vanishing mean oscillation, where G(z) is the usual transfer function. However, he does not consider the important problem of finite-dimensional approximations for nonrational transfer functions, which was the main motivation in [5].

For convenience we collect several known facts on discrete-time systems in section 2 and in section 3 we present the main results on balanced realisations from [5]. The conclusions are in section 3.

2. Representations of Discrete-time Systems

For convenience we summarize some basic facts concerning discrete-time infinite-dimensional systems from [3], [7], [12], which we shall using in this paper.

We shall be concerned with discrete-time linear systems in input-output form

$$(2.1)\quad y(k) = \sum_{i=0}^{k-1} h(i)u(k-i-1)$$

where $h(i) \in R^{k \times m}$ for each $i \geq 0$ is the *impulse response* of the system and defines the *Hankel operator* Γ given by

$$(2.2)\quad (\Gamma u)(j) = \sum_{i=0}^{\infty} h(i+j)u(j) \qquad \text{for } u \in \ell_2^m$$

Under very general conditions on h, it is known that systems in input-output form (2.1) possess representations in the *state space* form

$$(2.3)\quad z_{k+1} = Az_k + Bu_k$$

$$(2.4)\quad y_k = Cz_k$$

The state space H is a real, separable Hilbert space and $A \in L(H)$, $B \in L(R^m, H)$ and $C \in L(H, R^k)$. With zero initial condition, (2.3), (2.4) has as its input-output map

(2.5) $$y_k = \sum_{i=0}^{k-1} CA^i Bu(k-i-1)$$

which agrees with (2.1) if and only if

(2.6) $$h(i) = CA^i B \;;\; i = 0,1,2,\ldots$$

For the state space realisation (2.3), (2.4), we use the following concepts from [7].

Definition 2.1.

(i) The reachability operator $\mathcal{B}: \ell_2^m \longrightarrow H$ is given by

(2.7) $$\mathcal{B}u = \sum_{i=0}^{\infty} A^i Bu_i \qquad \text{for } u \in \ell_2^m$$

(ii) The observability operator $\mathcal{C}: H \longrightarrow \ell_2^k$ is given by

(2.8) $$\mathcal{C}h = \{CA^i h\}_{i=0}^{\infty} \qquad \text{for } h \in H$$

(iii) If $\mathcal{B} \in L(\ell_2^m, H)$, then $P = \mathcal{B}\mathcal{B}^* \in L(H)$ is the controllability gramian and satisfies

(2.9) $$P = \sum_{i=0}^{\infty} A^i BB^* (A^*)^i$$

(iv) If $\mathcal{C} \in L(H, \ell_2^k)$, then $Q = \mathcal{C}^* \mathcal{C} \in L(H)$ is the observability gramian and satisfies

(2.10) $$Q = \sum_{i=0}^{\infty} (A^*)^i C^* CA^i$$

Now (2.3) is reachable if $\overline{\text{range } \mathcal{B}} = H$, ([7]), and from (iii) if $\mathcal{B}$ is bounded, this is equivalent to the controllability gramian, P, being strictly positive. Similarly (2.3), (2.4) is observable if $\text{Ker}\mathcal{C} = \{0\}$ and this is equivalent to the observability gramian, Q being strictly positive.

(2.1) has infinitely many possible realisations, (C,A,B), which satisfy (2.6), but here we are interested in one special type.

Definition 2.2.

Given a system in input-output form (2.1), we say that (C,A,B) is a s t a b l e r e a l i s a t i o n for (2.1) with state space H if H is a separable Hilbert space, $A \in L(H)$, $B \in L(R^m, H)$, $C \in L(H, R^k)$, the reachability and observability operators $\mathcal{B}$ and $\mathcal{C}$ defined by (2.7) and (2.8) are bounded and (2.6) holds. (C,A,B) is a b a l a n c e d r e a l i s a t i o n if it is a stable realisation and its controllability gramians are both equal to the same positive, self adjoint operator. (2.1) can equally well be defined by the Z-transform of h, G(z), which is termed its t r a n s f e r f u n c t i o n. If (C,A,B) realizes (2.1), we have the relationship

$$(2.11)\quad G(z) = D + C(zI - A)^{-1}B$$

which is analytic for $|z| > \|A\|$.
It turns out that the key to balanced realisations are the singular values and Schmidt vectors of the Hankel operator, Γ, give by (2.2)

If $h \in \ell_1$, then $\Gamma^*\Gamma$ is compact, self adjoint and postive with eigenvalues $\sigma_1^2 > \sigma_2^2 > \sigma_3^2 > \ldots$ and eigenvectors $v_i, i \geq 1$. $\Gamma\Gamma^*$ has the same eigenvalues but different eigenvectors $w_i, i \geq 1$.

Definition 2.3.

σ_i are the singular values of Γ and (v_i, w_i) are its Schmidt pairs, $i \geq 1$. We recall that the Hankel operator is H i l b e r t S c h m i d t if the following holds

$$(2.12)\quad \sum_{i=1}^{\infty} \sigma_i^2 < \infty$$

and n u c l e a r if

$$(2.13)\quad \|\Gamma\|_1 := \sum_{i=1}^{\infty} \sigma_i < \infty$$

In [5] the following relationships were establshed under the assumption that $h \in \ell_1^{k \times m}$

$$(2.14)\quad \begin{cases} \Gamma v_i = \sigma_i w_i & ;\ i = 1,2,\ldots \\ \Gamma^* w_i = \sigma_i v_i & \end{cases}$$

and if $\sigma_i \neq \sigma_j$

(2.15) $<w_i, w_j> = \delta_{ij} = <v_i, v_j>$

(2.16) $h(k+\alpha) = \sum_{i=1}^{\infty} w_i(k)\sigma_i v_i(\ )$

where $<\cdot,\cdot>$ denotes the inner product in any ℓ_2^r space.

3. Balanced Realisations

In [5] the following explicit balanced realisation for Γ was found, assuming that $h \in \ell_1$.

(3.1) $A_{ij} = (\sqrt{\sigma_i/\sigma_j}) < Sv_i, v_j >$

(3.2) $C = (\sqrt{\sigma_i} w_1(0), \ldots\ldots, \sqrt{\sigma_i} w_i(0), \ldots\ldots)$

(3.3) $B^* = (\sqrt{\sigma_1} v_1(0), \ldots\ldots, \sqrt{\sigma_i} v_i(0), \ldots\ldots)$

(3.1) - (3.3) is a well defined realisation for (2.1) on the state space ℓ_2, and A is a contraction. The realisation (3.1) - (3.3) suggests approximating the infinite dimensional system by its truncations, that is, by the system (C^n, A^n, B^n) defined by

(3.4) $A^n_{ij} = A_{ij}$; $i,j = 1,\ldots,n.$

(3.5) $C^n = C_i$; $i = 1,\ldots,n.$

(3.6) $B_n = B_i$; $i = 1,\ldots,n.$

The system (C^n, A^n, B^n) has an impulse response h^n defined by

(3.7) $h^n(i) = C^n(A^n)^i B^n$

and a corresponding Hankel operator Γ defined by (2.2). In [5] several convergence results for these truncated systems were proved.

Lemma 3.1.

If Γ is Hilbert Schmidt then

(a) $(\Gamma - \Gamma_n)u \longrightarrow 0$ as $n \longrightarrow \infty$ for all $u \in \ell_2^m$

(b) $\| h - h_n \|_2 \longrightarrow 0$ as $n \longrightarrow \infty$

(c) $G_n(z) = C_n(zI - A_n)^{-1}B_n \longrightarrow G(z) = C(zI - A)^{-1}B$

as $n \longrightarrow \infty$ for all z in $|z| > 1$.

To obtain stronger convergence results one needs the extra assumption of nuclearity.

Lemma 3.2.

If Γ is nuclear and $h \in \ell_1$, then

(a) $\| h - h_n \|_1 \longrightarrow 0$ as $n \longrightarrow \infty$

(b) $\| G - G_n \|_\infty \longrightarrow 0$ as $n \longrightarrow \infty$

4. Conclusions

In [5] it was shown that infinite-dimensional discrete-time systems whose impulse response is in ℓ_1 possess a balanced realisation with state space ℓ_2. This realisation is given explicitly in terms of the singular values and the Schmidt pairs of the Hankel operator and it can be calculated numerically. Truncations of this balanced realisation yield a sequence of finite-dimensional impulse responses, which converge in ℓ_2 to the impulse response. This only implies pointwise convergence of the transfer functions. To prove L_∞ convergence of the transfer functions the assumption that Γ is nuclear or $\sum_{i=1}^{\infty} \sigma_i < \infty$ was introduced. In this case we can claim that the truncations will provide useful approximations to the original infinite-dimensional system.

1. V.M. Adamjan, D.Z. Arov and M.G. Krein, Analytic properties of Schmidt pairs for a Hankel Operator and the Generalized Schur-Takagi problem, Math. USSR Sbornik, 15, p. 31-73, 1971.

2. R.F. Curtain and A.J. Pritchard, "Infinite Dimensional Linear Systems Theory", Lecture Notes in Control and Information Sciences, Vol. 8, 1978, Springer Verlag.

3. R.F. Curtain, Sufficient conditions for Infinite-rank Hankel Operators to be Nuclear, J. Math. Control and Info. (to appear).

4. R.F. Curtain and K. Glover, Balanced Realisations for Infinite-Dimensional Systems, Workshop on Operator Theory and its Applications. June 3 - 7, 1985,

Amsterdam, the Netherlands.

5. R.F. Curtain, Finite-dimensional Approximations for Discrete-time Infinite-dimensional Systems, (submitted).

6. R.F. Curtain and K. Glover, Realisation and Approximation for infinite dimensional systems. (in preparation).

7. P.A. Fuhrmann, Linear Systems and Operators in Hilbert Space, 1981, McGraw Hill.

8. K. Glover, All Optimal Hankel-norm approximations of linear multivariable systems and their L_∞-error bounds. Int. J. Control, 1984, 39, pp. 1115-1193.

9. S.-Y. Kung and D.W. Lin, Optimal Hankel-Norm Model Reductions: Multivariable Systems, I.E.E.E. Trans. Aut. Control, 26, 1981, p. 832 - 852.

10. B.C. Moore, Principal Component Analysis in Linear Systems: controllability, observability and model reduction, I.E.E.E. Trans. Aut. Control, Vol. AC-26, 1981, p. 17 - 32.

11. L. Pernebo and L.M. Silverman, Model Reduction via Balanced State-space Representation, I.E.E.E. Trans. Aut. Control, Vol. AC-27, 1982, pp. 382 - 387.

12. S.C. Power, Hankel Operators in Hilbert space, Research Notes in Mathematics, 64, 1982, Pitman Advanced Publishing Program.

13. K.M. Przyluski, The Lyapunov equations and the problem of stability for linear, bounded discrete-time systems in Hilbert space, Appl. Math. Optim., 1980, 6, pp. 97 - 112.

14. N. Young, Balanced Realisations, ibid [4].

Ruth F. Curtain
Rijksuniversiteit Groningen
Mathematisch Instituut
Postbus 800
9700 AV GRONINGEN
The Netherlands

NONLINEAR PARAMETER ESTIMATION BY GLOBAL OPTIMIZATION: COMPARISON OF LOCAL SEARCH METHODS IN RESPIRATORY SYSTEM MODELLING

T. Csendes, B. Daróczy and Z. Hantos
Kalmár Laboratory of Cybernetics, József Attila University,
H-6720 Szeged, Hungary

1. Introduction

The optimization techniques applied for the identification of respiratory mechanical models from forced oscillatory impedance data are variable [4,6,8] and, in general, have not been described in sufficient detail to establish whether their reproducibility and accuracy and their reliability for finding the global minimum are suitable for the particular model and data used. In this paper a versatile identification method, allowing specification of a great variety of models, is presented briefly. The ability to recover original values of model parameters from generated impedance data, as well as the computational efficiency, are compared in the cases of three local search procedures.

For the identification of model parameters we used the following objective function:

$$F(x) = \{1/m \sum_{i=1}^{m} |z(f_i)-\tilde{z}(f_i,x)|^2\}^{1/2} \qquad (1)$$

where z and $\tilde{z}$ are the measured and modelled complex impedances, respectively, x is the vector of model parameters and the f_i-s are frequency points. The constraints are represented by simple bounds:

$$a_j \le x_j \le b_j \ ; \quad a_j,\ b_j \in R, \quad j = 1,2,\dots,n. \qquad (2)$$

The problem specified by (1)-(2) is of a least-mean-squares type. In general, the objective function has many, even substantially different, local minima. Apart from the simplest models [10], where the identification is straightforward, the parameter estimation has been accomplished by using the methods of multiple local searches [3] or the local minimization following a thorough search for a proper starting point [4].

2. The parameter estimation procedure

The models of the respiratory mechanics are generally formulated by using the analogy between mechanical and electric systems, and $\tilde{z}$ is given in the form of rational fractions [1]. We allowed for specification of a great variety of models by building them from two-port networks connected in cascade, and $\tilde{z}$ is computed from the chain matrices [7].

The applied global optimization procedure [2] is a combination of sampling, local search and clustering. The starting points for the local minimization are supplied by the sampling: it generates the initial sample, a certain number of random points in the region of interest with uniform distribution. At these points the objective function values are evaluated and the best of them are selected to produce the transformed sample. The clustering procedure is then applied to decide whether a starting point belongs to the region of attraction of an already known local minimum. If it does not, then the algorithm starts a local search from this point. The clustering step makes the global optimization procedure substantially more efficient. The steps of sampling, local search and clustering are repeated until a new local minimum is found. The smallest local minimum is regarded as the global one.

We examined three local minimization methods. The random walk algorithm UNIRANDI [9] is a simple robust method occupying a small amount of core space. The Levenberg-Marquardt procedure [5] is designed specifically to solve least-squares problems; it utilizes the known particular structure of this objective function. The third local search algorithm is a quasi-Newton method with the DFP update [5]. All three routines are gradient-free techniques, i.e. they do not require the partial derivatives of the objective function.

For each parameter to be optimized an interval (2) has to be specified, in which the corresponding parameter is expected to allow a good fit. We note, however, that the local search procedure is able to leave this domain, and thus the given intervals do not mean bounds in a strict sense for the problem. Negative parameters, which are inconsistent with passive linear models, may appear; these indicate the inadequacy of the particular model used.

From the numerical point of view, it is desirable to optimize parameters of about the same order of magnitude, which, unfortunately, is not the case for our identification problem. Hence, to improve the accuracy and speed of the optimization the variables are scaled [5]:

the parameter estimation problem is transformed so that all variables of the new problem are searched in the interval (-1,1), and then transformed back according to the original intervals. The number of model parameters to be optimized is limited to 15, and the values of those already estimated by some other methods can be fixed. The fitting is carried out using at most 100 frequency points, which need not be equidistantly spaced.

The program is written in FORTRAN IV-H for the IBM System/370 compatible Robotron R55M computer (GDR) running the OS MVT operating system with 2 megabytes of main storage, and it needs 120 kilobytes of core space. The required computation time depends mainly on the number of parameters to be identified, but for most problems (5-7 parameters) it does not exceed 10 minutes.

3. Test results and discussion

We examined the accuracy of our program by using theoretical impedance values. These were generated by a five-parameter model of respiratory mechanics (Model 1 in refs. [1] and [4]) with the parameter values listed in Table 1, at frequencies 2.5, 3.0,...,10.5 Hz. As the procedure is stochastic, the program was run with the three local minimization algorithms seven times each.

Table 1 shows the mean parameter values recovered and the coefficients of variation (c.v.). Although the accuracy of each method is satisfactory, the quasi-Newton method is by far the best. It should be stressed here that a small c.v. indicates the accuracy of a procedure only in the event of an accurate theoretical impedance, and it does not when the measured impedance values are noisy [4]. However, the fact that the c.v. of parameter C2 is about an order of magnitude higher than that of the others may provide an explanation for the relatively high variations found for C2 in real impedance data [4].

Table 2 summarizes the results concerning the efficiency of the procedures studied. Here F^* denotes the objective function value belonging to the global minimizer; CPU and NFE stand for the required CPU time in seconds and for the needed number of objective function evaluations, respectively. All the figures in the first three columns are average values from the same seven independent runs as in Table 1. As regards all indices, the procedure with the quasi-Newton technique is clearly the best, but it is difficult to decide which of the other two methods is the more efficient. The numbers in the last two columns, which can be regarded as proportional to the efficiency of the

TABLE 1

PARAMETER REPRODUCTION OF THREE PARAMETER ESTIMATION ALGORITHMS USING THE ORIGINAL IMPEDANCE VALUES

local search		parameter				
		Rc	Lc	C1	R2	C2
original values:		4.00	0.02	0.005	7.00	0.1
random walk	mean	4.0335	0.01944	0.005082	6.9738	0.1105
	c.v.	0.0041	0.0144	0.0092	0.0019	0.0760
Levenberg-Marquardt	mean	4.0195	0.01969	0.005055	6.9838	0.1044
	c.v.	0.0128	0.0381	0.0218	0.0044	0.1073
quasi-Newton	mean	4.0003	0.02000	0.005001	6.9998	0.1001
	c.v.	0.0003	0.0007	0.0004	0.0001	0.0023

TABLE 2

EFFICIENCY OF THE THREE STUDIED PARAMETER ESTIMATION PROCEDURES

local search	100 F^*	CPU(s)	NFE	$\frac{1}{CPU*F^*}$	$\frac{1}{NFE*F^*}$
random walk	0.8100	646.7	10957	0.00191	0.000113
Levenberg-Marquardt	1.2676	209.0	3632	0.00377	0.000217
quasi-Newton	0.0219	124.7	2324	0.36690	0.019693

method, confirm the superiority of the quasi-Newton algorithm for our identification problem and show that the parameter estimation procedure is somewhat more efficient with the Levenberg-Marquardt local search than with the random walk.

For comparison we have found only one paper providing data on the efficiency of parameter estimation procedures for respiratory mechanics. In that study [4] more than 47,000 function evaluations were performed to identify the five-parameter model from 31 impedance data. In contrast, our identification procedure with the quasi-Newton method used less than 10,000 function evaluations in the majority of computations with measured impedance data.

Acknowledgement:

This study was supported by Hungarian Ministry of Education Grant No. 588/81.

References:

[1] Avanzolini, G. and Barbini, P., Comment on "Estimating respiratory mechanical parameters in parallel compartment models", IEEE Trans. Biomed. Eng., BME-29(1982) pp. 772-774.

[2] Boender, C.G.E., Rinnooy Kan, A.H.G., Timmer, G.T. and Stougie, L., A stochastic method for global optimization, Math. Prog., 22(1982) pp. 125-140.

[3] Dixon, L.C.W. and Szegő, G.P., The global optimisation problem: an introduction, in: Towards global optimisation 2, Dixon, L.C.W. and Szegő, G.P.(eds.), North-Holland, Amsterdam, 1978, pp. 1-15.

[4] Eyles, J.G. and Pimmel, R.L., Estimating respiratory mechanical parameters in parallel compartment models, IEEE Trans. Biomed. Eng., BME-28(1981) pp. 313-317.

[5] Gill, P.E., Murray, W. and Wright, M.H., Practical optimization, Academic Press, London, 1981.

[6] Gillespie, J.R., Gender, J.P., Sagot, J.C. and Bouverot, P., Impedance of the lower respiratory system in ducks measured by forced oscillations during normal breathing, Respir. Physiol., 47(1982) pp. 51-68.

[7] Guillemin, E.A., Synthesis of passive networks, John Wiley, New York, 1957.

[8] Peslin, R., Duvivier, C. and Gallina, C., Total respiratory input and transfer impedances in humans, J. Appl. Physiol., 59 (1985) pp. 492-501.

[9] Törn, A.A., A search-clustering approach to global optimization, in Towards Global Optimization 2, Dixon, L.C.W. and Szegő, G.P., (eds.) North-Holland, Amsterdam, 1978, pp. 49-62.

[10] Tsai, M.J., Pimmel, R.L., Stiff, E.J., Bromberg, P.A. and Hamlin, R.L., Respiratory parameter estimation using forced oscillatory impedance data, J. Appl. Physiol., 43(1977) pp. 322-330.

THE "BENCHMARK" CHRONOLOGICAL SIMULATION MODEL: A NEW TOOL FOR BETTER UNDERSTANDING THE ECONOMIC AND PHYSICAL BEHAVIOUR OF ELECTRICAL GENERATING SYSTEMS

J.K. Delson, EPRI, Palo Alto, California, USA
R.T. Jenkins, Tennessee Valley Authority, USA
W.L. Stillinger, Northeast Utilities, USA
R. Beune and N. Halberg, N.V. SEP, The Netherlands

SUMMARY

With some recently developed software, operators and planners can simulate chronologically the hourly commitment and dispatch of a power supply system. This allows them to see if the system is physically capable of supplying the expected load and to estimate the cost of production. Dispatch simulation based on load duration curves is simpler and quicker than chronological simulation, however these curves do not capture the full effect of commitment decisions and technological constraints-- for example the limited reservoir capacity of pumped-storage plants and the ramp-rate limitations of base-load thermal units. If such constraints are overlooked, the role of peaking units, storage units, or dispatcher-controlled load management may be estimated incorrectly. The effects of commitment decisions and technological constraints can now be analyzed more precisely with a new computer program called BENCHMARK that will be released this year by the Electric Power Research Institute. As the name signifies, the program is intended to calibrate or benchmark calculations based on load duration curves.

LOAD DURATION CURVES

Dispatch techniques have evolved along with multigenerator power supply systems, which appeared at the turn of the century. An initial approach was average cost loading, which, strictly speaking, means taking the generators in order, loading one fully before the next is brought on line. The units are selected according to their cost per kilowatt-hour at full output. This type of dispatch does not depend on the chronological sequence of the hourly loads. It could be readily represented by a load duration curve, which is formed by sorting the loads by magnitude. The area under the load duration curve can then be divided into horizontal bands corresponding to the output of the individual units, with the base load units in the lower bands, the intermediate units further up, and the peaking units in the uppermost bands. The calculation is simple and quick, but does not necessarily minimize production cost.

In the 1920s, dispatch techniques based on marginal cost were introduced. In this approach increments of power are apportioned to all the units in operation, with output adjusted so that all units operate at the same marginal cost. In unconstrained systems, this technique minimizes operating cost for the set of units in operation. The startup and shutdown of units becomes a separate commitment and decommitment problem, which cannot be represented directly with load duration curve dispatch.

EQUIVALENT LOAD DURATION CURVES

An issue that arose in analysis of operation contingencies concerned the representation of forced outages. In the electric utility industry, operating records were analyzed in order to define statistically forced outage rates for various classes of generators. The utilization of this information first led to the use of pseudo deratings in load duration curves: for example, a 100 MW unit with a 10% forced outage rate was represented as a 90 MW unit. Although this practice was designed to give the correct energy output, it does not give either the correct number of operating hours or the correct power output. Such load duration curves, therefore, cannot assess the adequacy of a supply system.

The analytical problem was overcome in the 1960s through the concept of the equivalent load duration curve. The vertical axis was presented as a probability axis, and the horizontal axis was utilized to represent generator capacities in loading order sequence. Normally, the sum of these capacities would exceed the peak load. Again, energy bands could be drawn for each generator, but the equivalent load duration curve is constructed so that the area of a band is the energy that a generator will deliver, taking into account the probability that it and the other generators will be subject to forced outages. The equivalent load duration curve is thus a graphical representation of probabilistic production costing. It includes an estimate of loss of load probability, which can be used to replace various rules of thumb for specification of required reserve margins.

PUMPED STORAGE PLANTS

Although forced outages could thus be taken into account, another limitation important for some systems concerns dispatch of pumped-storage hydroelectric plants. The basic problem with using load duration curve in this situation is that the hourly sequence is lost, only information on load magnitude remains. To the degree that load magnitude is correlated with time of occurrence, a simple economic rule can be assumed: pump at low load levels, which probably occur at night, and generate at high loads, which probably occur during the day. Nighttime pumping would be followed by daytime generating, and a reservoir of limited capacity would suffice.

This simple picture may break down for several reasons. The load duration curve could cover a period of many days, in a generation expansion study it often represents a full year. The peak loads of such a duration curve would refer to the highest loads for the entire period and the analysis could miss the daytime pumping for days when the peak load was considerably lower than the period peak. Other roles of a pumped storage plant that take advantage of its quick response time would also be overlooked in such a simplified analysis. Dispatch of pumped storage plants based on a load duration curve is therefore inherently limited by the loss of chronological detail.

HYBRID MODELS

To reintroduce chronological constraints in production cost analyses a hybrid technique arose. It replaces a single load duration curve with a set of load duration curves and in this way preserves some chronological features. For example, the PROMOD computer program, developed by Energy Management Associates of Atlanta, Georgia, treats a week at a time but takes all peak load hours as one composite load, all off-peak hours Monday through Friday as a second load, and weekend hours as a third load. This arrangement permits representation, at least approximately, of energy storage at night and weekends and utilization of the stored energy during peak periods. PROMOD helps analyze production costs and loss of load probability but does not specify the expansion plan.

An example of a computer program that treats the installation and retirement of generating units as decision variables is EGEAS, the Electric Generating Expansion Analysis System developed by the Electric Power Research Institute. It also allows the user to subdivide time periods into intervals such as the daytime hours for a group of days, but for a multiyear analysis this can be a substantial computational burden. Neither PROMOD nor EGEAS are designed to represent inflexibility in generator operation, particularly the inability to ramp or change output rapidly from one hour to the next.

CHRONOLOGICAL MODELS

The hourly chronology is maintained and chronological constraints (minimum up and down times, ramp rates, pumped hydro reservoir constraints) are simulated in POWRSYM, a computer program originally developed at the Tennessee Valley Authority. POWRSYM treats forced outages stochastically by a statistical method or by a Monte Carlo method using weekly random draws.
The statistical method for approximating the capacity outage distribution relies on the central limit theorem of mathematics and statistics. The central limit theorem can be used to prove that the capacity outage distribution will approach a normal distribution.
The degree to which it becomes normal is dependent on the number of units and the forced outage rates. This method has been found to be an acceptable approximation for systems composed of ten or more near identical units so long as no single unit (with significant forced outage rate) is greater than 10 percent of the total capacity of other units having significant outage rates.
POWRSYM should therefore not be used for systems which do not meet the above criteria. The error introduced by failure of the normal approximation will tend to reduce the energy expected of intermediate loaded units. To overcome these problems POWRSYM may be optionally executed in a Monte Carlo mode rather than using the probabilistic reserve dispatch. Even then some conceptual limitations remain, such as:

- each week is executed independently without foresight or hindsigth in relation to other weeks
- the pumped storage optimization based on a peak shaving/valley filling technique may not lead allways to the lowest total system production cost.

The production cost of additional pumping and the production savings for additional pumped storage generation is computed by executing the unit commitment, dispatch, and reserve dispatch again only for the affected hours. This assumes independence of the hours and is not exact due to hourly coupling by minimum up time, minimum down time, and ramp rate constraints on the thermal units. An exact, but computationally infeasible, solution would be to recompute the thermal system unit commitment, dispatch and reserve dispatch for each increment of pumped-storage dispatch. The new detailed pumped storage dispatch routines developed for the BENCHMARK-program, lead however to results quite comparable to those to be obtained with an ideal model.

EXAMPLE OF RAMP LIMITS

Consider the case of nuclear or coal-fired baseload units that were not designed to follow load. Their operating inflexibility could prevent the generation from tracking load and lead to temporary capacity or load shortages, where load rises or drops faster than output of these units can be changed. Ramp rate limitations may also cause economic dispatch to increase the output of some units while the total load is decreasing. Here the operator would purposely not dispatch units at equal marginal cost. Further, to avoid ramp rate problems, an operator may start to unload efficient units a few hours before load decreases, which is another example of why dispatch is a dynamic problem that requires consideration of more than one hour at a time.

Three ramp-rate situations are indicated diagrammatically in Figure 1. In hours 3 and 4 there are capacity shortages, and load would either be shed or there would be an inadvertent inflow of power over the tie lines. In hours 11 and 15, load shortages occur because the units are unable to ramp down quickly enough. This is reported as an inadvertent outflow over the tie lines. In hours 18 and 19, the load drops below the minimum permissible output of the two units, leading to another inadvertent outflow. Hours 24 and 26 show an example of the output of the inexpensive unit increasing while the total load is dropping off. Instead of such a dispatch, the operator could have looked ahead to hours 11 to 15 and ramped down the inexpensive unit several hours earlier. Load shortages in hours 18 and 19 are due to minimum operating limits on the generators and therefore could not be avoided without shutting down a unit.

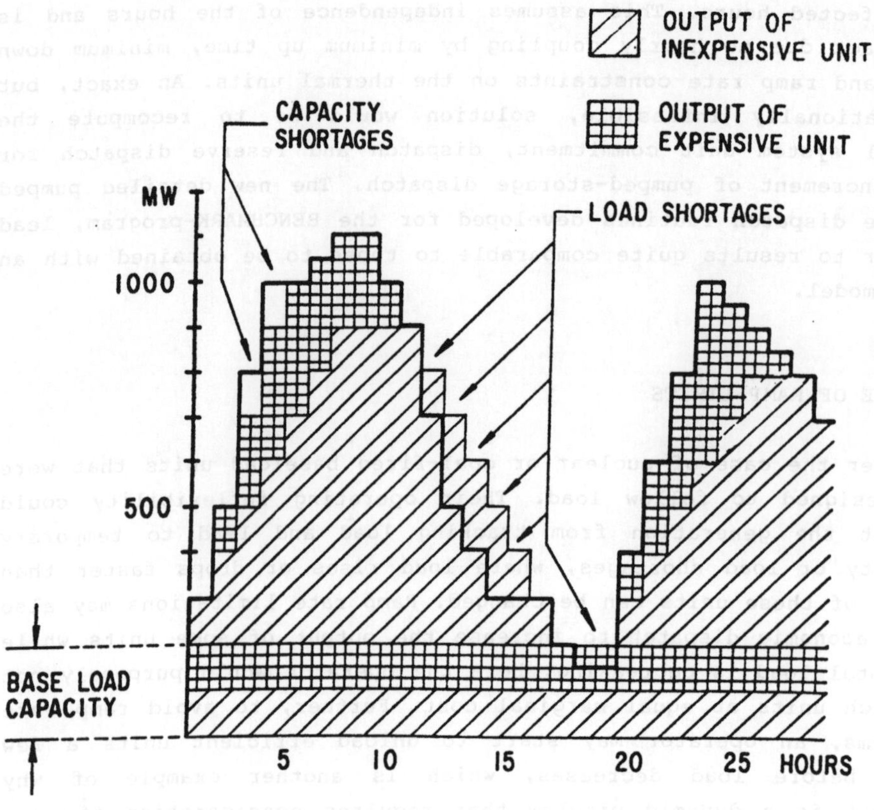

Figure 1
In this simulation of various ramp rate effects, load is indicated by the heavy black line and unit output by the shaded areas. Unit operating limits can lead to mismatches between load and generating output. Thus in hours 3 and 4 demand would exceed output in hours 11 to 15 and hours 18 and 19, output would exceed demand.

CHARACTERISTICS OF THE BENCHMARK COMPUTER PROGRAM

In 1980, an advisory committee of EPRI, the Electric Power Research Institute, decided to sponsor research on production simulation, particularly for evaluating interaction between generator operation and load management. This lead to development of the BENCHMARK computer program, a joint undertaking of the following universities and electric power utilities: Commonwealth Edison, Duke Power, Northeast Utilities, N.V. SEP (The Netherlands), Southern Company, Tennessee Valley Authority, Union Electric, Ohio University and University of Tennessee at Chattanooga.

The decision to write a new computer program was not made immediately. The initial effort was to compare hourly hybrid and conventional simulation based on equivalent load duration curves. It was decided, however, to develop a new chronological simulator that could show sequences of events and account for forced outages. A Monte Carlo analysis was chosen based on sequences of one-week simulations to match pumped storage cycles. In the program, availability or partial availability of each generating unit during a week is treated as an uncertain event that can be represented as a random number.

The program progresses in stages. To carry out a run, draws are taken to determine the availability of the generating units, an initial multiweek dispatch of thermal units is carried out to determine the most favorable hours for using pondage hydroelectric energy, and the thermal dispatch is updated. This is followed by pumped storage plant dispatch with the objective of eliminating infeasibilities (inadvertent tie-line flows), followed by another pumped storage dispatch, this time to minimize operating cost, and then a final thermal dispatch. The commitment and dispatch process is integrated in the sense that ramprate limits effect the commitment decision. Prespecified minimum operating and minimum down times are treated as constraints, and the program incorporates a look-ahead feature to mimic the action of operators in avoiding anticipated load-tracking problems.

The user has available optional graphical outputs, including progress reports referred to as probe points that report on intermediate results at each of the major updates in the dispatch.

The user can override the Monte Carlo feature and specify deterministically the availability of generating units to make a run, for example, that would replicate a historical operating record.

In the Monte Carlo mode, expected values as well as distribution of outcomes are reported, and measures are presented indicating the precision of results. Experiments have been performed that show that antithetical sampling is an effective means for variance reduction, i.e., to reduce the required number of calculations. Typical run time on an IBM 3081 mainframe is 2 or 3 CPU seconds per draw-week of simulation, i.e., a simulation for a 10-week period requiring eight draws would require 160 CPU seconds. Large systems require more calculating time, and if energy storage plants are not involved, the calculation time is considerably reduced.

LOAD TRACKING AND SPINNING RESERVE

The BENCHMARK production simulator can be applied to analysis of both physical and economic behavior of power systems. Load tracking is an example of a physical problem that can be studied through deterministic simulation, Here, the question is whether the system can serve the load, not the question of expected cost or of quantity of fuel required. The Monte Carlo technique can also be applied to this type of problem, but the purpose would be to define statistically the distribution of possible outcomes rather than to determine an expected value that would be based on the average of all possible outcomes.

The physical behavior of a system would also be the focus of studies on cycling of nuclear or other base load units in order to estimate stress that could shorten service life. In particular, operators may wish to study nighttime dispatch periods where the minimum permissible loading level of individual generating units may require undesirable shutdowns.

Recent interest in intermittent generation suggests that load-following and spinning reserve penalties must be considered as part of the economic evaluation of such technologies. Increasing penetration of intermittent generation sources such as wind can impose more severe duties with respect to ramping and minimum loading on the other generators. This will also apply to cogeneration as increasing cost of energy stimulates new interest in the use of cogeneration in industry and for district heating projects. Careful assessment of non-dispatchable and cogeneration technologies require determination of their impact on the operation and dispatch of the whole generating system. If new spinning reserve requirements are established, BENCHMARK in its role as an hourly chronological simulator can measure the cost of increasing spinning reserve. In contrast to a short-cut simulation based on load duration curves, the chronological approach in BENCHMARK is particularly suited to studies of cogeneration in order to show the effect on the entire power supply system of serving independent patterns of heat load and electric power load.

PRODUCTION COST STUDIES

Instead of physical behavior, economic performance is the purpose for calculation of production cost. Here, the Monte Carlo technique can be applied in order to obtain a probabilistic or expected value that will take into account random forced outages. The precision required depends on the application.

If the study period covers several decades and requires many scenarios, simplified calculations are all that may be possible, even with the highest speed computers. It may be possible, however, to calibrate or "benchmark" the procedures. For example, dispatch based on load duration curves could be compared to a corresponding sample of detailed simulations in order to define generalized correction functions that could be used to interpret the bias of a load duration curve dispatch.

The operating detail and chronology of the BENCHMARK production simulator can be applied to estimating expected energy expenses, particularly for the short term, for periods of several weeks to one year. The Monte Carlo simulation in BENCHMARK addresses the uncertainty of the availability of generating equipment but not the uncertainty of unexpected changes in fuel prices or load. This might call for sensitivity studies, for example with low, reference, and high values of fuel price and of load, altogether a combination of 3x3=9 sets of input. If needed, the results could be fitted in a single, polynomial multi-variate regression to obtain production cost as a function of the fuel cost and load variables. The Monte Carlo process used in BENCHMARK would provide the dispersion of production cost outcomes attributable to combinations of unscheduled full and partial generating unit outages.

Other studies in which both physical and economic behavior of the supply system are important include:

- The effect of interruptions of fuel supply or of water supply for hydroelectric plants.
- The possible impact of inter-utility power purchases or sales, either short term of long term.
- The effect on the supply system of load management such as dispatcher control of water heaters.
- The cost or benefit of an adjustment of generator maintenance if unexpected opportunities or needs arise.
- The benefit of a change in operating characteristics of equipment--for example, retrofit of boiler controls in order to lower the minimum operating level of the boiler.

CONCLUSION

Attention to the chronological or time-linked behavior of a power supply system becomes imperative when the role of storage units is considered, when ramp rate limitations affect system response, or other dispatch considerations such as minimum up time or minimum down time are brought into play. Dispatch based on load duration curves suppresses chronology and therefore cannot strictly represent commitment decisions and will produce biased results. Ramp rate limitations may require peaking units such as combustion turbines or pumped-storage units to be operated whenever base-load units cannot respond rapidly enough to sudden increases in load. If ramp rate limitations are overlooked, the usage of the peaking units or the advantages of dispatcher-controlled load management would then be underestimated. Actual operation of storage units requires tracking the amount of stored energy available, i.e., the water level in a pumped-storage plant. This is readily accomplished in a chronological simulation but not in the load duration domain.

AN ALGORITHM IN CONSTRAINED OPTIMIZATION

NADA I. DJURANOVIĆ-MILIČIĆ

Department of Mathematics
Faculty of Technology and Metallurgy
University of Belgrade, Yugoslavia

ABSTRACT. In this paper we cosider a constrained version of a modification of the Curry-Altman step-size algorithm. This modification of the Curry-Altman algorithm (see [3]) is based on so called "forcing functions". It is proved that this modified algorithm is well-defined, then a proof of the convergence of the obtained sequence of points to an optimal solution of the problem of constrained optimization is given.

1. INTRODUCTION

In this paper we are concerned with the nonlinear programming problem

$$\min \{\phi(x) \mid x\in X\}, \text{ where } X=\{x\in D \mid f_j(x)\leq 0,\ j\in I_o=\{1,\dots,m\}\}, \qquad (1)$$

and the functions ϕ, $f_j: D\subset R^n \to R$, $j\in I_o$, are continuously differentiable on the open convex set D.

We shall make the following assumptions:

1. $\text{int}(X) = \{x\in D \mid f_j(x)<0, \text{ for all } j\in I_o\}\neq\emptyset$;
2. every point of the set $\{x\in X \mid f_j(x) = 0 \text{ for some } j\in I_o\}$ is an accumulation point of $\text{int}(X)$;
3. $L^o\cap X$ is closed and $\text{int}(L^o\cap X) \neq \emptyset$;
4. ϕ is bounded below on $L^o\cap X$,

where by L^o we denote the connected component of the level set $L=\{x\in D \mid \phi(x)\leq(x_o)\}$ containing x_o (x_o is any point in X).

We shall consider iterative algorithms generating sequences of points $\{x_k\}$ of the form

$$x_{k+1}=x_k-\alpha_k p_k,\quad x_{k+1}\in X,\quad \phi(x_k)>\phi(x_{k+1}),\quad k=0,1,\dots, \qquad (2)$$

with $p_k\in R^n$, $\alpha_k\geq 0$ to solve (1).

Under the certain assumptions about the direction vector p_k we shall separately concern a step-size algorithm for finding a step-size α_k.

2. PRELIMINARIES

We begin first with a few preliminaries.

D e f i n i t i o n 2.1. (see [4]) A mapping $\sigma:[0,\infty)\times\dots\times[0,\infty)\subset R^m \to R$ is a forcing function of m variables if for any m sequences $\{t_i^k\}\subset[0,\infty)$, $i = 1,\dots,m$

$$\lim_{k\to\infty} \sigma(t_1^k,\dots,t_m^k) = 0 \qquad \text{implies} \qquad \lim_{k\to\infty} t_i^k = 0$$

for at least one i, $1 \leq i \leq m$.

At first, before we formulate the constrained version of the modified Curry-Altman step-size algorithm, we recall that by the modified Curry-Altman step-size algorithm for unconstrained case (see [3]) the step α'_k is defined in the following way:

$$\alpha'_k = 0 \quad \text{if} \quad \langle \nabla\phi(x_k), p_k \rangle = 0; \quad \text{otherwise}$$

$\alpha'_k \in I_k$, where I_k is the first interval of positive solutions of the inequality

$$\langle \nabla\phi(x_k - \alpha p_k), p_k \rangle \geq \sigma(\langle \nabla\phi(x_k), p_k \rangle),$$

where $\sigma : [0,\infty) \to [0,\infty)$ is a forcing function such that $\sigma(t) \leq \delta t$ for every $t \geq 0$ and some $0 < \delta < 1$ and $\alpha'_k \geq q\bar{\alpha}_k$ for some $0 < q < 1$ and every k, where $\bar{\alpha}_k$ is the smallest positive solution of the generalized Curry-Altman equation

$$\langle \nabla\phi(x_k - \alpha p_k), p_k \rangle = \sigma(\langle \nabla\phi(x_k), p_k \rangle).$$

(Recall that the original Curry-Altman algorithm (see [1], [2]) defines α_k in the following way:

$$\alpha_k = 0 \quad \text{if} \quad \langle \nabla\phi(x_k), p_k \rangle = 0; \quad \text{otherwise}$$

$$\alpha_k = \min\{\alpha \geq 0 \,|\, \langle \nabla\phi(x_k - \alpha p_k), p_k \rangle = \mu \langle \nabla\phi(x_k), p_k \rangle\},$$

where $\mu \in [0,1)$ is a fixed number).

We use in the sequel the following notation. For any $x \in X$ and $\varepsilon \geq 0$ we define the index set

$$I(x,\varepsilon) = \{j \in I_o \,|\, -\varepsilon \leq f_j(x) \leq 0\}$$

and use the notation $|I(x,\varepsilon)|$ for the cardinality of $I(x,\varepsilon)$. For any index set $I \subset I_o$ we denote by I^N or I^L the subsets of indices $j \in I$ for which f_j is a nonlinear or linear function, respectively.

D e f i n i t i o n 2.2. (see [4]) The constraint set X is said to be regular if for each $x \in X$ rank $(\nabla f_j(x)^T \,|\, j \in I(x,0)) = q < n$, whenever $q = |I(x,0)| > 0$.

We shall describe certain minimal conditions which must be satisfied by the direction $p_k \in R^n$.

Let $\zeta_j : [0,\infty) \to [0,\infty)$, $j \in I_o^N$, be given forcing functions with $\zeta_j(0) = 0$, and for any $x \in X$, index set $I \subset I_o$ and $\nu \in [0,\infty)$, consider the set

$$K(x,I,\nu) = \{p \in R^n \,|\, \|p\| = 1, \ \langle \nabla f_j(x), p \rangle \geq \zeta_j(\nu), \ j \in I^N, \ \langle \nabla f_j(x), p \rangle \geq 0, \ j \in I^L\}.$$

Following Rauch ([4]) we shall assume that $L^o \cap X$ is compact and that at given point

$x_k \in L^o \cap X$ a direction

$$p_k \in K(x_k, I(x_k, \varepsilon_k), \nu_k), \quad <\nabla\phi(x_k), p_k> \geq 0 \tag{3}$$

has been chosen for some $\nu_k > 0$ and $\varepsilon_k > 0$.

We will also need the following three lemmas.

L e m m a 2.1. (see [4]) Suppose that $p_k \in R^n$ satisfies (3) at $x_k \in L^o \cap X$. If

$$\tilde{\alpha}_k = \sup\{\alpha \geq 0 \mid x_k - tp_k \in L^o, \text{ for all } t \in [0,\alpha)\} \text{ and}$$

$$\alpha_k^* = \sup\{\alpha \geq 0 \mid x_k - tp_k \in X, \text{ for all } t \in [0,\alpha)\}, \tag{4}$$

then $\alpha_k^* \geq \min\{\tilde{\alpha}_k, \Psi(\nu_k, \varepsilon_k)\}$, where $\Psi : [0,\infty) \times [0,\infty) \subset R^2 \to [0,\infty)$ is a forcing function of two variables which depends only on $L^o \cap X$.

L e m m a 2.2. (see [4]) Suppose that $p_k \in R^n$ satisfies (3) at $x_k \in L^o \cap X$ with $<\nabla\phi(x_k), p_k> > 0$. Then $L^o \cap X$ contains an open interval $(x_k, x_k - \alpha_o p_k)$ with some $\alpha_o > 0$. Moreover, if $\alpha_1 > 0$ is any number such that

$$\phi(x_k - \alpha p_k) < \phi(x_k) \text{ for all } x_k - \alpha p_k \in (x_k, x_k - \alpha_1 p_k] \cap L^o \cap X, \tag{5}$$

then $[x_k, x_k - \alpha_1 p_k] \cap X \subset L^o \cap X$.

L e m m a 2.3. (see [4]) Suppose that X is regular and that $L^o \cap X$ is compact. Let $\{x_k\} \subset L^o \cap X$ be any sequence which satisfies

$$\lim_{k\to\infty} \| P_{q_k}(x_k) \nabla\phi(x_k)^T \| = 0, \quad q_k = | I(x_k, \varepsilon_k) |$$

for some given positive sequence $\{\varepsilon_k\}$ such that $\lim_{k\to\infty} \varepsilon_k = 0$. If the set

$$\Omega = \{x \in L^o \cap X \mid \| P_q(x) \nabla\phi(x)^T \| = 0, \ q = | I(x,0) | \}$$

consists of only one point, or if Ω is finite and $\lim_{k\to\infty} \| x_k - x_{k+1} \| = 0$, then

$$\lim_{k\to\infty} x_k = x^*, \quad \| P_q(x^*) \nabla\phi(x^*)^T \| = 0, \ q = | I(x^*,0) |.$$

Moreover, all of the limit sets I^* of the sequence $\{I(x_k, \varepsilon_k)\}$ are contained in $I(x^*, 0)$. (By $P_q(x)$ we denote the projection matrix

$$P_q(x) = E - N_q(x) [N_q^T(x) N_q(x)]^{-1} N_q^T(x), \text{ where}$$

$$N_q(x) = (n_{i_1}(x), \ldots, n_{i_q}(x)), \quad n_i(x) = \nabla f_i(x), \ i \in I_o.)$$

3. THE CONSTRAINED VERSION OF THE MODIFIED CURRY-ALTMAN STEP-SIZE ALGORITHM

The main results dealing with the algorithm are given in the following two theorems. In those theorems the results of [3] and [4] are generalized.

Theorem 3.1. Suppose that $\phi : D \subset R^n \to R$ is a continuously differentiable function on the open convex set D. Assume that $L^o \cap X$ is compact and that $p_k \in R^n$ satisfies (3) at $x_k \in L^o \cap X$. Define α_k by

$$\alpha_k = \min\{\alpha_k', \alpha_k^*\} \tag{6}$$

where α_k' is defined by the modified Curry-Altman algorithm (see [3]) and α_k^* is given by (4). Then:

1. $x_{k+1} = x_k - \alpha_k p_k \in L^o \cap X$;
2. $\phi(x_k) \geq \phi[\lambda x_k + (1-\lambda)x_{k+1}] \geq \phi(x_{k+1})$ for all $\lambda \in [0,1]$;
3. $\phi(x_k) - \phi(x_{k+1}) \geq F(<\nabla\phi(x_k), p_k>, \nu_k, \varepsilon_k)$, where F is a forcing function of three variables which depends only on $L^o \cap X$.

Proof. We proceede by induction and assume that $x_k \in L^o \cap X$. If $<\nabla\phi(x_k), p_k> = 0$, then $x_{k+1} = x_k$; hence, assume that $<\nabla\phi(x_k), p_k> > 0$. Then, by Lemma 2.2, $(x_k, x_k - \alpha_o p_k) \subset L^o \cap X$ for some $\alpha_o > 0$. From Lemma 2 in [3] it follows that $\alpha_k' > 0$.

If $x_k - \alpha p_k \in (x_k, x_k - \alpha_k' p_k] \cap L^o \cap X$, then, by the mean value theorem

$$\phi(x_k) - \phi(x_k - \alpha p_k) = \alpha <\nabla\phi(x_k - \hat{\alpha} p_k), p_k> \text{ for some } \hat{\alpha} \in (0,\alpha).$$

From this relation and the proof of Theorem 1 in [3], it follows

$$\phi(x_k) - \phi(x_k - \alpha p_k) \geq \alpha\sigma(<\nabla\phi(x_k), p_k>) > 0.$$

Thus it follows that (5) holds with $\alpha_1 = \alpha_k'$. Thus, by Lemma 2.2 $[x_k, x_k - \alpha_k' p_k] \cap X \subset L^o \cap X$.

Further on, since $\alpha_k \leq \alpha_k^*$, it follows that $[x_k, x_k - \alpha_k p_k] \cap X = [x_k, x_k - \alpha_k p_k] \subset L^o \cap X$ and $x_{k+1} = x_k - \alpha_k p_k \in L^o \cap X$.

Further, using $\alpha_k \leq \alpha_k'$ and Lemma 2 in [3], we have $\phi(x_k) \geq \phi[\lambda x_k + (1-\lambda)x_{k+1}] \geq \phi(x_{k+1})$ for all $\lambda \in [0,1]$.

By the mean value theorem, using again $\alpha_k \leq \alpha_k'$, Lemma 2 and Theorem 1 in [3], we obtain

$$\phi(x_k) - \phi(x_{k+1}) = \alpha_k <\nabla\phi(x_k - \alpha_s p_k), p_k> \geq \alpha_k \sigma(<\nabla\phi(x_k), p_k>), \quad \alpha_s \in (0, \alpha_k). \tag{7}$$

Moreover, from Lemma 2.1, (6) and the proof of Theorem 1 in [3], it follows that

$$\alpha_k = \min\{\alpha_k', \alpha_k^*\} \geq \min\{\alpha_k', \min\{\alpha_k', \hat{F}(\nu_k, \varepsilon_k)\}\} = \tag{8}$$

$$= \min\{\alpha_k', \hat{F}(\nu_k,\varepsilon_k)\} \geq \min\{ qs[(1-\delta)<\nabla\phi(x_k),p_k>], \hat{F}(\nu_k,\varepsilon_k)\},$$

where by $s:[0,\infty)\to[0,\infty)$ we denote the reverse modulus of continuity of $\nabla\phi$ (see [3]).

Together (7) and (8) yield

$$\phi(x_k)-\phi(x_{k+1}) > \min\{qs[(1-\delta)<\nabla\phi(x_k),p_k>], \hat{F}(\nu_k,\varepsilon_k)\}\sigma(<\nabla\phi(x_k),p_k>) \tag{9}$$
$$= F(<\nabla\phi(x_k), p_k>, \nu_k,\varepsilon_k),$$

where F is a forcing function of three variables given by

$$F(t_1,t_2,t_3) = \min\{ qs[(1-\delta)t_1], \hat{F}(t_2,t_3)\} \sigma(t_1).$$

T h e o r e m 3.1. Consider the iteration (2), where at $x_k \in L^o \cap X$ ε_k is chosen by an algorithm for ε (see [4]), and $p_k \in R^n$ is any direction such that with α_k obtained from (6) we have that $x_{k+1} \in L^o \cap X$ and that relation (9) holds with

$$\nu_k = \| P_{q_k}(x_k)\nabla\phi(x_k)^T \|, \qquad q_k = | I(x_k,\varepsilon_k)|.$$

Then, if the sequence $\{p_k\}$ has the property that $\| P_{q_k}(x_k)\nabla\phi(x_k)^T\| \to 0$ if $<\nabla\phi(x_k),p_k>\to 0$, $k\to\infty$ and if the set $\{x\in L^o\cap X | \| P_q(x)\nabla\phi(x)^T\| = 0,\ q = |I(x,0)|\}$ consists of only one point x^*, it follows that $x_k\to x^*$, $k\to\infty$.

P r o o f. From our assumptions it follows that

$$\phi(x_k) - \phi(x_{k+1}) \geq F(<\nabla\phi(x_k),p_k>, \| P_{q_k}(x_k)\nabla\phi(x_k)^T\|, \varepsilon_k), \quad q_k = |I(x_k,\varepsilon_k)|,$$

where F is a forcing function of three variables.

By Lemma 2.3 and Definition 2.1, it follows that either

$$<\nabla\phi(x_k),p_k>\to 0, \ \| P_{q_k}(x_k)\nabla\phi(x_k)^T \| \to 0 \ \text{or}\ \varepsilon_k \to 0,\ k\to\infty.$$

By the algorithm for ε (see [4]) it follows that $\lim_{k\to\infty} \varepsilon_k = 0$ if and only if $\lim_{k\to\infty} \| P_{q_k}(x_k)\nabla\phi(x_k)^T\| = 0$, while $\lim_{k\to\infty} <\nabla\phi(x_k),p_k> = 0$ implies, by assumption, $\lim_{k\to\infty} \| P_{q_k}(x_k)\nabla\phi(x_k)^T\| = 0$.

Therefore, in any case we have that $\lim_{k\to\infty} \| P_{q_k}(x_k)\nabla\phi(x_k)^T\| = 0$ holds and, by Lemma 2.3, $\lim_{k\to\infty} x_k = x^*$.

REFERENCES

1. Altman M., Generalized Gradient Methods of Minimizing a Functional, Bull. Acad. Polon. Sci. Ser.Sci. Math. Astronom. Phys. 14, pp. 313-318, 1966.

2. Curry H., The Method of Steepest Descent for Nonlinear Mnimization Problems, Quart. Appl. Math., 2, pp 258-261, 1944.

3. Miličić N., O jednoj modifikaciji step-size algoritma, Zbornik radova SYM-OP-IS'83, pp 267-274, Herceg-Novi, 1983.

4. Rauch S.W., Convergence Theory For a Class of Nonlinear Programming Problems, SIAM J. Numer. Anal., vol. 10, No.1, 1973.

AN EFFICIENT ALGORITHM FOR THE OPTIMAL OPERATION OF A CASCADE REACTOR

A. Dourado
Departamento de Engenharia Electrotecnica, Universidade de Coimbra, P-3000 COIMBRA, PORTUGAL

J. L. Calvet
Laboratoire d'Automatique et d'Analyse des Systèmes, 7 Avenue du Colonel Roche, F-31077 TOULOUSE CEDEX, France

ABSTRACT

The optimal start-up and regulation of a cascade reactor for ethanol fermentation is faced. By reducing the dimension of the model and based on the phase-plane analysis, the conditions ensuring the asymptotic stability of the system are stated. They are determined by the overall specific dilution rate and the initial conditions. The hamiltonian formulation of minimum-time start-up and regulation of the production steady-state is analysed. This time-optimal problem is approched by solving the two-point boundary value optimization problem with a quadratic criterium that accelarates the transient behaviour of the system and with decreasing final time. Results show that important savings in time can be obtained by optimal control.

Keywords: biotechnological systems; optimal control; hierarchical control.

INTRODUCTION

The industrial production of ethanol by fermentation of sugars is actually object of a large atention. The ethanol used as fuel is a credible and renewable source of energy. The high efficiency in the industrial production is required in order to increase the energy balance and to low the production costs. In a previous work(Dourado, 1983) the optimal configuration of the cascade, from the steady state point of view, was studied. The optimal regulation of the steady-state is the object of the present work. The importance of these questions lies in the fact that this system is caracterized by a very slow evolution, and then any savings in time becomes important.

Several studies appear in the litterature concerning the dynamic optimization of different kinds of fermentation processes. Constantinides(1979) gives an interesting review of them. More recent works are those of Gallegos and Gallegos(1982) who developed an explicit closed-loop control law for a certain class of nonlinear systems where the manipulated variables appear linearly on the state equation; they applied it to the fermentation of biomass obtaining a robust control. The case of a cascade reactor can be profitably faced by decomposition methods. In this way Fawzy and Hinton(1982) applied to a three reactors cascade a method of hierarchical decomposition developped by Hassan and co-workers(1978); in the lower level each sub-system is a linearized model of each reactor. On the upper level a mixte coordination with direct iteration is implemented.

In this paper a successive approximations method is presented. It takes profit, of the stair-case structure of the cascade reactor.

MODEL OF THE CASCADE REACTOR

Based on biological considerations and on mass-balance principles the following equations constitute the nonstructured model of the cascade reactor, ilustrated in fig.1, in the case of head-feeding (D^f_1)=0, D^f_i=0, i>1)

$$X_i(t)= \mu_i X_i(t) - D^f(t)X_i(t)+ D^f(t)X_{i-1}(t) \quad (1)$$

$$S_i(t)= \nu_{si}X_i(t)-D^f(t)S_i(t)+D^f(t)S_{i-1}(t)+D^f(t)S^f(t) \quad (2)$$

$$P_i(t)= \nu_{pi}(t)-D^f(t)P_i(t)+D^f(t)P_{i-1}(t) \quad (3)$$

with:

$$\mu_i= \mu_0 \cdot \frac{S_i}{K_s+S_i+\frac{S_i^2}{K_{si}}} \frac{P_i}{K_p+P_i+\frac{P_i^2}{K_{pi}}} (1 - P_i/P_1)^R \quad (4)$$

$$\nu_{pi}= \nu_0 \frac{S_i}{K_s+S_i+\frac{S_i^2}{K_{si}}} \frac{P_i}{K_p+P_i+\frac{P_i^2}{K_{pi}}} (1-P_i/P_1{}')^{R'} \quad (5)$$

$$\nu_{si}= - \nu_{pi}/Y_p \quad (6)$$

where $\left[u_0, K_s, K_{si}, K_p, K_{pi}, P_1, R, Y_p, {}_0, K_s', K_{si}', K_p', K_{pi}', P_1', R'\right]^T$ is the parameter vector. We refer to Dourado(1983) for a detailed description of them.

STABILITY

The conditions for the stability of the steady-states of the model describing the overall cascade are reduced to the conditions for the stability of each sub-system (corresponding to each reactor) by the application of the first principle of Lyapounov (Csáki, 1972). Linearizing the overall model on the steady-state, the resulting evolution matrix is quasi- block diagonal inferior and bounded. If the block diagonal elements correspond to an asymptoticaly stable system (if their eigenvalues are all negative) the overall system is asymptotically stable (Gantmacher, 1964). It should be noted that for time varying systems this condition is not sufficient but is also necessary (Michel, 1983).

Each diagonal block of the linearized system matrix corresponds to an isolated subsystem, which is described in state form by a linearized model of order 3. By reducing the dimension of the model of each reactor from three to two variables, and by analyzing the phase-plane trajectories, we can easily deduce the conditions assuring the stability of each reactor. The reduced model is obtained by a changment of variables in the state space representation: $x_1=X_i$, $x_2=S_iY_p$, $x_3=P_i$, $x_2^f=Y_pS^f$. The model of each reactor is rewritten in a simpler manner (7), (8) and (9)

$$\dot{x}_1=\mu(x_2,\ x_3)x_1 - ux_1 \tag{7}$$

$$\dot{x}_2=-\nu_p(x_2,x_3)x_1-ux_2+ux_2^f \tag{8}$$

$$\dot{x}_3=\nu_p(x_2,\ x_3)x_1 - ux_3 \tag{9}$$

where u is the specific dilution rate (D^f) at the head of the cascade. Adding (8) and (9), (10) is obtained,

$$\dot{x}_2+\dot{x}_3= u(x_2^f-x_2-x_3) \tag{10}$$

and by integrating (10) between t and t_f we obtain (11),

$$x_2^f-x_2(t_f)-x_3(t_f)=\left[x_2^f-x_2(t)-x_3(t)\right]e^{-a(t_f-t)} \tag{11}$$

with $a(t_f-t) = \int_0^t u(\tau)d\tau$.

Under the assumptions (i) $a(t_f-t)$ is bounded and (ii) the differential equation (10) is asymptotically stable one gets (12),

$$x_2^f - x_2(t) - x_3(t) = 0, \quad t \in [o, t_f] \tag{12}$$

Then equation (12) allows to reduce the model from three to two variables. We can choose the state variables $X_i = x_1$ and $S_i = x_2/Y_p$. If S^f is considered constant, D^f is the single input variable. Then the stability may be analyzed in the phase-plane.

For a given value of D^f, there exists two possible steady-states of the state equations. One is E^p defined by $\mu = D^f$, it is the production state; the other is E^1 defined by $X_i = 0$, $S_i = S^f$, it is the washout state.

In fig.2 are shown the trajectories in the phase-plane for the case, $D^f = 0.45\ h^{-1}$, $S^f = 100 g/l$. If the initial state lies in the region A, the steady-state will be the production state, E^p, asympotically stable. If the initial state lies in the zone B, then the washout state, E^1, is reached and it is also asympotically stable. Fig. 3 shows the case where $D^f = 0.58\ h^{-1}$, $S^f = 100\ g/l$. It can be seen that whatever the initial state is the washout state is always reached.

Resuming, we can say that for fixed values of D^f and S^f, there are two possible equilibrium states. If $D^f \leqslant D^{max}$ the two both are asympotically stables, one is the production state and the other one is the washout state; the initial condition determines which is reached. If $D^f > D^{max}$, the washout state is the only equilibrium one and it is asympotically stable. It can be shown (Dourado, 1983) that

$$D^{max} = \mu_0 \frac{\sqrt{K_p K_{pi}}}{2K_p + \sqrt{K_p K_{pi}}} \cdot \frac{S^f}{K_s + S^f + (S^f)^2 / K_{si}} \tag{13}$$

OPTIMAL CONTROL OF THE CASCADE

Known the desired steady-state (which is found by economi cal considerations (Dourado,1983)), we are interested to reach it from the initial condition (start-up problem) and to control it with respect to possible perturbations (regulation problem). Both should be solved for minimum time. However, the number of variables makes impossible to face the problem in this way. We consider then the optimization problem with a quadratic criterium over a fixed final time and we solve successive optimization problems with decreasing final time. The resulting optimal control accelerates the transient behaviour of the system.

Let the criterium be stated as

$$J=1/2\left\{\|x(t_f)-x\|_T+\int_0^{t_f}\left[\|x(t)-\bar{x}\|_Q+\|u(t)-\bar{u}\|_R\right]dt\right\} \qquad (14)$$

where $x = (X_1, S_1, P_1, \ldots, X_N, S_N, P_N)^T$

$\bar{x}$: steady-state, $\bar{u}$: steady input

Q, T positive semidefinites matrices, R:positive definite

$\|x\|_A$ denotes the quadratic form $x(t)^T Ax(t)$.

By a convenient choice of Q, the state x(t) is forced to approach x. The ponderation R over the control means an energetic limitation.

The stationnarity conditions of the Hamiltonian associated with this problem leads to the two-point boundary value problem composed by the state equations (1) to (3) and the costate equations (18),

$$\dot{\lambda}_i(t)=g_{i1}(\lambda_i(t),\lambda_{i+1}(t),\psi_i(t),\gamma_i(t),X_i(t),S_i(t),P_i(t))$$

$$\dot{\psi}_i(t)=g_{i2}(\lambda_i(t),\psi_i(t),\psi_{i+1}(t),\gamma_i(t),X_i(t),S_i(t),P_i(t)) \qquad (15)$$

$$\dot{\gamma}_i(t)=g_{i3}(\lambda_i(t),\psi_i(t),\gamma_i(t),\gamma_{i+1}(t),X_i(t),S_i(t),P_i(t))$$

$$i=1, \ldots, N$$

where λ_i, ψ_i, γ_i are the costate variables.

The final conditions on the adjoint variables are given by $p(t_f) = T\left[x(t_f)-\bar{x}\right]$. Moreover, one must consider the stationnarity conditions with respect to the control variable, $\mathcal{H}_u=0$.

Hierarchical method of sequential decomposition

The solution of the two-point boundary value problem is carried out by a hierarchical method of sequential decomposition that can be resumed as follows:

1. The interval $[0, t_f]$ is discretized into M identical intervals.

2. We choose and memorize an initial approximmation of the control trajectory $u^{(0)}(t)$ (u (j), j=1,..., M after discretization). Do k, the iteration indice equal to 0.

3. With the control $u^{(k)}$, the state equations are integrated from 0 to t_f, with the initial condition $x(0)=x^{(0)}$.

The serial structure of the differential equations allows to solve sequentially the state equations in a spatial decomposed form (16),

$$x_i^{(k)}(j+1) = f_i(x_i^{(k)}(j),\ x_{i-1}^{(k)}(j+1),\ u^{(k)}(j)) \qquad (16)$$
$$i = 1 \text{ to } N \qquad j = 1 \text{ to } M$$

4. Starting from the final conditions on the costate variables, the costate equations are then integrated backward.
In a dual manner, this integration is done sequentially resulting of the spacial decomposed form of the costate equation in the reverse sense (17),

$$p_i^{(k)}(j-1) = g_i(p_i^{(k)}(j),\ p_{i+1}(j-1),\ x_i^{(k)},\ u_i^{(k)}(j)) \qquad (17)$$
$$i = N \text{ to } 1,\quad j = M \text{ to } 1$$

Let us note that the funtions g_i are linear with respect to the costate variables.

5. From the trajectories $x^{(k)}(j)$, $p^{(k)}(j)$, j=1 to M, the control trajectory is updated as follows (18),

$$u_i^{(k+1)}(j)=u_i^{(k)}(j)-H_{k+1}\left[h(u^{(k)}(j),x^{(k)}(j),p^{(k)}(j))\right] \qquad (18)$$
$$i = 1 \text{ to } N,\quad j = 1 \text{ to } M$$

Substitute into the memory $u^{(k)}(j)$ by $u^{(k+1)}(j)$, j=1 to M. Do k=k+1 and go to phase 3 if the coordination task in not satisfied.

This algorithm leads to a calculation structure shown in fig.2. For deeper study of such kind of methods, see Calvet (1983).

RESULTS

Let us consider a cascade of 4 reactors, with head feeding. The optimal steady-state is defined by the table 1.

Let us suppose a perturbation of 20% of this steady-state, resulting in the initial perturbed state is shown table 2.

The natural evolution of the system ($u=\overline{u}$), is shown in fig. 5. The optimal control for a horizon-time of 20 h is shown fig.6. It can be seen that after 10h the system is already in the vicinity of the steady-state (with some overshoot).

The following values of Q, R ant T are compatible with the relative values of state and input variables:

$Q=diag[0.01,0.001.0.001.0.01,0.001,0.001,0.01,0.001,0.001,0.1,0.01,0.01]$

$R=1000.0$, $T=0$.

The convergence of the algorithm is ilustrated in fig.7. After 140 iterations the test of convergence $\|u^{(k+1)}(j)-u^{(k)}(j)\|<10^{-3}$ is satisfied.

Let it be remarqued that convergence is sensible to the optimization horizon-time. If it is too long, due to the instability of the adjoint variables, the algorithm may diverge. In this example such a situation is found for $t\geq 30$ h.

CONCLUSION

Due to the stair case structure, the stability of the cascade reactor is reduced to the stability of each reactor. Moreover, the stability of each reactor is easily solved by phase-plane analysis after a model order reduction.

The optimal regulation of the cascade has been solved by a hierarchical method of sequential decomposition. The results show that important savings in time may be obtained by optimal control. Moreover, the sub-problems solved are of dimension 3, which allows to face the problem of on-line control by microprocessors.

The optimal start-up raises serious problems of convergence, if a too long optimization horizon-time is needed to solve the problem. These numerical problems are not completely fulfilled until now.

REFERENCES

Calvet,J.L.(1984). Methodologies iteratives et partitionnement. Application aux problemes de commande. These de Docteur d'Etat, Univ. Paul Sabatier, Toulouse, France, 1984.

Constantinides, Alkis(1979).Application of rigurous optimization methods to the control and operation of fermentation processes. Annals of the New York Academy of Sciences, Biochemical Engineering, vol 326, 1979.

Csáki, F. (1972). Modern Control Theories, Akedemi Kiadó,Budapest, 1972.

Dourado, A(1983). Ethanol production in a cascade reactor:contribution a la modelization, a l´identification et a l´optimization.These de Docteur-Ingenieur, Universite Paul Sabatier, Toulouse, France, 1983.

Fawzy, A.R. and O.R. Hinton(1982).Hierarchical dynamical optimization of the fermentation process for producing Gramacidin S, using microprocessors. Intern. J. Systems Science, vol.13, p193-208, 1982.

Gallegos, J.A. and J.A. Gallegos(1982).Optimal control of a class of nonlinear systems. Application to a fermentation process. Trans. of the ASME, vol.104, p212, 1982.

Gantmacher, G.(1964). Theorie des matrices.Tome 1. Dunod, Paris.

Hassan, R. Hurteau, M.G. Singh, A. Titli (1978). A three-level costate prediction method for continuous dynamical systems.Automatica, vol.14, p177-181, 1978.

Michel, A.N.(1983). On the status of stability of interconnected systems. IEEE trans. on AC, vol AC-28, p639-653, 1983.

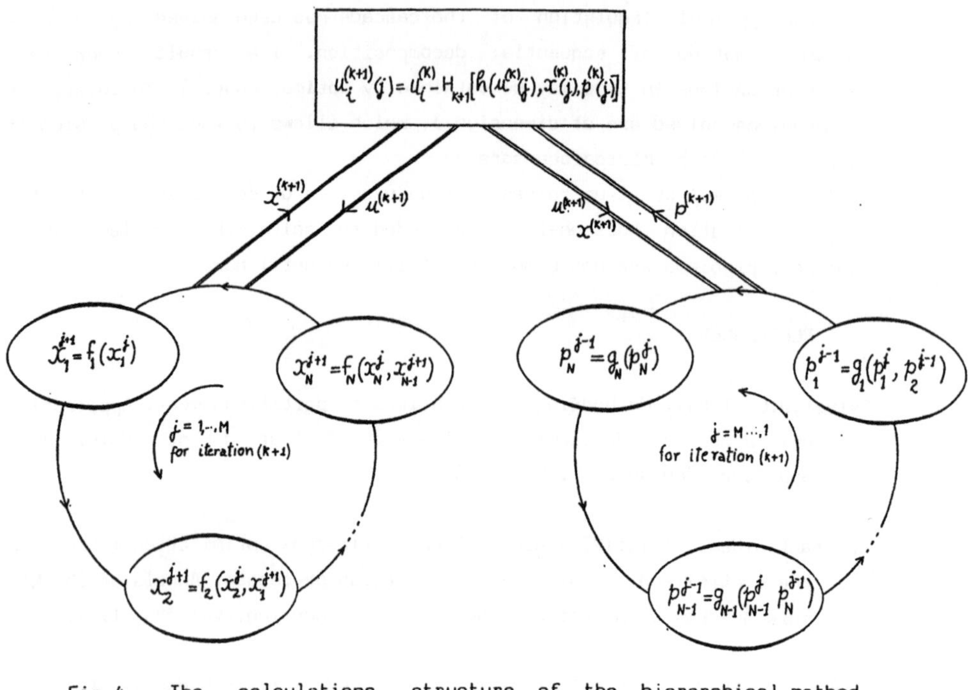

Fig.4. The calculations structure of the hierarchical method of sequential decomposition

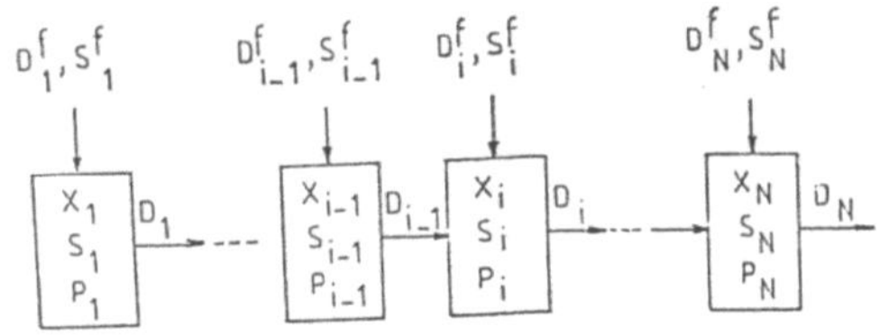

Fig. 1

A cascade reactor representation

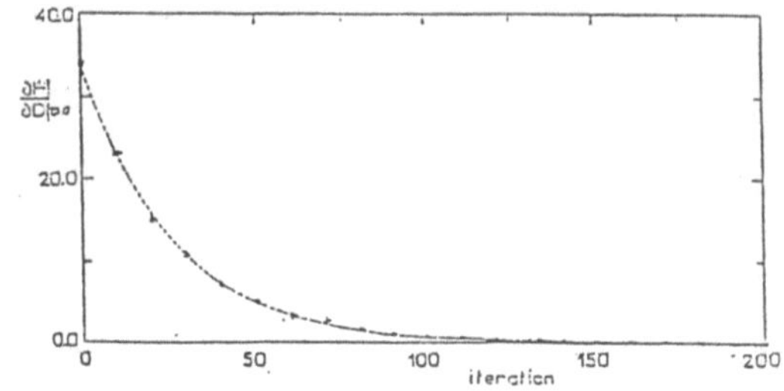

Fig. 7

First derivation of the Hamiltonian function with respect to the control D^f versus the number of iterations (at t=0)

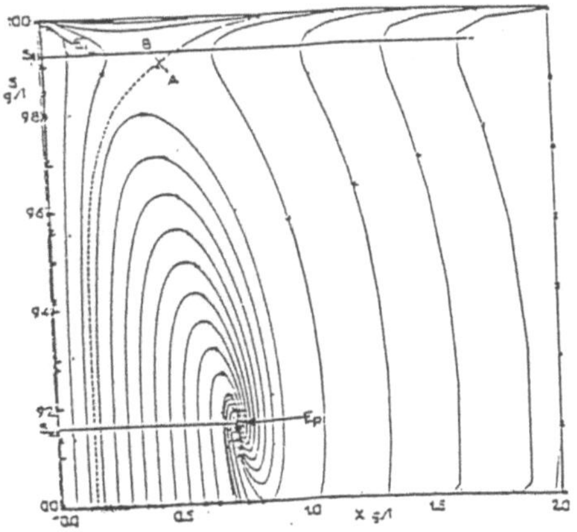

Fig. 2

Phase plane trajectories for the case D^f=0.45 h^{-1}, S^f= 100 g/l

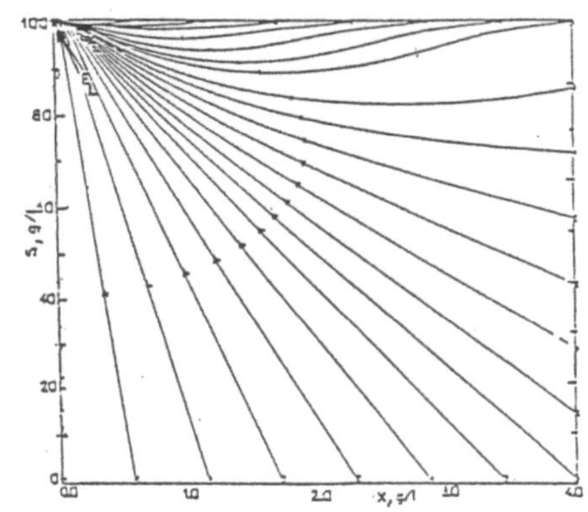

Fig. 3

Phase plane trajectories for the case D^f=0.58 h^{-1}, S^f=100 g/l

Table 1. The optimal steady-state.

i	$\bar{D}^f$ (h^{-1})	$\bar{S}^f$ (g/l)	$\bar{X}_i$	$\bar{S}_i$	$\bar{P}_i$ (g/l)
1	0.287	85.0	1.03	70.6	6.9
2			2.21	48.6	17.5
3			3.40	23.8	29.4
4			4.48	1.2	40.2

Table 2. The perturbed state.

i	X_i (0)	S_i (0)	P_i (0) (g/l)
1	0.83	73.5	5.5
2	1.77	55.9	14.0
3	2.72	32.1	23.5
4	3.57	17.9	32.2

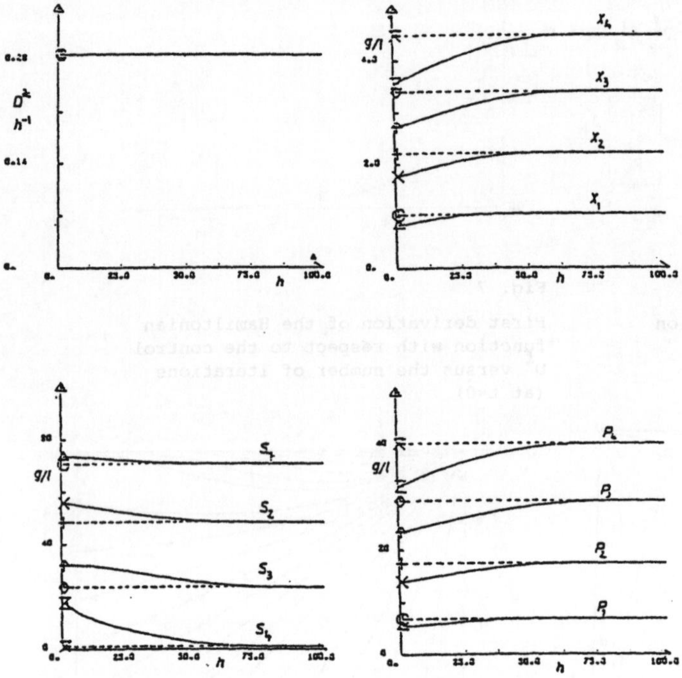

Fig. 5. Response of the uncontrolled perturbed system.
--- :steady-state, ––– natural trajectory.

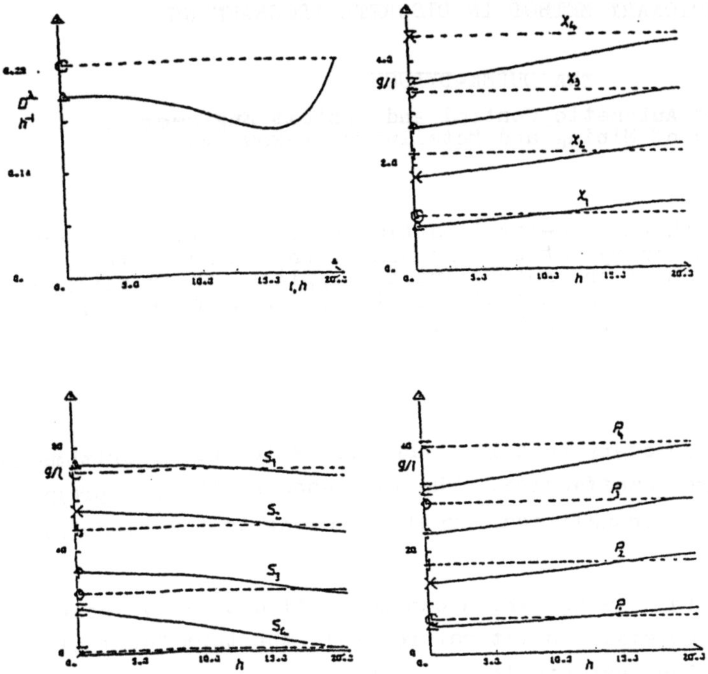

Fig.6. Optimal control over an horizon of 20h. ---:steady state, ---- optimal trajectory. Note the different time-scale with respect to fig.5

EVOLUTIONARY METHOD IN DISCRETE PROGRAMMING

EWA DUDEK-DYDUCH

Institute of Automatic Control and Systems Engineering,
University of Mining and Metallurgy, Kraków, Poland.

ABSTRACT

The paper deals with the two-level partially stochastic optimization method named evolutionary method. The aim of the paper is to present a formal description of the method and its application in discrete programming. The algorithm for planning of investments in chemical industry, based on the method, has been presented.

INTRODUCTION

At present it is necessary to solve tasks for which no algorithms exist or the existing are noneffective. The large-scale discrete programming, scheduling and some nonanalitical problems belong to the class of such tasks.

The paper deals with a method which can be applied to solve some of above mentioned problems. In literature [3] the term "evolutionary" refers to more or less stochastic algorithms. In this paper this term refers to the exactly determined class of partially stochastic algorithms and because of it the formal description of the method is presented.

The algorithm based on the method has been proposed for the planning of investment in chemical industry. In the paper a conception of the algorithm is described.

EVOLUTIONARY METHOD

The method can be applied to optimization tasks which has been transformed to the form suitable for two-level algorithms.
Thus the task: to find $\hat{v} \in V$ minimizing function f :

$$f(\hat{v}) = \min_{v \in V} f(v)$$

where V is a subset of a linear space V', can be replaced by : to find the pair $(\hat{m}, \hat{w}) \in M \times W = V$ such that

$$f(\hat{m}, \hat{w}) = \min_{(m,w) \in V} f(m,w) = \min_{m \in M} [\min_{w \in W_m} f(m,w)]$$

where $W_m \subset W$ is a set determined by means of fixed m .

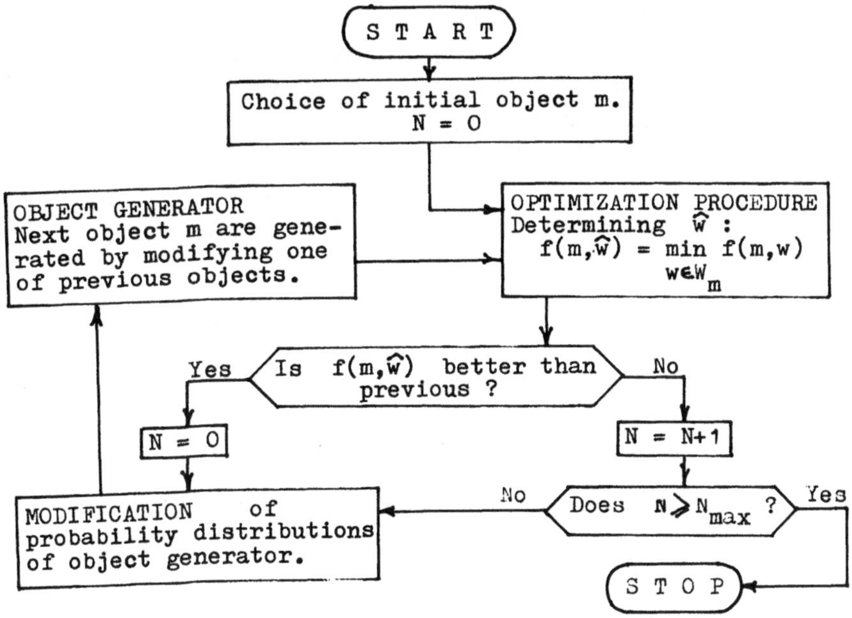

Fig. 1. The block schema of the method.

The values of m are generated by stochastic procedure on higher level. Then variables w are computed on a lower level as a result of a deterministic procedure, where m are parameters. Parameters of probability distributions are modified after each iteration on a basis of information obtained from earlier ones.

The main problems which should be solved to apply the method are :

a) to determine the proper division of searched variables into each level
   - to solve the low level problem in a possibly short time,
   - to obtain from the low level problem solution the suitable information to correct parameters of a stochastic generator,

b) to determine a way of utilizing the accumulated information to give the best direction to searching.

OPTIMIZATION TASK

During the planning of chemical industry development there was a need to solve following static optimization task. There are given :

- $\bar{b}$ - vector of interior demand for each chemical product in a fixed period of time,
- I - set of existing plants numbers,
- J - set of numbers of plants considered to be built,

h: (I∪J)→$\mathbb{R}$ - function giving maximum production level for each plant,

$\overline{c}_1$, $\overline{c}_2$ - respectively, vector of export and import prices,

$\overline{c}_3$ - vector of unit cost of production for each plant,

Q - value of investment outlay,

A - matrix where a_{ij} is coefficient of utilization or production for i-th product in j-th plant, dim A = dim $\overline{\overline{b}}\times(\overline{\overline{J}}+\overline{\overline{I}})$, matrix represents all plants, existing and those which can be built.

We assume a linear relation between vector of production level of plants $\overline{z}$ and vector $\overline{d}$ determining value of production for each product:

$A\,\overline{z} = \overline{d}$

We must compute: the optimal arrangement of investments, vector of production level $\overline{z}$, exportation and importation vectors $\overline{y}$, $\overline{x}$ which assure fulfilling the interior demand in terms of minimal social cost. The role of export and import is auxiliary in this task i.e. we assume that it is possible to sell or buy all products the production of which is greater or less then interior demand.

A cost of an investment depends on the arrangement of investments which it belongs to. These costs are given in tables. Let $\overline{v}$ denote the vector for which $v_i = 1$ if i-th investment will be built and $v_i = 0$ in opposite case. dim $\overline{v} = \overline{\overline{J}}$. The function $g(i,\overline{v})$ formally represents the cost of i-th investment in arrangement defined by $\overline{v}$.

The task can be formally described as a discrete programming one as follows. Vectors $\overline{z}, \overline{x}, \overline{y}, \overline{v}$ are to be found, which minimize function

$$f = \overline{c}_2\overline{x} - \overline{c}_1\overline{y} + \overline{c}_3\overline{z} \qquad \overline{c}_1 < \overline{c}_2 \qquad (1)$$

and satisfy constrains:

$$A\,\overline{z} + \overline{x} - \overline{y} = \overline{b} \qquad (2)$$

$$z_i \leq v_i\, h(i) \quad \text{for } i \in J \qquad (3)$$

$$z_i \leq h(i) \quad \text{for } i \in I \qquad (4)$$

$$\sum_{j \in J} v_j\, g(j,v) \leq Q \qquad (5)$$

$$\overline{x}, \overline{y}, \overline{z} \geq 0 \qquad v_i \in \{0,1\} \qquad (6)$$

The relation $\overline{c}_1 < \overline{c}_2$ causes that univocal and limited solution exists. Discussion of this model has been given in [1].

Appart from constrains (2)-(6) other nonlinear and logical constrains exist in real task. They result from taking into account modernization, alternative production and other reasons. These constrains do not affect the conception of algorithm. They have been omitted for the sake of simplicity.

The nonlinearities and large dimension make impossible to apply discrete programming method to solve the task. Therefore the evolutionary algorithm has been proposed.

Let us denote:

$L \subset J$ - a subset of coordinates numbers for which $v_i = 1$ (named arrangement of investments),

$\bar{p} : 2^J \to \mathbb{R}^n$, $\bar{q} : 2^J \to \mathbb{R}^n$, $n = \bar{\bar{J}}$, - vector functions defined as follows:

$$p_k(L) = \begin{cases} h(k) & \text{for } k \in L \\ 0 & \text{for } k \in (J \setminus L) \end{cases}$$

$$q_i(L) = \begin{cases} g(i,L) & \text{for } i \in L \\ 0 & \text{for } i \in (J \setminus L) \end{cases}$$

The task has been transformed to the following form: vectors $\bar{x}$, $\bar{y}$, $\bar{z}$ and set $L \subset J$ are to be found which minimize function:

$$f = \bar{c}_2 \, \bar{x} - \bar{c}_1 \, \bar{y} + \bar{c}_3 \, \bar{z} \qquad \bar{c}_1 < \bar{c}_2 \tag{1'}$$

and satisfy constrains:

$$A \, \bar{z} + \bar{x} - \bar{y} = \bar{b} \tag{2'}$$

$$z_i \leqslant p_i(L) \qquad i \in J \tag{3'}$$

$$z_i \leqslant h(i) \qquad i \in I \tag{4'}$$

$$\sum_{i=1}^{n} q_i(L) \leqslant Q \tag{5'}$$

$$\bar{x}, \bar{y}, \bar{z} \geqslant 0 \tag{6'}$$

EVOLUTIONARY ALGORITHM

It is easy to notice that when L is fixed the rest of searched variables $\bar{x}$, $\bar{y}$, $\bar{z}$ can be computed by means of linear programming procedure. Because of that on the upper level a group of new plants L which satisfy the constrains (5´) is determined by a stochastic procedure. Then on the lower level the linear programming task (1´) ÷ (4´),(6´) is solved.

Which information obtained from this solution can be utilized and in which way to determine direction of L changes (or changes of vector $\bar{q}$, what is equivalent) ?

Let us notice that after solving linear programming task also the dual prices are known. This information is utilized in algorithm.

Utilizing computed from tables vector of mean investment cost $\overline{q}_m$ we estimate the relation between vectors $\overline{p}$ and $\overline{q}$ as follows:

$$p_i \approx k_i \cdot q_i \quad \text{where} \quad k_i = h(i)/q_{mi} \tag{7}$$

Assuming for a while that vectors $\overline{p}$ and $\overline{q}$ can change continuously we can estimate gradient $\frac{\partial f}{\partial \overline{q}}$

$$\frac{\partial f}{\partial \overline{q}} = \frac{\partial f}{\partial \overline{p}} \frac{dp}{d\overline{q}} = \overline{\lambda} \cdot \frac{dp}{d\overline{q}} \tag{8}$$

$$(7),(8) \rightarrow \frac{\partial f}{\partial q_i} \approx \lambda_i \; k_i = e_i \tag{9}$$

If value of indicator e_i is big this indicates that i-th plant has big influence on the performance index and it suggests that this investment should be left in new set L.

Due to discrete set of vectors $\overline{q}$ there is no possibility of changing them in every direction and with any short step.

Let us denote $K(L^o)$ - neighbourhood of L^o.

$$K(L^o) = \{L \subset J_Q : \|\overline{q}(L) - \overline{q}(L^o)\|_1 \leqslant \varrho\}$$

where J_Q is the admissible set $J_Q \subset 2^J$ and ϱ is a parameter.

$$J_Q = \{L \subset 2^J : \sum_{i=1}^{n} q_i(L) \leqslant Q\}, \qquad \left(\|\overline{q}\|_1 \overset{df}{=} \sum_{i=1}^{n} |q_i| \right)$$

The new set L^{n+1} is generated as follows:

1. Probabilistic choice of plants numbers that are to be removed from arrangement L^n. The probability that i-th investment will stay in L^{n+1} is proportional to $s_i(L^n)$ where vector $\overline{s}$ is defined:

   $\overline{s}(L^n) = (1-r)\,\overline{s}(L^{n-1}) + r\,\overline{e}(L^n)$, r $(0,1)$ is a parameter

   $\overline{s}(L^1) = \overline{e}(L^1)$

2. Probabilistic choice of new investments. The probability is proportional to $s_i(L^n)$. Set L^{n+1} must belong to the neighbourhood $K(L^n)$

3. Solving of linear programming task. Computing value of function f denoted as $f(L^{n+1})$.

4. If $f(L^{n+1})$ is better than previous ones then set L^{n+1} replaces L^n and algorithm goes to 1. If performance index did not improve the sampling is repeated.

The searching is stopped after a fixed number of iterations, when the best value of performance index does not change.

The indicator $\bar{s}(L)$ is used to improve the stability of the procedure.

CONCLUSION

The presented method enables to take into account any other limitations refering to arrangements of investments what could be impossible or difficult when solved analitically.

Large scale of matrix A implies necessity of decomposition the linear programming task. It is much easier due to the fact that restriction (5´) is regarded on upper level.

The evolutionary algorithms are heuristic ones. They should be applied:

- when approximate solution is good enough,
- when the problem is very complicated and it must be divided into smaller ones solved by means of different algorithms or methods (for example partial simulation),
- we may use the information about previous solutions to modify parameters of stochastic generator.

REFERENCES

1. Dudek-Dyduch, E. An Evolutional Optimization Method of Industry Branch. Zeszyty Naukowe AGH No 146 pp.83-91, Kraków 1981
2. Dudek-Dyduch, E. On Evolutionary Algorithm. Proc.1-st IASTED Symp. on Applied Informatics vol.III p.37, Lille France 1983
3. Wala, K. Symulacyjne metody optymalizacji dyskretnych procesów produkcyjnych. Zeszyty Naukowe AGH z.21, Kraków 1979

ENVIRONMENTAL SYSTEMS MODELLING WITH RESPECT TO THE FUTURE

A. FARKAS
Department of Industrial Engineering,
Technical University, H-1521 Budapest,
Hungary

Introduction

During the last decades environmental systems modelling achieved a considerable progress in the field of technological forecasting. Only in the recent years have more systematic approaches and scientific tools been developed and applied to any great extent, Alter [1], Duperrin and Godet [2], Mitchell and Tydeman [7] and Porter et al. [8] .

In this paper we present a multiperiod, event-type, combined operational research and probability theory based model, together with a reliability systems application. Due to the nature of its phenomena the procedure concerns the future describing the multivariable, uncertain environment in which much of decision making takes place. Considering the above facts our principal goals are the following:

- to establish a flexible model structure in which quantitative and qualitative methods could be applied as well;
- to ensure the internal consistency of the model as much as possible;
- to connect theory and practice adequately;
- to make it clear that environmental systems modelling generally has to be considered as an individual approach of problem solving.

Problem formulation

Environmental systems and their constituent events are increasingly interdependent. In these type of problems the future states of the system cannot be determined relying upon only the events of the past. The presented approach utilizes both quantitative data and subjective judgments on the events of the studied system. The event assessment procedure requires that for each event, each expert has to provide the probability estimates of the events which will occur within the predetermined time intervals. Similar process has to be accomplished for the expected impacts on the system, caused by the occurances of the external events. In a time-dependent manner different cross-impacts will be emerged as a consequence of the responses of the system.

The numerical values of the estimated cross-impacts are the primary inputs for the scenario generation. This procedure seeks all of the possible future states of the studied system. As the systems are assumed to be a finite set of the possible future states, thus all of the mutually exclusive and collectively exhaustive scenarios can be identified.

Model structure

The model starts with three types of inputs, denoting as follows:

(i) The event set of the studied system, whose elements should be clear defined, nonrecurring, discrete events only,

$$A:=\{E_1, E_2, \ldots, E_i, \ldots, E_j, \ldots, E_n\}.$$

(ii) The time-dependent cumulative probabilities P, of each event E_i, through each interval t_k,

$$P(E_i, t_k), \qquad i=1,\ldots,n, \quad k=1,\ldots,m.$$

(iii) The quadratic type cross-impact matrix $\underline{\underline{C}}$, elements of which express the effects of the occurences, or nonoccurances on each other,

$$C_{E_i E_j}, \qquad i=1,\ldots,n, \quad j=1,\ldots, n.$$

Let us introduce the steps of a finite algorithm, which is used during the computations and could give us a substancial insight into the nature of the process itself.

Step (1) The input cumulative probabilities of events are converted to transition probabilities in each interval, setting into action the modifying effects of the cross-impacts.

Step (2) Using random numbers a computer simulation of the event pairs is run, in each interval, to attain the final probability distributions.

Step (3) The mutual entropy of the event pairs is computed as a measure of the similarity between the events.

Step (4) The maximalization of the correlation coefficient is made between each of the possible subsystem and their constituent events.

Step (5) A clustering technique is used to decompose the original system into closely connected subsystems.

Step (6) A scenario generation is made for each decomposed subsystem by applying linear programming methods.

Step (7) The whole systems scenarios are computed, after cardinal ranking of the scenario probabilities of each subsystem.

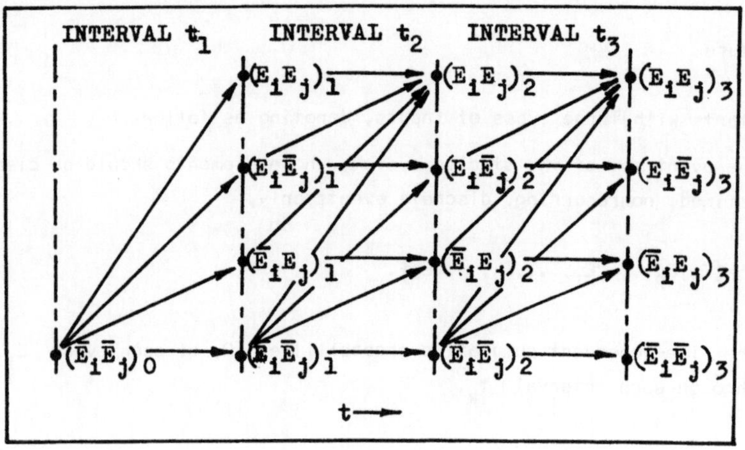

Figure I. Graph representation of the probability process.

The time-dependent probability process is illustrated in Figure I., through the example of two binary events E_i and E_j, assuming a problem having of three time intervals. The whole process could be modeled by a connected, directed graph. The depicted graph has a specific accessibility property between its origin-destination node pairs, which will meet our preliminary assumption about the occurances of the events. The origin node represents the starting nonoccurance state of the events while the destination nodes signify the possible probability outcomes at the required future time. To obtain these joint probability values, first we have to determine each chain of the graph leading from the origin node to the destination nodes. Depending on the k time intervals, there are a total of $(k+1)^2$ feasible chains. E.g. one of these chains could be adequately given by the sequence of the nodes

$$(\bar{E}_i,\bar{E}_j)_0,(E_i,E_j)_1,(E_i,E_j)_2,(E_i,E_j)_3.$$

The directed arcs indicate the transition probabilities between the appropriate joint event pairs. Their values could be determined according to Enzer [3], using the balancing law

$$p(E_i,t_k)=p(E_i,t_k|E_j)P(E_j,t_{k-1})+p(E_i,t_k|\bar{E}_j)[1-P(E_j,t_{k-1})].$$

Given n binary events, $N = 2^n$ different system scenarios can be identified. E.g. the h-th scenario is defined as the following state function

$$S_h(t_k) = (E_1, E_2, \ldots, \overline{E}_i, \ldots, \overline{E}_j, \ldots, \overline{E}_n), \qquad h=1,\ldots,N.$$

Each state $S_h(t_k)$, has an unknown probability p_h. The coefficients of the event probabilities and the joint probabilities

$$a(i) = \begin{cases} 1, & \text{if } E_i \in S_h(t_k) \\ 0, & \text{otherwise} \end{cases} \qquad i=1,2,\ldots,n;$$

$$a(ij) = \begin{cases} 1, & \text{if } E_i \text{ and } E_j \in S_h(t_k) \\ 0, & \text{otherwise} \end{cases} \qquad i=1,2,\ldots,n-1; \quad i<j,$$

$$\vdots$$

$$a(ij\ldots n) = \begin{cases} 1, & \text{if } E_i \text{ and } E_j \text{ and } \ldots \text{ and } E_n \in S_h(t_k) \\ 0, & \text{otherwise} \end{cases} \qquad i=1.$$

The entire event system may be described in terms of N pieces of linear equations:

$$\sum_h a(i)\,p_h = P(i),$$
$$\sum_h a(ij)\,p_h = P(ij),$$
$$\vdots$$
$$\sum_h a(ij\ldots n)\,p_h = P(ij\ldots n).$$

Subject to the constraint

$$\sum_{h=1}^{N} p_h = 1, \qquad p_h \geq 0 \quad \text{for all } h.$$

The maximum possible probability values of the scenarios are obtained with the objective function

$$\max Z = \max\left(\sum_{h=1}^{N} p_h\right).$$

The final computational results of the model are the list of all the possible event combinations and the most probable future states of the system together with the probability values of the expected occurances.

Limitations and the accuracy of the model

Let us interpret the limitations and the accuracy of the model:

- The model assumes that the event set holds the theorem of the total probabilities. Since the problem relates to the future, this condition is considered as an abstract idea which could hardly be realized in the practice.
- If the input data were not arised from measuring, or experimental results, then the quality of the expert assessments obviously could exert a doubtful effect on the output.
- While the event probabilities are time-dependent, the cross-impact values are constant in time.
- Although the method produces a cardinal ranking of the scenarios, the scenario probabilities cannot be uniquely determined. Since that any set of n events have 2^n scenarios, at least the same amount of information is required in the linear programme to find solution for the 2^n unknowns. However, it may obtain at the most, $\frac{1}{2}n(n+1)+1$ linearly independent equations, and therefore all linear combinations of the basic feasible solutions may also be solutions, see e.g. Kelly [5] and Mitchell and Tydeman [6]. The existence of the multiple solutions is a consequence of the lack of information for the higher order interrelationships. That is the reason, why the model cannot satisfy the overall consistency criteria.

Application

We introduce a reliability systems application for a tower crane, as an illustration of the model. The outlined results provide primary information for the preventive maintenance of the crane.

The reliability circuit diagram of the crane is shown in Figure 2. It can be seen, that 10 serially connected subsystems were defined, signed by Roman numbers. In this study the failures of the structural elements were called "defect-events". The cross--impacts between the elements cause a change in the probability of failures. In structural reliability problems the probability of failures may only increase during operation.

The failure data of 50 cranes were then analysed. These data were recorded during an 8 years time-period of service. The cross-impact values were estimated by an expert group, having six members of engineers. The completed cross-impact matrix is illustrated in Figure 3. It contains qualitative cross-impact values with three levels of intensity. Note, that the matrix is not symmetrical, because of the different intensity and direction of the impacts on the defect-events. In the attached three columns of the matrix, three external events are also generated. These are the additional wind loading (01), the chemical effects (02) and the inaccurate assembling effects (03). They represent those environmental effects, which have the strongest impacts on the crane.

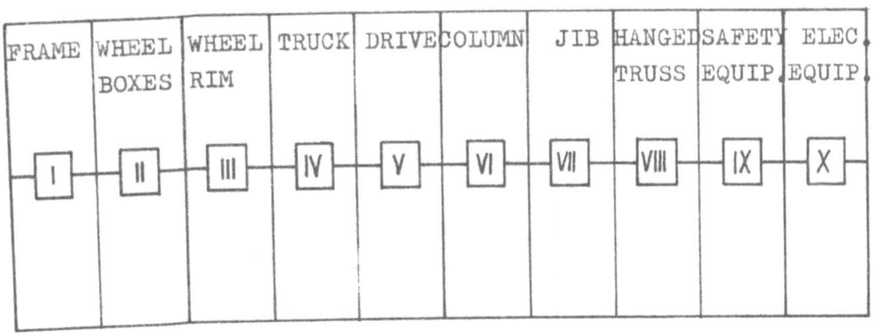

Figure 2. Reliability circuit diagram of the crane.

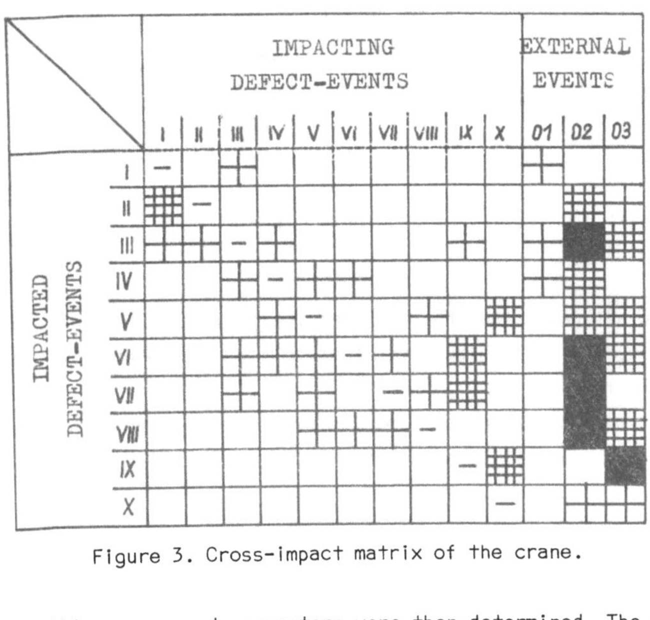

Figure 3. Cross-impact matrix of the crane.

The usual reliability curves and parameters were then determined. The results were significantly different before and after the cross-impacts have been activated. For a detailed description about these topics, see Farkas [4].

In our case the number of the internal and external defect-events are rather sizable. Even though, considering the combinatorial features of the problem and hence the computations, it needs to decompose the system into closely connected subsystems, using the similarity measures. The results of clustering is displayed in Figure 4., whereas a two-dimensional plot is used to group the similar events. This figure implies, that those events form a cluster, which are located close together.

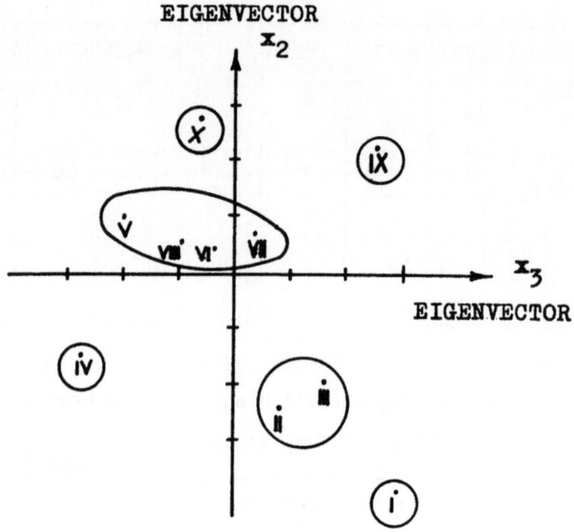

Figure 4. The result of clustering of the crane subsystems.

In the next computational phase the subsystems, belonged to the same cluster have to be treated jointly, while the singleton clusters may be regarded as failure--independent subsystems.

Then the whole systems scenarios are computed. The most probable future states, accompanied by their probabilities give the systems reliability over the investigated 960 hours of service time, as we illustrate below:

Ranking	EI	EII	EIII	EIV	EV	EVI	EVII	EVIII	EIX	EX	p_h
1.	0	0	0	0	1	0	0	0	0	0	0.340
2.	0	0	0	0	0	0	0	0	0	1	0.247
3.	0	0	0	0	1	0	0	0	0	1	0.210
4.	0	0	0	0	0	0	0	0	0	0	0.138
⋮											⋮
1024.	1	1	1	1	0	1	1	1	1	1	0.000

The first column contains the scenario ranking of the states of the crane system. The defect-events are indicated in the upper row. In the table 1's denote that the given subsystems will be failing, while 0's denote that they won't. The joint probability values of the most likely systems scenarios are listed in the last column.

Conclusions

A general method is not currently available to forecast the uncertain future of complex systems. The discussed model is partly based on judgmental data and is thus inevitably subjective. However, we believe that the flexibly combined methods are useful to accomplish the various aspects of problems of environmental systems. Although there are still difficulties and limitations to this type of analysis, continuation of present developments will surely increase the scope and need for its applications.

References

[1] Alter, S.: The Evaluation of Generic Cross-Impact Models. Futures, Vol.11, (4), 1979,pp.132-150.

[2] Duperrin, J.C. and Godet, M.: SMIC 74 - A Method for Constructing and Ranking Scenarios, Futures, Vol.7, (8), 1975,pp.302-312.

[3] Enzer, S.: Cross-Impact Techniques in Technology Assessment. Futures, Vol.4,(3), 1972,pp.30-51.

[4] Farkas, A.: Reliability of Mechanical Systems - An Attempt to Applying Cross-Impact Method. Periodica Polytechnica, Mech. Eng. Vol.30,(3-4) 1986, pp.155-168. (in press)

[5] Kelly,P.: Further Comments on Cross-Impact Analysis. Futures, Vol.8,(8), 1976, pp.341-345.

[6] Mitchell, R.B. and Tydeman, J.: A Note on SMIC 74. Futures, Vol.8,(2), 1976, pp.336-340.

[7] Mitchell, R.B. and Tydeman, J.: Subjective Information Modelling. Operational Research Quarterly, Vol.28, (1), 1977,pp.1-19.

[8] Porter, A.L. et al.: A Guidebook for Technology Assessment and Impact Analysis. North Holland Publishing Company, Amsterdam-Oxford, 1982.

ON THE CONNECTIONS BETWEEN MATHEMATICAL PROGRAMMING AND DISCRETE OPTIMAL CONTROL

José A. Soêiro Ferreira *)

René Victor Valqui Vidal **)

*) Department of Electronical Eng., Faculty of Engineering, Oporto, Portugal.

**) The Institute of Mathematical Statistics and Operations Research, The Technical University of Denmark, DK-2800 Lyngby, Denmark.

ABSTRACT

This paper gives a short review of different theories and topics in the fields of Mathematical Programming and Discrete-time Optimal control Theory. Well-known and new results (for instance, the generalized Maximum Principle) are presented within a unified framework. Emphasis is placed in the trends of interplay. New research areas are also identified.

1. Introduction

This paper gives a brief account of different theories and topics in the fields of mathematical programming (section 2) and discrete optimal control theory (section 3). In particular we mention the recent contributions to discrete optimal control - the upper boundary approach. This work has deep roots in mathematical programming, a reason for making its introduction in section 2. Finally, in section 4, we will point out some connections between those fields. In general, the contribution of mathematical programming to the field of discrete optimal control is better recognized than the reverse case. The same happens with the development of the upper boundary approach.

This paper is a condensed version of research report [1], that can be obtained from the authors. In [1] a complete list of references are also presented. Moreover in [2] a survey of the discrete-time maximum principle is presented.

To provide a common basis for the discussion in the remainder of this paper, specific forms of the problems are now stated.

The mathematical programming (MP) problem is:

$$\begin{aligned} &\text{Maximize} && r(u) \\ &\text{subject to} && f(u) = \bar{x} \\ & && u \in U \end{aligned}$$

where $r(.)$ is a real function defined on E^n, $f(.)$ is a function from E^n into E^m, $u \in U \subset E^n$, and $\bar{x}$ is a given vector in E^m. Note that the stated MP problem is quite general. Inequality constraints can be transformed into equality constraints by adding slack variables. Moreover, we do not specify, for the moment, how the U is defined.

The discrete optimal control (DOC) problem is:

$$\text{Maximize} \quad R = \sum_{i=0}^{N-1} r_i(x_i, u_i)$$

$$\begin{aligned} \text{subject to } x_{i+1} &= f_i(x_i, u_i) \\ u_i &\in U_i \subset E^n, && i = 0,1,\dots,N-1 \\ x_i &\in X_i \subset E^m, && i = 0,1,\dots,N \\ x_0 &= \bar{x}_0 \end{aligned}$$

where $r_i(.,.)$ is a real function defined on E^{m+n}, $f_i(.,.)$ is a function from E^{m+n} into E^m, for $i = o,1,\dots,N-1$, and $\bar{x}_0$ is a constant vector.

2. Mathematical Programming

2.1. Differential Optimization

Differentiability of the functions involved in the MP problem is required. In this way we have a basic distinction relative to the upper-boundary approach presented in section 2.2.

Necessary optimality conditions are characterized, in geometric terms, by an empty intersection of two sets, a tangent cone and an open half space defined by gradients. The convertion of this (or similar) geometric conditions into more usable algebraic statements involving equations, can be achieved by introducing mild assumptions - constraint qualifications. Multipliers are then introduced, in the context of differentiable functions, and are used to produce constrained stationary points of the Lagrangian function. This is well known from Kuhn-Tucker conditions, which represent the basic results in mathematical programming theory (the set U in the <u>MP</u> problem formulation usually defined by inequality constraints).

The use of conical approximations to unspecified sets ($U' = \{u: u \in U, f(u) - \bar{x} = 0\}$) to define a class of perturbations around the optimal solution, permits to obtain necessary conditions of optimality in a very general form. In particular, that formulation is suitable for application to optimal control theory (Canon et. al.).

2.2 Upper Boundary Approach

In the upper boundary approach multipliers or multiplier functions appear as a technique whose goal is maximization, rather than location of stationary points (e.g. Kuhn-Tucker conditions), of a generalized Lagrangian function. This viewpoint gives no restrictions such as continuity or differentiability of the function to be maximized and the domain can be any set.

The upper boundary approach has its origin in Everett's work. In that line, extended states (perturbation set) are defined, over which we construct an upper-boundary function (perturbation function) or then an upper boundary (set). The analysis is then carried in the payoff-resource space where general supporting functions to the upper boundary function are defined, together with a generalized Lagrangian function (which is constructed with those supports). This one may be interpreted as measuring the symmetric of the distance between points of the extended states and points on the supports. That is why each time the Lagrangian is maximized (this is an unconstrained problem), a point on the upper boundary is obtained. But the particular problem we are solving is not known in advance (it is a modified <u>MP</u> problem) but arises in the course of solution and it is, in fact, the problem whose constraints equal the "resources expended by the strategy" that

solved the unconstrained problem. This technique (generalized Everett's theorem) is quite attractive, because there are no conditions to verify in advance. Moreover the risk of wrong conclusions seems to be very small, because in that case no $u^* \in U$ maximizing $r(u)$ will be found, such that $f(u^*) = \bar{x}$.
The information obtained by this procedure can be used either to approximate the solution of the MP problem or to do a parametric study on the right-hand side. Although the MP problem may have a solution, the solution may not be obtained by maximizing the Lagrangian (relative to a specific type of supporting function). The situation can be such one that the desirable right-hand side cannot be generated, just because a support (of that type) to the upper boundary function, at the point $\bar{x}$, cannot be found - we have a gap.

The importance of considering linear supports is well known. Concavity of the upper boundary functions is a sufficient condition for the existence of supporting hyperplanes, and this property can be derived from special characteristics of the given r, f and U in the MP problem.

Saddle-point theory also provides conditions for the existence of a linear support. Its existence is guaranteed if a saddle-point to the Lagrangian linear in the multipliers exists.

3. Discrete Optimal Control

3.1 Maximum Principle

A point of contact between mathematical programming and discrete optimal control appeared along with the attempts, taken by both mathematical programming and control theories, to generate a parallel result of the continuous-time maximum principle for discrete optimal control.

Although the maximum principle caused no big difficulties when applied to the optimization of continuous systems it became apparent the infeasibility to obtain, for the optimization of discrete-time systems, as strong results as in the continuous-time case. In fact, an equivalent formulation in discrete-time is, in general, neither necessary nor a sufficient condition for optimality.

The available results from mathematical programming such as Kuhn-Tucker conditions, were able to yield conditions in the form of a maximum principle, but only stationarity of the Hamiltonian resulted as necessary conditions.

A great progress in understanding the role of the maximum principle in optimization of discrete-time systems was done by Halkin and Propoi. From a control theories viewpoint (i.e. by explicit consideration of trajectories) they have shown that the maximum principle is valid when the set consisting of all possible values of states and criteria in each stage is convex. This condition was then weakened by Holtzman and Halkin to directional convexity.

It has been shown (Canon et.al.) that the necessary conditions for optimality, in the form of the maximum principle (with convexity or directional convexity assumption), can be derived by recasting the DOC problem in the form of a MP problem and basically using Kuhn-Tucker theory combined with the useful conical approximations to special sets.

3.2 Upper Boundary Approach

A Generalized Maximum Principle

Nahorski, Ravn and Vidal developed a new approach to the DOC problem - the upper boundary approach to discrete optimal control. A generalized maximum principle is obtained as a result of the generalization of the (classical) Hamiltonian.

The idea behind this framework is the extension of Everett's work to the optimization of dynamic systems and the argumentation relies heavily on the definitions, concepts and results of the static optimization theory (see the upper boundary approach to mathematical programming - section 2.2). The reason is that, as we look for necessary conditions the overall optimization can be split into a finite number of subproblems to which that theory can be applied. It is clear in the new approach the influence of the generalization and extension of concepts, (generalized supports, generalized Lagrangian, upper-Boundary functions and sets,...), carried out in mathematical programming.

A critical assumption for the derivation of the maximum principle was the directional convexity property - which may be translated i terms of existence of supporting hyperplanes to the upper boundary functions (at each stage) at the optimal state. The generalized maximum principle offers, a priori, more chances by allowing more general supporting functions and, therefore, generalized Hamiltonian functions. The generalized maximum principle basically affirms that if the DOC problem has an optimal solution then supports to the upper-boundary functions exist so that the respective Hamiltonian (at every stage) is maximized by the optimal solution.

It is possible to distinguish in the necessary conditions of the maximum principle (on the linear or generalized version) a "maximum principle" part and an adjoint equations part. Different assumptions may be taken for each part, still keeping the validity of the main results. However the "maximum principle" part without the adjoint equations part has no practical relevance unless there is a way of choosing the right supporting surfaces to the upper boundary functions. This is in straight relation with the possibility of the upper boundary approach to deal with nonsmooth problems.

Halkin's principle of optimal evolution stems naturally from the upper boundary approach. Some work has been done to connect, in some way, this principle with Bellman's principle of optimality, in the spirit of the current approach. Therefore we can see dynamic programming (as applied to discrete optimal control) naturally placed within the upper boundary approach.

4. Mathematical Programming and Discrete Optimal Control

Along with the discussion around the different techniques to cope with the MP problem and the DOC problem, made in the previous sections, we have already touched the subject of the connections between mathematical programming and discrete optimal control. The equivalence of the MP and DOC problems immediately suggests the possibility of contacts and of an interplay between those theories. Moreover, we can expect that existing connections translate, in some way, the inner life of each of those fields - we mean the relations relative to the different approaches developed a) within mathematical programming and b) within discrete optimal control.

a) In what concerns mathematical programming we can say that the upper boundary approach (section 2.2) based on Everett's work, is more general than the differential optimization (section 2.1), in the sense that it permits non-linear multiplier functions and neither continuity nor differentiability are assumed. Basically it provides sufficient optimally conditions. There is a parallel situation in duality theory. The classical approach was related to diffenrentiability in an essential way. The generalized approach does not need such assumption.

Kuhn-Tucker conditions is a well known and essential result of what we called differential optimization. By using linear approximations to sets related to the feasible sets it basically provides, under some assumptions (constraint qualification, differentiability,...), necessary optimality conditions.

Therefore, we have been referring different theories of multipliers. However, if we work in differentiable space a very important class of <u>MP</u> problems exists for which both theories are applicable - the concave programming problems verifying a constrain qualification. Saddle-points exists, as well as linear supports to the upper boundary functions which, in case of being tangent to these functions, determine the multipliers uniquely. Thus, multipliers in the saddle point conditions are multipliers in Kuhn-Tucker conditions and multipliers for the linear supports. Moreover, the optimal dual variables for the dual problem are precisely the multipliers for Kuhn-Tucker, or saddle point or tangent linear support conditions. Linear programming problems are the best known member of that class of <u>MP</u> problems.

b) In relation to discrete optimal control the maximum principle with linear Hamiltonian fits smoothly into the upper boundary approach. Its proof (not covering the case of a <u>DOC</u> problem with state-space constraints) through this approach is in fact, very simple. The critical assumption attached to this maximum principle is the existence of supporting hyperplanes to the upper boundary functions at the optimal states. This condition may be stated in terms of directional convexity of certain sets of reachable states a condition already appearing in the earlier control approach. Remind that this directional convexity assumption is removed in the generalized maximum principle.

If linear supporting functions exists, then using linear Hamiltonian conditions arising from the different types of approaches, referred in section 3, give the same information.

Retaking, more closely the initial subject about the connections between mathematical programming and discrete optimal control, we observe that the search for necessary conditions of optimality, resulting in the formulation of the maximum principle (linear or generalized version) permitted us to suppose that the optimal values of the states x_i and x_{i+1} (see the statement of the DOC problem) are known, for any i, therefore obtaining static optimization problems (interrelated) relative to the control variables u_i . Mathematical programming becomes, then, applicable; the upper boundary approach to discrete optimal control is a clear example of that, in the way that it is a natural extension of concepts and methods used in static optimization theory (see section 2.2). Saddle-point theorems, results of duality theory in mathematical programming have their own role and interpretation in discrete optimal control. For instence, in the case of the maximum principle (linear) it is clear a relation between the separated application (stage by stage) of that principle and duality theory. Maximization of the Hamiltonian, for given x_i and x_{i+1} , corresponds to determining a dual function (regardless constant parcels). Dual variables appear as slopes of supporting hyperplanes (tangent) to the upper boundary sets.

The maximum principles, by dividing a problem in a series of interrelated ones, to which mathematical programming may be applied, give a good insight to the DOC problem. This profundity in analysis is an advantage of these principles relatively to dynamic programming. Adjoint equations, relating multipliers, are the connections among the interrelated problems mentioned.

We have already mentioned, in section 3.1., that the maximum principle (linear) may also be obtained only by mathematical programming theory, - not considering techniques using control theory.

In conclusion we observe that we have different tools to cope with the MP problem and the DOC problem. Multipliers in linear or non-linear forms, appearing in different contexts, always played an essential role. However classes of MP problems and DOC problems were identified for which a correspondence (equality in some cases) could be found among the different types of multipliers.

On the other hand we would like to remind that control problems often serve as convenient sources of large dimensional test problems for new algorithms in mathematical programming.

Finally we present a diagram reflecting some lines of our discussion. It is not a comprehensive one, neither has been the discussion about those broad and potential subjects. More could have been said and there is room for the development of new interactions.

REFERENCES

[1] Ferreira, J.A., and Vidal, R.V.V., On the connections between Mathematical Programming and Discrete Optimal Control - a survey and some new results, IMSOR, The Technical University of Denmark, Research Report Nr. 2, 1985, pp. 25.

[2] Nahorski, Z., Ravn, H., and Vidal R.V.V., The Discrete-time Maximum Principle - a survey and some new results, International Journal of Control, Vol. 40, Nr. 3, 1985, pp. 533-554.

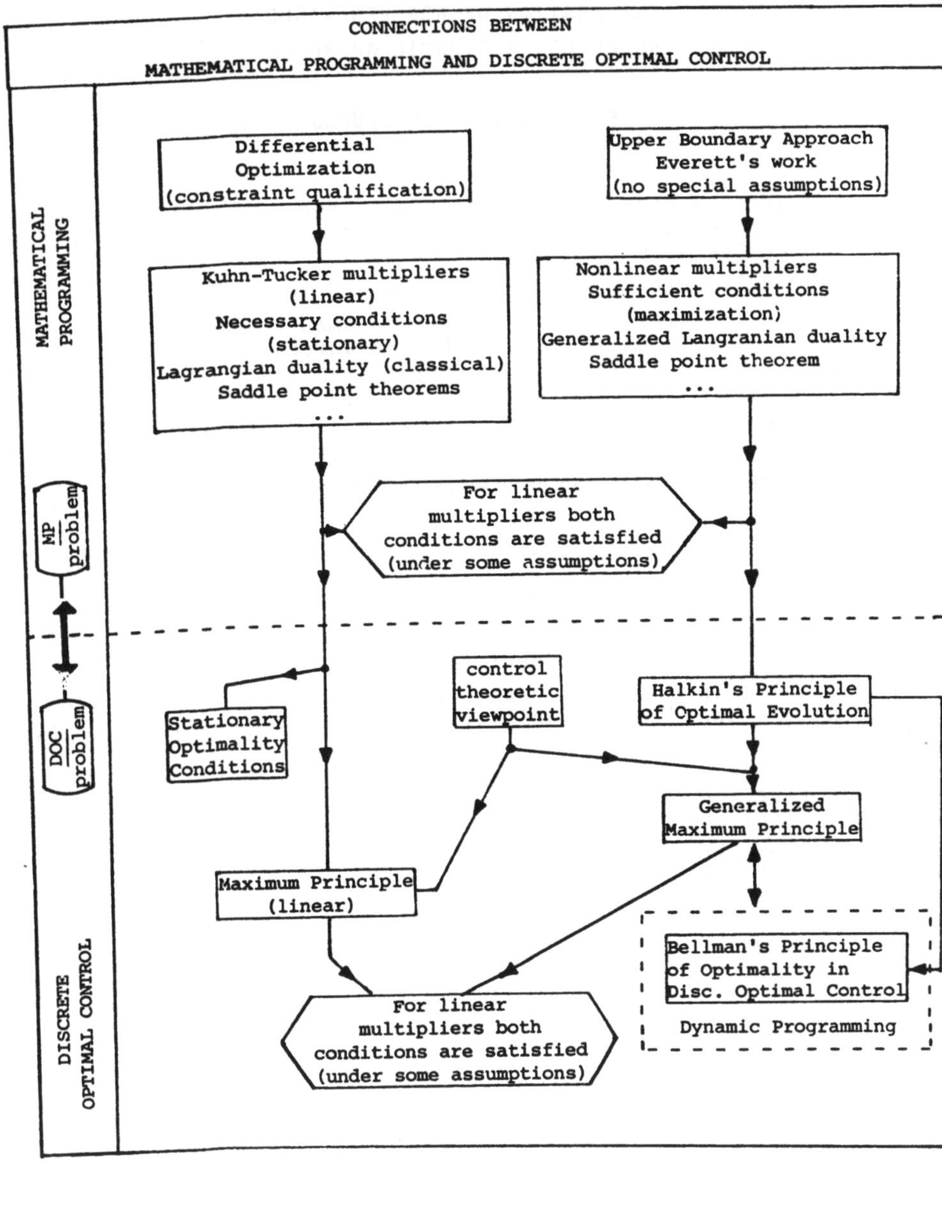
CONNECTIONS BETWEEN
MATHEMATICAL PROGRAMMING AND DISCRETE OPTIMAL CONTROL
MATHEMATICAL PROGRAMMING
Differential Optimization (constraint qualification)
Upper Boundary Approach Everett's work (no special assumptions)
Kuhn-Tucker multipliers (linear) Necessary conditions (stationary) Lagrangian duality (classical) Saddle point theorems ...
Nonlinear multipliers Sufficient conditions (maximization) Generalized Langranian duality Saddle point theorem ...
MP problem
For linear multipliers both conditions are satisfied (under some assumptions)
DOC problem
Stationary Optimality Conditions
control theoretic viewpoint
Halkin's Principle of Optimal Evolution
Generalized Maximum Principle
Maximum Principle (linear)
Bellman's Principle of Optimality in Disc. Optimal Control
Dynamic Programming
For linear multipliers both conditions are satisfied (under some assumptions)
DISCRETE OPTIMAL CONTROL

EXPLORATION OF NON RENEWABLE RESOURCES A DYNAMIC APPROACH

SAUL FUKS - COPPE/UFRJ

ABSTRACT

Arrow studies a model of exploration of non renewable natural resources, in continuous time (see |4|), where, the quantity of resources, to be found is a random variable.

In our model there is a T planning period to go.

If society consumes, in one period, an amount c of non renewable resource, it receives a utility u(c). New resources can be found through exploration, α reflecting the effort applied to get them, spending p per unit of effort.

The number of units found when effort α is employed is given by a random variable w_α, having a Poisson distribution with parameter $\lambda\alpha$.

Society is then faced with the problem of maximizing the descounted total utility:

$$V^T(y) = \text{Max}\left[u(\alpha) - p\alpha + \delta E\, V^{T-1}(y - c + w_\alpha)\right]$$

Subject to: $o \leq c \leq y \quad , \quad \alpha \geq 0$

where y is the stock of resources available at the begining of the initial period and δ is the discount factor.

The new fact in the above problem, in relation with similar ones in literature, is that the probability distribution is defined by choice variable.

Concavity plays a major role in thes kind of model, it can be proved that, under some conditions [related to decreasing risk aversion for u'(c) and u"(c)], V(y) is a concave function.

The main result is that there exists $\bar{y}$ such that for $y > \bar{y}$, $\alpha=0$ (there is no exploration).

The Model

We will assume that the utility that Society obtains for consumption is expressed by $u(c)$, where c is the amount of the non renewable resource used. New resources can be found by exploration.

If α is the effort applied in order to obtain new resources, then the number of units found is given by a random variable w_α that has a Poisson distribution with mean $\lambda\alpha$, where λ is a scale parameter and the probability of finding m units in given by.

$$P\left[w_\alpha = m\right] = \frac{(\lambda\alpha)^m e^{-\alpha m}}{m!}$$

The society has a T-period planning to go and has to decide at each period how much to consume and how much effort to apply looking for new resources, in order to maximize the total utility accumulated in the T periods.

The problem thus is to Max $E \sum_{t=0}^{T} \delta^{T-t}\left[u(c) - p\alpha\right]$

where: $y_{t-1} = x_t + w_\alpha^t$

$$y_t = x_t + c_t, \; c_t \geq 0 \quad , \quad x_t \geq 0 \text{ and } \alpha_t \geq 0$$

We will denote:

$V^T(y)$ = best that can be done (indirect utility) if we have T periods to go and we begin with y (resources).The dynamic Programming formulation is:

$$V^T(y_t) = \text{Max}\left[u(c) - p\alpha + \delta E\, V^{T-1}(y_T - c_T + w_{\alpha_T})\right]$$

$$0 \leq c_T \leq y_T \quad , \quad \alpha \geq 0$$

We are assuming that $u(c)$ is strictly concave, differentiable , strictly increasing and bounded.

In addition we will assume that $V^T(y)$ is concave (see in | 9 | sufficient conditions for that).

Resources are assumed to be distributed randomly throughout a relevant area and they can be discovered by costly exploration.

It can easily be proved that the random variable w_α is stochastically increasing with α, i.e. if we increase the effort α, the probability to get a quantity of resources greater than k increases.

The cost of exploration p will be supposed constant and independent of t and y.

In pratice the exploration cost may rise as reserves decline, although the exact relation between the two may be complex (see |8|).

The optimality equations are:

$$u'\left[c(y)\right] \leq \delta\, E\, V^{T-1}(x+w_\alpha) = V'^{T}(y) = p^t(y) \qquad \text{(a)}$$

with equality if $0 < c(y) < y$

$$P > \delta\lambda\, \Delta V^{T-1}(x,\alpha), \text{ with equality if } \alpha > 0 \qquad \text{(b)}$$

where $\Delta V^{T-1} = E\left[V^{T-1}(x + w_\alpha + 1) - V^{T-1}(x + w_\alpha)\right]$

The economic interpretation of the equality case in (a) is that the shadow price of consumption must equal the shadow price of holding reserves; and in (b) is that the unit cost of exploration (left hand side) must equal the discounted gain in total utility (a finite increment) due do finding one more unit of the resource (measured by ΔV^{T-1} which is equal the difference in expected value of V(y) when y is increased by one unit).

T h e o r e m 1: ΔV^{T-1} is a decreasing function of x

P r o f:

$$EV^{T-1}(x + w_\alpha) = \sum_{i=0}^{\infty} \frac{(\lambda\alpha)^i c^{-\lambda\alpha}}{i!} V^{T-1}(x + 1) \qquad (1)$$

$$EV^{T-1}(x+w_\alpha+1) = \sum_{i=0}^{\infty} \frac{(\lambda\alpha)^i e^{-\lambda\alpha}}{i!} V^{T-1}(x+i+1)$$

subtracting (1) from (2) we get

$$\Delta V^{T-1} = \sum_{i=0}^{\infty} \frac{(\lambda\alpha)^i e^{-\lambda\alpha}}{i!} \left[V^{T-1}(x+i+1) - V^{T-1}(x+i)\right]$$

We will denote

$V^{T-1}(x+i+1) - V^{T-1}(x+i)$ by A.

Since $V^{T-1}(y)$ is supposed concave, A is decreasing with x and so is ΔV^{T-1}.

Q.E.D.

So the economic intuition is correct, with the increasing of the reserves, the marginal difference ΔV^{T-1} must drop, (we will see later that x increases with y).

Theorem 2: $c_t(y)$ is an increasing function of y.

Prof:

Let $y' \geq y$.

We will show that $c_t(y') \geq c_t(y)$.

[The proof depends on the signs of the values that $c_t(y')$ assumes ; for simplicity we will consider only one case; the other cases have similar proofs]

Let $c_t(y') \geq 0$ and $c_t(y) \geq 0$, from the optimality conditions it follows:

$$p^t(y') = u'\left[c_t(y')\right]$$

$$p^t(y) = u'\left[c_t(y)\right]$$

Then by monotonicity of $p^t(y)$,

$$u'\left[c_t(y')\right] \leq u'\left[c(y)\right]$$

So $c_t(y') \geq c_t(y)$, by monotonicity of $u'(c)$. Q.E.D.

Theorem 3: $x_t(y)$ is an increasing function of y.

Proof:

$$V^T(y) = \text{Max}\left[u(c) - p\alpha + \delta EV^{T-1}(x+w_\alpha)\right] = g\left[y, x(y), \alpha(y)\right]$$

$$x+c=y \ ; \ c \geq 0, \quad \alpha \geq 0 \quad x \geq 0$$

$$g_1 = \frac{\partial g}{\partial y} = u'(y-x)$$

$$g_2 = \frac{\partial g}{\partial x} = -u'(y-x) + \delta EV^{T-1}(x+w_\alpha)$$

$$g_3 = \frac{\partial g}{\partial \alpha} = -p + \delta\lambda \ EV^{T-1}$$

$$g_{11} = \frac{\partial g_1}{\partial y} = u''(y-x) \qquad g_{21} = \frac{\partial g_2}{\partial y_1} = -u''(y-x)$$

$$g_{12} = \frac{\partial g_1}{\partial x} = -u''(y-x) \qquad g_{22} = \frac{\partial g_2}{\partial x^2} = u''(y-x) + EV''^{T-1}(x+w_\alpha)$$

$$g_{13} = 0 \qquad g_{23} = \frac{\partial g_2}{\partial \alpha} = \delta\lambda E \left[V'^{T-1}(x+w_\alpha+1) - V'^{T-1}(x+w_\alpha) \right]$$

$$g_{31} = \frac{\partial g_3}{\partial y} = 0$$

$$g_{32} = g_{23} = \frac{\partial g_3}{\partial x}$$

$$g_{33} = \frac{\partial g_3}{\partial \alpha} = \lambda^2 \delta E \left[V^{T-1}(x+w_\alpha+2) - V^{T-1}(x+w_\alpha+1) - (V^{T-1}(x+w_\alpha+1) - V^{T-1}(x+w_\alpha)) \right]$$

$$g_{11} = -g_{21} \qquad g_{22} = -g_{21} + EV''^{T-1}(x+w_\alpha)$$

Using the envelope theorem

$$(1) \qquad \begin{aligned} g_{12} + g_{22}\frac{dx}{dy} + g_{23}\frac{d\alpha}{dy} &= 0 \\ g_{31} + g_{32}\frac{dx}{dy} + g_{33}\frac{d\alpha}{dy} &= 0 \end{aligned}$$

Solving system (1) we get:

$$\frac{dx}{dy} = \begin{bmatrix} g_{22} & g_{23} \\ g_{32} & g_{33} \end{bmatrix}^{-1} \begin{bmatrix} g_{21} \\ g_{31} \end{bmatrix} \text{ since } \begin{bmatrix} g_{22} & g_{23} \\ g_{32} & g_{33} \end{bmatrix}^{-1} = -\frac{1}{\Delta} \begin{bmatrix} g_{33} & -g_{23} \\ -g_{23} & g_{22} \end{bmatrix}$$

$$\frac{dx}{dy} = -\frac{1}{\Delta}(g_{33}\ g_{21} - g_{23}\ g_{31})$$

where $\Delta = g_{22}\ g_{33} - (g_{23})^2 > 0$

$$\frac{dx}{dy} = -\frac{g_{33}\ g_{21} - g_{23}\ g_{31}}{\Delta} \gtrless 0$$

since $g_{31} = 0$

$$\frac{dx}{dy} = -\frac{g_{33}\ g_{21}}{\Delta}$$

and $g_{33} \lesseqgtr 0$

$$\frac{dx}{dy} \gtreqless 0$$ Q.E.D.

Both Theorems (2) and (3) are economically intuitive; they say that as resources increase, consumption and also saving will increase (i.e. $\Delta y \geq 0$ implies $\Delta c \geq 0$ and $\Delta x \geq 0$.

It can be shown that if u(c) is strictly concave, increasing and bounded then $\lim_{c\to\infty} u'(c) = 0$

Our main result is to obtain a bound for y such that if $y > \bar{y}$ there is no exploration (i.e., $\alpha = 0$) stated in | 4 |(for the continuous case).

The following steps are:

1) If $y \to \infty$ then $c_t(y) \to \infty$ and $x_t(y) \to \infty$

2) If $y \to \infty$ then $\Delta V^{T-1} \to 0$

3) There exists $\bar{y}$ s.t. if $y > \bar{y}$ then $\alpha = 0$

Lemma 1 - If u (c) is bounded then $V^T(y)$ is bounded

Proof:

$$V^T(y) = \sum_{T=0}^{T} u(c) - p\alpha, \ \alpha \geq 0$$

Since u(c) is bounded the result follows immediately. If $T \to \infty$ it is also true, since if B is a bound on u(c), for $0 < \delta < 1$

$$V^T(y) < B(1+\delta+\delta^2+\ldots) \quad \text{so}$$

$$V^T(y) < B\frac{1}{1-\delta}.$$ Q.E.D.

Lemma 2 - If u(c) is bounded then $EV^T(y)$ is bounded.

Proof:

If u(c) is bounded then $V^T(y)$ is bounded (by Lemma 1). Let B be a bound for $V^T(y)$

$$EV^T(y) < \sum_{i=0}^{\infty} P_i B < B \sum_{i=0}^{\infty} P_i < B$$

Since $\sum_{i=0}^{\infty} P_i = 1$ (where P_i is the probability to find one mine with i units of resources). Q.E.D.

Lemma 3 - If $y_t \to \infty$ and $c_t(y) \to \bar{c}(y)$ then $\alpha(y) \to \bar{\alpha}(y)$

Proof:

$$V^T(y_t) = \max \left[u(c_t) - p\alpha_t + EV^{T-1}(y_t+w) \right]$$

$$x_t + c_t = y_t$$

$$c_t \geq 0, x_t > 0, \alpha_t > 0$$

$y_t \to \infty$ and suppose $c_t \to \bar{c}$ then $u(c_t) \to u(\bar{c})$ by continutity of $u(c)$. If $y_t \to \infty$ then $V^T(y_t) \to A$ since $V^T(y_t)$ is bounded and a non decreasing function of y_t.

If $y_t \to \infty$ then $EV^{T-1}(x_t+w) \to B$ by Lemma 2 since $y_t \to \infty$ and $c_t(y) \to \bar{c}(y)$ implies $x_t \to \infty$ (since $EV^{T-1}(x)$ is increasing with x), so in the limit

$$A = u(\bar{c}) - \lim \alpha_t p + \delta B$$

$$\lim p\alpha_t = u(\bar{c}) - A + \delta B$$

$$\text{Let } u(\bar{c}) - A + \delta B = K$$

$$\text{so } p \lim \alpha_i = k$$

$$\text{and } \lim \alpha t = \frac{k}{p} = \bar{\alpha}$$

Q.E.D.

Lemma 4 - If $x_t \to \infty$ then $\Delta V^{T-1} \to 0$ where

$$\Delta V^{T-1} = EV^{T-1}\left[x_t+1+w\right] - EV^{T-1}\left[x_t+w\right]$$

Proof:

$$V^{T-1}(x_t+1) - V^{T-1}(x_t) \geq V^{T-1}(x_t+1+w) - V^{T-1}(x_t+w)$$

by concavity of $V^{T-1}(y)$

$$EV^{T-1}(x_t+1+w) - EV^{T-1}(x_t+w) \leq \sum_{j=0}^{\infty} \beta_j \left[V^{T-1}(x_t+1) - V^{T-1}(x_t)\right]$$

Since $\sum_{j=0}^{\infty} \beta_j = 1 \leq V^{T-1}(x_t+1) - V^{T-1}(x_t)$

but since $u(c)$ is strictly concave, strictly increasing and bounded, and $V^{T-1}(x)$ is bounded and concave, and $x_t \to \infty$ then $\Delta V^{T-1} \to 0$. Q.E.D.

Theorem 4: If $u(c)$ is strict concave, stricti increasing and bounded and if $y \to \infty$ then $c(y) \to \infty$.

See Proof in |9|.

Theorem 5: If $y_i \to \infty$ then $x_i(y) \to \infty$ [$u(c)$ is supposed bounded].

Proof: $V(y_i) = u(c_i) - P_i + EV^{T-1}(x_i + w)$

by absurd suppose $x_i \nrightarrow \infty$

So since (by Theorem 3) x_i is non decreasing with y_i this implies that

$$\lim_{y_i \to \infty} x_i = \bar{x}$$

Since $u(c)$ is supposed bounded.

$$\lim_{y_i \to \infty} V(y_i) = u(\infty) - p\alpha_i + EV^{T-1}(x+w)$$

Since $V^{T-1}(y_i)$ is bounded

$$\lim_{y_i \to \infty} V^{T-1}(y_i) = T, \; u(\infty) = K$$

$T = K - p\alpha_i + EV^{T-1}(\bar{x}+w)$

$\phi(\alpha_i) = EV^{T-1}(\bar{x}+w)$ is a function of α_i.

Let $T-K = A$

$A = \lim \phi\ (\alpha_i) - \lim p\alpha_i$

Since $EV^{T-1}(y)$ is strictly concave then $\phi(\alpha_i) - \alpha_i p$ is strictly concave and there exists a unique $\lim_{y \to \infty} \alpha_i = \bar{\alpha}$.

Since $V^T(y)$ is strictly increasing then $EV^{T-1}(y)$ is strictly increasing.

Let the solution of $V^T(y_i)$ be $\{c_i, x_i, \alpha_i\}$.

I will obtain a feasible solution better than the above.

We will fix Δc_r; $\Delta c_r < c_r$ for a given r, y_i such that for it

$\{c.$ $V^T(y_i) = u(c_i) - p\alpha_i + \delta EV^{T-1}(x_i+w)$

$y_i \to \infty$ for the feasible solution $(c_i - \Delta c_r, x_i + \Delta c_r, \alpha_i)$

the value of the function is

$u(c_i - \Delta c_r) - p\alpha_i + \delta EV'(x_i+\Delta c_r+w)$

Since $y_i \to \infty$, $x_i \to \bar{x}$ and since $V^T(y)$ is strictly increasing

$\lim EV^{T-1}(x_i+\Delta c_r+w) - EV^{T-1}(x_i+w) > \varepsilon\ (\varepsilon > 0)$

but

$u(c_i) - u(c_i - \Delta \dot{c}_r) < \varepsilon$, $\forall$ given ε

(where if $y \to \infty$ then $c_i \to \infty$) for $c_i > \bar{c}_i$ by continuity of u(c).

T h e o r e m 6: If $y_i \to \infty$ then $\Delta V^{T-1} \to 0$.

P r o o f:

Since $y_i \to \infty$ implies $x_i \to \infty$ (Theorem 5) and by previous Lemma $x_i \to \infty$ and $\Delta V^{T-1} \to 0$. Q.E.D.

L e m m a 6 - If y = 0 then $\alpha > 0$

P r o o f:

$$V^T(0) = \max_{\substack{x+c=0 \\ x>0,x>0,\alpha>0}} \left[u(c) - p\alpha + EV^{T-1}(x+w) \right]$$

Since y = 0 implies x = 0 and c = 0.

We will prove that $\alpha = 0$ cannot be part of the solution; for $\alpha = 0$, we get a feasible solution

$x = 0, \quad c = 0, \quad \alpha = 0$

The value of the function is

$u(0) - 0 + 0 = u(0)$.

We will obtain one value of α that give a better value of the function (with x = 0 and c = 0)

$u(0) - p\alpha + EV^{T-1}(0+w)$

This is a value of α such that

$- p\alpha + EV^{T-1(0+w)} > 0$

$EV^{T-1}(0+w) > p\alpha$

$EV^{T-1}(0+w) > V^{T-1}(0)$ since $V^{T-1}(y)$ is not decreasing sith y

$V^{T-1}(0) > p\alpha$, $\alpha < \dfrac{V^{T-1(0)}}{p}$ Q.E.D.

T h e o r e m 7: Exixts $y = \bar{y}$ such that if the reserves are greater than $\bar{y}$ exploration must cease (i.e. $\alpha = 0$).

P r o o f:

From the optimality equations:

1) $p > \delta\lambda\Delta V^{T-1}$ for $\alpha \geq 0$

2) $p = \delta\lambda\Delta V^{T-1}$ for $\alpha > 0$

We know that

3) $\frac{\partial y}{\partial \alpha} < 0$ (see |9|).

(that means if reserves increase the rate of exploration decrease) we know that (Lemma 4 and Theorem 5).

$y \to \infty$ implies $\to \infty$ and $\Delta V^{T-1} \to 0$

So for y sufficient large ΔV^{T-1} is sufficient small and $\alpha = 0$ by 3) and Lemma 6 there must exist $y = \bar{y}$, such that if $y < \bar{y}$ then $\alpha > 0$ and if $y > \bar{y}$ then $\alpha = 0$ ($\bar{y} \neq 0$).

Note that using 3) if $y \to \infty$ then α decreases and

$\lim_{y\to\infty} \alpha = \bar{\alpha} > 0$ Q.E.D.

BIBLIOGRAPHY

|1| Josef Stoer and Christoph Witzgall, Convexity and Optimization in Finite Dimensions, Springer-Verlag, 1970.

|2| R.T. Rockafellar, Convex Analysis, Princenton University Press, 1972.

|3| Funchell, Convex Cones, Sets and Functions, Princton University Notes, 1953.

|4| Kenneth Arrow, Optimal Pricing: Use and Exploration of Uncertain Natural Resource Stokes, first draft, Conference on National Resource, Princing, Trail Lake (1977).

|5| D. Gale, "Non Linear Duality and Qualitatie Properties of Optimal Growth in Integer and Non-Linear Programming", (with J. Abadie), Ed.Amsterdam, 1970.

|6| D. Gale and W.R. Sutterland, "Analysis of a one good Model of Economic Development", Math. of Decicion Science, G. Dantzig and F. Veinott), Part. 2, American Math, Society, 1968.

|7| Jack Schechtman, "Competitive Prices, Dynamic Programming under Uncertainty, a non Stationary Case", Operations Research Center, Berkeley, University of California, 1976.

|8| Robert S. Pindick, "The Optimal Exploration and Production of Nonrenewable Resources", Journal of Political Economy.

|9| S. Fuks, As Some Results about Exploration and Use of uncertain Natural Resource Stocks - Ph.D. Dissertation - Uc. Berkeley.

|10| K.J. Arrow and S. Chang, "Optimal Pricing, Use and Exploration of Uncertain Natural Resources Stocks: Harvard University Technical Report nº 31.

DYNAMIC AND STOCHASTIC OPTIMIZATION OF TRANSPORT NETWORKS

A. Garliauskas, A. Burkauskas, G. Lesinskas, R. Vygovskaya *

Institute of Mathematics and Cybernetics of the Lithuanian SSR Academy of Sciences

*Computing Centre of the USSR Academy of Sciences

Abstract

Optimization of complex developing transport networks is considered. For a deterministic dynamic problem a three-stage approach with iteration recounting is used. The dialogue system of optimization is realized by EC-computers. For a stochastic case an additional qualitative analysis is made. The conditions for the initial stochastic data are given to guarantee the convexity of constraints.

1. Introduction

A great number of objects in nature is presented in the form of complex systems. To them refer: transport-energetic, computer, informational, medico-biological, organizational-economic, behaviouristic, etc., systems.

As a rule, all these networks develop in the conditions of stochastic variation of parameters.

General algorithms for optimizing flows in static networks are reviewed in papers[1-3]. The problems of developing transport networks are presented in[4,5]. Analogous questions for specialized systems were solved in papers[6-10]. Papers[8,10] investigate networks taking into account reliability and uncertainty of a priori information. To the same class of problems papers[11,12] belong, where effective algorithms and programs for optimizing transport-energetic and other complex networks are proposed.

Naturally, a lot of papers is devoted to computer networks, informational, biological structures and medical systems. It is practically impossible to enumerate all of them.

All the problems connected with networks are devided into five general groups: conceptual, statical, stochastic, dynamic and the problems of formation and control of data bases (Fig. 1).

From the methodological point of view the conceptual problems can be solved on the basis of construction of different algebras, such as: algebra of networks sets, algebra of fuzzy networks sets, structural-functional, logical-heuristic and other approaches.

Statical problems can be different: deterministic for obtaining a flow, for choosing constructive parameters, stochastic with the properties of uncertainty, adaptivity and reliability.

Many of the problems mentioned are already solved, though stochastic aspects are generally considered for simple objects only. In this case an important question of information means, i.e. validity and reliability of statistical data arises. The methodology for the solution of problems with uncertainty conditions is far from being worked out.

The dynamic problems in their reliable and stochastic version appear to be especially difficult to solve in the case of complex networks.

The class of problems dealing with the formation and control of data bases is especially important for the solution of concrete problems in practice.

We shall concentrate mainly on setting and solving of the dynamic and stochastic problems of complex networks optimization.

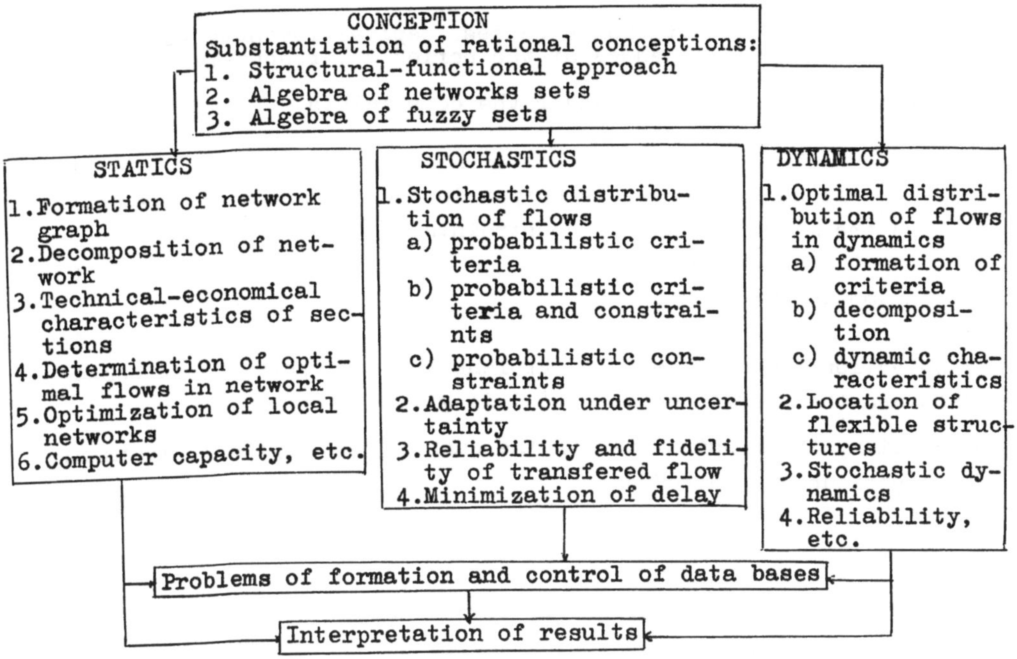

Fig. 1

2. Dynamics

For the deterministic dynamic problem a structure graph with surplus nodes of network, the number of calculation levels in dynamics, demands at the nodes, economical and some technological parameters are given. The objective is a discounted cost functional. While solving the nonlinear dynamic problem a three-stage approach with iteration recounting is used. At the first stage the dynamic equivalizing characteristics are determined, at the second one the optimization is proceeded, and at the third stage optimal constructive parameters are reproduced.

The minimization of the following nonlinear functional is considered:

$$\Sigma_{k\in K} (\Sigma_{i\in L} (E_i^k(I_i^k, P_i^k)+ C_i^k(\Delta P_i^k)+ \Sigma_{j\in V} D_j^k(d_j^k)) \tag{1}$$

where K set of optimization periods

L set of network sections

V set of network nodes

I_i^k flow through the i th section at the period k

P_i^k capacity of the i th section at the beginning of the period k

$\Delta P_i^k = P_i^k - P_i^{k-1}$ increment of the capacity

d_j^k selection of the j th node of the output at the period k

E_i^k, C_i^k, D_j^k exploitation expenses, capital investments and output expenses at the period k, respectively.

The functional is minimized under the constraints

$$I_i^k \leqslant P_i^k, \quad d_j^k \leqslant Q_j^k \tag{2}$$

where Q_j^k is maximal possible output in the j th node at the period k, and the material balance constraints

$$\Sigma_i \alpha_i (I_i^k) I_i^k = Q_j^k \tag{3}$$

for consumption nodes and

$$\Sigma_i \alpha_i (I_i^k) I_i^k + d_j^k = 0 \tag{4}$$

for selection nodes. A detailed explanation concerning these constraints and the coefficients $\alpha_i(I_i^k)$ is given below.

In addition, the constraints for the quantity of capital investments at the periods are introduced:

$$\Sigma_{i\in L} C_i^k(\Delta P_i^k) \leqslant B^k \tag{5}$$

In a general case the functional (1) for most cases of transport networks is not only nonlinear but it also consists of concave functions of equivalizing characteristics. However, due to the fact that the range of flow at every discrete step is not large and within small ranges the functions are convex, dynamic problems can be considered as piecevise convex and can be solved by linear programming methods jointly with iterative procedures. The convexity of characteristics is intensified by the fact that in constructing them there is a limited set of diameters or carrying capacities,and this smooths down the effect of technological progress expressed by the concave functions of reduced expenses.

In order to solve the above problem a dialogue-automated system is constructed. It is used for the analysis of system characteristics of complex network structures, decomposition of active networks, optimization of parameters and flows in statics and dynamics, and for solving stochastic, reliability, adaptation, statistical data accumulation problems.

In a general case at a given period of work the system deals with the following problems:

a) optimization of constructive parameters for the main-lined sections of transport networks in statics when five-year or long-term plans are drawn up.
b) optimization at the periods indicated in a) of parameters for the complex transport sections which develop in stages, and construction of dynamic characteristics for global network optimization in dynamics.
c) analysis and synthesis of reliability indices for the multichannel sections of a network with heterogeneous initial reliability indices of section elements.
d) optimization of single-product flows for a global transport network in prospect of seven discrete-time levels for a common planning period.
e) optimization of multi-product flows in analogy with d).
f) description and coding of the primary technico-economic information by sections and of common system indices with the purpose of construction of a system for data bases control.
g) organization of the dialogue part of the system containing display control, semantic processing of consumer's answers, processing of erroneous situations, correction of the sequence of process stages.

In future this system will be supplemented with conceptual problems of networks development, stochastic problems, the problems of planning under uncertainty conditions, adaptive problems, reliability problems, problems of structural flexibility, vital capacity, etc.

As an example, the results of optimization in dynamics for conditional developing gas-producing networks at five planning periods are presented in the scheme (Fig. 2); a) b) c) give the initial data, and d) e) present the numerical results. Numbers in columns correspond to one node or one section in five planning periods.

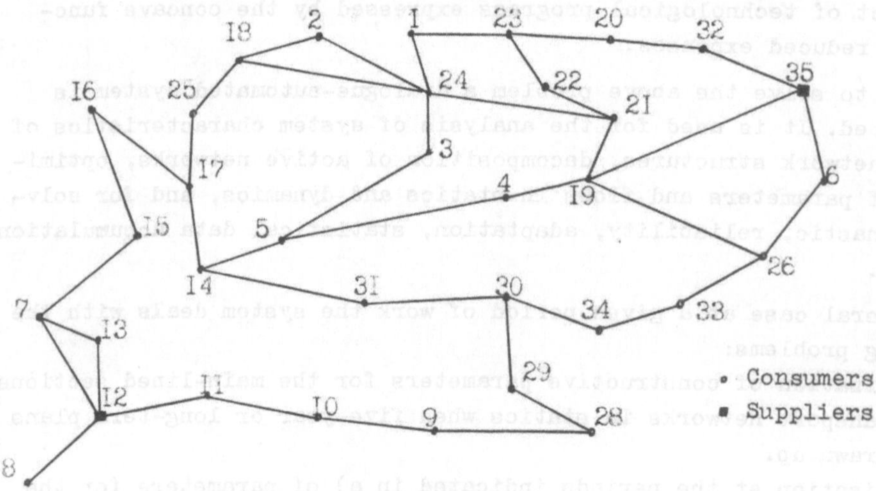

a) Network graph

I	2	3	4	5	6	7	8	9	I0	II	I2
I.3	2.5	0.3	2.0	0.4	39.4	5.9	66.4	9.0	I.0	0.6	-I.6
I.5	3.4	0.3	2.I	0.5	4I.6	6.2	76.7	9.4	I.2	0.6	-I.7
I.7	3.4	0.5	2.2	0.8	42.3	6.5	83.0	9.9	I.2	0.7	-I.6
2.0	3.7	0.7	2.3	0.9	43.7	6.5	85.I	I0.6	I.3	0.7	-I.4
2.4	3.8	0.8	2.7	I.0	45.8	6.7	87.3	I0.8	I.3	0.7	-I.3

I3	I4	I5	I6	I7	I8	I9	20	2I	22	23
I.2	3.8	2.6	I.4	0.0	3.9	4.0	0.6	0.0	3.0	I2.3
4.3	3.7	2.6	I.4	0.0	4.3	3.5	0.8	0.0	3.0	I2.8
4.3	4.3	2.8	I.5	0.0	4.9	3.6	0.8	0.0	3.2	I3.7
4.2	4.6	2.8	I.6	0.0	5.4	3.7	I.2	0.0	3.2	I4.0
4.0	5.2	3.0	I.6	0.0	5.8	3.8	I.5	0.0	3.3	I4.2

24	25	26	28	29	30	3I	32	33	34	35
0.3	2.I	2.3	II.0	0.8	0.5	I.3	3.7	I.8	2.3	-I95.
0.4	2.3	2.5	II.I	0.8	0.6	I.4	3.9	2.2	2.5	-2I5.
0.5	2.4	2.6	II.6	0.9	0.6	I.6	4.I	2.3	2.6	-230.
0.5	2.6	2.6	I2.I	0.9	0.6	I.7	4.6	2.3	2.7	-240.
0.6	2.7	2.8	I2.2	I.0	0.6	I.7	5.I	2.5	2.7	-250.

b) Demands at nodes

300 000 300 000 300 000 300 000 300 000

c) Investment upper bounds

7-I2	II-I2	8-I2	7-I5	I2-I3	I5-I6	I6-I7	4-I9	4-5	5-I4
55.80	I5.00	68.00	62.00	2.00	66.80	67.80	77.45	75.00	74.00
63.42	I5.00	77.45	75.95	2.00	79.65	8I.09	92.66	90.0I	89.0I
76.I3	I5.00	88.I5	88.86	2.00	93.I5	94.79	I07.67	I05.02	I04.02
76.I3	I5.00	88.I5	88.86	2.00	93.I5	94.79	I08.35	I05.40	I04.02
76.I3	I5.00	88.I5	88.86	2.00	93.I5	94.79	II2.I5	I08.78	I06.56

I4-I7	I7-25	I4-I8	2-I8	I8-24	3-5	3-24	I-24	I-23	22-23	2I-22	20-23	2I-24
70.00	3.50	0.I3	2.00	3.30	0.30	0.00	0.50	I.I0	0.00	4.50	I4.00	6.99
84.25	3.50	0.92	2.00	3.30	0.30	0.00	0.50	I.87	0.00	4.50	I5.0I	6.99
98.49	3.50	I.62	2.00	3.30	0.5I	0.00	0.50	I.87	0.00	4.50	I5.26	6.99
98.49	3.50	2.55	2.00	3.30	0.7I	0.00	0.50	I.9I	0.00	4.50	I6.28	6.99
98.49	3.50	2.55	2.00	3.30	0.8I	0.00	0.50	I.9I	0.00	4.50	I6.49	8.35

I9-2I	I8-25	I4-3I	30-3I	29-30	28-29	6-35	9-28	9-I0	I0-II	7-I3
I0.00	3.50	I.70	I.50	36.00	35.00	90.00	22.00	22.00	22.00	6.50
I0.0I	3.50	I.70	I.50	36.00	35.00	90.00	22.00	22.00	22.00	6.50
I0.0I	3.50	I.70	I.50	36.00	35.00	90.00	22.00	22.00	22.00	6.50
I0.2I	3.50	I.70	I.50	36.00	35.00	90.00	22.00	22.00	22.00	6.50
II.66	3.50	I.70	I.50	36.00	35.00	92.39	22.00	22.00	22.00	6.50

6-26	2-24	20-32	32-35	I9-26	30-34	33-34	26-33	I9-35
44.50	3.30	I5.60	I9.00	0.00	38.70	40.50	42.00	9I.50
44.50	3.54	I5.87	I9.83	0.00	38.70	40.50	42.00	I06.5I
44.50	3.54	I6.I2	20.28	0.00	38.70	40.50	42.00	I2I.52
45.62	3.72	I7.55	22.22	0.00	38.72	40.50	42.72	I22.65
56.22	3.82	I8.06	23.23	0.00	38.72	40.53	43.I2	I28.02

d) Carrying capacities

7-I2	II-I2	8-I2	7-I5	I2-I3	I5-I6	I6-I7	4-I9	4-5	5-I4
54.62	II.09	-67.05	-6I.74	0.00	-65.24	-66.67	-77.45	74.99	74.00
63.42	I0.62	-77.45	-75.95	-2.0	-79.65	-8I.09	-92.66	90.0I	88.86
72.97	9.54	-83.8I	-83.80	0.00	-87.8I	-89.35	-I04.42	I0I.60	99.9I
75.60	9.23	-85.93	-86.33	0.00	-90.38	-92.02	-I08.35	I05.40	I03.39
76.I3	9.03	-88.I5	-88.86	-2.0	-93.I5	-94.79	-II2.I5	I08.78	I06.56

I4-I7	I7-25	I4-I8	2-I8	I8-24	3-5	3-24	I-24	I-23	22-23	2I-22	20-23
70.00	2.63	0.I3	0.00	-3.30	-0.30	0.0	-0.2I	-I.I0	0.0	3.0I	I3.7I
84.25	2.32	0.92	0.I2	-3.30	-0.30	0.0	0.35	-I.87	0.0	3.0I	I5.0I
93.78	3.50	I.62	0.00	-2.27	-0.5I	0.0	-0.50	-I.2I	0.0	3.2I	I5.26
95.93	2.96	2.55	0.00	-2.57	-0.7I	0.0	-0.I0	-I.9I	0.0	3.2I	I6.28
98.49	2.72	2.55	0.00	-3.30	20.8I	0.0	-0.50	-I.9I	0.0	3.3I	I6.49

2I-24	I9-2I	I8-25	I4-3I	30-3I	29-30	28-29	6-35	9-28	9-I0	I0-II
6.42	9.43	-0.5I	0.00	I.3I	-33.98	-33.0I	-82.48	-2I.87	I2.78	II.73
6.99	I0.0	0.00	-0.09	I.50	-34.2I	-33.24	-86.03	-22.00	I2.5I	II.26
6.79	I0.0	-I.08	0.II	I.50	-34.32	-33.25	-87.I5	-2I.5I	II.52	I0.27
6.99	I0.2	-0.34	0.22	I.50	-35.32	-34.24	-89.68	-22.00	II.3I	9.97
8.35	II.7	0.00	0.22	I.50	-35.52	-34.34	-92.39	-22.00	II.II	9.77

7-I3	6-26	2-24	20-32	32-35	I9-26	30-34	35-34	26-33	I9-35
I.20	42.75	-2.52	-I4.37	-I8.I3	0.0	-35.88	38.29	40.I8	-9I.I8
6.3I	44.09	-3.54	-I5.87	-I9.83	0.0	-36.40	39.0I	4I.30	-I06.5I
4.3I	44.50	-3.42	-I6.I2	-20.28	0.0	-36.5I	39.22	4I.6I	-II8.4I
4.2I	45.62	-3.72	-I7.55	-22.22	0.0	-37.5I	40.33	42.72	-I22.65
6.0I	46.22	-3.82	-I8.06	-23.23	0.0	-37.7I	40.53	43.I2	-I28.02

e) Flows in sections

Fig. 2

3. Stochastics

In practical situations the definiteness of initial data is often doubtful. Considering their origin, we see that the parameters are the results of measurement, estimation, forecasting, i.e. they are sooner random variables by their nature. Taking it into account, in the sense of numerical solution of the problem, we need a certain decision concept for the formalization of the problem in terms of mathematical programming.

Despite of a concrete network structure of the problem, let us consider a general situation of the linear constraints

$$x \geqslant 0, \ Ax \leqslant b \tag{6}$$

with random matrix A and vector b . One of the most natural approaches is to require system functioning (i.e., satisfying the random constraints (6)) with the probability no less than a certain given reliability level p . Then we obtain a deterministic equivalent of the stochastic problem with the probabilistic constraints of the type

$$x \geqslant 0, \ P(Ax \leqslant b) \geqslant p \tag{7}$$

In this case we loose the linearity and, in general, the convexity of constraints. The latter, however, is necessary for a satisfactory application of numerical solution algorithms.

The aim of the following is to investigate the constraints set (7), particularly, the establishment of generalized concavity properties of the constraint function

$$F(x) = P(Ax \leqslant b) \tag{8}$$

The following considerations may serve as the basis of the approach to the solution of the problem:

1. A certain role may be played by the relative magnitude of location parameters (expectations) and scale parameters (variances) of random variables. In the limiting case, when the dispersions vanish, the random variables degenerate and the constraints set becomes a convex polyhedra.
2. Despite of the "bad" shape (in the sense of concavity) of the constraints function (8), the latter, having certain unimodality properties, may determine the convex level sets
$$x \geqslant 0, \ F(x) \geqslant p$$
provided the reliability level p is high enough.
3. These presumptions do not seem unnatural because in practical situations the absolute magnitude of random parameters is usually relatively larger than the measure of their dispersion, and the reliability level p is high, as a rule (usually close to unity).

Let us pass on to the formulation of the results (proofs and details can be found in[13]). Let the rows of random matrix (A,b) be independent normal vectors (a_i^T, b_i), $i=1,\dots,m$, with the numerical characteristics

$$m_i=Ea_i,\ \mu_i=Eb_i,\ Q_i=E(a_i-m_i)(a_i-m_i)^T,\ \sigma_i^2=Db_i,\ i=1,\dots,m .$$

Define $(n+1)\times(n+1)$-matrices in the following way:

$$\hat{M}_i=\begin{matrix} m_i m_i^T & -\mu_i m_i \\ -\mu_i m_i^T & \mu_i \end{matrix} \qquad \hat{Q}_i=\begin{matrix} Q_i & 0 \\ 0^T & \sigma_i^2 \end{matrix}, \quad i=1,\dots,m .$$

<u>Theorem 1</u>. If there exist positive numbers α_i such that the matrices $\hat{M}_i-\alpha_i^2\hat{Q}_i$ are copositive [1)], then the function (8) is logconcave on the set

$$\bigcap_{i=1}^{m} \{x \geqslant 0,\ x^T Q x \leqslant h(\alpha_i)\sigma_i^2\}$$

where $h(z)=z^2+z\varphi(z)/\Phi(z)$, $\Phi(z)$ and $\varphi(z)$ are normal distribution and density functions, respectively.

If determination of the logconcavity region of the constraints function in terms of dispersion characteristics is undesirable, one can use

<u>Theorem 2</u>. If there exist positive numbers α_i such that the matrices $\hat{M}_i-\alpha_i^2\hat{Q}_i$ are copositive, then the function (8) is logconcave on the set

$$\bigcap_{i=1}^{m} \{x \geqslant 0,\ m_i^T x \geqslant \mu_i-\sigma_i\alpha_i(1+h(\alpha_i))^{\frac{1}{2}}\}$$

In the case, when $\mu_i=0$, the copositivity of the matrices cannot be achieved. Then the following statement can be applied:

<u>Theorem 3</u>. If there exist $\alpha_i>1$ such that the matrices $m_i m_i^T-\alpha_i Q_i$ are copositive, then the function (8) is logconcave on the set

$$\bigcap_{i=1}^{m} \{x \geqslant 0,\ x^T Q_i x \leqslant \sigma_i^2(\alpha_i-1)\}$$

<u>Note</u>. To determine the logconcavity region of the constraints function, one can use a combination of Theorems 1-3, according to the numerical characteristics of each row-vector of random matrix (A,b) selecting such a characterization of the logconcavity region which gives the most complete intersection of the sets.

Finally, for arbitrary numerical characteristics of random variables we formulate the following statement:

<u>Theorem 4</u>. The function (8) is logconcave upon the level $F(x)=p_0$, where

$$p_0=\Phi(d^{-1}(\bar{\alpha})),\ d(z)=(h(z)+1)/(z^2(h(z)-3)),$$
$$\bar{\alpha}=\min_{1\leqslant i\leqslant m}\{\bar{\alpha}_i\},\ \bar{\alpha}_i=\max\{\alpha_i>0,\ Q_i-\alpha_i m_i m_i^T \text{ is p.s.d.}\}$$

---

1) A matrix A is copositive if $x^T A x \geqslant 0$ for $x \geqslant 0$.

<u>Note</u>. Similar concavity statements for the constraints function can be proved in the case of stable distributions of random variables, when the characteristic parameter α is between unity and two.

A possible application of the statements to the solution of problems with given numerical characteristics of random variables is presented in the flow-chart (Fig.3).

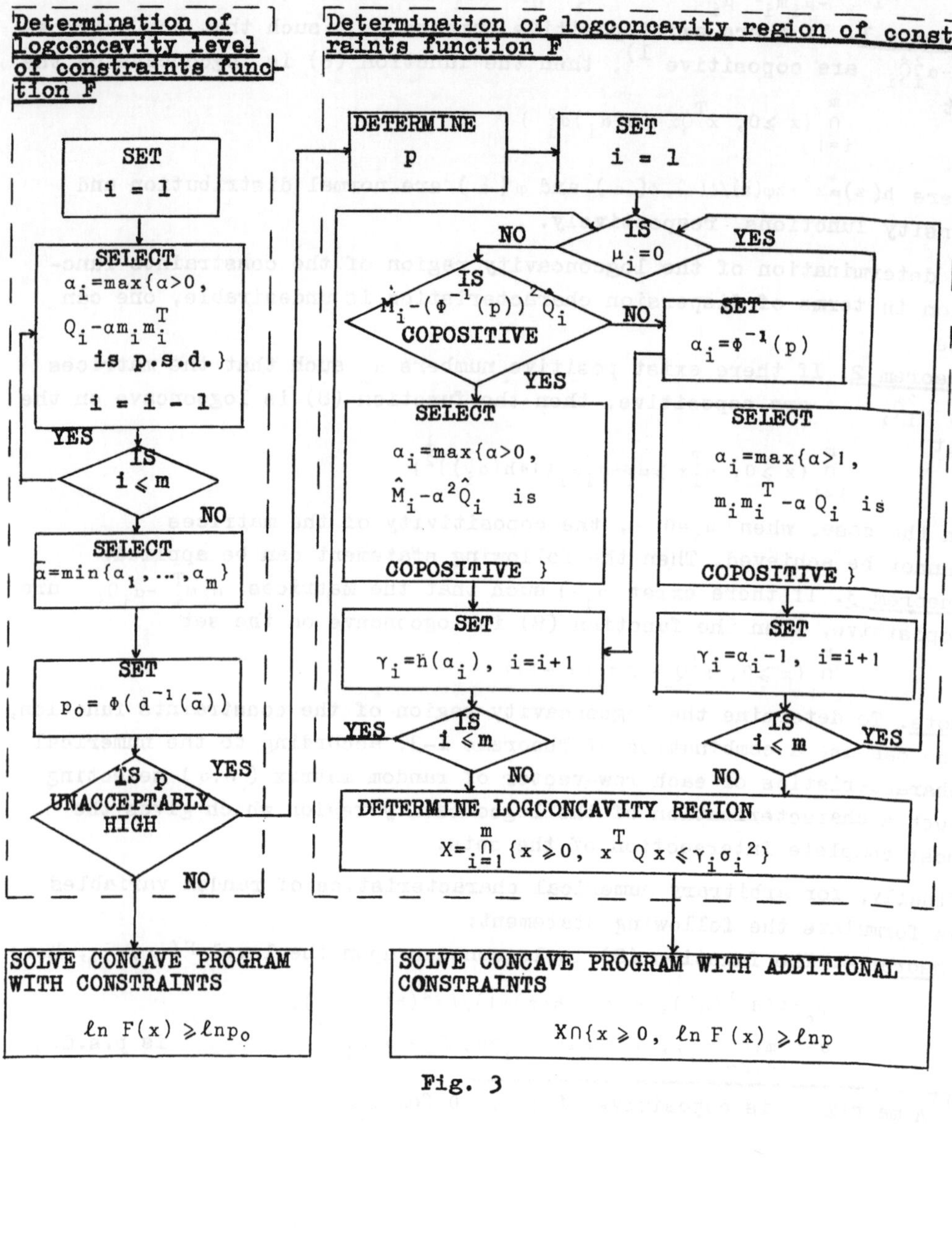

Fig. 3

References

1. Ford, L.R., Fullkerson, D.R., Flows in Networks, Princeton Univ. Press, Princeton, N.Y., 1962.
2. Garliauskas, A.I., Kalika, V.I., Farfel, S.Ya., Solution of Perspective Development Problems of Gas Supply Systems with Uncertain Initial Data (in Russian), Moscow, 1978.
3. Setting and Methods of Solution of Pipe-Line Transport Problems (in Russian), Ufa, 1978.
4. Tsvirkun, A.D., Foundations of Complex Systems Structure Synthesis (in Russian), Moscow, 1982.
5. Ivanishchev, V.V., Krasnoshchekov, A.D., Control of Networks with Varying Structure (in Russian), Moscow, 1982.
6. Garliauskas, A.I., Freygin, V.I., Dynamic Economico-Mathematical Models of Optimization of Integrated Gas Supply System (in Russian), Moscow, 1975.
7. SPOR System for Optimal Five-Year Planning of Gas Flows in ESG. General Description (in Russian), VNIIEgasprom, 1983.
8. Volski, E.L., Garliauskas, A.I., Gerchikov, S.V., Reliability and Optimal Reserves of Gas Extractions and Turnpike Gas-Lines (in Russian), Moscow, 1979.
9. Livshits, V.N., Optimal Decision Selection in Technological-Economical Calculations (in Russian), Moscow, 1971.
10. Garliauskas, A.I., Unanyan, L.A., Problems of Optimization of Regional Gas Supply Systems, in: Electronic Modelling, Kiev, 1981, No.3 (70-74).
11. Belyaeva, L.V., Biletski, V.I., Shor, N.Z., On Decomposition Algorithm for Selection of Optimal Railroad Profile (in Russian), Kibernetika, 1983, No.3 (76-79).
12. Mikhalevich, V.S., et al., Package of Applied Programs for Solution of Large-Scale Planning of Producing and Transporting (PLANER) (in Russian), Kibernetika, 1983, No.3 (57-71).
13. Burkauskas, A.A., On Convexity of Chance-Constraints in the Case of Normal- and Stable with $1 < \alpha < 2$ Laws (in Russian), LitNIITI, No.1376-85.

SIMULATION OF BIOMEDICAL SYSTEMS

A. Garliauskas, A. Garliauskienė

Institute of Mathematics and Cybernetics
Lithuanian SSR Academy of Sciences
Vilnius, USSR

Research Institute of Mother and Child
Care Lithuanian SSR Ministry of Health
Vilnius, USSR

ABSTRACT

Problems of simulation of systems from molecular level up to ecosystem of lake are considered. The way of modelling the immune system as a distributed one is proposed. The attempt of modelling complex processes of sepsis and rejection in kidney transplantation is carried out. The ecosystem of lake-cooler of nuclear station is modelled with a help of submodel system. Examples of numeral solution of problems of biomedical systems are presented.

1. INTRODUCTION

Biomedical systems belong to the class of systems which are especially complex and diffieult for formalize. The features as multilinkedness, inertness, dynamism, nonlinearity, stochastic, systematization are characteristic to such systems.

Mathematical modelling is of great importance in this case.

On the first stage it is used an auxiliary instrument in processing results of observations and experiments. On the second stage - for obtaining correlation. On the third - as a mean for analysis and synthesis of complex structural formation. On the fourth stage- for forecasting new biological regularities and unexplained phenomena.

The comparison of mathematical modelling results and real observations is the important problem in practice. There are feed back tasks of degree of reliability of received hypothesis, regulation of physiological and biological processes, stability, adaptivity, optimization, systemology and etc. However, for decision these problems it is necessary to integrate the forces of many specialists of different areas of sciences.

The more detailed description of biomedical systems is presented in following levels:

1) molecular level is studied on the grounds of a vast range of physical-chemistry, 2) cellular and clonal level is investigated on the basis of bioorganic chemistry, genetic engineering and bioelectrical chemistry, 3) level of tissues is explored with a help of bioenergetic chemistry and fermentative kinetics, 4) level of organs and functional system is synthesized with a help of cybernetic methods, methods for processing sensor information and controling visceral systems, 5) level of organisms is explored basing on the methodology of physiology of the higher nervous activity, theory of control, ecological physiology and human ecology, 6) level of species is studied on the grounds of mathematical genetics, theoretical ecology, ecological physiology, monitoring methodology, 7) level of biosphere is investigated with a help of systemology, biogeno - cenosis, general biology and sociology.

It is natural to use the vast range of mathematical methods in all mentioned above levels.

Hereafter, we shall pay attention mainly to the problems from the first four levels with the exception of the first two levels for the case of immune system of organism's regulation.

General problems of theoretical biology are considered in the work [1]. Notable contribution to the field of mathematical modelling of biomedical phenomena is carried in by Marchuk G.I. and his school [2-4]. Problems in the fields of ichtyology and ecology of lakes are dealt in the works [5-7], in the fieds of medicine and ecology - [8-15]. It is only few works mentioned above but they stimulate the expansion of future investigation in the field of applying mathematics in biomedical systems considerably.

2. MODELLING OF PROCESSES OF RECOGNIZING ANTIGENS IN IMMUNE SYSTEM

Today the immune system is generally modelled in a point representation.

Even on this level of representation of the immune system, many problems are to be solved, yet, i.e. simulation of processes of recognizing antigens by antibodies, sensibilization of lymphocytes, modelling of clonal formations, extention of the investigation of regulation mechanism and links with other systems and etc.

In our opinion, it is necessary to go over from point mathematical models to the extended description of all organs of the immune system and modelling from the position of network structures.

This allows us to have the modelling results much more exact, imitate processes of circulation between immuno- protective clones and antigens , eliminate centres of infection, prognose the immune response and tolerance.

The mechanism for recognizing heterologenous cells and molecules is explained, but not to the end. Five versions of recognition are known, but we look only into two principially different versions: the direct and double recognition of antigens . The definition of direct recognition is relative because we have a multi-step process.

In double recognition T_h - lymphocytes have receptors of two types: one of them finds "their own", i.e. identical according to reconsiliation of H-2 antigen, and the other recognizes virus if the first type of recognition took place (see fig.1).

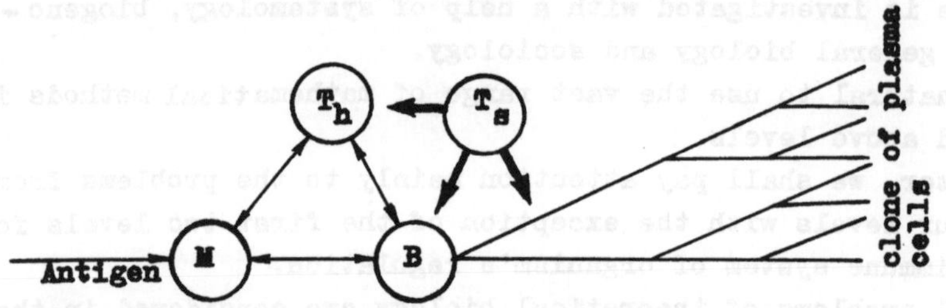

Fig.1. The double recognizing principle of antigens. M-macrophage, T_h-helper, T_s-supressor, B- lymphocytes.

Recognizing process is complex for mathematical modelling. In general case this process could be solved with a help of recognizing methods or nonlinear wave equations.

However, we can mention some of the approaches the modelling of basic and fully explored processes.

Concentration of B - lymphocytes in blood and lymph is changeable, therefore the diffusive processes are possible. And proliferation of lymphocytes corresponds to their reproduction. Besides, there are inductors of immunopoeze, i.e. outside powers have an effect uppon the process of proliferation.

Then dynamics of the generation process of plasmatic cells can be presented in a limited size in a form of the following equation:

$$\frac{\partial L(x,t)}{\partial t} = \frac{D}{C} \Delta L(x,t) + \alpha L(x,t) \qquad (1)$$

where L - concentration of B - lymphocytes in a set Ω at the moment t, D - diffusion coeficient, C - constant of porosity of boundaries, Δ - operator of second partial derivatives, α - reproduction coefficient.

If $\alpha = 0$, we have a pure diffusion.

If $\alpha < 0$, the equation (1) describes the diffusion process of lymphocytes, when their fusion rate is proportional to their concentration.

If $\alpha > 0$, the equation corresponds with the diffusion, if only the chain reaction takes place.

Futher, it is concretely ascertained in immunology that while recognizing the determinant of an antigen, a network of lymphocytes and antibodies is formed [5]. An apperance an antigen modulates the network.

There are data consisting the quantitive threshold values when an antigen sensitizes lymphocytes, i.e. it awakens net and suppresses the process of immunoglobulin producing.

Though it is a complex problem of presenting these processes in a mathematical form, but, obviously, it demands more attention from mathematics for the accelerated investigation of the theory of random graphs with elements threshold excitations, according to the type of representation of neuronal nets.

In a view that the elements of cooperation and competition between clones in their simultaneous and random interaction display, it is apparently, sensible to search for mathematical methods from the field of coalitional games in mixed stratagies and apply integral equations [15].

3. MATHEMATICAL MODELLING OF DISTRIBUTED PROCESSES OF BLOOD CIRCULATION

In our point of view together with the extention of investigation field it is necessary to work out a mathematical discription of the immune system in detail and move on to the modelling of the distributed dynamic systems of immune reactivity.

All necessary promises and a vast range of modern mathematical methods and powerful computers are available in this case.

More favourable conditions in comparison with the general problem of immune modelling are characterized to the mathematical modelling of blood circulation and flow of lymph, but there are many common, complex and unexplained to the end aspects.

Therefore, mathematical modelling can become of essential stimulus for clearing up all the details of action mechanisms of the organs from immune system.

Great quantity of works dealing with mechanics and hydrodynamics of blood circulation are generally systematized in the works[16,17].

Mechanics and hydrodynamics of blood circulation differs from analogical technical systems in a number of features. Elasticity of vessels which modifies the size, curvity, branching and compressing, influence of surrounding tissues are related to them.

The elasticity laws of Hook and Yung differ because of their nonlinearity, i.e. module of Yung is variable. The impulsivness of blood pumping is characteristic also, and as a result the waves of pressure and flow, reflection waves in branching places arise.

The complete mathematical discription of blood hydrodynamics on arteries when the area of cross-section and the strain of arteries change sharply is possible only if nonlinear effects would be taken into account.

Then the equations simultaneously into consideration the distributed outflow through vascular wall, are presented in a form:

equation of quantity of motion

$$\frac{\partial U}{\partial t} + U\frac{\partial U}{\partial x} + \left(\frac{1}{\rho}\right)\frac{\partial p}{\partial x} = f \qquad (2)$$

equation for mass conservation

$$\frac{\partial \varphi}{\partial t} + \frac{\partial U\varphi}{\partial x} + \delta Q = 0 \qquad (3)$$

function of properties of vascular wall

$$\varphi = \varphi\,(\bar{p}, \bar{p}_t, x) \qquad (4)$$

where f - a viscous power of inhibition depending on the gradient of pressure in arbitrary point x; δ - δ-function of the distributed outflow Q:

$$\delta = \begin{cases} 0 \text{ if } x < x_1 \\ 1 \text{ if } x_1 \leq x < x_2 \\ 0 \text{ if } x \geq x_2 \end{cases} \qquad (5)$$

where x_1, x_2 - fixed points of spatial coordinates, $\bar{p}$ - transmural pressure

$$\bar{p} = p - p_o\,(x,t) \qquad (6)$$

U (x,t) - rate of blood movement (it is constant in transverse section of a vessel);

p (x,t) - excess pressure, ρ -density,

φ (x,t) - area of transverse section of a tube.

For solving the system of equations initial conditions U(x,0), p(x,0) and boundary conditions, for example at the point x=0 corresponding to the outlet of aorta of heart or pulmonary artery, are set.

Motion processes in the main lymphoidic system can be described in analogy to the blood movements with a help of mathematical expressions (2)-(6).

More complicated from mathematical point of view are the processes in capillary and lymphatic vessels. Here, the transmission of mass bases upon processes of diffusion and convection and on transport through the boundary of phases: liquid-membrane-liquid. The motive powers for mass-transmission are gradients of concentration, pressure(an osmotic one, also), temperature and viscousness.

It is necessary to note that a number of objective laws of movement of substance stated here occur entirely in all other organs of the immune system, mentioned above in cellular and molecular levels.

Therefore, the distributed mathematical description of the immune system and using of mathematics in applied fields hold much of unknown, unexplained to the end.Though the first fragmental mathematical presentation is formulated; it is necessary to expand, improve and develop it.

Of great interest and importance appear applied fields of immunology in oncology, allergology, transplantology and etc.

We shall consider the problem of modelling immune processes in renal transplantation.

4. PROGNOSTICAL MATHEMATICAL MODELS IN RENAL TRANSPLANTATION

The system of mathematical models is created for solving different problems of renal transplantation: models of histocompatibility, models of sepsis - rejection, models of preoperation state, dynamic models of rejection mechanism and models of date bases.

The most important is the first group of models. It contains models for choosing recipient, i.e. which has the best compatibility with donor's kidney. True, an important role can play the order of priority, which is determined according to the patient's state,i.e. it is important if the patient is able to wait for next suitable case of compatibility.

Mathematical model of histocompatibility must evaluate a series of factors: inspection of indicators from the systems ABO and R_h, compatibility according to the system HLA. When the system HLA is checking, it is important to take into account not only simple inspection of compatibility between recipient's and donor's antigens , but to trace what preceeding antibodies(what antigens generated them) possesses recipient, what immune status he has, how many haemotransfusions were carried out, which transplantation is in turn.

These factors together with those of sex, transplantat's state, result of cross-serologic checking and etc., complicate the problem itself becomes unsurveyable.

This problem could be solved only with a help of probabilistic-combinative mathematical methods interpreted as a system of models.

As a big experience in renal transplantation is gained, so we shall dispose of apriori information about transplantat's survival in general and when depending on separate features.

However, for finding out the influence of the whole feature set to the survival, information, usually, is not sufficient, because the number of transplantations in one centre is limited to few hundreds.

In the case of independence of various features the predictive survival curves can be constructed on the basis of the Bayesian methods.

We shall have a look at two events:

1) the time T of transplantat's survival is longer(or equal) then the fixed time t;

2) and it will be rejected, if $T < t$.

The evaluation of probability for the first event in general case and independence with separate features also, can be found from statistical data:

$$F(t) = P(T \geqslant t) \tag{7}$$

$$q_{z_i} = P(T \geqslant t / z_i), \quad i = 1,2,\ldots, n \tag{8}$$

Using the Bayesian method we shall have finally formula:

$$q_{z_1,z_2\ldots,z_n} = \left[1 + \left(\frac{1}{F(t)} - 1\right) f_{z_1} f_{z_2} \cdots f_{z_n}\right]^{-1} \tag{9}$$

where

$$f_{z_i} = \frac{F(t)(1 - q_{z_i})}{[1 - F(t)]\, q_{z_i}}, \quad i = 1,2,\ldots,n;$$

z_i - features of compatibility, i - its number.

The transplantation centre of kidney in Vilnius had done over 450 operations. Prognosing survival curves is presented in fig.2 and 3. They got after proceeding of statistical data of the transplantation

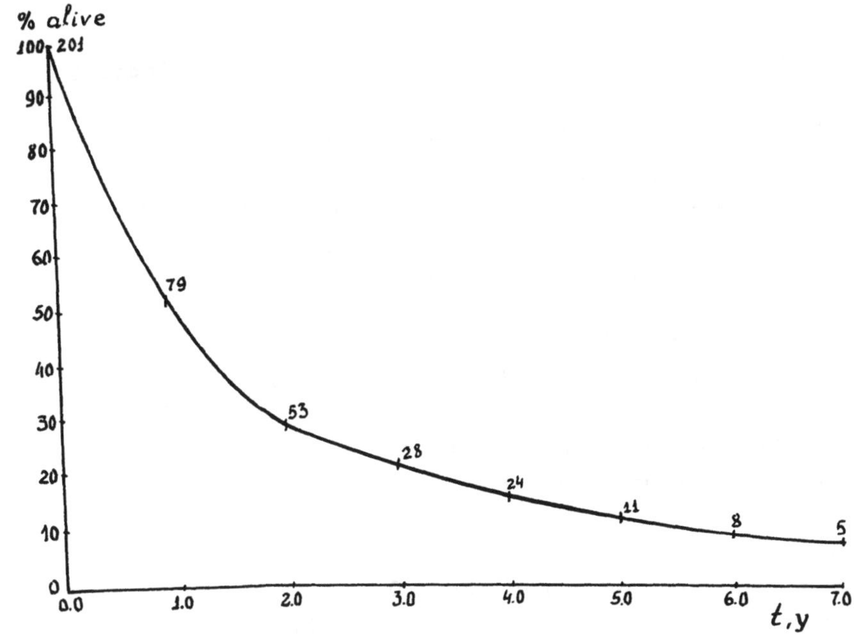

Fig. 2. General survival curve.
(Number above curve-quantity of alives).

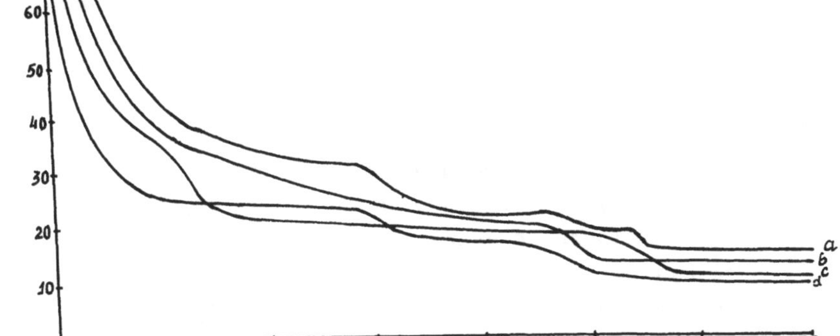

Fig.3. Prognosing survival curves on the different features. a-donor and recipient-men, groups of blood 0. b-men as a, donor-0, recipient-A. c-men as a, donor -A, recipient -0. d-number of HLA-2 antigens.

centre.

Table 1

Comparison of prognosing values q with real data

Group of features	Prognostic	Real
Donor and recipient - man, group of blood 0	0,53	0,61
Donor and recipient - man; donor-group of blood 0, recipient - group of blood A	0,6	0,54
Donor and recipient - man; donor - group of blood A, recipient - group of blood 0	0,48	0,52
Number of antigens of compatibility for groups A and B - 3; donor-group of blood 0, recipient - man	0,63	0,67

The comparison of real probabilities with prognosing ones for different features shows satisfactory results.

5. MATHEMATICAL MODELLING OF SEPSIS - REJECTION

Relying on the theory of immunology, morphological description of rejection mechanism, etiology and morphology of sepsis, some mathematical models are formed.

It is known that the infection processes arise when a microbe-antigen invades into organism. For "fighting" with them lymphocytes emerge, and after concrete differentiation they specialize on a certain antigen producing plasmatic cells. Forming of these cells takes place with delay because the cells are maturing. Analogically, transplantat's cells in the case when his compatibility is not full (ideally compatible are only the organs of monozygotic turns) become exactly like the heterologenous agents which are the objects of fight for the immune system.

These two phenomena can be linked with a help of nonlinear differential equations with delay.

It is necessary to note that equations of the type Hutchinson [10] or Voltera [11] are well known in mathematical biology.

The principle "predator - victim" according to its biological essence is not applied here, for in this case the predator perishes together with its victim. Therefore, in our opinion, "bee principle" occurs according to which a bee perishes (together with its victim) also. This principle, not speaking directly about it, is widelly used in the mathematical immune theory [2-5,9,21].

On the basis of this principle we describe processes of transplantat's rejection and sepsis as interconnected ones in a form of following system of differential equations:

$$\begin{aligned} \dot{V}(t) &= \alpha_1 V(t) - \alpha_2(a)L(t-\tau)V(t) - \mu_V V(t) \\ \dot{L}(t) &= \beta_1 E(t)L(t) - \beta_2(b)L(t-\tau)W(t) - \beta_3(b)L(t-\tau)V(t) - \mu_L L(t) \\ \dot{E}(t) &= \vartheta_1 \delta(L)E(t) - \vartheta_2 V(t) - \mu_E \delta(L)E(t) \\ \dot{W}(t) &= \gamma_1 E(t)W(t) - \gamma_2 \beta_2(b)L(t-\tau)W(t) \end{aligned} \qquad (10)$$

where the first member from the right side of the first equation means the reproduction rate of antigens V(t) when the growth coefficient is α_1; the second member - the member of "meetings" between antibodies L(t) and antigens V(t) with a certain probabilistic coefficient $\alpha_2(a)$, which depends upon the parameter of control a that imitates the dose of antibiotics; the third member is the rate of perishing of antigens arbitrary or on the basis of influence of other nonspecific protective mechanisms of immunity.

The second equation from system (10) characterizes the change rate that depends upon the number of erythrocytes E(t) (the first member from the right), number of transplantat's cells W(t) (the second member), number of antigens and on its proper perishing; coefficients $\beta_2(b)$ and $\beta_3(b)$ depend upon the dose of immune - inhibitor b.

The third equation from that system shows the change of quantity of erythrocytes in a transplantat depending on the quantity of antibodies (the first member), where δ- function imitates vessel occlusion, moreover:

$$\delta(L) = \begin{cases} a_1 & \text{if } L \leq L_c \\ -a_2 & \text{if } L > L_c,\ a_2 > a_1 \end{cases} \qquad (11)$$

where L_c - critical number of antibodies.

The second and the third members mean the reduction of receiving erythrocytes in the transplanted organ with the dependence on the number of antigens and on the proper death; ϑ_1, ϑ_2 and μ_E - corresponding coefficients of growth and death.

The last equation from system (10) shows the change of the number of transplantat's cells depending on the quantity of erythrocytes (the first member from the right) and of antibodies (the second one), γ_1, γ_2 - coefficients of growth and reduction, respectively.

Besides, the value $L(t - \tau)$ signifies the number of lymphocytes which mature in the corresponding antibodies. Therefore, the delay τ is put into for restricting the reduction rate of corresponding components.

This system constructs two mechanisms of rejection. One of them is linked up with the cessation of receiving blood in transplantat, and the other one characterizes the spontaneity of falling of transplantat's cell complexies - antibody. Both mechanisms are presented by the curves E(t) and W(t), respectively (see fig.4).

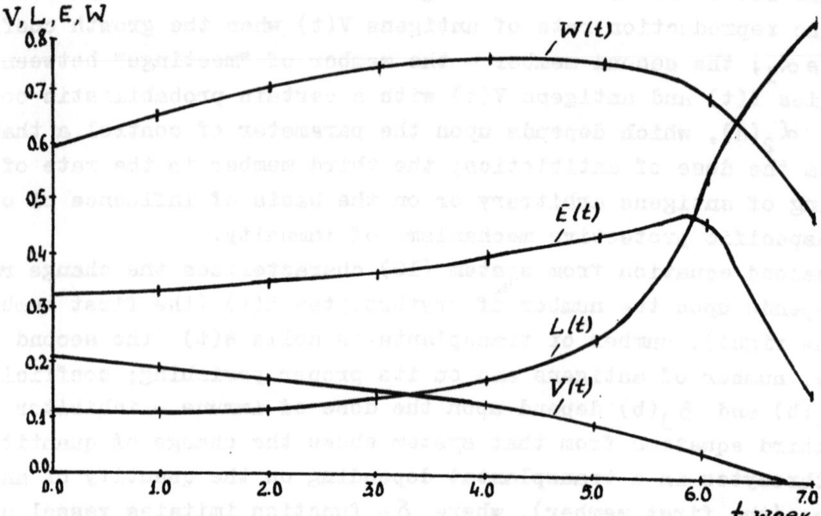

Fig. 4. Curves of dynamics of transplantat's rejection

After reaching certain critical level of antibodies the fast descent of transplantat's complexies of cells(curve W(t),fig.4) and thrombosis of erythrocytes (curve E(t)) occur. Fast rejection of a transplantat takes place at the moment $t > 6$. There is also shown that conditions for development of sepsis are unfavorable (curve

V(t)).

On the whole, the qualitative analysis of the most complex interactive mechanisms of the transplantating organs takes place.

In future, the quantitive comparison of the real course of postoperative process with the results of numerical simulation will be accomplished. This will help to construct complete system of mathematical prognosing of processes of transplantation of kidneys, vessels, skin and other tissue formations of human organism.

6. PROBLEMS OF MODELLING ECOSYSTEMS

A number of works in different levels of generalization, estimation of great number of links, system characteristics, influence of exactness of information are dedicated to the questions of modelling global and local ecosystems [6,18,19,20] .

Our aim is to look through some of arising problems of modelling ecosystems of a nature lake - cooler of nuclear electric station.

Problems of imitative modelling of a cooler of an electric station are analysed in the papers [7,22] . Quality models [22,23,24] , models concerning the change of temperature [25,26], stochastic models [20,27] and models of the second generation [28,29,30] are proposed also.

In order one could control ecosystem of region of nuclear station in different trophic levels and under effect of radiational, thermal and chemical influence, it is necessary to create a technically complex monitoring system. There is a block of models of radiational and chemical pollution of atmosphere, a block of analogical influence upon land by the lake of that region, a block of thermal effect of ecosystem of reservoir, a block of models of trophic interaction of biological populations. In future, a block of monitoring of rational heat employment, a general genetic monitoring, block of parisite complex, accumulation of radionuclides of plants and etc. will be developed.

The block of modelling of hydrothermal regime of reservoir-lake consists of models of a distribution of temperature in area and in depth when the boundary and initial conditions are set. Although the presented problem is not simple, especially in a spatial case, because of a complex configuration of the lake but it can be solved without playing attention to the fact that it is described as a system of differential equations with partial derivatives.

It is presented in a following form:

$$\left.\begin{aligned}
&\frac{\partial H\upsilon}{\partial t}+\frac{\partial H\upsilon^2}{\partial x}+\frac{\partial H V\upsilon}{\partial y}=-gH\frac{\partial\xi}{\partial x}-\frac{g}{\varrho_0}\frac{\partial\varrho}{\partial x}\frac{H}{2}-g\frac{\upsilon(\upsilon^2+V^2)^{1/2}}{c^2}+\frac{\tau_x}{\varrho_0}\\
&\frac{\partial HV}{\partial t}+\frac{\partial H\upsilon V}{\partial x}+\frac{\partial HV^2}{\partial y}+fHV=-gH\frac{\partial\xi}{\partial y}-\frac{g}{\varrho_0}\frac{\partial\varrho}{\partial y}\frac{H}{2}-\frac{gV(\upsilon^2+V^2)^{1/2}}{c^2}+\frac{\tau_y}{\varrho_0}\\
&\frac{\partial\varrho}{\partial t}+\frac{\partial H\upsilon}{\partial x}+\frac{\partial HV}{\partial y}=0\\
&\frac{\partial H\bar{P}}{\partial t}+\frac{\partial H\upsilon\bar{P}}{\partial x}+\frac{\partial HV\bar{P}}{\partial y}=\frac{\partial}{\partial x}\left(HD\frac{\partial\bar{P}}{\partial x}\right)+\frac{\partial}{\partial y}\left(HD\frac{\partial\bar{P}}{\partial y}\right)+[K]H\bar{P}+H\bar{F}\\
&\varrho=\varrho_0\,(1+\sigma_p\cdot 10^{-3})
\end{aligned}\right\}\quad(12)$$

σ_p = 28,14 - 0,0735 T - 0,00469 T^2 + (0,802 + 0,002 T) (S-35)
where T- temperature, S- salinity %, $\bar{P}=\begin{Bmatrix}P_1\\P_2\end{Bmatrix}$, P_1, P_2-averaged concentration values, h(x,y) - depth, ξ(x,y,t) - level of free surface W =(u, υ)-rate, H =h + ξ , $\upsilon=\frac{1}{H}\int_h^{\xi}u(x,y,z,t)dz$, $V=\frac{1}{H}\int_h^{\xi}\upsilon(x,y,z,t)dz$

f - Koriolise parameter, g - acceleration of the gravity force, ϱ_0 -average density, (τ_x , τ_y) = $\bar{\tau}$ -tension of wind friction, ϱ - density, c - coefficient by Shezies, $D=\begin{pmatrix}D_1&0\\0&D_2\end{pmatrix}$, D_1, D_2 - coefficients of turbulent diffusion,

$[K]=\begin{pmatrix}K_{11}&K_{12}\\K_{21}&K_{22}\end{pmatrix}$ - matrix of interaction,

$F=\begin{pmatrix}F_1\\F_2\end{pmatrix}$ - functions of sources or outflows.

The model of hydrothermal regimes is bounded with a block of ecosystems of reservoir through biogens and temperature. The model estimates also the gradient of wind, motion of surficial stratum, rate of mixing of water and other boundary conditions of the change of temperature.

The mathematical model of biosystem of the lake is described as a system of nonlinear differential equations:

$$\left.\begin{aligned}
&\dot{x}_i=b_{in}x_ix_n-x_i\sum_{j\in K_i}(1+e_{ij})b_{ij}x_j-g_ix_i^2-d_ix_i\,,\quad i=\overline{1,m}\\
&\dot{x}_i=x_i\sum_{j\in P_i}b_{ij}x_j-x_i\sum_{j\in K_i}(1+e_{ij})b_{ij}x_j-g_ix_i^2-r_ix_i-d_ix_i\,,\quad i=\overline{m+1,n-2}\\
&\dot{x}_{n-1}=\sum_{j=1}^{n-2}(g_jx_j^2+d_jx_j)-x_{n-1}\sum_{j\in K_{n-1}}b_{n-1,j}x_j-\alpha x_{n-1}+\sum_{i,j}e_{ij}b_{ij}x_ix_j
\end{aligned}\right\}\quad(13)$$

$$\dot{x}_n = \sum_{i=m+1}^{n-1} r_i x_i + \beta x_{n-1} - \sum_{j=1}^{m} b_{nj} x_n x_j$$

The first group of equations ($i=\overline{1,m}$) from (13) belongs to producers (macrophytes and phytoplanktons). The second group ($i=\overline{m-1, n-2}$) - to consumers (zooplankton,benthos), the third one (i=n-1) - to detritus and the fourth group (i=n) belongs to nutrients. In a real model the number of detritus and biogens is bigger than one.

In equations (13) following meanings are used: b_{ij}-coefficients of assimilated materials, e_{ij}-specific coefficients of methabolic excretion, r_i-coefficients of respiration, d_i-constant of natural mortality, g_i-coefficients of "interspecies" overcrowding, P_i- a set of species which are food for i ones, K_i- a set of species for which j^{th} ones are food, α -coefficient of loss of detritus, β -coefficient of decomposition of detritus to biogens.

Fragments of curves according to the ecological submodel (13) are shown in fig.5.

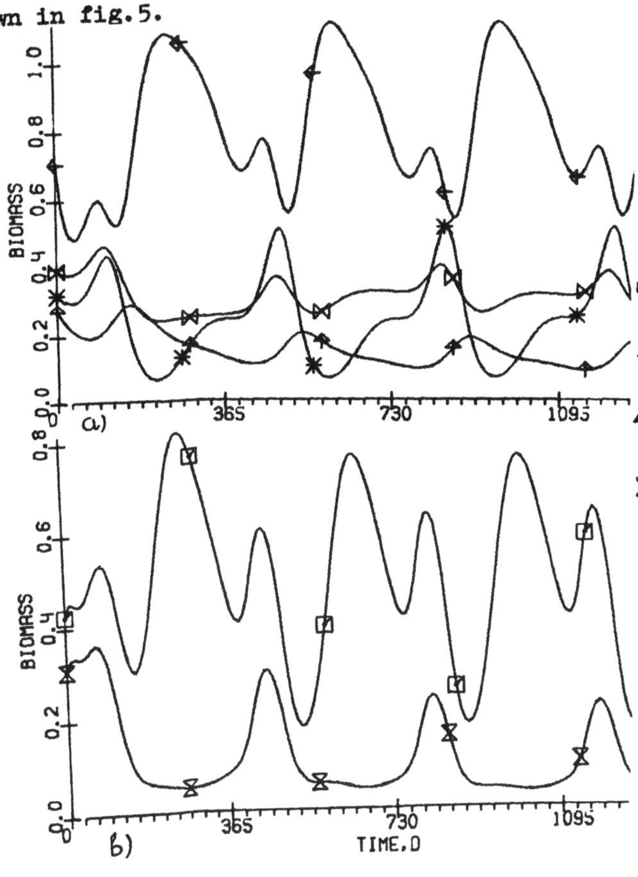

Fig.5a and b. Simulated seasonal changes in the ecosystem.
- bacteriaplancton,
- protozooplancton,
- blue phytoplancton,
- non-predator zooplancton,
- predator zooplancton,
- other phytoplanctons

The two maximums (in spring and autumn) for biomass are cleary seen from given curves (fig.5).

Next submodel of ecosystem of the lake is an ichtyo-model which includes two species of nonpredatory, two species of predatory fishes and four young species (fipes) of fishes:

$$
\left.\begin{aligned}
\dot{x}_1 &= a_{11}x_1 - a_{12}x_1\sum_{s=1}^{m_1}\beta_{1s}x_1(t-h_1-s) - a_{13}x_3x_1 - a_{12}x_1x_2 - a_{14}x_4x_1 - \\
&\quad - a_{15}x_1\sum_{s=1}^{m_1'}\gamma_{1s}x_1(t-h_1-s) - a_{16}\sum_{s=1}^{m_1''}\alpha_{1s}x_1(t-h_1-s) \\
\dot{x}_2 &= a_{21}x_2 - a_{22}x_2\sum_{s=1}^{m_2}\beta_{2s}x_1(t-h_2-s) - a_{23}x_3x_2 - a_{12}x_1x_2 - a_{24}x_2x_4 - \\
&\quad - a_{25}x_2\sum_{s=1}^{m_2'}\gamma_{2s}x_2(t-h_2-s) - a_{26}\sum_{s=1}^{m_2''}\alpha_{2s}x_2(t-h_2-s) \\
\dot{x}_3 &= a_{31}x_1x_3 + a_{32}x_2x_3 - a_{33}x_3\sum_{s=1}^{m_3}\beta_{3s}x_3(t-h_3-s) - a_{34}\frac{x_3x_4}{C_0} - \\
&\quad - a_{35}x_3\sum_{s=1}^{m_3'}\gamma_{3s}x_3(t-h_3-s) - a_{36}\sum_{s=1}^{m_3''}\alpha_{3s}x_3(t-h_3-s) - a_{37}\frac{x_3^2}{C_0} \\
\dot{x}_4 &= a_{41}x_1x_4 + a_{42}x_2x_4 - a_{43}x_4\sum_{s=1}^{m_4}\beta_{4s}x_4(t-h_4-s) - a_{44}\frac{x_3x_4}{C_0} - \\
&\quad a_{45}x_4\sum_{s=1}^{m_4'}\gamma_{4s}x_4(t-h_4-s) - a_{46}\sum_{s=1}^{m_4''}\alpha_{4s}x_4(t-h_4-s) - \\
&\quad - a_{47}\frac{x_4^2}{C_0}, \qquad C_0 = x_3 + x_4 \qquad \left.\begin{aligned}
&\sum_{s=1}^{m_i}\beta_{is} = 1 \\
&\sum_{s=1}^{m_i'}\gamma_{is} = 1 \\
&\sum_{s=1}^{m_i''}\alpha_{is} = 1
\end{aligned}\right\} i=\overline{1,4}
\end{aligned}\right\} \quad (14)
$$

Equations (14) are given only for adult fishes. Components of equations of this submodel are: members of growth of biomasses with coefficients a_{11}, a_{21}, a_{31}, a_{32}, a_{41}, and a_{42} ; members of maturing dependence on age with coefficients a_{12}, a_{22}, a_{33}, a_{43}, maturing ages h_i (i=1,4) and per cent of fishes of the same age (β_{is}); members of adult mortality due to fishing with a_{15}, a_{25}, a_{35}, a_{45} and γ_{is}; members of natural mortality of fishes with a_{16}, a_{26}, a_{36}, a_{46} and α_{is}; members of equations expressing the competition between victims due to overcrowding with coeffients a_{13}, a_{14}, a_{23}, a_{24}, a_{34}, a_{37}, a_{44}, a_{47} and a_{12}.

Fragments of simulated seasonal curves of submodel (14) are presented in fig.6.

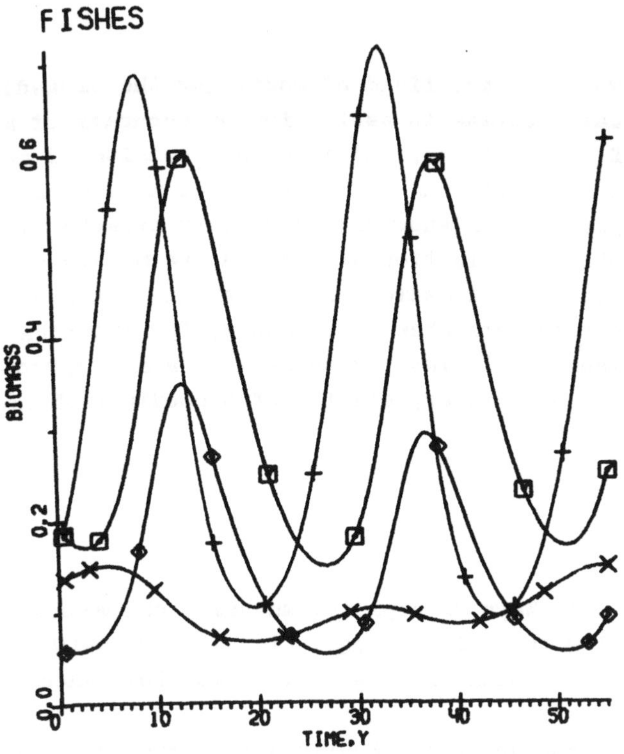

Fig.6. Simulated year changes of biomass of fishes. + -roach, x - bream, ⊡ - perch, ◇ - pike.

One can easily notice the periodicity, delay, interspecific competition and other important factors of local interspecific relations in dynamics of certain species of biomass.

At present the system of models is developed further by increasing the number of submodels and,also,considering all possible factors - informations gathering in purpose of comparison the results of modelling with real data.

7. CONCLUSION

Theoretical works in the field of modelling the biomedical systems raise actual problems in estimation of accuracy of models and accumulation of observations, and also in precision of measurement of experimental data. Testing of mathematical models on real statistical data gives final answers about their expediency. Even formulation of problems, selection of qualitative models and their preliminary usage in real biomedical processes gives a good outset for researches in various branches of sciences, for interchanging ideas between physicians, biologists, mathematicians and systemologists that leads to the best interpretation of biomedical phenomena in future.

REFERENCES

1. Mitropolski J.A. Ways of applying mathematical methods in biology.-Mathematical methods in biology.-Kiev,1977,p.84-106(in Russian).
2. Marchuk G.I. Mathematical methods in immunology.-Moscow,1980(in Russian).
3. Nisevich N.I.,Marchuk G.I., Zubikova I.I.,Pogozhev I.B.Mathematical modelling of viral hepatitis.-Moscow,1981(in Russian).
4. Marchuk G.I.Mathematical modelling in problem of environments.-Moscow,1982 (in Russian).
5. Garliauskas A.Mathematical modelling of biomedical systems.Mathematical models in biology and medicine.-Vilnius,1985,N.1,p.5-29 (in Russian).
6. Menshutkin V.V. Mathematical modelling of populations and communities of aquatic animals.-Leningrad,1971 (in Russian).
7. Park R.A. et al. Simulation,1974,vol.3,No 2,p.33-50.
8. Romanovski J.M., Stepanova N.V.,Chernavski D.S. Mathematical modelling in biophysics.-Moscow,1975 (in Russian).
9. Klein P., Hraba T., Doležal J. The use of immunological tolerance to investigate B lymphocyte replacement kinetics in chickens.-Math.Biology, 1983,vol.16,p.131-140.
10. Hutchinson G.E. Circular causual systems in ecology.-Ann.N.Y. Acad. Sci.,1948,vol.50,p.221-246.
11. Volterra V. Mathematical theory of struggle for existence.- Moscow,1976(in Russian).
12. Beili N. Mathematics in biology and medicine.-Moscow,1970(in Russian).

13. Marri G. Nonlinear differential equations in biology.-Moscow, 1983 (in Russian).
14. Antomonov J.G. Principles of neurodynamics.-Kiev,1974 (in Russian).
15. Glushkov V.M.,Ivanov V.V., Yanenko V.M. Modelling of developing systems.-Moscow,1983(in Russian).
16. Pedley T.J. The fluid mechanics of large blood vessels.-Cambridge Univ. Press, 1980.
17. Caro C.G., Pedley T.J., Schroter R.C., Seed W.A. The mechanics of the circulation.-Oxford Univ. Press, 1978.
18. Beliaev V.I. Theory of complex geosystems.-Kiev,1978(in Russian).
19. Mesarovich M. Theoretist's point of view.-System theory and biology.-Moscow,1971 (in Russian).
20. Fleishman B.S. Fundamentals of systemology.-Moscow,1982(in Russian).
21. Garliauskas A.,Griškevičiūtė A. Lesinskas G.,Šekštelienė N.Some aspects of prognosing the survival rate of a renal transplant.-Statistical problems of control.-Vilnius,1984,No 64,p.24-38 (in Russian).
23. Kellar H.N. Metabolism and model of an estaurine bay ecosystem affected by a coastal power plant."Ecol. modell.",1977,vol.3, No 2,p.85-118.
23. Smith G.M. Models in ecology.-Moscow,1976(in Russian).
24. Alekseev V.V. Achievments of physical sciences.-Moscow,1976, vol. 120, p. 647-676 (in Russian).
25. Krishev I.I. Mathematical models of dynamics of ecosystem of reservoir-cooler.- Experimental and field researches of biological grounds of productivity of lakes.-Leningrad,1979,p.131-146 (in Russian).
26. Babkin N.A. About dynamics of lake ecosystem.-Mathematical models in ecology.-Gorkyi,1980,p.3-17 (in Russian).
27. Tiwari J.L. at all. Some stochastic differential equation of an aquatic ecosystem.-Ecological Modelling.-Amsterdam,1978,No4, p.3-27.
28. Straškraba M. Natural control mechanisms in models of aquatic ecosystems.- Ecol. Modell.,1979,vol.6,No 4,p.305-321.
29. Fedra K. Modelling biological processes in the aquatic environment.- Working paper 79-20 (IIASA),1979.
30. Jorgenson S.E., Mejer H. A holistic approach to ecological modalling.- Ecol.Modell.,1979,vol.7,No3,p.169-189.

STABILITY OF SEMILINEAR SYSTEMS IN HILBERT SPACES

PIOTR GRABOWSKI
Inst.Contr.Engg.Syst.Sci.Telecom.
Academy of Mining and Metallurgy
30-059 Kraków,Mickiewicza 30,POLAND

1.Abstract semilinear systems.

Several problems of mathematical physics,control and circuit theories lead to the following semilinear problem

$$\dot{x}(t) = Ax(t) + B\,F[x(t)] \tag{1.1}$$

where $x(t)\in X$ for every fixed $t\geqslant 0$, X is a real Hilbert space with scalar product $\langle .,.\rangle_X$; $A : (\mathcal{D}(A)\subset X)\longrightarrow X$ is a linear operator,which is the generator of a linear C_0-semigroup on X ; U is another real Hilbert space with scalar product $\langle .,.\rangle_U$; $B\in\mathcal{L}(U,X)$.

It follows from the earlier results due to Segal [1],Pazy [2] and Ball [3] that for every initial condition $x_0\in X$ there exists a unique weak (mild) solution of (1.1) prolongable on its right maximal interval of existence $[0,t_{max}(x_0))$,provided that $F:X\longrightarrow U$ is a locally Lipschitz function.

The aim of this paper is to give sufficient conditions for global (as well as global uniform) asymptotic stability of the equilibrium $0\in X$, independently of F from the prescribed subclass of locally Lipschitz functions from X into U,vanishing at 0.

2.Main results.

The main results will be formulated as theorems and remarks.

Theorem 1.

Let $\mathfrak{M}$ be a class of functions such that :

(i) $\mathfrak{M}\subset\{$ P: P is a locally Lipschitz mapping from X into U, $P(0)= 0\}$,

(ii) There exist operators $Q\in\mathcal{L}(U,X)$, $M=M^*\in\mathcal{L}(X)$, $L\in\mathcal{L}(U,X)$, $K=K^*\in\mathcal{L}(U)$ such that :

(H1) $\forall F \in \mathfrak{M}$: QF is a gradient type operator,

(H2) $\forall x \in X$, $\forall F \in \mathfrak{M}$:

$$\langle x, -Mx\rangle_X + \langle x, LF(x)\rangle_X + \langle F(x), L^* x\rangle_U + \langle F(x), -KF(x)\rangle_U \geqslant 0 \qquad (2.1),$$

(H3) There exist $H=H^* \in \mathcal{L}(X)$ and a real positive ε such that

$$\langle Ax, Hx\rangle_X + \langle x, HAx\rangle_X + \langle x, -Mx\rangle_X + \langle x, HBu\rangle_X + \frac{1}{2}\langle u, Q^* Ax\rangle_U +$$
$$+ \frac{1}{2}\langle Q^* Ax, u\rangle_U + \langle u, B^* Hx\rangle_U + \langle x, Lu\rangle_X + \langle u, L^* x\rangle_U + \frac{1}{2}\langle u, Q^* Bu\rangle_U + \qquad (2.2),$$
$$+ \frac{1}{2}\langle u, B^* Qu\rangle_U + \langle u, -Ku\rangle_U \leqslant -\varepsilon[\|x\|_X^2 + \|u\|_U^2] \quad \forall (x,u) \in \mathcal{D}(A) \times U$$

(H4) If $F \in \mathfrak{M}_L = \mathcal{L}(X,U) \cap \mathfrak{M}$ then $A + BF$ generates an exponentially stable semigroup,

(H5) $\forall F \in \mathfrak{M}$, $\forall x \in X$, $x \neq 0$ $\exists \varkappa = \varkappa(x,F) \in \mathfrak{M}_L$:

$$\int_0^1 \langle x, Q[F(sx) - \varkappa sx]\rangle_X \, ds = 0 ,$$

Then the equilibrium $0 \in X$ is globally asymptotically stable (GAS) for every $F \in \mathfrak{M}$.

<u>Skech of the proof.</u>

Let us consider a continuously Fréchet-differentiable functional

$$V(x) = \langle x, Hx\rangle_X + \int_0^x \langle dy, QF(y)\rangle_X \quad \left(= \langle x, Hx\rangle_X + \int_0^1 \langle x, QF(sx)\rangle_X ds \right) \quad (2.3).$$

The second term in (2.3) is the antiderivative of the gradient operator QF. The assumptions (H1), (H2), (H3) allow us to prove the following inequalities :

$$\langle Ax + BF(x), V'(x)\rangle_X \leqslant -\varepsilon[\|x\|_X^2 + \|F(x)\|_U^2] \quad \forall x \in \mathcal{D}(A), \forall F \in \mathfrak{M} \qquad (2.4),$$

where $V'(x)$ denotes the gradient of V at x, and

$$V[x(t,x_0)] - V(x_0) \leqslant -\varepsilon \int_0^t \{\|x(\tau)\|_X^2 + \|F[x(\tau)]\|_U^2\} d\tau \quad \forall x_0 \in X, \quad \forall t \in [0, t_{max}), \forall F \in \mathfrak{M} \qquad (2.5).$$

If V is nonnegative on X for every $F \in \mathfrak{M}$ then we can prove that $t_{max} = +\infty$ and

$$\varepsilon \int_0^{+\infty} \{\|x(t)\|_X^2 + \|F[x(t)]\|_U^2\} dt \leqslant V(x_0) \quad \forall x_0 \in X, \forall F \in \mathfrak{M} \qquad (2.6).$$

The conditions assuring that $V(x) \geqslant 0$ $\forall x \in X, \forall F \in \mathfrak{M}$ can be derived using the linear comparison system technique. The linear comparison system arises from (1.1) by substitution of $\mu\in\mathcal{M}_L$ instead of F. Using the results known in the stability theory of linear systems defined on Hilbert space, it can be proved that $V(x)\geq 0$ $\forall x\in X$, $\forall F\in\mathcal{M}$ under the assumptions (H4), (H5). Now, (2.6) and the variation-of-constants formula yield the following estimation valid for all weak solutions :

$$\| x(t,x_0)\|_X \leq \bar{M}\|x_0\|_X\left[1+\|B\|\sqrt{\tfrac{2}{\varepsilon}}\max\{1,\|\mu\|\}\sqrt{\|H\|+\tfrac{1}{2}\|Q\|\,l_F(\|x_0\|)}\right],$$

where $\bar{M}\geq 1$ is a constant, $l_F : \mathbb{R}^+\longrightarrow \mathbb{R}^+$ is a continuous, strictly increasing function, appropriately constructed due to the fact that F is a locally Lipschitz mapping, $\mu\in\mathcal{M}_L$ is such that A+BF generates an exponentially stable semigroup (2.7).

The global uniform stability follows from (2.7) and the global attractivity follows from some basic properties of convolution operators in L^p spaces (it suffices to use p=2 plus a corollary from the Paley - - Wiener Theorem).

It turns out that we must restrict $\mathcal{M}$ to the class $\mathcal{M}^c$, described in the theorem given below, in order to obtain results concerning the global uniform asymptotic stability of the equilibrium $0\in X$. □

Theorem 2.

Let $\mathcal{M}^c$ be a class of functions such that :

(i) $\mathcal{M}^c\subset\{P$: P is a locally Lipschitz mapping from X into U, P maps every weakly convergent sequence in X into a strongly convergent sequence in U, $P(0)=0\}$,

(ii) There exist operators $Q\in\mathcal{L}(U,X)$, $M=M^*\in\mathcal{L}(X)$, $L\in\mathcal{L}(U,X)$, $K=K^*\in\mathcal{L}(U)$ such that :

(H1) $\forall F\in\mathcal{M}^c$: QF is a gradient type operator,

(H2) $\forall x\in X$, $\forall F\in\mathcal{M}^c$:

$$\langle x,-Mx\rangle_X+\langle x,LF(x)\rangle_X+\langle F(x),L^*x\rangle_U+\langle F(x),-KF(x)\rangle_U\geq 0 \qquad (2.8),$$

(H3) There exist $H=H^*\in\mathcal{L}(X)$ and a positive real ε such that

$$\begin{aligned}&\langle Ax,Hx\rangle_X+\langle x,HAx\rangle_X+\langle x,-Mx\rangle_X+\langle x,HBu\rangle_X+\tfrac{1}{2}\langle u,Q^*Ax\rangle_U+\\ &+\tfrac{1}{2}\langle Q^*Ax,u\rangle_U+\langle u,B^*Hx\rangle_U+\langle x,Lu\rangle_X+\langle u,L^*x\rangle_U+\tfrac{1}{2}\langle u,Q^*Bu\rangle_U+ \qquad (2.9),\\ &+\tfrac{1}{2}\langle u,B^*Qu\rangle_U+\langle u,-Ku\rangle_U\leq-\varepsilon\left[\|x\|_X^2+\|u\|_U^2\right]\quad\forall(x,u)\in\mathcal{D}(A)\times U\end{aligned}$$

(H4) If $F\in \mathfrak{M}^c_L = \mathfrak{M}^c \cap \mathcal{L}(X,U)$ then $A + BF$ generates an exponentially stable semigroup ,

(H5) $\forall F\in\mathfrak{M}^c, \forall x\in X, x\neq 0\ \exists \varkappa = \varkappa(x,F)\in \mathfrak{M}^c_L : \int_1^0 \langle x, Q[F(sx)-\varkappa sx]\rangle_X ds = 0$,

Then the equilibrium $0\in X$ is globally uniformly asymptotically stable (GUAS) for every $F\in\mathfrak{M}^c$.

Remark 1.

It can be proved that if the assumptions of Theorems 1 and 2 are satisfied with $Q = 0$ then even global exponential stability (GEXS) of the equilibrium $0\in X$ takes place.

Remark 2.

If A is the generator of an exponentially stable semigroup then the statements in Theorems 1 and 2 remain valid after eliminating the term "$-\varepsilon\|x\|^2_X$" from the right-hand side of (2.2) or (2.9).

Special problems of the form (1.1) have been considered in earlier papers [5-finite dimensional case], [6-time delay systems with $Q=0$], [7-the case $Q=0$], [8-the case $Q\neq 0$, but with A,B,F of very particular forms. The next theorem allows us to obtain more precise conditions for (GAS) or (GUAS), or (GEXS) ,provided that some additional assumptions are satisfied.

Theorem 3.

Let $\mathfrak{M}$ be a class of functions such that :

(i) $\mathfrak{M}\subset\{F : F$ is a locally Lipschitz mapping from X into $U, F(0)=0\}$,

(ii) There exist operators $Q\in\mathcal{L}(U,X), M=M^*\in\mathcal{L}(X), L\in\mathcal{L}(U,X), K=K^*\in\mathcal{L}(U)$ such that :

(H1) $\forall F\in\mathfrak{M}$: QF is a gradient type operator,

(H2) $\forall x\in X, \forall F\in\mathfrak{M}$:

$$\langle x,-Mx\rangle_X + \langle x, LF(x)\rangle_X + \langle F(x), L^*x\rangle_U + \langle F(x), -KF(x)\rangle_U \geqslant 0 \qquad (2.10),$$

(H3') There exists a triplet of operators (H,G,V), $H=H^*\in\mathcal{L}(X)$, G is a linear, A-bounded operator from X into U, $V\in\mathcal{L}(U)$ such that

$$\begin{cases} \langle Ax,Hx\rangle_X + \langle x,HAx\rangle_X + \langle x,-Mx\rangle_X = -\langle Gx,Gx\rangle_U & \forall x\in\mathcal{D}(A) \quad (2.11a),\\ B^*Hx + 0.5Q^*Ax + L^*x = -V^*Gx & \forall x\in\mathcal{D}(A) \quad (2.11b),\\ K - 0.5[Q^*B + B^*Q] = V^*V & (\text{in } \mathcal{L}(U)) \quad (2.11c),\end{cases}$$

(H6) There exists a positive real α such that for every $x \in X$ and $F \in \mathfrak{M}$ we have the following inequality :

$$V(x) \stackrel{df}{=} \langle x, Hx \rangle_X + \int_0^1 \langle x, QF(sx) \rangle_X \, ds \geqslant \alpha \, \| x \|_X^2 \qquad (2.12),$$

(H7) The semigroup $\{S(t)\}_{t \geqslant 0}$,generated by A,is either compact for $t > t_0 \geqslant 0$ or is exponentially stable and each F belonging to $\mathfrak{M}$ is compact,

(H8) For every $F \in \mathfrak{M}$ the null solution is the only one mild solution of (1.1),defined for all $t \in \mathbb{R}$,on which $V[x(t,x_0)] = V(x_0)$ $\forall t \in \mathbb{R}$.

Then the equilibrium $0 \in X$ is (GAS) for every $F \in \mathfrak{M}$.

<u>Sketch of the proof</u>.

In the same way as it was done in the proof of Theorem 1,we can show that the following inequalities are valid for the functional V defined by (2.3) :

$$\langle Ax + BF(x), V'(x) \rangle_X = - \| Gx + VF(x) \|_U^2 - \langle x, -Mx \rangle_X - \langle F(x), L^* F(x) \rangle_U -$$
$$- \langle x, LF(x) \rangle_X - \langle F(x), -KF(x) \rangle_U \leqslant 0 \quad \forall x \in \mathcal{D}(A), \forall F \in \mathfrak{M} \qquad (2.13),$$

where $V'(x)$ denotes the gradient of V at x and

$$V[x(t,x_0)] - V(x_0) \leqslant 0 \quad \forall t \in [0, t_{max}), \forall F \in \mathfrak{M}, \forall x_0 \in X \qquad (2.14).$$

Now,by (H7) and the local Lipschitz property for F,

$$\| x(t,x_0) \| \leqslant \sqrt{\tfrac{1}{\alpha} V(x_0)} \leqslant \sqrt{\tfrac{1}{\alpha} [\| H \| + \tfrac{1}{2} \| Q \| \ell(\| x_0 \|)} \; \| x_0 \| \quad \forall t \in [0, t_{max}),$$

$\forall F \in \mathfrak{M}, \forall x_0 \in X$,where $\ell = \ell(\sigma)$ is a continuous,strictly increasing function from $\mathbb{R}^+$ into $\mathbb{R}^+$, $\ell(0) = 0$ (see also (2.7)) (2.15).

The unbounded prolongability to the right of all mild solutions and the global uniform stability of the equilibrium $0 \in X$ follow from (2.15) independently of $F \in \mathfrak{M}$.In the case of compactness of $\{S(t)\}$ for all $t > t_0 \geqslant 0$,in (H8),one can apply the generalization of Pazy's criterion for precompactness of bounded trajectories,to prove that all trajectories have precompact ranges,independently of $F \in \mathfrak{M}$ .Originally,this criterion was stated for $t_0 = 0$ [2,p.236] or [3,Lemma 5.3] but after a modification of its proof it is possible to handle with $t_0 \geqslant 0$.

If the second case in (H8) holds, the same conclusion can be deduced from Webb's criterion for precompactness of bounded trajectories [4]. Now, by LaSalle Invariance Principle, all trajectories tend to the largest strongly-invariant set contained in the set $\{x_0 \in X: V[x(t,x_0)] = V(x_0)\ \forall t \in \mathbb{R}\}$, but, since (H9) holds, this set is equal to $\{0\}$. Thus, $0 \in X$ is globally attractive, which together with global uniform stability implies (GAS). □

Remark 3.

If the term "$-\varepsilon\langle x,x\rangle_X$" can be added in the right-hand side of (2.11a) where ε is a positive real, then the function $v(t) = V[x(t)]$ satisfies the following differential inequality

$$\dot v(t) \leqslant -\varepsilon\{\beta^{-1}[v(t)]\}^2 \text{ for almost all } t \geqslant 0,\ v(0) = V(x_0) \qquad (2.16),$$

where $\beta(\sigma) = [\|H\| + \frac{1}{2}\|Q\|\,l(\sigma)]\sigma^2$ (see (2.15)). Since $\alpha\|x(t)\|^2 \leqslant v(t)\ \forall t \geqslant 0$ then (2.16) frequently allows to obtain an estimate for the rate of decay of solutions or even to convert Theorem 3 into a result concerning (GUAS) or (GEXS) (the last holds f.i. if $Q=0$ or F satisfies a global Lipschitz condition).

Remark 4.

If in (H7) the semigroup $S(t)$ is compact for all $t>0$, then it suffices to assume only $\mathfrak{M} \subset \{P: P$ is continuous and maps bunded sets in X into bounded sets in $U, P(0)=0\}$ in (i) (see [2, p.193]).

3. Example.

Let us consider the electronic circuit shown in Fig.1a. The dynamics of this circuit is governed by the following system of equations

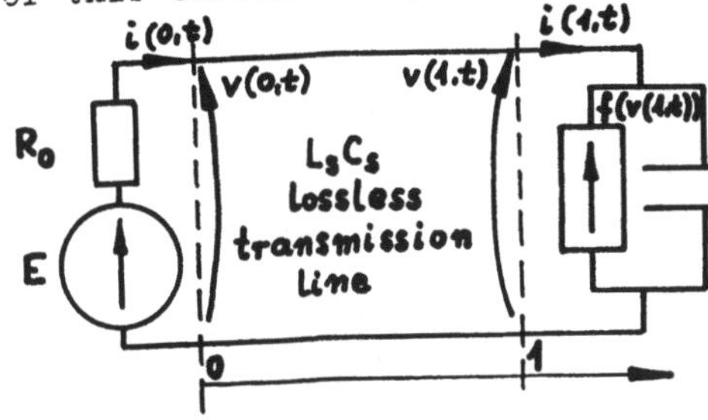

Fig.1a.

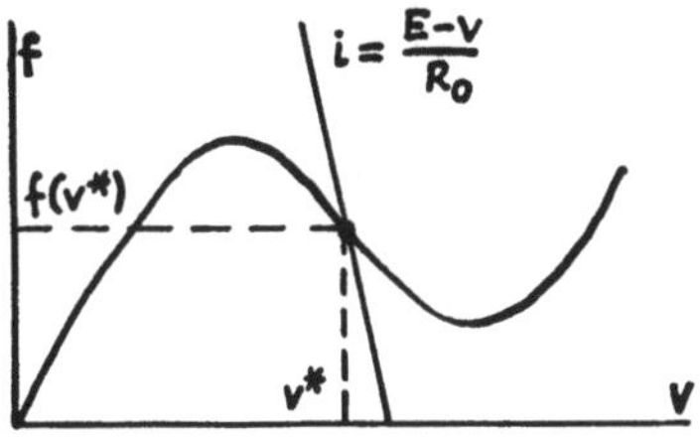

Fig.1b.

$$-L_s\frac{\partial I}{\partial t} = \frac{\partial V}{\partial x},\quad -C_s\frac{\partial V}{\partial t} = \frac{\partial I}{\partial x} \qquad 0 \leqslant x \leqslant 1,\ t>0 \qquad (3.1),$$

$$0 = V(0,t) + I(0,t)R_0, \quad -C\frac{dV(1,t)}{dt} = -I(1,t) + g[V(1,t)] \qquad (3.2),$$

where $g(y) = f(y+v^*) - f(v^*)$, f is the characteristic of the nonlinear active element (e.g. tunnel diode), v^* is defined in Fig.1b. The system (3.1),(3.2) can be reduced to a simple NFDE in the same way, as it was done by O.Lopes [9],

$$C\frac{d}{dt}\left[\xi(t) + K\xi(t-r)\right] = -g\left[\xi(t) + K\xi(t-r)\right] + \frac{K}{z}\xi(t-r) - \frac{1}{z}\xi(t) \quad t>0$$

where $K = (R_0 - z)\cdot\frac{1}{R_0+z}$, $z = \sqrt{L_s\cdot\frac{1}{C_s}}$, $r = \frac{2}{\sqrt{L_sC_s}}$ (3.3).

The system (3.3) can be interpreted as the abstract semilinear system (1.1). To do this, we take

$X = \mathbb{R}\times L^2(-r,0)$ with scalar product $\langle x_1,x_2\rangle = v_1v_2 + \int_{-r}^{0}\psi_1(\theta)\psi_2(\theta)d\theta$,

$x_i = \begin{pmatrix} v_i \\ \psi_i \end{pmatrix} \in X, i=1,2$; $U = \mathbb{R}$, $A\begin{pmatrix} v \\ \psi \end{pmatrix} = \begin{pmatrix} -\frac{1}{zC}v + \frac{2K}{zC}\psi(-r) \\ \psi' \end{pmatrix}$, $\mathcal{D}(A) = \left\{\begin{pmatrix} v \\ \psi \end{pmatrix} \in\right.$ $\left. X : \psi\in AC[-r,0], \psi' \in L^2(-r,0), \psi(0) + K\psi(-r) = v\right\}$, $Bu = \begin{pmatrix} -\frac{1}{C} \\ 0 \end{pmatrix}u$, $F\begin{pmatrix} v \\ \psi \end{pmatrix} = g(v)$

It can be proved that A is the infinitesimal generator of a linear C_0-semigroup on X (see [10] for details) while $B\in\mathcal{L}(U,X)$. Now it is quite natural to consider a class of mappings

$D = \left\{F : F\begin{pmatrix} v \\ \psi \end{pmatrix} = g(v), g:\mathbb{R}\longrightarrow\mathbb{R} \text{ is a locally Lipschitz function}, g(0)=0\right\}$.

Notice that if $F\in D$ then $F(0)=0$ and F maps every weakly convergent sequence $\{x_n\}_{n\in\mathbb{N}}$ in X into a strongly convergent sequence of values of F, $\{F(x_n)\}$ in U. Indeed, it is obvious that $F(0)=0$ for $F\in D$. By the definition of weak limit we have

$X \ni \begin{pmatrix} v_n \\ \psi_n \end{pmatrix} = x_n \longrightarrow x = \begin{pmatrix} v \\ \psi \end{pmatrix}\in X$ iff $\left\langle\begin{pmatrix} v_n \\ \psi_n \end{pmatrix},\begin{pmatrix} w \\ \phi \end{pmatrix}\right\rangle = v_nw + \int_{-r}^{0}\phi(\theta)\psi_n(\theta)d\theta \ \forall\begin{pmatrix} w \\ \phi \end{pmatrix}\in$ $\in X$. Thus, in particular, for $w=1$ and $\phi=0$ one obtains $v_n\longrightarrow v$. Now by the continuity of g and since $F\in D$ we have $F(x_n) = g(v_n)\longrightarrow g(v) = F(x)$.

Let R be an arbitrary real positive number. If $x_i = \begin{pmatrix} v_i \\ \psi_i \end{pmatrix}$ $i=1,2$ belong to the ball $\{x\in X : \|x\|\leqslant R\}$ then v_i, $i=1,2$ belong to the ball $\{v\in\mathbb{R} : |v|\leqslant R\}$ and hence by the fact that g is a locally Lipschitz function

there exists $l(R) > 0$ such that
$\| F(x_1) - F(x_2) \|_U = | g(v_1) - g(v_2) | \leqslant l(R) |v_1 - v_2| \leqslant l(R) \| x_1 - x_2 \|_X$ and
therefore the local Lipschitz property for F takes place. It follows from the above facts that if $\mathfrak{M}^c$ is any subset of D then the assumption (i) in Theorem 2 is satisfied.

Let us take $Qu = \begin{pmatrix} q \\ 0 \end{pmatrix}, q \in \mathbb{R}$. Then the operator QF is of gradient type for every $F \in D$ having the antiderivative $q \int_0^v g(s)\,ds$. Hence (H1) holds if $\mathfrak{M}^c$ is any subset of D. Assume that $|K| < 1$, ε is a sufficiently small positive real and take the operators M,K,L of the forms :

$$M\begin{pmatrix} v \\ \psi \end{pmatrix} = \begin{pmatrix} -\frac{1-|K|}{zC(1+|K|)} v + \varepsilon v \\ 0 \end{pmatrix}, Lu = \begin{pmatrix} \left(\frac{1}{2C} + \frac{\varepsilon zr}{2(1-K^2)}\right) u \\ 0 \end{pmatrix}, \quad Ku = \varepsilon u .$$

Then (H2) holds iff $\frac{g(v)}{v}$ lies between or at the roots of the following quadratic equation

$$Cy^2 - y\left(\frac{1}{\varepsilon} + \frac{Czr}{1-K^2}\right) - \left(\frac{1-|K|}{\varepsilon z(1+|K|)} - C\right) = 0 \qquad (3.4).$$

It is not difficult to check that if $|K|<1, q=0$ and ε is a sufficiently small positive real, then the assumption (H3) is satisfied, since for

$$H\begin{pmatrix} v \\ \psi \end{pmatrix}(\theta) = \begin{pmatrix} \left(\frac{1}{2} + \frac{Czr\varepsilon}{2(1-K^2)}\right) v \\ \left(\frac{|K|}{zC} + \varepsilon\left(\frac{r}{1-K^2} + \theta\right)\right)\psi(\theta) \end{pmatrix}$$

inequality (2.9) is fulfilled for every $(x,u) \in \mathcal{D}(A) \times U$. Now, in order to have the assumptions (H1),(H2),(H3) satisfied, we assume that $|K| < 1$ and define the class $\mathfrak{M}^c$ is Theorem 2 as follows

$$\mathfrak{M}^c = \left\{ F \in D : y_1(\varepsilon) \leqslant \frac{g(v)}{v} \leqslant y_2(\varepsilon) \;\; \forall v \neq 0, y_1 y_2 \text{ are the roots of } (3.4) \right\}$$

We claim that (H4) and (H5) also hold in this case. Indeed, by definitions of $\mathfrak{M}^c_L$ and $\mathfrak{M}^c$, $F \in \mathfrak{M}^c_L$ iff $F(x) = \mu v$, where $\mu \in [y_1, y_2]$. Thus, for $F \in \mathfrak{M}^c_L$ we have

$$(A + BF)\begin{pmatrix} v \\ \psi \end{pmatrix} = \begin{pmatrix} \left(-\frac{1}{zC} + \frac{\mu}{C}\right) v + \frac{2K}{zC}\psi(-r) \\ \psi' \end{pmatrix}, \quad \mathcal{D}(A + BF) = \mathcal{D}(A).$$

A+BF generates an exponentially stable semigroup, independently of the delay r, iff the spectrum of A+BF lies in the open left complex half-

-plane and $|K|<1$. If $|K|<1$, then by Pontriagin's theorem this holds for $\mu\in\left(-\frac{1-|K|}{z(1+|K|)},+\infty\right)$. As $y_1>-\frac{1-|K|}{z(1+|K|)}$ (H4) is fulfilled. (H5) holds evidently, since Q=0. Now by Theorem 2 and Remark 1 we establish that if $|K|<1$ then the equilibrium $0\in X$ is (GEXS) for every locally Lipschitz function $g:\mathbb{R}\longrightarrow\mathbb{R}$ such that $y_1\leqslant\frac{g(v)}{v}\leqslant y_2$ $\forall v\neq 0$, $g(0)=0$, where y_1,y_2 are the roots of (3.4).

In the case of the tunnel diode characteristic f (Fig.1b) the fulfilment of the above conditions depends on the behaviour of f in a neighbourhood of v^* (in this case the lower constraint for $\frac{g(v)}{v}$ becomes active) and on the rate of growth of g for large $|v|$ (in this case the upper bound for $\frac{g(v)}{v}$ becomes active). To obtain the best possible conditions for (GEXS) one therefore should take ε as a very small number for we have $y_1(\varepsilon)\searrow-\frac{1-|K|}{z(1+|K|)}$, $y_2(\varepsilon)\nearrow+\infty$ as $\varepsilon\longrightarrow 0+$. If for large $|v|$ g behaves as, say, v in the power 3 then the upper bound for $\frac{g(v)}{v}$ can be reached for any $\varepsilon>0$.

Now, we see the need for removing the upper bound for $\frac{g(v)}{v}$ from our conditions for (GEXS). To do this we use Theorem 3 and Remark 3, taking X,U,A,B,Q,M,L,H as previously and

$$\mathfrak{M}=\left\{F : F\begin{pmatrix}v\\ \psi\end{pmatrix}=g(v);\ g:\mathbb{R}\longrightarrow\mathbb{R}\text{ is a locally Lipschitz function,}\right.$$

$$\left. g(0)=0,\ \frac{g(v)}{v}\geqslant-\frac{\frac{1-|K|}{z(1+|K|)}}{\frac{Czr\varepsilon}{1-K^2}+1}\ \left(\underset{\varepsilon\longrightarrow 0+}{\longrightarrow}-\frac{1-|K|}{z(1+|K|)}\right)\ \forall v\neq 0\right\},$$

$K=0\in\mathcal{L}(U)$ in Theorem 3. It can be readily seen that the assumption (i) and (H1),(H2) in (ii) hold. If one takes the following linear, A-bounded functional on X

$$G\begin{pmatrix}v\\ \psi\end{pmatrix}=\sqrt{\frac{1-|K|}{Cz}}\left[\sqrt{\frac{|K|}{1+|K|}}\ v-(\operatorname{sign}K)\sqrt{|K|(1+|K|)}\,\psi(-r)\right]\text{ and}$$

$V=0\in\mathcal{L}(U)$ then

$$\langle Ax,Hx\rangle+\langle x,HAx\rangle+\langle x,-Mx\rangle=-\langle Gx,Gx\rangle-\varepsilon\langle x,x\rangle\quad\forall x\in\mathcal{D}(A),$$

$HB+L=0$ (in $\mathcal{L}(U,X)$), $K=V^*V$ (in $\mathcal{L}(U)$).

Thus the strengthened version of (H3'), mentioned in Remark 3, holds.

As for $\theta\in[-r,0]$: $\theta+\frac{r}{1-K^2}\geqslant\frac{rK^2}{1-K^2}>0$ then $\langle x,Hx\rangle\geqslant\alpha\|x\|_X^2$ $\forall x\in X$, where $\alpha=\min\left\{\frac{1}{2}+\frac{\varepsilon Czr}{2(1-K^2)},\ \frac{|K|}{zC}+\frac{\varepsilon rK^2}{1-K^2}\right\}>0$, so (H6) is fulfilled.

Let us recall that the semigroup generated by A is exponentially stable and every F belonging to $\mathfrak{M}$ is compact, therefore (H7) holds.

In our example, we can take (2.14) with $-\varepsilon\int_0^t \|x(\tau,x_0)\|^2 d\tau$ in the right-hand side, which yields (H8).

By Theorem 3 and Remark 3, the equilibrium $0 \in X$ is (GEXS) for every g determined by $\mathfrak{M}$, provided that $|K| < 1$. This result is equivalent to that given by O. Lopes [9, Theorem 3.5], who assumed that $C[-r,0]$ is the state space for (3.3).

More sophisticated considerations lead to the additional result conserning (GAS) of the equilibrium $0 \in X$. The (GAS) takes place for all locally Lipschitz functions $g: \mathbb{R} \longrightarrow \mathbb{R}$ such that $\frac{g(v)}{v} > -\frac{1-|K|}{z(1+|K|)}$ $\forall v \neq 0, g(0) = 0$.

4. References.

[1] I.Segal.Non-linear semigroups.Annals of Math.1963.78.3.339-364.

[2] A.Pazy-Semigroups of Linear Operators and Applications to PDEs. Springer.N.Y.-Berlin.1983.

[3] J.A.Ball-On the as.behavior of generalized processes with applications.J.Diff.Eqs.1978.27.2.224-265.

[4] G.F.Webb-Compactness of bounded trajectories of dynamical systems in infinite dimension spaces.Proc.Roy.Soc.Edinb.1979.A84.19-33.

[5] C.A.Desoer,M.Vidyasagar-Feedback Systems.Ac.Press.N.Y.1975.

[6] L.O.Barsuk,V.A.Brusin-Stability,unboundedness and dichotomy for solutions of FDE.Differencjalnyje Uravn.1977.13.9.1547-1557.

[7] Likhtarnikov A.L-A criterion for absolute stability of nonlinear operator eqs.Izvestija ANSSSR,s.Mat.1977.41.5.1064-1083.

[8] D.Wexler-On frequency-domain stability for evolution eqs. in H-spaces via the ARE.SIAM.J.Math.Anal.1981.11.6.969-983.

[9] O.Lopes-Stability and forced oscillations.J.Math.Anal.Appl.1976. 55.3.686-698.

[10] P.Grabowski-A Lyapunov functional approach to a parametric optimization problem for a class of infinite-dimensional control systems.Elektrotechnika.1983.2.3.207-232.

A DECOMPOSITION ALGORITHM FOR THE DETERMINATION OF OPTIMAL BUS FREQUENCIES

WOJCIECH GREGA
Institute of Automatic Control, System Enginnering and Telecommunication
Technical University of Mining and Metallurgy
PL-30-059 Kraków, Al.Mickiewicza 30/B1

1. INTRODUCTION

The problem to be discussed can be described in the following general terms: given the matrix of origin-destination passengers´ flows on a urban public transportation network. How optimally allocate an assumed fleet of vehicles among these routes in order to minimize travel time for passengers?

Passenger flows on routes are a function of vehicle allocation due to different travel possibilities between some node pairs. Hence, origin destination flows affects the allocation of vehicles and allocation of vehicles affects the passenger flows (assignment of passengers to routes).

The main requirement of an optimization method for this problem is the ability to handle system of high dimensionality with non-convex objective function. It should be noted that single calculation of the objective function value requires an expensive computational effort due to allocation-assignment relationships. For some major transfer points in the network algorithms have succesively assign origin-destination matrix to each route, usually for the whole network.

Various decomposition methods for the solution of assignment-allocation problem have been proposed over the last decade [1],[2]. The main factors affecting the efficiency of the methods are on one hand problem´s order and structure, on the other hand the properties of the optimization algorithms. Current papers are examining decomposition methods for the purpose of implementing hierarchical algorithms on multicomputer systems [3], but there are only a few papers reporting about the application to large scale transportation systems.

The minimization problem formulated in the next section was previously discussed [5], and several solution methods were proposed. In this paper, the additive separability of the objective function with respect to some decision variables is exploit in order to introduce local optimization problems. Each of them deals with local objective function $P_{i,j}$ and describes a subsystem for node "i" and travel demand from "i" to "j". Based on the solutions of the local problems, the coordination problem is defined. A penalty function method is used for coordination of local solutions. With all well-known weaknesses of this method, convergence can be proved for a broad class of decomposed objective functions. The applicability conditions turn out to be weaker that in any other two level method.

2. FORMULATION OF THE PROBLEM

Let us study the urban transportation network during a time period t, using the following definition and notations:

$\mathcal{L}$ - set of routes in the network,

$\mathcal{I}$ - set of nodes in the network,

$n = (n_1,\dots,n_l,\dots,n_L)$

n_l - number of buses on route l (allocation),

N - number of vehicles in the fleet (resources),

$n \in \mathcal{N} \subset R^L$

n_{lo} - minimum allocation on route l.

Demand matrix. We suppose given a demand matrix:

$$\{O_{i,j}\}$$

$$O_{i,j} = \mathcal{I} \times \mathcal{I} \rightarrow R^+$$

We shall think $O_{i,j}$ as the steady demand for travel from node i to node j.

Modified demand matrix.

$$Q_{i,j} = \mathcal{I} \times \mathcal{I} \rightarrow R^+$$

$$Q_{i,j} \in \mathcal{T}_{i,j} \subset R^+$$

$Q_{i,j}$ may be thought as the number of direct trips from node i to node j. The node where a transfer takes place is treated as the destination and the new origin.

Assignment function. The variable flow of passengers between node pair i,j, which wil be shared among a set of competing routes, can be characterized by an assignment function:

$$G_{i,j} : \mathcal{I} \times \mathcal{I} \times \mathcal{N} \times \quad \times \mathcal{T}_{i,j} \rightarrow R^1$$

Local objective function.

$$P_{i,j} : \mathcal{T}_{i,j} \times \mathcal{N} \rightarrow R^+$$

$P_{i,j}$ - average travel time for passengers travelling from node i to node j.

Basic assumptions.

- Set of routes is given.
- Set of passengers' paths in the network is given and is not affected by the allocation of buses.
- Minimum allocation to routes is given in the sense, that no excess capacity is allowed on the peak load link of any route. It can be shown [4] that n_{lo} is the lower bound of allocation.

Optimization problem. The following optimization problem will be considered:

$$\min_{\substack{n \in \mathcal{N},\ Q_{i,j} \in \mathcal{T}_{i,j}}} \sum_{i,j \in I} P_{i,j}(n, Q_{i,j}) \tag{1}$$

where:

$$n \in \mathcal{N} = \left\{ n : \sum_{l \in \mathcal{L}} n_l = \bar{N}, n_l \geqslant n_{lo};\quad l = 1,2,\ldots,L \right\} \tag{2}$$

$$Q_{i,j} \in \mathcal{T}_{i,j} = \left\{ Q_{i,j} : G_{i,j}(Q_{i,j}, n) = 0;\ i,j \in I \right\} \tag{3}$$

3. DECOMPOSITION OF THE PROBLEM

The model given in section 2 is of the type:

$$\min p(v,m) = \min \psi\{p_1(v,m_1), \ldots, p_N(v,m_N)\} \tag{4}$$

$$v \in V \subset R$$

$$m_i \in M_i \subset R^r$$

where:

$$V = \{ v : g_0(v) \leqslant 0,\ h_0(v) = 0 \} \tag{5}$$

$$M_i = \{ m_i : g_i(v,m_i) \leqslant 0,\ h_i(v,m_i) = 0 \} \tag{6}$$

$$p_i : V \times M_i \rightarrow R^1$$

$$g_i : V \times M_i \rightarrow R^{l_i}$$

$$h_i : V \times M_i \rightarrow R^{k_i}$$

$$\psi : \Omega \rightarrow R^1$$

ψ is strictly preserving order on Ω .

Let us consider the vectors v of separable parts of objective function (4) as independent variables and put v_i instead of v (in general $v_i \neq v$). Therefore, we introduce local variables v_i and we define locals problems:

$$\min\{p_i(v_i,m_i) + K_{i,k}(v - v_i)\} \tag{7}$$

$$v_i, m_i \in W_i$$

$$W_i = \{ (v_i,m_i) : q_i(v_i,m_i) \leqslant 0,\ h_i(v_i,m_i) = 0 \} \tag{8}$$

where v is a vector of coordinating (global) variables, not constrained on W_i .

$K_{i,k}$ is a penalty function for i-local and k-iteration.

$$K_{i,k}: R^r \rightarrow R^1 , \quad i = 1,2,\dots,N, \quad k = 0,1,\dots,K$$

Coordination problem can be formulated as follows:

$$\min_{v \in V} \psi\{p_{1,k}(v),\dots,p_{N,k}(v)\} \tag{9}$$

where:

$$p_{i,k} = p_i(v_i,m_i) + K_{i,k}(v_i - v)$$

Following basic result was proved in [1]:

Suppose that:

I. Functions p_i are continuous on W_i.

II. W_i are non-empty and compact sets.

III. If solutions of the problem (4) exist they create compact and separable sets $\hat{W}$.

IV. Exists $k = k_0$ such that for $k \in k_0$, $v \in V$ solutions of the local problems (7) are global.

then:

I. Problem (4) has at least one solution.

II. For every set of solutions $\hat{W}$ exists the set $\tilde{W}$ such that $\hat{W} \in \mathrm{int}\tilde{W}$. Furthermore, for sufficiently large "k" exists unconditional minimum of the problem (9) in the points of the set $\tilde{V}$, where:

$$\tilde{V} = \{ v \in R^r : \exists m \in R^N, (v,m_1,m_2,\dots,m_N) \in \tilde{W}$$

III. Exists at least one convergent subsequence of the sequence: $(\hat{v}_k,\hat{m}_{i,k})$ with the limit in $\hat{W}$, where $\hat{v}_k$ and $\hat{m}_{i,k}$ are solutions of problems (9) and (7) for k-iteration.

$$\lim_{k' \rightarrow \infty} \hat{v}_{i,k'} = \lim \hat{v}_{k'}$$

IV. $$\lim_{k \rightarrow \infty} \psi(p_{1,k},\dots,p_{N,k}) = \hat{\psi}$$

where $\hat{\psi}$ is the solution (not necessaey global!).

Now, we apply the algorithm to a large-scale bus allocation-assignment problem by putting:

ψ - additive operator

$p_i = P_{i,j}$

$v = n$; $r = L$; $N = |J|^2$

where: $|J|$ - cardinality of the set J

$m_i = Q_{i,k}$

$$\sum_{l \in \mathcal{L}} n_l - \bar{N} = g_0(n)$$

$$h_i(\cdot) = G_{i,j}(\cdot)$$

4. CONCLUSIONS

The network consisting of two routes and three stops nodes was considered as a sample problem. It was assumed that:

- Passengers arrive randomly at the stops and that they catch the first bus arriving which takes them to their destination.
- All passengers are able to board the bus.
- The objective function (1) includes the average waiting time for all groups of passengers. Assuming that bus arrivals at each route are Poisson process, the average waiting time for passengers travelling from node "i" to node "j" was defined by the formula:

$$P_{i,j}(n, Q_{i,j}) = Q_{i,j} \Big/ \sum_{l \in \mathcal{L}} d_{i,j,l} \cdot n_l$$

$$d_{i,j,l} = \begin{cases} 1/y_l & \text{- if passenger travelling from node "i" to uses route "l".} \\ 0 & \text{- otherwise.} \end{cases}$$

where:

y_l - travel time on route l .

The assignment function was defined as:

$$Q_{i,j} = \sum_{k \in I} \left\{ \sum_{l \in \mathcal{L}} (n_l \cdot d_{k,i,l}) \Big/ \sum_{l \in \mathcal{L}} (n_l \cdot d_{k,j,l}) \right\} \cdot Q_{k,j}$$

Results of numerical experiments were presented during the Conference.

The following conclusions can be drawn:

1. Limited numerical experience shows, that the number of coordinative iterations needed for the solution of the decomposed problem is not greater than square of the number of subsystems.
2. Convergence can be proved in more convenient way than in the case of a central treatment of the overall problem.

REFERENCES

1. Findeisen W., et al. - Control and Coordination in Hierarchical Systems, Wiley, New York 1980.
2. Jörnsten K.O. - Adaptation of Decomposition Methods to Some Transportation Problems, Linkoping Studies in Science and Technology, Dissertations, No. 45, 1980.
3. Papageorgiou M., Schmidt G. - Implementation of Hierarchical Optimization Algorithm on a Multimicrocomputer System, IEEE Transactions on Systems, Man and Cybernetics, vol.SMC-13, No.1, pp.11-18.

4. Han A.F., Wilson H.M. - The Allocation of Buses in Heavily Utilized Networks with Overlapping Routes, Transportation Research, pp.221 -232, 1982.
5. Grega W., Werewka J. - Verification of Public Transportation Models by Computer Simulation Methods, Proc. of 11 IMACS World Congress, Oslo 1985.

THE EFFECT OF POINT-LIKE NOISE SOURCES ON CHEMICAL DISTRIBUTED PARAMETER SYSTEMS

K.M. HANGOS, T. VIRÁG*

Computer and Automation Institute, Hungarian Academy of Sciences,
H-1502 Budapest, POB. 63, Hungary
* Research Institute for Technical Chemistry of the Hungarian Academy of Sciences,
H-1502 Budapest, POB. 98, Hungary

Abstract

The modelling of stochastic distributed parameter systems has a great importance in chemical engineering from the viewpoint of fault detection, measurement device placing and control. In the following paper an integral equation has been chosen for the stochastic model of a one-dimensional chemical system in transient state. The source of the fluctuations is assumed to be a steady-state process with continuous trajectories w.p. 1. Under weak assumptions it can be shown that the state variable process of a steady-state system is stationary, its correlation function fulfils an explicit operator equation which is closely related to the operator equation of the solution. The proposed method can be easily extended to more complicated noise sources, too.

Introduction

The modelling of stochastic distributed parameter systems has a great importance in chemical engineering from the viewpoint of fault detection, measurement device placing and control. For chemical engineering systems there are various deterministic mathematical models in use. These models can have a form of a set of partial differential equations, algebraic equations or a set of mixed type equations depending on the nature of mixing in the modelled equipment.

Serious mathematical problems arise introducing some noise source into a model of partial differential equation-type, that is when stochastic variation of some parameters or variables in the equation is assumed. Therefore, a unified model of the chemical distributed parameter system has been chosen which has a form of an integral equation.

This model fits very well to one of the favourite concepts of modelling of stochastic processes in Hilbert-spaces. The theory of stochastic processes in Hilbert-spaces is well developed [1,2] and its applications have already appeared in various fields (physical, geophysical, etc).

The stochastic model

Let us denote the Hilbert-space of the square integrable functions on a finite $[0,L]$, $L \in \mathbb{R}$ interval by $\mathcal{L}_2(0,L)$ with a scalar product of the form:

$$\langle f,g \rangle = \int_0^L f\,g \quad , \quad f,g\ \mathcal{L}_2(0,L).$$

The one-dimensional model of a distributed parameter chemical engineering system is assumed to be in the following form:

$$\Theta(t) = \Theta_I(t) + (\mathbb{K}\phi)(t) - \frac{d}{dt}\mathbb{K}\,\Theta(t)$$
$$\Theta(0) = \Theta_o \;, 0 \leq t \qquad (1)$$

and

$$\mathbb{K} : \mathcal{L}_2(0,L) \rightarrow \mathcal{L}_2(0,L)$$
$$\mathbb{K}\, g = \int_0^L K(.,u)g(u)du \qquad (2)$$

In the above equations the following notations are used:

t	time
$\Theta(t) \in \mathcal{L}_2(0,L)$	state function of the model (for example energy density, concentration etc.)
$\phi(t) \in \mathcal{L}_2(0,L)$	source function of the model (for example energy or mass source)
$\Theta_o \in \mathcal{L}_2(0,L)$	initial condition
$K(.,.)$	kernel function of the operator $\mathbb{K}$ characterizing the mixing conditions in the modelled system
$\Theta_I(t) \in \mathcal{L}_2(0,L)$	constant function on [0,L] is equal to the input value.

It can be shown (see e.g. [3] that the integral equation model has a rather general form and the well-known models for various mixing conditions (ordinary differential equations for perfectly stirred systems, partial differential equations for ideal tubular systems or for diffusion type systems or even cascade models) can be regarded as special cases of equations (1)-(2). In these cases the K(.,.) kernel function is equal to Green's function of the given boundary value problem.

In the following an extension of the above deterministic model will be derived for the stochastic case. For the sake of easy readability the stochastic version of the state function will be denoted by T while that of the source function will be called f.

There is a great family of possible disturbances to the system which appear in the source function, say disturbances in the heating conditions or in the environment's energy density or concentration. In order to model these effects, let us assume that the source (f) as well as the model state (T) can be modelled with appropriate stochastic processes in the given Hilbert-space, e.g.

$$f(t,\omega), T(t,\omega) \in \mathcal{L}_2(0,L), \; \forall t,\omega \;.$$

It is fairly not a trivial and solved problem how to model noises with given properties. One of the most simple cases is when the noisy source function is assumed to be in the neighbourhood of a given deterministic source function in the mean sense, i.e.:

$$f(t,\omega) = \phi(t) + F^o(t,\omega)$$
$$E\{f\} = \phi \qquad (3)$$

where $E\{,\}$ is the expectation operator.

In the above equation F^{o} can be regarded as a noise source with zero mean. For the mathematical simplicity it is assumed that F^{o} has continuous trajectories w.p.1.

In the following the concept of the "point-like noise source" will be used in order to have results in an analytical form. A noise source F^{o} is called point-like if it has the following form:

$$[F^{o}(t,\omega)](l) = f^{o}(t,\omega)\delta_{\ell\xi}, \quad l \in [0,L], \quad \xi \in [0,L] \text{ fixed} \tag{4}$$

where

$f(.,.)$ a real valued stochastic process with the covariance function $B_{f^{o}}(.,.)$ and with continuous trajectories w.p. 1

$\delta_{\ell\xi}$ Kronecker-delta function (in space).

With help of the above noisy source function the stochastic model can be formulated analogously to the deterministic model (1)-(2) in the form:

$$T(t,\omega) = T_I(t) + (\mathbb{K}\, f)\,(t,\omega) - \frac{d}{dt}\;\mathbb{K}\, T(t,\omega)$$
$$T(0,.) \equiv T_o(.),\ t \geq 0 \tag{5}$$

where

T_o initial condition, $E\{T_o\} = \Theta_o$

and the operator $\mathbb{K}$ is defined by equation (2).

As $\mathbb{K}$ is a bounded linear operation on $\mathcal{L}_2(0,L)$ it is easily seen that the expectation of equation (5) gives the same equation for the mean values as the deterministic model equation (1), i.e.:

$$E\{T(t,,.)\} = \Theta(t) \tag{6}$$

This fact gives the possibility to construct a model equation only for the fluctuation $\Delta(t,.)$ which is defined by the following equation:

$$\Delta(t,\omega) = T(t,\omega) - \Theta(t)\ . \tag{7}$$

Subtracting the expectation of equation (5) from itself the following model equation is obtained for the fluctuation:

$$\Delta(t,\omega) = \mathbb{K}\, F^{o}(t,\omega) - \frac{d}{dt}\ \mathbb{K}\ \Delta(t,\omega)$$
$$\Delta(0,.) \equiv \Delta_o(.),\ E\{\Delta_o\} = 0. \tag{8}$$

Equations for the covariance operator

The covariance operator ($\mathbb{B}_\Delta(.,.)$) of the fluctuation can be determined according to the definition

$$\mathbb{B}_\Delta(t,u) = E\{\Delta(t,.)\circ\Delta\ (u.,)\} \tag{9}$$

where the symbol $\circ$ denotes the tensor product on $\mathcal{L}_2(0,L)$. For this purpose the explicit solution of Eq. (8) is needed which has the following form:

$$\Delta(t,\omega) = \int_0^t \mathbb{Z}(t-\tau)\, F^{o}(\tau,\omega)d\tau \tag{10}$$

where operators $\mathbb{Z}(.)$ form a continuous semigroup on $\mathcal{L}_2(0,L)$, i.e.:

$$\mathbb{Z}(t): \mathcal{L}_2(0,L) \to \mathcal{L}_2(0,L)$$
$$\mathbb{Z}(0) = \mathbb{I},\ \mathbb{Z}(t)\ \mathbb{Z}(u) = \mathbb{Z}(t+u) \tag{11}$$

and $\mathbb{I}$ is the identity operator, $\mathbb{I}\ g = g$.

There is an explicit relation between the operators $\mathbb{K}$ and $\mathbb{Z}(t)$ which permits also the numerical computation of $\mathbb{Z}(t)$ in a recursive manner [4]:

$$\mathbb{Z}(t)g = \lim_{n\to\infty} [(\mathbb{K} - \frac{t}{n}\mathbb{I})^{-1}\ \mathbb{K}]^{n}\ g\ ,\ g\in\mathcal{L}_2(0,L) \tag{12}$$

Eq. (12) is only one side of the mutually unique relation between operators $\mathbb{K}$ and $\mathbb{Z}(t)$.

The covariance operator of the fluctuation can then be determined from an explicit operator equation using the definition (9) and the explicit solution equation (10) with the operator semigroup (11):

$$\mathbb{B}_{\Delta}(t,u) = \int_0^t\int_0^u \mathbb{Z}(t-\tau)\ \mathbb{B}_{F^o}(\tau,\nu)\ \mathbb{Z}^{*}(u-\nu)d\nu d\nu \tag{13}$$

where $\mathbb{B}_{F^o}$ denotes the covariance operator of F^o.

Eq. (13) gives only a formal solution together with Eq. (12). Therefore, in practical applications some dicretized (both in time and in space) approximation of the above equations is needed. There are only a few special cases when explicit analytical solution is possible. These cases will be shown in the next section, together with the method of the numerical computations for the discretized case.

Special cases and computational method

1. Continuously stirred tank reactor (CSTR)

As a first special case, the stochastic model of a CSTR was chosen, however, a CSTR is not a distributed parameter system. If the volume of the tank is denoted by V and the inlet volume rate by w, the kernel function of the operator K takes the following special form:

$$K(x,\xi) = L/V,\ 0 \le x,\ \xi \le L\ . \tag{14}$$

In this case the explicit form of the operator semigroup $\mathbb{Z}(t)$ can also be given

$$[\mathbb{Z}(t)q\ (.,\tau)](x) = \exp\ (-\frac{t}{\bar{t}}) \int_0^L q(\xi,\tau)d\xi \tag{15}$$

where $\bar{t}=V/w$ is the mean residence time. With the above equation the explicit solution of the fluctuation equation is the following:

$$\Delta(.,t,\omega) = \int_0^t \exp\left(\frac{t-\tau}{\bar{t}}\right) \int_0^L F^o(z,\tau,\omega)\, dz d\tau. \tag{16}$$

It can be seen from Eq. (16) that $\Delta(.,.,\xi)$ does not depend on space.

Assuming that F^o is point-like (Eq. (4)), the covariance operator takes a nice simple form:

$$B_\Delta(.,.,t,u) = \int_0^t \int_0^u \exp\left(\frac{t-\tau+u-\nu}{\bar{t}}\right) B_{f^o}(\tau,\nu)\, d\nu d\tau \tag{17}$$

2. Cell models

In this case the state function and the source function will be projected into a subspace of $\mathcal{L}_2(0,L)$ generated by a given $\{\Psi_i\}_{i=1}^m$ finite orthonormal system on $\mathcal{L}_2(0,L)$. The corresponding chemical engineering system is a set of CSTR-s with various set of recirculations, feedbacks and feedforwards of the material in them. The elements of the finite orthonormal system are then in the form above a given $\Gamma^{(m)}=\{z_i = \frac{iL}{m}$, $i=0,\dots,m\}$ equidistant grid:

$$\Psi_i(z) = \begin{cases} \frac{m}{L}, & z_{i-1} \le z < z_i \\ 0 & \text{otherwise} \end{cases} \tag{18}$$

while the operator $\mathbb{K}$ is characterized by an m×m matrix:

$$\mathbb{K}\, g = \sum_{i=1}^m \sum_{j=1}^m a_{ij}\, \Psi_i \langle \Psi_j, g\rangle, \quad g \in \mathcal{L}_2(0,L) \tag{19}$$

The elements of the operator semigroup $\mathcal{Z}(t)$ are also determined by an m×m matrix with time dependent elements:

$$\mathcal{Z}(t) g = \sum_{j=1}^m \sum_{j=1}^m \zeta_{ij}(t)\, \Psi_i \langle \Psi_j, g\rangle \tag{20}$$

and the ζ_{ij} functions can be computed according to Eq. (12).

In the case of point-like noise source, the projection of F^o to this finite subspace has the following simple form:

$$F^o(t,\omega) = f^o(t,\omega) \Psi_\ell \langle \Psi_\ell\, , F^o\rangle \tag{21}$$

and the covariance operator of F^o is also very simple

$$\mathbb{B}_{F^o}(t,u)\; g = B_{f^o}(t,u) \Psi_\ell \langle \Psi_\ell\, ,g\rangle \tag{22}$$

If parameters a_{ij}, $i,j=1,\dots,m$ are given, the above equations give us the possibility to compute numerically the parameters of the covariance operator $\mathbb{B}_\Delta(t,u)$ according to Eq. (13) and the computation can be done in a recursive manner using ordinary matrix operations.

Conclusion

In the paper a unified model of stochastic distributed parameter chemical engineering system was shown which is in the form of stochastic differential equation in the Hilbert-space.

For practical industrial cases the parameters of the operator $\mathbb{K}$ in Eq. (19) can be obtained from direct measurements [5]. It gives the possibility to calculate these parameters of the covariance operator of the fluctuations (Eq. (13)) applying also Eq. (12) and the point-like noise property (Eq. (22)).

The proposed method can easily be extended to the case of more independent point-like noise sources.

References

[1] Tsafestas, G. (ed.): Distributed parameter control systems. Theory and applications.
Pergamon Press, Oxford-New York-Toronto-Sydney-Paris-Frankfurt, 1982.

[2] Arnold, X., Curtain, R.F., Kotelenez, P.: Nonlinear Stochastic Evolution Equations in Hilbert Space Report.
Universität Bremen, No. 17, Bremen, 1980.

[3] Halász, G., Virág, T.: A numerical method for heat and mass transfer processes. Computer and Chemical Engineering, 6 (1982) 177-179.

[4] Martin, L.H.: Nonlinear operators and differential equations in Banach spaces. John Wiley and Sons, New York-London-Sydney-Toronto, 1976.

[5] Alsaidy, A., Blickle, T., Halász, G.: Investigation of axial mixing in chemical reactors by measuring the kernel function.
Hung. J. of Industrial Chemistry, 12 (1984) 369-383.

Optimal Energy Management

P.G. Harhammer (IBM Österreich)

A. Schadler (STEWEAG)

1. Introduction

The Decision Support System "Energy Management" (figure 1) /1,2/ can be applied to economically plan the operation of any single or combined energy system (electric power system, district heating system and its combinations) for a week and/or a day in advance. Its development was based on two main items:

- Correct modelling of the highly non-linear energy system yielding practice acceptable results.
- CPU-time effective optimization procedures.

2. Problem Definition

The Decision Support System "Energy Management" was developed based on the STEWEAG power system (figure 2) consisting of 5 hydro storage plant, 4 run-of-river plants with small storage capacity (Enns river) including 3 weir power plants, 6 thermal units, 1 energy contract and run-of-river energy (Mur river) /3/. The result are cost-optimal schedules of all components of the given power system for the planning period of a week including an underlying day with 6, 12, 24 or 48 time steps per day. The components are specified by their constructive and operational data, supplemented by the load curve of the energy systems (electric power system, district heating system) and the resources available for the planning period (quantities of primary and electric energy).

3. Decision Support System "Energy Management"

The Decision Support System /4/ consists of 4 basic modules: problem data input, matrix generator (MIP model), optimization (MPSX/370 with MIP/370 extension and performance accelerating tools), color graphic and/or numeric output of results. It is characterized by two main items:

- Correct and timely solution of the given highly non-linear problem

- User orientation by menu driven data input and output of resulting schedules.

4. Model System

The operation planning procedure is based on a hierarchical model system with two levels. The weekly optimization with 42 time steps (6 time steps a day) plans the available resources (stored water, coal, oil, gas) while the underlying daily optimization prepares the cost effective schedules of all power system elements involved in the operation.

5. The Mixed-Integer Programming (MIP) Model

The main goal in effectively modelling electric power generation systems are reasonable execution times together with a general applicability to all power systems achieved by its modular structure. The MIP model is derived from the nonlinear dynamic process by discretization in time and piecewise linear approximation. The most important aspect to yield short execution times are submodels with a small number of integer variables summarized in special ordered sets if possible /5/.

The planning period comprises up to seven days with 6, 12, 24 or 48 time steps each. The number of system elements to be modelled is more or less unlimited.

6. Matrix Generator

The matrix generator produces the MPSX data input files out of the given data set of the power generation system and its preassigned energy resources. The matrix generator has a modular structure and each module consists of an input and a checking subroutine, a printing subprogram to produce a review and of one or more subroutines to generate the input files for the optimization program by applying the mathematical model (MIP formulation).

All input data are plausibility checked and converted according to the desired number of time steps per day. If necessary detailed error messages and hints for corrections are produced. The present version of the matrix generator accepts models up to 30.000 variables. All generated MPSX input files are directly applicable except the COLUMNS - section, being rearranged by an additional subroutine called "SORTER". All MPSX names for rows and columns have a unique structure including type, system element, day and time step number.

The general applicable modules of the matrix generator are the base to solve successfully the economic scheduling problem in any power generation systems for the planning period.

7. Optimization

The highly non-linear models are optimized by the already mentioned program package (MPSX/370 with MIP/370 extension) supplemented by a number of CPU-time decreasing measures to achieve the predefined targets of execution times based on IBM system 4341-2.

- Daily optimization (30 minutes of CPU-time)
- Weekly optimization including the daily one (60 minutes of CPU-time).

The accelerating measures are described in connection with the daily optimization although used extensively in different combinations for the hierarchical model system (week - day) too.

7.1 Model Analysis Techniques

A large number of tests with models of different power systems proved the solution of both the unit commitment problem and the resource allocation one as well being practically independent of the time step width (figure 3). Based on this knowledge a special model analysis technique was developed to reduce the model size by removing all variables including especially the CPU-time consuming integer ones not needed to represent the respective power unit mix of the energy system under investigation.

Therefore, the daily optimization problem is solved at first with 6 time steps requiring 1 or 2 minutes of CPU-time. Afterward its result is automatically analyzed by program to remove from the model energy conversion units being not in operation and those with constant power output during the planning period. The output of the analysis program is a so called REVISE file which is called by the MPSX control program when starting the optimization run to reduce the size of the model with 48 time steps a day as required.

The matrix of the model representing the given STEWEAG power system was reduced from some 3.000 rows and 7.800 variables including 1.000 integer ones to 1.900 rows and 5.300 variables including 490 integer ones by applying the described analysis techniques to the model of the investigated day saving in this case 40 % (range: 30 to 70 %) of CPU-time.

7.2 Starting Solution

The standard program package MPSX/370 allows for the definition of an LP (Linear Programming) starting basis to be activated by one of the MPSX-control languages (MPSCL - Mathematical Programming System Control Language, ECL - Extended Control Language).

Therefore, the result of a previous planning period (week, day) is used as an LP-starting basis for each new optimization run saving up to 70 % CPU-time necessary for the LP solution, the first step of solving an MIP optimization problem CPU-time effectively.

7.3 Preoptimization Techniques

Preoptimization is an accelerating method to improve the performance of the Branch and Bound algorithm being the second step of an MIP optimization procedure to solve the IP (Integer Programming) part of the problem. Two techniques are included as integral part of the preoptimizer (PROPTIM) in order to generate a good IP-starting solution.

- Insertion of IP solutions /6/
- Heuristic starting solutions /7/

The insertions of IP starting solutions to the Branch and Bound process is not a standard feature of the MIP/370 program. Therefore, a new algorithm was developed using the ECL (Extended Control Language - PL/I superset) of the MPSX/370 program package.

Heuristic starting solutions are found by fixing certain integer variables to integer values not being integer after the LP solution. A rounding technique is used to assign 1 to certain integer variables depending on their continuous values or to their corresponding costs. Presently a two stage process is implemented. At a first stage to assign certain values grater than 0.5 are assigned to 1 (all other basic variables remain unfixed) and if this measure fails another trial is made assigning values greater than 0.8 to 1.

At first the PROPTIM algorithm attempts to find a feasible solution of the given optimization problem by inserting one of these IP starting solutions. If the search is successful but the corresponding solution fails to meet the termination creterion the previously fixed integer variables are freed again and the standard method (OPTIMIX) is called to seek an IP solution. If the search is not successful another IP starting solution is put in place and the PROPTIM algorithm starts again.

The preoptimizer allows a number of parameters to be defined in advance in order to adapt its possibilities closely to the structure and the requirements of the given optimization problem.

Another 10 to 30 % of CPU-time can be saved in the IP solution stage by applying the preoptimization techniques.

7.4 Termination Criteria

A number of termination criteria (e.g. desired number of integer solutions, maximum number of iterations, maximum CPU-time allowed) can be defined to stop the optimization process.

Two termination criteria (figure 4) were found by analyzing a large number of test cases of different models.

- Maximum allowed deviation of the IP solution from the LP solution (GAP)
- Gradient criterion.

One of the main advantages of all LP solution techniques of optimization problems is the possibility to know in quantified figures how good an IP solution is when comparing it to its LP solution or to the best functional value of the waiting nodes of the Branch and bound search. Therefore, it is a reasonable stopping rule to end the optimization process if the deviation of the IP solution from the LP solution is smaller than a previously defined gap (e.g. 1 %).

The quotient out of the difference between the values of two consecutive objective functions of IP solutions and the respective CPU-times can also be used as termination criteria if the so defined gradient is smaller than a predefined value.

8. Results in CPU-time

The implementation of all accelerating measures described above resulted in a 77 % decrease of CPU-time (optimization procedure) from 32.00 to 7.24 minutes (IBM System 3033).

The CPU-time of an IBM System 3033 multiplied by 3.2 yield the CPU-time of an IBM System 4341-2. 23.17 minutes would be the appropriate execution time of an IBM System 4341-2 being below the predefined target of 30 minutes for the daily optimization.

The weekly optimization including an underlying day results in some 60 minutes CPU-time as average value taking four different weeks and seven underlaying days into account. The respective target of approximately 60 minutes in average is therefor met again. It should be mentioned that the achieved CPU-times of the investigated weeks varies strongly (summer week: 15 min, transition week: 74 min, winter weeks: 75 min).

A benchmark between an IBM System 4341-2 and an IBM System 4361-5 with the transition week resulted in 24 % better performance of the latter one.

9. Summary

This report describes the optimization of large scale MIP models to plan economically the operation of energy systems. A number of measures were implemented to accelerate the optimization procedures included in the program package "Energy Management System". This program package is designed to economically plan the operation of any single and combined energy system (electric power system; district heating and its combinations) a week or a day in advance. Between 1 and 2 % savings in operation costs (fuel and start up costs of thermal units and costs for contracted energy) can be expected based on the experience up to now.

The user oriented program package runs under control of VM/CMS and is menue driven as far as both the input of system and operation data and the color graphic output of optimization results (power and cost schedules, water economics) via terminals (displays, printers) are concerned.

10. References

/1/ Steinbauer E. et al
Kraftwerkseinsatzoptimierung
ÖZE, 1985, No1, Pages 1-7

/2/ Harhammer P.G.
Project Reports "Kraftwerkseinsatzoptimierung - Tag", "Kraftwerkseinsatzoptimierung Woche-Tag"
unpublished, May 1983 and June 1984.

/3/ Steinbauer E.
Kraftwerke an der Steirischen Enns - Betriebsführung der Kraftwerkskette
ÖZE, 1973, No. 5, Pages 185 - 191

/4/ Steinbauer E. et al
Menue Driven Program Package
"Operation Planning of Energy Systems"
PSCC VIII Report, Helsinki, August 1984

/5/ N.N.
IBM Mathematical Programming System Extended/370
Program Reference Manual, 1975

/6/ Nimführ W.
Ganzzahlige Startlösungen für das Programmsystem MPSX/370 mit MIP/370 Zusatz
Diplomarbeit, TU-Wien, unpublished, 1984

/7/ Suhl U.
Solving Large Scale Mixed-Integer Programs with Fixed Charge Variables
Research Report, IBM Thomas J. Watson Research Center,
RC 10266 (No. 45712), 11/83

Prof. Dr. Peter Harhammer
IBM Österreich
Obere Donaustraße 95
A-1020 Wien, Austria

Dipl.-Ing. Alois Schadler
STEWEAG
Leonhardgürtel 10
A-8011 Graz

FIGURE 1

MENUE DRIVEN PROGRAM PACKAGE

ENERGY MANAGEMENT SYSTEM

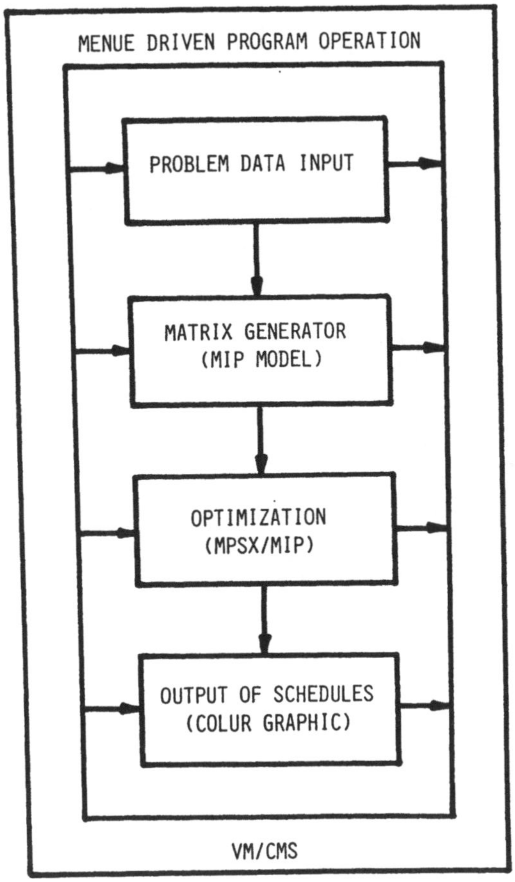

STEWEAG POWER SYSTEM

FIGURE 2

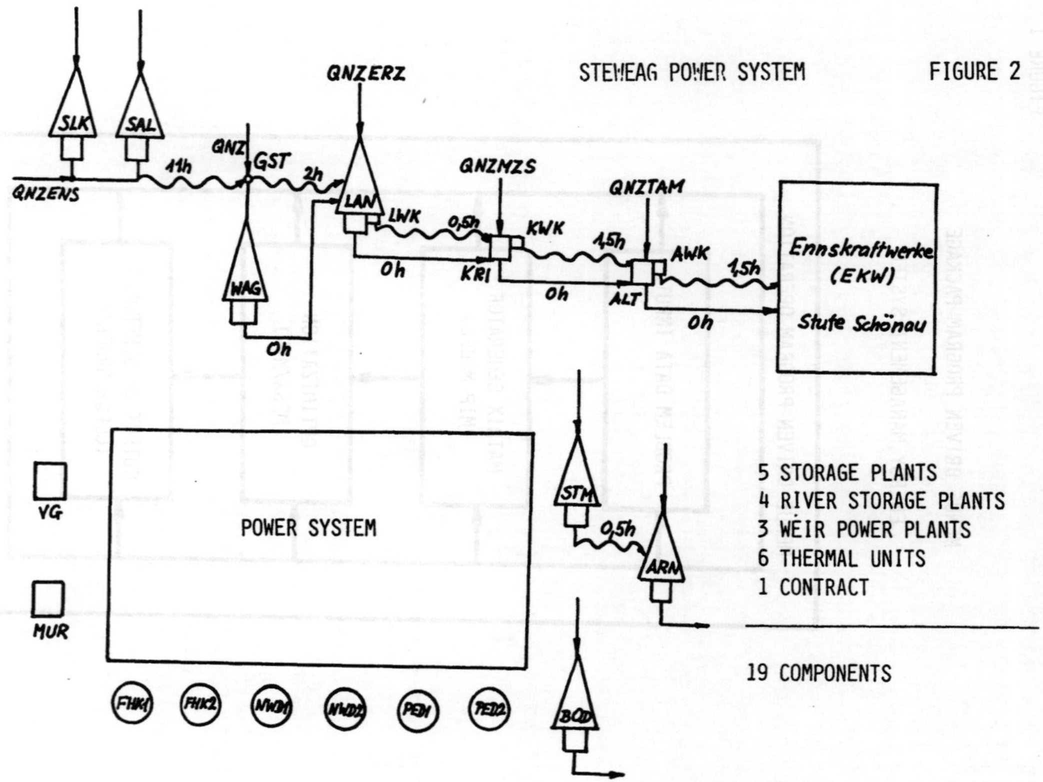

UNIT COMMITTMENT — FIGURE3

	1	2	3	4	5	6	7	8	9	10	11	12
U1							✔					
U2	✔	✔	✔	✔	✔	✔	✔	✔	✔	✔	✔	✔
U3												
U4	✔	✔			✔	✔	✔	✔			✔	✔
TS	1	2	3	4	5	6	7	8	9	10	11	12

U1													✔	✔										
U2	✔	✔	✔	✔	✔	✔	✔	✔	✔	✔	✔	✔	✔	✔	✔	✔	✔	✔	✔	✔	✔	✔	✔	✔
U3																								
U4	✔	✔	✔	✔					✔	✔	✔	✔	✔	✔	✔					✔	✔	✔	✔	✔
TS	1	2	3	4	5	6	7	8	9	10	11	12	13	14	15	16	17	18	19	20	21	22	23	24

U1																								✔
U2	✔	✔	✔	✔	✔	✔	✔	✔	✔	✔	✔	✔	✔	✔	✔	✔	✔	✔	✔	✔	✔	✔	✔	✔
U3																								
U4	✔	✔	✔	✔	✔	✔	✔	✔									✔	✔	✔	✔	✔	✔	✔	✔
TS	1	2	3	4	5	6	7	8	9	10	11	12	13	14	15	16	17	18	19	20	21	22	23	24

U1	✔	✔	✔	✔																				
U2	✔	✔	✔	✔	✔	✔	✔	✔	✔	✔	✔	✔	✔	✔	✔	✔	✔	✔	✔	✔	✔	✔	✔	✔
U3																								
U4	✔	✔	✔	✔	✔	✔									✔	✔	✔	✔	✔	✔	✔	✔	✔	✔
TS	25	26	27	28	29	30	31	32	33	34	35	36	37	38	39	40	41	42	43	44	45	46	47	48

FIGURE 4

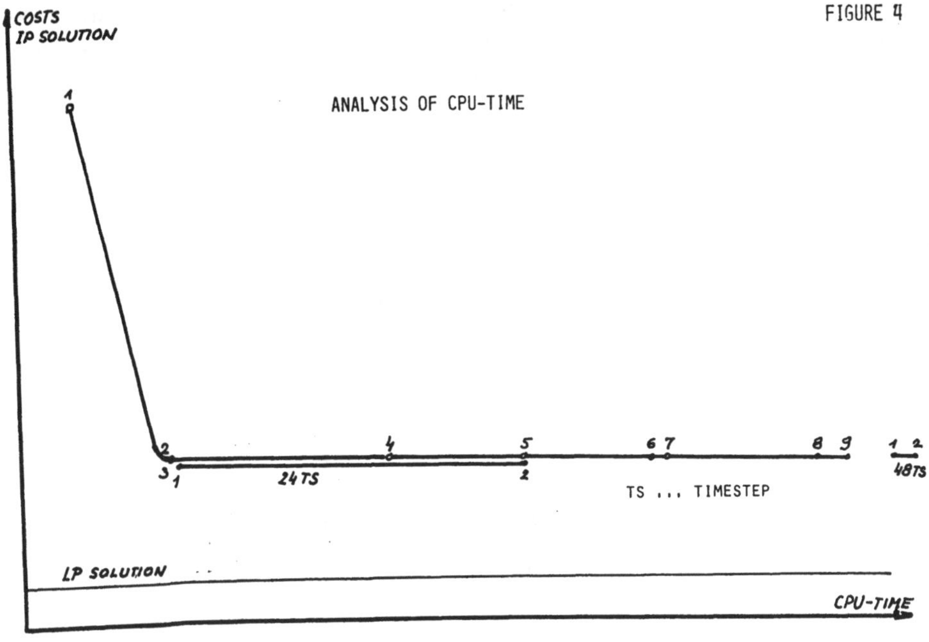

EFFECTS OF POWER-CONVERTERS, COMPENSATION EQUIPMENT AND FILTER CIRCUITS ON THE VOLTAGE TRAJECTORIES IN ELECTRICAL DISTRIBUTION SYSTEMS

HENDRICH CH., DRAXLER B.
University of Federal Defence Munich
Department of Electrical Engineering
D-8014 Neubiberg, FRG

INTRODUCTION: By means of modern power converters an optimal and dynamic control of the power flow in electrical distribution systems can be achieved. Reactive power flow and network losses are minimized by means of elctronical or mechanical operated compensation equipment. On the other hand more and more low voltage electronical equipment is used for control, protection, calculation, research and others. But both power converters and the power compensation equipment generate upper harmonics and under-/overvoltages which may disturb each other or damage electronical equipment. In this paper measurements and calculations are presented, which show the effect of power converters, compensation equipment and filter circuits on system voltages and line currents.

1. Power converters: The time function of the 20 kV-terminal voltage at the feeding bus bar of an industrial power system was measured by means of a high quality fast fourier transient analyser. The figures 1 and 2 show the frequency spectra of two selected measurements.

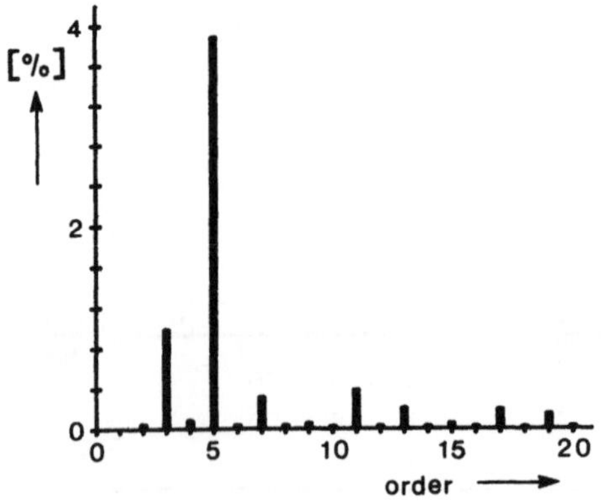

Fig. 1: Frequency spectrum of terminal voltage. Typical working-day.

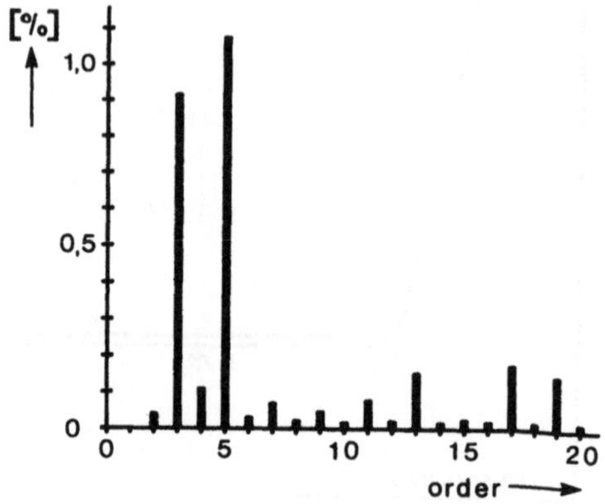

Fig. 2: Frequency spectrum of terminal voltage. Dynamic operation of one dc-engine.

Figure 1 was measured at a typical working-day. We can see that there are installed six-pulse bridges in the network, which generate the characteristic harmonics with the order 5, 7, 11 and 13. The fifth harmonic amounts up to 4 percent of the fundamental harmonic. The third harmonic may be generated by asymmetrical three-phase currents in the system.

Figure 2 show the frequency spectrum in case of the dynamic operation of a dc-engine, which is supplied by a six-pulse-bridge rectifier. Based on the conventional rule for rectifiers the non-characteristic upper harmonics should not exist. But figure 1 and especially figure 2 show that there are generated non-characteristic harmonics with the order 2, 4 and 6.

For getting some more information about the principle behaviour of the upper harmonics and their qualitative dependence on system parameters and operation mode of engines a lot of transient calculations were done by means of time-step-method. The investigated configuration consists of a feeding 20 kV-power system, a smoothing inductance and filter circuit, a 20 kV/520 V transformer (1,2 MVA), a 6-pulse bridge rectifier converter (460 kW) and a controlled dc-engine (400 kW).

In figure 3 the amplitudes of line current harmonics are given for different filter circuits in case of a start up procedure of the engine. In this dynamic operation mode all characteristic harmonics, figure 3a, are much higher than in the steady state operation and non-characteristic harmonics occur, figure 3b.

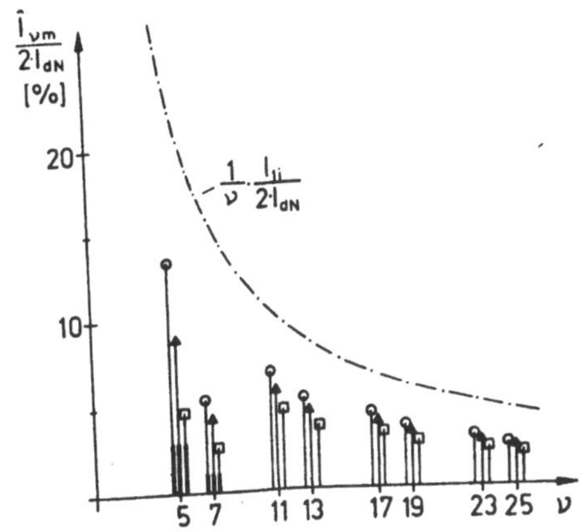

Fig. 3a: Characteristic harmonics of line current.

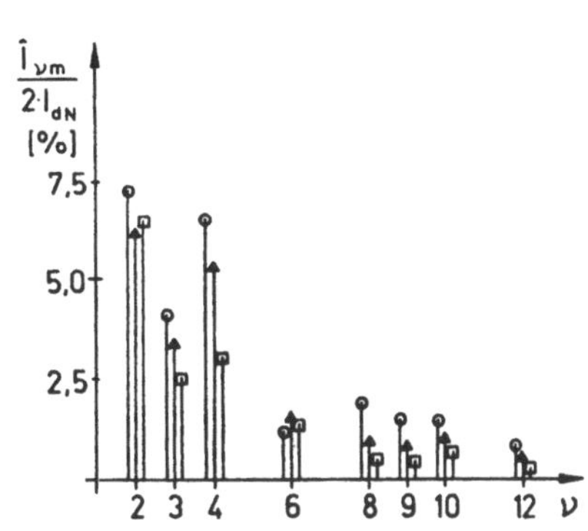

Fig. 3b: Non-characteristic harmonics of line current.

In figure 4 and 5 the time functions of the harmonics of alternating line current and terminal voltage are presented. It is obvious that the non-characteristics only appear in case of dynamic operation. In case of the line current the critical harmonic with order 4 reaches up to 7 % of the fundamental harmonic.

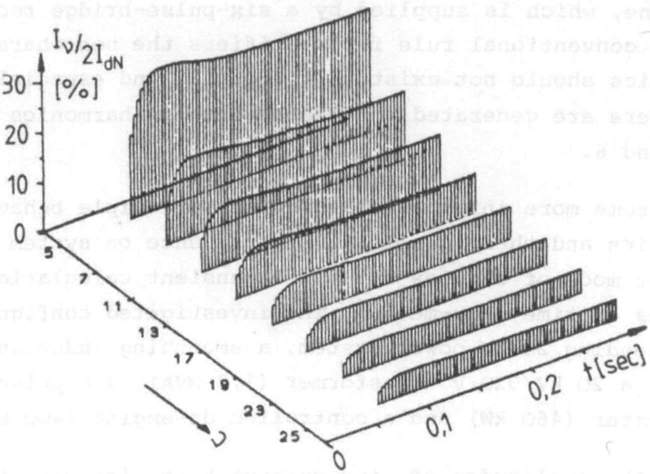

Fig. 4a: Characteristic harmonics of line current.

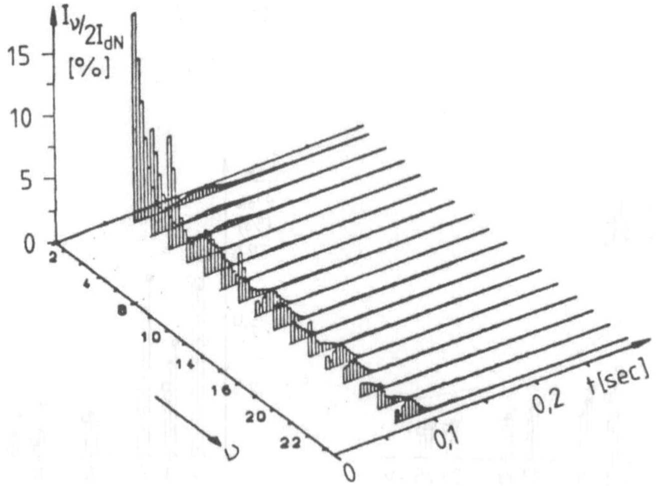

Fig. 4b: Non-characteristic harmonics of line current.

In case of the terminal voltage the 4th and 5th harmonics amount up to 2,5 % and 7,5 % of the nominal value.

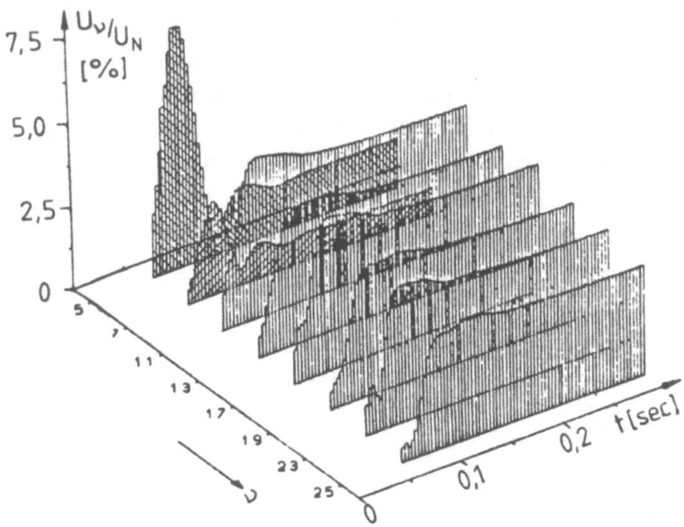

Fig. 5a: Characteristic harmonics of terminal voltage

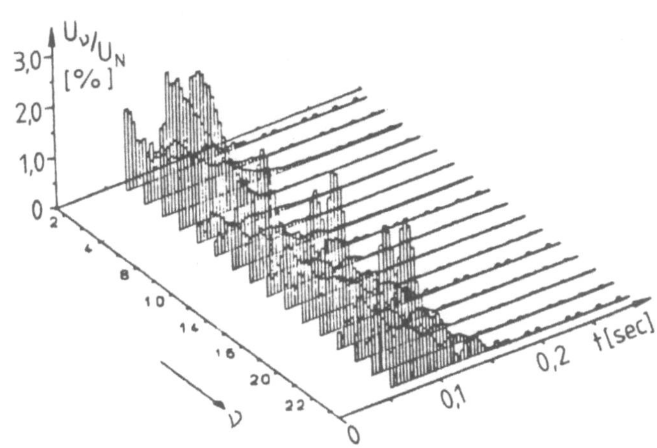

Fig. 5b: Non-characteristic harmonics of terminal voltage

The relation between the operation mode of the engine and the harmonics some more clearly can be explained in the figures 6 and 7. Figure 6a shows the time behaviour of the direct current of the engine. First starting up and than constant speed, slowing down, lower speed and speed up again. By means of laying figure 6a on the figures 6b and 6c, it can be demonstrated that the time behaviour of the characteristic harmonics depends on the time function of the direct current of the engine and the time behaviour of the non-characteristic harmonics depends on the time function of the gradient of the direct current. The maximal gradient occurs at the start up procedure and there we have the maximal non-characteristic harmonics, too.

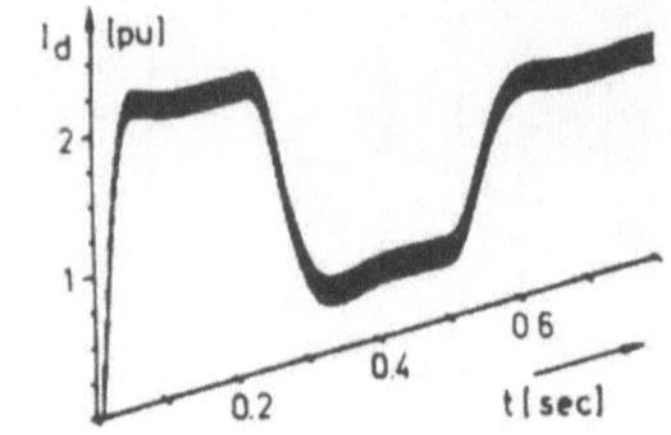

Fig. 6a: Direct current. No filter circuits.

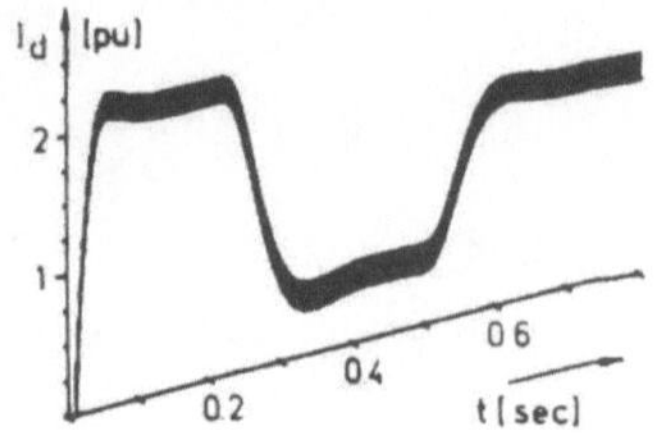

Fig. 7a: Direct current. Filter circuits.

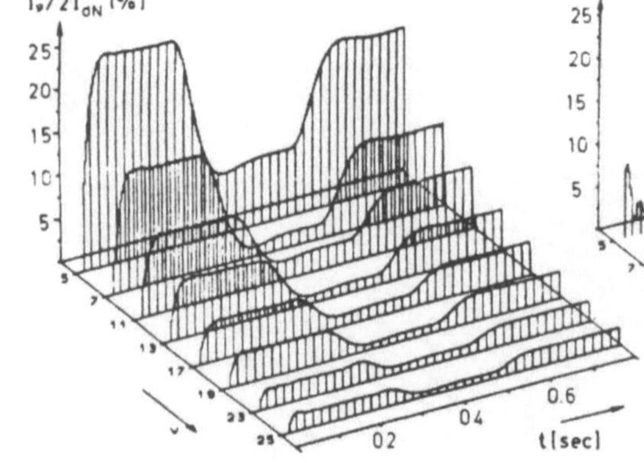

Fig. 6b: Characteristic harmonics of alternating current. No filter circuits.

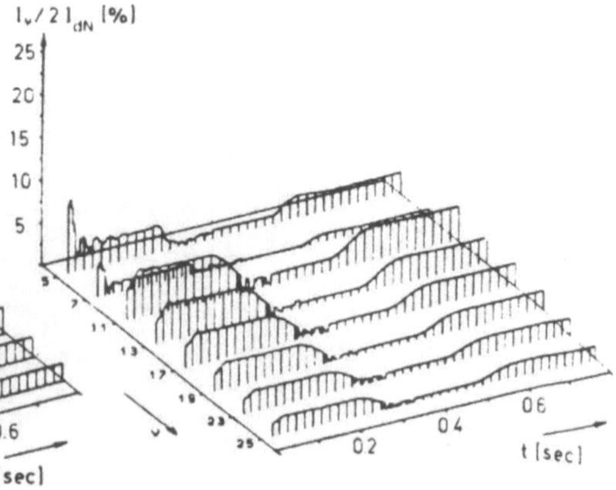

Fig. 7b: Characteristic harmonics of alternating current. Filter circuits.

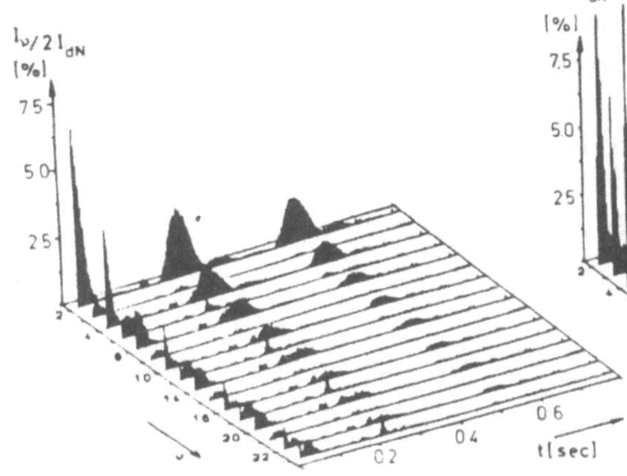

Fig. 6c: Non-characteristic harmonics of alternating current. No filter circuits.

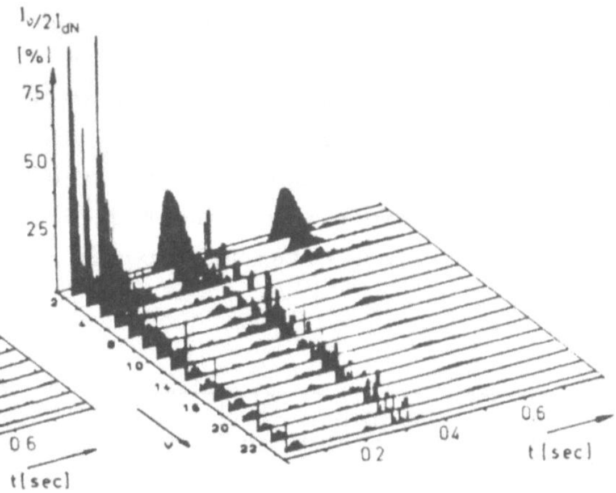

Fig. 7c: Non-characteristic harmonics of alternating current. Filter circuits.

2. Filter circuits: For the compensation of the upper harmonics in the a-c-currents filter circuits must be installed. The planning of these filter circuits in general is based on the conventional ideal rule for the harmonics, which means that only characteristic harmonics occur. For example in case of a six-pulse bridge converter, only filter circuits for the 5th and 7th harmonics are considered. But chapter 1 shows that in the dynamic operation mode not only these characteristic but also non-characteristic upper harmonics of the order 2 and 4 occur in the alternating current, which have remarkable amplitudes.

The figures 7a, b, c show the time behaviour of the direct current, characteristic and non-characteristic harmonics in the case of installed filter circuits for the 5th and 7th harmonic. Comparing the figures 6b and 7b the absorbing effect of filter circuits on the characteristic harmonics can be demonstrated. But comparing the figures 6c and 7c it is obvious that the non-characteristic harmonics are not reduced but amplified by means of the filter circuits.

This disagreeable amplification of non-characteristic harmonics is caused by the fact that an installed filter circuit together with the inductance of the feeding network will have one more lower resonance frequency. The figure 8a shows that in case of an installed filter circuit for the 5th harmonic another frequency near to the 4th harmonic appears. This effect should be regarded because of the d-c-engines in an industrial area often are operating in a dynamic mode.

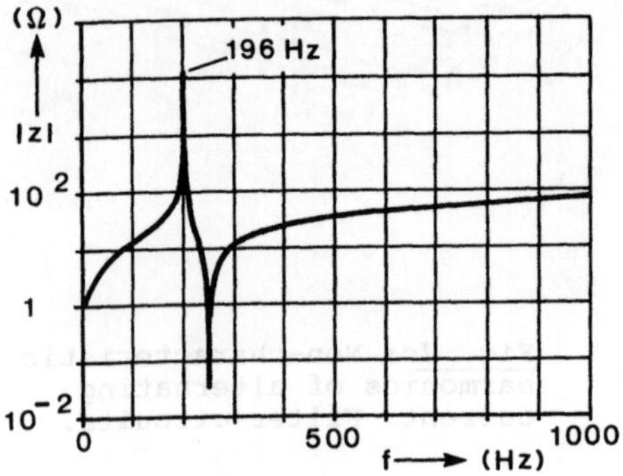

Fig. 8a: Frequency response of an industrial distribution system. Filter circuits for the 5th harmonic.

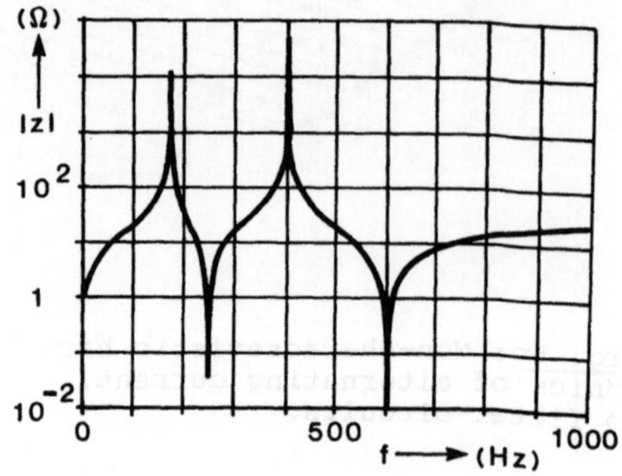

Fig. 8b: Frequency response of an industrial distribution system. Filter circuits for the 5th harmonic and series rejector circuit.

If there are installed not only parallel filter circuits but also series rejector circuits we have 2 additional resonance frequencies, as shown in the figure 8b.

3. Compensation equipment: For the fundamental harmonics the filter circuits operate capacitive. This must be considered in the case of dimensioning the reactive power compensation equipment. If the compensation of reactive power is realized by means of controlled capacitors, under- and overvoltages occur in case of switching, which may be dangerous for electronical equipment. The values of these under-/overvoltages depend on the ratio of the switched capacity to the capacity, which is already connected to the system. Considering this rule prohibited voltage deviations can be avoided.

Figure 9a shows the time function of the system voltage in case of switching the first compensation capacitor on the power system assuming the worst case that there is no capacity already connected to the system. The voltage deviations are equal to the system voltage. In figure 9b the second stage is switched on, which has the same capacity as the first one. The voltage deviation amounts only half the value above.

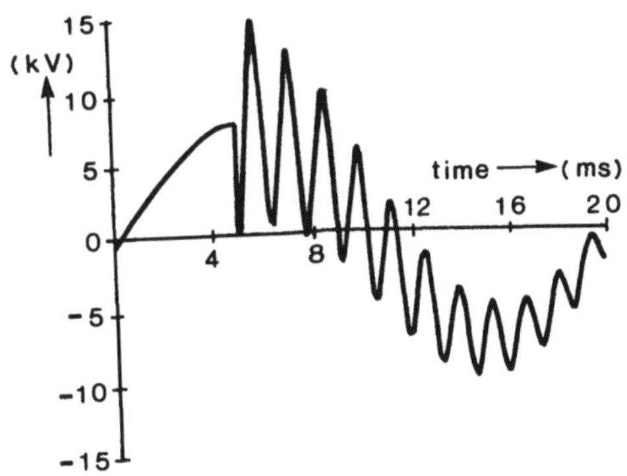

Fig. 9a: Switching the first stage.

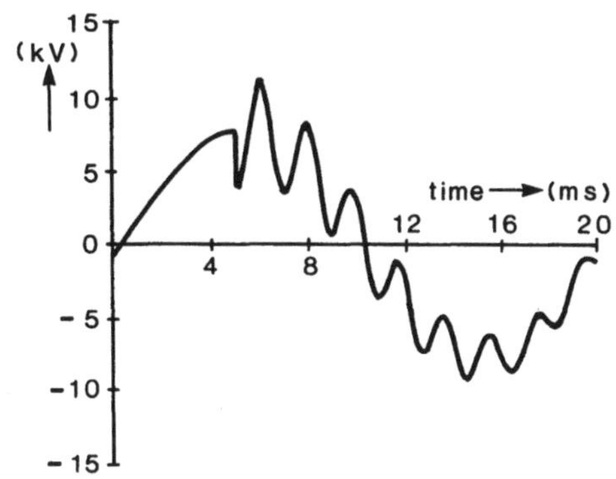

Fig. 9b: Switching the second stage.

For estimation of the maximal amount of the voltage deviation ΔV at the switching point a very simple rule can be used:

$$\Delta V \leq \frac{C_S}{C_S + C_F} \sqrt{2} \cdot V_o ,$$

V_o = nominal system voltage against neutral point

C_S = Capacity switched on or off

C_F = Capacity, which is already connected to the system.

If, for example, voltage deviations of maximal 20 percent of the nominal voltage are allowed, the switched capacity C_S should be less or equal 25 % of the capacity C_F, which is already at the system.

<u>CONCLUSION</u>: Undue upper harmonics or aperiodical voltage distortions in industrial power systems should be avoided in order to reach an undisturbed operation of electronical equipment.

On principle the interaction of power converters, power compensation equipment, filter circuits and low voltage electronical equipment should be taken into account already in case of planning and studied by means of transient analysis.

ORELIA
A PROBABILISTIC MODEL FOR THE EXPANSION PLANNING OF THE EHV ELECTRIC POWER SYSTEM

by A. HERTZ - EDF-DER - 92141 CLAMART - FRANCE

Abstract : ORELIA is a tool for the long-term scheduling of the electric energy power system.

It seeks the localization of new means of generation and network reinforcements which, when carried out, will result in the lowest investment and operating cost. The random factors liable to affect the system, are taken into account. In order to solve the stochastic optimization problem, two methods are used concurrently :

- the "stochastic gradient" method which provides results expressed in terms of continuous variables (the tuning of this method - a little delicate in practice - has been partially automatized),
- an heuristic method whose results are integers of lines, or plants, to be constructed.

In addition, for both methods, a technique for reducing the variance improves the relevancy of the results.

1 - INTRODUCTION

1.1 - The search for an electric power target system

One of the tasks of engineers in charge of long-term scheduling studies for the main grid, is to propose the appropriate localization of future conventional-thermal or nuclear power stations which are scheduled to be constructed. This must be done while taking into account the outcomes upon the network : the line investments which are undertaken each year depend very much on this localization.

The problem of installing generation facilities on the new sites is presented in the following way at Electricité de France :

- owing to preliminary studies, the overall volume of generating equipment which has to be put into service each year in order to satisfy the demand for electricity, is known (this knowledge includes each type of generation facility : nuclear, conventional thermal, gas turbines, etc.),
- in addition, there is a precise list of the possible sites suitable for the installation ; the list considers each site for each type of generation, and the maximum power which it seems possible to install.

From the set of localizations which satisfy the two types of the above-mentioned constraints, the planner will seek the one which will allow for a total minimum cost. This cost includes :

- the cost of the transmission lines which have to be built,
- the investment cost of the new generation facilities, (the cost of a power station is variable, depending on its geographical location which, among other things, determine the type of cooling process required),
- the mathematical expectation of the annual management cost, taking into account all the types of uncertainties which affect the power system (unavailability of generating units, line failures, random factors affecting the load).

This management cost is divided into two parts :

- the first part corresponds to the cost of generating the energy actually used,
- the second part corresponds to the cost of the loss-of-load energy.

Until recent years, an appropriate solution to the problem was obtained using the deterministic model TASSILI [1,2] (in this model, the values of random variables are replaced by mathematical expectations). The results of TASSILI were usually validated through the use of the probabilistic model ORTIE [3], but with fixed localization assumptions for the units. The increasingly wider use on the French system, of nuclear units with large unitary dimensions, and the greater discrepancy of thermal generation costs (nuclear, coal, fuel) led to the design of a probabilistic model which would better account for the "interconnection function" of the electric power system.

1.2. - The ORELIA software

The ORELIA software was thus designed so as to "optimize" simultaneously the localization of the new generation facilities and the development of the transmission system, while taking into account the random aspect of the problem. The use of the "stochastic subgradient" method has made this possible [4,5,6,7].

Theoretically, this method converges. In practice, too. However, determining the series of coefficients used to multiply the gradient (or subgradient) turned out to be a delicate operation. Thus, it will be briefly described how this was partially automatized in the ORELIA model.

In addition, the "stochastic subgradient" method provides solutions in terms of continuous variables, i.e., quantities of MW for the units and lines to be built.

It was thus complemented by an heuristic method, but in integers (of unitary plants or lines). The principle of this second method consists in narrowing, in successive stages, the range of feasible solutions : with each stage, this narrowing action is guided by using the same mechanism as that of the "stochastic gradient".

The two methods programmed in ORELIA, use a succession of situations of the electric power system which are selected at random, and on which certain functions are computed (in particular, the gradient of the economic function): this paper will explain how a method for reducing the variance - applicable to the Monte-Carlo methods - is implemented in ORELIA.

2 - MODELLING (AND SIMPLIFICATION) OF THE PROBLEM

Before describing the two optimization methods use in ORELIA, the problem being handled should be described.

This problem may be summarized by the formula :

$$\min_{y \in D} \left[a.y + Ef(y,s) \right]$$

y is the vector of unknown investments (per possible site and generation type, or per link between regions).

D is the set in which y must be found. It is the intersection of an hyperbox and hyperplanes. The hyperbox reflects the bound constraints which each of the potential investments must respect: site constraints for the generation units, corridor constraints for the transmission lines. Each hyperplane reflects the possible knowledge of the volume of investment for a given type of equipment (e.g., total number of 600 MW thermal coal units, number of MW x KM of lines to be built).

a is the vector of the unitary costs of investment (in annuities): it is thus assumed that the costs of investments vary linearly as a function of the quantities of investment as expressed by y.

s is a random vector reflecting the possible situations on the electric power system : greater or lesser availability of generation and transmission equipement, and the demand for power at the system nodes.

Actually, for practical reasons (available data), and also for greater efficiency, the situations are stratified a priori into subsets designated as "hourly points" (about ten).

Each hourly point corresponds to a period of the year (winter, summer), and a certain level of demand. From one hourly point to the next, the availability rates of the units may be differentiated in order to reflect the variation in the maintenance rates in the course of a year. The number of selected situations s is the same in each subset (for each "hourly point"): the variables computed for a situation s must thus be weighted as a function of the subset to which s belongs. In fact this weighting may be seen as the duration (in hours) of the period of the year that the hourly point corresponding to s represents.

f is the function of the management cost for a certain value of the vector of investments y and a random situations s. Its computation is a priori that of an "economic dispatch". Actually, both for practical reasons (CPU time), and theoretical reasons (the stochastic gradient method converges only if f is a convex function in respect to y), this economic dispatch is modeled in the form of a minimal cost flow problem : in other words, KIRCHHOFF's second law, from the direct current approximation, is neglected (to compensate for this simplification of electrical laws, the capacities of the system lines are brought down ; in addition, the cost of the losses may be taken into account in an approximate manner by indicating a linear cost of energy transmission.

E designates the mathematical expectation on the probability space of s. If S is the set (assumed to be discrete) of the possible situations s and p(s) the probability matched with a situation s,

$$E\, f(y,s) = \sum_{s \in S} f(y,s)\; p(s)$$

3 - THE STOCHASTIC GRADIENT METHOD

3.1. - Brief description

The principle of this method was the subject of two papers [6,7] in 1981 and 1982. Since then, it has had some degree of success in actual use. However, it will be briefly described again here, in order to point out how some difficulties inherent in the method were skirted, and also because the method serves as a basic mechanism for the integer method.

The method consists in the construction of a series of values for y.

$$y^{n+1} = P\left[D, y^n - r_n g(y^n, s^n)\right]$$

n varies from 0 to N-1, and, in practice, y^N is selected as the "best solution".

P (D,y) is the projection of y on set D.

r_n is the element of a series in $\frac{a}{b+n}$.

g(y,s) designates the gradient (or, if there is no gradient, a pseudogradient) of the function a.y + f (y,s) whose mathematical expectation we wish to minimize.

This gradient is the sum of the vector a and a vector inferred from the dual solution of the minimal cost flow problem (which is used as an "economic dispatch").

s^n is a series of vectors, selected at random, which represent a series of random situations for the electric power system.

3.2. - Convergence of the method

With a certain number of assumptions (see [5,6]) among which, in practice, the most restrictive are the convexity of the set D of y and of the function f(y,s) (for any s, in respect to y), a certain number of theoretical results ensure a probabilistic convergence (e.g., the convergence in a quadratic mean for y^n, an almost certain convergence for E $f(y^n,s)$).

On an experimental basis, the method converges at the price of tens of thousands of situations selected at random (For 130 unknowns in y, this means 30 minutes CPU IBM 3081). This CPU time looks fairly big, but we would point out that this duration is about ten times shorter than for ORTIE [3], when seeking optimal network reinforcements for a given localization of units. For that purpose, ORTIE solves a huge linear program which expresses simultaneously a thousand system random situations.

When using the stochastic gradient method, two difficulties arose :

- the convergence is probabilistic ; y^N is, in fact, a random variable whose value is more or less close to the optimum. Therefore, in order to ensure a certain stability in the results, N has to be chosen fairly large. The user of the model must, however, exercise a certain degree of caution, and not hesitate to look for several "solutions" for the same set of data (e.g., by changing the random situations sequence). This is because no indicator for the appropriate or inapropriate convergence of the method has been found which is sufficiently independent of the problems presented...
- to ensure the best time/convergence compromise for the algorithm, the coefficients a and b of the series $r_n = \frac{a}{b+n}$ and the total number of iterations N must be well adjusted.

Thus, r_n values which are too low and result in a very slow convergence make it necessary to carry out a large number of iterations ; r_n which are too high will make the y^n values move too much. In addition, the CPU time is practically proportional to N.

Lastly, the optimal a, b, and N values (i.e., ensuring the best time/accuracy compromise) depend to a considerable extent on the data of a problem (e.g., b may be as large as y^o is close to the solution being sought).

3.3. - Semi-automatic determination of the coefficients r_n

Describing the complete method would be too lengthy a task. Further, in spite of everything, it is linked to the problem presented. The description will thus be limited to giving an idea of the three conditions used to determine a, b and N.

- 1st condition : at the beginning (small n values) the movements of y^n must be large enough to make the choice of y^o indifferent (more or less for from the optimal solution).

  This is a condition concerning the value of r_m, m being app. 1000, or 2000, in current practice.

- 2nd condition : unlike in the first condition, the movements of y^n must be sufficiently low when n is close to N.

This is a condition reflected in ORELIA for :

$$\frac{r_N}{r_m} = \frac{b+N}{b+m}$$

- 3rd condition : the situations (selected at random) which are the most catastrophic for the electric power system are those situations characterized by negative gradient components and big absolute values (due to the high cost of the loss-of-load energy). These situations result in the sharp variations of certain components of the vector y (which increase). For a system in which the investments are close to the optimum, such situations have a probability which is low, but not a zero probability. The number of iterations must therefore be high enough to ensure that the effects of a "catastrophic" situation may be offset, if necessary, by the effects of a sufficient number of "ordinary" situations. This condition is reflected in the relation between r_m and r_N which takes into account the decrease of the series r_n.

Conditions 2 and 3 thus make it possible, in principle, to compute b and N and, then, using condition 1, to compute a.

In reality, the mechanism is a little more complicated. In particular, it is the user who proposes two possible values of N and the program chooses the lowest value between them which satisfies both conditions 2 and 3.

3.4 - Computing the cost of a solution

It should be noticed that the "stochastic gradient" method yields a solution ($y = y^N$) without the need of computing E f(y,s). Besides, this computation is impossible during the optimization process, due to y variations. But, most of the time, the user needs estimates of the cost and of some other random variables values, dependent on the solution (y). So, in ORELIA, an optimization process is usually followed by an estimation process, computing estimates of those random variables for the y which was selected. This is made, using a MONTE-CARLO method and selecting, here again, a large number of system random situations s (usually 5000).

4 - PRINCIPLE OF THE INTEGER METHOD

In certain respects, this method is similar to the "cut method" used for integer programming. In regard to the mechanism of the algorithm, it is an iterative method, whose each iteration includes the execution of the stochastic gradient algorithm.

It should be remembered that the latter constructs a series :

$$y^{n+1} = P\left[D, y^n - r_n \ g(y^n, s^n)\right]$$

for n varying between 0 and N-1.

For the integer method, N is chosen much lower (currently 1000) and r_n may be thought as a constant (fixed with the help of a condition such as the first one of 3.3).

In addition, between each execution of the "stochastic gradient" the set D of y is shrunk. Actually, some of the bounds, between which the y_i values are to stay, are brought closer together.

Obviously, the method stops when D is reduced to a point, in practice, this happens after a little more than ten calls of the "stochastic gradient".

What is the rule followed to bring together (possibly) the upper and lower bounds $\overline{y_i}$ and $\underline{y_i}$ of a component of y with a suffix i ?

This rule uses two results provided by an execution of the gradient method :

- the final value y^N of y,
- a vector h, the mean on the N iterations of the values taken by the gradient $g(y^n, s^n)$.

It is more or less expressed by stating $z = y^N - exh$:

if $h_i > 0$ then lower $\overline{y_i}$ to $I_i^+(z)$

if $h_i < 0$ then raise $\underline{y_i}$ to $I_i^-(z)$

$I_i^+(z)$ designates the lowest multiple of the unitary size of investment i, higher than the value z_i (e.g., if i designates a nuclear site for units of 1400 MW, and if $z_i = 1800$, $I_i^+(z) = 2800$).

$I_i^-(z)$ designates the greatest multiple of the unitary size of investment i, lower than z_i (e.g., whith the preceding assumptions $I_i^-(z) = 1400$).

e is a scalar value : its value is zero at the start of the integer method and could be termed an extrapolation coefficient : indeed e multiplies the tendency embodied in vector h, to give a vector z extrapolated from y^N.

The value of e is modified at each iteration, so that the shrinkage of D is neither too fast, nor too slow. If the shrinkage were too fast, the method could be equated to a simple rounding-off method, which generally provides poor results. If the shrinkage is too gradual, too much CPU time will be needed.

Actually, for an iteration of the integer method, e is determined to ensure one move at least of a bound $\underline{y_i}$ or $\overline{y_i}$, while taking into account the number of bounds modifield at the preceding iteration compared to the number of components of y which are still "free" ($\underline{y_i} \neq \overline{y_i}$). The guiding idea behind this determination is to try to execute, at each iteration, only the most natural "bringing together" (of bounds), and to carry out bolder extrapolations only when the convergence proves too slow. This method, in theory rather simple, provides nevertheless quite good solutions (in integers), compared with the solutions in terms of continuous variables (obtained with the stochastic gradient method).

5 - UTILIZATION OF A VARIANCE REDUCTION TECHNIQUE IN ORELIA

5.1. - Why try to reduce variance ?

When implementing a MONTE-CARLO method, means are often used to reduce the variance of the results being sought.

Let it be remembered that a typical execution of the ORELIA program includes two phases : an optimization phase to seek an optimal value of the vector y of investments, followed by an "estimation" phase to compute, among other elements, the value of the economic function a.y + E f(y,s) for the y which was selected. It is thus natural to try to reduce the variance of the results.

This is made clearer when we consider that, for instance, with a line failure rate of q = 1%, an incident affecting two given lines supplying the same node, will be selected with a probability of only 1/10000. Then, out of 5000 selections, there is only one chance in two for the simultaneous failure of two given lines to be chosen.

Now, this type of incident can obviously have very serious economic consequences !

5.2. - Implementation of variance reduction method

A conventional method, and one which is rather easy to implement, consists in simply selecting a line failure with a modified probability u > q, and then correcting adequately the values of the random variables, computed for a situation.

If v(s) designates such a variable, p(s) the probability of occurrence of s, and t(s) the probability of its being selected, then v(s) must be replaced by :

$$\frac{v(s) \times p(s)}{t(s)} \quad \text{because :}$$

$$\sum_s \frac{v(s) \times p(s)}{t(s)} \times t(s) = \sum_s v(s) \times p(s)$$

For the rest, it will be assumed that the components of a situation vector s are values of independent random variables, the probabilities of which values are $p(s_i)$. Consequently, $p(s) = \prod_i p(s_i)$.

In addition, for purposes of simplification, interest will be focused on only the s_i, each of which represents the state of a single line : $p(s_i)$ is thus worth q or 1-q, and $t(s_i)$ is worth u or 1-u.

The whole problem consists in determining u in order to truly reduce the variance. It has been observed how sensitive this variance is to the choice of u : the optimal value of u varies considerably as a function of the problem being processed (e.g., from 1,2 % to more than 2 % for q = 1 %).

Moreover, whith a well-adjusted u, the standard deviation of the economic function is divided by a factor of the order of 2 for most of the problems...

In the final analysis, the solution for determining u consisted in taking the greatest value yet guaranteeing a sufficient relative representation of the situations without line failures.

If m is the number of lines, this means that a certain value, greater than one is given to the probability ratio of these situations, which is expressed as follows :

$$\left(\frac{1-u}{1-q}\right)^m$$

This procedure does not always lead to the best possible determination of u, but, up till now, it has always allowed for the effective reduction of the variance of the different results of ORELIA.

5.3. - Use of the variance reduction technique during the optimization process

If s is selected with the probability t(s), instead of p(s), a.y + E f(y,s) may be written $\sum_s \left[a.y + f(y,s)\right] \times \frac{p(s)}{t(s)} \times t(s)$

It so happens that the gradient, in relation to y, of the function

$$\left[a.y + f(y,s)\right] \times \frac{p(s)}{t(s)} \quad \text{is} \quad g(y,s) \times \frac{p(s)}{t(s)}$$

which is also the expression to be used in order to compute the mathematical expectation of the gradient, using the MONTE-CARLO method with modified selection probabilities).

Consequently, for the stochastic gradient method, nothing speaks against using a selection probability of t(s), instead of p(s). Obviously, the above gradient expression must be used instead of g(y,s). In the final analysis, this is what was done in ORELIA, with t(s) being determined in the same way thus observed. To obtain the same accuracy for the final solution without modifying the selection probability, it seems that approximately twice the number of iterations are needed.

6 - CONCLUSION

The strenght of the stochastic gradient method makes it possible to suggest localization for generation units and network reinforcements, which minimize investments and management cost, while taking into account the random factors liable to affect the electric power system.

However, this method is a little difficult to use, and its "tuning' has been partially automatized to better guarantee the convergence.

Moreover, the method provides results in terms of continuous variables. It has thus been complemented by an integer method which is heuristic, but which provides integer solutions for lines and units to be constructed. These solutions, which are more meaningful, are, in respect to the economic function, rather close to the solutions provided by the stochastic gradient method.

REFERENCES

[1] PERSOZ H., Relations entre choix des sites de production et volume du réseau de transport". Revue Générale de l'Electricité, Tome 88, n°3 (mars 1979), pp. 210-214.

[2] DODU J.C. et MARECHAL P., "Un modèle pour la détermination de la localisation optimale des moyens de production : le modèle TASSILI". EDF Bulletin de la Direction des Etudes et Recherches, Série C, n°2 (1980).

[3] DODU J.C. and MERLIN A. "A dynamic model for long term expansion planning studies of power transmission systems : the Ortie model". Proc. PSCC 6 (1978).

[4] POLYAK B.T. "Pseudogradient adaptation and training algorithme". Automatika i Telemekhanica, n°3 (1973).

[5] POLYAK B.T. and TSYPKIN Y.Z. "Convergence and convergence rate of iterative stochastic algorithms". Automatika i Telemekhanica, n°12 (1976).

[6] DODU J.C. - HERTZ A., "Un nouveau modèle probabiliste pour la planification à long terme des réseaux électriques THT utilisant une méthode du gradient stochastique" P SCC 81.

[7] DODU J.C., GOURSAT M., HERTZ A., QUADRAT J.P., VIOT M., "Méthodes de gradient stochastique pour l'optimisation des investissements dans un réseau électrique" EDF. Bulletin de la Direction des Etudes et Recherches série C Mathématiques, informatique n°2 (1981).

SENSITIVITY ANALYSIS OF AN OPTIMAL CONTROL PROBLEM ARISING FROM AIR QUALITY CONTROL IN URBAN AREA

Piotr Holnicki, Jan Sokołowski and Antoni Żochowski

Systems Research Institute
Polish Academy of Sciences
ul. Newelska 6
01 - 447 Warszawa, Poland

1. Introduction

The paper deals with an optimal control problem for the atmospheric pollutant dispersion in the urban scale.

The computer forecasting model of the system [2] is used for the control purposes. The model is based on two-dimensional advection - diffusion equation, which is numerically solved by a combined FE - characteristic method [3]. The convex control constrained, optimal control problem is formulated.

The method of the sensitivity analysis of constrained optimization problems [8-11] is applied to the problem under consideration.

The directional derivative of an optimal control with respect to the parameter is derived in the form of an optimal solution to an auxiliary optimal control problem. The numerical results for an example are presented.

2. Forecasting model

We consider the urban-scale forecasting model [2] predestinated for short-term prediction of air pollution in a large city. The horizontal scale of simulated dispersion process is 20-40 km, while the time horizon of model's prediction is 1-3 days. The input data can be divided, in general, into the following three groups: i) independent of time structural data, ii) meteorological forecast, iii) characteristics of emission sources.

The physical process of pollutant dispersion in the atmosphere is considered in a sequare domain $\Omega = (L \times L)$ (see Fig.1). The process is discribed by two-dimensional, averaged over the mixing height H, advection - diffusion equation of the following form [1, 6]

$$\frac{\partial c}{\partial t} + \underline{w} \cdot \nabla c - K_H \Delta c + \gamma c = \overline{Q} + (E - v_d c)/H \quad \text{in} \quad \Omega \times (0,T) \qquad (2.1)$$

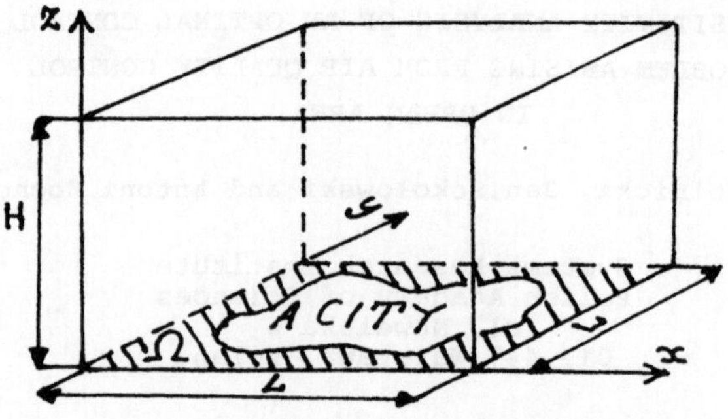

Fig. 1. Domain of simulation

along with the boundary conditions

$$\frac{\partial c}{\partial \underline{n}} = 0 \quad \text{on} \quad S^{+}, \tag{2.2}$$

$$c = 0 \quad \text{on} \quad S^{-} \tag{2.3}$$

and the initial condition

$$c(0) = y^{o} \quad \text{in} \quad \Omega, \tag{2.3}$$

where

$$S^{+} = \{(\underline{x},t) \mid \partial\Omega \times (0,T) / \underline{w} \cdot \underline{n} \geq 0\},$$

$$S^{-} = \{(\underline{x},t) \mid \partial\Omega \times (0,T) / \underline{w} \cdot \underline{n} < 0\}.$$

Here we use the following notation:

- c - pollutant concentration in $[\mu g/m^3]$,
- $\underline{w} = col(u,v)$ - wind vector in $[m/s]$,
- K_H - horizontal diffusion coefficient in $[m^2/s]$,
- $\overline{Q}$ - averaged over height H pointwise emission field in $[\mu g/m^3 s]$,
- E - area emission field in $[\mu g/m^2 s]$,
- v_d - dry deposition coefficient in $[m/s]$,
- γ - wet deposition factor depending on the precipitation intensity.

The values of the wind field vector $\underline{w}(x,y,t)$ in domain $\Omega \times (0,T)$ are predicted by a special procedure [2] as

$$\underline{w} = \underline{w}_o + \underline{w}_t + \underline{w}_\theta, \tag{2.4}$$

where $\underline{w}_o$ is the averaged value, $\underline{w}_t$ is the topographical correction, $\underline{w}_\theta$ is the thermal correction.

The model generates a sequence of forecasts of pollutant concentration within a period - T, which is discretized with time interval - δT. It's length is determined by the frequency of introducing meteorological data (in our case $\delta T=6$ hrs). Each time interval is segemented with discretization step-τ of a numerical procedure solving the advection - diffusion equation.

All time - dependent data at the inner points of the interval δT are linearly interpolated. The initial-boundary value problem (2.1)-(2.3) is numerically solved by an effective combination of the method of charcteristics with the finite element procedure [3].

The forecasting model has been tested on real data for Warsaw and Krakow areas [2].

3. Control problem

Basing on the forecasting model, the real-time emission control problem for the system of sources covering the area was formulated. A general idea of controlling consists in minimizing environmental impacts by redistributing the production (emission) among the set of selected sources, according to the meteorological situation.

In order to define an optimal control problem, we introduce the state equation, the cost functional and the set of sdmissible controls of the form:

State equation:

find concentration $c=c(\underline{u};\underline{x};t)$, for a given vector function $\underline{u}\in L^2(0,T;R^N)$, $(\underline{x},t)\in\Omega\times(0,T)$, which satisfies the parabolic equation

$$\frac{\partial c}{\partial t}+\underline{w}\cdot\nabla c-K_H\Delta c+\gamma c=Q+\sum_{i=1}^{n}\chi_i F_i(u_i), \text{ in } \Omega\times(0,T) \tag{3.1}$$

with the boundary conditions (2.2) and the initial conditions (2.3). Here $\underline{u}=\mathrm{col}(u_1,\dots,u_N)$ denotes the control.

Cost functional:

$$J(\underline{u})=\frac{1}{2}\int_o^T\int_\Omega r\max^2(0,c(\underline{u})-c_d)\,d\underline{x}\,dt+\frac{\alpha}{2}\int_o^T\sum_{i=1}^{N}\beta_i(u_i-\tilde{u}_i)^2dt, \tag{3.2}$$

where α,β_i, $i=1,\dots,N$ are given constants such that $\alpha\geq 0$, $\beta_i>0$; $r\in L^\infty(\Omega\times(0,T))$ is a given function, $r(\underline{x},t)\geq 0$, $(\underline{x},t)\in\Omega\times(0,T)$.

Set of admissible controls:

$$U_{ad}=\{\underline{u}\in L^2(0,T;R^N)\mid \underline{\underline{u}}_i(t)\leq u_i(t)\leq\bar{\bar{u}}_i(t),\quad i=1,\dots,N, \tag{3.3}$$

$\sum_{i\in N_j} a_{ij}u_i(t) \geq b_j,\ j=1,\dots,N,\ N_j \subset \{1,\dots,N\},$ for a.e. $t \in (0,T)\}$.

The functions $\chi_i(\underline{x})$ in (3.1) describe the location of controlled sources; $F_i(u_i)$ relate emission to the production level. The factor $r(\underline{x},t)$ in (3.2) is a region weight function, $c_d(\underline{x})$ denotes the admissible level of pollution. The second term in (3.2) constitutes the cost of deviation of production levels u_i from the desired economic values $\tilde{u}_i$. The control constraints represent technological limitations, and reflect the demand requirements b_j imposed on homogeneous groups of palnts N_j.

4. Differential stability of optimal controls

In this section we derive the form of the right-derivative of an optimal control u^ε with respect to the parameter ε at $\varepsilon=0$. To this end we will define the control constrained optimal control problem (P_ε), $\varepsilon \in [0,\delta)$.

We assume that the set of admissible controls U_{ad} is given by (3.3). Let $\varepsilon \in [0,\delta)$ be parameter, we denote

$$d^\varepsilon = (K_H^\varepsilon, \gamma^\varepsilon), \tag{4.1}$$

where

$$K_H^\varepsilon = K_H + K_H' + o(\varepsilon), \tag{4.2}$$

$$\gamma^\varepsilon = \gamma + \varepsilon\gamma' + o(\varepsilon); \tag{4.3}$$

here $\qquad |o(\varepsilon)|/\varepsilon \to 0 \quad$ with $\quad \varepsilon \downarrow 0$.

Let us denote by $c_\varepsilon = c_\varepsilon(\underline{u})$ a unique solution of the state equation:

$$\frac{\partial c_\varepsilon}{\partial t} + \underline{w} \cdot \nabla c_\varepsilon - K_H \Delta c_\varepsilon + \gamma^\varepsilon c_\varepsilon = Q + \sum_{i=1}^{N} \chi_i F_i(u_i) \quad \text{in} \quad \Omega \times (0,T) \tag{4.4}$$

with boundary conditions (2.2) and initial condition (2.3).

Let us consider the following optimal control problem:

Problem (P_ε)

Find an element $\underline{u} = \underline{u}(d^\varepsilon) \in L^2(0,T;R^N)$ which minimizes the cost functional

$$J_\varepsilon(\underline{u}) = \frac{1}{2}\int_0^T\int_\Omega r \max{}^2(0, c_\varepsilon(\underline{u}) - c_d)\,d\underline{x}\,dt + \frac{\alpha}{2}\sum_{i=1}^{N}\int_0^T \beta_i(u_i - \tilde{u}_i)^2 dt \tag{4.5}$$

over the set U_{ad}.

We assume for simplicity that

$$F_i(r) = r, \quad \forall r \in R, \quad i=1,\dots,N. \tag{4.6}$$

It can be shown that the optimal control is uniquely determined by the following optimality system:

Find $(\underline{u}^{\varepsilon}, c^{\varepsilon}, p^{\varepsilon})$ such that

$$\left\{\begin{array}{ll}
\dfrac{\partial c^{\varepsilon}}{\partial t} + \underline{w}\cdot\nabla c - K_H^{\varepsilon}\Delta c^{\varepsilon} + \gamma^{\varepsilon}c^{\varepsilon} = Q + \sum_{i=1}^{N}\chi_i u_i \quad \text{in} \quad \Omega \times (0,T), & (4.7)\\
\dfrac{\partial c^{\varepsilon}}{\partial n} = 0 \quad \text{on} \quad S^{+}, & (4.8)\\
c^{\varepsilon} = 0 \quad \text{on} \quad S^{-}, & (4.9)\\
c^{\varepsilon}(0) = y^{o} \quad \text{on} \quad \Omega, & (4.10)
\end{array}\right.$$

p^{ε} satisfies the adjoint state equation

$$\left\{\begin{array}{ll}
-\dfrac{\partial p}{\partial t} - \mathrm{div}(\underline{w}p^{\varepsilon}) - K_H^{\varepsilon}\, p^{\varepsilon} + \gamma^{\varepsilon}p^{\varepsilon} = r\max\{0, c^{\varepsilon}-c_d\} \quad \text{in} \quad \Omega \times (0,T), & (4.11)\\
p^{\varepsilon} = 0 \quad \text{on} \quad S^{-}, & (4.12)\\
K_H^{\varepsilon}\dfrac{\partial p^{\varepsilon}}{\partial n} + \underline{w}\cdot\underline{n}\, p^{\varepsilon} = 0 \quad \text{on} \quad S^{+}, & (4.13)\\
p^{\varepsilon}(x,T) = 0 \quad \text{on} \quad \Omega, & (4.14)
\end{array}\right.$$

and the following optimality condition is verified

$$\underline{u}^{\varepsilon} \in U_{ad}$$

$$\sum_{i=1}^{N} \left\{ \int_{o}^{T}\int_{\Omega} \chi_i p^{\varepsilon}(u_i - u_i^{\varepsilon})\,d\underline{x}dt + \alpha \int_{o}^{T} \beta_i (u_i^{\varepsilon} - \tilde{u}_i)(u_i - u_i^{\varepsilon})dt \right\} \geq 0, \quad \forall \underline{u} \in U_{ad}. \tag{4.15}$$

Theorem 1

For $\varepsilon > 0$, ε small enough

$$\underline{u}^{\varepsilon} = \underline{u}^{o} + \varepsilon\underline{q} + o(\varepsilon) \quad \text{in} \quad L^2(0,T;R^N), \tag{4.16}$$

where

$$\|o(\varepsilon)\|_{L^2(0,T;R^N)}/\varepsilon \to 0 \quad \text{with} \quad \varepsilon \downarrow 0.$$

The element $\underline{q} \in L^2(0,T;R^N)$ is given by a unique solution of the following optimal control problem:

Problem (Q):

Find an element $\underline{q} \in L^2(0,T;R^N)$ which minimizes the cost functional

$$I(\underline{u}) = \frac{1}{2}\int_0^T\int_\Omega r(\theta_o \max{}^2\{\theta,z\}+\theta_1 z^2)\,d\underline{x}dt - \int_{S^+} K'_H \frac{\partial p^o}{\partial \underline{n}}\, z\, d\Sigma + \frac{\alpha}{2}\int_0^T \beta_i u_i^2 dt \quad (4.17)$$

over the set of admissible controls (4.21), subject to state equation (4.18)-(4.20) with $\underline{q}=\underline{u}$.

The elements θ_o, θ_1 are given by (4.25), (4.26), respectively. The proof of Theorem 1 is given in [4].

We use here the following notation for problem (Q).

State equation:

$$\frac{\partial z}{\partial t} + \underline{w}\cdot\nabla z - K_H\Delta z - K'_H\Delta c^o + \gamma^o z + \gamma' c^o = \sum_{i=1}^N \chi_i u_i \quad \text{in} \quad \Omega\times(0,T), \qquad (4.18)$$

$$\frac{\partial z}{\partial n} = 0 \quad \text{on} \quad S^+, \quad z = 0 \quad \text{on} \quad S^-, \qquad (4.19)$$

$$z(x,0) = 0 \quad \text{on} \quad \Omega. \qquad (4.20)$$

Set of admissible controls:

$$C = \{\underline{u}\in L^2(0,T;R^N)\ |\ u_i(t)\geq 0 \ \text{ a.e. on } \Xi_1^i,\ u_i(t)\leq 0 \ \text{ a.e. on } \Xi_2^i,$$

$$\sum_{i\in N_j} a_{ij}u_i(t)\geq 0 \ \text{ a.e. on } \Xi_3^j,\ j=1,\dots,M, \qquad (4.21)$$

$$\sum_{i=1}^N \Big[\int_\Omega\int_0^T \chi_i p^o u_i\,d\underline{x}dt + \alpha\int_0^T \beta_i(u_i^o - u_i)u_i dt\Big] = 0\}.$$

Here

$$\Xi_1^i = \{t\in(0,T)\ |u_i^o(t) = \underline{\underline{u}}_i(t)\}, \qquad i=1,\dots,N, \qquad (4.22)$$

$$\Xi_2^i = \{t\in(0,T)\ |u_i^o(t) = \overline{\overline{u}}_i(t)\}, \qquad i=1,\dots,N, \qquad (4.23)$$

$$\Xi_3^j = \{t\in(0,T)\ |\sum_{i\in N_j} a_{ij}u_i^o(t) = b_j\},\ j=1,\dots,M, \qquad (4.24)$$

$$\theta_o(x_1,t) = \begin{cases} 1, & c^o(x_1,t) = c_d(x_1,t), \\ 0, & c^o(x_1,t) \neq c_d(x_1,t), \end{cases} \qquad (4.25)$$

$$\theta_1(x_1,t) = \begin{cases} 1, & c^o(x_1,t) > c_d(x_1,t), \\ 0, & c^o(x_1,t) \leq c_d(x_1,t). \end{cases} \qquad (2.26)$$

5. Numerical results

In order to ilustrate the results of previous paragraphs, a somewhat simplified control problem was studied. The most important simplification concerned the horizon of control, which was taken as equal

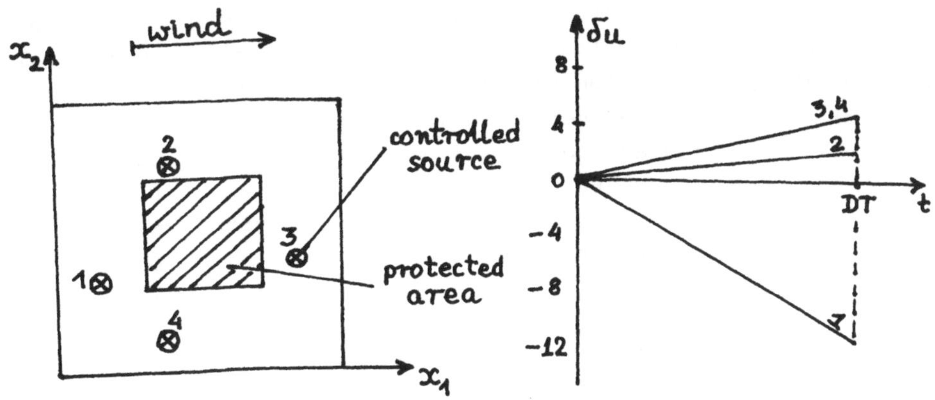

Fig.2. The control for K_H=300, α=2.

Table 1

source number		1	2	3	4
complementary case K_H=300					
K'_H = 100, sensitivity		0	-0,27	0,12	0,15
$u(K_H$=200)	approx.	-12.00	2.79	4.61	4.61
	exact	-12.00	2.76	4.60	4.64
$u(K_H$=100)	approx.	-12.00	3.06	4.49	4.46
	exact	-12.00	3.08	4.44	4.48
noncomplementary case K_H=300					
K'_H = 200, sensitivity		1.70	-0.82	-0.46	-0.42
$u(K_H$=500)	approx.	-10.30	1.95	4.14	4.21
	exact	-10.13	2.10	4.01	4.02
K'_H = -200, sensitivity		0.0	0.66	-0.30	-0.36
$u(K_H$=100)	approx.	-12.00	3.43	4.30	4.27
	exact	-12.00	3.25	4.36	4.39

to δT, see Section 2.In each of the intervals of the length δT the controls were approximated by linear functions, analogically to meteorological data. Thus, the decision variables (discrete) were reduced to increments of u_i, $i=1,...,N$ over the period δT, denoted δu_i.

The emission field and topographical data correspond to real neighbourhood of Warsaw. Four artificial controlled sources with emmission 35 each (1 unit ≈ 57,6 kg/sec) were added. The numerical values of other parameters were $\alpha=2$, $T=\delta T=6h$; $\underline{\underline{u}}_i=20$, $\overline{\overline{u}}_i=50$, for $i=1,...,4$. Only the city center was considered as protected area , what means that $r(\underline{x},t)$ was zero outside the shaded region in Fig.2. The value of K_H was 300. Only one demand constraint was imposed

$$\sum_{i=1}^{4} u_i \geq 140 .$$

It means that only transfer of power from one source to another was permissible.

The results of computation for $F_i(u_i) \equiv u_i$ and $\beta_i=1$, $i=1,...,4$ are shown in Fig.2.

In the next step the directional derivatives of u_i corresponding to the change of K_H in the direction K_H' were calculated.

Two cases were considered: one, when the strict complementarity condition connected with constraint imposed on δu_1 holds (corresponding dual variable positive) and the second, when the strict complementarity condition is not satisfied. In order to achieve the approximately noncomplementary situation the value of α was changed to $\alpha=2,55$.

The difference between both cases lies in the fact, that in the noncomplementary case the sensitivity had to be calculated twice, separately for $K_H' > 0$ and $K_H' < 0$.

The results are summarized in Table 1. The approximated values of controls δu_i are obtained using formula (5.8), while exact numbers correspond to the solution of original problem with new K_H.

All optimization problems encountered is this work have been solved with quickly convergent version [3] of the linearisation algorithm [7].

6. Concluding remarks

In the paper some theoretical and numerical results for the sensitivity analysis of an optimal control problem arising in air pollution control are presented.

We prove that an optimal control is locally stable (Lipschitz continuous) with respect to the perturbations of the coefficients of the state equation.

Our numerical results confirm that the formulae for the right-derivatives of an optimal control can be used in order to approximate the increments of the optimal control with respect to the perturbations of data.

References

[1] G. Fronza and P. Melli (eds.): Mathematical Models for Planning and Controlling Air Quality, Proceedings of IIASA Workshop, Pergamon Press, 1982.

[2] P. Holnicki and al.: An urban-scale computer model for short - term prediction of air pollution, Arch. Autom. i Telemech. (to appear).

[3] P. Holnicki and A. Żochowski: Numerical methods in forecasting and controlling air pollution problems, Systems Research Institute, Report No. ZTS-15-7/84, Warsaw, 1984, (in Polish).

[4] P. Holnicki, J. Sokołowski and A. Żochowski: Differential stability of solutions to air quality control problems in urban area , (to appear).

[5] J.L. Lions: Controle optimal de systemes gouvernes par des equations aux derivees partielles. Dunod, Paris, 1968.

[6] G.I. Martchuk: Mathematical Modelling in Enviromental Problems, Nauka, Moscov 1982 (in Russian).

[7] B.N. Pshenitchny: Linearization Method, Nauka, Moscov 1983, (in Russian).

[8] J. Sokołowski: Differential stability of solutions to constrained optimization problems, Applied Mathematics and Optimization, 13, 97-115 (1985).

[9] J. Sokołowski: Differential stability of control constrained optimal control problems for distributed parameter systems, Proceedings of 2nd International Conference on Control Theory for Distributed Parameter Systems and Applications, Springer Verlag, (to appear).

[10] J. Sokołowski: Sensitivity Analysis and Parametric Optimization of Optimal Control Problems for Distributed Parameter Systems, Zeszyty Naukowe Politechniki Warszawskiej, seria Elektronika,(to appear).

[11] J. Sokołowski: Differential stability of solutions to boundary optimal control problems for parabolic systems, in the present proceedings.

ON MATHEMATICAL MODEL OF IMMUNOLOGICAL TOLERANCE

T. HRABA[1], J. DOLEŽAL[2]

[1]Institute of Molecular Genetics
Czechoslovak Academy of Sciences
142 20 Prague, Czechoslovakia

[2]Institute of Information Theory and Automation
Czechoslovak Academy of Sciences
182 08 Prague, Czechoslovakia

1. Introduction and Previous Results

In the last decade our experimental work has been concentrated on analyzing the mechanisms responsible for the specific inhibition of antibody formation in chickens rendered tolerant to human serum albumin (HSA) shortly after hatching [1, 2]. Our findings suggested that the mechanism operative in the studied system is probably the unresponsiveness at the B lymphocyte level induced by the administration of tolerizing dose of antigen.

It is difficult and costly to obtain sufficient amount of experimental data giving thus deeper insight of respective mechanisms. Therefore a mathematical model of this tolerant state was successively developed [3, 4, 5] with the intention to propose more efficient experimental schemes. The model assumed that the tolerance is induced by irreversible inactivation of B cells by antigen and that the escape from tolerance was effected by differentiation of new B cells (lymphocytes) reactive to HSA from the stem cells after the elimination of antigen from the organism. However, the observed recovery from tolerance in chickens challenged with HSA was much slower than that calculated according to this model.

At the time, when our mathematical model was formulated, we observed that anti-HSA antibody production induced in tolerant chickens by a cross-reactive antigen (bovine serum albumin - BSA) was much more intensive than after the challenge with the tolerated antigen. The rate of faster recovery from tolerance to HSA observed after BSA challenge agreed well with the calculated values.

In the first experiments tolerance was induced by four injections of HSA administered on 1st, 4th, 7th and 11th day after hatching. The computed duration of tolerance to HSA induced by one or four injections did not reveal any principal difference. The highest difference between the simulated B cell recovery from tolerance induced by these two injection schedules was at the age of four weeks. This prediction was tested experimentally and the obtained results agreed reasonably well with the calculated values [5].

Experimentation with the mathematical model motivated us to test whether lower doses of HSA are able to induce tolerant state in hatched chickens. We found that also doses 100 time lower than those used in our earlier experiments induced detactable tolerance [6]. The model was also compared with the experimental data of other authors concerning the tolerance in adult mice [7]. The kinetics of B cell tolerance to an antigen with fast elimination - sheep red blood cells - and to an antigen with slow elimination - human gamma globulin - were used for this comparison. In both cases a good correspondence was observed between the experimental a calculated values.

These findings suggest that our model can exhibit B cell tolerance well. However, the discrepancy between the experimentally observed escape from tolerance to HSA in chickens and the values calculated by the means of the model clearly indicated that some additional mechanism to B cell tolerance is operative in this tolerant state. Based on experimental results we concluded that neither T lymphocytes not macrophages played any important role. Thus we tested in our model the assumption that anti-HSA antibody production can occur only when two B cell clones reactive to two different determinants of the HSA were present. The computed results indicated that the recovery from tolerancewas not much slower than earlier and thus thes assumption cannot explain the noticed difference [8].

The mathematical model was recently modified by taking into the account also T helper (T_h) lymphocytes. Preliminary observations in this respect we summarized in [8]. Here we present an augmented version of our model which includes not only T helper cell compartment, but also takes into the account possible T suppressor (T_s) lymphocytes. The values predicted by this model are compared with experimental data collected on chickens and mice.

2. Augmented Mathematical Model of Tolerance

In this version of our model three lymphocyte categories specifically reactive to HSA are considered:

(i) B lymphocytes - precursors of antibody producing cells.

(ii) T_h lymphocytes - their interaction with HSA specific B cells is considered as necessary condition for antibody production.

(iii) T_s lymphocytes - their suppressive activity is assumed to be induced by the tolerizing dose of antigen.

In B and T_h cell categories two developmental compartments are anticipated:

(i) The immature immunocompetent cell compartment - I cells in the case of B lymphocytes and J cells in the case of T_h lymphocytes.

(ii) The mature immunocompetent cell compartment - X cells in the case of B lymphocytes and U cells in the case of T_h lymphocytes.

Sizes of I and X cell compartments of B cells are described by the following differential equations with the given initial values:

$$dI(t)/dt = t_I[I_E - I(t)] - M\,a(t)\,I(t), \quad I(0) = I_0, \tag{1}$$

$$dX(t)/dt = t_I I(t) - t_X X(t) - a(t)\,X(t), \quad X(0) = X_0, \tag{2}$$

where I(t) and X(t) are the numbers of I and X cells at time t, t_I and t_X are the rates of maturation of I into X cells and death of X cells, respectively. The quantity a(t) is the rate of irreversible inactivation of X cells by antigen and analogously Ma(t) that of I cells with $M > 1$; $a(t) = a_o \exp(-\lambda t)$, where a_o depends on the amount of antigen injected and λ is the rate constant of its non-immune elimination. From the steady-state considerations in the absence of antigen, i.e. $a_o = 0$, it simply follows that $t_i I_E = t_X X_E$ with index E denoting the steady-state values.

Denote $X_c(t)$ the number of X cells in the controls at time t, which is obtained as the solution of the model equations (1)-(2) with $a_o = 0$. Then

$$r_0(t) = 100\,[X(t)/X_c(t)] \tag{3}$$

is the percent measure of X cell recovery from tolerance.

The sizes of J and U cell compartments are quite analogously described by the following differential equations:

$$dJ(t)/dt = t_J[J_E - J(t)] - P\,a(t)\,J(t), \quad J(0) = J_0, \tag{4}$$

$$dU(t)/dt = t_J J(t) - t_U U(t) - N\,a(t)\,U(t), \quad U(0) = U_0, \qquad (5)$$

where J(t) and U(t) are the numbers of J and U cells at time t, t_J and t_U the rates of maturation of J into U cells and death of U cells. respectively, Again, in the steady-state one has $t_J J_E = t_U U_E$ and also now we assumed $P > 1$ and $N > 1$. The percent recovery from tolerance was measured as

$$r_1(t) = 100\,[X(t)/X_c(t)]\,[U(t)/U_c(t)], \qquad (6)$$

where $U_c(t)$ is defined by equations (4)-(5) by setting $a_o = 0$.

Because the escape from tolerance of B cells is much faster than that of T_h cells, the latter becomes the limiting factor in the recovery of immune reactivity to the tolerated antigen. When B cells escaped already from tolerance and T_h cells are not yet reactive to the tolerated antigen, the challenge of the tolerant animal with a cross-reactive antigen can detect the recovered B cell reactivity. This is due to the fact that T_h cells reactive to epitopes of the cross-reactive antigen, which are different from those shared with the tolerated one, can cooperate with B cells reactive to the tolerated antigen and help them to produce antibodies.

As there were described at least two types of antigen specific T_h cells collaborating with B cells in antibody production to the same antigen, the mathematical model was further modified by the assumption of the activity in anti-HSA antibody formation of two T_h cell populations of the same size and lifespan. In this case the percent recovery from tolerance is measured as

$$r_2(t) = 100\,[X(t)/X_c(t)]\,[U(t)/U_c(t)]\,[U(t)/U_c(t)]. \qquad (7)$$

The recovery from tolerance is further slowed down and, consequently, the same effect is achieved for considerably shorter T_h cell lifespans with respect to r_1. On the other hand, if only one T_h cell compartment is involved, it is necesary to admit some additional mechanism which, on the contrary, can be of suppressive type. The effect of such mechanism prolongs the escape from tolerance of T_h cells. If antigen specific T_s cells are present, the tolerizing dose of antigen will activate them and the resulting suppressive effect will again slow down the recovery from tolerant state.

At this place only the so-called suppressive activity is introduced explicitely not taking into the account the underlying cell dynamics. It is assumed that the suppressive activity on T_h cells is a relative factor of the form

$$s_a(t) = \begin{cases} \sigma, & 0 \leqslant t < 3\lambda, \\ 1 - (1-\sigma)\exp(\gamma(t-3\lambda)), & t \geqslant 3\lambda. \end{cases} \qquad (8)$$

Also now it is assumed that the tolerizing dose of antigen is being administered at time t = 0. As σ, $0 < \sigma < 1$, is denoted the residual T_h cells activity (corresponds to maximal suppression) and γ is the decline rate of the suppressive activity on T_h cells. Maximal suppression is assumed to last over the period of persistence of tolerizing dose of antigen. The above used value of 3λ corresponds to 95 % of antigen elimination. Then the percent recovery from tolerance can be measured as

$$r_3(t) = 100\,[X(t)/X_c(t)]\,[U(t)/U_c(t)]\,s_a(t), \qquad (9)$$

i.e., $r_3(t) = r_1(t)\,s_a(t)$. From the practical point of view the values of the residual level σ and the duration of maximal suppression 3λ are not crutial as in the inital phase the tolerizing dose eliminates both B and T_h cells.

3. Tolerance in Chickens

In our model we anticipate that the higher susceptibility of immature B and T_h cells (I and J cells) is responsible for the relatively easy induction of tolerance to HSA in newly hatched chickens. We assume that at the time of hatching only immature immunocompetent cells (ICC), both B and T_h, i.e., I and J cells, are present in chickens and that mature ICC differentiate only after hatching. The presence of mature ICC, which are less susceptible to tolerance induction, is assumed to be the reason of the resistance to tolerance induction observed in older chickens. The transition from the immature to the mature cell compartment and the differentiation of immature lymphocytes from precursor cells lacking immunocompetence are spontaneous processes which are not dependent on action of antigen.

The following values of various parameters were used during the simulation runs: $I_0 = I_E = 50$, $X_0 = 0$, $X_E = 100$, $M = 5$, $a_0 = 2.4$, $\lambda = 0.72$, $t_I = 0.2$, $t_X = 0.1$, $J_0 = 100$, $U_0 = 0$, $U_E = 100$, $P = 50$, $N = 5$, and $t_J = 0.2$. Various t_U together with according to the steady-state equation corresponding values of J_E were used in each case. All rate constants are in day^{-1}. The fairly high value of J_0 was assumed to describe the relatively fast maturation of T_h cell reactivity shortly after hatching. The r_0 curves characterizing the escape of B cells from tolerance are always included for the sake of comparison in Figs. 1 - 3.

The dynamics of tolerance computed according to the present model were compared with the relative numbers of anti-HSA antibody forming cells in spleens of tolerant chickens challenged with HSA at the age of 2, 4, or 6 weeks. In these experiments tolerance was induced by one injection of 100 mg of HSA on the day of hatching [9]. These experimental values are compared with the computed r_1 curves of recovery from tolerance in Fig. 1. To see the complete picture of the recovery from tolerance the time interval of 300 days is always used, although, because of the available data, the trial-and-error fitting was performed in more appropriate scales. Curves a, b, and c depict $r_1(t)$ time courses for U cell lifespans t_U = 0.01, 0.007, and 0.004 (day^{-1}) and for the corresponding values of J_E = 5, 3.5, and 2. The experimental values depicted as black circles agree best with the curve b for the U cell lifespan of 143 days. This value, however, seems too high, especially for the young individuals.

The curves a, b, and c depicted in Fig. 2 give the $r_2(t)$ time courses for U cell lifespan t_U = 0.025, 0.02, and 0.015 (day^{-1}) and the corresponding values of J_E = 12.5, 10, and 7.5. Recovery of tolerance c mputed according to this modification of the model was substantially slower than that given by $r_1(t)$. Experimental data fitted best the curve b for the corresponding U cell lifespan 50 days.

For the case of suppressive activity $s_a(t)$ the following values

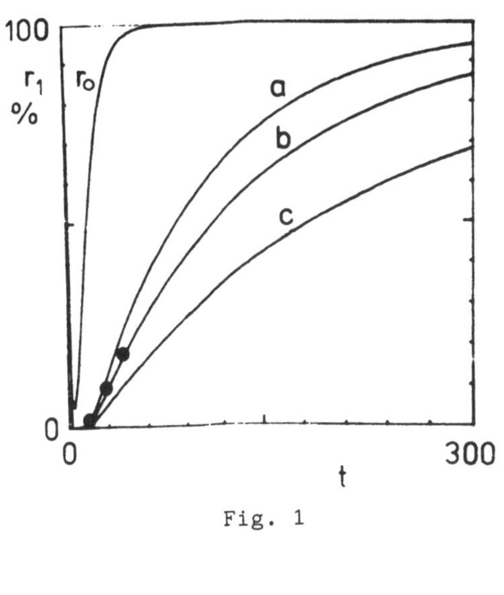

Fig. 1

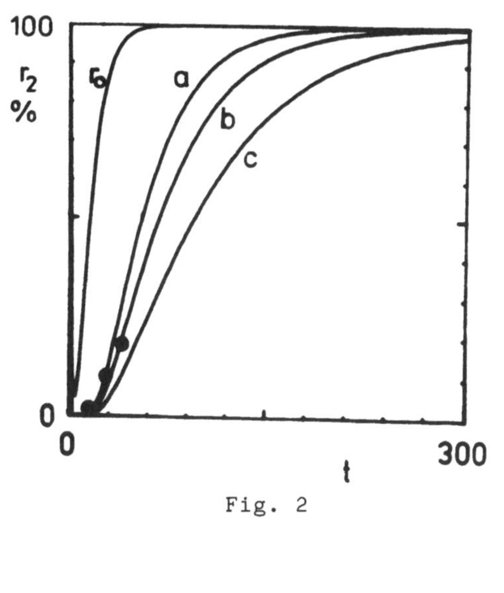

Fig. 2

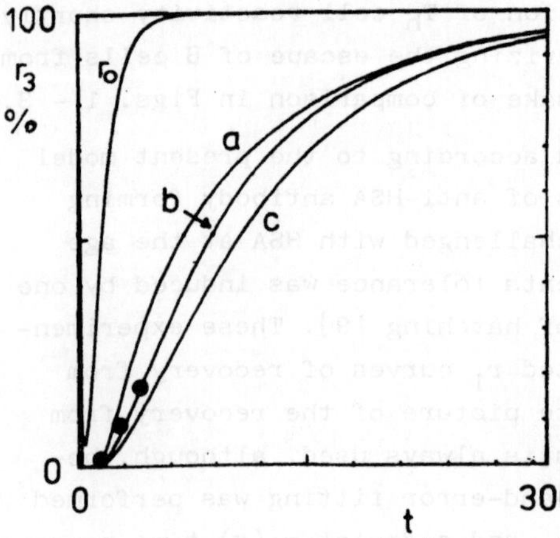

Fig. 3

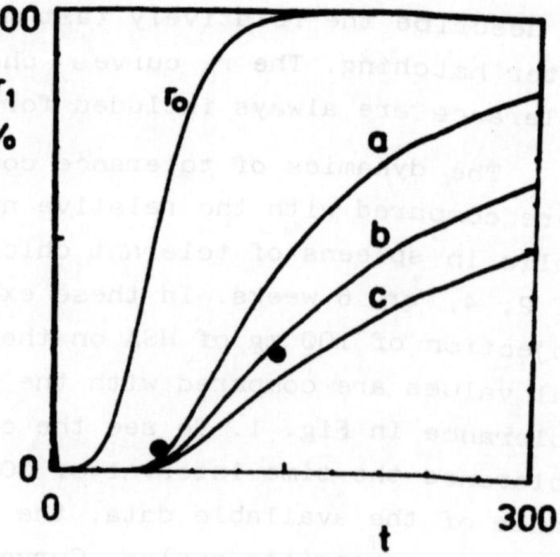

Fig. 4

were used: $\sigma = 0.1$, $\gamma = 0.01$ (day^{-1}). Fig. 3. depicts the $r_3(t)$ time courses (curves a, b, and c) for U cell lifespans $t_U = 0.05$, 0.025, and 0.007, and the corresponding values of $J_E = 25$, 12.5, and 3.5. The best approximating curve b corresponds to $t_U = 0.025$, i.e., the U cell lifespan of 40 days.

4. Tolerance in Mice

The mathematical model was further compared with experiments of other authors [10, 11] on tolerance to human gamma globulin (HGG) in mice. This antigen administered in deaggregated, soluble form induces tolerance even in adult mice. The duration of B and T_h cell tolerance is well known in this case. In our previous paper [7] the kinetics of B cell tolerance observed experimentally was compared with the values calculated according to our model.

The similar analysis as in the case of chickens is illustrated in Fig. 4 - 6. Again, the corresponding $r_0(t)$ curve representing the pure B cell tolerance is always included for the sake of comparison. The following set of model parameters was used during the simulation runs: $I_0 = I_E = 25$, $X_0 = X_E = 100$, $M = 5$, $a_o = 2.4$, and $\lambda = 0.1032$ for the B cell compartment, and choosing the best fit according to [7] with $t_I = 0.1$ and $t_X = 0.05$. In addition, for the U cell compartment: $U_0 = U_E = 100$, $P = 50$, $N = 5$, and $t_J = 0.2$. The value of t_U was varied for the best fit and the corresponding values according to the steady-state equation were set for $J_0 = J_E$. All rate constants are in day^{-1}.

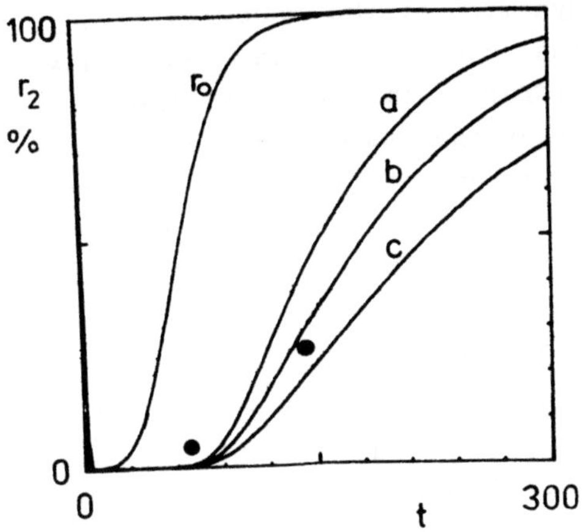

Fig. 5

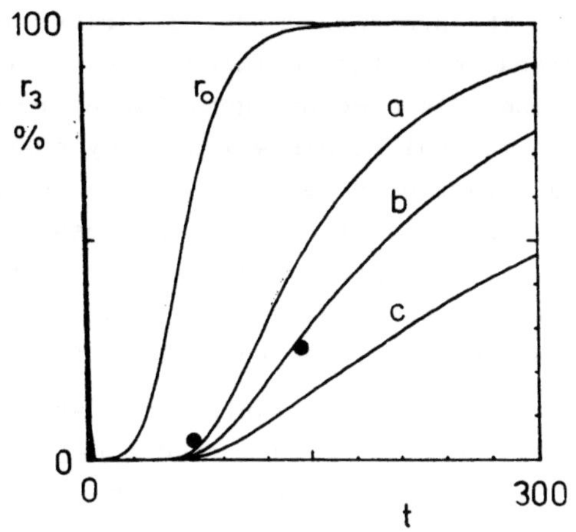

Fig. 6

In Fig. 4 the $r_1(t)$ dynamics of recovery from tolerance for U cell lifespans t_U = 0.008, 0.0045, and 0.003 (day^{-1}) and the corresponding values of $J_0 = J_E$ = 4, 2.5, and 1.5 are compared with the experimental data of [10]. From the resulting curves a, b, and c the best approximating curve b was obtained for the U cell lifespan of 222 days.

The $r_2(t)$ time courses for U cell lifespans t_U = 0.015, 0.011, and 0.008 (day^{-1}) and the corresponding $J_0 = J_E$ = 7.5, 5.5, and 4 are depicted as curves a, b, and c, respectively, in Fig. 5. The curve b for the U cell lifespan 91 days fitted best the experimental data.

Finally, for the suppressive activity $s_a(t)$ the same values as in the case of chickens were used in (8). Fig. 6 depicts the $r_3(t)$ time courses (curves a, b, and c) for U cell lifespans t_U = 0.015, 0.007, and 0.003 (day^{-1}) and the corresponding values $J_0 = J_E$ = 7.5, 3.5, and 1.5. The best approximating curve b corresponds to the U cell lifespan of 143 days.

5. Discussion

The problems related with the study of mechanisms of tolerance to HSA in chickens called for the formulation of a mathematical model which, in turn, gave better insight into the involved celular processes and stimulated new experimental work. The close relation and interaction with experimental work led to several modifications of the original model, some of them being quite promising.

The presented model describes the dynamics of the recovery from tolerance at the cellular level, i.e., the return of immune reactivity to the tolerated antigen. Our approach differs from the most other mathematical models which study quantitative aspects of immune response. They start with mature immunocompetent cells and study their differentiative and proliferative processes after challenge with antigen. In our case the preceding phase - the maturation of immunocompetent cells in tolerized animals - is studied from the point of view of the qualitative aspects of the transition from tolerance to immune reactivity.

This paper summarizes our attempts to incorporate T helper cells in the originally suggested model of neonatal tolerance to HSA in chickens and to extend it further by the inclusion of T suppressor cell activity. This augmented model is then confronted with the experimental data on the kinetics of immunological tolerance to HGG induced in adult mice. In both cases no clear evidence for any of the suggested alternatives is available due to the lack of sufficient experimental data. On the other hand, cellular mechanisms of immune reactions were more extensively studied in mice and it seems probable that it would be easier to reach a more definite conclusions there in the future. This being the case, it would help also to better understand the situation in chickens.

References

[1] Hraba, T., Karakoz, I., Madar, J.: Mechanisms of immunological tolerance to a xenogeneic serum protein in chickens. In: "Cellular and Molecular Mechanisms of Immunological Tolerance." (Eds. T. Hraba and M. Hašek), Marcel Dekker, Inc., New York 1981, pp. 361-366.

[2] Hraba, T., Karakoz, I., Madar, J.: Immunological unresponsiveness to HSA in chickens. Ann N.Y. Acad. Sci. 392: 47, 1982.

[3] Klein, P., Doležal, J., Hraba, T.: Compartmental model of immunological tolerance to HSA in chickens. Folia biol. (Prague) 25: 345, 1979.

[4] Klein, P., Doležal, J., Hraba, T.: Compartmental models of immunological tolerance. Kybernetika 19: 285, 1980.

[5] Klein, P., Hraba, T., Doležal, J.: The use of immunological tolerance to investigate B cell replacement kinetics in chickens. J. Math. Biology 16: 131, 1983.

[6] Hraba T., et al.: Induction of tolerance in chickens by low doses of human serum albumin administered after hatching. Folia biol. (Prague) 30: 276, 1984.

[7] Klein, P., Hraba, T., Doležal, J.: Mathematical model of B lymphocyte replacement kinetics: its application to the recovery from tolerance in adult mice. Math. Biosci. 73: 228, 1985.

[8] Doležal, J., Hraba, T.: T helper cell inclusion into the mathematical model of immunological tolerance. Folia biol. (Prague) 30, 281, 1984.

[9] Hraba, T., Madar, J., Karakoz I.: Antibody-producing cells in chickens tolerant to human serum albumin. Folia biol. (Prague) 30: 281, 1984.

[10] Chiller, J.M., Weigle, W.O.: Termination of tolerance to human gamma globulin in mice by antigen and bacterial lipopolysacchride (endotoxin). J. Exp. Med. 137: 740, 1973.

[11] Parks, D.E., Weigle, W.O.: Maintance of immunological unresponsiveness to human γ-globulin: evidence for irreversible inactivation in B lymphocytes. J. Immunol. 124: 1230, 1980.

PROBLEM OF OPTIMAL CONTROL FOR INTEGRAL DYNAMIC MODELS WITH CONTROLLABLE PREHISTORY

V.V. IVANOV, Ju.P. JACENKO

V.M. Glushkov Institute of Cybernetics
Academy of Sciences of the Ukrainian SSR
252207 Kiev 207 USSR

The class of dynamic models [1-3] suggested by academician V.M. Glushkov turned out to be the promising tool for modelling and optimization of system development in various applications: economics, ecology, biology, etc. Similar models in economics are known as models with materialized technical progress and disaggregated funds. The mathematical apparatus of this class of models is the integral equations of the Volterra type with controllable previous history of the form

$$x_i(t)=\sum_{j=1}^{m}\int_{z_j(t)}^{t} K_{ij}(\tau,t)\,y_{ij}(\tau)\,x_j(\tau)\,d\tau + f_i(t), \quad i=\overline{1,n}, \tag{1}$$

$$0 \le y_{ij}(\tau) \le 1, \quad -\infty < z_j(t) < t, \tag{2}$$

where some of variables x_i, y_{ij}, z_j are unknown. The models (1)-(2) are the generalization of the well-known models of linear nonstationary systems of automatic control for the case of transients with variable (or unknown) durations $t-z_j(t)$ (the so-called "previous histories" of processes). If at least one of the functions z_j is unknown then the corresponding mathematical problems are nonlinear and require the development of special methods for their investigation.

The most interesting problems from application and mathematical points of view are the problems of optimal control for models (1)-(2) which are reduced to the definition of n_1 unknown functions from the set $\{x_i, y_{ij}, z_j\}$, $n_1 > n$, satisfying the relations (1)-(2) and providing the extremum of some functional

$$I = I(x_i, y_{ij}, z_j) \rightarrow max\,(min). \tag{3}$$

The existence of the solution of problems (1)-(3) is proved on the basis of paper [4] by using the generalized control functions (the generalized Young curves). The technique of qualitative investigation of problems is based on the derivation of the gradient of the functional (3) under constraints (1) by using the Lagrange functional and on the investigation of the integro-functional equations obtained above.

Until the present time a number of meaningful problems of optimal control for economic systems with one and two control functions have been obtained. There have been investigated the qualitative behaviour and the structure of optimal control functions, the uniqueness conditions for solutions as well as the number of switching points of control functions and other related problems.

The research carried out shows that even for simplest "classical" figures of merit the optimization problems even in cases of one- and two-product models $(n=2;\ m=1,2)$ have solutions of very complex structure.

Main properties of problems of optimal control in the two-product macroeconomic model are given below.

Description of the model. The considered two-production model describes the interaction of two groups of production in macroeconomics: A-production of means of production (new work places - WP), B-production of consumption goods. It includes the following system of equations and inequalities [1-3, 5] :

$$m(t)=\int_{a_1(t)}^{t} \alpha(\tau,t)\,y(\tau)\,m(\tau)\,d\tau, \tag{4}$$

$$c(t)=\int_{a_2(t)}^{t} \beta(\tau,t)\,[1-y(\tau)]\,m(\tau)\,d\tau, \tag{5}$$

$$P(t)=\int_{a_1(t)}^{t} y(\tau)m(\tau)\,d\tau+\int_{a_2(t)}^{t}[1-y(\tau)]\,m(\tau)\,d\tau, \tag{6}$$

$$N(t)=c(t)+\varphi(t),\quad \varphi(t)=\lambda(t)m(t),\quad t\geqslant t_0, \tag{7}$$

$$0\leqslant y(\tau)\leqslant 1,\quad a_i(t)<t,\quad a_i'(t)\geqslant 0,\quad i=1,2, \tag{8}$$

with initial conditions

$$a_i(t_0)=a_i^0\geqslant \tau_0,\quad i=1,2,$$
$$y(\tau)\equiv y_0(\tau),\quad m(\tau)\equiv m_0(\tau),\quad \tau\in[\tau_0,t_0]. \tag{9}$$

It is assumed that functions $\alpha(\tau,t)$, $\beta(\tau,t)$, $c(t)$, $\lambda(t)$, $\varphi(t)$, $N(t)$, $P(t)$, $y_0(\tau)$, $m_0(\tau)$ are continous, positive and satisfy (4)-(9) for $t=t_0$. Economic meaning of model functions see in [1-3, 6].

I. Problem of maximization of the output of useful product under the given labour resources is formulated in the following way [1-3]: under given $\alpha, \beta, P, y_0, m_0$ define functions y, a_1, a_2, m, c, providing the extremum

$$I_1 = \int_{t_0}^{T} c(t)\,dt \to max \tag{10}$$

under constraints (4)-(6), (8) and initial conditions (9).

The following results are valid.

Theorem 1. There exists the moment Θ, $t_0 \leq \Theta < T$, such that solutions $y(t)$, $a_2(t)$ are minimally permissible $y_m(t)$, $a_{2m}(t)$ by virtue of conditions (8) on the interval $(\Theta, T]$. If $\beta'_\tau \leq 0, \alpha'_\tau \leq 0$, then $\Theta = t_0$.

Theorem 2. If $\alpha'_\tau > 0$, $\beta'_\tau > 0$, and on some interval $[t_0, \Theta)$ the inequalities

$$\frac{\alpha(u,t)}{\beta(u,t)} \leq \frac{\alpha(v,t)}{\beta(v,t)}, \quad 0 \leq u \leq v \leq t \leq T, \tag{11}$$

$$\beta(0,t)/\alpha(0,t) < \int_t^T [\beta(t,\tau) - \beta(0,\tau)]\,d\tau \tag{12}$$

are valid, the solutions $y(t)$, $a_1(t)$, $a_2(t)$ are greater than $y_m(t)$, $a_{1m}(t)$, $a_{2m}(t)$ on some subinterval Δ of interval $[t_0, \Theta)$.

Condition (12) is fulfilled if the rate of technical progress in group B exceeds the rate of possible reduction of productivity of all WP because of factors of current year; condition (11) is fulfilled if the rate of technical progress in group A is not less than in group B.

Thus, for problem 1 the result is valid which was proved earlier in [2,3] for this problem for $a_1 \equiv a_2$ about increasing a relative part of WP which go in production of means of production at the initial part of the planning interval $[t_0, T]$ in the case of large $T - t_0$ and the presence of technical progress.

Theorem 3. If $\beta(\tau,t) \equiv K(t)\,\alpha(\tau,t)$ and $\beta'_\tau > 0$, then $a_1(t) \equiv a_2(t)$ at least on the intervals where solutions $a_1(t)$, $a_2(t)$, $y(t)$ are inner in region (8).

Theorem 4. Let $\frac{\partial}{\partial \tau}[\alpha(\tau,t)/\beta(\tau,t)] \not\equiv 0$, $\alpha'_\tau > 0$, $\beta'_\tau > 0$ and solution $y(t)$ is differed from boundary $y_m(t)$. Then $a_1(t)$ and $a_2(t)$

are not identically equal, moreover, near the right boundary $t=T$ of interval $[t_0,T]$, $a_1(t)<a_2(t)$ for $\partial/\partial\tau[\alpha/\beta]<0$ and $a_1(t)>a_2(t)$ for $\partial/\partial\tau[\alpha/\beta]>0$.

Hence, the introduction of different moments of elimination of obsolete WP $a_1(t)$ and $a_2(t)$ into the model (4)-(9) is efficient in the case when the rates of technical progress in groups A and B are different. In addition, accelerated renovation of WP is optimal in that group where the rate of progress is higher.

There exists the following theorem on a structure of solutions of problem I which develops the results of papers [6,7].

Theorem 5. In problem I for $\beta(\tau,t)=K(t)\alpha(\tau,t)$ there exists some "best" function $\tilde{a}(t)$, $t\in[t_0,T]$, which does not satisfy the conditions (8)-(9) in the general case on which the gradient of functional (10) $\partial I_y\equiv 0$ for $t\in[t_0,\theta)$, $\theta<T$, $\partial I_y(t)<0$ for $t\in(\theta,T]$. Depending on the length of the planning interval $T-t_0$ and values of disagreement $\Delta_a=|\tilde{a}(t_0)-a(t_0)|$, $\Delta'_a=|\tilde{a}'(t_0)-a'(t_0)|$ the solutions $y(t)$, $a(t)=a_1(t)\equiv a_2(t)$ can have the following versions of behaviour:

1) $y(t)\equiv y_m(t)$, $a(t)\equiv a_m(t)$;

2) $y(t)=\begin{cases}1, & t\in[t_0,\tau_1],\\ y_m(t), & t\in(\tau_1,T],\end{cases}$ $a(t)$ does not intersect $\tilde{a}(t)$;

3) $y(t)=\begin{cases}1, & t\in(\tau_{i-1},\tau_i],\\ y_m(t), & t\in(\tau_i,\tau_{i+1}],\end{cases}$ $i=\overline{1,n}$, $\tau_0=t_0$, $\tau_{N+1}=T$,

$a(t)$ intersects $\tilde{a}(t)$ N times, $N\geq 1$.

4) $y(t)$ has an infinite denumerable set of isolated switching points $\{\tau_i\}_{i=1}^{\infty}$, $\tau_i\to\tau^*$, $a(t)$ intersects $\tilde{a}(t)$ infinite number of times, $a(t)\to\tilde{a}(t)$, $a'(t)\to\tilde{a}'(t)$ for $t\to\tau^*$, $a(t)\equiv\tilde{a}(t)$ for $t\in[\tau^*,T]$ (such behaviour corresponds to small values Δ_a, Δ'_a as compared to $T-t_0$);

5) $a(t)\equiv\tilde{a}(t)$, $t\in[t_0,T]$, $y(t)$ has no switching points and is inner for $t\in[t_0,\theta]$ (if $\Delta_a=\Delta'_a=0$).

Problem I may be investigated similarly in the general case (for $\alpha\not\equiv K(t)\beta$) in which the analogous theorem will have even two "best" functions $\tilde{a}_1(t)$ and $\tilde{a}_2(t)$.

Theorem 5 is close to the theorems of mathematical economics which describe the asymptotics of optimization problems for large terms of planning. By some modification and detalization of it we obtain the following statement.

Theorem 6. Let conditions

1) $\beta(\tau,t)=k(t)\alpha(\tau,t),\ \beta'_\tau(\tau,t)>0;$

2) $\int_\tau^\infty [\beta(\tau,u)-\beta(0,u)]du > k(\tau),\ \tau \geq t_0$

be fulfilled. Then in problem I there is some best function $\tilde{a}(t)$, moreover, for all $\Theta > t_0$ there is $\bar{T}(\Theta)$ such that for all $T \geq \bar{T}$ the behaviour of $\tilde{a}(t)$ on interval $[t_0,\Theta)$ does not depend on value $T-t_0$ and is defined only by functions $\beta(\tau,t)$ and $\alpha(\tau,t)$. For large $T-t_0$ the solution $a(t)\equiv\tilde{a}(t)$ for $t\in[\tau^*,T],\ \tau^*\geq t_0$, only if $0\leq y(\tau)\leq 1$ for $t\in[\tau^*,T]$ .Here the duration of the initial "transient process" τ^*-t_0 does not depend on values $T-t_0$.

Theorem 7. Let $\beta(\tau,t)\equiv K(t)\cdot\alpha(\tau,t)$. Then on interval $[t_0,\Theta)$ defined by theorem 5 the function $\tilde{a}(t)$ has the following properties:

1) if $\beta'_\tau(\tau,t)$ increses (decreases) with respect to τ , then $t-\tilde{a}(t)$ decreases (increases);

2) if $\beta'_\tau(\tau,t)$ does not depend on τ , and $\alpha(\tau,t)$ and $\beta(\tau,t)$ decrease (increase) with respect to t , then $t-\tilde{a}(t)$ increases (decreases);

3) if $\beta(\tau,t)$ and $\alpha(\tau,t)$ do not depend on t and are linear with respect to t_0 , then $\tilde{a}(t)\equiv t-const$.

Thus, if the rate of technical progress increases ($\beta''_{\tau\tau}>0$) then the optimal period of functioning of WP $t-\tilde{a}(t)$ decreases and vice versa. If the rate of progress is constant ($\beta''_{\tau\tau}\equiv 0$) and there is no decrease of WP productivity because of current factors ($\beta'_t\equiv\equiv\alpha'_t=0$), the period of functioning of WP is also constant.

II. Problem of maximization of the output of national income under given labiour resources: under given $\alpha,\ \beta,\ \lambda,\ P,\ y_0,\ m_0$ define functions $y,\ a_1,\ a_2,\ m,\ N$ providing the extremum

$$I_2=\int_{t_0}^{T}N(t)\,dt\to max \tag{13}$$

under constraints (4)-(9).

Qualitative picture of the behaviour of solutions of this problem is similar to problem I: theorems 2-6 in the general case remain for problem II. However theorem I is valid for problem II only in the case $\lambda(t)\cdot\alpha(\tau,t)<\beta(\tau,t)$. In the case when $\lambda(t)\cdot\alpha(\tau,t)>\beta(\tau,t)$ and the technical progress exists $(\alpha'_\tau>0,\ \beta'_\tau>0)$, solutions $y(t)$, $a_1(t)$ and $a_2(t)$ are greater than minimally permissible on the whole interval $[t_0,T]$ (moreover, here $y(t)\equiv 1$. If α and β are independent of τ, and $\lambda(t)\cdot\alpha(t)\equiv\beta(t)$, then the value of functional I_2 does not depend on unknown y, a_1, a_2, m and $I_2=\int_{t_0}^{T}\beta(t)\,P(t)\,dt$.

III. Problem of maximization of the output of useful product under given dynamics of bringing into being capital funds: under given α, $\beta, \lambda, \varphi, y_0, m_0$ define functions y, a_1, a_2, m, C providing the extremum (10) under constraints (4), (5), (7)-(9).

The behaviour of solutions of this problem is qualitatively distinct from preceding ones, and theorems 2-7 are not valid. And exactly for any length of the interval of planning $T-t_0$ and arbitrary $\alpha(\tau,t)$ and $\beta(\tau,t)$ the behaviour of solutions of problem III is trivial: for $t\in[t_0,T]$ solutions $y(t)$ and $a_2(t)$ coincide with minimally permissible functions $y_m(t)$ and $a_{2m}(t)$ by virtue of constraints (8), and $a(t)$ and $m(t)$ are determined from equations (4) and (7).

A comparative analysis of problems I-III shows that qualitative and structural properties of solutions of optimization problems in model (4)-(9) similar to theorems 2-6 are defined to a greater extent by given functions than by the form of the functional.

IV. Problem of minimization of labour expenditure under given dynamics of output of consumption product: under given $\alpha, \beta, C, y_0, m_0$ define functions y, a_1, a_2, m, P providing the extremum

$$I_3=\int_{t_0}^{T}P(t)\,dt\to\min \tag{14}$$

under constraints (4)-(6), (8), (9).

This problem is in a certain sense inverse to problem I and, as investigation shows, problems I and IV are equivalent at the qualitative level: all theorems 1-5 are valid for problem IV too. However this does not mean that their solutions coincide. More detailed asymptotic properties of $\tilde{a}(t)$ (and, hence, of solutions a_1, a_2, y)

are defined by other functional equations and are found to be different in problem I.

V. Problem of maximization of the output of useful product under given dynamics of labour resources and national income [5] : under given $\alpha, \beta, \lambda, P, N, y_0, m_0$ define functions y, a_1, a_2, m, c providing the extremum (10) under constraints (4)-(9).

In this and next problems one more constraint (for the national income N) is added that results in sufficient distinctions in the qualitative behaviour of solutions.

In particular, solutions y and a_2 of problems I-IV cannot be inner in region (8) on the whole interval of planning $[t_0, T]$ by virtue of theorem I (or analogous to it).

Theorem 8. Under natural assumptions connected with constraints of the rate of increase of given functions $P(t)$ and $N(t)$ solutions y, a_1, a_2 of problem V can be inner on the whole interval $[t_0, T]$.

Theorem 9. If functions α and β are independent of τ and $\lambda(t)\cdot\alpha(t) \neq \beta(t)$ then the value of functional I in problem V is independent of the behaviour of unknown functions y, a_1, a_2, m and is equal to:

$$I_1 = \int_{t_0}^{T} \frac{\beta(t)}{\beta(t) - \lambda(t)\alpha(t)} \cdot [N(t) - \lambda(t)\alpha(t) P(t)]\,dt.$$

If $\beta(t) \equiv \lambda(t)\cdot\alpha(t)$ then problem V coincides with problem I; if $\beta(t) \not\equiv \lambda(t)\cdot\alpha(t)$ but the equality can be valid for some t then problem V is incorrect: small variations of control $y(t)$ may cause very great variations of I_1.

Then, for problem V theorems 3 and 4 are found to be valid. Four unknown functions : $a_1 \equiv a_2$, y, m and c in the case of theorem 4 are defined from equations (4)-(7) and their behaviour depends on correlations between given $P(t)$ and $N(t)$.

The structure of solutions of problem V in the general case is also defined just as theorem 5 since $\tilde{a}(t)$ does not satisfy the initial conditions (8) as in previous problems.

VI. Problem of minimization of labour expenditure under the given output of national income and consumption product: under given α, $\beta, \lambda, N, c, y_0, m_0$ define functions y, a_1, a_2, m, P providing the extremum (14) under constrains (4)-(9).

At a qualitative level the behaviour of solutions of problem VI is equivalent to problem V, in particular, theorems 3-6,8,9 are valid for it too (in theorem 9 $I_3 = \int_{t_0}^{T} [m(t)/\alpha(t) + c(t)/\beta(t)]dt$).

Thus, the qualitative behaviour of solutions of all considered problems is defined first of all by the rates α'_τ and β'_τ of increase of productivity α and β of new WP. In particular, if the productivity of newly created WP remains constant then for given resources the optimization reserves at the expense of redistribution of new WP between groups A and B and liquidation of obsolete WP are highly limited (theorem 1) or are absent in general (theorem 9).

Investigation of some problems of optimal control (1)-(3) in multiproduct integral economic models [6] confirms this conclusion. One of such problems - the problem of reorganization of production structure of multiple-industry economic system in minimal time under given labour resources of branches is studied in [7].

REFERENCES

1. Glushkov V.M. On the class of dynamic macroeconomic models. Upravljajushchie sistemy i mashiny, 1977, N2, p. 3-6.
2. Glushkov V.M., Ivanov V.V., Jacenko Ju.P. Analytical investigation of the class of dynamic models. Kibernetika, 1980, N2, p.1-12, N3, p.112-119.
3. Glushkov V.M., Ivanov V.V., Janenko V.M., Modelling of developing systems. M., Nauka, 1983, 380 p.
4. Warga J. Optimal control of differential and functional equations. - New York: Academic Press, 1982, 620 p.
5. Ivanov V.V., Jacenko Ju.P., Galiev U.E. Decision of the problem of maximization of the output of external product of system on the basis of V.M. Glushkov model. - Doklady AN USSR.Seria A, 1983, N3, p. 69-72.
6. Ivanov V.V., Jacenko Ju.P. Integral macroeconomic models for planning and predicition of the development of national economy. -All-unia conference "Problemy ekonomicheskogo modelirovanija", (Odessa, 1984). Part 1. Kiev, 1984, p.10-14.
7. Jacenko Ju.P. Investigation ofintegral model of the management of structure of economic systems. - Vestnik Kievskogo universiteta. Modelirovanie i optimizacija slozhnykh syste. Kiev, 1985, N4, p.107-111.

ONE-MACHINE SCHEDULING PROBLEMS WITH RESOURCE CONSTRAINTS

ADAM JANIAK

Institute of Eng. Cybernetics
Techn. University of Wrocław
Wrocław, Poland

ABSTRACT. One-machines scheduling problems with maximum lateness and maximum tardiness criteria are generalized on the case with allocation of continuously-divisible constrained nonrenewable resource. Models of operation are assumed to be duration versus resource amount linear functions. For the problems with maximum lateness and the maximum tardiness criteria under equal release dates (i.e. moments at which operations are available for processing) polynomial-time algorithms are found. It is shown that the problem with maximum lateness criterion under unequal release dates is NP-hard. For this problem a branch-and-bound algorithm (utilizing properties shown) is outlined.

1. INTRODUCTION. PROBLEM FORMULATION

One-machine scheduling problems arise in those production processes in which it appears only one critical machine and other machines are practically non-bottleneck ones (i.e. of infinite capacity). The processing times of operations often are not constant but depend on an amount of resources, e.g. energy, fuel, oxygen, catalyst, raw materials (or money). These problems also arise in theoretical context of computing lower bounds for the flow-shop and job-shop problems with resource constraints [3,4].

One-machine scheduling problems with allocation of constrained resource can be formulated as follows. There are n jobs $J_1, J_2, \ldots, J_n$ that are to be processed on one machine. Each job J_j consists of one operation O_j. The machine can handle only one job at a time. The operation O_j corresponds to the processing of the job J_j on the machine during an uninterrupted processing time p_j. We shall assume that $p_j \triangleq (p_j(u_j)) \triangleq b_j - a_j u_j$, where $a_j > 0$, $b_j > 0$ are known parameters and u_j is the amount of resource alloted to the operation O_j. We will say that a resource allocation $u = [u_1, \ldots, u_n]$ is admissible for the problems under consideration if the following constraints are satisfied:

$$\alpha_j \leq u_j \leq \beta_j, \quad j = 1,2,\ldots,n, \quad \sum_{j=1}^{n} u_j \leq U',$$

where α_j, β_j are known parameters, $0 \leq \alpha_j \leq \beta_j < b_j/a_j$ and U' is the global amount of resource alloted to realization of all operations O_j, $j = 1,2,\dots,n$. The set of all admissible resource allocations will be denoted by U.

For each job J_j the release date r_j, i.e. the moment at which the job J_j is available for processing and the due date d_j, i.e. the moment by which the job J_j should be completed, may be given. There may exist precedence constraints between jobs that can be conveniently represented by sets $A_j = \{ i | J_j < J_i \}$, $j = 1,2,\dots,n$, where $J_j < J_i$ means that the processing of J_i cannot start before J_j has been completed. Denote by Π the family of all schedules (permutations of job indices) satisfying given precedence constraints.

Given a feasible schedule $\pi \in \Pi$ and resource allocation $u \in U$, we can compute for each J_j its completion time $C_j(\pi,u)$, its lateness $L_j(\pi,u) = C_j(\pi,u) - d_j$ and its tardines $T_j(\pi,u) = \max\{ 0, C_j(\pi,u) - d_j \}$. The schedul π with the resource allocation $u \in U$ can be evaluated according to various criteria; in the presented paper we seek to minimize the maximum lateness $L_{max} \triangleq L_{max}(\pi,u) \triangleq \max_{1 \le j \le n} \{ L_j(\pi,u) \}$ and the maximum tardiness $T_{max} = T_{max}(\pi,u) = \max_{1 \le j \le n} \{ T_j(\pi,u) \}$.

The formulated above one-machine scheduling problems with resource constraints have not been considered in the literature until now.

For the first time in the sequencing problems the presented above models of operations were introduced in [3,4] for the job-shop and flow-shop problems. Scheduling problem for splittable operations with the models in the form of performing speed-resource (renewable resource) amount functions was considered in [8] for the case of identical parallel machines. Resources allocation problems in which operation resource requirements are discrete, i.e. concern resource amount belonging to given finite sets were investigated in [7] under multicriteria approach. Some of the sequencing and scheduling problems with discrete resource requirements were also considered in [1].

For the problems with the maximum lateness and the maximum tardiness criteria under equal release dates (considered in Section 2 and 3) efficient (polynomial-bounded) [2] algorithms are found. Computational time (or a number of computational steps) of such an algorithm is bounded by a polynomial w(n) in the size of the problem considered (i.e. a number of jobs - n). In the following, we will say that computational complexity of such an algorithm (problem) is O(w(n)). The problem with the maximum lateness criterion under unequal release dates (Section 2) has been shown to be NP-hard [2]. It is thereby equivalent to a host of notorious combinatorial problems, in the sense that a polynomial-bounded method for its solution would provide similar algorithms for all other NP-hard problems as well.

2. MINIMIZATION OF THE MAXIMUM LATENESS

We shall prove the following.

Theorem 1

If $r_j = 0$, $d_j \geqslant 0$, $A_j \neq \emptyset$, $j = 1,2,\ldots,n$. Then the maximum lateness L_{max} is minimized (in $O(n^2)$ steps) by ordering jobs and by resource allocation obtained in agreement with the following Procedure 1 and Procedure 2, respectively:

Procedure 1

Step 1. Set $S:=\{1,2,\ldots,n\}$, $k:=n$ and go to Step 2;

Step 2. Identify the set $Q:=\{h | h \in S,\ A_h \cap S = \emptyset\}$ and find an $i \in S$ satisfying $d_i = \max_{h \in Q} \{d_h\}$, then set $\pi^*(k):=i$, $k:=k-1$, $S:=S-\{i\}$ and go to Step 3;

Step 3. If $S \neq \emptyset$, then go to Step 2, otherwise Stop – π^* is the optimal schedule.

Procedure 2

Step 1. Set $u_j^*:=0$ for any $j = 1,2,\ldots,n$ and $l:=n$. Find the completion times of jobs in the following way:

$$C_{\pi^*(j)} := \sum_{i=1}^{j} b_{\pi^*(i)},\ j=1,2,\ldots,n \text{ and go to Step 2;}$$

Step 2. Find the smallest index k, $1 \leq k \leq l$ that satisfies the following equation

$$C_{\pi^*(k)} - d_{\pi^*(k)} = \max_{1 \leq i \leq l} \left[C_{\pi^*(i)} - d_{\pi^*(i)}\right].$$ Then find the set $P := \{\pi^*(j) \mid 1 \leq j \leq k,\ u^*_{\pi^*(j)} = 0\}$. If the set P is empty or $U' = 0$, then Stop – u_j^*, $j=1,2,\ldots,n$ is the optimal resource allocation, otherwise go to Step 3;

Step 3. Find an index t, that satisfies $a_{\pi^*(t)} := \max_{\pi^*(j) \in P} a_{\pi^*(j)}$ and then set $h:=\left[(C_{\pi^*(k)} - d_{\pi^*(k)}) - \max_{1 \leq i < t} (C_{\pi^*(i)} - d_{\pi^*(i)})\right]$, $z:=\min\{\beta_{\pi^*(t)},\ U',\ h/a_{\pi^*(t)}\}$.

Next bring up to date resource allocation and completion times by setting: $u^*_{\pi^*(t)}:=u^*_{\pi^*(t)}+z$, $U':=U'-z$, $C_{\pi^*(i)}:=C_{\pi^*(i)} - z \cdot a_{\pi^*(t)}$ for $t \leq i \leq n$. Finally set $l:=k$ and go to Step 2.

Proof. The proposed ordering rule is self-evident since it has been used to solve the classical problem with maximum lateness criterion (i.e. under constant processing times of jobs) and the yielded schedule in this case has been independent of the actual processing times [6]. It is obvious that the global amount of resource should be allocated among the jobs composing the path for which $\sum_{i=1}^{i} p_i(u_i)-d_j$ achieves the maximal value (for j=1,...,n). The maximal deterioration of the length of this path is obtained by assignment of feasible resource amounts to the jobs with the maximal parameters a_j values.

Remark. Note that for any fixed π the resource allocation generated by the Procedure 2 gives the minimal value $L_{max}(\pi,\cdot)$.

Theorem 2

If $r_j \geqslant 0$, $d_j \geqslant 0$ $(A_j = \emptyset)$, j=1,2,...,n, then the one machine scheduling problem with resource constraints under the maximum lateness criterion is NP-hard.

Proof. The considered problem under assumption that p_j = const, for j=1,2,...,n (i.e. the classical one machine scheduling problem under the maximum lateness criterion with $r_j \geqslant 0$) is NP-hard [5] and it is reducible to the problem under consideration since the first one can be considered as a special case of the second one.

This complexity result strongly suggests that an enumerative solution approach is essentially unavoidable. We therefore restrict ourselves to outlining of an algorithm of the branch-and-bound type.

This problem may have an equivalent statement which is more convenient for analysis. Let M_1 and M_3 be nonbottleneck machine of infinite capacity and M_2 be a bottleneck machine processing only one job at a time. Each job J_j has to be processed on M_1, M_2 and M_3, in that order, and has to spend a time from 0 to r_j on M_1, a time $p_j(u_j)$ from S_j to $C_j = S_j + p_j(u_j)$, $(S_j \geqslant r_j)$, on M_2, and a time q_j from C_j to $C_j + q_j$ on M_3 where $q_j = d - d_j$, $d = \max_{1 \leqslant j \leqslant n} d_j$. The objective is to find a processing order on bottleneck machine M_2 and a resource allocation $u \in U$ that minimize the time of the last job completion on machine M_3, $\max_{1 \leqslant j \leqslant n} (C_j + q_j)$. Denote by $C_{max}(\pi, u_\pi^*)$ the time of the last job completion on M_3 for schedule $\pi \in \Pi$ under the optimal resource allocation $u^* \in U$ for this schedule π (i.e. the resource allocation minimizing the value $\max_{1 \leqslant j \leqslant n} (C_{\pi(j)}(\pi,u) + q_{\pi(j)})$). It is obvious that

$$C_{max}(\pi, u_\pi^*) = \max_{1 \leqslant j_1 \leqslant j_2 \leqslant n} \left(r_{\pi(j_1)} + \sum_{j=j_1}^{i_2} p_{\pi(j)}(u^*_{\pi(j)}) + q_{\pi(j_2)}\right). \quad (1)$$

A subschedule of jobs $\langle J_{\pi(j_1)},\dots,J_{\pi(j)},\dots,J_{\pi(j_2)}\rangle$ on M_2 for which (1) holds will be called a critical path in π under $u^*_\pi \in U$. Note that there may be several critical paths in π under $u^*_\pi \in U$. A partial schedule of jobs $\langle J_f,\dots,J_j,\dots,J_l\rangle$ on M_2 containing the maximal number of jobs belonging to the same critical paths will be called a section in π under $u^*_\pi \in U$. Let k_π be the number of sections in π. Analogically to Theorem 1 [4] the followings theorem may be proved.

Theorem 3

For each $\pi \in \Pi$, if the schedule π' has been obtained from π by interchange of a performance order of operations and if $C_{max}(\pi',u^*_{\pi'}) <$ $C_{max}(\pi,u^*_\pi)$, then in π at least one job J_j of the k-th section of π has been moved before the first J_f (or after the last J_l) job of this section, for k=1 or 2 or,..., or k_π. At the same time, if the job J_j has been moved before the first job J_f (or after the last job J_l) of the k-th section, then $r_j - r_f < 0$ (or $q_j - q_l < 0$) has been held for the job J_j in this section.

This property implies the enumerative scheme and branching rule which will be used while solving the problem under consideration by a branch-and-bound procedure with mixed strategy of bound.

Starting with any initial schedule π_1 (e.g. obtained by some heuristic method) a sequence of schedules π_r is generated. For each π_r from this sequence the optimal allocation of resource $u^*_{\pi_r}$ is computed and next the critical paths and all sections are identified. Each new schedule π_s is obtained from a preceding schedule π_r of the sequence by moving one job J_j of same section in π_r before the first job J_f (or after the last one J_l) of this section. Only such jobs J_j, for which $r_j - r_f < 0$ ($q_j - q_l < 0$) holds, are moved before the first J_f (after the last J_l) job of the section considered. After a movement of a job J_j certain set of precedence constraints F_s is fixed in π_s. For each π_r from the generated sequence a lower bound on the value $C_{max}(\pi_s,u^*_{\pi_s})$ for every π_s, which can be generated from the schedul π_r, is found. We can eliminate considered currently schedule π_r and backtrack to the immediate predecessor π_p of π_r if one of the following conditions holds:

- the lower bound on the value $C_{max}(\pi_s,u^*_{\pi_s})$ of all the possible descendents π_s of π_r is not less than the current upper bound (i.e. the minimal value $C_{max}(\pi,u^*_\pi)$ found so far);
- no section consists of the job J_j with $r_j - r_f < 0$ or with $q_j - q_l < 0$;
- no job with $r_j - r_f < 0$ or $q_j - q_l < 0$ can be moved after the first

and before the last job of any section without violating the fixed precedence constraints F_r.

The algorithm of finding the resource allocation $u^*_\pi \in U$ minimizing the value $C_{max}(\pi,\cdot)$ for fixed schedule π is similar to Procedure 2.

The branch-and-bound procedure outlined above is similar to one used to solve the job-shop problem with resource constraints [4].

3. MINIMIZATION OF THE MAXIMUM TARDINESS

The following theorem holds.

Theorem 4

If $r_j = 0$, $d_j \geqslant 0$, $A_j \neq \emptyset$, $j=1,2,\dots,n$, then a schedule $\pi^* \in \Pi$ and a resource allocation $u^* \in U$ that are optimal with respect to the maximum lateness criterion L_{max}, i.e. $\min_{\pi\in\Pi} \min_{u\in U} L_{max}(\pi,u) = L_{max}(\pi^*,u^*)$, are also optimal with respect to the maximum tardiness criterion T_{max}, i.e. $\min_{\pi\in\Pi} \min_{u\in U} T_{max}(\pi,u) = T_{max}(\pi^*,u^*)$.

Proof. It is easy to notice that if the resource allocation u^*_π minimizes the value of the maximum lateness criterion $L_{max}(\pi,\cdot)$ for a fixed schedule π, then u^*_π minimizes also the value of the maximum tardiness criterion $T_{max}(\pi,\cdot)$ for this schedule π. (In the case when $L_{max}(\pi,u^*_\pi) < 0$, then there may also exist other resource allocations $\hat{u}_\pi$, for which $T_{max}(\pi,u^*_\pi) = T_{max}(\pi,\hat{u}_\pi)(=0)$). Given two scheduls π and π' with the value of the maximum lateness criterion $L_{max}(\pi,u^*_\pi)$ and $L_{max}(\pi',u^*_{\pi'})$ under the optimal resource allocation u^*_π and $u^*_{\pi'}$, respectively. Given also the value of the maximum tardiness criterion $T_{max}(\pi,u^*_\pi)$ and $T_{max}(\pi',u^*_{\pi'})$. We have that $L_{max}(\pi,u^*_\pi) \le L_{max}(\pi',u^*_{\pi'})$ implies that

$$T_{max}(\pi,u^*_\pi) = \max_j\{\max(0,L_j(\pi,u^*_\pi))\} = \max(0,L_{max}(\pi,u^*_\pi) \leqslant$$
$$\leqslant (0,L_{max}(\pi,u^*_\pi)\} \leqslant \max(0,L_{max}(\pi',u^*_{\pi'})) = T_{max}(\pi',u^*_{\pi'}) .$$

Remark. Note that L_{max} and T_{max} are not equivalent: a schedule π with a resource allocation u for which $T_{max}(\pi,u) = 0$ may be suboptimal with respect to the maximum lateness criterion.

Concluding the optimal schedule and the optimal allocation of resource with respect to the maximum tardiness criterion, may be obtained due to the Procedure 1 and 2 or by special procedures constructed in like maner as mentioned ones.

It may be well to add that the Procedure 1 and 2 give the optimal solution to the problem of finding a schedule of jobs on single machine with a resource allocation (subject to the given precedence constraints) that will cause each job to be completed by its dedline (due date) d_j, if such a schedule with resource allocation exists. If the schedule with resource allocation obtained by Procedure 1 and 2 is not successful in this regard, no other schedule (with resource allocation) is.

4. CONCLUSION

It would be a good thing to consider other one-machine scheduling problems with resource constraints e.g. problems with such criteria as: the maximum completion time, the maximum cost, the total weighted completion time, the total weighted tardiness, the total cost. Most of them are probably NP-hard.

REFERENCES

[1] J. Błażewicz, J.K. Lenstra and A.H.G. Rinnooy Kan, Schduling subject to resource constraints: Classification and complexity. Discrete Appl. Math, 11-24, 1983.

[2] M.R. Garey and D.S. Johnson, Computers and Intractability: A Guide to the Theory of NP-Completeness. Freeman, San Francisco, 1979.

[3] J. Grabowski and A. Janiak, Sequencing problem with resource constraints in:Cybernetics and Systems Research 2, Elsevier Science Publishers B.V. (North-Holland), 329-333, 1984.

[4] J. Grabowski and A. Janiak, On job-shop scheduling with resource constraints in:Lecture Notes in Control and Information Sciences 59, Springer-Verlag, 193-200, 1984.

[5] B.J. Lageweg, J.K. Lenstra and A.H.G. Rinnooy Kan, Minimizing maximum lateness on one machine: computational experience and some applications, Statistica Neerlandica 30, 25-41, 1976.

[6] L. Lawler, Optimal sequencing of a single machine subject to precedence constraints, Manag. Science 19, 544-546, 1973.

[7] R. Słowiński, Multiobjective network scheduling with efficient use of renewable and nonrenewable resources, Europ. J. Opl. Res. 7, 265-273, 1981.

[8] J. Węglarz, Project scheduling with discrete and continuous resources, IEEE Trans. SMC-9, 644-650, 1979.

Multiobjective structural optimization

STEFAN JENDO & WOJCIECH MARKS
Institute of Fundamental Technological Research
Polish Academy of Sciences
Warsaw, Poland

Abstract

The paper deals with an application of multicriteria optimization to optimum structural design. A brief formulation of multicriteria optimization problem is presented. The paper contains two examples. The first example is concerned with application of utility function in optimum design of prestressed concrete I beam. The second example illustrates the application of metric function in izostatic truss optimization. Both, the compromise and preferable solutions are founded.

1. Introduction

With this paper an attempt to fill the gap between the theory of multiobjective optimization and its application in optimum structural design will be presented. The paper deals with the formulation of multiobjective optimization problems and generating the compromise sets as well as selecting the preferable solutions. The optimization of building and machine structures usually involves a number of requirements that should be met at the same time to obtain the fully useful design. In the case of single criterion optimization, one of the requirements is selected as the criterion, while the remaining ones are met by including them into the constraints set. But with such an approach, it is necessary to determine a priori the bounds which these requirements should fulfill. Multiobjective optimization enables us to take into account numerous criteria that are often mutually conflicting. It is then possible to find the compromise and preferable solution which - althrough non of the criteria involved attains its extremum - quarantees meeting all the requirements in the best way possible. The criteria appearing most frequently in the theory of structures are as follows [2]: i) minimum volume or weight of the structures, ii) minimum potential energy or maximum structural stiffness, iii) minimum displacements at selected points or regions of the structure, iv) maximum critical force, v) maximum frequency of free vibration, vi) maximum safety or reliability, vii) maximum moment of inertia. It should be emphasized that to satisfy any of the criteria mentioned above, it is necessary to find the extremum of an objective function taking into account the related constraints. In the theory of structures, the most important constraints are related with: i) permissible stresses or safety factors under all possible loading conditions, ii) permissible displacements for a given structure, iii) minimum and maximum sizes that are available in view of services and constructional reasons.

2. Formulation of optimization problem

The problem of multicriteria optimization can be formulated as follows:

$$\min_{\underset{\sim}{x} \in \Omega} \underset{\sim}{f}(\underset{\sim}{x})$$

where $\underset{\sim}{f}: \Omega \to R^k$ is objective function vector given by

$$\underset{\sim}{f}^T(\underset{\sim}{x}) = \{f_1(\underset{\sim}{x}), f_2(\underset{\sim}{x}), \ldots, f_k(\underset{\sim}{x})\}$$

and $\Omega \subset R^n$ is a feasible domain defined by the equality and inequality constraints, i.e.

$$\Omega = \{\underset{\sim}{x} \in R^n \mid \underset{\sim}{h}(\underset{\sim}{x}) = 0,\ \underset{\sim}{g}(\underset{\sim}{x}) \leq o\}.$$

The components $f_i : \Omega \to R$, $i=1,2,\ldots,k$, are called the partial criteria of optimization and $\underset{\sim}{x}$ is the vector of design variables.

3. Elements of solution of optimization problem

The solution obtained after meeting a few criteria mentioned above has usually the form of the compromise set. The optimal structure should be selected from this set on the basis of a global criterion which can take various forms in different optimization problems. Its choice is particulary vital because it will be decisive for results of the entire optimization problem. The global criteria appearing most frequently are as follows: i)utility function which often has a form of a cost function, ii)metric function which describes the distance between ideal solution and compromise set. It is proposed that the solution of multicriteria optimization problem should contain the following elements: i)set of compromise solutions, ii)ideal solution, iii)preferable solution i.e. $\tilde{f}^{pr}$ and corresponding $\tilde{x}^{pr}$, iv)width of compromise set, i.e. $\Delta f_i=\bar{f}_i-f_i^{id}$, $i=1,2,\ldots,k$, v)sensitivity vector $\frac{\partial F}{\partial f_i}\Big|_{f^{pr}}$, $i=1,2,\ldots,k$, where F describes the global criterion.

4. Examples of multicriteria structural optimization problems

4.1. Application of utility function in optimum design of prestressed concrete I beam

The optimum design problem of prestressed concrete I beam can be formulated as follows. Find the dimensions of the cross-section area of prestressed concrete I beam as well as the eccentricity of prestressing force according to two criteria: minimum volume of concrete and prestressing steel. Two extremal loading conditions represented by the bending moments M_1 and M_2 are considered. The vector of design variables $\tilde{x}^T=\{x_i\}$ $i=1,2,\ldots,6$ contains the following quantities: x_1 -beam depth, x_2-flange width, x_3 -flange thickness, x_4 -web thickness, x_5 and x_6 -the values of prestressing force and its eccentricity, respectively (Fig.1). The feasible domain Ω is defined by the following constraints:

- stress constraints in two loading conditions

$$\frac{x_5}{A}-\frac{x_5x_6}{I}\frac{x_1}{2}+\frac{M_1}{I}\frac{x_1}{2}\leqslant\bar{\sigma}_1,$$

$$\frac{x_5}{A}+\frac{x_5x_6}{I}\frac{x_1}{2}-\frac{M_1}{I}\frac{x_1}{2}\geqslant\underline{\sigma}_1,$$

$$\frac{x_5}{A}-\frac{x_5x_6}{I}\frac{x_1}{2}+\frac{M_2}{I}\frac{x_1}{2}\geqslant\underline{\sigma}_2,$$

$$\frac{x_5}{A}+\frac{x_5x_6}{I}\frac{x_1}{2}-\frac{M_2}{I}\frac{x_1}{2}\leqslant\bar{\sigma}_2;$$

- side constraints

$$x_1\leqslant x_1,\quad x_2\leqslant x_2,\quad x_3\geqslant x_3,\quad x_4\geqslant x_4,\quad x_6\leqslant\frac{x_1}{2}-d.$$

The optimization criteria can be written as follows

$$\min_{\underset{\sim}{x}\in\Omega} A = 2x_2x_3 + x_4(x_1 - 2x_3),$$

$$\min_{\underset{\sim}{x}\in\Omega} S = x_5 .$$

The segment of curve given by

$$x_5 = \frac{\bar{\sigma}_1 - \frac{M}{I}\frac{\bar{x}}{2}1}{\frac{1}{A} + \frac{1}{I}\left(\frac{\bar{x}}{2}1 - d\right)\frac{\bar{x}_1}{2}},$$

contained between two following curves

$$x_5 = \frac{\bar{\sigma}_2 + \frac{M_2}{I}\frac{\bar{x}_1}{2}}{\frac{1}{A} + \frac{1}{I}\left(\frac{\bar{x}}{2}1 - d\right)\frac{\bar{x}_1}{2}} \quad \text{and} \quad x_5 = \frac{\underline{\sigma}_1 + \frac{M_1}{I}\frac{\bar{x}_1}{2}}{\frac{1}{A} + \frac{1}{I}\left(\frac{\bar{x}}{2}1 - d\right)\frac{\bar{x}_1}{2}}$$

is the set of compromise solutions [4].

The preferable solution is selected on the basis of utility function

$$F = A + kS, \qquad k = \frac{c_s}{c_c}\frac{\gamma}{\bar{\sigma}_s}[m^2/kN],$$

where k is the ratio of unit cost of prestressing steel over concrete unit cost. The following notation is used: c_c - unit cost of 1 cubic meter of concrete, c_s - unit cost of 1 kg of prestressing stell, γ -bulk density of prestressing steel, $\bar{\sigma}_s$ -allowable tension stress for prestressing steel. The preferable solution occurs at one of the end points of compromise set or at the tangent point between compromise set and indifference lines of utility function F=A+kS=const depends on the value of k. *Numerical example:* The following data are taken: M_1=320kNm,M_2=100 kNm, $\bar{x}_1$=0.70 m, $\bar{x}_2$=0.40 m, $\underline{x}_3$=0.10 m, $\underline{x}_4$=0.08 m, d=0.07 m, $\bar{\sigma}_1 = \bar{\sigma}_2$=10000 kN/m, $\underline{\sigma}_1=\underline{\sigma}_2$=-500 kN/m and k=0.000054 m^2/kN. The solution is presented on Figure 1 and contains the following information:

i) the coordinates of the end points of compromise set

- point No 1 A= 0.13286m^2, S= 750.00 kN;
- point No 2 A= 0.13797 m^2; S= 655.20 kN;

ii) ideal solution (point No 3) A^{id}= 0.13286m^2,S^{id}=655.20 kN;

iii) preferable solution (point No 4) A^{pr}=0.13600m^2, S^{pr}=690.45 kN

and corresponding design variables vector x_1^{pr}= 0.70m, x_2^{pr}= 0.40m,

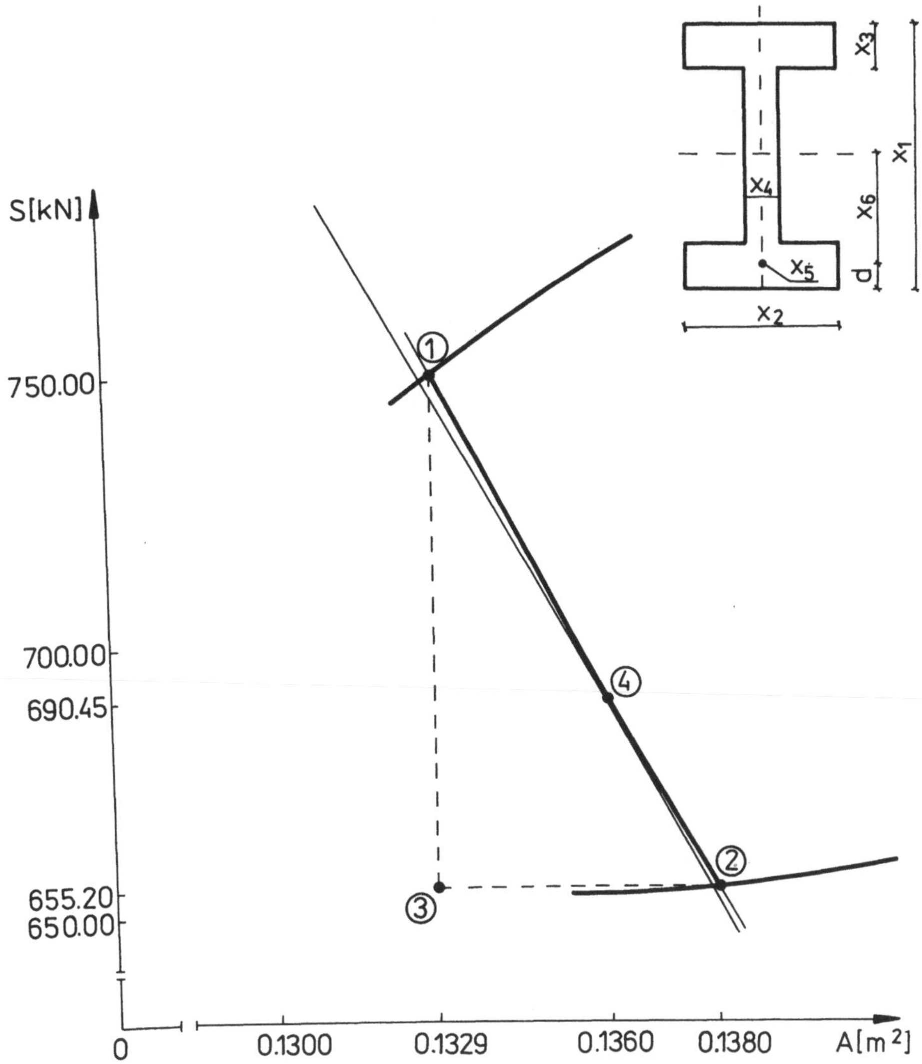

Fig.1. Representation of objective space for Example 4.1.

$x_3^{pr} = 0.125m$, $x_4^{pr} = 0.08m$, $x_5^{pr} = 690.45kN$, $x_6^{pr} = 0.28m$;

iv) width of compromise set

$$\Delta A = 0.00511m , \qquad \Delta S = 94.80 \text{ kN};$$

v) sensitivity vector of global criterion F at the preferable solution

$$\frac{\partial F}{\partial A} = 1, \qquad \frac{\partial F}{\partial S} = 0.000054.$$

4.2. Application of metric function in isostatic truss optimization

The multicriteria optimization problem of isostatic steel truss structures can beformulated as follows. Find the cross-section areas of truss bars $\tilde{x}^T=\{x_i\}$, i=1,2,...,7 according to two criteria - minimum volume of truss and minimum vertical displacement at point B (Fig.2). The feasible domain of solutions is defined by stress and side constraints i.e.

$$\Omega = \{x_i \ R^n | \underline{\sigma} \leq \sigma_i \leq \bar{\sigma} , \qquad x_i \leq \bar{x}_i , \qquad i=1,2,\ldots,7\}$$

where $\underline{\sigma}$ and $\bar{\sigma}$ denote the allowable compression and tension stresses, respectively; and $\sigma_i = N_i/x_i$.

The optimization criteria of truss structure can be written in the following form

$$\min_{x_i \in \Omega} V = \sum_{i=1}^{7} x_i \ell_i$$

$$\min_{x_i \in \Omega} \delta_B = \sum_{i=1}^{7} \frac{\partial}{\partial P}\left(\frac{N_i^2 \ell_i}{2Ex_i}\right)$$

The optimization problem is solved using dynamic programming. The compromise set has the following form

$$V = \frac{1}{32}\frac{P}{E}\frac{(5l_1 + 3l_2 + 3l_3 + 5l_4)}{\delta_B} \tag{1}$$

and it is contained between $\underline{V}$ and $\bar{V}$ in the coordinate system (V,δ_B).The minimal feasible volume of truss $\underline{V}$ is found on the basis of allowable stresses in truss bars

$$V = 2l_1\frac{5}{8}\frac{P}{\underline{\sigma}} + 2l_2\frac{3}{8}\frac{P}{\bar{\sigma}} + 2l_3\frac{5}{8}\frac{P}{\bar{\sigma}} + l_4\frac{3}{4}\frac{P}{\underline{\sigma}} = \underline{V}.$$

The maximal feasible volume of truss $\bar{V}$ is found on the basis of side constraints $x_i = \bar{x}_i = \bar{x}$ i.e.

$V \leq (2l_1 + 2l_2 + 2l_3 + l_4)\bar{x} = \bar{V}$.

Since there is no preferences concern with objective functions V and σ_B the solution is found according to minimization of metric function [3]. Three solutions based on different forms of metric function are found.

Solution No 1. The metric function in dimension form is taken as follows

$$F^{(2)} = \sqrt{(V-V^{id})^2 + \mu^2(\delta_B-\delta_B^{id})^2}, \qquad \mu = 1m^2. \tag{2}$$

The ideal solution has the following coordinates

$$V^{id} = \frac{1}{4}P(5\frac{l_1}{\underline{\sigma}} + 3\frac{l_2}{\bar{\sigma}} + 5\frac{l_3}{\bar{\sigma}} + 3\frac{l_4}{\underline{\sigma}})$$

$$\delta_B^{id} = \frac{1}{32}\frac{P}{E}\frac{(5l_1 + 3l_2 + 5l_3 + 3l_4)^2}{\bar{x}(2l_1 + 2l_2 + 2l_3 + l_4)} \tag{3}$$

Substituting (1) and (3) into (2) we have

$$F^{(2)} = \{[V-\frac{1}{4}P(5\frac{l_1}{\underline{\sigma}} + 3\frac{l_2}{\bar{\sigma}} + 5\frac{l_3}{\bar{\sigma}} + 3\frac{l_4}{\underline{\sigma}})]^2 +$$

$$+ \mu^2(\frac{P}{32E})^2(5l_1 + 3l_2 + 5l_3 + 3l_4)^4[\frac{1}{V} - \frac{1}{\bar{x}(2l_1+2l_2+2l_3+l_4)}]\}^{\frac{1}{2}}$$

The necessary condition for extremum gives the following equation

$$V - \frac{1}{4}P(5\frac{l_1}{\underline{\sigma}} + 3\frac{l_2}{\bar{\sigma}} + 5\frac{l_3}{\bar{\sigma}} + 3\frac{l_4}{\underline{\sigma}})V^3 +$$

$$+\mu^2(\frac{1}{32}\frac{P}{E})^2\frac{(5l_1+3l_2+5l_3+3l_4)^4}{\bar{x}(2l_1+2l_2+2l_3+l_4)}V - \mu^2(\frac{1}{32}\frac{P}{E})^2(5l_1+3l_2+5l_3+3l_4)^4 = 0,$$

which allows to determine the preferable solution $\underline{V}<V^{pr}<\bar{V}$.

Solution No 2. Taking the metric function in dimensionless form we have

$$F^{(2)} = \left\{\left(\frac{V-V^{id}}{V}\right)^2 + \left(\frac{\delta_B-\delta_B^{id}}{\delta_B}\right)^2\right\}^{1/2}$$

Introducing the following notation: $\nu = \frac{V}{\bar{V}}$, $\eta = \frac{\delta_B}{\bar{\delta}_B}$, $\beta = \frac{\underline{V}}{\bar{V}}$ and using analogical procedure as above we can found the following equation

$$\nu^4 - \beta\nu^3 + \beta^2\nu - \beta^2 = 0,$$

which allows to determine the preferable solution ν^{pr}.

Solution No 3. Taking the metric function in another dimensionless form i.e.

$$F^{(2)} = \left[\left(\frac{V - V^{id}}{\bar{V}}\right)^2 + \left(\frac{\delta_B-\delta_B^{id}}{\ell}\right)^2\right]^{1/2}$$

and introducing analogical notation as above

$$\nu = \frac{V}{\bar{V}},\ \xi = \frac{\delta_B}{\ell},\ \alpha = \frac{1}{32}\frac{P}{E}(5l_1 + 3l_2 + 5l_3 + 3l_4)^2$$

the following equation is found

$$\bar{V}^2\nu^4 - \bar{V}\underline{V}\nu^3 + \frac{\alpha^2}{\ell^2}\nu - \frac{\alpha^2}{\ell^2} = 0$$

which allows to determine the preferable solution ν^{pr}. *Numerical example.* The following data are taken: P=2000 kN, E=200kN/mm, $\underline{\sigma}=\bar{\sigma}$=250MPa, $l_1=l_3=l_5=l_7$=5.00m, $l_2=l_4=l_6$=6.00m, $\bar{x}$=40000mm. Three solutions of optimal truss are presented in Table 1 and they are shown on Figure 2. The compromise sets and ideal solutions are independent of the form of metric function. Unfortunately, the preferable solution and sensitivity vector essentially depend on the form of metric function.

5. Conclusions

i) Multicriteria optimization approach allows to determine the compromise solution according to a few conflicting criteria and gives much more information about the optimal solution than the single criterion optimization.

ii) The best compromise solution can be determined on the basis of user preferences or according to utility function. If there is not preferences and it is difficult to build the utility function then the metric function can be used to found the preferable solution.

iii) The compromise set is an objective notion but the preferable solution depends on global criterion. In the case of metric function the preferable solution depends also on the form of dimensionless metric function.

References

[1] P.H.Farquhar, *A survey of multiattribute utility theory and applications* (in) M.K. Starr & M.Zeleny (eds), **Multiple criteria decision making**, North Holland, New York 1977, pp.59-90.

[2] S.Jendo, W.Marks and G.Thierauf, *Multicriteria optimization in optimum structural design*, (in) **Large Scale Systems**, North Holland, Amsterdam, 1985.

[3] J.Koski, *Truss optimization with vector criterion*, Tampere University of Technology, Tampere, 6, 1979.

[4] W.Marks, *Optimum design of prestressed bending elements* (in Polish), **IFTR Reports**, 52/1978.

Table 1. List of solutions of multicriteria optimization of isostatic truss

	Solution No 1	Solution No 2	Solution No 3
Equation of compromise set	$V[m^3]=\frac{0.002311}{\delta_B[m]}$	$\nu=\frac{0.113137}{\eta}$	$\nu=\frac{0.0001267}{\xi}$
Coordinates of end points of compromise set - point No 1	$V= 1.52\ m^3$ $\delta_B= 0.00152\ m$	$\nu= 1$ /$V =1.52\ m^3$/ $\eta= 0.113$ /$\delta_B =0.00152m$/	$\nu=1$ /$V=1.52\ m^3$/ $\xi=0.000127$ /$\delta_B=0.00152m$/
- point No 2	$V= 0.172m^3$ $\delta_B= 0.0134\ m$	$\nu= 0.113$ /$V =0.172m^3$/ $\eta= 1$ /$\delta_B =0.0134\ m$/	$\nu=0.1132$ /$V =0.172\ m^3$/ $\xi=0.001119$ /$\delta_B=0.0134m$/
Ideal solution /point No 3/	$V^{id}= 0.172\ m^3$ $\delta_B^{id}= 0.00152\ m$	$\nu^{id}= 0.113$ /$V^{id}=0.172m^3$/ $\eta^{id}= 0.113$ /$\delta_B^{id}=0.00152m$/	$\nu^{id}=0.113$ /$V^{id}=0.172\ m^3$/ $\xi^{id}=0.000127$ /$\delta_B^{id}=0.00152\ m$/
Preferable solution /point No 4/	$V^{pr}= 0.1729\ m^3$ $\delta_B^{pr}= 0.0134\ m$	$\nu^{pr}= 0.336$ /$V^{pr}=0.5107\ m^3$/ $\eta^{pr}= 0.336$ /$\delta_B^{pr}=0.00453\ m$/	$\nu^{pr}=0.1132$ /$V^{pr}=0.172\ m^3$/ $\xi^{pr}=0.001119$ /$\delta_B^{pr}=0.0134\ m$/
Design variables $x_1[m^2]$	0.005034	0.014846	0.0050
$x_2[m^2]$	0.003017	0.008907	0.0030
$x_3[m^2]$	0.005026	0.014846	0.0050
$x_4[m^2]$	0.006030	0.017814	0.0060
Width of compromise set	$\Delta V= 1.348\ m^3$ $\Delta\delta_B= 0.0119\ m$	$\Delta\nu= 0.887$ /$\Delta V= 1.348\ m^3$/ $\Delta\eta= 0.887$ /$\Delta\delta_B=0.0119m$/	$\Delta\nu=0.887$ /$\Delta V =1.348\ m^3$/ $\Delta\xi=0.000993$ /$\Delta\delta_B=0.0119\ m$/
Sensitivity vector	$\frac{\partial F}{\partial V}= 0.0757$ $\frac{\partial F}{\partial \delta_B}=0.9971$	$\frac{\partial F}{\partial \nu} = 0.7071$ $\frac{\partial F}{\partial \eta} = 0.7071$	$\frac{\partial F}{\partial \nu} = 0$ $\frac{\partial F}{\partial \xi} = 1$

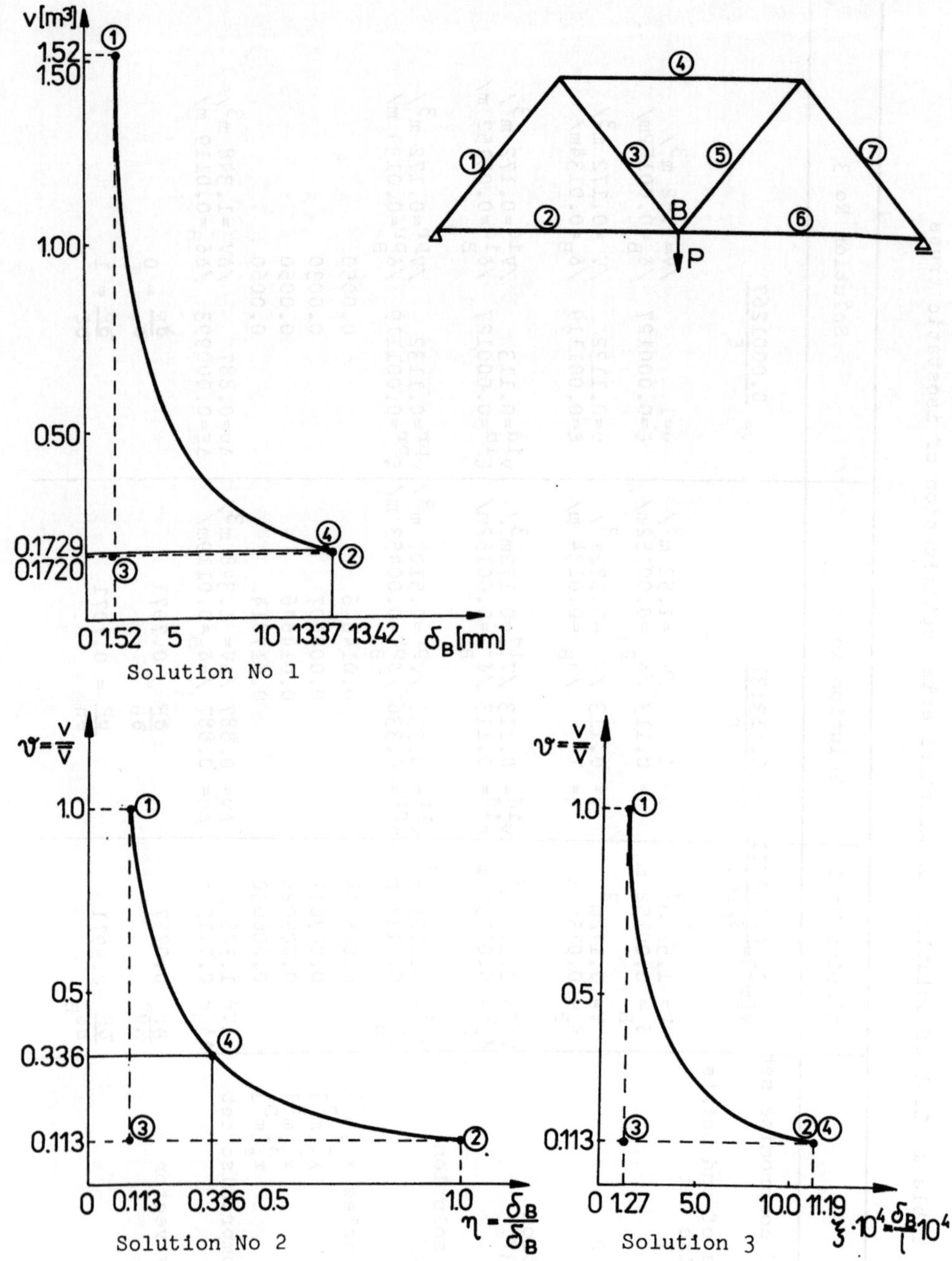

Fig.2. Representation of objective space for Example 4.2.

LINEAR APPROXIMATION OF SURFACES WITH A CRITERION "MINIMUM OF AVERAGE QUADRATIC TIME" TREATING OPTIMAL CONTROL PROBLEMS

J.I.KALEV, L.AT. GUNCHEV*

NPP "Techenergo", Kozloduj Str.14, Sofia 1202, Bulgaria
*Higher Institute of Mechanical and Electrical Engineering (VMEI "Lenin" Dept. of Automatica, bl.2) Sofia - 1156, Bulgaria

SUMMARY. The paper discusses a new criterion for evaluation the precision of linear approximation when treating optimal control problems. The latter relates the deviations of the functional of quality of motion from its optimal value to the error of approximation. The system of quadratic equations for determing the vector of parameters "c" of the approximating plane is derived. Application of this norm is considered involving an ascillatory second order plant. The synthesized near-optimal controllers were computer simulated and the corresponding time responses are given.

1. INTRODUCTION

Treating optimization problems with systems of higher order n ($n>4$), no matter what approach we use, we come accross a major obstacle, named by R.Bellman "the doom of dimension". This is expressed in setting impossible requirements (even for modern powerful computers) for computer memory capacity. It is due to the need of storing the values of functions of many variables in the computer, and that is necessary both for optimal control design and for its technical realization. For example, if the design is carried out using the dynamic programming method, then for each optimization stage it is necessary to memorize the values of the Bellman functions for the current and the previous stages. On its turn, the optimal control structure involves at least one functional transformer of many variables.

An effective approach to overcoming those difficulties is the approximation of the surfaces representing geometrically the corresponding functions. Some peculiarities of the topology of these surfaces [1] make piecewise-linear approximation especially suitable for the case.

After choosing the class of approximating functions one has to define a criterion for evaluating the precision of approximation. The essence of the suggested in this paper method of approximating is in the following: the deviations of the functional of quality of motion from its optimal value are related to the time for "erroneous motion" By "erroneous motion" one should understand the motion of the system for control, differing from the optimal one. For better understanding of this term let us assume that in the state space there is a part of a surface which we would like to approximate with a plane segment. For illustration, let us consider a (second order) two-dimentional case.

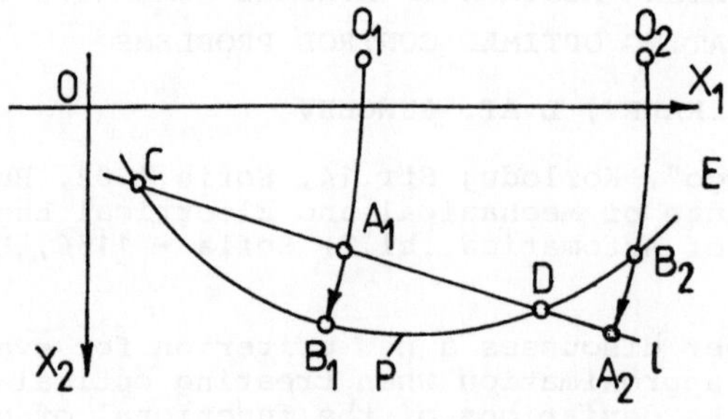

Fig. 1

On Fig.1, O_iB_i (i=1,2) are the trajectories of the near-optimal motion of the plant. Curve "p" is the optimal control switching line, and "l" is the switching straight line. Let Δt_{AiBi} is the time for motion of the plant-A_i to B_i on the phase plane. Then Δt_{A1B1} is the time during which the strictly optimal system should move with the old control, and Δt_{A2B2} is the time during which the near-optimal system should move with the new control. Δt_{AiBi} is the time during which the system moves with a control different from the optimal one.

The validity of the above statement for a time-optimal control system is obvious. In this case the error of approximation of the control switching surface is exactly the time for non-optimal motion of the system, i. e. Δt_{AiBi}. If we apply the dynamic programming method, and an approximation of the isochrones is carried out, then the following expression is valid:

$$T_i = T_{ai} \pm \Delta t^{(i)} \tag{1}$$

where T_i is the minimum time that takes the system to get to a given final state (pole of the isochrone), T_{ai} - the time during which the system moves along the optimal trajectory from a point on the approxiating segment to the defined final state. For $T_i \approx T_{ai}$ it is necessary that $\Delta t^{(i)}$ be minimum.

For a quality criterion

$$I = \int_{t_o}^{t_f} f_o\,(x,\ u,\ t)\,dt$$

the error caused by non-optimal control for a time interval $\Delta t^{(i)}$ of motion of the system, is expressed by:

$$\Delta I = \int_{t_o}^{t_o \pm \Delta t^{(i)}} f_o\,(x,\ u,\ t)\,dt \tag{2}$$

If $f_o(.)$ is constrained, then for a sufficiently small $\Delta t^{(i)}$ we get:

$$\Delta I = f_o[x(t_o),\ u(t_o),\ t_o]\ \Delta t^{(i)} \tag{3}$$

Therefore, we can draw the conclusion that based on the intervals of erroneous motion of the system, a function can be formed, from which a criterion for evaluation of the precision of approximation can be derived.

2. STATING THE APPROXIMATION PROBLEM

Let the motion of the system be defined by the following system:

$$\dot{x} = f(x, u, t), \ x \in R^n, \ u \in R^r \tag{4}$$

where f(.) is a n-dimentional function. For f(.) are fulfilled the conditions for existence and singularity of solution of equation (4).

For small enough Δt, we can get, using (4):

$$x(t_o + \Delta t) = x(t_o) + f[x(t), u(t), t] \cdot \Delta t \tag{5}$$

We suppose that "s" points from the approximated surface are given, which we take for initial points of the systems motion at $t_o = 0$. We define

$$z^{(i)} \triangleq f(x^{(i)}, u^{(i)}, 0), \ i=1, \ldots\ldots, s \tag{6}$$

where we consider $u^{(i)}(t) = \text{const}$, for $0 \leq t \leq \Delta t_{BiAi}$

Taking into account the location of points A_i and B_i, as related to one another, we can write

$$x^{(i)}(t_o \pm \Delta t^{(i)}) = x^{(i)}(t_o) + f[x^{(i)}(t_o), u^{(i)}(t_o), t_o](\pm \Delta t^{(i)}) \tag{7}$$

where $\Delta t^{(i)} = \Delta t_{AiBi}$

The "+" sign corresponds to the location of B_i and A_i, similar to that on Fig. 1 for $A_i \in CD$, and the " - " sign - to that of $A_i \notin CD$.

We assume that $x^{(i)}(t_o \pm \Delta t^{(i)})$ are points on the approximating plane with the equation

$$c^T x^{(i)}(t_o \pm \Delta t^{(i)}) = \alpha \tag{8}$$

where $c \in R^n$ is a vector of parameters of the approximating surface; α is a constant; $c^T \triangleq [c_1 \ldots c_n]$ is a vector.

Let $x^{(i)}(t_o)$ be points in the approximated surface. We denote $x^{(i)}(t_o) \triangleq x^{(i)}$, i. e. points B_i are with coordinates $x^{(i)}$, and points A_i are with coordinates $x^{(i)}(t_o \pm \Delta t^{(i)})$.

We multipy (7) by c^T. Then, from (6), (7) and (8) we obtain

$$c^T x^{(i)}(t_o \pm \Delta t^{(i)}) = c^T . x^{(i)} + c^T . z^{(i)} . (\pm \Delta t^{(i)}) = \alpha \tag{9}$$

Therefore:

$$\pm \Delta t^{(i)} = \frac{\alpha - c^T x(i)}{c^T z(i)} \tag{10}$$

As a criterion for the precision of the approximation we choose the function:

$$J_t(c) = \sum_{i=1}^{s} (\Delta t^{(i)})^2 = \sum_{i=1}^{s} \frac{(\alpha - c^T x^{(i)})^2}{(c^T z^{(i)})^2} \tag{11}$$

The parameters of the approximating straight line "c" are

determined from the minimization condition:

$$J_t^o = \min_c J_t \tag{12}$$

In analogy with the error criterion "Minimum of average quadratic deviation", the criterion (12) will be named "minimum of average quadratic time and denoted "L_{2t}".

3. DETERMINING THE PARAMETERS OF THE APPROXIMATING PLANE

The optimum C_{opt} of (12) is found using the equation:

$$\frac{\partial J}{\partial c} = 0 \tag{13}$$

$$\frac{\partial J}{\partial c} = \sum_{i=1}^{s} \frac{-2[\alpha - c^T x^{(i)}] x^{(i)} (c^T z^{(i)})^2 - 2(c^T z^{(i)}) [\alpha - c^T x^{(i)}]^2}{[c^T z^{(i)}]^4}$$

After dividing eacg side of eq.(13) by $C^T z^{(i)}$ using condition

$$c^T z^{(i)} \neq 0 \tag{14}$$

we get the equation:

$$[\sum_{i=1}^{s} (\alpha x^{(i)} z^{(i)^T} - 2\alpha z^{(i)} x^{(i)^T})]c + \sum_{i=1}^{s} [c^T x^{(i)}]\{[z^{(i)} x^{(i)^T} - x^{(i)} z^{(i)^T}]c\} = -\alpha^2 \sum_{i=1}^{s} z^{(i)} \tag{15}$$

When deriving (15), the following identity was used:

$$(c^T z) x = (x z^T) c \tag{16}$$

In order to give a more compact expression of (15), we make the following denotations:

$$\text{a)}\ E \triangleq \sum_{i=1}^{s} (x^{(i)} z^{(i)T} - 2z^{(i)} x^{(i)T}) \triangleq \begin{bmatrix} l_1 \\ \vdots \\ l_k \\ \vdots \\ l_n \end{bmatrix}, \tag{17}$$

where $E = [l_{mn}]$ is n x n matrix, and l_k is the k- line of E;

$$\text{b)}\ g \triangleq \sum_{i=1}^{s} z^{(i)} \triangleq \begin{bmatrix} g_1 \\ \vdots \\ g_k \\ \vdots \\ g_n \end{bmatrix} \tag{18}$$

where g is a n-dimensional vector;

$$\text{c)}\ F^{(i)} \triangleq z^{(i)} x^{(i)T} - x^{(i)} z^{(i)T} \triangleq \begin{bmatrix} h_1^{(i)} \\ \vdots \\ h_k^{(i)} \\ \vdots \\ h_n^{(i)} \end{bmatrix}, \tag{19}$$

where $h_k^{(i)}$ is the k- line of the n x n matrix $F^{(i)} = [f_{mn}]$.

With these denotations equation (14) becomes:

$$\alpha E c + \sum_{i=1}^{s} (c^T x^i) F^i c = -\alpha^2 g \tag{20}$$

The equation that corresponds to the k-line of (20) is:

$$\alpha l_k c + \sum_{i=1}^{s} (c^T x^{(i)}) h_k^{(i)} c = -\alpha^2 g_k, \quad k = 1,\dots n \qquad (21)$$

For a fixed k, a n x n matrix D_k is introduced:

$$D_k \triangleq \sum_{i=1}^{s} x^{(i)} h_k^{(i)} = [d_{lmk}]; \quad l,m = 1,\dots,n \qquad (22)$$

Finally, from (21) is derived the following system of quadratic equations for determining the vector of parameters c of the approximating plane:

$$\alpha l_k c + c^T D_k c = -\alpha^2 g_k, \quad k = 1,\dots,n \qquad (23)$$

The parameters of D_k and l_k of the system (23) are expressed directly from (17), (18), (22).

If a 3-dimensional matrix is used

$$D = [d_{lmk}], \quad l,m,k = 1,\dots,n \qquad (24)$$

the system (22) can be presented as

$$\alpha Ec + c^T Dc = -\alpha^2 g \qquad (25)$$

For further considerations some of the properties of the matrix D are used, namely:

(I) The diagonal columns of D are of zero elements. That property follows directly from (22), since when m=k, it follows

$$d_{lkk} = 0, \quad l = 1,\dots,n \qquad (26)$$

As a consequence, the k-column of D_k is of zero elements.

(II) The columns of D, symmetrical to the diagonals, have elements of equal absolute values and with opposite signs. That property follows from (21).

Using properties (I) and (II) considerably shortens the amount of computational work when determining the elements of D. Instead of n^3 elements, the number of elements to be computed is $n^2(n-1)/2$.

For the design of near optimal control structures one must carry out approximation of surfaces passing through the origin. The same requirement is valid for the approximating surface. For that reason, near the origin the surface is approximated by a plane passing through the origin.

The equation for determining the vector of parameters of the approximating plane passing through the origin, is derived from (23), respectively (25) putting $\alpha = 0$, i.e:

$$c^T Dc = 0 \text{ or } c^T D_k c = 0, \quad k = 1,\dots,n \qquad (28)$$

4. A PROGRAMME FOR DETERMINING THE PARAMETERS OF THE OPTIMAL APPROXIMATING STRAIGHT LINE

A programme in FORTRAN IV computes the matrix parameters of eq.(25) i. e. mattrices E, D and g respectively through (17), (22) and (18), taking into account the above-stated properties (I) and (II) of matrix D.

In the general case, computations are carried out for a linear stationary system of order n. The n x s matrix Z is defined with vector columns $z^{(i)}$. Matrix Z is computed using the equation:

$$Z = AX + BU, \qquad (29)$$

where:

X is a n x s matrix, the "s" columns of which are points on the approximated surface in n-dimensional state space.

U is a r x s matrix, the "s" columns of which are r-dimensional control vectors in the above-defined points on the approximated surface for the time-interval Δt. We admit that the control remains constant during Δt.

From (23), and for n=2, we get a polynomial for C_2. The programme computes the coefficients and the zeroes of the polynomial using programme POLRT of the package SSP. Then the value of c_1 corresponding to the real value of c_2 and the parameters of the approximating straight lines are computed.

$$x_2 = kx_1 + b; \quad k = -\frac{c_1}{c_2}\ ; \ b = \frac{1}{c_2} \ \text{for } \alpha = 1 \qquad (30)$$

$$x_2 = kx_1 \ , \ k = -\frac{c_1}{c_2} = \frac{d_{221}}{d_{121}} = \frac{d_{212}}{d_{112}} \ \text{for } \alpha = 0 \qquad (31)$$

5. EXAMPLE

Consider an oscillating plant, described by the equation:

$$\begin{bmatrix} \dot{x}_1 \\ \dot{x}_2 \end{bmatrix} = \begin{bmatrix} 0 & 1 \\ -1 & -0.4 \end{bmatrix} \begin{bmatrix} x_1 \\ x_2 \end{bmatrix} + \begin{bmatrix} 0 \\ 1 \end{bmatrix} u \qquad (32)$$

The parametric equation of the optimal switching line is given in [1]. Using the above-described programme, a piecewise-linear approximation of the optimal switching line is carried out, for m = 4 and m = 3 approximating segments. For comparison, for the same initial data were determined the parameters of the approximating straight lines with a nearness norm "minimum of average quadratic deviation" L_2. The corresponding programme in FORTRAN IV for a Gram-Schmidt regression was used for the purpose [3]. The obtained results are in Table 1.

In Table 1 the approximated part of the curve is defined through the definition interval of x_1. Here for values of x_1 are given the corresponding parameters of the approximating straight lines for nearness norms L_2 and L_{2t}, and m = 3, m = 4. Figure 3 shows the optimal switching line for this plant (the smooth curve). In continuous line are shown the approximating straight lines for norms L_{2t}, and in dotted line - for norms L_2.

For the considered oscillating plant (32) was synthesized a near-optimal control device. The structures of the controller were of the following type:

1) P - structure

$$V_4^p(x_1x_2) = \mu_4 \text{sign}[\mu_3 \text{sign}(\mu_2 \text{sign } W_1 + W_2) + W_3] + W_4 \quad (33)$$

$$W_i = c_{i1}x_1 + x_2, \; i=1,\dots 4, \; \mu_i = c_{io}$$

2) P_1 - structure

$$V_4^{p1}(x_1x_2) = v_4 \text{sat}\{v_3 \text{sat}[v_2 \text{sat}(r_1x_1) + r_2x_1] + r_3x_1\} + r_4x_1 + x_2 \quad (34)$$

The equations for determining the parameters in eqs. (33) and (34) are given in [5]. The value of the coefficients for number of approximating elements of the control device m = 3 and m = 4 are given in Table 1.

Figure 2 shows the transient processes of the thus synthesized control system.

6. CONCLUSIONS

1) The synthesized near-optimal controllers for a nearness norm L_{2t} guarantee a minimum (in the average-statistic sense) of the time of motion of the system with non-optimal control.

2) The error function J_t expresses a direct connection between the error of approximation and the character of motion of the system. In that sense, the introduced nearness norm is dynamic. For one and the same approximated surface and different dynamic properties of the plant we get different parameters of the approximating straight lines.

3) The regions where the phase portrait is more strongly curved (the trajectories near the approximating surface have a greater radius of the curve), the parameters of the approximating straight lines for L_2 and L_{2t} nearness norms differ more strongly.

4) The greater the part of a given surface to be approximated the more the parameters of the approximating planes differ for the two norms.

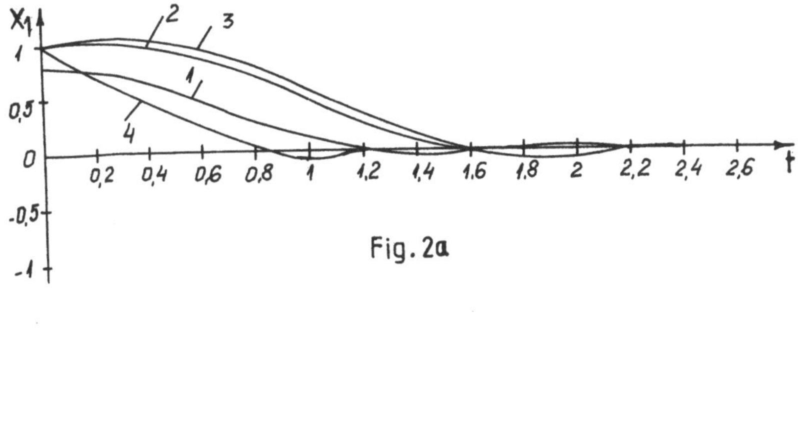

Fig. 2a

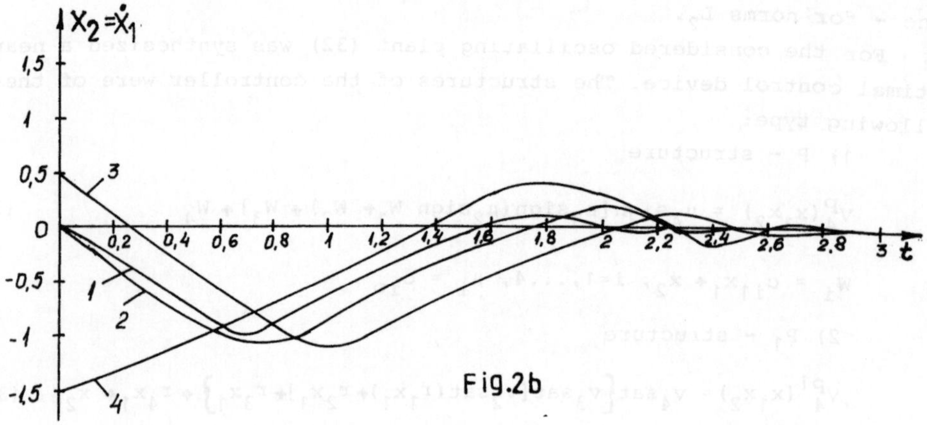

Fig.2b

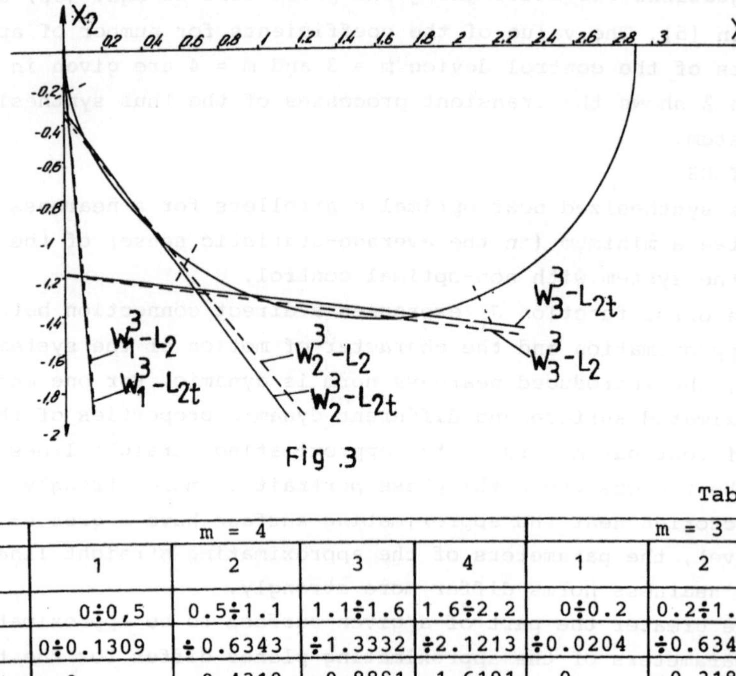

Fig 3

Table 1

		m = 4				m = 3		
i		1	2	3	4	1	2	3
τ_1		0÷0,5	0.5÷1.1	1.1÷1.6	1.6÷2.2	0÷0.2	0.2÷1.1	1.1÷2.2
x_1		0÷0.1309	÷0.6343	÷1.3332	÷2.1213	÷0.0204	÷0.6343	÷2.1213
L_{2t}	b=-cio	0	-0.4319	-0.8881	-1.6191	0	-0.3185	-1.1787
	k=-ci1	-5.0003	-1.1516	-0.4181	0.1226	-12.4617	-1.4128	-0.1261
L_2	b=-cio	0	-0.4454	-0.8958	-1.601-	0	-0.3437	-1.1830
	k=-ci1	-4.3733	-1.1104	-0.4095	0.1125	-11.8073	-1.2980	-0.1210
L_{2t}	r_i	8.9108	0.8259	0.1824	0.1226	34.684	1.0916	0.1261
	v_i	0	0.4863	0.5485	1.6191	0	0.2702	1.1787
L_2	r_i	7.32524	0.7823	0.3260	-0.1125	30.5747	0.9948	0.1210
	v_i	0	0.4971	0.5595	1.6010	0	0.2905	1.1830

REFERENCES

1. Gunchev L.At., Kalev J.I and L.K.Temelkov (1985). A modified algorithm of the dynamic programming method for optimal control design. Proceedings of the 11-th IMACS World Congress, Oslo.
2. Gunchev L.At., Kalev J.I. and L.K.Temelkov, W.Kr.Radonov (1985) Conditions of optimality and algorithms for design determined through the quickest alternation of the optimal Lyapunov function.
Jubilee Session of the HIMEE, Sofia.
3. Farebrother R.W. (1981), Gram-Schmidt Regression. Collected Algorithm from ACM.
4. Seber G.A.F.(1977). Linear regression analysis.
5. Gunchev L.At. (1967). On approximation of the sign-functions of several variables. Report of the 2-nd Congress of the Bulgarian mathematicians, Varna.
6. Gunchev L.At., Temelkov L.K. and J.I.Kalev (1984). An algorithm of modified dynamic programming. Tenth International Conference on Nonlinear Oscillations, Varna.

VECTOR LYAPUNOV FUNCTIONS METHOD FOR DIFFERENCE-DIFFERENTIAL SYSTEMS AND ITS APPLICATIONS TO IMMUNOLOGY

N.A.KARATUEVA, V.M.MATROSOV
Irkutsk Computer Center of the Siberian Branch of the USSR Academy of Sciences

Vector Lyapunov functions(VLF) method is elaborated in the analysis of stability, positive invariance, periodical solutions of nonlinear difference-differential equations. For this class of systems Lyapunov functionals and functions method is developed by N.N.Krassovskii [I], B.S.Razumikhin [2], S.N.Shimanov, L.E. El'sgol'ts [3], Kolmanovsky V.B., Nosov V.R. [4], G.K.Hale [5]. Vector Lyapunov functions [6] were applied to the systems with delay in works by P.S.Gromova [7] and authors [8]. We obtained the comparison principle with VLF for derivation of theorems on different dynamical properties of differencedifferential equations and the means of VLF construction in problems of exponential stability of complex bilinear systems.

In this paper there are presented the theorems on stability, positive invariance, existence of periodical solutions, where Lyapunov-Krassovskii functionals are chosen as VLF components, and B.S.Razumikhin's condition is taken into account. Algorithms of VLF construction with components of quadratic forms, modulus of linear forms [9] for interconnected systems with polynomial right-hand sides are suggested. On the base of VLF method some investigation of immunological models, developed by G.I.Marchuk [10] and R.V.Petrov [11] is produced.

I. A functional-differential equation is considered

$$\dot{x}(t)=F(t, x_t), \qquad (1)$$

with following notations

$$x\in R^n, \; t\in \mathfrak{J}=[t_0, +\infty), \; x_t=x(t+s), -\tau\leq s\leq 0, \; \tau\text{-const}>0,$$

$$\varphi: [-\tau,0]\to R^n,$$

$$\varphi^n= C[[-\tau,0], R^n], \; \varphi\in\varphi^n, \; \|\varphi\| = \sup_s |\varphi(s)|,$$

$$\rho\text{-const} > 6, \; C_\rho=\{\varphi\in C^n: \|\varphi\|<\rho\}.$$

$F(t,x(t),x_t)$ is n-vector, whose components are piecewise-continuous functional by t, satisfying Lipschitz condition in x_t [I].

Let $V\in C[[-\tau,\infty)\times C_\rho, R^m_+]$, $\varphi\in C_\rho$ V be vector comparison function, satisfying the differential inequality:

$$\bar{D}_+ V(t,x_t)\leq g(t,V(t,x_t)), \qquad (2)$$

where $\bar{D}_+$ is the upper right derivative by time, $x_t \in \Omega_j$, $j=\overline{0,2}$. g is quasimonotonously increasing function (Wazewski condition)

$g \in C$, $g \in W$: $g_i(t,u) \leq g_i(t,\bar{u})$, $u_i=\bar{u}_i$, $u_j \leq \bar{u}_j$, $(j \neq i)$, $i,j=\overline{1,m}$.

$$\Omega_0 \triangleq \{\varphi \in C_\rho : |V_t^i|_0 = V^i(t, \varphi(0)), i=\overline{1,m}, t \in \mathcal{J}\}$$

$$\Omega_1 \triangleq \{\varphi \in C_\rho : \sum_{i=1}^m V^i(t+s, \varphi(s)) < f(\sum_{i=1}^m V^i(t, \varphi(0))), t \in \mathcal{J}\}$$

$$\Omega_2 \triangleq \{\varphi \in C_\rho : V^i(t+s, \varphi(s)) < f^i(V^i(t, \varphi(0))), t \in \mathcal{J}\}$$

Here $\Omega_2 \subset \Omega_1$, the norm $|V_t^i| = \sup_s V(t + s, \varphi(s))$, $i=\overline{1,m}$;

$f(u)$ is continuous on R_+, non-decresing function by u, $f(u) > u$, (respectively, $f^i(u_i) > u_i$, $i=\overline{1,m}$). Comparison system(CS) is introduced

$$\dot{u}=g(t,u) \tag{3}$$

The specification of derivative estimations from $\bar{D}_+ V(t,x_t)$ by the sets Ω_j showed, that CS is obtained in the class of ordinary differential equations in R^m, the investigation of which is much more simple.

L e m m a .

Let $V \in C[[-\tau,\infty) \times C_\rho, R_+^m]$ satisfies the differential inequality (2) for $\varphi \in \Omega_j$, $j=0 \vee 1 \vee 2$, $x(t_0,\varphi_0)(t)$ is the solution of the equation(I), sup V $(t_0,\varphi_0(s)) \leq u_0$. Then $V(t,x(t_0,\varphi_0)(t)) \leq \bar{u}(t,t_0,u_0)$, for $t \in [t_0,t_1)$, $t_1 = \min[\tau(x), \tau(u)]$. Here $\bar{u}(\cdot, t_0, u_0)$ is upper solution of the equation (3), $\tau(x)$ ($\tau(u)$) is the maximum interval of the solution existence for the equation (I) (equation(3)).

Using the general scheme of comparison principle [8] and known sufficient conditions of stability(instability, asymptotic stability), existence of periodical solutions and positive invariance of ordinary differential equations for CS(in terms of their right-hand sides), also the theorems of conversion [1,5], we shall state the following theorems.

T h e o r e m I. For stability(corresponding to uniform stability with respect to t_0) of solution x=0 of the system(I) $(F(t,0,0)=0)$ the existence of VLF $v(t,x_t)$ in some region C_ρ is necessary and sufficient for which $\sum_{i=1}^m v^i(t,x_t)$ is positive definite(respectively, and $v^s(t,x_t)$, $s=\overline{1,m}$ admits infinitesimal higher limit), the differential inequality holds.

$$\bar{D}_+ V(t,x_t) \leq P(t)v(t,x_t), \quad (t,x_t) \in \mathcal{J} \times C_\rho, x_t \in \Omega_0 \tag{4}$$

where P(t) is the real continuous positive $(P_{s\nu}(t) \geq 0$ for $\nu \neq s)$

$m \times m$ matrix and

$P(t) = A(t) + B(t)$, where $A(t), B(t)$ are real continuous matrices, $\int_{t_0}^{\infty} \|B(t)\|\, dt < +\infty, t_0 \in [0, +\infty)$ and $A(t)$ satisfies one of the conditions:

a) for the system $\dot{y} = A(t)y$ Caushy matrix $K_A(t,t_o) = Y_A(t)Y_A^{-1}(t_o)$ is bounded for $0 \le t_o \le t$, $t < +\infty$;

b) $A(t)$ is the w-periodical matrix, $A(t+w)=A(t)$, for which the matrix of monodromy $Y_A(w)$ has eigenvalues $\rho_1, \ldots, \rho_k$ (multipliers), located at the unique circle ($|\rho_s| \le 1$, $s=\overline{1,m}$), besides, to those of them that lie on the circumference ($|\rho_s| = 1$), correspond to the simple elementary divisors.

c) $A(t)=A + C(t)$, where A is the constant real matrix, that has eigenvalues $\lambda_s(A)$ with non-positive parts ($Re \lambda_s(A) \le 0$, $S=\overline{1,m}$), besides, or $C(t)\equiv 0$ or with simple eigenvalues with zero real parts, $\|C(t)\| \to 0$ for $t \to 0$, $\int^{+\infty} \left\|\frac{dC(t)}{dt}\right\| dt < +\infty$, $Re\lambda_s(A+C(t)) \le 0$, $s=\overline{1,m}$, $t \in [0,+\infty)$.

T h e o r e m 2. (on instability). For instability of the solution x=0 of the system (I) the existence of the bounded, Lipschitz condition by x_t VLF $v(t,x_t)$ in any region $C_{\rho_0} \subset C_\rho$ is necessary and sufficient the derivative of which satisfies the differential inequality

$$\mathcal{D}_+ v(t,x_t) \ge Pv(t,x_t), \quad (t,x_t) \in J \times C_{\rho_0}, \quad x_t \in \Omega_o \tag{5}$$

P is the real positive constant $m \times m$ matrix, possessing the various eigenvalues $\lambda_1 \ldots \lambda_m$, having maximum eigenvalue $\lambda_1 > 0$, besides in any amount of small neighbourhood of the motion x=0 for any t_o there is found x_o^*, such that for any $i = \overline{1,m}$

$$c_1^i \det C \sum_{j=1}^{m} A_{j1} v^j (t_o, x_o^*) > 0,$$ here c_1^i is i-th component of the eigenvector C_1, corresponding to λ_1, A_{j1} $(j=\overline{1,m})$ the algebraic complements of elements of the first column of determinant C, composed of eigenvectors, $c_s = \underset{i}{col}\, c_s^i$, $s=\overline{1,m}$.

In the following theorem, analogous to Baley's one for the systems with delay [7], the expediency of the set input Ω_2 is shown on a level with Ω_0, Ω_1.

Let the system (I) be consisted of m interconnected subsystems and be presented in the form

$$\dot{x}_i(t)=F_i(t,x_i(t), x_i(t-\tau_i)) + \sum_{j=1}^{m} (R_{ij}(x_t,t)x_j(t) + R_{ij}(x_t,t)x_j(t-\tau_j)) \tag{6}$$

where for any fixed $t \ge t_o$, $x_i \in R^{n_i}$; $\sum_{i=1}^{n} n_i = n$; $x_t = (x_1(t), \ldots, x_m(t), x_1(t-\tau_1) \ldots x_m(t-\tau_m))$.

The isolated subsystems are considered simultaneously

$$\dot{x}_i(t)=F_i(t,x_i(t),x_i(t-\tau_i)) \tag{7}$$

Let for every isolated subsystem(7)one can construct Lyapunov function $V_i(t,x_i)$with estimations,characteristic of quadratic forms

$$C_{i2}\ \|x_i\|^2 \leq V_i(t,x_i) \leq C_{i1}\ \|x_i\|^2,\quad \dot{V}_i \leq -\varkappa_i\ V(t,x_i), \tag{8}$$

where $\varkappa_i > 0$, $C_{ij} > 0$, $i=\overline{1,\ m}$, $j=1,2$, besides (8) is carried out on indispensable curves φ, satisfying the condition $\varphi \in \Omega_2$.

Theorem 3. If the following condition is satisfied: by the system(6)

$$\dot{v}^k(t,x) \leq \langle P^k, v\rangle + V^T(t,x)Q^k v(t,x),$$

where P,Q^k are $m \times m$ constant matrices, P^k- k-th string of the matrix P, $P_{s\nu} \geq 0$, $s \neq \nu$, $(-1)^s|P_{\mu\nu}|_1^s > 0$, $q^k_{ij} > 0$, $i \wedge j \neq k$, $s=\overline{1,m}$, then the trivial solution of the system(I) is exponentially stable.

Theorem 4. The set $\{V(t,x_t) \geq 0\}$ by the system(1)is positive invariant, if there exists continuous vector function $V(t,x_t)$,satisfying the differential inequality $\underline{D}^+V(t,x_t) \geq g(t,V(t,X_t))$ for x_t $\bar{\Omega}_0(\bar{\Omega}_1)$,where $g_i(t,u) \geq 0$ for $u_i=0$, $u_j \geq 0$, $j \neq i$, $g \in W$.

Theorem 5. Let for (1) $(F(t,x_t)=F(t+T,x_t))$, $F(t,x_t)$ maps bounded sets of $R \times C$ into bounded sets of R^n, and there exists VLF:$V:R \times C \to R^m$ $V \in C$, $\sum\limits^m V_i(t,\varphi) > 0$, for $\varphi \in K$, where K is positively invariant for (1) convex subset of C, $V(t,\varphi) \xrightarrow[t]{} \infty$, as $\|\varphi\| \to \infty$, V maps bounded set of $R \times C$ into bounded sets of R^m and for $k > 0$, $k \in R^m$, the set $\{\varphi \in K,\ V(t_0,\varphi) \leq k\}$is closed convex set in C.

Then, if $g \in w$, $\dot{V}(t,x_t) \leq g(t, V(t,x_t))$, for $x_t \in \Omega_1$, and CS $\dot{u} = g(t,u)$ has upper positive solution such that $u(t_0+T) \leq \bar{u}(t_0)$ for any $t_0 \in \mathcal{J}$, the system (1) has periodical solution.

2.At present the central problem in Lyapunov function method is LF and VLF construction. The general means and algorithms for Lyapunov function constructions are not found for nonlinear systems. In 8 the ways of VLF constructions with components of quadratic forms kind $V_i=x_i^TC_ix_i$ for interconnected bilinear systems with delay were presented.

Let us consider the interconnected polynomial system,consisting of subsystems:

$$\dot{x}_K = A_{KK}x_K(t) + B_{KK}x_K(t-\tau_K) + \sum_{i=1}^{m}(A_{Ki}x_i(t) + \tag{9}$$

$$B_{Ki}x_i(t-\tau_i)) + G^K(x_K(t)) + \Phi^K(x_K(t-\tau_K)) + \bar{G}^K + \bar{\Phi}^K$$

$$(i \neq K)$$

where $x=(x_1,\dots,x_m)^T$, $x_k\in R^{n_k}$,$k=\overline{1,m}$,

$\sum_{k=1}^{m} n_k=n$, $A_{kk},B_{kk}(A_{ki},B_{ki})$ are the $n_k\times n_k$ $(n_k\times n_i)$ blocks of matrices A and B of the linear part of the system, we see that polynomial nonlinearities are defined by the vectors:

$G^k(x_k)=\text{colon}\ G_i^k\ (x_k);\ \bar{G}^k(x_1,\dots,x_{k-1},\ x_{k+1},\dots,x_m) = \text{colon}\ G_i^k(\dots)$,

$\Phi^k=\text{colon}\,\Phi^k_i\ (x_k(t-\tau_k))$,

$\bar{\Phi}^k=\text{colon}\bar{\Phi}^k_i\ (x_1(t-\tau_1),\dots,x_{k-1}(t-\tau_{k-1}),x_{k+1}(t-\tau_{k+1}),\dots,x_m(t-\tau_m))$.

Isolated subsystems of the system(9)

$$\dot{x}_k=A_{kk}x_k + B_{kk}x_k(t-\tau_k) + G^k(x_k) + \Phi^k(x_k(t-\tau_k)) \qquad (10)$$

To simplify calculations and taking into account the pecularities of applied problems [10], we restrict ourselves to the trilinear terms in polynomial parts of the systems (9)-(10):

$$G_i^k=\sum_{s,\tau=1}^{n_k} g^{ki}_{s,\tau}x_k^s(t)x_k^\tau(t)+\sum_{j,s,\tau}^{n_k}\tilde{g}^{ki}_{j,s,\tau}x_k^j(t)x_k^s(t)x_k^\tau(t)$$

$$\bar{G}_i^k=\sum_{\substack{\mu=1\\ \mu\neq k}}^{m}\Big[\sum_{s,\tau=1}^{n_\mu}\bar{g}^{\mu i}_{s\tau}x_\mu^s(t)x_\mu^\tau(t)+\sum_{j,\tau,s}^{n_\mu}\bar{g}_{j,s,\tau}x_\mu^j(t)\cdot x_\mu^\tau(t)\cdot x_\mu^s(t)$$

Analogously Φ_i^k, $\bar{\Phi}_i^k$.

Vector-function $V=(V_1\dots V_m)$, whose components are the quadratic forms $V_k=x_k^T(t)C_kx_k(t)$, $C_k^T=C_k$ allows for the case of trilinear nonlinearities to obtain CS vector Ricatti equation type:

$$\dot{u}^k\leq\langle P^k,u\rangle+u^TQ^ku,$$

where $k=\overline{1,m}$, P and Q^k are the matrices of dimension $m\times m$, whose elements are defined by the formulae, obtained by majorizing the derivative of VLF. Algorithms of vector-function and CS constructions are realized by a programm.

For the system(9) we shall take the algorithm of vector Lyapunov function construction with components of the linear form modulus. Advantages of VLF construction of the module of linear form, as it is known [9] , are determined by the possibility and simplicity of estimation construction for initial systems by CS estimations; this is based on the experience of investigation of ordinary differential equations [9]. Let $V^k(x_k) = W_k(x_k$), where $W_k(x_k)=|T^kx_k|$, $\Lambda=T^k(A_k+B_k)T^{-1}$.

Consider the linearized isolated systems

$$\dot{x}_k(t) =A_kx_k(t) + B_kx_k(t-\tau_k) \qquad (11)$$

Defining the set

$$\Omega_0=\{x(t):\ V_j(x(s))\leq V_j(x(t)),\ s\in[-\tau,\ 0],\ \ j=\overline{1,m}\ \}$$

we estimate the right derivative of V_k by the system (II) on the set Ω_o:

$$\overset{+}{D} V_k \leq (\Lambda + \| B_k \| \cdot (E + |T_K^{-1}| \, |T_K|^{-1}) \, |T_K^{-1}| \cdot V_k$$

The CS may be written as $\dot{u}_k = P_k \cdot u_k$, where p_k are diagonal elements of the matrix $P=(P_{ij})_1^m$, that is $P_{kk}=P_k$.

Majorization of the derivative of V is produced by the complete system (9), taking into account the connections

$$V \leq |T| \, |x| \quad , |x| \leq |T^{-1}| V.$$

There is given the algorithm of m-dimensional comparison system construction in the class of trilinear ordinary differential equations. The presented algorithms of VLF construction are implemented.

Estimations of the attraction regions and others, obtained by VLF and CS are strict, but may be rough. The attraction regions boundary construction for (9) is realized by combination of VLF method with multiple numerical integration (9) analogously [8].

3. Consider the model, which describes the immune reaction of organism:

$$\begin{aligned} dv(t)/dt &= (\beta - \gamma F(t)) v(t) \\ dF(t)/dt &= \rho c(t) - \mu_f F(t) + \eta \gamma V(t) F(t) \\ dc(t)/dt &= \alpha V(t-\tau) F(t-\tau) - \mu_c (C(t) - C^*) \\ dm(t)/dt &= \sigma V(t) - \mu_m m(t) \end{aligned} \qquad (12)$$

with initial conditions $V(t)=0, t < t^o$, $V(t^o)=V^o$, $c(t^o)=c^o$, $f(t^o)=F^o$, $m(t_o)=m^o$.

(The simplest immunological model by G.I.Marchuk [10]). In system (12) there are two stationary regimes. The first one: $V_I = 0$, $F_I = \gamma \rho c^*/\mu_f$, $C_I = C^*$, $m_I = 0$, is interpreted as the state of healthy organism. Stability condition $\beta < \gamma F^*$ holds. The attraction region boundary of the first stationary regime is constructed (Fig.1). It should be noted, that for numerical values of parameters, the instability of the first regime excludes the stability of the second one. The second stationary regime [10] is interpreted as the

$$\begin{aligned} V_{2cT} &= \mu_c(\mu_f \beta - \gamma \rho c^*)/\beta(\alpha \rho - \mu_c \eta \gamma), \quad F_{2cT} = \beta/\gamma, \\ C_{2cT} &= (\alpha \mu_f \beta - \eta \gamma^2 \mu_c c^*)/\gamma(\alpha \rho - \mu_c \eta \gamma), \quad m_{2cT} = \frac{\sigma}{\mu_m} V_{2cT} \end{aligned} \qquad (13)$$

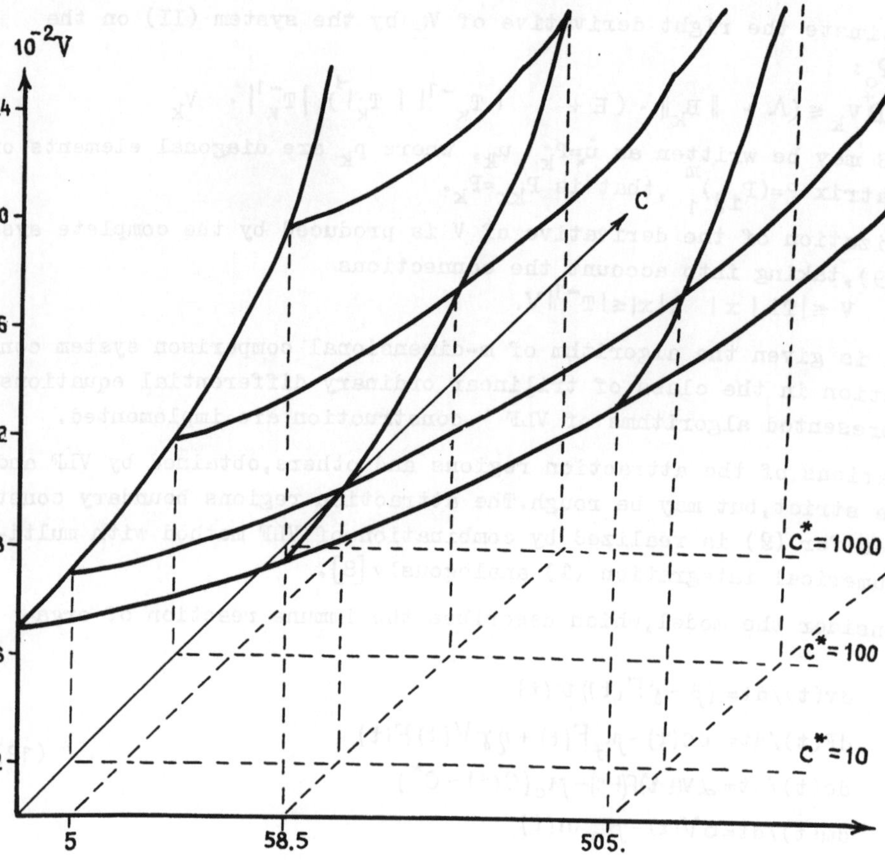

Fig.1

state of non-healthy organism, corresponding to the chronic desease. G.I.Marchuk obtained the sufficient condition of positivity for the second stationary and the sufficient condition of stability for (13). In our work for arbitrary the following sufficient conditions are obtained

$$a(b-g) + (d-f) > 0$$
$$d > g^2/4a + af^2/g$$

where $a = \mu_c + \mu_f + \eta\gamma V_2$, $d = \eta\gamma\mu_c\beta V_2$,

$b = \mu_c \cdot (\eta\gamma V_2 + \mu_f) - \eta\gamma\beta V_2$, $g = \alpha\rho V_2$, $f = \alpha\rho V_2 \beta$.

For investigation of the influence of parameters change on the location of quasipolynomial roots :

$$W_2 = -\lambda^3 - a\lambda^2 - b\lambda + d + g\lambda e^{-\lambda\tau} - f e^{-\lambda\tau}$$

of the system (12), linearized in the neighbourhood of second stationary solution the D-splitting (see Fig.2) in the plane of parameters $k_1 = \gamma \rho c^*$, $k_2 = \alpha\rho/\eta\gamma$ is constructed.

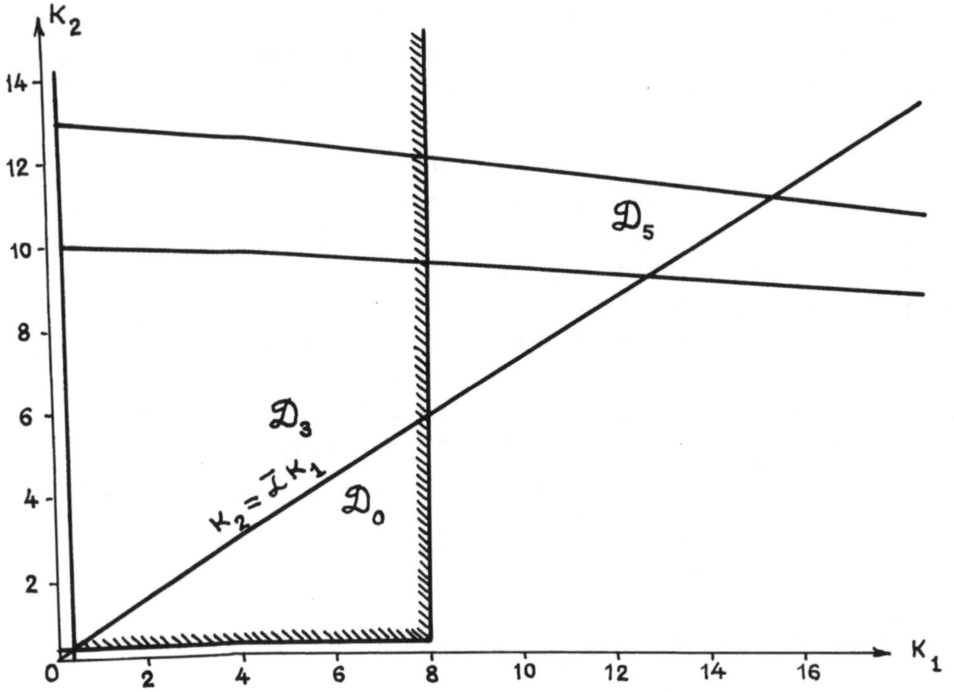

Fig 2

As well Marchuk as the authors by numerical modelling with (12) we obtain the periodical regimes.

There is obtained the sufficient condition of the periodical regime absence in the neighbourhood of stationary solution for arbitrary τ:

$$d > q^2/4a + af^2/q^2$$

If we take into account, that stationary periodical regimes are interpreted as chronic desease of the organism, then the obtained condition is tracted as the possibility to avoid now chronic deseases (excluding the second stationary), that is of interest from the medical point of view.

For $w=\beta$ for $w_2(\lambda), \lambda=jw$ in explicit form the first harmonics of parametrized family of periodical solutions of the system (12) are described

$$v(t) = -v_0 \sin(wt + \psi)$$

$$f(t)=f_0\cos(wt+\psi),\quad f_0=v_0 w/\gamma V_2;$$
$$c(t)=f_0(\mu_f+\eta\gamma V_2)\cos(wt+\psi)-v_0(\eta\gamma F_2/\rho+w^2/\gamma\rho V_2)\sin(wt+\psi).$$

The family depends on two arbitrary parameters ψ, v_0 which are determined by the initial conditions of Caushy problem. For the case $T = 2\pi/\beta$ the system of equations has one parametric family of periodical solutions, parametrized by the parameter β (Hopf theorem). Besides there do not exist another periodical solutions in sufficiently small neighbourhood.

Construction of autooscillations is realized by Ju.S.Kolessov's method. In Fig.2 it is obviously, that for $k_2=\bar{\alpha}k_1$ for the small parameter (in notations [11] one can take $E=k_2-\bar{\alpha}k_1$, as for $\varepsilon=0$ $W_2(\lambda)$ has the pair of purely imaginary roots, for the parameter μ the vector$(\gamma, \gamma, \alpha)^T$ is taken.

The modified model of virose desease [11] represents the trilinear system of differential equations with 4 delays: $\tau_B>0, \tau_H>0, \tau_H^B>0, \tau_E>0$ we represent it as three interconnected subsystems($n=10$, $n_1=5$, $n_3=2$, $n_2=3$).

$$\dot{V}_f=n b_e f(\ell)C_v E+\rho b_m C_v-\gamma_m V_f M_v-\gamma_f f(\ell)V_f F-\kappa\sigma V_f(C_v+\delta-m)$$
$$\dot{F}(t)=\rho\Pi-\eta_f\gamma_f f(\ell)V_f F-\alpha_f F$$
$$\dot{C}_v(t)=\sigma V_f(\ell-m)-b_E f(\ell)C_v E-b_m C_v$$
$$\dot{m}(t)=\mu b_E f(\ell)C_v E+\eta b_m C_v-\lambda m$$
$$\Pi(t)=b_n\xi(m,\ell)\rho_B M_v H_B B|_{t-\tau_B}+\alpha_\Pi(\Pi^*-\Pi)$$
$$\dot{M}_v(t)=\gamma_M V_f M_v-\alpha_M M_v-\sigma_M M_v E$$
$$\dot{H}_E(t)=b_H[\xi(m,\ell)\rho_H M_v H_E|_{t-\tau_H}-M_v H_E]-b_p M_v H_E E+\alpha_H(H_E^*-H_E)$$
$$\dot{E}(t)=b_p[\xi(m,\ell)\rho_E H_E M_v E|_{t-\tau_E}-M_v H_E E]-b_E C_v E+\alpha_E(E^*-E)$$
$$\dot{H}_B(t)=b_H^{(B)}[\xi(m,\ell)\rho_H^{(B)}M_v H_B|_{t-\tau_H^{(B)}}-M_v H_B]-b_p^{(B)}M_v H_B B+\alpha_H^{(B)}(H_B^*-H_B)$$
$$\dot{B}(t)=b_p^{(B)}[\xi(m,\ell)\rho_B M_v H_B B|_{t-\tau_B}-M_v H_B B]+\alpha_B(B^*-B)$$

For the condition $\xi\rho_B>1$, $\xi\rho_H^B>1$, $\xi\rho_E>1$, $\xi\rho_H>1$ the system possesses the property of positive invariance(theorem4,VLF), $v=(V_f(t)F(t),C_v(t), m(t), \Pi(t), M_v(t), H_E(t), E(t), H_B(t), B(t))$. The complex system has the stationary solution, which corresponds to the state of healthy organism:
$M_v^*(t)=C_v^*(t)=m^*(t)=0$, $B_{IcT}=B^*$, $E_{IcT}=E^*$, $H_{IcT}=H_E^*$, $H_{BIcT}=H_B^*$, $\Pi_{IcT}=\Pi^*$

and another stationaries, for which $M_v^*>0$, $C_v^*>0$, $m^*>0$.

The sufficient condition of asymptotic stability of the first stationary has the form:

$$b_m((1+\mu)\delta E^*+1)(\kappa\sigma\delta+\gamma_f(1+\mu)\delta E^*)-\sigma\delta(\beta_m\eta+\mu(1+\mu)\beta_E\delta E^*)>0.$$

The implementation of described algorithms in principle,allows to obtain the results by construction of the attraction regions of stationary points of the complex model.

LITERATURE

1. Krassovskii N.N. Some problems of the theory of stability. M.,Physmathgiz.,1959.
2. Razumikhin B.S. On stability of systems with delay.- PMM.,1956,20, N 4,p.500-512.
3. Mishkiss A.D.,Shimanov S.N.,Els'gol'ts L.E, Stability and vibrations of systems with delay. Proceedings of the international symposium on nonlinear vibrations., P.,1963,p.242-267.
4. Kolmanovskii V.B.,Nosov V.R. Stability and periodic solutions of controlled systems with aftereffects.-Moscow,Nauka,1981.
5. Hale Y.K. Functional differential equation. N.Y.Springer,1971.
6. Matrosov V.M. To the theory of stability of motion. PMM, 1962,XXVI, issue 6.
7. Gromova P.S. Vector Lyapunov functions method for systems with deviating argument. In book: Direct method in the theory of stability. M.,Science,1981,p.46-54.
8. Matrosov V.M.,Karatueva N.A. Vector Lyapunov functions method and its applications to immunology. In:Mathematical modelling in immunology and medicine.,North-Holland,1983, pp.175-186.
9. Vachonina G.S.,Kozlov R.I.,Malikov A.I. Algorithms of investigation of nonlinear controlled systems by vector Lyapunov functions method.- In book:Theory of stability and its applications.Novosibirsk,Nauka, 1979,p.14-24.
10. Marchuk G.I. Mathematical models in immunology. M.,Science,1980.
11. Marchuk G.I.,Petrov R.V. Virose organ infection and immune-physiological reactions of immunology.
12. Zemlyakov A.S. To problem of comparison system construction. Proceedings of KAI. Issue 4., Kazan,1972.

A NEW GLOBAL OPTIMIZATION TECHNIQUE FOR SOLVING PARTIAL DIFFERENTIAL EQUATIONS

I.KARPOUZAS , Y.CHERRUAULT

ABSTRACT

We propose to use an original global optimization method for solving functional equations and specially partial differential equations. We know that a lot of biological or physical phenomena can be represented by such equations and thus a new method for their resolution will be useful.

1. INTRODUCTION

Let us consider a general functional equation having the form :

$Au(x,t) = f(x,t)$ with initial and boundary conditions. A being a known linear or nonlinear differential operator. The function $u(x,t)$ is also known and $u(x,t)$ will be calculated. Several numerical methods allow to approach its solutions.

For example the operator A is approximated using the finite differences method.

On the other hand, with the help of the finite elements method, we keep A unchanged but we are looking for the solution in an approached set, that is to say we choose an approximation for $u(x,t)$ [3].

The proposed method can be related to the second possibility when the operator A remains unchanged. The basic idea is very simple and consists to choose a particular approximated form for representing the unknowns functions.

To do that, we use a transformation which has been developped about global optimization methods [2].

2. DESCRIPTION OF THE METHOD

First we explain the global optimization method : ALIENOR.
Then we give the description of our method for solving partial differential equations.

2.1 Global optimization method : ALIENOR

With the help of the generalized Archimedean spiral, we can reduce the unknown variables to a single one. Then we obtain an optimization problem according to a single variable θ, [2], generally $\theta \geqslant 0$.

First, we make the Alienor transformation described below. Then the densification principle is applied. We calculate a relative minimum using a probabilistic technique. Afterwards, we advance, on the length, from $\theta = 0$ and we look for successive minima more and more small. The lowest obtained minimum is the desired approximation. We have thus calculated a good approximation of the absolute minimum. Finally, we find the absolute minimum of the function by a sub-routine using local variations method [7].

For example, with 4 variables x_i, $i = 1,\dots, 4$ the transformation formulae are the following :

$$\theta_0 = \frac{\theta\cos\theta}{2\pi} \;, \; \theta_1 = \frac{\theta\sin\theta}{2\pi} \;, \; \theta_2 = \frac{\theta_o\cos\theta_o}{2\pi} \;, \; \theta_3 = \frac{\theta_o\sin\theta_o}{2\pi}$$

Then $Y_1 = \dfrac{\theta_2\cos\theta_2}{2\pi}$, $Y_2 = \dfrac{\theta_2\sin\theta_2}{2\pi}$,

$$Y_3 = \frac{\theta_3 \cos\theta_3}{2\pi} \quad , \quad Y_4 = \frac{\theta_3 \sin\theta_3}{2\pi}$$

are the transformed variables.

Densification principle.

Let m_i, M_i be the lower and upper bound of the variables x_i, then the principle of densification is such that :

if $0 < Y_i < 1/2$ then $z_i = m_i + (M_i - m_i) \cdot Y_i \cdot 2$ otherwise

$z_i = m_i + (M_i - m_i) \cdot (2.Y_i - \mathrm{INT}(2.Y_i))$ where $i = 1, \ldots, 4$.

2.2 Our technique for solving partial differential equations

Using a transformation related to the Archimedean spiral (space filling curve [1]) and considering the following relations :

$$\left.\begin{matrix} x = a\theta\cos\theta \ , \ t = a\theta\sin\theta \\ \theta \geqslant 0 \end{matrix}\right\} \Rightarrow \theta = \frac{1}{a} \sqrt{(x^2 + t^2)}$$

we can reduce the variables x,t of the function u(x,t) to a single one θ involving a function G(θ) deduced from u(x,t). This transformation can be easily generalized to n variables ($n \geqslant 2$).

This function G(θ) can be approximated for instance, by a polynomial function :

$$G(\theta) = \alpha_1 \theta + \ldots + \alpha_n \theta^n$$

If A_1 is the initial conditions, and A_2 the boundary conditions associated to the equation Au = f, we get the following penalized function [4], [6] :

$$J_\varepsilon = ||Au(x,t) - f(x,t)||^2_{L^2(\Omega)} + \frac{1}{\varepsilon_i} ||A_1||^2_{L^2(\Omega_1)} + \frac{1}{\varepsilon_2} ||A_2||^2_{L^2(\Omega_2)} \qquad (2.2.a)$$

with $\Omega = \{(x,t) \ / \ Au(x,t) = f(x,t)\}$, $\Omega_1 = \{(x,t) \ / \ u(x,t) \text{ satisfies } A_1\}$

$\Omega_2 = \{(x,t) \ / \ u(x,t) \text{ satisfies } A_2\}$.

In (2.2.a), we can associate the following optimization problem :

Find α_i , $i = 1, \ldots, n$, such that :

$$J_\varepsilon = ||AG(\theta) - F(\theta)||^2_{L^2(\Omega)} + \frac{1}{\varepsilon_1} ||A_1||^2_{L^2(\Omega_1)} + \frac{1}{\varepsilon_2} ||A_2||^2_{L^2(\Omega_2)} \quad \text{be minimum}$$

The terms $\frac{1}{\varepsilon_i} ||\ \ ||$ are the penalty terms, with $\{\varepsilon_1, \varepsilon_2\} \to 0$

A discretisation leads easily to :

$$\underset{\alpha_i}{\operatorname{Min}} \left\{ \sum_j \left(A\left(\frac{\alpha_1}{a} (x_j^2+t_j^2)^{1/2}+\ldots+\frac{\alpha_n}{a^n}(x_j^2+t_j^2)^{n/2}\right)- f(x_j,t_j)\right)^2 + \frac{1}{\varepsilon_1} \sum_k A_{1,k}^2 + \right.$$

$$\left. + \frac{1}{\varepsilon_2} \sum_m A_{2,m}^2 \right\} , \quad i = 1,\ldots, n.$$

If the operator A is linear, the functional problem is transformed into the linear algebraic system :

$$\frac{\partial J}{\partial \alpha_\tau} = 0 \ , \ \tau = 1, \ \ldots, \ n$$

If the operator A is nonlinear we can find the absolute minimum of the function J using the global optimization method : ALIENOR. We can reduce the unknown variables $\alpha_1,\ldots, \alpha_n$ to a single one θ . Thus we minimize the functional J in relation to a single variable θ.

The generalization of our method is not difficult. For example for three variables x,y,z we have :

$$x = a\theta_1\cos\theta_1$$

$$y = a\theta_1\sin\theta_1$$

$$\theta_1 = a\theta\cos\theta$$

$$z = a\theta\sin\theta \quad , \quad \theta > 0$$

and conversely we obtain :

$$\theta = \frac{1}{a} \sqrt{z^2 + \frac{1}{a^2} (x^2+y^2)}$$

The function $G(\theta)$ can be also approximated by :

- a logarithmic development
- an exponential development
  or other convenient approximation.

3. APPLICATION OF THE METHOD

The heat equation.

This method was tested on the heat equation :

$$\frac{\partial u}{\partial t} = u_0 \frac{\partial^2 u}{\partial x^2} \quad \text{with } u_0 = 1 \text{ and with the boundary conditions :}$$

$$u(0,t) = u(\pi,t) = 0 \quad \text{and the initial conditions :}$$

$$u(x,0) = 2 \sin x \ , \quad 0 \leq x \leq \pi \ .$$

Considering the linear differential operator :

$$A = \frac{\partial}{\partial t} - u_0 \frac{\partial^2}{\partial x^2}$$

We replace u(x,t) by :

$$\alpha_0 + \alpha_1 [\, x^2 + t^2 \,]^{1/2} + \ldots + \alpha_q [\, x^2 + t^2 \,]^{q/2}$$

Then we minimize the following penalized functional [6] for finding the α_i :

$$J_\varepsilon = \sum_{j=1}^{N} \left[Au(x_j,\, t_j) \right]^2 + \frac{1}{\varepsilon_1} \sum_{j=1}^{M_1} \left[\sum_{i=0}^{q} \alpha_i \,(x_j^2)^{\,i/2} - 2 \sin x_j \right]^2$$

$$+ \frac{1}{\varepsilon_2} \sum_{j=1}^{M_2} \left[\sum_{i=0}^{q} \alpha_i \,(x_j^2 + t_j^2)^{\,i/2} \right]^2$$

according to α_i , $i = 0, \ldots, q$.

4. LAGRANGE METHOD

Consider the functional [5] :

$$J = \sum_{i=1}^{N} \left[Au(x_i, t_i) \right]^2 + \sum_{k=1}^{N_1} \lambda_1^k A_{1k} + \sum_{\ell=1}^{N_2} \lambda_2^\ell A_{2\ell} + \sum_{m=1}^{N_2} \lambda_3^m A_{3m}$$

with $A_{1k} = 2 \sin x_k - \sum_{j=0}^{M} \alpha_j \left(\sqrt{x_k^2}\, \right)^j$

$$A_{2\ell} = \sum_{j=0}^{M} \alpha_j \left(\sqrt{t_\ell^2}\, \right)^j$$

$$A_{3m} = \sum_{j=0}^{M} \alpha_j \left(\sqrt{\pi^2 + t_m^2}\, \right)^j$$

We need to :

$$\min_{\substack{\alpha_0,\ldots,\alpha_M \\ \lambda_1^k,\, \lambda_2^{\ell},\, \lambda_3^m}} J(\alpha_0,\ldots,\alpha_M,\lambda_1^k,\lambda_2^{\ell},\lambda_3^m)$$

This implies the resolution of a linear system :

$$\frac{\partial J}{\partial \alpha_i} = 0 \quad i = 0,\ldots, M.$$

$$\frac{\partial J}{\partial \lambda_1^k} = \frac{\partial J}{\partial \lambda_2^{\ell}} = \frac{\partial J}{\partial \lambda_3^m} = 0 \quad k = 1,\ldots, N_1,\ \ell,\ m = 1,\ldots, N_2.$$

For accelerating this process, a good choice of the parameters λ_1^k, λ_2^{ℓ}, λ_3^m is necessary. We found them by an empirical technique.

5. RESULTS

Actually we have only considered 2 values of t (except $t = 0$) have only been used and we advanced step by step (with a step equal to 2 units of time).

The previous methods were numerically tested on microcomputers (Apple, Commodore). In all cases, we can calculate an exact solution. This fact allows to calculate exactly the errors.

For the heat equation, with boundary and initial conditions, the penalty technique with $q = 12$, a step of space equal to 1/10 and a step of time = 0,05, has given small relative errors (10^{-3} in mean with some peaks at 10^{-2}).

The same precisions were obtained with the Lagrange functional.

6. CONCLUSIONS

This technique has all the advantages of the finite elements methods. But on top of that, this method is simple, fast and can be used on microcomputers because we don't need a large memory space.

This method can also be used for approximating functions with several variables.

REFERENCES

[1] V. ALEKSANDROV, N.D. GORSKY, A.O. POLYAKOV. 'Recursive algorithms of data representation and processing'. Leningrad, 1979.

[2] Y. CHERRUAULT, A. GUILLEZ. 'A simple method for optimization'. Kybernetes, Vol. 12, Number 1, pp 59-63, 1983.

[3] P. CIARLET. 'The finite element method for elliptic problems'. North Holland, 1979.

[4] A. FIACCO, G. Mac CORNICK. 'Nonlinear programming : Sequential unconstrained minimization techniques'. J. Willey, 1968.

[5] M. FORTIN, R. GLOWINSKI. 'Méthodes de Lagrangien augmenté. Applications à la résolution numérique de problèmes aux limites'. Méthodes mathématiques de l'informatique, 9, Dunod, Paris, 1982.

[6] J.L. LIONS. 'Contrôle optimal de systèmes gouvernés par des équations aux dérivées partielles'. Etudes Mathématiques, Ed. Dunod, Paris, 1968.

[7] J. VIGNES. 'Méthodes numériques d'optimisation d'une fonction de plusieurs variables'. Chimie Industrie - génie chimique, Vol. 97, n° 8, avril 1967, pp 1264-1276.

Y. CHERRUAULT, I. KARPOUZAS
Université PARIS VI - MEDIMAT
15, rue de l'Ecole de Médecine
75270 PARIS Cédex 06 - FRANCE

MODELLING AND OPTIMIZATION OF BUFFER STOCKS IN A PRODUCTION LINE

PÉTER KELLE

Computer and Automation Institute,
Hungarian Academy of Sciences,

Budapest, Hungary

Abstract

Stochastic models are formulated for the optimal allocation of the safety stocks in raw material, in internal stocks on the different production levels and in finished goods. Simple approximate solutions are derived, too.

1. Introduction

The production-inventory system considered has N stages of production. The raw material is processed successively at the N facilities before reaching the costumer. There is a final product store S_{N+1} and an internal store S_i before each stage of processing $(i=1,2,\dots,N)$ including raw material store S_1. Buffer stocks are planned for each store as initial stocks of a production cycle to ensure the continuous supply for the whole production period. These safety stocks are necessary because of the uncertainties in demand and in production which may often be disturbed by random factors such as machine failures, faulty products, breakdowns etc. In this case there is great difficulty in providing for continuous production and demand satisfaction with reasonably low levels of buffer stocks.

The uncertainty in material requirement planning systems was considered and different buffering policies were given by Berry and Whybark [2], Whybark and Williams [18], Miller [12] and New [13]. The effect of random demand in internal stocking was analysed recently in the papers of Schmidt and Nahmias [16], Lambrecht et. al. [11] and De Bodt and Graves [4]. The optimal safety stock policy for a continuous deterministic production process and Poisson demand was derived by Axsäter and Lundell [1].

We consider a multi-stage production system where both the demand and the production process may have random factors. We derive a mathematical method for planning the buffer stocks for this system. First we describe the different stages of this disturbed production process by random processes including a new one which we have constructed for this problem. In Section 3 stochastic programming models are formulated for the optimal allocation of the buffer stocks among the stores S_i $(i=1,\dots,N+1)$. The solution of this problem is rather sophisticated, so for the practical application simple approximate solutions are derived. These are described shortly in Section 4. In Section 5 we refer to the practical application.

2. Models of processing for the different production stages

MODEL I is typical for the raw material delivery. Here the total amount Q processed in the production period $[0,T]$ is delivered with a random lead-time L, characterized by a known distribution function $H(L)$.

MODEL II is applied for the processing in fixed batch size where the periodicity is disturbed by machine failures, breakdowns which are random. Let n denote the number of batches in $[0,T]$. The setup times $t_1 < t_2 < \ldots < t_n$ are independent, uniformly distributed random numbers arranged in increasing order. The cumulative amount processed until time t $(0 \leq t \leq T)$ is the empirical distribution function of the uniform distribution:

$$g_t = \begin{cases} 0 & \text{if} \quad 0 \leq t \leq t_1 \\ kQ/n & \text{if} \quad t_k < t \leq t_{k+1} \quad (k=1,\ldots,n-1) \\ Q & \text{if} \quad t_n < t \leq T . \end{cases} \tag{1}$$

MODEL III is constructed for the processing in random batch sizes. It is typical for the production stages where often not the whole amount of a batch can be immediately processed at the next stage because of quality problems which have to be corrected. Afterwards the repaired quantity is added to the next batch. The random batch size β_i/n is approximated by using the statistical data of the quality control results. The cummulative amount processed is expressed by

$$g_t = \begin{cases} 0 & \text{if} \quad 0 \leq t \leq t_1 \\ \frac{1}{n} \sum_{i=1}^{k} \beta_i & \text{if} \quad t_k < t \leq t_{k+1} \quad (k=1,\ldots,n;\ t_{n+1} \equiv T) \end{cases} \tag{2}$$

assuming setup times $t_1 < t_2 < \ldots < t_n$, which may also be random as in Model II.

MODEL IV is the Poisson process which may serve an approximation for the cumulative demand of final product.

MODEL V is the continuous demand with known or random rate α. The random rate is typical in the phase of planning since at the time of decision making only rough estimation of demand is available.

MODEL VI is the Wiener process serving an approximation for continuous processing with random fluctuation resulted by the interaction of many random influences.

3. Optimization models for the allocation of buffer stocks

For production managers it is a great problem to provide the continuous supply of costumers and production stages with reasonably low level of raw material, in-process and final product stocks. We intend to allocate these buffer stocks for the system of $N+1$ stores in such a way that they should jointly provide for a high service level of the whole production line with possibly small investments in stocks. The service level is measured by the probability of non-interruption in supply of processing and in demand satisfaction.

In the first model a constrained investment capacity K has to be allocated in buffer stocks among the $N+1$ stores in such a way that they guarantee the maximal service level:

$$\begin{aligned} &\text{maximize} \quad P(g_t^i \leq M_i + g_t^{i-1} \ , \ 0 \leq t \leq T \ , \ i=1,2,\dots,N+1) \\ &\text{subject to} \sum_{i=1}^{N+1} d_i M_i \leq K \ , \end{aligned} \tag{3}$$

where g_t^i denotes the cumulative amount processed in the period $[0,t]$ on level i. Here g_t^1 means the delivery of raw material and g_t^{N+1} means the final product demand until time t. A unit of safety stock M_i on level i has the investment cost d_i.

In the second model a prescribed high service level $1-\varepsilon$ of the production-supply system should be ensured with minimal investment in safety stocks:

$$\begin{aligned} &\text{minimize} \quad \sum_{i=1}^{N+1} d_i M_i \\ &\text{subject to} \ P(g_t^i \leq M_i + g_t^{i-1} \ , \ 0 \leq t \leq T \ , \ i=1,2,\dots,N+1) \end{aligned} \tag{4}$$

The first model is an allocation model, which can be used for safety stock planning in a static case where the total investment is prescribed. In case of considerable changes in demand, production or raw material supply system the second model should be preferred to safety stock planning. The service level can be fixed by using the results of the static model applied to the previous planning periods. On the joint application and analysis of the above two optimization models a decision support system of buffer stock planning can be built.

4. Exact and approximate solutions

The first step in the solution of the above stochastic optimization problems is to derive the probability of continuous supply in connection with two consecutive production levels and given buffer stocks between them. Here we can describe it only for the most important models of delivery, batch processing and demand.

By choosing appropriate units of time and amount we may assume further in this paper that $T=1$ and $Q=1$ for the sake of simple notations. For a delivery according to Model I and processing at the first stage according to Model II the initial stock $k/n \leq M_1 < (k+1)/n$ ensures the continuous raw material supply with probability

$P(t_k \leq L) = n\binom{n-1}{k-1} \int_0^{\infty} \int_0^{L} t^{k-1}(1-t)^{n-k} \, dt\,dH(L)$. For the in-process stores S_i $(i=2,\dots,N)$ both input (as the processing of level $i-1$) and the demand (as the processing of level i) can be described by Model II of batch processing. For the number of lots processed during the planning period the relation $n_i = kn_{i-1}$ holds for the optimal lot sizes with an integer k. Then we get the following probability using the result of Koroliuk [10]:

$$p(M_i) = P(g_t^i - g_t^{i-1} \leq M_i\ ,\ 0 \leq t \leq 1) =$$

$$= 1 - \frac{c}{\binom{n_i+n_{i-1}}{n_i}} \sum_{\substack{m=c \\ m\equiv c(\mathrm{mod}\,(k+1))}}^{n_i-n_{i-1}-\left[\frac{c}{k}\right]} \frac{1}{m} \binom{m}{\frac{m-c}{k+1}} \binom{n_i+n_{i-1}-m}{n_i-\frac{m-c}{k+1}} \tag{5}$$

where the constant c is the smallest integer for which $c \geq M_i n_i^2/n_{i-1}$. For the general case of arbitrary integers n_i, n_{i-1} the exact probability is not known. Here an approximation can be derived based on the asymptotic distribution of Smirnov [17] in the form

$$p(M_i) \approx 1 - \exp\left[-\frac{2n_i n_{i-1}}{n_i+n_{i-1}} M_i^2\right] . \tag{6}$$

If two consecutive processing levels with continuous processing are approximated by Model VI the theorem of Baxter and Donsker [2] can be applied as it is published in Kelle [7]. Similarly for Model IV the results of Pyke [15] are applicable.

For the final product store S_{N+1} we assume a continuous demand by Model V with random α which has a known distribution function $G(\alpha)$. The setup times of the final processing level are uniformly distributed as in Model II but the random batch sizes are not necessarily uniformly distributed. Using the statistical data of final quality control we fit appropriate distributions. Let $H_k(x)$ denote the distribution function characterizing the cumulative amount of perfect final products of the first k batches, i.e. $\sum_{i=1}^{k} \beta_i$ in Model III according to (2). We have derived the following exact distribution for fixed $\alpha (>0)$ and g_t defined by (2), when the random vector $\beta=(\beta_1,\dots,\beta_n)$ is interchangeable (see Kelle [6], [8])

$$p(M,\alpha) = P(\alpha t - g_t \leq M \ , \ 0 \leq t \leq 1) =$$

$$= 1 - \left(1 - \frac{M}{\alpha}\right)^n - \frac{M}{\alpha} \sum_{k=1}^{n} \binom{n}{k} \int_0^{n(\alpha - M)} \left(\frac{M+x/n}{\alpha}\right)^{k-1} \left(1 - \frac{M+x/n}{\alpha}\right)^{n-k} dH_k(x) . \tag{7}$$

The probability of continuous supply for random demand intensity α can be expressed in the form

$$P(\alpha t - g_t \leq M \ , \ 0 \leq t \leq 1) = \int_0^{\infty} p(M, \alpha \mid \alpha = x) \ dG(x) \ . \tag{8}$$

If α has normal distribution with parameters m and s $(s < 1/\sqrt{n})$ the following approximation has been derived based also on an asymptotic distribution (Kelle [5]):

$$P(\alpha t - g_t \leq M \ , \ 0 \leq t \leq 1) \approx 1 - \exp\left[-2nM(M+1-m-nMs^2)\right] . \tag{9}$$

The solution of the stochastic optimization problems (3) and (4) can be reduced in such a way to deterministic nonlinear programming problems for independent joint constraint or objective function. This formulation is simple for the above cases using the exact distribution of continuous supply on each two subsequent processing levels. The consideration of stochastic dependence among the subsequent processing levels leads to a lot of difficulties in solution and data requirement. Most of these problems can be solved only by using simulation technique and the statistical data of previous periods (see e.g. Prékopa and Kelle [14]).

The approximate distributions described above enable us to construct a very simple, effective solution based on Lagrange multiplier method. The algorithm is detailed in the paper Kelle [9]. The error of the approximate solution has been analysed compared with the exact solution in the cases when the letter was available, too. For the most part, especially when the number of batches is above 20 during the production cycle, the approximation is satisfactory. The relative error is less than 5%. In other cases the approximation can serve as a starting point for the solution of the nonlinear programming problem defined by the exact distributions.

5. Application

The stochastic programming models formulated for the optimal allocation of the investment in raw material, in-process and final product safety stocks could be reduced to simple deterministic problems in some important cases. These results have made it possible for us to solve these problems effectively and have given an efficient tool for decision makers in planning the safety stock allocation. The above models and methods have been tested on real production processes and have partly been introduced in practice in a rolling mill in Hungary.

References

1. Axsäter, S. and Lundell, P.: In-Process Safety Stocks, Working Paper, Linköping Institute of Technology, 1983.
2. Baxter, G. and Donsker, M.D.: On the Supremum Functional for Processes with Stationary Independent Increments, Transactions of the American Math. Soc. 1957. 73-87.
3. Berry, W.L. and Whybark, D.C.: Buffering Against Uncertainty in Material Requirement Systems, Discussion Paper No. 82, School of Business, Indiana University, 1977.
4. De Bodt, M.A. and Graves, S.C.: Continuous-Review Policies for a Multi-Echelon Inventory System with Stochastic Demand, Working Paper 82-18, Katholicke Univ. Leuven, 1982.
5. Kelle, P.: Chance Constrained Inventory Models and their Application, Methods of OR 44. Athenaum/Hain, 1981, 607-616.
6. Kelle, P.: Reliability-Type Inventory Models for Random Delivery Process, in: Chikán, A. /ed./: Proc. First Int. Symp. on Inventories, Budapest, Elsevier Publ.Co. 1980, II. 385-395.
7. Kelle, P.: An Inventory Control Program System Based on Reliability Type Models, in: Chikán, A. /ed./: Proc. Second Int. Symp. on Inventories, Budapest, Elsevier Publ. Co. 1982, 491-497.
8. Kelle, P.: On the Safety Stock Problem for Random Delivery Process, European Journal of OR. 17(1984) 191-200.
9. Kelle, P.: Safety Stock Planning in a Multi-Stage Production-Inventory System, Engineering Costs and Production Economics, 9(1985) 231-237.
10. Koroliuk, D.: On the Difference of Two Independent Empirical Distribution Functions, Isvestia Akad. Nauk SSSR, Ser.Math. 19(1955) 81-96. (in Russian)
11. Lambrecht, M.R., Muckstadt, J.A. and Luyten, R.: Protective Stocks in Multi-Stage Production Systems, Onderzocksrapport No. 8215, Katholicke Univ. Leuven, 1982.
12. Miller, J.G.: Hedging the Master Schedule, in: Ritzman et.al. /ed./: Disaggregation: Problems in Manufacturing and Service Organizations, Martinus Nijhoff Publ., Boston, 1979.

13. New, C.: Safety Stocks for Requirements Planning, Production and Inventory Management, 2nd quarter, 1975.
14. Prékopa, A. and Kelle, P.: Reliability Type Inventory Models Based on Stochastic Programming, Math. Programming Study 9(1978) 43-58.
15. Pyke, R.: The Supremum and Infimum of the Poisson Process, Annals of Math. Stat. 30(1959) 568-575.
16. Schmidt, C.P. and Nahmias, S.: Optimal Policy for a Two Stage Assembly System Under Random Demand, Working Paper, Univ. of Santa Clara, California, 1981.
17. Smirnov, N.V.: Approximation of Random Amount by Empirical Data, Uspechi Math. Nauk (1944) 179-206.
18. Whybark, D.C. and Williams, J.C.: Material Requirements Planning Under Uncertainty, Decision Sciences, 7(1976).

SIMULATING EXPERT JUDGMENT: ACTUARIAL MODELS AND APPLICATIONS

Arshad M. Khan
Box 19017
University of Texas, Arlington, Texas 76019

INTRODUCTION

If uncertainty is the essence of decision, then the modeling of human decision making, at its core, must model uncertainty. Given historical perspective, the uncertainty can be translated into probability distributions that can form the basis for future predictions. On the other hand, a lack of explicit data leads to a situation where the expert decision maker is likely to exercise intuitive judgment. This decision environment can be modeled by simulating the expert decision maker's judgment. And such models fall within the domain of a body of knowledge called behavioral decision theory that has developed in the last two dozen years.

PREDICTING JUDGMENT

The lens model proposed by Brunswik [4] presents a graphic conceptual framework: The decision situation consists of an actual event separated from the decision maker by time (usually), and the judgment or decision anticipating the event, which is made by the decision maker on the basis of a set of factors (variables) that are less than perfect predictors of the event. Thus the predictor variables serve as a lens through which the actual event is observed.

Dawes and Corrigan [7] note that as early as 1923 Henry Wallace, Vice-President of the United States under Roosevelt had proposed a linear regression model to simulate the predictions of experts judging corn quality [19]. His idea apparently did not prosper. Several decades later however, Hoffman [8] independently proposed an actuarial model to simulate human judgment. And this time the idea was received with great enthusiasm. Hoffman called his model "a paramorphic representation" because he did not wish to imply that the model represented the actual human cognitive process, but that it could be used to simulate the final decisions resulting from the cognitive process. Since then several different types of models have been identified as successful simulators of human judgment.

Types of Models: Judgment models can be categorized into scoring models and screening models. Scoring models aggregate the effect of all attributes and are therefore compensatory, that is the decrease of one attribute value can be compensated by an increase in another. There are three types of scoring models: the simple additive process or linear model, the configural process or linear model with interactions between attributes included, and the curvilinear process or linear model with higher order functions of attributes.

Screening models tend to be noncompensatory. Decision alternatives are not scored; instead they are considered acceptable or screened out as unacceptable. Examples of screening models are the conjunctive process, the disjunctive process, and the lexicographic model. The conjunctive model establishes a minimum evaluation function, and an acceptable alternative must satisfy the required minimum for each attribute. On the other hand, a disjunctive model has a maximum evaluation function, and an acceptable alternative has to surpass just one attribute maximum regardless of other attribute levels. In the lexicographic model, the attributes are listed in order of importance, and the alternative scoring highest on the most important attribute is selected. If there is a tie, then the next important attribute is considered.

Examples: Numerous studies from varied decision environments are now in the literature. Thus Bowman [3] and Kunreuther [13] have modeled production planning, Naylor and Wherry [16] job performance, Newton [17] and Dawes [6] student performance, Little [14] marketing, Ashton [1] and Joyce [9] internal control judgments of public accountants, Khan [10] venture capital investments, Connell, Adam and Moore [5] health care, and Khan and Shih [11] personality characteristics.

MODEL PERFORMANCE

The predictive success of these models has been hypothesized earlier (Meehl [15]). Yet it is still surprising: Wherever actual outcomes have become available, the models have surpassed the human decision maker in performance. And this paradox whereby the model of human judgment performs better than the human himself has been dubbed "bootstrapping".

A common feature in many studies has been the particular success of the linear model, both in representing judgment, as measured by the correlations, and in its importance as a predictor, as evidenced by the bootstrapping reported. Several studies (for e.g. Ashton [1] and Joyce [9] have indicated that nonlinear terms do not add any significant increment to the variance accounted for by the linear model in the context of human judgment. And Yntema and Torgerson [20] have shown earlier that a linear model can approximate even nonlinear functions extraordinarily well -- they are able to predict 94% of the variance of a truly configural model from a linear combination of the main effects alone.

Dawes and Corrigan [7] reason that the conditionally monotone relationship prevailing between predictor variables and the criterion, in most decision situations, favors the linear model. Thus the function tends to become more linear as error increases in the predictor variables. In addition, the relative weights computed for the predictors are not affected by errors in the criterion. Törn [18] has demonstrated the robustness of the linear model by obtaining a correlation coefficient of 0.5 with a linear model of a highly nonlinear (90° angle) curve. His only precondition is conditional monotonicity.

APPLICATIONS

Country Risk: The first application models the judgments of experts who assess country risk and is taken from the author's own collaborative research (Khan and Shih [12]). The tremendous growth during recent years in international trade and investment has developed the need for forecasting country risk. And, as might be expected, there are now major consulting firms plus individual experts providing risk assessment services.

In this study, experts were used first to ascertain relevant risk factors. These were determined to be political stability, inflation rate, exchange rate stability, funds transferability, current account balance, and per capita gross domestic product. For these variables, statistical and descriptive data on 17 countries, that had been selected to cover a broad cross-section of economic strata, were complied. Five experts next evaluated the predictive factors on a scale of 1 to 5 for each country and also provided an assessment of overall risk on a 1 to 10 scale. In each case, 1 represented "good" meaning low risk or desirable quality and vice versa. Intermediate steps on the scales were also given to achieve scale uniformity among the experts.

The first concern in a study of this kind, where the criterion cannot be measured explicitly, was to demonstrate that expertise does in fact exist. One measure of expertise has been the consistency and reliability of the expert's judgments. Thus a few months later a second set of data were provided to the same experts. This time 20 country profiles were used and among them were six countries repeated from the previous data set. These were then used to measure expert consistency over time. The results showed that the judgments of the experts were, in general, quite consistent. With the exception of one expert, consistency in judging the predictor variables, as measured by the correlations between the two evaluations of the six repeat countries, varied from 0.79 to 0.93. The criterion (overall country risk) had slightly lower consistency with correlations from 0.47 to 0.90 for the different experts. This was not unexpected because assessment of the criterion involves the additional step of combining the individual predictors.

Linear compensatory models were fitted to the evaluations of each of the experts for the initial data set of 17 countries. The resulting correlations ranged from 0.86 to 0.97 with a composite judgment model of all five experts yielding 0.91. The most significant variable was political risk, followed by exchange rate stability, and inflation rate. Cross-validation of the pooled composite model on the second data set of 14 countries (20 minus the six repeats) showed shrinkage down to 0.83.

The variables were also tested for multicollinearity and a problem was observed between exchange rate stability and inflation rate. Consequently, a second composite model was constructed excluding inflation rate which was the less important of the two variables. Cross-validation of this second model showed slightly less shrinkage down to 0.84. And this model was then used to develop risk assessments for the countries in the study. These assessments ranged from a low risk of 1.25 for Japan to a high of 8.13 for Argentina. Some of the countries in between were Algeria (4.34), Philippines (5.88) and Brazil (6.85).

Capital Budgeting: The second application is taken from a study that has been conducted by Bao [2]. It utilizes the attribute-dynamic attitude model (ADAM) and compares the results with a linear model. Proposed by Zeleny [22], ADAM purports to overcome some potential problems with linear regression models of judgment.

As delineated by Bao [2], these problems are (i) that causality is not clear, as cognitive process is not expressed, (ii) that the linear model provides judgments, not actual choices which might be different, (iii) that it assumes fixed preferences between alternatives whereas inconsistent decision-maker behavior has often been observed, and (iv) that it assumes fixed relative weights for the predictors.

Yet the crux of the problem is simulating final judgment, not cognitive process, and the bootstrapping that is universally reported appears to confirm the success of judgment models. The remaining three reasons distill down to inconsistent decision-maker behavior. And it merely serves to reinforce the argument for a consistent actuarial model - an argument that is empirically sustained by actual performance.

Bao's setting is capital budgeting and his experts are fifteen accountants (company controllers, financial analysts, and corporate accountants) who assume the role of company controller in deciding preferences for 10 independent sets of capital budgeting alternatives. Each set contains three capital projects that have to be ranked.

Following a procedure outlined by Zeleny [21], an ADAM model is developed. The model is based on the assumption that predictor weighting is a combination of a stable weight reflecting the decision maker's own background and a context-dependent weight reflecting variations in predictor importance over different decision settings. This model is then compared with a linear model of judgment that is developed in the customary manner.

Since three projects have to be ranked, yielding six possible rankings, a random assignment of choice should yield 16.67 percent correct choices. Both models exceed this figure comfortably with the linear model marginally better in performance (56 percent correct choices versus 54.7 percent for ADAM). Next the predictive correlation coefficients of both models against actual expert ratings are compared. Here the linear model is a better predictor for seven of the experts while ADAM performs better for three experts. Clearly the evidence supporting ADAM does not appear to be inordinately strong. At the same time the case for modeling expert judgment is sustained.

CONCLUSIONS

Models simulating the judgments of experts work, and particularly successful are linear models. The success of linear models is due to two major factors: Firstly, most situations in which these models are applied involve a conditionally monotone relationship between the predictors and the criterion. Secondly, measurement errors in the variables tend to make the functions more linear.

Of the two studies presented (country risk assessment and capital budgeting), country risk involves a relatively vaguely defined criterion compared with capital budgeting. Yet the experts in both studies are modeled successfully. A linear model is used in the country risk study, and the ADAM model is compared with a linear model in the capital budgeting study.

ADAM employs the concept that the weights assigned to predictor variables can be subject to change over time/decision settings, and should therefore consist of a combination of a stable weight reflecting the decision maker and a context-dependent weight recognizing variations in predictor importance over different decision settings. The results of the comparison with a linear model do not at present indicate that ADAM can supplant the linear model.

REFERENCES

1. Ashton RH (1974) An experimental study of internal control judgments, Journal of Accounting Research, 12:143-157.
2. Bao DH (1985) An empirical validation of ADAM in a capital budgeting context, Decision Sciences, 16:265-283.
3. Bowman EH (1963) Consistency and optimality in managerial decision making, Management Science, 9:310-321.
4. Brunswik E (1952) The Conceptual Framework of Psychology, University of Chicago Press, Chicago.
5. Counell BC, Adam EE, Moore AN (1984) Aggregate planning in health care foodservice systems with varying technologies, Journal of Operations Management, 5:41-55.
6. Dawes RM (1971) A case study of graduate admissions: application of three principles of human decision making, American Psychologist, 26:180-188.
7. Dawes RM, Corrigan B (1974) Linear models in decision making, Psychological Bulletin, 81:95-106.
8. Hoffman PJ (1960) The paramorphic representation of clinical judgment, Psychological Bulletin, 57:116-131.
9. Joyce EJ (1976) Expert judgment in audit program planning, Studies on Human Information Processing in Accounting, supplement to Journal of Accounting Research, 14:29-60.
10. Khan AM (1983) The analysis of investments in new ventures, Sixth European Congress on Operational Research, Vienna.
11. Khan AM, Shih CT (1984) Linear, conjunctive and disjunctive models in spousal selection, working paper, University of Texas, Arlington.

12. Khan AM, Shih CT (1985) On country risk: some preliminary results, working paper, University of Texas, Arlington.

13. Kunreuther H (1969) Extensions of Bowman's theory of managerial decision making, Management Science, 15B:415-439.

14. Little JDC (1970) Models and managers: the concept of a decision calculus, Management Science, 16B:466-485.

15. Meehl PE (1954) Clinical versus Statistical Prediction: A Theoretical Analysis and Review of the Literature, University of Minnesota Press, Minneapolis.

16. Naylor JC, Wherry RJ (1965) The use of simulated stimuli and the "JAN" technique to capture and cluster the policies of raters, Educational and Psychological Measurement, 25:969-986.

17. Newton JR (1965) Judgment and feedback in a quasi-clinical situation, Journal of Personality and Social Psychology, 1:336-342.

18. Törn AA (1980) Correlation coefficients of linear regression models of human decision making, Omega, 8:393-394.

19. Wallace HA (1923) What is in the corn judge's mind? Journal of the American Society of Agronomy, 15:300-304.

20. Yntema DC, Torgerson WS (1961) Man-computer cooperation in decisions requiring common sense, IRE Transactions of the Professional Group on Human Factors in Electronics, HFE-2(1):20-26.

21. Zeleny M (1975) The theory of the displaced ideal, in M. Zeleny (Ed), Multiple Criteria Decision Making, Springer-Verlag, New York.

22. Zeleny M (1976) The attribute-dynamic attitude model (ADAM), Management Science, 23:12-26.

CORPORATE PLANNING AND THE VALUE OF INFORMATION

HANNU KIVIJÄRVI and MARKKU TUOMINEN
Helsinki School of Economics
Runeberginkatu 14 - 16
SF-00100 Helsinki, Finland

University of Oulu
Linnanmaa
SF-90570 Oulu, Finland

Abstract

In this research, the strategies of information systems development are created at the corporate level. A corporate model is usually aimed to support general corporate planning and it describes simultaneously the relations between the finance, marketing and production functions of the firm.

A multiattribute and multicriteria approach is used to assess the value of information at the corporate level. Delay, inconsistency, underestimation, and overestimation are used as the quality attributes of information and total profits, cash balance, market share, or the value of plant may be used as evaluation criteria.

The main contribution of this study is to show that there are no conceptual, mathematical, computational or other unsolvable difficulties inherent in applying our approach in realistic situations. The information value, information attributes, and information costs are readily observed, understood, and the results are directly interpretable in terms of common corporate planning.

1. Introduction

Because decisions are generally based on information and because decisions have valuable effects, the 'value of information' concept is always at least indirectly present in every decision making situation. Sometimes even an explicit (secondary) decision is made in order to produce the needed information. For instance, decisions concerning market surveys, credit investigations, quality controls and especially, development of an information system are examples of such decisions which are primarily based on the potential value of information.

The concept 'value of information' has been studied for instance through statistical decision theory (Bayesian theory) or information economics, control theory, system dynamics and gaming (Kleijnen, 1980).

From a point of view of the financial and organizational scope perhaps the most important utilization area of the 'information value' concept is the development of an information system. Epsten and King (1982) even argue that all aspects of information system development are related to the information value. "Virtually every controllable aspect of any information system (IS), its design and implementation, are effected by how information is valued. For instance, information that is valued relatively higher than other information will be chosen over the other to be incorporated into the data base, to be made relatively more accessible to more users, to be incorporated into a report, etc."

Generally, every information system development project is initiated or should be initiated with an activity called information requirements analysis. Typically such a requirement analysis is accomplished by direct methods just by interviewing decision makers and asking their conceptions about the relevancy and importance of a certain information unit. In this way knowledge about user-perceived relevancy of the single information units is created.

This direct, individual oriented approach may be relevant in very early phases of the assessment process but when searching and selecting new alternatives this approach is not sufficient enough. Ackoff (1967) puts it as follows: "One cannot specify what information is required for decision making until an explanatory model of the decision process and the system involved has been constructed and tested. Information systems are subsystems of control systems. They cannot be designed adequately without taking control into account. ... A decision model identifies what information is required and, hence, what information is relevant."

Hence, it seems that the best information system for the whole organization is not necessarily created by means of the individual, user-perceived information requirements. Also the indirect and intermediate (chained) effects of decisions must be valued, not only the direct effects on an individual decision maker. When the approach is extended from the individuals to the whole organization some conceptual and practical problems are met. The first difficulty is the definition of the attributes (state set, characteristics, measurement scale) that describe the quality of information. It is quite unrealistic to define the state of information only dicotomically (information - no information, or perfect information - imperfect information). Instead, information may be measured and described by several attributes, such as accuracy, timeliness, and aggregation. These characteristics, of course, represent directly the nature of proposed information system alternatives.

The second difficulty is how to measure the utility or benefit that would accrue from the proposed information system, that is, the difficulty to select the evaluation criterion: the point of the whole system where the influence of information is measured.

The entire approach adopted in this study can be characterized by following three key concepts: systemic, multiattribute, multicriteria. The object system is assumed to be a firm which is studied as a whole, at the corporate planning level. The final value of information is measured on multiple criteria as well as the quality of information is measured on multiple attributes.

We have used a methodology based on an optimal control approach to optimize the selected criteria. After the integration of some features of systems dynamics to the optimal control approach we have developed a useful tool to determine the value of information in realistic decision situations.

Next, the method is described shortly and then some features of a corporate model are presented. Then the information quality attributes and evaluation criteria are defined. Finally, the results of the optimization experiments are evaluated.

2. Methodology

Methodologically, in this study the corporation is described by a dynamic corporation model, where the essential relations inside and between financial, production and marketing activities are defined. In the model, the different components of the whole feedback system are explicitely defined, namely, the object system (real world), control system (or decision making system), decision information and feedback information. Technically, the model is developed as a system dynamics model by Dysmap simulation language, which is connected with a heuristic optimization algorithm. The developed method is called the Dysmap-Optimizer (Keloharju, 1983).

In the Dysmap-Optimizer package the dysmap simulation language and the Search Decision Rule algorithm are integrated into a single interactive computer system. SDR is developed by Taubert (1968) on the basis of the pattern search algorithm that was developed by Hooke and Jeeves. In the Dysmap-Optimizer, the heuristic optimization procedure starts with the nomination of the objective function (criteria for performance). The optimization is based on the final value the objective function has at the end of the simulation period. Then the parameters to be optimized must be selected. Finally, some technical parameters must also be defined to guide the searching process.

In the Dysmap-optimizer there are relevant tools available to model optimal control problems. As it is generally known, the optimal control problem is to choose control functions u(t), $t \in (0,T)$ to

$$\max J = \int_0^T F(x(t), u(t), t)dt$$

subject to

$$x'(t) = f(x(t), u(t), t).$$

The values of the control vectors may be restricted to a defined subset U, i.e., $u(t) \in U$, and an initial (and/or a terminal) state may be given by $x(0) = xo$.

The most fundamental idea is that an optimal control problem may be conceptualized, formulated, solved and the results interpreted with substantially minor mathematical requirements, see the more detailed discussion in Kivijärvi and Tuominen (1985). Such a flexible system is needed when complicated corporate models with several relationships are modelled and solved.

Also Politzer and Wilmes (1977) have used an optimal control approach to design a management information system. However, due to the mathematical complexity they had to use a very simple HMMS model. And the process of abstraction and simplification may eliminate important variables and relations.

The approaches for determining information requirements are generally (for instance Srinivasan and Kaiser, 1981) classified into two groups: decision analysis and data analysis. The decision analysis approach concentrates on decisions, using them for determining indirectly the information requirements. The data analysis approach, on the other hand, rather focuses on data (information) flows in an organization. Traditional system dynamics has an advantage over these two approaches. It is a hybrid technique combining both approaches. Namely, when we describe an organization with the principles of system dynamics we are forced to identify explicitely the major decision situations and rules. On the other hand, in system dynamics, information flows are explicitely isolated from the other system elements. When this methodological advantage is joined with the technical optimization ability, we have a tool with high potentials for analyzing available information values.

3. A corporate model

According to Naylor (1979) a corporate model is "an attempt to describe the complex interrelationships among a corporation's financial, marketing, and production activities in terms of a system of mathematical and logical relationships which have been programmed into a computer." For the purposes of this study a ready corporate model is adopted from Coyle (1977), Management system dynamics.

The example firm operates on metal industry and it is a subsidiary of a parent company. It comprises semifabrication and manufacturing divisions. The semifabrication division takes metal from the parent company or buys it outside the company. A part of the semifabricated products is sold outside the company another part to the manufacturing. The firm's market share for semifabricated metal is roughly 25 % and much less for the metal used in final products. The metal demand for industry is initially 400000 tons per year but it increases to 480000 tons during the fifth year (10 year planning horizon). The firm is financed by retained earnings, long term borrowings and original share capital.

The example model is relatively complicated. It includes all aspects of a corporate model in general, namely financial, production and marketing functions. The main contribution of this study is to show that there are no conceptual, mathematical, computational or other unsolvable difficulties inherent in applying this approach in realistic situations. The information value, information attributes, and information costs are readily observed, understood, and the results are directly interpretable in terms of common corporate planning.

We must emphasize that according to our approach the value of information is analyzed by a corporate model that is originally aimed to support general corporate planning. Some minor alternations must be made but because no new model is needed, the costs of the analysis are low, and the analysis can be performed within a short time period.

4. Information quality attributes and evaluation criteria

System dynamics has often been used or suggested to be used as a tool for analyzing the dependence of information values on information attributes (Swanson, Kleijnen, Verhelst, Martin). For instance the following attributes are used in those studies: error, distortion, delay, and sampling.

In this study the effects of the following attributes on the value of information are analysed: delay, inconsistency, overestimation, and underestimation. In general, a delay is the time interval between an actual event and the successive decision. A delay may be constituted of the following elements (Keijnen 1980): update processing delay, update interval, retrieval delay, and decision delay. In system dynamics delays are implemented by special macro functions, which have different kinds of transfer effects between inputs and outputs.

In this case, a delay is introduced into the order book information. The following three new variables are defined in the following manner. The degree of the delay is described by a decision parameter DP1. DP1 is a decision variable in the sense that we may select its value; if DP1 is fixed to be one then perfect information is assumed, but if DP1 is zero then delayed information is used. Every value between zero and one may also be selected.

By inconsistent information is meant such kind of bias where the same information is delivered to different decision situations with different contents. A bias of this kind may arise for instance from a defective data management system, where the same data is stored in different places. In such a case the data may be updated in a place but remain unchanged in another place.

In the corporate model, the information concerning semifabricate capacity is used in four decision situations. Now we assume, that decision concerning the semifabricated production that is retained for own manufacturing is misled by an additive permanent deviation from the real semifabricate capacity. The decision variable two DP2 describes the degree of the bias. On the contrary, the other equations using semifabricate capacity information will remain unchanged.

Lots of efforts have been made to develop more and more accurate forecasts, especially demand forecasts. Unfortunately, the performance of the whole system (corporation) may not inevitably be very sensitive to the accuracy of the forecasts (Coyle, 1977). In this case, the demand forecasts are used to count the indicated semifabrication capacity (ISFCAP). In the original model perfect forecasts are generated. Here, a multiplicative bias element is inserted into the demand forecast equation. DP3 describes the accuracy of forecasts. If DP3 = 1, then accurate (perfect) forecasts are generated, if $0 \leq$ DP3 < 1, then demand is underestimated and if $1 <$ DP3 ≤ 2 then demand is overestimated.

In sum, three parameters (DP1, DP2, DP3) are defined to describe the attributes of information. If values close to ones are assigned to the parameters, then perfect information is assumed. On the other hand, a great deviation from ones indicates a great deviation from perfect information.

Because it is not easy to show one particular, unambiguous and accepted criterion for corporate planning in general, it is not less easy to determine only one measuring criterion on information value. As a potential criterion we may select the best one for instance from the following set of criteria: cumulative profits (CUPROF), value of plant (VP), cash reserve (CASH), and the market share of manufactured metal (UMS). Of course, in the model there are several other potential evaluation criteria, for instance production capacity, return on investments, working capital, etc.

In this case, all criteria variables are integrals over a planning period (120 months) and all of them are going to be maximized. In the systemic view, any information unit and any information attribute has at least indirect effects on all those criteria.

5. The optimizing experiments and the results

Although it is argued for (Marschak, 1969, Feltham, 1968) a simultaneous selection of a set of decision rules and an information system, we assume in this study that the decision rules are given and they will remain relatively unchanged despite the changes in the contents of information.

In experiments each evaluation criterion is maximized by turns with respect to the information attributes (decision parameters) and the results are summarized in Table 1. The reported results were achieved after twenty iterations.

Evaluation criterion	Perfect information Criterion value	Degradated information DP1	DP2	DP3	Criterion value
CUPROF	.10495+9	1	1	0	.11483+9
VP	.48170+8	.993	1	.84	.50220+8
CASH	.23782+6	1	1	0	.75705+8
UMS	.25769	.833	.994	.045	.31368

Table 1: Results of optimization (In perfect information DP1=DP2=DP3=1, DP1 - delay, DP2 - inconsistency, DP3 - under/overestimation)

At first sight the results might seem to be at least to some degree surprising. They indicate that perfect information is not leading to the best results but better criterion values are reached by degradated information. They indicate that if the goal of the corporation is to maximize the value of plant (VP) or the average level of semifabricate throughput (LTPUT) then the information very close to the perfect information would lead to the best results (slight deviation in the delay of the order book information and in the accuracy of forecasts). But if the goal of the corporation is to maximize the cumulative profit over a planning period or the ending cash then the information system producing heavily underestimated demand forecasts would lead to the best result! On the top, the maximization of the market share seems to lead to relatively confusing results.

How can the achieved results be explained? There is only one general explanation for the odd deviation from perfect information, namely due to the conflicting set of joined criteria there cannot be such a set of decision rules that would maximize all criteria by the same value set of decision parameters. This is true even with perfect information! Thus, with fixed decision rules, the value of information may be negative with respect to one specific evaluation criterion. And consequently, in the multicriteria, multigoal decision making situations there is no single, ideal information system that would dominate all other information system alternatives.

When analysing concretely the dependency of the optimal information characters on the given decision rules it appeared for instance that the firm is not a very profitable business at all. And if the decision rules are fixed to keep an unprofitable business going according to the demand forecasts then the underestimated forecasts would lead to larger cumulative profits than the accurate forecasts. This explanation can be easily verified by increasing the product prices. If the prices are increased by 50 % at the time 12 and the cumulative profit is maximized, then after 22 iterations the cumulative profit is .19826+9 and the decision parameters will achieve the following values: DP1=.945, DP2=1, DP3=.99. Thus in these business conditions, (nearly) perfect information leads to the best results on given decision rules.

After some additional iterations it was found that the market share function is very flat, that is, the same market share (31 %) is achieved by a great variety of parameter values.

In Table 2 the final values of the five criteria are listed. As expected, the table indicates that a criterion reaches its best values when that particular criterion is maximized.

Criterion to be maximized	Value of criterion			
	CUPROF	VP	CASH	UMS
CUPROF	115+6	837+4	757+5	.314
VP	102+6	502+5	44	.291
CASH	115+6	837+4	757+5	.314
UMS	114+6	101+5	705+5	.314

Table 2: The values of criteria

For reasons of space, only the plotted time series over the planning period are shown in Figures 5.1. In each figure the respective four evaluation criteria are plotted. As a whole, it seems obvious that the dynamic modes of the system heavily depend on the criterion having been optimized. In other words, the results are Pareto optimal, which means that if the value of a criterion is increased then the value of another criterion would decrease.

Another set of experiments was also performed. For this set of experiments an information cost function was defined. It is a function from the set of attributes (decision parameters) so that the closer to the ones the parameters are the more costs are also accrued. The costs of information are directly substracted from the profits. It appeared for instance that the inconsistency in the information concerning the semifabricate capacity is most dangerous and it is worth money to decrease this bias to its minimum.

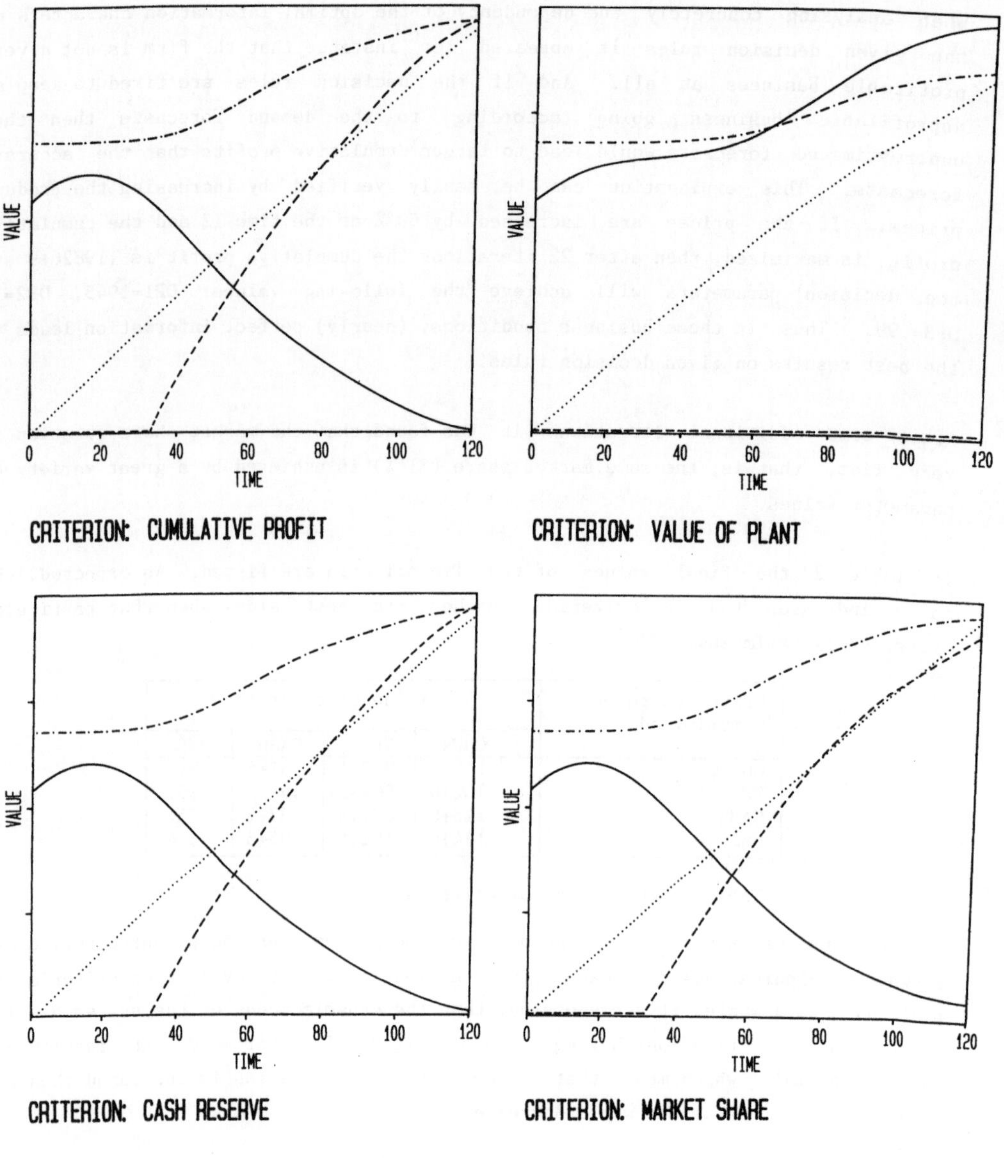

FIGURE 5.1 MAXIMIZATION WITH RESPECT TO DIFFERENT CRITERIA

········	CUMULATIVE PROFIT	(0, 12E7)
———	VALUE OF PLANT	(80E5, 51E6)
– – – –	CASH RESERVE	(0, 77E6)
–·–·–	MARKET SHARE	(0.1, 0,32)

6. Concluding remarks

At the theoretical level we have argued in this paper that the final value of information in a firm should be measured at the corporate level by multiple criteria and the state of information should be described by multiple attributes. At the practical, operational level a systemic, multicriteria and multiattribute approach means, for instance, that the best information system for the whole organization is not developed by means of the individual, user-perceived information requirements but also the joint effects of decisions must be evaluated.

From the technical point of view we have introduced a flexible and user oriented optimization tool to determine information values and to develop information systems without complicated mathematical manipulations.

The primary finding is that in a multicriteria corporate planning situation with fixed decision rules there is no single, ideal information system that would dominate all other information system alternatives. Probably the clearest message to the design of information systems is the care which must be given in defining the evaluation criterion.

Of course, the results with a single corporate model cannot establish the validity of the approach but they would have been able to discredit it totally. In this case this operational and practical approach performed very well suggesting a potential for further studies.

REFERENCES

1. ACKOFF, RUSSELL L., Management Misinformation Systems, Management Science, December, 1967

2. COOPER, RANDOLPH B. and SWANSON, E. BURTON, Management information requirements assessment: the state of the art, Data Base, Fall 1979

3. COYLE, R.G., Management System Dynamics, John Wiley & Sons, 1977

4. EPSTEIN, BARRY J. and KING, WILLIAM R., An Experimental Study of the Value of Information, Omega, Vol. 10, No. 3, 1982

5. FELTHAM, GERALD A., The Value of Information, The Accounting Review, October 1968

6. GALLAGHER, CHARLES A., Perceptions of the Value of a Management Information System, Academy of Management Journal, March 1974

7. KELOHARJU, RAIMO, Relativity Dynamics, Publications of the Helsinki School of Economics, A-40, Helsinki 1983

8. KING, WILLIAM R. and EPSTEIN, BARRY JAY, Assessing the value of information, Management Datamatics, Vol. 5, No. 4, 1976

9. KIVIJÄRVI, HANNU and TUOMINEN, MARKKU, The value of information and corporate financial, marketing and production planning, Helsinki School of economics, Working Paper, F-129, 1985

10. KLEIJNEN, JACK P. C., Computers and Profits: Quantifying Financial Benefits of Information, Addison-Wesley, 1980

11. LUENBERGER, D. G., Introduction to dynamic systems, John Wiley and Sons, 1979

12. MARSCHAK, JACOB, Economics of Inquiring, Communicating, Deciding, American Economic Review, Vol. 58, No. 2, 1968

13. MARTIN, MERLE P., Determining Information Requirements for DSS, Journal of Systems Management, December 1982

14. NAYLOR, THOMAS H., Corporate Planning Models, Addison-Wesley Publishing Company, 1979

15. POLITZER, J. P. and WILMES, Ph., Optimal design of management information systems through control theory, Universite de Sherbrooke and Universite de Louvain, Working Paper 77-5

16. SRINIVASAN, ANANTH and KAISER, KATE M., The Role of Information Accessibility in Information Requirements Analysis, Systems, Objectives, Solutions, No. 4, 1984

17. SWANSON, C. V., Evaluating the Quality of Management Information, Working Paper, Sloan Management School, 1971

18. TAUBERT, WILLIAM H., A Search Decision Rule for the Aggregate Scheduling Problem, Management Science, Vol. 14, No. 6, 1968

19. VERHELST, MAURICE, On possible approaches and techniques for determining financial benefits of organizational information, In Economics of Informatics edited by A. B. Frielink, North-Holland Publishing Company, 1975

Duality in probabilistic constrained linear programming

Eva Komaromi
National Management Development Centre, Budapest, Hungary

A b s t r a c t. In this paper a probabilistic constrained linear programming pair of problems is introduced. We establish conditions under which the bilinear function involved in the objective functions of both problems has a saddlepoint.

<u>Introduction.</u> In this paper we introduce and investigate the following pair of problems:

(P) Minimize $\sup \{ uAX : u \in U \}$ over

$$X = \{ x \in R^n : x \geq 0, \; \mathcal{P}(Ax \geq \beta) \geq p \},$$

(D) Maximixe $\inf \{ uAX : x \in X \}$ over

$$U = \{ u \in R^m : u \geq 0, \; \mathcal{P}(uA \leq \gamma) \geq q \},$$

where A is a given $m \times n$ dimensional deterministic matrix; $\mathcal{P}$ denotes probability; β is a m-dimensional and $-\gamma$ is a n-dimensional random vector with given joint continuous probability distribution functions F_β and $F_{-\gamma}$, respectively; p and q are prescribed reliability levels, $0<p<1$, $0<q<1$; x and u are decision vectors.

The formulation of this pair of problems is an extension of the linear programming duality concept [1]. Its interpretation is similar to the game theoretical interpretation of the primal-dual pair of linear programs.

We will also require that the support sets supp $F_{-\gamma}$ and/or supp F_β of the probability distributions $F_{-\gamma}$ and F_β be bounded. (The support of a probability distribution is the smallest closed subset of R^n the probability measure of which equals 1.) This requirement seems to be quite natural since in practical problems the support of the probability distribution of the random variables in question is necessarily bounded. However, the support of the approximating theoretical distribution function is often R^n, like for the multivariate normal distribution. That is the reason why this boundedness can not be taken for granted.

We begin by formulating the deterministic equivalents of (P) and (D), in Section 1. Then, in Section 2, we compare the optimality sets of (P) and (D) to the ones of the somewhat simpler problems

(P′) Minimize $\sup \{ cx : \mathcal{P}(\gamma \geq c) \geq q \}$ over X

and (D′) Maximize $\inf \{ ub : \mathcal{P}(\beta \leq b) \geq p \}$ over U,

respectively. Finally we establish convexity and regularity conditions under which uAx has a saddlepoint (with respect to minimizing over X and maximizing over U), in Section 3.

1. Equivalent reformulations of (P) and (D).

The formulation of the deterministic equivalent of (P) is based on the following observations: (i) $x \in X$ if and only if there exists $b \in R^m$ such that x and b satisfy the constraints $Ax \geq b$, $x \geq 0$, $\mathcal{P}(\beta \leq b) \geq p$; (ii) $Ax \geq b$ implies that $Ax \geq b'$ for all $b' \leq b$ therefore b can be chosen in $\operatorname{supp} F_\beta$; (iii) $\mathcal{P}(\beta \leq b) = F_\beta(b)$ by the continuity of F_β. Similar observations are valid for the set U . Define the sets B , X(b) for a given $b \in R^m$, C, U(c) for a given $c \in R^n$ to be

$$B = \{ b \in \operatorname{supp} F_\beta : F_\beta(b) \geq p \} ,$$

$$X(b) = \{ x \in R^n : Ax \geq b , x \geq 0 \} ,$$

$$C = \{ c \in \operatorname{supp} F_{-\gamma} : F_{-\gamma}(-c) \geq q \} ,$$

$$U(c) = \{ u \in R^m : uA \leq c , u \geq 0 \} .$$

C and B are nonempty because $q<1$ and $p<1$ and they are closed by the continuity of $F_{-\gamma}$ and F_β. If $\hat{x}$ is feasible to (P) then, by thr monotonicity of the probability distribution, there exists $\hat{b} \leq A\hat{x}$ such that $F_\beta(\hat{b}) \geq p$ and $\hat{b} \in \operatorname{supp} F_\beta$ (possibly, $\hat{b} = A\hat{x}$ might do). Thus $(\hat{b},\hat{x})$ is feasible to the following problem:

$$\text{(P1) Minimize} \quad \sup_{c\in C} \{ \sup_{u\in U(c)} uAx \} \quad \text{over}$$

$$Z = \{ (b,x) : b \in B,\ x \in X(b) \} .$$

Conversely, if $(\hat{b},\hat{x})$ is feasible to (P1) then $\hat{x}$ is feasible to (P).

Similarly, (D) is equivalent to

$$\text{(D1) Maximize} \quad \inf_{b\in B} \{ \inf_{x\in X(b)} uAx \} \quad \text{over}$$

$$S = \{ (c,u) : c \in C,\ u \in U(c) \} .$$

2. Nonequivalent reformulations of (P1) and (D1).

Let $b \in B$, $x \in X(b)$ be fixed. By the linear programming (LP) duality theorem $\sup_{u\in U(c)} uAx = \inf_{x'\in X(Ax)} cx'$ for any fixed $c \in R^n$. If $U(c) = \emptyset$ then $\inf cx' = -\infty$ because $X(Ax) \neq \emptyset$, if $U(c) \neq \emptyset$ then both the inf and the sup are attained. Hence

$$\sup_{c\in C} \{ \sup_{u\in U(c)} uAx \} = \sup_{c\in C} \{ \inf_{x'\in X(Ax)} cx' \} \tag{1}$$

and both sides are $-\infty$ if $\{ c\in C : U(c) \neq \emptyset \} = \emptyset$.

Furthermore, by Rockafellar's result in [2,Ch.37] the saddlevalue of cx' exists (with respect to minimizing over $X(Ax)$ and maximizing over C) if $C \subset \operatorname{supp} F_{-\gamma}$ is bounded because cx' is convex in c, concave in x' and continuous on $C \times X(Ax)$. Thus

$$\sup_{c\in C} \{ \inf_{x'\in X(Ax)} cx' \} = \inf_{x'\in X(Ax)} \{ \sup_{c\in C} cx' \} \tag{2}$$

Moreover, $\{ x' \geq 0 : \exists\ x \in X(b) : Ax' \geq Ax \} = X(b)$ so that

$$\inf_{x\in X(b)} \{ \inf_{x'\in X(Ax)} \{ \sup_{c\in C} cx' \}\} = \inf_{x\in X(b)} \{ \sup_{c\in C} cx \}$$

where both sides are defined, as usual, to be $+\infty$ if $X(b) = \emptyset$.

This argumentation and a similar one lead to the problems

(P2) Minimize $\sup\{ cx : c \in C \}$ over Z,

(D2) Maximize $\inf\{ ub : b \in B \}$ over S .

If $(\hat{b},\hat{x})$ is optimal to (P2) then it is optimal to (P1) because otherwise $\sup\limits_{c\in C}\{ \sup\limits_{u\in U(c)} uAx \} < \sup\limits_{c\in C}\{ \sup\limits_{u\in U(c)} uA\hat{x} \}$ would hold for a $x \in X(\hat{b})$ which, by (1) and (2), would imply the contradiction

$$\inf_{x'\in X(Ax)}\{ \sup_{c\in C} cx' \} < \sup_{c\in C} c\hat{x}.$$

However, if we want to solve (P1) via solving (P2) we should also make sure that not all the optimal solutions of (P1) are lost during this transformation to (P2). It is established in the following theorem under additional assumptions.

T h e o r e m 1. Suppose C is compact, $\operatorname{int}\{ c \in C : U(c) \neq \emptyset \} \neq \emptyset$, suppose $F-\gamma$ is quasiconcave on C. If for $b \in R^m$ $X(b) \neq \emptyset$ then cx has a saddlepoint with respect to minimizing over $X(b)$ and maximizing over C .

P r o o f. $\inf\{ cx : x \in X(b) \}$ as a function of c is concave on R^n by definition and its domain $\{ c \in R^n : U(c) \neq \emptyset \}$ is closed as it was first shown by Williams in [3]. Thus the continuous function $\inf\{ cx : x \in X(b) \}$ achieves its maximum over the nonempty compact set $\{ c \in C : U(c) \neq \emptyset \}$. Then the statement follows from the convexity and regularity assumptions as it was proved by Komaromi in [4].

Thus, under the assumptions of Theorem 1 and by (1), for an optimal solution $(\hat{b},\hat{x})$ of (P1) there exists $x' \in X(A\hat{x}) \subset X(\hat{b})$ such that $(\hat{b}, x')$ is optimal to (P2) . Moreover,

C o r o l l a r y 1. Suppose the assumptions of Theorem 1 are fulfilled. If $\hat{b}$ is an optimal solution of

(P3) Minimize $\inf\limits_{x\in X(b)}\{ \sup\limits_{c\in C} cx \}$ over B

then the infimum is attained at a $\hat{x} \in X(\hat{b})$ so that $(\hat{b},\hat{x})$ is optimal to both (P1) and (P2).

Similar argumentation leads to the following statements.

T h e o r e m 1'. Suppose B is compact, int $\{ b \in B : X(b) \neq \emptyset \} \neq \emptyset$, suppose F is quasiconcave on B. If for $c \in R^n$ $U(c) \neq \emptyset$ then ub has a saddlepoint with respect to minimizing over B and maximizing over $U(c)$.

C o r o l l a r y 1'. Suppose the assumptions of Theorem 1' are fulfilled. If $\hat{c}$ is an optimal solution of

$$\text{(D3)} \quad \text{Maximize} \quad \sup_{u \in U(c)} \{ \inf_{b \in B} ub \} \quad \text{over } C$$

then the supremum is attained at a $\hat{u} \in U(\hat{c})$ so that $(\hat{c},\hat{u})$ is optimal to both (D1) and (D2).

3. Duality.

Clearly, $\inf_{b \in B} \{ \inf_{x \in X(b)} \{ \sup_{c \in C} cx \}\} \geq \sup_{c \in C} \{ \inf_{b \in B} \{ \inf_{x \in X(b)} cx \}\}$. Furthermore, $\inf \{ cx : x \in X(b) \} = \sup \{ ub : u \in U(c) \}$ by the LP duality theorem. Finally, $\sup_{c \in C} \{ \inf_{b \in B} \{ \sup_{u \in U(c)} ub \}\} \geq \sup_{c \in C} \{ \sup_{u \in U(c)} \{ \inf_{b \in B} ub \}\}$.

Therefore, if $\sup \{ c'x : c' \in C \} = \inf \{ ub' : b' \in B \}$ for a $(b,x) \in Z$ and a $(c,u) \in S$ then (b,x) and (c,u) are optimal for (P2) and (D2), respectively.

In the following theorem we assume that $F_{-\gamma}$ is strictly logarithmic concave and continuously differentiable on int supp $F_{-\gamma}$. We refer to Prekopa's results on logarithmic concave measures [5] in order to indicate the richness of this class of probability distributions. Under this assumption $\sup cx$ is attained at a unique point of the compact C for any $x \geq 0$, $x \neq 0$. Denote this point by $c(x)$, and $\sup cx$ by $xc(x)$ even at $x = 0$. It was shown in [4] that $c(x)$ is continuous on $\{ x : x \geq 0, x \neq 0 \}$; $xc(x)$ is convex continuous on $\{ x : x \geq 0 \}$, it is differentiable on $\{ x : x > 0 \}$ with gradient $c(x)$, and it is strictly convex on $\{x \geq 0 : \Sigma x_i = 1\}$. Furthermore, $xc(\delta x + x')$ is a strictly increasing function of δ for

$\delta > 0, \quad x \geq 0, \quad x \neq 0, \quad x' \geq 0, \quad x' \neq 0, \quad x/|x| \neq x'/|x'|$ with $\lim_{\delta \to \infty} xc(\delta x + x') = xc(x)$.

T h e o r e m 2. Suppose F is quasiconcave on B, C is compact, $F_{-\gamma}$ is strictly logarithmic concave and continuously differentiable on int supp $F_{-\gamma}$. Suppose $\text{int}\{b \in B : X(b) \neq \emptyset\} \neq \emptyset$. If $(\hat{b},\hat{x})$ is optimal to (P2), $\hat{x} \neq 0$, then for $c(\hat{x}) = \text{Arg max } \{c\hat{x} : c \in C\}$ there exists $\hat{u} \in U(c(\hat{x}))$ such that $\hat{b} \in \text{Arg min } \{\hat{u}b : b \in B\}$ so that $(c(\hat{x}), \hat{u})$ is optimal to (D2).

P r o o f . The proof is based on the Kuhn-Tucker saddlepoint necessary optimality theorem [6, Ch.5]. According to this theorem if $Z^{\circ} = \{ (b,x) : x \geq 0, b \in B \}$ is convex,
if there exists $(b',x') \in Z^{\circ}$ such that $Ax' > b'$, and
if $(\hat{b},\hat{x})$ minimizes $xc(x)$ over $Z = \{ (b,x) \in Z^{\circ} : Ax \geq b \}$
then there exists $\hat{u} \geq 0$ such that

$$\hat{u}(\hat{b} - A\hat{x}) = 0 \tag{3}$$

$$\hat{x}c(\hat{x}) + u(\hat{b} - A\hat{x}) \leq \hat{x}c(\hat{x}) + \hat{u}(\hat{b} - A\hat{x}) \leq xc(x) + \hat{u}(b - Ax)$$

hold for all $u \geq 0$ and $(b,x) \in Z^{\circ}$. (The hypotheses are among the assumptions of the theorem.) The second inequality implies that

$$\hat{b} \in \text{Arg min } \{ \hat{u}b : b \in B \} \tag{4}$$

and $$\hat{x}c(\hat{x}) - \hat{u}A\hat{x} = \min_{x \geq 0} (c(x) - \hat{u}A)\, x = 0 \tag{5}$$

If $c(\hat{x}) \geq \hat{u}A$, that is, if $\hat{u} \in U(c(\hat{x}))$ then $(c(\hat{x}), \hat{u})$ is optimal to (D2) by (3),(4) and (5). Thus we prove the theorem by showing that

$$\min_{x \geq 0} (c(\hat{x}) - \hat{u}A)\, x = 0 .$$

Suppose, on the contrary, that there exists $x' \geq 0$ such that $(c(\hat{x}) - \hat{u}A)\, x' < 0$. Then $x' \neq 0$, $x' \neq \hat{x}$. Since $x'c(\delta x' + \hat{x})$ is a strictly increasing function of δ $(0 \leq \delta)$ which takes any value between $x'c(\hat{x})$ and $x'c(x')$ hence there exists $0 < \delta_o$ such that $x' [c(\delta_o x' + \hat{x}) - \hat{u}A] < 0$ still holds, while $\hat{x} [c(\delta_o x' + \hat{x}) - \hat{u}A] < \hat{x} [c(\hat{x}) - \hat{u}A] = 0$ also holds by the definition of $xc(x)$. Then for $x'' = \delta_o x' + \hat{x}$: $x'' [c(x'') - \hat{u}A] < 0$ which contradicts (5) and completes the proof.

The following theorem is proved similarly.

T h e o r e m 2′. Suppose $F_{-\gamma}$ is quasiconcave on C, B is compact, F_β is strictly logarithmic concave and continuously differentiable on int supp F_β. Suppose $\text{int}\{c \in C : U(c) \neq \emptyset\} \neq \emptyset$. If $(\hat{c},\hat{u})$ is optimal to (D2), $\hat{u} \neq 0$, then for $b(\hat{u}) = \text{Arg min } \{\hat{u}b : u \in U\}$ there exists $\hat{x} \in X(b(\hat{u}))$ such that $\hat{c} \in \text{Arg max } \{c\hat{x} : c \in C\}$ so that $(b(\hat{u}),\hat{x})$ is optimal to (P2).

The conclusion is summarized in the last two statements:

C o r o l l a r y 2. Suppose the assumptions of Theorem 2 are fulfilled. If $(\hat{b}, \hat{x})$ is optimal to (P2), $\hat{x} \neq 0$, then there exists $\hat{u}$ such that $(\hat{x},\hat{u})$ is the saddlepoint of uAx with respect to minimizing over X and maximizing over U.

C o r o l l a r y 2′. Suppose the assumptions of Theorem 2′ are fulfilled. If $(\hat{c}, \hat{u})$ is optimal to (D2), $\hat{u} \neq 0$, then there exists $\hat{x}$ such that $(\hat{x},\hat{u})$ is the saddlepoint of uAx with respect to minimizing over X and maximizing over U.

References

[1] G.B.Dantzig,"Linear programming and extensions" (Princeton University Press, Princeton, N.J., 1963).

[2] R.T.Rockafellar,"Convex analysis" (Princeton University Press, Princeton, N.J., 1970).

[3] A.C.Williams,"Marginal values in linear programming",J.Soc.Indust. Appl.Math.11(1963)82-94.

[4] É.Komáromi,"A dual method for probabilistic constrained problems" in print in the Mathematical Programming Study on Stochastic Programming.

[5] A.Prékopa,"On logarithmic concave measures and functions", Acta Sci.Math.34(1973)335-343.

[6] O.L.Mangasarian,"Nonlinear programming"(Mc Graw-Hill,New York,1969).

NONLINEAR ESTIMATION PROBLEMS OF POISSON CLUSTER PROCESSES

FRANZ KONECNY
Universität für Bodenkultur - Wien
Institut für Mathematik u. Angewandte Statistik
A - 1180 Wien, Gregor Mendel-Straße 33
Austria

Summary. Doubly stochastic Poisson processes whose unobservable intensity is a shot-noise process with random amplitude arise when each event of a primary Poisson process generates a random number of subsidiary events. We derive a stochastic partial differential equation for the unnormalized conditional moment generating function. This equation can be used for recursive compution of the minimum variance estimator of the unobservable intensity as well as the likelihood ratio with respect to the reference measure, on the basis of point process observations.

1. Introduction

Cluster point processes are mathematical models for series of events, which occur in groups distributed around some centers. When the centers follow a Poisson process, we speak of a Poisson cluster process.

In this paper we are concerned with Poisson cluster processes of Neyman-Scott (N-S) type. Originally invented for the description of the spatial distribution of galaxies, N-S processes have become important in geophysics (earthquake aftershocks) and hydrology (rainfall and flood models). Such a process consists of two components. These are (1) the Poisson process of cluster centers (T_n) and (2) a random number S of following subsidiary events, whose distances to the center are i.i.d. as well as the cluster size S. The superposition of all subsidiary events constitutes the N-S process $N = (N_t)$.

Bartlett [1] and Vere-Jones [9] established the following important representation for a subclass of N-S processes:
Assume

i) (T_n) are the occurrence times of a homogeneous Poisson process with the intensity c.

ii) The common distribution of the cluster sizes S is compound Poisson with the mixing distribution q(du))

iii) The distances from each cluster center to its members have the density function h(t).

Then the N-S process (N_t) can be represented as a doubly stochastic Poisson process with the shot-noise process

$$X_t = \sum_{T_n<t} U_n h(t-T_n) \tag{1.1}$$

as stochastic intensity. The random amplitudes (U_n) are i.i.d. according to $q(du)$. When $h(t) = \beta e^{-\beta t}$, $\beta > 0$, then (X_t) is a Markov process.

Usually, the intensity of a doubly stochastic Poisson process is unobservable. Therefore the optimal estimation of X_t based on observations of the point process up to time t is an important task.

We adopt an approach which was successfully applied by Boel and Beneš [2] on the estimation of a diffusion of known structure. By the method of reference probability a stochastic partial differential equation for the conditional moment generating function of the stochastic intensity will be derived, which reduces between the observed events to a linear deterministic p.d.e. of first order.

2. Estimation theory of point processes

On a stochastic basis $(\Omega, \underline{\underline{F}}, P, (F_t; t \geq 0))$, where $(\underline{\underline{F}}_t)$ is non-decreasing family of sub-σ-algebras of $\underline{\underline{F}}$, we assume the following processes are defined:

(i) a Markov process X of the type (1.1) with respect to $(\underline{\underline{F}}_t;\ t \geq 0)$

(ii) An increasing family of random variables $0 < \tau_1 < \ldots < \tau_n < \ldots$ together with the associated counting process $N = (N_t)$:

$$N_t = \sum_{n=1}^{\infty} 1_{(\tau_n \leq t)} \ ,$$

which we assume to be conditional Poisson for $t \geq 0$, given $(X_s; s \leq t)$.

The stochastic intensity or the rate of the point process can be defined as follows:

$$\lambda_t = \lim_{h \downarrow 0} (1/h) P(N_{t+h} - N_t > 0 \mid \underline{\underline{F}}_t).$$

In our situation $\lambda_t = \psi(X_t)$, where ψ is an deterministic function.

One approach to provide schemes for recursive computation of estimates of the form $\hat{\phi}(X_t) = E(\phi(X_t) \mid N_s; s \leq t)$ is the method of reference probability, cf. [3]. With t fixed, the computation can be accomplished by means of a measure transformation, which yields

$$\hat{\phi}(X_t) = \frac{\sigma_t(\phi)}{\sigma_t(1)} \ , \tag{2.1}$$

where

$$\sigma_t(\phi) = E(\phi(X_t)\exp\{\int_0^t \log \lambda_s dN_s + \int_0^t (1-\lambda_s)ds\}).$$

The expectation on the R.H.S. is with respect to X.

The following stochastic equation has been derived for the unnormalized conditional expectation $\sigma_t(\phi)$:

$$\sigma_t(\phi) = \sigma_o(\phi) + \int_0^t \sigma_{s-}\ (\phi(\lambda_{s-}\ -1))d(N_s-s), \tag{2.2}$$

where L is the generator of X. Using the notation of stochastic differentials this equation takes the form:

$$d\sigma_t(\phi) = \sigma_t(L\phi)dt + \sigma_{t-}(\phi(\lambda_{t-}-1))d(N_t-t). \tag{2.2'}$$

3. The estimation of the shot-noise process

Let q(du) denote the common distribution of the U_n and c be the intensity of the Poisson process (T_n). The shot-noise process (1.1) with $h(t) = \beta e^{-\beta t}$ satisfies the stochastic differential equation

$$dX_t = -\beta X_t dt + \int u\nu(dt, du), \qquad X_o = 0, \tag{3.1}$$

where $\nu(.,.)$ is a Poisson measure on $[0,T]\times(0,\infty)$, $T>0$, with $E\nu(dt,du) = cq(du)dt$. The generator of the Markov process (X_t) is

$$L\phi(x) = -\beta x\phi'(x) + c\int[\phi(x+u) - \phi(x)]q(du),$$

cf. [8, p.187].

We assume the stochastic intensity of the counting process N is of the form

$$\lambda_t = \mu + X_t, \tag{3.2}$$

with $\mu > 0$.

The conditional moment generating function (c.m.g.f.) of (3.2)

$$e^{s\lambda_t} = E(e^{s\lambda_t}|N_r;\ r \le t)$$

specify the optimal estimate $\hat{\lambda}_t$ as follows:

$$\hat{\lambda}_t = \frac{d}{ds}\, e^{s\lambda_t}\ \Big|_{s=0}.$$

The evolution of the c.m.g.f. can be obtained from equation (2.2):

$$d\sigma_t(e^{s\lambda_t}) = \sigma_t(-\beta s(\lambda_t-\mu)e^{s\lambda_t} + \rho(s)e^{s\lambda_t})dt$$
$$+ \sigma_{t-}((\lambda_{t-}-1)e^{s\lambda_{t-}})d(N_t-t), \qquad (3.3)$$

with $\sigma_o(e^{s\lambda_o}) = e^{s\mu}$ and

$$\rho(s) = c\int [e^{su}-1]q(du).$$

Setting $g(s,t) = \sigma(e^{s\lambda_t})$ we get

$$dg(s,t) = \{-(1+\beta s)\frac{\partial g}{\partial s} + (1+\beta\mu s+\rho(s))g(s,t)\}dt$$
$$+ \{\frac{\partial g}{\partial s}(s,t-) - g(s,t-)\}dN_t. \qquad (3.4)$$

Between the events (τ_n) of the point process (N_t) the stochastic equation (3,4) reduces to a deterministic linear p.d.e. of first order:

$$\frac{\partial g}{\partial t} + (1+\beta s)\frac{\partial g}{\partial s} = (1+\beta\mu s+\rho(s))g(s,t). \qquad (3.5)$$

At the occurrence times (τ_n) we obtain the following updating relation:

$$g(s,\tau_n) = \frac{\partial}{\partial s}g(s,\tau_n-). \qquad (3.6)$$

Thus, the integration of the p.d.e. (3.5) together with the relation (3.6) yields an expression for the c.m.g.f. of the stochastic intensity

$$\widehat{e^{s\lambda_t}} = \frac{g(s,t)}{g(0,t)},$$

cf. Boel-Beneš [2].

Most important in applications are N-S processes, where the distribution of the cluster sizes is a Poisson distribution with the parameter α. In this case the directing shot-noise process is of the form

$$X_t = \alpha\beta\sum_{T_n<t} e^{-\beta(t-T_n)}$$
$$= \alpha\beta\int_0^t e^{-\beta(t-r)}dN_r^C, \qquad (3.7)$$

and $\rho(s) = c(e^{s\alpha\beta}-1)$. Setting $g_n(s,t) = g(s,t)1_{(\tau_n\le t\le\tau_{n+1})}$ we found in [11] the solution of (3.4) as follows:

$$g_n(s,t) = \phi_n\left(\frac{1}{\beta}(1+\beta s)e^{-\beta(t-\tau_n)}-1\right)$$

$$\times \exp \left\{\frac{ce^{-\alpha}}{\beta} \int_{e^{-\beta(t-\tau_n)}}^{1} \frac{e^{\alpha(1+\beta s)x}}{x}\, dx + (1-c-\mu)(t-\tau_n)\right.$$

$$\left. + \frac{\mu}{b}(1+\beta s)(1-e^{-\beta(t-\tau_n)})\right\}$$

with

$$\phi_n(s) = \begin{cases} e^{s\mu} & n = 0 \\ \frac{\partial}{\partial s} g_{n-1}(s,\tau_n) & n \geq 1. \end{cases}$$

4. Doubly stochastic Poisson process with a linear intensity process

The intensity process (3.7) is a special case of a class of processes which was introduced in a paper of Ogata and Akaike [10]:

$$\lambda_t = \mu + \int_0^t h(t-r)\, dN_r^c\,, \tag{4.1}$$

where $h(t) = e^{-\beta t}(\alpha_0 + \alpha_1 t + \ldots + \alpha_n t^n)$, with suitable restrictions on α_k and β to ensure that

$$\sum_{k=0}^{n} \alpha_k t^k \geq 0 \quad \text{for} \quad t \geq 0$$

and

$$\int_0^\infty h(t)\, dt < 1.$$

It was reported that in applications to earthquakes occurrences these parametrisation produces more realistic models than (3.7).

For the description of the c.m.g.f. of (4.1) we consider a piecewise deterministic Markov process given by the linear stochastic differential system

$$dX_t = AX_t + b\, dN_t^c\,, \quad X_0 = 0, \tag{4.2}$$

where $X_t = (X_t^0,\ldots,X_t^n)'$, $b = (b^0,\ldots,b^n)'$ and A denote the matrix $(a_{ij};\ i,j = 0,1,\ldots,n)$. Assume the intensity of the counting process N under consideration is of the form

$$\lambda_t = \mu + \theta' X_t ,$$

$\theta \in \mathbb{R}^{n+1}$, such that $\theta' X_t \geq 0$ for $t \geq 0$.

Substituting $e^{v' X_t}$ into (2.2), $v \in \mathbb{R}^{n+1}$, we get the following equation for the unnormalized c.m.g.f. of X_t:

$$\begin{aligned} d\sigma_t(e^{v'X_t}) &= \sigma_t(v'AX_t e^{v'X_t} + \rho(v) e^{v'X_t})dt \\ &+ \sigma_{t-}(e^{v'X_{t-}}(\theta' X_{t-} + \mu - 1))d(N_t - t) \\ &= \sigma_t((v'A-\theta')X_t e^{v'X_t} + (\rho(v) + 1 - \mu) e^{v'X_t})dt \\ &+ \sigma_{t-}(\theta' X_{t-} e^{v'X_t} + (\mu - 1) e^{v'X_{t-}})dN_t \qquad (4.3) \end{aligned}$$

Setting $g(v,t) = \sigma_t(e^{v'X_t})$ and $\frac{\partial g}{\partial v} = \text{grad}_v g$, the stochastic equation (4.3) reduces to the linear p.d.e between the events τ_n of (N_t):

$$\frac{\partial g}{\partial t} = (v'A - \theta') \frac{\partial g}{\partial v} + (\rho(v) + \mu - 1) g(v,t)$$

together with the updating relations

$$g(v,\tau_n) = \theta' \frac{\partial g}{\partial v}(v,\tau_n -) + \mu g(v,\tau_n -).$$

When λ_t is given by (4.1), we have

$$A = \begin{pmatrix} -\beta & 1 & 0 & 0 & \dots & 0 \\ 0 & -\beta & 1 & 0 & \dots & 0 \\ \dots & \dots & \dots & \dots & \dots & \dots \\ 0 & \dots & \dots & 0 & -\beta & 1 \\ 0 & \dots & \dots & \dots & 0 & -\beta \end{pmatrix} \qquad b = \begin{pmatrix} \alpha_0 \\ 1!\alpha_1 \\ \vdots \\ n!\alpha_n \end{pmatrix}$$

and $\lambda_t = \mu + X_t^o$.

The c.m.g.f. of λ_t is

$$e^{s\lambda_t} = e^{s\mu} e^{se_o' X_t} = e^{s\mu} \frac{g(se_o,t)}{g(0,t)} ,$$

with $s \in \mathbb{R}$, $e_o = (1,0,\dots 0)'$ and the minimum variance estimator of λ_t is given by the expression

$$\hat{\lambda}_t = \mu + \hat{X}_t^o = \mu + \frac{\frac{\partial}{\partial s} g(se_1,t)|_{s=0}}{g(0,t)} .$$

REFERENCES

[1] Bartlett, M.S. (1964). The spectral analysis of two-dimensional point processes. Biometrika 51, 299-311.

[2] Boel, R.K. and Benes, V.E. (1980). Recursive Nonlinear Estimation of a Diffusion Acting as the Rate of an Observed Poisson Process. IEEE Trans.Inform.Theory IT-26/5, 561-575.

[3] Bremâud, P. (1981). Point Processes and Queues, Martingale Dynamics. Springer-Verlag, Berlin.

[4] Lawrence, A.J. (1972). Some models for stationary series of univariate events. In: P.A.W. Lewis (ed.), Stochastic Point Processes, Wiley, New York.

[5] Kavvas, M.L. and Delleur, J.W. (1981). A Stochastic Cluster Model of Daily Rainfall Sequences. Water Resour.Res.17/4, 1151-1160.

[6] v. Schuppen, J.H. (1977). Filtering, prediction and smoothing for counting process observations, a martingale approach. SIAM J.appl.Math 32/3.

[7] Smith, J.A. and Karr A.F. (1984). Statistical inference for point process models of rainfall. Water Resour.Res., to appear.

[8] Snyder, D.L. (1975). Random Point Processes. Wiley, New York.

[9] Vere-Jones, D. (1970). Stochastic models for earthquake occurrence (with discussion). J.R. Statist.Soc. B 32, 1-62.

[10] Ogata, Y. and Akaike, H. (1982). On the Linear Intensity Models for Mixed Doubly Stochastic Poisson and Self-exciting Point Processes. J.R. Statist.Soc.B 44/1, 102-107.

[11] Konecny, F. (1985). On the estimation of the stochastic intensity and the parameters of Neyman-Scott trigger processes (submitted to Mathem. Operationsfoschung und Statistik).

On the Manifold of Control Processes in Lagrange Problems

ANDRÁS KÓSA
Dept. of Applied Analysis, Eötvös Loránd University
1088 Budapest, Muzeum krt. 6-8.

The Morse theory treated in the book [1] turned out to be a highly efficient tool in the global study of finite dimensional extremum problems. In [2] an extension of this theory has been developed to find geodetic curves. In order to set up a similar theory for general Lagrange problems of the calculus of variations, first of all, an appropriate differential manifold structure must be given on the set of the admissible processes.

In this talk, to simplify the calculations, we consider the special case when the control differential equation is linear. We note, however, that the method applied generalizes considerably.

In what follows, an imbedded manifold structure will be constructed on the set of the admissible processes provided that the control differential equation has certain properties of reachability in system-theoretical sense. As an application of this construction to Lagrange problems, in [3] the homotopy type of the manifold of the admissible processes has been calculated. In [4] the same construction has been applied to system identification problems.

1. The set of the admissible processes

Suppose we are given $m,n \in \mathbb{N}$; $t_0, t_1 \in \mathbb{R}$, $t_0 < t_1$; $\xi_0, \xi_1 \in \mathbb{R}^n$ and the square-integrable functions

$$A:[t_0,t_1] \to \mathbb{R}^{n\times n} \quad , \quad B:[t_0,t_1] \to \mathbb{R}^{m\times n} \quad .$$

Denote simply by H_k^1 , resp. L_k^2 the Sobol'ev space $H_k^1[t_0,t_1]$, resp. the space $L_k^2[t_0,t_1]$ of the square-integrable functions with values in $\mathbb{R}^k$ $(k \in \mathbb{N})$. Then define the set M of the admissible function as follows:

$$M:=\{(x,u) \in H_n^1 \times L_m^2 \mid \dot{x}=Ax+Bu \ , \ x(t_0)= \xi_0, \ x(t_1)= \xi_1\}$$

If we define the mapping

$$F:H_n^1 \times L_m^2 \to L_n^2 \times \mathbb{R}^n \times \mathbb{R}^n \quad ,$$

$$(x,u) \mapsto (\dot{x}-Ax-Bu \ , \ x(t_0)- \xi_0 \ , \ x(t_1)- \xi_1) \quad ,$$

we have

$$M=\{(x,u) \in H_n^1 \times L_m^2 \mid F(x,u)=0\} .$$

2. Introduction of the manifold structure

Denote by $\Phi:[t_o,t_1] \times [t_o,t_1] \to \mathbb{R}^{n\times n}$ the fundamental matrix of the equation $\dot{x}=Ax$ and define the Kalman matrix

$$W:= \int_{t_o}^{t_1} \Phi(t_1,\cdot)\, BB^T\, \Phi^T(t_1,\cdot)$$

Theorem. Suppose that the matrix W is invertible. Then $M \subset H_n^1 \times L_m^2$ is an imbedded manifold.

Proof. According to the standard construction well-known in differential geometry, the proof consists of the following steps:

1. Verifying the conditions

a/ the mapping F is continuously differentiable,

b/ for each $(x,u) \in M$ the range of the operator $F'(x,u)$ coincides with the whole space $L_n^2 \times \mathbb{R}^n \times \mathbb{R}^n$.

2. Giving a continuous projection of the space $H_n^1 \times L_m^2$ onto the kernel $N[F'(x,u)]$ explicitely.

Now we verify the given conditions.

1° Condition a/ can be verified immediately. For the proof of b/ consider the derivative

$$F'(x,u)(h,v)=(\dot{h}-Ah-Bv\ ,\ h(t_o),\ h(t_1)) \qquad ((x,u) \in H_n^1 \times L_m^2\ ).$$

/Observe that the derivative $F'(x,u)(h,v)$ does not depend on the process (x,u)./ Pick $(\ell,\eta_o,\eta_1) \in L_n^2 \times \mathbb{R}^n \times \mathbb{R}^n$ arbitrarily. Then the solvability of the equation

$$F'(x,u)(h,v)=(\ell,\eta_o,\eta_1)$$

is equivalent to the fact that the boundary value problem

$$\dot{h}=Ah+Bv+\ell \quad ,\ h(t_o)=\eta_o\ ,\ h(t_1)=\eta_1$$

can be solved for same $v \in L_m^2$. Denote by h_o the solution of the initial value problem

$$\dot{h}=Ah+\ell \quad ,\ h(t_o)=\eta_o \ .$$

Then, for the difference $h_1:=h-h_o$ we have

$$\dot{h}_1=Ah_1+Bv\ ,\ h_1(t_o)=0\ ,\ h_1(t_1)=\eta_1-h_o(t_1)=:\eta \quad , \tag{1}$$

that is, condition b/ is satisfied if for any $\eta \in \mathbb{R}^n$ there exists a control $v \in L_m^2$ such the boundary value problem (1) can be solved for h_1. The latter condition is equivalent to the global reachability of the system $\dot{h}_1=Ah_1+Bv$ from zero over the interval $[t_o,t_1]$. It is well-known /see [5]/ that for this a necessary and sufficient condition is the invertibility of the Kalman matrix W. So, under the hypothesis of our theorem, condition b/ also satisfies.

2° For the construction of a projection onto the kernel

$$N:=N[F'(x,u)]$$

observe that any pair $(h,v) \in H_n^1 \times L_m^2$ belongs to the kernel N if and only if

$$\dot{h}=Ah+Bv \ , \ h(t_o)=0 \ , \ h(t_1)=0 \ . \tag{2}$$

Using the fundamental matrix Φ define a linear continuous operator

$$R:L_m^2 \to H_n^1 \ , \ (Rv)(t):=\int_{t_o}^{t} \Phi(t,\cdot) \ Bv \qquad (t \in [t_o,t_1]).$$

Then, for the solution h of the problem (2) corresponding to the control v, we have

$$h=Rv \ , \ h(t_1)=\int_{t_o}^{t_1} \Phi(t_1,\cdot) \ Bv=0 \ .$$

Thus, a function $v \in L_m^2$ is an admissible control for the problem (2) if and only if it is orthogonal to every row of the matrix $\Phi(t_1,\cdot) \ B$. Denote by L_1 the subspace spanned by these rows in L_m^2. It is clear that the Kalman matrix W is nothing else that the Gram matrix of the column vectors of the matrix $\Phi(t_1,\cdot) \ B$. This way, the invertibility of the matrix W implies that $\dim L_1=n$.

Now we construct a continuous projection P of $H_n^1 \times L_m^2$ onto N. To this end define the continuous linear operator

$$Q:L_m^2 \to L_m^2 \ ,$$
$$v \mapsto v-B^T\Phi^T(t_1,\cdot) \ W^{-1} \int_{t_o}^{t_1} \Phi(t_1,\cdot) \ Bv \ .$$

Denote by $L_1^{\perp}$ the orthogonal complement of L_1. It is easy to see that range $Q=L_1^{\perp}$ and the relations

$$Q(v)=v \quad (v \in L_1^{\perp}) \ , \quad Q(v)=0 \quad (v \in L_1)$$

hold. Define now P in the following way

$$P:H_n^1 \times L_m^2 \to H_n^1 \times L_m^2 \quad , \quad (h,v) \mapsto (RQv, \ Qv) \ .$$

Since, by the definition of the operator R,

$$N=\{(h,v) \in H_n^1 \times L_m^2 \mid h=Rv \ , \ v \in L_1\} \ ,$$

it is easy to see that P is the required projection.

Let us construct now a local chart at the fixed point (x,u).

Define the continuously differentiable mapping

$$\Psi : H_n^1 \times L_m^2 \to N \times (L_n^2 \times \mathbb{R}^n \times \mathbb{R}^n) \ , \ (h,v) \mapsto (P(h-x,\ v-u),\ F(h,v)) .$$

Clearly, $G(x,u)=0$. We shall prove that the operator

$$\Psi' : H_n^1 \times L_m^2 \to N \times (L_n^2 \times \mathbb{R}^n \times \mathbb{R}^n)$$

is an isomorphism. By the Banach isomoprhism theorem it is enough to prove that $N[\Psi'(x,u)] = 0$ and the operator is onto. Since $\Psi'(x,u) = (P, F'(x,u))$, from the properties of the projection P , the injectivity of the derivate $\Psi'(x,u)$ follows. Choose now an arbitrary

$$((y,w),\ (\ell,\eta_0,\eta_1)) \in N \times (L_n^2 \times \mathbb{R}^n \times \mathbb{R}^n) \quad .$$

Then, by the condition b/ there exists a pair $(\bar{h},\bar{v}) \in H_n^1 \times L_m^2$ such that $F'(x,u)(\bar{h},\bar{v}) = (\ell,\eta_0,\eta_1)$. Put $(z,s) := (y,w) - P(\bar{h},\bar{v})$ then

$$\Psi'(x,u)(\bar{h}+z,\ \bar{v}+s) = ((y,w),(\ell,\eta_0,\eta_1)) \ .$$

Now apply the inverse function theorem to the mapping Ψ at the point (x,u). Then there exist a neighbourhood U of the point (x,u) in the space $H_n^1 \times L_m^2$ and a neighbourhood $V \times W$ of the point $(0,0) \in N \times (L^n \times \mathbb{R}^n \times \mathbb{R}^n)$ for which the mapping

$$\Psi|_U : U \to V \times W$$

is a diffeomorphism. Let

$$p := \Psi|_{U \cap M} \quad ,$$

then

$$p(h,v) = (P(h-x,\ v-u),\ 0) \qquad ((h,v) \in U \cap M) \ .$$

Identify now the points

$$((h,v),\ 0) \in N \times (L_n^2 \times \mathbb{R}^n \times \mathbb{R}^n)$$

and

$$(h,v) \in N \ .$$

Then p assumes its values from the kernel N of operator $F'(x,u)$. From the definition of the mappings Ψ and p it follows that $p(U \cap M)$ is open in N and the map p is a homeomoprhism. So we obtained that

$$(U \cap M \ ,\ p)$$

is a local chart at the point (x,u) with the coordinate space N .

Since the point $(x,u) \in M$ was arbitrary, we have defined a chart at each point of M .

Fix now two arbitrary points (x_1, u_1) and (x_2, u_2) of M and carry out the above construction in both cases. Denote the resulting objects by the same symbols as above distinguishing them by indices. So we obtain charts

$$(U_1 \cap M\ ,\ p_1\ ),\ (U_2 \cap M, p_2) \tag{3}$$

at the points (x_1, u_1) and (x_2, u_2) respectively. Suppose that $U_1 \cap U_2 \neq \emptyset$. The mapping

$$\Psi_1 \circ \Psi_2^{-1}$$

is a diffeomorphism and so is the mapping

$$p_1 \circ p_2^{-1}\ .$$

Indeed, $p_1 \circ p_2^{-1}$ can be obtained from the diffeomorphism $\Psi_1 \circ \Psi_2^{-1}$ of type

$$N \times (L_n^2 \times \mathbb{R}^n \times \mathbb{R}^n) \to N \times (L_n^2 \times \mathbb{R}^n \times \mathbb{R}^n)$$

by restriction to a function of type

$$N \to N$$

while differentiability is preserved. Therefore the charts (3) are compatible.

Theorem is completely proved.

References

[1] J. Milnor, Morse Theory, Princeton University Press /Princeton, R. J., 1963/

[2] J. T. Schwartz, Nonlinear Functional Analysis, Gordon and Breach /New York-London-Paris, 1969/

[3] A. Kósa, On the homotopy type of some spaces occurring in the calculus of variations, Acta Sci. Math. Szeged, 45 /1983/, 273-277.

[4] A.S.A.R. Al-Janabi, On Volterra equations and identification of systems, Zeitschrift für Angewandte Mathematik und Mechanik, 63, No 4/5, 1983, P. 396-397.

[5] R. E. Kalman, P. L. Falb., M. A. Arbib, Topics in mathematical system theory, McGraw-Hill Book Company /New York-San Francisco--St. Louis-Toronto-London-Sydney, 1969/

On the Numerical Solution of Certain Time-Minimal Control Problems of Second Order

W. Krabs, Fachbereich Mathematik der Technischen Hochschule Darmstadt,
Schlossgartenstr. 7, 6100 Darmstadt, West Germany

ABSTRACT

This paper is concerned with the problem of time-minimal null-controllability of vibrating systems described by an abstract wave equation via distributed controls of bounded L^2-norm.

For the approximate solution of this problem Galerkin's method is applied and numerically demonstrated for the problem of time-minimal null-control of a rectangular vibrating plate.

1. THE PROBLEM AND BASIC RESULTS

As in [1] and [3] the following abstract process of vibrations is investigated: Let $y : [0,T] \to H$, for any $T > 0$, be a function that describes the deviation of a vibrating medium from the position of rest as a function of time t with values in a (finite- or infinite-dimensional) Hilbert space H. We assume y to satisfy an abstract wave equation of the form

$$\ddot{y}(t) - Ay(t) = f(t), \quad t \in (0,T), \tag{1.1}$$

where $\dot{y}$ denotes the derivative with respect to t, A is a self adjoint positive definite linear operator defined on a dense domain D(A) in H and $f(t) \in H$ for almost all $t \in [0,T]$, $\|f(.)\|_H$ is measurable and satisfies $\int_0^T \|f(t)\|_H^2\, dt < \infty$ where $\|.\|_H$ denotes the norm in H. The space of all (classes of) such functions is called $L^2([0,T], H)$. Let $N = \dim H$. In addition to the above requirements we assume that A has a complete sequence $(\varphi_j)_{j=1,\dots,N}$ of orthonormal eigenelements $\varphi_j \in D(A)$ and a corresponding sequence $(\lambda_j)_{j=1,\dots,N}$ of eigenvalues λ_j of finite multiplicity with $0 < \lambda_1 \le \lambda_2 \le \dots$ and $\lim_{j\to\infty} \lambda_j = \infty$, if $N = \infty$. Then it follows that

$$D(A) = \{ v \in H \mid \sum_{j=1}^{N} \lambda_j^2 \langle v,\varphi_j\rangle_H^2 < \infty \}$$

and

$$Av = \sum_{j=1}^{N} \lambda_j \langle v,\varphi_j\rangle_H \, \varphi_j \quad \text{for all } v \in D(A).$$

Furthermore there exists a unique "square root" $A^{1/2}$ of A with domain

$$D(A^{1/2}) = \{ v \in H \mid \sum_{j=1}^{N} \lambda_j <v,\varphi_j>_H^2 < \infty \}$$

which is defined by

$$A^{1/2} v = \sum_{j=1}^{N} \lambda_j^{1/2} <v,\varphi_j>_H \varphi_j \quad \text{for all } v \in D(A^{1/2}).$$

Let $y_0 \in D(A^{1/2})$ and $\dot{y}_0 \in H$ be given. Then we require initial conditions for $t = 0$ to be given by

$$y(0) = y_0 \quad \text{and} \quad y(0) = y_0 \tag{1.2}$$

We put $V = D(A^{1/2})$ provided with the scalar product

$$<v,w>_V = \sum_{j=1}^{N} \lambda_j <v,\varphi_j>_H <w,\varphi_j>_H, \quad v, w \in V.$$

Then V becomes a separable Hilbert space which is continuously and densely imbedded into H because of $D(A) \subset V \subset H$ and

$$\|v\|_V \geq \lambda_1^{1/2} \|v\|_H \quad \text{for all } v \in V.$$

The dual space V^* of V consists of all linear functionals $v^* : V \to \mathbb{R}$ such that

$$\sum_{j=1}^{N} \frac{1}{\lambda_j} v^*(\varphi_j)^2 < \infty$$

and

$$v^*(v) = \sum_{j=1}^{N} <v,\varphi_j>_H v^*(\varphi_j), \quad v \in V.$$

If we identify H with its dual space, we obtain the following chain of continuous and dense imbeddings: $V \subset H \subset V^*$.

If we define a linear mapping $\tilde{A} : V \to V^*$ by

$$(\tilde{A}v)(w) = <A^{1/2} v, A^{1/2} w>_H = <v,w>_V$$

for all $v, w \in V$, then we have

$$(\tilde{A}v)(v) = \|v\|_V^2 \quad \text{for all } v \in V$$

and all the assumptions of Theorem 1.1 in Chapter IV of [5] are satisfied. This in connection with Remark 1.3 loc. cit. implies the existence of exactly one weak solution $y : [0,T] \to H$ of (1.1), (1.2) for every choice of $f \in L^2([0,T], H)$, $y_0 \in V =$

$D(A^{1/2})$ and $\dot{y}_0 \in H$ in the following sense:

1) $y \in C([0,T],V)$, $\dot{y}(t)$ exists for all $t \in [0,T]$ in the strong sense and $\dot{y} \in C([0,T],H)$ where, for an arbitrary Hilbert space Z, $C([0,T],Z)$ is the space of continuous functions from $[0,T]$ into Z.

2) $\lim_{t\to 0+} \|y(t) - y_0\|_V = 0$ and $\lim_{t\to 0+} \|\dot{y}(t) - \dot{y}_0\|_H = 0$.

3) The second derivative $\ddot{y}$ in the sense of distributions can be identified with a function in $L^2([0,T],V^*)$ and

$$\ddot{y}(t)(v) + \tilde{A}y(t)(v) = \langle f(t),v\rangle_H \tag{1.3}$$

is satisfied for all $v \in V$ and almost all $t \in (0,T)$. This weak solution can be explicitly expressed with the aid of sine and cosine operators (see [1] and [3]).

The main concern of this paper is the

PROBLEM OF TIME-MINIMAL NULL-CONTROLLABILITY

Given $(y_0,\dot{y}_0) \in V \times H$ and some $M > 0$.

a) Does there exist a time $T > 0$ and a control function $f \in L^2([0,T],H)$ with

$$\|f\|_{2,T} = \left(\int_0^T \|f(t)\|_H^2\, dt\right)^{1/2} \le M \tag{1.4}$$

such that the weak solution $y : [0,T] \to H$ of (1.1), (1.2) satisfies

$$y(T) = \Theta_V \text{ and } \dot{y}(T) = \Theta_H \tag{1.5}$$

with Θ_V and Θ_H being the zero element of V and H, respectively?

b) If this is possible, then the infimum $T(M)$ of all such times $T > 0$ is well defined and the question arises whether there exists a time-minimal control function $f \in L^2([0,T(M)],H)$ with (1.4) for $T = T(M)$ such that the corresponding weak solution $y : [0,T(M)] \to H$ satisfies (1.5) for $T = T(M)$.

A further question is the uniqueness of time-minimal control functions and how they can be characterized.

In [3] we have proved the following results:

1) For every $T > 0$ there exists exactly one control function $f_T \in L([0,T],H)$ with least norm such that the corresponding weak solution $y : [0,T] \to H$ of (1.1), (1.2) satisfies (1.5). Moreover, the norm of f_T can be estimated in the form

$$\|f_T\|_{2,T}^2 \le \frac{8\sqrt{\lambda_1}}{T\sqrt{\lambda_1}-1}\left(\|y_0\|_V^2 + \|\dot{y}_0\|_H^2\right), \tag{1.6}$$

if $T > 1/\sqrt{\lambda_1}$. This estimate implies

$$\lim_{T\to\infty} \|f_T\|_{2,T} = 0$$

so that part a) of the problem of time-minimal null-controllability has a solution.

2) For every $M > 0$ the least norm control $f_{T(M)} \in L^2([0,T(M)],H)$ in 1) is the unique time-minimal control function and is characterized by

$$\|f_T\|_{2,T} = M \iff T = T(M). \tag{1.7}$$

2. GALERKIN'S METHOD

From the results in [3], in principle, for each $T > 0$ the least norm control $f_T \in L^2([0,T],H)$ that makes the corresponding weak solution $y : [0,T] \to H$ of (1.1), (1.2) satisfy (1.5) can be explicitly computed with the aid of the eigenvalues and eigenelements of the operator A so that $\|f_T\|_{2,T}$ can also be computed. Hence the determination of the time-minimal control $f_{T(M)}$ requires the solution of the left equation in (1.7).

In most cases, however, the eigenvalues and eigenelements of A are not readily available so that this method cannot be applied. We therefore use Galerkin's method in order to obtain approximate time-minimal control functions. This method has also been used in [5] for proving the existence of weak solutions of (1.1), (1.2) in a more general situation than described above.

In order to apply Galerkin's method we choose an orthonormal base $(v_j)_{j=1,\dots,N}$ in V (which could be $(\varphi_j)_{j=1,\dots,N}$, if known explicitly) (observe that N = dim H) we define

$$Y_n = \{y_n = \sum_{j=1}^{n} y_j^n v_j \mid y_j^n \in H^2[0,T],\ \ j = 1,\dots,n\}$$

which is an n-dimensional subspace of $H^2([0,T],V)$. For a given $n \in \{1,\dots,N\}$ we replace the weak equation (1.3) by

$$\langle \ddot{y}_n(t),v^n\rangle_H + \langle A^{1/2}y_n(t),\ A^{1/2}v^n\rangle_H = \langle f(t),v^n\rangle_H \tag{1.3_n}$$

for all

$$v^n \in V_n = \{\sum_{j=1}^{n} c_j^n v_j \mid c_j^n \in \mathbb{R},\ \ j = 1,\dots,n\}$$

and all $t \in (0,T)$ which is equivalent to

$$\ddot{y}_j^n(t) + \sum_{k=1}^{n} a_{jk}^n y_j^n(t) = f_j^n(t) \tag{2.1}$$

for all $j = 1,\dots,n$ and all $t \in (0,T)$ where

$$a_{jk}^n = \langle A^{1/2}v_j, A^{1/2}v_k\rangle \quad \text{for all } j, k \in \{1,\dots,n\}$$

and

$$f_j^n(t) = \langle f(t), v_j\rangle_H \text{ for all } j \in \{1,\dots,n\}.$$

The initial conditions (1.2) are replaced by

$$y_j^n(0) = y_{0j}^n \quad \text{and} \quad \dot{y}_j^n(0) = \dot{y}_{0j}^n \quad \text{for all } j = 1,\dots,n \tag{2.2}$$

where

$$y_{0j}^n = \langle y_0, v_j\rangle_H \text{ and } \dot{y}_{0j}^n = \langle y_{0j}, v_j\rangle_H$$

for $j = 1,\dots,n$.

In vector-matrix notation (2.1) and (2.2) can be rewritten as

$$\ddot{y}^n(t) + A^n y^n(t) = f^n(t), \quad t \in (0,T), \tag{1.1$_n$}$$

and

$$y^n(0) = y_0^n \text{ and } \dot{y}^n(0) = \dot{y}_0^n \tag{1.2$_n$}$$

where $A^n = (a_{jk}^n)$ is a symmetric and positive definite matrix which defines a self adjoint positive definite linear operator A^n mapping $D(A^n) = \mathbb{R}^n$ onto $\mathbb{R}^n$ and $f^n \in L^2([0,T], \mathbb{R}^n)$. The Operator A^n has a complete sequence $(\varphi_j^n)_{j=1,\dots,n}$ of orthonormal eigenvectors $\varphi_j^n \in \mathbb{R}^n$ and corresponding eigenvalues λ_j^n with $0 < \lambda_1^n \le \lambda_2^n \le \dots \le \lambda_n^n$. So we are in the same situation as at the beginning of Section 1 with $A = A^n$, $H = \mathbb{R}^n$ and $\dim H = N = n$. In this case we have $V = V^* = H = \mathbb{R}^n$ and weak solutions of $(1.1)_n$ are vector functions $y : [0,T] \to \mathbb{R}^n$ such that $\dot{y}(t)$ exists for all $t \in [0,T]$, the vector function $\dot{y} : [0,T] \to \mathbb{R}^n$ is absolutely continuous and $(1.1)_n$ holds for almost all $t \in (0,T)$. Using the results of [3] for $H = \mathbb{R}^n$, for every choice of $f^n \in L^2([0,T], \mathbb{R}^n)$ and $y_0^n, \dot{y}_0^n \in \mathbb{R}^n$, the unique weak solution of $(1.1)_n$, $(1.2)_n$ can be written down as

$$y^n(t) = C(t)y_0^n + S(t)\dot{y}_0^n + \int_0^t S(t-s)f^n(s)\, ds, \tag{2.3}$$

$t \in [0,T]$, where, for every $v \in \mathbb{R}^n$,

$$C(t)v = \sum_{j=1}^{n} \cos \sqrt{\lambda_j^n}t <v,\varphi_j^n>_{\mathbb{R}^n} \varphi_j^n$$

and

$$S(t)v = \sum_{j=1}^{n} \frac{1}{\sqrt{\lambda_j^n}} \sin \sqrt{\lambda_j^n}t <v,\varphi_j^n>_{\mathbb{R}^n} \varphi_j^n$$

are the so called cosine and sine operators of $\mathbb{R}^n$ into $\mathbb{R}^n$, respectively. The strong derivative of $y^n = y^n(t)$ reads

$$\dot{y}^n(t) = - AS(t)y_0^n + C(t)\dot{y}_0^n + \int_0^t C(t-s)f^n(s)ds. \tag{2.4}$$

The problem of time-minimal null-controllability as being formulated in Section 1 is formulated here in the same way with (1.1) and (1.2) being replaced by $(1.1)_n$ and $(1.2)_n$, respectively, (1.4) by

$$\|f^n\|_{2,T} = (\int_0^T <f^n(t),f^n(t)>_{\mathbb{R}^n} dt)^{1/2} \leq M \tag*{$(1.4)_n$}$$

and (1.5) by

$$y^n(T) = \Theta_n \text{ and } \dot{y}^n(T) = \Theta_n \tag*{$(1.5)_n$}$$

with Θ_n being the zero vector of $\mathbb{R}^n$.

Let $T_n(M)$ be the infimum of all times $T > 0$ such that a control function $f^n \in L^2([0,T], \mathbb{R}^n)$ with $(1.4)_n$ exists whose corresponding weak solution (2.3) of $(1.1)_n$, $(1.2)_n$ satisfies $(1.5)_n$. If $f^n_{T_n(M)} \in L^2([0,T_n(M)], \mathbb{R}^n)$ is the least norm control function such that the corresponding weak solution (2.3) of $(1.1)_n$, $(1.2)_n$ satisfies $(1.5)_n$ we have, by the results given in Section 1, $\|f^n_{T_n(M)}\|_{2,T_n(M)} = M$ and we take $T_n(M)$ as an approximation of $T(M)$ and

$$f_{T_n(M)}(t) = \sum_{j=1}^{n} <f^n_{T_n(M)},v_j>v_j \tag{2.5}$$

as an approximation of the time-minimal control $f_{T(M)} \in L^2([0,T(M)],H)$ we are looking for.

In order to find $f^n_{T_n(M)}$ we have to determine, for an arbitrary time $T > 0$, the least norm control function $f_T^n \in L^2([0,T], \mathbb{R}^n)$ such that the corresponding weak solution (2.3) of $(1.1)_n$, $(1.2)_n$ satisfies $(1.5)_n$.

According to [3] we can proceed as follows: We assume f_T^n to be of the form

$$f_T^n(t) = S(T-t)f_1 + C(T-t)f_2 \tag{2.6}$$

and have to determine

$$f_1 = \sum_{j=1}^{n} x_j^1 \varphi_j^n \quad \text{and} \quad f_2 = \sum_{j=1}^{n} x_j^2 \varphi_j^n, \tag{2.7}$$

$x_j^1, x_j^2 \in \mathbb{R}$ for $j = 1,\dots,n$, such that the equations

$$\int_0^T S(T-t)f^n(t)dt = C(T)y_0^n - S(T)\dot{y}_0^n,$$

$$\int_0^T C(T-t)f^n(T)dt = AS(T)y_0^n - C(T)\dot{y}_0^n$$

are satisfied which are equivalent to $(1.5)_n$. This leads to the solution n linear 2×2-systems from which x_j^1 and x_j^2 for $j = 1,\dots,n$ can be obtained in terms of explicit formulas. Insertion into (2.7), (2.6) then gives the desired least norm control $f_T^n \in L^2([0,T], \mathbb{R}^n)$. Moreover we obtain the estimate (see (1.6))

$$\| f_T^n \|_{2,T}^2 \le \frac{8\sqrt{\lambda_1^n}}{T\sqrt{\lambda_1^n}-1} \sum_{j=1}^{n} \lambda_j^n \langle y_0^n, \varphi_j^n \rangle_{\mathbb{R}^n}^2 + \langle \dot{y}_0^n, \varphi_j^n \rangle_{\mathbb{R}^n}^2, \tag{2.8}$$

if $T > 1/\sqrt{\lambda_1^n}$.

In order to determine the time-minimal control $f_{T_n(M)}^n$ to be inserted into (2.5) we have to find $T = T_n(M)$ such that $\| f_T^n \|_{2,T} = M$.

In view of $(1.4)_n$ and (2.8) we can assume that $T_n(M) \in [0, T_0^n]$ with

$$T_0^n = \frac{1}{\sqrt{\lambda_1^n}} \left[\frac{8\sqrt{\lambda_1^n}}{M} \left(\sum_{j=1}^{n} \lambda_j^n \langle y_0^n, \varphi_j^n \rangle_{\mathbb{R}^n}^2 + \langle \dot{y}_0^n, \varphi_j^n \rangle_{\mathbb{R}^n}^2 \right) + 1 \right]. \tag{2.9}$$

For finding the solution of the equation

$$g(T) = \| f_T^n \|_{2,T}^2 - M^2 = 0$$

in $[0, T_0^n]$ it is reasonable to use the secant method beginning with $T_0 = T_0^n$, $T_1 = \alpha T_0^n$ for some $\alpha \in (0,1)$.

In concluding this section we remark that Galerkin's method has also been applied to parabolic time optimal control problems (see [2] and [4]).

3. APPLICATION TO A VIBRATING PLATE

Let $\Omega \subset \mathbb{R}^2$ be a bounded domain with boundary $\partial\Omega$. Then $\bar{\Omega} = \Omega \cup \partial\Omega$ is considered as the rest configuration of a vibrating plate whose deviation is given as a function $y = y(x,t)$, $x \in \bar{\Omega}$, $t \in [0,T]$ for any $T > 0$ and is assumed to be governed by the wave equation

$$y_{tt}(x,t) + \Delta^2 y(x,t) = f(x,t), \qquad x \in \Omega, \quad t \in (0,T) \tag{3.1}$$

where $\Delta^2 = \Delta \circ \Delta$, Δ = Laplacian operator, $f \in L^2(\bar{\Omega} \times [0,T])$ and submitted to boundary conditions

$$y(x,t) = \Delta y(x,t) = 0, \qquad x \in \partial\Omega, \quad t \in [0,T]. \tag{3.2}$$

If we define

$$D(A) = \{z \in H_0^1(\Omega) \mid \Delta z \in H_0^1(\Omega), \quad \Delta^2 z \in L^2(\Omega)\}$$

and

$$A : D(A) \to H = L^2(\Omega) \quad \text{by} \quad Az = \Delta^2 z,$$

then we obtain a self adjoint, positive definite operator that possesses a complete sequence $(\varphi_j)_{j \in \mathbb{N}}$ of orthonormal eigenfunctions $\varphi_j \in D(A)$ and corresponding sequence $(\lambda_j)_{j \in \mathbb{N}}$ of eigenvalues λ_j of finite multiplicity with

$$0 < \lambda_1 \le \lambda_2 \le \ldots \quad \text{and} \quad \lim_{j \to \infty} \lambda_j = \infty.$$

The wave equation (3.1) and the boundary conditions can be combined to one abstract wave equation of the form (1.1) and all the requirements at the beginning of Section 1 are met.

In addition the square root of A has the domain

$$D(A^{1/2}) = \{v \in H_0^1(\Omega) \mid \Delta v \in L^2(\Omega)\}$$

and is defined as $A^{1/2}v = -\Delta v$. Initial conditions are therefore given in the form

$$y(x,0) = y_0(x), \quad y_t(x,0) = \dot{y}_0(x), \quad x \in \Omega, \tag{3.3}$$

where $y_0 \in H_0^1(\Omega)$ with $\Delta y_0 \in L^2(\Omega)$ and $\dot{y}_0 \in L^2(\Omega)$.

A NUMERICAL EXAMPLE

Let $\Omega = (0,1) \times (0,1)$. Then the eigenvalues of $A = \Delta^2$ read

$$\lambda_{i,k} = \pi^4 (i^2 + k^2)^2 \quad \text{for } i, k \in \mathbb{N}$$

and the corresponding normalized eigenfunctions are given by

$$\varphi_{i,k}(x_1,x_2) = 2 \sin i\pi x_1 \cdot \sin k\pi x_2$$
$$x_1, x_2 \in [0,1], \quad i, k \in \mathbb{N}.$$

We assume initial conditions (3.3) with

$$y_0(x_1,x_2) = x_1 x_2 (1-x_1)(1-x_2),$$
$$\dot{y}_0(x_1,x_2) = 0 \quad \text{for} \quad x_1, x_2 \in (0,1).$$

As an orthonormal base for $V = D(A^{1/2})$ in Galerkin's method we choose first

$$v_j(x_1,x_2) = \varphi_{i,k}(x_1,x_2) \quad \text{with}$$
$$j = i + (k-1) \cdot m \quad \text{for} \quad i = 1,\dots,m, \quad k = 1,\dots,m$$

with $m \in \mathbb{N}$. Then $n = m^2$.

The following table gives the values of $T_n(M)$ for n = 1, 4, 9, 16 and M = 0.1, 1, 10, 100:

M \ n	1	4	9	16
0.1	86.33869	86.33869	92.25277	92.25277
1	0.87079	0.87079	0.90327	0.90327
10	0.04823	0.04823	0.04841	0.04841
100	0.01096	0.01096	0.01097	0.01097

Second we choose the orthonormal sequence of polynomials $v_j = v_j(x_1,x_2)$, $j = 1,\dots,n = m^2$, for some $m \in \mathbb{N}$ generated from the sequence

$$w_{i,k}(x_1,x_2) = x_1^i x_2^k (1-x_1)^i (1-x_2)^k,$$
$$i,k = 1,\dots,m,$$

by the method of Gram-Schmidt.

The following table gives the values of $T_n(M)$ for n = 1, 4, 9, 16 and M = 0.1, 1, 10, 100:

M \ n	1	4	9	16
0.1	97.75499	97.78098	97.78848	97.77813
1	0.97680	1.00054	1.00030	1.00058
10	0.04799	0.04860	0.04858	0.04859
100	0.01097	0.01097	0.01097	0.01097

These results show that the second choice of $(v_j)_{j=1,\ldots,n}$ gives a better approximation of T(M) by $T_{16}(M)$ than the first choice, in particular for small values of M. For $M \geq 10$ the difference of the results is acceptable.

REFERENCES

[1] Fattorini, H.O.: The Time Optimal Problem for Distributed Control of Systems Described by the Wave Equation. In: Aziz, A.K., Wingate, J.W., Balas, M.J. (eds.): Control Theory of Systems Governed by Partial Differential Equations. Academic Press, New York, San Francisco, London, 1977.

[2] Knowles, G.: Finite Element Approximation of Parabolic Time Optimal Control Problems. SIAM J. of Control and Optimization 20 (1982), 414-427.

[3] Krabs, W.: On Time-Minimal Distributed Control of Vibrating Systems Governed by an Abstract Wave Equation. Appl. Math. and Optim. 13 (1985), 137-149.

[4] Lasiecka, I.: Ritz-Galerkin Approximation of the Time Optimal Boundary Control Problem for Parabolic Systems with Dirichlet Boundary Conditions. SIAM J. of Control and Optimization 22 (1984), 477-500.

[5] Lions, L.J.: Optimal Control of Systems Governed by Partial Differential Equations. Springer Verlag: Berlin, Heidelberg, New York, 1971.

NASH GAME WITH REGULAR POLYNOMIAL PERTURBATION

Maciej Krawczak

Systems Research Institute
Polish Academy of Sciences
ul. Newelska 6
01 - 447 Warsaw, POLAND

1. Introduction

Very little work has been reported about approximation techniques for determining sub-optimal strategies of differential game problems. The Nash equilibrium solution of a two-person, nonzero-sum differential game can be expressed in terms of the solution of a nonlinear two-point boundary value problem. In [6], the solution of the Riccati matrix differential equations, associated with a linear-quadratic Nash game, is assumed to be an analitic function of a small parameter. In [11], the Newton-Raphson method in function space and the gradient optimization of the Hamiltonian function with respect to strategies in an almost linear pursuit-evasion game (a case of Nash game) with a quadratic criterion is investigated. In other papers e.g. [5] and [7] one can find solutions of only very specialized games and first of all for the open-loop case.

In this paper we consider a continuous-time, two-person, deterministic differential game with a nonlinear state equation and two quadratic cost functionals. The nonlinearity of the state equation appears as a regular perturbation. The perturbation term is a polynomial in the state of order higher than one. The presented method is based on the classical perturbation method [2]. It is assumed that two co-states, appearing in the two-point boundary value problem, can be expanded in the form of power series in the small perturbation parameter. The method requires to solve the coupled Riccati matrix differential equations, known from linear-quadratic Nash differential games, and a sequence, up to some order, of quasi-linear partial differential equations. Under the assumption that the perturbation term is a polynomial the problem of solving these quasi-linear equations is reduced to find coefficients of appropriate polynomials. These coefficients are solutions of a series of corresponding sets of ordinary differential equations.

2. Problem statement

There is a two-person, nonlinear differential game described by

$$\dot{x} = A(t) + \varepsilon f(x,t) + B_1(t)u_1 + B_2(t)u_2 \tag{1}$$

$$x(0) = x_o$$

where x is an n-vector, u_1 and u_2 are m_1- and m_2-vector strategies of Player 1 and Player 2, respectively. $f(x,t)$ is a vector-valued polynomial in x such that $f(0,0)=0$. The matrices $A(t)$, $B_1(t)$, $B_2(t)$ are piecewice continuous of consistent dimensions. ε is a small, the so-called, perturbation parameter. The review of differential game models [5] justifies the choice of the considered model (1).

The objective for Player i, i=1,2, is to choose the strategy $u_i(x,t)$ in such a way to minimize the following performance index

$$J_i(u_1,u_2) = \frac{1}{2}x'(T)Q_{if}x(T) + \frac{1}{2}\int_o^T [x'Q_i(t)x + u_1'R_{ij}(t)u_1 + u_2'R_{ij}(t)u_2]dt \tag{2}$$

where $i,j=1,2$, $i \neq j$, the terminal time T is fixed, Q_{if} is a positive semidefinite symmetric matrix, Q_i is a positive - semidefinite symmetric matrix with elements piecewise contonuous in t, R_{ii} is a positive - definite and R_{ij} $(i \neq j)$ a positive-semidefinite symmetric matrix with elements piecewise continuous in t.

In a game, where cooperation between players is impossible to be accepted the players are interested in solution which have the Nash equilibrium property [12]:

<u>Definition 2.1</u>

<u>A strategy set</u> $\{u_1^*, u_2^*\}$ <u>is called a Nash equilibrium strategy set if</u>

$$J_i(u_i^*,u_j^*) \le J_i(u_i,u_j^*) \tag{3}$$

$i,j=1,2$, $i \neq j$.

The Nash solution is secure against any attempt by one player to unilaterally change his strategy, since that player can only lose by deviating from his equilibrium strategy.

The necessary conditions for the feedback Nash strategies have been derived [12] using an extension of the variational methods used in the optimal control theory. These conditions can be obtained [9] in an easier way using the minimum principle. The necessarv conditions are given by the following nonlinear two-point boundary value problem

$$\dot{x} = Ax + \varepsilon f - B_1R_{11}^{-1}B_1'p_1 - B_2R_{22}^{-1}B_2'p_2 \ . \tag{4}$$

$$\dot{p}_1 = -Q_1x - A'p_1 - \varepsilon(\frac{\partial f}{\partial x})'p_1 + (\frac{\partial p_2}{\partial x})'B_2R_{22}^{-1}(B_2'p_1 - R_{12}R_{22}^{-1}B_2'p_2) \quad (5)$$

$$\dot{p}_2 = -Q_2x - A'p_2 - \varepsilon(\frac{\partial f}{\partial x})'p_2 + (\frac{\partial p_1}{\partial x})'B_1R_{11}^{-1}(B_1'p_2 - R_{21}R_{11}^{-1}B_1'p_1) \quad (6)$$

with the boundary conditions

$$x(0) = x_o$$

$$p_i(T) = Q_{if}x(T) \quad \forall x, \quad i=1,2. \quad (7)$$

While the strategies of the players can be written as follows:

$$u_i^*(x,t;\varepsilon) = -R_{ii}^{-1}B_i'p_i(x,t;\varepsilon). \quad (8)$$

3. Recurrence relations

It is difficult to apply known numerical methods to solve the two -point boundary value problem (4)-(7) due to appearing partial derivatives and to the fact that the boundary conditions (7) hold for any x. Therefore, it is reasonable to assume that the co-state p_i, i=1,2, can be expanded in the form

$$p_i(x,t;\varepsilon) = \sum_{k=0}^{\infty} \varepsilon^k p_i^{(k)}(x,t). \quad (9)$$

It is assumed that the series (9) converges for $t \in [0,T]$ and ε is sufficiently small.

3.1. Series of quasi-linear equations

Substituting (9) into (4)-(6) and equating powers of the perturbation parameter ε one can obtain: For k=0 we have a matrix Riccati differential equation

$$\dot{P}_i = -P_iA - A'P_i - Q_i + P_iS_{ii}P_i + P_iS_{jj}P_j + P_jS_{jj}P_i - P_jS_{ij}P_j, \quad P_i(T) = Q_{if} \quad (10)$$

which solution (a symmetric matrix) gives the solution of a linear-quadratic game (ε=0); For k=1,2,3,... we have the following quasi-linear partial differential equation

$$\frac{\partial p_i^{(k)}}{\partial t} + \frac{\partial p_i^{(k)}}{\partial x} Dx + D'p_i^{(k)} + (\frac{\partial p_j^{(k)}}{\partial x})'E^{ij}x + (E^{ij})'p_j^{(k)} = h_i^{(k)} \quad (11)$$

$$p_i^{(k)}(x,T) = 0 \qquad \forall x$$

where

$$h_i^{(k)} = \begin{cases} -(\frac{\partial f}{\partial x})' P_i x - P_i f & \text{if} \quad k=1 \\ -(\frac{\partial f}{\partial x})' p_i^{(k-1)} - \frac{\partial p_i^{(k-1)}}{\partial x} f \\ \quad + \sum_{l=1}^{k-1} \{ \frac{\partial p_i^{(l)}}{\partial x} [S_{11} p_1^{(k-l)} + S_{22} p_2^{(k-l)}] \\ \quad\quad + (\frac{\partial p_j^{(l)}}{\partial x})' [S_{ii} p_i^{(k-l)} - S_{ij} p_j^{(k-l)}] \} & \text{if} \quad k=2,3,\ldots \end{cases} \qquad (12)$$

with

$$S_{ij} = B_j R_{jj}^{-1} R_{ij} R_{jj}^{-1} B_j'$$

$$D = A - S_{11} P_1 - S_{22} P_2$$

$$E^{ij} = S_{ij} P_j - S_{jj} P_i \qquad (13)$$

for $i,j=1,2,\ i \neq j$.

3.2. Linear matrix equations

Under the assumption that the perturbation term $f(x,t)$ is a polynomial in x of degree q, then $h_i^{(1)}$ in (12) is also a polynomial in x of degree q. Now the following theorem is established.

Theorem 3.1 [9]

If $f(x,t)$ in (1) is a polynomial in x of degree q, then the solution $p_i^{(k)}$ of (11), for $i,j=1,2,\ i \neq j,\ k=1,2,\ldots$, can be written in the form a polynomial in x of degree $k(q-1)+1$.

The proof of the above theorem is based on the theory of quasi-linear partial differential equations with the so-called "same principal part" [4].

We can easily notice that equation (11) has the same form for any $k=1,2,3,\ldots$ Therefore, let us discuss the way of solving the recurrence equation (11) by considering the following equation

$$\frac{\partial z_i(x,t)}{\partial t} + \frac{\partial z_i(x,t)}{\partial x} Dx + D' z_i + [\frac{\partial z_j(x,t)}{\partial x}]' E^{ij} x + (E^{ij})' z_j = v_i \qquad (14)$$

$$z_i(x,T) = 0 \quad \forall x, \quad i,j=1,2, \quad i \neq j.$$

It has been shown [9] that (14) or (11) can be recognized as a system of quasi-linear partial differential equations with "the same principal part". Next, according to [4] integration of the system of partial differential equations with the same principal part is equivalent to integration of the system of ordinary differential equations, the so-called characteristic equations. For (14), these characteristic equations can be written in the matrix form

$$\dot{x} = Dx$$
$$\dot{z}_i = -D'z_i - \left(\frac{\partial z_j}{\partial x}\right)'E^{ij}x - (E^{ij})'z_j + v_i \tag{15}$$
$$i,j=1,2, \quad i \neq j.$$

According to Theorem 3.1 the solution of (14) can be written in the form (for details of the used summation convention see Appendix A)

$$z_{il} = Z_{(il)\alpha_1\alpha_2\ldots\alpha_r} x_{\alpha_1} x_{\alpha_2} \ldots x_{\alpha_r} \tag{16}$$

(the subscript l denotes the l-th component of the vector z_i).

The derivative of (16) w.r.t. time t can be written in the following matrix form

$$\frac{dz_i}{dt} = \dot{Z}_{(i)\alpha_1\alpha_2\ldots\alpha_r} x_{\alpha_1}, x_{\alpha_2} \ldots x_{\alpha_r} + \frac{\partial z_i}{\partial x}\frac{dx}{dt}$$
$$= \dot{Z}_{(i)\alpha_1\alpha_2\ldots\alpha_r} x_{\alpha_1}, x_{\alpha_2} \ldots x_{\alpha_r} + [Z_{(i)\beta\alpha_2\ldots\alpha_r} x_{\alpha_2} \ldots x_{\alpha_r} + \ldots +$$
$$Z_{(i)\alpha_1\alpha_2\ldots\alpha_{r-1}\beta} x_{\alpha_1}, x_{\alpha_2} \ldots x_{\alpha_{r-1}} x_\beta] Dx \tag{17}$$

with $\alpha_1, \alpha_2, \ldots, \alpha_r, \beta = 1,2,\ldots,n$.

On substituting (16) into (15) and then equating (15) to (17) one can obtain the following linear differential equation for the coefficients of the polynomial being the solution of (14)

$$\dot{Z}_{(il)\alpha_1\alpha_2\ldots\alpha_r} = -D_{\beta l} Z_{(i\beta)\alpha_1\alpha_2\ldots\alpha_r}$$
$$-D_{\beta\alpha_1} Z_{(il)\beta\alpha_2\ldots\alpha_r} - \ldots - D_{\beta\alpha_r} Z_{(il)\alpha_1\alpha_2\ldots\beta}$$
$$-E^{ij}_{\beta l} Z_{(j\beta)\alpha_1\alpha_2\ldots\alpha_r}$$
$$-E^{ij}_{\beta\alpha_1} Z_{(jl)\beta\alpha_2\ldots\alpha_r} - \ldots - E^{ij}_{\beta\alpha_r} Z_{(il)\alpha_1\alpha_2\ldots\beta}$$
$$+V_{(il)\alpha_1\alpha_2\ldots\alpha_r} \tag{18}$$

$(i,j=1,2, \quad i \neq j, \quad l,\beta,\alpha_1,\alpha_2,\ldots,\alpha_r=1,\ldots,n)$

where $D_{\beta l}$ is the (β,l)-element of the matrix D and $E^{ij}_{\beta l}$ is the (β,l)-element of E^{ij}, while $Z_{(il)\alpha_1\alpha_2\ldots\alpha_r}$ is the l-th component for the i-th player's co-state vactor; provided that the right hand side of (14) expressed by (12) has the following polynomial form

$$v_{il} = V_{(il)\alpha_1\alpha_2\ldots\alpha_r} x_{\alpha_1} x_{\alpha_2} \ldots x_{\alpha_r} \tag{19}$$

In the above equations the order of polynomials is described by r and due to Theorem 3.1 is equal to $k(q-1)+1$.

By virtue of Appendix B equation (18) can be written in the following matrix form

$$\sigma[\dot{Z}_{(i)\alpha_1\alpha_2\ldots\alpha_r}] = -\bigoplus_{\gamma=1}^{r+1} D'_\gamma \sigma[Z_{(i)\alpha_1\alpha_2\ldots\alpha_r}]$$
$$-\bigoplus_{\gamma=1}^{r+1} (E^{ij}_\gamma)' \; [Z_{(j)\alpha_1\alpha_2\ldots\alpha_r}] + \sigma[V_{(i)\alpha_1\alpha_2\ldots\alpha_r}] \tag{20}$$
$$\sigma[Z_{(i)\alpha_1\alpha_2\ldots\alpha_r}(T)] = 0.$$

where $i,j=1,2,\ i\neq j,\ \alpha_1\alpha_2,\ldots,\alpha_r=1,2,\ldots,n.$

The equation (20) looks rather complicated, but in fact this is a simple set of ordinary differential equations satisfying the Lipschitz conditions for ordinary differential equations.

Thus successive solution of (11) by increasing index k through solution of equations like (20) can determine two co-states p_i, i=1,2, (9). In the perturbation problems one is satisfied in cpomputation only a few first terms of the expansion (9). The strategies for two players will then be of the form of truncated series

$$u^*_i(x,t;\varepsilon) = -R^{-1}_{ii} B'_i \sum_{k=0}^{w_i} \varepsilon^k p^{(k)}_i(x,t) \tag{21}$$

where w_i, i=1,2, is the order of approximation.

4. Conclusion remarks

In this paper a method for suboptimal strategies for the two-person, nonlinear Nash differential games has been developed. The nonlinearity of the game model is assumed to be a polynomial in x. The suboptimal strategies can be obtained by solving a truncated series of sets of ordinary differential equations. The method has been outlined only due to a very limited space. Results of the required further investigation will be presented in forthcoming papers.

References

[1] J.F. Baldwin, J.H. Sims Williams: The Use of a Method of Perturbations in the Synthesis of Closed-Loop Optimal Control Laws for Nonlinear Systems, Automatica 5 (1969).

[2] R. Bellman: Perturbation Techniques in Mathematics Physics and Engineering, Holt, N-Y, 1964.

[3] R. Bellman: Introduction to Matrix Analysis, Mc Graw-Hill, N-Y, 1960.

[4] R. Courant, D. Hilbert: Methods of Mathematical Physics, Interscience Publishers, N-Y, 1962.

[5] G. Feichtinger, S. Jorgensen: Differential Game Models in Management Science, European J. Oper. Research 14 (1983).

[6] H. Khalil: Approximation of Nash strategies, IEEE Trans. Autom. Control, AC-25 (1980).

[7] M. Krawczak, K. Mizukami: On the Application of Delayed Sampled Data Observes to some Interception Problems, Int. J. Systems Sci., 14 (1983).

[8] M. Krawczak: Regularly Perturbed Zero-Sum Diff. Games, Polish-DDR Conference, Warsaw, May 1984.

[9] M. Krawczak: Suboptimal Strategies for Nash Nonlinear Differential Games, in: Analysis and Algorithms for Solution of Optimization Problems, K. Malanowski, K. Mizukami (eds.), to be published in Lecture Notes in Control and Information Sciences, Springer-Verlag.

[10] H. Lass: Elements of Pure and Applied Mathematics, Mc Graw-Hill, N-Y, 1957.

[11] V.H. Quintana, E.J. Davison: Two Numerical Techniques to Solve Differential Game Problems, Int. J. Control, 16 (1971).

[12] A.W. Starr, Y.C. Ho: Nonzero-Sum Diff. Games, JOTA, 5 (1969).

Appendix A: The Summation Convention

In this paper we use a convenient summation convention. A Einstein notice that a polynomial

$$w = \sum_{\alpha=1}^{n} a_\alpha x_\alpha \tag{A1}$$

can be written as

$$w = a_\alpha x_\alpha \ , \quad \alpha=1,2,\dots,n \tag{A2}$$

provided it is understood that whenever an index occurs as a subscript in neighbouring a coefficient (a_α) and a variable (x_α) over a full range of definition.

Es an example, let us consider the following polynomial

$$S = a_{\alpha\beta}x_{\alpha}x_{\beta}, \quad , =1,2,3.$$

$$= a_{1\beta}x_1x_\beta + a_{2\beta}x_2x_\beta + a_{3\beta}x_3x_\beta$$

$$= a_{11}x_1x_1 + a_{12}x_1x_2 + a_{13}x_1x_3$$

$$+ a_{21}x_2x_1 + a_{22}x_2x_2 + a_{23}x_2x_3$$

$$+ a_{31}x_3x_1 + a_{32}x_3x_2 + a_{33}x_3x_3$$

Appendix B: Kroncker Product and Sum

L t A and B be two matrices of dimension m and n respectively.

Definition B.1

The Kronecker product of A and B is defined as

$$A \otimes B = (a_{ij}B) \, , \quad i,j=1,2,\dots,m \tag{B1}$$

Definition B.2

The Kronecker sum of two matrices A and B is defined by

$$A \oplus B = A \otimes I_n + I_m \otimes B \tag{B2}$$

where I_m and I_n are respectively the m and n dimensional identity matrices.

Now, we will show the way in which the Kronecker product and sum are applied in our consideration.

Let x satisfies

$$\frac{dx}{dt} = Ax \tag{B3}$$

and consider the linear transformation induced in z, the m^2 dimensional vector

$$z = \sigma[x_\alpha x_\beta] \, , \qquad \alpha,\beta=1,2,\dots,m \tag{B4}$$

where $z[\cdot]$ denotes the "stacking operator" piling up the summands of the sum $x_\alpha x_\beta$ (refer to Appendix A). Then, it can be shown that

$$\frac{dz}{dt} = (A \oplus A)\, z. \tag{B5}$$

In the same mood, if we consider

$$w(x) = W_{\alpha_1\alpha_2\dots\alpha_r}\, x_{\alpha_1}x_{\alpha_2}\dots x_{\alpha_r} \tag{B6}$$

and define

$$z = \sigma[Z_{\alpha_1\alpha_2\ldots\alpha_r} x_{\alpha_1} x_{\alpha_2} \ldots x_{\alpha_r}] \tag{B7}$$

then, it can be inductively shown that

$$\frac{dz}{dt} = \bigoplus_{i=1}^{r} A_i z \tag{B8}$$

or

$$\frac{d}{dt} \sigma[Z_{\alpha_1\alpha_2\ldots\alpha_r}] = \bigoplus_{i=1}^{r} A_i \sigma[W_{\alpha_1\alpha_2\ldots\alpha_r}] \tag{B9}$$

where $A_i \equiv A$, and the symbol $\bigoplus_{i=1}^{r}$ denotes the Kronecker sum repeated r times.

A WORST CASE ANALYSIS FOR THE OPTIMAL GRADIENT METHOD

I. Kun

§1. Introduction

It is wellknown in the theory of unconstrained convex programming methods that to prove an at least linear rate of convergence requires that the function is twice continuously differentiable and the Hessian matrix is invertible in the optimum, hence also in some neighbourhood of the optimum. In the present paper we give an example which demonstrates that this requirement is necessary, because dropping it leads to the theoretical possibility of sublinear convergence. Sublinear convergence is however hardly experienced in practice, even in the case of singular Hessian. The reason lies in the effect of numerical errors. If the domain of bad convergence properties is of a small relative measure, then random numerical errors help to stay out of this domain, while they hardly slow down convergence.

Let $f(\underline{x})$ be convex and twice continuously differentiable on an open convex set $A \subseteq R_n$ and

$$\min_{\underline{x} \in A} f(\underline{x}) = f(\underline{x}^*)$$

for some $\underline{x}^* \in A$. Suppose for the sake of simplicity that $\underline{x}^*$ is unique. We are interested in the behaviour of optimum seeking algorithms, more precisely, in gradient-related algorithms, where for the given point $\underline{x}_k$ of the trajectory and the direction vector $\underline{s}_k$ we have

$$\frac{|(\operatorname{grad} f(\underline{x}_k))\, \underline{s}_k|}{\|\operatorname{grad} f(\underline{x}_k)\| \, \|\underline{s}_k\|} \geq \alpha > 0 \tag{1.1}$$

As far as steplength is concerned, we restrict ourselves to the line search method

$$f(\underline{x}_{k+1}) = f(\underline{x}_k + \lambda_k \underline{s}_k) = \min_{\lambda} f(\underline{x}_k + \lambda\, \underline{s}_k) \tag{1.2}$$

because many other steplength algorithms behave similarly.

§2. On the possibility of sublinear convergence

We know that (1.1) together with the other conditions implies

$$|f(\underline{x}^*) - f(\underline{x}_k)| = O(1/k) \tag{2.1}$$

(see e.g. Karmanov [2]). It is however not trivial to present a case where - under the above assumptions - instead of (2.1)

$$|f(\underline{x}^*)-f(\underline{x}_{k+1})| \ / \ |f(\underline{x}^*)-f(\underline{x}_k)| \longrightarrow 1 , \quad k \to \infty \tag{2.2}$$

holds, i.e. convergence is really sublinear.

As an example, consider the function on R_2

$$z = z(x,y) = \exp\{C[q(x,y)]^m\} \tag{2.3}$$

where $q = q(x,y)$ satisfies the implicit relation

$$\begin{aligned} &A(q)x^2 + 2B(q)xy + A(q)y^2 = 1 \quad \text{for } xy \neq 0 \\ &q(0,0) = 0 \end{aligned} \tag{2.4}$$

with

$$A(q) = \frac{1}{q^2-g(q)^2} \qquad B(q) = -\frac{g(q)}{q^2-g(q)^2} \tag{2.5}$$

where

$$g(q) = q - q^4/2$$

and m, C are finite positive numbers with the following roles: if m is large enough then $[q(x,y)]^m$ is twice continuously differentiable in some neighbourhood of the origin with zero-valued first and second partial derivatives in the origin. This can be checked by a lengthy but elementary computation. It is easy to see that $xy \neq 0$ implies $q(x,y) > 0$, therefore A, B, $A^2-B^2 > 0$ due to the first row of (2.5); i.e. each level set is a non-degenerate ellipse, which means that $[q(x,y)]^m$ is strictly quasiconvex. Now we can apply Theorem 2. of Gerencsér [1] which states that on a convex compact subset of the domain there is a real number $0 < C < +\infty$ for which $z=z(x,y)$ is convex.

Now let us launch the optimal gradient (or, synonymously, steepest descent) method from a point $(x_o, y_o=g(x_o))$, where x_o is small enough to ensure that this point is inside the convexity domain. Then the gradient is parallel to the x axis and the line $(x_o,y_o)^T + \lambda[\text{grad } z(x_o,y_o)]^T$ is tangential to another ellipse at the point $(g(g(x_o)), g(x_o))$ etc. Recursively

$$(x_{i+1}, y_{i+1}) = \begin{cases} (g(y_i), y_i) & \text{if } i \text{ is even} \\ (x_i, g(x_i)) & \text{if } i \text{ is odd} \end{cases} \tag{2.6}$$

The function value is $z = z(q,g(q)) = z(g(q),q) = \exp[Cq]^m$ and the next function value is $z = \exp[Cg(q)]^m$. As $z(0,0) = 1$, it is easy to check that the given trajectory produces $\lim\, [z(x_{i+1},y_{i+1})-1]/[z(x_i,y_i)-1] = 1$. At the same time $\lim\, (x_i,y_i) = (0,0)$, becasuse $g(q) < q$ for $q > 0$. So sublinear convergence is proved for function values. As for the arguments, sublinear convergence is an almost trivial corollary.

It is interesting to illustrate how such a situation is connected with the singularity of the Hessian at $\underline{x}^*$.

Let us apply (1.2) with $\underline{s}_k = -[\text{grad } f(\underline{x}_k)]^T$:

$$f(\underline{x}_k)-f(\underline{x}_{k+1}) \geq f(\underline{x}_k)-f(\underline{x}_k-\lambda[\text{grad } f(\underline{x}_k)]^T) = \tag{2.7}$$

(using Taylor expansion with an integral mean remainder term)

$$= \lambda \| \text{grad } f(\underline{x}_k)\|^2 - \lambda^2 \int_0^1 (1-t)[\text{grad } f(\underline{x}_k)] H(\underline{x}_k - t\lambda[\text{grad } f(\underline{x}_k)]^T) \cdot \\ \cdot [\text{grad } f(\underline{x}_k)]^T dt$$

Choose

$$\lambda = \frac{c[f(\underline{x}_k)-f(\underline{x}^*)]}{\|\text{grad } f(\underline{x}_k)\|^2} \tag{2.8}$$

where $c > 0$ is to be determined later.

$$\frac{f(\underline{x}_k)-f(\underline{x}_{k+1})}{f(\underline{x}_k)-f(\underline{x}^*)} \geq c - c^2 \frac{f(\underline{x}_k)-f(\underline{x}^*)}{\|\text{grad } f(\underline{x}_k)\|^2} \int_0^1 (1-t) \frac{\text{grad } f(\underline{x}_k)}{\|\text{grad } f(\underline{x}_k)\|} \cdot \tag{2.9}$$

$$\cdot H(\underline{x}_k - \lambda[\text{grad } f(\underline{x}_k)]^T) \frac{[\text{grad } f(\underline{x}_k)]^T}{\|\text{grad } f(\underline{x}_k)\|} dt = c - c^2 D$$

If in (2.9) the coefficient of c^2 is bounded along the trajectory, then by choosing a sufficiently small c it seems that $f(\underline{x}_k) \rightarrow f(\underline{x}^*)$ linearly. Now we analyze this coefficient D. Using again Taylor expansion for grad $f(\underline{x}_k)$ around $\underline{x}^*$, since grad $f(\underline{x}^*) = \underline{0}^T$

$$[\text{grad } f(\underline{x}_k)]^T = \int_0^1 H(\underline{x}^*+t(\underline{x}_k-\underline{x}^*))(\underline{x}_k-\underline{x}^*)\, dt \tag{2.10}$$

Further, because of convexity

$$\frac{f(\underline{x}_k)-f(\underline{x}^*)}{\|\text{grad } f(\underline{x}_k)\|} \leq \|\underline{x}_k-\underline{x}^*\| \tag{2.11}$$

and so

$$|D| \leq \frac{\| \underline{x}_k-\underline{x}^* \|}{\| \int_{1-\delta} H(\underline{x}^*+t(\underline{x}_k-\underline{x}^*))(\underline{x}_k-\underline{x}^*)\, dt \|} \int_0^1 \frac{\text{grad } f(\underline{x}_k)}{\|\text{grad } f(\underline{x}_k)\|} \cdot H(\underline{x}_k - t\lambda[\text{grad } f(\underline{x}_k)]^T) \frac{[\text{grad } f(\underline{x}_k)]^T}{\|\text{grad } f(\underline{x}_k)\|}\, dt \tag{2.12}$$

If δ and c are appropriately small then both arguments of H in (2.12) are close to $\underline{x}_k$ and the continuity of the Hessian guarantees that the change of eigenvalues of H is small in both integration domains. Now if we suppose that the quotient of the maximal and minimal eigenvalues (or, equivalently, the condition number) of H is bounded in the neighbourhood of $\underline{x}^*$, then D is bounded from above and so by choosing a sufficiently small c the left hand side of (2.9) is bounded from below by a positive number.

R e m a r k 2.1: If D is not bounded, we may still get linear convergence. Consider e.g. the set of all unit direction vectors belonging to any point of any trajectory producing sublinear convergence. If the accumulation points of this set can be covered by a set of very small measure (relative to the surface of the unit sphere) then sublinear convergence is in fact only a local singularity.

R e m a r k 2.2: In order to give some idea on the situation mentioned in the previous remark, suppose that $\|\text{grad } f(\underline{y})\|^2$ is demonstrated in the form of a multivariate Taylor polynomial where for given directions the sum of the lowest order terms disappears. This means an equation for the coordinates of the unit direction vector. Its solutions form a surface whose intersection with the unit sphere surface has certainly measure 0 relative to the latter. Outside the previous intersection $\|\text{grad } f(\underline{x})\|^2$ retains its original order, if $\underline{x}/\|\underline{x}\|$ keeps away from the intersection.

§3. On the effect of random errors

Suppose that we have an iteration algorithm defined by the function G: A→A. Suppose further that

$$\frac{\|G(\underline{x})-\underline{x}^*\|}{\|\underline{x}-\underline{x}^*\|} \leq \begin{cases} r<1 & \underline{x} \in K-K_\delta \\ 1 & \underline{x} \in K_\delta \end{cases} \tag{3.1}$$

where K is compact, $K \subset A$ and $\underline{x}^* \in K$; further, let

$$\mu(K_\delta) \leq \delta \tag{3.2}$$

where μ is the Lebesgue measure. That is, the algorithm works well on K, but it may converge slowly on K_δ. We are interested in proving R-linear convergence (as defined in Ortega-Rheinboldt [3]):

$$\limsup_{k \to \infty} \sqrt[k]{\frac{\|\underline{x}_k - \underline{x}^*\|}{\|\underline{x}_0 - \underline{x}^*\|}} < 1 \tag{3.3}$$

(It is easy to see that if the function value converges sublinearly then the same holds for the argument.) By means of the arithmetic-geometric mean inequality and telescope expansion

$$\sqrt[k]{\frac{\|\underline{x}_k - \underline{x}^*\|}{\|\underline{x}_0 - \underline{x}^*\|}} \leq \frac{1}{k} \sum_{i=1}^{k} \frac{\|\underline{x}_i - \underline{x}^*\|}{\|\underline{x}_{i-1} - \underline{x}^*\|} \tag{3.4}$$

The right hand side of (3.4) has a form which is typical in probability theory, more precisely, in limit theory of sums of random variables. So if the quotients on the right hand side have some regular random nature then we immediately have at hand tools to prove (3.3). Perhaps the strongest among them is the law of the iterated logarithm (LIL).

In order to have a random process, we allow the presence of random errors (e.g. in an actual computation random roundoff errors):

$$\underline{x}_{i+1} = G(\underline{x}_i) + \xi_i$$

where ξ_i has the following properties for each i=0,1,2... :

(i) $E\,\xi_i = \underline{0}$

with the conventions $\underline{x}^* = \underline{0}$ and $S(\underline{z}, d) = \{\underline{y} \mid \|\underline{y} - \underline{z}\| \leq d\}$

(ii) there exist positive constants ε_0, $\mu_1 < \mu_2$ such that if $B \subseteq S(G(\underline{x}_i), \varepsilon_0 \|\underline{x}_i\|)$ is Lebesgue measurable then

$$\mu_1 \frac{\mu(B)}{\mu(S(G(\underline{x}_i), \varepsilon_0 \|\underline{x}_i\|))} \leq P(G(\underline{x}_i) + \xi_i \in B) \leq \mu_2 \frac{\mu(B)}{\mu(S(G(\underline{x}_i), \varepsilon_0 \|\underline{x}_i\|))}$$

i.e. $\underline{x}_{i+1}$ has positive density in $S(G(\underline{x}_i), \varepsilon_0 \|\underline{x}_i\|)$.

(iii) $\mu(K_\delta \cap S(G(\underline{x}_i), \varepsilon_0 \|\underline{x}_i\|)) \leq \delta\, \mu(S(G(\underline{x}_i), \varepsilon_0 \|\underline{x}_i\|))$

i.e. K_δ is almost uniformly scattered in K

(iv) there exists a constant $0 < P_0 < 1$ such that $P\big(G(\underline{x}_i) + \xi_i \in S(G(\underline{x}_i), \varepsilon_0 \|\underline{x}_i\|)\big) \geq 1 - P_0$

(v) there exists $\varepsilon_1 > 0$ such that $\varepsilon_1 \|\underline{x}_i\| \geq E(\|\underline{\xi}_i\|) - \varepsilon_0 \|\underline{x}_i\|$

T h e o r e m 3.1: Let $\mathcal{F}_i$ be the σ-field generated by $\underline{x}_0, \underline{x}_1, \ldots \underline{x}_i$. Suppose that with probability 1 (i)-(v) also hold if P(.) and E(.) are substituted by $P(.|\mathcal{F}_{i-1})$ and $E(.|\mathcal{F}_{i-1})$ respectively. (i.e. if we know the trajectory up to $\underline{x}_{i-1}$.) Suppose that there are positive constants $\varepsilon_{21} < \varepsilon_{22}$ such that for all i with probability 1

$$\varepsilon_{21} < \sqrt{E\left\{\left[\frac{\|x_{i+1}\|}{\|x_i\|} - E\left(\frac{\|x_{i+1}\|}{\|x_i\|}|\mathcal{F}_{i-1}\right)\right]^2 \middle| \mathcal{F}_{i-1}\right\}} \leq \varepsilon_{22}$$

and

$$\sum_{i=0}^{\infty} P\left(\sqrt{\frac{\log\log(i\,\varepsilon_{22}^2)}{i}}\ \frac{\|\underline{\xi}_i\|}{\|\underline{x}_i\|} < \frac{1}{2\sqrt{2}}\varepsilon_{21}\right) < \infty$$

Then with probability 1

$$\limsup_{k\to\infty} \sqrt{\frac{k}{\log\log(k\,\varepsilon_{22}^2)}}\left\{\frac{1}{k}\sum_{i=0}^{k} \frac{\|\underline{x}_{i+1}\|}{\|\underline{x}_i\|} - r_1\right\} \leq 2\sqrt{2}\,\varepsilon_{22}$$

where $r_1 = r + [P_0 + 1 + (\mu_2 - \mu_1)\delta](\varepsilon_0 + \varepsilon_1) + P_0 + \delta(\mu_2 - \mu_1 r)$

The main step of the proof is to deduce from the assumptions that $E((\|\underline{x}_{i+1}\|/\|\underline{x}_i\|)/\mathcal{F}_{i-1}) \leq r_1$ with probability 1 for i=1,2,... hence $\|\underline{x}_{i+1}\|/\|\underline{x}_i\| - r_1$ is a supermartingale difference sequence with respect to $\mathcal{F}_{i-1}$. Then a LIL for supermartingales can be applied (see Stout [5], Theorem 5.4.1.).

R e m a r k 3.2: The assumptions of the theorem give only bounds for the relative fluctuation of $\underline{\xi}_i$ with respect to $\|\underline{x}_i - \underline{x}^*\|$.

R e m a r k 3.3: The theorem states that in spite of the role of K_δ R-linear convergence holds with probability 1, for the whole set K.

R e m a r k 3.4: The R_1-factor of the convergence (in case of small P_0, ε_0, ε_1, δ) is hardly worse after an appropriate number of iterations than on $K-K_\delta$ in the deterministic case.

C o n c l u s i o n 3.5: Although linear convergence is guaranteed for gradient-related methods only if $H(\underline{x}^*)$ is positive definite, sublinear convergence would hardly occur in the course of an actual computation since random errors prevent entering K_δ too many times.

C o n c l u s i o n 3.6: An iterative process with linear convergence is in general not very attractive in itself. It may serve however as the basis of multistep methods (e.g. discretized Newton methods, among others Steffensen's method; see Ortega and Rheinboldt [3], Ch. 7.2.). The latters are proved to converge quadratically if (again) $H(\underline{x}^*)$ is not singular, but the reasoning exposed here may carry over. This kind of convergence acceleration is described theoretically in one dimension in Stoer-Bulirsch [4] Ch. 5.10.

References:

[1] Gerencsér, L., "On a close relation between quasiconvex and convex functions and related investigations", Math. Operationsforschung u. Stat. 4(1973), No.3. pp. 201-211.

[2] Karmanov, V., Programmation Mathématique, Mir, Moscou, 1977.

[3] Ortega, J.M., and Rheinboldt, W.C., Iterative solution of nonlinear equations in several variables, Academic Press, New York-London, 1970.

[4] Stoer, J., and Bulirsch, R., Introduction to numerical analysis, Springer, New York - Heidelberg - Berlin, 1980.

[5] Stout, W.F., Almost sure convergence, Academic Press, New York - Heidelberg - London, 1974.

István Kun
Computer and Automation Institute, Hungarian Academy of Sciences
H-1502 Budapest XI.
Kende u. 13/17

Energy Modelling Approach

at the

Swiss Federal Institute for Reactor Research

S. Kypreos
Swiss Federal Institute for Reactor Research (EIR)
Division of Prospective Studies
CH - 5303 Würenlingen, Switzerland

The submodels MACRO, SMEDE and MARKAL are linked to an integrated model for energy systems analysis.

1 INTRODUCTION

Recent energy systems analysis for Switzerland concentrates on energy demand projections using the econometric model of Geneva University, and on the analysis of Energy supply using the model MARKAL (1, 2, 3).

Although the price effects on energy demand have been taken into account by econometrics, the influence of energy prices to the economic growth is not assessed by a generalized economic framework.

The ideal situation is therefore to establish a general framework which takes the partial equilibrium problem between demand and prices into account as well as the general equilibrium problem between economic growth, demand and prices.

This objective could be obtained by a set of interlinked models:

- The macroeconomic model MACRO which describes the economic equilibrium among demand and supply and the substitution effects among capital, labour and energy.
- The model SMEDE which simulates energy demand in consistency to economic activity and population growth, and
- The energy allocation model MARKAL which optimizes interfuel substitution potential and technology implementation.

The objective of the proposed model is to indogenize the effect of the energy prices to the economic growth and improve the internal consistency of the analysis. The method should be simple and easily applied and it should be able to reduce the uncertainties in generating potential demand and supply projection for Switzerland. A short description of the three submodels and their links follows.

2 THE ENERGY SUPPLY MODEL MARKAL

2.1 General Structure of the Model

MARKAL is a time-phased L.P. model which is structured around a representation of the energy system of a country.

The variables of the model characterize the various forms of energy from the sources of supply to the end use devices, the sources of the energy system and the installed capacities of the different technologies.

The equations (or constraints) of the model describe the balance of energy flows either on an annual base or on a seasonal and diurnal base.

The capacity built-up, the load management for electricity and district heat and of course the different objective functions constitute the other parts of the model.

The discounted energy system cost is the most commonly used objective function. (Other objectives could also be specified such as minimizing of oil imports, environmental emissions, fossil fuel consumption, energy imports, total foreign capital transfer for energy and technology investments, etc.)

2.2 Technologies and Sources of Supply

The model distinguishes among technologies and the system sources of supply.

Sources of supply cover all possibilities by which energy can enter or leave the system. This includes mining of resources, imports, exports and stockpiling. Stockpiling allows that energy resources produced in one period to be consumed in a subsequent period of time.

Technologies are divided into three subgroups:

- Process-Technologies
  Process technologies include load independent systems which convert one or more energy forms into others. Coal liquifaction/gasification and oil refining are typical examples of such systems.

- Demand Technologies
  Usually energy demand is disaggregated to different demand sectors like industry, residential, commerical, transport and agriculture.

  Each of these sectors is further disaggregated to submarkets of useful energy demand like space heating, water heating, industrial processes, road transport, etc. Many devices compete to satisfy each submarket. These devices are called demand technologies.

- Convertion Technologies
  This category of technologies includes electrical plants, heating plants and coupled production plants. The load characteristics of demand, forces the use of seasonal and diurnal production variables to specify the installed capacity necessary to operate the electricity or district heating network.

2.3 System Variables

It follows a mathematical description of model constraints. An exact description of the model could be found in Ref. (4) and (5). The objective of this paper is to define the most important characteristics of the model without getting lost in details. Table 2.1. is a list of the model variables.

Table 2.1: System Variables

a) Capacity Variables

- X_{rt}^{t} installed capacity at time period t and technology r.
- XN_r added capacity (investments) in period t and technology r.

Capacity variables which are expressed in PJ/Yr could also represent annual energy flows.

b) Activity Variables

- Y_p^t Annual total production of process p at time period t.
- $YE_e^{t,\Delta t}$ Electricity production of plant e, at period t, and time division Δt (there are 6 time divisions in a year).
- $YM_e^{t,\Delta s}$ Maintenance of plant e, at period t, and season Δs (there are three seasons in a year).

- $YH_{cp}^{t,\Delta t}$ Heat production of coupled production plant cp, at period t, time division Δt.
- $YH_{hr}^{t,\Delta s}$ Heat production of heating plant hp, at period t and season Δs.

c) Sources of Energy Supply

- $S_{i,en,k}^{t}$ Annual energy flow of energy carrier <u>en</u>, Index <u>k</u>, related to the source S

Sources are: imports, exports, mining and stockpiling (e.g. i=1,4). The index k is used to express the linear steps of the price/resource availability function for each energy carrier <u>en</u>.

Each of the indexes used above represents a set of quantities defined by the user. The technology index r refers to all the technologies used in the model.

Although the user could change the duration of the period and the number of periods per time horizon, usually 6 - 9 periods of 5 years duration have been used in the different studies.

2.4 Simplified Description of the Model Constraints

2.4.1 Balance Constraints

- Demand Balance

$$\sum_{dd_k} X_{dd_k}^{t} \geq d_k^{t}$$

One constraint for each time step t, and demand market K.

The sum of annual useful energy flows through demand devices dd_k, competing to satisfy demand market K, should be higher or equal to the total demand of the market d_k.

- Fuel Balance Equations (except electricity and heat)

The sum of imports, mining and production of an energy carrier is greater than or equal to the total consumption.

One equation for each time step and energy carrier, <u>en</u>:

$\sum_{i,k} S^{t}_{i,en,k} \cdot tr_{en}$ — Net production of <u>en</u> by all sources of supply i and subindex k.

$\sum_{p} out_{en} \; Y_p^t \cdot tr_{en} \geq$ — Production of <u>en</u> by processes p.

$\sum_{p} inp_{en} \cdot y_p^t$ — Consumption of <u>en</u> by a process p.

$+ \sum_{dd} ma_{en} \frac{1}{n_{dd}} \cdot X_{dd}^{t}$ — Consumption by demand devices dd.

$+ \sum_{e} \sum_{\Delta t} inp \cdot YE_e^{t,\Delta t}$ — Consumption by electrical plants e.

$+ \sum_{hp} \sum_{\Delta S} inp_{en} \cdot YH_{np}^{t,\Delta s}$ — Consumption by heating plants hp.

$+ \sum_{cp} \sum_{\Delta t} inp_{en} \cdot YH_{cp}^{t,\Delta t}$ — Consumption by coupled production plants corresponding to heat produced.

tr_{en} is the transmission efficiency of energy carrier <u>en</u> to the consumers.

out_{en} is the factional output of energy carrier <u>en</u> per unit of process output.

inp_{en} are the units of energy carrier <u>en</u> per unit of output Y , etc.

ma_{en} is the fractional input of energy carrier en per unit of useful output X

n_{dd} is the efficiency of demand device dd.

The exact formulation of energy carrier balance is more complex since "lead" and "lag" times for production and consumption are taken explicitly into account. This option is important for the nuclear fuel cycle management.

- Electricity (Heat) Balance

This constraint ensures that production of electricity is greater than or equal to the consumption of electricity. It is expressed per time division and time step.

2.4.2 Load Management Constraints (for electricity or district heat)

- Plant Utilization Constraint: It ensures that the sum of plant maintenance is less than or equal to the plant unavailability.

- Seasonal Plant Availability (or electricity production): This constraint ensures for each electrical plant, that the electricity production in each time division is less than or equal to the available capacity.

- Peaking Plant Constraint

This constraint ensures that enough capacity exists to meet peaking requirements at the time division of the greatest load.
A very simplified representation of this equation is as follows:

$$\frac{\Sigma X_e^{t,\Delta t} \cdot tr}{1 + res} \geq \sum_k \sum_{dd_k} \frac{fr(\Delta t)_k \cdot ma_{el} \cdot Y^t_{dd_k}}{dr(\Delta t) \cdot n_{dd_k}}$$

where dd_k are the demand devices competing for the demand market k.

The factor (1+res) takes the reserve capacity required to meet the electric load at the time division with the greatest load into account. The model calculates an average load (right hand side of equation) for the season of highest demand. The (1+res) factor should compensate the model limitations and the forced and scheduled maintenance.

2.4.3 Capacity Transfer Constraint

This constraint ensures that the installed capacity of a technology is equal to the residual capacity plus the investments made in previous periods within the life time of a technology.

$$X_r^t = Res_r^t + \sum_{k=t-L_r+1}^{t} XN_r^k$$

where L_r is the life time of technology r in terms of time periods.

2.4.4 Bounds on Technologies and Resources

The user could insert any bound to limit:

- the annual growth of a technology,
- cummulative resource use, and
- the penetration of a set of linked technologies.

These bounds modify the solution considerably and should be used only for reflecting real world situations not explicitly formulated by other system constraints.

2.5 The Objective Function(s) of the Model

The most commonly used objective function "price" refers to the total discounted system cost which is a linear combination of three functions:

INVEST : Capital investment cost of newly installed capacities in period t.

ANNCOST : The annual operation and maintenance cost of all installed capacities, which could include costs proportional to the capacity (fixed) or proportional to the annual production (variable).

SALVAGE : The terminal corrections at the end of the time horizon, due to the unused fraction of the life time of a technology.

All these cost components are discounted back to the beginning of the time horizon. Other functions used are:

SECURITY: Total oil import
QSLOPE: A linear combination of "price" and "security".
FOSSIL: total fossil energy use.
RENEW: Total renewable use
ENV(env): Total emission inventories for pollutant (env)

A very interesting function is "QSLOPE" which has been used to define the trade-off among the energy system cost and the oil imports.

All these objective functions are minimized in the process of optimization, with the exception of RENEW which is maximised.

3 THE ENERGY DEMAND SIMULATION MODEL SMEDE*

3.1 Generalities

SMEDE is, in principle, an accounting framework of energy demand.

The function of this model is to generate the disaggregated demand constraints for MARKAL. The model follows the scenario approach and it is driven by a set of demand indicators and their specific energy consumption.

The scenario parameters of SMEDE could be grouped in two broad categories: the socioeconomic data and the technical data.

The socioeconomic group of data describes the economic growth and the structural changes in the industry and the service sector, the evolution of population and its structure and the shifts in the consumer preferences.

The technical group of data describes the improvements in the energy intensiveness for the industry, the service sector and the households.

Conservation and substitution activities are explicitely formulated as scenario parameters by implementing new technological options.

It follows a short description of the industrial, residential and the transport sectors of the model.

3.2 The Industrial Sector

Four economic sectors are included in the Swiss model; manufacturing industries, construction, services and agriculture.

----

*) Swiss Model of Energy Demand. The model is based on MEDEE-2 (6)

The manufacturing industry sector is subdivided into four subsectors. Durables or capital goods; consumer goods (non-durables); basic materials and other industries (including mining and energy).

There are two options for taking the structural changes of the industry into account:

- Either by an exogenous disaggregation of the GDP based on economic studies of the problem, or
- by econometric analysis based on the macroeconomic variables of the MACRO model.

Once the economic activities per industrial sector is known, the next parameter required to generate projections, concerns the evolution of energy intensiveness per industrial subsector. An assessment of the most energy intensive industries and the trends on the evolution of intensiveness is necessary for that. The energy consumption in industry is further disaggregated to space and water heat, process heat (low temperature, steam medium temperature (up to 300 C) and high temperature) as well as mechanical power. Although SMEDE could consider substitution effects for fossil fuels by district heat, cogeneration, solar energy and electric heat pumps, this job is left for MARKAL. The improved energy statistics for the Swiss industrial sectors is a great help for running the model.

3.3 The Residential/Commercial Sector

The energy use in the residential/commercial sector is disaggregated according to:

Space heating / Water heating
Air conditioning
Cooking and house appliances
Process heat and air conditioning for commercial sector
and small industry.

Different house types are considered like single family houses, apartements and commercial buildings. Each house type is described according to its energy consumption per year.

The evolution of housing stock is driven by the average family size and population growth. Conservation in buildings and interfuel substitution will be estimated by the optimization model.

The electricity consumption for house appliances could in principle be estimated based on a saturation curve and the specific annual energy consumption per appliance. However, it is difficult to have reliable estimations on appliance ownership and specific energy consumption. Thus as an alternative approach an econometric relation based on the personal consumption per household has been introduced. Another problem concerns the shifts in consumer preferences among single family houses and apartements.

3.4 The Transport Sector

The transport sector includes the following categories; international/military transport which is correlated to GNP; freight transport which is correlated to industrial production; and the passenger transport.

Passenger transport is divided into urban and intercity transport by private or public modes of transport.

Car-ownership is estimated by a logistic curve while the average distance travelled per person is assessed as an exogenous variable.

Finally the load factors (passengers per car) and the fuel economy are taken into consideration to generate energy demand projections.

4 THE MACROECONOMIC MODEL MACRO

The concept of constant elasticity of substitution and a simple macroeconomic model for Switzerland has been adopted as the structure which generates the macroeconomic environment for SMEDE.

This chapter will give a short description of the procedure adopted. The model is going to be evaluated in cooperation with the St. Galler Zentrum für Zukunftsforschung (SGZZ) and it is now in a preliminary stage. The model is based on the work of H.H. Rogner of IIASA (8).

A possibility exists to develop a disaggregated I/O structure for the industrial and service sectors once a more satisfactory I/O table for Switzerland will be available.

4.1 The Production Function and the Constant Elasticity of Substition (CES)

The vehicle to analyse factors of production and substitution is the production function. The production function used in the MACRO-model is a Cobb-Douglas function for the input factors capital and labor, and a CES function among the factor Energy and the aggregate output of capital and labor:

$$(4.1)\qquad Y = \gamma\ \underbrace{[aE^{-\beta} + b\,\underbrace{(\theta\, K^{1-\alpha} .\ L^{\alpha})^{-\beta}}_{\text{Cobb-Douglas}}]^{-1/\beta}}_{\text{CES}}$$

K is the gross capital formation of the economy
L is the labor force
E is the final energy use
and are time dependent functions which express the increase of the economic output over time with constant primary input factors, due to the technical progress. The constants a, b, α , β and the functions θ , γ should be estimated by regression analysis.

The prices are correlated to the production function assuming profit maximization conditions.

$$(4.2)\qquad E^{R} = [P_{E}\ (1 + \varepsilon_{1})\ .\ \gamma^{\beta}\ .\ \alpha^{-1}]^{-\sigma}\ .\ Y$$

Equ. 4.2 specifies the energy demand (E) as a function of GDP, (Y) and the energy price, P . is the price elasticity of energy demand.

Similar relations could be written for the capital and the labor demand.

$$(4.3)\qquad K^{D} = (1 - \alpha)\ \gamma^{-\beta}\ .\ V^{-\beta}\ \ Y^{\beta+1}\ .\ P_{K}^{-1}$$

$$(4.4)\qquad MH^{D} = \alpha\ .\ \gamma^{-\beta}\ .\ b\ .\ V^{-\beta}\ .\ Y^{\beta+1}\ .\ P_{L}^{-1}$$

The production function defines maximum potential economic output at full utilization of production factors labor and capital and it also assumes that energy is not a constraint to economic growth.

The supply of the primary production factors , capital and labour must be expressed separately by another set of macroeconomic equations.

The macroeconomic model should include at least a capital formation and labour sector, the government sector, the trade sector and the price sector, together with the principal GDP identities.

This procedure defines a system of non-linear simultaneous equations on demand and

supply of primary production factors and their prices.

The system is solved by the Gauss-Seidel algorithm to "clear the market" and to specify the equilibrium prices. If a new price structure is assumed, the model will specify the substitution among the production factors and it will estimate the share of primary inputs which establishes a new equilibrium.

4.2 Linking MACRO to SMEDE

Fig. 1 explains the overall flow of information among the three submodels and their exogenous variables in a schematic way.

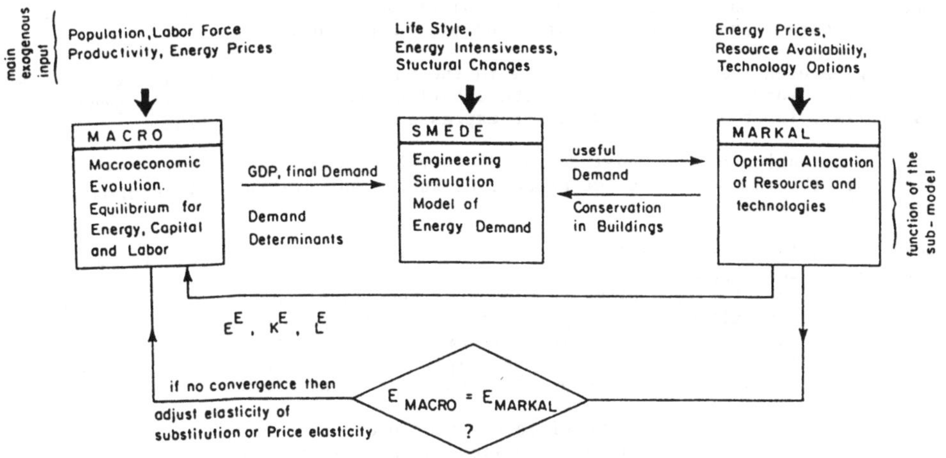

Fig. 1: Overall loop of interlinked models

SMEDE drives energy demand projections based on the time evolution of some determinants. The value added in the main industrial sectors, the evolution of residential and commercial floor area and the evolution of the passenger and freight transport are the most important ones.

The time dependent values of those determinants times their corresponding energy intensiveness result in useful (or final) demand projections.

The aggregate macro-economic variables must therefore be correlated to the demand determinants of the industry and the other sectors. The use of judgemental parameters is in any case unavoidable, due to the degree of disaggregation in SMEDE and the specific structure of this model.

4.3 Linking SMEDE to MARKAL

The interface of SMEDE to MARKAL is straight forward since most of the simulation results are in terms of useful energy. Conservation options, substitution effects and technology implementation is then estimated in MARKAL for the fuel price schedules assumed in the scenario specifications.

Iterations between SMEDE and MARKAL are necessary to obtain the same energy intensiveness in both models.

4.4 Linking MARKAL to MACRO

Energy use and energy prices, imports, investments and labor force in the energy sector are the results of MARKAL which need to be introduced in MACRO and close the loop.

It is a necessary (but not sufficient) condition for consistency that the disaggregated projections of final energy consumption, as estimated by SMEDE/MARKAL is similar to MACRO results, for the assumed price development.

5 CONCLUDING REMARKS

The SMEDE/MARKAL approach makes the demand projections and the analysis of the energy supply system more transparent and understandable than econometrics.

The coupling between these energy models with the macroeconomic framework will make the projections more consistent, since the substitution effects among the production factors capital, labor and energy would be expressed as function of price.

Although the extension of the model to an I/O formalism seems appropriate, it is not feasible for the moment due to the absence of sufficient statistical information to estimate an I/O table.

The model in its full development could be applied for long term strategic analysis of energy demand and supply.
It could also be used for technology ranking and specification of R/D priorities. The effectiveness of different energy policies like energy taxes and subsidies could be analysed and their macroeconomic implications could be shown.
Environmental control technologies have been implemented in MARKAL and the trade-off among the cost of emission control and pollution could be studied.

Finally the implementation of the model in advanced microcomputers makes the direct feed-back with high level executives and policy makers feasible, when performing analysis on the national or regional level.

However, it should never be forgotten that energy modelling is not a substitute for judgements. It is a useful instrument in the hands of experts to start from a set of assumptions and derive conclusions in a consistent way.
Whether the input assumptions and the way the economy is modelled are reasonable or not, it must be decided for every case by the working team.

Energy modelling provides background information and defines boundary conditions to discuss the energy problem. It is neither a substitute for discussion, nor does it provide for solutions ready to be applied.

References

1) F. Carlevaro et C. Spierer: Nouvelles perspectives d'evolution de la demande d'energie en Suisse jusqu'a l'an 2000. Aout 1983, Uni Geneve.

2) B. Giovannini et A. Delfosse: Influence sur la consommation d'energie des scenarios de politique energetique en Suisse.
Septembre 1983, Centre Universitaire d'etude des problemes de l' energie.

3) S. Kypreos, P. Kesselring: Kostenoptimierte Energieversorgungsszenarien für die Schweiz und deren politische Randbedingungen. EIR Reports Nr. 516 and Nr. 558, 1984/85, Würenlingen

4) L. Fishbone et al: User's Guide for MARKAL: A multi period, linear programming model for energy systems analysis. BNL 51701 / KFA Report, 1983

5) Fishbone L., Abilock H.: A technical description of MARKAL (to be published in the International Journal of Energy Research).

6) A.M. Khan, A. Hölzl: Evolution of Future Energy Demand. RR-82-14, April 1982.

7) H.H. Rogner: A long term macroeconomic equilibrium model for the European Community. RR-82-18, April 1982.

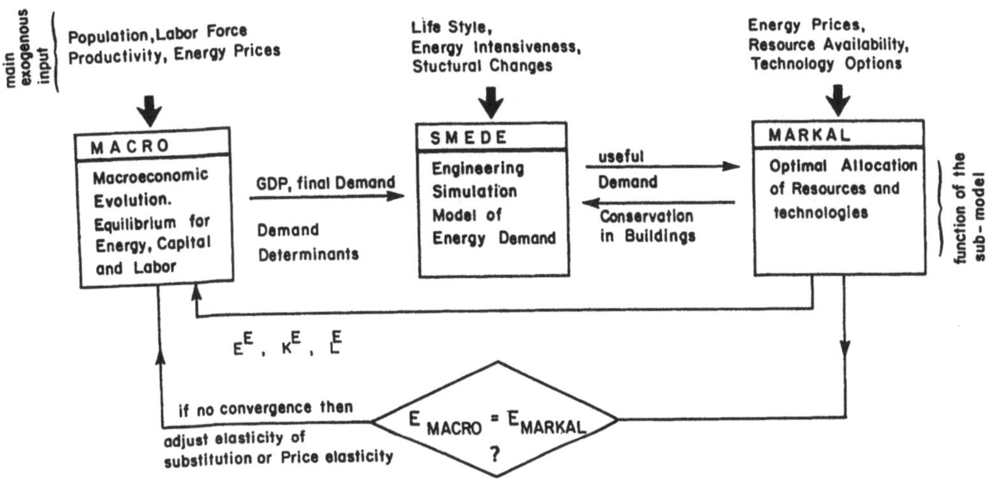

Fig. 5.1 Overall loop of interlinked models

ANALYSIS OF DNA DISTRIBUTIONS FROM FLOW CYTOMETRY BY MEANS OF AN OPTIMIZATION PROCEDURE*

F. Lampariello, S. Lucidi
Istituto di Analisi dei Sistemi ed Informatica del CNR,
Viale Manzoni 30, 00185 Roma, Italy

ABSTRACT

In this paper the problem is considered of FCM histogram processing in the case of mixtures of two or more cell populations, which appears of great interest in diagnosis and treatment of tumor diseases.

A suitable mathematical model is first developed in order to represent the pattern of DNA and fluorescence distribution in the samples. An optimization procedure is then proposed for the automatic estimation of the unknown model parameters, which employs a version of Newton's method.

1. INTRODUCTION

The problem of estimating the fractions of a population corresponding to the various phases of the cell cycle from the DNA content distribution obtained by flow cytometry is an important problem in the diagnosis and treatment of tumor diseases.

The application of flow cytometry (FCM) to the study of human malignancies reveals very often an abnormal DNA content consisting in the presence of abnormal stemlines (aneuploidy) with respect to the normal diploid cell population. Recently a graphical method for estimating the fraction of cells in S-phase from clinical tumor samples containing aneuploid cell populations has been proposed [1].

However a qualitative or semi-quantitative analysis of DNA histograms may lead to somehow unreliable results particularly when the different components in the mixture exhibit a high overlapping degree or when the aneuploid component fraction is relatively small, as may be in the so-called precancerous conditions [2]. On the other hand it appears of great interest to give a quantitative evaluation of aneuploidy as a tool of predicting to some extent the clinical behaviour of the neoplastic disease or for confirming a possible suspicion of malignant lesion.

In this paper we present an automatic procedure for recovering the unknown model parameters from samples containing mixtures of two or more cell populations.

At first we propose some modeling assumptions for DNA and fluorescence distributions. Then we present a parameter estimation procedure based on the minimization of an error index (the chi-square or, alternatively, the maximum likelihood index) resulting from the comparison between the theoretical and the experimental histograms. In order to take constraints on the parameters into account, we introduce suitable changes of variables so that the original minimization problem is transformed into an unconstrained one. The solution is attained by means of a Newton-type method which offers, as known, fast convergence properties. As regards the line search technique, we adopt the "non monotone" steplength selection rule suggested in [3], which allows a considerable saving in computations.

The performance of the procedure has been tested against simulated data and the results obtained show its reliability. The procedure has been also applied to some clinical samples with satisfactory results as far as the final fit is concerned.

2. THE MATHEMATICAL MODEL

In this section we introduce the fundamental assumptions about the DNA distribu-

---

* This work was supported by Progetto Finalizzato Oncologia, CNR, under grant n. 56825.

tion and the fluorescence distribution which our mathematical model is based on.

As already mentioned in the introduction, we consider the case of a mixture of two or more cell populations.

For each of them, the DNA content distribution will feature a G_o/G_1 and a G_2+M peak, suitably located, and an S-phase continuously distributed in between.

In order to take into account the intrinsic property of the DNA synthesis mechanism, we assume that, for each i-th population ($i = 1,2,\ldots,\nu$) the DNA content $y_2^{(i)}$ of the G_2+M peak is twice that one $y_1^{(i)}$ of the G_o/G_1 peak:

$$y_2^{(i)} = 2y_1^{(i)}, \quad i = 1,2,\ldots,\nu$$

Moreover the S-phase is assumed to have a piecewise constant distribution density in $[y_1^{(i)}, y_2^{(i)}]$; thus the number n of intervals of the partition of $[y_1^{(i)}, y_2^{(i)}]$ with constant S-phase density corresponds to the number of parameters employed for reconstructing the S-phase shape.

Denoting by y the DNA content, the corresponding global distribution density can be given the form:

$$g(y) = \sum_{i=1}^{\nu} g^{(i)}(y) = \sum_{i=1}^{\nu} [\lambda_1^{(i)}\delta(y-y_1^{(i)})+\lambda_2^{(i)}\delta(y-2y_1^{(i)}) + \sum_{k=1}^{n} a_k^{(i)}\varphi_k^{(i)}(y)] \qquad (1)$$

where the first and second terms in brackets are two impulse functions located at $y_1^{(i)}$ and $2y_1^{(i)}$ which represent the $G_o/G_1(\lambda_1^{(i)})$ and $G_2 + M(\lambda_2^{(i)})$ phase fractions respectively and the last term describes the S-phase distribution density with:

$$\varphi_k^{(i)}(y) = \begin{cases} \dfrac{n}{y_1^{(i)}}, & y_1^{(i)}(1 + \frac{k-1}{n}) \le y < y_1^{(i)}(1 + \frac{k}{n}) \\ 0, & \text{otherwise} \end{cases} \qquad i = 1,2,\ldots,\nu \qquad (2)$$

Since g is a distribution, the parameters introduced in (1) are subjected to the following constraints:

$$\lambda_1^{(i)} \ge 0,\ \lambda_2^{(i)} \ge 0,\ a_k^{(i)} \ge 0, \quad k = 1,2,\ldots,n, \quad i = 1,2,\ldots,\nu$$
$$\sum_{i=1}^{\nu} [\lambda_1^{(i)} + \lambda_2^{(i)} + \sum_{k=1}^{n} a_k^{(i)}] = 1 \qquad (3)$$

Furthermore we assume the cell populations pairwise not fully overlapping, for this clearly helps in detecting each of them out of the mixture. By suitably reordering the ν populations with respect to the DNA content, this amounts to assume:

$$y_1^{(i+1)} - y_1^{(i)} > \varepsilon^{(i)}, \quad i = 1,2,\ldots,\nu-1 \qquad (4)$$

for fixed positive $\varepsilon^{(i)}$. As a matter of fact, should this constraint be removed allowing for $y_1^{(i+1)} = y_1^{(i)}$ for some i, then separation of the corresponding populations may be acomplished only if they exhibit a sufficiently different behaviour in the flow cytometric analysis.

When processed by FCM techniques, a given cell sample gives rise to a fluorescence distribution whose density is obviously related to the DNA distribution density g(y). To this connection two main effects are to be taken into account.

a) the fluorescence spreading

b) the background caused by cell disintegration.

As regards a) we assume it is essentially the result of two independent effects, the first one accounting for the dispersion in the staining process, and the second one for additional instrumental dispersion. As usually done in the literature, we use a gaussian distribution for the description of both effects, the first one with mean y and variance $\gamma^{(i)}y$ and the second one with mean·0 and variance δ^2y^2. As a result, denoting by x the fluorescence intensity and by $p^{(i)}(x|y)$ the fluorescence density for a cell of the i-th population with DNA content y, which describes the fluorescence spreading, we have:

$$p^{(i)}(x|y) = \frac{1}{\sqrt{2\pi v^{(i)}(y)}} e^{-\frac{(x-y)^2}{2v^{(i)}(y)}} \qquad (5)$$

where

$$v^{(i)}(y) = \gamma^{(i)}y + \delta^2 y^2 \;, \; \gamma^{(i)} \geq 0, \;\; i = 1,\ldots,\nu, \;\; \delta \geq 0 \qquad (6)$$

Notice that the variance of the first effect is allowed to be different in each population for a possible different stoichiometric behaviour with respect to the staining process. On the other hand, the instrumental dispersion is likely to be the same for all populations.

The background effect is essentially due to the cell fragmentation process caused by the cell pretreatment procedure necessary for the FCM measurement.

A correct approach to the study of this effect should rely on a statistical analysis of the breaking mechanisms, connecting the DNA content distribution of cell fragments to the original DNA distribution. In this sense we refer to [4].

Here, for simplicity, we assume for the description of the fluorescence distribution density due to DNA fragments an exponential expression:

$$b(x) = \sigma e^{-\sigma(x-\alpha)} \;, \; \sigma > 0, \text{ for } x \geq \alpha \qquad (7)$$

which is coherent with most pertinent literature [5] and usually supported by experimental evidence.

However, we remark that the assumed structure (7) for the background effect is part of our mathematical model. Therefore, differently from the widely adopted procedure of preliminarily filtering out the background by an exponential interpolation of the initial part of the fluorescence histogram, we prefer here to estimate the unknown parameter σ together with the other unknown parameters in the model.

In conclusion, denoting by η the fraction of the cell sample due to the background effect, the fluorescence distribution density model takes the form:

$$f(x;\theta) = \eta b(x) + (1-\eta) \sum_{i=1}^{\nu} \int_{-\infty}^{\infty} p^{(i)}(x|y) g^{(i)}(y) dy \qquad (8)$$

where θ denotes the $[\nu(n+4)+3]$-dimensional parameter vector:

$$\theta = [y_1^{(1)} \lambda_1^{(1)} a_1^{(1)} \ldots a_n^{(1)} \lambda_2^{(1)} \gamma^{(1)} \Big| \ldots \Big| y_1^{(\nu)} \lambda_1^{(\nu)} a_1^{(\nu)} \ldots a_n^{(\nu)} \lambda_2^{(\nu)} \gamma^{(\nu)} \Big| \delta \; \eta \; \sigma]$$

Let $[j_i, j_f]$ (j_i = initial channel, j_f = final channel) be a prefixed interval of the fluorescence axis, where data are to be processed.

By integrating (8) over the interval corresponding to the j-th channel and by normalizing to unit area over the content of all channels in $[j_i, j_f]$, we get:

$$F_j(\theta) = \int_{j-0.5}^{j+0.5} f(x;\theta)dx, \;\; H_j(\theta) = \frac{F_j(\theta)}{\sum_{j=j_i}^{j_f} F_j(\theta)} \;, \; j \in [j_i, j_f] \qquad (9)$$

which defines the theoretical histogram to be compared with the measurements.

Let E_j , $j \in [j_i, j_f]$ be the experimental data, that is the cell (or fragment) number registered in the j-th channel and let $N = \sum_{j=j_i}^{j_f} E_j$ be the total cell number in $[j_i, j_f]$. We define: $\tilde{H}_j = E_j/N$ and we can write:

$$\tilde{H}_j = H_j(\theta) + R_j$$

where R_j is an error term which accounts for the statistical fluctuations due to the finiteness of the cell sample and which will implicitely be taken into account by means of the estimation procedure.

3. THE ESTIMATION PROCEDURE

An optimal estimate of the parameter vector θ can be obtained by minimizing an error index which results from the comparison between $H_j(\theta)$ and $\tilde{H}_j$.

Noting that $H_j(\theta)$ can be interpreted as the conditional probability of finding a cell in the j-th channel relative to the hypothesis that this cell is registered in $[j_i, j_f]$, we can consider the *minimum chi-square estimation* which corresponds to determine θ so as to minimize:

$$\chi^2(\theta) = \sum_{j=j_i}^{j_f} \frac{[E_j - NH_j(\theta)]^2}{NH_j(\theta)} \tag{10}$$

or, alternatively, the *maximum likelihood estimation* which corresponds to determine θ so as to minimize:

$$L(\theta) = -\log \prod_{j=j_i}^{j_f} [H_j(\theta)]^{E_j} = N \log \sum_{j=j_i}^{j_f} F_j(\theta) - \sum_{j=j_i}^{j_f} E_j \log F_j(\theta) \tag{11}$$

As regards the statistical properties of (10) and (11) we refer to [6], [7].

As already mentioned, our objective is to implement, for the minimization of (10) or (11), a version of Newton's method which requires, as well known, the analytical expressions for first and second order derivatives of the functional.

In order to perform the integration in (8), we divide each interval of length $y_1^{(i)}/n$ of the partition of $[y_1^{(i)}, 2y_1^{(i)}]$ into m equal parts and to each one of them apply the Simpson integration rule. Consequently, recalling (1),(2), we obtain for $f(x;\theta)$ the following expression:

$$f(x;\theta) = \eta b(x) + (1-\eta) \sum_{i=1}^{\nu} \Bigg\{ (\lambda_1^{(i)} + \frac{a_1^{(i)}}{6m}) p^{(i)}(x|y_1^{(i)}) + \tag{12}$$

$$+ \frac{1}{6m} \sum_{k=1}^{n} [(a_{k+1}^{(i)} - a_k^{(i)}) p^{(i)}(x|(1+\frac{k}{n})y_1^{(i)}) + a_k^{(i)} \sum_{\ell=1}^{2m} \{3-(-1)^{\ell}\} p^{(i)}(x|(1+\frac{k}{n}-h)y_1^{(i)})] \Bigg\}$$

where $h = \frac{2m-\ell}{2nm}$ and $a_{n+1}^{(i)} \triangleq 6m\lambda_2^{(i)}$.

Applying again the Simpson integration rule we obtain for (9):

$$F_j(\theta) = \frac{1}{6}[f(j-0.5;\theta) + 4f(j;\theta) + f(j+0.5;\theta)] \tag{13}$$

Now, using (12) and (13) in (10) or (11) we have an approximation for the chi-square or maximum likelihood error index and their first and second order derivatives with respect to θ can be easily obtained into an analytical form, thus avoiding further numerical integrations.

Of course, in the minimization of (10) or (11) we must take constraints on θ into account. These constraints are:

a) a nonnegativity constraint for all components of θ

b) the normalization constraint in (3) which, being $0 \leq \eta < 1$ and letting, for $i = 1,2,\ldots,\nu$:

$$\hat{\lambda}_1^{(i)} = (1-\eta)\lambda_1^{(i)}, \hat{\lambda}_2^{(i)} = (1-\eta)\lambda_2^{(i)}, \hat{a}_k^{(i)} = (1-\eta)a_k^{(i)}, \quad k = 1,2,\ldots,n \tag{14}$$

can be rewritten in the form:

$$\eta + \sum_{i=1}^{\nu} [\hat{\lambda}_1^{(i)} + \hat{\lambda}_2^{(i)} + \sum_{k=1}^{n} \hat{a}_k^{(i)}] = 1 \tag{15}$$

By introducing suitable changes of variables, we transform the original constrained minimization problem into an unconstrained one.

The normalization constraint (15) allows us to eliminate one of the parameters to be estimated, which is reasonably chosen to be $\hat{\lambda}_1^{(1)}$. In fact, it is preferable to choose a parameter whose value is not close to zero, since in this case some of the other transformed variables would become badly scaled.

Then, by reordering the parameter vector as follows:

$$\hat{\theta} = [\,\eta\hat{a}_1^{(1)} \ldots \hat{a}_n^{(1)}\hat{\lambda}_2^{(1)} \,\Big|\, \hat{\lambda}_1^{(2)}\hat{a}_1^{(2)} \ldots \hat{a}_n^{(2)}\hat{\lambda}_2^{(2)} \,\Big|\, \ldots \,\Big|\, \hat{\lambda}_1^{(\nu)}\hat{a}_1^{(\nu)} \ldots \hat{a}_n^{(\nu)}\hat{\lambda}_2^{(\nu)} \,\Big|$$

$$y_1^{(1)} \ldots y_1^{(\nu)}\gamma^{(1)} \ldots \gamma^{(\nu)}\,\delta\sigma\hat{\lambda}_1^{(1)}]$$

constraint (15) will be automatically satisfied by defining the new variables z_j, $j = 1,2,\ldots,\nu(n+2)$, such that:

$$\hat{\lambda}_1^{(1)} = \frac{c}{c + \sum_{j=1}^{\nu(n+2)} e^{-sz_j}}$$

$$\eta = \frac{\hat{\lambda}_1^{(1)}}{c} e^{-sz_1}, \quad \hat{a}_k^{(1)} = \frac{\hat{\lambda}_1^{(1)}}{c} e^{-sz_{k+1}}, \quad k = 1,\ldots,n, \quad \hat{\lambda}_2^{(1)} = \frac{\hat{\lambda}_1^{(1)}}{c} e^{-sz_{n+2}}$$

$$\hat{\lambda}_1^{(i)} = \frac{\hat{\lambda}_1^{(1)}}{c} e^{-sz_{(i-1)(n+2)+1}}, \quad \hat{a}_k^{(i)} = \frac{\hat{\lambda}_1^{(1)}}{c} e^{-sz_{k+(i-1)(n+2)+1}}, \quad k = 1,\ldots,n \tag{16}$$

$$\hat{\lambda}_2^{(i)} = \frac{\hat{\lambda}_1^{(1)}}{c} e^{-sz_{i(n+2)}}, \quad i = 2,\ldots,\nu$$

where s and c are scale factors.

As regards a), in order to obtain a better numerical stability, we add for the components $y_1^{(i)}, \gamma^{(i)}, \delta, \sigma$ some boundedness constraints, allowing these parameters to vary only in a prefixed range.

More specifically we assume:

$$\zeta < y_1^{(1)} < \mu^{(1)}, \; y_1^{(i)} + \varepsilon^{(i)} < y_1^{(i+1)} < \mu^{(i+1)} \qquad i = 1,2,\ldots,\nu-1$$

$$0 < \gamma^{(i)} < 1, \quad i = 1,2,\ldots,\nu \tag{17}$$

$$0 < \delta < 1, \quad 0.01 < \sigma < 1$$

and denoting by ℓ_j and u_j these lower and upper bounds respectively, we use for the corresponding variables the following transformation:

$$\hat{\theta}_j = \ell_j + \frac{u_j - \ell_j}{1 + e^{-s_j z_j}}, \quad j = \nu(n+2)+1,\ldots,\nu(n+4)+2 \tag{18}$$

where z_j is the new variable associated with $\hat{\theta}_j$ and s_j is a scale factor.

Thus, redefining $\chi^2(\hat{\theta})$ (or $L(\hat{\theta})$) as a function of the new $[\nu(n+4)+2]$-dimensional vector z of unknown parameters, with $\hat{\theta} = \psi(z)$ given by (16) and (18), the original estimation procedure becomes an unconstrained minimization problem of χ^2 (or L) with respect to z.

4. THE MINIMIZATION ALGORITHM

To achieve the minimization of the functional J(z) given by (10) or (11) with respect to z, an algorithm is constructed which implements a version of Newton's method. At each step, the new search direction in the parameter space is computed using the analytical expressions for first and second order derivatives of J, and the new estimate is given by:

$$z(k+1) = z(k) + \alpha_k d_k, \quad k = 0,1,\ldots$$

$$d_k = -[\nabla^2 J(z(k))]^{-1} \nabla J(z(k))$$

where ∇J is the gradient of J, $\nabla^2 J$ is the Hessian matrix and α_k is the stepsize.

Denoting by:

$$C_j \triangleq \left(\frac{E_j \Sigma F_j}{N F_j}\right)^2$$

by simple calculations we have for the chi-square index (10):

$$\frac{\partial \chi^2(z)}{\partial z_h} = \frac{N}{\Sigma F_j}\left\{\Sigma(1-C_j)\frac{\partial F_j}{\partial z_h} - \frac{\Sigma F_j(1-C_j)}{\Sigma F_j}\,\Sigma\frac{\partial F_j}{\partial z_h}\right\}$$

$$\frac{\partial^2 \chi^2(z)}{\partial z_h \partial z_\ell} = \frac{N}{\Sigma F_j}\left\{\Sigma(1-C_j)\left(\frac{\partial^2 F_j}{\partial z_h \partial z_\ell} - \frac{F_j}{\Sigma F_j}\Sigma\frac{\partial^2 F_j}{\partial z_h \partial z_\ell} + \frac{2F_j}{(\Sigma F_j)^2}\Sigma\frac{\partial F_j}{\partial z_h}\Sigma\frac{\partial F_j}{\partial z_\ell}\right) + \right.$$

$$+ 2\Sigma C_j\left(\frac{1}{F_j}\frac{\partial F_j}{\partial z_h}\frac{\partial F_j}{\partial z_\ell} + \frac{F_j}{(\Sigma F_j)^2}\Sigma\frac{\partial F_j}{\partial z_h}\Sigma\frac{\partial F_j}{\partial z_\ell}\right) -$$

$$\left. - \frac{1}{\Sigma F_j}\Sigma(1+C_j)\left(\frac{\partial F_j}{\partial z_h}\Sigma\frac{\partial F_j}{\partial z_\ell} + \frac{\partial F_j}{\partial z_\ell}\Sigma\frac{\partial F_j}{\partial z_h}\right)\right\}$$

and for the maximum likelihood index (11):

$$\frac{\partial L(z)}{\partial z_h} = \frac{N}{\Sigma F_j}\Sigma\frac{\partial F_j}{\partial z_h} - \Sigma\frac{E_j}{F_j}\frac{\partial F_j}{\partial z_h}$$

$$\frac{\partial^2 L(z)}{\partial z_h \partial z_\ell} = \frac{N}{\Sigma F_j}\left(\Sigma\frac{\partial^2 F_j}{\partial z_h \partial z_\ell} - \frac{1}{\Sigma F_j}\Sigma\frac{\partial F_j}{\partial z_h}\Sigma\frac{\partial F_j}{\partial z_\ell}\right) -$$

$$- \Sigma\frac{E_j}{F_j}\left(\frac{\partial^2 F_j}{\partial z_h \partial z_\ell} - \frac{1}{F_j}\frac{\partial F_j}{\partial z_h}\frac{\partial F_j}{\partial z_\ell}\right)$$

$$h = 1,2,\ldots,\nu(n+4)+2, \quad \ell = h,h+1,\ldots,\nu(n+4)+2$$

where the indices of all summations range from j_i to j_f. The expressions of $\partial F_j/\partial z_h$ and $\partial^2 F_j/\partial z_h \partial z_\ell$ can be found in [8] for the case $\nu = 2$.

As regards the stepsize, the usual line search techniques along the Newton's direction guarantee a monotonical decrease of the objective function in order to ensure global convergence towards local minima. However, enforcing monotonicity of the function values by selecting a stepsize not equal to unity can considerably slow the rate of convergence. Therefore we adopt the steplength selection rule suggested in [3] which, allowing an increase in the function values without affecting the convergence properties, may realize a considerable saving both in the number of line searches and in the number of function evaluations.

This aspect is of relevant importance in the present application because of the great amount of calculations involved at each iteration.

In particular the minimization algorithm implements the following model (for more details see [3]):

Data. $z(0)$, $\omega \in (0,1)$, $\tau \in (0,1)$, $T > 0$, M (integer) ≥ 0, $c_1 > 0$, $c_2 > 0$.

Step 1. Set $k = 0$, $m(0) = 0$ and compute $J_o \triangleq J(z(0))$.

Step 2. Compute $\nabla J(z(k))$; if $\|\nabla J(z(k))\| \leq T$ stop, else compute $\nabla^2 J(z(k))$ and d_k by solving the linear system equation

$$\nabla^2 J(z(k))d_k = -\nabla J(z(k))$$

(Should this system be ill-conditioned, then several modification procedures for the Hessian matrix can be devised thus obtaining an approximation of Newton's direction). If $|d_k' \nabla J(z(k))| < c_1\|\nabla J(z(k))\|^2$ or $\|d_k\| > c_2\|\nabla J(z(k))\|$ set $d_k = -\nabla J(z(k))$, $m(k) = 0$ and go to step 4.

Step 3. If $d_k^{\prime}\nabla J(z(k)) > 0$, set $d_k = -d_k$.

Step 4. Set $\alpha = 1$.

Step 5. Compute $J_\alpha = J(z(k)+\alpha d_k)$. If

$$J_\alpha \leq \max_{0\leq j\leq m(k)} J_{k-j} + \omega\alpha d_k^{\prime}\nabla J(z(k))$$

set $J_{k+1} = J_\alpha$, $z(k+1) = z(k)+\alpha d_k$, $k = k+1$, $m(k) = \min[m(k-1)+1, M]$ and go to step 2.

Step 6. Set $\alpha = \tau\alpha$ and go to step 5.

The values chosen for the parameters are: $\omega = 10^{-3}$, $\tau = 0.5$, $M = 5$, $c_1 = 10^{-5}$, $c_2 = 10^5$.

Of course, convergence is guaranteed to a local minimum; therefore particular care should be used in the choice of the initial point $z(0)$. Some comments on this issue will be made in the next section.

The choice of n and m deserves some remarks. The number n of intervals with constant S-phase density is taken quite low ($n = 4$). This allows to have some information of the shape of the S-phases with a limited number of parameters. At the same time it avoids ill-conditioning problems and yields a sufficient stability and speed of convergence. To further improve these last features, the n value is initially set to 1 and then switched to 2 as soon as the minimization trajectory gets close enough to a local minimum (which is checked by comparing $\|\nabla J\|$ to the fixed threshold T); a similar procedure is subsequently implemented to switch from $n = 2$ to $n = 4$.

As regards m we assume $m = 32/n$ so that the numerical integration in (12) is performed with a step which remains constant during all minimization stages.

5. PRELIMINARY DATA PROCESSING

Before data are taken by the minimization program, it is necessary to preliminarily process them in order to overcome possible drawbacks due to instrumental biasedness, to select the most suitable fluorescence interval in which data are to be considered and restrictions for the allowed parameter ranges and to properly choose the initial value for the parameter vector.

We did not find convenient to automatically perform these tasks, both because of their complexity due to high variety of possible experimental configurations and because of the advantages that experience and skilfulness of the operator may offer in these circumstances.

In the model we assumed $y_2^{(i)} = 2y_1^{(i)}$; therefore a similar behaviour should be followed by the location of peaks in the experimental histogram. Should this one exhibit significant deviation from this pattern, a lack in the instrument calibration is likely to have occurred. In such a case the operator may decide to properly shift the whole experimental histogram by a suitable number of channels.

As regards the interval $[j_i, j_f]$, the operator should select the initial channel j_i such that possible data on the left tail, which clearly disagree with the exponentially decreasing behaviour assumed for $b(x)$, are discarded and the final channel j_f after $2y_1^{(\nu)}$ so as to discard data on the right tail too close to zero.

Finally, the choice of the bounds $\zeta, \varepsilon^{(i)}, \mu^{(i)}$ in (17) and the initial guess for $\hat{\theta}$ are left to the experience of the operator who obviously will choose them on the basis of the experimental histogram. Because of the great importance of the starting point with respect to the problem of convergence to the global minimum, the procedure includes a preliminary interactive phase in which the "theoretical" histogram is displayed corresponding to each choice the operator makes for $\hat{\theta}$. Comparing this histogram to the experimental one should help in improving the choice itself.

6. DATA PROCESSING AND CONCLUSION

The procedure described has been tested against simulated data (obtained by means of a Monte Carlo technique) for various cases of mixture of two cell populations.

Such a test had a twofold purpose: a) to check capability of the procedure in the separation of the two populations, and b) to verify its robustness against statistical noise, background and sphape of the S-phases.

a) Two series of data were simulated and processed: in the first one the proportion of the two populations is changed, while in the second one the difference $y_1^{(2)} - y_1^{(1)}$ between the locations of the G_o/G_1 peaks is progressively decreased.

The procedure was able to properly separate the two populations by recovering the unknown parameters within few percents of the actual values, up to an unbalance in their relative weight of 8:1 and to a degree of nonoverlapping of one channel only (in this last case, however we took different values for $\gamma^{(1)}$ and $\gamma^{(2)}$; notice that the simulated histogram did not exibit any evidence of the presence of a second population).

b) The effect of statistical noise was checked by simulating and processing a series of data with a decreasing total number of cells N. Some instability begins to arise when N decreases below 7500.

Several sets of data were also simulated in which a background was added and a non uniform distribution density for S-phases was taken. Processing these data led always to a complete identification of the background term (and of the other parameters as well) and to an approximate reconstruction of the S-phase shape. We add that the relatively largest deviations from the actual values occurred in the S-phase estimation, as one should have expected.

The numerical results relative to all these experiments can be found in [9].

The procedure has been also applied to some clinical samples taken from the digestive tract.

In figs. 1-2 the experimental data corresponding to already established neoplastic diseases are reported, along with its best fit. The optimal estimates obtained for the phase fractions are also shown.

Fig. 3a refers to a case of a chronic atrophic gastritis (a so-called precancerous condition). From the histogram there is no evidence of a second population while the numerical result reveals a non-zero value for the $\lambda_1^{(2)}$ peak. In fig. 3b where another sample from the same biopsy having more cells than the previous one is reported, the presence of a second population is also graphically confirmed.

In conclusion, it appears that the proposed procedure can improve the evaluation of the results obtained by flow cytometric measurements, as referred to the magnitude, the ploidy level and the cell cycle phase percentages of abnormal stemlines. Some of these data may help in predicting the clinical behaviour of the neoplastic disease. Moreover, the possibility of detecting and evaluating with accuracy an abnormal population even when this one is not clearly identifiable by a qualitative analysis of the histogram looks very promising in the application of FCM data processing to early diagnosis of neoplasia. Finally, an extensive application of the procedure to a variety of experimental cases is needed in order to help in classifying different disease patterns and isolating suitable markers with diagnostic and prognostic significance.

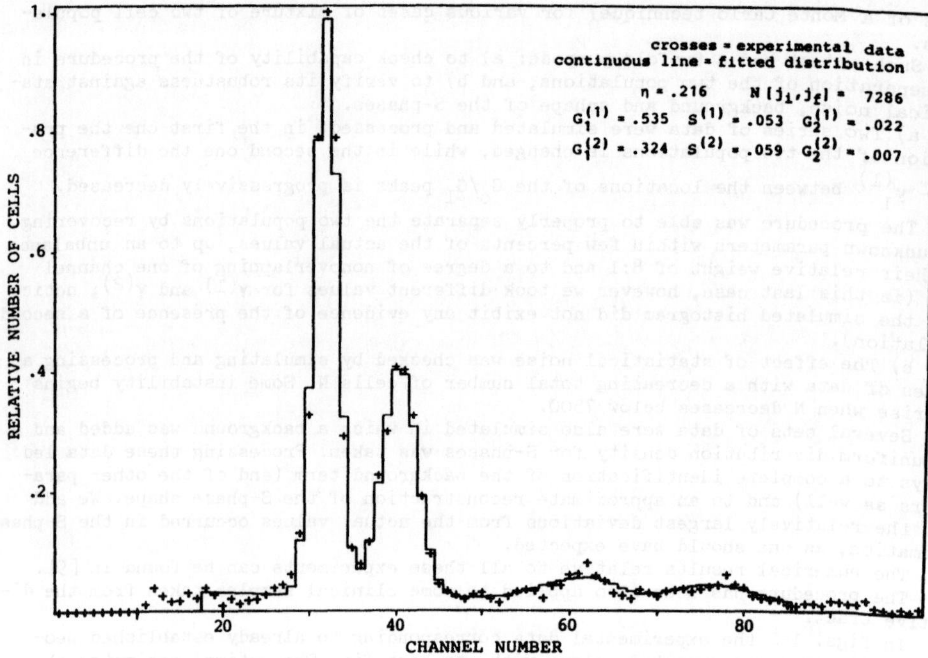
crosses = experimental data
continuous line = fitted distribution
η = .216 N [j₁,j₂] = 7286
G$_1^{(1)}$ = .535 S$^{(1)}$ = .053 G$_2^{(1)}$ = .022
G$_1^{(2)}$ = .324 S$^{(2)}$ = .059 G$_2^{(2)}$ = .007
RELATIVE NUMBER OF CELLS
CHANNEL NUMBER

Fig. 1

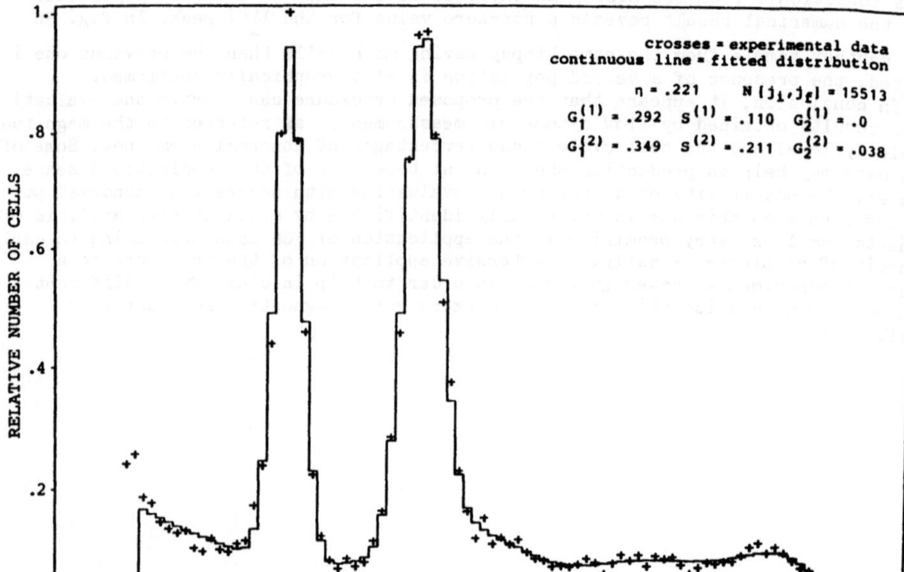
crosses = experimental data
continuous line = fitted distribution
η = .221 N [j₁,j₂] = 15513
G$_1^{(1)}$ = .292 S$^{(1)}$ = .110 G$_2^{(1)}$ = .0
G$_1^{(2)}$ = .349 S$^{(2)}$ = .211 G$_2^{(2)}$ = .038
RELATIVE NUMBER OF CELLS
CHANNEL NUMBER

Fig. 2

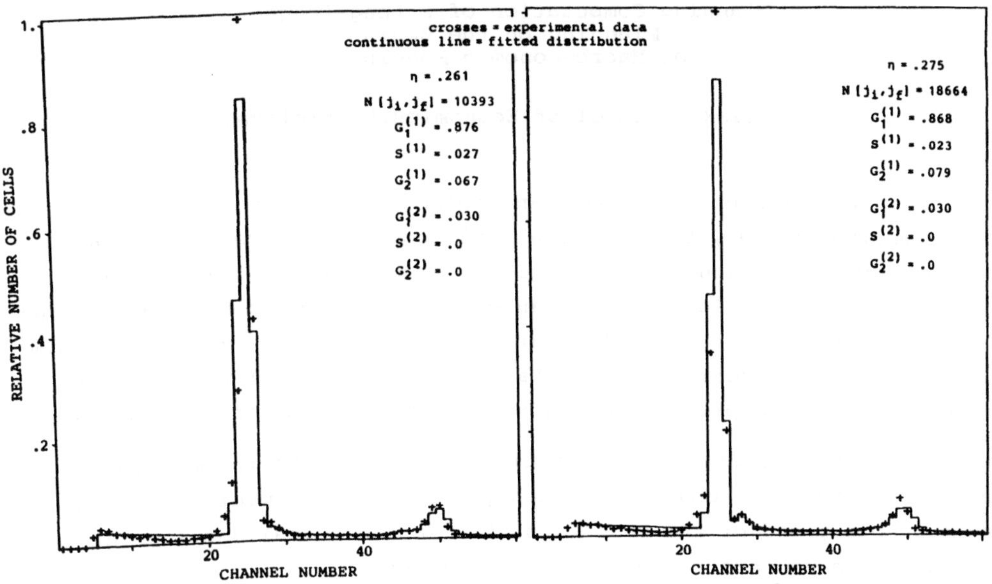

Fig. 3a Fig. 3b

REFERENCES

[1] P.S. RITCH, S.E. SHACKNEY, W.H. SCHUETTE, T.L. TALBOT, C.A. SMITH: *A Practical Graphical Method for Estimating the Fraction of Cells in S in DNA Histograms from Clinical Tumor Samples Containing Aneuploid Cell Populations.* Cytometry, 4, pp. 66-74, 1983.

[2] L. TEODORI, L. CAPURSO, E. CORDELLI, R. DE VITA, M. KOCH, M. TARQUINI, F. PALLONE, F. MAURO: *Cytometrically Determined Relative DNA Content as an Indicator of Neoplasia in Gastric Lesions.* Cytometry, 5, pp. 63-70, 1984.

[3] L. GRIPPO, F. LAMPARIELLO, S. LUCIDI: *A Non Monotone Line Search Technique for Newton's Method.* IASI Tech. Rept., R.100, Nov. 1984. (to appear on SIAM J. on Numerical Analysis).

[4] W. SCHMIDT: *The Problem of Cell Debris in Pulse-Cytophotometry: A Probability Theoretic Model.* J. Theor. Biol., 103, pp. 581-599, 1983.

[5] H.P. BECK: *Evaluation of Flow Cytometric Data of Human Tumours. Correction Procedures for Background and Cell Aggregations.* Cell Tissue Kinet., 13, pp. 173-181, 1980.

[6] H. CRAMER: *Mathematical Methods of Statistics.* Princeton University Press, Princeton, 1963.

[7] R.C. MANN, R.E. HAND Jr., G.R. BRASLAWSKY: *Parametric Analysis of Histograms Measured in Flow Cytometry.* Cytometry, 4, pp. 75-82, 1983.

[8] F. LAMPARIELLO, S. LUCIDI: *Analysis of DNA Distributions from Flow Cytometry by means of an Optimization Procedure.* IASI Tech. Rept. R.113, Feb. 1985.

[9] C. BRUNI, L. CAPURSO, G. KOCH, M. KOCH, F. LAMPARIELLO, S. LUCIDI, L. TEODORI: *Automatic Analysis of Flow Cytometrically Determined DNA Distributions in the Presence of Abnormal Stemlines.* IASI Tech. Rept. R.114, March 1985. (to appear on J. of Math. Modeling).

Planning Computation of a Long-Range Planning on the Basis of Macroeconomic Models

A.Lasciak - School of Economics,Bratislava,CSSR

The essential mission of a long-range planning is an elaboration of comprehensive social, technological and economic forecasts of the development conceptions of the national economy that form the main basis for determination of long-range political and economic goals and fixing of the principal development directions of the national economy in the given Five-Year Plans so that a more complete utilisation of national and socio-productive conditions of the economy are obtained.

The purpose of the long-range planning /LP/ is to put through the necessary structural changes, to apply the new achievments of science and technology and of the results of the technological development with regard to the cooperation of the CMEA countries.

In the present stage of the activities concerning the long-range planning in the CSSR we start from the fact that we deal with a formation of the economic structure in a period when hitherto application of the achievments of classical physics on which the contemporary industrial production is based will continually be replaced by a new technological basis enabling the realisation of new production technologies based on the last achievments of nuclear physics, quanto mechanics, chemistry, molecular biology and applied cybernetics and using cybernetics microelectronics for automation of production and management process. The forecast of the development of science and technology and the information of conditions for a utilisation of their achievments has a decisive importance for the development of our economy. The relevance of a given long-range planning can be presented step by step by straightforward methods:

Subject approach	Form approach	Methodological instruments	Research tree of mission phases
1° Population and its projection ↓	Tree of goals	Forecasts Morphpologic. analysis	1° Research project ↓
2° Standard of living ↓	Complex development programs	Economic projections /Scenarios/	2° Basic science ↓
3° Production, investments, capital funds ↓	Frames of long-range planning	System of DISPLAN models	3° Technological capability and economic potential ↓
4° Labour forces, employment ↓	Proposal of long-range outlook	National economy balances	4° Programs /technology creation/ ↓
5° Natural resources ↓	Proposal of the Five-Year Plan		5° Product of technology
6° Technical progress ↓			

In planning of the long-range outlook there are multidimensional aspects of economic rationality often of a different and conflict nature. Three categories of planning procedures can be distinguished here: algorithmic ones which represent the DISPLAN sphere,
heuristic ones and
fuzzy approaches /or uncertainty in the development/.

The DISPLAN sphere in long-range planning will represent a dialogue planning system on the basis of a complex economic, mathematical, statistical and technical means bounded with the data base which will be formed on line and in which the user will handle the extra prepared language or instructions.

The task of the DISPLAN will be to secure the management by the parameters of the formed balance system, the choice procedures, the goal programming and resource approach in a long-range planning on the basis of forecasting and economic projections. The function of this system in a broader sense of the term will be a demonstration of mutually tuned quantificated conceptions of the development of the national economy and balancing of complex programs of the economic development on the basis of formulated goals and effectiveness criteria.

Consequently, the human factor in the system of construction of the long-range planning has a dominant position for the present. The process of interconnection of human and machine components of the system is not only directed to remove the heuristic approaches but gradually to learn and to improve up to heuristic programs. As the user of the system is not usually acquainted with programming, all the existing applicable algorithms will be concentrated in the method and model bank in the user's version.

At the same time, the single moduls, after being determined by the user, send their requirements to the data fund to provide problems with data. The generator of outputs gives the presentation of results.

New Approaches in Long-Range Planning

The long-range planning represents a complex program of the economic development in the form of comprehensive programs of a complex development, aggregate models and socio-economic indicators of the economic development. The complex program of the economic development represents a quantification of the goal tree. The goal tree is a certain classification of the needs of society and population formed to a certain degree of differentiation and that are realised by means of complex programs of the economic development.

To every goal of a certain level and tree branch a number is assigned which determines the importance of the goals that means a norm of the goals is carried out.

The starting point of the complex programs of development are the demographic projections of population and dynamics of macroeconomic constrains. Their parameters will be expressed in the balances of the national economy and in different structural models by which the planning structure of the basic macro-ties is projected in the following construction procedure:

- goal tree
- forecasts

} goal programming approach

- complex programs
- economic projections

} goal-resource approach

By a strategy of economic development we mean a transformation of basic political, social and economic goals into the set of quantificated problems , programs, characterizing development and economic goals of the society in the area of sectoral structure and economy resources. The network structure of the goal-resource arrangement of processes and programs can be mathematically expressed by aset of multidimensional matrices or multidimensional graphs.

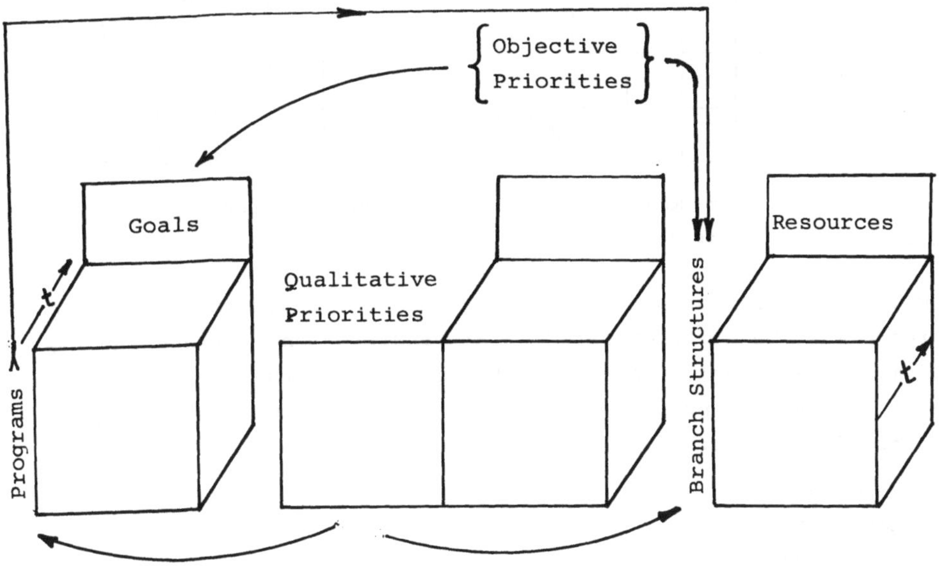

To establish the relevance of specific research projects to technological capabilities the k x m and m x n matrices above can be multiplied together to produce a single composite k x n matrix relating projects directly to technologies:

$$\left[\underbrace{\}\; k \text{ projects}}_{m \text{ sciences}} \right] \times \left[\underbrace{\}\; m \text{ sciences}}_{n \text{ technologies}} \right] = \left[\underbrace{\}\; k \text{ projects}}_{n \text{ technologies}} \right]$$

Similarly we can form a matrix set for the research tree:

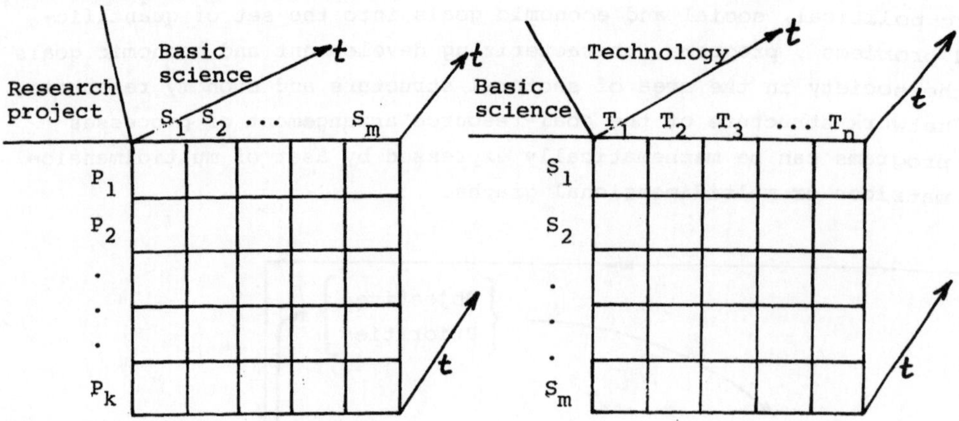

See * Project-science matrix Science-technology matrix

From the viewpoint of structuralizing the goal programming research approach must be differentiated by alternatives and variants of the economic development bounded with the meta-goal, the leading goals, goal criterias and preferences. In the same case we must differentiate 3 sorts of programs:

- development
- attenuation and
- liquidation ones.

From the viewpoint of space organization we must keep in view the international /CMEA/, national, sectoral and regional level. With the development programs also the development programs of science and technology are closely connected.

The task of forecasts in the long-range planning is to investigate and analyse the qualities of the dynamics of national economy; the qualities of which will be basically different from the state when the forecast has been set up. In a special case the forecast investigates the behaviour of the aggregate economy objects in a changing or stable environment in junction, goal or fraction situations. We can set up the following forecasts:

* R.U.Ayres: Technological Forecasting and Long-Range Planning, McGraw-Hill Comp., New York 1969
Optimum, Prices and Effectiveness, Svoboda Prague 1975

1. Demographic forecasts
2. Prognoses of natural resources and state of the living environment
3. Scientific and technological forecast
4. Social forecast
5. Forecast of the foreign and economic development
6. Army and strategy forecast
7. Economic forecast

Forecasting involves both the motion of relevant resources as well as the dynamics of social development.

Economic projections represent various techniques of illustrations scenarios, variants and alternatives of a long-range planning /as well as of a medium-range planning/ in the form of multistage management processes and planning of economic dynamics and its equilibrium. In a narrow sense the multistage processes represent multilevel planning with one or more effectiveness criteria. Economic projections are based on the principle of a historical road and the principle of Markov's road.

A general procedure in the area of economic projections is the following:

1° The construction of hypotheses, formulation of goals, forecast of national economy, forecast of scientific and technological development, control parameters, effectiveness criteria.

2° Construction of complex programs of the economic development leading into models or system of models on the basis of adaptive indicators for the given stages and volume of the material problems.

3° Solving of programs and models according to selected algorithms or heuristic approaches.

4° Analysis of solution. Formulation of conceptions of the long-range planning.

5° Proposals of the long-range planning.

6° Analysis of the development of the national economy.

7° New programs and models.

Economic projections represent the searching phase for making of different development ways.

Balances of national economy belong to the set of classical tools of socialist planning. We can imagine the traditional process of planning as a gradual information processing about tasks of the plan in order to obtain an equilibrium economic growth in its resources with needs /sector-resources approach/. The individual stages of information processing on the planning tasks represent a subject of a discussion between the supplier and carrier of the problem. The way of argumentation is understandable to both sides. The models in such a procedure remained a long time outside the planning process and were used only as a confrontation with the results of traditional balance procedures.

There is a need of elaborating also the traditional balance approaches in the areas of their dynamization.

From the methodological point of view dynamization is supported by macroeconomic analysis of production that is by an analysis of production factors in the changing conditions and proportions for the given production object.

System of Models of Long-Range Planning

The system of models for the needs of planning was developed at three levels so that they could be used for long-range planning, medium range planning and annual performance plan.

a. System of models at a horizontal level which describes various sides of planning activities at the level of the central planning authority.
b. System of models with a vertical aspect which describes the functioning of the planning process of different organisation stages in hierarchical relation. It is a process of multilevel planning.
c. System of models with a dynamic and a long-range planning.

When we construct a system of models we can follow two ways. The first one is based on building up of a system of models on the objects of national economy leading to an economic analysis. It is a traditional procedure. The second one deals with assumptions nearer to the planning reality and it can be described at three levels:
- at the first level processes are modelled which can appear in economic dynamics and consequently the basic strategies of decision on the basis of economic analysis are derived. A system of models on

objects with multicriteria effectiveness corresponds to this level.

- at the second level social and economic alternatives of preferences of the planning subjects are modelled which take part in the decision process concerning planning problems,
- at the third level the decision process is modelled.

The second way has its origin in multicriteria analysis as a means of evaluation of strategies and scenaria of long-range planning. This multicriteria decision process is carried out in fact on all planning stages where a coordination of partial, structural, local goals with global social interest is necessary. The above principles are elaborated in the system of models especially of 1st and 3rd level. The models of the 2nd level are only at the beginnings. For the needs of the Slovak planning organs for a long-range planning an open system of models was proposed, experiments on which have been carried out in 1967-79. The following system of models is involved here:

1. Production functions for four basic sectors of national economy
2. Production functions for ten sectors
3. Production functions for sixteen sectors
4. Model of extensive and intensive growth factors
5. Aggregate forecasting model of development of social product and national income /GNPNI/
6. Sectoral model of development of the social product
7. Projection of components of final production on the basis of input-output analysis
8. Dynamic optimalisation model for projection of the structure of basic sectors on the basis of various effectiveness criteria
9. Model of production and investments
10. Model of production and consumption
11. Model of incomes and expenditures of the population
12. Model of personal consumption
13. Model of social consumption
14. Model of stocks
15. Model of import and export
16. Model of investments

The above system of models has two parts. The first part is represented by aggregate macroeconomic models /5,6,7,8,9/. The second part consists of analytical models which have a partial character /3,4,10, 11,12,13,14,15,16/. The models express both the demand the supply aspect of the economy.

Mutual Ties between Individual Models

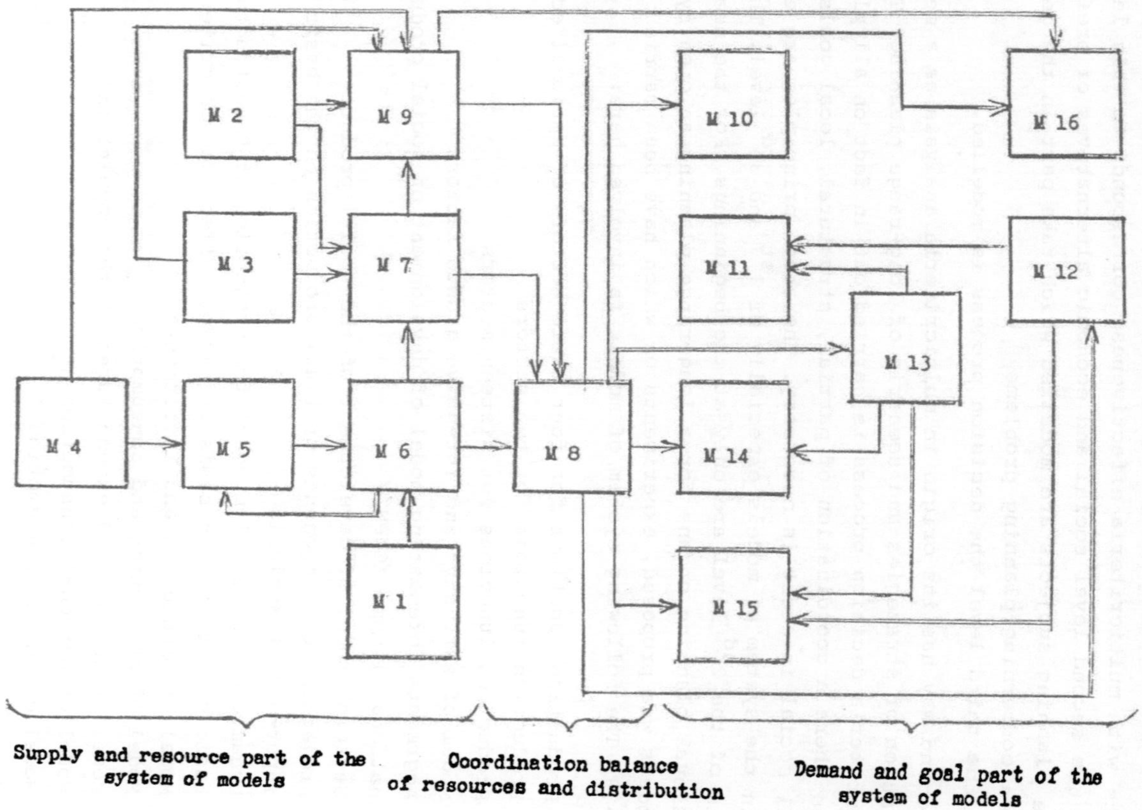

Analytical Instruments of Automatized System of Planning Calculations - for Long-range Outlook

The Part of Models

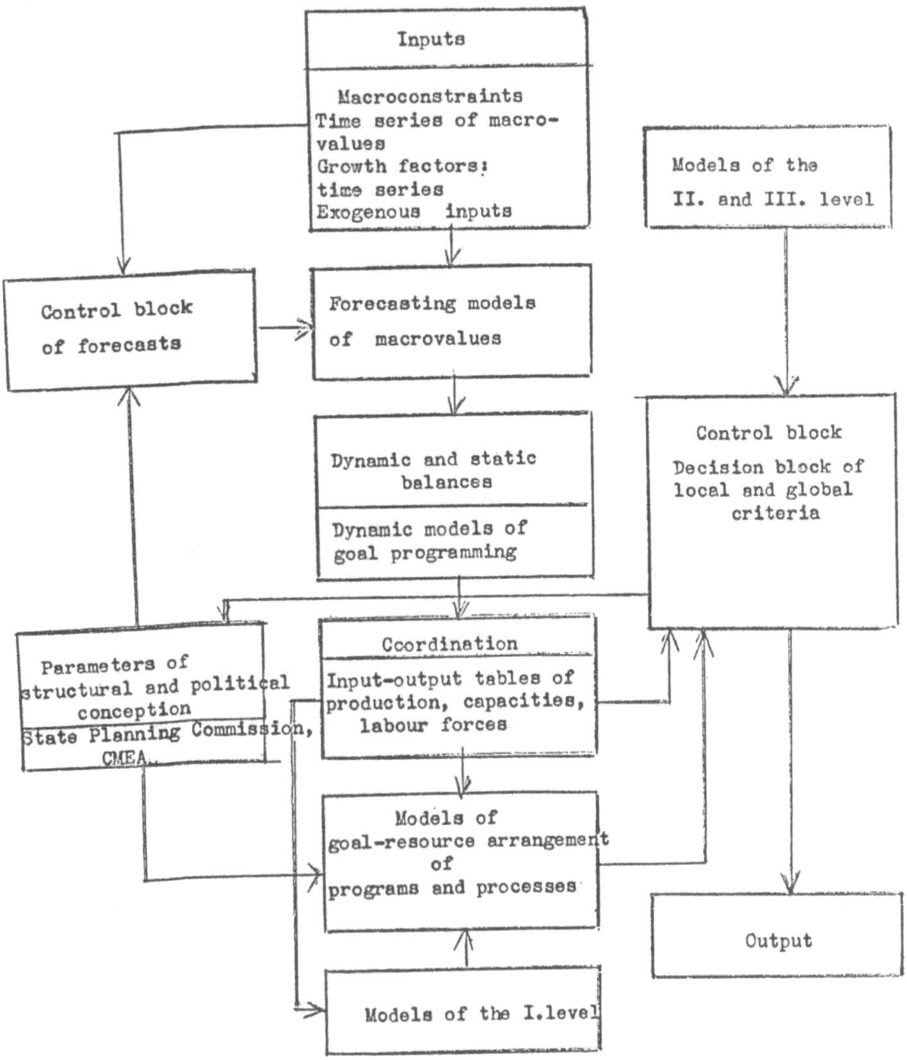

Figure

The construction of complex models for the needs of planning had to respect certain methodological and material requirements.

The system of experimented models was based on the balance methodology so as to secure the balance equilibrium between the resources and their use. The programming character had to respect qualitatively different parameters and restrictions which entered the individual Five-Year Plans and the " maturity " of development of the Czechoslovak economy. The Czechoslovak national economy also achieved a certain stage of complexity of labour division and at the same time the limiting factors of this development / energy, raw materials, labour forces / started to work. On the other hand we can see some possibilities to combine new technologies and to apply structural changes not only by means of investment but also non-investment factors /scientific and technological progress, management level, rationalisation degree, organisation level etc./.

The process of this new balancing in the form of a system of models is an interactive and iterative process. The system of models has to be organically included into the planning process so that the system of models accepts the above goal approach. The goals can be divided into two groups here:

a/ goals that will be in the system of scenarios /models of 1st and 2nd level/

b/ goals that will be expressed by strategies /models of 3rd level/.

The given case deals with an application of long-range searching with choice procedures that means to have a possibility of making decisions according to changing situations in the process of the solution of plans.

R e f e r e n c e s

1.Laščiak A.: Dynamické modely /Dynamic Models/,Alfa,Bratislava 1985
2.Bouška a kol.:Systém modelů pro střednedobé plánování,Akadémia,Praha 1978
3.J.de Mongolfier - P.Bertier:Approche multicritère des problèmes de décision, Édition Hommes et Techniques, Paris 1978
4.Martin K. Starr - Milan Zelený: Multiple Criteria Decision Making North-Holland Publ. Comp.,Amsterdam 1977

Boundary stabilization problems for hyperbolic dynamics - a review

IRENA LASIECKA
Mathematics Department
University of Florida
Gainesville, Florida 32611

1. Introduction

1.0. Preliminaries

The past several years have seen a surge of research activity focused on the general problem of feedback stabilization for hyperbolic partial differential equations, particularly in the case where the feedback action is exercised on the boundary Γ of the bounded spatial domain Ω in R^n. There have been various factors which have contributed to accelerate what was already a natural process of evolution on stabilization or stabilizability questions - from the finite dimensional theory (of linear, ordinary differential equations) of the late sixties to present day research on boundary control systems modeled by partial differential equations. Among them a non negligible role was apparently played by newly formed interest in the study of flexible structures [B-4]. Our emphasis here will be on boundary control problems rather than on distributed control problems. Our choice needs hardly be justified or defended. Physically, it is of greater interest and appeal to have only the boundary Γ of Ω available to external manipulations. For instance, there might be technical limitations that force one to place actuators and perhaps also sensors on the boundary Γ and not in the interior Ω. Mathematically, the questions raised in the boundary control case are far more challenging than the counter-part questions in the distributed control case, and - moreover - they are of interest in their own rights. Thus, in short, the present paper intends to review (selected portions of) the theory presently available in the area of boundary feedback stabilization problems for hyperbolic partial differential equations, with particular emphasis on the work of the past few years. We make no claim of completeness in our review and we regret that because of space limits we are forced to omit a number of interesting results.

The present review paper is loosely based on a talk given at IFIP Conference in Budapest Sept. 2-6, 1985. However, of the various topics touched in the talk, we are forced here by space constraints to focus only on the problem of boundary feedback stabilization of hyperbolic equations.

To begin with, we shall generally consider only canonical situations (involving the Laplacian operator, the bi-harmonic operator, etc.) and refer to the literature for more general elliptic operators, as appropriate. One reason of our choice of presentation is that the intrinsic difficulties of the problems in question are already present at the level of the canonical cases.

1.1. Canonical hyperbolic problems

We shall consider three fundamental cases of hyperbolic dynamics: 1) a second order scalar equation (wave equation) with Dirichlet or Neumann (or Robin) boundary conditions; 2) a fourth order scalar equation (bi-harmonic equation) with different boundary conditions; 3) a first order hyperbolic system (e.g. Maxwell equation). Henceforth in this paper, Ω will be an open bounded domain in R^n with smooth boundary Γ. We set, as usual $Q_T = (0,T) \times \Omega$ and $\Sigma_T = (0,T) \times \Gamma$. The case $T = \infty$ will also be included.

Second order scalar equation (wave equation)

$$\begin{cases} \dfrac{\partial^2 y}{\partial t^2} = \Delta y & \text{in } Q_T \\ y(0,\xi) = y_o(\xi) & \xi \in \Omega \\ y_t(0,\xi) = y_1(\xi) & \end{cases} \tag{1.1}$$

and boundary condition either of Dirichlet type

$$y(t,\sigma) = u(t,\sigma) \qquad \text{in } \Sigma_T \tag{1.1.D}$$

or else of Neumann type

$$\frac{\partial y}{\partial \nu} = u(t,\sigma) \qquad \text{in } \Sigma_T \tag{1.1N}$$

Fourth order scalar equation (beam, plate equation, etc.)

$$\begin{cases} \dfrac{\partial^2 y}{\partial t^2} = -\Delta^2 y & \text{in } Q_T \\ y(0,\xi) = y_o(\xi) & \\ y_t(0,\xi) = y_1(\xi) & \xi \in \Omega \end{cases} \tag{1.2}$$

with either boundary conditions (simply supported or hinged B.C.)

$$\begin{cases} y = 0 & \text{in } \Sigma_T \\ \Delta y = u(t,\sigma) & \end{cases} \tag{1.2.a}$$

or else (free B.C.)

$$\begin{cases} \Delta y = u_1(t,\sigma) & \text{in } \Sigma_T \\ \frac{\partial}{\partial \nu} \Delta y = u_2(t,\sigma) & \end{cases} \tag{1.2.b}$$

First order hyperbolic system

$$\begin{cases} \frac{\partial y}{\partial t} = A(x,\partial)y & \text{in } Q_T \\ y(0,\xi) = y_o(\rho) & \text{in } \Omega \end{cases} \tag{1.3}$$

the boundary conditions

$$My(t,\sigma) = u(t,\sigma) \qquad \text{in } \Sigma_T \tag{1.3a}$$

where $y(x)$ is a k-vector, $A(x,\partial)$ is strictly hyperbolic operator of the following form

$$A(x,\partial)y = \sum_{j=1}^{n} A_j(x)\frac{\partial y}{\partial x_j} + B(x)y$$

with coefficients A_j, B smooth $k \times k$ matrices. The boundary operator M is given by an $\ell \times k$ matrix valued function $M(x)$ (ℓ-stands for the number of negative eigenvalues of $A^N(x) \equiv \sum_{j=1}^{n} A_j N_j(x)$ $\vec{N} = (N_1,\dots,N_n)$ inward unit normal). We assume that Rank $M(x) = \ell$.

In all the above cases, the control function u acts in the boundary conditions. A qualitative statement of the boundary feedback stabilization problem may now be given at this point in the following terms with reference to the second order problems (1.1)-(1.1.D-N) or else (1.2)-(1.2.a-b) [respect. to the first order problem (1.3)], seek if possible a linear generally unbounded operator F (feedback operator) acting on the state (y,y_t) [respect. on the state y] from a suitable space based on Ω or even on Γ, into the space $L_2(\Sigma_\infty)$, $\Sigma_\infty = (0,\infty) \times \Gamma$, i.e. $F\begin{vmatrix} y \\ y_t \end{vmatrix} \in L_2(\Sigma_\infty)$ [respct. $Fy \in L(\Sigma_\infty)$], such that with $u = F\begin{vmatrix} y \\ y_t \end{vmatrix} \in L_2(\Sigma_\infty)$ inserted in (1.1.D-N) or else in (1.2.a-b), [respect. with $u = Fy \in L_2(\Sigma_\infty)$ inserted in (1.3a)], the corresponding closed loop feedback system possesses the following two properties:

(i) it is well posed in the variables (y,y_t) [respect., in the variable y] in some suitable underlying space (for instance, it generates a s.c. semigroup on such underlying space); (ii) the feedback solutions $\left|\begin{matrix} y(t;y_o,y_1) \\ y_t(t;y_o,y_1)\end{matrix}\right|$ [resp. $y(t,y_o]$, converge to zero in the weak, strong or - more desirably - uniform operator norm of the underlying space as $t \to \infty$, for all initial data (y_o,y_1) [resp. y_o] in it.

2. Semigroup models for hyperbolic boundary problems, Regularity of the solutions.

2.0. Preliminaries

Several approaches and techniques have been used to solve problems of the type reviewed in this article: energy or multiplier methods, functional analytic or operator theoretic methods, pseudo-differential operator techniques, etc. Indeed, often the final solution has emerged as a combination of a few different techniques. An illustrative case in point is paper [L-T.1], of whose content we shall have to say more below. Thus, while the successful technique appears to be dependent upon, and subordinated to, the particular problem under study, one may also make the point that it has been found advantageous to provide abstract operator theoretic models for the original P.D.E.'s problems (1.1)-(1.1.D-N); (1.2)-(1.2.a-b), (1.3) and others. These operator models posses some convenient features, including the conceptual one of allowing for a reformulation of the original P.D.E.'s problems (and other dynamics as well) in a unified manner and in a language (formalism) which closely resembles the classical finite dimensional theory (of linear ordinary differential equations). Thus, justified by the preceding comments, we shall report below operator theoretic models of the dynamics introduced in section 1.1. Because of space restriction we shall confine ourselves only to the treatment of Dirichlet model 1.1 while for other dynamics introduced in section 1 we refer to [Ch-L], [L-T-1], [L-T-4].

2.1. Wave equation with Dirichlet - boundary control

Problem (1.1)-(1.1.D) admits the following operator model [T-1], [L-T-2]

$$y(t) = C(t)y_o + S(t)y_1 + (L_D u)(t) \tag{2.1}$$

$$(L_D u)(t) = A \int_0^t S(t-z)Du(z)dz \tag{2.2}$$

$$y_t(t) = -AS(t)_{yo} + C(t)y_1 + (\tfrac{d}{dt}L_D u)(t) \tag{2.3}$$

$$(\tfrac{d}{dt}L_D u)(t) = A \int_0^t C(t-z)Du(z)dz \tag{2.4}$$

Here we have

$$-A : L_2(\Omega) \supset \mathcal{D}(A) \equiv H_o^1(\Omega) \cap H^2(\Omega) \to L_2(\Omega) \tag{2.5}$$

is the generator of a s.c. cosine operator C(t) on $L_2(\Omega)$ with $S(t)y = \int_o^t C(z)y\,dz$.

D is the Dirichlet map defined by (2.6)

$$Dv = h \iff \begin{cases} \Delta h = 0 & \text{in } \Omega \\ h = v & \text{on } \Gamma. \end{cases}$$

Elliptic theory gives [L-M], [N-1], [F-1]

$$D : \text{continuous } L_2(\Gamma) \to H^{1/2}(\Omega) \subset H^{1/2-2\varepsilon}(\Omega \equiv \mathcal{D}/A^{1/4-\varepsilon}) \tag{2.7}$$

the operator

$$\mathcal{A} \equiv \begin{pmatrix} 0 & I \\ -A & 0 \end{pmatrix}: \mathcal{D}(\mathcal{A}) = \mathcal{D}(A^{1/2}) \times L_2(\Omega) \tag{2.8}$$

generates a s.c. unitary group $e^{\mathcal{A}t}$ on the space

$$E \equiv L_2(\Omega) \times H^{-1}(\Omega).$$

$$B : L_2(\Gamma) \to E \tag{2.9}$$

$$Bu = \begin{vmatrix} 0 \\ ADu \end{vmatrix} \qquad \text{on } [\mathcal{D}(\mathcal{A})]' \ (*)$$

Notice that $\mathcal{B}u \in E \Rightarrow u = 0$ (since $ADu \in H^{-1}(\Omega) \Rightarrow u = 0$). Thus $\mathcal{B}$ in (2.9) is intended as $\mathcal{B} : L_2(\Gamma) \to [\mathcal{D}(\mathcal{A})]'$.

With the above notation (2.1) and (2.2) can be rewritten as the first order equation.

$$\frac{d}{dt}\begin{bmatrix} y \\ y_t \end{bmatrix} = \mathcal{A}\begin{bmatrix} y \\ y_t \end{bmatrix} + \mathcal{B}u \qquad \text{on } [\mathcal{D}(\mathcal{A})]' \tag{2.10}$$

$$\begin{bmatrix} y(0) \\ y_t(0) \end{bmatrix} = \begin{bmatrix} y_o \\ y_1 \end{bmatrix} \in E \tag{2.11}$$

Model (2.10), while reminiscent at finite dimensional or distributed control system (where the "B" - operator is bounded on the basic space) displays the pathologies that now occur in the modeling of the Dirichlet boundary control problem (1.1.1-11.1.D) through the operator $\mathcal{B}$; i.e. the domain of AD as an operator $L_2(\Gamma) \to H^{-1}(\Omega)$

(*) $[\mathcal{D}(\mathcal{A})]'$ stands for the fual space to $\mathcal{D}(\mathcal{A})$.

is trivial ; and likewise, the domain of $\mathscr{B}$ as an operator from $L_2(\Gamma) \to E$ is also trivial. Thus, the corresponding equation (2.10) has to be regarded in enlarged space $[\mathscr{D}(\mathscr{A})]' \supsetneq E$. All this accounts for considerable extra difficulties over the distributed control case of "B" bounded on the basic space, since our final results are sought after in the basic space, say E for problem (1.1)-(1.1.D). In this context, one should also justify the choice of the "natural state space" - $E \equiv L_2(\Omega) \times H^{-1}(\Omega)$. This problem is closely related to the regularity of the operators L_D and $\frac{d}{dt} L_D$ as acting on $L_2(\Sigma)$. The one dimensional case, dim $\Omega = 1$, admits an ad hoc direct (and simple) analysis, based on the physical part that the "input" u, say applied only at one endpoint of Ω, travels with speed equal to one and is reflected at the other endpoint of Ω, as to satisfy the zero B.C. This way one can write a direct formula for the solution y(t), which - in turn - readily reveals that $y(t) \in C[0T;L_2(\Omega)]$ when dim $\Omega = 1$. The case dim $\Omega > 1$ is considerably more difficult. Here, sharp - in fact optimal regularity results have been obtained only recently

<u>Theorem 2.1</u> [L-T-2], [L-T-6], [L] [L-L-T]
We have for the operators in (2.2), (2.4)

$$L_D : \text{continuous } L_2(\Sigma_T) \to C(0T;L_2(\Omega)] \tag{2.12}$$

$$\frac{d}{dt} L_D : \text{continuous } L_2(\Sigma_T) \to C[0T;H^{-1}(\Omega)].$$

Thus the map

$$\{y_o,y_1,u\} \in E \times L_2(\Sigma_T) \to \begin{bmatrix} y \\ y_t \end{bmatrix} \in C[0T;E] \tag{2.13}$$

is continuous. Moreover

$$\left.\begin{matrix} D^*A^*S^*(\cdot) \\ D^*A^{*\,1/2}\,C^*(\cdot) \end{matrix}\right\} : \text{continuous } L_2(\Omega) \to L_2(\Sigma_T) \tag{2.14}$$

equivalently

$$\left\|\frac{\partial\phi}{\partial\gamma}\right\|_{L_2(\Sigma_T)} \leq C_T\|\{\phi_o,\phi_1\}\|_E$$

for the problem

$$\begin{cases} \phi_{tt} = \Delta\phi & \text{in } Q_T \\ \phi|_\Gamma = 0 \\ \phi|_{t=0} = \phi_o;\ \phi_t|_{t=0} = \phi_1 \end{cases} \tag{2.15}$$

The above regularity result justifies the choice of the space E as the appropriate state space for the problem (1.1)-(1.1.D).

Remark

Similar approach can be applied to all hyperbolic problems described in section 1. The semigroup model for the wave equation with Neumann Boundary condition is presented in [L-T-2], [L-T-7], [L-T-8]. Operator models for bi-harmonic equation are given in [L-T-4]. First order hyperbolic systems are discussed in [Ch-L].

3. Stabilization - general approach

3.1. General considerations

In Section 2 we have seen that all our four prototypes of boundary control problems for hyperbolic dynamics share the following features. They can be written as an abstract operator equation

$$\dot{z} = \mathscr{A}z + \mathscr{B}u \qquad \text{on Hilbert space Z} \tag{3.1}$$

$$z(0) = z_o \in H \qquad H \not\supseteq Z$$

where:

a) $\mathscr{A}$ is the generator of a s.c. semigroup (group) on the basic Hilbert space H (which was $L_2(\Omega) \times H^{-1}(\Omega)$ for problem (1.1)-(1.1.D); $H^1(\Omega) \times L_2(\Omega)$ or $L_2(\Omega) \times (H^1(\Omega))$, for problem (1.1)-(1.1.N);

b) $\mathscr{B}$ as an operator $U(\equiv L_2(\Gamma)) \to H$, has trivial domain {0}, while - in fact - $\mathscr{B} \in \mathscr{L}(U,Z)$ where, however Z is another Hilbert space strictly larger than H : $Z \supsetneq H$, and with weaker topology than H; we have seen that we can take $Z = [\mathscr{D}(\mathscr{A})]'$. The pathology of point b) is an intrinsic feature of the boundary nature of the hyperbolic dynamics.

We then say that (3.1) is strongly stable on H, [respectively; uniformly (exponentially) stable on H], in case: there is a linear feedback operator F, u = Fz, such that

(i) $\mathscr{A} + \mathscr{B}F$ generates a s.c. semigroup on H, denoted by $e^{(\mathscr{A}+\mathscr{B}F)t}$;

(ii) $e^{(\mathscr{A}+\mathscr{B}F)t}$ converges strongly to zero on H, as $t \to \infty$

$$\|e^{(\mathcal{A}+\mathcal{B}F)t} z_o\|_H \to 0 \text{ as } t \to \infty \;\; \forall z_o \in H \tag{3.2}$$

[respectively $e^{(\mathcal{A}+\mathcal{B}F)t}$ satifies

$$\|e^{(\mathcal{A}+\mathcal{B}F)t}\|_{\mathcal{L}(H)} \leq M\, e^{-\delta t}, \quad t > 0, \; \delta > 0 \tag{3.3}$$

It may be worth at this point to analyze the problem of stabilization in the hyperbolic case, by pointing out relevant differences with the corresponding parabolic case. At the outset, we may remark that while weak, strong and uniform stability coincide for parabolic dynamics where $e^{\mathcal{A}t}$ is compact, [B.1], this is not so for hyperbolic dynamics, where one may have strong stability without achieving uniform stability. Let us analyze the differences between parabolic and hyperbolic dynamics more closely.

If $\mathcal{A}$ is an unstable generator of a s.c., analytic semigroup on H, as in the parabolic case on a bounded domain $\Omega \subset R^n$, its spectrum $\sigma(\mathcal{A})$consists of isolated eigenvalues in a triangular sector as in Fig. 3.a

Thus, there are only finitely many eigenvalues in Re $\lambda \geq 0$. The problem of stabilization is then to seek linear operator F such that $\mathcal{A} + \mathcal{B}F$ generates a s.c., analytic semigroup on H, which moreover satisfies (see Fig. 3.b)

$$\sup \operatorname{Re} \sigma(\mathcal{A} + \mathcal{B}F) < \delta, \qquad \delta > 0 \tag{3.5}$$

so that the uniform decay (3.3) is then assured. This is a special case of semigroups which satisfy the spectrum determined growth condition (SDGC) [T.2].(3) Thus, in the parabolic case, the function of the feedback operator F is to remove the finitely many eigenvalues of $\mathcal{A}$ in Re $\lambda \geq -\delta$ in Fig. 3.a and achieve the eigenvalue placement (3.5), $\delta > 0$ pre-assigned in Fig. 3.a. This guarantees then the uniform decay (3.3) on H. Moreover, the well-established perturbation theory of analytic semigroups $\mathcal{A}$ on H can be invoked to establish that $\mathcal{A} + \mathcal{B}F$ is likewise a generator of an analytic semigroup on H [K.1]. The situation is drastically different in the hyperbolic case. To begin with, if $\mathcal{A}$ is the generator of a unitary s.c. group on H, non dissipative perturbations P of the type

(3) The SDGC is satisfied by analytic semigroups, differentiable semigroups, positive semigroups.

naturally encountered in boundary feedback problems may destroy semigroup generations; e.g. one may construct an operator P which is one dimensional range, $\mathcal{A}^{\varepsilon}$-bounded for any $\varepsilon > 0$ yet $\mathcal{A}$ and P does not generate a s.c. semigroup, see [T-3], also [L-T-4]. Thus, to prove generation of $\mathcal{A} + \mathcal{B}F$ with $\mathcal{A}$ generator of a unitary group (wave equation) and a feedback operator F coming from the boundary, ad hoc devices have to be used, see e.g. [L-T-4, L-T-5]. Second, the spectrum location of the homogeneous wave equation (which gives rise to a unitary group generator $\mathcal{A}$ on H) consists now of an infinite sequence of eigenvalues of the imaginary axis. To obtain strong stability, a necessary condition is that these infinite eigenvalues be all removed by feedback action F to Re $\lambda < 0$; while to obtain uniform stability with upper bound $Me^{-\delta t}$, $\delta > 0$ as in (3.3) a necessary condition is that these infinite eigenvalues be all removed by feedback action to Re $\lambda < -\delta$. However, in these cases the SDGC cannot be invoked directly and the above conditions are only necessary and this is the third difficulty, in the hyperbolic case a serious one; to ascertain the uniform decay (3.3) once the spectrum location (3.5) has been achieved, one has to work further in the hyperbolic case. One very desirable way is to establish simultaneously that the corresponding new eigenvalues of $\mathcal{A} + \mathcal{B}F$ form a Riesz basis in H. This program was carried out e.g. in [L-T-4] in every general setting, with applications which include parabolic, hyperbolic second order equations with or without damping, and fourth order hyperbolic equations.

Thus, the problem of uniform stabilization is much more delicate and demanding in the hyperbolic case over the parabolic case. The pre-assigned form of the operator F is also an issue. We shall see in the next section that uniform stabilization of the wave equation with feedback in the Neumann or Dirichlet B.C. was achieved only in recent years. Moreover, these results require some geometrical conditions to be satisfied by the bounded domain Ω. Techniques to show uniform stabilization are mostly different from those that may be used to show strong stabilization see e.g. [L-T-1] strong stabilization - unlike uniform stabilization does not require geometric conditions on Ω. Spectral analysis techniques are helpful in showing strong stabilization, but have not been fruitful so far in showing uniform stabilization of the wave equation.

3.2. Negative results: lack of uniform (exponential) stabilization

In the case of parabolic equations, weak and uniform stabilization coincide [B-1], since the semigroup is compact (on bounded domains Ω). By contrast, hyperbolic dynamics may be strongly stabilizable without being uniformly (exponentially stabilizable). This situation occurs for example when the free dynamics generates a group of contractions, (as in the case of the wave equation or the biharmonic equation) and the feedback operator is compact.

Theorem 3.2.1. [R.1] *

Let $\mathscr{A}$ generate a s.c. group G(t) on the Hilbert space H with $\|G(-t)\| \leq 1$, $t \geq 0$. Let K be compact $U \to H$. Then

$$\|e^{(A+K)t}\| \geq 1, \quad t \geq 0$$ □

Thus, by the above Theorem, uniform (exponential) stabilization for the dynamics described by the examples in section 1, however with distributed rather than boundary control (B bounded), and, say - with a finite rank bounded (compact) feedback operator F is out of question, as K = BF is compact. Finite rank feedback operators are attractive as they correspond to finitely many actuators. In veiw of this, it is natural then to resort to boundary control which do not give rise to a feedback BF compact. However, even in this situation, a counterpart of Theorem 3.2.1 still holds true; it concerns (unbounded) $\mathscr{B}^*$ subject to a certain L_1-condition, and finite rank F.

Theorem 3.2.2 [L-T-4]

Let A be a generator of a group e^{At} on H with $\|e^{-At}\| \leq 1$, $t \geq 0$. Let the feedback operator F be of the form $Fy = \sum_{j=1}^{N} (y,w_j)_H b_j$, with $w_j \in H$ and $b_j \in U$. Assume the following property on the operator $\mathscr{B}^*$ (possibly unbounded boundary feedback cases)

$$\int_0^T |(B^*e^{A^*t}y_o,b_j)_U| dt \leq C\|y_o\|_H\|b_j\|_U \tag{3.1.1}$$

Then, A + BF generates a s.c. semigroup $e^{(A+BF)t}$ on H which satisfies

$$\|e^{(A+BF)t}\| \geq 1, \quad t \geq 0$$

Remark 3.1. (i) The first claim of Theorem 3.2.2, i.e. that A + BF is a generator of a s.c. semigroup should not be under estimated. Counterexamples more recently constructed [T-3] [L-T-4] showing that when A is a unitary group generator on H, on unbounded additive perturbation P which is A-bounded and finite rank (therefore unclosable and with A-bound equal to zero) may well destroy the property of generation, in the sense that A and P may well for $\mathscr{A}_o$ be the generator of a s.c. semigroup, in the absence - (typical of boundary feedback problems) - of dissipativity on P. See also [D-5] for a different counterexample.

Remark 3.2 It can be shown that assumption (3.1.1) is typically satisfied for classes of hyperbolic dynamics with a feedback acting on the boundary (in the L_2-sense, in fact). In fact in case of wave equation with Dirichlet-boundary

* A result of similar nature was reproved later in [Gibson] [G-1]

conditions this is a consequence of (2.14). Thus, on the basis of Theorem 3.2.2, two strategies may be delineated in achieving stabilization of boundary feedback dynamics. One is to use finite rank operators F (appealing in their simple physical implementation) at the price of giving up on exponential stabilization, and settle only for strong stabilization as an acceptable substitute. The other strategy is to escape the class covered by the negative result of Theorem 3.2.2 and still strive for exponential stabilization. These two situations will be examined next. Section 4 deals with strong stabilization and section 5 with uniform stabilization via boundary feedbacks.

4. Strong stabilization

4.1 Dissipative feedbacks

We begin, as a motivation, with the case of distributed control, i.e. with B bounded $U \to H$. If the original s.c. semigroup generated by A is contraction on H, then a natural candidate for the feedback action is to take the operator $-\mathcal{B}^*$ which preserves dissipativity. This way, $A - BB^*$ is still a generator of a contraction semigroup. To show strong stabilization, a controllability assumption is imposed. Several proofs are now available of the following result.

Theorem 4.1 [S-1], [B-2]
Let A generate a s.c. contraction semigroup e^{At} on H, let $B \in \mathcal{L}(U,H)$ and assume that the system (A,B) is approximately controllable. Then, $e^{(A-BB^*)t}$ is weakly stable. Moreover, it is strongly stable, if in addition the resolvent $R(\lambda,A)$ is compact □

The same idea, with the appropriate new technicalities, works also to obtain dissipative candidates for stabilization in the boundary control case. This is, if one likes, a bonus of using abstract operators models (3.1) to describe boundary and weak dynamics.

For instance, let us return to the wave equation into Dirichlet boundary control, (1.1), (1.1.D), whose abstract model was provided in (2.1)-(2.10). We recall that $E \equiv L_2(\Omega) \times H^{-1}(\Omega)$, $H^{-1}(\Omega) = [\mathcal{D}(A^{1/2})]$ and also the definition (2.11) of $\mathcal{B} : L_2(\Gamma) \supset \mathcal{D}(\mathcal{B}) \to E$. Then, if $(\mathcal{B}u,v)_E = (u,\mathcal{B}^*v)_{L_2(\Gamma)}$ with u,v, in appropriate domains we have

$$(\mathcal{B}u,v)_E = (ADu,v_2)_{[\mathcal{D}(A^{1/2})]'} = (Du,v_2)_{L_2(\Omega)} = (u,D^*v_2)_{L_2(\Gamma)} \tag{4.1}$$

Thus

$$\mathcal{B}^*v = D^*v_2 = D^*A\,A^{-1}v_2 = +\frac{\partial}{\partial\nu}(A^{-1}v_2) \tag{4.2}$$

with $v = [v_1, v_2] \in \mathcal{D}(\mathcal{B}^*)$. Thus, the hyperbolic feedback dynamics corresponding to the operator $\mathcal{A} - B\mathcal{B}^*$, dissipative on E, is

$$\begin{cases} \frac{\partial^2 y}{\partial t^2} = \Delta y & \text{on } Q_\infty = (0,\infty) \times \Omega \\ y_o, y_1 \in L_2(\Omega) \times H^{-1}(\Omega) & \\ y = -\frac{\partial}{\partial \nu}(A^{-1} y_t) & \text{on } \Sigma_\infty \neq (0,\infty) \times \Gamma \end{cases} \tag{4.3}$$

One can then verify that the "range condition" in Lummer-Phillips theorem also holds and conclude that $\mathcal{A} - \mathcal{B}\mathcal{B}^*$, i.e. (4.3), generates a contraction semigroup $[y(t), y_t(t)] \in e^{(\mathcal{A}-\mathcal{B}\mathcal{B}^*)t}$ on E. Moreover, using the Nagy-Foias-Fogel decomposition theory, see [L-2] for an excellent account, combined with spectral analysis to exclude the existence of eigenvalues of $\mathcal{A} - \mathcal{B}\mathcal{B}^*$ on the imaginary axis, one also obtain strong stabilization of (4.3) via an operator-theoretic proof. No assumption on the geometry of Ω is required except the usual of the boundary.

Theorem 4.2 [L-T-1].
Let Ω be an open bounded domain in R^n, with sufficiently smooth boundary. For each initial conditions y_o, $y_1 \in L_2(\Omega) \times H^{-1}(\Omega) \equiv E$, the (semi-group) solutions of the feedback system (4.3) are strongly stable, in E, i.e.

$$\left\| \begin{vmatrix} y(t) \\ y_t(t) \end{vmatrix} \right\|_E = \left\| e^{(\mathcal{A}-\mathcal{B}\mathcal{B}^*)t} \begin{vmatrix} y_o \\ y_1 \end{vmatrix} \right\|_E \to 0 \text{ as } t \to \infty$$

A similar approach applied to the Neumann problem for the wave equation (1.1)-(1.1.N), leads to the following system

$$\begin{cases} \frac{\partial^2 y}{\partial t^2} = \Delta y & \text{in } Q_\infty \\ y_o,\ y_1 \in H^1(\Omega) \times L_2(\Omega) & \\ \frac{\partial y}{\partial \nu} = -y_t & \text{in } \Sigma_\infty \end{cases} \tag{4.5}$$

The same operator-theoretic proof given for the Dirichlet case above permits to demonstrate [L-T-1] the following result, originally proved through different techniques by various authors.

Theorem 4.3 [S-1] [Q-R] [Z-1]
With Ω a general open bounded domain in R^n, let the initial conditions y_o, $y_1 \in Z \equiv H^1(\Omega) \times L_2(\Omega)$. Then the (semi-group) solutions of the feedback system (4.5) are

strongly stable, i.e.

$$\left\| \begin{vmatrix} y(t) \\ y_t(t) \end{vmatrix} \right\|_Z = \| e^{(\mathcal{A}-\mathcal{B}\mathcal{B}^*)t} \begin{vmatrix} y_o \\ y_1 \end{vmatrix} \|_Z \to 0 \text{ as } t \to \infty$$

Remark 4.1. We shall see in the next section that under additional geometrical conditions on Ω, the feedback systems (4.3) and (4.5) are, in fact, uniformly (exponentially) stable in the norms of $L_2(\Omega) \times H^{-1}(\Omega)$ and $H^1(\Omega) \times L_2(\Omega)$, respectively. Moreover the boundary feedback functions are in $L_2(0,\infty,L_2(\Gamma))$ result important in relation to the regulator theory of these dynamics. The proofs of this results are, however, much more complicated and require the contributions of other techniques (multiplier methods), as mentioned in section 2.

4.2. Feedbacks of finite rank

Another class of feedback operators, of interest in its own right, is represented by finite rank (not necessarily bounded) operators. (There is no space in this review paper to even mention dynamic compensators). Although, under the assumptions of Theorem 3.3.2, they may at best yield strong stabilization - a much weaker conclusion - their appeal is in their simple structure, which may be thought of as being easily implementable by means of finitely many actuators. Mathematically, they present serious difficulties of their own, and require special ad hoc treatments. Generally, they do not yield dissipative feedbacks, i.e. contraction feedback systems, and this is a feature both interesting and also revealing of the difficulties involved. We begin with the wave equation with feedback in the Dirichlet B.C.

$$\begin{cases} \dfrac{\partial^2 y}{\partial t^2} = \Delta y & \text{in } Q_\infty \\ y_o, y_1 \in L_2(\Omega) \times H^{-1}(\Omega) & \\ y|_\Sigma = u & \text{in } \Sigma_\infty \\ u = F(y,y_t) = [y(t),w_1) + y_t(t),w_2)]g & \\ g \in L_2(\Gamma),\ (\ ,\ ) = L_2(\Omega)\text{-inner product} & \end{cases} \tag{4.6}$$

Since the natural state space for this problem is the space $L_2(\Omega) \times H^{-1}(\Omega)$, see the regularity result in Thm. 2.1, the stabilization problem can then be formulated as follows: given a suitable vector $g \in L_2(\Gamma)$ find suitable $w_1 \in L_2(\Omega)$ and $w_2 \in H_o^1(\Omega)$ such that the corresponding feedback problem generates a s.c. semigroup on $L_2(\Omega) \times H^{-1}(\Omega)$which is strongly stable.

Before starting the strong stabilization result, we remark that the preliminary question of feedback generation (wellposedness) is far from being clear. Indeed, it is known by now (see [T-3], [L-T-4], [D-Sch]) that unbounded perturbation of unitary groups (which is certainly the case in 4.6) may destroy generation. However it turns out, the operator situation which arises from (4.6) does not fall into the pathology of the counterexamples provided in the above stated references and indeed (4.6) generates a s.c. semigroup. This follows from the following abstract result:

Theorem 4.4 [L-T-7], [L-T-4]
Let $\mathcal{A}$ be a s.c. semigroup generator on H (generally unbounded). Assume that the operator $B : U \to H$ satisfies

$$\int_0^T |B^* e^{\mathcal{A}^* t} x|_U^2 dt \leq C|x|_H^2 \qquad \forall x \in H. \tag{4.7}$$

Then, for any $F \in \mathcal{L}(H \to U]$ we have that A + BF generates s.c. semigroup on H. □
Theorem 4.4 applies in particular to problem 4.6. Indeed, it turns out that with $H = L_2(\Omega) \times H^{-1}(\Omega)$ we have $\mathcal{B}^* e^{\mathcal{A}^*} x = D^*A^*S^*(t)x_1 + D^*A^{1/2} C^*(t)A^{-1/2} x_o$ and (4.7) is a simple consequence of the trace regularity given by (2.14).
Similar trace regularity" results for hyperbolic first order" systems are given in [L-T-4] and [Ch-L].
Notice that condition (4.7) when applied to boundary system reflects certain behavior of the traces of the homogenous solutions to hyperbolic dynamics which is usually "better" than it would follow from the interior regularity coupled with the standard trace theory. This is a very specific feature of our hyperbolic problems which makes possible our treatment of boundary control problems.
Having settled the issue of wellposedness for (4.6) we turn our attention to the stabilization. Here, again our stability result will follow from more general abstract result which gives certain conditions on the finite rank in case when A is similar to normal perturbation (possibly unbounded) P in order to guarantee two very desirable conditions:

(i) the difference $|\alpha_m - \lambda_m|$ between the new eigenvalues α_m of the perturbed operator A and P and "old" ones λ_m of A can be asymptotically arbitrary preassigned.

(ii) the new eigenvectors of A + P form a Riesz basis

This result is more properly a spectrum allocation and Riesz basis assignment type of result. It encompasses, in particular, a large variety of stabilization problems for many dynamics (including the feedback system 4.6), applications to parabolic hyperbolic systems of the second order with or without damping, and hyperbolic systems of the fourth order - see [L-T-4]).
Below we shall state the above mentioned result, although not in the full form (for sake of clarity of exposition) and for the technical details we refer the reader to

[L-T-4].

Theorem 4.5 [L-T-4]
Let A be similar to normal operator, which generates a s.c. semigroup on H. Let A have compact resolvent. Denote by λ_n and ϕ_n the eigenvalues of A, assumed simple for simplicity, and the corresponding eigenvalues forming an orthonormal basis in H. Let the feedback operator be of the form: $A_F x = Ax + Px$ where

$$Px = (Ax,a)b; \; a, \; b \in H.$$

Assume $(a,\phi_n) \neq 0$ (approximate controllability assumption).
Let the nonzero constant $\{\varepsilon_n\}$ $n = 1,2,\dots$ be assigned subject to certain conditions (see L-T-4) stating that (ε_n) should be asymptotically "small" enough. Then given a vector $a \in H$, there is a vector $b \in H$ such that A_F generates a s.c. semigroup, such that

(i) the eigenvalues of A_F are given precisely by $\alpha_n = \lambda_n + \varepsilon_n$

(ii) the corresponding eigenvectors of A_F form a Riesz basis.

Remark Result in the spirit of Thm. 4.5 has been proved in [S-2] for the case where P is bounded finite rank operator and the eigenvalues λ satisfy a so called "gap condition."
Our Theorem 4.5 when specified to our feedback problem 4.3 leads to the following result.

Theorem 4.6 [L-T-4]
Suppose that the eigenvalues of A are simple. Let $(g,\frac{\partial}{\partial\eta}\phi_n)_{L_2(\Gamma)} \neq 0$ where ϕ_n are eigenvectors of A. Then there exist suitable vectors $(w_1, w_2) \in L_2(\Omega) \times H_o^1(\Omega)$ such that the corresponding semigroup generated by the feedback system (4.6) is strongly stable (and can never be made exponentially stable).

Remark
The proof of the Thm. 4.5 is constructive. One can determine explicitly the stabilizing vectors w_1 and w_2 in Thm. 4.6.

Remark
Similar approach to the one used in the proof of Thm. 4.6 can be applied to study wave equation with a small damping. Here the problem (motivated by engineering application) is to increase at will the margin of stability of an arbitrary finite number of eigenvalues, while preserving the small decay (introduced originally by the damping) of all solutions. A priori it is not clear that this can be done. Higher eigenvalues may be effected as to stabilize the system. Application of Thm

4.5 (after some modification), to the above problem leads to the positive result i.e. we can construct finite feedback operator acting on the boundary with the properties that finitely many modes are arbitrary preassigned, while the overall decay of the solution is preserved. For details see [L-T-4]. Numerical results for this problem are presented in [B-3].

Remark 4.4.
Through a different approach - rather than as application of Thm. 4.5 - one may obtain strong stabilization of the semigroup generated by the feedback system (4.6) simply with $w_1 = 0$, i.e. using only the velocity vector y_t (damping) in the feedback. In fact we have the following variation

Theorem 4.7 [L-T-5]
Let $w \in L_2(\Omega)$ and $g \in L_2(\Gamma)$ be vectors satisfying the following two assumptions

$$(w,\phi_n)_\Omega (g,\frac{\partial\phi_n}{\partial\nu})_\Gamma < 0 \tag{4.8}$$

$$\frac{(w,\phi_n)_\Omega}{(g,\frac{\partial\phi_n}{\partial\nu})_\Gamma} \sim \frac{1}{\lambda_n^{3/2+\varepsilon}},\ \varepsilon > 0 \tag{4.9}$$

where $\{\phi_n\}$ are as in them 4.6 and $(\ ,\ )_\Omega$, $(\ ,\ )_\Gamma$ denotes L_2-norms. Then the feedback system (4.6) with $w_1 = 0$ and $w_2 = w$ generates a s.c. group on $L_2(\Omega) \times H^{-1}(\Omega)$ which is strongly stable here. Moreover, for any $w \in L_2(\Omega)$ and $g \in L_2(\Gamma)$, such feedback system can never be uniformly stable, by theorem 3.2.2 □

Remark 4.5
The proof of theorem 4.8 faces the following two difficulties: (1) to show the spectrum of the feedback system as strictly on the left half of the complex plane; (ii) to establish that the feedback semigroup is uniformly bounded . Assumption (4.10) yields (i). To show (ii), one applies a similarity transformation (explicitly defined by virtue of assumption (4.11)) which makes the resulting semigroup contraction in a larger space with weaker topology. Strong stabilization is first proved in this larger space (by either applying here the Nagy-Foias-Fogel decomposition of contraction semigroups, or else by employing the property that the closure of the span of the generalized eigenvectors of the feedback semigroup). Next, strong stabilization is proved in a smaller space and finally it is extended in between these two spaces by interpolation. When viewed on $L_2(\Omega) \times H^{-1}(\Omega)$, the feedback semigroup need not be dissipative □

Remark A result similar to the one of Theorem 4.7 was proved in [S-3] in the case of Neumann B.C. i.e.

$$\frac{\partial y}{\partial \nu} = (y_t, w)_\Omega g, \qquad \text{in } \Sigma_\infty$$

$w \in L_2(\Omega)$, $g \in H^{1/2}(\Gamma)$ with strong stabilization asserted in $H^1(\Omega) \times L_2(\Omega)$, under a variety of technical conditions including a gap conditions for the eigenvalues of the free dynamics □

In case of Neumann boundary conditions a mathematically more challenging problem - which is physically more appealing as well - arises if one considers the feedback system

$$\begin{cases} \dfrac{\partial^2 y}{\partial t^2} = \Delta y & \text{in } Q_\infty \\ y_o, y_1 \in H^1(\Omega) \times L_2(\Omega) & \\ \dfrac{\partial y}{\partial \nu} = (y_t, w)_{L_2(\Gamma)} g & \text{in } \Sigma_\infty \end{cases} \tag{4.12}$$

g, $w \in L_2(\Gamma)$, involving boundary observation . To begin with, (4.12) may or may not be well-posed. The techniques used in proving theorem 4.8 had to be replaced by an ad hoc treatment for problem (4.12). However, use is still made of the idea of introducing a similarity transformation of the candidate of the semigroup generator to show generation by it of a contraction semigroup in a larger space see Remark 4.5. In any case one can show

<u>Theorem</u> 4.8 [L-T-8]
Let g, $w \in L_2(\Gamma)$ satisfy

$$g_n w_n < 0, \text{ where } g_n = (g, \phi_n)_\Gamma, \; w_n = (w, \phi_n)_\Gamma \tag{4.13a}$$

$$0 < c < \left|\frac{g_n}{w_n}\right| < C \qquad \forall n = 1,2,\ldots \tag{4.13b}$$

Then, the feedback system (4.12) generates a s.c. semigroup on $H^1(\Omega) \times L_2(\Omega)$ which is strongly stable here, but not uniformly stable

<u>Remarks</u>

(i) Using abstract result of Thm. 4.5 one also obtains results on strong boundary stabilization of biharmonic models (sec 1.2). The details are given in [L-T-4].

(ii) Results on strong stabilization of beam equations using nonlinear feedbacks operators are given in [M-S-1], [S-4].

5. <u>Exponential (uniform) stabilization</u>

The most desirable type of stability is of course uniform (exponential) stability, where we have a specified rate of convergence to zero of all solutions. Unfortunately, this type of stability is much more difficult to obtain, if at all possible (see Theorem 3.3.2 for negative results for group generators). For this reason, as already mentioned before, boundary controls offer an attractive alternative, as they lead intrinsically to unbounded perturbations. Also, if the feedback operators are unbounded (this way conditions of Theorem 3.3.2 are violated) we may have some chances to obtain exponential stability - provided however that the whole feedback system is well posed. On the other hand wellposedness of feedback system usually fails with unbounded feedbacks (condition (4.7) of Theorem does not hold!), unless the feedback operator is dissipative. Therefore, one way to look for exponentially stabilizable feedbacks for hyperbolic dynamics is to look for unbounded dissipative feedback operators acting on the boundary. Below we shall present several unbounded, dissipative feedback operators producing exponential stability for hyperbolic dynamics. We start with the wave equation with Dirichlet boundary control. We consider feedback system

$$\begin{cases} y_{tt} = \Delta y \\ y|_{\Gamma} = -\frac{\partial}{\partial\eta}A^{-1}y_t \\ y(0) = y_o;\ y_t(0) = y_1;\ y_o,\ y_1 \in E \equiv L_2(\Omega) \times H^{-1}(\Omega) \end{cases} \tag{5.1}$$

Our result is:

Theorem 5.1 [L-T-1]
Assume that the domain Ω is strictly convex open bounded set in R^n.
Then there exists $c > 0$; $w > 0$ such that the solution y of (5.1) satisfies

$$\left|\begin{matrix} y(t) \\ y_t(t) \end{matrix}\right|_E < Ce^{-wt}\left|\begin{matrix} y_o \\ y_1 \end{matrix}\right|_E$$

Remark 5.1
Theorem 5.1 was proved in [L-T-1] under more general assumptions imposed on Ω

Remark 5.2
Notice that (5.1) can be written in an abstract form as (notation of sect 2.1)

$$\frac{d}{dt}\binom{y}{y_t} = \mathcal{A}y - \mathcal{B}\mathcal{B}^*y \tag{5.2}$$

where

$$(\mathcal{B}^*\binom{y}{y_t},v)_U = (\binom{y}{y_t},\mathcal{B}v)_E = (y_t,ADv)_{H^{-1}(\Omega)} = (A^{-1/2}y_2,A^{1/2}Dv)_{L_2(\Omega)} =$$

$$(D^*y_t,v)_{L_2(\Gamma)} = (D^*AA^{-1}y_t,v)_{L_2(\Gamma)} = (\frac{\partial}{\partial\eta}A^{-1}y_t,v)_{L_2(\Gamma)}$$

Thus, feedback operator $F : E \to U = L_2(\Gamma)$ given by

$$F\binom{y}{y_t} = \mathscr{B}^*\binom{y}{y_t} = \frac{\partial}{\partial\eta}A^{-1}y_t$$

is clearly unbounded from $E \equiv L_2(\Omega) \times H^{-1}(\Omega) \to U \equiv L_2(\Gamma)$. (With $y_t \in H^{-1}(\Omega) = D(A^{1/2})'$ we have $A^{-1}y_t \in H_o^1(\Omega)$ and $\frac{\partial}{\partial\eta}A^{-1}y_t$ is in $H^{-1/2}(\Gamma)$!)

The next result deals with exponential stabilization for the wave equation with Neumann boundary conditions. We consider

$$\begin{cases} y_{tt} = \Delta y & \text{in } \Omega \\ \frac{\partial y}{\partial\eta}\Big|_\Gamma = -y_t & \text{on } \Gamma \\ y(0) = y_o;\ y_t(0) = y_1 & (y_o,y_1) \in H^1(\Omega) \times L_2(\Omega) \end{cases} \tag{5.3}$$

Theorem (5.2) [Ch-1], [L-3]
Assume that the domain Ω is star-shaped (see Ch-1 and L-3 for precise definition). Then $\exists$ c,w > 0 such that solution y,y_t of (5.3) satisfies

$$\left|\begin{matrix} y \\ y_t \end{matrix}\right|_{H^1(\Omega) \times L_2(\Omega)} \leq Ce^{-wt}\left|\begin{matrix} y_o \\ y_1 \end{matrix}\right|_{H^1(\Omega) \times L_2(\Omega)}.$$

Remark 5.3
Theorem 5.2 was proved in [Ch-1], [L-3] for more general domains their star-shaped. Also, the case where the control acts only on a position of the boundary has been discussed.

Remark 5.4
Similarly as in Dirichlet case, our model 5.3 leads to the dissipative feedback of the form $-\mathscr{B}\mathscr{B}^*$ where feedback operator $F\binom{y}{y_t} = -y_t|_\Gamma$ is unbounded when acting from $H^1(\Omega) \times L_2(\Omega)$ sub $L_2(\Gamma)$.
The next result an exponential stabilization will deal with the one dimensional version of biharmonic equation (1.2.2). Consider

$$\begin{cases} \frac{\partial^2 y}{\partial t^2} + \frac{\partial^2 y}{\partial x^2} = 0 & . \ 0 < x < L \\ y(0,t) = 0; \ y_x(0,t) = 0 & \\ \frac{\partial^2 y}{\partial x^2}(L,t) = -K, \frac{\partial^2 y}{\partial t\, \partial x} & \\ \frac{\partial^3 y}{\partial x^3} = K_o \frac{\partial y}{\partial t} & \end{cases} \tag{5.4}$$

$y(0) = y_o$; $y_t(0) = y_1$ $y_o, y_1 \in E$ where $E \equiv \{u \in H^2(\Omega);\ u(0) = u_x(0) = 0\} \cap L_2(\Omega)$

Theorem 5.3 [Ch-D-K-P]

If $K > 0$ and $k_o > 0$ then $\exists\, C > 0$, $u > 0$ such that the solution $\binom{y}{y_t}$ of (5.4) satisfies:

$$\left|\begin{matrix} y \\ y_t \end{matrix}\right|_E < Ce^{-wt} \left|\begin{matrix} y_o \\ y_1 \end{matrix}\right|_E$$

Remarks

(i) In a similar manner as before, one can show that feedback operator defined in (5.9) in unbounded and dissipative

(ii) It was observed numerically that if $K_1 \neq 0$ the spectrum of the feedback system is contained in some triangular sector. Thus, the system behaves "parabolic like". This is an evidence of the so-called structural damping.

(iii) A result on exponential stabilization for some hyperbolic systems with the special nature of boundary conditions is provided in [Q-R].

(iv) Notice that the results on exponential stability provided by Theorems (5.1) and (5.2) and (5.3) lead a fortiori to the exact controllability through Russell's "controllability via stabilization" approach. However, at the present, is a direct proof of exact controllability, which in some cases (for example Dirichlet case) does not require any geometrical assumptions. Using there exact controllability results as the starting point of the analysis of the algebraic Riccati equations, one can show like in [L-T-9] that the feedback operator

$$Fy = -\mathscr{B}^*Py$$

with P being an appropriate Riccati operator will produce exponential stability without any geometrical condition (at least in the Dirichlet case).

Because of the space constraints problems related to the issue of boundary "stability via Riccati operator" shall not be discussed here. Interested readers may consult [L-T-9], [D-L-T].

References

[B-4] A. V. Bakalarischman, A Mathematical Formulation of a Large Space Control Problem. Proceedings 24IEEE Conference on Decision and Control Ft. Lauderdale 1985.

[B-1] A. V. Bakalrischman, Applied Functional Analysis Springer Verlag 1976

[B-2] C. D. Benchimol, A note on weak stabilization of contraction semigroups SIAM J. Control Optimiz. 16 (1978) 373-379

[B-3] J. Bartolomeo

[Ch-L] S. Chang, I. Lasiecka, Riccati equations for nonsymmetric and nondissipative hyperbolic systems to appear in J. Math. Analysis and Applications in 1986.

[Ch-1] C. Chen, Energy decay estimates and exact boundary value controllability for the wave equation in a bounded domain. J. Math. Pures Appl. 58 (1979) 249-273.

[Ch-D-K-P] G. Chen, M. Delfour, A. M. Krall, G. Payre Modeling, Stabilization and control of serially connected beams. Proceedings of the 23rd IEEE Conference on Decision and Control, Las Vegas 1984.

[D-Sch] W. Desh - W. Schappracher, On Relatively Bounded Perturbations of Linear C_o-Semigroups Anneli della Suida Normale Superiore di Pisa vol. XI, 2, 1984 (1327-341).

[D-L-P] R. Datko, J. Lagnese, M. Polis, An Example of the effect of time delays in boundary feedback stabilization of wave equations SIAM J. on Control vol. 24 1986 pp. 152-157.

[D-L-Sch] W. Desh, I. Lasiecka, W. Schappacher, Finite dimenisonal boundary feedback control problems for linear infinite dimensional systems. Israel Journal of Mathematics, vol. 51, No. 3, 1985 (177-207).

[D-L-T] G. De Prato, I. Lasiecka, R. Triggiani, A direct study of Riccati equations arising in hyperbolic boundary control problems. To appear in Journal of Differential Equations.

[F-2] H. O. Fattorini, Ordinary differential equations in linear topological spaces I, II, J. Differential Equations 5 (1968) 72-105; 6 (1969) 50-70.

[F-1] D. Fujiwara, Concrete characterization of the domain of fractional powers of some elliptic differential operators of the second order. Proc. Japan. Acad. Ser. A. Math. Sci. 43 (1967) 82-86.

[G-1] S. Gibson, A note on stabilization of infinite dimensional linear ascillators by compact feedback. SIAM J. Control Opt. 18 (1980) 311-316.

[K-1] T. Kato, Perturbation Theory of Linear Operators, Springer Verlag 1966.

[L-1] I. Lasiecka, Sharp Regularity Results for mixed hyperbolic problems of second order. Proceedings of International Conference on Abstract Differential Equations. Bologna, July 1986.

[L-2] N. Levan, The stabilization problem: A Hilbert space operator decomposition approach, IEEE Trans. Circuits and Systems CAS-25 (9) (1978) 721-727.

[L-3] J. Lagnese, Decay of solutions of wave equation in a bounded region with boundary dissipation J. Diff. Eq. (46) (1983).

[L-4] J. L. Lions, Controle des systemes distribués singuliers. Gouthier Villars Paris 1983.

[L-M] J. L. Lions, E. Magenes, Nonhomogenous Boundary Value Problems and Applications. vol. I, II, Springer Verlag 1972.

[L-T-1] I. Lasiecka, R. Triggiani, Uniform exponential energy decay in a bounded region with $L_2(0,\infty;L_2(\Gamma))$ feedback control in the Dirichlet Boundary conditions to appear in Journal Diff. Eq.

[L-T-2] I. Lasiecka, R. Triggiani, A cosine operator approach to modeling $L_2(0T;L_2(\Gamma))$ boundary input hyperbolic equations. Appl. Math. and Optimization vol. 7, (35-83) 1981.

[L-L-T] I. Lasiecka, J. L. Lions, R. Triggiani, Nonhomogenous boundary value problems for second order hyperbolic value problems for second order hyperbolic operators to appear in Journal de Math. Pure et Applique.

[L-T-3] I. Lasiecka, R. Triggiani, Sharp regularity results for hyperbolic equations of second order, in prepartion.

[L-T-4] I. Lasiecka, R. Triggiani, Finite rank, relatively bounded perturbations of C_o-semigroups. Part II. Spectrum allocation and Riesz basis in parabolic and hyperbolic feedback systems to appear in Ann. Matem Pure et Applic.

[L-T-5] I. Lasiecka, R. Triggiani, Dirichlet Boundary stabilization of the wave equation with damping feedback of finite range. Journal of Math. Anal. and Appl. vol 97, No. 1, (112-130) 1983.

[L-T-6] I. Lasiecka, R. Triggiani, Regularity of hyperbolic equations under $L_2(0t:L_2(\Gamma))$ Dirichlet boundary terms. Appl. Math. Optimiz. 10, (1983) 275-286.

[L-T-7] I. Lasiecka, R. Triggiani, Finite rank, relatively bounded perturbations of C_o-semigorups. Part I: Welleprosduen and boundary feedback hyperbolic dynamics to apear in Annali Scuola Normale Superiore di Pisa.

[L-T-8] I. Lasiecka, R. Triggiani, Nondissipative boundary stabilization of hyperbolic equations with boundary observation Journal de Mathematique Pure et Appl. 63(1984) 59-80.

[L-T-9] I. Lasiecka, R. Triggiani, Riccati equations for hyperbolic partial differential equations with $L_2(0T;L_2(\Gamma))$ Dirichlet boundary terms to appear in SIAM Journal an Control.

[M-S-1] J. R. McLaughlin and M. Slemrod, Scanning Control of a Vibrating String Proceeding 24 IEEE Conference on Decision and Control. Ft. Lauderdale 1985.

[N-1] J. Necas, Les methodes directes an Theore des Equations Elliptiques. Mosson et Cie Paris 1967.

[Q-R] J. P. Quinn and D. L. Russell, Asymptotic stability and energy decay rates for soltuions of hyperbolic equations with boundary damping. Proc. Roy Soc. Edinburgh. Ser. A 77 (1977/78) 97-127.

[R-1] D. L. Russell, Decay rates for weekly damped systems in Hilbert space obtained via control - theoretic methods. J. Diff. Eq. 19 (1975) 344-370.

[R-2] D. L. Russell, Canonical forms and spectral determination for a class of hyperbolic distributed parameter control systesm J. Math. Anal. and Applic. 62, (1978) pp. 186-225.

[R-3] D. L. Russell, Differential delay equation as canonical forms for controlled hyperbolic systems with application to spectral assignment. Control theory of systems governed by partial differential equations. Academic Press pp. 119-150.

[S-1] M. Slemrod, Stabilization of boundary control systems J. Diff. Eq. 22(1978) 402-415.

[S-2] S. Shun-Hua, On spectrum distribution of completely controllable linear systems. SIAM J. Control. Optimiz. 19(1981) pp. 730-743.

[S-3] S. Shun-Hua, Boundary stabilization of hyperbolic systems with no dissipative conditions. SIAM J. on control vol. 20 No 6 1982 (862-884).

[S-4] M. Slemrod, Feedback stabilization of $\frac{du}{dt} = Au + \beta f$ in Hilbert space. Proceedings 24 IEEE Conference on Decision and Control Ft. Lauderdale 1985.

[T-1] R. Triggiani, An approach to modeling boundary input problems for hyperbolic equations, Springer-Verlag Lectures Notes in Control (1978) 380-390.

[T-2] R. Triggiani, On the stabilization problem in Banach space. J. Math. Anal. Appl. 52 383-403 (1975).

[T-3] R. Triggiani, A^{ε} finite rank perturbations of a s.c. group generators A: counterexample to generation and another condition for well-poisedness Lecture notes in Mathematics Springer-Verlag 1984.

[T-4] R. Triggiani, Improvement of stability properties of hyperbolic damped wave equation via boundary feedack, Lecture notes in Control and Information #75, Springer Proceedings 2nd International Conference in Vorau (Austria) 400-410.

[Z-1] J. Zabczyk, Stabilization of boundary control systems Lecture Notes in Control and Information Sciences, #14, Springer-Verlag 1979.

UNCERTAIN SYSTEMS: ROBUSTNESS OF ULTIMATE BOUNDEDNESS CONTROL WITH RESPECT TO NEGLECTED DYNAMICS

G. Leitmann
Department of Mechanical Engineering, University of California
Berkeley, California CA 94720, USA.

E.P. Ryan
School of Mathematics, University of Bath
Claverton Down, Bath BA2 7AY, UK.

A. Steinberg
Department of Aeronautical Engineering
TECHNION - Israel Institute of Technology, Haifa, Israel.

Abstract Feedback control of a class of imperfectly known dynamical systems is considered. On the basis of known functional properties and bounds relating to the uncertain elements of the generic system and initially neglecting actuator and sensor dynamics, a feedback structure is first determined (using established Lyapunov-based techniques [1],[2]) which guarantees uniform ultimate boundedness of all motions of the *reduced-order system* (i.e. the uncertain system in the absence of actuator and sensor dynamics). Robustness properties of the full system are subsequently derived (related questions of robustness are also addressed in e.g., [3]-[6]). In particular, it is shown that the feedback controlled system is structurally stable in the sense that the features designed for the reduced-order system are qualitatively retained in the presence of sufficiently fast actuator or sensor dynamics. Moreover, explicit threshold measures of "fastness" of these dynamics are obtainable.

1. Introduction

The generic system considered consists of a dynamical process

$$\left.\begin{aligned} \dot{x}(t) &= Ax(t) + \bar{B}(t)u(t) + F(t,x(t)), \quad x(t) \in \mathbb{R}^n \\ \bar{B}(t) &= B + \Delta B(t) \end{aligned}\right\} \tag{1}$$

which is to be controlled via an appropriately determined feedback operator Φ, operating on the output $y(t)$ of a dynamical state sensor

$$\left.\begin{aligned} \mu_s\dot{y}(t) &= \bar{D}(t)\,[y(t) - x(t)], \quad y(t) \in \mathbb{R}^n \\ \bar{D}(t) &= D + \Delta D(t), \quad \sigma(D) \subset \mathbb{C}^- \text{ (the open left half complex plane)} \end{aligned}\right\} \tag{2}$$

and generating $u(t) = \Phi(y(t)) \in \mathbb{R}^m$ at the input of a dynamical control actuator

$$\left.\begin{aligned} \mu_a\dot{z}(t) &= \bar{C}(t)\,[z(t) - u(t)], \quad z(t) \in \mathbb{R}^m \\ \bar{C}(t) &= C + \Delta C(t), \quad \sigma(C) \subset \mathbb{C}^-. \end{aligned}\right\} \tag{3}$$

Here, A, B, C and D are known constant matrices; in (2) and (3) σ denotes spectrum (so that C and D are asymptotically stable) and $\mu_a \ge 0$, $\mu_s \ge 0$ are real parameters, the values of which reflect (inversely) the "fastness" of the actuator and sensor dynamics. For example, the gas turbine model of [7] is of the above form. The overall system is subject to uncertainty: the functions $\Delta B: \mathbb{R} \to \mathbb{R}^{n\times m}$, $\Delta C: \mathbb{R} \to \mathbb{R}^{m\times m}$, $\Delta D: \mathbb{R} \to \mathbb{R}^{n\times n}$ and $F: \mathbb{R} \times \mathbb{R}^n \to \mathbb{R}^n$ are unknown members of known uncertainty classes $\mathcal{B}$, $\mathcal{C}$, $\mathcal{D}$ and $\mathcal{F}$, respectively, defined below (via assumptions A1 - A7). The classes $\mathcal{B}$ and $\mathcal{F}$ are such that, in the absence of actuator and sensor dynamics, the feedback operator

Φ can be determined so as to yield satisfactory performance in the sense that, for arbitrary realizations $\Delta B \in \mathcal{B}$ and $F \in \mathcal{F}$, global uniform ultimate boundedness with respect to an acceptable neighbourhood of the zero state is guaranteed. The main question then addressed is that of robustness with respect to neglected dynamics, viz. how does the presence of actuator and sensor dynamics affect the performance of the feedback controlled uncertain system? Before considering this question, some preliminary assumptions are introduced in §2 which enable the determination of a suitable operator Φ for the reduced-order system. The analysis underlying the ensuing sections is essentially that of [8]; for the sake of brevity, proofs of all lemmas are omitted here but can be found in [8].

2. Reduced-order system: Feedback control guaranteeing global uniform ultimate boundedness

The following assumptions are first introduced.

A1: (A,B) is a stabilizable pair; $\operatorname{rank} B = m \le n$.

A2: There exists a C^1 function (unknown) $E: \mathbf{R} \to \mathbf{R}^{m\times m}$ such that for all $t \in \mathbf{R}$

(i) $\Delta B(t) = BE(t)$, (ii) $\|E(t)\| \le \kappa_1 < 1$, (iii) $\|\dot{E}(t)\| \le \kappa_2$,

where κ_1 and κ_2 are known constants.

The above assumptions implicitly define $\mathcal{B}$.

A3: For some subspace $S \subset \mathbf{R}^n$ complementary to $\operatorname{Im} B$ (i.e. $S \oplus \operatorname{Im} B = \mathbf{R}^n$), every unknown function F admits a decomposition

(i) $F = f + Bg$

where $f: \mathbf{R} \times \mathbf{R}^n \to S$ and $g: \mathbf{R} \times \mathbf{R}^n \to \mathbf{R}^m$ are (unknown) Carathéodory functions [1] with, for all $(t,x) \in \mathbf{R} \times \mathbf{R}^n$,

(ii) $\|f(t,x)\| \le \kappa_3\|x\| + \kappa_4$, (iii) $\|g(t,x)\| \le \kappa_5\|x\| + \kappa_6$,

where κ_3, κ_4, κ_5 and κ_6 are known constants.

The above assumption, together with assumption A4 to be introduced below, implicitly defines $\mathcal{F}$.

Remarks: (i) The Euclidean inner product $\langle x, y\rangle = x^T y$ and norm $\|x\| = \langle x, x\rangle^{\frac{1}{2}}$ on $\mathbf{R}^q$ (q determined by context), with associated matrix norm $\|M\| = [\sigma_{max}(M^T M)]^{\frac{1}{2}}$ where σ_{max} denotes maximum eigenvalue, is assumed throughout. (ii) In the terminology of [9] - [11], ΔB and Bg collectively model the *matched* uncertainty while f models the *unmatched* or *residual* uncertainty in the reduced-order system (i.e. the system without actuator/sensor dynamics).

The primary objective is now to characterize a continuous feedback operator $\Phi: \mathbf{R}^n \to \mathbf{R}^m$ such that, for each $\Delta B \in \mathcal{B}$ and $F \in \mathcal{F}$, the feedback controlled reduced-order system

$$\dot{x}(t) = G(t,x(t)), \quad G(t,x) := Ax + \overline{B}(t)\Phi(x) + F(t,x) \tag{4}$$

is globally uniformly ultimately bounded with respect to some suitable set N containing the zero state, in the sense of the following definition.

Defn. 1 Global uniform ultimate boundedness with respect to N

System (4) is globally uniformly ultimately bounded w.r.t. N iff:-

(i) Existence and continuation of solutions: For each $(t_o, x^o) \in \mathbf{R} \times \mathbf{R}^n$, there

exists a (local) solution $x: [t_o, t_1) \to \mathbb{R}^n$ with $x(t_o) = x^o$ and every such solution can be extended into a solution on $[t_o, \infty)$.

(ii) Uniform boundedness: For each $r \geq 0$, there exists $R(r) \geq 0$ such that, for each $t_o \in \mathbb{R}$, $\|x(t)\| \leq R(r)$ for all t on every solution $x: [t_o, \infty) \to \mathbb{R}^n$ with $\|x(t_o)\| \leq r$.

(iii) Uniform ultimate boundedness w.r.t. N: For each $r \geq 0$, there exists $\tau(r) \geq 0$ such that, for each $t_o \in \mathbb{R}$, $x(t) \in N$ for all $t \geq t_o + \tau(r)$ on every solution $x: [t_o, \infty) \to \mathbb{R}^n$ with $\|x(t_o)\| \leq r$.

A1 ensures the existence of a unique, positive-definite, symmetric solution $K > 0$ of the algebraic Riccati equation

$$KA + A^TK + I - KBB^TK = 0 . \tag{5}$$

The following assumption, which (together with A3) completes the characterization of the uncertainty class F, is now made:

A4: $\kappa_3 < \frac{1}{2}\|K\|^{-1}$.

Suppose that a desired set N of ultimate boundedness is specified as the closed ball of radius $\bar{\rho} > 0$ in $\mathbb{R}^n$, i.e.

$$N = \Sigma(\bar{\rho}) := \{x \in \mathbb{R}^n: \|x\| \leq \bar{\rho}\} . \tag{6}$$

Defining

$$\nu := 1 - 2\kappa_3\|K\| > 0 \tag{7}$$

and introducing the parameter $\rho > 0$ with the property

$$\rho = \bar{\rho} - 2\nu^{-1}\kappa_4\|K\|[\|K\|\|K^{-1}\|]^{\frac{1}{2}} \quad \text{if} \quad \bar{\rho} > 2\nu^{-1}\kappa_4\|K\|[\|K\|\|K^{-1}\|]^{\frac{1}{2}} \tag{8}$$

then, analogous to the control proposed in [2], the feedback operator $\Phi: \mathbb{R}^n \to \mathbb{R}^m$ is taken to be *linear* and is given by

$$\Phi(x) = \Phi x, \quad \Phi := -\tfrac{1}{2}(1+\gamma)B^TK \tag{9a}$$

where

$$\gamma := (1-\kappa_1)^{-1}\left[\kappa_1 + \nu^{-1}\left(1 + 2\nu^{-1}\kappa_4\|K\|\rho^{-1}[\eta\|K\|\|K^{-1}\|]^{\frac{1}{2}}\right)^{-1}\left(\kappa_5 + \hat{\kappa}_6\rho^{-1}[\eta\|K\|\|K^{-1}\|]^{\frac{1}{2}}\right)^2\right] \tag{9b}$$

$$\hat{\kappa}_6 := 2\nu^{-1}\kappa_4\kappa_5\|K\| + \kappa_6 \tag{9c}$$

and $\eta > 1$ is an arbitrary constant.

Lemma 1: For every uncertainty realization $\Delta B \in B$ and $F \in F$, the feedback controlled system defined by (4) & (9) is globally uniformly ultimately bounded with respect to the closed ball $\Sigma(r) \subset \mathbb{R}^n$ of radius

$$r = \begin{cases} \bar{\rho} & ; \text{ if } \bar{\rho} > 2\nu^{-1}\kappa_4\|K\|[\|K\|\|K^{-1}\|]^{\frac{1}{2}} \\ 2\nu^{-1}\kappa_4\|K\|[\|K\|\|K^{-1}\|]^{\frac{1}{2}} + \rho & ; \text{ otherwise} \end{cases} \tag{10}$$

Proof: See [8].

Remarks: The above lemma implies that, if the residual uncertainty is such that $\kappa_4 = 0$, then global uniform ultimate boundedness with respect to *any* prescribed ball $N = \Sigma(\bar{\rho})$ can be achieved in principle (but in practice at the expense of high gain since $\gamma \sim \rho^{-1} = \bar{\rho}^{-1}$). If the residual uncertainty is such that $\kappa_4 \neq 0$, then a *desired* set N is a *feasible* ultimate boundedness set if it contains the closed ball of radius $2\nu^{-1}\kappa_4\|K\|[\|K\|\|K^{-1}\|]^{\frac{1}{2}}$ strictly in its interior.

3. Global uniform ultimate boundedness in the presence of actuator dynamics

The effects on performance of the introduction of actuator dynamics will now be considered. In this case, the governing equations are

$$\left.\begin{array}{l} \dot{x}(t) = Ax(t) + \bar{B}(t)z(t) + F(t,x(t)), \quad \bar{B}(t) = B + \Delta B(t), \quad x(t) \in \mathbf{R}^n \\ \mu_a \dot{z}(t) = \bar{C}(t)[z(t) - \Phi x(t)], \; \mu_a > 0, \quad \bar{C}(t) = C + \Delta C(t), \quad z(t) \in \mathbf{R}^m \end{array}\right\} \qquad (11)$$

with Φ defined by (9). Writing $e(t) = z(t) - \Phi x(t)$, then (11) may be expressed as

$$\left.\begin{array}{l} \dot{x}(t) = \Gamma(t,x(t),e(t)), \quad \Gamma(t,x,e) := G(t,x) + \bar{B}(t)e \\ \dot{e}(t) = \mu_a^{-1}\bar{C}(t)e(t) - \Phi\Gamma(t,x(t),e(t)) \end{array}\right\} \qquad (12)$$

with G defined as in (4). Recalling that $\sigma(C) \subset \mathbf{C}^-$, there exists unique $P > 0$ such that

$$PC + C^TP + I = 0\,. \qquad (13)$$

The following assumption on the actuator uncertainty ΔC is now made, implicitly defining the uncertainty class $\mathcal{C}$:

A5: $\Delta C\colon \mathbf{R} \to \mathbf{R}^{m\times m}$ is continuously differentiable with (i) $\|\Delta C(t)\| \le \kappa_7 < \frac{1}{2}\|P\|^{-1}$, (ii) $\|\Delta\dot{C}(t)\| \le \kappa_8$, for all $t \in \mathbf{R}$, where κ_7 and κ_8 are known constants.

Prop.1: For each $\Delta C \in \mathcal{C}$, there exists a continuously differentiable function $\bar{C}^{-1}\colon \mathbf{R} \to \mathbf{R}^{m\times m}$ with, for all $t \in \mathbf{R}$, (i) $\bar{C}^{-1}(t) = [C + \Delta C(t)]^{-1}$, (ii) $\|\bar{C}^{-1}(t)\| \le [\frac{1}{2}\|P\|^{-1} - \kappa_7]^{-1}$, (iii) $\|\dot{\bar{C}}^{-1}(t)\| \le \kappa_8[\frac{1}{2}\|P\|^{-1} - \kappa_7]^{-2}$.

Proof: See [8].

The function $W\colon \mathbf{R} \times \mathbf{R}^n \times \mathbf{R}^m \to [0,\infty)$ (Lyapunov function candidate) is now introduced, where

$$W(t,x,e) := \frac{1}{2} \left\langle \begin{bmatrix} x \\ \mu_a^{\frac{1}{2}} e \end{bmatrix}, M_{\mu_a}(t) \begin{bmatrix} x \\ \mu_a^{\frac{1}{2}} e \end{bmatrix} \right\rangle \qquad (14a)$$

$$M_{\mu_a}(t) := L_{\mu_a}^T(t) \begin{bmatrix} K & 0 \\ 0 & \mu_a P \end{bmatrix} L_{\mu_a}(t) \qquad (14b)$$

$$L_{\mu_a}(t) := \begin{bmatrix} I & 0 \\ -P^{-1}[\bar{C}^{-1}(t)]^T \bar{B}^T(t)K & \mu_a^{-\frac{1}{2}} I \end{bmatrix} \qquad (14c)$$

It is readily verified that the (uncertain) function $M_{\mu_a}(\cdot)$ is positive and continuously differentiable with C^1 inverse $M_{\mu_a}^{-1}(\cdot)$. Moreover, the uncertain functions $t \mapsto \|M_{\mu_a}(t)\|$ and $t \mapsto \|M_{\mu_a}^{-1}(t)\|$ are bounded; in particular, there exist calculable constants (expressible in terms of μ_a and the known quantities $\kappa_1, \ldots, \kappa_8$, B, C, K and P; the dependence on μ_a only will be explicitly indicated) $\bar{m}_{\mu_a}$ and $\bar{\bar{m}}_{\mu_a}$ such that, for all $(\Delta B, \Delta C) \in \mathcal{B} \times \mathcal{C}$,

$$\|M_{\mu_a}(t)\| \le 2\bar{m}_{\mu_a}\,, \qquad \|M_{\mu_a}^{-1}(t)\| \le \tfrac{1}{2}\bar{\bar{m}}_{\mu_a} \qquad \text{for all } t \qquad (15)$$

which, when combined with (14), yield the inequalities

$$(\bar{\bar{m}}_{\mu_a})^{-1}[\|x\|^2 + \mu_a\|e\|^2] \le W(t,x,e) \le \bar{m}_{\mu_a}[\|x\|^2 + \mu_a\|e\|^2] \qquad (16)$$

valid for all $(t,x,e) \in \mathbf{R} \times \mathbf{R}^n \times \mathbf{R}^m$. Now, by considering the behaviour of W along solutions $(x(\cdot),e(\cdot))$ of (12), the following lemma may be established.

Lemma 2: There exists a calculable† constant (threshold value) $\mu_a^* > 0$ (expressible in terms of $\kappa_1, \ldots, \kappa_8$, B, C, K and P) such that, for each $\mu_a \in (0,\mu_a^*)$ and for all $(\Delta B,\Delta C,F) \in \mathcal{B} \times \mathcal{C} \times \mathcal{F}$, the feedback controlled system (12) is globally uniformly ultimately bounded with respect to the ellipsoid

$$\{(x,e) \in \mathbb{R}^n \times \mathbb{R}^m:\ \|x\|^2 + \mu_a\|e\|^2 \le \bar{m}_{\mu_a}\bar{\bar{m}}_{\mu_a}\tilde{\rho}^2_{\mu_a}\}$$

where $\tilde{\rho}_{\mu_a}$ is a calculable† constant (expressible in terms of μ_a, $\kappa_1, \ldots, \kappa_8$, η, B, C, K and P).

Proof: See [8].

Remarks: The above lemma essentially establishes structural stability with respect to neglected actuator dynamics insofar as the property of global uniform ultimate boundedness designed for the reduced-order system is qualitatively preserved in the presence of actuator dynamics provided that $\mu_a < \mu_a^*$.

4. Global uniform ultimate boundedness in the presence of sensor dynamics

In this case, the governing equations are

$$\dot{x}(t) = Ax(t) + \bar{B}(t)\Phi y(t) + F(t,x(t)), \quad \bar{B}(t) = B + \Delta B(t), \quad x(t) \in \mathbb{R}^n \tag{17a}$$

$$\mu_s\dot{y}(t) = \bar{D}(t)[y(t) - x(t)],\ \mu_s > 0, \quad \bar{D}(t) = D + \Delta D(t), \quad y(t) \in \mathbb{R}^n \tag{17b}$$

Let S_o denote the subspace *span* $\begin{bmatrix} 0 \\ \Phi \end{bmatrix} \subset \mathbb{R}^n$. The following assumption is now made which, together with A7 below, implicitly defines the uncertainty class $\mathcal{D}$.

A6: The subspace $S_1 := S_o^{\perp}$ (the orthogonal complement of S_o) is both D - invariant ($DS_1 \subset S_1$) and $\Delta D(t)$-invariant ($\Delta D(t)S_1 \subset S_1$) for all $t \in \mathbb{R}$.

Remarks: Assumption A6 is imposed here to facilitate direct application of the results of §3; a fully independent analysis of the case of sensor dynamics may possibly enable weakening of this assumption.

Introducing

$$S = \begin{bmatrix} S_1 \\ \Phi \end{bmatrix} \quad \text{with inverse} \quad S^{-1} = [\,T_1 \ \vdots\ \Phi^T(\Phi\Phi^T)^{-1}\,] \tag{18}$$

where S_1 is such that *span* $\begin{bmatrix} S_1 \\ 0 \end{bmatrix} = S_1 = S_o^{\perp}$, then, under the sensor output transformation

$$\begin{bmatrix} \hat{y} \\ \hat{z} \end{bmatrix} = \begin{bmatrix} S_1 y \\ \Phi y \end{bmatrix} = Sy \tag{19}$$

(17b) becomes

$$\mu_s \begin{bmatrix} \dot{\hat{y}}(t) \\ \dot{\hat{z}}(t) \end{bmatrix} = \hat{D}(t) \begin{bmatrix} \hat{y}(t) - S_1x(t) \\ \hat{z}(t) - \Phi x(t) \end{bmatrix}, \quad \hat{D}(t) = \hat{D} + \Delta\hat{D}(t) \tag{20a}$$

---

† Using appropriate estimates, it is straightforward to calculate both μ_a^* and $\tilde{\rho}_{\mu_a}$; however, the resulting expressions in terms of the known system parameters are notationally cumbersome and are not given here.

with, in view of A6,

$$\hat{D} = S\,D\,S^{-1} = \begin{bmatrix} \hat{D}_1 & \hat{D}_2 \\ 0 & \hat{C} \end{bmatrix}, \quad \Delta\hat{D}(t) = \begin{bmatrix} \Delta\hat{D}_1(t) & \Delta\hat{D}_2(t) \\ 0 & \Delta\hat{C}(t) \end{bmatrix} \tag{20b}$$

where

$$\hat{D}_1 = S_1 D\, T_1, \quad \hat{D}_2 = S_1\, D\Phi^T(\Phi\Phi^T)^{-1}, \quad \hat{C} = \Phi\, D\, \Phi^T(\Phi\Phi^T)^{-1} \tag{20c}$$

$$\Delta\hat{D}_1(t) = S_1 \Delta D(t) T_1, \quad \Delta\hat{D}_2(t) = S_1 \Delta D(t)\Phi^T(\Phi\Phi^T)^{-1}, \quad \Delta\hat{C}(t) = \Phi\,\Delta D(t)\Phi^T(\Phi\Phi^T)^{-1} \tag{20d}$$

Recalling that $\sigma(D) \subset \mathbb{C}^-$, it follows that $\sigma(\hat{D}_1) \subset \mathbb{C}^-$ and $\sigma(\hat{C}) \subset \mathbb{C}^-$; hence, the Lyapunov equations

$$\hat{P}_1\hat{D}_1 + \hat{D}_1^T\hat{P}_1 + I = 0 \tag{21}$$

$$\hat{P}\,\hat{C} + \hat{C}^T\,\hat{P} + I = 0 \tag{22}$$

admit unique positive solutions $\hat{P}_1 > 0$ and $\hat{P} > 0$. The final assumption is now made:

A7: $\Delta D\colon \mathbf{R} \to \mathbf{R}^{n\times n}$ is continuously differentiable with, for all $t \in \mathbf{R}$,

(i) $\|\Delta\hat{D}_1(t)\| \le \kappa_9 < \tfrac{1}{2}\|\hat{P}_1\|^{-1}$, (ii) $\|\Delta\dot{\hat{D}}_2(t)\| \le \kappa_{10}$,

(iii) $\|\Delta\,\hat{C}\,(t)\| \le \kappa_{11} < \tfrac{1}{2}\|\,\hat{P}\,\|^{-1}$, (iv) $\|\Delta\dot{\hat{C}}(t)\| \le \kappa_{12}$,

where κ_9, κ_{10}, κ_{11} and κ_{12} are known constants.

Using (21) and A7(i)-(ii), it is readily verified that the sensor sub-system

$$\mu_s\dot{\hat{y}}(t) = [\hat{D}_1 + \Delta\hat{D}_1(t)]\,[\hat{y}(t) - S_1x(t)] + [\hat{D}_2 + \Delta\hat{D}_2(t)]\,[\hat{z}(t) - \Phi x(t)] \tag{23}$$

is bounded-input, bounded-output stable (i.e. for bounded $x(\cdot)$ and $\hat{z}(\cdot)$, $\hat{y}(\cdot)$ is bounded). Hence, it may be concluded that the original system (17) is globally uniformly ultimately bounded if and only if the associated system

$$\left.\begin{aligned} \dot{x}(t) &= A\,x(t) + \bar{B}(t)\hat{z}(t) + F(t,x(t)) \\ \mu_s\dot{\hat{z}}(t) &= [\,\hat{C} + \Delta\hat{C}(t)]\,[\,\hat{z}(t) - \Phi\,x(t)] \end{aligned}\right\} \tag{24}$$

is globally uniformly ultimately bounded. Clearly, (24) is structurally equivalent to (11); moreover, on making the identifications: $\mu_a \sim \mu_s$, $C \sim \hat{C}$, $P \sim \hat{P}$, $\kappa_7 \sim \kappa_{11}$, $\kappa_8 \sim \kappa_{12}$, assumption A5 is seen to hold. Hence, the results of §3 may be applied directly to conclude the existence of a calculable threshold value $\mu_s^* > 0$ such that (24) (and thus (17)) is globally uniformly ultimately bounded with respect to a (calculable) compact neighbourhood of the origin for all $\mu_s \in (0,\mu_s^*)$.

5. Conclusion

Robustness properties of ultimate boundedness control with respect to neglected dynamics have been established for a restricted class of uncertain systems. However, the linear forms assumed for the actuator and sensor dynamics and the linear† structure assumed for the feedback are somewhat special; generalization of the analysis, by

---

† *Nonlinear* feedback (continuous and discontinuous) control of uncertain systems, in the absence of neglected dynamics, is considered in, for example, [1],[10],[12]-[15].

weakening these assumptions, is warranted. Tentative studies in this direction suggest that, if a continuous *nonlinear* feedback operator Φ of a form similar to that of [1] is adopted, then (under smoothness assumptions on Φ) structural stability with respect to neglected linear actuator and sensor dynamics can be established by demonstrating the *existence* of threshold values $\mu_a^* > 0$ or $\mu_s^* > 0$; however, in contrast to the case of a linear feedback operator Φ, these values may *not* be explicitly *calculable* in general.

6. References

[1] Corless, M. and Leitmann, G.: Continuous state feedback guaranteeing uniform ultimate boundedness for uncertain dynamic systems, *IEEE Trans. Automatic Control, AC-26,* 1139-1144, 1981.

[2] Barmish, B.R., Corless, M and Leitmann, G.: A new class of stabilizing controllers for uncertain dynamical systems, *SIAM J. Control & Optim., 21,* 246-255, 1983.

[3] Khalil, H.K.: On the robustness of output feedback control methods to modeling errors, *IEEE Trans. Automatic Control, AC-26,* 524-526, 1981.

[4] Khalil, H.K.: A further note on the robustness of output feedback control methods to modeling errors, *IEEE Trans. Automatic Control, AC-29,* 861-862, 1984.

[5] Kokotovic, P.V.: Recent trends in feedback design: an overview, *Automatica, 21,* 225-236, 1985.

[6] Vidyasagar, M.: Robust stabilization of singularly perturbed systems, *Systems & Control Letters, 5,* 413-418, 1985.

[7] Porter, B. and Manganas, T.: Design of robust digital controllers for gas turbines with explicit actuator and sensor dynamics, *AIAA J. Guidance, 8,* 644-649, 1985.

[8] Leitmann, G., Ryan, E.P. and Steinberg, A.: Feedback control of uncertain systems: robustness with respect to neglected actuator and sensor dynamics, *Int. J. Control,* to appear.

[9] Barmish, B.R. and Leitmann, G.: On ultimate boundedness control of uncertain systems in the absence of matching conditions, *IEEE Trans. Automatic Control, AC-27,* 153-157, 1982.

[10] Ryan, E.P. and Corless, M.: Ultimate boundedness and asymptotic stability of a class of uncertain dynamical systems via continuous and discontinuous feedback control, *IMA J. Math. Control & Info., 1,* 223-242, 1984.

[11] Petersen, I.R.: Structural stabilization of uncertain systems: necessity of the matching condition, *SIAM J. Control & Optim., 23,* 286-296, 1985.

[12] Gutman. S.: Uncertain dynamical systems - a Lyapunov min-max approach, *IEEE Trans. Automatic Control, AC-24,* 437-443, 1979.

[13] Leitmann, G.: Guaranteed asymptotic stability for some linear systems with bounded uncertainties, *ASME J. Dynam. Systems, Meas. & Control, 101,* 212-216, 1979.

[14] Leitmann, G.: On the efficacy of nonlinear control in uncertain linear systems, *ASME J. Dynam. Systems, Meas. & Control, 103,* 95-102, 1981.

[15] Gutman, S. and Palmor, Z.: Properties of min-max controllers in uncertain dynamical systems, *SIAM J. Control & Optim., 20,* 850-861, 1982.

DISCRETE STABILITY OF STOCHASTIC PROGRAMMING PROBLEMS WITH RECOURSE

R. Lepp

Institute of Cybernetics
Estonian Academy of Sciences
Akadeemia tee 21
SU-200108 Tallinn, USSR

1. Introduction. In this paper we propose a way of solving approximately the stochastic programming problem with recourse (two-stage stochastic programming problem). However, because of space limitations we can only sketch the approach. Let us consider the dynamic (canomic) formulation of the problem, defined recursively in finite-dimensional space: minimize

$$h(x) + \int_S Q(x,s)m(ds) \tag{D}$$

over all $x \in X$, where

$$Q(x,s) = \inf_y \{ f(s,x,y) \mid y \in Y,\ g(s,x,y) \leq 0 \} .$$

Here $x \in R^r$, $y \in R^t$, $h: R^r \to R^1$, $f: R^1 \times R^r \times R^t \to R^1$, $g: R^1 \times R^r \times R^t \to R^k$, $X \subset R^r$, $Y \subset R^t$ and s is a random vector with the known distribution. It is well-known that under quite natural assumptions the function $Q(x) = E\, Q(x,s)$ is Lipschitz continuous. Moreover, if the distribution of a random vector s has a density, then the function $Q(x)$ is differentiable [3].

Despite of these promising properties it is hard to solve the problem (D) by mathematical programming methods due to difficulties arising in numerical evaluation of $Q(x)$ and its derivatives. Besides, there exists another, static formulation of the problem: minimize

$$h(x) + \int_S f(s,x,y(s))\, m(ds) \tag{S}$$

over all vectors $x \in X$ and vector-valued functions $y(s)$ satisfying conditions: for almost all (a.a.) $s \in S$

$$y(s) \in Y,$$

$$g(s,x,y(s)) \leq 0.$$

Here $y \in L^p(S,B,m;R^t)$, $1 < p < \infty$, B is the sigma-algebra of Borel sets in $S \subset R^l$ and m is the probability measure on B determined by the random vector $s = s(\omega)$.

In a series of papers Rockafellar and Wets considered $y(s)$ as an essentially bounded function, $y \in L^\infty$, Eisner and Olsen as an element of L^p, $1 < p < \infty$ (see for instance [2], [5], [6]). They present conditions that guarantee the equivalence of the two statements (D) and (S) in the sense that x minimizes the problem (D) if and only if the pair $[x,y(s)]$ minimizes the problem (S). In this pair x denotes a vector and $y(s)$ a function. Due to this equivalence we can conclude that difficulties which arise solving two-stage stochastic programming problems are the same as those which arise solving an optimization problem in Banach spaces.

As pointed out by Olsen [5], the dynamic formulation of the problem is more fruitful for charaterizing solutions where the static formulation is more computationally tractable and can be settled in some cases by solving a sequence of finite-dimensional "discretizations".

One possibility to solve the problem (S) consists in its replacing by a more simple, finite-dimensional problem. Among all approximations the discrete one seems to be more natural, especially in multidimensional L^p-spaces. In order to avoid singularities let us consider only the case of relatively complete recourse, i.e. for every $x \in X$ and almost all $s \in S$ there exists an y such that $y \in Y$ and $g(s,x,y) \leq 0$. In the case of relatively complete recourse the space L^∞ has no advantages compared to L^p-spaces, $1 < p < \infty$, but for analysis of discrete stability the (L^p, L^q) - framework, $1/p + 1/q = 1$, is more convenient than $(L^\infty, (L^\infty)^*)$ - framework.

2. Discretization of the problem. In order to replace the problem (S) by a simpler problem in finite-dimensional space we have to start with the weak convergence of a sequence of discrete measures to our initial measure m. Let us suppose that there exists a sequence of discrete measures (m_n) $(n \in N) = \{1,2,3,\dots\}$, $m_n = (m_{1n},\dots,m_{nn})$, such that

$$\sum_{i=1}^{n} h(s_{in})\, m_{in} \rightarrow \int_S h(s)\, m(ds) \qquad \text{as} \quad n \rightarrow \infty$$

for every bounded and continuous on S function $h(s)$ and for some sequence of points $s_n = (s_{1n},\dots,s_{nn}) \in S$. Now we are able to consider instead of the problem (S) the following finite-dimensional problem (S_n) in space $R^r + l_n^p$ with the norm $\| y_n \|_n = (\sum_{i=1}^{n} |y_{in}|^p m_{in})^{1/p}$ in space l_n^p : minimize

$$h(x) + \sum_{i=1}^{n} f(s_{in},x,y_{in})\, m_{in} \qquad (S_n)$$

with respect to

$$x \in X$$

$$y_{in} \in Y, \; i= 1,2,\dots,n,$$

$$g(s_{in},x,y_{in}) \leq 0, \; i=1,2,\dots,n.$$

Since the problems (S) and (S_n) are defined in different spaces $R^r \times L^p$ and $R^r \times l_n^p$ respectively, we have to introduce a system of connection operators $p_n : L^p \rightarrow l_n^p$ $(n \in N)$. In L^p-spaces we are obliged to use a piecewise integral connection operators in the form

$$p_n y = (m(A_{in}))^{-1} \int_{A_{in}} y(s)\, m(ds), \quad i=1,2,\dots,n). \qquad (1)$$

To guarantee the discrete stability of problems (S) and (S_n) we have to restrict the choice of sets A_{in} from the initial Borel sigma-algebra B: suppose that

1) $\bigcup_{i=1}^{n} A_{in} = S$, 2) $A_{in} \cap A_{jn} = \phi$ as $i \neq j$,

3) $s_{in} \in A_{in}$, 4) diam $A_{in} \to 0$ as $n \to \infty$,

5) $\max_{1 \leq i \leq n} |m_{in}\, m\,(A_{in})^{-1} - 1| \to 0$ as $n \to \infty$,

6) $m\,(A^{\circ}_{in}) = m\,(A_{in}) = m\,(\overline{A}_{in})$,

where A° denotes interior and $\overline{A}$ closure of a set A.

Let us explain briefly the sense of restrictions 1)-6). The first four restrictions are quite natural. The fifth restriction means that we have to choose sets A_{in} in connection with weak convergence of a sequence (m_n) ($n \in N$). The sixth condition restricts our initial probability measure to the certain collection of sets from B. This collection of sets A_{in} with property 6) constitutes only an algebra. For instance, the sixth condition is fulfilled if the measure m has a density, and for example, if we deal with the Lebesgue measure on $[0,1]$, then integrability over so restricted measure means simply integrability in Riemann sense.

3. Conditions of discrete stability. In this section we present conditions that guarantee discrete stability of problems (S) and (S_n) with respect to cost functions and optimal solutions. Under discrete convergence of elements we mean the following: a sequence (y_n) ($n \in N$) converges discretely (or p_n-converges)

to an element y if $\| y_n - p_n y(s) \|_n \to 0$ as $n \to \infty$.

Let f^* and f_n^* be optimal values, $[x^*, y^*(s)]$ and $[x_n^*, y_n^*]$ optimal solutions of problems (S) and (S_n) respectively. Suppose that

1) the function $h(x)$ is convex and continuous,

2) the functions $f(s,x,y)$ and $g(s,x,y)$ are continuous in (x,y) convex and differentiable in y for all x and a.a. s and let the m-measure of their discontinuity points be zero,

3) the measure m has a density and bounded support S,

4) there exists a point $[\tilde{x}, \tilde{y}(s)]$ such that the strong inequality

$$g(s,\tilde{x},\tilde{y}(s)) < 0$$

holds for a.a. $s \in S$,

5) the sets X and Y are bounded.

<u>Theorem</u>. Suppose that a sequence of discrete measures $(m_n)(n \in N)$ converges weakly to the probability measure m and that the conditions 1)-5) are fulfilled. Then

$$f_n^* \to f^* \quad \text{as } n \to \infty$$

and there exists a subsequence $([x_n^*, y_n^*])$ $(n \in N' \subset N)$ of the sequence of extremum elements of problems (S_n) converging discretely to the solution $[x^*, y^*(s)]$ of the problem (S).

<u>Remark.1</u>. It seems quite natural to use instead of connection system (p_n) $(n \in N)$ the following connection system (p_n') $(n \in N)$

$$p_n' y(s) = (y(s_{in}), \; i = 1,2,\dots,n)$$

(see for instance [4]). But, unfortunately, this system suits only for continuous functions since changing values of an integrable function on a set of measure zero we can vary the value of sums in our problem so that they are equal to zero for all n N but the value of the integral does not change. So, for

L^p- spaces only a piecewise integral connection operators are admissible.

Remark 2. For discrete approximation of the problem (S) we cannot directly use Daniel's scheme about the approximate minimization of functionals [1], since a piecewise integral connection operator does not need to transform the set of admissible solutions of the problem (S) to the set of admissible solutions of discretized problems. This may also happen with restoration of points taken from the set of admissible solutions of discretized problems.

Remark 3. For the same reason we cannot replace our initial problem (S) by a problem in a class of piecewise constant functions. As piecewise constant functions belong to L^p- space, so it seems that we do not need the discretization theory at all. In order to prove the convergence $f_n^* \to f^*$ we must show that $f^* \leq \liminf f_n^*$ and $f^* \geq \limsup f_n^*$. However it is impossible to prove the first inequality even in the case of very simple linear functions using piecewise constant approximations.

REFERENCES

1. Daniel J., On the approximate minimization of functionals. Math. Comput. 23 (1969), pp. 573-581.

2. Eisner M.J., Olsen P., Duality for stochastic programming interpreted as LP in L_p space. SIAM J. Appl. Math., 28 (1975), pp. 779-792.

3. Kall P., Stoyan D., Solving stochastic programming problems with recourse including error bounds. Math. Operationsforsch. Statist Ser. Optimization, 13 (1982), pp. 431-447.

4. Olsen P., Discretizations of multistage stochastic programming problems. Math. Programming Stud., 6 (1976), pp.111-124.

5. Olsen P., Multistage stochastic programming with recourse as mathematical programming in an L_p space. SIAM J. Control Optimiz., 14 (1976), pp. 528-537.

6. Rockafellar R.T., Wets R.J-B., Stochastic convex programming: Basic duality. Pacific J. Math., 62 (1976), pp.173-195.

Time optimal boundary controllability of a viscoelastic beam

G. Leugering, Fachbereich Mathematik der Technischen Hochschule Darmstadt,
Schlossgartenstr. 7, 6100 Darmstadt, West Germany

ABSTRACT

It is shown that the vibrations of a viscoelastic beam can be steered to rest in minimal time using a $L_2(0,T)$-boundary control realizing a prescribed norm-bound.

1. INTRODUCTION

There is currently considerable interest in controlling the vibrations of flexible structures which are usually modelled by structurally damped beam equations. The choice of the damping mechanism is related to the problem of finite-rank-feedback-stabilization where one is interested to improve the decay rate of finitely many lower modes of the underlying structure. The uncontrolled modes should automatically decay to zero at an exponential rate. The models used, however, are of rate type damping and give rise to analytic semigroups; [5], [7], [13]. The frequency spectrum, therefore, is not confined in a strip parallel to the imaginary axis which is the case in most problems involving vibrations. There is a very well established theory, namely the theory of linear viscoelastic solids, which reveals a very subtle damping mechanism, due to the action of internal parameters, and which is felt to give a more adequate and more appealing picture of reality. We do not claim to give a feedback stabilization device here but what we want to do is to turn the attention onto the necessary background associated with this problem i.e., the question of admissible controllability and time optimal control. Let us consider a vibrating beam of length ℓ which is taken to satisfy the model equation

$$u_{tt}(t,x) + u_{xxxx}(t,x) - \int_0^t a(t-s)u_{xxxx}(s)ds = 0 \quad \text{on } (0,T)\times(0,\ell) \tag{1.1}$$

with initial conditions

$$u(0,x) = u_0(x), \quad u_t(0,x) = v_0(x) \quad \text{on } (0,\ell) \tag{1.2}$$

and appropriate boundary conditions, say,

$$\begin{aligned} u(t,0) = u(t,\ell) = 0 \\ u_{xx}(t,0) = 0, \quad u_{xx}(t,\ell) = f(t) \end{aligned} \quad \text{on } (0,\ell) \tag{1.3}$$

see Gaiduk [6] for special choices of a. For physical and technical reasons, of course, the function f, our control instrument, has to be restricted in size. The control problem then is to specify an admissible phase space such that the "trajectories" satisfying the initial conditions (1.2) can be steered to some given final state under the action of an admissible boundary control f. Without the restriction on f this problem was recently solved by the author ([9]). As was pointed out by Narukawa [11] the problem of admissible controllability is intimately connected with the stability of motion. In [1] Ahmed gives a necessary and sufficient criterion for admissible controllability, which, incidently, is a direct generalization of Marzollo's results obtained for finite dimensional systems (see [10]), but which appears to be difficult to establish in concrete cases. Since we are interested in the behavior of solutions of (1.1), (1.2), (1.3) for eventually large times we have to specify the relaxation kernel a(.). Because of the limited spaces availlable we decide to choose the approach by Dafermos [3] rather than to extend the more recent theory of Desch and Grimmer [4] to questions of asymptotic stability.

2. EXISTENCE AND UNIQUENESS RESULTS

Let A be defined in $H := L_2(0,\ell)$ by $An = u_{xxxx}$ and $D(A) = \{u \in W_{2,4}(0,\ell) \mid u(0) = u(0) = u''(0) = u''(\ell) = 0\}$ where $W_{p,m}(0,\ell)$ is the familiar Sobolev space of order m associated with $L_p(0,\ell)$. Then A is a positive definite selfadjoint operator in H with discrete spectrum $\{\lambda_j\}_{j\in\mathbb{N}}$ $0 < \lambda_1 < \lambda_2 < \dots < \lambda_j < \lim_{j\to\infty} \lambda_j = \infty$, and orthonormal eigenbase $\{\varphi_j\}_{j\in\mathbb{N}}$, i.e. $\lambda_j = (\frac{j\pi}{\ell})^4$, $\varphi_j = \frac{2}{\ell} \sin \frac{j\pi}{\ell} (.)$. Let $\alpha = W_{2,2}(0,\ell) \cap \overset{0}{W}_{2,1}(0,\ell)$ denote the form domain of A. Define $w(x,\zeta,t) := u(x,t-\zeta)$, $0 \le \zeta \le t$ and zero otherwise, that is, if u(x,t) is the given displacement of the viscoelastic beam $w(x,\zeta,t)$ denotes its history which is itself trivial before t = 0. We put $u_t = v$ in order to transform Equation (1.1) into the system

$$u_t = v$$

$$v_t = - u_{xxxx} + \int_0^\infty a(\zeta) w_{xxxx}\, d\zeta \qquad (2.1)$$

$$w_t = - w_\zeta$$

As in [3], we define an energy space E by

$$E := Q \times H \times L_2([0,\infty),a,Q)$$

where $L_2([0,\infty),a,Q)$ is a space of fading memory i.e. $L_2([0,\infty),a,Q)$ is the space of a-weighted Q-valued L_2-distributions. We provide W with the energy product:

$$<(u,v,w),(\hat{u},\hat{v},\hat{w})> = \int_0^\infty \{u_{xx}\,\hat{u}_{xx} + v\hat{v} + \int_0^\infty a(\zeta)[u_{xx} - w_{xx}] \,.$$

$$\cdot\, [\hat{u}_{xx} - \hat{w}_{xx}]d\zeta\, dx \tag{2.2}$$

We hasten to state the following assumption on a(.):

(A1) $\quad a(\zeta) \geq 0, \quad a'(\zeta) \leq 0, \quad \int_0^\infty a(\zeta)d\zeta < 1$

which is familiar in its mechanistic interpretation. Obviously, the function $p(x) := \frac{1}{6}(\frac{x^2}{\ell} - \ell)x$ solves the boundary value problem $u_{xxxx} = 0$, $u(0) = u(\ell) = u_{xx}(\ell) = 0$, $u_{xx}(\ell) = 1$. Let z be the solution of the elastic beam equation

$$z_{tt} + z_{xxxx} = 0 \quad \text{in} \quad (0,\ell)\times(0,T) \tag{2.3}$$

with boundary conditions (1.3) and initial conditions (1.2). Following the lines of [8] it is then shown in [9] that z, which can be decomposed as $z = z + p\,.\,f$ (p given above) is in $C(0,T,H)$ if f is in $L_2(0,T)$. Define $\varphi = u - z$, then

$$\varphi_{tt}(t) + \varphi_{xxxx}(t) - \int_0^t a(t-s)\varphi_{xxxx}(s)ds = + \int_0^t a(t-s)z_{ss}(s)ds \tag{2.4}$$

Obviously, φ satisfies homogeneous boundary conditions (1.3) $(f \equiv 0)$. Defining $h(t) = \int_0^t a(t-s)z_{ss}ds$, $\varphi_t = \psi$, $\eta(t,s) = \varphi(t-s)$, (2.4) as (1.1) can be put into the form (2.1), that is

$$\begin{aligned} \varphi_t &= \psi \\ \psi_t &= -A\varphi + \int_0^\infty a(s)A\varphi(s)ds + h(t) \\ \eta_t &= -\eta_s \end{aligned} \tag{2.5}$$

Define the system operator $\mathcal{A}$ in E by $\mathcal{A}(u,v,w) = (v, -Au + \int_0^\infty a(s)ds, -ws)^t$, $D(\mathcal{A}) = \{(u,v,w)\in E|\ v\in Q,\ w_s\in L_2([0,\infty),a,Q),\ w(.,0) = u(.),\ Au - \int_0^\infty a(s)Au(s)ds\in H\}$. It is shown by Dafermos [3] that $\mathcal{A}$ is a generator of a contractive semigroup S(t) in E. Thus, if we denote the state (φ,ψ,η) by ρ and the inhomogenity h(t) imbedded in E by $\chi(t)$, (2.5) can be reformulated as

$$\rho_t = \mathcal{A}\rho + \chi(t) \tag{2.6}$$

In order to solve (2.6) by the variation of constants formula we need some assumption on a(.)

(A2) $\quad a \in C^2([0,\infty)),\ \ a(0) = 0.$

Integrating by parts h(t) is then given by

$$h(t) = a'(0)z(t) - a(t)z_t(0) - a'(t)z(0) + \int_0^t a''(t-s)z(s)ds \tag{2.7}$$

Since z(.), the elastic part, is in C(0,T,H), we conclude that h is in C(0,T,H). This obviously implies $\chi(.) \in C(0,T,E)$. By the standard semigroup argument, ρ, provided with the initial condition $\rho(0) = 0$, is given by $\rho(t) = \int_0^t S(t-s)\chi(s)ds$ and is in fact the unique mild solution of (2.6). That is, $g \in C(0,T,E)$. Therefore, the first two components, $(u-z,\ u_t-z_t)(T)$, of g(T) satisfy $(u-z,u_t-z_t)(T) \in Q\times H \to H\times Q^*$, the embedding being compact. Thus, we have seen that combining the semigroup approach by Dafermos with the existence and uniqueness of weak solutions of the reference problem (2.3) proved in [9] Theorem 2.1 we can uniquely solve the original problem (1.1), (1.2), (1.3) in a weak sense:

THEOREM 1: Let $u_0 \in Q$, $v_0 \in H$ and $f \in L_2(0,T)$ be given. Then there exists a unique weak solution u of (1.1), (1.2), (1.3) such that $u \in C^1(0,T,H) \cap C(0,T,Q^*)$. Furthermore at t = T : $(u,u_t)(T,u_0,v_0,f) = (\varphi,\varphi_t)(T,u_0,v_0,f) + (z,z_t)(T,u_0,v_0,f)$ where φ is given by (2.4) - (2.7) and z is the unique weak solution of (2.3), (1.2), (1.3). As a map from $L_2(0,T)$ to $H\times Q^*$, $(\varphi,\varphi_t)(T,u_0,v_0,..)$ is compact. □

From [3] we adopt:

THEOREM 2: Let $u_0 \in Q$, $v_0 \in H$ be given. Then the free solution, u, and its derivative, u_t, tend to zero as t tends to infinity.

PROOF: [3] Section 6. The arguments given there do not depend on the order of the operator A. □

REMARK 1: We have shown in [9] that the shifted energy space $H\times Q^*$ can be replaced by $Q_1\times H_1$ where Q_1 is given by $Q_1 := \overset{0}{W}{}^{1}_{2,1}(0,\ell) = \{u\in H \mid \sum_j \lambda_j^{1/2}\ |(u,\varphi_j)|^2 < \infty\}$, and $H_1 = Q_1^*$. Therefore, even the weak solution satisfies the geometric boundary conditions. Since $Q\times H$ is compactly embedded in $Q_1\times H_1$ as well, all conclusions of Theorem 1 hold true in the case $Q_1\times H_1$ instead of $H\times Q^*$. □

3. OPTIMAL CONTROLLABILITY

As we have seen by Theorem 1, the solution of Equation (1.1), (1.2), (1.3) can be decomposed into a part solving the reference problem (2.3), (1.2), (1.3) being responsible for the propagation of discontinuities coming in from the boundary and another part revealing the damping mechanism due to the viscoelastic behavior of the material. As we have shown in [9], the elastic reference system is exactly controllable in $Q_1 \times H_1$ by $L_2(0,T)$-boundary controls, that is, the operator mapping the L_2-boundary data into the phase-space $Q_1 \times H_1$, given in Remark 1, is onto. By Theorem 1 and Remark 1, the whole process is a compact perturbation of this surjective map and is, hence, governed by a semi-Fredholm operator. The same procedure as in [9] applies to the extend that the kernel of the adjoint solution operator can be shown to be trivial. This is proved by Laplace-transform techniques and a well-known Abelian result on the asymptotic behavior of such transforms.

In this note we are mainly concerned with admissible and time-optimal controllability of the system (1.1).

THEOREM 3: Let $u_0 \in Q$, $v_0 \in H$ be given. Then for each $M > 0$ there exists a time $T_0 > 0$ and a boundary control f_0 such that the solution, $u(.,u_0,v_0,f_0)$, of (1.1), (1.2), (1.3) satisfies $u(T,u_0,v_0,f_0) = u_t(T,u_0,v_0,f_0) = 0$, and moreover $\|f_0\|_{L_2(0,T)} \leq M$.

PROOF: Let $T > 0$ be given, By Theorem 1 and Remark 1 the operator mapping $L_2(0,T)$ into $Q_1 \times H_1$ is bounded. Define, as in [11], the closed subspace N of $L_2(0,T)$ as $N := \{f \in L_2(0,T) \mid u(T,0,0,f) = u_t(T,0,0,f) = 0\}$. By Theorem 3.2 in [ ], to u_0, v_0 given above there exists a control $\tilde{f} \in L_2(0,T)$ such that $u(T,u_0,v_0,\tilde{f}) = u_t(T,u_0,v_0,\tilde{f}) = 0$. Let F denote the map transferring (u_0,v_0) into the equivalence class of $\tilde{f}$ in $L_2(0,T)/N$. By the well-posedness of (1.1), (1.2), (1.3) established in Theorem 1, F is a closed and hence bounded operator, thus mapping bounded sets into bounded sets. Therefore, if we are given u_0, v_0 as above, by Theorem 2, all we have to do is to wait until the energy norm $\|u(T,u_0,v_0,0), u_t(T,u_0,v_0,0)\|_{Q\times H}$, of the free solution has become sufficiently small. Hence, if $M > 0$ is arbitrarily given, there exists a time $T > 0$ (possibly large) such that for each $T_0 > T$ we have a boundary control, f_0, with $f_0(t) = 0$ on $[0,T]$ a.e. satisfying $u(T_0,u_0,v_0,f_0) = u_t(T_0,u_0,v_0,f_0) = 0$ and $\|f_0(.)\|_{L_2(0,T_0)} \leq M$. □

THEOREM 4: For each $u_0 \in Q$, $v_0 \in H$ and $M > 0$, there exists a minimal transition time $T_M > 0$ and a boundary control f_M in $L_2(0,T_M)$ such that $u(T_M,u_0,v_0,f_M) = u_t(T_M,u_0,v_0,f_M) = 0$ and furthermore $\|f_M\|_{L_2(0,T_M)} = M$.

PROOF: Since we have arranged the phase space $Q_1 \times H_1$ in such a way that the boundary-input operator is bounded, the existence of a time-minimal control, f_M, is proved by standard means, (cf. [1]); the proof of the extremum principle, namely $\|f_M\|_{L_2(0,T_M)} = M$, follows basically the lines of Narukawa's proof in Theorem 3 [11]. This is because, by [3] we are provided with a semigroup, which can easily be extended to the energy space E_1 where Q and H are replaced by Q_1 and H_1, respectively, by density arguments outlined in [8]. □

REMARK 2: A theorem similar to Theorem 3 has been shown by Narukawa, [12], for second order operators, different space setting and, more important to note, for kernels which are small together with their derivative at each point $t \in [0,\infty)$. □

REFERENCES:

1. N.N. Ahmed, Finite-time null controllability for a class of linear evolution equations on a Banach space with control constraints, *JOTA*, 47/2, 129 - 158, (1985).

2. A.V. Balakrishnan, *Applied functional analysis*, Springer-Verlag 1976.

3. C.M. Dafermos, Contraction semigroup and trend to equilibrium in continuum mechanics, in *Applications of Functional Analysis to Problems in Mechanics* (Springer, 1976).

4. G.W. Desch and R.C. Grimmer, Initial boundary value problems for integro-differential equations, *J. of Int. Eq.*, to appear.

5. R.F. Curtain, Finite dimensional compensators for some hyperbolic systems with boundary control; *Control theory for distributed parameter systems and applications*, Lecture Notes in Control and Information Sciences 54, Springer (1983), 77 - 92.

6. S.I. Gaiduk, Some problems related to the theory of the action of a transverse impulse on a rod, *Differential Equations*, 13/II, 854 - 861 (1977).

7. J.S. Gibson, An analysis of optimal modal regulation: convergence and stability; *SIAM J. Control and Opt.* 19/5, 686 - 707 (1981).

8. I. Lasiecka and R. Triggiani, A cosine operator approach to modelling $L_2(0,T,L_2(\Gamma))$-boundary hyperbolic equations, *J. Appl. Math. & Opt.*, 7, 35 - 93 (1981).

9. G. Leugering, Boundary controllability of a viscoelastic beam, *Applicable Analysis*, to appear.

10. A. Marzollo, *Controllability and optimization*, CISM Lecture No. 17, Udine 1969, Springer Verlag (1972).

11. K. Narukawa, Exact and admissible controllability of viscoelastic systems with boundary controls, *Preprint*.

12. K. Narukawa, Admissible null controllability and optimal time control, *Hiroshima Math. J.*, 11/3, 533 - 551 (1983).

13. Y. Sakawa, Feedback control of second order evolution equations with damping; *SIAM J. Control and Opt. 22/3*, 343 - 361 (1984).

STABILIZATION BY THE FEEDBACKS $-B^*$ AND $-B^*P$

N. Levan
Department of Electrical Engineering
7732 Boelter Hall
University of California, Los Angeles, CA 90024, USA

1. INTRODUCTION

Let A be the generator of a C_o semigroup of bounded linear operators $T(t)$, $t \geq 0$, over a Hilbert space H --with inner product $[.,.]$ and norm $||.||$-- and let B be a bounded linear operator from a second Hilbert space U to H. Denote by (A,B) the "distributed" system

$$\dot{x}(t) = A\, x(t) + B\, u(t), \quad t \geq 0. \tag{1-1}$$

This note will study feedback stabilization of (A,B) by the state feedback operators $-B^*$ and $-B^*P$ --where P is to be specified. In other words, what we would like to know is when do the closed loop systems

$$\dot{x}(t) = (A - BB^*)\, x(t), \quad t \geq 0, \tag{1-2}$$

and

$$\dot{x}(t) = (A - BB^*P)\, x(t), \quad t \geq 0, \tag{1-3}$$

stable?

Let $S(t)$, $t \geq 0$, be the "closed" loop semigroup, i.e, the one with generator $A - BB^*$ or $A - BB^*P$. Then the question is "what are the necessary and sufficient conditions for $S(t)x \to 0$, (in a suitable sense), $t \to \infty$, for all x in H. These will be explored in section 2. We begin with the class of contraction semigroups, then we generalize to the class of "quasi-affine" transforms of contraction semigroups. It is interesting to note that the latter is precisely the class of closed loop semigroups resulting from stabilizing the system, using a non-negative solution of the Steady State Riccati Equation (SSRE). We shall also discuss the problem of stability enhancement by state feedback.

2. MAIN RESULTS

We begin by recalling the following definitions of stability.

Definition

A semigroup $T(t)$, $t \geq 0$, over H is: (i) e(exponential)-stable if $||T(t)|| \leq Me^{-\omega t}$ for some $M \geq 1$ and some $\omega > 0$; (ii) s(strong)-stable if, for every x ($\neq 0$): $||T(t)x|| \to 0$, $t \to \infty$; and (iii) w(weak)-stable if for every x and y: $[T(t)x,y] \to 0$, $t \to \infty$.

Now let $T(t)$, $t \geq 0$, be a semigroup of contractions --i.e., $||T(t)|| \leq 1$, $t \geq 0$-- over H, and let A be its generator, then we know that A is "maximal" dissipative:

$$\text{Re.}[Ax,x] \leq 0, \quad \text{for } x \text{ in } \mathcal{D}(A) \text{ (the domain of } A).$$

Let $F: H \to U$, $u = Fx$, be a state feedback then the closed loop semigroup is generated by $A + BF$. It then follows that if both A and BF are dissipative then so is $A + BF$. This suggests that we can take $F = -B^*$ so that $A - BB^*$ generates the contraction semigroup $S(t)$, $t \geq 0$. One then uses results from "structure" theory of Hilbert space contractions to find necessary and sufficient conditions for $S(t)$, $t \geq 0$, to be stable, i.e., for the pair (A,B) to be stabilizable by the feedback $-B^*$.

We state the following results.

Theorem 1 [1]

Let $T(t)$, $t \geq 0$, be a contraction semigroup with generator A in H, and define

$$W(T) = \{x \text{ in } H: \ T(t)x \to 0, \text{ weakly}, \ t \to \infty\}. \tag{2-1}$$

Let $W(T)^{\perp}$ denote the orthogonal complement in H of $W(T)$. Then the semigroup $S(t)$, $t \geq 0$, generated by $A - BB^*$ is w-stable if and only if $W(T)^{\perp}$ is controllable for the pair (A,B), i.e.,

$$W(T)^{\perp} \subset M_c(A,B), \tag{2-2}$$

where $M_c(A,B)$ is the controllable subspace of (A,B),

$$M_c(A,B) = \bigcup_{t\geq 0} T(t)BU. \tag{2-3}$$

The above results refine earlier results of Slemrod [2]. We must note that the proof of Theorem 1 relied on the canonical decomposition of contractions due to Nagy and Foias [3], and another decomposition of contractions due to Foguel [4].

We now prove the following results for stabilizability of unitary or isometric semigroups.

Theorem 2

Let $T(t)$, $t \geq 0$, be a unitary or an isometric semigroup with generator A in H, and suppose that the resolvent of A is compact. Then $T(t)$, $t \geq 0$, is s-stabilizable by the feedback $-B^*$ if and only if the pair (A,B) is controllable.

Proof

Let $W(T)$ be as in Theorem 1 then according to Foguel [4] --for a contraction semigroup-- $W(T)$ is a closed subspace, $W(T) = W(T^*)$ hence it is reducing, and the part of the semigroup on $W(T)^\perp$ is unitary. This together with the canonical decomposition of Nagy and Foias [3] tell us that the space H admits the unique orthogonal decomposition

$$H = H_{cnu}(T) \oplus W_u(T) \oplus W(T)^\perp, \tag{2-4}$$

where $H_{cnu}(T)$ is reducing and the part of the semigroup on it is completely nonunitary, and

$$H_{cnu}(T)^\perp = H_u(T) \text{ (say)} = \{x \text{ in } H: ||T(t)x|| = ||x|| = ||T(t)^*x||,\ t \geq 0\}, \tag{2-5}$$

is the "maximal" unitary subspace of the semigroup. Moreover

$$H_u(T) = W_u(T) \oplus W(T)^\perp, \tag{2-6}$$

and

$$W(T) = H_{cnu}(T) \oplus W_u(T). \tag{2-7}$$

We also note that $T(t)|W_u(T)$ is absolutely continuous unitary and $T(t)|W(T)^{\perp}$ is singular unitary. For details on these decompositions we refer to [5].

Now if $T(t)$, $t \geq 0$, is a unitary semigroup (or a unitary group) then it is plain that $H_{cnu}(T) = \{0\}$, hence

$$H = W(T) \ (= W_u(T)) \oplus W(T)^{\perp}. \tag{2-8}$$

In addition if the resolvent of A is compact then, clearly,

$$\text{for } x \text{ in } W(T): \ ||T(t)x|| \to 0, \ t \to \infty.$$

But this is not possible since the semigroup is unitary. Therefore, in this case, (2-8) becomes

$$H = W(T)^{\perp},$$

i.e., the semigroup --or equivalently, the system (A,B)-- is weakly unstable on all of H.

Next, since the resolvent of $A - BB^*$ is also compact, we have for the semigroup $S(t)$, $t \geq 0$,

$$H = H_{cnu}(S) \ (= W(S)) \oplus H_u(S) \ (= W(S)^{\perp}). \tag{2-9}$$

It then follows from this decomposition that the semigroup $S(t)$, $t \geq 0$, is w-stable if and only if $W(S)^{\perp} = \{0\}$. For this we only have to note that, for x in $H_u(S) \cap \mathcal{D}(A)$:

$$\text{Re.}[(A - BB^*)x,x] = 0 = -\ ||B^*x||^2.$$

Therefore, it is plain that

$$H_u(S) \subseteq \ker.B^*. \tag{2-10}$$

More is true. It follows from this relation that the two semigroups $T(t)$, $t \geq 0$, and $S(t)$, $t \geq 0$, are identical on $H_u(S)$. Therefore $H_u(S)$ is also reducing for the former. It then follows from this and from (2-10) that

$$H_u(S) \subset M_{uc}(A,B),$$

where

$$M_{uc}(A,B) = M_c(A,B)^{\perp} = \{x \text{ in } H: B^*T(t)^*x = 0, \ t \geq 0\} \tag{2-11}$$

is the uncontrollable subspace of (A,B). Therefore

$$M_{uc}(A,B) = \{0\} \implies H_u(S) = \{0\},$$

and one half of the Theorem is proven.

For the other half, we first note that

for x in $M_{uc}(A,B)$: $S(t)^*x = T(t)^*x, \quad t \geq 0$.

Now suppose that $S(t)$, $t \geq 0$, is weakly stable and $M_{uc}(A,B)$ is not trivial. Then it follows from the Foguel result that

for x in $M_{uc}(A,B)$: $T(t)x \to 0, \ t \to \infty$.

This is not possible since we have seen that $H = W(T)^{\perp}$, unless $M_{uc}(A,B)$ is trivial. This completes the proof of the Theorem, keeping in mind that, weak stability implies strong stability as soon as the resolvent of A is compact, [1], while exactly the same arguments hold for the case of an isometric semigroup.

We must note that if, for a unitary semigroup, the resolvent of A is not compact, then controllability of the subspace $W(T)^{\perp}$ is necessary and sufficient for the semigroup to be w-stabilizable by the feedback $-B^*$. This has not been noticed before. Infact, [4] requires the system to be controllable for weak stabilizability of a unitary semigroup. The results of Theorem 2 were given in [2] --using a different approach and were only sufficient. For strong stabilizability of contraction semigroups --even when the resolvent of A is not compact, we refer to [6].

We now turn to the feedback $-B^*P$ where P is a self-adjoint non-negative operator on H. First, let us take P to satisfy the Steady State Riccati Equation (SSRE) [7] --which resulted from a Linear Quadratic Regulator (LQR) Problem associated with the system (A,B),

$$[PAx,x] + [x,PAx] - [PBB^*Px,x] + [Rx,x] = 0, \quad x \text{ in } \mathcal{D}(A), \tag{2-12}$$

where R is a given non-negative operator on H. It follows that, for x in $\mathcal{D}(A)$:

$$2\text{Re.}[P(A - BB^*P)x,x] = - ||B^*Px||^2 - [Rx,x] \leq 0. \tag{2-13}$$

Again let $S(t)$, $t \geq 0$, denote the semigroup generated by $A - BB^*P$ --i.e., the closed loop semigroup, then it is easy to see from (2-13) that

$$[PS(t)x,S(t)x] \leq [Px,x], \quad \text{for } x \text{ in } H. \tag{2-14}$$

Let $Q^2 = P$ then (2-14) becomes

$$||QS(t)x|| \leq ||Qx|| \quad \text{for } t \geq 0 \text{ and all } x \text{ in } H. \tag{2-15}$$

If the self-adjoint operator P is positive, $P > 0$, then so is its positive square root Q. Then since the range $R(Q)$ is dense in H we can define a contraction semigroup $C(t)$, $t \geq 0$, (say) over H by

$$QS(t)x = C(t)Qx, \quad \text{for all } x \text{ in } H.$$

We therefore conclude that

Lemma 1

If $P > 0$ satisfies the SSRE (2-12) then $A - BB^*P$ generates the semigroup $S(t)$, $t \geq 0$, which is a quasi-affine transform [3] of a contraction semigroup over H.

The above suggests that one can use the decompositions of contractions to obtain sufficient conditions for stabilizability in this case. Thus the key question is "when does the quasi-affine transform of a contraction semigroup stable?"

To proceed, we define

$$M(S) = \{x \text{ in } H: [PS(t)x,S(t)x] = [Px,x],\ t \geq 0\}, \qquad (2\text{-}16)$$

then it is plain from (2-14) that $M(S)$ is a closed subspace of H. This clearly is the analog of the unitary subspace $H_u(T)$ --equation (2-5). One then expects that weak stability of $S(t)$, $t \geq 0$, will be achieved as soon as $M(S)$ is trivial. This, unfortunately, is not quite the case since we can only have

Proposition 1

Let $S(t)$, $t \geq 0$, be a C_o semigroup over H and, at the same time, it is a quasi-affine transform of a contraction semigroup over H. Then if its subspace $M(S) = \{0\}$ then the semigroup is weakly stable on a dense subspace of H.

Proof

For details of the proof we refer to [8].

Combining Lemma 1 and Proposition 1 we have

Theorem 3

Let $P \geq 0$ be a solution of the SSRE (2-12) then controllability of the pair (A^*,R) implies that the semigroup $S(t)$, $t \geq 0$, generated by $A - BB^*P$ is weakly stable on a dense subspace, i.e., the open loop system (A,B) is weakly stabilizable on a dense subspace by the feedback $-B^*P$.

Proof

We have from (2-13), for x in H and for $t \geq 0$:

$$[PS(t)x,S(t)x] - [Px,x] = -\int_0^t ||B^*PS(\sigma)x||^2\, d\sigma - \int_0^t [RS(\sigma)x,S(\sigma)x]\, d\sigma.$$

Therefore it is clear that

$$\text{for } x \text{ in } M(S):\ B^*PS(t)x = 0 = RS(t)x,\quad t \geq 0. \qquad (2\text{-}17)$$

Moreover, the subspace $M(S)$ is also invariant for $S(t)$, $t \geq 0$. Hence it is easy to see that $S(t)x = T(t)x$ for x in $M(S)$. This implies that $M(S)$ is also invariant for $T(t)$, $t \geq 0$. We therefore conclude from (2-17) that

$M(S) \subseteq \{x \text{ in } H: RT(t)x = 0,\ t \geq 0\} = M_{uc}(A^*,R)$ --the uncontrollable subspace of the pair (A^*,R). More is true. We also have $\ker.P \subseteq M(S)$. Therefore $\ker.P \subseteq M(S) \subseteq M_{uc}(A^*,R)$. This completes the proof of the Theorem.

Note that if the resolvent of A is compact, in addition, then the semigroup $S(t)$, $t \geq 0$, is strongly stable , again only on a dense subspace.

3. CONCLUSION AND DISCUSSIONS

We have in this note discussed stabilizability by the feedback $-B^*$ and $-B^*P$ --where P is a positive solution of the SSRE. We now close the note with the following cases:

(i) In view of the decompositions of Hilbert space contractions --which resulted in stabilizability of contraction semigroups by $-B^*$, we would like to know "when is the generator $A - BB^*K$ dissipative?" Here A is the generator of a C_o semigroup -- A needs not be dissipative, and K is a non-negative operator. The reason for this is that if such an operator K can be found then the semigroup $S(t)$, $t \geq 0$, generated by $A - BB^*K$ is certainly contractive. Hence sufficient conditions for it to be stable can be found. It turns out that if the semigroup $T(t)$, $t \geq 0$, generated by A is such that $||T(t)|| \leq e^{\omega t}$ for $t \geq 0$, then the operator K must satisfy the Lyapunov equation

$$K(BB^*) + (BB^*).K = V$$

where V is a self-adjoint strictly positive operator on H such that, for x in H: $2\omega\,||x||^2 \leq [Vx,x]$. For details and related results we refer to [9].

(ii) Stability Enhancement Problem. Suppose that the open loop semigroup $T(t)$, $t \geq 0$, is s-stable: $||T(t)x|| \to 0$, $t \to \infty$. What we ask is that is it possible to find a state feedback so that the closed loop semigroup $S(t)$, $t \geq 0$, is e-stable: $||S(t)|| \leq Me^{-\alpha t}$ for some $M \geq 1$ and some $\alpha > 0$? Now let us take $-B^*$ be the state feedback, then according to a result of Datko [10], the semigroup $S(t)$, $t \geq 0$, generated by $A - BB^*$ is e-stable if and only if there is a self-adjoint positive operator $P > 0$ such that, for x in $\mathcal{D}(A)$: $2\text{Re}.[P(A - BB^*)x,x] = -||x||^2$. It then follows that, for x in H:

$$[PT(t)x,T(t)x] - [Px,x] = -\int_0^t [(I - W)T(\tau)x,T(\tau)x]\, d\tau,$$

where

$$W = PBB^* + BB^*P.$$

Letting $t \to \infty$ and using the fact that the semigroup $T(t)$, $t \geq 0$, is s-stable we obtain:

$$[Px,x] = \int_0^\infty [(I - W)T(t)x,T(t)x]\, dt > 0.$$

Thus it is enough to require that $I - W > 0$. Hence the operator P must satisfy the operator inequality $PBB^* + BB^*P < I$. Results for this case and for the general case of a feedback of the form $-B^*K$, as well as related problems are being investigated and will be reported elsewhere.

4. REFRENCES

[1] C.D. Benchimol, A Note On Weak Stabilizability of Contraction Semigroups, SIAM J. Contr. and Optimi., 16 (1978), 373-379.

[2] M. Slemrod, A Note On Complete Controllability and Stabilizability For Linear Control Systems In Hilbert Space, SIAM J. Contr. and Optimi., 12 (1974), 500-508.

[3] B. Sz-Nagy & C. Foias, Harmonic Analysis of Operators On Hilbert Space, American Elsevier, New York, 1970.

[4] S.R. Foguel, Powers of a Contraction in Hilbert Space, Pacific J. Math., 13 (1963), 551-562.

[5] P.A. Filmore, Notes On Operator Theory, Van Nostrand, New York, 1970.

[6] N. Levan & L. Rigby, Strong Stabilizability Of Linear Contractive Control Systems On Hilbert Space, SIAM J. Contr. and Optimi., 17 (1979), 23-35.

[7] A.V. Balakrishnan, Applied Functional Analysis, Springer Verlag, 1981.

[8] N. Levan, Approximate Stabilizability Via The Algebraic Riccati Equation, SIAM J. Contr. and Optimi., 23 (1985), 153-160.

[9] N. Levan, Stabilization By Dissipative Feedback Systems, Proc. 23rd CDC Conference, Dec. 12-14, 1984, Las Vegas, Vol.2, 1176-1179.

[10] R. Datko, Extending A Theorem Of A.M. Liapunov to Hilbert Space, J. Math. Anal. and Appl., 32 (1970), 610-616.

THE STOCHASTIC MODELS FOR ESTIMATION OF FUNCTIONING QUALITY OF COMMUNICATIONS AND THEIR APPLICATION

Y.V. Livanov. Computing Centre of the USSR Academy of Sciences

Stochastic models for estimation of communications average productivity in stationary and working models during time T are considered. These models are used for designing oil and gas communications.

1. Description

When designing communications it is necessary to estimate their efficiency in case of equipment failure. A network, which may suffer some of its components failure during operation, is convenient to be described by probability graph.

During operation, a i-branch of communications ($i=\overline{1,N}$) at an arbitrary moment may be in one of a number of conditions α_i ($\alpha_i=\overline{1,m_i}$) which are characterized by the maximum possible productivity. Further on in this paper we shall omit the wording "maximum possible" for the sake of simplification.

The communications condition shall be defined through the condition of their branches.

Let the condition of i-branch be α_i, then the communications condition is

$$(\alpha_1, \dots, \alpha_N)$$

Let us write the communications productivity in a condition $(\alpha_1,\dots,\alpha_N)$ as $q(\alpha_1,\dots,\alpha_N)$.

The general number of the communications conditions is

$$K_0 = \prod_{i=1}^{N} m_i \tag{1}$$

In every condition and with technological limitations observed, the communications productivity is assumed as known.

2. Model 1

Let us consider the communications operation in a stationary mode. The stationary probability of the branch i being in the condition α_i will be written as $P(\alpha_i)$.

Here it is obvious that

$$\sum_{\alpha_i=1}^{m_i} P(\alpha_i) = 1, \quad \alpha_i = \overline{1, m_i} \tag{2}$$

The communications productivity at an arbitrary moment is a random variable with a distribution law defined by the set of probabilities.

$$P(\alpha_1, \ldots, \alpha_N) = \prod_{i=1}^{N} P(\alpha_i) \tag{3}$$

The probabilities $P(\alpha_1, \ldots \alpha_N)$ satisfy the relation

$$\sum_{\alpha_1=1}^{m_1}, \ldots, \sum_{\alpha_N=1}^{m_N} P(\alpha_1, \ldots, \alpha_N) = \sum_{(\alpha_1, \ldots, \alpha_N)} P(\alpha_1, \ldots, \alpha_N) \tag{4}$$

Average productivity of the communications is expressed as mathematical expectation of a random variable $q(\alpha_1, \ldots, \alpha_N)$

$$M[q] = \sum_{\alpha_1=1}^{m_1}, \ldots, \sum_{\alpha_N=1}^{m_N} q(\alpha_1, \ldots, \alpha_N) P(\alpha_1, \ldots, \alpha_N) =$$

$$= \sum_{(\alpha_1, \ldots, \alpha_N)} q(\alpha_1, \ldots, \alpha_N) P(\alpha_1, \ldots, \alpha_N) \tag{5}$$

With independent processes of failures and reconditionings in the communications branches, the probability $P(\alpha_1, \ldots, \alpha_N)$ is found out from the relation

$$P(\alpha_1, \ldots, \alpha_N) = \prod_{i=1}^{N} P(\alpha_i) \tag{6}$$

With a large value of N, it is difficult to determine the average productivity analytically using formula (5). Therefore, it is usually determined by statistical modelling.

3. Model II

In Model II the communications operation during time T is considered. Due to independent failures and reconditionings of the communications branches Model II is based on operation models for each branch.

Let us write the time evolution of states for branch i as $Y_i(t)$ and describe it by the Markov homogeneous chain with a finite number of conditions m_i. The process $Y_i(t)$ can change its conditionsonly at moments t_ν, when $t_\nu = \nu \cdot \Delta t \ (\nu = \overline{1, l})$, i.e. inside the interval Δt there will be no condition change.

The Markov homogeneous chain is defined by [1] :

1) probabilities distribution at the initial moment

$$P(Y_i(t_o)) = P(Y_i(t_o) = 1), \ldots, P(Y_i(t_o) = m_i) \tag{7}$$

when conditions

$$\sum_{\alpha_i=1}^{m_i} P(Y_i(t_o) = \alpha_i) = 1, P(Y_i(t_o) = \alpha_i) \geqslant 0 \tag{8}$$

are satisfied.

2) matrix of transitions probabilities

$$P_i = \{p_{ikj}\} \tag{9}$$

where

$$p_{ikj} = P(Y_i(t_\nu) = j / Y_i(t_{\nu-1}) = k), \; j,k = \overline{1, m}. \tag{10}$$

Here the relations

$$\sum_{j=1}^{m_i} p_{ikj} = 1, \; p_{ikj} \geqslant 0 \tag{11}$$

are satisfied.

To determine the mean communications productivity, one has to find out the $P(Y_i(t_\nu) = \alpha_i)$, i.e. the probability of branch being in condition α_i at the moment t_ν. This probability is found out from recurrence relation

$$P(Y_i(t_\nu) = j) = \sum_{k=1}^{m_i} P(Y_i(t_{\nu-1}) = k) \cdot p_{ikj} \tag{12}$$

From (12) average product volume supplied through the communications within the time $[t_{\nu-1}, t_\nu]$ will be

$$Q_\nu = \Delta t \cdot \sum_{(\alpha_1, \ldots, \alpha_N)} q(\alpha_1, \ldots, \alpha_N) \cdot \prod_{i=1}^{N} P(Y_i(t_{\nu-1}) = \alpha_i) \tag{13}$$

Hence, average product volume supplied through the communications within the time $[0, T]$ will be

$$Q(T) = \sum_{\nu=1}^{l} Q_\nu \tag{14}$$

Average communications productivity within the time T is determined from the relation

$$M[q(t_o, t_1)] = \frac{Q(T)}{T} \tag{15}$$

The quantity $M[q(t_0,t_1)]$ is determined through the probability values $P(Y_1(t_\nu)=\alpha_1)$, which can be found out by both analytical methods and statistical modelling of the process of the communications branches failures and reconditionings.

Average productivity based on the above examined models is an index of the communications efficiency.

4. Estimation of the plan fulfilment probability

Let us consider the second index of the communications efficiency, which is the fulfilment of the plan.

Each curve of the communications condition change in time will have a correspondent W - i.e. the product supply to the customer within the estimated time T.

The supply volume is a random variable W owing to failures and reconditionings in the communications. A random variable W can be realized by means of modelling failures and reconditionings process within the time T. These realizations can be used to formulate the random variable distribution law for W.

Distribution laws of the random variables of failure-free time and of reconditioning time are used as initial data. Models that ensure effective mode control and equipment repair are employed.

Thus, the probability of the plan fulfilment is determined as

$$P(W \geqslant W^{\circ}) = \gamma \qquad (16)$$

where W° is the planned volume of supply to the customer.

5. Models application

The above described stohastic models are applied in designing oil and gas supply communications. Such a design with provided reliability could be formulated as a two-stage design. The first stage envisages communications ignoring reliability. The second stage provides correction of the result obtained at the first stage.

The communications design with reliability provided could be formulated in the following generalized form.

To find such a vector of the parameters X_1 for the designed communications that will ensure minimum costs for the estimated time T:

$$C(X_1)=\min_{X \in \Omega}(\sigma K(X) + U(X)) \tag{17}$$

where under the variant X: K(X) is capital expenditures, U(X) is working costs, σ is known coefficient, Ω is set of variants, $X_1 \in \Omega$ is vector of an optimum variant parameters.

Here the following limitations are observed:

$$Z_r(X) \geqslant 0, \quad r = \overline{1,h} \tag{18}$$

Apart from technological, technical and economic limitations, included in (18) are also those associated with reliability. They are written as:

$$d_2 q^\circ \leqslant M[q] \leqslant d_1 q^\circ \tag{19}$$

$$P(W(X) \geqslant W^\circ) \geqslant \gamma^\circ \tag{20}$$

where d_1, d_2 are known coefficients provided by solution of the top priority problem, q°, W° are planned productivity and planned supply to the customers, accordingly, γ° is the preset probability of the plan fulfilment.

Wall thickness and diameter of the pipeline, equipment types and locations, amount of standby equipment and its distribution, standby reservoir capacity are components of objective variables vector X_1.

Thus, every communications variant has a correspondent vector X .

Limitations (5), (15), (16) associated with communications efficiency indices make the problem (17) difficult to solve.

Algorithm of the communications design with provided reliability considered in [2] is based on solution of the following problems: 1) communications design ignoring reliability [3]; 2) estimation of the variant using efficiency index (5), (15), (16); 3) optimum reserve distribution; 4) correction of the communications variant with regard for reliability.

The stochastic models considered in this paper are used in designing oil and gas fields construction [4, 5] and oil pipelines [6].

References

1. Shiryaev A.N.: "Probability". Nauka Publishing House, Moscow, 1980, 575 p. (In Russian).
2. Livanov Y.V.: "Iterating search for an optimum variant of a dendritic network with regard for reliability", the sizes for papers presented at XXVI Conference of the Lithnanian Mathematical Society, Vilnius Institute of mathematics and cybernetics, Academy of Sciences of Lithnanian Soviet Socialist Republic, 1985, p.p. 158-159. (In Russian).
3. Khachaturov V.R., Arzhanov F.G., Astakhov N.D. et al.: System for computerized designing general oil field construction and its application experience, Review information of All-Union Research Institute for Organization, Management and Economics of Petroleum Industry, Moscow, 1980, 68 p.(In Russian).
4. Livanov Y.V., Miroshnichenko V.I., Pisarenko T.V.:"Problem of Optimum Equipment Reserve in Designing Oil-field Construction", Review Papers: "Oil Field Production and Oil Transport", All-Union Research Institute for Organization, Management and Economics of Petroleum Industry, Moscow, 1985, N 7, pp. 51-53. (In Russian).
5. Livanov Y.V., Miroshnichenko V.I., Pisarenko T.V.: "Imitation Modelling Used for Reliability Estimation of an Oil-field Production System", Review Papers of All-Union Research Institute for Organization, Management and Economics of Petroleum Industry, Moscow, 1985, N 6, pp. 10-13. (In Russian).
6. Livanov Y.V.: "Reliability Factor in Computerized Oil Pipeline Designing", Review Papers: "Automatization and Telemetry in Oil Industry", All-Union Research Institute for Organization, Management and Economics of Petroleum Industry, Moscow, 1983, N 12, pp. 7-9. (In Russian).

MDS: AN INSTRUMENT FOR THE CONSTRUCTION OF MODELS FOR COMPANY PLANNING USING THE CONCEPTS OF SYSTEM DYNAMICS

Francesco De Lorenzo, ENI, Roma
Vincenzo Gervasio, ENIDATA, Milano
Paolo Verrecchia, TEMA, Bologna

1. THE METHODOLOGY FOR COMPANY STRATEGIC DECISIONS MODELLING (MDS).

Within a company, two different categories of observable phenomena may be distinguished:

a) those which refer to single points in time (for example, the level of bank deposits on a given day) which are thus not linked to the passage of time, i.e. having "atemporal" characteristics.

b) those which refer to intervals of time (e.g. the flow of interest matured on a deposit account from one day to another) and which are thus linked to the passage of time, i.e. having "temporal" characteristics.

The "temporal" phenomena correspond to the ACTIVITIES of the company whereas the "atemporal" ones correspond to the STATES in which the company finds itself due to its ACTIVITIES. Thus, it can be said that the "atemporal" phenomena are the EFFECTS of which the "temporal" ones are the CAUSES, in the chain of events occurring which constitutes the company in its entirety.
A fundamental characteristic of the company as a system is that of presenting open or closed chain reactions between ACTIVITIES and STATES, CAUSES and EFFECTS.
The theory of System Dynamics (1) provides a suitable representation of this characteristic, by assuming:

1. that the variations in the "atemporal" quantities (levels) are determined by the values of the "temporal" ones (flows);

2. that the values of the flows are determined by the values of the levels unless they are exogenous.

For these reasons, this approach is thus particularly suitable for representing positive or negative feedback phenomena which are the determinants of the company's "real dynamics".
Company phenomena can thus be examined using simulation models based upon System Dynamics.
The MDS methodology (Modelli per Decisioni Strategiche Aziendali - Models for Company Planning and Strategies) (2) is of this type, possessing the following characteristics:

a) the company is represented as a system and is thus composed of parts which mutually interact through relationships which represent the regularity of company phenomena; the the system is broken down into flows and levels.

b) the system is an open one and, therefore, there are (represented by exogenous variables) both influences deriving from the context in which the company operates as well as the effects of important decisions made by the management.

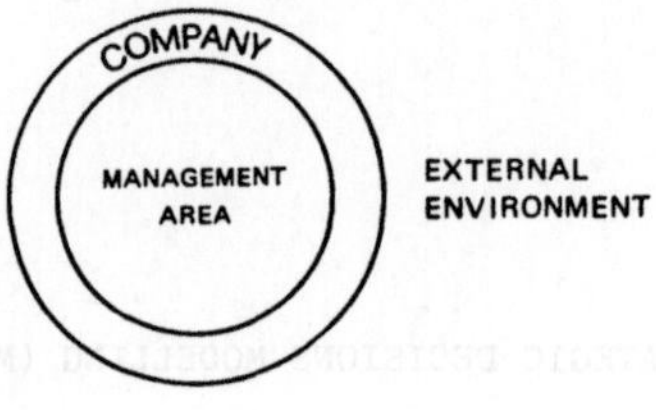

Fig. 1

This concept is shown schematically in Fig. 1, where the loop called COMPANY represents the domain of the known relationships and the origin of the principles of irregularity. The COMPANY loop is immersed in an EXTERNAL CONTEXT from which the exogenous influences derive with active centres distant from the company. Within this loop the MANAGEMENT AREA is represented and is not considered to belong to the system being simulated as it can not be associated with deterministic principles nor with phenomena showing a certain regularity.

c) The regular structure of the company described by the model consists of three subsystems (Economic/financial, Production and Market), which interact with the mutual exchange of information.
The Economic/financial subsystem may be considered as a "projection" of production and market phenomena, measured in monetary terms, within a single structure which comprises both accounting principles and the laws of financial methematics. This structure is, fundamentally, the same for any type of company.

In order to obtain an MDS model for company simulation, the Production and Market subsystems must be built and the structure of the Economic/financial subsystem adapted to the particular case being considered.
From the above, it can be seen:

1. An MDS model, however, does not, by itself, give any indication regarding optimum decisions, or the way forward in given company situations. There is, in fact, no optimizing algorithm.

2. Rather than predicting the future, the MDS model allows the decision-maker to construct "plausible histories" by combining particular scenarios regarding the external context with given hypotheses of strategic choices.
For each scenario and each hypothetical strategic choice one can obtain the simulation of a plausible company future, from given initial conditions. Each simulation can then be evaluated (e.g. with respect to economic and financial

trends). In this way one can assess a range of possible company futures, on the basis of the evaluation criteria used and arrive at a choice.

The MDS model is thus to be considered not as an automatic mechanism but as an instrument in the hands of company management, as a "decisional aid".

2. COMPANY MODEL

It is possible to describe the company as a dynamic system in a brief but precise manner using MDS with the aim of showing:

a) the aspects of instability, originating from the nature of financial phenomena (important in the life of the company, especially when money costs are high);

b) the stabilizing effect which may be obtained through given control policies actuated by the shareholders.

It will be shown that, under fairly general conditions and given the assumption that the shareholders behave in a standard manner, the company may be considered as a system with two stable states, which are:

1. the state which corresponds to the ability of the company to distribute dividends to the shareholders;

2. the state which corresponds to the shareholders' need to sustain the company.

It will also be shown that the change from one state to another does not take place in a continuous manner, but through "transitions" where the system is not in equilibrium. The basic structure used for the present study is shown in Fig. 2.

ENVIRONMENT OF COMPANY EVOLUTION

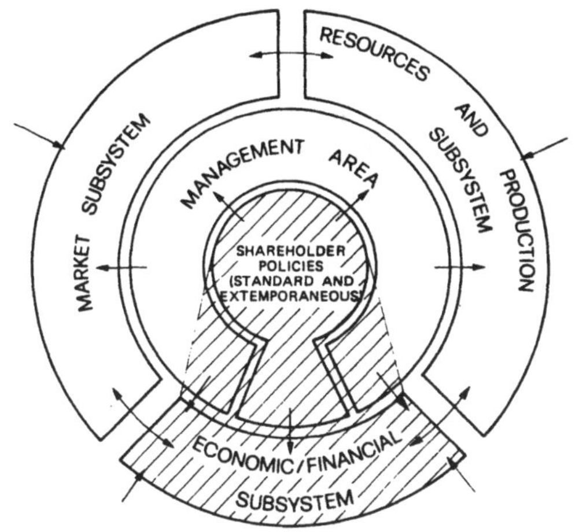

GENERATION OF BOUNDARY CONDITIONS
AND SCENARIO PARAMETERS

Fig. 2

The point of view considered here is that of the shareholder, so that in the overall company system there will be an open subsystem, represented in Fig. 2 by the shaded area, which includes the economic/financial subsystem, the area of management for which it is responsible and the policies of the shareholders. In particular, keeping in mind the objectives to be reached, it will be sufficient to define the state of the company subsystem examined, by means of a single variable (net capital). The "upstream" variables which influence the subsystem, e.g. scenario variables, interaction with production, management styles and accounting policies, lead to certain trends in the results of the economic management (excluding financial effects) and in balance sheet items such as credits, debits, surpluses, capital expenditures.
It will be shown that a particular linear combination of these functions may be used as a "control variable" of the company subsystem.
During the course of this treatment particular hypotheses about the control variables will not be necessary except that, over an interval sufficiently long for a complete representation of the evolution of company phenomena to be studied, an analytical description must be possible using an "average value plus variations". While this statement is extremely general in nature, it is necessary to represent a relatively stable scenario over the interval considered: a possible instability in the company which becomes evident during that interval can thus not be considered, in a strict sense, as having been imported from the exterior.

3. INSTABILITY PHENOMENA AND COMPANY CONTROL SYSTEMS

The company, considered as a dynamic system, presents characteristics of instability. In order to understand this, one can consider the economic/financial subsystem of the company which is represented using the MDS symbology of flows and levels shown in Fig. 3.

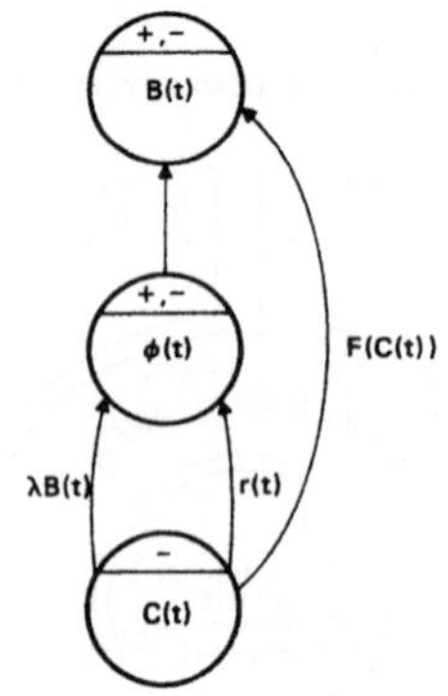

Fig. 3

At each point in time t, B(t) is the level of short-term deposits (or exposure): $\phi(t)$ is a sum level of the credit, surpluses, capital expenditure levels, less the sum of the various debit and fund levels; C(t) the level of net capital (given a negative sign); r(t) the flow of returns (positive or negative) from the industrial activity, λ B(t) the flow of financial returns (or costs) from short-term deposits (or exposure), where λ the short-term cost of money referred to the unit time interval; $F(C(t))$ is the financial flow deriving from the company shareholders' policies.

The following two cases will be examined: A and P - where A is the total absence of shareholder policies and P their presence.

CASE A - Absence of shareholder policies.

If, during the interval dt there are no variations in the stock (registered) capital, and neither ordinary nor extraordinary interventions regarding the latter then $F(C(t))=0$, and from a flow balance carried out around the level $C(t)$ over an interval dt, one has:

$$dC(t) = [r(t) + \lambda B(t)]\,dt \tag{1}$$

If the capital is given a negative sign, following the MDS convention, Eq. 1 becomes:

$$dC(t) = -[r(t) + \lambda B(t)]\,dt \tag{2}$$

and thus:

$$dC(t)/dt = -r(t) - \lambda B(t) \tag{3}$$

Recalling that, at each point in time, t, in the company's balance sheet:

ASSETS (t) = LIABILITIES (t)

And in this case (see Fig. 3):

$$\text{ASSETS } (t) = B(t) + \phi(t) \tag{4}$$

$$\text{LIABILITIES} = -C(t) \tag{5}$$

So that, from Eqs. 4 and 5:

$$B(t) + \phi(t) + C(t) = 0 \tag{6}$$

Management aims to reach a given positive result r_0 (i.e. a net profit) and to exercise control over the level of credits, debits, surpluses and capital expenditure

A virtually stationary situation, as far as management are concerned, may be characterised by the trends in $r(t)$ and $\phi(t)$ (excluding inflation) which oscillate very little with respect to r_0 and ϕ_0, the average values of the variables $r(t)$ and $\phi(t)$ over a sufficiently long interval.

It will be seen that, even under the conditions described here, assuming that the short-term cost of money takes a given value λ, the overall trend of the company can not be considered as "stable" in the normal sense of the word.

Combining Eqs. 3 and 6:

$$dC(t)/dt = -r(t) + \lambda[C(t) + \phi(t)] \tag{7}$$

and therefore,

$$d\,C(t)/dt = \lambda\, C(t) - [r(t) - \lambda\, \phi(t)] \tag{8}$$

The binomial $(r(t) - \lambda\phi(t))$, from the above reasoning, is a function which depends upon management trends and may be considered as a control function of the "company system" taken as an economic/financial structure.
Given that

$$v(t) = r(t) - \lambda\,\phi(t) = r_0 - \lambda\,\phi_0 + \Delta v(t) = v_0 + \Delta v(t) \tag{9}$$

Eq. 8 becomes:

$$d\,C(t)/dt = \lambda\, C(t) - (v_0 + \Delta v(t)) \tag{10}$$

To integrate Eq. 10 both sides are multiplied by $e^{-\lambda t}$ giving:

$$C(t) = \frac{v_0}{\lambda} + \left(C_0 - \frac{v_0}{\lambda}\right) e^{\lambda t} - e^{\lambda t} \int_0^t \Delta v(\tau)\, e^{-\lambda \tau}\, d\tau \tag{11}$$

Where C_0 is the value of C(t) for t=0.

In the particular case where $\Delta v(t) = 0$ for each value of t, (which corresponds to a "stationary" company management in a strict sense), Eq. 11 becomes:

$$C(t) = \frac{v_0}{\lambda} + \left(C_0 - \frac{v_0}{\lambda}\right) e^{\lambda t} \tag{12}$$

The stationary solution clearly corresponds to:

$$C_0 - \frac{v_0}{\lambda} = C_0 - \frac{r_0}{\lambda} + \phi_0 = -B_0 - \frac{r_0}{\lambda} = 0 \tag{13}$$

where B_0 is the value of "short-term deposits (liabilities)" for t=0.
If however $\lambda B_0 \neq - r_0$, a solution is obtained such that the net capital C(t) and, consequently also the level B(t) (having assumed $\phi(t)$ constant $= \phi_0$), varies exponentially with time (Eq. 12).
It follows that each different initial condition, even for very small adjustments from that corresponding to the equality in Eq. 13 gives rise to a time evolution which tends to move the system continually away from a stationary situation.

CASE P - Presence of a shareholder policy.

Some suitable mechanisms for controlling an inherently unstable system such as a company where there is a complete absence of policies on the part of the shareholders, will now be examined.
Taking account of reality, the following assumptions are made:

a) the "net capital", C(t), is the variable to be controlled;

b) no action is taken while C(t) remains within the interval $C_a - C_b$ (see Fig. 4).

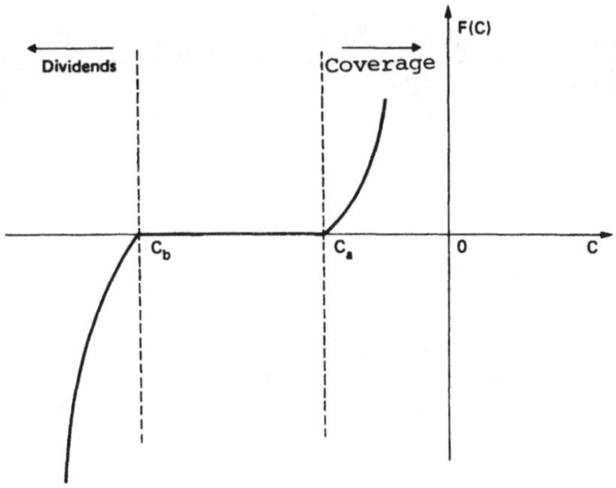

Fig. 4

"Standard control" is defined as the measures taken by the shareholders (loss coverage for $C_a \leq C(t) \leq 0$; distribution of dividends for $C(t) \leq C_b$) in the following manner:

A) the flow of coverage is, as absolute value, an increasing function of the difference $C(t) - C_a$;

B) the flow of dividends is, as absolute value, an increasing value of the difference $C_b - C(t)$

On the basis of these assumptions, from a balance of flows carried out at a level $C(t)$ in Fig. 3, over an interval dt, one has:

$$dC(t)/dt = -r(t) - \lambda B(t) - F(C(t)) \tag{14}$$

where the shape of the function $F(C(t))$ is given schematically in Fig. 4.
Before further development, Eq. 14 is transformed by introducing a dimensionless variable $y(t)$* related to the net capital by the following linear relationship:

$$C(t) = -|C_0| - y(t)|\Delta C_0| \tag{15}$$

with $|C_0| = \frac{|C_b| + |C_a|}{2}$, $|\Delta C_0| = \frac{|C_b| - |C_a|}{2}$

Substituting Eq. 15 in Eq. 14 and recalling Eq. 6:

$$-\frac{dy(t)}{dt}|\Delta C_0| = -r(t) + \lambda\left(\phi(t) - |C_0| - y(t)|\Delta C_0|\right) - F(C(t)) \tag{16}$$

---

* The meaning of $y(t)$ can be shown as follows:
(a) for $-1 < y(t) < +1$ the company is not subject to the standard control of the shareholders and tends to move towards one of the extremes of the interval;
(b) the standard control of the shareholders tends to keep the level of the company from going below $y(t) = -1$; or to keep $y(t)$ from going above +1 (apart from small variations outside of the interval itself).

and also:

$$\frac{dy(t)}{dt} = \frac{r(t)}{|\Delta C_0|} + \lambda\frac{(|C_0| - \emptyset(t))}{|\Delta C_0|} + \lambda y(t) + \frac{F(C(t))}{|\Delta C_0|} \tag{17}$$

Transforming the dimensional function F (C(t)) using:

$$F(C(t)) = -\lambda|C_0| f(y(t)) \tag{18}$$

where f(y(t)) is a dimensionless function.

Eq. 17 thus becomes:

$$\frac{dy(t)}{dt} = \frac{r(t)}{|\Delta C_0|} + \lambda\frac{(|C_0| - \emptyset(t))}{|\Delta C_0|} + \lambda y(t) - \lambda\frac{|C_0|}{|\Delta C_0|} f(y(t)) \tag{19}$$

If:

$$\begin{cases} \alpha = |C_0|/|\Delta C_0| & \text{dimensionless} \\ x(t) = r(t)/|\Delta C_0| + \lambda\,(|C_0| - \emptyset(t))/|\Delta C_0| \end{cases} \tag{20}$$

x(t) can be defined as a normalised control function and Eq. 19 thus becomes:

$$\frac{dy(t)}{dt} = \lambda y(t) - \lambda\alpha f(y(t)) + x(t) \tag{21}$$

Eq. 21 relates the state variable of the system y(t) to the normalised control function x(t).

An investigation of Eq. 21 should identify the possible steady states.

Assuming that dy(t)/dt=0 (assuming that in each case λ, which is the indicative parameter of the short-term cost of money, $>$ 0), one has:

$$\alpha f(y(t)) - y(t) = \frac{x(t)}{\lambda} \tag{22}$$

which may be satisfied, for each constant value of x(t), by values of y(t) corresponding to the locus A'ABB' in Fig. 5.

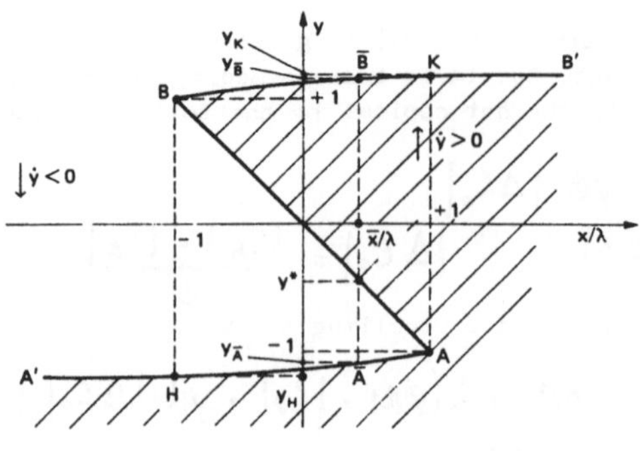

Fig. 5

The "S-curve" shown in Fig. 5 is thus a locus of the equilibrium points of the system, and divides the plane into two regions: the shaded region, corresponding to values of (dy/dt)$>$ 0, and the unshaded one where (dy/dt)$<$0.

The AB branch of the curve, characterized by zero values of f(y(t)), represents a

locus of unstable equilibrium points, as outlined for case A.
The branches BB' and AA' are, on the other hand, loci of stable equilibrium points. It can also be seen that:

a) for each constant value of $x(t)/\lambda > 1$ there is only one stable equilibrium state characterised by a value of y(t) which is very close to +1;

b) for each constant value of $x(t)/\lambda < 1$ there is only one state of stable equilibrium with a value of y(t) very close to -1;

c) for each constant value of $\bar{x}(t)/\lambda$ over the interval +1, -1 there are three values of y(t) corresponding to a possible equilibrium. Two of these, $y_{\bar{A}}$ and $y_{\bar{B}}$ correspond to stable equilibrium states:

one has $y_H \leq y_{\bar{A}} \leq -1$ and $1 \leq y_{\bar{B}} \leq y_K$ with y_H and y_K, respectively, the value of y on AA' for $x/\lambda = -1$ and on BB' for $x/\lambda = +1$. Whereas the third value y^*, between -1 and +1 represents a state of unstable equilibrium.

Before continuing, it is worthwhile examining the meaning of the normalised control function x(t).
Each value of x(t) (see Eq. 20) is given by the sum of the two contributions: $r(t)/|\Delta C_0|$ which is a measure of the profitability of the industrial activity, referred to a capital $|\Delta C_0|$, $\lambda[|C_0| - \phi(t)]/|\Delta C_0|$ which is the interest (referred to $|\Delta C_0|$ of a capital $(|C_0| - \phi(t))$ utilized at the interest rate of λ. Clearly, both r(t) and $(|C_0| - \phi(t))$ may be either positive or negative.
High values of r(t) contribute towards creating high values of x(t), whereas increases in the value of ϕ(t), without corresponding increases in r(t) (as, for example, in the case of investments with long-term profitability), tend to decrease the value of x(t).
Having clarified the meaning of the "normalised control function", and recalling that, both r(t) and ϕ(t) are the results of a given type of management, the process of evolution is examined, as that illustrated in Fig. 5. Starting from an equilibrium state $\bar{A}\{\bar{x}/\lambda, y_{\bar{A}}\}$ and assuming that the value of x(t) undergoes changes which are so slow that the "company system" can re-locate its equilibrium position. Starting, for example, from a situation in which stability is guaranteed by shareholder support, a gradual increase in x(t) will lead the system to $A\{x/\lambda = 1$ and $y = -1\}$; further increases in x(t) will correspond to equilibrium states described by the upper branch of the "S-curve". In other words, even for a very small increase in x(t) beyond the λ value, there will be a transition phase which breaks the succession of equilibrium states, (which are possible from $\bar{A}$ to A), leading the system to a new state represented by K.
Under these conditions, the company is profitable and can also distribute dividends to the shareholders (who, as assumed here, adopt a "standard control" policy).
Following x(t) further shows the company system to be in a "positive" situation ($y(t) \simeq 1$), also for values of $x(t)/\lambda$ below the initial value of $\bar{x}/\lambda$, and even for negative values of x(t) : down to $x(t) = -\lambda$.
At this point, even a slight further decrease in x(t) causes a transient which leads towards conditions which require coverage by the shareholders in order to guarantee the company's survival.

4. CONCLUSIONS

This paper has demonstrated the possibility of describing company phenomena which are important from the shareholders' point of view, concisely and with the necessary accuracy. The MDS methodology thus allows an extension of the use of dynamic company models from the specific field of management support instruments to that of top level strategic projects.

As an example of the use of the models described here, one may cite the preparation of "business plans" for particular initiatives which require the temporary support of possible shareholders or sponsors before becoming profitable or taking off as a company.

Formalizing shareholders' policies has allowed the observation of two types of possible stable steady states during the life of a company, and of the mechanisms which enable the transition between them.

This provides an untraditional view of company phenomena for the shareholders: for example, it is possible to represent the trends of a company using a "trajectory" in an x/λ ,y plane (such as that in Fig. 5). Comparisons can also be made between the performance of different companies, given that the variables used are normalised.

REFERENCES

1. Forrester,J.W. "Industrial Dynamics" 1961, MIT Press, Cambridge, Massachussets.

2. Bartezzaghi,E., Mariotti,S. (a cura di) "Modelli per le decisioni strategiche aziendali. Concetti e metodologie di base. Il sistema MDS- Tema (ENI)", Franco Angeli, Milano, Italia, 1983.

MATHEMATICAL SIMULATION MODEL OF POWER TRANSFORMER FOR ELECTRICAL POWER SYSTEM PROTECTIVE SCHEMES

S u m m a r y

The conventrional methods of electromagnetical circuit analysis do not provide means of comprehensive taking of the magnetic network topology into account. This article suggests an approach employing two novel circuit elements - electromagnetic gyrator /EMG/ and magnetic loss /K/ - which enable the process of mathematical model building to be done faster . Besides , the use of these elements is presented in building a mathematical model of a power transformer. The model is employed for simulating the performance of the transformer by means of digital simulation languages .

1. Introduction

While performing a simulation investigation of electromagnetic systems, it is really significant to take both electrical network topology and magnetical network topology as well as the basic properties of magnetic materials /i.e.magnetizing characteristic, eddy-current loss, hysteresis, etc./ into consideration .

There were many papers referring to electromagnetic systems and, in particular, transformer's transient current and voltage waveforms,most of them being written in the last 40 years .

These authors most often used circuit analysis methods for the electrical and magnetical waveforms and employed the term of mutual inductance M and, in consequence, equivalent circuits with transverse non--linear inductance branch were obtained . The authors of this paper suggest an approach with the use of two novel circuit elements which enable the electric and magnetic circuits to be combined in a smart way. The second section in this paper presents an electromagnetic gyrator /EM/ which is to be the basic circuit element for interfacing the electric circuits and the magnetic circuits. The third section discusses the other element - magnetic loss K - which is responsible for the eddy-current losses in the magnetic parts . The fourth section presents a circuit diagram and a mathematical model of a power transformer with the use of elements mentioned above. The concluding section presents a digital simulation model and examples of the results obtained by means of the numerical computation.

2. Electromagnetic Gyrator

The general discussion of the gyrator can be found in most publica-

tions referring to the circuit theory, e.g. [1]
The gyrator equations are as follows :

$$\begin{bmatrix} u_1 \\ i_1 \end{bmatrix} = \begin{bmatrix} 0 & \pm r_1 \\ \pm \frac{1}{r_2} & 0 \end{bmatrix} \cdot \begin{bmatrix} u_2 \\ i_2 \end{bmatrix} \qquad (1)$$

where : r_1 and r_2 are gyration constants .
Fig. 1 shows a graphical symbol of the gyrator .

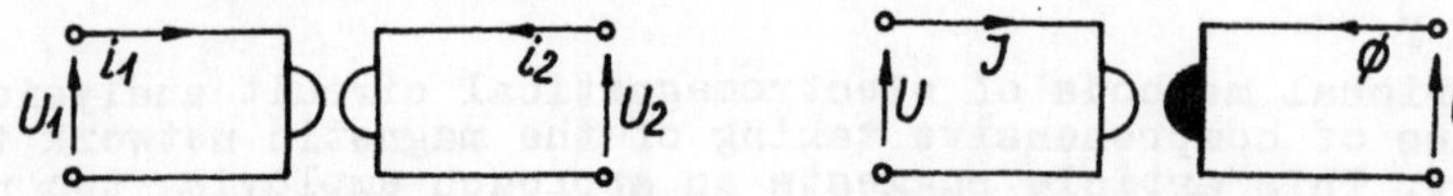

Fig.1.a/Conventional Gyrator Symbol . b/Electromagnetic G.S.
A simple electromagnetic circuit /e.g. a coil wound on a torroidal core, the winding resistance being negligible/ can be presented in the following way :

$$u = - n d\phi/dt \; , \qquad i = \Theta/n \; , \qquad (2)$$

where : ϕ is a magnetic flux /equivalent to electric current/ ,
Θ is a magnetomotive force, n is number of turns .
The equations can be presented in the matrix form

$$\begin{bmatrix} u \\ i \end{bmatrix} = \begin{bmatrix} 0 & -nd/dt \\ \frac{1}{n} & 0 \end{bmatrix} \cdot \begin{bmatrix} \Theta \\ \phi \end{bmatrix} \qquad (3)$$

$$\begin{bmatrix} U(s) \\ I(s) \end{bmatrix} = \begin{bmatrix} 0 & -n\,s \\ \frac{1}{n} & 0 \end{bmatrix} \cdot \begin{bmatrix} \Theta\,s \\ \Phi(s) \end{bmatrix} . \qquad (4)$$

The Laplace transformation representation being valid for the zero initial conditions. A four-pole described by the above equations provides conversion of the magnetic into electric quantities. The reverse conversion is provided by the reversed equations. From a formal point of view, the equatrions /4/ are similar to the commonly known gyrator equatrions. It is suggested however to give a name of "electromagnetic gyrator" /EMG/ to the gyrator equations and to use a symbol shown in fig. 1b as this gyrator perform some specific functions in the building of magnetic network model .

Properties of Electromagnetic Gyrator :

A. The use of gyrators provides means of clear combinations of electric and magnetic circuits. The combination pronciples are presented in figs. 2a and 2b .

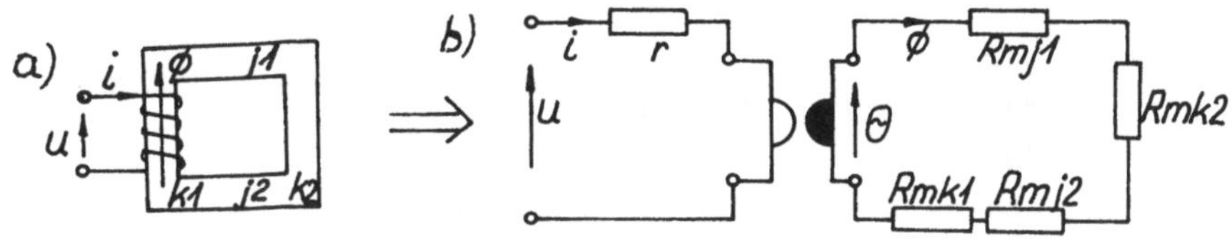

Fig.2. Transformer Circuit Diagram and Equivalent Diagram Employing the EMG Gyrator .

B. Input impedance /as seen from the electrical side/ is

$$z_{wej} = \frac{n^2 s}{z_{wyj}} , \tag{5}$$

where : $z_{wyj} = \frac{\Theta}{\Phi}$ /equivalent reluctance of the magnetic circuit/ .

C. Series combination of reluctances in a magnetic circuit coresponds to a parallel combination of inductances in an electric circuit , while a parallel combination of reluctances corresponds to a series combination of inductances .

D. All the properties of a general gyrator become the properties of the suggested gyrator EMG.

3. **Magnetic Loss**

The eddy-current losses in magnetic circuit in the linear range with respect to the B = f/H/ characteristics are directly proportional to the rate of change of magnetic flux. It is suggested to introduce such an element K - called a magnetic loss hereafter - into the circuit diagram, the magnetomotive force /MMF/ across which is proportional to the magnetic loss and the first derivative of time-varying flux Φ. Thus, the following equation can be written

$$\Theta_k = K\, d\Phi/dt . \tag{6}$$

If a toroidal core is used, by substituting

$$\Theta = H.l \tag{7}$$

$$\text{and} \quad \Phi = B \cdot A, \tag{8}$$

into Eq./6/ , the follwing relation with the magnetizing force H is obtained :

$$H = \frac{K . A}{l} \cdot \frac{dB}{dt} , \tag{9}$$

where : K is magnetic loss in the magnetic circuit, l is the length of magnetic path, and A is the cross-sectional area of the core .

Let us discuss a response with sinusoidal excitment. If

$$B = B_m \cdot \sin (wt) \tag{10}$$

then

$$H = kw \cdot \sin(wt). \qquad (11)$$

As the power loss per unity of volume is expressed by

$$L' = \int B\, dH, \qquad (12)$$

then, from the above relations, the following equation can be obtained

$$p = \frac{L'}{t} = k\, w^2 B_m^2 /2, \qquad (13)$$

where : $p = p_m \cdot \gamma$, p_m - eddy-current loss per unity of mass, γ - material specific density, w - angular frequency .

The last equation provides means of determining the value of magnetic loss coefficient

$$k = 2\, p_m / w^2 B_m^2 , \qquad (14)$$

which can be expressed by

$$K = kl/A \qquad (15)$$

for a specific part .

Properties of Magnetic Loss Coefficient :

- Value K is constant for a specific material and does not depend on the frequency or the value of flux density at which it is determined /in the linear section of the magnetizing characteristic/.
- By introducing the element K into a magnetic circuit diagram and a mathematical model, means of taking the eddy-current loss in the magnetic circuit into consideration are provided even in the excitment is a non-sinusoidal one .
- The element K in a magnetic circuit is an analog of the inductance in an electric circuit .

4. Mathematical Model of Three-Phase Transformer

While developing the mathematical model of transformer, the following simplifications were made : the hysteresis loop was neglected, it was assumed that the parameters were lumped, linear and stationary, no interaction of leakage fluxes was taken into account, no inter-coil or inter-winding capacitance was taken into account, the reluctances of winding leakage fluxes were determined by means of simplified Rogowski's formula .

As the scope of this paper is limited, only a mathematical model which can be quickly translated into a digital simulation model /using any of type CEMMA, SCOP, CSMP languages/, as well as an equivalent diagram of the transformem and examples of simulation test results are presented in this paper .

Below presented are all the algebraic and differential equations /in their matrix form/ which represent the electric and magnetic circuits of a three-phase transformer. The eq./16 is valid for the transformer primary side

$$\underline{U}_k' = \underline{U}' - \underline{R}_u' \cdot \underline{I}', \tag{16}$$

where : $\underline{U}_k'$ - column matrix of voltages across the primary magnetic circuit, the matrix dimension being 3x1 ; $\underline{U}'$ column matrix of phase voltages across the transformer terminals, the matrix dimension being 3x1 ; $\underline{I}'$ column matrix of primary phase current, the matrix dimension being 3x1 ; $\underline{R}_u'$ diagonal matrix of transformer primary winding resistances, the matrix dimension being 3x3 .

The matrix equations for the secondary side are

$$\underline{U}_k'' = \underline{U}'' - \underline{R}_u'' \cdot \underline{I}'', \tag{17}$$

where the /"/ indices indicate the transformer secondary side .
The matrix dimensions are same as those in eq. /16, respectively .
The total flux equations in their matrix form are

$$\underline{\Phi}' = \frac{1}{s} \underline{N}' \underline{U}_k' + \underline{\Phi}'(0). \tag{18}$$

The particular matrices are as follows : $\underline{\Phi}'$ - /3x1/ column matrix of the total primary fluxes, $\underline{N}'$ - /3x3/ diagonal matrix with the diagonal elements equal to the primary-turn numbers, $\underline{U}_k'$ - as that in eq./16/ , $\underline{\Phi}'/0/$ - /3x1/ column matrix of the initial values of fluxes in the primary circuit, $\frac{1}{s}$ - integral operator .
The relation for the secondary circuit is

$$\underline{\Phi}'' = \frac{1}{s} \underline{N}'' \underline{U}_k'' + \underline{\Phi}''(0). \tag{19}$$

The magnetic fluxes in legs were determined from the magnetomotive force /MMF/ drops across the ohmic parts . These were expressed by

$$\underline{\Phi}_k = \frac{1}{s} \underline{K}_k \underline{\Theta}_s + \underline{\Phi}_k(0), \tag{20}$$

where : $\underline{\Phi}_k$ - /3x1/ column matrix of the leg fluxes, $\underline{K}_k$ - /3x3/ diagonal matrix whose diagonal with parameters k_k is responsible for the eddy-current losses, $\underline{\Theta}_s$ - /3x1/ column matrix of MMF drops across the ohmic parts of the legs, $\underline{\Phi}_k(0)$ - /3x1/ column matrix of the initial fluxes in the legs .
The leakage fluxes are calculated by subtracting the leg fluxes from the total fluxes algebraically

$$\underline{\Phi}_r' = \underline{\Phi}' - \underline{\Phi}_k, \tag{21}$$

$$\Phi_r'' = \Phi'' - \Phi_k, \tag{22}$$

where : Φ_r' and Φ_r'' - /3x1/ column matrices of the primary and secondary leakage fluxes, respectively . For the directions of vectors of the magnetic fluxes in the legs and yokes assumed in this example , the leakage fluxes to the tank are /see fig. 3a / :

$$\Phi_{10} = \Phi_{k1} - \Phi_{12}, \tag{23}$$

$$\Phi_{20} = \Phi_{k2} + \Phi_{12} + \Phi_{23}, \tag{24}$$

$$\Phi_{30} = \Phi_{k3} - \Phi_{23}. \tag{25}$$

The values of the magnetic fluxes in the yokes were calculated in the same way as those in the legs .
From the equations for MMF drops across the ohmic parts of yoke it is seen that

$$\Phi_{ji} = \frac{1}{s} \; \underline{k}_j \; \underline{\Theta}_{sji} + \phi_{ji}(0), \tag{26}$$

where : Φ_{ji} - /2x1/ column matrix of magnetic fluxes in the yokes, K_j - /2x2/ diagonal matrix with the values of parameters k responsible for the losses in the yokes, $\underline{\Theta}_{sji}$ - /2x1/ column matrix of MMF drops across the ohmic parts of the yokes, $\phi_{ji}(0)$- /2x1/ column matrix of the initial fluxes in the yokes .
Having determined the magnetic fluxes, the MMF drops across the reluctances of the yokes as well as those of air gaps for the leakage fluxes between the particular legs and the tank were calculated .
The MMF drops across the reluctances of the legs and yokes can be simply read from the magnetizing characteristics of the laminations employed for the transformer core . The characteristics expressed by the $H = f(B)$ function should be translated into $H \cdot l_{AVERAGE} = f(B \cdot A)$ function, i.e. $\Theta = f(\phi)$. The $l_{AVERAGE}$ and A quantities express the average length of the magnetic path in the leg /yoke/ and the cross-sectional area of the leg /yoke/, respectively .
The MMF drops across the reluctances of the legs are $\Theta_{k1} = f_1(\phi_{k1})$, $\Theta_{k2} = f_1(\phi_{k2})$, $\Theta_{k3} = f_1(\phi_{k3})$ and of the yokes are $\Theta_{j12} = f_2(\phi_{j12})$ $\Theta_{j23} = f_2(\phi_{j23})$.
The MMF drop across the reluctance of air gap for the leakage flux between the legs and the tank is computed by means of matrix equation

$$\underline{\Theta}_{10} = \underline{R}_{m10} \; \Phi_{10}, \tag{27}$$

where : $\underline{\Theta}_{10}$ - /3x1/ column matrix of the MMF drops , R_{m10} - /3x3/ diagonal matrix with the values of the reluctances of air gaps for the leakage fluxes between the legs and the tank - positioned along the diagonal , Φ_{10} - /3x1/ column matrix of the leakage fluxes to the tank .

For the equivalent diagram of the transformer refer to fig. 3b .

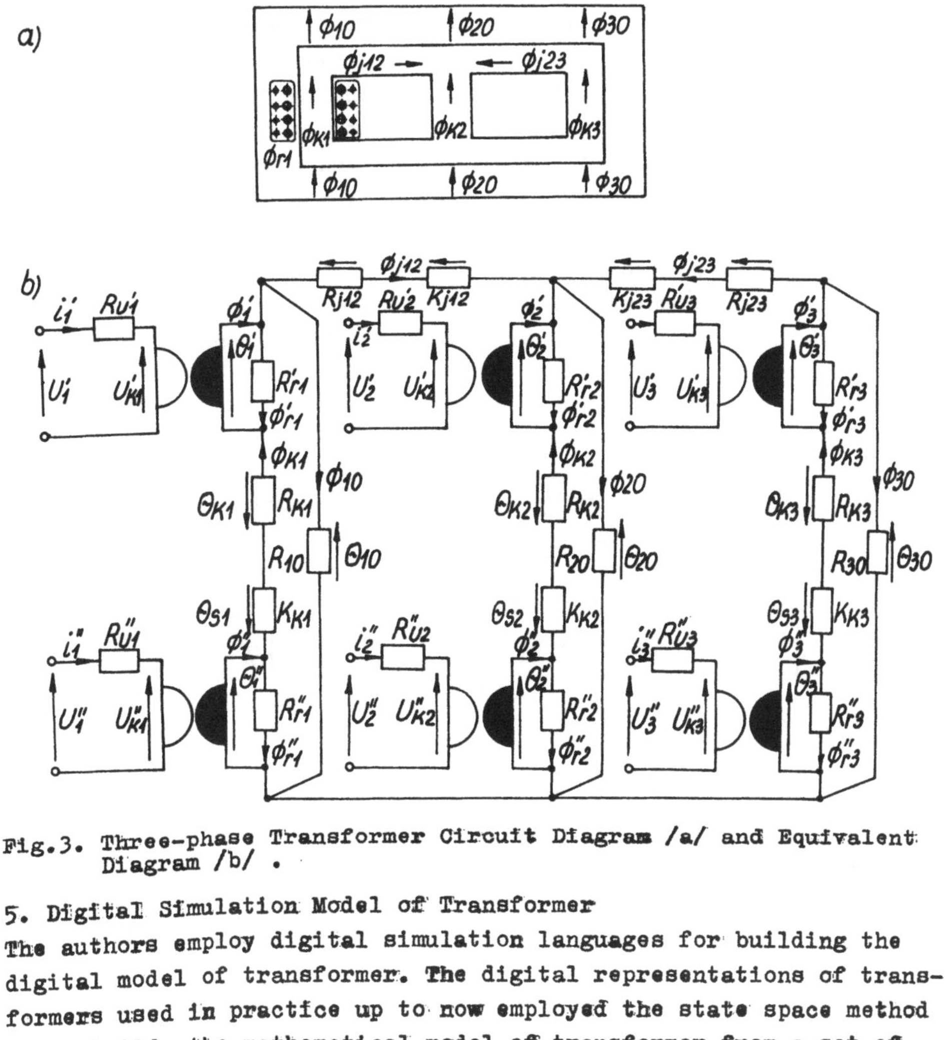

Fig.3. Three-phase Transformer Circuit Diagram /a/ and Equivalent Diagram /b/ .

5. Digital Simulation Model of Transformer

The authors employ digital simulation languages for building the digital model of transformer. The digital representations of transformers used in practice up to now employed the state space method which builds the mathematical model of transformer from a set of differential equations with the use of an algorithm solving these equations by means of matrix analysis method .

Such a digital representation is very difficult, time-consuming and not capable of determining the currents and voltages at freely selected points of the modxlled system in a simple and quick way .

While building the digital simulation model, the authors first built an analog block diagram /structural analog model/ of the system under investigation and then employed this diagram to write a program which took the system structure of the modelled system into account by describing the individual subassembly functions and their interrelations . Both analytic model-building and structural model-building techniques were used for creating the analog block diagram . The former took the present mathematical model of the system under investigation /i.e. the set of differential and algebraic equations/ as the starting point , while the latter used the system structure for building individual component models of the real system and involving them into a structural scheme .
The analog block diagram of the transformer, which represents the diagram shown in fig. 3b and equations /16/-/27/ is presented in fig. 4.

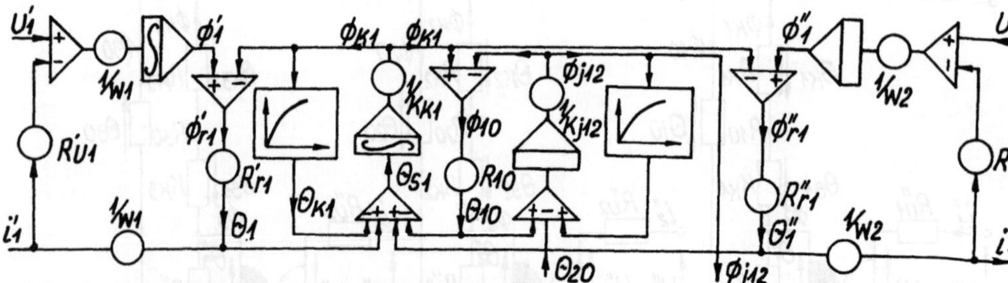

Fig.4. One-phase Analog Block Diagram of Electric and Magnetic Circuits of three-phase Transformer .

The transformer is not a stand-alone unit in an electric power system and each its state is to be analysed together with the powering system and the load with various connection schemes of both transformer sides taken into account. Fig.5 shows a diagram of the transformer linked with an electric power system which was used in further analysis.
The analog block diagram of the power system-transformer combination /figs. 4/ was used to write a program in the SCOP simulation language [6], which provided means of determining the current waveforms for the transformer /110/15 kV, 16 MVA, Yd11/ being energized in its non-load mode .

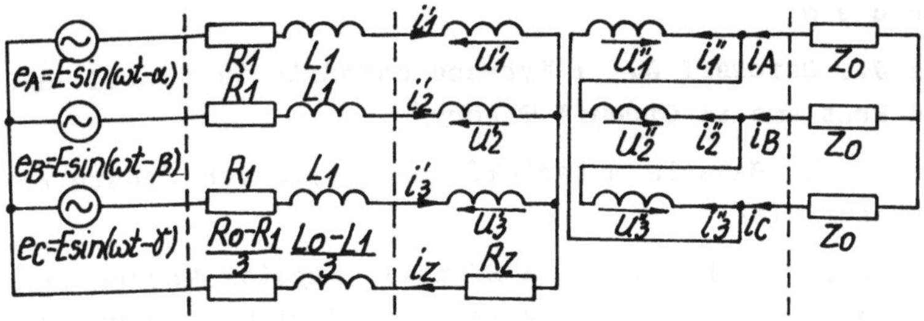

Electricity Souce **Power System** **Transformer** **Load**

Fig.5. Diagram of Power System-Transformer Combination

The required accuracy of computation was obtained by using the Runge-Kutt's integration method /of fourth order/ with a step of $\Delta t = 0.01$ milliseconds .

In spite of this integration step, the total integration time was rather fast as from the transformer-protection technique standpoint the waveform duration of interest was not longer than 100-200 milliseconds .

The computation time would be even faster in the authors' opinion if variable-step numerical integration methods are used .

Fig.6 shows an example of inrush current waveform obtained from the numerical computations for a phase angle of 90 degrees at time $t = 0$ /refer to e_B voltage in fig. 5/ .

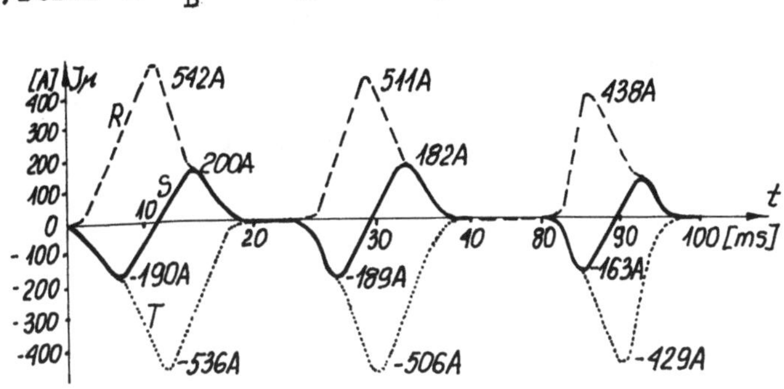

Fig.6. Inrush Current Waveforms Obtained by Means of Digital Simulation Computations

R e f e r e n c e s

1. Kudrewicz J., Osiowski J. : Wybrane zagadnienia teorii obwodów /Selected Problems in Circuit Theory/

2. Kuryłowicz J. : Badania materiałów magnetycznych /Tests of Magnetic Materials/

3. Tripathy S., Yusuf Khan M. : A Digital Computer Method for Electromagnetic Transients in Multiphase Nonlinear Power Systems. IEEE PES Summer Meeting, 1975

4. Wright A., Bradley K. : The Computer Modelling of Protective Schemes . IV International Conference „Problems of Power System Automation and Control" , Gliwice, 1985

5. Yamashita H., Nakamae E., Hamman M., Wakisho H. : A Program fo Analize Transient Phenomena of Circuits Including Precisely Presented Transformers. IEEE PES Summer Meeting, 1975

6. Język symulacyjny SCOP na m.c. ODRA 1305. Centrum Informatyki Energetyki i Energii Atomowej . Warszawa, 1977 . /SCOP Simulation Language for the ODRA 1305 Digital Computer . Information Science Centre for Power Systems and Atomic Energy/.

dr ing ANDRZEJ MAGDZIARZ
dr ing ZBIGNIEW ŻAGAŃ
INSTITUTE OF ELECTRICAL POWER SYSTEMS
WARSAW TECHNICAL UNIVERSITY

TIME-OPTIMAL CONTROL OF ACCELERATION CONSTRAINED MECHANICAL SYSTEMS

Pencho MARINOV, Petko KIRIAZOV
Institute of Mechanics & Biomechanics
Bulgarian Academy of Sciences

Abstract

The subject of this work is the problem for time-suboptimal point-to-point control in the presence of control and acceleration constraints. A direct method for control synthesis of the corresponding two-point boundary-value problems, combined with a parametric optimization procedure is proposed. Control laws are obtained on the base of the relay principle - the motion of the mechanical system is such that in every samplinh interval, either the control functions or the dynamic constraints have their limit values. Using backward integration in time, we can solve the problem when deceleration constraints are imposed only in some fixed final intervals - which case is of special practical interest. To be more illustrative, the dynamical model of a d.c. manipulator with cylindrical coordinates is taken into consideration.

Introduction

It is well known that in the present status of tecgnology, many mechanical systems work as positioning devices. The path between motion end points is not specified and depends on the control algorithm and hardware. The limitations of the strains and vibrations of the mechanical systems in the case of higher operation speeds and havier manipulation loads have to be taken into considerations [1,2]. So, the subject of this work is the problem for time-suboptimal point-to-point control of a class of mechanical systems in the presence of control and acceleration constraints.

A direct method for control synthesis of the corresponding two-point boundary-value problem (TPBVP), combined with a parametric optimization procedure is proposed. This method is an extension of the earlier proposed method [3], where acceleration constraints have not been considered. Control laws are obtained on the base of the relay principle - the motion of an N-degrees of freedom mechanical system is such that in every sampling interval, either the control functions or the dynamic constraints have their limit values. The TPBVP reduces to the solution of a system of N-shooting equations with N-unknowns (switching values). The optimization procedure is performed over the set of all feasible solutions with respect to the initial times.

As an illustrative example, a dynamic model of manipulator with cylindrical coordinates is taken into consideration. As dynamical constraints, the limitations on radial and centrifugal parts of the gripper's accelerations have been imposed.

1. Statement of the Problem

Using Lagrangian formulation, the mathematical model of a dynamical system with N-degrees of freedom can be written as:

$$M(q)\ddot{q} + L(q,\dot{q}) = u \qquad (1)$$

where: M is an inertia matrix, diagonal for our case: $M(q) = (m_{ii})$ $i = 1,\dots,N$, $q(t)$ - vector of N-generalized (joint) coordinates, $L(q,\dot{q})$ - vector of centrifugal, Coriolis, gravity or friction forces, u - vector of control inputs.

- boundary conditions:

$$q(t^o) = q^o,\ \dot{q}(t^o) = 0 \text{ - initial state} \qquad (2)$$

$$q(t^f) = q^f,\ \dot{q}(t^f) = 0 \text{ - final state} \qquad (3)$$

A. Control constraints:

$$|u_i(t)| \le u_{i\,max},\ i = 1,\dots,N \qquad (4)$$

B. Acceleration constraints:

$$|h_i| = |f_i(t,q,\dot{q})\ddot{q} + g_i(t,q,\dot{q})| \le c_i,\ i = 1,\dots,N \qquad (5)$$

where: f_i and g_i are some functions, $f_i > 0$.

C. Let the signs of the general velocities are constant during the whole motion:

$\text{sign}\ \dot{q}_i(t) = \text{sign}(q_i^f - q_i^o) = s_i$, $i = 1,\dots,N$ and each joint motion has one switching point from acceleration to deceleration.

The initial times t_i^o for the joint motions may be different and so as to the final times t_i^f, $i = 1,\dots,N$. Therefore $t^o = \min t_i^o$, $t^f = \max t_i^f$, $i = 1,\dots,N$, and the problem consist in minimization of the movement execution time $T = t^f - t^o$ for the mechanical system (1) from initial state (2) to the final state (3) under constraints A, B and C.

Having an algorithm for control synthesis of the TPBVP (1 ÷ 3) under above mantioned constraints, one can obtain a sufficiently rich set of feasible solutions varying initial times. Over this set, a parametric optimization can be easily performed, applying, e.q., steepest descent method.

So, the problem in the following, is to elaborate control synthesis algorithm, generating feasible solutions. The control laws will be derived in accordance with the relay principle, as time-optimal control concept. A general appearance of the requested control functions and corresponding joint motions is given on Fig. 1.

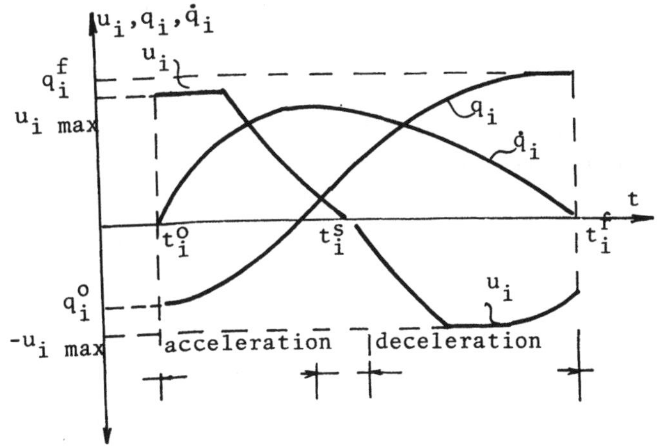

Fig. 1. Constrained joint motion

2. Control Synthesis Algorithm

Let take the equality sign in the acceleration constraints (5):

$$f_i(t,q,\dot{q})\ddot{q} + g_i(t,q,\dot{q}) = (\text{sign } u_i)c_i, \quad i = 1,\ldots,N \tag{6}$$

where: $\text{sign } u_i = s_i, \quad t\in(t_i^o, t_i^s)$

$$\text{sign } u_i = -s_i, \quad t\in(t_i^s, t_i^f) \tag{7}$$

So, to follow the time-optimal concept, under the constraints (4) and (5), the control functions u_i should have the form:

$$|u_i| = \min\{u_{i\ max};\ |m_{ii}(q)((\text{sign } u_i)c_i - g_i(t,q,\dot{q})/f_i(t,q,\dot{q}) + L(q,\dot{q})|\} \tag{8}$$

In the off-line sence, the control synthesis algorithm, solving the TPBVP (1 ÷3) with the control laws (8) and the observation also of (C) condition, one consist in performing several simulation test mevements from the given initial state to some terminal states, converging to the required one.

Each such movement means that the system of differential equations (1) with initial values (2), given starting times t_i^o and some approximate switching times t_i^s, is integrated until satisfying the terminal condition $(3)_2$: $\dot{q}_i(t_i^f) = 0$. Each joint motion is settled down after the moment t_i^f with a terminal value of the corresponding coordinate $q_i(t_i^f) = F_i$.

The N-dimensional vector F only depends on the switching times t_i^s, $i = 1,\ldots,N$, if the values of all other parameters are fixed. So, in order to satisfy the final condition $(3)_1$, one has to solve the following shooting equation:

$$F(t^s) = q^f \tag{9}$$

Solving this equation by Newton's or bisections methods [4], one obtains a feasible motion q(t) with the corresponding control input u(t) and movement execution time T. Varying the values of initial times t_i^o, we can find the best solution over this class of feasible solutions.

3. Numerical Example

To be more illustrative, the proposed method is applied to a dynamic model of manipulator with cylindrical coordinates as a typical example of a mechanical system performing point-to-point operation, Fig. 2. Including direct current actuator systems and taking some specified numerical values of the model parameters, the equations of the two coupled joint motions are written as [3]:

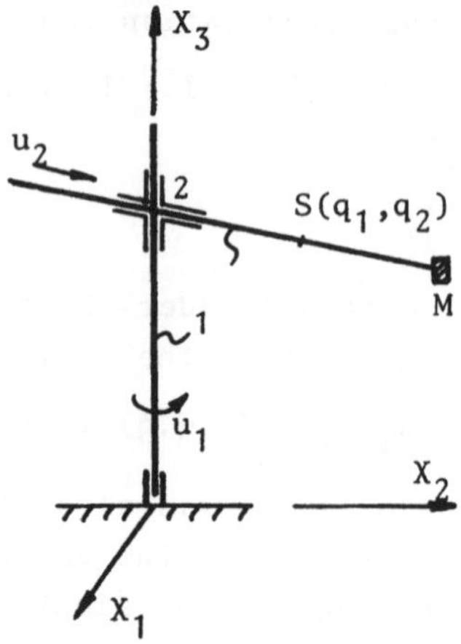

Fig. 2. Manipulator with cylindrical coordinates

$$\ddot{q}_1 = (u_1-6.8\dot{q}_1-0.34(10.0q_2+5.0(q_2+1.0))\dot{q}_1\dot{q}_2/$$
$$(0.4+0.17(20.0+10.0q_2+5.0(q_2+1.0)^2))$$
$$\ddot{q}_2 = (u_2-6.8\dot{q}_2+0.17(10.0q_2+5.0(q_2+1.0))\dot{q}_1^2/2.95 \tag{10}$$

where: $u_{i\ max} = 15$ V, $u_{2\ max} = 10$ V, $q_1^o = 0.4$ rad, $q_1^f = 0.0$ rad,
$q_2^o = 0.4$ m, $q_2^f = 0.0$ m.

As dynamical constraints, the following inequalities for the radial and tangential parts of the gripper's acceleration are taken:

$$|h_1| = |\ddot{q}_1(q_2+a) + 2\dot{q}_1\dot{q}_2| \leq c_1$$
$$|h_2| = |\ddot{q}_2 - \dot{q}_1^2(q_2+a)| \leq c_2 \tag{11}$$

where: $c_1 = 3.0$ rad.m/s^2 and $c_2 = 3.0$ m/s^2.

In the previous work [5], which have been reported in Udine, a numerical example is given for sucg a dynamical model without being considered actuators effects and the dynamics constraints are imposed through the whole motion.

In the present example, these constraints will be imposed in some specified final time interval $\Delta t^f = t^f - t^a = 0.4$ s. Since the final moment t^f is not known apriori, we apply in this case backward in time integration from q^f to q^o and the acceleration constraints will act in the beginning. The obtained suboptimal solution is shown on Fihs. 3 and 4.

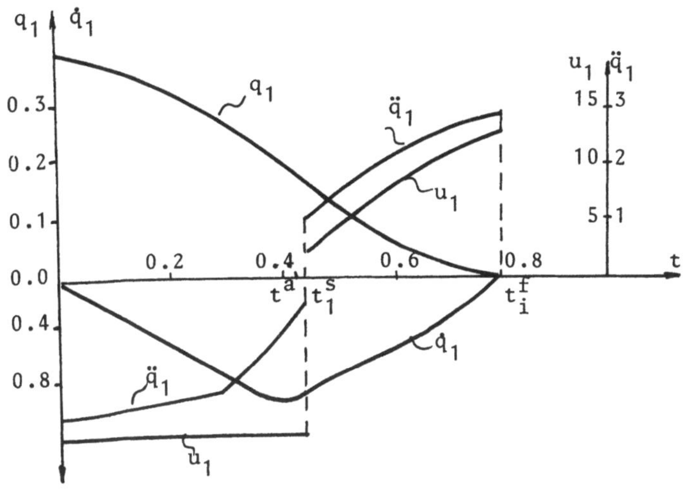

Fig. 3. First joint motion

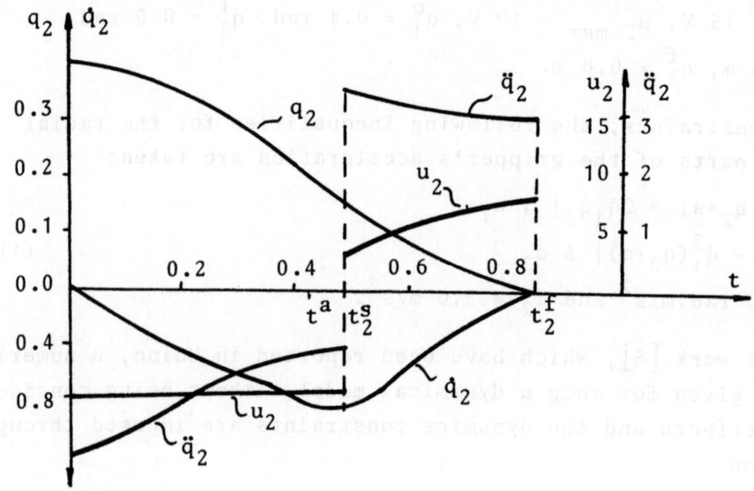

Fig. 4. Second joint motion

4. Conclusions

A direct method for time-soboptimal control synthesis of a class of mechanical system in the presence of control and acceleration constraints has been proposed. In case the letter constraints are imposed in some specified final interval, this method has been verified on a dynamic model of manipulator with direct current actuators.

References

1 Truckenbrodt, A., Dynamics and Control Methods for Moving Flexible Structures and their Application to Industrial Robots, Proceding of the Fifth Congress on the Theory of Machines and Mechanisms-1979, Publ. by ASME, 345 E, v. 1, 1980, pp. 831-834

2 Bremer, H., Dynamics of Flexible Hybrid Satellites - Evaluation and Computation of a Symbolic Formalism, Proc. of 3th VPI/AIAA Symp. Dynamics and Controw of Large Flexible Space-craft, 1981, Blacksburg/USA, pp. 409-423.

3 Marinov, P. and P. Kiriazov, Synthesis of Time-Optimal Control for Manipuwator Dynamics, Theor. Appl. Mech., Publ. House Bulg. Acad. Sci., (1), 1984, pp. 13-19

4. Stoer, J. and R. Bulirsh, Introduction to Numerical Analysis, Springer-Verlag, New York - Heidelberg - Berlin, 1980

5 Kiriazov, P. and P. Marinov, A method for time-optimal control of dynamically constrained manipulators. Prepr. of 5th CISM-IFToMM RoManSy'84. June 26-29, Udine, Italy, 1984, pp. 131-138

A Relational View of a Linear Optimization Model and its Consequence for the Modelling Process.

MASCHTERA ULRIKE SCHNEIDER WALTER
Institut f. Informatik, J.-K.-Universität
Altenbergerstr. 69 A-4040 Linz

The modelling process consists of a series of abstraction levels deriving a formulation suitable for the solution process from observations of the real system. A relational view is typical for the real system. By applying it to the target formulation a shift of the application of conceptual knowledge from the modeller to the computer becomes possible. The general methodology is shown at linear optimization models and mixed integer models. Furthermore this approach is compared with recent modelling aids, such as matrix generators and modelling languages, and the advantages of this approach especially with respect to the model life cycle are shown.

Introduction

The modelling process consists of a series of abstraction levels deriving a formulation suitable for the solution process from real system observations. In the first abstraction step - discerning the system elements and their relationships - the (factual) knowledge of the system is gathered. By using his conceptual knowledge with respect to the target formulation the modeller maps the factual knowledge into such a target formulation.

In case of linear optimization problems the target formulation is a set of linear equations together with a (set of) optimization functions. Figure 1 shows the abstractions done during the linear optimization modelling enterprise. The starting point is a set of relationships representing the factual knowledge of the real system. Then the linear optimization model (called LOM in the following) is derived by applying the (well known) conceptual knowledge of linear modelling (CK means conceptual knowledge in the following):

(CK1) The minimal information units are relationships between exactly two -jects (subjects/objects, system elements). They are called RATOMs in the following.

(CK2) The relationships can be expressed in two r-dual forms (see ex1)

(CK3) The relationships are quantitatively variable, but structurally constant.

⟹ a) the system elements become variables

b) the (constant) semantics of the relationships is mapped to 'ENHANCE', 'DIMINISH' or 'GAIN', which are the only ways of influence representable within linear equations

c) the coupling is strictly coordinative ('AND', but not: 'EITHER...OR', 'IF... THEN...ELSE..', etc.)

ex. 1: relationships: PRODUCES(machine,product_A)

PRODUCES(machine,product_B)

system elements (-jects): machine,product_A,product_B

RATOMs(primal form): machine produces maximal x units of product_A;

machine produces maximal y units of product_B;

(dual form): product_A is produced by maximal x (capacity)units of machine

product_B is produced by maximal y (capacity)units of machine

mapping: produces → ENHANCE

coupling: machine produces x units of product_A and y units of product_B.

(subject: machine; objects: product_A, product_B)

linear optimization (RATOMs): machine: product_A $\leq$ x [units_A]

machine: product_B $\leq$ y [units_B]

(coupled): machine: $\frac{1}{x}$ product_A + $\frac{1}{y}$ product_B ≤ 1 [x units_A ≡ y units_B]

Derivation of a linear optimization model

Computer aided linear modelling means a shift of the application of the conceptual knowledge from the modeller to the computer. This has to be done step by step (see figure 2).

Given a system description in (1:1) relationships, these relationships have to be transformed into combinable, conceptually valid information units (here RATOMs) in the first step. According to CK3a) the identifiers of sub- and object of each RATOM have to be stored and due to CK3b) the semantics of the relationship - describing part of the constant structure of the system - has to be found out. Regarding the names of the -jects as identifiers (strings) only, the semantics is allocated solely at the praedicate. This means, that the praedicate has to be mapped to one of the LOM-specific synonyms (ENHANCE/DIMINISH/GAIN). The mapping praedicate → synonym is usually constant (eg. produce → ENHANCE) and is stored in a base to be available for further usage. However, depending on the context of the input, a user may associate different meanings with one and the same praedicate (see ex2,ex3). Thus the responsibility for allocation of a synonym at a praedicate rests solely at the user. It reflects the user's individual context (ex2)/perception (ex3). Therefore, whenever there is a mapping in the base for the actual praedicate, this mapping is suggested to the user for confirmation. If the suggested allocation does not suit his context, he may allocate the praedicate multiply.

ex. 2: relationship: BELONGS_TO(member,set)

When quantifying membership the praedicate BELONGS_TO has to be mapped to ENHANCE/DIMINISH depending on whether the inclusion of a member enhances or diminishes the quantum, which is represented by the set?

eg. set...... set of products_stored

member... products_sold　　　　products_sold DIMINISH products_stored

member... products_produced　　　　products_produced ENHANCE products_stored

Note that multiple allocation of BELONG_TO shows inhomogeneous semantics of the described set.

ex. 3: bottling_machine fills(ENHANCES) bottles.

bottling_machine fills(DIMINISHES) wine.

Obviously, the user used the identifier 'bottles' for 'full_bottles'. Correctly, filling always has the semantics of diminishing, because only empty_bottles can be filled. However, the individual perception (bottles are full_bottles) and the individual style (full_bottles are filled) of a user should be preserved.

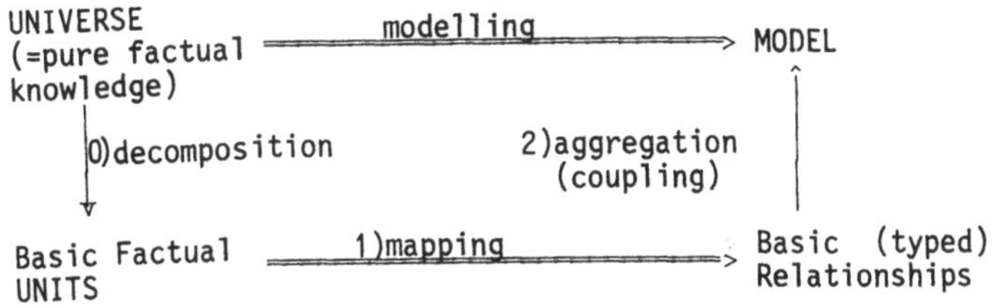

fig.2: steps of the modelling process

Given a set of (mapped) RATOMs, these have to be coupled in the next step. The result is a set of constraints and a (set of) optimization function(s), where each coefficient is decomposed into its sign (which is already fixed according to the semantics) and its absolute value (which can be a variable in case of a general model). In linear optimization modelling there are two coupling operations which can be applied to RATOMs with common subjects: SHARED (for RATOMs with equal type: both ENHANCE or DIMINISH) or EXPAND (for RATOMs with unequal type: ENHANCE and DIMINISH), and one coupling operation which is applied to RATOMs with different subjects: ADD representing the logical 'AND' by which constraints are combined to form a model (/3/). Note that the result of the coupling operations are aggregated RATOMs, these are RATOMs where either the objects are aggregated (in case of SHARED/ EXPAND) or sub- and objects are aggregated (in case of ADD), see fig. 3. Furthermore note in case SHARED/EXPAND that the subject of the aggregated RATOM is the subject which is common to both component RATOMs; this is the slack variable of the constraint. With respect to distributive modelling or novel users, it is not obvious which system elements will become subjects. Therefore partitioning into sub- and objects has to be done in this step by using CK2. The result of this step is a hierarchical system element structure which is one of the characteristics of conceptual modelling knowledge (compare /4/).

The next steps are well-known transformations of a general LOM into a specific LOM.

fig. 1: abstraction steps during the linear optimization modelling process (CALM ... Computer Aided Linear Modelling, the name of the modelling support presented in this paper)

modelling process	view of the model	abstraction steps	model formulation	knowledge	interface MG	interface ML	interface CALM
target formulation	set of equations+ set of opt.functions		$\Sigma_j a_{ij} x_j + x_i = b_i \quad \forall i$ $\Sigma_j p_{kj} x_j \rightarrow opt \quad \forall k$	←— conceptual knowledge —→	←— modeller responsible for MG —→	←— modeller≡user —→	←— CALM —→|
		4)transforming inequalities to equalities					
	set of inequalities+ set of opt.functions		$\Sigma_j a_{ij} x_j \underline{<>} b_i \quad \forall i$ $\Sigma_j p_{kj} x_j \rightarrow opt \quad \forall k$				
		3)uniting amount and direction of influence within the coefficients					
	LOM composed of RATOMs		$\Sigma_j sign_{ij} \lvert a_{ij} \rvert x_j \underline{<>} b_i \quad \forall j$ $\Sigma_j sign_{kj} \lvert p_{kj} \rvert x_j \rightarrow opt \quad \forall k$				
		2)coordinative coupling of RATOMs (including partitioning of -jects)		CK2 CK3c			
	set of RATOMs		$\{x_i : x_j \underline{<>} b_i / a_{ij}\} \cup$ $\{z_k : x_j = p_{kj};\ z_k \rightarrow opt\};\forall i,j,k$				
		1)mapping the praedicates to the conceptual way of expressing semantics		CK3a CK3b			|←— user + CALM —→|
system description	set of (1:1) relationships		(machine$_1$ produces maximal a_{ij} products$_A$),(bottling-machine fills maximal a_{ij} bottles),...				
		0)splitting of relationships in more than two -jects		CK1			
system observations	set of relationships		any sentence	factual			|←user

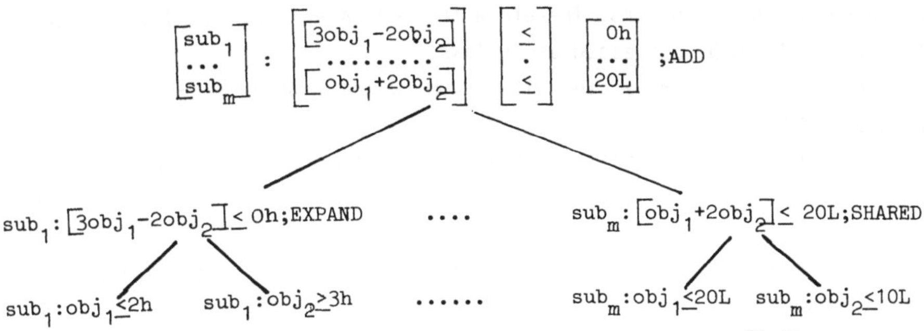

fig.3: coupling operations of linear optimization modelling; [,] represent aggregated system elements

Thus the abstraction steps of linear optimization modelling have been shifted from the modeller to the computer. However, the zero'th abstraction step is allocated at the user: he has to decompose and input his system observations in simple sentences which 1) have exactly one subject and one object

2) express the semantics solely within the praedicate

3) have to be a valid, consistent and complete description of the modelled system. Step 0 and step 1 are the major source of problems and errors. In order to achieve a valid and tight model a help function (for splitting the system knowledge into (1:1)relationships) and model (in)validation aids are applied in these steps. The model (in)validation aids can be divided into three groups: definitional ones (checking typing and similar errors), numerical ones (checking the range of system elements) and structural ones (checking semantic model integrity and model connectivity) (see /3/). Tightness of the target formulation is achieved by reformulating models according to their structure and by adjusting the coefficients according to their measures (see /3/, where the LOM-knowledge of a mixing problem: $\Sigma_j a_{ij}=1$ ($\forall j$) with a_{ij}..percentage of ingredient$_i$ used for mixture$_j$, is derived in this manner).

Logical requirements

Adding logical requirements to a LOM renders a mixed integer problem (abbreviated MIP in the following). A summary of logical requirements (called LR in the following) is given in Gabriel(/2/). He distinguishes three classes of LRs:

(1) selection of alternatives

(2) implications

(3) complex combined logic requirements.

The first two of the above classes are distinguished by the fact that every -ject occurs at most once in the set of LRs. Therefore the necessary conceptual knowledge for adding LRs to a LOM can be expressed as follows:

CK4) a) Any LR couples one or more states with a LOM (where a RATOM is considered as minimal LOM).

b) Any LOM expresses the variablility of a relationship. It is transformed into a state by multiplying the (non-zero) right hand side with a decision variable $y_i \varepsilon \{0,1\}$).

c) The sum of the decision variables is bounded (this results in a socalled decision constraint).

Therefore the following extensions are necessary in order to tackle LRs of type 1 and type 2:

A) step 1: acceptance of factual knowledge has to be extended

a) by accepting distributive coordination of input sentences such as

(1) choose exactly/at most/at least k elements out of a set of variables/constraints.

(2) if..then..else..; either..or; .. only if..; .. if and only if ..

b) by including a blocking feature like the BEGIN..END of programming languages and by including logic operations (AND,OR,NOT) for the coupling of partial LOMs and states.

B) step 2 has to be extended by distributive coupling operations. Their basic features are

1) the definition of decision variables

2) the modification of involved RATOMs

3) the generation of decision constraints.

They require upper and/or lower bounds of the -jects (CK4b) and give different results when applied for subject (constraint) or object (variable). Therefore partitioning of concerned -jects has to be fixed before distributive coupling operations can be applied.

Note that the necessity of formulating alternatives can be detected by the model (in)validation aids. On the one hand, the user can define two RATOMs with equal semantics and different numerics, or two RATOMs with equal -jects, but different semantics. The real situations - which are described by such two RATOMs - cannot exist at the same time. Thus either the user has to eliminate one of the contradicting RATOMs (the one which describes reality in a wrong way) or he has to view the two RATOMs as alternatives, that means he has to couple them distributively.
On the other hand, whenever the user defines two conceptually different subjects (eg. a bottling machine and a cleaning machine) as specializations of one system element (eg. machine), then he has to decide whether transformation of one specialization into the(an) other one (eg. setting up of the machine) is relevant to the model (rendering a MIP) or not (rendering a LOM).

The conceptual knowledge necessary to tackle with LRs-where the sets of involved -jects are not disjunct- has to be expanded for:

CK5) LRs for the same -ject can be combined via an exclusive OR ('elseif') or an inclusive OR ('and if'). In case of elseif the LRs can be coupled by the coordinative coupling operation SHARED: (where x_1 is plays the role of 'the numeral')

eg: if $x_1 \geq 5$ then model1
elseif $x_1 \geq 10$ then model2
elseif $x_1 \geq 17$ then model3 .....

$\Rightarrow x_1 \geq 5y_{11} + 10y_{12} + 17y_{13} + \ldots\ldots\ldots$ with $\Sigma_i\, y_{1i} = 1$
$y_{1i} \varepsilon \{0,1\}$
model_i dependant on y_{1i}

In case of 'and if' LRs for the decision variables are implied:

eg: if $x_1 \geq 5$ then model1
and if $x_1 \geq 10$ then model2
and if $x_1 \geq 17$ then model3 .....

$\Rightarrow x_1 \geq 5y_{11} + (10-5)y_{12} + (17-10)y_{13} + \ldots\ldots$ with $\Sigma_{i=1}^{j}\, y_{1i} \geq j \cdot y_{1,j+1}$ $(\forall j)$ (1)
$y_{1i} \varepsilon \{0,1\}$
model_i dependant on y_{1i}

(1) is the LR: $(y_{1j}=1) \Rightarrow (\forall k<j)\ (y_{1k}=1),\ \forall j$

CK6) Nested ifs have to be resolved (see /5/,p27) and should be checked whether they can be transformed to the selection of alternatives.

Comparison to recent modelling aids

Recent modelling aids can be grouped into matrix generators (abbreviated MG) and modelling languages (abbreviated ML). While MGs define general standard models and therefore have the semantics incorporated into their formulas, 'a modeling language ... expresses the modeler's form of a linear program that a computer system can interpret' (/1/) by providing means to input a LOM. Therefore the user of a MG has no influence on the semantics and can do neither changes nor extensions of the predefined model; on the other hand, the user of a ML has to know the LOM before starting to work with the ML. So in both cases the conceptual knowledge is applied by a person (see fig. 1). Therefore changes and extensions have to be done by the/a modeller too. This implies the need of suitable documentation tools. Furthermore changes and extensions have to be adjusted to the hierarchic system element structure of the existing model. This means that either the one of the two possible r-dual views (compare CK2) is preserved which has been chosen when the model has been defined initially, or that the whole model has to be reformulated. The reason for that is that MG and ML are based on applying the equation oriented view of the target formulation to the real system.

However, by applying the relational view - which is typical for the real system - to the target formulation, the abstraction steps of modelling are exposed and the application of conceptual knowledge can be shifted from the modeller to the computer. Furthermore system semantics can be defined in a user-oriented form by extracting the mapping of the 'user's vocabulary' interactively. Therefore such a dialog serves as a documentation. It can be stopped and resumed by any person at any time, making modelling a stepwise, distributive process instead of a singulary, solitary enterprise (compare fig. 4).

	matrix generators	modelling languages	modelling support described above
objective	general standard models	aid for input of any (predefined) model	derivation of models from factual knowledge interactively
application of conceptual knowledge	by the modeller who defined the MG	by the modeller who uses the ML	computer + user, who checks validity of derived knowledge
changes/extensions	NOT by the user	by the modeller (new input)	computer + user (same way as above)
documentation	separately	separately	≡dialog computer-user
view	equation-oriented	equation-oriented	relational
modelling process	singulary,solitary	singulary,solitary	stepwise,distributive

fig. 4: comparison of described modelling support to recent modelling aids

References:

/1/ R. Fourer: 'Modeling Languages Versus Matrix Generators For Linear Programming', ACM Transactions on Mathematical Software, Vol.9, No.2, June 83,143-183

/2/ R. Gabriel: Optimierungsmodelle bei logischen Verknüpfungen, Minerva München 1982

/3/ U. Maschtera, W. Schneider: 'Structuring of Linear Optimization Models', in: Modelling and Simulation Methodology in the Artificial Intelligence Era, ed: M.S.Elzas, T.I.Ören, B.P.Zeigler, North Holland

/4/ U. Maschtera: 'Duality Concepts in Discrete Event Simulation', Applied Informatics 1984, ed. E.H.Hamza, Acta Press 1984, p.29-32

/5/ Neta Amit: 'A Different Solution for Improving the Readability of Deeply Nested IF-THEN-ELSE Control Structure, Sigplan Notices V19(1), Jan.84,24-29

AN ADAPTIVE COORDINATION ALGORITHM FOR THE INTERACTION BALANCE METHOD WITH FEEDBACK

Maciej Michniewicz

1. Introduction.

The problem of large-scale system control has been recently intensively investigated. The Interaction Balance Method with Feedback (see ref. 1) is one of the most fundamental methods created for large-scale system optimization. The selection of a proper optimization algorithm for that method is the essential problem of the control system designer. The elaboration of the on line upper level algorithm applicable to the stochasticaly disturbed system seems to be of great importance. An interesting and very simple algorithm has been proposed by Ruszczynski(see ref. 2). His concept requires however suffitiently small difference between the real system and its mathematical model. The algorithm fails as that difference increases to excess because of varying disturbances.

Two years ago at the 11-th IFIP Conference on System Modelling and Optimization in Copenhagen the author presented the paper dealing with the on line optimization of the objective function having a drifting minimal point (see ref. 3). The algorithm described in that paper is applicable as an upper level optimization algorithm of the Interaction Balance Method with Feedback. Thereby the disturbances could change very much if they only drift slowly enough. The proposed below algorithm is an extension of the one suggested by Ruszczynski.

2. Description of the control system.

Let us consider a system composed of many interconnected subsystems (see example in fig. 1). Each subsystem is supplied with stabilizing regulators, thus we can omit dynamical effects and consider outputs y as functions of inputs u, control variables c and disturbances z:

$$y = \Phi(u,c,z). \qquad (1)$$

We assume that function Φ is differentiable and may be locally approximated as follows:

$$y = Au + Bc + Cz + d, \tag{2}$$

where:

$$A = \frac{\partial\Phi(u_o, c_o, z_o)}{\partial u} \quad ; \quad B = \frac{\partial\Phi(u_o, c_o, z_o)}{\partial c};$$

$$C = \frac{\partial\Phi(u_o, c_o, z_o)}{\partial z} \quad ; \quad d = \Phi(u_o, c_o, z_o).$$

The subsystem interconnections will be writen down in the following form:

$$u = My, \tag{3}$$

where M is a certain matrix consisting of 0 and1.
From (2) and (3) we can calculate outputs y as functions of control variables and disturbances:

$$y = (I-AM)^{-1}Bc + (I-AM)^{-1}Cz + (I-AM)^{-1}d. \tag{4}$$

The existance of $(I-AM)^{-1}$ is equivalent to the controlability of the system.

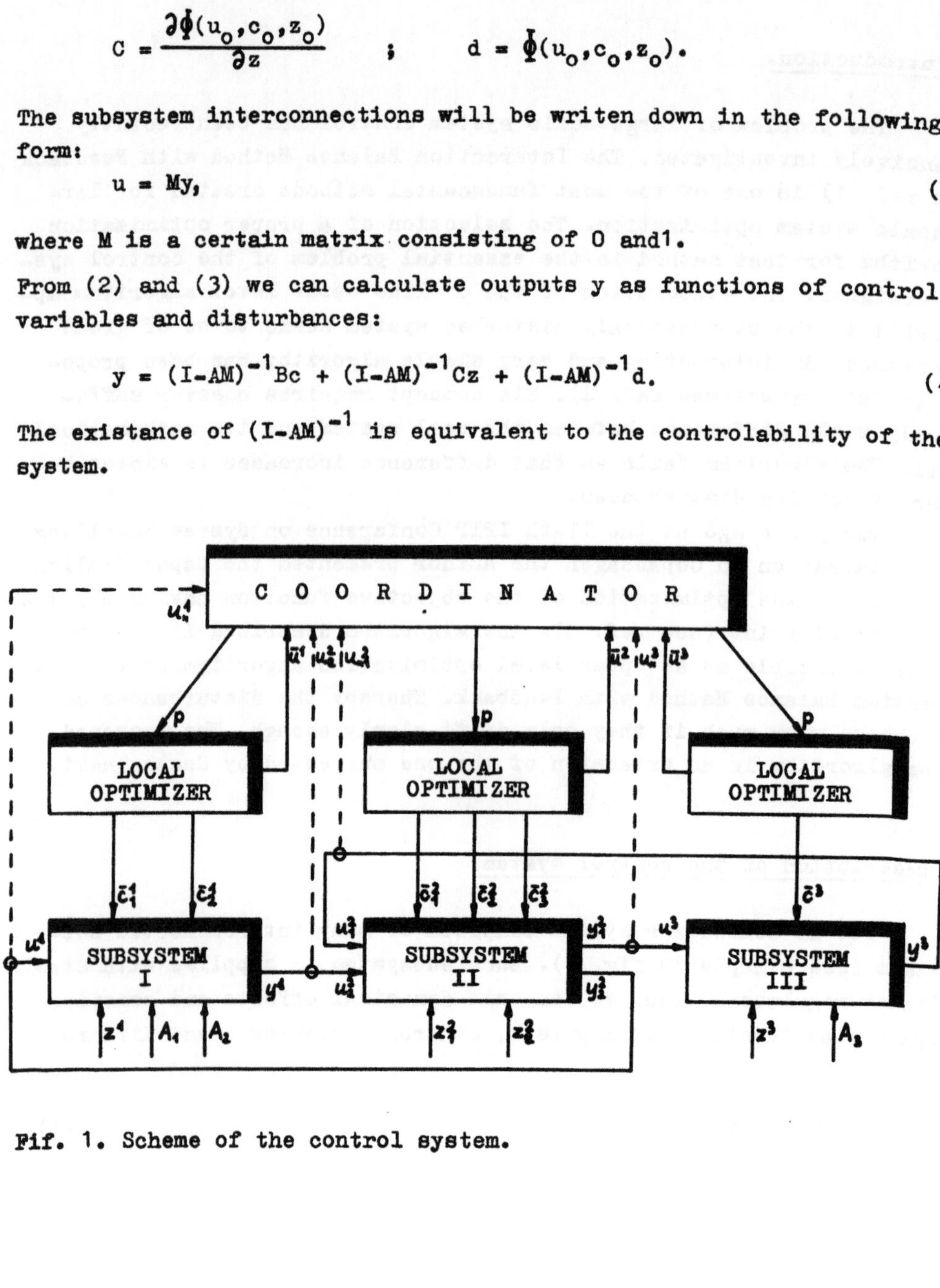

Fif. 1. Scheme of the control system.

Assume we have the linear model of the system in the form:

$$\bar{y} = \bar{A}u + \bar{B}c + \bar{d}. \tag{5}$$

The lower level of the Interaction Balance Method (local optimizers in fig. 1) calculates the values $\bar{u}(p)$, $\bar{c}(p)$ so as to minimize the Lagrangian:

$$L(u,c,p) = Q(u,c) + p^T(u-My) \tag{6}$$

with constraint (5) having regard to coordination variables p. The objective function Q(u,c) is usually assumed to be the sum of the subsystem objective functions. Thus the problem of Lagrangian minimization may be divided into many easier problems assigned to the local optimizers.

Assume that Q(u,c) is a quadratic function:

$$Q(u,c) = \frac{1}{2}u^T Su + \frac{1}{2}c^T Rc + r^T u + w^T c + q, \tag{7}$$

where S, R are positive definite matrices, and there are no constraints in the system. The Lagrangian minimization problem can be solved immediately now:

$$\bar{u}(p) = S^{-1}(M\bar{A}-I)^T p - S^{-1}r, \tag{8}$$

$$\bar{c}(p) = R^{-1}\bar{B}^T M^T p - R^{-1}w. \tag{9}$$

Applying the calculated control values $\bar{c}(p)$ to the real system (2) and (3) we obtain the following values of the real inputs $u^*(p,z)$:

$$u^*(p,z) = (I-MA)^{-1}(MBR^{-1}\bar{B}^T M^T p+MCz+Md-MBR^{-1}w, \tag{10}$$

and the so colled discoordination (the difference between $\bar{u}(p)$ and $u^*(p,z)$ may be writen down in the form:

$$D^*(p,z) = Fp + Gz + k, \tag{11}$$

where:

$$F = S^{-1}(M\bar{A}-I)^T - (I-MA)^{-1}MBR^{-1}\bar{B}^T M^T, \tag{12}$$

$$G = -(I-MA)^{-1}MC, \tag{13}$$

$$k = -(I-MA)^{-1}M(d-BR^{-1}w) - S^{-1}r. \tag{14}$$

The upper level problem (coordinator in fig. 1) consist in appointing $\hat{p}(z)$ which minimizes the discoordination norm:

$$\hat{p}(z) = \text{argmin} \| D^{*}(p,z) \|^2 = -F^{-1}(Gz+k). \tag{15}$$

From (11) and (15) we obtain:

$$\hat{p}(z) = p - F^{-1} D^{*}(p,z). \tag{16}$$

The inverse of matrix F does exist if there exists solution $\hat{p}(z)$ for any feasable value z. Note that the discoordination $D^{*}(p,z)$ is measured. Thus from the formula (16) we obtain a very simple coordination algorithm:

$$p' = p - F^{-1} D^{*}(p,z) \tag{17}$$

Ruszczynski has shown how to evaluate matrix F^{-1} basing on the model of the system in nonlinear case with constraints. He has also proved the convergence of the algorithm (17) provided that the difference between the model and the system is small enough.

It is worth noting that algorithm (17) is analogous to the variable metric algorithm considered in the authors paper (ref. 3). Discoordination $D^{*}(p,z)$ plays the role of a measured gradient, F is analogous to the Hessian matrix and may be estimated in the same way described in ref. 3. Consequently all results from the mentioned above paper are applicable to our coordination algorithm. Let us rewrite those results in the suitable form now.

3. Coordination algorithm and its properties.

The coordination algorithm is composed of two different steps made in turn: the identification step from an a'priori given set, and the optimization step like in the Newton method.

Initial parameters:

- $p_1 \in R^n$ - initial upper level control value,
- $\{s_1, s_2, \ldots, s_n\}$ - set of linearly independent identification steps,
- F_o^{-1} - initial approximation of the inverse of matrix F^{-1} (e.g. Ruszczynski's operator).

0° Put $i = 1$, $k = 0$.

1° Measure the discoordination $D^{*}_{nk+i} = D^{*}(p_{nk+i}, z_{nk+i})$.

2° Put $p'_{nk+i} = p_{nk+i} + s_i$ (i d e n t i f i c a t i o n s t e p).

3° Measure the discoordination $(D^{\ast}_{nk+i})' = D^{\ast}(p'_{nk+i}, z'_{nk+i})$ and calculate $r_{nk+i} = (D^{\ast}_{nk+i})' - D^{\ast}_{nk+i}$.

4° Update F^{-1}_{nk+i}, the evaluation of matrix F^{-1} satisfying the formula:

$$F_{nk+i}\, s_j = \begin{cases} \dfrac{1}{k+1} \displaystyle\sum_{l=0}^{k} r_{nl+j} & \text{for } j = 1,2,\dots,i \\ \dfrac{1}{k} \displaystyle\sum_{l=0}^{k-1} r_{nl+j} & \text{for } j = i+1, i+2, \dots, n \end{cases} \tag{18}$$

5° Put $p_{nk+i+1} = p_{nk+i} - F^{-1}_{nk+i} D^{\ast}_{nk+i}$ (o p t i m i z a t i o n s t e p).

6° Put $i = i + 1$. If $i > n$ then $i = 1$ and $k = k + 1$. Go to 1.

We have denoted:

k - iteration number,
i - step number in any iteration,
p_i - the control value calculated at time t_i,
z_i - the value of disturbances at time t_i, $(z_i = z(t_i))$.

Note that the inverse of matrix F_i is available immediately. For instance the well known on line Least-Squeres Estimation algorithm can be used (see ref. 3).

Let us adapt now the basic assumptions which have been made in reference 3:

Assumption 1. The considered large-scale system is linear, unconstrained with a quadratic objective function. Both the upper level optimization problem and the lower level one have solution for any feasable value of disturbances.

Assumption 2. The interval of time between two succeding modifications of the control variables is constant and equal to T.

Assumption 3. The drift of the optimal point is bounded as follows:

$$\| \hat{p}(z'_i) - \hat{p}(z_i) \| \leqslant vT$$

$$\| \hat{p}(z_{i+1}) - \hat{p}(z'_i) \| \leqslant vT \qquad \text{for all } i.$$

Drift rate may be evaluated either experimentaly or on the basis of (15) using the model estimation of matrices F, G and the characteristics of disturbances z.

We denote the distance between a given point p and the R-neighbourhood of the mooving optimal point $\hat{p}$ as follows:

$$\mathrm{dist}(p,\hat{p})_R = \max(\ \|p-\hat{p}\| - R\ ,\ 0\ ). \tag{19}$$

The theorems proved in ref. 3 may be rewriten now in the following form:

Theorem 1. **Let the assumptions 1 ÷ 3 be satisfied and:**

$$\sigma = \sqrt{n}\cdot vT[\ s_1\ ,\ s_2\ ,\ \ldots\ ,\ s_n\ ]^{-1} \leqslant 1/2. \tag{20}$$

Then:

$$\mathrm{dist}(p_{i+1},\hat{p}(z(t_{i+1})))_R \leqslant q\cdot \mathrm{dist}(p_i,\hat{p}(z(t_i)))_R$$

where:

$$q = \sigma/1-\sigma) \tag{21}$$

$$R = 2vT/(1-q) \tag{22}$$

Theorem 2. **Let the assumptions 1 ÷ 3 hold and random values $(z_i'-z_i)$ for i = 1,2,... be independent and have zero mean.**

a) Then:

$$F_i \xrightarrow[i\to\infty]{} F \qquad \text{with probability 1}$$

and

$$\mathrm{dist}(p_i,\hat{p}(z(t_i)))_R \xrightarrow[i\to\infty]{} 0 \qquad \text{with probability 1}$$

$$R = 2vT \tag{23}$$

b) Choose $q_o, \vartheta > 0$ and k_o such that:

$$k_o \gg n\sigma^2(1+q_o)^2/\vartheta q_o^2 \tag{24}$$

and eliminate the identification steps beginning from the iteration number k_o with $F_j = F_{k_o n}$ for all $j > k_o n$. Then with probability ϑ :

$$\text{dist}(p_{j+1}, \hat{p}(z(t_{j+1})))_R \leqslant q_o \cdot \text{dist}(p_j, \hat{p}(z(t_j)))_R \,,$$

$$R \leqslant vT/(1-q_o) \qquad \text{as} \quad j > k_o n \tag{25}$$

Theorem 1 shows that the upper level control variables generated by the algorithm linearly converge to the R-neighbourhood of their mooving optimal values. Tracking efficacy parameters R and q are evaluated. Note that any small value of those parameters is available if identification steps are long enough. But we want them to be as small as possible. In theorem 2 the assumption of small σ is eliminated. Assuming independance of disturbances increments we obtain the convergence of generated p_i to the R-neighbourhood of $\hat{p}_i$ with probability 1. The evaluation of R is the best possible. Furthermore any fixed quality of the convergence is available with any fixed probability provided that identification steps have been made during the sufitiently long time and then eliminated. Both the theorems may be proved in the same way as shown in ref. 3.

The described algorithm is able to estimate matrix F which is constant in the considered linear-quadratic case without constraints. If there are constraints or nonlinearities in the system the operator F becomes nonlinear and depends on the value of the disturbances. In that case our coordination algorithm appoints an average of operators F(z,p) for p and z appearing in the real system. Consequently the estimation of operator F ought to be periodicaly restarted because the drifting disturbances may change significantly if the considered period of time is long. In that case the model of the system used by the lower level optimization algorithm also ought to be adapted. But that adaptation is a separate problem and one of the well known methodes could be applied. The described above modification allows us to apply that adaptive algorithm to the more general case provided that rate v is small enough.

4. Computational experiments.

The tested system shown in fig. 1 is described as follows:

SYSTEM EQUATIONS

Subsystem I $\quad y^1 = c_1^1 + c_2^1 + 2u^1 + A_1(c_1^1)^2 + A_2(c_1^1 - c_2^1 - 2)u^1 + z^1$

$$\begin{aligned} &\textbf{Subsystem II} \quad y_1^2 = c_1^2 - c_2^2 + u_1^2 - 3u_2^2 + z_1^2 \\ &\qquad\qquad\qquad\quad y_2^2 = 2c_2^2 - c_3^2 - u_1^2 + 2u_2^2 + z_2^2 \end{aligned} \qquad (26)$$

Subsystem III $\quad y^3 = c^3 - 4u^3 + A_3c^3u^3 + z^3$

EQUATION OF SUBSYSTEM CONNECTION

$$u = My \qquad M = \begin{bmatrix} 0 & 1 & 0 & 0 \\ 1 & 0 & 0 & 0 \\ 0 & 0 & 0 & 1 \\ 0 & 0 & 1 & 0 \end{bmatrix} \qquad (27)$$

where: $u = [\ u^1\ , u_1^2\ , u_2^2\ , u^3\ ]^T\ ,\quad y = [\ y^1\ , y_1^2\ , y_2^2\ , y^3\ ]^T$

CONSTRAINTS

$$\begin{aligned} &\textbf{Subsystem I} \quad c_1^1 + u^1 \leqslant 1.006 \\ &\textbf{Subsystem III} \quad c^3 - u^3 \leqslant 0.5 \end{aligned} \qquad (28)$$

OBJECTIVE FUNCTION

$$\begin{aligned} &\textbf{Subsystem I} \quad Q_1 = (u^1-1)^2 + (c_1^1)^2 + (c_2^1-2)^2 \\ &\textbf{Subsystem II} \quad Q_2 = 2(c_1^2-2)^2 + (c_2^2)^2 + 3(c_3^2)^2 + 4(u_1^2)^2 + (u_2^2)^2 \\ &\textbf{Subsystem III} \quad Q_3 = (c^3+1)^2 + (u^3-1)^2 \end{aligned} \qquad (29)$$

$$Q = Q_1 + Q_2 + Q_3$$

Note that the system is nonlinear with constraints. The equations of the model have been obtained from system equations with disturbances A_i and z_j^k equal to zero.

Let us now compare two cases of the system coordination. One of them is the case of the constant F^{-1} obtained from (12) using only the model matrices. In the second one the described adaptive coordination algorithm is applied. Fig. 2 shows that the coordination algorithm using the on line modification of matrix F works even when the first algorithm diverges. We have put:

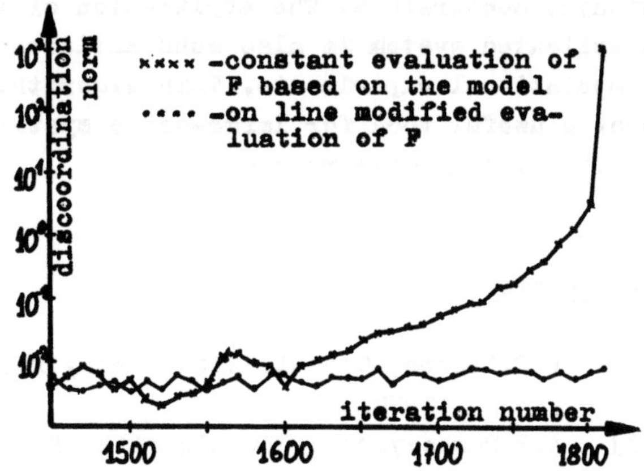

Fig. 2. Illustration of superiority of adaptive coordination algorithm.

$$vT = 0.0005$$

$$[\, s_1 \,,\, s_2 \,,\, s_3 \,,\, s_4 \,] = 0.001 \cdot I \qquad (30)$$

The estimation of matrix F has been restarted after any 400 iterations.

In fig. 3 the case of a single significant variation of the disturbances is illustrated. The first algorithm does not converge while the second one linearly converges with rate q equal to about 0.61.

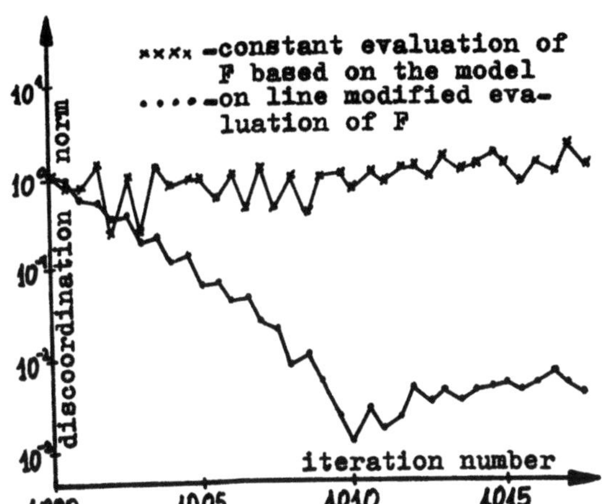

Fig. 3. Consequence of a single significant disturbances variation.

5. Conclusions.

The paper shows the convergence properties of the proposed algorithm in the case of linear system without constraints. The application of the algorithm to control more complicated system is also substantiated and then illustrated by the computational experiments. This algorithm therefore seems to represent a useful tool for large-scale system coordination in the presence of drifting disturbances.

References.

[1] L.S.Lasdon, Optimization Theory for Large Systems, Mac Millan, 1970.

[2] A.P.Ruszczynski, "An Algorithm for Real System Coordination", Large Scale System Theory and Applications, Proc. of IFAC Symp. Udine 1976.

[3] M.M.Michniewicz, "Application of Variable Metric Method in Tracking the Minimal Point of Nonstationary Objective Function", presented at the 11-th IFIP Conference on System Modelling and Optimization in 1983, prepared for publication in Mathematical Programming.

dr Maciej MICHNIEWICZ
Institute of Aviation
The Lukasiewicz Technical University
ul. Wincentego Pola 2
35-959 Rzeszów
POLAND

A STOCHASTIC MULTIOBJECTIVE DYNAMIC PROGRAMMING METHOD WITH APPLICATION TO ENERGY MODELLING

by

Sándor Molnár[1] and Ferenc Szidarovszky[2]

Abstract

A stochastic multiobjective dynamic programming model is described for the optimal utilization of mineral resources for satisfying the energy demand of the national economy. The model is based on a special block-structure, where each block is characterized by its input and output vectors and state variables. The whole system is considered as a superposition of the subsystems defined by the individual blocks. Input-output and state transition relations are defined for each block, and the overall objectives of the whole system are defined by investment cost, production cost, manpower demand and environmental effect.

A special stochastic multiobjective dynamic programming algorithm is developed for the numerical solution of the model, which is the common generalization of deterministic single or multiobjective dynamic programming algorithms and Bayesian decision methods.

A case study illustrates the model and the solution methodology.

1. Introduction

Optimal satisfaction of long-range demands within the national economy requires the use of primarily domestic resources. Therefore the task is twofold: first, to meet the demand on final products; and secondly, to assure the optimal exploitation and use of domestic resources which depend on the possibilities within the national economy and the dynamism of foreign trade. The main objective of the development structure is the minimization of expenditures needed for functioning of the system under consideration. This development structure should guarantee production with at least normative efficiency. The model also compares the expenses of different products depending on the economical level of processing and the expenses incurred by importing certain products.

---

[1]Dept.head, Mining Development Institute, Budapest III. Mikoviny u. 2-4, H-1037, Hungary

[2]Prof. and head, Dept. of Math. and Comp. Sci., Univ. of Horticulture, Budapest XI. Villányi ut 29-35, H-1118, Hungary

The individual phases of each activity in the national economy can be characterized by units which are usually independent technologically, technically, and economically. These units are called technological blocks. The products produced by these technological blocks can be either processed further, utilized domestically, or sold in the foreign market.

The export structure has been created for three reasons: the avoidance of exploiting certain domestic resources; the substitution of certain phases of processing; and the counter balancing of imports of the same value. This last reason is the most important.

In the model, domestic and import expenses are separated. A further distinction will be made between the ruble and the dollar-based expenses. The foreign exchange rate for converting import expenses into domestic currency /forints/ will be not used; instead the demand on foreign currencies in import /machinery, equipments, materials, etc./ will be replaced by expenditures needed to produce a sufficient quantity of exportable products in export-oriented enterprises. The export products mentioned above may also influence the total import demand.

Two types of technological blocks, will be distinguished in the model. These elements are connected with each other and with the technical-economical environment through their input and output parameters, which will be called signals.

Technological blocks are production units which manufacture certain products for the direct satisfaction of domestic and/or export demands.

The technological blocks are not necessarily organizationally independent units. Their capacities are characterized by the yearly quantity of products manufactured by them. An activity which results in the same main product as output, but has different technology and consequently different technical-economic parameters and characteristics, is considered to be a different block. The export-producing blocks must also include all of the economic activities which must be performed in the export-import business. These technological blocks are assumed to be given, and their parameters are assumed to be known. The alternatives examined by the model are based on the same set of technological blocks; they differ through the time periods when these blocks actually enter the system.

The mathematical description of the model will be presented in the next section.

2. The mathematical description of the model

Introduce the following notation:

N = number of technological blocks,

J = number of export blocks producing rubles,

I = number of export blocks producing dollars,

r = total number of products,

T = length of the time interval,

T_{max} = maximal possible lifetime of technological blocks.

Thus, the model considers N+J+I different kind of technological blocks.

(A) The state variables of the system are given as:

$s(t,k,tt)$ = the number of technological blocks of the type k in year t which entered the system in year t-tt. Here, $1 \leq t \leq T$, $1 \leq k \leq N + J + I$, $1 \leq tt \leq T_{max}$.

(B) The decision variables are as follows:

$\delta(k,t)$ = the number of technological blocks of the type k entering the system in year t;

$W_r(k)$ = capacity of the ruble-producing block of the type k;

$W_d(k)$ = capacity of the dollar-producing block of the type k;

$a_{1l}(t)$, $a_{2l}(t)$, $a_{3l}(t)$ = the export-import rate of product in ruble, dollar, and forint markets.

(C) The parameters of the models are the following:

(a) Technological blocks:

$B_k(tt)$ = forint investment cost of block of type k in year tt;

$BD_k(tt)$ = dollar investment cost of block of type k in year tt;

$BR_k(tt)$ = ruble investment cost of block of type k in year tt;

$U_k(tt)$ = forint production cost of block of type k in year tt;

$U_{Dk}(tt)$ = dollar production cost of block of type k in year tt;

$UR_k(tt)$ = ruble production cost of block of type k in year tt;

$M_k(tt)$ = manpower need of block k in year tt;

$P_{jk}(tt)$ = need for product j of block k in year tt [in the case of demand, $P_{jk}(tt) < 0$, and[in the case of production, $P_{jk}(tt) > 0$];

$d_l(t)$, $r_l(t)$, $f_l(t)$ = unit price of product l in year t given in dollars, rubles, and forints, respectively;

$UT_k(tt)$ = utility for environmental damage caused by block k in year tt;

(b) Export-import blocks:

$b_k(tt)$ = forint investment cost per unit capacity of block k in year tt;

$bd_k(tt)$ = dollar investment cost per unit capacity of block k in year tt;

$br_k(tt)$ = ruble investment cost per unit capacity of block k in year tt;

$u_k(tt)$ = forint production cost per unit capacity of block k in year tt;

$ud_k(tt)$ = dollar production cost per unit capacity of block k in year tt;

$ur_k(tt)$ = ruble production cost per unit capacity of block k in year tt;

$m_k(tt)$ = manpower demand per unit capacity of block k in year tt;

$P_{jk}(tt)$ = demand on product j per unit capacity of block k in year tt;

$ut_k(tt)$ = utility for environmental damage per unit capacity of block k in year tt.

(c) Discounting factors:

QF(t) = discount factor of forint, in year t;

QD(t) = discount factor of dollar, in year t;

QR(t) = discount factor of ruble, in year t;

QU(t) = discount factor of environmental utility;

(d) $\Delta_l(t)$ = demand of the national economy on product   in year t.

(D) The output variables are as follows:

S (t) = net quantity of product   in year t;

BE (t) = total forint investment cost in year t;

BER (t) = total ruble investment cost in year t;

BED (t) = total dollar investment cost in year t;

UZ(t) = total forint production cost in year t;

UZR(t) = total ruble production cost in year t;

UZD(t) = total dollar production cost in year t;

ML(t) = total manpower demand in year t;

COF(t) = total forint cost in year t;

COD(t) = total dollar cost in year t;

COR(t) = total ruble cost in year t;

UTT(t) = total environmental utility in year t.

(E) The state-transition relations are given next. The variables s (0,k,tt) express the initial block structure of the subsystem, which can be considered to be given. Then obviously

$$s(t+1,k,tt) = \begin{cases} s(t,k,tt-1) & \text{if } tt > 1, \\ \delta(k,t) & \text{if } tt = 1 \end{cases} \tag{1}$$

for all values of t,k, and tt. The first case can be interpreted as follows: the existing blocks become one year older, and the number of new blocks equals the number of blocks just entering the system.

(F) The output functions can be determined in the following manner. Consider first the total quantity of product in year t. Then one can easily verify that

$$S_l(t) = \sum_{tt=1}^{T_{max}} \left\{ \sum_{k=1}^{N} s(t,k,tt)\,P_{lk}(tt) + \sum_{k=N+1}^{N+J} s(t,k,tt)\,P_{lk}(tt)\,W_r(k) + \sum_{k=N+J+1}^{N+J+I} s(t,k,tt)\,P_{lk}(tt)\,W_d(k) \right\}, \qquad (2)$$

where the summations correspond to technological blocks and ruble- and dollar-producing blocks, respectively. We can similarly construct the forms of BE(t), BED(t), BER(t), BED(t), UZ(t), UZR(t), UZD(t), ML(t). Furthermore

$$COF(t) = BE(t) + UZ(t) - \sum_{l=1}^{L} [S_l(t) - \Delta_l(t)]\,a_{3l}(t)\;f_l(t). \qquad (3)$$

In the last equation the export income is subtracted from the sum of the investment and production costs. In the case of import costs, the last sum is negative. Similarily,

$$COR(t) = BER(t) + UZR(t) - \sum_{l=1}^{L} [S_l(t) - \Delta_l(t)]\,a_{1l}(t)\;r_l(t)$$

$$- \sum_{k=N+1}^{N+J} \sum_{H=1}^{T_{max}} s(t,k,tt)\,P_{N+1,k}(tt)\,W_r(k), \qquad (4)$$

$$COD(t) = BED(t) + UZD(t) - \sum_{l=1}^{L} [S_l(t) - \Delta_l(t)]\,a_{2l}(t)\,d_l(t)$$

$$- \sum_{k=N+J+1}^{N+J+I} \sum_{tt=1}^{T_{max}} s(t,k,tt)\,P_{N+2,k}(tt)\,W_d(k), \qquad (5)$$

and

$$UTT(t) = \sum_{H+1}^{T_{max}} \left\{ \sum_{k=1}^{N} s(t,k,tt)\,UT_k(tt) + \sum_{k=N+1}^{N+J} s(t,k,tt)\,ut_k(tt)\,W_r(k) + \sum_{k=N+J+1}^{N+J+I} s(t,k,tt)\,ut_k(tt)\,W_d(k) \right\}. \qquad (6)$$

G The objective functions of the model are the following:

(a) Minimize the total dollar based investment cost:

$$\text{minimize } Z_1 = \sum_{t=1}^{T} BED(t)QD(t). \tag{7}$$

(b) Minimize the dollar based production cost:

$$\text{minimize } Z_2 = \sum_{t=1}^{T} UZD(t)QD(t). \tag{8}$$

(c) Minimize the average manpower demand:

$$\text{minimize } Z_3 = \frac{1}{T} \sum_{t=1}^{T} ML(t); \tag{9}$$

(d) Total costs in forints, rubles, and dollars have to be minimized:

$$\text{minimize } Z_4 = \sum_{t=1}^{T} COF(t)QF(t),$$

$$\text{minimize } Z_5 = \sum_{t=1}^{T} COD(t)QD(t), \tag{10}$$

$$\text{minimize } Z_6 = \sum_{t=1}^{T} COR(t)QR(t).$$

(e) In minimizing the total environmental damage, we get

$$\text{minimize } Z_7 = \sum_{t=1}^{T} UTT(t)QU(t). \tag{11}$$

(H) The set of constraints can be described as follows:

(a) Upper bounds for investment costs are given in yearly basis as

$$\begin{aligned} BE(t) &\leq K_1(t), \\ BER(t) &\leq K_2(t), \quad (\forall t) \\ BED(t) &\leq K_3(t), \end{aligned} \tag{12}$$

(b) Manpower limitations:

$$ML(t) \leq K_4(t). \quad (\forall t) \tag{13}$$

(c) Ruble and dollar balances on a yearly basis:

$$\begin{aligned} K_5(t) &\leq COR(t) \leq K_6(t), \\ K_7(t) &\leq COD(t) \leq K_8(t). \end{aligned} \quad (\forall t) \tag{14}$$

Note that by choosing $K_5(t) = K_6(t) = K_7(t) = K_8(t) \equiv 0$, exact balances are required.

(d) Export and import limitations can be considered by means of the feasible sets of the decision variables $a_{1l}(t)$, $a_{2l}(t)$, $a_{3l}(t)$.

(e) Bounds on the quantity of products to be exported or imported can be given as

$$\begin{aligned} K_9(l,t) &\leq [S_l(t) - \Delta_l(t)]\, a_{1l}(t) \leq K_{10}(l,t), \\ K_{11}(l,t) &\leq [S_l(t) - \Delta_l(t)]\, a_{2l}(t) \leq K_{12}(l,t). \quad (\forall l,t) \end{aligned} \tag{15}$$

(f) Bounds on the utility characterizing the environmental damage have the form

$$K_{k3}(t) \leq UTT(t) \leq K_{14}(t), \tag{16}$$

Observe that the mathematical formulation of the model satisfies the following properties:

(1) The state-transition relation has the usual form, that is, the present state depends on the previous state and the decision.

(2) All constraints either consist of additive functions /where the summations are made over the stages/, or give direct restrictions for the decision or state space.

(3) All objective functions are additive.

The above properties imply that the mathematical formulation of this model is a special case of the multiobjective dynamic-programming problem.

The parameters of the model are uncertain values in the case of a long-term optimization. This uncertainty can be modelled by considering the parameters as random variables. Their means are assumed to be given by their estimated values, and their variances are estimated subjectively by their uncertainties. Thus, the objective functions and the

constraints are stochastic. By taking expectations, the objective functions can be transformed into deterministic forms, and the constraints can be rearranged by using chance-constraint conditions.
By using thise well known ideas, a deterministic multiobjective programming problem is obtained, which can be solved by standard techniques [12], [13] .

4. An application of the model

The model described in the previous sections has been tested on a special subsystem of the Hungarian coal mining and related industry. The satisfaction of the national aconomy with regard to brown coal, electricity, and coke and the replacement of oil and gas by brown coal has been considered. Eighteen blocks and ten different products have been included in the model. In this case we have chosen $I = J = 1$, that is only one ruble and one dollar producing block have been assumed to belong to the subsystem under consideration.

The single objective function Z_4 was minimized under the constraints (12), for investment costs, (13) for manpower, and (15) for export and import.

In the constraints (14) we had

$$K_5(t) \equiv K_6(t) \equiv K_7(t) \equiv K_8(t) \equiv 0.$$

Since we have both integer and real variables, a mixed real-integer dynamic-programming algorithm [2,3] was used. The total computation on a type R20 computer /made in the Soviet Union/ of the University of Horticulture, Budapest, required 35 minutes.

There was no theoretical difficulty in applying the model for this small-dimensional case. But in more realistic cases, one of the following possibilities should be applied for the actual application of the model:

(1) apply regular nonlinear programming methods [5,4];

(2) use a suitable version of differential dynamic-programming [7,8];

(3) if more than one objective function have to be considered, use a multiobjective dynamic-programming method [11,10,9].

References

[1] L. Kapolyi, Systems Models of Mineral Resources Management /in Hungarian/, Akademiai Kiado, Budapest, 1981.

[2] R.Bellman and S. Dreyfus, Applied Dynamic Programming, Princeton U.P., Princeton, N.J., 1962.

[3] E.R. Larson and J.L. Casti, Principles of Dynamic Programming II, Marcel Dekker, New York, 1982.

[4] G.Hadley, Nonlinear and Dynamic Programming, Addison-Wesley, Reading, Mass., 1964.

[5] B. Martos, Nonlinear Programming, Theory and Methods, North Holland, Amsterdam, 1975.

[6] L. Lengyel, S. Molnar, L. Neumann, F. Szidarovszky, and A. Vancsa, A systems model for the utilization of domestic mineral resources /in Hungarian/, Report of the Central Mining Development Institute, Budapest, Hungary, 1981.

[7] D.Jacobson, and D. Mayne, Differential Dynamic Programming, Elsevier, New York, 1970.

[8] F. Szidarovszky, M. Gershon, and A. Bardossy, Application of multiobjective dynamic programming to regional natural resource management, presented, at CORS/ORSA/TIMS Joint National Meeting, Toronto, Ontario, 3-6 May 1981.

[9] G. Tauxe, R.Inman, and D.Mades, Multiobjective dynamic programming; A classic problem redressed, Water Resources Research 15(6): 1398-1402 /Dec. 1979/.

[10] F. Szidarovszky and L. Duckstein, A general numerical framework for dynamic multiobjective techniques, Working Paper #81-27, SIE, Univ. of Arizona, Tucson, 1981.

[11] F. Szidarovszky, Notes on multiobjective dynamic programming, Working Paper # 79-1, SIE, Univ. of Arizona, Tucson, 1979.

[12] F. Szidarovszky and S. Yakowitz, Principles and Procedures of Numerical Analysis, Plenum, New York, 1978.

[13] S. Molnar and F. Szidarovszky, Classical and Differential Games, Multiobjective Programming with Engineering Applications /in Hungarian/, to be published by Müszaki Könyvkiadó, Budapest.

OPERATION PLANNING OF POWER SYSTEMS

Dipl.-Ing. Dr. Michael MUSCHICK
Dipl.-Ing. Alois SCHADLER
AUSTRIA

ABSTRACT

This paper deals with economical optimization in electric utilities. After a short review which summarizes the optimization activities of the Styrian electric utility STEWEAG, a report is given about an attempt of IBM Austria and STEWEAG to practically implement a short term operation planning system which is universally applicable to electric power systems and district heating systems. Some of the highlights of this optimization model are:

- Complex water reservoir systems including reservoir networks, time delays for the water flow and water contracts to downstream plants
- Coupled energy systems with electric and heating power generation
- Sophisticated interchange contracts taking into account several types of contracts
- New acceleration strategies for solving the optimization problem

The solution of the optimization problem is based upon the mixed integer programming package MPSX and MIP. The optimization kernel is embedded in an extended control program containing a number of new acceleration measures which were necessary to achieve practically acceptable results within reasonable computing times.

OPTIMIZATION ACTIVITIES IN THE ELECTRIC UTILITY STEWEAG

Energy management in an electric utility is a very complex problem due to the fact that there are various technical and economical processes involved which influence each other to a high degree. The difficulty of determining optimal strategies for operation and expansion planning has created the need for systematic approaches.

The problems arising in economical optimization of electric power systems have been analysed in our utility in the sixties and early seventies. Concepts have been derived and a number of papers for operation and expansion planning have been published by Steinbauer and Lehner (see /1/, /2/, /3/, /4/).

Based on these concepts the Styrian electric utility developed the first optimization program about 1970 /5/. This program was used as a basic tool for operation planning as well as for expansion planning on a monthly and yearly basis.

The heart of this program system was a simple and partially heuristic algorithm based upon the optimization method of EULER-LAGRANGE. This method used the principle of equal incremental costs for the load dispatch. The unit commitment problem was determined using a heuristic strategy which took into account economical and technical considerations. Energy constraints were handled in an iterative manner.

Due to the simplicity of this method, this program was very fast, thus allowing the calculation of different model variations with minimum effort. But unfortunately - and this seems to be a characteristic of such algorithms - some conditions which were essential to this model changed during the oil crisis and it was not possible to adapt the program to the new situation.

Therefore investigations were made to get an improved algorithm with the ability to handle a wider range of constraints. It turned out that mixed integer programming seemed to satisfy the demands for the power system of the STEWEAG. However - despite of existing results of several research workers using mixed integer programming for the solution of similar problems - nobody was able to guarantee that this method could solve our model in reasonable time with acceptable results. The reason for this situation was that all similar previous applications of mixed integer programming have treated relatively small and less complex problems without taking into account things like generation of electric power coupled with district heating generation, complex reservoir networks and sophisticated interchange contracts (with various interchange possibilities).

Therefore in 1981 a joint study was started by IBM Austria and STEWEAG comprising the development of a program system for the short time optimization. This system calculates optimal operation schedules for time periods of one up to seven consecutive days and takes care of the technical, economical and operational constraints.

OPERATION PLANNING STRATEGY

The short term operation planning is executed in 2 steps:

- Weekly optimization
- Daily optimization

Based upon the computer assisted short term forecast the unit commitment of the whole (or the remaining) week is optimized with time steps of four hours. It is the main goal of the weekly optimization to deal with the start-up and shut-down decisions of thermal plants taking into account the different load patterns of the days, the weekend influence of the water reservoirs and the fuel restrictions. The results are global unit commitment, storage management and fuel management schedules. The uncertainty in the weekly forecast justifies the rough time discretisation.

Embedded in the weekly optimization the daily optimization is calculated on a 48 time step basis in order to get smooth schedules for the components of the energy system. The boundary constraints for the daily models are automatically transferred from the weekly model into the daily one.

In addition the solution of the weekly model is used as a starting solution for the daily optimization.

THE MATHEMATICAL MODEL

The mathematical model is generated automatically by universally applicable matrix generator consisting of modules for

- storage systems
- hydro plants
- thermal plants for electric and/or district heating generation
- logical operation conditions
- global fuel management constraints
- electric energy interchange contracts
- heating energy interchange contracts
- electric balances (power balance and/or energy balance)
- heating balances (power balance and/or energy balance)

General mixed integer linear programming problems have the following well-known form:

Minimize the objective function

$$F = c\,x + d\,y \qquad (P)$$

subject to the constraints

$$A\,x + B\,y \gtrless a$$

A ... r x n matrix
B ... r x m matrix

and the boundary conditions for the continuous and the integer variables

$$\underline{x} \leq x \leq \overline{x}$$

$$\underline{y} \leq y \leq \overline{y}$$

y ... integral

Each of the modules mentioned above generates a set of constraints, boundary conditions and contributions to the objective function specific to the type of plant, contract, reservoir or other element to be taken into account. It is neither necessary nor possible in the space provided to list every single one of them. As an example, however, it may be appropriate to give a more detailed, if simplified, outline of the element "extraction condensing turbine".

Extraction condensing turbines

The main difference to back pressure turbines is given by the fact that the heating generation can be chosen independently from the electric power output. Therefore this type of plant can be described by a series of generation curves with the heating output as parameter.

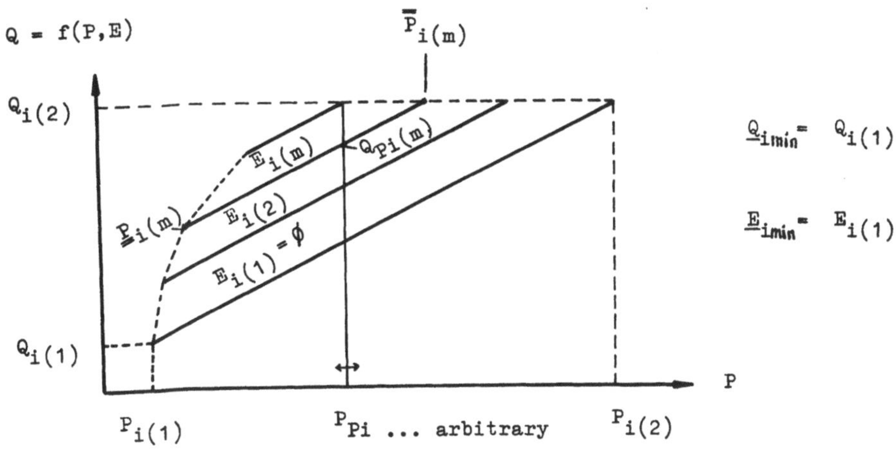

Simplifications:

It is assumed that the generation curves are piecewise linear and that they can be derived from the basic generation curve (the one without heating output, i.e. $E_{i(1)} = 0$) by parallel shifting. Furtheron it will be assumed that the characteristic curves are limited by linear segments as shown in the figure. For this case the relationship between primary power input, electric and heating power output can be expressed as:

$$Q_{imin} * U_{ik} + \sum_{s=1}^{S_i} \eta_{is} * P_{iks} + \sum_{m=1}^{M_i} \xi_{im} * E_{ikm} - \sum_{f=1}^{F_i} Q_{ikf} = 0$$

$$\eta_{is} = \frac{Q_{i(s+1)} - Q_{i(s)}}{P_{i(s+1)} - P_{i(s)}}$$

$$\xi_{im} = \frac{Q_{Pi(m+1)} - Q_{Pi(m)}}{E_{i(m+1)} - E_{i(m)}}$$

i .... plant index

k .... time step index

s .... segment index

f .... fuel index

m .... curve index

M_i .... number of characteristic curves

η_{is} .... electric power production coefficient

ξ_{im} .... heating power production coefficient

P_{iks} ...... electric power output variables

E_{ikm} ...... heating power output variables

Q_{ikf} ...... fuel input variables

U_{ik} .... operation state variable

Bounds for the district heating variables:

$$\emptyset \leq E_{ikm} \leq \max \left\{ 0 , \min \left\{ \bar{E}_{ik} - E_{ik(m)} , E_{ik(m+1)} - E_{ik(m)} \right\} \right\}$$

Restrictions for the electric power generation:

Upper limit:

$$- (\bar{P}_i - \underline{P}_i) * U_{ik} + \sum_{s=1}^{S_i} P_{iks} - \sum_{m=1}^{M_i} \bar{e}_{im} * E_{ikm} \leq \emptyset$$

Lower limit:

$$\sum_{s=1}^{S_i} P_{iks} - \sum_{m=1}^{M_i} \underline{e}_{im} * E_{ikm} \geq \emptyset$$

$$\bar{e}_{im} = \frac{\bar{P}_{i(m+1)} - \bar{P}_{i(m)}}{E_{i(m+1)} - E_{i(m)}} , \qquad \underline{e}_{im} = \frac{\underline{P}_{i(m+1)} - \underline{P}_{i(m)}}{E_{i(m+1)} - E_{i(m)}}$$

These equations take into account the reduction of the electric generation if the heating generation is increased (generation curve limitations, see figure).

Restriction for the heating generation:

$$- \bar{E}_i * U_{ik} + \sum_{m=1}^{M_i} E_{ikm} \leq \emptyset$$

In addition all the constraints for simple thermal plants (not shown here) are valid exept for the generation equation which must be replaced by the equations above.

OPTIMIZATION

Existing mixed integer programming software packages like MPSX/MIP are still not sophisticated enough to be treated as black boxes. Their efficiency can be considerably improved by a deep knowledge of their functioning and the possibilities they offer to the user. Furthermore it is necessary to take into account all the special knowledge that is available from the problem structure. Therefore a lot of extensions and modifications are necessary to get an efficient tool to solve large scale mixed integer problems like this one. Some of these extensions which are available in the control program shall be discussed. Note that usage of these possibilities is not compulsory. Experience has shown that a great part of the improvements is due

a) to speeding up the linear optimization part using a starting solution and

b) to tidying up the integer search.

Accelerating measures for the LP-solution:

The first step in the optimization process is the solution of the continuous problem. Although the simplex algorithm of MPSX has a very good technique to get a starting solution it is advantageous to use the results of previous planning periods as starting solution.

Accelerating measures for the IP-solution:

- preoptimization techniques
  - rounding heuristic
  - integer starting solutions
- directing the B&B search using priority lists for SOS, pseudo costs and weights for SOS variables
- different search strategies for the preoptimization phase and the final search phase
- analysis techniques for the elemination of units which will not be in operation in the optimal solution

RESULTS

The STEWEAG power system has a very complex nature because nearly all types of plants are present. The energy generation is characterized by a strong hydro generation, a relatively complex interchange contract and thermal power plants for electric power and district heating generation. Included are five hydro storage plants. Three of them are located at the Enns river, followed by a chain of downstream hydro plants.

Furtheron there are eight thermal units, two of them with extraction condensing turbines and one with a back pressure turbine. The electric power system is coupled with the district heating system as shown in the following figure. The heating demand is determined by the daily heating energy and the maximum heating power. Both depend on the average temperature.

The results obtained from the optimization program are cost-optimal schedules of the given power system. One example is shown on the following page. The weekly optimization including one underlying day resulted in about one hour of CPU time as an average value across four different weeks of the year.

The positive experience gained during the different test runs encouraged us to take up computer aided operation planning with this program package in our dispatching centre in autumn 1985. Furtheron we decided to extend the system to the yearly optimization. This includes flexible handling of planning scenarios and further extensions of the mathematical model and the optimization process.

PRESENTATION OF THE STEWEAG POWER SYSTEM FOR THE WEEKLY OPTIMIZATION

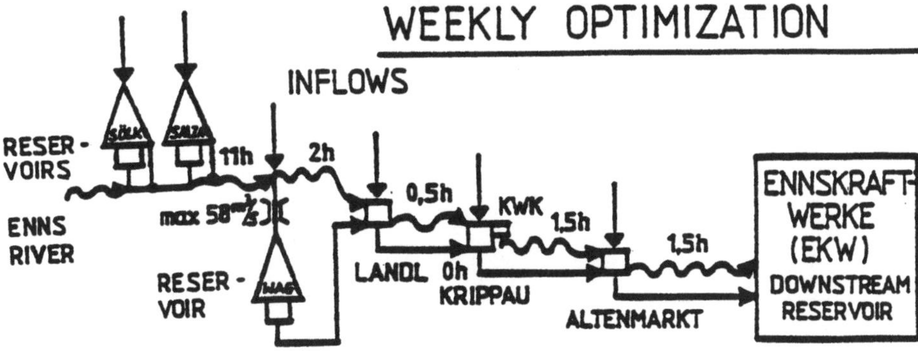

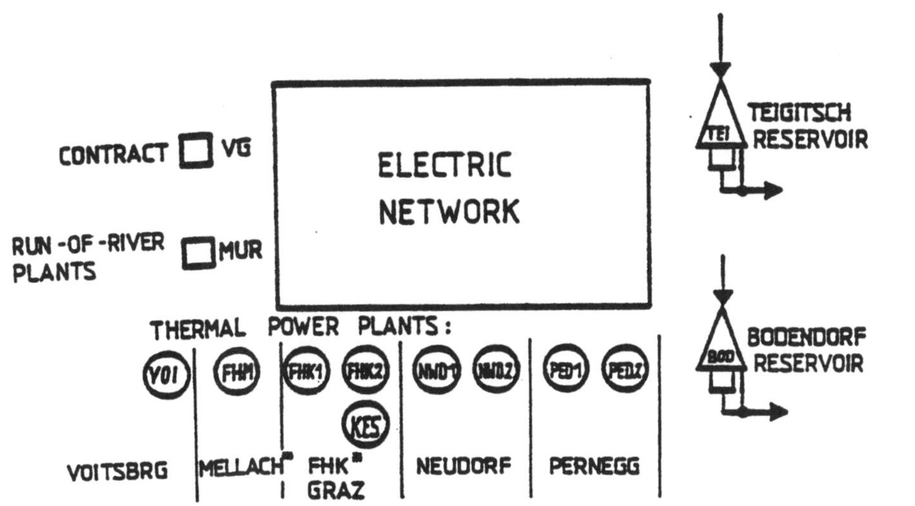

5 STORAGE POWER PLANTS
4 HYDRO POWER PLANTS
8 THERMAL UNITS FOR ELECTRIC POWER AND HEATING* GENERATION
1 BOILER
1 CONTRACT

19 UNITS

II B 19.6.1984 Kn

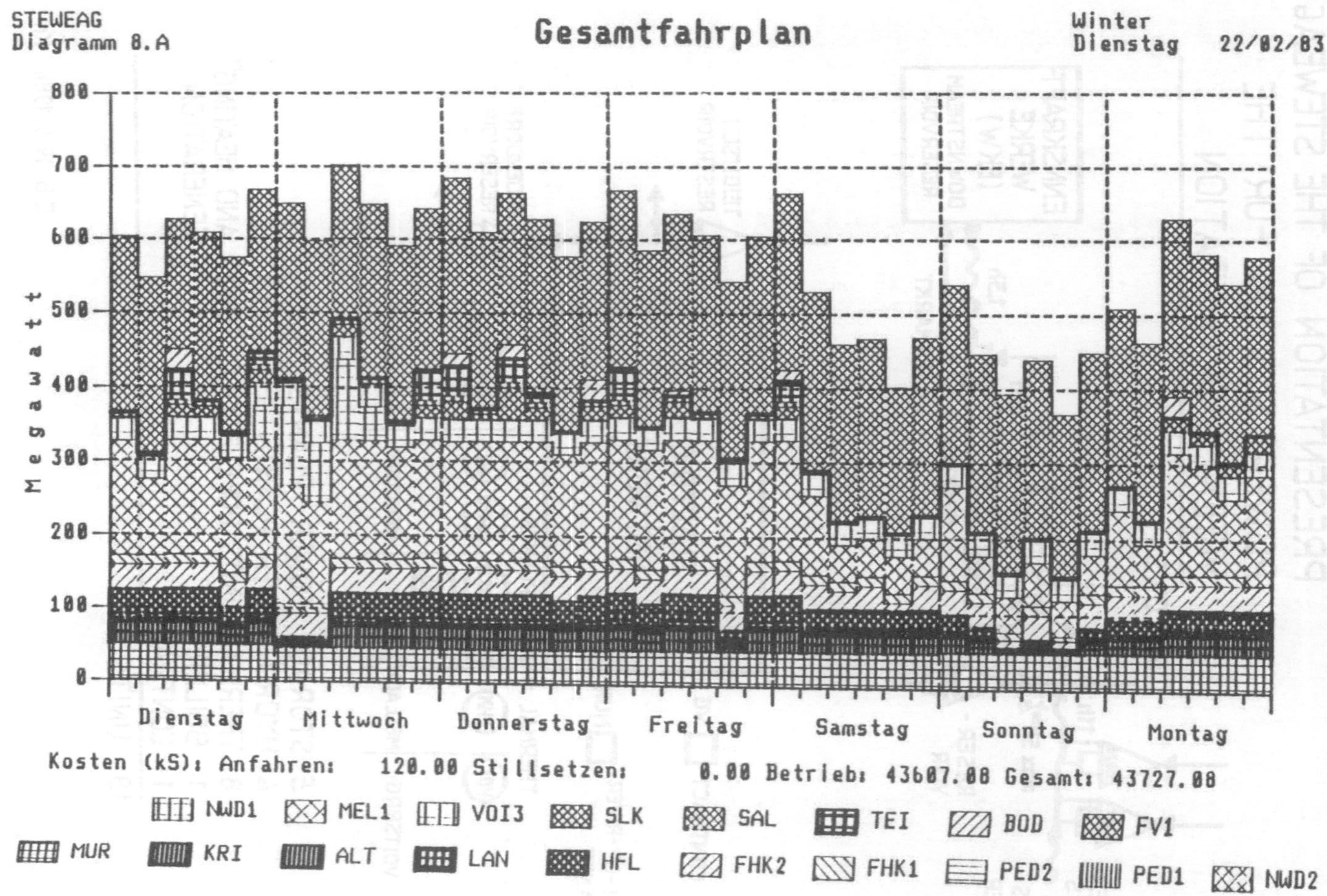
STEWEAG
Diagramm 8.A
Gesamtfahrplan
Winter
Dienstag 22/02/83
800
700
600
500
400
300
200
100
0
Megawatt
Dienstag
Mittwoch
Donnerstag
Freitag
Samstag
Sonntag
Montag
Kosten (kS): Anfahren: 120.00 Stillsetzen: 0.00 Betrieb: 43607.00 Gesamt: 43727.00
NWD1
MEL1
VOI3
SLK
SAL
TEI
BOD
FV1
MUR
KRI
ALT
LAN
HFL
FHK2
FHK1
PED2
PED1
NWD2

LITERATURE

/1/ E. Steinbauer — Planungs- und Betriebsoptimierung in der Elektrizitätswirtschaft mit Hilfe von Rechenanlagen. ÖZE, Jg. 26 (1973), Heft 6.

/2/ E. Steinbauer — Das wirtschaftliche Optimum in der Elektrizitätswirtschaft. EuM, 89. Jg. (1972), Heft 1, Seite 12.

/3/ E. Steinbauer — Suche nach optimalen Lösungen in der Energiewirtschaft - gestern, heute und für morgen. ÖZE, Jg. 34 (1981), Heft 2.

/4/ E. Steinbauer — Das wirtschaftliche Optimum in der Elektrizitätswirtschaft. Dissertation TH Wien, 1969.

/5/ N. Lehner — Methoden zur numerischen Berechnung der optimalen Lastverteilung im hydrothermischen Verbundbetrieb nach dem Prinzip der Euler-Lagrangeschen Variationsableitung. Dissertation TH Graz, 1971.

/6/ P.G. Harhammer — Wirtschaftliche Lastaufteilung auf Basis der gemischt-ganzzahligen Planungsrechnung. Dissertation TH Wien, 1974.

/7/ P.G. Harhammer — Optimization of large scale MIP models
M. Muschick, A. Schadler — Operation planning of energy systems

/8/ M. Muschick — Wochenoptimierung mit MPSX/MIP (unpublished)

/9/ E. Steinbauer — Techischer und wirtschaftlicher Einsatz hydraulischer und thermischer Kraftwerke ÖZE, Jg. 38 (1985), Heft 2.

MODELLING AND NUMERICAL SIMULATION OF WIND-GENERATED CIRCULATION AND MATTER TRANSPORT IN SHALLOW LAKES

Wolfdietrich Müller
Academy of Sciences of the G.D.R.
Karl Weierstrass Institute of Mathematics
Berlin, G.D.R.

1. Introduction

Permanently growing water demand and permanent decrease of water quality call for increasing investments in water quality management. Here the decision maker can no longer employ thumb rules but ought to make use of general computer models and program packages.

Circulation of water and distribution of pollutants in shallow lakes are significantly influenced by the wind. As a mathematical model of "medium" complexity, the Ekman model (derived from the Navier-Stokes equations by using the concept of turbulence viscosity and assuming hydrostatic pressure distribution and small circulation velocities) gives a globally good approximation to the three dimensional velocity field of circulation in shallow lakes. Horizontal velocity distributions in arbitrary depths can easily be calculated from a "stream function" which turns out to be the solution of a boundary value problem for a Poisson-like partial differential equation (with additional convection terms) in two space variables.

This partial differential equation is solved numerically (on an arbitrary two dimensional domain) by a specially adapted finite difference scheme on an equidistant rectangular grid. From the stream function, velocity vectors are computed by direct calculations. The obtained velocity field determines the coefficients in the ordinary differential equation of drifting body trajectories (solved numerically by a predictor-corrector method) and in the mass transport equation (a partial differential equation of diffusion-convection type, solved by a finite difference scheme closely related to the one mentioned above).

2. The circulation model

For a complete derivation of the equations of the Ekman model from the Navier-Stokes equations we refer to /6/ where also further references are given. The core of this model is the following: Assume a cartesian coordinate system (x,y,z) (where the z-axis is directed upward) and denote by $v = (v_x, v_y, v_z)$ the three dimensional flow velocity field. The volume stream density $u = (u_x, u_y)$ is obtained from v by a vertical integration

$$u_x(x,y) = \int_{-h(x,y)}^{\zeta(x,y)} v_x(x,y,z)\,dz, \quad u_y(x,y) = \int_{-h(x,y)}^{\zeta(x,y)} v_y(x,y,z)\,dz, \tag{1}$$

where h(x,y) denotes the (known) water depth and $\zeta(x,y)$ denotes the (unknown) water surface level. Then the components of u can be obtained as the partial derivatives of a scalar stream function $\phi(x,y)$,

$$u_x = \frac{\partial \phi}{\partial y}, \quad u_y = -\frac{\partial \phi}{\partial x}, \tag{2}$$

and ϕ is the solution of a boundary value problem for a Poisson-like partial differential equation in two space variables which - if for the vertical viscosity coefficient an empirical relation given by Barannik and Seljuk /1/ is used - reads

$$\Delta\phi - \frac{2}{h}\left(\frac{\partial h}{\partial x}\frac{\partial \phi}{\partial x} + \frac{\partial h}{\partial y}\frac{\partial \phi}{\partial y}\right) = k_A\left(a\frac{\partial h}{\partial x} - b\frac{\partial h}{\partial y}\right). \tag{3}$$

Here the abbreviations

$$a = w_{10}\sin\varphi, \qquad b = w_{10}\cos\varphi \tag{4}$$

are used, φ denotes the angle of the wind direction with respect to the positive x-axis, w_{10} denotes the wind velocity at 10 m above water surface, k_A is an empirical constant (e.g., $k_A = 0.03$).

Along the boundaries of the (eventually multiply connected) integration domain Ω the stream function is piecewise constant,

$$\phi = \text{const} \tag{5}$$

along each shore section and

$$\phi(P_j) - \phi(P_i) = Q_{ij} \tag{6}$$

along inflow or outflow sections, where Q_{ij} is the volume stream passing the boundary between points P_i and P_j.

If the stream function is computed by solving the boundary value problem (3)-(6), then all subsequent calculations remain on a comparatively elementary level. Volume streams u are obtained by (2), average stream velocities $w = (w_x, x_y)$ are given simply by

$$w_x(x,y) = u_x(x,y)/h(x,y), \quad w_y(x,y) = u_y(x,y)/h(x,y), \tag{7}$$

and from these values the Ekman model approach allows to calculate also the real stream velocity components in arbitrary depths z (which seemed to have been lost forever by the averaging procedure (1)) using the algebraic representations

$$v_x(x,y,z) = 2bk_A(1 + \frac{z}{h}) + \frac{3}{2}(1 - (\frac{z}{h})^2)(w_x - bk_A), \tag{8}$$
$$v_y(x,y,z) = 2ak_A(1 + \frac{z}{h}) + \frac{3}{2}(1 - (\frac{z}{h})^2)(w_y - ak_A)$$

(cf. /6/). The water surface level $\xi(x,y)$ can be computed be integrating the system

$$\frac{\partial \xi}{\partial x} = k_B w_{10} \frac{1}{h}(bk_A - w_x), \quad \frac{\partial \xi}{\partial y} = k_B w_{10} \frac{1}{h}(ak_A - w_y) \tag{9}$$

from a suitably chosen zero level point (x_0,y_0):

$$\xi(x,y) = \int_{(x_0,y_0)}^{(x,y)} (\frac{\partial \xi}{\partial x} dx + \frac{\partial \xi}{\partial y} dy) \tag{10}$$

(k_B is another empirical constant, e.g., $k_B = 15.3\ 10^{-6}$).
Trajectories of drifting bodies can be calculated by integrating a (nonlinear, ordinary) differential system

$$\frac{dx}{dt} = \widetilde{w}_x(x,y), \quad \frac{dy}{dt} = \widetilde{w}_y(x,y), \tag{11}$$

where $\widetilde{w}_x$, $\widetilde{w}_y$ denote the stream velocity components acting on the body. These components are easily obtained from (8). For instance, for a body drifting at constant depth,

$$\widetilde{w}_x(x,y) = \frac{1}{z_1 - z_2} \int_{z_2}^{z_1} g(z)\ v_x(x,y,z)\ dz$$
$$\widetilde{w}_y(x,y) = \frac{1}{z_1 - z_2} \int_{z_2}^{z_1} g(z)\ v_y(x,y,z)\ dz\ , \tag{12}$$

where z_1, z_2 denote the upper and lower depth coordinate, respectively, of the drifting body, and g(z) is a weight function representing the geometrical shape of the body.

3. The mass transport model

For shallow lakes (with average depths of less than 10 m) it can be assumed (cf. /1/, /4/) that, by turbulence effects, scalar quantities like temperature or concentrations of pollutants are homogeneously distributed on each vertical line. The horizontal spreading of admixtures is then essentially determined by the dispersion which results from the vertical stream velocity shearing. So, for stationary wind and stream situations, mass transport can be described by the following transport equation

$$h \frac{\partial c}{\partial t} + u_x \frac{\partial c}{\partial x} + u_y \frac{\partial c}{\partial y} = \frac{\partial}{\partial x}(h\, D_x \frac{\partial c}{\partial x}) + \frac{\partial}{\partial y}(h\, D_y \frac{\partial c}{\partial y}) - k_1\, h\, c + q \,. \tag{13}$$

Here $c = c(x,y,t)$ denotes the mass concentration or excess temperature, h, u_x, u_y have the same meaning as in the previous section ($h(x,y)$ water depth, (u_x,u_y) volume stream), $k_1 = 0$ is a decay coefficient, $q = q(x,y,t)$ is a source term. The dispersion terms represent the vertical shearing effects (cf. /2/, /3/), the dispersion coefficients D_x, D_y are assumed to be given by

$$\begin{aligned} D_x(x,y) &= \varepsilon_x + (\chi h^2(x,y)/\varepsilon_z) \left\langle (v_x(x,y,z)-w_x(x,y))^2 \right\rangle , \\ D_y(x,y) &= \varepsilon_x + (\chi h^2(x,y)/\varepsilon_z) \left\langle (v_y(x,y,z)-w_y(x,y))^2 \right\rangle , \end{aligned} \tag{14}$$

where $\langle \ldots \rangle$ denotes vertical averaging (from $z = 0$ to $z = -h(x,y)$), (w_x,w_y) and (v_x,v_y) are given by (7) and (8) respectively, ε_x is the horizontal turbulence viscosity coefficient, ε_z is the vertical turbulence viscosity coefficient, and χ is a numerical parameter whose value depends on the shape of the vertical velocity distribution ($\chi = 0.001 \ldots 0.1$).

A physically correct description of the behaviour of concentration values near the boundary is obtained in the following way: Second kind boundary conditions

$$\frac{\partial c}{\partial n} = 0 \qquad \text{(n: normal to the boundary)} \tag{15}$$

are prescribed in all boundary points, and the source term q is taken equal to the mass input per time unit in all inflow boundary points (and equal to zero in all other boundary points).

In addition, for a complete description of instationary problems initial conditions

$$c(x,y,0) = c_0(x,y) \tag{16}$$

have to be given. (Stationary concentration distributions may be computed by instationary imbedding for $t \rightarrow \infty$ or, more easily, by directly solving a modified equation (13) where the term $h\, \partial c/\partial t$

is omitted.)

By solving an instationary mass transport problem for a sufficiently large time horizon, the mean detention time of water in a lake crossed by a river may be calculated: Denote by $c_O(t)$ the mass concentration at the outflow caused by an impulse-like mass input $c_I(t) = \delta(t)$ at the inflow, then the mean detention time T_M is given by

$$T_M = \lim_{T \to \infty} \left[\int_0^T t\, c_O(t)\, dt \Big/ \int_0^T c_O(t)\, dt \right]. \qquad (17)$$

During the integration of equation (13) the quotient in (17) can be approximated numerically and so an estimate for T_M can be obtained.

4. Numerical methods

As already mentioned, the core of the circulation model is given by the partial differential equation (3) for the stream function ϕ. This equation is solved numerically by a finite difference scheme on an equidistant rectangular grid. This approach implies the approximation of the two dimensional integration domain Ω (which is given by the geometrical shape of the lake under consideration) by a discrete set $\bar{\Omega}$ which consists of all grid points of the above-mentioned grid that belong to Ω (or, eventually, lie in a certain neighbourhood of the boundary of Ω).

A complete description of the finite difference scheme is given in /6/. It has been developed by G. Stoyan and is a two dimensional version of the scheme for the one dimensional convection-diffusion equation described in /5/. In inner points of $\bar{\Omega}$ a five-point approximation is used which combines the standard approximation of the Laplacian with a weighted approximation of the convection terms, in boundary points Dirichlet conditions corresponding to (5) or (6) are prescribed. In this way a large linear system (of order N, where N is the number of points belonging to $\bar{\Omega}$) is obtained for the values of $\tilde{\phi}$ in the points of $\bar{\Omega}$ (where $\tilde{\phi}$ denotes the discrete approximation to the stream function ϕ). Since its coefficient matrix is sparse (in each equation at most five coefficients are different from zero), this system is solved by standard sparse matrix techniques.

The partial derivatives in (2) are replaced by central (in boundary points by one-sided) finite differences. The calculations in (7) and (8) are elementary. The calculation of water surface heigths according to (9), (10) offers no numerical problems, but may be non-trivial with respect to its organization: Starting at some zero level grid point (i_o,j_o), the integration procedure must reach each point of the (eventually multiply connected) point set $\overline{\Omega}$ once and only once.

For the system (11) of ordinary differential equations of drifting body trajectories, the right-hand sides have to be calculated numerically in each point where they are needed. In grid points the calculations are based on formulas (12), in intermediate points values are obtained by interpolation. The integration is performed using a predictor-corrector method with automatically adapted time step length.

As to the mass transport model and its "core", the instationary transport equation (13), it turned out that the structure of its elliptic part is essentially the same as that of equation (3). So a straightforward generalization of Stoyan's finite difference approach for the spatial derivatives (replacing the approximation of the Laplacian by an analogous approximation of the dispersion terms) and a backward difference approximation of the time derivative (assuming an equidistant or piecewise equidistant discretization of the time axis) were used to compose a two-level, fully implicit difference scheme for the parabolic equation (13). The obtained scheme is unconditionally stable for arbitrary ratios of spatial grid widths and time step lengths. It is mass balance preserving and, as careful investigations showed, its numerical diffusion is small compared to the really acting dispersions as described by (14).

5. Computational realization: Program LAKE

For the computational realization of the various numerical models described above, at the Karl Weierstrass Institute of Mathematics the computer program package LAKE has been created. LAKE (developed by G. Stoyan and the author together with H. Baumert from the Institute of Water Management, Berlin, G.D.R.) is a system of about 35 modules (all in FORTRAN, originally intended for running on the computer EC 1040, but fully portable to other machines) which perform particular tasks and can be combined by a "main program" according to a specific scenario. During a typical run of LAKE the following tasks have to be accomplished in turn:

- read-in of data (geometrical data of the lake, depth coordinates, boundary values of stream function, direction and intensity of wind, etc.),

- internal aggregation of data,
- set-up of system of discretized equations for stream function,
- solution by sparse-matrix techniques,
- calculation of volume streams,
- calculation of average stream velocities,
- calculation of stream velocities in prescribed depths,
- (if requested:) computation of certain drifting body trajectories,
- computation of coefficients of mass transport equation,
- set-up of system of discretized equations for mass transport,
- solution by sparse matrix techniques (repeatedly, for each time step),
- (if requested:) computation of estimate for mean detention time according to formula (17),
- output of computed results by printer (various printing regimes are available which arrange the computed data according to the geometric shape of the lake),
- output of computed results by plotter (equilines for stream function and mass concentrations, arrow fields for velocities),
- storing of computed results on external files for subsequent analysis.

The system LAKE is applicable to lakes with arbitrary geometry, arbitrary depth distribution, arbitrary distribution of inflow and outflow links. "Main programs" for various standard situations are available.

The package was handed over to the Institute of Water Management, Berlin, G.D.R., and is there, in different versions, available by licence.

6. Computational examples

The program LAKE has been applied to the numerical simulation of various circulation and mass transport scenarios in lakes belonging to the Spree river system as well as in shallow bays on the Baltic Sea. In order to give an impression of the efficiency of the system the technical data of two test runs will be reported below.

(i) A special (fictive) mass transport scenario has been computed for the Müggelsee (size: about 3 x 5 km, maximal depth: 8 m). Discretization with a grid width of 100 m led to a discrete model with 803 points (704 inner points, 99 boundary points). On the EC 1040, the computation of the stream function and subsequent calculation of some velocity profiles took about 1 min CPU time, 1250 time steps of integration of the mass transport equation (with a time step length of 12 min, such covering a time interval of 250 hours and confirming the well-known phenomenon of a global circulation with a period of 70 ... 90 hours) took about 20 min CPU time (required main storage: 460 KByte).

(ii) Estimates for mean detention times have been computed for the Schwielochsee assuming a (fictive) dislocation of the Spree inflow. For a discrete model with 1286 points (950 inner points, 336 boundary points) a run of the integration procedure with 504 time steps of 40 min (covering a time interval of 14 days) took a CPU time of about 9 min (required main storage: 650 KByte). The results (mean detention times between 4.5 and 13.2 days for different wind situations) clearly show the strong influence of wind-induced circulations on matter transport processes in shallow lakes.

References

/1/ W.A. Barannik, N.I. Seljuk, Flow calculations for cool waters under wind influence (Russian). Problemy ochrany wod 8 (1977) 49-63.

/2/ H. Baumert, H. Rische, Über Vermischungsprozesse in Fließgewässern unter besonderer Berücksichtigung der transversalen Vermischung. Acta Hydrophysica 26 (1981) 109-148.

/3/ H.B. Fischer et al., Mixing in inland and coastal waters. Academic Press, New York 1979.

/4/ M.A. Nikolsky, V.M. Sedov, The use of the numerical modelling method for the solution of problems of the environment protection from radioactive contaminations. Proc. Internat. Conference Num. Modelling of River, Channel and Overland Flow, Bratislava 1981.

/5/ G. Stoyan, On a monotone finite difference approximation for one dimensional partial differential equations (Russian). Differencial'nye Uravnenija 18 (1982) 1257-1270.

/6/ G. Stoyan, W. Müller, H. Baumert, Numerische Simulation von wind- und durchflußinduzierten Strömungen in Flachgewässern auf der Basis des Ekman-Modells (Rechenprogramm LAKE). Acty Hydrophysica 30 (1986) 51-67.

Results of Evolutional Random Search Procedure for Load Flow Optimization in Electric Networks

H. MÜLLER G. THEIL W. WALDMANN

Institute for Electrical Power Systems and High Voltage Technology,
Technical University of Vienna, Gusshausstrasse 25, A-1040 Vienna, Austria

Abstract

A derivation-free random search procedure following the principles of biological evolution was chosen to solve the constrained reactive optimal power flow in order to overcome convergence difficulties originating from the imposed constraints of different physical dimension (scaling problems!) and from the 'fissured' solution space caused by them. Starting from an arbitrary system state the solution process is decomposed into two steps, determining first a feasible and then optimal power flow. If the pregiven range of variation of normal control parameters is unsufficient to obtain a feasible state a further facility was included which allows additional load adjustments in some prespecified nodes.

The performance of higher developed evolutionary strategies was investigated with some example networks and two real systems. Results and experiencies about the proper choice of strategy parameters which provide best performance are reported.

Introduction and Problem Formulation

Load flow calculation is a standard element in transmission network analysis for planning purposes, where the load flow equations

$$P_i + j\,Q_i = \underline{V}_i \sum_{k=1}^{n} (\underline{Y}_{ik}^{*}\,\underline{V}_k^{*}) \quad , \quad i = 1 \ldots n \tag{1}$$

have to be solved for the voltages $\underline{V}_i$, the state variables of the system. Reactive power distribution is not well known in many planning cases, on the other hand the solution is rather sensitive to it. This is a particular difficulty, especially when the system planner has to compare and evaluate several variants. Computer aid for this task should offer some freedom with regard to the quantities to be predetermined, adjusting the values automatically, so that the solution provided is feasible according to imposed constraints and comparable with regard to a - preferable global - criterion. This automatic adjustment is yield by formulation as a control problem - the optimal load flow. For the sub-case dealt here - the optimal control of r e a c t i v e power flow (with real power inputs considered fixed) - Table 1 gives an overview over the involved quantities.

Table 1. Grouping of quantities for the reactive optimal power flow.

load flow problem			given quantities **z**		dependent unknowns
control problem			predetermined fixed parameters **c**	adjustable control variables **x**	(uncontrollable) **y**
slack			δ_i	V_i	P_i , Q_i
node type	P,V	(generator)	P_i	V_i	δ_i , Q_i
	P,Q	generator, reactive power source	P_i	Q_i	δ_i , V_i
		load	P_i , Q_i		δ_i , V_i
transmission elements			transformer and line constants topology	t_1 (taps of TCT)	branch flows (currents, power)

The multivariable nonlinear constrained optimization problem reads [1], [2]: Minimize

$$f(\mathbf{x},\mathbf{y},\mathbf{c}) \overset{!}{=} \text{Min} \tag{2}$$

subject to

$$\mathbf{e}(\mathbf{x},\mathbf{y},\mathbf{c}) = \mathbf{0} \tag{3}$$

and

$$\mathbf{x}_{min} \leqq \mathbf{x} \leqq \mathbf{x}_{max} \tag{4}$$

$$\mathbf{h}(\mathbf{x},\mathbf{y},\mathbf{c}) \leqq \mathbf{0} \tag{5}$$

The equality constraints (3) correspond to the load flow equations (1) and include the demand balance. Parametric inequality constraints (4) on control variables define the allowable ranges for the quantities listed in column '**x**' in Table 1, while the functional inequality constraints (5) impose allowable ranges or upper bounds to the quantities in column '**y**' of Table 1.

Possible objectives are 1) real power losses (as a natural economic criterion):

$$f_1(\mathbf{x},\mathbf{y},\mathbf{c}) = \sum_{i=1}^{n} P_i \qquad n \text{ ... number of nodes} \tag{2a}$$

- with real power injections considered fixed this is equal to

$$\tilde{f}_1 = P_{slack} \tag{2b}$$

(compare [3]) - or 2) the variance of bus voltages (when looking for a suitable voltage profile):

$$f_2(\mathbf{x},\mathbf{y},\mathbf{c}) = \frac{1}{m}\cdot\sum_{i=1}^{m} \left(\frac{1}{n_i}\cdot\sqrt{\sum_{k=1}^{n_i} (V_{ki} - \overline{V}_i)^2} \right) \tag{2c}$$

where

$$\overline{V}_i = \frac{1}{n_i}\cdot\sum_{k=1}^{n_i} V_{ki} \tag{2d}$$

m is the number of subnetworks and n_i the number of nodes in the i'th subsystem.

It is not as much the exact objective function value, which is in the center of interest. Yet it is meaningful to optimize the reactive power flow according to transmission losses or voltage profile in the sense of a systematical search process in presence of a certain number of degrees of freedom, since this adjustment of control parameters with regard to a well defined, global criterion ensures a solution, which is not only feasible with respect to specified technical limits, but also unique and comparable.

Numerous approaches to optimal load flow have been reported so far. With regard to gradient techniques an earlier review about the state of the art is found in [4] and a more recent one in [3], while second order methods are reviewed in [5]. Very comprehensive state of the art reviews are given in [1], [2]. However, as most high voltage power systems encounter problems in reactive power and voltage management for planning one has to be aware of the nonlinear and non-convex character of reactive power optimization: i.e. the nonlinear dependence between control variables and objective, but especially the discontinuity and nonlinearity caused by imposed constraints (e.g. by branch current security limits, due to the strong nonlinearity of currents versus voltage magnitude) which fissure the hypersurface (distort the hypercontours) of the objective! Since in addition the involved quantities are of different physical dimension the suitable weighting of penalty function terms (mutually and in relation to the primary objective) poses scaling problems to gradient techniques, whose convergence may be hampered by all these features - problems, which are only partly overcome by second order approaches. Contrary to system operation, where calculation can start from an existing feasible state, it is the typical planning situation that the starting point is not in the immediate vicinity of the final solution point and very often is unfeasible by a wide margin. This fact as well as the mentioned nonlinearity/non-convexity can be problematic for compact methods too (though these may be adequate for real time use).

Therefore a complete derivation-free search procedure has been tried: The performance of higher developed evolutionary strategies [6] was investigated [7], [8], motivated by experiences about the use of evolutionary random search in it's simplest, two-membered form for power flow optimization, reported in [9], and based upon own good results, obtained with multi-membered strategies for optimal thermal generation scheduling [10].

Evolutionary Strategic Search [6]

The scheme of the random search procedure following the principles of biological evolution is illustrated in Fig. 1 together with a simple example in Fig. 2. A second example in combination with load flow optimization is shown in Fig. 4.

By means of strategy parameters alternative options can be chosen in higher developed evolutionary strategies: 'Plus' and 'Comma' strategies indicate different survive-philosophies for parents (chance of survival by success or forced natural death by senility). The population size is determined by the number of parents n_p and the number of descendants n_d in each generation. The standard deviations vector σ of the (multidimensional) distribution for random variations of control variables is responsible for mutation rate (and consequently the step length). Starting from a chosen initial variance the adaption to the space curvature of the objective function hypersurface is possible (alternatively or in combination) by adjustment of σ according to 1/5 rule of success, by random variation of σ itself, by separate adjustment of individual σ_i's (resulting in a hyperelliptic contour of constant random hit probability), with correlations between σ_i's (revolution of the hyperellipsoid with random variation of rotation angles). Finally different forms of recombination (modelling of sexual reproduction) can be chosen.

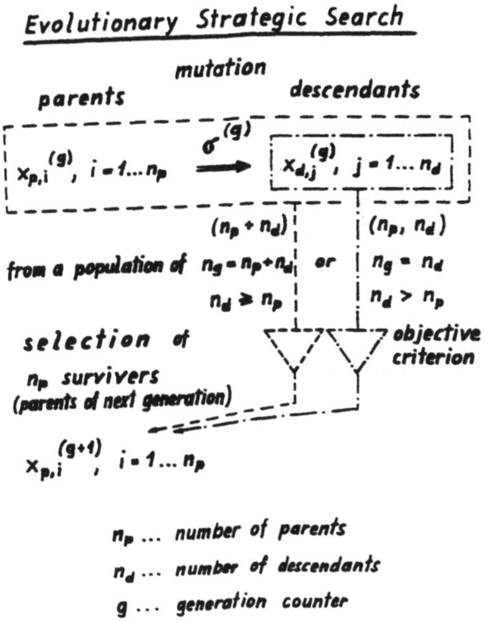

Fig. 1.

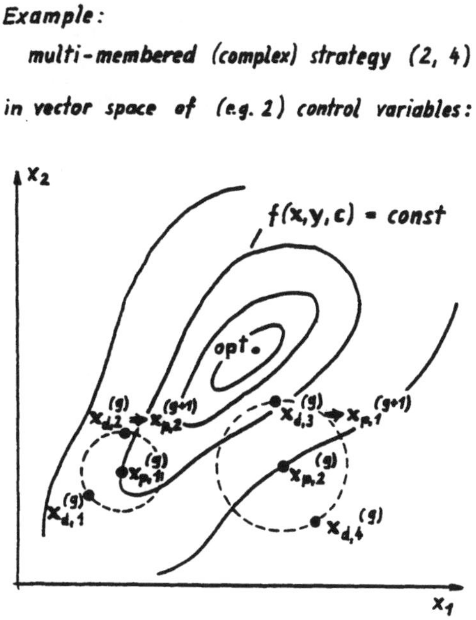

Fig. 2.

Solution Technique [8]

The computational approach is decomposed into two main steps as indicated in Fig. 3: First the search for a feasible region is performed, followed by optimization of the chosen primary objective. Beyond that the facility of load adjustments for security reasons was also included in the search process: Cases may occur, where constraints are violated and where these violations cannot be eliminated completly by adjusting the normal control parameters (reactive power injections, voltage magnitude at reactive power sources, tap sets of transformers). In a first sub-step load adjustments are determined in some prespecified nodes, which are additionally necessary to get

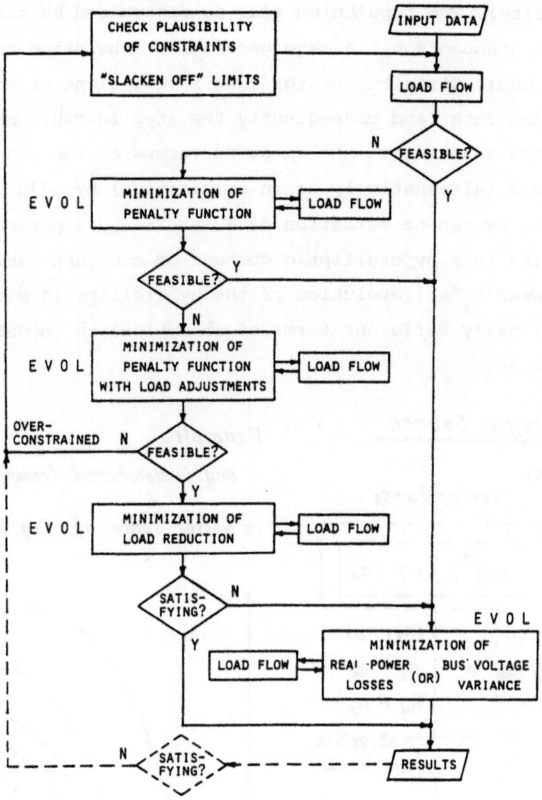

Fig. 3. Flow chart of solution procedure

the system state feasible. The necessary unavoidable amount of this load reduction is minimized then in a second sub-step (in combination with the other control adjustments).

The minimization procedures are done by complex evolutionary strategic search in all steps, wherein the assignment of objective function values to the generated individuals - as criterion for selection - is carried out via load flow calculations (using a fast load flow method with a constant solution matrix [11]).

As computational detail it is to note that some control variables (reactive power injections and tap sets of transformers) are scaled into the interval -1 to +1, while other control variables (voltage magnitudes) are scaled into the interval 0 to +1. During feasibility search penalty function values are given in linear dependence to the absolut amount of limit excess, normalized by the violated bound value.

Test Series and Results [7], [8]

Alternative combinations of strategy parameters were tested with regard to good performance by means of tests with two example networks (with sizes of 6 / 17 buses, 2 / 10 control variables, 25 / 80 bounds) and two real systems (25 / 62 buses, 9 / 15 control variables, 134 / 300 constraints). Briefly summarized the results [7], [8] were:

Fig. 4 a) - d) illustrates the movement of the parents (i.e. the best solutions within each generation) from the start to the optimum. The two voltage magnitudes V_1, V_2 of the two generator nodes of a 6 bus example network were the free control parameters in this case.

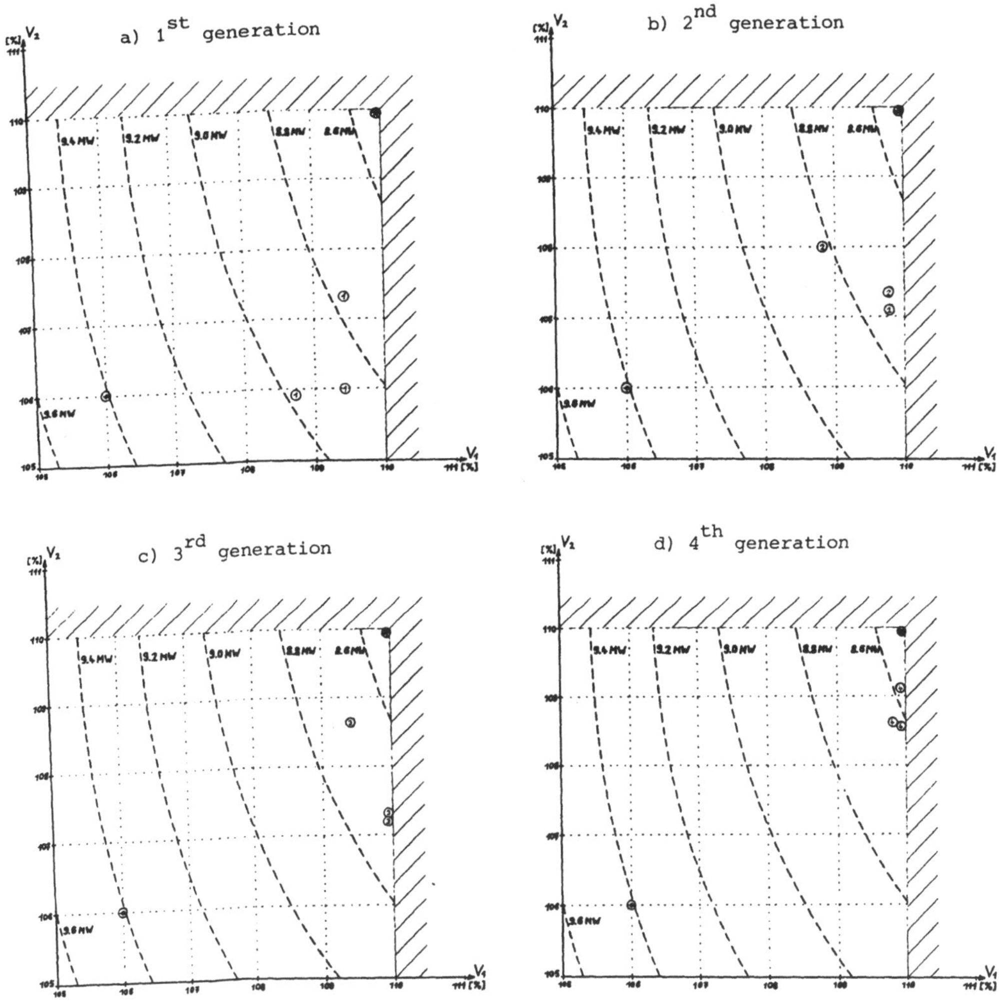

Fig. 4. Optimization progress with a (3,18)-strategy
6 bus example network
⊙ start ⊛ optimum ⓘ parents in ith generation

The improvement by using a more complex, higher developed evolutionary strategy becomes more significant for larger systems (with more control variables and limits). Yet it is disadvantageous to choose a too great population size since this slows down the adaption process to the problem structure. A (3,18)-strategy and allowance of recombinations turned out to be a good choice for both secure and fast convergence. This is indicated in Table 2 as well as the result that correlation (revolution of hyperellipsoid) together with a not too great number of preference axis (i.e. about 5 seperately and individually adjusted variances) was advantageous. 10% initial variance was best suited in most cases.

Table 2. Feasibility search. Variation of descendants number.
(Example system: 17 bus 110/30 kV network - Fig. 7,
10 control variables, 80 constraints.
Execution times are on basis of a SIEMENS 7.760 computer.)

Strategy	(3, 12)	(3, 18)	(3, 24)	(3, 30)	(3, 18) + Corr.
CPU-seconds	28.0	12.0	20.0	29.0	3.0
Iterations	151	65	107	157	15

Fig. 5 gives an example of the search performance with respect to the real 25 bus transmission network: The free control parameters were either the reactive power injections (a) or the voltage magnitudes (b) of 8 generator nodes. Fig. 6 shows the loss

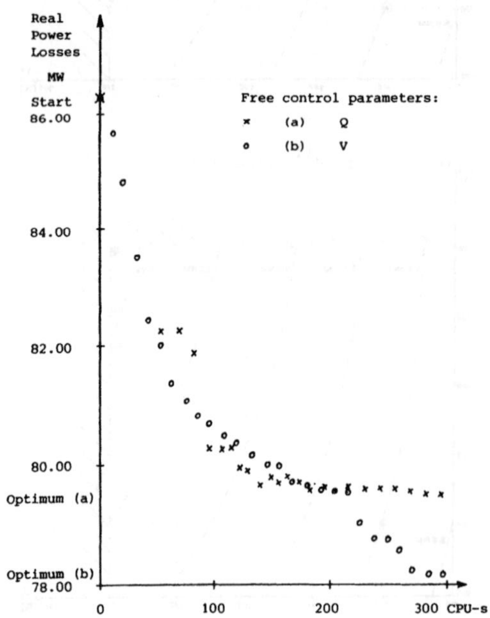

Fig. 5. Loss minimization
Real 25 bus network
(3,18)-strategy

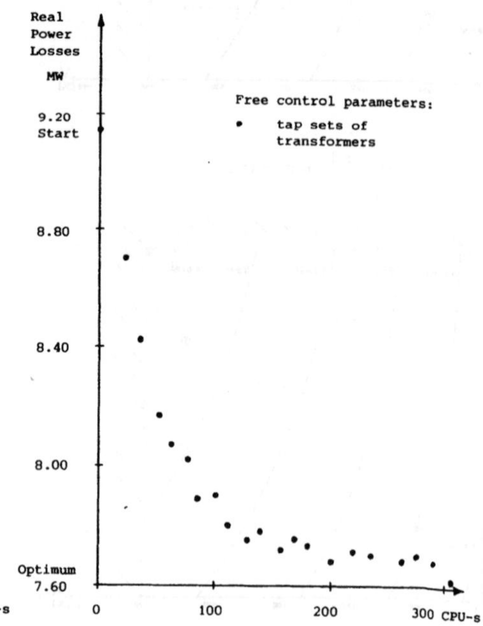

Fig. 6. Loss minimization
Real 62 bus network
(3,18)-strategy

minimization progress for the real 62 bus municipal network: In this case the free control parameters were the tap sets of 15 transformers. (In Fig. 5 and Fig. 6 the execution times are on basis of a VAX 11/780 computer.)

Only minor differences were found between minimization of losses and bus voltage variance optimization - Fig. 8 illustrates this with the results for the 17 bus 110/30 kV example network (Fig. 7).

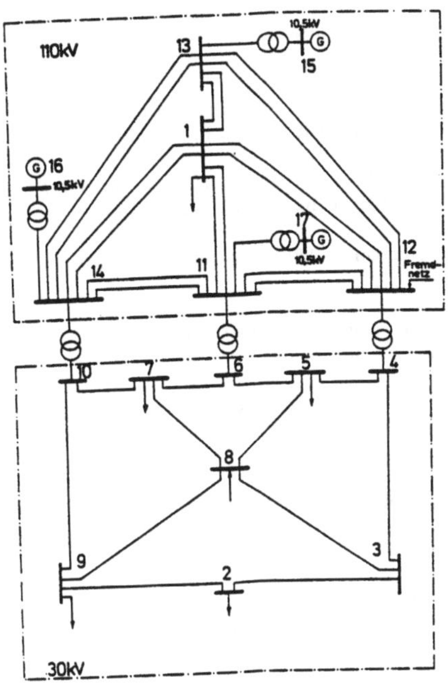

Fig. 7. Example system:
17 bus 110/30 kV network

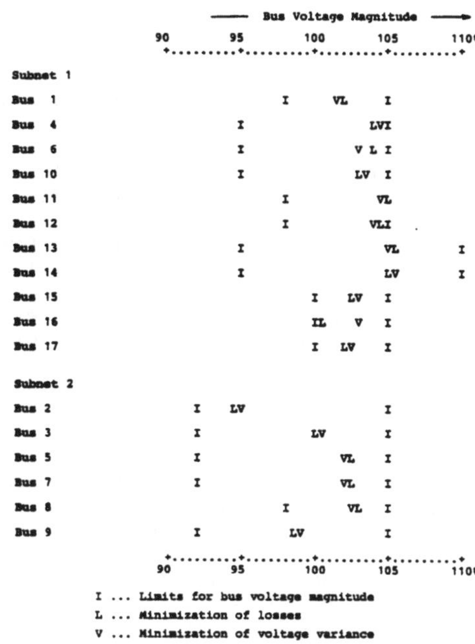

Fig. 8. Voltage profiles obtained with different objective functions (17 bus 110/30 kV example network)

Conclusions

Advantages of complex evolutionary search are: It poses no requirements on the objective function or (type of) constraints as it is a derivation-free method. It's good adaption to the curvature of objective hypersurface and it's feature of being able to follow more than one single trajectory towards the optimum makes it to a 'robust' algorithm with good convergence characteristic even in case of a 'fissured' hypersurface. The process needs no feasible initial state to start from and is suited to enter a feasible region relatively fast even in cases of rather restricted solution spaces.

A certain drawback is the computing time demand due to the many recalculations of objective function value. However, this can be reduced by an order of magnitude - when parallel processing is available [12] - with parallel treatment of the individuals within one generation.

Acknoledgement

The authors would like to thank the Austrian Science Foundation, which has sponsored the work to this report as part of the research project no. P5068 'Methods for Computer Aided Operation and Planning of Electric Power Systems'.

References

1. Carpentier, J.: Optimal power flows. Electrical Power & Energy Systems, vol.1 (1979) no.1, pp.3-15.

2. Carpentier, J.; Merlin, A.: Optimization methods in planning and operation of power systems. Electrical Power & Energy Systems, vol.4 (1982) no.1, pp.11-18.

3. Burchett, R.C.; Happ, H.H.; Vierath, D.R.; Wirgau, K.A.: Developments in Optimal Power Flow. IEEE Trans. Pwr.App.Syst., vol.PAS-101 (1982) no.2, pp.406-414.

4. Dommel, H.W.; Tinney, W.F.: Optimal Power Flow Solutions. IEEE Trans. Pwr.App. Syst., vol. PAS-87 (1968) no.10, pp.1866-1876.

5. Giras, T.C.; Talukdar, S.N.: Quasi-Newton method for optimal power flows. Electrical Power & Energy Systems, vol.3 (1981) no.2, pp.59-64.

6. Schwefel, H.-P.: Numerische Optimierung von Computer-Modellen mittels der Evolutionsstrategie. 1st edn. Basel, Stuttgart: Birkhäuser, 1977. (Interdisciplinary systems research series; ISR 26)

7. Müller, H.; Pollhammer, G.: Evolutionsstrategische Lastflussoptimierung. Research report FB 4/1983 of the Institute for Electrical Power Systems and High Voltage Technology, Technical University of Vienna.

8. Theil, G.; Waldmann, W.: Kurvenanpassung und Lastflussoptimierung mittels Evolutionsstrategie. Research report FB 2/1984 of the Institute for Electrical Power Systems and High Voltage Technology, Technical University of Vienna.

9. Fuchs, F.; Maier, H.A.: Optimierung des Lastflusses in elektrischen Energieversorgungsnetzen mittels Zufallszahlen. Archiv für Elektrotechnik, vol.66 (1983) no.2, pp.75-83 (part 1) and pp.85-94 (part 2).

10. Painz, R.: Wöchentliche Brennstoffaufteilung eines thermischen Kraftwerkssystems. Diploma graduation thesis at the Institute for Electrical Power Systems and High Voltage Technology, Technical University of Vienna, 1981.

11. Theil, G.: Schnelle Lastflussrechnungen zweiter Ordnung. Elektrotechnik und Maschinenbau (E und M), vol.101 (1984) no.9, pp.415-420.

12. Happ, H.H.: Parallel processing in power systems. Electrical Power & Energy Systems, vol.4 (1982) no.1, pp.37-41.

ON THE OPTIMAL COOLING OF THE STEEL DURING CONTINUOUS CASTING

P. Neittaanmäki
Department of Mathematics
University of Jyväskylä
Seminaarinkatu 15
SF-40100 Jyväskylä, Finland

1. Introduction

A considerable amount of the world's steel is now produced by the continuous casting process [4]. The complexity of this process is such that until recently innovations to machine design could only be tested by full scale runs on production machine. This is extremely costly and in many cases production cannot be delayed for such tests to occur. However, numerical models of the continuous casting process have recently been developed, [1, 3, 4, 5, 9, 10], which offer alternative means of testing new designs. In the continuous casting the cooling is used to accelerate steel solidification and to strenghten the solified shell. Improper water quantity or distribution may form rapid changes of temperature. The strand is to be cooled down according to the pattern which depends on steel quality, product size, casting speed and machine design.

The essential features of a continuous casting machine are shown in Figure 1.1.

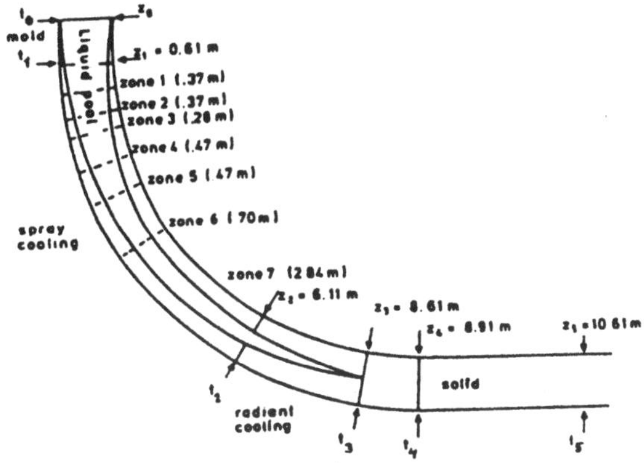

Figure 1.1.

The molten steel is fed from a reservoir into a water cooled mold. The product is supported by rolls after mold and cooled by water sprays during the secondary cooling (zones 1-7 in Figure 1.1) and through the radiant cooling. The unbending point is z_4 meters from meniscus ($z_4 - z_1$ meters from the bottom of the mold). The cutting point is z_5 meters from the meniscus.

In our mathematical model we shall assume that the casting speed V is constant (in fact it could be optimized as well). Consequently $t_i = z_i/V$, where t_i is time and z_i distance from the meniscus (see Figure 1.1). In order to define the water cooling strategy and spray system design the heat flow and solidification rate must be known. The problem for finding the optimal cooling strategy for water spray cooling leads us to a boundary control problem for heat transfer coefficients. The state is governed by a nonlinear parabolic-type equation (with phase changes) with nonlinear boundary conditions. By this model the heat transfer coefficients for the secondary cooling of the steel billets are determined in order to get a desirable surface temperature. Two different models are presented. The first method (chapter 3) is based on a combination of Dirichlet boundary data (with desired surface temperature) and of the Neumann boundary condition (2.2) in the secondary cooling. The second method (chapter 4), which is based on the optimal control theory, is more sophisticated [1, 5, 6, 9, 10]. It takes into account the metallurgical constraints as well. Numerical results on the use of both methods are given.

2. The state problem

During the continuous casting process the temperature field $T = T(x,t)$ in the cross-section Ω of the billet at time t is governed by the nonlinear equation of parabolic type

$$(2.1) \qquad \frac{\partial}{\partial t} H(T) = \operatorname{div} \operatorname{grad} K(T) \quad \text{in } \Omega \times (0,t_5) .$$

The heat transfer from the strand to the environment is assumed to happen by convection and by radiation:

$$(2.2) \qquad -k(T)\frac{\partial}{\partial n} T = \begin{cases} h_m(T - T_{mold}) & \text{on } \Gamma \times [0,t_1] , \\ h(T - T_{H_2O}) + \sigma\varepsilon(T^4 - T_{ext}^4) & \text{on } \Gamma \times (t_1,t_2] , \\ \sigma\varepsilon(T^4 - T_{ext}^4) & \text{on } \Gamma \times (t_2,t_5] . \end{cases}$$

For $t = 0$ we assume that the temperature distribution at the meniscus to be uniform and equal to the temperature of the incoming metal stream:

$$(2.3) \qquad T(x,t) = T_1 \quad \text{for } t = t_0 = 0 \text{ and } x \in \Omega .$$

In the equation (2.1) $H(T)$ and $K(T)$ are the enthalpy and the Kirchoff transformation, respectively. They are defined by the following equations:

$$(2.4) \qquad H(T) = \int_0^T \rho c - \rho L(T_l - T_s)^{-1}\, ds \;, \qquad K(T) = \int_0^T k(s)\, ds \,.$$

The list of symbols used is given in Table 2.1.

t	- time	$k(T)$	- thermal conductivity (see Figure 2.2)
T	- temperature	$c(T)$	- specific heat (see Figure 2.2)
T_{mold}	- ambient mold temperature	ρ	- density
T_{H_2O}	- spray water temperature	L	- latent heat
T_l	- liquidus temperature	σ	- Stefan-Bolzmann constant
T_s	- solidus temperature	ε	- emissity
$\frac{\partial}{\partial n}T$	- temperature gradient to strand surface	V	- casting speed
H	- enthalpy	x^+	- the positive part of x; $x^+ = x$, for $x \geq 0$ and $x^+ = 0$, for $x < 0$
h_m	- heat transfer coefficient in the mold	Ω	- cross-section of a billet
h	- water heat transfer coefficient	Γ	- boundary of Ω
w	- water flux		

Table 2.1. List of symbols.

The specific heat and the thermal conductivity of the steel are defined separately for three temperature regions, inside which they are assumed to be constant. These three regions are liquid region, the mushy region and the solid shell. The density of the steel is assumed to be independent on temperature. These assumptions imply by equation (2.4) that K and H are piecewise linear functions of T as indicated in Figure 2.2.

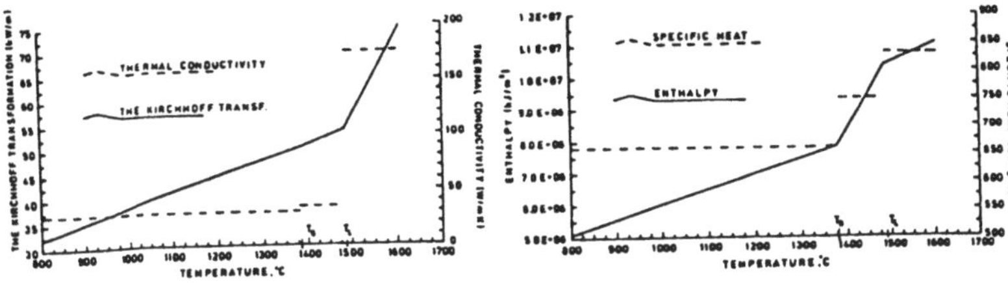

Figure 2.2. $K(T)$ and $H(T)$

The assumed heat transfer coefficients in the midface and corner of the mold are given in Figure 2.3.

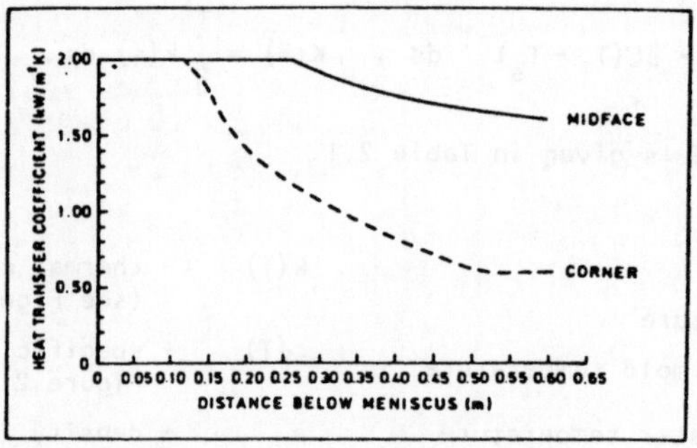

Figure 2.3. $h_m(\cdot,t)$, $t \in (0,t_1]$

Other values for $h_m(\cdot,t)$ are obtained by interpolation. The air gap formation reduces the heat transfer coefficient in the corner.

The problem (2.1) - (2.3) is multiphase Stefan problem. The solvability of similar system is discussed in [8] for example.

3. The billet with the desired surface temperature

Suppose that during the secondary cooling we want to have the surface temperature T_d $(= T_d(x,t))$. In order to find the water spray cooling which gives this surface temperature field we apply the boundary condition (2.2) and the relation between the spray heat transfer coefficients and the water flux. This relation is linear or nonlinear [2], depending on the model, (strand size, machine design). Suppose that

(3.1) $$h = 0.14\ w + 0.1\ , \quad ([w] = 1/m^2\,s)\ .$$

The algorithm for finding "optimal" h is given below:

Algorithm 3.1.

Step 1. For a given surface temperature T_d solve the problem (2.1) - (2.3) with the Dirichlet boundary data.

(3.2) $$T = T_d \qquad \text{on} \quad \Gamma \times (t_1,t_2]\ .$$

Step 2. Determine the water heat transfer coefficient h from the relation

$$(3.3) \qquad -k(T)\frac{\partial}{\partial n}T = h(T - T_{H_2O}) + \sigma\varepsilon(T^4 - T_{ext}^4) \qquad \text{on } \Gamma \times (t_1,t_2] .$$

Step 3. Determine the water flux w from the relation (3.1) (or from some other related relation).

We shall illustrate the usefulness of above algorithm by an example.

Example 3.1. Consider the continuous casting machine shown in Figure 1.1. Suppose the casting speed $V = 2.8$ m/min. The material and casting data used are given in Table 3.1.

Carbon content	0,5 %	ρ	7200 kg/m^3
$T_I = T_0$	1485 ^oC	L	272 kJ/kg
T_s	1377 ^oC	σ	$5{,}67 \times 10^{-8}$ W/m^2K^4
T_{mold}	80 ^oC	ε	0,8
T_{H_2O}	27 ^oC	Cross-section	100 × 100 mm
T_{ext} spray z.	97 ^oC	T_{ext} rad.z.	437 ^oC

Table 3.1. Material and casting data used.

The desired surface temperature T_d is given on the corner and midface by the equations

$$T_d(x,t) = 0.54\ t + 904.43 \qquad , t \in (t_1,t_4)$$

$$T_d(x,t) = -0.16 \cdot 10^{-3}\ t + 1100.03 \qquad , t \in (t_1,t_4) .$$

For the other points of Γ T_d is given by the parabolic fitting.

Figure 3.2 shows the results.

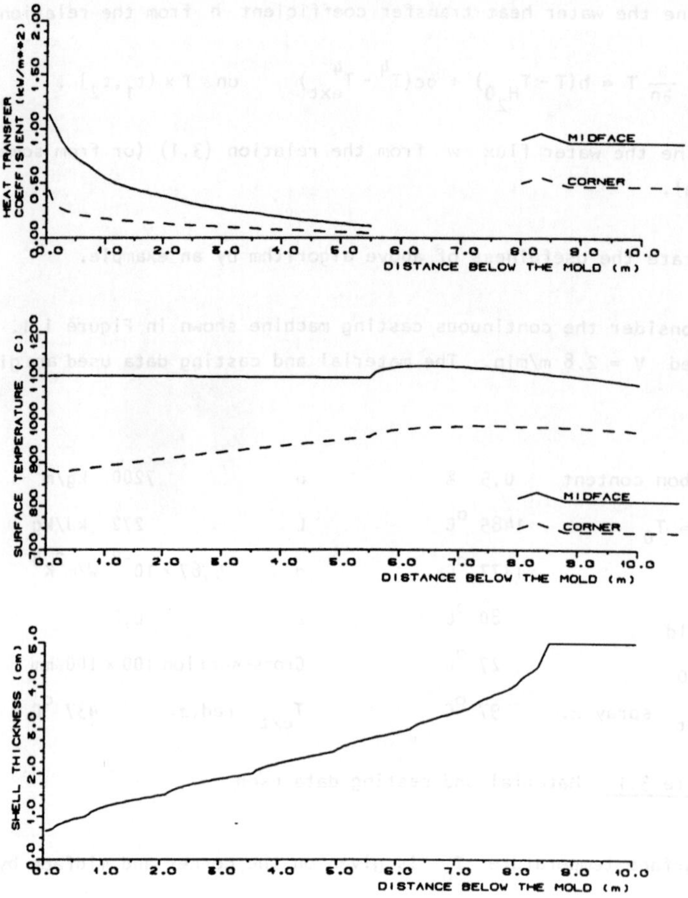

Figure 3.2.

4. Optimization of heat transfer coefficient

As we can see in Example 3.2 the metallurgical constraints are not satisfied. Suppose we want to have the average surface temperature as near as possible to given temperature T_d and moreover following metallurgical constraints must be satisfied:

(1) The surface temperature along the caster must be between the lower bound T_{min} and the upper bound T_{max}.

(2) The temperature variation along the surface must lie in range [CN,CL]; $CN < 0$ and $CL > 0$ some given constants.

For a constant casting speed above constraints give that the length of the liquid pool is less than z_3 : $T \leq T_s$ for $t \geq t_3$.

We are led to the control problem of finding optimal heat transfer coefficients in boundary condition (2.2). The relation (3.1) then gives information about corresponding water flux.

Let $\mathcal{U}_{ad}$ denote the set of admissible controls (heat transfer coefficients):

$$\mathcal{U}_{ad} = \{h \mid h^i_{min} \leq h^i \leq h^i_{max} \quad , \quad i = 1, \dots, 7\} \tag{4.1}$$

where h^i_{min} and h^i_{max} are the minimum and maximum possible heat transfer coefficients in the spray zone i. For the metallurgical constraints (1) - (2) we apply the penalty technique with the nonmeeting of the constraints. Thence our control problem is

$$\min_{h \in \mathcal{U}_{ad}} J(T(h);h) \tag{4.2}$$

where T solves the state problem (2.1) - (2.3). The cost function J is a sum of partial costs $J = J_0 + J_1 + J_2$, where

$$J_0 = \int_{t_1}^{t_4} \int_{\Gamma} [T(x,t) - T_d(x,t)]^2 \, dx \, dt \, ,$$

$$J_1 = C_1 \int_{t_1}^{t_5} \int_{\Gamma} [(T(x,t) - T_{max})^+]^2 + [(T_{min} - T(x,t))^+]^2 \, dx \, dt \, ,$$

$$J_2 = \frac{C_2}{2} \int_{t_1}^{t_5} \int_{\Gamma} \{[\frac{\partial}{\partial t} T(x,t) - CP]^+\}^2 + \{[CN - \frac{\partial}{\partial t} T(x,t)]^+\}^2 \, dx \, dt \, .$$

Above, C_i are constants which express the relative importance of criterion J_i with respect to other ones.

The optimal control problem (4.2) can be solved by the steepest descent method (in fact conjugate gradient version of it with subgradient).

Algorithm 4.1. (For solving problem (4.2).)

Step 1. Give an initial guess $h_0 \in \mathcal{U}_{ad}$ (possible the one obtained by Algorithm 3.1). Set $n = 0$.

Step 2. For a given $h = h_n$ solve the state problem (2.1) - (2.3) in order to obtain $T_n = T(h_n)$.

Step 3. Compute, if possible, the gradient of J and set

$$G_n = \nabla_{h_n} J(T(h_n)) .$$

Step 4. Set $h_{n+1} = h_n - \rho_n G_n$ for a real parameter $\rho_n > 0$ so that $h_{n+1} \in \mathcal{U}_{ad}$.

Step 5. Perform the terminal tests. If necessary set $n = n + 1$ and GO TO Step 2.

The convergence of above type algorithm is proved in [7]. For finding a numerical solution of (2.1) - (2.3) the discretization can be employed by the implicit method in time and by finite element method in spatial domain [5]. The computation of G_n leads to the adjoint state for (2.1) - (2.3), (4.2), which can be solved as well by finite difference (backward Euler) and finite element method ([5, 6, 9, 11]).

Example 4.2. Consider the same situation as in Example 3.1. The cooling criteria can be seen in Table 4.1.

Criterion	Aimed value
The desired surface temperature T_d	as in Example 3.1
Maximum reheating rate on the surface	+0.5 °C/sec
Minimum cooling rate on the surface	-0.5 °C/sec
Minimum surface temperature along the secondary cooling zones	900 °C
Maximum surface temperature along the secondary cooling zones	1200 °C

Table 4.1. Cooling criteria used.

The lower bounds h^i_{min} in constraints are 0.05 for $i = 1, \dots, 7$ and the upper bounds h^i_{max} are 1.2, 1.2, 1.2, 1.2, 1.2, 0.9 and 0.5 kW/m^2K. The lenghts of the cooling zones are 0.37, 0.37, 0.28, 0.47, 0.47, 0.70 and 2.84 meters.

Figure 4.2 shows the results.

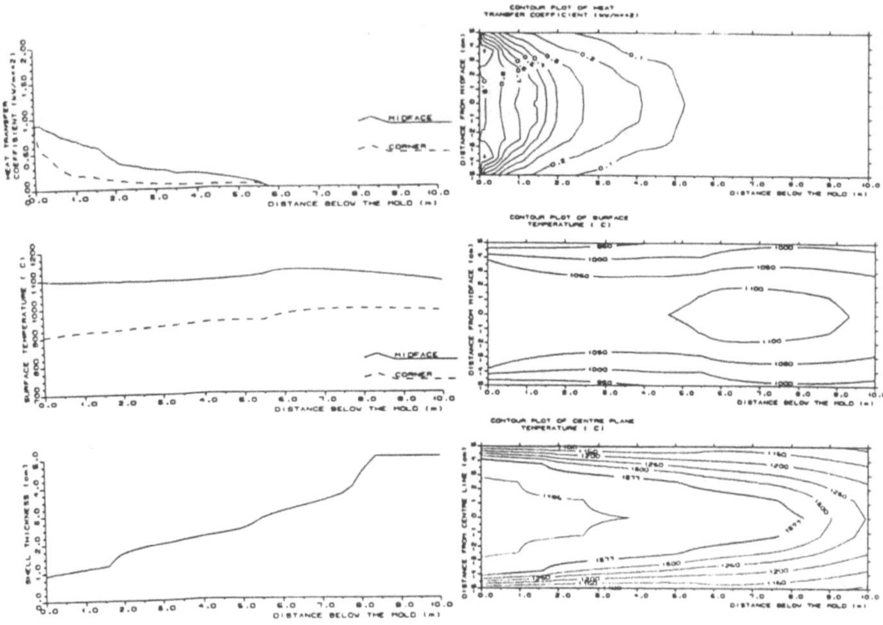

Figure 4.2.

We find that the surface temperature is very near to T_d as well in the radiant cooling zone. Moreover, the length of the liquid pool is less than 8.5 m and the surface temperature is more than 1000 ^{o}C (high temperature solution) at the unbending point. By the algorithm 3.1 it is not possible to control these facts. The idea of Algorithm 3.1 is just to give quickly a rough approximation for the control. Algorithm 4.1 makes it possible to improve this control if necessary.

Acknowledgements. This research was supported by the Academy of Finland and by TEKES. The author is indebted to to Mr. E. Laitinen for his assistance in numerical tests.

References

[1] Birat, J.P., Larrecq, M., Le Bon, A., Jeanneau, M., Poupon, M. and Senaneuch, D.: 'Control of secondary cooling in continuous casting of plate grades.' IRSID, ACI 83 RE 1009, Août 1983.

[2] Brimacombe, J.K., Samarasekera, I.V. and Lait, J.E.: Continuous Casting, Vol 2, Book Crafters, Ins., Chelsea, MI, 1984.

[3] Holappa, L., Louhenkilpi, S., Laitinen E. and Neittaanmäki, P.: 'Optimization of the secondary cooling in the continuous casting of billets', In Proc. of 24th Annual Conference of Metallurgists, 18.-21. August 1985, Vancouver.

[4] Morgan, K., Lewis, R.W. and Williams, J.R.: 'Thermal stress analysis of a novel continuous casting process', MAFELAP III, 1978, J. Whiteman (ed.) Academic Press 1979.

[5] Neittaanmäki, P.: 'On the control of cooling during continuous casting', In Proc. of the 4th Int. Confr. on Numerical Methods in Thermal Problems, R.W. Lewis and K. Morgan (eds.) Pineridge Press, Swansea, 1985, 240-249.

[6] Neittaanmäki, P. and Tiba, D.: 'On the finite element approximation of the boundary control for two-phase Stefan problems', Analysis and Optimization of Systems, Ed. Bensoussan, A. and Lions, J.L. Lecture Notes in Control and Information Sciences, Springer Verlag, 1984, 356-370.

[7] Neittaanmäki, P. and Tiba, D.: A steepest descent method for approximation of the boundary control in two-phase Stefan problem, Lappeenranta University of Technology, Dept. of Physics and Math., Research report 1/85.

[8] Niezgódka, M. and Pawlow, I.: A generalized Stefan problem in several space variables, Appl. Math. Optim. 9, 1983, 193-224.

[9] Saguez, C.: 'Contrôle optimal de systemès à frontiere libre', Thèse, Univ. Technologique Compiegne, 1980.

[10] Saquez, C.: 'Optimal control of free-boundary problems', in 12th IFIP-conference on system modelling and optimization, Budapest, September 2-6, 1985.

[11] Zlamal, M.: 'A finite element solution of a nonlinear heat equation', RAIRO, Num. Anal. 14, 1980, 203-216.

GOOD LATTICE POINTS FOR QUASIRANDOM SEARCH METHODS

HARALD NIEDERREITER
Mathematical Institute
Austrian Academy of Sciences
Dr. Ignaz-Seipel-Platz 2
A-1010 Vienna/Austria

1. Introduction

Random search is a familiar method in nondifferentiable optimization (see Rubinstein [12, Ch. 7]). It is based on random sampling from the domain on which the target function is defined. If random samples are replaced by well-chosen deterministic point sets, we arrive at a so-called quasirandom search method. Quasirandom search methods were studied numerically by Niederreiter and McCurley [8] and Niederreiter and Peart [9], and a systematic theory of such methods was developed by Niederreiter [5]. Work on quasirandom search methods was also carried out by Russian mathematicians (see e.g. Sobol' [13], [14]). Quasirandom search methods have already been applied to problems in science and engineering, see e.g. an application to radiation detection in Bandyopadhyay and Sarkar [1].

In this paper we study a special class of deterministic point sets and its application to quasirandom search methods. These point sets are easy to generate and with a proper choice of parameters they are well suited for quasirandom search methods. The structure of these point sets permits their use for continuous as well as for discrete optimization problems. In Section 2 we describe these point sets and establish a general existence theorem for a good choice of parameters ("good lattice points"). Section 3 is devoted to effective constructions of good lattice points. Applications to discrete optimization are discussed in Section 4.

We briefly review the basic elements of quasirandom search methods. Let the target function f be defined on a bounded subset X of $\mathbb{R}^s$, $s \geq 1$. We select deterministic points $x_1, x_2, \ldots$ in X and use the approximation

$$M = \sup_{x \in X} f(x) \approx m_N = \max_{1 \leq n \leq N} f(x_n).$$

The convergence of m_N to M as $N \to \infty$ is guaranteed whenever f is continuous and the sequence x_n, $n = 1,2,\ldots$, is dense in X. To obtain an effective error bound, we introduce the metric

$$d(x,y) = \| x - y \|_\infty \quad \text{for} \quad x,y \in X,$$

where $\|z\|_\infty$ denotes the maximum absolute value of the coordinates of $z \in \mathbb{R}^s$. Define the modulus of continuity

$$\omega(t) = \sup_{\substack{x,y\in X \\ d(x,y)\leq t}} |f(x) - f(y)| \quad \text{for } t \geq 0$$

and the dispersion

$$d_N = \sup_{x\in X} \min_{1\leq n\leq N} d(x,x_n). \tag{1}$$

Then by [5] we have the error bound

$$M - m_N \leq \omega(d_N).$$

Since $\omega(t)$ is a nondecreasing function of t, we see that $x_1,\ldots,x_N$ should be chosen in such a way that their dispersion d_N is small. We note that for any N points in X we have

$$d_N \geq c(X)N^{-1/s}, \tag{2}$$

where the constant $c(X) > 0$ depends only on X, and this lower bound is best possible (see [5]). The method described above is the crudest form of quasirandom search. More refined search strategies are obtained by using adaptive methods (see Niederreiter [6]) or the method of localization of search (see Niederreiter [6] and Niederreiter and Peart [10]). However, in all these techniques it is essential to work with point sets of small dispersion. Therefore, the problem of constructing low-dispersion point sets that are easy to implement is crucial in this area.

2. Good lattice points

We normalize the domain X by letting $X = [0,1]^s$, the s-dimensional unit cube. For other domains, low-dispersion point sets can be obtained by suitable transformations of low-dispersion point sets in $[0,1]^s$ (see Niederreiter and Peart [11]).

Let N be a large integer and let $g = (g_1,\ldots,g_s)$ be an s-dimensional lattice point, i.e. a point with integer coordinates. Then we define the points

$$x_n = (\{\tfrac{n}{N} g_1\},\ldots,\{\tfrac{n}{N} g_s\}) \in [0,1]^s \quad \text{for } 1\leq n\leq N, \tag{3}$$

where $\{t\} = t - \lfloor t\rfloor$ and $\lfloor t\rfloor$ is the greatest integer $\leq t$. Thus, the points x_n are obtained by forming multiples of the point $N^{-1}g$ and reducing them modulo 1. The suitability of the point set (3) for quasirandom search methods depends on the choice of the lattice point g. We call g a good lattice point (modulo N) if the points $x_1,\ldots,x_N$ have a small dispersion d_N.

The dispersion of the point set (3) can be bounded by the inequality in Theorem 1. We need some definitions and a lemma. Let

$$C(N) = \{h: -N/2 < h \leq N/2 \text{ and } h \text{ an integer}\},$$
$$C_s(N) = \{(h_1,\ldots,h_s): h_j \in C(N) \quad \text{for } 1\leq j\leq s\},$$

and let $C_s^*(N)$ be the set $C_s(N)$ with the origin deleted. For $h\in C(N)$ put

$$r(h,N) = \begin{cases} 1 & \text{if } h = 0, \\ \dfrac{1}{N \sin(\pi|h|/N)} & \text{if } h \neq 0, \end{cases}$$

and for $h = (h_1,\dots,h_s) \in C_s(N)$ put

$$R(h,N) = \prod_{j=1}^{s} r(h_j,N).$$

For real t we set $e(t) = e^{2\pi i t}$. Let $\langle u,v \rangle$ denote the standard inner product of $u,v \in \mathbb{R}^s$.

Lemma 1. Let $v_1,\dots,v_k$ be s-dimensional lattice points and $N \geq 2$ an integer. Then the dispersion d_k of the points $\frac{1}{N} v_1,\dots,\frac{1}{N} v_k$ (reduced modulo 1 to lie in $[0,1]^s$) satisfies

$$d_k \leq \frac{1}{N} + \Big(\sum_{h \in C_s^*(N)} R(h,N) \, \Big|\frac{1}{k} \sum_{n=1}^{k} e\Big(\frac{1}{N} \langle h,v_n \rangle\Big)\Big|\Big)^{1/s}.$$

Proof. Put $w_n = \frac{1}{N} v_n$ (reduced modulo 1), then from (1) and the compactness of $X = [0,1]^s$ it follows that there exists $y \in [0,1]^s$ with

$$d_k = \min_{1 \leq n \leq k} d(y,w_n).$$

Consequently, there exists an open cube C in $[0,1]^s$ of edge length $d/N \geq d_k$ and with the coordinates of all vertices being rationals with denominator N, say

$$C = \prod_{j=1}^{s} \Big(\frac{c_j}{N}, \frac{c_j+d}{N}\Big) \subseteq [0,1]^s,$$

such that C does not contain any point w_n. For $a_j \in \{0,1,\dots,N-1\}$ put

$$\psi\Big(\frac{a_1}{N},\dots,\frac{a_s}{N}\Big) = \psi_1\Big(\frac{a_1}{N}\Big)\cdots\psi_s\Big(\frac{a_s}{N}\Big),$$

where

$$\psi_j\Big(\frac{a_j}{N}\Big) = \begin{cases} 0 & \text{for } 0 \leq a_j \leq c_j \quad \text{or} \quad c_j + d \leq a_j \leq N-1, \\ 1 & \text{for } c_j < a_j < c_j + d. \end{cases}$$

By discrete Fourier expansion,

$$\psi_j\Big(\frac{a_j}{N}\Big) = \sum_{t_j=c_j+1}^{c_j+d-1} \frac{1}{N} \sum_{h_j \in C(N)} e\Big(\frac{1}{N} h_j(a_j - t_j)\Big)$$

$$= \sum_{h_j \in C(N)} F_j(h_j) \, e\Big(\frac{1}{N} h_j a_j\Big)$$

with

$$F_j(h_j) = \frac{1}{N} \sum_{t_j=c_j+1}^{c_j+d-1} e(-\frac{1}{N} h_j t_j).$$

Put $B = F_j(0) = (d-1)/N$ and $F(h) = F_1(h_1)\ldots F_s(h_s)$ for $h = (h_1,\ldots,h_s)$. Then

$$\psi(\frac{a_1}{N},\ldots,\frac{a_s}{N}) = \sum_{h\in C_s(N)} F(h)\, e(\frac{1}{N}\langle h, a\rangle)$$

$$= B^s + \sum_{h\in C_s^*(N)} F(h)\, e(\frac{1}{N}\langle h,a\rangle).$$

Now $\psi(w_n) = 0$ for all n, and so

$$0 = \sum_{n=1}^{k} \psi(w_n) = kB^s + \sum_{h\in C_s^*(N)} F(h) \sum_{n=1}^{k} e(\frac{1}{N}\langle h,v_n\rangle).$$

Consequently,

$$d_k \leq \frac{1}{N} + B \leq \frac{1}{N} + (\sum_{h\in C_s^*(N)} |F(h)|\, |\frac{1}{k}\sum_{n=1}^{k} e(\frac{1}{N}\langle h,v_n\rangle)|)^{1/s}.$$

For $h_j = 0$ we have $|F_j(0)| \leq 1 = r(0,N)$ and for $h_j \neq 0$ we have

$$|F_j(h_j)| \leq \frac{2}{N|e(h_j/N)-1|} = r(h_j,N),$$

hence the result of the lemma follows. □

Theorem 1. The dispersion d_N of the point set (3) satisfies

$$d_N \leq \frac{1}{N} + A(g,N)^{1/s},$$

where

$$A(g,N) = \sum_{\substack{h\in C_s^*(N) \\ \langle h,g\rangle \equiv 0 \bmod N}} R(h,N).$$

Proof. We apply Lemma 1 with $v_n = ng$ for $1 \leq n \leq N$. The desired result follows from the observation that

$$\frac{1}{N}\sum_{n=1}^{N} e(\frac{1}{N}\langle h,ng\rangle) = \frac{1}{N}\sum_{n=1}^{N} e(\frac{n}{N}\langle h,g\rangle) = \begin{cases} 1 & \text{if } \langle h,g\rangle \equiv 0 \bmod N, \\ 0 & \text{if otherwise.} \end{cases}$$ □

Theorem 2. For all $s \geq 1$ and $N \geq 2$ there exists an s-dimensional lattice point g such that the dispersion d_N of the corresponding point set (3) satisfies

$$d_N < cN^{-1/s} \log N$$

with an absolute constant $c > 0$.

Proof. The case $s = 1$ is trivial. For $s \geq 2$ the result is obtained from Theorem 1 and the following existence theorem shown in [7, Theorem 3]: if N is a prime power, there exists a g with

$$A(g,N) < \frac{1}{N}(0.9 + \log N)^s,$$

and if N is not a prime power, there exists a g with

$$A(g,N) < \frac{1}{N}[(0.9 + \log N)^s + (1.24)(1.09)^s s + (1.16)(1.17)^s s].$$

3. Effective construction of good lattice points

The proof of Theorem 2 is nonconstructive. For the numerical practice it is however important to know good lattice points explicitly. A detailed discussion of the case $s = 2$ was carried out by Larcher [4]. We show in this section that for any given s there are infinitely many N for which one can effectively construct an s-dimensional lattice point g such that the dispersion d_N of the corresponding point set (3) satisfies $d_N = O(N^{-1/s})$. According to (2) we thus attain the minimal order of magnitude for the dispersion.

The construction proceeds as follows. Let $1,\theta_1,\ldots,\theta_s$ be a basis of a real algebraic number field of degree $s + 1$ over the rationals. Let N be a positive integer for which there exist integers $g_1,\ldots,g_s$ such that

$$|N\theta_j - g_j| < K_1 N^{-1/s} \quad \text{for } 1 \leq j \leq s, \tag{4}$$

where the constant K_1 depends only on $\theta_1,\ldots,\theta_s$. By Dirichlet's theorem on simultaneous approximation (see [2, p. 14]) there are infinitely many N for which (4) holds with $K_1 = 1$. Effective constructions of such N and of the corresponding lattice points $g = (g_1,\ldots,g_s)$ for suitable values of K_1 are described in [3, Ch. 1] and tables are found in the appendix of [3]. Take such an N and let $g = (g_1,\ldots,g_s)$ be the corresponding lattice point obtained from (4). For $1 \leq n \leq N$ we have

$$|n\theta_j - \frac{n}{N} g_j| < K_1 n N^{-1-(1/s)} \leq K_1 N^{-1/s}. \tag{5}$$

Let d'_N be the dispersion of the points $(\{n\theta_1\},\ldots,\{n\theta_s\})$, $1 \leq n \leq N$. Then from [6, Theorem 3] it follows that

$$d'_N \leq K_2 N^{-1/s} \tag{6}$$

with a constant K_2 depending only on $\theta_1,\ldots,\theta_s$. Let $u = (u_1,\ldots,u_s) \in [0,1]^s$ be such that

$$(K_1 + K_2)N^{-1/s} \leq u_j \leq 1 - (K_1 + K_2)N^{-1/s} \quad \text{for } 1 \leq j \leq s. \tag{7}$$

Then by (6) there exists an n with $1 \leq n \leq N$ such that

$$|\{n\theta_j\} - u_j| \leq K_2 N^{-1/s} \quad \text{for} \quad 1 \leq j \leq s.$$

By (5) this implies

$$|\{n\theta_j\} - \{\tfrac{n}{N} g_j\}| < K_1 N^{-1/s} \quad \text{for} \quad 1 \leq j \leq s,$$

hence

$$|\{\tfrac{n}{N} g_j\} - u_j| < (K_1 + K_2) N^{-1/s} \quad \text{for} \quad 1 \leq j \leq s.$$

With x_n denoting the points in (3) corresponding to g, we have thus shown

$$\min_{1 \leq n \leq N} d(u, x_n) < (K_1 + K_2) N^{-1/s}.$$

If $u \in [0,1]^s$ does not satisfy (7), then we can find a point at a distance from u at most $(K_1 + K_2)N^{-1/s}$ for which (7) holds. Consequently,

$$\min_{1 \leq n \leq N} d(u, x_n) < 2(K_1 + K_2) N^{-1/s}.$$

Altogether we have shown that the dispersion d_N of $x_1, \dots, x_N$ satisfies

$$d_N \leq K N^{-1/s}$$

with a constant K depending only on $\theta_1, \dots, \theta_s$. Therefore g is a good lattice point modulo N.

4. Applications to discrete optimization

Suppose we want to maximize a given function f on the discrete grid G consisting of all lattice points $(a_1, \dots, a_s)$ with $0 \leq a_j \leq N - 1$ for $1 \leq j \leq s$, where N is a given positive integer. Let g be an s-dimensional good lattice point modulo N, let x_n be the corresponding points in (3), and put

$$y_n = N x_n \quad \text{for} \quad 1 \leq n \leq N. \tag{8}$$

Then all $y_n \in G$, and the points $y_1, \dots, y_N$ provide an efficient way of sampling from the N^s points of G. We use the approximation

$$M = \max_{a \in G} f(a) \approx T = \max_{1 \leq n \leq N} f(y_n).$$

The following error bound can be given. We introduce the modulus of continuity

$$\omega(t) = \max_{\substack{a,b \in G \\ d(a,b) \leq t}} |f(a) - f(b)| \quad \text{for} \quad t \geq 0$$

and the dispersion

$$D_N = \max_{a \in G} \min_{1 \leq n \leq N} d(a, y_n).$$

Then as in [5] one shows that

$$M - T \leq \omega(D_N).$$

If d_N is the dispersion of $x_1, \dots, x_N$, then it is seen immediately that $D_N \leq N d_N$, hence

$$M - T \leq \omega(N d_N).$$

If the sample $y_1,\dots,y_N$ is too sparse, one may repeat the procedure with other choices for g and take the largest value of T as an approximation for M.

Often we do not maximize over the whole grid G but over a subset, a typical case being the "simplex"

$$S = \{(b_1,\dots,b_s) \in G:\ b_1 \leq b_2 \leq \dots \leq b_s\}.$$

Here we can work with a discrete analog of a method in [11]. To describe the underlying general principle, we take arbitrary points $z_1,\dots,z_N \in G$ with dispersion

$$D_N = \max_{a\in G} \min_{1\leq n\leq N} d(a,z_n).$$

Apply the transformation

$$U:\ (a_1,\dots,a_s) \in G \rightarrow (b_1,\dots,b_s) \in S,$$

where $b_1 \leq b_2 \leq \dots \leq b_s$ is the nondecreasing arrangement of the numbers $a_1,\dots,a_s$. Then we get the points $U(z_n) \in S$ for $1 \leq n \leq N$. Consider the dispersion of these points in S, namely

$$E_N = \max_{b\in S} \min_{1\leq n\leq N} d(b,U(z_n)).$$

<u>Lemma 2.</u> For any $z_1,\dots,z_N \in G$ we have $E_N \leq D_N$.

<u>Proof.</u> It suffices to show that for all $b = (b_1,\dots,b_s) \in S$ and $w = (w_1,\dots,w_s) \in G$ we have

$$d(b,U(w)) \leq d(b,w). \tag{9}$$

We note that if the sequence $w_1,\dots,w_s$ contains an inversion, i.e. if we have $w_i > w_k$ for some $i < k$, then we can repair this inversion by interchanging w_i and w_k, thus obtaining the sequence $u_1,\dots,u_s$ with $u_i = w_k$, $u_k = w_i$, and $u_j = w_j$ for $j \neq i,k$. We claim that

$$\max_{1\leq j\leq s} |b_j - u_j| \leq \max_{1\leq j\leq s} |b_j - w_j| = d(b,w). \tag{10}$$

If $b_i \leq w_k$, then $b_i \leq w_k < w_i$, hence

$$|b_i - u_i| = |b_i - w_k| \leq |b_i - w_i| \leq d(b,w).$$

If $b_i > w_k$, then $w_k < b_i \leq b_k$, hence

$$|b_i - u_i| = |b_i - w_k| \leq |b_k - w_k| \leq d(b,w).$$

Similarly, one shows that

$$|b_k - u_k| \leq d(b,w)$$

by distinguishing the cases $b_k \leq w_i$ and $b_k > w_i$. Since $|b_j - u_j| = |b_j - w_j|$ for $j \neq i,k$, (10) is established. After finitely many steps all the inversions in $w_1,\dots,w_s$ can be repaired, so that we obtain the nondecreasing arrangement of $w_1,\dots,w_s$. Using (10) in each step, we get (9). □

Therefore, if $z_1,\dots,z_N$ is a low-dispersion point set in G, then $U(z_1),\dots,U(z_N)$ is a low-dispersion point set in S. In the present context we apply this principle with $z_n = y_n$ from (8).

References

1. T. Bandyopadhyay and P. K. Sarkar: A quasi random search technique for functional fitting of detector responses, Pramāna 24, 643-655 (1985).
2. J. W. S. Cassels: An Introduction to Diophantine Approximation, Cambridge Univ. Press, London, 1965.
3. L. K. Hua and Y. Wang: Applications of Number Theory to Numerical Analysis, Springer-Verlag, Berlin, 1981.
4. G. Larcher: The dispersion of a special sequence, preprint, Univ. of Salzburg, 1985.
5. H. Niederreiter: A quasi-Monte Carlo method for the approximate computation of the extreme values of a function, Studies in Pure Mathematics (To the Memory of Paul Turán), pp. 523-529, Akadémiai Kiadó, Budapest, 1983. (Submitted in 1977).
6. H. Niederreiter: Quasi-Monte Carlo methods for global optimization, Proc. Fourth Pannonian Symp. on Math. Statistics (Bad Tatzmannsdorf, 1983), D. Reidel, Dordrecht-Boston, to appear.
7. H. Niederreiter: Multidimensional numerical integration using pseudorandom numbers, Proc. Numerical Methods for Stochastic Optimization (Laxenburg, 1983), Math. Programming Study, North-Holland, Amsterdam, to appear.
8. H. Niederreiter and K. McCurley: Optimization of functions by quasi-random search methods, Computing 22, 119-123 (1979).
9. H. Niederreiter and P. Peart: A comparative study of quasi-Monte Carlo methods for optimization of functions of several variables, Caribbean J. Math. 1, 27-44 (1982).
10. H. Niederreiter and P. Peart: Localization of search in quasi-Monte Carlo methods for global optimization, SIAM J. Sci. Statist. Computing, to appear.
11. H. Niederreiter and P. Peart: Quasi-Monte Carlo optimization in general domains, preprint.
12. R. Y. Rubinstein: Simulation and the Monte Carlo Method, Wiley, New York, 1981.
13. I. M. Sobol': On the systematic search in a hypercube, SIAM J. Numer. Analysis 16, 790-793 (1979).
14. I. M. Sobol': On an estimate of the accuracy of a simple multidimensional search (Russian), Dokl. Akad. Nauk SSSR 266, 569-572 (1982).

A VISUAL INTERACTIVE SIMULATION MODEL FOR THE DESIGN OF A RAILWAY STATION

RUI CARVALHO OLIVEIRA

CESUR - TECHNICAL UNIVERSITY OF LISBON
CESUR/IST, Av. Rovisco Pais, 1000 Lisbon
PORTUGAL

ABSTRACT

Designing a railway station is a complex decision problem that calls for extensive investigation of alternative solutions. With the purpose of assisting the design studies of Campanhã Railway Station, a major node in the Portuguese Railway Network, a simulation model was developed enabling the planners to evaluate different lay-out solutions under various traffic load conditions. This model uses SIMUVI, an event-based visual interactive simulation system developed by the author that is also briefly described in the paper.

1.INTRODUCTION

A large program to improve railway facilities is taking place in the northern part of Portugal, including the renewal of Campanhã Railway Station. Campanhã is the main station in Oporto area, operating as a terminal for several long distance services (such as international trains and inter-city services to/from Lisbon) and also as a commuting station for several regional and suburban trains (heavily used for commuting purposes) connecting Oporto central urban area (served by S. Bento railway station) with several origins and destinations. It is a busy station (more than 440 trains daily), and its present track and platform layout and safety systems can hardly cope with the present traffic. With the construction of a new bridge over Douro river, which will eliminate a severe bottleneck for train circulation in this area, the situation is bound to worsen and a complete re-design of Campanhã railway station will become necessary.

Designing a railway station is a complex decision problem. First, there are multiple objectives to be attained, some contradictory, some ill-defined, and numerous constraints of various natures. This raises many modelling problems amplified by the fact that the performance of the station can not be easily related (by traditional engineering techniques) to the envisaged lay-out solution. Therefore, a fully analytical model could hardly be suitable to cope with the complexity and requirements of the problem.

Furthermore, the construction of new railway stations or the renewal of existing ones represent very high investment costs. And, as the lifetime of railway stations is very long (some 50 years) the conditions under which the station is going to operate are bound to vary in time. All this calls for extensive investigation of alternative lay-out solutions and it is crucial that planners can assess the variants under (uncertain) future scenarios.

These difficulties are common to many design problems, and computer simulation has proved to be a powerful decision-aid tool (eg |2|) namely to cope with the modelling difficulties and the unstructured features of decision-making. Simulation enables decision-makers to carry out extensive experimentations and to use their experience and judgement in generating and evaluating alternative solutions.

This paper describes a simulation model that was developed to assist the design studies of Campanhã Railway Station. It is part of an interactive computer-based system that is briefly presented in the next section. The model was constructed using SIMUVI, a Pascal based simulation system that has been developed by the author and is described in section 3. In the final section several conclusions are presented, namely discussing the importance of the visual and interactive characteristics of the model in improving communication and strengthning the confidence of end-users in the model.

2.OVERALL STRUCTURE OF THE COMPUTER SYSTEM

The simulation model is the core of a fully interactive computer-based system designed to ease the task of preparing the relevant data for the simulation experiments. The system comprises modules for:

i) Basic data editing;
ii) Traffic load definition;
iii) Lay-out solutions generation.

The first module includes a number of menu-driven routines to edit data on travel times, minimum stopping times, train/platform assignment preferences, alternative routes definition, etc.

The second module enables the user to define timetables reflecting projected traffic loads that will be imposed on the station.

The last includes a routine that enables the user to add or eliminate components of the instalation (e.g. switch points), depicting on the screen a diagram of the resulting lay-out. At the end of the session, an analysis is carried out to determine feasible routes and incompatibilities between routes. The results are then recorded in a file used as input for the simulation model.

3. AN OVERVIEW OF SIMUVI

The model for Campanhã railway station was constructed using SIMUVI, a visual interactive simulation system developed by the author. It is based on the concepts originally developed by Hurrion |3| and it is the result of a cooperative project with Oporto University (GEIN - Fac. of Engineering) where a similar system was developed for super mini-computers |5|.

SIMUVI is an event-based general purpose simulation system written in Pascal and designed to run on micro-computers. In addition to the common simulation facilities, SIMUVI also presents visual and interactive features: it depicts a dynamic representation of the system on a colour screen showing the changes of state as simulation progresses, and this appears to the user as a "silent movie". Furthermore, it enables the user to interrupt the simulation run and interact with the model via a set of commands. Figure 1 shows the structure of SIMUVI.

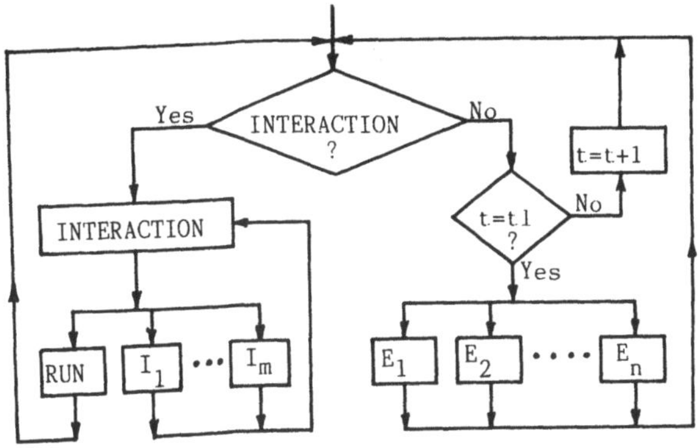

t - current simulation time
t1 - time of next event
E_i - event i

I_j - interaction j

Fig. 1 (Conceptual) Structure of SIMUVI.

SIMUVI includes all the required routines for simulation flow control. The boxes (E_i) on the right-hand side of Fig. 1 represent the events that must be supplied by the model-maker in Pascal. For this purpose, SIMUVI provides the model-maker with a comprehensive set of routines and functions that carry out all the usual manipulations of entities, queues and histograms. The model-maker can also write his own routines for particular manipulations not included in the basic-system, and for that he can use special functions provided by SIMUVI that make transparent all the internal data structure.

When the user interrupts the simulation run, control is passed to the interaction manager (left-hand side of Fig. 1) and the user can then maintain a dialogue with the model via a set of commands. SIMUVI provides a number of interactions:

i) To show the current contents of data elements, such as attributes of entities, contents of queues or histograms, scheduled events, etc;

ii) To alter the state of the model, such as modifying the value of an attribute, or removing an entity from a queue, etc.

iii) To make a dump (storing the current state of the model) and restoring a previously saved situation.

These are general purpose interactions and tend to use some simulation jargon, since they were primarily designed to be used by the model--maker. However, the model-maker can also supply his own Pascal written interactions specially designed for dealing with the particular needs of each application and constructed on a user-friendly basis.

4.THE SIMULATION MODEL

The simulation model was built with the SIMUVI facilities and emulates the operation of Campanhã railway station and its interactions with the neighbouring stations. The main inputs for the model are the data files generated by the facilities mentioned above which define:

i) a lay-out solution;

ii) a train timetable

The user must also specify the simulation period and whether random disturbances are to affect traffic generation. If not, trains enter the simulated system (when they are ready to depart from a neighbouring station to Campanhã) precisely at the times defined in the timetable. In the other case, randomly generated delays are added to these times. A statistical analysis was carried out to identify the distribution of these delays.

The diagram in Fig. 2 shows, in broad terms, the flow of trains through events and queues.

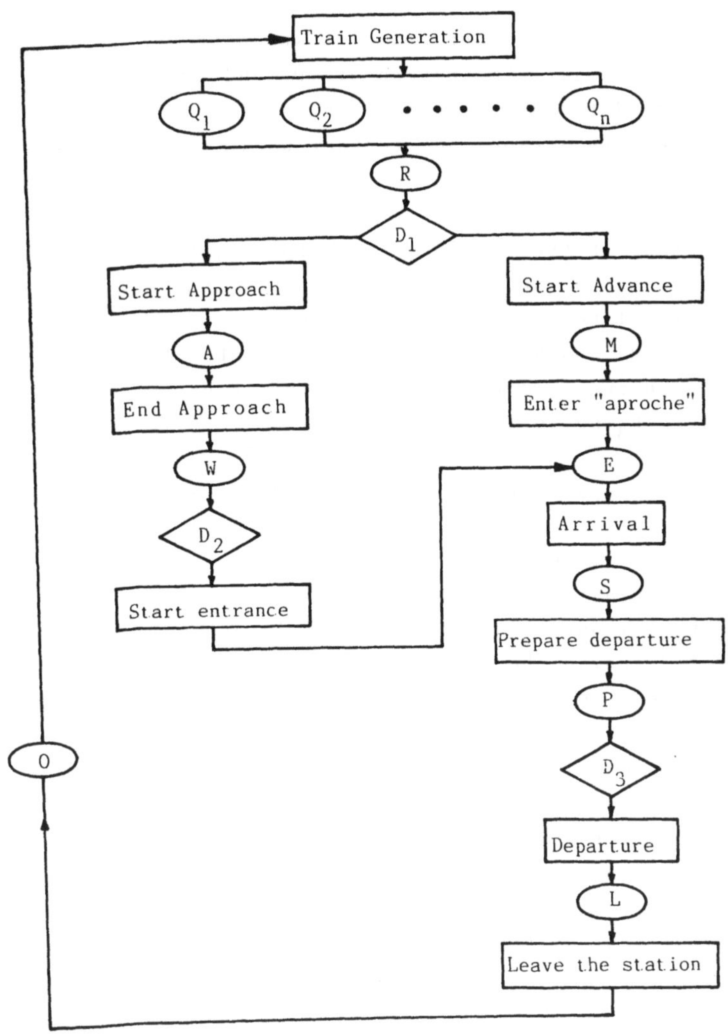

Fig. 2 Train Activity Cycle Diagram

At the first stage, the train enters the queue Q_i of the appropriate neighbouring station and, at the same time, it is placed in a common dummy queue R to which priority rules are applied. A first decision mechanism D_1 checks the conditions for train movement: if a platform and corresponding route are free, the train "advances" to the selected platform; if these conditions are not met at the time of departure from the neighbouring station but there are expectations that they will be met before the train reaches the entrance signal, a (conditioned) "approach" movement is selected; otherwise, the train remains in the neighbouring station.

The selection of the platform and route to which the train is assigned is based on a greedy algorithm invoked by D_1 and D_2 that takes into account preferences defined à priori. This decision can be overruled by the user via an interaction, and further work is being done to include an optimization procedure that will support user´s intervention.

Once the train stops at the platform in Campanhã, the departure time is defined taking into account the timetable and minimum stopping time, if it arrived late. The train joins then a common dummy queue P (with priority rules) and a decision mechanism D_3 checks the conditions for departure and selects the outcoming route. The remarks on D_1 and D_2 apply to D_3 as well.

The events routines were written using the SIMUVI facilities. The same applies to a set of display routines that are responsible for depicting on a colour screen all the train movements and changes of state. Fig. 3 is a diagram showing the visual representation depicted by the model. The lay-out solution, with its lines and switch points, is represented and inside the boxes associated with each neighbouring station is displayed the identification of trains that are stopped at or departing from those stations. In Fig. 3, trains no. 322 and 21 are leaving from platforms III and VII to S. Bento and Terminal, respectively; train no. 719 has departed from Gaia, while train no. 52124 is still stopped there; trains no. 6121 and 52023 are stopped at platforms XII and IX, respectively.

The dynamic visual representation, besides other aspects that will be discussed below, allows for a qualitative assessment of the solution performance since one can observe how and where congestion occurs.

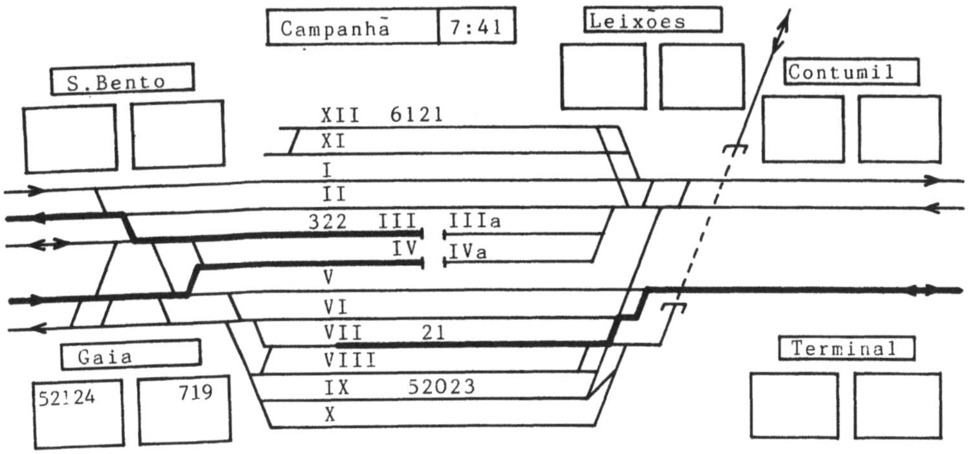

Fig. 3 The (dynamic) visual representation
of the station as it is depicted by the model

The model includes additional interaction facilities that complement those provided by SIMUVI, namelly:

a) To obtain additional information on:

i) static data used as input for the model, such as travel times, train/platform assignment preferences, alternative routes, incompatibilities between routes, etc;

ii) state of the system:

- trains (origin, destination, type, scheduled or actual arrival and departure times, etc)
- platforms and routes (status, scheduled time of release, etc)

iii) statistics and performance measures gathered up to the present simulation time;

b) To alter the state of the system by changing the state of a component of the instalation (such as disabling platforms or routes) or effecting train movements (such as forcing an early departure of a train).

During the execution the model collects a number of statistics and performance measures that are displayed, printed and/or recorded in a file at the end of the simulation run. These include:

a) Train delay statistics, for several levels of aggregation;
b) Train/platform assignment statistics;
c) Utilization of components of the instalation (platforms, routes, etc);
d) List of all train movements.

These elements constitute a basis for a quantitative assessment of the solution performance.

5.CONCLUSIONS

It is the author opinion that the success of a model developed to support decisions should be assessed on the basis of the contribution of the model to improve decisions. This assessment can not be made yet since the Railway Company staff responsible for the design of Campanhã station have just started using it and therefore the impact of the model in decision-making is still to be determined. However, the model is already regarded as a powerful instrument for analysis and an important decision support tool that decision-makers can understand and trust.

For this end-users satisfaction and confidence a very important contribution comes from the visual and interactive characteristics of the model.

It is generally recognised that one of the main reasons why results obtained with simulation models are not implemented derives from the fact that these models are not easily understood by non-simulation specialists. Quoting Lucke |4|, "If a manager can not understand a computer model, naturally he will not have full confidence in it. The obvious result: the manager will be reluctant to use the model for important decision-making. It is crucial, therefore, that future models shrung off the "black-box" image so that they will be more readily - even enthusiastically - accepted and used by management".

In a way, the visual representation depicted by the model can be viewed as an interface so that the non-simulation specialist can understand the model whose inner workings he will not necessarily know in detail. By enabling the user to observe how the model works, it provides a means of understanding (how results are obtained) and of verification (of the similitude between the model and the real system), leading to much greater user confidence. Furthermore, the interactive facilities that provide the user with full information and control of the simulation run create unparalleled means for validation, and contribute also for strengthning user confidence.

The visual and interactive facilities also played a very important role in improving the communication between the Railway Company staff and the author. The former had not been exposed to simulation concepts and felt some difficulties in specifying the model. A prototype was then developed: it was a rough representation of the existing station and its operation rules and it created conditions for the Railway Company staff to understand how the model could work and to become aware of the potentialities of the approach. The first specifications for the model were then produced, based namely on criticisms about the prototype. The modifications and additions that were made as the project progressed were typically sugested observing the dynamic representation depicted by the model, using frequently the interactive facilities to generate situations of particular interest. Without the visual and interactive facilities, end-users involvement and communication with the analyst would not presumably reach the level achieved in this project.

From the experience gained with this project it can be concluded that computer simulation represents a novel way of tackling the problem of designing railway stations. It presents several advantages over traditional engineering techniques, particularly by enabling extensive investigation of alternative solutions. Visual interactive simulation, as discussed above, enhances these advantages by creating conditions to overcome some motivational and communication problems. To quote Brooks |1|, "Moving images on a screen have great power to inform; images that move in response to one s manipulation seem to be perceived more as real things and studied more intently. Such systems achieve a high degree of transparency".

REFERENCES

|1| - Brooks, F.P. (1977) - "The Computer Scientist Toolsmith: Studies in Interactive Computer Graphics", in Gilchrist (Ed.), "Information Processing 77" IFIP, North Holland.

|2| - Fiddy, E. et al (1981) - "SEE-WHY Interactive Simulation on the Screen", Conference of the Inst. of Mech. Engineers, London.
|3| - Hurrion, R. D. (1976) - The Design, Use and Required Facilities of an Interactive Visual Computer Simulation Language to Explore Production Planning Problems", PhD Thesis (unpublished), Univ. of London (L.S.E.).
|4| - Lucke, M.C. (1974) - "Computer Models: "Black box" or Management Oriented", in Sanders (Ed) "Computers and Management in a Changing Society", McGraw-Hill.
|5| - Moreira da Silva, C. and Bastos, J.M. (1984) - "SIMVIS: An Event Based Visual Simulation System", Progress report n.2, GEIN, Univ. Porto.

SIMULTANEOUS SYNTHESIS AND ECONOMIC OPTIMIZATION OF THE POWER SYSTEM ON AN OIL/GAS PRODUCTION PLATFORM

ARILD OLSBU

The Foundation for Scientific and Industrial Research at the Norwegian Institute of Technology, 7034 Trondheim-NTH, Norway

PER A. LOEKEN

Division of Heat and Combustion Engineering, Norwegian Institute of Technology, 7034 Trondheim-NTH, Norway

IGNACIO E. GROSSMANN

Department of Chemical Engineering, Carnegie-Mellon University, Pittsburgh, PA 15213, U.S.A.

ABSTRACT

This paper is aimed at the selection of optimal configurations for power systems on oil platforms for generating electricity and satisfying power demands of process equipment. The variable power demands arising from the variable production profile are approximated through a finite sequence of time periods. A systematic procedure for the design of these flexible power systems is presented that is based on a multiperiod mixed-integer linear programming model. The model, which has as an objective to minimize the net present value of the investment and operating cost, can select among several types of gas turbines for satisfying the power demands, and it also considers the optimal assignment of drivers for rotary process equipment. The application of this model is illustrated on the power system on an oil platform in the North Sea. The problem consists in selecting gas turbine type as driver for the generator, and to decide what kind of driver, gas turbines or motor, is going to be used at the two selected equipment nodes. The optimal selection was obtained in approx 234 sec of CPU time on a DEC-20 computer. The type choosen is the Solar Mars gas turbine, which was the smallest of the gas turbines. The two nodes are both driven by electrical motors.

INTRODUCTION

An important part in the design of a production platform is the central power generation system which supplies electricity and power to meet demands from the processing utility and crew facilities. The basic process consists of an effective system to separate the oil feedstock from the wells into the component crude oil,gas and water phases.The separation is done through a single train system with three stages, high,medium and low pressure separations.In addition, systems for fuel gas,heating and cooling,water injection etc. are required.

To meet the demand of electricity from units such as compressors, pumps and from the crew facilities,a central power generation system has to be designed.In most cases,it will consist of a generator driven by gas turbine. The gas turbine can be of different types and sizes, and with different load ranges,efficiencies and costs. The driver package will therefore consist of one or more units, depending on load range and maximum demand.Instead of using electrical motors as drivers for the compressors,pumps etc., it can be more economical to use gas turbines for direct drive.It is also possible to put more than one rotary process equipment on the same gas turbine shaft.

An important feature in oil platforms is that the production rate varies with time.Therefore,the power system has to be designed to have the flexibility to meet variable electricity and power demands throughout the life of the platform.

The objective of this work is to develop a general model based on mixed-integer linear programming (MILP) for the optimal design of power systems on oil platforms. The model will choose the gas turbine type for generator drive and synthesize the driver system for different rotary process equipment so as to minimize the net present value of investment and fuel cost. An important point in this model is that variations of the production rate throughout the life of the platform are taken into account.Therefore,the model also has the capability to indicate how the power system could be run throughout the life time of the platform.

TIME-VARYING POWER DEMANDS

For the development of the proposed model it is convenient to divide the electricity and power requirement into two categories:

1. Constant electricity demand during the production period for crew facilities.

2. Variable electricity and power demand for processing facilities.

The electricity demand for the crew facilities,hvac and lighting is constant during the production period,and therefore it can easily be handled in a MILP-model.

On the other hand, the electricity and power demand for for the process units are functions of the production rate. The production rate is in turn dependent on the number of wellhead in production and the wellhead pressure. Since the power demand for the processing facilities are time-dependent variables,they can only be handled as discrete-time variables in a MILP-model.It is therefore necessary to divide the power demand curve into a finite number of representative discrete values. This can be accomplished by dividing the total production time into a finite number of periods,each of which involves an average power demand.If a reasonably number of periods are introduced a good approximation to the power profile can be obtained.Therefore, the number of periods to be selected depends on both the shape of the production profile and the availability of computing resources.It is clearly the user who ultimately has to decide how many time periods should be introduced in the model.

SYSTEMATIC ANALYSIS OF ALTERNATIVES

In order to determine the optimal design of the central power generation system and the configuration of equipment driver system,a general superstructure will be developed which has embedded all the alternative configurations that can be derived for the power system.To clarify the alternatives to be included in the superstructure,the following aspects must be considered:

1. Choice of gas turbine type as generator driver.

2. Choice of gas turbine type or electrical motor as driver for each rotary process equipment.

3. Synthesis of the equipment-driver system

Driver type for the generator.

The generator can be driven by one of the given types of gas turbines that must be be choosen.The number of units, generator and gas turbine, to be used depends on the electricity demand and maximum load on each unit.Also,each type of gas turbine must be described with min/max load,efficiencies and cost.

Driver type at process equipment.

The driver for each rotary process equipment can be an electrical motor or a gas turbine.For the latter alternative,several choices of types of gas turbines can be considered.

Synthesis of equipment-driver configuration.

A driver can either be used for one and only one rotary process equipment, or in some cases two or more rotary equipment can be put on the same shaft. If the drivers is a electrical motors,the common procedure is to use the first option. However, if a gas turbine is used as driver, it can be more economical to have more than one rotary equipment on the same gas turbine shaft.

The superstructure which includes all of the above alternatives is shown in fig 1. By deriving the appropriate equations, constraints and objective function for this superstructure, the proposed model has the possibility to choose the gas turbine type for the generator driver, as well as the gas turbine type or electrical motor as driver for the equipment. Therefore, the next step will address some general issues in the modelling of the superstructure.

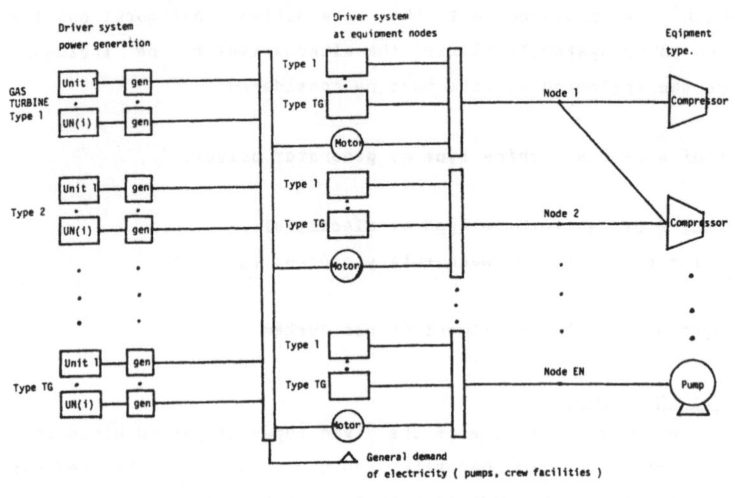

Fig. 1. General superstructure which has embedded all the alternative configurations.

COST MODELS

The costs to be considered in the proposed model are the following:

1. Investment cost
2. Running cost

Investment cost

The investment cost of gas turbine and electrical motor are given as a set of constant values,since standard equipment sizes are assumed. For an oil platform the cost must include the purchase cost, installation cost, as well as the weight and space cost.

Running cost.

The only running cost that is cosidered in the model, is the cost of fuel consumption in the gas turbines. The fuel consumption is calculated by the following equation:

$$F = W/H*\eta$$

where:

F :	fuel consumption	Stm3/s
W :	load	kW
H :	calorific value	kW/Stm3
η :	efficiency	-

This equation shows that F is a function of the load and the efficiency of the gas turbine.However,the efficiency is also a function of load and since this is a nonlinear function it can not be implemented as such in the model.Instead of directly approximating this function,it is better to use the relationship between load and fuel consumption. This function can be expressed in the following way:

$$W = \alpha *F + \beta$$

The coefficients α and β are calculated on the basis of fuel consumption and efficiencies at maximum and minimum output.

Economic objective function

As mention earlier the total production period has to be divided into a finite number of periods. The fuel consumption cost factor for each period is based on a given interest rate and the time from the middle of each period to a year before production start.

Since the power system is installed on the platform before production start, no correction for the direct investment cost is considered.

FORMULATION OF THE MILP-MODEL

Having discussed the development of superstructures for power systems, and the modelling of the cost and fuel consumption, the synthesis problem of power systems will be formulated as a multiperiod MILP problem. Binary variables that can take only 0-1 value will be assigned for logical decisions involving the possible choice of gas turbines and electrical motors, while non-negative continuous variables will be assigned to parameters such as fuel consumption. A value of 1 in the binary variables means that the potential choice is included, a value of 0 is when the choice is excluded. In the development of the model, it will be assumed that power demands are specified for TP periods af operation. The postulated superstructure will contain in general TG types of gas turbines for electricity generation. Each type of gas turbine may consist of several units running in paralell. Also, it is assumed that power must be supplied to TE types of rotary equipment. These can be assigned in EN equipment nodes where power can be supplied either by an electrical motor or by direct drive with any of the TG types af gas turbines.

The objective function can be formulated as:

$$\text{Minimize } \Sigma\ (I + A\ .\ F)$$

where the parameter I represent the total investment cost, F is the fuel consumption and A is discounted cost factor. A complete formulation of the objective function would be more complex regarding time periods, running/not running mode etc.

The most important constraints are:

Power balance on each power supply point.

Max/min. load constraints on each gas turbine.

To ensure that this constraints produce a power system that is implementable in an oil platform, also the following logical constraints must be included:

Constraints for selecting only one type of gas turbine as driver for the generators.

Constaints to ensure that only one type of gas turbine is selected.

Is should be emphasized that this MILP model could be generated automatically within a computer program for the synthesis of these power systems.

NUMERICAL EXAMPLE

An oil production platform in the North Sea was considered to demonstrate the application of the MILP model for optimal synthesis of the power system. The production profile for this field is shown in fig.2.

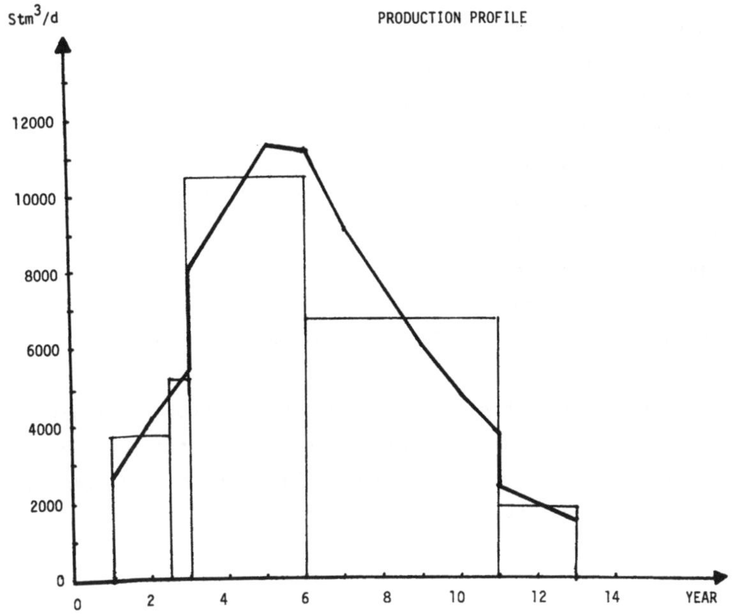

Fig. 2. Production profile for the oil/gas field.

In this paticular case, the oil/gas ratio and their respective composition are assumed constant during the production period. Based on these assumptions, the electricty and power demands were calculated on the basis of the production profile. From the flowsheet it can be seen that there are six different rotary process equipment where a gas turbine or electrical motor can be used as driver: three compressors for the separation process,a sales gas compressor,a lift gas compressor and the water injection pump.The remaining electricity demands are from general process equipment and the crew facilities.

As shown in fig. 2. the production period is divided into five different periods.Using the average production level in each period the power requirement for each equipment in each period is calculated. The electrical and power requirements together with the peak demands are presented in Tab.1.

The problem is then to select gas turbine type as driver for the generator, and to decide what kind of driver, gas turbines or motor, is going to be used at the to selected equipment nodes (compressor train and water injection pump) There are four types of gas turbine to choose from, and data for the different types are shown in Tab.2. The net present value of the fuel cost is calculated with a interest rate of 12%, and the time is calculated from the middle of each period to one year before production start.

		GAS TURBINE			
		LM 5000	LM2500-30	LM2500-20	SOLAR MARS
PEAK LOAD	MW	28.0	22.0	13.00	8.00
MAX LOAD	MW	21.0	16.5	9.75	6.24
EFFICIENCIES AT MAX LOAD	%	25.0	25.0	27.0	20.00
MIN LOAD	MW	8.4	6.6	3.9	2.4
EFFICIENCIES AT MIN LOAD	%	16.7	14.6	13.3	8.4
α	-	16.7	21.3	36.0	56.3
β	-	10.5	-15.0	-21.7	-25.2
γ	-	-31.5	-31.5	-31.5	-31.5
κ	-	18.9	21.6	25.6	27.6

Fig. 2. Gas turbine data.

The size of the MILP model is 55 binary variables, 135 continuous variables and 300 constraints. The optimal solution was obtained in approx. 234 sec. of CPU time on a DEC-20 computer using the branch and bound algorithm of the LINDO computer code. The resulting optimal configuration is shown in Fig 4.

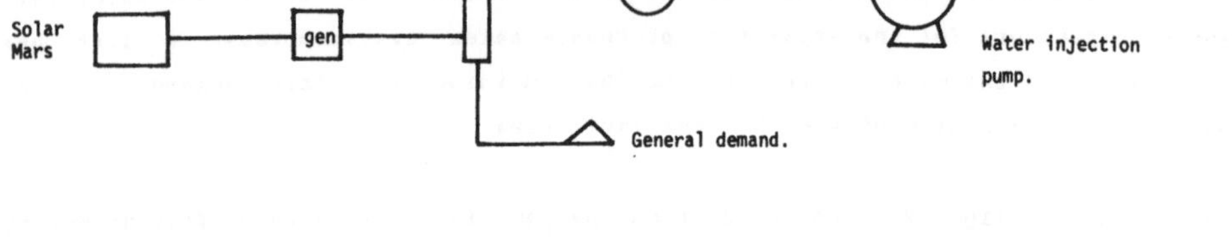

Fig. 4. Optimal configuration. There is four units in the central power generations system and the two nodes are both driven by electrical motors.

Since the power demands of the 2 and 3 stage compressors are relatively small, the three compressors together with the lift gas compressor were considered to be on the same shaft. The sales gas compressor is only driven by an electrical motor, while the water injection pump can be driven either by an electrical motor or by gas turbines. The resulting superstructure for the alternatives considered is shown in fig 3.

UNIT		PERIOD					
		A	B	C	D	E	PEAK
GENERATOR	MW	3.318	4.238	7.774	5.452	2.120	8.357
HP-COMP.	MW	1.421	1.945	3.958	2.636	0.738	4.293
MP-COMP.	MW	0.193	0.265	0.538	0.359	0.100	0.584
LP-COMP.	MW	0.129	0.177	0.360	0.239	0.067	0.390
SG-COMP.	MW	0.909	1.244	2.534	1.687	0.427	2.746
LG-COMP	MW	0.000	0.603	1.227	0.817	0.229	1.330
INJECTION WATER PUMP	MW	0.000	2.311	4.705	3.133	0.877	5.100

Tab. 1. Power demands.

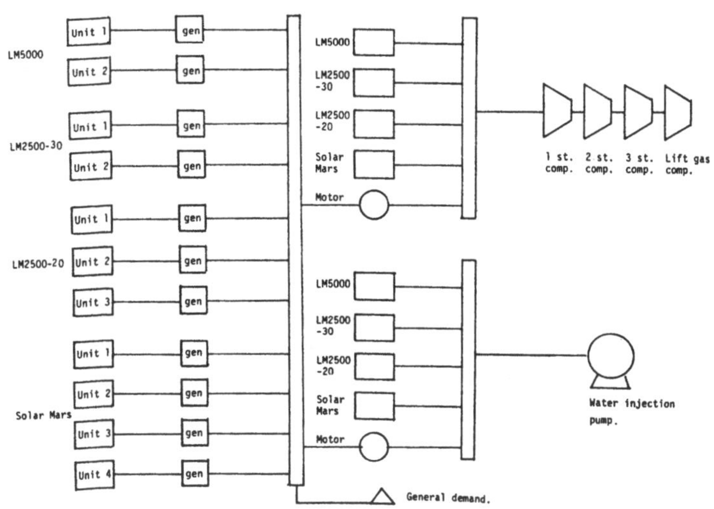

Fig.3. Resulting superstructure for the alternatives considered in the numercical example.

The type choosen is the SOLAR MARS gas turbine which was the smallest of the gas turbines.The two nodes are both driven by electrical motors. The net present value for minimum investment and fuel cost was 44.1 M$. In period A only one gas turbine has to run, in period B two turbines, in period C four, in period D three and in period E only one.

To compare the effect of selecting other types of gas turbines the MILP model was resolved by fixing the binary variables of the corresponding types of different gas turbines. The results are shown in Tab. 3.

GAS TURBINE TYPE	NUMBER OF UNITS	NUMBER RUNNING IN PERIOD A	B	C	D	E	DRIVER TYPE AT COMP.TRAIN	WAT.INJ	NET PRESENT VALUE
SOLAR MARS	4	1	2	4	3	1	MOTOR	MOTOR	44.1M $
LM2500-20	3	1	2	3	2	1	MOTOR	MOTOR	60.1M $
LM2500-30	2	1	1	2	1	1	MOTOR	MOTOR	53.0M $
LM-5000	2	1	1	2	1	1	MOTOR	MOTOR	59.0M $

Tab.3. Results for different selections of gas turbine types.

These results clearly indicate that large economic savings can be achieved by synthesizing an optimal power system for an oil platform with the proposed MILP model.

CONCLUSIONS

This paper has presented a multiperiod MILP model for the optimal synthesis of power systems in oil platforms. The model has the capability of accounting for the variable production profile in an oil platform. This model can be a very useful design tool for achieving significant economic savings as was shown in the example problem. The extension of this model for incorporating availability considerations and where the number of back-up units is optimized, is accomplished and will be presented later on.

REFERENCES

* Duran, M.A. and I.E. Grossmann: "An outher-approximation algorithm for a class of mixed-integer nonlinear programs. Part I: The algorithm and its properties", paper presented at the TIMS/ORSA meeting, Chicago (1983).
* Papoulios, S.A., Grossmann, I.E. "Optimal synthesis of Flesible Utility Systems, Proceedings of the 18th IECEC conference, 2066-2071, Orlando (1983).
* Schrage, L.E., "User Manual for Lindo". The scientific Press, Palo Alto (1981).

DC MOTOR DRIVE PARAMETERS OPTIMIZATION

MILUTIN M. OSTOJIĆ
University of Titograd
Yugoslavia

APSTRACT

This paper presents the procedure for optimum parameters adjustment of the separetly excited dc motor drive using sensitivity analysis, The dc motor drive, containing dc motor, thyristor, regulators of armature current and speed is described by the set of the first order linearised differential equation in the state space form. The modes of the regulators are assumed to be known. On the base of the mathematical model of the real system, sensitivity model of the system relative to the variable parameters is developed. The output of the system is the speed of the motor. The criterion has been chosen as an integral of the squared error of the motor speed variation. The parameters of the system that have been optimised are: the gains and time constants both current and speed regulators, gains of the current and speed feedbacks. For minimisation of the criterion are emploied both Rosen gradient procedure and Fletcher-Powell procedure. Starting point may be chasen in the stable region as well as in the unstabile one. An example is carried out to illustrate the procedure.

1. INTRODUCTION

Under assumption of small variations of all variables around an arbitrarily chosen refference operating state, DC motor drive can be linearised and discribed by set of first order differential equations with constant coeficients. For the parameters synthesis of DC Motor drive it is very important to chose a rotional structure of the regulators and the values of the parameters in order to obtain high quality of transients and small steady - state error. Meny papers have been published on this subject. Usualy parameters of the regulators are chosen on the base of known relationship between transients and logarithmic AF caracteristic of the chose loop system. In that case, optimal parameters values are obtained on the base of adjustment of logarithmic characteristices to the desired shape. Such a approach is very complicated, especially in the case when transient, for obtained parameters, have high quality coresponding to the desired speed signal, but coresponding to the disturbauce these parameters are not satisfacting. As autor is aware no attempts have been made to apply sensitivity theory for optimisation of DC motor drive. Therefore, this paper briefly examines application of the sensitivity theory to the particular DC motor drive in order to obtain optimum values of the system parameters that minimise the integral of the squared error criterion.

2. MATHEMATICAL MODEL OF THE SYSTEM

The hybrid diagram of the separetly excited DC motor drive system is shown in Fig.1. This diagram is a cross between a block diagram and an electromechanical network, which is quite convinient method to display interrelation of system variables in order to build a mathematical model. The essential parts of the system is a separetly exited motor, three phase full wave thyristor converter and current and speed regulators.

Matrix representation of the state equations that describe behaviour of the system may be written in the form:

$$Y = AY + FP;\ Y(t_o) = Y_o \qquad (1)$$
$$X = CY,$$

where:

Y - n - dimensional vector of states

X - k - dimensional disturbance vector

X - p - dimensional output vector

A - nxn - constant state distribution matrix

F - nxk - constant disturbance distribution matrix

C - pxn - constant output distribution matrix

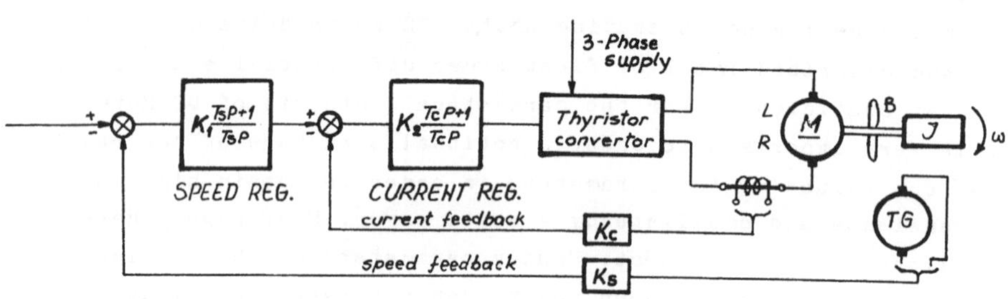

Fig.1. Hybrid diagram of the studied system

3. OPTIMUM CRITERION

Generally, in developing procedure for optimization we have to find variable system parameters which will minimize the performance index given by the integral at the squared errors in the form:

$$J = \int_{t_o}^{t_o+T} \left[Y(t,q)-Y_r(t)\right]^T \left[R\right] \left[Y(t,q)-Y_r(t)\right] dt \qquad (2)$$

where:

Y(t,q) - n - dimensional state vector as a function of the system parameters q_i;

$Y_r(t)$ - n-dimensional state vector of refference system,

R - nxn - positive-definite weighting matrix

T - time of integration.

The elements of the weighting matrix are the value which determine relative importance attached to the various errors. Time of integration is theoretically infinite but in practice a sufficiently long time. If we chose appropriately value for T,R and reference system outputs, then performance index becoms a function of the unknown system parameters vector q only i.e.

$$J = J(q) \tag{3}$$

4. OPTIMUM CONDITION

The partial derivatives of the performance index J with respect to the system parameters vector q give the optimum condition as

$$\frac{\partial J}{\partial q} = 0 \text{ or grad } J = \left\{\frac{\partial J}{\partial q_1}, \frac{\partial J}{\partial q_2} \cdots \frac{\partial J}{\partial q_e}\right\} = 0 \tag{4}$$

Having in mind this, the expression (2) may be written as:

$$\frac{\partial J}{\partial q} = \int_{t_0}^{t_0+T} \left[Y(t,q) - Y_r(t) \right]^T \left[R+R^T\right]\left[V(t,q)\right] dt \tag{5}$$

where

$$V(t,q) = \frac{\partial y(t,q)}{\partial q} \tag{6}$$

The vector function V(t,q) is sensitivity function an may be written as

$$V(t,q) = \left[\frac{\partial Y_i(t,q)}{\partial q_j}\right] l \times n \tag{7}$$

where

i=1,2 ... n - number of the system outputs and j=1,2...,1 number of unknown parameters.

5. SENSITIVITY FUNCTIONS

In order to obtain the gradient components of the criterion it is necessary to have the system sensitivity functions of all of the system outputs relative to the system parameters which are supposed to be optimize.

The sensitivity functions may be determined by definition, but number of equations can be significantly redused if we apply procedure described here. To apply this procedure, we have to know block diagram of the system. If we assume that transfer function of the current and speed regulators are respectivly:

$$W_c(p) = K_2 \frac{T_2 p+1}{T_2 p} \text{ and } W_s(p) = K_1 \frac{T_1 p+1}{T_1 p} \quad . \tag{8}$$

and knowing differential equations of the separetly excited dc motor with constant field exoitation

$$U_f = R_f F_f$$
$$U_a = \omega C_e + (R_a + L_a p) L_a \qquad (9)$$
$$T = C_e C_a = T_L + \omega (B + Jp),$$

We can draw block diagram of the system as shown in Fig.2.

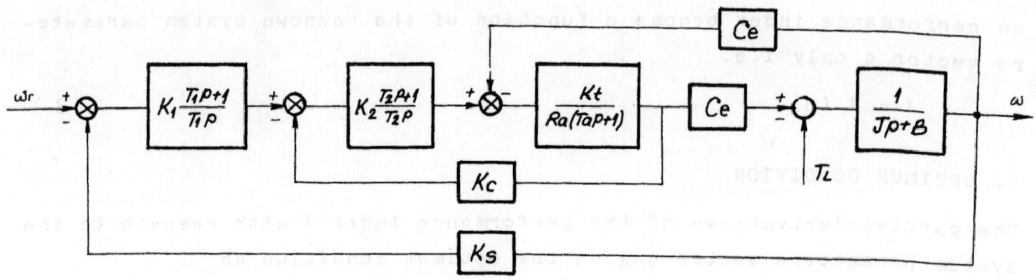

Fig.2. Block diagram of the system

This block diagram can be easyly modified to get more convinient shape for obtaining sensitivity functions. The modified block diagram is shown in Fig.3.

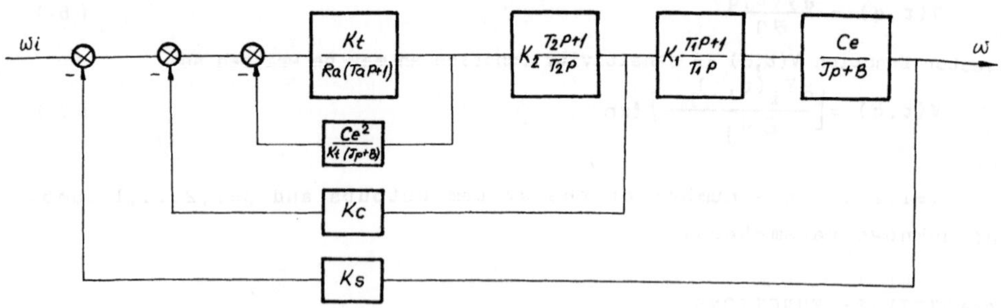

Fig.3. Modified block diagram

We consider the motor speed as the output of the system. To obtain first derivatives of the speed relative to the system parameters $q = T_1$, T_2, K_1, K_2, K_c, K_s , we have to develope system sensitivity block diagram and to chose sensitivity points. If is very easy to obtain sensitivity block diagram [3] , because if is the same as the basic system block diagram, but the input to the sensitivity block diagram is output of the basic one, and it is shown in Fig.4. Now we can chose sensitivity points S_i, $i=1,\dots,6$. Multiplying signals in the sensitivity points with transfer functions $T_i(p)$, we obtain sensitivity functions of the system.

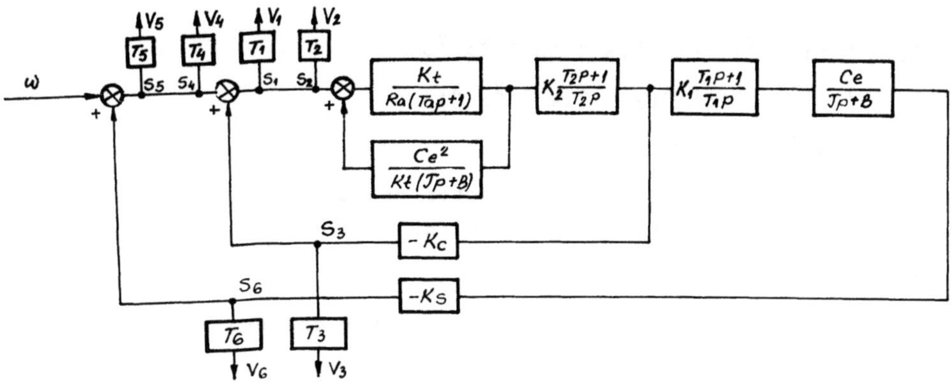

Fig.4. The system sensitivity block diagram

Generally, the T functions are defined as:

$$T_i(p) = \frac{q_i}{W_i(p)} \quad \frac{\partial}{\partial q_i}\left[W_i(p)\right] \qquad (10)$$

In this case T functions are:

$$T_1(p) = \frac{1}{T_2 p+1}, \quad T_2(p)=T_s(p)=T_5(p)=T_6(p)=1 \qquad (11)$$

$$T_4(p) = -\frac{1}{T_1 p+1}$$

Having in mind Fig 3 and 4, the basic system and system sensitivity models are:

- Basic system model

$$\begin{aligned} Y_1 &= (-BY_1 + C_e Y_2)/J \\ Y_2 &= K_1 X_3/T_1 \\ Y_3 &= K_2 Y_4/T_2 \\ Y_4 &= (K_a K_t x 5 - Y_4)/T_a \\ Y_5 &= (-C_e^2 Y_4 + BK_t Y_5)/JK_t \end{aligned} \qquad (12)$$

where

$$\begin{aligned} X_7 &= -K_s Y_1, \quad X_3 = Y_3 + Y_4 K_2 \\ X_2 &= Y_2 + K_1 X_3, \quad X_6 = X_7 - K_c X_3 \\ X_5 &= Y_5 + X_6 \end{aligned} \qquad (13)$$

- System sensitivity model

$$\begin{aligned} Y_6 &= (-BY_6 + C_e X_9)/J \\ Y_7 &= K_1 X_{1o}/T_2 \\ Y_8 &= K_2 Y_9/T_2 \\ Y_9 &= (K_a K_t X_{12} - Y_9)/T_a \\ Y_{1o} &= (-C_e^2 Y_9 + B K_t Y_{1o})/ K_t \end{aligned} \qquad (14)$$

Where

$$X_{14} = Y_1 - K_s Y_6; \quad X_{1o} = Y_8 + K_2 Y_9$$

$$X_9 = Y_7 + K_1 X_{1o}; X_{13} = X_{14} - K_c X_{1o} \tag{15}$$

$$X_{12} = Y_{1o} + X_{13}$$

The performance index depends on the motor speed variation only and the following equation is to be added to the equations (12) and (14):

$$J = Y_{11} = \frac{1}{2} (-K_s Y_1)^2 \tag{16}$$

The sensitivity function, that are needed for the criterion minimization are:

$$V_1 = Y_{12}, V_2 = X_{13}, V_3 = -K_e X_{1o},$$

$$V_4 = Y_{13}, V_5 = X_{14}; V_6 = -K_s Y_6, \tag{17}$$

where Y_{12} and Y_{13} (see Fig.4) are solutions of the following equations:

$$Y_{12} = -(Y_{12} + X_{13})/T_2$$

$$Y_{13} = -(Y_{13} + X_{14})/T_1 \tag{18}$$

The gradient components of the criterion $J=Y_{11}$, i.e. $\partial J/\partial qi$ may be obtained solving following equations:

$$\frac{d}{dt} \left(\frac{\partial J}{\partial q_i} \right) = \frac{dy_i}{dt} = [\omega - K_s Y_1)K_s V_j/q_j$$

$$\frac{d}{dt} \left(\frac{\partial J}{\partial k_s} \right) = \frac{dy_{19}}{dt} = [\omega - K_s Y_1] [Y_1 + V_6]$$

where i = 14,15...18 i j=1,2...5.

Solwing set of 19 differential equations we ostain all system variable, sensitivity functions, value of the criterion J and its first determitives with respect to the parameters q.

6. MINIMIZATION OF THE CRITERION

If we have a point defined as $q_{opt} = q_o$ in which is $\partial J/\partial q=0$ or sufficient small, then point q_{opt} we may call as an optimum point. If this condition is not satisfied we shall change value of the parameters in a starting point for on increment Δq. Chosing initial step sise s and knowing $\partial J/\partial q_i$ we can evaluate direction vector components D_i by using Rosen [6] or Fletcher - Powell constrained methods. Then we can locate new point as

$$q_{(new)} = q_{(old)} + S_i D_i$$

This process is repeated until convergence is achieved.

An implementation of computation is given by block diagram shown in Fig.5. The computation procedure is as follows:

1. - The set of differential equations which describe system model and sensitivity model is solved by using Runge-Kutta Adams-Moulton predictor-corrector method.
2. - The responce $Y(t,q)$ of the system and responce $Y_r(t)$ of the reference model give the error $Y(t,q)-Y_r(t)$.
3. - On the base of this error and weighting matrix R we calculate value of the criterion $J(q)$ and partial derivatives $\partial J/\partial q$ at any value of the parameters.
4. - With these data we supply on optimization routine in order to obtain optimal parameter values.
5. - With optimal parameter values we enter into basic sistem model to obtain behavior of the optimized system.

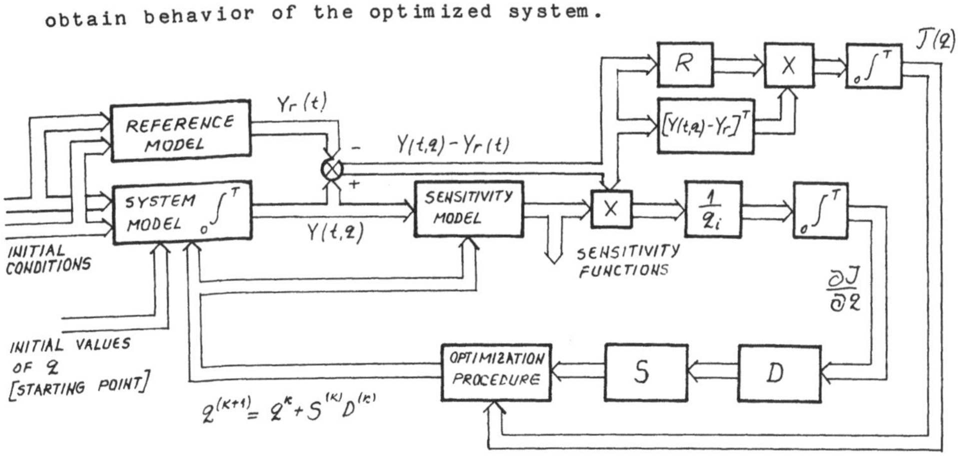

Fig.5. Structural interpretation of optimization procedure

7. NUMERICAL EXAMPLE

The foregaing theory was applied for parametar optimization of the DC motor drive which is characterized with following parameters.

$P=118okW$, $V=5ooV$, $Ia=25ooA$, $R_a=o,o8412$, $L_a=o,oo31H$

$C_e=3,92$ V/rad/s, $T_m=o,1$ s and $K_t = 97$.

The input to the system is the reference value of the motor speed in the form of step function. The results of calculation are shown in Table 1 and Fig.6. for stable and unstable starting points using both

Fletcher-Powell and Rosen gradient procedure. In the Table time of the calculation is included in order to compare computer time. Calculation time needed by Rosen gradient procedure is taken as 1. Starting from the some starting point, we have been obtained, using both methods, very chose final suboptimal solutions. For different starting points we found different suboptimal solutions i.e. different final values of the parameters, but final value of the criterion was almost the same.

Table 1.

	T_1	T_2	K_1	K_2	K_c	K_s	J	t
Unstable star.point.	o.o17	o.3o24	38.623	o.434	o.o82	1	11943.98	
Rosen	o.o94	o.o712	28.o67	o.271	o.o17	1.ol	3.617E-o3	1
Fletcher--Powell	o.o3o	o.o737	28.o66	o.654	o.oo6	o.988	3.75E-o3	2.2
Stable star.point	o.o16	o.o756	28.o67	o.434	o.o41	1	1.5988	
Fletcher--Powell	o.378	o.o164	28.o68	o.374	o.olo	1.o28	3.23E-o3	o.95
Rosen	o.232	o.o736	28.o67	o.357	o.oll	o.997	3.39E-o3	o.45
Rosenb-rock	o.339	o.o113	28.o68	o.363	o.o12	1.o44	5.14E-o3	3.5

For minimization of the criterion we applied Rosenbrock hillclimb procedure which has been proved meny times as very efficient in solwing different minimization problems when first derivatives of the cost function are not known. However, in this case the procedures which require first derivatives have been appeared as more efficient regarding computer time and converqence. It has happen meny times when starting point is in unstable region that converqence has not been achieved or it has been achieved for very large value of the criterion. The results that are shown in Table 1. using Rosenbrock hillclims algoritham are obtained after many trials, chosing different starting points in stable region. When starting point is in the stable area and in the neighborhood of the optimum, then Rosenbrock on the euad, it should be noted that suboptimal solution can be improved if we use obtained optimal parameters (using one procedure) as starting points for another one. Doint so we obtained curve 3 shown in Fig 6, using alternatively Rosem-Fletcher Powell-Rosen procedures.

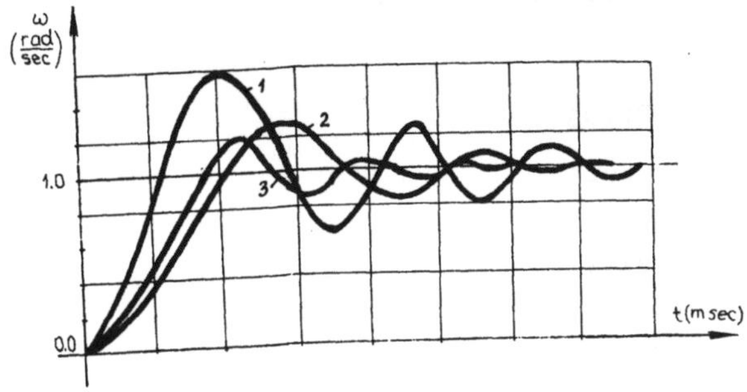

1. Fletcher-Powell: Unstable starting point.
2. Rosen gradient procedure: Unstable starting point
3. Rosen - Fletcher - Powell - Rosen.

Fig.6. Time response of the speed

CONCLUSION:

In this paper, an approach to the optimization dc motor drive based on the system sensitivity analysis is developed. The dynamic performance index is optimised by varing the parameters of the system. The number of the parameters theoretically is not limited but the computation fime depends on that, and also on the applied algorithm fcr optimization and capacity of the computer.

An numerical example has been carried out to illustrate the feasibility of the procedure.

REFERENCES:

1. V.M. Nikitin, A.D.Pozdeev and J.A.Sabinin, "Analysis and systhesis of dc motor drive", Elektrocthestvo, N° 3, 1974. (in Russian).

2. J.A.Sabinin, V.P.Andreev: "Basic electric drives", Gosenergoizdat, Moskow 1963. (in Russian).

3. R.Tomović, "Sensitivity analysis of dynamic Systems", Mc Graw-Hill Company, New York, 1963.

4. M.M.Ostojic, "The Automatic control synthesis of a winding nachine dc motor drive", Etan, 1976, Opatia (in Serbian).

5. H.H.Rosenbrock, "An automatic method for finding the greatest or least value of a function", computer J. 3, 196o.

6. J.B.Rosen, The Gradient projection method for nonlinear programing. Part I, Linear constraints", J.Soc. Indust.Appl. Math., 8, pp 181-217, 196o.

7. R.Fletcher and M.J.D.Powell "A rapidly convergent desent method for minimization", computer J.,, 6, 1963.

Optimizing Simulated Markov Processes

Georg Ch. Pflug

Math. Institute
Arndtstraße 2
JL University
D-6300 Gießen

1. Introduction

Consider a stochastic system, which is controllable by a control vector x. Suppose that the performance of the system is measured by the expected value of a real function H under the steady-state law. The problem is to find the vector x within some fearible set S, which optimizes the performance of the system.

To be more specific, we assume that $\{Z_n\}$ is a discrete-time continuous-state Markovian process on $\mathbb{R}^m$ with transition

$$Z_{n+1} = K(Z_n, x, \xi_n) \tag{1}$$

where ξ_n is an i.i.d. sequence of random vectors on a probability space $(\Omega, \mathcal{A}, P)$. Note that (1) is a kind of canonical representation of the transition kernel

$$P_x(z,A) := P\{K(z,x,\xi) \in A\} \tag{2}$$

Every transition kernel (2) may be represented by a transition function $K(\cdot,\cdot,\cdot)$ and uniformly $[0,1]^k$-distributed r.v.'s ξ_n, since all other distributions can be obtained by a transformation from the uniform.

Suppose that μ_x ist the unique invariant measure of (1), i.e.

$$z_{n+1} \sim \mu_x \quad (\sim \text{means: "is distributed according to"})$$

$$\text{if} \quad z_n \sim \mu_x$$

Or, more formally

$$\mu_x(A) = (P_x \cdot \mu_x)(A) := \int P_x(z,A)\, d\mu_x(z) \,, \quad A \in \mathcal{A}.$$

μ_x is the distribution of the steady-state situation. Let $H(z,x)$ be a real function describing the behavior of the system at state z under control x. The performance F of the system is defined as the expected value of H under the steady-state situation, i.e.

$$F(x) := \int H(z,x)\, d\mu_x(z) \qquad (3)$$

The problem is to minimize $F(\cdot)$ over S, i.e.

$$F(x) = \min! \qquad (4)$$
$$x \in S$$

If the invariant measure μ_x, as well as the integral in (3) can be calculated analytically, then the optimality problem reduces to a (typically) nonlinear deterministic constrained optimization problem, as is treated in OR-books.
However, if an analytic solution is impossible, the system has to be investigated by simulation. Then the connection between simulation and optimization is the basic problem. There are two groups of methods described in the literature

(i) Response surface techniques

A grid is defined in the fearible set S. For each point x_i in the grid, a simulation study is conducted to get an estimation $\hat{F}(x_i)$ of $F(x_i)$. A curve is fitted using these estimates and finally the optimal point of this curve is taken as a solution of the problem.

(ii) Experimental design techniques

This method is borrowed from the design of statistical experiments. First, simulation experiments are conducted only at few parameter values x . Based on the result, a "direction of descent" is defined and further observations are made only at points lying in this direction. The geometric pattern of the design points may be cubic, hexagonal etc.... For a detailed description see e.g. Kleijnen [1] or Mihram [4].

It is the purpose of this paper to show that simulation experiments are not simply statistical experiments. In fact, since the randomness in simulation stems from generated random numbers we are often much better off: More information is available to us in a simulation study.

In particular, estimates of the gradient of F(x) are available, as we shall see. These gradient estimates can be used to define a iterative algorithm for the solution of the optimality problem. This algorithm is characterized by a close interplay between the simulation and optimization, in contrast to the previous methods, where simulation is used to replace the objective-function call in an optimization package.

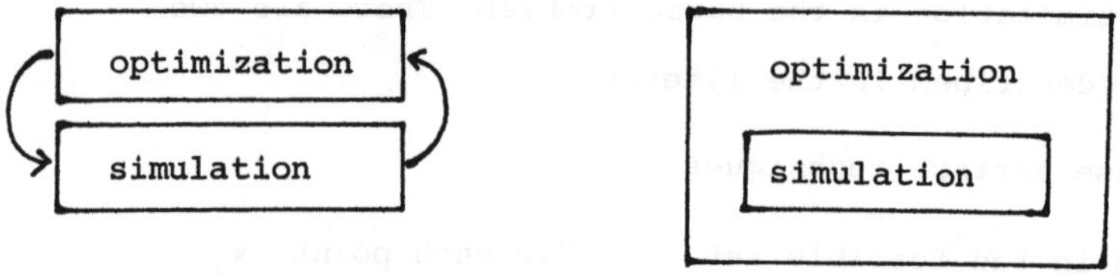

stochastic gradient technique

response surface technique
experimental design technique

2. The algorithm

We assume that the control variable x belongs to a closed, convex set $S \subseteq \mathbb{R}^k$. The projection operator onto S is denoted by Π_S. The evolution of the Markovian process $Z_n \in \mathbb{R}^m$ under control x is

$$Z_{n+1} = K(Z_n, x, \xi_n) .$$

We assume that $(z,x) \mapsto K(z,x,\xi)$ is boundedly L^1-differentiable in the sense that there is a bounded $(m \times n)$-matrix $J_z(z,x,\xi)$ and a bounded $(m \times k)$ matrix $J_x(z,x,\xi)$ such that

$$\int \|K(z,x,\xi)-K(z_o,x_o,\xi)-J_z(z,x,\xi)(z-z_o)-J_x(z,x,\xi)(x-x_o)\|dP(\xi)=0(\|x-x_o\|+\|z-z_o\|)$$

The performance generating function is denoted by $H(z,x)$ and it is assumed that the gradients w.r.t. x and z, namely $\nabla_x(H(z,x)$ and $\nabla_z H(z,x)$, and all second derivatives exist and are bounded.

The algorithm approximates the optimal x by a stochastic iteration $\{X_n\}$:

$$X_{n+1} = \Pi_S (X_n - a_n Y_n) \tag{5}$$

where

$$Y_n = \nabla_x H(X_n, Z_n^{(0)}) + \\ + \sum_{l=0}^{k_n} \nabla_z H(X_n, Z_n^{(l)})' \prod_{j=1}^{l} J_z(Z_n^{(j)}, X_n, \xi_n^{(j)}) J_x(Z_n^{(0)}, X_n, \xi_n) \tag{6}$$

Here $\{\xi_n\}$ and $\{\xi_n^{(1)}, \xi_n^{(2)}, \dots, \xi_n^{(k_n)}\}$ $n \geq 1$ are i.i.d. random variables on $(\Omega, \mathcal{A}, P)$. The sequence $Z_n^{(0)}, Z_n^{(1)}, \dots$ is an intermediate simulation of the Markovian process with fixed control X_n:

$$z_n^{(0)} = z_{n-1}^{(k_{n-1})}$$

$$z_n^{(j+1)} = K(z_n^{(j)}, X_n, \xi_n^{(j+1)}). \qquad (7)$$

The different steps are summarized by the following diagram

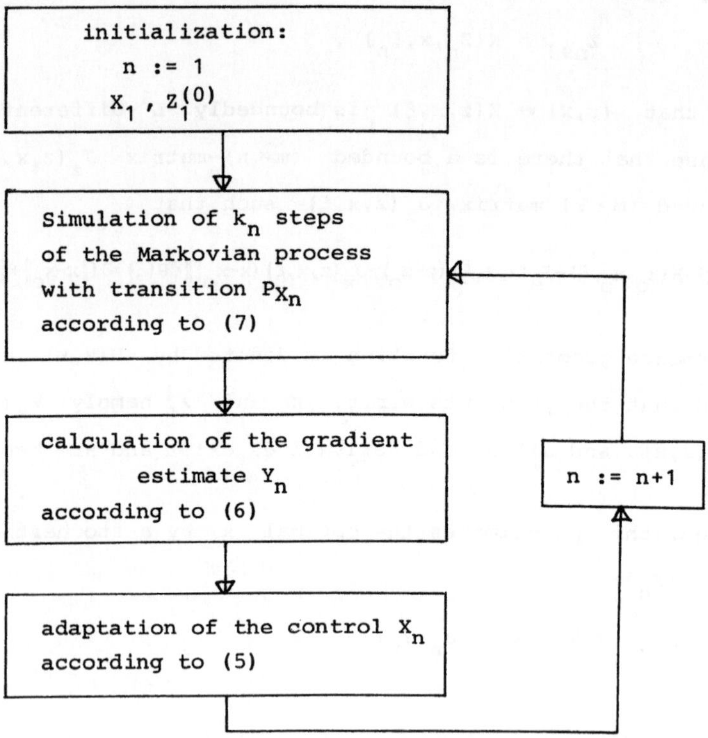

The following theorem exhibits the convergence properties of the approximation process $\{X_n\}$.

Theorem.

Let the following assumptions be fulfilled

(i) There is a constant ρ , $0 < \rho < 1$, such that for all $x \in S$

$$\int \|K(z_1,x,\xi) - K(z_2,x,\xi)\| dP(\xi) \le \rho \|z_1 - z_2\|$$

(ii) There is a constant C, such that for all $z \in \mathbb{R}^n$

$$\int \|K(z,x_1,\xi) - K(z,x_2,\xi)\| dP(\xi) \leq C \|x_1 - x_2\|$$

(iii) Let μ_x be the invariant measure of P_x, which exists according to (i) and $F(x) := \int H(z,x) d\mu_x(z)$.

Let furthermore x^* be the solution of $F(x)=\min!$ under the constraint $x \in S$, which fulfills

$$(\nabla F(x))'(x - x^*) \geq \delta \quad \|x - x^*\|^2 \qquad \text{for a} \quad \delta > 0$$

$$x \rightarrow \|\nabla F(x)\| \quad \text{is continuous.}$$

(iv) $a_n \geq 0$, $\sum a_n = \infty$, $\sum a_n^2 k_n^2 < \infty$ $k_n \rightarrow \infty$, $\sum_{i=1}^{n} \rho^{k_i} k_i a_i = 0(a_n)$.

Then the iteration $\{X_n\}$ converges a.s. to x^*, the solution of the optimization problem.

The major steps of the proof are:

1.) We show that $d(\mu_{x_1},\mu_{x_2}) \leq \frac{C}{1-\rho} \|x_1 - x_2\|$

where $d(\mu_1,\mu_2)$ is the Wasserstein-distance

$d(\mu_1,\mu_2) = \inf\{E\|x_1 - x_2\|$ s.t. the marginal distribution of x_1 and x_2 is μ_1 and μ_2 respectively$\}$.

2.) We show that $E(Y_n | X_1,\ldots,X_n) = \nabla_x < H(\cdot,x),\mu_x > |_{x=X_n} + 0(\rho^{k_n}) + 0(K_n^2 a_n)$

3.) We use known results (cf. e.g. Kushner & Clark [2] or Ljung [3]) to infer from 2 together with $\sum a_n \rho^{k_n} < \infty$, $\sum k_n^2 a_n^2 < \infty$ the desired convergence of the algorithm.

Details are omitted.

3. An example

Consider the following network of service systems

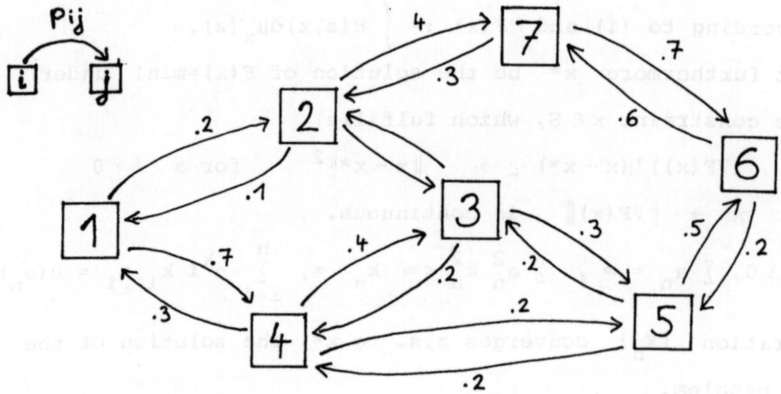

We assume that the service is done in discrete steps and that the amount of work is measured by a continuous variable. At each step, a maximal amount of work x_i can be accomplished by service unit i; i=1,...,7. After leaving the node i , the amount of work just finished splits randomly to reach the next node. A certain proportion leaves the system. At the same time, a random amount of work η enters newly the system. To describe it formally, let $Z_n^{(i)}$ the amount of work waiting for service at node i at time n . Then

$$Z_{n+1}^{(i)} = (Z_n^{(i)} - x_i)^+ + \sum_{j \neq i} \pi_j \; i \;\; \min(Z_n^{(j)}, x_j) + q_i \eta \qquad (8)$$

where π_{ij} are random proportions, such that $E(\pi_{ij}) = p_{ij}$. The mean transition proportions $\{p_{ij}\}$ are indicated in the above figure. The quantity $1 - \sum_j p_{ij}$ represents the proportion of work leaving the system.

The equation (8) shows the transition of the system in terms of the control $x = (x_1, \ldots, x_7)$ and the random quantities $\{\pi_{ij}\}$ and η. The performance function is the mean unfinished work, i.e.

$$E(H(Z^{(1)}, \ldots, Z^{(7)})) = E(\sum_{i=1}^{7} Z^{(i)})$$

E(H) should be minized for x under the budget constraint

$$\sum_{i=1}^{7} x_i \leq C$$

and the nonnegativity constraints

$$x_i \geq 0.$$

The chosen specifications were

$\eta \sim$ Exponential with expectation 10.0

$$\pi_{ij} = \frac{V_{ij}}{\sum_j V_{ij}}$$ with $V_{ij} \sim$ Exponential with expectation p_{ij}

$C = 80.$

$q_i = \frac{1}{7}$ for all i .

The gradients needed for step (6) of the algorithm are easily calculated as

$$\nabla_x H(\cdot,\cdot) = 0$$
$$\nabla_z H(\cdot,\cdot)' = (1,1,\ldots,1)$$

$$(J_z(z,x,\xi))_{ij} = \begin{cases} 1_{\{z \geq x\}} & \text{if} \quad i = j \\ \pi_{ji} \, 1_{\{z \leq x\}} & \text{if} \quad i \neq j \end{cases}$$

$$(J_x(z,x,\xi))_{ij} = \begin{cases} -1_{\{z > x\}} & \text{if} \quad i = j \\ \pi_{ji} \, 1_{\{z > x\}} & \text{if} \quad i \neq j \end{cases}$$

We remark here that the method works also with numerical derivatives, if the values of the exact derivatives are difficult to obtain. However taking numerical gradients in more dimensions is rather time-consuming.

The size of the "inner loop" k_n (step (7)) was held constant $k_n = 10$. Although the theory says that these constants should slowly increase to infinity, we found practically no difference in the results for a larger inner loop.

The stepsizes a_n were chosen not adaptively as $a_n = \frac{30}{10+n}$. There is a large literature on adaptive stepsize control (for a survey see Pflug [5]). In fact, adaptive stepsize control can lead to sharper bounds for the solution by cleverly exploiting the information gathered during the approximation. The optimal stepsize problem is however beyond the scope of this paper.

The optimal allocation of the budget 80 to the 7 service stations in our problem was found in 500 steps to be

$$x_1=6.8 \quad x_2=13.5 \quad x_3=10.0 \quad x_4=9.8 \quad x_5=9.1 \quad x_6=15.8 \quad x_7=15.0$$

In order to demonstrate the accuracy we have also calculated the optimal solution for a perfectly circular symmetric transition matrix $\{p_{ij}\}$. By symmetry, the optimal solution must be $x_1=x_2=\ldots=\ldots=x_7$. The algorithm found as solution (500 steps)

$$x_1=11.18 \quad x_2=11.41 \quad x_3=11.50 \quad x_4=11.23 \quad x_5=11.65 \quad x_6=11.70 \quad x_7=11.26$$

which is rather close to the true solution $x_i \equiv 11.43$, compared to the uncertainity inherent in the simulation. For both simulation-optimization runs the starting value was $(0,0,\ldots,0)$.

5. Concluding remarks

We have presented a method of "on line" optimization of simulated Markov processes. In the presentation we have restricted ourselves to a discrete-time continuous-state process, for which the needed derivatives can be calculated analytically. However, if numerical derivatives are considered to be sufficient, the method can easily be generalised to a continuous-time process, which is halted at regular time intervals to adapt the new control x . The performance of the method was demonstrated by a numerical example. It can be hoped that in the future an optimization environment can be created (e.g. in the framework of a simulation language) such that every simulation model can be automatically optimized (by preprogrammed structures) with respect to some specified control parameters.

References

[1] J.P.C. Kleijnen. Design of simulation experiments. in: Encyclopedia of Systems and Control, Pergamon Press, Oxford, 1982

[2] H. Kushner and D.S. Clark. Stochastic approximation for constrained and unconstrained systems. Appl. Math. Sci., vol. 26. Springer Verlag, 1978

[3] L. Ljung. Analysis of recursive stochastic algorithms. IEEE Trans. Autom. Control, vol. AC-22, No. 2, 551-575 (1977)

[4] G.A. Mihram. Simulation: Statistical foundations and methodology. Academic press, 1972

[5] G. Pflug. Stepsize rules, stopping times and their implementation in stochastic quasigradient algorithms. in: Numerical Methods of Optimization (R. Wets, editor), Springer-Verlag (1986)

GLOBALLY CONVERGENT EXACT PENALTY ALGORITHMS FOR CONSTRAINED OPTIMIZATION

G. Di Pillo *, L. Grippo ** and S. Lucidi **

* Dipartimento di Informatica e Sistemistica
Università di Roma "La Sapienza"
Via Eudossiana 18, 00184 Roma - Italy

** Istituto di Analisi dei Sistemi ed Informatica
Consiglio Nazionale delle Ricerche
Viale Manzoni 30, 00185 Roma - Italy

ABSTRACT

In this paper we define two classes of algorithms for the solution of constrained problems. The first class is based on a continuously differentiable exact penalty function, with the additional inclusion of a barrier term. The second class is based on a similar modification performed on a continuously differentiable exact augmented Lagrangian function. In connection with these functions, an automatic adjustment rule for the penalty parameter is described, which ensures global convergence, and Newton-type schemes are proposed which ensure an ultimate superlinear convergence rate.

1. INTRODUCTION

In recent years an increasing attention has been devoted to exact penalty methods for the solution of constrained problems by means of unconstrained minimization techniques (see, e.g. [1]).

A common feature of existing penalty methods is that "exactness" can only be established with reference to some compact set S containing the problem solutions, so that the threshold value of the penalty coefficient depends on S. This causes, in principle, an inherent difficulty in the unconstrained minimization, since the level set corresponding to the penalty parameter and to some given initial point, even if compact, need not be contained in S. As a consequence, global convergence results for exact penalty methods, employing automatic procedures for the adjustment of the penalty parameter, require the additional assumption that the sequence constructed by the algorithm is bounded.

In this paper we define two classes of algorithms which avoid these difficulties under mild regularity and compactness assumptions on the problem. The first class is based on the continuously differentiable exact penalty function considered in [2], with the additional inclusion of a barrier term on the boundary of S [3]. The second class is based on a similar improvement performed on a continuously differentiable exact augmented Lagrangian function [4]. For both functions, a complete equivalence can be established between the solution of the constrained problem and the unconstrained minimization of a continuously differentiable function whose global and local minimizers are contained in the interior of S. On this basis globally convergent algorithms, employing an automatic adjustment rule for the penalty parameter, are defined which cannot produce unbounded sequences, thus overcoming the main drawback of existing exact penalty function methods. For both classes, Newton-type schemes are described, which ensure an ultimate superlinear convergence rate.

Numerical results are reported.

2. PROBLEM FORMULATION

The problem considered is the nonlinear programming problem:

Problem P minimize $f(x)$, $x \in R^n$
subject to $g(x) \leq 0$

where the functions $f: R^n \to R^1$ and $g: R^n \to R^m$ are assumed to be three times continuously differentiable on R^n.

We denote by $L(x,\lambda) \triangleq f(x)+\lambda' g(x)$ the Lagrangian function for Problem P and by $A(x) \triangleq \frac{\partial g(x)}{\partial x}$ the Jacobian matrix of the constraints.

Let $\alpha \in R^m$ be a given vector of positive numbers α_i, $i = 1,\dots,m$ and define the set:

$$S \triangleq \{x \in R^n: g(x) \leq \alpha\},$$

with interior $\overset{\circ}{S}$ and boundary ∂S.

Given any $x \in S$, define the index sets $I_o(x) \triangleq \{i: g_i(x) = 0\}$, $I_\pi(x) \triangleq \{i: g_i(x) \geq 0\}$, $I_\nu(x) \triangleq \{i: g_i(x) < 0\}$.

We assume that the following hypotheses are satisfied:

Assumption A1. The set S is compact.

Assumption A2. For any $x \in S$, the gradients $\nabla g_i(x)$, $i \in I_o(x)$ are linearly independent.

Assumption A3. For any $x \in S$, the system

$$\sum_{i=1}^{n} \frac{\partial g_j(x)}{\partial x} z_i > 0, \quad j \in I_\pi(x)$$

has a solution $(z_1,\dots,z_n)' \in R^n$.

A discussion of these assumptions can be found in [3].

We consider two classes of algorithms for the solution of Problem P. The first is based on the exact penalty function proposed in [3], the second is based on the exact augmented Lagrangian function studied in [4], which constitutes a recent development of the approach introduced in [5],[6].

Let $\lambda(x)$ be the multiplier function introduced in [7] and defined by:

$$\lambda(x) \triangleq -M^{-1}(x)A(x)\nabla f(x)$$

with

$$M(x) \triangleq A(x)A(x)' + \gamma^2 G^2(x)$$

$$G(x) \triangleq \mathrm{diag}[g_i(x)], \; i = 1,\dots,m,$$

and let $B(x)$ be the diagonal matrix with elements $b_i(x) \triangleq \alpha_i - g_i(x)$, $i = 1,\dots,m$.

Then, we can define on $\overset{\circ}{S}$ the exact penalty function:

$$Z(x;\varepsilon)=f(x)+\lambda(x)'(g(x)+Y(x;\varepsilon)y(x;\varepsilon))+\frac{1}{\varepsilon}(g(x)+Y(x;\varepsilon)y(x;\varepsilon))'B^{-1}(x)(g(x)+Y(x;\varepsilon)y(x;\varepsilon))$$

where $\varepsilon > 0$ and:

$$Y(x;\varepsilon) \triangleq \mathrm{diag}[y_i(x;\varepsilon)], \; i = 1,\dots,m$$

$$y(x;\varepsilon) = (y_i(x;\varepsilon)), \quad i = 1,\dots,m$$

$$y_i(x;\varepsilon) \triangleq \{-\min[0, g_i(x) + \frac{\varepsilon}{2} b_i(x)\lambda_i(x)]\}^{1/2}.$$

Moreover we can define over $\overset{\circ}{S} \times R^m$ the exact augmented Lagrangian function:

$$\begin{aligned} W(x,\lambda;\varepsilon) = {} & f(x)+\lambda'(g(x)+V(x,\lambda;\varepsilon)v(x,\lambda;\varepsilon)) \\ & + \frac{1}{\varepsilon}(g(x)+V(x,\lambda;\varepsilon)v(x,\lambda;\varepsilon))'B^{-1}(x)(g(x)+V(x,\lambda;\varepsilon)v(x,\lambda;\varepsilon)) \\ & + \eta\|A(x)\nabla f(x)+M(x)\lambda\|^2, \end{aligned}$$

where $\eta > 0$ and:

$$V(x,\lambda;\varepsilon) \triangleq \mathrm{diag}[v_i(x,\lambda;\varepsilon)], \; i = 1,\dots,m$$

$$v(x,\lambda;\varepsilon) \triangleq (v_i(x,\lambda;\varepsilon)), \; i = 1,\dots,m$$

$$v_i(x,\lambda;\varepsilon) \triangleq \{-\min[0, g_i(x) + \frac{\varepsilon}{2} b_i(x)\lambda_i]\}^{1/2}.$$

It can be observed that the last term of (2) satisfies

$$\|A(x)\nabla f(x) + M(x)\lambda\|^2 = \|M(x)(\lambda(x) - \lambda)\|^2,$$

so that it can be viewed as a penalty term on the difference between the multiplier function $\lambda(x)$ and the current value λ of the multiplier.

In particular, if we substitute $\lambda(x)$ for λ in the expression of $W(x,\lambda;\varepsilon)$ we obtain $W(x,\lambda(x);\varepsilon) \equiv Z(x;\varepsilon)$.

In the sequel we will refer to any pair $(\bar{x},\bar{\lambda}) \in R^n \times R^m$ satisfying the Kuhn-Tucker necessary conditions $\nabla_x L(\bar{x},\bar{\lambda}) = 0$, $G(\bar{x})\bar{\lambda} = 0$, $g(\bar{x}) \leq 0$, $\bar{\lambda} \geq 0$, as a Kuhn-Tucker (K-T) pair for Problem P.

3. UNCONSTRAINED MINIMIZATION ALGORITHMS FOR $Z(x;\varepsilon)$

We report first the properties of $Z(x;\varepsilon)$ which are more relevant for the construction of unconstrained minimization algorithms. The proofs of Propositions 1-6 can be found in [5].

PROPOSITION 1. *For any* $\varepsilon > 0$:

(a) *the function* Z *is continuously differentiable for all* $x \in \mathring{S}$, *with gradient*

$$\nabla Z(x;\varepsilon) = \nabla f(x)+A(x)'\lambda(x) + \frac{\partial\lambda(x)'}{\partial x}(g(x)+Y(x;\varepsilon)y(x;\varepsilon))$$
$$+ \frac{2}{\varepsilon} A(x)'B^{-1}(x)(g(x)+Y(x;\varepsilon)y(x;\varepsilon))$$
$$+ \frac{1}{\varepsilon} A(x)'[G(x)+Y^2(x;\varepsilon)]B^{-2}(x)(g(x)+Y(x;\varepsilon)y(x;\varepsilon));$$

(b) *for any* $x_o \in \mathring{S}$, *the level set* $\{x \in \mathring{S}:Z(x;\varepsilon) \leq Z(x_o;\varepsilon)\}$ *is compact so that* Z *admits a global minimum point on* $\mathring{S}$. □

PROPOSITION 2.

(a) *Let* $(\bar{x},\bar{\lambda})$ *be a* K-T *pair for Problem* P; *then for any* $\varepsilon > 0$, $\bar{x}$ *is a stationary point of* Z, $\lambda(\bar{x}) = \bar{\lambda}$ *and* $Z(\bar{x};\varepsilon) = f(\bar{x})$;

(b) *there exists an* $\varepsilon^* > 0$ *such that, for all* $\varepsilon \in (0,\varepsilon^*]$, *if* $\bar{x} \in \mathring{S}$ *is a stationary point of* Z, *the pair* $(\bar{x},\lambda(\bar{x}))$ *satisfies also the* K-T *conditions for Problem* P.□

PROPOSITION 3.

(a) *If* $\bar{x} \in \mathring{S}$ *is an isolated local minimum point of Problem* P, *then there exists an* $\varepsilon^* > 0$ *such that, for all* $\varepsilon \in (0,\varepsilon^*]$, $\bar{x}$ *is a local unconstrained minimum point of* Z;

(b) *there exists an* $\varepsilon^* > 0$ *such that, for all* $\varepsilon \in (0,\bar{\varepsilon}]$, *if* $\bar{x} \in \mathring{S}$ *is a local unconstrained minimum point of* Z, $\bar{x}$ *is a local minimum point of Problem* P *and* $\lambda(\bar{x})$ *is the associated* K-T *multiplier.* □

PROPOSITION 4. *There exists an* $\varepsilon^* > 0$ *such that for all* $\varepsilon \in (0,\varepsilon^*]$, *any global minimum point of Problem* P *is a global minimum point of* Z *on* $\mathring{S}$ *and conversely.* □

In order to state second-order optimality results we need some additional notation. More specifically, we introduce the index sets:

$$I_+(x) \triangleq \{i:2g_i(x)+\varepsilon b_i(x)\lambda_i(x) > 0\}, \quad I_-(x) \triangleq \{i:2g_i(x)+\varepsilon b_i(x)\lambda_i(x) \leq 0\};$$

accordingly, we define:

$$g_+(x) \triangleq [g_i(x)]_{i\in I_+(x)} \qquad g_-(x) \triangleq [g_i(x)]_{i\in I_-(x)},$$
$$\lambda_+(x) \triangleq [\lambda_i(x)]_{i\in I_+(x)} \qquad \lambda_-(x) \triangleq [\lambda_i(x)]_{i\in I_-(x)},$$
$$A_+(x) \triangleq \left[\frac{\partial g_i(x)}{\partial x}\right]_{i\in I_+(x)} \qquad A_-(x) \triangleq \left[\frac{\partial g_i(x)}{\partial x}\right]_{i\in I_-(x)},$$

and we refer to g, λ and A reordered in the form:

$g(x) = [g_+(x)'g_-(x)']'$ $\lambda(x) = [\lambda_+(x)'\lambda_-(x)']'$ $A(x) = [A_+(x)'A_-(x)']'$.

A similar convention is adopted for the reordering of g, λ and A induced by the index sets $I_o(x)$ and $I_\nu(x)$. Then, we have:

PROPOSITION 5. *Let $(\bar{x},\bar{\lambda})$ be a K-T pair for Problem P and assume that strict complementarity holds at $(\bar{x},\bar{\lambda})$.*

Then, for any $\varepsilon > 0$ the function Z is twice continuously differentiable in a neighbourhood of $\bar{x}$. □

PROPOSITION 6.

(a) *Let $(\bar{x},\bar{\lambda})$ be a K-T pair for Problem P, assume that strict complementarity holds at $(\bar{x},\bar{\lambda})$ and that $\bar{x}$ is an isolated local minimum point for Problem P satisfying the second order sufficiency condition:*

$$x'\nabla_x^2 L(\bar{x},\bar{\lambda})x > 0 \text{ for all } x : A_o(\bar{x})x = 0,\ x \neq 0;$$

then, there exists an $\varepsilon^* > 0$ such that for all $\varepsilon \in (0,\varepsilon^*]$, $\bar{x}$ is an isolated local minimum point for Z and the Hessian matrix $\nabla^2 Z(\bar{x};\varepsilon)$ is positive definite.

(b) *Suppose that strict complementarity holds at any K-T pair $(\bar{x},\bar{\lambda})$ of Problem P; then, there exists an $\varepsilon^* > 0$ such that, for all $\varepsilon \in (0,\varepsilon^*]$, if $\bar{x} \in \mathring{S}$ is a local unconstrained minimum point of $Z(x;\varepsilon)$, with positive definite Hessian $\nabla^2 Z(\bar{x};\varepsilon)$, $\bar{x}$ is an isolated local minimum point of Problem P, satisfying the second order sufficiency condition.* □

On the basis of the preceding results, it is possible to define Newton-type algorithms, employing consistent approximations of the Newton's direction, which avoid the evaluation of third order derivatives.

In the sequel we shall confine ourselves to state the consistency of these algorithms. The proofs of Propositions 7 and 8 can be found in [8].

A first possibility is that of replacing $\nabla^2 Z(x;\varepsilon)$ with the approximating matrix $H(x;\varepsilon)$ defined by:

$$H(x;\varepsilon) = \nabla_x^2 L(x,\lambda(x)) + \frac{\partial\lambda_+(x)'}{\partial x} A_+(x) + A_+(x)'\frac{\partial\lambda_+(x)}{\partial x} + \frac{2}{\varepsilon} A_+(\bar{x})'B_+^{-1}(x)A_+(\bar{x}) - \frac{\varepsilon}{2}\frac{\partial\lambda_-(x)'}{\partial x} B_-(x) \frac{\partial\lambda_-(x)}{\partial x},$$

where $B_+(x) = \text{diag}[b_i(x)]_{i\in I_+(x)}$ and $B_-(x) = \text{diag}[b_i(x)]_{i\in I_-(x)}$. In fact, we have:

PROPOSITION 7. *Let $(\bar{x},\bar{\lambda})$ be a K-T pair of Problem P, satisfying the assumptions of Proposition 6(a). Then we have $H(\bar{x};\varepsilon) = \nabla^2 Z(\bar{x};\varepsilon)$. Moreover, there exists an $\varepsilon^* > 0$ such that for all $\varepsilon \in (0,\varepsilon^*]$, the matrix $H(x;\varepsilon)$ is positive definite in a neighbourhood of $\bar{x}$.* □

By employing the approximation H of $\nabla^2 Z$ we can define the following Newton-type algorithm:

ALGORITHM 1

$\hat{x} = x + \alpha d$

$H(x;\varepsilon)d = -\nabla Z(x;\varepsilon)$

where $x,\hat{x}$ are respectively the present and the next iterate, d is the search direction and α is the stepsize.

A different Newton-type algorithm for the minimization of Z can be derived from the solution $d \triangleq [d_x'\ d_+'\ d_-']'$ of the following system:

$$\begin{bmatrix} \nabla_x^2 L(x,\lambda(x)) & A_+(x)' & A_-(x)' \\ A_+(x) & 0 & 0 \\ 0 & 0 & I \end{bmatrix} \begin{bmatrix} d_x \\ d_+ \\ d_- \end{bmatrix} = - \begin{bmatrix} \nabla_x L(x,\lambda(x)) \\ g_+(x) \\ \lambda_-(x) \end{bmatrix}. \qquad (1)$$

In fact, it can be shown that the component d_x of d is a consistent approximation of the Newton's direction for Z. More specifically, we can state the following proposition.

PROPOSITION 8. *Let* $(\bar{x},\bar{\lambda})$ *be a K-T pair for Problem* P *and assume that strict complementarity holds at* $(\bar{x},\bar{\lambda})$. *Then, there exists a neighbourhood* $\Omega \subset \mathring{S}$ *of* $\bar{x}$ *such that, for all* $x \in \Omega$, *if* $d \triangleq [d_x' \; d_+' \; d_-']'$ *is a solution of system* (1), *we have*:

$$\hat{H}(x;\varepsilon)d_x = -\nabla Z(x;\varepsilon),$$

where $\hat{H}(x;\varepsilon)$ *is a continuous matrix defined on* Ω *and satisfying* $\hat{H}(\bar{x};\varepsilon)=\nabla^2 Z(\bar{x};\varepsilon)$. □

Thus, letting $z = d_+ - \lambda_+(x)$, we can define the following algorithm.

ALGORITHM 2

$\hat{x} = x + \alpha d_x$,

where d_x is obtained by solving the system:

$$\begin{bmatrix} \nabla_x^2 L(x,\lambda(x)) & A_+(x)' \\ A_+(x) & 0 \end{bmatrix} \begin{bmatrix} d_x \\ z \end{bmatrix} = - \begin{bmatrix} \nabla f(x) \\ g_+(x) \end{bmatrix}.$$

It can be observed that Algorithm 2 does not require second order derivatives in the r.h.m., so that, by using a Quasi-Newton approximation of $\nabla_x^2 L(x,\lambda(x))$, it is possible to derive Quasi-Newton algorithms for the minimization of Z, employing only first order derivatives of the problem functions.

The Newton-type algorithms considered before ensure an ultimate superlinear convergences rate.

As regards global convergence, an important point is that, because of the barrier term $B^{-1}(x)$, the level set $\{x \in \mathring{S}, \; Z(x;\varepsilon) \leq Z(x_o;\varepsilon)\}$ is compact for any given $\varepsilon > 0$ and $x_o \in \mathring{S}$. Therefore, making use of the algorithms considered in association with some suitable stabilization scheme, it is possible to define an iteration map I: $\mathring{S} \to 2^{\mathring{S}}$, such that the sequence produced by I remains in $\mathring{S}$ and admits limit points whichare stationary points of Z.

The convergence towards K-T points of the constrained problem can be ensured by employing the automatic adjustment rule for the penalty parameter ε, described in the following algorithm, where $\{\varepsilon_j\}$ is a preselected sequence such that $\varepsilon_{j+1} < \varepsilon_j$ and $\varepsilon_j \to 0$ as $j \to \infty$.

ALGORITHM EPSZ

Initial guess: $z_o = x_o \in \mathring{S}$.

Step 0: Set $j = 0$.

Step 1: Set $k = 0$ and $x_o = z_j$.

Step 2: If $\nabla Z(x_k;\varepsilon_j) = 0$ go to step 3; else go to step 4.

Step 3: If $g(x_k)+Y(x_k;\varepsilon_j)y(x_k;\varepsilon_j) = 0$ stop; else go to step 6.

Step 4: If $\|\nabla Z(x_k;\varepsilon_j)\|^2 + \|A(x_k)\nabla Z(x_k;\varepsilon_j)\|^2 \geq \|g(x_k)+Y(x_k;\varepsilon_j)y(x_k;\varepsilon_j)\|^2$ go to step 5; else go to step 6.

Step 5: Compute $x_{k+1} \in I[x_k]$, set $k = k+1$ and go to step 2.

Step 6: Set $z_{j+1} = x_k$, $j = j+1$ and go to step 1.

The convergence of the preceding algorithm is stated in the following proposition, whose proof is given in [3].

PROPOSITION 9. *Suppose that for every* $\varepsilon > 0$ *and* $x_o \in \mathring{S}$, *any accumulation point of the sequence* $\{x_k\}$ *generated by the iteration map* I *is a stationary point of*

Z *belonging to* $\overset{\circ}{S}$. *Then, either the algorithm terminates at some* $x_\nu \in \overset{\circ}{S}$ *and* $(x_\nu,\lambda(x_\nu))$ *is a* K-T *pair for Problem* P, *or the algorithm produces an infinite sequence* $\{x_k\} \subseteq \overset{\circ}{S}$ *such that every limit point* $\bar{x}$ *yields a* K-T *pair* $(\bar{x},\lambda(\bar{x}))$ *for Problem* P. □

It is important to observe that the algorithm above cannot produce unbounded sequences and this constitutes a relevant improvement with respect to algorithms for the automatic adjustment of the penalty parameter defined up to now. (See, for instance [7], [9] and [10]).

4. UNCONSTRAINED MINIMIZATION ALGORITHMS FOR $W(x,\lambda;\varepsilon)$

In this section we report the main properties of the exact augmented Lagrangian $W(x,\lambda;\varepsilon)$ and we describe unconstrained minimization algorithms for this function on the product space $\overset{\circ}{S} \times R^m$. We refer to [4] for the proofs of Propositions 10-15.

PROPOSITION 10. *For any* $\varepsilon > 0$:

(a) *the function* W *is continuously differentiable with respect to* x, λ *on* $\overset{\circ}{S} \times R^n$, *with gradient given by*

$$\nabla_x W(x,\lambda;\varepsilon) = \nabla f(x)+A(x)'\lambda + \frac{2}{\varepsilon} A(x)'B^{-1}(x)[g(x)+Y(x,\lambda;\varepsilon)y(x,\lambda;\varepsilon)]$$
$$+\frac{1}{\varepsilon} A(x)'[G(x)+Y^2(x,\lambda;\varepsilon)]B^{-2}(x)[g(x)+Y(x,\lambda;\varepsilon)y(x,\lambda;\varepsilon)]$$
$$+N(x,\lambda)[A(x)\nabla f(x)+M(x)\lambda]$$

where

$$N(x,\lambda) \triangleq 2\eta[\nabla_x^2 L(x,\lambda)A(x)' + \sum_{i=1}^{n} \nabla_x^2 g_i(x)\nabla_x L(x,\lambda)e_i' + 2\gamma^2 A(x)'G(x)\Lambda],$$

e_i *denotes the* i^{th} *column of the* $m \times m$ *identity matrix and* $\Lambda \triangleq \mathrm{diag}(\lambda_i)$, $i = 1,\ldots,m$;

$$\nabla_\lambda W(x,\lambda;\varepsilon) = g(x)+Y(x,\lambda;\varepsilon)y(x,\lambda;\varepsilon)+2\eta M(x)(A(x)\nabla f(x)+M(x)\lambda);$$

(b) *for any* $(x_o,\lambda_o) \in \overset{\circ}{S} \times R^m$ *the level set*

$$\{(x,\lambda): x \in \overset{\circ}{S},\ \lambda \in R^m,\ W(x,\lambda;\varepsilon) \le W(x_o,\lambda_o;\varepsilon)\}$$

is compact, so that the function W *admits a global minimum point on* $\overset{\circ}{S} \times R^m$. □

PROPOSITION 11.

(a) *Let* $(\bar{x},\bar{\lambda})$ *be a* K-T *pair for Problem* P; *then for any* $\varepsilon > 0$, $(\bar{x},\bar{\lambda})$ *is a stationary point of* W *and* $W(\bar{x},\bar{\lambda};\varepsilon) = f(\bar{x})$;

(b) *there exists an* $\varepsilon^* > 0$ *such that, for all* $\varepsilon \in (0,\varepsilon^*]$, *if* $(\bar{x},\bar{\lambda}) \in \overset{\circ}{S} \times R^m$ *is a stationary point of* W, *the pair* $(\bar{x},\bar{\lambda})$ *satisfies also the* K-T *conditions for Problem* P. □

PROPOSITION 12.

(a) *If* $\bar{x} \in \overset{\circ}{S}$ *is an isolated local minimum point of Problem* P *and* $\bar{\lambda}$ *is the associated* K-T *multiplier, then there exists an* $\varepsilon^* > 0$ *such that, for all* $\varepsilon \in (0,\varepsilon^*]$, *the pair* $(\bar{x},\bar{\lambda})$ *is a local unconstrained minimum point of* W;

(b) *there exists an* $\varepsilon^* > 0$ *such that for all* $\varepsilon \in (0,\varepsilon^*]$, *if* $(\bar{x},\bar{\lambda}) \in \overset{\circ}{S} \times R^m$ *is a local unconstrained minimum point of* W, $\bar{x}$ *is a local minimum point of Problem* P *and* $\bar{\lambda}$ *is the corresponding* K-T *multiplier.* □

PROPOSITION 13. *There exists an* $\varepsilon^* > 0$ *such that for all* $\varepsilon \in (0,\varepsilon^*]$ *if* $\bar{x}$ *is a global minimum point of Problem* P *and* $\bar{\lambda}$ *is the corresponding* K-T *multiplier, the pair* $(\bar{x},\bar{\lambda})$ *is a global minimum point of* W *on* $\overset{\circ}{S} \times R^m$ *and conversely.* □

In connection with function W we introduce the following index sets

$I_+(x,\lambda) \triangleq \{i:2g_i(x)+\varepsilon b_i(x)\lambda_i > 0\}$, $I_-(x,\lambda) \triangleq \{i:2g_i(x)+\varepsilon b_i(x)\lambda_i \leq 0\}$;

and we adopt for the indexed submatrices and subvectors the same conventions introduced in the preceding section.

Then we have:

PROPOSITION 14. *Let $(\bar{x},\bar{\lambda})$ be a K-T pair for Problem P and assume that strict complementarity holds at $(\bar{x},\bar{\lambda})$. Then, for any $\varepsilon \geq 0$ the function W is twice continuously differentiable in a neighbourhood of $(\bar{x},\bar{\lambda})$.* □

PROPOSITION 15.

(a) *Let $(\bar{x},\bar{\lambda})$ be a point satisfying the K-T conditions for Problem P; assume that strict complementarity holds at $(\bar{x},\bar{\lambda})$ and that $\bar{x}$ is an isolated local minimum point for Problem P satisfying the second-order sufficiency conditions*

$$x'\nabla_x^2 L(\bar{x},\bar{\lambda})x > 0, \quad \forall x: A_o(\bar{x})x = 0,\ x \neq 0;$$

then there exists an $\varepsilon^* > 0$ such that, for all $\varepsilon \in (0,\varepsilon^*]$, $(\bar{x},\bar{\lambda})$ is an isolated local minimum point for W and the Hessian matrix $\nabla_x^2 W(\bar{x},\bar{\lambda};\varepsilon)$ is positive definite.

(b) *Suppose that strict complementarity holds at any K-T pair $(\bar{x},\bar{\lambda})$ of Problem P; then, there exists an $\varepsilon^* > 0$ such that, for all $\varepsilon \in (0,\varepsilon^*]$, if $(\bar{x},\bar{\lambda}) \in \mathring{S} \times R^m$ is a local unconstrained minimum point of W with positive definite Hessian $\nabla_x^2 W(\bar{x},\bar{\lambda};\varepsilon)$, $\bar{x}$ is an isolated minimum point of Problem P satisfying the second-order sufficiency condition.* □

We can now define Newton-type algorithms for the unconstrained minimization of W on $\mathring{S} \times R^m$, along the same lines followed in the preceding section. For the proofs of Propositions 16 and 17 we refer to [8].

Let $H(x,\lambda;\varepsilon)$ be the symmetric matrix defined by $H(x,\lambda;\varepsilon) = (H_{ij})$ $i,j=1,2,3$, where the submatrices H_{ij} are given by:

$$H_{11} = \nabla_x^2 L(x,\lambda) + \frac{2}{\varepsilon} A_+(x)'B_+^{-1}(x)A_+(x)+2\eta[\nabla_x^2 L(x,\lambda)A(x)'A(x)\nabla_x^2 L(x,\lambda)]$$

$$H_{22} = 2\eta\, A_+(x)A(x)'A(x)A_+(x)'$$

$$H_{33} = -\frac{\varepsilon}{2} B_-(x)+2\eta(A_-(x)A(x)'A(x)A_-(x)'+\gamma^4 G_-^4(x)+\gamma^2 A_-(x)A_-(x)'G_-^2(x)+\gamma^2 G_-^2(x)A_-(x)A_-(x)')$$

$$H_{12} = A_+(x)'+2\eta\nabla_x^2 L(x,\lambda)A(x)'A(x)A_+(x)'$$

$$H_{13} = 2\eta\nabla_x^2 L(x,\lambda)(A(x)'A(x)A_-(x)'+\gamma^2 A_-(x)'G_-^2(x))$$

$$H_{23} = 2\eta(A_+(x)A(x)'A(x)A_-(x)'+\gamma^2 A_+(x)A_-(x)'G_-^2(x)).$$

Then we can state the following proposition:

PROPOSITION 16. *Let $(\bar{x},\bar{\lambda})$ be a K-T pair of Problem P, satisfying the assumptions of Proposition 15 (a). Then, we have $H(\bar{x},\bar{\lambda};\varepsilon) = \nabla^2 W(\bar{x},\bar{\lambda};\varepsilon)$. Moreover, there exists an $\varepsilon^* > 0$ such that, for all $\varepsilon \in (0,\varepsilon^*]$, the matrix $H(x,\lambda;\varepsilon)$ is positive definite in a neighbourhood of $(\bar{x},\bar{\lambda})$.* □

Therefore we can define the following algorithm on $\mathring{S} \times R^m$.

ALGORITHM 3

$$\begin{bmatrix} \hat{x} \\ \hat{\lambda} \end{bmatrix} = \begin{bmatrix} x \\ \lambda \end{bmatrix} + \alpha d$$

$$H(x,\lambda;\varepsilon)d = -\nabla W(x,\lambda;\varepsilon)$$

where $d \in R^{n+m}$.

It can be shown that also the solution $d = [d_x' \; d_+' \; d_-']'$ of the system:

$$\begin{bmatrix} \nabla_x^2 L(x,\lambda) & A_+(x)' & A_-(x)' \\ A_+(x) & 0 & 0 \\ 0 & 0 & I \end{bmatrix} \begin{bmatrix} d_x \\ d_+ \\ d_- \end{bmatrix} = - \begin{bmatrix} \nabla_x L(x,\lambda) \\ g_+(x) \\ \lambda_- \end{bmatrix} \tag{2}$$

is a consistent approximation of the Newton's direction for $W(x,\lambda;\varepsilon)$.

Note that if we set $\lambda = \lambda(x)$ in system (2) and in the index sets $I_+(x,\lambda)$, $I_-(x,\lambda)$, system (2) reduces to system (1) considered in the preceding section.

The next proposition establishes the consistency of the search direction d.

PROPOSITION 17. *Let* $(\bar{x},\bar{\lambda})$ *be a* K-T *pair for Problem* P *and assume that strict complementarity holds at* $(\bar{x},\bar{\lambda})$. *Then there exists a neighbourhood* $\Omega \times \Lambda \subseteq \overset{\circ}{S} \times R^m$ *of* $(\bar{x},\bar{\lambda})$ *such that, for all* $(x,\lambda) \in \Omega \times \Lambda$, *if* $d \triangleq [d_x' \; d_+' \; d_-']'$ *is a solution of system* (2), *it results*:

$$\hat{H}(x,\lambda;\varepsilon)d = -\nabla W(x,\lambda;\varepsilon)$$

where $\hat{H}(x,\lambda;\varepsilon)$ *is a continuous matrix defined on* $\Omega \times \Lambda$ *and satisfying*:

$$\hat{H}(\bar{x},\bar{\lambda};\varepsilon) = \nabla^2 W(\bar{x},\bar{\lambda};\varepsilon). \; \square$$

Then we can define a Newton-type algorithm based on the consistent approximation of the Newton's direction obtained by solving system (2).

ALGORITHM 4

$$\begin{bmatrix} \hat{x} \\ \hat{\lambda} \end{bmatrix} = \begin{bmatrix} x \\ \lambda_+ \\ \lambda_- \end{bmatrix} + \alpha \begin{bmatrix} d_x \\ d_+ \\ d_- \end{bmatrix}$$

where d_x, d_+, d_- are obtained by solving the equations

$$\begin{bmatrix} \nabla_x^2 L(x,\lambda) & A_+(x)' \\ A_+(x) & 0 \end{bmatrix} \begin{bmatrix} d_x \\ z \end{bmatrix} = - \begin{bmatrix} \nabla f(x) \\ g_+(x) \end{bmatrix}$$

$$d_+ = z + \lambda_+$$

$$d_- = -\lambda_-.$$

By using a Quasi-Newton approximation of $\nabla_x^2 L(x,\lambda)$ we can also derive Quasi-Newton algorithms for the minimization of W employing only first order derivatives of the problem functions.

Algorithms 3 and 4 define Newton-type iterations on $R^n \times R^m$, which ensure an ultimate superlinear convergence rate.

As regards global onvergence, we recall that the level set $\{(x,\lambda), \; x \in \overset{\circ}{S}, \; \lambda \in R^m: W(x,\lambda;\varepsilon) \leq W(x_o,\lambda_o;\varepsilon)\}$ is compact for any given $\varepsilon > 0$ and $(x_o,\lambda_o) \in \overset{\circ}{S} \times R^m$, as stated in (b) of Proposition 10. Hence, making use of the Newton-type algorithms considered before, in conjunction with some suitable stabilization scheme, it is possible to introduce an iteration map Q: $\overset{\circ}{S} \times R^m \to 2^{\overset{\circ}{S} \times R^m}$ such that the sequence produced by Q remains in $\overset{\circ}{S} \times R^m$ and admits limit points which are stationary points of W.

Then we can introduce an algorithm employing an automatic adjustment rule for the penalty parameter ε, which ensures convergence towards K-T points of the constrained problem. In this algorithm we make use of a preselected sequence $\{\varepsilon_j\}$, with $\varepsilon_{j+1} < \varepsilon_j$, $j = 0,1,\ldots$, and $\varepsilon_j \to 0$ as $j \to \infty$.

ALGORITHM EPSW

Initial guess: $z_o = (x_o,\lambda_o) \in \overset{\circ}{S} \times R^m$

Step 0: Set $j = 0$.

Step 1: Set $k = 0$ and $(x_o,\lambda_o) = z_j$.

Step 2: If $\nabla_x W(x_k,\lambda_k;\varepsilon_j) = 0$ and $\nabla_\lambda W(x_k,\lambda_k;\varepsilon_j) = 0$ go to step 3; else go to step 4.

Step 3: If $g(x_k)+Y(x_k,\lambda_k;\varepsilon_j)y(x_k,\lambda_k;\varepsilon_j) = 0$ stop; else go to step 6.

Step 4: If $\|\nabla_\lambda W(x_k,\lambda_k;\varepsilon_j)\|^2 + \|\nabla_x W(x_k,\lambda_k;\varepsilon_j)\|^2 + \|A(x_k)\nabla_x W(x_k,\lambda_k;\varepsilon_j)\|^2$

$\geq \|g(x_k)+Y(x_k,\lambda_k;\varepsilon_j)y(x_k,\lambda_k;\varepsilon_j)\|^2$ go to step 5; else go to step 6.

Step 5: Compute $(x_{k+1},\lambda_{k+1}) \in Q(x_k,\lambda_k)$, set $k = k+1$ and go to step 2.

Step 6: Set $z_{j+1} = (x_k,\lambda_k)$, $j = j+1$ and go to step 1.

As regards the convergence of the preceding algorithm we can state:

PROPOSITION 18. *Suppose that for every* $\varepsilon > 0$ *and* $(x_o,\lambda_o) \in \mathring{S} \times R^m$ *any accumulation point of the sequence* $\{x_k,\lambda_k\}$ *generated by the iteration map* Q *is a stationary point of* W *belonging to* $\mathring{S} \times R^m$. *Then, either the algorithm terminates at some* $(x_\nu,\lambda_\nu) \in \mathring{S} \times R^m$ *and* (x_ν,λ_ν) *is a* K-T *pair for Problem* P *or the algorithm produces an infinite sequence* $\{x_k,\lambda_k\} \subseteq \mathring{S} \times R^m$ *such that every limit point* $(\bar{x},\bar{\lambda})$ *yields a* K-T *pair for Problem* P. □

The proof of Proposition 18 is given in [4].

5. NUMERICAL RESULTS

The algorithms introduced in the preceding sections have been tested by solving two standard test problems (TP).

TP1. (Rosen and Suzuki)

Minimize $f(x) = -5(x_1+x_2)+7(x_4-3x_3)+x_1^2+x_2^2+2x_3^2+x_4^2$ subject to:

$$\left(\sum_{i=1}^{4} x_i^2\right) + x_1 - x_2 + x_3 - x_4 - 8 \leq 0$$

$$x_1^2 + 2x_2^2 + x_3^2 + 2x_4^2 - x_1 - x_4 - 10 \leq 0$$

$$2x_1^2 + x_2^2 + x_3^2 + 2x_1 - x_2 - x_4 - 5 \leq 0.$$

Solution: $x^* = (0,1,2,-1)'$ with $f(x^*) = -44$.

Starting point: $x_o = 0$.

TP2. (Wong)

Minimize $f(x) = (x_1-10)^2+5(x_2-12)^2+x_3^4+3(x_4-11)^2+10x_5^6+7x_6^2+x_7^4-4x_6x_7-10x_6-8x_7$

subject to:

$$2x_1^2 + 3x_2^4 + x_3 + 4x_4^2 + 5x_5 - 127 \leq 0$$

$$7x_1 + 3x_2 + 10x_3^2 + x_4 - x_5 - 282 \leq 0$$

$$23x_1 + x_2^2 + 6x_6^2 - 8x_7 - 196 \leq 0$$

$$4x_1^2 + x_2^2 - 3x_1x_2 + 2x_3^2 + 5x_6 - 11x_7 \leq 0.$$

Solution: $x^* = (2.33050, 1.95137, -0.47754, 4.36573, -0.62448, 1.03813, 1.594)'$

with $f(x^*) = 680.630$.

Starting point: $x_o = (1,2,0,4,0,1,1)'$.

For both problems, it was taken $\gamma = 1$ in the expression of the matrix $M(x)$. The starting values of the multipliers in the minimization of W was $\lambda_o = 0$. As regards the barrier term it was taken $\alpha_i = 1$ in all cases.

It can be easily verified that TP1 has a compact feasible set; as regards TP2, since the feasible set is not compact, it was added the constraint:

$$\|x\|^2 \le 50.$$

In Table 1 we report the results obtained by employing Algorithms 2 and 4. More precisely, we show, for various values of ε, the number LS of line searches and the number NF of function evaluations needed to attain the solution with an accuracy on the objective function of the order 10^{-6}. In Algorithm 4 it was taken $\eta = 10^{-3}$ for TP1 and $\eta = 10^{-4}$ for TP2.

	ε	ALG2 LS	ALG2 NF	ALG4 LS	ALG4 NF
TP1	1	5	6	10	12
	0.1	11	19	9	12
	0.01	15	27	13	24
TP2	1	18	22	12	16
	0.1	19	23	12	16
	0.01	23	35	13	20

Table 1

It can be observed that, as usual with penalty methods, both LS and NF increases for decreasing values of ε, below the threshold value.

We remark that, if we take the values indicated for ε in table 1 as starting values in ALGZ and ALGW, the solution is attained without changing ε. In order to see the effect of the automatic adjustment of ε, it was necessary, for both problems, the adoption of larger starting values for ε. In this connection it was experienced that when the adjustment of ε is due to a transfer to step 6 from step 3, the best policy is that of restarting the algorithm from the given initial point or, if available, from the best feasible point produced in the preceding iterations.

REFERENCES

[1] R. FLETCHER: *Penalty functions* in: Mathematical Programming. The State of the Art, A. Bachem, M. Grötschel and B. Korte eds., Springer Verlag, Berlin, 1983.

[2] G. DI PILLO and L. GRIPPO: *A continuously differentiable exact penalty function for nonlinear programming problems with inequality constraints*, SIAM Journal on Control and Optimization, Vol. 23, 1985, pp. 72-84.

[3] G. DI PILLO and L. GRIPPO: *An exact penalty method with global convergence properties for nonlinear programming problems*, IASI-CNR Techn. Rep. R99, 1984 (to appear in Mathematical Programming).

[4] S. LUCIDI: *New results on a class of exact augmented Lagrangians*, IASI-CNR Techn. Rep. R116, 1985.

[5] G. DI PILLO and L. GRIPPO: *A new class of augmented Lagrangians in nonlinear programming*, SIAM J. on Control and Optimization, Vol. 17, 1979, pp. 618-628.

[6] G. DI PILLO and L. GRIPPO: *A new augmented Lagrangian function for inequality constraints in nonlinear programming*, J. of Optimization Theory and Applications, Vol. 36, 1982, pp. 495-519.

[7] T. GLAD and E. POLAK: *A multiplier method with automatic limitation of penalty growth*, Mathematical Programming, Vol. 17, 1979, pp. 140-155.

[8] G. DI PILLO, L. GRIPPO and S. LUCIDI: *Globally convergent exact penalty algorithms for constrained optimization*, IASI-CNR Techn. Rep. R 128, 1985.

[9] H. MUKAI and E. POLAK: *A quadratically convergent primal-dual algorithm with global convergence properties for solving optimization problems with equality constraints*, Mathematical Programming, Vol. 9, 1975, pp. 336-349.

[10] D.Q. MAYNE and N. MARATOS: *A first order, exact penalty function algorithm for equality constrained optimization problems*, Mathematical Programming, Vol. 16, 1979, pp. 303-324.

GLOBAL OPTIMIZATION ALGORITHMS: THEORY AND SOME APPLICATIONS

J. PINTÉR and J. SZABÓ

Research Centre for Water Resources Development
(VITUKI), H-1453 Budapest, P.O. Box 27., Hungary

1. INTRODUCTION

In the course of solving different decision problems, one often has to determine a number of parameters "optimally". As examples from the field of water resources management, planning and operation of reservoir systems or water quality treatment plants, identification and calibration of different descriptive models, hydrologic time-series analysis etc. can be mentioned. As a rule, the optimality criterion is expressed by a (real-valued) objective function, while the feasible domain for parameter selection is usually given by a number of constraints. Hence, the mentioned type of tasks frequently leads to a mathematical programming problem of the following form:

$$\min f(x) \qquad (1)$$
$$g_i(x) \leq 0 \quad i = 1, \ldots, m.$$

Here the vector x represents the decision variables (we suppose that x belongs to the n-dimensional Euclidean real space R^n), while f and g_i i = 1, ..., m are the objective function and the constraints, respectively.

As it is well-known, under fairly elementary (compactness and continuity) assumptions one can prove that problem (1) has a (not necessarily unique) global optimizer, i.e. such vector x^* for which there holds

$$g_i(x^*) \leq 0 \quad i = 1, \ldots, m,$$

and for all other feasible vectors x we have the inequality

$$f(x^*) \leq f(x).$$

However, most numerical methods in mathematical programming are capable to find only local optimizers. In other words, unless problem (1) is specified a priori to have only a single local (thus global) optimum, there may be significant differences between local optima, depending on the starting point of the applied locally convergent algorithm. This fact has motivated studies in the field of global optimization, where we seek for a globally optimal solution of (1), while it may have several local optima. Due to some inherent theoretical and numerical difficulties, this research became more intensive only recently, cf. e.g. the works of Archetti (1980), Boender et al. (1982), Boender (1984), Dixon and Szegő (1975, 1978), Pintér (1984a), Strongin (1978), Timmer (1984) or Zilinskas (1982, 1985).

In the sequel, instead of the general problem (1), first its specified form will be investigated:

$$\min_{a \le x \le b} f(x) \qquad f \in F[a,b], \tag{2}$$

where a,b are n-dimensional (finite) vectors, and f belongs to the class $F[a,b]$ of those Lipschitz-continuous real functions which have only a finite number of global optima on the n-dimensional interval [a, b]. In other words,

$$F[a,b] = \bigcup_{L>0} F_L[a,b] ,$$

where $F_L[a,b]$ denotes the set of those functions f (with a finite set X^* of global minima), for which there holds

$$|f(x_1) - f(x_2)| \le L \, ||x_1 - x_2|| \ ; \ a \le x_1, x_2 \le b.$$

Note that the special constraint-set of (2) is typical in many applications.

In Pintér (1983) we defined a general class of globally convergent one-dimensional (directional) optimization procedures for solving problem (2). The examples, presented there (different known and some new algorithms), indicated the applicability of the described axiomatic approach. Different extensions of this result are given in Pintér (1984b, 1985a,b).

Section 2 summarizes the theory introduced in the mentioned papers. A global optimization algorithm and two numerical examples are briefly presented in Sections 3,4: for more computational details see (Pintér, 1985a), (Pintér, Szabó and Somlyódy, 1985).

2. BASIC CONCEPTS AND THEORETICAL RESULTS

Assume that the sequence of points $\{x^k\}$, generated for finding optimal solution(s) $x^* \in X^*$ to (2), is defined by the following type of algorithm, which gradually divides [a,b] into adaptively constructed n-dimensional subintervals.

GA(n) conceptual n-dimensional global optimization algorithm

Denote by ℓ, $m = m(\ell)$ and $k = k(\ell)$ the current number of iteration cycle, generated subintervals and points (vertices of subintervals), respectively. The lexicographically ordered vertices of the i-th subinterval will be denoted by

$$x(i) = \{x(i,1), x(i,2), \ldots, x(i,2^n)\},$$

($x(i)$ is a matrix of $(n \times 2^n)$ size). The vector of respective function values is denoted by

$$z(i) = \{z(i,1), z(i,2), \ldots, z(i,2^n)\}$$

$$z(i,j) = f(x(i,j)) \qquad j = 1, \ldots, 2^n$$

(hence, $z(i)$ is a vector, consisting of 2^n components). Further on, we assume that a real-valued interval selection function

$$R(i) = R(x(i), z(i))$$

and a real vector-valued point selection function

$$S(i) = S(x(i), z(i))$$

can be evaluated for each generated interval, respectively. Applying the above notations, the algorithm-scheme GA(n) consists of the following steps:

Step 0: Initialization. Set $\ell:=1$, $m:=1$, $k:=2^n$. The initial search information is represented by the vertices and corresponding function values of $[a,b]$.

Step 1: Interval sequencing. This can be accomplished e.g. by lexicographically ordering the "lower left corners" (i.e. the vertices with component-wise minimal coordinates) of the subintervals, generated up to the ℓ-th cycle.

Step 2: Interval selection. Choose an (arbitrary) index t, for which there holds

$$R(t) = \max_{1 \le i \le m} R(i) \qquad m = m(\ell).$$

Step 3: Point selection. Choose a new point from interval $x(t)$:

$$x^{k+1} = S(t)$$

(unless some termination criteria are fulfilled). Then x^{k+1} defines the partitioning of $x(t)$ to 2^n new subintervals: these are generated by the intersections of the boundary of $x(t)$ and the hyperplanes which contain x^{k+1} and are parallel to the boundary hypersurfaces of $x(t)$. The algorithm continues at Step 1 of the next iteration cycle ($\ell := \ell+1$).

The above algorithm-scheme is of fairly general character to cover a rather broad class of direct (derivative-free) global optimization procedures: some essentially GA(n)-type univariate algorithms are collected in Pintér (1983). Obviously, the particular realizations of GA(n) methods depend primarily on the choice of the decision functions R and S. Below we quote some theoretical results concerning GA(n) algorithms.

Theorem 1. (Pintér, 1984b). Assume that the functions R and S satisfy the following conditions:

A1(n). R(i) is continuous with respect to (w.r.t.) the vertices of interval i.

A2(n). R(i) is translation invariant w.r.t. x(i) (i.e. the position of interval i on [a,b]).

A3(n). R(i) is a strictly monotonously decreasing function of the components of z(i) (i.e. the function values at the vertices of interval i).

A4(n). $R(i) > R(x,f(x)) = \lim_{\substack{\ell\to\infty \\ x(i)\to x}} R(x(i), z(i))$

for all points of interval i (the right-hand side limit function is supposed to be existing and continuous).

A5(n). $x^{k+1} = S(t)$ belongs to $x_c(t)$, where $x_c(t)$ denotes the central contraction of interval t by a fixed factor $0<c<1$.

Then, for arbitrary $f \in F[a,b]$, the GA(n) algorithm defined by the above R and S generates a sequence of points $\{x^k\}$ such that the set of its accumulation points is identical to X^*, the solution set of the global optimization problem (2).

Besides Theorem 1, an efficiency estimate was given which shows that GA(n)-type methods may generate only a bounded number of points on subintervals of [a,b], which do not contain global optimizer(s). Both mentioned results are specified also for the case, when $f \in F_L[a,b]$ and the constant L (or its upper estimate) is known. The details of these theoretical investigations are described in Pintér (1983, 1984b).

The general character of the mentioned results shows that, in principle, one has a great freedom when constructing GA(n)-type algorithms. On the other hand, there are some obvious difficulties, connected with the numerical realization of such algorithms: namely, the usual dimensionality and computational complexity problems. Therefore it is important to define such GA(n)-type algorithms, which have relatively small storage requirements and easily computable decision functions R and S. E.g. it is possible to construct diagonal-type interval selection functions of the form

$$R_d(i) = R_d(||x(i,1)-x(i,2^n)||, z(i,1), z(i,2^n)) \quad (3)$$

which depend only on the information represented by the main diagonal of interval i (i.e. the "lower left", "upper right" vertices of x(i) and the corresponding function values).

Theorem 2. (Pintér, 1985a). Assume that $R_d(i)$ is continuous w.r.t. $x(i,1)$ and $x(i,2^n)$; continuous, symmetrical and strictly monotonously decreasing w.r.t. $z(i,1)$ and $z(i,2^n)$; finally, that it satisfies requirement A4(n) of Theorem 1. Then $R_d(i)$ can be applied in the GA(n) algorithm-scheme. In other words, the method defined by the diagonal characteristic function $R_d(i)$ and a suitable function S(t) will generate such a sequence $\{x^k\}$, the accumulation point set of which is identical with the solution set X^* of (2).

The main numerical advantage of applying diagonal-type decision functions R, S lies in the fact that their use permits to reduce the storage requirements: viz. instead of all the 2^n vertices of x(i) (plus respective function values) only two vertices and function values are to be stored.

The extension of the above results to the case of convex feasible sets is summarized below. Suppose now that instead of (2), we face a more general global optimization problem of the form

$$\min_{x \in H} f(x) \quad (4)$$

(cf. (1)), where H is the closure of a (non-empty), bounded, open convex set in R^n. Assume that H is imbedded into a (known) n-dimensional interval [a,b]. It is shown in (Pintér, 1985b), that if the objective function f is Lipschitz-continuous on H, then it can be extended to be Lipschitz-continuous on [a,b] (with a suitably defined Lipschitz-constant). Hence, the above cited results permit to solve multiextremal optimization problems on convex (compact) feasible sets. Some numerical aspects of realizing this approach are treated in (Pintér, 1985b).

3. A GA(n)-TYPE ALGORITHM

As an example, consider the global optimization method, defined by the following decision functions:

$$R_d(i) = K\,||\,x(i,2^n) - x(i,1)\,|| + \frac{[z(i,2^n)-z(i,1)]^2}{K\,||x(i,2^n)-x(i,1)||} - 2[z(i,1) + z(i,2^n)], \quad (5)$$

where

$$K=K(\ell) = (4 + \frac{C}{\ell}) \cdot \max_{1 \le i \le m} \frac{|z(i,2^n) - z(i,1)|}{||x(i,2^n)-x(i,1)||}, \quad m=m(\ell) \quad (C>0);$$

$$S(t) = Q\{x(t,1), \frac{1}{2}[x(t,1)+x(t,2^n)], \; x(t,2^n)\}. \tag{6}$$

We remark that (5) is a diagonal-type generalization of Strongin's univariate interval-selection function (Strongin, 1978). It is easy to verify that R is continuous w.r.t. x(i) and z(i); translation-invariant w.r.t. x(i), symmetrical and strictly monotonously decreasing in z(i,1) and $z(i,2^n)$ and - with proper parameterization - its value is greater than the respective value of it for arbitrary degenerated (one-point) subinterval of x(i). Basically, these properties are sufficient to assure global convergence of the algorithm, defined by (5) and (6) (cf. the assertions of Theorem 1 and 2). We note that formula (6) symbolizes a quadratic interpolation on the main diagonal of x(t). It is worth to point out that the function S has only a minor role in terms of theoretical convergence; however, it may significantly influence the local efficiency of the global search algorithm. (Actually, S is supplemented with some technical modification, in order to maintain convergence.)

Our computational algorithm combines global and local optimization phases: the global search procedure outlined above is changed for local search, when some stopping criterion is fulfilled (e.g. the main diagonal of the subinterval chosen at Step 2 is less, than a fixed threshold value). If one determines a priori the number M of (global or local) minima to be found, then M global and local search phases will be iteratively accomplished before the optimization procedure is ended. The applied local search consists of several cycles of a conjugated gradients type method (Fletcher and Reeves, 1964), in which gradients are estimated by finite differences. (It is felt important that the algorithm should be applicable also in absence of higher order analytical information, as in many technical problems the analytical form of the figuring functions is unknown or can not be used to calculate derivates: see e.g. the examples below.)

The combined global-local algorithm has been applied first to some known univariate (n=1) and multidimensional (n = 2,3,4) test problems, given in the works of Shubert (1972), Strongin (1978),Dixon and Szegő, eds. (1978), respectively. According to the numerical results, presented in (Pintér, 1985a), the above described algorithm is fairly stable (in all test problems each global optimizer was found) and efficient (the number of function evaluations, necessary to locate the optimal solution with prescribed accuracy, was in general significantly less,than in the results surveyed by the above mentioned references).

4. APPLICATIONS

For illustrating the "real-world" applicability of the proposed methodology, we mention two examples from water resources engineering.

Calibration of a streamflow forecasting model

The National Water Monitoring Service (Budapest, Hungary) computes daily streamflow forecasts. These forecasts are based on an adequate discretized version of the so-called Kalinin-Milyukov-Nash cascade model (for details cf. e.g. Szöllősi-Nagy (1982)). The respective discrete state-space model is described by the vector equations

$$\underline{x}_t = F_t(\Delta t)\ \underline{x}_{t-\Delta t} + G_t(\Delta t)\ \underline{u}_t \quad (\underline{x}_o \text{ is known}), \tag{7}$$

$$\underline{y}_t = H\ \underline{x}_t,$$

where the input process $\underline{u}(t)$ and the output process $\underline{y}(t)$ are sampled in equidistant time-moments $t = \Delta t, 2\Delta t, \ldots, N\Delta t$; further on, $\underline{x}_t$ is the system state, $F_t(\Delta t)$ is the state-transition matrix, $G_t(\Delta t)$ is the input-transition matrix, H is the output generator matrix. (In the considered special application H is a vector, thus $\underline{y}_t = y_t$ is a scalar.) Our aim is to find an optimal parameterization of the model (7), which minimizes the deviation between observed $\{\hat{y}_t\}$ and computed $\{y_t\}$ output time-series in the sense

$$\min_{\substack{n_{min}\le n\le n_{max}\\ k_{min}\le k\le k_{max}}} f(n,k) \qquad f(n,k) = \sum_{t=\Delta t}^{N\Delta t} (y_t-\hat{y}_t)^2, \qquad y_t=y_t(n,k). \tag{8}$$

The parameters n,k explicitly figure in the definition of the matrices $F_t(\Delta t)$ and $G_t(\Delta t)$: hence, (8) can be regarded as a special case of the global optimization problem (2). (Note that - as illustrated by Pintér, Szabó and Somlyódy (1985) - the parameter estimation problem (8) is typically multiextremal, if the values y_t are defined by a nonlinear function of the model parameters.) From the physical interpretation of the model (7) we know that n is a positive integer, while k is a continuous variable. Consequently, for each feasible n, we have to solve a univariate optimization problem for finding k. In our case the bounds $1\le n\le 10$, $0\le k\le 10$ were given; the solution of each univariate problem needed only some tens of steps (evaluations of f(n,k)).

Calibration of a sediment-water interaction model for shallow lakes

Wind-induced sediment-water interaction has a considerable impact on the artificial eutrophication of shallow lakes by two phenomena:

a) the change in the concentration of particulate and dissolved phosphorus fractions, due to particle resuspension and possible sorption;
b) the variation of light transparency properties influencing algal growth, due to the change of suspended solids (SS) concentration.

From the viewpoint of both influences, the determination of the dynamics of SS is of primary importance. As shown by Somlyódy (1982), the temporal variation of volume-averaged SS concentration in a lake (or its spatially homogeneous subbasin) can be approximately described by the differential equation

$$\frac{dy}{dt} = - b_1 y + b_2 w^n, \tag{9}$$

where t is time, $y = y(t)$ denotes the concentration value, $w = w(t)$ is the absolute value of wind speed; finally, b_1, b_2 and n are model parameters to be optimized.

Assume that the value of w is constant between t and $t + \Delta t$. Then, for $t = 0, \Delta t, 2\Delta t, \ldots N\Delta t$, the discretized solution of (9) can be explicitly calculated from

$$y_{t+\Delta t} = y_t \, e^{-b_1 \Delta t} + \frac{b_2}{b_1} w_t^n (1 - e^{-b_1 \Delta t}) \tag{10}$$

(the initial value y_o is supposed to be known). Using the expression (10) for computing the values y_t - given b_1, b_2 and n - and knowing the observed values $\hat{y}_t$, the parameter estimation problem can be formulated as

$$\min f(b_1,b_2,n) \qquad f(b_1,b_2,n) = \sum_{t=\Delta t}^{N\Delta t} (y_t - \hat{y}_t)^2, \tag{11}$$

$$b_{1min} \le b_1 \le b_{1max}$$

$$b_{2min} \le b_2 \le b_{2max} \qquad y_t = y_t \; (b_1,b_2,n).$$

$$n_{min} \le n \le n_{max}$$

Again, this problem can be regarded as a special case of (2): now we have three continuous parameters, which are to be optimized. The feasible domain was given by the constraints $10^{-20} \le b_i \le 2$ $i=1,2$, $0 \le n \le 6$. The yielded three-dimensional global optimization problem was solved computing some 450 function values of $f(b_1, b_2, n)$; for a detailed analysis of the results see Pintér, Szabó and Somlyódy (1985).

5. CONCLUDING REMARKS

In this paper the global optimization problem (2) and a class of methods for its solution were briefly described. Some basic results and their extensions were surveyed, a GA(n) method was given and some numerical results were briefly presented.

Of course, a small number of experimental runs does not justify statements about the general efficiency and robustness of any particular method. We think, however, that - according to the obtained numerical results - the proposed method is reliable and rather fast.What is probably more important that the general framework of GA(n)-type methods makes possible to construct flexible procedures to solve global optimization problems, which are inherently more difficult than local optimization problems of similar size.

ACKNOWLEDGEMENT

All experimental runs, mentioned or described in this paper, were accomplished on an MO8X personal computer and an R10 computer of the Research Centre for Water Resources Development (Budapest, Hungary).

REFERENCES

1. Archetti, F. (1980) "Analysis of stochastic strategies for the numerical solution of the global optimization problem", in: Archetti,F. and Cugiani, M. (eds.), Numerical Techniques for Stochastic Systems (North Holland, Amsterdam), pp. 275-295.
2. Boender,C.G., Rinnooy Kan, A.H.G., Stougie,L. and Timmer, G.T. (1982) "A stochastic method for global optimization", Mathematical Programming 22, 125-140.
3. Boender, C.G.E. (1984) "The generalized multinomial distribution. A Bayesian analysis and applications", Ph.D. dissertation,Erasmus University, Rotterdam.
4. Dixon, L.C.W. and Szegő, G.P. (eds.) (1975, 1978) Towards Global Optimization, Vol. 1-2. (North-Holland, Amsterdam).
5. Fletcher,R. and Reeves,C.M. (1964) "Function minimization by conjugate gradients", Computer Journal 7, 149-154.
6. Pintér,J. (1983) "A unified approach to globally convergent one-dimensional optimization algorithms", Technical report IAMI-83.5, Istituto per le Applicazioni della Matematica e dell'Informatica CNR, Milan.

7. Pintér, J. (1984a) "Convergence properties of stochastic optimization procedures", Mathematische Operationsforschung und Statistik, Series Optimization 14, 405-427.

8. Pintér, J. (1984b) "Globally convergent methods for n-dimensional multiextremal optimization", Mathematische Operationsforschung und Statistik, Series Optimization (to appear).

9. Pintér, J. (1985a) "Extended univariate algorithms for n-dimensional global optimization", Computing (to appear).

10. Pintér, J., Szabó, J. and Somlyódy, L. (1985) "Multiextremal optimization for calibrating water resources models", Environmental Software (to appear).

11. Pintér, J. (1985b) "Global optimization on convex sets" (submitted for publication).

12. Shubert, B.O. (1972) "A sequential method seeking the global maximum of a function", SIAM Journal on Numerical Analysis 9, 379-388.

13. Somlyódy, L. (1982) "Water quality modelling: a comparison of transport-oriented and ecology-oriented approaches", Ecological Modelling 17, 183-207.

14. Strongin, R.G. (1978) Numerical methods for multiextremal problems (in Russian), (Nauka, Moscow).

15. Szöllősi-Nagy, A. (1982) "On the discretization of the continuous linear cascade by means of the state-space analysis", Journal of Hydrology 58, 223-236.

16. Timmer, G.T. (1984) "Global optimization: A stochastic approach", Ph.D. dissertation, Erasmus University, Rotterdam.

17. Zilinskas, A. (1982) "Axiomatic approach to statistical models and their use in multimodal optimization theory", Mathematical Programming 22, 104-116.

18. Zilinskas, A. (1985) "Axiomatic characterization of a global optimization algorithm and investigation of its search strategy", Operations Research Letters 4, 35-39.

PERIODIC SOLUTIONS OF AN INFINITE DIMENSIONAL RICCATI EQUATION

G. Da Prato

Scuola Normale Superiore
Piazza dei Cavalieri 7
56100 PISA, Italy

1. INTRODUCTION

We shall be concerned with the problem:

Minimize

$$J(u) = \frac{1}{2} \int_0^\tau [\langle My,y\rangle + \langle Nu,u\rangle]\, dt \qquad (1)$$

over all $u \in L^2(0,\tau;U)$ subject to

$$y' = Ay + f + Bu \quad , \quad y(0) = y(\tau) \qquad (2)$$

where the period τ is given.

We shall denote by H and U two Hilbert spaces and by $\Sigma^+(H)$ and $\Sigma^+(U)$ the set of all positive symmetric operators in H and U respectively.

We shall assume:

$$\left.\begin{array}{l} \text{i) } A \text{ is the infinitesimal generator of a } C_0\text{-semigroup} \\ \quad e^{tA} \text{ on } H,\ B \in L(U;H). \\ \text{ii) } f:\mathbb{R} \to H,\ M:\mathbb{R} \to \Sigma^+(H),\ N:\mathbb{R} \to \Sigma^+(U) \text{ are } \tau\text{-periodic} \\ \quad \text{and } N(t) \geq \varepsilon > 0,\ t \in \mathbb{R} \\ \text{iii) } 1 \text{ belongs to the resolvent set of } e^{\tau A} \end{array}\right\} \qquad (3)$$

Assumption iii, assures that problem (2) has a unique solution for any periodic control u.

Problem (1)-(2) have been extensively studied when H has finite dimension (see for instance [1], [4], [6]).

We look for τ-periodic solutions of the Riccati equation

$$Q' + A^{\star}Q + QA - QBN^{-1}B^{\star}Q + M = 0 \quad . \tag{4}$$

A strongly continuous mapping $Q:\mathbb{R} \to \Sigma^{+}(H)$ is called a <u>solution</u> of (4) if

$$Q(t)x = e^{(\tau-t)A^{\star}}Q(\tau)e^{(\tau-t)A}x + \tag{5}$$

$$+ \int_t^{\tau} e^{(s-t)A^{\star}}(M(s) - Q(s)BN^{-1}(s)B^{\star}Q(s))e^{(s-t)A}x, \; t\in[0,\tau], \; x \in H.$$

If Q is a periodic solution of (4) it is not difficult to show (see [3]) that the optimal control $u^{\star}$ is related to the optimal state $y^{\star}$ by the synthesis formula

$$u^{\star} = -N^{-1}B^{\star}(Qy^{\star} + r) \tag{6}$$

where r verifies

$$r' = -(A^{\star} - QBN^{-1}B^{\star})r - Qf \quad , \quad r(0) = r(\tau) \quad . \tag{7}$$

Moreover the optimal cost is given by

$$J(u^{\star}) = \int_0^{\tau}[\langle f,r\rangle - \frac{1}{2}\|B^{\star}r\|^2]\,dt \quad . \tag{8}$$

In [3] we proved the existence and uniqueness of a periodic solution Q of (4) under suitable hypotheses of stabilizability and detectability.

In this paper we shall assume $M(t) = M$, $N(t) = N$; then any solution of the algebraic Riccati equation

$$A^{\star}Q + QA - QBN^{-1}B^{\star}Q + M = 0 \tag{9}$$

is clearly τ-periodic; however we are interested in finding τ-periodic solutions of (4) which are not constant. This problem has been studied in [6] when H is two dimensional. We present here some preliminary results for a general Hilbert space.

2. PERIODIC SOLUTIONS

In all this section we assume (3) and moreover $M(t) = M$, $N(t) = N$, $t \in \mathbb{R}$.

Setting $P(t) = Q(-t)$, $K = BN^{-1}B$, Riccati equation becomes:

$$P' = A^{\star}P + PA - PKP + M \tag{10}$$

If (A,B) and $(A^{\star},\sqrt{M})$ are stabilizable it is well known (see for instance [5]) that any solution of (10) converges to the unique solution of the algebraic Riccati equation (9). Thus in this case there are not periodic non-constant solutions of (10). It is convenient, for the following discussion, to consider the solution $\bar{P}$ of (10) such that $\bar{P}(0) = 0$. It is well known that $\{\bar{P}(t)\}_{t \geq 0}$ is an increasing sequence in $\Sigma^{+}(H)$; moreover if (A,B) is stabilizable then $\{\bar{P}(t)\}$ is bounded and converges to a positive solution P_{∞} of (9). P_{∞} is called the <u>minimal solution</u> of (9).

<u>Proposition 1</u>. Assume that equation (9) has no solutions in $\Sigma^{+}(H)$; then equation (10) has not periodic solutions.

<u>Proof</u>. If (9) has no solutions in $\Sigma^{+}(H)$, $\{\bar{P}(t)\}$ is not bounded in $\Sigma^{+}(H)$; by the Uniform Boundedness Theorem there exists $x \in H$ such

that $\langle \bar{P}(t)x,x\rangle \to +\infty$, thus no periodic solution can exist ∎

Now, in looking for periodic solutions of (10) it is natural to assume:

$$\left.\begin{array}{l} \text{i) } (A,B) \text{ is stabilizable} \\ \text{ii) } (A^\star,\sqrt{M}) \text{ is not stabilizable} \end{array}\right\} \qquad (11)$$

Proposition 2. Assume (3) and (11), let P be a τ-periodic solution of (10). Then we have $P(t) \geq P_\infty$, $t \in \mathbb{R}$.

Proof. Fix x in H , then, for any $\varepsilon > 0$ there exists $t_\varepsilon > 0$ such that $\langle P_\infty x,x\rangle - \langle \bar{P}(t)x,x\rangle < \varepsilon$ for $t \geq t_\varepsilon$.
If $t \geq t_\varepsilon$ it follows

$$\langle P(t)x,x\rangle \geq \langle \bar{P}(t)x,x\rangle \geq \langle P_\infty x,x\rangle - \varepsilon .$$

Since P is periodic and ε is arbitrary the conclusion follows ∎

By Proposition 2 it follows that P is a periodic solution of (10) if and only if $S = P - P_\infty$ is a positive periodic solution of the equation

$$S' = L^\star S + SL - SKS \qquad (12)$$

where $L = A - KP_\infty$.

Lemma 3. Assume (3) and let $X \in \Sigma^+(H)$. Then the solution S of (12) such that $S(0) = X$ is given by:

$$S(t) = e^{tL^\star} X(1 + \int_0^t e^{sL}Ke^{sL^\star}ds X)^{-1} e^{tL} \qquad . \qquad (13)$$

Proof. We know that there exists a unique solution S of (12) such that $S(0) = X$; moreover $S(t)x = \lim_{n\to\infty} S_n(t)x$ uniformly in $t \in [0,\tau]$ where S_n is the solution of the approximating problem

$$S_n' = L_n^\star S_n + S_n L_n - S_n K S_n \quad , \quad S_n(0) = X + \frac{1}{n} \tag{14}$$

where $L_n = A_n - KP_\infty$ and $A_n = nA(n-A)^{-1}$ (see [2]). Setting $V_n = S_n^{-1}$ problem (14) becomes

$$V_n' = -L_n V_n - V_n L_n^\star + K \quad , \quad V_n(0) = (X + \frac{1}{n})^{-1} \tag{15}$$

whose solution is given by

$$V_n(t) = e^{-tL_n}(X + \frac{1}{n})^{-1} e^{-tL_n^\star} + \int_0^t e^{-sL_n} K e^{-sL_n^\star} ds \tag{16}$$

from which

$$S_n(t) = e^{tL_n^\star}(X + \frac{1}{n})\,[1 + \int_0^t e^{sL_n} K e^{sL_n^\star} ds\,(X + \frac{1}{n})]^{-1} e^{tL_n} \tag{17}$$

and the conclusion follows as $n \to \infty$ ∎

The following proposition is clear:

<u>Proposition 4</u>. Assume (3) and (11). Then P is a τ-periodic solution of (10) if and only if $X = P(0) - P_\infty$ verifies the following equation:

$$X = e^{\tau L^\star} X (1 + \int_0^\tau e^{sL} K e^{sL^\star} ds X)^{-1} e^{\tau L} \quad . \tag{18}$$

The study of equation (18) seems in general not very easy; we shall consider a special situation and we shall give an application to wave equation.

Let Π and Ξ be orthogonal projectors on H such that $\Pi + \Xi = 1$.

Set $H_0 = \Xi H$, $H_1 = \Pi H$. Assume:

$$\left.\begin{array}{l} \text{i) } AH_i \subset H_i,\ BH_i \subset H_i,\ MH_i \subset H_i,\ NH_i \subset H_i \qquad i = 0,1 \\ \text{ii) Setting } A_i = A/H_i,\ B_i = B/H_i,\ M_i = M/H_i, \\ \quad N_i = N/H_i,\ K_i = K/H_j \text{ we have} \\ \qquad A_0^{\star} + A_0 = \lambda\Xi \quad , \quad \lambda > 0 \\ \qquad M_0 = 0 \end{array}\right\} \tag{19}$$

Then Riccati equation (10) splits into

$$P_0' = A_0^{\star}P_0 + P_0A_0 - P_0K_0P_0 \tag{20}$$

and

$$P_1' = A_1^{\star}P_1 + P_1A_1 - P_1K_1P_1 + M_1 \tag{21}$$

and algebraic equation (9) into

$$A_0^{\star}P_0 + P_0A_0 - P_0K_0P_0 = 0 \tag{22}$$

and

$$A_1^{\star}P_1 + P_1A_1 - P_1K_1P_1 + M_1 = 0 \quad . \tag{23}$$

Assume, moreover

$$\left.\begin{array}{l} (A,B) \text{ is stabilizable} \\ (A_1^{\star},\sqrt{M_1}) \text{ is stabilizable} \end{array}\right\} \tag{24}$$

then equation (23) has a unique solution $P_{1\infty} \in \Sigma^+(H)$ and $P_{1\infty}$ is also the unique periodic solution of (21). Now Proposition 5 follows immediately from Proposition 4.

Proposition 5. Assume (3), (19) and (24). Then P is a 2π-periodic solution of (10) if and only if $P = P_0 + P_{1\infty}$ where P_0 is a solution of (20) and $X = P_0(0)$ is a solution of:

$$(e^{2\pi\lambda} - 1)X = XZX \tag{25}$$

where

$$Z = \int_0^{2\pi} e^{\lambda t} e^{tA_0} K_0 e^{tA_0^\star} dt \tag{26}$$

Example 6. (WAVE EQUATION)

Let F be a separable Hilbert space, let $\{e_K\}_{K \in \mathbb{N}}$ be an orthonormal basis in F. Let Λ be a positive self-adjoint operator in F such that

$$\Lambda e_K = \lambda_K e_K \quad , \quad K \in \mathbb{N} \quad , \quad \lambda_K > 0 \quad . \tag{27}$$

Consider the following state equation in F (with control $u \in L^2(0,2\pi;F)$)

$$\xi'' = (\Lambda - \frac{\alpha^2}{4})\xi + \alpha\xi' + u \quad , \quad \alpha \in [0,1[\quad . \tag{28}$$

For $\alpha = 0$ equation (28) reduces to wave equation. Setting $\eta = \xi' - \frac{\alpha}{2}\xi$ equation (28) transforms in the system

$$\left.\begin{aligned} \xi' &= \frac{\alpha}{2}\xi + \eta \\ \eta' &= -\Lambda\xi + \frac{\alpha}{2}\eta + u \end{aligned}\right\} \tag{29}$$

which we consider as an equation in the Hilbert space $H = D(\sqrt{\Lambda}) \oplus F$

(endowed with the inner product $\left\langle \begin{pmatrix} \xi \\ \eta \end{pmatrix}, \begin{pmatrix} w \\ z \end{pmatrix} \right\rangle = \langle \sqrt{\Lambda}\xi, \sqrt{\Lambda}w \rangle + \langle \eta, z \rangle$).

Thus we shall write (29) in the form

$$\begin{pmatrix} \xi \\ \eta \end{pmatrix}' = A \begin{pmatrix} \xi \\ \eta \end{pmatrix} + Bu \tag{30}$$

where

$$A = \begin{pmatrix} \alpha & 1 \\ -\Lambda & \alpha \end{pmatrix}, \quad Bu = \begin{pmatrix} 0 \\ u \end{pmatrix}, \quad BB^\star = \begin{pmatrix} 0 & 0 \\ 0 & 1 \end{pmatrix}$$

(A,B) is stabilizable since $A - pBB^\star$ is stable for $2\alpha < p < 2$.

Consider the cost function

$$J(u) = \frac{1}{2} \int_0^{2\pi} \left[\sum_{i=1}^{n} |\langle \sqrt{\Lambda}\xi, \sqrt{\Lambda}e_i \rangle|^2 + |\eta|^2 + |u|^2 \right] dt \tag{31}$$

where $n = 1,2,\ldots,\infty$ is fixed.

Denote now by Π_n and Ξ_n the orthogonal projectors in H defined by

$$\Pi_n \begin{pmatrix} \xi \\ \eta \end{pmatrix} = \sum_{i=1}^{n} \begin{pmatrix} \langle \sqrt{\Lambda}\xi, \sqrt{\Lambda}e_i \rangle e_i \\ \langle \xi, e_i \rangle e_i \end{pmatrix}, \quad \Xi_n = 1 - \Pi_n .$$

If $n = +\infty$ which corresponds to the cost

$$J(u) = \frac{1}{2} \int_0^{2\pi} [|\sqrt{\Lambda}\xi|^2 + |\eta|^2 + |u|^2] \, dt$$

we have $\Pi_n = 1$, $M = 1$ so that Riccati equation has not non-trivial periodic solutions. In the sequel we shall assume $n < +\infty$. In this case all the assumptions of Proposition 5 are fulfilled. We will show now that equation (20) has non-trivial periodic solutions if $\alpha > 0$.

By representing e^{tA_0} in terms of the basis $\{e_k\}$ we can write

$$e^{tA_0} = e^{\frac{\lambda}{2}t} \begin{pmatrix} \cos\sqrt{\lambda_k}t & \frac{1}{k}\sin\sqrt{\lambda_k}t \\ -k\sin\sqrt{\lambda_k}t & \cos\sqrt{\lambda_k}t \end{pmatrix}, \quad k \geq n .$$

and by elementary computations we find

$$Z = \frac{e^{2\pi\alpha} - 1}{\alpha(\alpha^2 + 4\lambda_k^2)} \begin{pmatrix} 2\lambda_k^2 & -\alpha \\ -\alpha\lambda_k^2 & \alpha^2 + 2\lambda_k^2 \end{pmatrix}, \quad k \geq n \ , \ \alpha \neq 0$$

and equation (25) has the unique solution $X = 0$ if $\alpha = 0$ whereas if $\alpha > 0$ we have non-trivial solutions

$$X = \frac{\alpha(\alpha^2 + 4\lambda_k^2)}{2\lambda_k^2} \begin{pmatrix} 1 & 0 \\ 0 & 0 \end{pmatrix}, \quad X' = \frac{\alpha(\alpha^2 + 4\lambda_k^2)}{\alpha^2 + 2\lambda_k^2} \begin{pmatrix} 0 & 0 \\ 0 & 1 \end{pmatrix}$$

as easily checked the corresponding periodic solutions of (20) are not constant. Thus in this case Riccati equation (10) has no trivial periodic solutions. ∎

REFERENCES

[1] Bittanti S., Locatelli A., Maffezzoni C., Periodic Optimization under small perturbations. Periodic Optimization Vol. II, edited by A. Marzollo, Udine, Springer-Verlag, 1972, 183-231.

[2] Da Prato G, Quelques résultats d'existence, unicité et regularité pour un problème de la théorie du contrôle, J. Math. Pures et Appl. 52, 1973, 353-375.

[3] Da Prato G., Synthesis of optimal control for an infinite dimensional periodic problem, submitted to SIAM Jour. on Control and Optimization.

[4] Kalman R.E., Falb P.L. and Arbib M.A., Topics in Mathematical System Theory, Mc Graw Hill, 1960.

[5] Pritchard A.J. and Zabczyk J., Stability and stabilizability of infinite dimensional systems, SIAM Review 23, 1981, 25-52.

[6] Sernik E. and Arapostathic A., Contributions to the study of the Riccati equation, Proceedings of the 23rd Conference on Decision and Control, Las Vegas, 1984.

COMPUTER AIDED EXPANSION PLANNING OF COMBINED ELECTRIC POWER AND DISTRICT HEATING SYSTEMS - A MIXED INTEGER PROGRAMMING MODEL -

G. Rabensteiner[1)]

Abstract: A Mixed-Integer-Programming-Model is presented within this paper for long-term expansion and operation planning of combined electric power and heat systems. The mathematical formulation of the monetary objective and the power and heat demand restraints is explained at first. The restraints for the long-term operation of a typical extraction-condensing power plant are shown afterwards. The application of the planning-model to a practical power and heat system is illustrated by some diagrams.

1. Introduction

The expansion planning problem in an electricity and heat supply system is characterized by the complex of questions arising from time, location and way of expanding or modification the existing system to satisfy the power and heat demand in the future considering technological, economical and last not least ecological restraints. Type and installed capacity of powerstations, way of heat decoupling, type and topological expansion of the power and the heat transmission network, reserve requirements and alternative contracts are some topics of interest to be handled during the planning process. This r e a l - problem represents a stochastic nonlinear optimization problem in the mathematical sense, including a great number of equations, inequations, variables and coefficients so that computer aided planning methods become very important and helpful /1,2,3/. The Mixed-Integer-Programming (MIP)-Model, described in this paper, is a long-term planning model for the solution of the above mentioned tasks. It is part and an expansion of a MIP-planning package for long-, mid- and short-term optimization tasks, that means expansion and operation planning, for hydrothermal electric power systems in original /3/.

2. MIP-Model

2.1 Objective

The standard form of the monetary objective as a minimum function of the total amount of the present values of fixed and variable costs over the planning period for both the electric power and heat generation and transmission system is presented in (2-1). Fig. 1 and 2 show the basic investment cost data for hard coal fired condensation power plants and the additional plant costs for the heat decoupling as an example for a certain time status /8/.

$$\min Z = \sum_{t \in \mathbf{T}} \{ \sum_{k \in \mathbf{K}'} [c_{BP,kt} \cdot \frac{1}{\eta_{P,k}} (p_{EG,k} \cdot I_{K,kt} + P_{KV,kt})] +$$

$$+ \sum_{k \in (\mathbf{K}'_Q \cup \mathbf{K}'_{PQ})} [c_{BQ,kt} \cdot \frac{1}{\eta_{Q,k}} (q_{EG,k} \cdot I_{KQ,kt} + Q_{KV,kt})] +$$

$$+ \sum_{l \in \mathbf{L}'} [c_{BP,lt} \cdot I_{L,lt}] + \sum_{l \in \mathbf{L}'_Q} [c_{BQ,lt} \cdot I_{LQ,lt}]\} +$$

$$+ \sum_{\bar{t} \in \bar{\mathbf{T}}} \{ \sum_{k \in \mathbf{K}} \sum_{\bar{d} \in \bar{\mathbf{D}}_{\bar{t}}} [c_{BA,k\bar{t}\bar{d}} (\frac{1}{\eta_{P,k}} A_{K,k\bar{t}\bar{d}} + \frac{1}{\eta_{Q,k}} \cdot f_{W,k} \cdot W_{K,k\bar{t}\bar{d}})] +$$

$$+ \sum_{k \in \mathbf{K}_Q} [\frac{1}{\eta_{Q,k}} \sum_{\bar{d} \in \bar{\mathbf{D}}_{\bar{t}}} c_{BW,k\bar{t}\bar{d}} \cdot W_{K,k\bar{t}\bar{d}}]\} \qquad (2\text{-}1)$$

$\mathbf{K}$	set of existing electric and combined electric and heat power stations, projects and electric power supply contracts at t=1
$\mathbf{K}'$	set of electric power stations, back-pressure and extraction-condensing power plants for combined heat and electric power generation and electric power contracts in the project catalogue at t=1 ($\mathbf{K}' \subset \mathbf{K}$)
$\mathbf{K}'_{PQ}$	set of back-pressure and extraction-condensing power stations for combined heat and electric power generation in the project catalogue at t=1 ($\mathbf{K}'_{PQ} \subset \mathbf{K}'$)
$\mathbf{K}_Q$	set of existing heat production power plants, peak heat boilers, heat supply contracts and projects at t=1
$\mathbf{K}'_Q$	set of heat production power plants, peak heat boilers and heat supply contracts in the project catalogue at t=1
$\mathbf{L}'$	set of electric power transmission lines in the project catalogue at t=1
$\mathbf{L}'_Q$	set of heat transmission lines in the project catalogue at t=1
$\mathbf{T}$	set of moments (beginning of each year) in the optimization horizon
$\bar{\mathbf{T}}$	set of time intervals (years) in the optimization horizon
$\bar{\mathbf{D}}_{\bar{t}}$	set of sub-time-intervals (summer, winter) in the year $\bar{t}$
$A_{K,k\bar{t}\bar{d}}$	net output electric energy of unit k over the sub-time-interval $\bar{d}$ in the year $\bar{t}$; (MWh)
$c_{BA,k\bar{t}\bar{d}}$	present value of the specific total variable costs for the electric operation of unit k over the sub-time-interval $\bar{d}$ in the year $\bar{t}$; (MU/MWh)
$c_{BP,kt}$	present value of the specific fixed costs referred to the rated electric power over the optimization horizon for project k and first commercial start-up at t; (MU/MW)
$c_{BP,lt}$	present value of the fixed costs over the optimization horizon for the electric power transmission line project l and first operation at t; (MU)
$c_{BQ,kt}$	present value of the specific fixed costs referred to the rated heat power over the optimization horizon for project k and first commercial start-up at t (additional costs to the electric power generation for heat decoupling in case of extraction-condensing power plants); (MU/MW)
$c_{BQ,lt}$	present value of the fixed costs over the optimization horizon for the heat transmission line project l and first operation at t; (MU)

$c_{BW,k\bar{t}\bar{d}}$ present value of the specific total variable costs for the heat production of unit k over the sub-time-interval $\bar{d}$ in the year $\bar{t}$; (MU/MWh)

$f_{W,k}$ factor for additional variable costs in case of heat decoupling from combined electric and heat power plants; (-)

$I_{K,kt}$ {0,1}-decision-variable for the commercial start-up of unit k at t; (-)

$I_{KQ,kt}$ {0,1}-decision-variable for the commercial start-up of heat unit k at t; (-)

$I_{L,lt}$ {0,1}-decision-variable for the construction of power transmission line l;(-)

$I_{LQ,lt}$ {0,1}-decision-variable for the construction of heat transmission line l;(-)

$p_{EG,k}$ net maximum capacity for the basic completion of power station k; (MW)

$P_{KV,kt}$ net capacity exeeding the basic value $p_{EG,k}$; (MW)

$q_{EG,k}$ net maximum heat capacity for the basic completion of station k; (MW)

$Q_{KV,kt}$ net heat capacity exeeding the basic value $q_{EG,k}$; (MW)

$W_{K,k\bar{t}\bar{d}}$ net heat output of unit k over the sub-time-interval $\bar{d}$ in the year $\bar{t}$; (MWh)

$\eta_{P,k}$ efficiency with regard to the auxiliary electric power supply for unit k; (-)

$\eta_{Q,k}$ efficiency with regard to the net heat generation of unit k; (-)

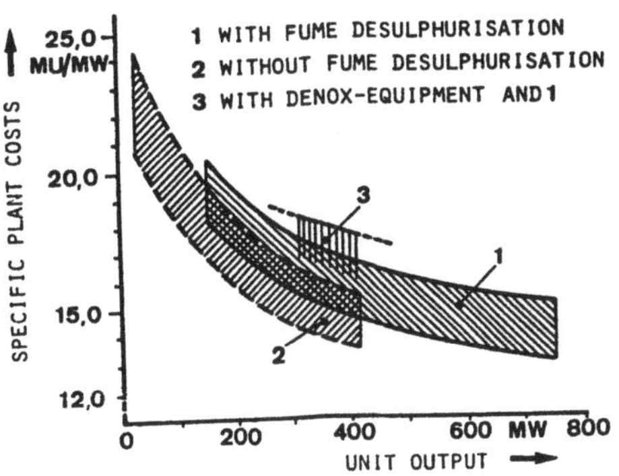

Fig. 1: Specific plant costs for hard coal condensation power plants (status 1980)

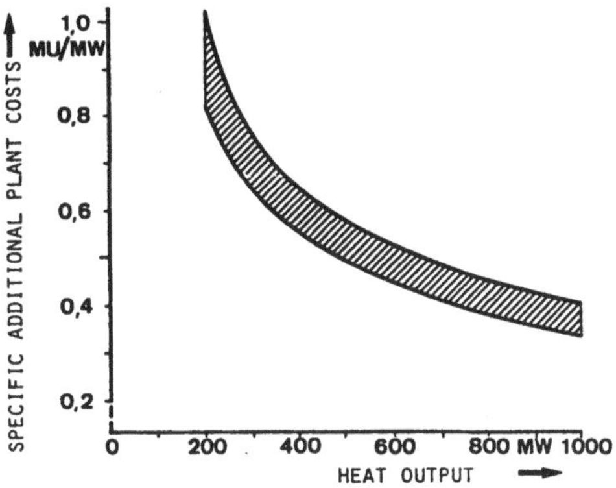

Fig. 2: Specific additional plant costs for heat decoupling in condensation power plants (status 1980)

2.2 Restraints

The system of restraints for the digital modelling of the long-term behaviour of the combined electric power and heat production system contains three groups:

- Demand restraints
- power station and contract restraints
- transmission network restraints

Only some of these constraints within the mentioned three groups will be discussed in this chapter, to show the MIP-modelling procedure.

2.2.1 Demand restraints

The input-parameters for the demand restraints are calculated out of prognosticated duration curves for the electric power and heat demand. Each year within the optimization period is subdivided into a summer and a winter halfyear according to the original electric power system expansion planning model for hydrothermal generation systems /3/. The demand restraints for electricity and heat are formulated as power and energy balance equations for the total demand or the topological demand nodes for summer and winter. The peak heat power demand constraint is given by (2-2) for the total district heating system including the transmission losses. The prognosticated demand is represented by a lower bounded variable within the model as shown in (2-3). The same procedure is to be done for the heat energy demand in each sub-time-interval as shown in (2-4) and (2-5).

Heat peak-load demand

$$\sum_{k\in(\mathbf{K}_{PQ}\cup\mathbf{K}_{Q})} Q_{K,k\bar{t}\bar{d}} - Q_{N,\bar{t}\bar{d}} = 0 \quad ; \quad \forall(\bar{t},\bar{d}) \qquad (2\text{-}2)$$

$$\text{LB: } Q_{N,\bar{t}\bar{d}} \geqq q_{NH,\bar{t}\bar{d}} \quad ; \quad \forall(\bar{t},\bar{d}) \qquad (2\text{-}3)$$

Heat energy demand

$$\sum_{k\in(\mathbf{K}_{PQ}\cup\mathbf{K}_{Q})} W_{K,k\bar{t}\bar{d}} - W_{N,k\bar{t}\bar{d}} = 0 \quad ; \quad \forall(\bar{t},\bar{d}) \qquad (2\text{-}4)$$

$$\text{LB: } W_{N,k\bar{t}\bar{d}} \geqq w_{NH,\bar{t}\bar{d}} \quad ; \quad \forall(\bar{t},\bar{d}) \qquad (2\text{-}5)$$

$Q_{K,k\bar{t}\bar{d}}$ net peak-load heat power output of unit k in the sub-time-interval $\bar{d}$ of the year $\bar{t}$; (MW)

$Q_{N,\bar{t}\bar{d}}$ peak-load heat demand model variable for the total heat supply system; (MW)

$q_{NH,\bar{t}\bar{d}}$ prognosticated total peak-load heat demand during the sub-time-interval $\bar{d}$ of the year $\bar{t}$ for the total heat supply system; (MW)

$W_{K,k\bar{t}\bar{d}}$ net heat energy output of unit k over $\bar{t}\bar{d}$; (MWh)

$W_{N,k\bar{t}\bar{d}}$ total heat demand model variable; (MWh)

$w_{NH,\bar{t}\bar{d}}$ prognosticated total heat demand over $\bar{t}\bar{d}$; (MWh)

LB lower bound

2.2.2 Power station and contract restraints

The power station and contract restraints are subdivided into two groups, the expansion planning constraints and the long-term operation planning restraints. It should be mentioned here that the application of the model constraints can be done for the generation units of power stations in detail, for the power station as a whole or for groups of power stations. Two examples for expansion planning restraints are listed below. The fixed coupling between electric power and heat generation in case of back-pressure plants is taken into consideration by (2-6).

$$I_{K,kt} - I_{KQ,kt} = 0 \quad ; \qquad \forall t \tag{2-6}$$

This equation has to be modified for extraction-condensing power plants as shown in (2-7).

$$I_{KQ,kt} - I_{K,kt} \leqq 0 \quad ; \qquad \forall t \tag{2-7}$$

It is the task of the long-term operation simulation of power plants by the formulation of restraints within this MIP-model to get information on the power output in case of peak-load and on the energy output of the power station for each time- or sub-time-interval within the optimization period to catch the prognosticated electric power and heat demand. The possible long-term operation of power plants for heat generation on basis of duration curves is shown in fig. 3 as an example.

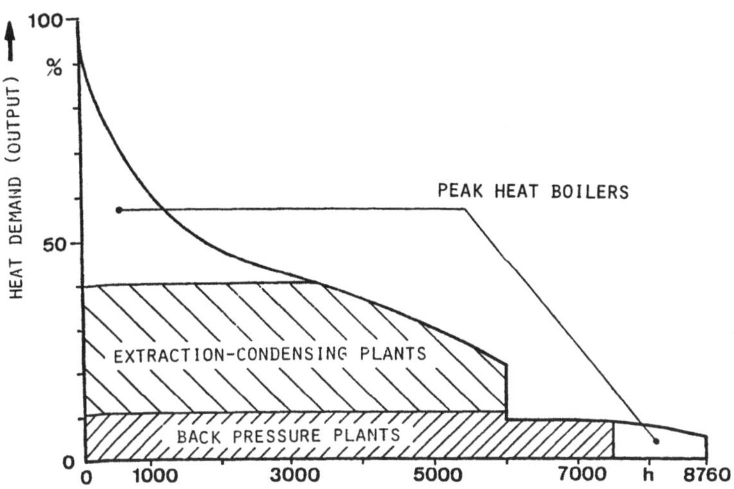

Fig. 3: Example of a power station long-term-operation (heat-production) plan on basis of annual duration curves

As described in /4/ the maximum electric power output of a power station or unit k is bounded by a spinning reserve part and an expected availability rate of the electric power output. For the maximum heat output of unit k the following restraint (2-8) has been formulated.

$$Q_{K,k\bar{t}\bar{d}} - e_{QV,k\bar{t}\bar{d}} \sum_{\substack{t \in \mathbf{T} \\ t \leq \bar{t}}} (q_{EG,k} \cdot I_{KQ,kt} + Q_{KV,kt}) \leq 0 \; ; \qquad \forall(\bar{t},\bar{d}) \;\; k \in \mathbf{K}'_{PQ} \tag{2-8}$$

$e_{QV,k\bar{t}\bar{d}}$ expected availability rate of heat output for unit k; (-)

The relation between electric power and heat generation for an extraction-condensing power plant is shown by the PQ-diagram in fig. 4.

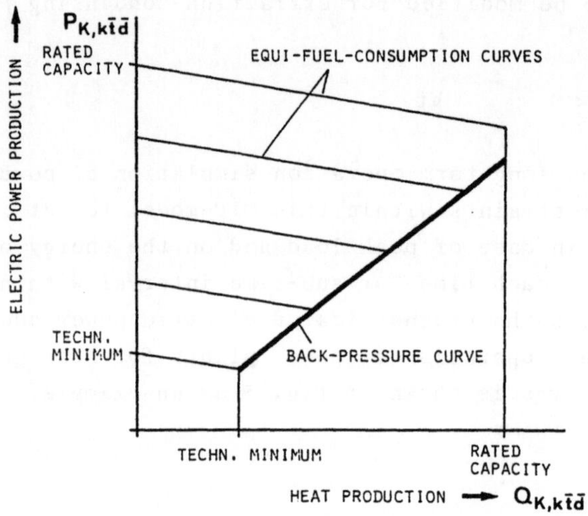

Fig. 4: Linearized PQ-diagram for extraction-condensing power plants

The simplified modelling of the possible operation range of extraction-condensing power stations within this model is given by the restraints (2-9) and (2-10).

$$P_{K,k\bar{t}\bar{d}} - f_{Q1,k} \cdot Q_{K,k\bar{t}\bar{d}} \geq 0 \quad ; \qquad \forall(\bar{t},\bar{d}) \tag{2-9}$$

$$P_{K,k\bar{t}\bar{d}} - (1 - r_{k\bar{t}\bar{d}}) \, e_{PV,k\bar{t}\bar{d}} \cdot \sum_{\substack{t \in \mathbf{T} \\ t \leq \bar{t}}} (p_{EG,k} \cdot I_{K,kt} + P_{KV,kt}) + $$
$$+ f_{Q2,k} \cdot Q_{K,k\bar{t}\bar{d}} \leq 0 \quad ; \qquad \forall(\bar{t},\bar{d}), k \in \mathbf{K}'_{PQ} \tag{2-10}$$

$f_{Q1,k}$ linearized back-pressure constant; (-)
$r_{k\bar{t}\bar{d}}$ spinning reserve part; (-)
$e_{PV,k\bar{t}\bar{d}}$ expected availability rate of electric power for unit k; (-)
$f_{Q2,k}$ equi-fuel-consumption curve slope; (-)

The back-pressure constant can be calculated as a function of the thermodynamic parameters of the process, e.g. forward/return temperature of the district heating water a.s.o. as described in /6/ for example.

The maximum heat energy output of unit k over one sub-time-interval is bounded by restraint (2-11).

$$W_{K,k\bar{t}\bar{d}} - e_{VZ,k\bar{t}\bar{d}} \cdot Q_{K,k\bar{t}\bar{d}} \leq 0 \quad ; \qquad \forall(\bar{t},\bar{d}) \qquad (2\text{-}11)$$

$e_{VZ,k\bar{t}\bar{d}}$ expected time duration of availability of unit k during the sub-time-interval $\bar{t}\bar{d}$; (h)

2.2.3 Transmission network restraints

The power flow problem within the planning model for combined electric generation and transmission expansion /5/ is based upon the case of peak-load for each sub-time-interval and the energy flow is calculated for the duration of each sub-time-interval, modelled as a transportation problem under consideration of linearized network losses. Similar aspects are valid for the inclusion of the heat transmission network topology into this MIP-model. Type and site of heating power stations, peak heat boilers and heat storage facilities are to be seen in close connection to type and right-of-way of heat transmission lines, transformer and pump stations within the planning procedure. The restraints for the heat transmission network operation describe a transportation problem including the heat power flow and energy flow variables. The transmission losses are considered by efficiency factors as a function of the transmission distance.

3. Application example

The diagrams and tables on the following pages give an overview on the application of the MIP-model to a practical example. The existing electric power generation and 220 kV-transmission system and the projects can be seen from fig. 5. The thermal power station projects 18 and 19 include extraction-condensing turbine units alternatively to meet the prognosticated mid-load heat demand from the middle of the optimization period on. Computational parameters for the application of the one-node model can be seen from tab. 1. The electric power and heat expansion plans over the optimization period are drawn up in fig. 6. The heat load duration curves for power station k=18 can be seen from fig. 7 for a winter interval.

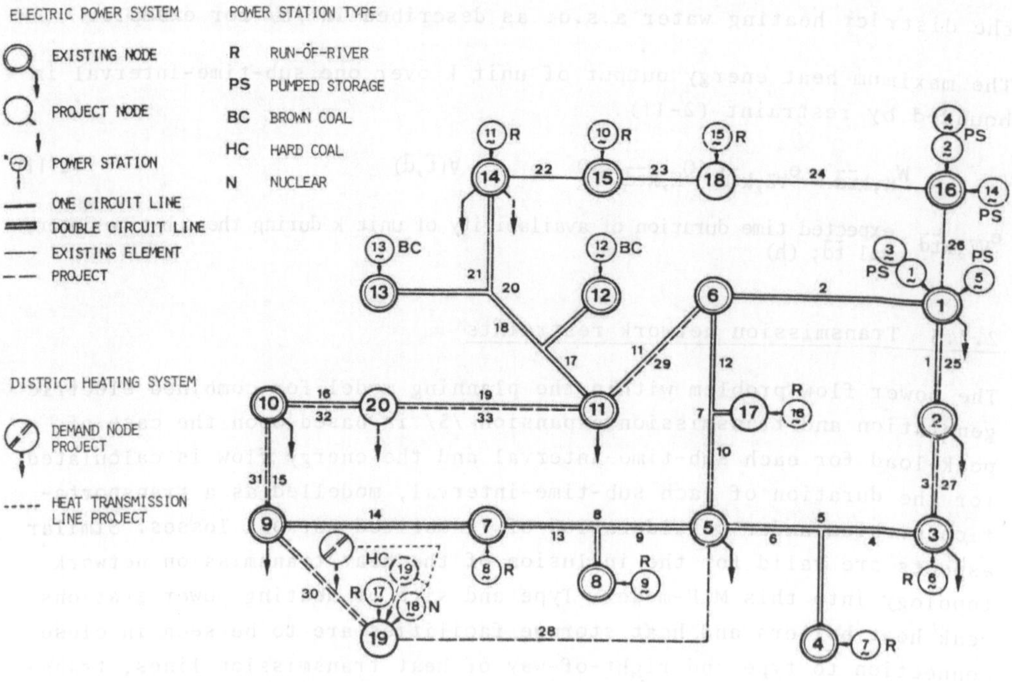

Fig. 5 : Electric power and heating system

MATRIX STATISTICS	
ROWS.....	1084
COLUMNS.....	1137
RHS.....	1
DENSITY.....	.38
MAX-COL-NZ'S	66
ELEMENTS.....	4631
LARGEST.....	.448500+007
SMALLEST.....	.462000-005
ZEROES.....	0
MAJOR ERRORS	0
MINOR ERRORS	0
INTEG. COLS.	80

SOLUTION	OBJECTIVE MU	OBJECTIVE P.U.	CPU-TIME MIN
LP	7336.59	1.0	0.97
1.MIP	8717.85	1.188	1.67
2.MIP	8396.23	1.144	4.27
MIP-OPTIMAL	8396.23	1.144	5.01

Tab. 1: Computational parameters (UNIVAC 1100/81) /7/

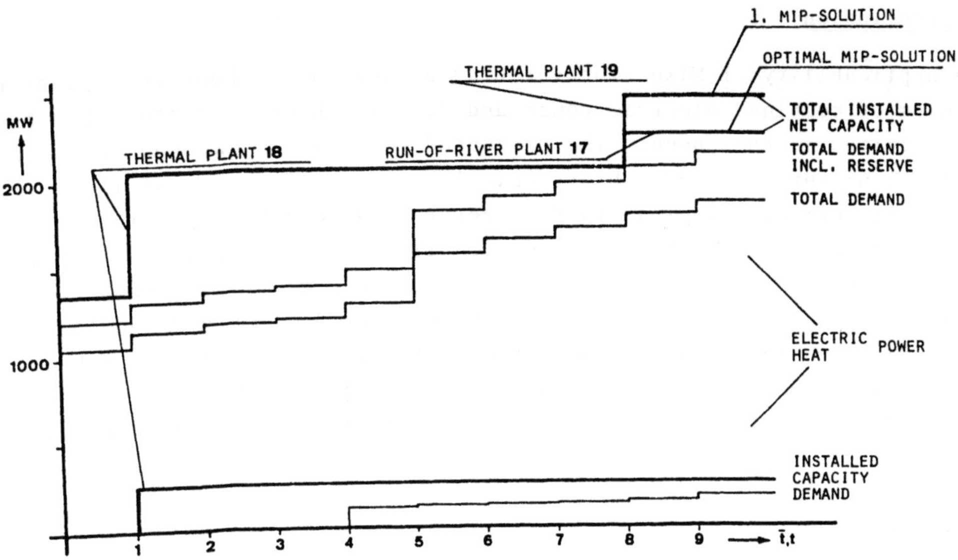

Fig. 6: Power development over the optimization period; run-of-river plants considered with the expected value of the firm capacity

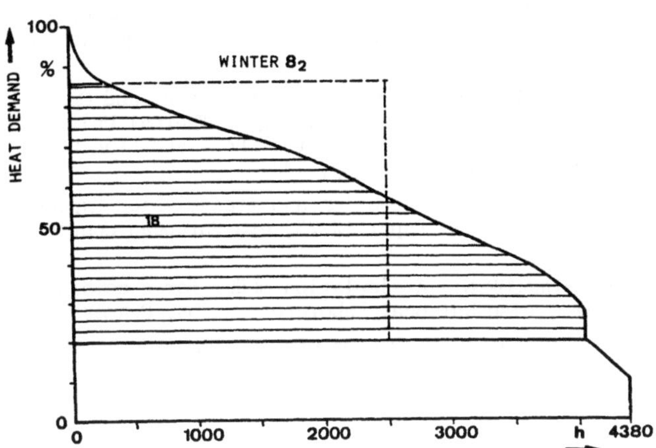

Fig. 7: Heat load duration curve for power station k=18

4. Conclusion

The applicability of Mixed-Integer-Programming to the long-term planning problem of combined electric power and district heating systems has been demonstrated in this paper. The presented model is an expanded version of the long-term part of a hierarchical structured package of long-, mid- and short-term planning models for hydrothermal electric power systems. Important requirements for this applicability are powerful MIP-software and computers, problem and practice oriented MIP-models and supporting software for efficient datahandling. Some advantages of MIP-models are their modular structure and their flexibility, which allow a quick application and conversion from the mathematical model to the practical problem and make easily surveyable modifications possible in a short time.

References

/1/ SAWEY, R.M.; ZINN, C.D.: A Mathematical Model for Long Range Expansion Planning of Generation and Transmission in Electric Utility Systems. IEEE Transactions on Power Apparatus and Systems. Vol. PAS-96, 1977

/2/ MUCKENHUBER, R.: Die gemischt-ganzzahlige lineare Optimierung- ein modernes Instrument zur Ausbauplanung elektrischer Energieversorgungssysteme. E & M, Jg. 102, H. 1, 1985

/3/ RABENSTEINER, G.: MIP-Planning Models for Expansion and Operation of Hydrothermal Electric Power Systems. Conference Proceedings, Verlag B.G. Teubner, Stuttgart, 1985

/4/ RABENSTEINER, G.: MIP-Planning Models for Expansion and Operation of Hydrothermal Electric Power Systems - Modelling of Thermal Units and Applications. Seminar on Modern Methods in Power System Short and Long Range Planning. Techn. University of Denmark, Lyngby, 1984

/5/ RABENSTEINER, G.: Combined Generation and Transmission Expansion Planning for Hydrothermal Power Systems. Paper IX/23, 6th European Congress on Operations Research, Wien, 1983

/6/ LARSEN, H.V.: SIMULACHRON - A simulation model for a combined heat and power production system. Riso National Laboratory R-508, Denmark, 1984

/7/ N.N.: Functional Mathematical Programming System (FMPS). FMPS Level 9R1, Programmer Reference, SPERRY CORPORATION USA, 1981

/8/ N.N.: Wärmetransport über große Entfernungen. Generalbericht zum Unichal-Kongreß 1983, Lahti

[1] G. Rabensteiner, Dipl.-Ing. Dr.techn.

Electric Power Systems Institute
Technical University of Graz
Inffeldgasse 18, A-8010 GRAZ (AUSTRIA)

CONVEX PROGRAMMING ON RIEMANNIAN MANIFOLDS

T. RAPCSÁK

Computer and Automation Institute
Hungarian Academy of Science
Budapest

A b s t r a c t. The paper presents a short survey of arcwise-connectedness and arcwise-convexity, then the convex programming will be extended to Riemannian manifolds.

1. I n t r o d u c t i o n. Both convexity and generalized convexity are important in nonlinear programming because any local minimizer to a convex and certain generalized convex programming problem is a global minimizer, therefore the algorithms are substantially more efficient than in other cases.

In the field of mathematical programming many authors have dealt with the generalization of the convexity concept. A survey of some classes of generalized convex functions can be found in [1]. These classes include quasi-convex, strictly quasi-convex, strongly quasi-convex, pseudo-convex, strictly pseudo-convex functions etc. A common property of these functions is that their level sets are convex.

The convex and generalized convex functions as well as the convex programming problems are defined on convex sets and this fact limits the use of this concept. In general, this property does not hold in nonlinear programming. For instance, the feasible set of the nonlinear programming problems where the constraints are nonlinear equalities representing a differentiable manifold is not convex, therefore the convex programming concepts cannot be on it.

The purpose of this paper is to extend the concepts of convexity and convex programming to differentiable manifolds and to connect some mathematical programming problems with deep mathematical, differential geometrical questions and results.

I wish to express my thanks to professor J. Szenthe for his help and valuable advice.

2. A r c w i s e - c o n n e c t e d n e s s and a r c w i s e - c o n v e x i t y in m a t h e m a t i c a l p r o g r a m m i n g.

Ortega and Rheinboldt [2] extended the families of quasi-convex and strongly quasi-convex functions by considering continuous arcs instead of line segments in the definitions. Arcwise-convex and arcwise-connected functions are defined and some of their properties are shown in [1]. In the paper of Avriel and Zang [3] the generalized arcwise-connected functions are defined and characterized with respect to the local-global minimum properties. A more general version of the local-global properties is derived in Martin's works [4], [5], [6]. Geodesically convex surfaces of (n-1)-dimensions are investigated in [7]. The second-order necessary and sufficient conditions are derived for functions defined on surfaces in [8], [9].

Following the preceding ideas we have the definitions:

D e f i n i t i o n 2.1. A set $C \subset R^n$ is said to be arcwise-connected if for every pair of points $\underline{x}_1 \in C$, $\underline{x}_2 \in C$ there is at least one continuous vector valued function $\underline{x}(t)$, called an arc, defined on the unit interval $[0,1] \subset R$ for which

$$\underline{x}(t) \in C, \ \forall t \in [0,1] \quad \text{and} \quad \underline{x}(0) = \underline{x}_1, \ \underline{x}(1) = \underline{x}_2 \ . \tag{1}$$

D e f i n i t i o n 2.2. Let $C \subset R^n$ be an arcwise-connected set. A subset of the connecting arcs is called a family of the feasible arcs or shortly, feasible arcs if for every pair of points in C there exists a connecting arc belonging to the selected subset.

D e f i n i t i o n 2.3. Let $C \subset R^n$ be an arcwise-connected set and F a family of the feasible arcs. The real valued function $f(\underline{x})$, defined on the set C, is then said to be arcwise-convex (for the set F) if for every pair of points $\underline{x}_1 \in C$, $\underline{x}_2 \in C$ and a feasible arc $\underline{x}(t)$ we have

$$f(\underline{x}(t)) \leq (1-t) f(\underline{x}_1) + t \, f(\underline{x}_2) \ , \quad 0 \leq t \leq 1 \ . \tag{2}$$

The first statement which results e.g. from the more general theorems of [3] shows the importance of the arcwise-convexity.

L e m m a 2.1. If $f(\underline{x})$ is arcwise-convex (for a set of functions F) on C and $\underline{x}_0$ is a local minimum, then $\underline{x}_0$ is also a global minimum point.

A principal question, however, with respect to the arcwise-connectedness is the following:

What restrictions the property of arcwise-connectedness provides for a set and do there exist arcwise-connected sets really interesting in nonlinear programming?

In what follows, the methods of investigations and their results are closely related to topological and differential geometrical ones.

3. Arcwise-connectedness in topological spaces.

In topological concept an arc is a set which is homeomorphic to the closed interval $0 \leq t \leq 1$ and a topological space is arcwise-connected if every two points can be joined by an arc contained in the space.

For the topological characterization of the arcs we need the following two definitions:

Definition 3.1. A topological space M is connected if for every decomposition $M = A \cup B$, $A \neq \emptyset$, $B \neq \emptyset$, the condition $A \cap B \neq 0$ holds.

Definition 3.2. A topological space is said to be locally connected if every point has a connected neighborhood base.

Theorem 3.1. [10] The topological space M is an arc if and only if

a/ M is a connected space

b/ M is a locally connected space

c/ M is metrizable

d/ there exist two points $a, b \in M$ so that every point $x \in M$ separates these points.

On the basis of this theorem it is possible to derive the Mazurkiewicz-Moore-Menger theorem which gives sufficient conditions for the arcwise-connectedness.

Theorem (Mazurkiewicz-Moore-Menger)[11]. Every connected, locally connected, complete metric space is arcwise-connected.

This proves that there are really interesting arcwise-connected sets and it is possible to introduce the concept of arcwise-convex functions.

The following theorem, due to Mazurkiewicz [12], will show the significance of the differentiability.

Theorem (Mazurkiewicz)[12]. Every connected, locally connected, compact space is a continuous image of the closed interval $0 \leq t \leq 1$.

It follows from this theorem that the property of having a continuous parametric representation over a closed interval does not **apply** to curves only. This fact shows that the topological definition of the arc is better in mathematical programming than the one in the Definition 2.1. and the hypothesis of differentiability is essential from this point of view. On the other hand, differentiability seems to be useful **in** characterizing the feasible arcs and the convexity property of functions, as well. Consequently, it should be worthwhile to combine an arcwise-connected metric space with a differentiable structure. The differentiable manifolds are general differentiable structures, for this reason the following question is given (e.g. [13]):

What are the necessary and sufficient conditions which imply that a metric space should be induced for instance by a Riemannian manifold?

This problem seems to be a very difficult open one which **has** been solved in some interesting special cases (e.g. J.Szenthe [13], [14]).

4. Arcwise-connectedness and convexity in differential geometry.

In case we choose the converse direction and consider a connected differentiable manifold (every differentiable manifold is locally connected) then for applying the Mazurkiewicz-Moore-Menger theorem, we need to introduce a metric because in general a differentiable manifold is not a metric space.

Thus this problem leads to deep results of differential geometry, the first of this kind was published by K. Menger [15] who carried out the systematic study of metric theory of curves.

By using the results of Menger and other authors [16] - [20], a metric on a Riemannian manifold can be introduced in the following way: let M be a Riemannian manifold and let $\varrho(x,y)$ be the infinum of the length of those piecewise differentiable continuous curves which join $x,y \in M$. Then ϱ is a distance function on M.

This metric does not depend on the differentiable structure and the function $\varrho(x,y)$ is equivalent to a geodesic of minimal length belonging to the Riemannian manifold, connecting the two points $x,y \in M$ [21].

It turns out that a Riemannian manifold is arcwise-connected where the feasible arcs are the geodesics characterized by a system of differentiable equations.

Now it is possible to introduce the concept of the convex function (or arcwise-convex or geodesically convex function) on a Riemannian manifold and the concept of the convex set (or arcwise-connected set).

It is said in differential geometry that a function $f:M \rightarrow R$ is convex (or geodesically convex) if its restrictions to every geodesic belonging to M is convex [22] and a set $A \subset M$ is convex (or geodesically convex) if one of the shortest geodesics or the shortest geodesic joining every two points of A belongs to A [23].

A survey of the convexity with respect to the Riemannian manifolds can be found in [24], [25].

5. Convex programming on Riemannian manifolds.

To define the convex programming on Riemannian manifolds, from nonlinear point of view it seems to be better to modify the definitions of convex sets and convex functions.

Definition 5.1. It is said that a set $A \subset M$ is (geodesically) convex if any two points of A are joined by a geodesic belonging to A.

Definition 5.2. It is said that a function $f:M \rightarrow R$ is (geodesically) convex if its restrictions to a family of the feasible geodesic arcs belonging to M are convex.

The difference of the two convex set definitions is shown by the following example and figure.

If we consider a sphere and the arc-metric on the sphere then a part of the sphere which is greater than a hemisphere is not geodesically convex in the former sense, but is geodesically convex in the latter sense.

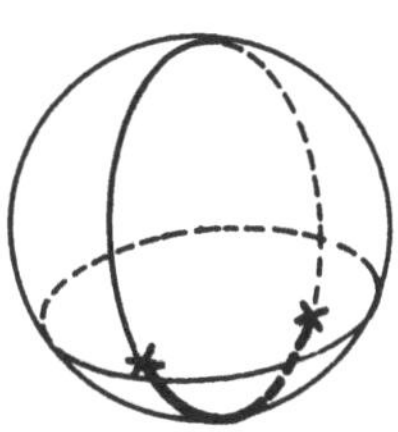

Figure 1.

In the definition of convex functions it is not necessary to require the consideration of every geodesic but a family of the feasible geodesic arcs only.

Now we have the following convex programming problem on Riemannian manifolds:

$$\min f(\underline{x}) \qquad \underline{x} \in A \subset M \subset R^n \tag{3}$$

where R^n is the Euclidean n-space
M is a Riemannian manifold
A is a geodesically convex set
$f(\underline{x})$ is a geodesically convex function on A .

The second order characterization of the geodesically convex functions is given in the next theorem.

T h e o r e m 5.1. [9] Let $M \subset R^n$ be a Riemannian manifold or $A \subset M$ be an open geodesically convex set. Then the twice continuously differentiable function $f(\underline{x})$ is geodesically convex on M or on A if and only if the following matrix is positive semidefinite in every point:

$$H_{|TM} + |\nabla f(\underline{x})_N| \cdot B_{\nabla f_N} \tag{4}$$

where $H_{|TM}$ is the Hesse-matrix of the function $f(\underline{x})$ restricted to the tangent space of M in point $\underline{x}$,

$B_{\nabla f_N}$ is the second fundamental form of M in the normal direction of the vector $\nabla f(\underline{x})$ in point $\underline{x}$.

We remark that without the convexity properties the optimization problem (3) is a generalization of nonlinear programming problems, namely it is known [26] that a general nonlinear programming problem, given in the form

$$\begin{aligned} &\min f(\underline{x}) \\ &h_i(\underline{x}) = 0 , \quad i=1,\dots,p \qquad \underline{x} \in R^n \\ &g_j(\underline{x}) \geq 0 , \quad j=1,\dots,m \end{aligned} \tag{5}$$

where f, h_i, $i=1,\dots,p$, g_j, $j=1,\dots,m$ are twice continuously differentiable functions and assuming the linear independence of the vectors

$$\nabla h_i, \ i \in \{1,\dots,p\} , \quad \nabla g_j, \ j \in \{j \mid g_j(\underline{x}) = 0\} \tag{6}$$

in each point, can be considered as a differentiable manifold with generalized boundary.

R e f e r e n c e s

1. Avriel, M., Nonlinear programming: Analysis and methods, Prentice Hall, Inc. New Yersey, 1976.
2. Ortega, J.M., and Rheinboldt, W.C., Iterative solution of nonlinear equations in several variables, Academic Press, New York, 1970.
3. Avriel, M., and Zang, I., Generalized arcwise-connected functions and characterizations of local-global properties, Journal of Optimization Theory and Applications, Vol. 32, No.4, pp. 407-425, 1980.
4. Martin, D.H., Connected level sets and uniqueness in optimization - Part I, National Research Institute for Mathematical Sciences, Pretoria, CSIR Special Report WISK 276, September 1977.
5. Martin, D.H., Connected level sets and uniqueness in optimization - Part II, National Research Institute for Mathematical Sciences, Pretoria, CSIR Special Report WISK 277, October 1977.
6. Martin, D.H., Connected level sets, minimizing sets and uniqueness in Optimization, Journal of Optimization Theory and Applications, Vol. 36, No. 1, pp. 71-93, 1982.
7. Nozicka, F., Spherical mappings of convex sets and convex parametrical optimization, Survey of Mathematical Programming Proceedings of the 9th International Mathematical Programming Symposium, Vol. 1, Edited by Prékopa, A., Akadémiai Kiadó, Budapest, pp. 299-307, 1979.
8. Rapcsák, T., On the arcwise-convexity, Alkalmazott Matematikai Lapok No. 10, pp. 115-123, 1984. (in Hungarian)
9. Rapcsák, T., Arcwise-convex functions on surfaces, Publicationes Mathematicae (to appear)
10. Császár, Á., Introduction to general topology, Akadémiai Kiadó, Budapest, 1970. (in Hungarian)
11. Kuratowski, C., Topologie I-II., Warszawa, 1958.
12. Kuratowski, K., Introduction to set theory and topology, Pergamon Press, New York, 1961.
13. Szenthe, J., A metric characterization of homogeneous Riemannian manifolds, Acta Scientiarum Mathematicarum, Vol. 33, No. 1-2, pp. 137-151, 1972.
14. Szenthe, J., A metric characterization of symmetric spaces, Acta Mathematica Academiae Scientiarum Hungaricae, Vol. 20, No. 3-4, pp. 303-314, 1969.
15. Menger, K., Untersuchungen über allgemeine Metrik I,II,III. Math. Annalen, Vol. 100, pp. 75-163, 1928.
16. Busemann, H., The geometry of geodesics, New York, 1955.
17. Busemann, H., and Mayer, W., On the foundations of the calculus of variations, Trans. Amer. Math. Soc., Vol. 49, pp. 173-198, 1941.

18. Rinow, W., Die innere Geometrie der metrischen Räume, Berlin, 1961.
19. Szenthe, J., On intrinsic metric homogeneous spaces, Magyar Tud. Akadémia Közleményei, Vol. 13, No. 2, pp. 125-132, 1963. (in Hungarian)
20. Szenthe, J., On the metric theory of Euclidean space curves I., Periodica Mathematica Hungarica, Vol. 3, No. 3-4, pp. 319-337, 1973.
21. Kobayashi, S., and Nomizu, K., Foundations of differential geometry, New York, 1963.
22. Greene, R.E., and Shiohana, K., Convex functions on complete noncompact manifolds: Topological structure, Inventiones Mathematicae, Vol. 63, pp. 129-157, 1981.
23. Spivak, M., A comprehensive introduction to differential geometry, Publish or Perish, Berkeley, 1979.
24. Busemann, H., Convex surfaces, Interscience Publishers, New York, 1958.
25. Shiohana, K., Convexity in Riemannian Geometry, Proceedings of the Beijing symposium on differential geometry and differential equation, Beijing, Aug. 18.-Sept. 21. 1980. Vol. 1-3, New York, 1982.
26. Jongen, H.Th., Jonker, P., and Twilt, F., Nonlinear Optimization in R^n, Verlag Peter Lang, Frankfurt am Main, 1983.

INTERACTIVE MULTICRITERIA GOAL PROGRAMMING BY REFERENCE OBJECTIVES

Nenad Rijavec

UNIVERSITY COMPUTING CENTRE
Engelsova bb, 41000 Zagreb, Yugoslavia

1. Introduction

In 1975. Wallenius |6| has made a comparative evaluation of several methods for interactive multicriteria programming. He tested several well-known methods on a large sample of students and experienced managers.

For comparison purposes, among methods tested was something he called "Unstructured approach", which was actually a simple guessing procedure, where decision maker (DM) proposed values for objective functions, and algorithm just checked whether this combination was feasible.

Wallenius found that this simple and unsophisticated method performed very well in comparison to other methods being tested. It was judged to be easiest to understand and use by the large majority of decision makers and results obtained were usually as close to efficient as for the other methods.

There were, however, some drawbacks. First, it was tought that decision makers would have difficulties in hitting feasible solutions in problems where there was a large number of objective functions. Furthermore, there was no guarantee that obtained solutions would even remotely approach efficient solution (though that proved to be the case during the testing).

The algorithm presented here adresses this shortcomings while attempting to preserve the simplicity and the ease to use of this method.

2. The Algorithm

We will want to solve standard multicriteria optimization problem

$$\max\ (f_1,\dots,f_K)$$
$$x \in X. \qquad (1)$$

The algorithm consists of following steps:

0. Solve K problems

$$\max\ f_k \qquad k = 1,\dots,K$$
$$x \in X \qquad (2)$$

to get the ideal solution of (1), denoted by $w_1^1,\dots,w_k^1$, where w_k^1 is the optimum of f_k. If this solution is feasible, there is no need for interaction and algorithm can terminate. Otherwise, denote $\hat{f}_k = f_k/w_k^1$, $u_k^1 = 1$, $k = 1,\dots,K$. Set iteration counter $i = 1$.

1. Solve problem

$$\min \sum_{k=1}^{K} (y_{2k-1} - \varepsilon\, y_{2k})$$

$$\hat{f}_k + y_{2k-1} - y_{2k} = u_k^i \qquad k = 1,\dots,K$$

$$x \in X,\quad y_k \geq 0,\quad k = 1,\dots,2K \tag{3}$$

and denote obtained values of $\hat{f}_k$ by $\hat{f}_k^0$, $k = 1,\dots,K$. Let perturbation distance be

$$d = \Big(\sum_{k=1}^{K} |u_k^{i\,2} - f_k^{0\,2}|\Big)^{1/2}$$

2. For $s = 1,\dots,K$ solve perturbed problems

$$\min \sum_{k=1}^{K} (y_{2k-1} - \varepsilon\, y_{2k})$$

$$\hat{f}_k + y_{2k-1} - y_{2k} = u_k^i + \delta_{sk}\, d \qquad k = 1,\dots,K$$

$$x \in X,\quad y_k \geq 0,\quad k = 1,\dots,2K. \tag{4}$$

Obtained values of $\hat{f}_k$ denote by $\hat{f}_k^s$, $k = 1,\dots,K$, $s = 1,\dots,K$.

3. For $s = 0,\dots,K$, $k = 1,\dots,K$ calculate $f_k^s = w_k^1\, \hat{f}_k^s$.

4. Present all w_k^i, $\hat{f}_k^s$, f_k^s to the decision maker. If for some s solution f_k^s are satisfactory, terminate the algorithm. Otherwise, go to 5.

5. Let $i = i+1$ and ask decision maker to set new reference objective $w_1^i,\dots,w_K^i$ of desirable (or satisfactory) values for f_k. Calculate $u_k^i = w_k^i/w_k^1$, $k = 1,\dots,K$ and go to step 1.

3. Discussion

In each step DM is specifying desired values for objective functions. The algorithm then tries to attain that values or, at least, come as close as possible (in the sense of $\|\ \|_1$) to it. If proposed point is feasible but not efficient, the algorithm will return a majorising efficient solution.

In step 2. the neighborhood of obtained efficient solution is explored. According the ideas in |7|, the perturbation distance is chosen so that the neighborhood contracts as the proposed point comes nearer to efficient surface.

All internal calculations are done with scaled objective functions, to eliminate different scales of objective functions. In step 3. DM sees both original and scaled values, where scaled values can be interpreted as the attained percentage of the ideal solution calculated in step 0.

Objective functions in (3) and (4) mean that we minimize the underachievement and maximize overachievement of each reference objective.

The factor ε means that we would much rather come as close as possible to reference objective with all objective functions than overshoot it by one or two functions at the expense of others. Ideally, ε should be as small as possible, but in practice, since too small ε would affect the stability of the problem, we should choose ε to be just small enough.

Let ε_0 be the lower bound to ε. Let

$$\varepsilon_k = \max_{x \varepsilon X} \|\nabla \hat{f}_k(x)\| \qquad k = 1,\ldots,K. \tag{5}$$

Let, further, be

$$\varepsilon_{k^*} = \max \left\{\varepsilon_k | k = 1,\ldots,K\right\} \tag{6}$$

and

$$\varepsilon_{k^{**}} = \min \left\{\varepsilon_k | k = 1,\ldots,K\right\}. \tag{7}$$

We can now define

$$\varepsilon = 0.9 \frac{\varepsilon_{k^{**}}}{\varepsilon_{k^*}} \quad . \tag{8}$$

If $\varepsilon < \varepsilon_0$, let $\varepsilon = \varepsilon_0$.

Factor ε obtained in this way is small enough for our purposes, but large enough not to cause numerical instability of problems (3) and (4), provided that lower bound ε_0 is chosen well.

Wierzbicki in |7| has given the theoretical background and algorithms for the use of reference objectives and scalarizing functions in interactive multicriteria programming. It can easily be proved that algorithm obtained here by such a pragmatical approach is an implementation of one of the algorithms proposed in |7|, which gives our approach necessary theoretical background.

4. Conclusion

The algorithm presented here can be used for optimization of a broad variety of functions. The conditions which objective functions must fulfill will depend of the choice of methods for solving problems (3) and (4).

However, regardless of type of objective functions, the algorithm presents an unified front to decision maker - no matter how (3) and (4) are solved, DM can always solve his problem in the same familiar way.

* * *

References

|1| Benayoun, R., de Montgolfier, J., Tergny, J. and Laritchev, O., *Linear Programming with Multiple Objective Functions: STEP-Method (STEM)*, Mathematical Programming, Vol. 1, No. 3 (December 1971), pp. 366-375.

|2| Dyer, J.S., *Interactive Goal Programming*, Management Science, Vol. 21, No. 12, August 1975.

|3| *** , *An Empirical Investigation of Man-Machine Interactive Approach to the Solution of the Multiple Criteria Problem*, in J. Cochrane and M. Zeleny (eds), *Multiple Criteria Decision Making*, University of South Carolina Press, 1973.

|4| Geoffrion, A.M., Dyer, J.S., Feinberg, A., *An Interactive Approach for Multi--Criterion Optimization with an Application to the Operation of an Academic Department*, Management Science, Vol. 19, No. 4, December, Part I, 1972.

|5| Hemming, T., *A New Method for Interactive Multiobjective Optimization: A Boundary Ranking Method*, in H. Thiriez and S. Zionts (eds), *Multiple Criteria Decision Making*, Jouy-en-Josas, Proceedings, Springer-Verlag, 1975.

|6| Wallenius, J., *Comparative Evaluation of Some Interactive Approaches to Multicriterion Optimization*, Management Science, Vol. 21, No. 12, August, 1975.

|7| Wierzbicki, A.P., *The Use of Reference Objectives in Multiobjective Optimization*, in *Multiple Criteria Decision Making: Theory and Application*, Springer-Verlag, 1979.

* * *

HYBRID METHODS IN NONLINEAR PROGRAMMING

by C. RICHTER* and K. SCHÖNEFELD**

* Abt. Mathematik/Rechentechnik der Ingenieurhochschule Köthen
DDR-4370 Köthen, Bernburger Straße 52-57

** Sektion Mathematik der Technischen Universität Dresden
DDR-8027 Dresden, Mommsenstraße 13

To construct robust and efficient optimization methods it is advantageous to combine a globally convergent algorithm and a locally superlinearly convergent one. Many of the proposed hybrid methods can be regarded as realizations of a general principle. For this class of methods (examples included) a general convergence result is presented.

1. INTRODUCTION

For the numerical determination of Kuhn-Tucker points $z^*=(x^*,u^*)\in R^n\times R^m$ of the nonlinear programming problem

$$f(x)=\min! \quad \text{subject to} \quad x\in G \quad \text{with} \qquad (1)$$
$$G=\{x\in R^n \mid g_i(x)\leq 0,\ i\in I\}\ ,\ I=\{1,\dots,m\}$$

locally superlinearly convergent methods are known [8],[13],[16],[11]. To obtain global convergence too it is possible to combine a globally convergent method I and a locally superlinearly convergent method II, for example the following combinations had been considered:

a penalty method - Robinson's method in [1],
a penalty method - Wilson-type methods in [7],[4],[6],
a method of feasible direct. - Wilson-type methods in [15],[2],[6],
a method of linearization - Wilson-type methods in [10],[5],[6],
a generalized meth. of feasible dir. - a Wilson-type method in [9].

These hybrid methods differ from other combinations (for instance modifications of Powell's dog leg strategy [6]) in that sense, that they can be regarded as realizations of the following general principle: After having generated the iterates $z^k=(x^k,u^k)$ by one step of the method I respectively II we decide by use of the convergence tests I respectively II wether the generation of $z^{k+1}\in R^n\times R^m$ by method II is advantageous or not.

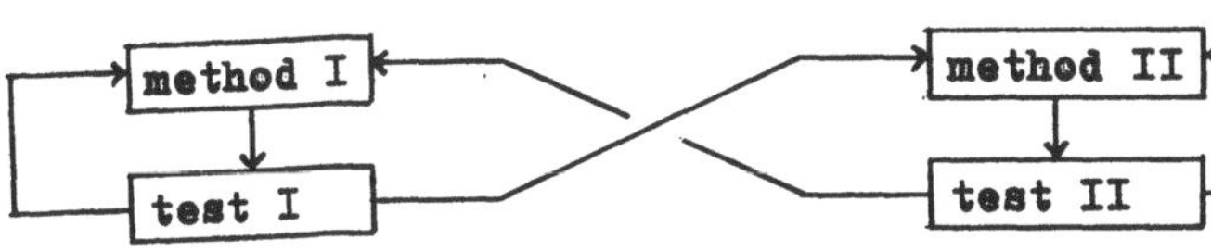

(2)

The aim of the paper is to present a general convergence result for the class of hybrid methods constructed by the principle above. The locally convergent methods II considered base on perturbation

principles described for example in [14],[11],[7],[4]. As method I algorithms can be used for which it is possible to show global convergence, with respect to both primal and dual iterates. We consider the following examples: a penalty method, a method of feasible directions and a modification of Pšeničnyj's linearization method.
For the so arising hybrid methods it has to be shown that after a finite number of iterations, roughly speaking, an infinite cycle in the right hand part of the diagram above occurs.
Now we want to distinguish two cases: If approximations I_k of the index set $I(x^*,0)$ of constraints active at x^* are used to switch over to method I respectively to method II (see [1],[10],[7]) this coupling is called to be of type B, otherwise of type A (see [9]). In the following we discuss the parts of hybrid methods described by diagram (2).

2. CONSTRUCTION OF HYBRID METHODS

According to diagram (2) the hybrid methods consist of a first phase (method I and test I) and a second phase (method II and test II).

method I(1) We start with the known primal iterate x^k of z^k and compute the corresponding dual iterate u^k. In case B, additionally, we determine the index set I_k.

test I Test of a merit number defined in dependence of the special algorithm. If the test is satisfied, then we switch over to method II, otherwise we perform the steps of the second part of method I.

method I(2) We compute the new primal iterate x^{k+1}.

method II With an appropriate perturbation p^k depending on $z^k=(x^k,u^k)$ and further informations (comp. [7] or [6]) we form the subproblem

$$\begin{aligned}&\bar f(x,p^k)=\min! \quad \text{subject to}\\ &\text{case A: } \bar g_i(x,p^k)\leq 0 \quad (i\in I) \quad \text{or}\\ &\text{case B: } \bar g_i(x,p^k)=0 \quad (i\in I_k),\ (u^k)_i=(u^{k+1})_i=0 \quad (i\in I\setminus I_k).\end{aligned} \tag{3}$$

We determine z^{k+1} as that Kuhn-Tucker point of (3) that possesses the smallest Euclidean distance to z^k. By use of the following approximations $\bar f$ respectively $\bar g_i$ we obtain from (3) the methods of Levitin/Poljak, of Robinson and of Wilson-type:

$$\begin{aligned}&\bar f:\ \nabla f(x^k)^T(x-x^k)+0.5(x-x^k)^T\nabla^2 f(x^k)(x-x^k) \qquad \bar g_i:\ g_i(x)\\ &\bar f:\ f(x)+(u^k)_1[g_1(x)-\bar g_1(x^k,p^k)]+\ldots+(u^k)_m[g_m(x)-\bar g_m(x^k,p^k)]\\ &\bar f:\ \nabla f(x^k)^T(x-x^k)+0.5(x-x^k)^T H_k(x-x^k)\\ &\bar g_i \text{ for the last both methods: } g_i(x^k)+\nabla g_i(x^k)^T(x-x^k),\end{aligned} \tag{4}$$

where H_k is a suitable approximation of the Hessian M_k of the Lagrangian with respect to x at (x^k,u^k).

test II Let z^{k+1} be generated by method II and let $j\leq k$ denote the greatest iteration index at which a jump from method I to method II took place:

If in case A: $\|z^{k+1}-z^k\| \le (1/2)^{k-j+1}$ (5)

or if in case B: (5) and $\bar{g}_i(x^{k+1},p^k) \le 0$ $(i \in I \setminus I_k)$, $(u^{k+1})_i \ge 0$ $(i \in I_k)$ (6)

hold then goto method II, otherwise goto method I(1).

3. CONVERGENCE

Corresponding to the first phase of the hybrid methods we define the index set $J := \{k \in N \mid x^{k+1}$ is computed by method I(2)$\}$.

T h e o r e m: Let the hybrid algorithm according to chapter 2 generate an infinite sequence of iterates $z^k=(x^k,u^k)$ as approximations of a Kuhn-Tucker point $z^*=(x^*,u^*) \in R^n \times R^m$ of (1). Let $\nabla^2 f$, $\nabla^2 g_i (i \in I)$ be Lipschitz-continuous in a neigbourhood of x^*, let the vectors $\nabla g_i(x^*)$ $(i \in I(x^*,0))$ be linearly independent , let the second order sufficiency conditions and the strict complementary slackness condition be fulfilled at z^*.

Furthermore the following conditions are required to be satisfied:

If the index set $J \subset N$ is infinite, then it holds:

a) There exists an infinite subset $K \subset J$, such that $z^k \to z^*$ $(k \in K)$.

b) Test I leads to method II for sufficiently large k K.

c) In case B, additionally, it holds for sufficiently large $k \in K$
$I_k = I(x^*,0)$, $(u^k)_i = 0$ $(i \in I \setminus I_k)$.

Then the index set J is finite and the sequence $\{z^k\}$ converges Q-superlinearly to z^*.

R e m a r k to the proof:

Suppose that J is infinite. Then the assumptions a), b) and in the case B, additionally, c) ensure for some sufficiently large k_o that the algorithm switches over to method II and local convergence theorems (comp. [14],[1],[7],[4],[6]) guarantee that method II is used exclusively for all $k \ge k_o$, contradiction. Using the first assertion we obtain the convergence from the test (5) and the Kuhn-Tucker conditions for the subproblem (3). Under the required assumptions superlinear convergence can be shown (comp. [14],[7],[4]).

4. EXAMPLES

We consider now globally convergent methods used for combination.

1. A penalty method ([1],[7]): We use a penalty-function $S | R^n \times R^m \to R^1$:

$$S(x,\mu)=f(x)+\mu_1 \max^4\{0,g_1(x)\}+\ldots+\mu_m \max^4\{0,g_m(x)\} \quad ,$$

a feasible point $x^+ \in G$ and numbers $\beta > 0, \gamma \in (0,1)$.

We start with x^k and compute

| method I(1) | case A: $(u^k)_i := 4(\mu^k)_i \max^3\{0,g_i(x^k)\}$ | (7) |

case B:(7) and $I_k := \{i \in I \,|\, |g_i(x^k)| < (u^k)_i\}$. (8)

test 1 — Set $\mu^{k+1} := 10\mu^k$.

If in case A: $\max\{g_i(x^k) \mid i \in I\} \leq \beta$ (9)

or if in case B:(9) and card $I_k \leq n$ (10)

hold then $\beta := \gamma\beta$, goto method II, otherwise goto method I(2).

method I(2) — Starting from x^k compute x^{k+1} satisfying

$$S(x^{k+1},\mu^{k+1}) \leq S(x^+,\mu^{k+1}) \text{ and } \|\nabla_1 S(x^{k+1},\mu^{k+1})\| \leq (1/2)^{k+1} \quad (11)$$

and goto method I(1).

R e m a r k: The required properties a),b)(case A) of the theorem above can be shown under the following assumptions:

1. There is a number τ such that $f(x) \geq \tau$ for all $x \in R^n$.
2. $G \neq \emptyset$, and $\max\{g_i(x) \mid i \in I\} \to \infty$ as $\|x\| \to \infty$.
3. For every accumulation point $\bar{x}$ of $\{x^k, k \in J\}$ it holds: The vectors $\nabla g_i(x)$ $(i \in I(\bar{x},0))$ are linearly independent.

To show also c)(case B), additionally, the strict complementary slackness condition at x^* is required (comp.[1],[7]).

2. A method of feasible directions([15],[3],[2]): Let for $\varepsilon \geq 0$ and $x \in G$ $I(x,\varepsilon) := \{i \in I \mid -\varepsilon \leq g_i(x) \leq 0\}$ denote the index set of constraints ε-active at x. Let be $\gamma \in (0,1)$. We start with $\beta > 0, x^k \in G$ (or a projection of x^k on G), $\theta_k > 0, \varepsilon_k > 0$ and set $I_k := I(x^k,\varepsilon_k)$.

method I(1) — Solve the problem

$$\sigma = \min! \quad \text{s.t.}$$

$$g_i(x^k) + \nabla g_i(x^k)^T s \leq \sigma \quad (i \in I_k) \quad (12)$$

$$\nabla f(x^k)^T s \leq \theta_k \sigma \quad \text{and} \quad (\sigma, s^T)^T \in S_k$$

(S_k compact with $0 \in \text{int } S_k$ appropriately chosen), i.e., determine a Kuhn-Tucker point $((\sigma_k,(s^k)^T)^T,(v^k)_i \ (i \in \{0\} \cup I_k)) \in R^{n+1}\ R^{1+\text{card}(I_k)}$ of (12) and compute $u^k \in R^m$ with $(u^k)_i = 0$ $(i \in I \setminus I_k)$ and $(u^k)_i = (v^k)_i/(v^k)_0$ $(i \in I_k)$.

test I — If in case A: $\sigma_k \leq \beta$ (13)

or if in case B: (13) and $(u^k)_i > -g_i(x^k)$ $(i \in I_k)$ (14)

hold then set $\beta = \gamma\beta$ and goto method II, otherwise goto method I(2).

method I(2) — Set $\varepsilon_{k+1} := \min\{\varepsilon_k, -\sigma_k\}$ in case B. Compute $\alpha_k \geq 0$ such that $x^{k+1} = x^k + \alpha_k s^k \in G$ and the difference $f(x^k) - f(x^{k+1})$ sufficiently large. Goto method I(1). If in the case B the set S_k is chosen appropriately then σ_k, s^k and u^k can be given explicitely (comp.[3]).

R e m a r k: The required properties a),b) (case A) of the theorem above can be shown under the following assumptions

1. There is a number τ such that $f(x) \geq \tau$ for all $x \in G$.
2. $G \neq \emptyset$, and the level set $\{x \in G \mid f(x) \leq f(x^0)\}$ is bounded.
3. The vectors $\nabla g_i(x)$ $(i \in I(x,\varepsilon_0))$ are linearly independent for all $x \in G$.

To show also c)(case B), additionally, the strict complementary slackness condition at x^* is required(comp. [15],[2]).For the following method the required properties can be shown under similar assumptions.

3.A method of linearization(comp.[5],[1o])

We use a penalty function $F(x)=\max\{0,g_1(x),\dots,g_m(x)$,an index set $\tilde{I}(x,\delta)=\{i\in I\mid F(x)-\delta\le g_i(x)\}$ and start with $\delta_k>0,x^k,N,\beta>0,\gamma\in(0,1)$.

method I(1) Solve the problem

$$\begin{aligned}&\nabla f(x^k)^T s+0.5 s^T T_k^T T_k s = \min! \quad \text{s.t.}\\ &g_i(x^k)+\nabla g_i(x^k)^T s = 0 \quad (i\in I(x^k,\delta_k))\end{aligned} \tag{15}$$

with a nonsingular (n,n)-matrix T_k, i.e., determine a Kuhn-Tucker point $(s^k,(u^k)_i(i\in\tilde{I}(x^k,\delta_k)))$ of (15) and a merit number $\varrho_k>0$.

Set $(u^k)_i:=0(i\in I\setminus\tilde{I}(x^k,\delta_k))$ and determine in case B the set I_k as in (8).

test I If in case A: $\varrho_k\le\beta$ (16)

or if in case B: (16) and card $I_k\le n$ (17)

hold then set $\beta:=\gamma\beta$ and goto method II, otherwise goto method I(2).

method I(2) Compute $\alpha_k\ge 0$ such that for $x^{k+1}=x^k+\alpha_k s^k$ the difference $B_N(x^k)-B_N(x^{k+1})$ is sufficiently large (with a function $B_N(x):=f(x)+NF(x)$. Goto method I(1).

If the matrix T_k is chosen appropriately then s^k,u^k and ϱ_k can be given explicitely(comp.[5]).

5.REMARKS

1.Compared to B, the coupling principle A is more simple and the numerical effort in method I can be reduced. The more complicated coupling principle B leads to equality constrained subproblems (3) for method II.This is in general an essential simplification, e.g. for Wilson-type methods systems of linear equations occur exclusively [7].

2.Other hybrid methods considered in this paper by coupling methods I with methods II and combinations with further methods are in detail discussed in [12], also with respect to convergence and numerical behaviour.

6.REFERENCES

[1] Best,M.J.;Bräuninger,J.;Ritter,K.;Robinson,S.M.; A globally and quadratically convergent algorithm for general nonlinear programming problems. Computing 26, 141-153 (1981).

[2] Ishutkin,V.S.;Schönefeld,K.; A globalization of Wilson-type methods ... , Proceedings X.IKM, Weimar 1984, Vol.4, 69-72.

[3] Ishutkin,V.S.;Kleinmichel ,H.;Verfahren der zulässigen Richtungen unter Benutzung reduzierter Gradienten für nichtlineare Optimierungsprobleme. Optimization 16, 373-390 (1985).

[4] Kleinmichel,H.; Zur Anwendung von Quasi-Newton-Verfahren in der

nichtlinearen Optimierung. Dissertation B, Technische Universität Dresden, 1982.

[5] Kleinmichel, H., Überlinear konvergente Verfahren der nichtlinearen Optimierung. Proceedings X.IKM, Weimar 1984, Vol.4, 73-76.

[6] Kleinmichel, H., Koch, W., Richter, C., Schönefeld, K., Überlinear konvergente Verfahren der nichtlinearen Optimierung.Preprints 84-07-20 bis 84-07-24,84-07-29. Technische Universität Dresden, 1984.

[7] Kleinmichel, H., Richter, C., Schönefeld, K., On the global and superlinear convergence of a discretized version of Wilson's method. Computing 29, 289-307 (1982).

[8] Levitin, E.S., Poljak, B.T., Metody minimizacii pri nalicii ogranicenii. Z.Vycisl.Mat.i Mat.Fiz. 6, 787-823 (1966).

[9] Polak, E., Mayne, D.Q., A robust secant method for optimization problems with inequality constraints. JOTA 33, 463-477 (1981).

[10] Pšeničnyj, B.N., Sobolenko, L.A., Uskorenije shodimosti metoda linearizacii dlja zadaci uslovnoi minimizacii. Z.Vycisl. Mat. i Mat.Fiz. 20, 605-614 (1980).

[11] Richter, C., Über Mehrschrittverfahren der nichtlinearen Optimierung. ZAMM 60, 129-136 (1980).

[12] Richter, C., Schönefeld, K., Hybrid methods for nonlinear programming problems (In preparation).

[13] Robinson,S.M.: A quadratically convergent algorithm for general nonlinear programming problems.Math. Programming 3, 145-156(1972).

[14] Robinson,S.M.: Perturbed Kuhn-Tucker-Points and rates of convergence for a class of nonlinear programming problems. Math.Pro - gramming 7, 1-16 (1974).

[15] Schönefeld,K.:A hybrid method for linearly constrained optimi - zation problems. to appear in Optimization ,1986.

[16] Wilson,R.B.: A simplicial algorithm for concave programming. Ph.D.Dissertation, Harward University, Cambridge, Mass.,1963.

Bundle-Based Decomposition: Description and Preliminary Results

STEPHEN M. ROBINSON

University of Wisconsin-Madison*

1. Introduction. This is a preliminary report on a computational method for solving the separable convex programming problem

$$\text{minimize}\{\Sigma_{i=1}^{n} f_i(x_i) \mid \Sigma_{i=1}^{n} A_i x_i = a\} , \tag{1}$$

where for each i f_i is a proper convex function from $\mathbf{R}^{n_i}$ to $(-\infty,+\infty]$ and A_i is a linear transformation from $\mathbf{R}^{n_i}$ to $\mathbf{R}^m$, and where $a \in \mathbf{R}^m$.

The problem (1) models an economic process in which n activities (subsystems) compete for m shared resources. The total amounts of resources available are specified by the m-vector a. The i-th activity is controlled by n_i parameters, represented by the vector $x_i \in R^{n_i}$; its resource requirements are $A_i x_i$ and its cost (or the negative of its profit contribution) is $f_i(x_i)$. Solving (1) then amounts to determining how to set the parameters $x_1,\ldots,x_n$ in order to minimize the total system cost while consuming only the available resources a. Evidently the case in which consumption is only restricted to be less than or equal to a is included in (1), since we can always add a dummy (slack) activity with zero cost.

The fact that the f_i may take values of $+\infty$ is important, since it permits us to include subsystem constraints implicitly. An example of a case in which this capability is particularly useful is the familiar block-angular linear programming problem:

$$\begin{array}{lrcl}
\text{minimize} & \langle c_1,x_1\rangle + \langle c_2,x_2\rangle + \cdots + \langle c_n,x_n\rangle & & \\
\text{subject to} & B_1x_1 \qquad\qquad\qquad\qquad & = & b_1 \\
& B_2x_2 \qquad\qquad\quad & = & b_2 \\
& \ddots \qquad\qquad & & \vdots \\
& B_nx_n & = & b_n \\
& A_1x_1 + A_2x_2 + \cdots + A_nx_n & = & a \\
& x_1,x_2,\ldots,x_n \geq 0 . & &
\end{array} \tag{2}$$

To model (2) in the form (1) we simply set

---

†Sponsored by the US National Science Foundation under Grant No. DCR-8502202.

*Department of Industrial Engineering, 1513 University Av., Madison, WI 53706, USA.

$$f_i(x_i) := \begin{cases} \langle c_i, x_i\rangle & \text{if } B_i x_i = b_i \text{ and } x_i \geq 0 , \\ +\infty & \text{otherwise} , \end{cases} \tag{3}$$

for $i = 1,\ldots,n$. The problem (2) occurs in multistage planning models as well as in two-stage stochastic linear programming (where it occurs as a result of discretizing the probability distribution of the second-stage model).

In this paper we deal with the case in which evaluation of the individual f_i does not present great difficulty, but in which the direct solution of (1) is impractical. Efforts have been made in the past to decompose (1) into smaller optimization problems, the best known of these being probably the decomposition method of Dantzig and Wolfe for (2) [Dantzig (1963), Ch. 23]. These algorithms have had mixed success. We propose here an approach applying convex duality theory directly to (1) to produce a nonsmooth convex dual problem in R^m, then taking advantage of recent progress in nonsmooth optimization to solve this dual in a way that provides approximate solutions to both the dual problem and the primal problem (1).

2. Duality analysis of (1). We begin by introducing perturbations $p \in R^m$ into (1) to yield the function

$$F(x_1,\ldots,x_n,p) := \begin{cases} \Sigma_{i=1}^n f_i(x_i) & \text{if } \Sigma_{i=1}^n A_i x_i = a - p \\ +\infty & \text{otherwise} \end{cases} \tag{4}$$

By standard perturbational duality methods [Rockafellar (1970), Rockafellar (1974)] we obtain from (4) the dual objective function

$$g(y) := \langle a,y\rangle - \Sigma_{i=1}^n f_i^*(A_i^* y) ,$$

where A_i^* is the adjoint of A_i and

$$f_i^*(x_i^*) := \sup_{x_i}\{\langle x_i^*, x_i\rangle - f_i(x_i)\} .$$

Under the assumption

$$a \in \Sigma_{i=1}^n A_i(\text{ri dom } f_i) \tag{5}$$

we know that the function g attains a maximum equal to the infimum in (1); here $\text{dom } f_i$ is the set $\{x_i | f_i(x_i) < +\infty\}$ and "ri" denotes relative interior. If we make the further assumption that there is some y_0 such that for each i

$$A_i^* y_0 \in \text{ri dom } f_i^* \tag{6}$$

then the subdifferential mapping ∂g is given by the formula

$$\partial g(y) = a - \Sigma_{i=1}^n A_i x_i(y) ,$$

where for each i the point $x_i(y)$ solves

$$\text{minimize}\{f_i(x_i) - \langle A_i^* y, x_i\rangle\} . \tag{7}$$

Note that solving (7) requires minimization only in n_i variables. Also, note that the solutions of (7) may not be unique, so that $\partial g(y)$ may be multivalued. Of course, if the f_i are strongly convex this will not happen, and in that case ∂g, which is then the Fréchet derivative of g, is actually Lipschitzian. In our analysis we do not assume any strong or strict convexity condition.

The assumption (5) amounts to a constraint qualification on the primal problem (1), whereas (6) says, roughly speaking, that the effective domains of f_i^* are large enough. This condition always holds, for example, if the functions f_i take the value $+\infty$ off some bounded set. In the case of the linear programming problem (2), this amounts to assuming that $\{x_i | B_i x_i = b_i, x_i \geq 0\}$ is bounded for each i. Of course, (6) can also hold under many other assumptions (e.g., that each f_i is strongly convex).

Thus, if we assume we can solve the subproblems (7) for given values of y, then for such values we are able to compute $g(y)$ and one element of $\partial g(y)$. This capability suggests the use of one of several methods currently under active investigation for optimization of nonsmooth functions. We shall concentrate here on the so-called "bundle method," which is discussed in detail in [Lemaréchal et al. (1981)]. In the next section we summarize this method and discuss its application to the maximization of g.

3. Applying the bundle method to g. The bundle method, very briefly described, involves computing for a sequence of points $y_1, \ldots, y_k$ a set of subgradient values $v_1, \ldots, v_k$, with $v_j \in \partial g(y_j)$ for each j. Then a convex combination of $v_1, \ldots, v_k$ is computed to approximate 0 as well as possible (with certain weighting restrictions). If the approximant thus found is not very close to 0, a new point y_{k+1} is determined and one of its subgradients, v_{k+1}, is then computed. The method terminates with y_k when, for previously prescribed positive values of ε and δ, there is a vector d_k with $\|d_k\| < \delta$ such that $-d_k \in \partial_\varepsilon g(y_k)$: thus, y_k is an approximate ε-maximizer of g, since the ε-subdifferential ∂_ε is defined by

$$\partial_\varepsilon g(\hat{y}) := \{v \mid \text{for each } y,\ g(y) \leq g(\hat{y}) + \langle v, y - \hat{y}\rangle + \varepsilon\} .$$

The vector d_k is obtained from the v_i by setting

$$d_k = -\Sigma_{j=1}^{k} \lambda_j^k v_j , \tag{8}$$

where the λ_j^k are convex coefficients (non-negative and summing to 1) subject to certain constraints.

The main practical difficulty in using a nonsmooth optimization method to solve (1) by maximizing g lies in the fact that even if we knew an exact maximizer $\bar{y}$ for g we might not be able to find a primal optimal solution $\bar{x}_1, \ldots, \bar{x}_n$. This is the case because, even though we know that $0 \in \partial g(\bar{y})$, we can readily compute only one element of $\partial g(\bar{y})$. We might therefore find $\hat{x}_1(\bar{y}), \ldots, \hat{x}_n(\bar{y})$ such that the point $w := a - \Sigma_{i=1}^{n} A_i\hat{x}_i(\bar{y})$ belonged to $\partial g(\bar{y})$ but was far from zero; then the $\hat{x}_i(\bar{y})$ would not constitute a primal solution because they would be infeasible.

The bundle algorithm very nicely circumvents this difficulty through the manner in which the approximate zero subgradient is constructed. If in (8) we substitute the expression

$$v_j = a - \Sigma_{i=1}^{n} A_i x_i(y_j) \qquad (j = 1,\ldots,k)$$

then we obtain

$$-d_k = a - \Sigma_{i=1}^{n} A_i x_i^k$$

where

$$x_i^k := \Sigma_{j=1}^{k} \lambda_j^k x_i(y_j) .$$

The linearity of the constraint in (1) therefore causes $x_1^k,\ldots,x_n^k$ to be nearly primal feasible if the norm of d_k is nearly zero, and this avoids the difficulty just mentioned.

In a (longer) forthcoming paper we shall examine the degree to which $x_1^k,\ldots,x_n^k$ will be approximately primal optimal, and we shall also present detailed computational results. However, in the next section we present some preliminary results for the block-angular linear programming problem (2), showing that the method can be implemented to compute approximate solutions for such problems.

4. Preliminary computational results. The results given here were obtained by generating linear programming problems of the form (2), then solving their duals by the bundle method described in Section 3. This method was implemented by using the code M1FC1, kindly made available by Dr. Claude Lemaréchal; the linear programming subproblems (7) were solved using the IMSL linear programming code ZX3LP. Computations were carried out on a DEC VAX 11/780. For these preliminary results, no attempt was made to program the method for speed, and therefore no timing results are provided. The purpose was simply to determine whether the method would work reliably on medium-sized linear programming problems of the type often attacked with the Dantzig-Wolfe method.

To generate the test problems, values of m, n, and the block sizes in (2) were first selected. The matrices B_i and A_i were then generated by drawing random numbers from uniform distributions on specified intervals; for most matrix elements the intervals [-8,9] or [-8,6] were used but in each matrix one positive row was generated by drawing from an interval such as [1,6] or [5,13]. The purpose of this positive row was to ensure boundedness of the resulting feasible region so that the effective domain of g would be $\mathbb{R}^m$, since the bundle code M1FC1 cannot handle constraints.

After the matrices had been generated, random non-negative vectors x_i were generated using the uniform distribution on [0,4], and cost coefficients using [-7,7]. The matrices, together with the points x_i, then determined the right-hand sides.

Results of the computations are given in Table 1. Column headings are generally self-explanatory, except that for "subproblem calls", which records the number of times that sets of subproblems of the form (7) had to be solved. Tolerances for the computation were set to be rather tight, as indicated by the small primal infeasibility (in the coupling constraints) and the close agreement between primal and dual objective functions. If these tolerances had been relaxed it is possible that significantly fewer calls would have had to be made.

Problem	Rows	Columns	n	m	Max. Primal Infeasibility	Primal Objective	Dual Objective	Subproblem Calls
A	150	255	25	5	$< 10^{-6}$	-1,173.99	-1,174.00	42
B	150	255	19	5	$< 10^{-6}$	-942.95	-942.96	88
C	200	350	26	5	1×10^{-6}	-1,552.0	-1,552.1	42
D	250	450	30	5	$< 10^{-6}$	-2,022.2	-2,022.3	38
E	300	550	41	5	1×10^{-6}	-2,540.15	-2,540.17	36
F	350	650	44	7	1×10^{-6}	-3,506.0	-3,506.1	52
G	450	850	57	5	2×10^{-6}	-4,188.0	-4,188.1	45

TABLE 1.

5. Acknowledgment. All computations reported here were carried out by Mr. Deepankar Medhi, Research Assistant in the Engineering Experiment Station of the University of Wisconsin-Madison, using the facilities of the Computer Systems Laboratory of the Computer Sciences Department.

REFERENCES

1. G. B. Dantzig (1963), Linear Programming and Extensions. Princeton University Press, Princeton, NJ.

2. C. Lemaréchal, J. J. Strodiot, and A. Bihain (1981), "On a bundle algorithm for nonsmooth optimization," in: O. L. Mangasarian, R. R. Meyer, and S. M. Robinson, eds., Nonlinear Programming 4, Academic Press, New York.

3. R. T. Rockafellar (1970), Convex Analysis. Princeton University Press, Princeton, NJ.

4. R. T. Rockafellar (1974), Conjugate Duality and Optimization. CBMS Regional Conference Series in Applied Mathematics No. 16, Society for Industrial and Applied Mathematics, Philadelphia, PA.

ON THE OPTIMIZATION OF A SHORT-RUN MODEL OF ENERGY PRODUCTION SYSTEMS

R. GONZALEZ* - E. ROFMAN**

1 - INTRODUCTION

In this paper, we apply the general procedure presented in [2] to the optimization of a short-run model of an energy production system. The numerical data have been provided by Electricity of France (EDF) : they describe a forecast of the French system for the year 2000.

The characteristics of the model allow us to introduce several improvements in the general method. These modifications produce a significant reduction of the time of computation, reduction due to the simplicity of the algorithms obtained.

The aim of our paper is to obtain the optimal cost V and the optimal production policy P of the hydraulic and thermopower (nuclear, fuel, coal and gas) plants of the system.

The method of solution is based on the characterization of the optimal cost function V as the maximum element of a suitable set of subsolutions of the Hamilton-Jacobi equation associated to the control problem. Here we develope an approximate method of numerical solution that uses new special discretization schemes for the partial derivatives $\frac{\partial V}{\partial t}$, $\frac{\partial V}{\partial x}$. This method has the following properties :

a) A discrete maximum principle is verified and, in consequence, the discrete problem has an unique solution that converges to the solution of the original problem as the discretization is refined.

b) The discrete problem is solved using fast algorithms for dynamic programming problems on graphs and simple algorithms of convex optimization.

We show, in the final paragraph of this paper, some numerical results and the times of computation employed.

2 - MODELING OF THE SYSTEM

2.1. General description

The energy production system consists of μ hydraulic plants, one nuclear plant,

* Electronic Dept., Univ. of Rosario, Argentine
** INRIA, Le Chesnay, France

one gas powered plant, ν thermic (coal or fuel) plants and one external (or fictitious) plant. We denote P_ℓ^h $(\ell=1,2,\ldots,\mu)$, P^n, P^g, $P_r^\theta (r=1,2,\ldots,\nu)$, P^e their respective levels of production and c_ℓ^h, c^n, c^g, c_r^θ, c^e the corresponding costs/KWh.

If D(t) is the demand of energy we have :

$$(1) \quad D(t) = \sum_{\ell=1}^{\mu} P_\ell^h(t) + \sum_{r=1}^{\nu} P_r^\theta(t) + P^n(t) + P^g(t) + P^e(t).$$

The function D(.) is a step function in the interval of time [0,T] considered in our problem.

2.2. Characteristic of operation of the plants

2.2.1. Hydraulic plants

The evolution of the stock of energy S_ℓ^h (measured in KWh) of each hydraulic plant is considered independent of the remaining stocks. Taking into account pumped hydraulic plants this evolution will be described by :

$$(2) \quad \frac{d}{dt} S_\ell^h = -(P_\ell^h(t))^+ + \eta_\ell \, (P_\ell^h(t))^- + A_\ell \; ; \; S_\ell^h \in [0, S_{\ell,max}^h]$$

P_ℓ^h will be positive if the plant deliver energy to the production system ; on the other hand it will be negative in the pumping mode of operation : it takes energy from the system and it storages that energy in the corresponding dam.

η_ℓ is the pumping-efficiency ; A_ℓ is the inflow of energy (rain, incoming rivers,...) to the 1th dam.

P_ℓ^h verifies the restrictions :

$$P_{\ell,min}^h \le P_\ell^h(t) \le P_{\ell,max}^h \text{ , with } 0 \in [P_{\ell,min}^h, P_{\ell,max}^h]$$

$$(3) \quad P_\ell^h(t) \le A_\ell \quad \text{if } S_\ell^h = 0$$

$$P_\ell^h(t) \ge A_\ell \text{ if } S_\ell^h = S_{\ell,max}^h$$

The cost of operation $c_\ell^h(S_\ell^h(t),t)$ of the hydraulic plants is related to the variation of the stock of energy. Actually we shall use "shadow" prices obtained from expected values given by stochastic models.

2.2.2. Thermic (coal and fuel) plants

For these plants we consider a start-up cost k_r equivalent to 6 hours of operation with maximum power :

$$(4) \quad k_r = 6 \, c_r^\theta \, P_{r,max}^\theta$$

The restrictions for the production level of an operational plant are, $\forall r = 1,\nu$:

(5) $\alpha_r^\Theta \cdot P_{r,max}^\Theta = P_{r,min}^\Theta \leq P_r^\Theta(t) \leq P_{r,max}^\Theta$, $\alpha_r^\Theta > 0$

2.2.3. Nuclear, gas and external plants

The production levels of these plants must verify :

$$0 < P_{min}^n \leq P^n(t) \leq P_{max}^n$$

(6) $0 \leq P^g \leq P_{max}^g$

$$0 \leq P^e < \infty$$

2.3. The optimization problem ($\mathcal{P}$).

Let us introduce the functional J that measures the total cost of operation of the system in the interval [0,T] for a given production policy :

(7) $$J = \int_0^T [\sum_{\ell=1}^{\mu} c_\ell^h(S_\ell^h(s),s)\ ((P_\ell^h(s))^+ - \eta_\ell(P_\ell^h(s))^- - A_\ell)]\ ds$$
$$+ \int_0^T [\sum_{r=1}^{\nu} c_r^\Theta\ P_r^\Theta(s) + c^n\ P^n(s) + c^g\ P^g(s) + c^e\ P^e(s)]\ ds + \sum_{r=1}^{\nu} k_r\ n_r^\Theta$$

where n_r^Θ gives the number of start-ups of the rth thermic plant in the interval [0,T].

($\mathcal{P}$) : Our problem is to find a production policy, in the frame of 2.1 and 2.2 that minimizes the functional J.

3 - THE DYNAMIC PROGRAMMING (DP) APPROACH AND THE ASSOCIATED SYSTEM OF QUASI-VARIATIONAL INEQUALITIES (IQV)

3.1. Parametrization of the problem

To apply the DP methodology let us define in (8), (9), (10) a set of parameters that identify the internal (and initial) state of the system :

(8) the time $t \in [0,T]$;

(9) the vector of hydraulic stocks $x \in \Omega = \prod_{\ell=1}^{\mu} ([0,S_{\ell,max}^h] \subset \mathbb{R}^\mu$

(10) the thermic state of operation $q \in \Gamma(\nu) = \{0,1,2,\dots, 2^\nu-1\}$ used to identify those thermic plants being operating. We use the following definitions :

(11) The rth plant is in operation iff $q_r(q) = 1$, where $q_r(q)$ are the (unique) coefficients of the binary expression of q :

(12) $q = \sum_{r=1}^{\nu} q_r(q)\ 2^{r-1}$, $q_r \in \{0,1\}$, $q \in \Gamma(\nu)$.

We denote $P_{ad}(x,t,q)$ the set of production policies P(.) verifying (1), (3), (5), (6) and given $q \in \Gamma(\nu)$, $(x,t) \in Q = \Omega \times [0,T]$ we consider the following functional

(13) $$J_q(x,t;P(.)) = \int_t^T [\sum_{\ell=1}^{\mu} c_\ell^h\ (S_\ell^h(s),s)\ ((P_\ell^h(s))^+ - \eta_\ell\ (P_\ell^h(s))^- - A_\ell)]\ ds +$$

$$+ \int_t^T [\sum_{r=1}^{\nu} c_r^{\theta} P_r^{\theta}(s) + c^n P^n(s) + c^g P^g(s) + c^e P^e(s)]\, ds + \sum_{r=1}^{\nu} k_r \eta_r^{\theta}(t)$$

where $\eta_r^{\theta}(t)$ is the number of start-ups of the rth thermic plant in the interval $[t,T]$ and S_ℓ^h is given by (2) with (x_ℓ,t) as initial conditions.

Let us define, depending of the given parameters, the optimal cost functions

(14) $V_q(x,t) = \inf_{P(.)\in P_{ad}(x,t,q)} J_q(x,t;P(.))$, $\forall q\in\Gamma(\nu)$, $\forall (x,t)\in Q$.

3.2. The equivalent problem

Using (14) we will solve ($\mathcal{P}$) after introducing an auxiliary equivalent problem. In fact THEOREM 1 gives us a characterization of V(x,t), optimal cost for our system as a function of the time and the hydraulic state. The knowledge of V(x,t) allow us (see [8], [6]) to obtain the optimal production policy.

THEOREM 1 :

Let be

(15) $\mathcal{W} = \{w_q : Q \to \mathbb{R} \,/\, w_q \in W^{1,\infty}(Q)$, $\forall q \in \Gamma(\nu)$, w_q verifies (16), (17), (18)$\}$

(16) $$\frac{\partial w_q}{\partial t}(x,t) + \min_{P(.)\in P_{ad}(x,t,q)} \{\sum_{\ell=1}^{\mu} \frac{\partial w_q}{\partial x_\ell}(x,t)\, (-(P_\ell^h(t))^+ + \eta\, (P_\ell^h(t))^- + A_\ell) +$$
$$+ \sum_{\ell=1}^{\mu} c_\ell^h(x_\ell,t)\, ((P_\ell^h(t))^+ - \eta_\ell (P_\ell^h)^- - A_\ell) + \sum_{r=1}^{\nu} c_r^{\theta} P_r(t) + c_n P^n(t) + c^g P^g(t) +$$
$$+ c^e P^e(t)\} \geq 0$$

(17) $w_q(x,t) \leq w_p(x,t) + k_q^p$, $\forall p \neq q$

(18) $w_q(x,T) = 0$

where $k_q^p = \sum_{r=1}^{\nu} (p_r(p) - q_r(q))^+ k_r$ is the cost associated to move the system from the state q to the state p.

Then :

(19) $V(x,t) \in \mathcal{W}$

(20) $V(x,t) \geq w_q(x,t)$ $\forall w_q \in \mathcal{W}$, $\forall (x,t) \in Q$

For the proof of THEOREM 1, see [6].

Remark : THEOREM 1 says us that our unknown V(x,t) is the maximum element (after introducing a suitable partial order) of the set $\mathcal{W}$. Based in this property we will discuss an approximation procedure to compute V.

In the following we will consider the equivalent problem :

($\mathcal{P}_e$) : Find V(x,t) such that (20) **and** (19) hold.

4 - NUMERICAL COMPUTATION OF THE OPTIMAL COST V(x,t)

4.1. Discretization of ($\mathcal{P}_e$)

4.1.1. Discretization of Q

The set Q is approximated with a net of nodes Q^a which are uniformly distributed :

$$
\begin{aligned}
&Q^a = \Omega^a . T^a\\
&T^a = \{t^m/t^m = m\,\Delta t^a,\ \Delta t^a = T/(NT)^a,\ m = 0,1,\ldots,(NT)^a\}\\
(21)\quad &\Omega^a = \{x^j \in \mathbb{R}^{\mu}/x^j = (x^{j_1},\ldots,x^{j_\mu}),\ x^{j_\ell} = j_\ell . \Delta S_\ell^a\}\\
&\Delta S_\ell^a = S^h_{\ell,max}/(NS)^a_\ell\ ,\ j_\ell = 0,1,\ldots,(NS)^a_\ell\}
\end{aligned}
$$

The integers $(NT)^a$, $(NS)^a_\ell$ and $\ell = 1,\ldots,\mu$ identify the discretization Q^a.

4.1.2. Discretization of the partial derivatives and formulation of a "discrete maximum principle" (DMP)

To approximate the partial derivatives that appear in (16) we use explicit discretization schemes depending of the points (x^j,t^m). They are defined in such a way that a DMP is verified (cfr. [4]). This DMP plays an essential rol in what follows.

We introduce the following definitions :

$$
(22)\quad Z^1_\ell(x^j) = \begin{cases} (x^{j_1},\ldots,x^{j_\ell+1},\ldots,x^{j_\mu}) & \text{if } 1 < j_\ell < (NS)^a_1 \\ (x^{j_1},\ldots,x^{j_\ell},\ldots,x^{j_\mu}) & \text{if } 1 = j_\ell \text{ or } j_\ell = (NS)^a_\ell \end{cases}
$$

$$
(23)\quad Z^2_\ell(x^j) = \begin{cases} (x^{j_1},\ldots,x^{j_\ell-1},\ldots,x^{j_\mu}) & \text{if } 1 < j_\ell < (NS)^a_\ell \\ (x^{j_1},\ldots,x^{j_\ell},\ldots,x^{j_\mu}) & \text{if } j_\ell = 1 \text{ or } j_\ell = (NS)^a_\ell \end{cases}
$$

$$
(24)\quad F^+_\ell(x^j) = \begin{cases} (x^{j_1},\ldots,x^{j_\ell+1},\ldots,x^{j_\mu}) & \text{if } j_\ell < (NS)^a_\ell \\ (x^{j_1},\ldots,x^{j_\ell},\ldots,x^{j_\mu}) & \text{if } j_\ell = (NS)^a_\ell \end{cases}
$$

$$
(25)\quad F^-_\ell(x^j) = \begin{cases} (x^{j_1},\ldots,x^{j_\ell-1},\ldots,x^{j_\mu}) & \text{if } j_\ell > 1 \\ (x^{j_1},\ldots,x^{j_\ell},\ldots,x^{j_\mu}) & \text{if } j_\ell = 1 \end{cases}
$$

$$
(26)\quad \frac{\partial w_q}{\partial t}(x^j,t^m) = \frac{1}{\Delta t^a} . \{\frac{1}{2\mu}[\sum_{\ell=1}^{\mu}(w_q(Z^1_\ell(x^j),\ t^{m+1}) + w_q(Z^2_\ell(x^j),\ t^{m+1}))] - w_q(x^j,t^m)\}
$$

$$
(27)\quad \frac{\partial w_q}{\partial x_\ell}(x^j,t^m) = [w_q(F^+_\ell(x^j),\ t^{m+1}) - w_q(F^-_\ell(x^j),\ t^{m+1}] \,/\, ||F^+_\ell(x^j) - F^-_\ell(x^j)||
$$

These descretization schemes verify the following DMP :

THEOREM 2 :

If it is satified the inequality :

(28) $(\Delta S)^a_\ell \geq \Delta t^a.\mu.\max\left(\left|P^h_{\ell,max} - A_\ell\right|, \left|A_\ell - \eta_\ell P^h_{\ell,min}\right|\right) \ \forall \ell = 1,\mu$

then, it is valid a DMP in this form :

$w_q(x^j,T) = 0 \ \forall x^j \in \Omega^a$ and

(29) $$\frac{\partial w_q}{\partial t}(x^j,t^m) + \min_{P\in P_{ad}(x^j,t^m)} \left[\sum_{\ell=1}^{\mu} \left(\frac{\partial w_q}{\partial x_\ell}(x^j,t^m)\left(-(P^h_\ell)^+ + \eta_\ell (P^h_\ell)^- + A_\ell\right)\right)\right] \geq 0$$
$$\forall x^j \in \Omega^a, \ \forall t^m < T$$

imply that

(30) $w_q(x^j,t^m) \leq 0 \ \forall x^j \in \Omega^a, \ \forall t^m < T.$

4.1.3. Formulation of the discretized problem $(\mathscr{P}_e)^a$

We introduce $(\mathscr{P}_e)^a$ as a natural discretized version of $(\mathscr{P}_e)$:

$(\mathscr{P}_e)^a$: Given

$\mathscr{W}^a_a = \{w^a_q : Q^a \to \mathbb{R} \,/\, q \in \Gamma(\nu),\ w_q$ verifies (31), (32), (33)$\}$

(31) $$\frac{\partial w^a_q}{\partial t}(x^j,t^m) + \min_{P(.)\in P_{ad}(x^j,t^m,q)} \left[\sum_{\ell=1}^{\mu}\left(c^h_\ell(x^{j\ell},t^m) - \frac{\partial w^a_q}{\partial x_\ell}(x^j,t_m)\right)\left((P^h_\ell)^+ - \eta_\ell (P^h_\ell)^- - A_\ell\right) + \sum_{r=1}^{\nu} c^\theta_r P^\theta_r + c^g P^g + c^n P^n + c^e P^e\right] \geq 0,$$
$$\forall (x^j,t^m) \in Q^a, \ \forall q \in \Gamma(\nu)$$

(32) $w^a_q(x^j,t^m) \leq \min_{p\neq q}\left(w^a_p(x^j,t^m) + k^p_q\right) \ \forall q \in \Gamma(\nu), \ \forall (x^j,t^m) \in Q^a, \ t^m < T$

(33) $w^a_q(x^j,T) = 0 \quad \forall x^j \in \Omega^a, \quad \forall q \in \Gamma(\nu),$

find $\bar{w}^a \in \mathscr{W}^a$ such that

$\bar{w}^a(x^j,t^m) \geq w^a_q(x^j,t^m) \qquad \forall w^a_q \in \mathscr{W}^a, \quad \forall (x^j,t^m) \in Q^a.$

Using definitions (22), (23) and (26) we transform inequality (31) in the explicit restrictions :

(34) $w^a_q(x^j,t^m) \leq \rho_q(x^j,t^m)$

where $\rho_q(x^j,t^m) = \frac{1}{2\mu} [\sum_{\ell=1}^{\mu} (w_q^a(Z_\ell^1(x^j), t^{m+1}) + w_q^a(Z_\ell^2(x^j), t^{m+1})] +$

$$+ \Delta t^a \min_{P(.)\in P_{ad}(x^j,t^m,q)} [\sum_{\ell=1}^{\mu} (c_\ell^h(x^{j_\ell},t^m) - \frac{\partial w_q^a}{\partial x_\ell}(x^j,t^m)) ((P_\ell^h)^+ - \eta_\ell (P_\ell^h)^- - A_\ell) + \sum_{r=1}^{\nu} c_r^\theta P_r^\theta + C^n P^n + c^g P^g + c^e P^e]$$

As (32) and (34) can be written in the unique inequality :

(35) $w_q^a(x^j,t^m) \le \min [\rho_q(x^j,t^m) , \min_{p\neq q} (w_p^a(x^j,t^m) + k_q^p)]$,

to verify (33) and (35) implies to belong to $\mathscr{W}^a$.

Concerning existence, unicity and characterization of $\bar{w}^a$ we have

THEOREM 3 :

If (28) is satisfied then there is only one solution $\bar{w}^a$ of $(\mathscr{P}_e)^a$; furthermore this solution is characterized by the condition :

(36) $(\bar{w}^a)_q (x^j,t^m) = \min [\rho_q(x^j,t^m), \min_{p\neq q} (\bar{w}^a)_p(x^j,t_m) + k_q^P)]$

Before showing how to solve numerically the nonlinear fixed point problem (36) let us give the convergence result :

THEOREM 4 :

When the mesh of the discretization (21) tends to zero the solutions $\bar{w}^a$ converges uniformly to V.

The proofs of THEOREMS 3 and 4 can be seen in [6]. They are mainly based in the DMP shown in THEOREM 2.

4.2. Numerical solution of $(\mathscr{P}_e)^a$

The computation of $\bar{w}^a$ is recursive, i.e. with the values $(\bar{w}^a)_q (x^j,t^{m+1})$ it is possible to compute $(\bar{w}^a)_q (x^j,t^m)$.

The numerical task can be divided in two principal steps :

α) Computation of ρ_q

β) Solution of the non linear fixed point problem :

$\bar{w}^a(x^j,t^m) = G(\bar{w}^a) (x^j,t^m)$

where $G : \mathbb{R}^{2^\nu} \longrightarrow \mathbb{R}^{2^\nu}$ is defined by

$$(37) \quad (G(w))_q = \min (\rho_q, \min_{p\neq q} (w_p + k_q^p)).$$

4.2.1. Computation of ρ_q

To compute ρ_q it is necessary to solve the linear programming problem (LP) that appears implicity in the definition (34) of ρ :

$$(38) \quad \min_{P\in P_{ad}(x^j,t^m,q)} [\sum_{\ell=1}^{\mu} (c_\ell^h(x^{j\ell},t^m) - \frac{\partial w_q^a}{\partial x_\ell}(x^j,t^m)) ((P_\ell^h)^+ - \eta_\ell (P_\ell^h)^- - A_\ell) + \sum_{r=1}^{\nu} c_r^\theta P_r^\theta + c^n P^n + c^g P^g + c^e P^e]$$

Because of the choice of the discretization schemes (22)-(27) this LP problem has a particular structure that avoid the use of the Simplex method. We apply a special algorithm that computes (38) in a finite numbers of steps without necessity of introducing a discretization of P(.). For the definition of this algorithm and proof of its convergence see [6].

4.2.2. Solution of the nonlinear fixed point problem (36)

To solve (36) we introduce an algorithm based in the definition of k_q^p given after (18) and some properties of the dynamic programming theory on a graph.

We have presented this algorithm in [1] where we have shown that it complexity is measured by the number of aditions and comparison employed, these numbers being $n_{ad} = n_{comp} = \nu.2^\nu$ with ν = number of thermic plants.

5 - SOME NUMERICAL RESULTS

5.1. The EDF problem

We have applied the method described in the previous paragraphs to the optimization of a model of the operation of a system during one week. The numerical data have been provided by Electricity of France (EDF).

The system comprises 3 hydraulic plants (2 of pumped type) and 7 thermic plants (1 nuclear, 2 of coal, 2 of fuel, 1 gas powered and 1 external plant). We have obtained the optimal policy of operation using the value of discretization Δt^a = 1 hour, $(NT)^a$ = 168.

The others parameters are : $(NS)_1^a = 4$, $(NS)_2^a = 5$, $(NS)_3^a = 6$.

In consecuence, Ω^a has 120 points and Q^a has 20.280 points. As we have $\nu = 4$, the total number of values of $w_q(x^j,t^m)$ computed is 324.480. The program has been processed in a Vax 11/780 and the time of computation was 18' 47".

Using the values of the function $\bar{w}^a$, it is possible to find the policies of operation as a feedback P(x,t,q). This policy is defined in the following form :

for any (x,q), we apply $\forall$ $t \in [t^m, t^{m+1})$ the control P that produces the minimum of problem (38), where the x^j that appears there, is the least distant point of x that belongs to Ω^a.

The evolution of the hydraulic stocks are obtained using the above defined control P in the integration of the differential equations (2).

In the following figures we show the data, the optimal policies and the evolution of the stock of the hydraulic plants. The optimal cost V was 1.397 MF.

5.2. Final remarks

a) We have also obtained an optimal production policy for the 21 periods in which the week was divided by E.D.F. With $(NS)_1^a = 4$, $(NS)_2^a = 5$, $(NS)_3^a = 3$ the time of computation was 1'10" with the optimal cost increased, related to the first one, in 7 %.

b) the number of start-ups n_r^θ are close related with the values of $P_{r,min}^\theta$. In [6] we discuss this problem.

6 - BIBLIOGRAPHY

[1] M.C. Báncora-Imbert, R. González, J.C. Miellou, E. Rofman : Numerical Optimization of Energy-Production Systems. Rapport de Recherche n°306, INRIA, Rocquencourt, France (1984) and International Journal of Energy Systems, Vol. 4, n°1, pp. 27-29, 1984.

[2] R. González, E. Rofman : On deterministic control problems : An approximation procedure for the optimal cost. Part I and II. Siam J. Control and Optimization, Vol. 23, n°2, 1985, pp. 242-266 and 267-285.

[3] A. Bensoussan, J.L. Lions : Contrôle impulsionnel et inéquations quasi variationnelles. Dunod, Paris, 1982.

[4] P.G. Ciarlet, P.A. Raviart : Maximum principle and uniform convergence for the finite element method. Computer Methods in Applied Mechanics and Engineering, Vol. 2, pp. 17-31, 1973.

[5] R. González : Sur la résolution de l'équation de Hamilton-Jacobi du contrôle déterministe. Thèse 3ème Cycle, Université de Paris IX, 1980 ; Cahiers de Math. de la Décision, Cérémade, 8029 and 8029 bis.

[6] R. González, E. Rofman : Remarks on the optimization of energy production systems, Rapport de Recherche INRIA, 1986, to appear.

[7] W. Fleming, R. Rishel : Optimal deterministic and stochastic control. Springer Verlag, New York, 1975.

	MONDAY			TUESDAY			WEDNESDAY			THURSDAY			FRIDAY			SATURDAY			SUNDAY		
PERIODS	1	2	3	1	2	3	1	2	3	1	2	3	1	2	3	1	2	3	1	2	3
DURATION (H)	10	10	4	8	10	6	8	10	6	8	10	6	8	10	6	4	11	9	4	10	10
DEMAND (GW)	84.1	78.5	73.0	87.6	83.0	79.5	88.8	84.3	80.9	92.1	87.6	84.3	85.3	80.9	77.9	77.2	68.4	62.8	82.7	75.4	67.8

THERMIC CHARACTERISTICS

PLANT		DAILY ADMISSIBLE MAXIMUM POWER (GW) MONDAY	TUESDAY	WEDNESDAY	THURSDAY	FRIDAY	SATURDAY	SUNDAY	COST (FFCENTS/KWH)	DURATION OF HEATING BEFORE START-UP (H)
1	NUCLEAR	64.24	64.61	66.14	67.09	64.27	65.41	66.37	5.15	-
2	COAL "A"	5.37	6.24	7.70	5.76	7.07	7.03	5.57	22.0	6
3	COAL "B"	3.62	3.92	4.48	3.88	3.87	4.80	3.55	22.2	6
4	OIL "A"	3.86	3.65	4.06	4.12	3.44	3.68	3.29	50.1	6
5	OIL "B"	1.84	1.12	1.66	1.72	1.29	0.92	0.85	56.5	6
6	GAS	2.11	2.23	2.11	2.37	2.00	1.86	2.10	111.00	-
7	ADDITIONAL SOURCE	∞	∞	∞	∞	∞	∞	∞	200.00	-

HYDRAULIC CHARACTERISTICS

PLANT	STOCK BETWEEN: (GWH)	VALUE OF KWH IN STOCK (FFCENTS/KWH) MONDAY	TUESDAY	WEDNESDAY	THURSDAY	FRIDAY	SATURDAY	SUNDAY	MAXIMUM CAPACITY (GWH)	MAX. POWER INS. TURBINE (GW)	MAX. POWER INS. PUMP (GW)	INPUTS (GW)	INITIAL LEVEL (GWH)
1	244.1 AND S_{MAX}	78	74	70	68	65	66	67	7 322	6.58	0	1.06	287.8
	0 AND 244.1	160	155	150	148	144	144	145					
2	72.2 AND S_{MAX}	74	70	65	63	69	60	61	361	2.24	1.83	0.01	88.3
	0 AND 72.2	145	140	130	120	92	100	110					
3	30 AND S_{MAX}	58 64 68	50 60 69	55 66 70	54 55 59	28 40 39	28 28 30	35 50 53	60	2.50	2.50	0.00	47.5
	0 AND 30	140 130 140	113 132 148	122 138 158	120 125 141	93 73 68	48 49 60	56 78 90					

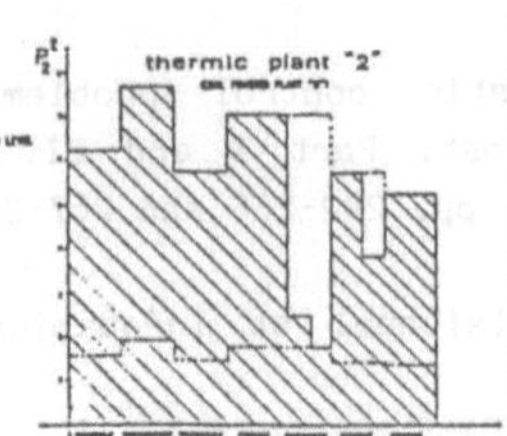

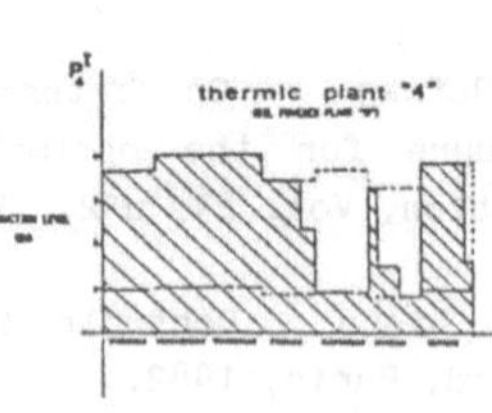

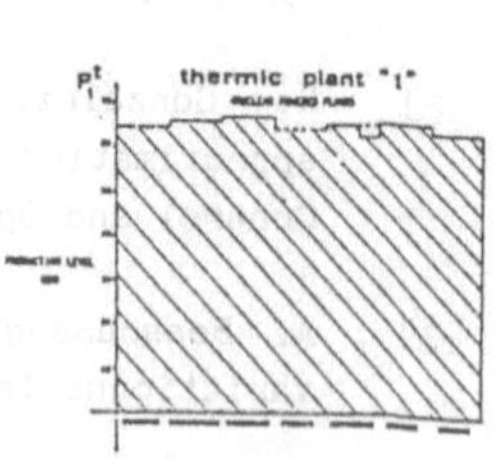

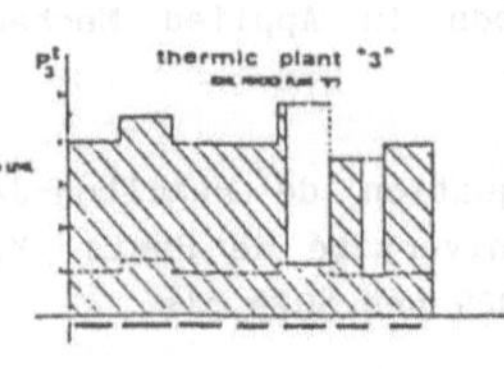

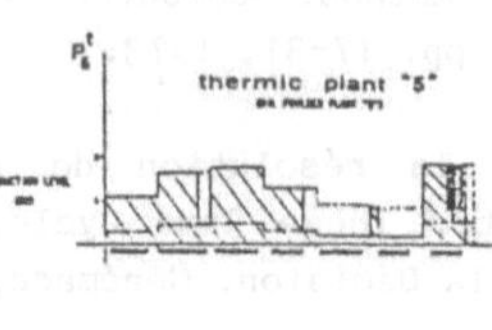

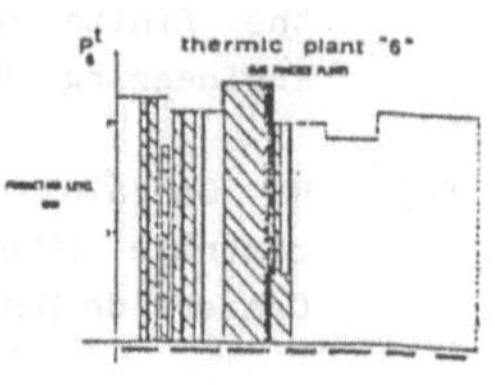

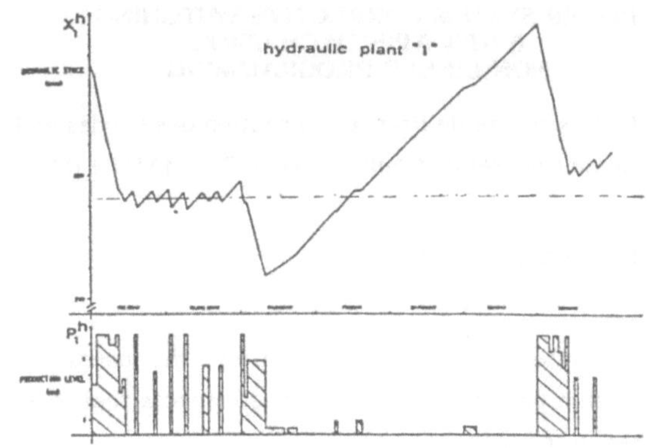
X_1^h
hydraulic plant "1"
P_1^h

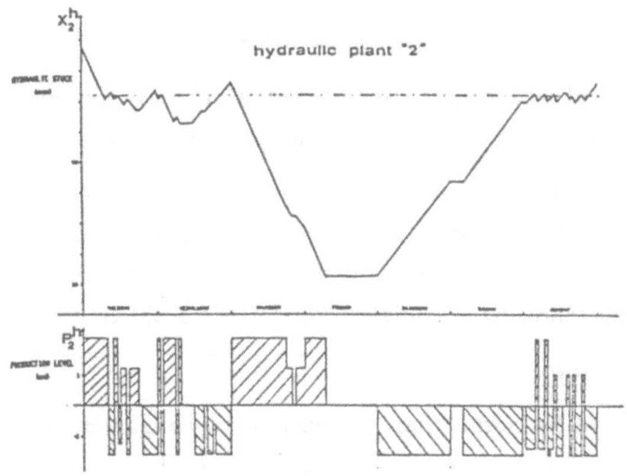
X_2^h
hydraulic plant "2"
P_2^h

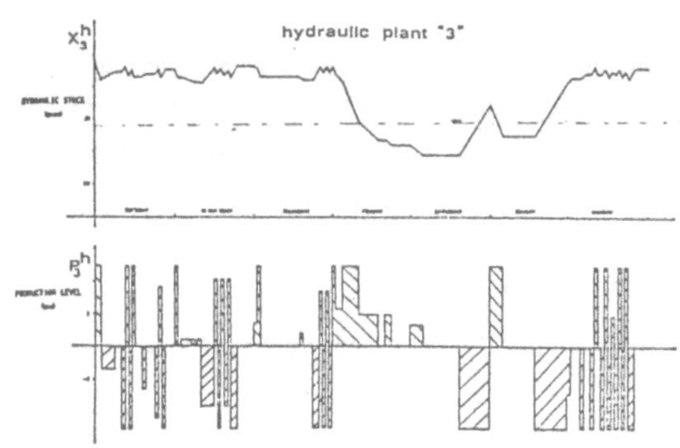
X_3^h
hydraulic plant "3"
P_3^h

POWER SYSTEM CORRECTIVE SWITCHING : A NEW APPROACH USING NON-LINEAR PROGRAMMING

J. RYCKBOSCH - Electricité de France - Direction des Etudes et Recherches
1, Avenue du Général de Gaulle - 92141 CLAMART FRANCE

1 - Corrective switching following an incident

1.1 Origin of the problem

To enable the power system to be operated safely, the operator must, at any given moment, be able to ensure that power flows on the lines respects their bounds. To achieve this, he may use two main types of actions :

. modify the generation schedule of the power stations,

. modify the grid itself, i.e. the topology of the system.

Although the first type of actions is extremely useful, it must be recognized that the time needed to implement these actions makes them impractical when a prompt reaction is required.

On the other hand, if, for example, the accidental loss of a transmission line, or a thermal unit, induces power flows on the network which can be supported for only a short moment, corrective switching can very quickly restore the situation to normal conditions : this is what is called "corrective switching following an incident " and is the problem dealt with in this paper.

1.2 Mathematical point of view

To begin with, it should be noted that appropriate modelization can always provide for corrective switching on a power system by adding or removing a certain number of lines (eventually fictitious). Thus, the first idea which comes to mind is to associate, with each of these mobile elements, a bivalent variable worth 1 or 0, depending on whether the elements are present or absent.

Unfortunately, the resulting problem includes a very large set of discrete and continuous variables such that no mathematical tools exist to solve it. Moreover, devising an heuristic method which would select a priori the most probable switching actions (and thus reduce substantially the size of the combinational set) seems particularly delicate : a power system is, in fact, an indivisible entity and a local switching action can very often have totally unforeseeable consequences on the entire system.

Such ideas, in spite of their simplicity, must thus be discarded.

2 - A new modelization

A new modelization of the problem will now be presented which makes it possible to bypass the mathematical difficulties mentioned previously.

In the direct current approximation, which consists in linearizing the equations of the system, the two Kirchhoff's laws are used.

1rst Kirchhoff's law written in each node	$\forall i \in I$	$\sum_{l \in \{\text{lines from } i\}} t_l = p_i - c_i$
2nd Kirchhoff's law written on each cycle of the network	$\forall j \in J$	$\sum_{l \in \{\text{cycle } j\}} x_l\, t_l = 0$

where I is the set of the nodes of the network

J the set of the cycles of the network

x_l, t_l reactance and power flow on line l

p_i, c_i generation, consumption at node i.

Let it be assumed that, following a switching action by the operator, line k is no longer connected with the network. Then $t_k = 0$. The first Kirchhoff's law remains exact : the contribution of line k to the energy balance in each node is zero ($t_k = 0$), which is logical, since it no longer exists.

On the other hand, the 2nd Kirchhoff's law becomes inexact : for all the cycles including line k, it is absurd to impose $\sum x_l\, t_l = 0$, since these are the cycles which no longer exist. This is why a deviation variable u_l will be introduced into the equations which will make it possible to write nevertheless :

$\sum x_l\,(t_l + u_l) = 0$, the variable u_l taking an appropriate value.

The system equations will thus be stated as follows :

$$\forall i \in I \quad \sum t_l = p_i - C_i$$

$$\forall j \in J \quad \sum_{l \notin \Omega} x_l\, t_l + \sum_{l \in \Omega} x_l\,(t_l + u_l) = 0$$

where Ω designates a set of lines whose connection and disconnection is modeled.

Thus, if t_l is not zero, u_l must be zero (the 2nd Kirchhoff's law is then seen again) ; if u_l is not zero, t_l must be zero (line l no longer exists).

By adopting a more synthetic formalization, the corrective switching can thus be presented in the form of the following system (S) :

(S)			
	Kirchhoff's laws	$A = \begin{bmatrix} t \\ u \end{bmatrix} = B$	A : matrix, B : vector — linear equations
	exclusion constraints	$\{ t_i\, u_i = 0 \quad \forall i \in \Omega$	non-linear equations
	bound contraints	$\{ \lvert t \rvert \leqslant \overline{t}$	inequations, expressing that the system is reliable

3 - Mathematical resolution

3.1 Penalty method

The constraints $t_i \cdot u_i = 0$ are difficult to handle, since they define a non-convex set ; thus, through the use of a penalty method, they will be eliminated by transforming the system (S) into a problem (P) of optimization :

$$(P) \quad \left[\begin{array}{l} \text{Min } \psi = \sum_{i \in \Omega} u_i^2 \, t_i^2 \\ \text{with : } A \begin{bmatrix} t \\ u \end{bmatrix} = B \quad \text{and} \quad \left\{ |t| \leqslant \bar{t} \right\} \end{array} \right.$$

Since the solution to S is the only solution being sought, we will, when resolving (P), stop as soon as any one of the local minima of ψ has been reached verifying $u_i t_i = 0, \ \forall i \in \Omega$; thus, the non-convexity of P is not an obstacle, since, unlike the usual procedure for optimization methods, the global minimum of the function ψ is not being sought.

However, it seems advisable not to be satisfied with any local minimum but, rather, to seek the local minimum which, for example, corresponds to a minimum number of switching actions. This is possible by choosing a start point suitable to the optimization.

3.2 Introduction of disturbances

Based on the description given up until now, the method might not supply the desired results. This is because some of the local minima which might be obtained are undesirable, as they do not verify $u_i \, t_i = 0 \ \ \forall i \in \Omega$.

This may be illustrated by a simple (and purely) mathematical example with one dimension :

$$\left\{ \begin{array}{l} \text{Min } \ u^2 \, t^2 \\ |t| \leqslant \bar{t} \\ u = u_o - at \end{array} \right.$$

Several figures are possible, depending on the value of $\bar{t}$.

• α]

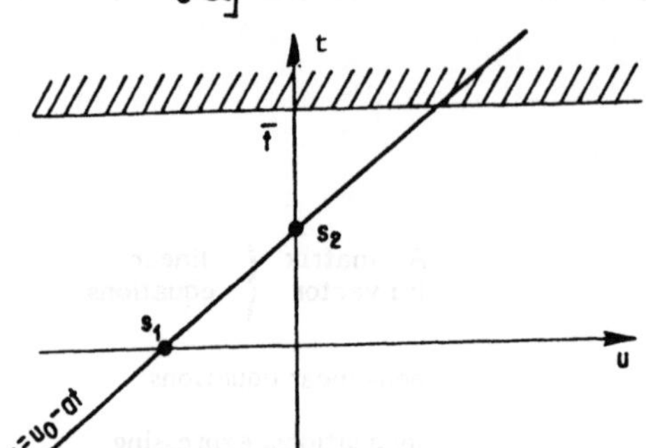

space u,t

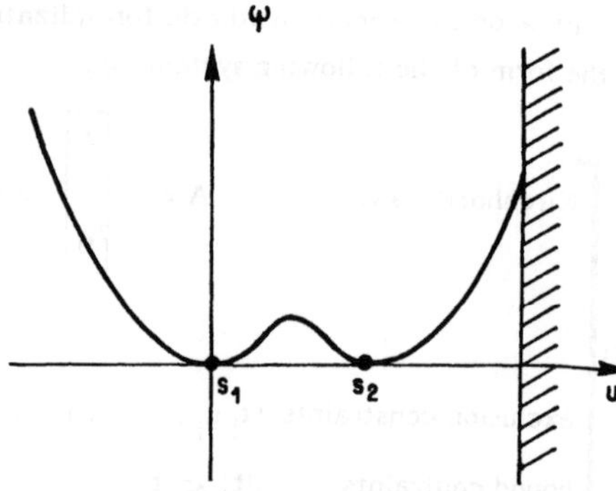

graph of the function to be minimized depending on $u = u_o - at$

In this case, the problem allows for two minima, which, indeed, verify the equation ut = 0.

• β]

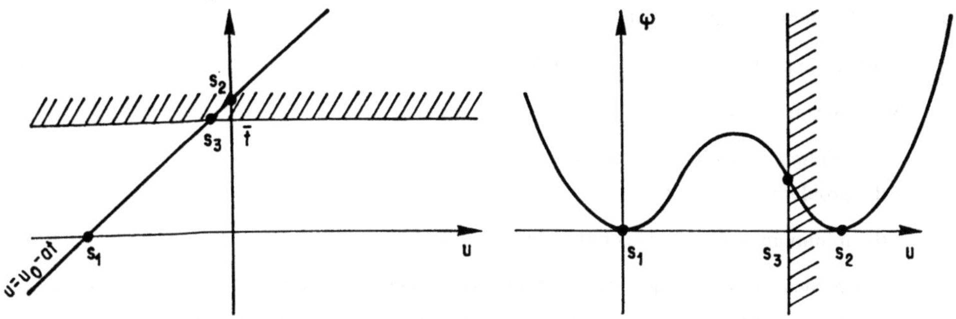

Here, the value S_3 is, indeed, a local minimum of our problem, but does not verify ut = 0.
On the other hand, by multiplying the economic function by a disturbance g, such as :

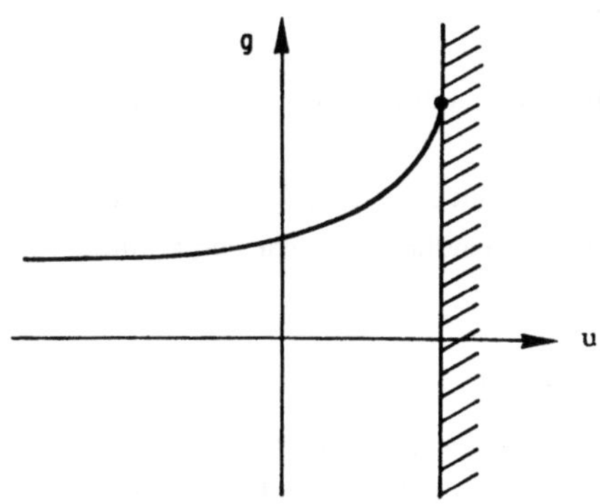

the shape of the new function to be minimized will be :

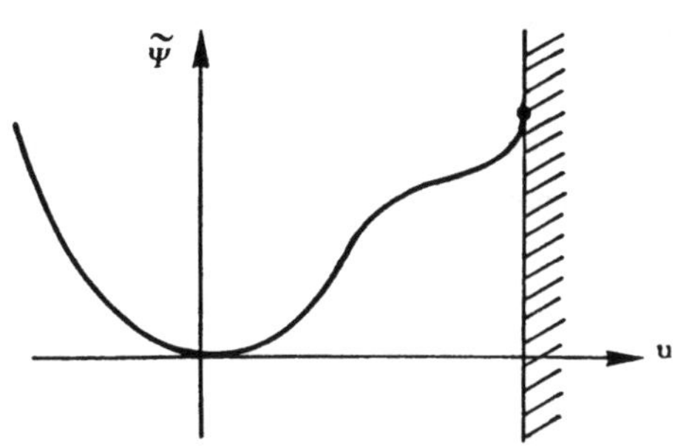

and it is seen that the undesirable minimum has disappeared.

This idea can be generalized rather easily for multiple dimensions.

The following algorithm was thus adopted, based on the observation that each undesirable minimum is, in fact, caused by a constraint of inequality which is binding.

$\underline{\alpha}$ * Seek a feasible start point

$\underline{\beta}$ * Solve the problem (P) of optimization.

Two cases are then possible :

. the minimum obtained satisfies $u_i^2 t_i^2 = 0$
the problem is solved ;

. the minimum obtained does not satisfy $u_i^2 t_i^2 = 0$;
this means that an inequality constraint has reached its bound ; the objective function is, in this case, multiplied by a very high penalty on this constraints ; this penalty decreases to 1 as the constraint is far from its bound. A new problem (P') is thus obtained, and we return to step β.

By using this procedure, the disturbances are introduced only one by one : the problem will thus remain numerically stable.

Nota Bene : a more precise algorithm is described in the appendix.

Lastly, it should be noted that successive optimization problems are solved by a reduced gradient method, well-suited to the handling of linear constraints.

4 - Performances of the model

The development of a computerized prototype enabled us to successfully test our method.

The first test consisted in processing disturbed situations on the IEEE test system which has 35 lines and 23 nodes ; the validity of the method was thus verified by comparing the results with the results of other tests performed on this subject throughout the world.

Next, in order to deal with real data, our computations were carried out for several situations which actually occurred on the French national grid (Figures 1 and 2 provide an example).

Lastly, it should be noted that the computational time is satisfactory (a few seconds on an IBM 30.81) and could be greatly improved by developing an optimization code specific to this problem.

FIGURE 1

February 10, 1984, French Southwest : initial topology

The problem is as follows :

After the tripping on the Marquis-Fléac line, Pessac-Montguyon is overload.

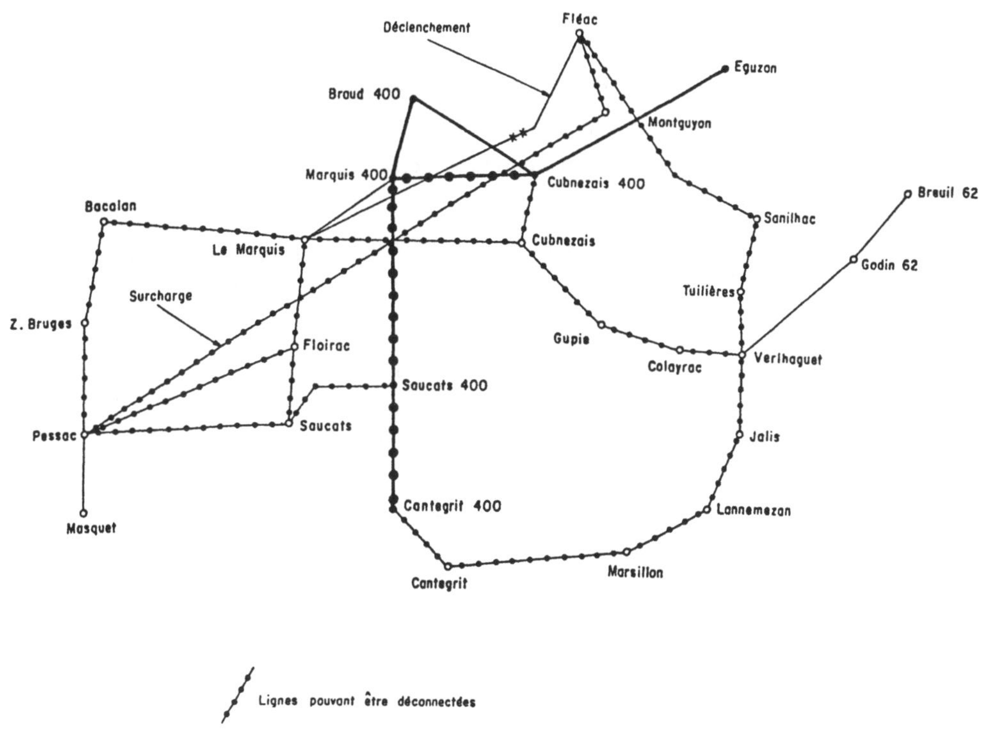

FIGURE 2

February 10, 1984, French Southwest : final topology

In order to lift the constraint, only 3 lines need to be disconnected :

- Floirac-Saucats
- Colayrac-Verlhaguet
- Jalis-Verlhaguet.

A few remarks concerning the way the model is running :

- Initial solution : constraint in Montguyon-Pessac
- First iteration : the constraint on Montguyon-Pessac remains active, a new disturbance is introduced
- Second iteration : disappearance of the Montguyon-Pessac constraint ; appearance of a constraint on Tuilières-Verlhaguet ; a new disturbance is thus introduced
- Third iteration : disappearance of all constraints and convergence of the model.

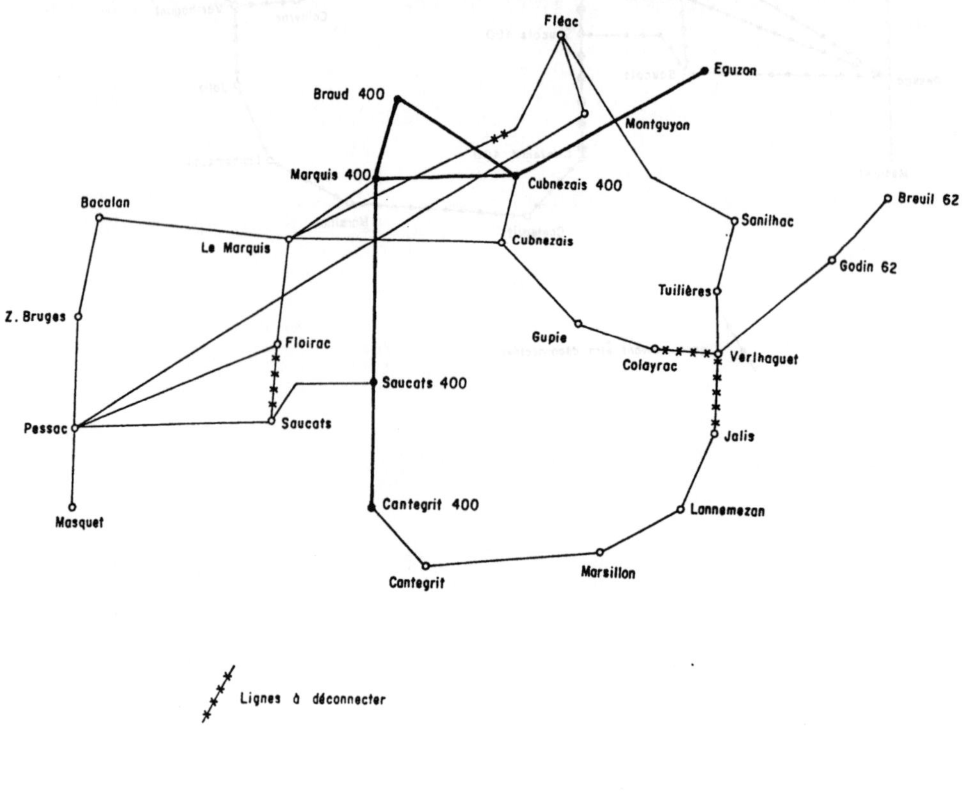

1 $L1 = \emptyset$; $L2 = \emptyset$; $\mu(i) = 0 \ \forall \ i \in L$;

L1 designates the set of the active constraints at a given moment ;

L2 is the set of the constraints which were active at least once during past or present stages ;

2 finished : = false ;

3 as long a (finished = false) perform.

<u>beginning</u>

3.1 Give the function to be minimized the value :

$$\Psi = \left(\sum_{\Omega} u_i^2 t_i^2\right) \times \prod_{L2} g(j, \mu(j), t_j)$$

$$\left\{ g(j, \mu(j), t_j) = 1 + e^{\mu(j) \left[(t_j - t_j) / t_j + t_j\right]} \right.$$

if constraints j is $t_i \leqslant t_j$

$$g(j, \mu(j), t_j) = 1 + e^{\mu(j) \left[(t_j + \bar{t}_j) / \bar{t}_j + t_j\right]}$$

if constraint j is $-\bar{t}_j \leqslant t_j$;

3.2 Solve

Min Ψ

$$A \begin{bmatrix} t \\ \\ u \end{bmatrix} = B \qquad \left\{ |t| \leqslant \bar{t} \right.$$

{ The list of active constraints at the optimum provides us L1. }

3.3 $L2 := L2 \cup L1$;

3.4 $(\forall i \in \Omega, |u_i t_i| < \epsilon)$

end ;

4 - <u>Stop</u>

OPTIMAL CONTROL OF FREE BOUNDARY PROBLEMS

C. SAGUEZ
SIMULOG - 3, avenue du Centre
78182 ST QUENTIN YVELINES CEDEX

ABSTRACT

Optimal control problems of systems governed by parabolic variational inequalities are studied. After a presentation of different kinds of functionals, two methods for proving necessary optimality conditions are presented. The paper is illustrated by the optimal control problem of the secondary cooling system of a continuous caster.

INTRODUCTION

In this paper, we present some aspects of the optimal control of free-boundary systems, when the state is solution of a parabolic or elliptic variational inequality. Such mathematical formulation appears in many physical problems, multiphase problems (solidification, freezing, ...), filtration problems (dam-problem,...), continuous mechanics problems (elasto-plasticity,...).

The optimal control of such systems is an important class of nonlinear optimal control problems. The two main difficulties, we have to solve, are the obtention of necessary optimality conditions and the implementation of efficient numerical algorithms.

In a first part, different classes of such optimal control problems of a parabolic variational inequality are presented. In a second part, different results of regularity of the state and some useful results of dependance of the solution with respect to the control are given. The third part is devoted to the obtention of necessary optimality conditions. In the last part, the results are illustrated for the industrial problem of the optimization of the secondary cooling system of a continuous casting process.

I. OPTIMAL CONTROL PROBLEMS

Let Ω be an open subset of $\mathbf{R}^n$ ($n \leqslant 3$). We consider the following parabolic variational inequality :

To find $y \in L^2(0,T;V)$; $\frac{\partial y}{\partial t} \in L^2(0,T;L^2(\Omega))$ s.t.

$$(1.1) \quad \begin{cases} (\frac{\partial y}{\partial t}, \phi - y) + a(y, \phi - y) \geqslant (f + v, \phi - y) & \forall \phi \in K \\ y(.,t) \in K \; ; \; y(0) = y_0 \end{cases}$$

with $\quad V = H_0^1(\Omega)$; $K = \{\phi \in V \mid \phi \geqslant 0 \quad \text{a.e. in } \Omega\}$

a the continuous bilinear form defined on $V \times V$ by

$$a(y,\phi) = \int_\Omega (\text{grad } y \;\; \text{grad } \phi + y\phi)d\Omega$$

f a given function in $L^2(Q)$ (or $L^p(Q)$)

v is the control variable (we suppose $v \in \mathcal{U} = L^2(Q)$ (or $L^p(Q)$)).

Remark 1.1 : In this paper, a distributed control in the second member is considered, but most of the results, which will be presented, can be extended to other types of control, in particular for boundary controls and for more general bilinear form. But important difficulties may appear, when other convex sets K are considered. □

Remark 1.2 : Optimum design problems may be also studied for such problems (I. Sokolovsky - J.P. Zolesio [9]) □

For a system governed by the variational inequality (1.1), different classes of functional may be introduced.

Case 1 : We observe the state y (or a linear transformation of the state). In this case, the cost function is of this type :

$$(1.2) \qquad J_1(v) = \|y - z_d\|^2_{L^2(Q)} + \nu\|v\|^2_{\mathcal{U}}$$

where z_d is a given function of $L^2(Q)$ and ν a real positive number.

Case 2 : Frequently for free boundary problems, the free boundary (or a domain directly associated with the free boundary) may be observed. For that, the domain $F \subset R^{n+1}$ is defined as

$$F(v) = \{(x,t) \mid y(v)(x,t) = 0\}$$

and the following criteria is introduced :

$$(1.3) \qquad J_2(v) = \|\chi_{F(v)} - \chi_d\|^2_{L^2(Q)} + \nu \|v\|^2_{\mathcal{U}}$$

where : χ_F denotes the characteristic function of the domain F

and χ_d the characteristic function of a given domain $F_d \subset Q$

Remark 1.3 : Sometimes it is interesting to introduce some other distances between F and F_d, for example the Hausdorff metric □

Case 3 :Let us consider Y_{ad} a non empty, closed, convex subset of V. We define, the set $\mathcal{U}_{ad}$ by :

$$\mathcal{U}_{ad} = \{v \mid y(.,t) \in Y_{ad} \ \forall\, t\}$$

we suppose that $\mathcal{U}_{ad}$ is empty. Then we introduce the set I(v), defined as follows :

$$I(v) = \{t \in [0,T] \mid y(v)(.,t) \notin Y_{ad}\}$$

and the cost function :

$$(1.4) \qquad J_3(v) = \mathcal{M}(I(v)) + \nu\|v\|^2$$

where $\mathcal{M}$ denotes the Lebesgue measure

Remark 1.4 :Sometimes, instead of the Lebesgue measure, we consider the diameter of the set I(v) □

Remark 1.5 : As example, for the continuous casting process, this functional corresponds to the following problem. Sometimes during transient period, there exists intervals of time for which it is impossible to verify all the quality constraints and defaults are appearing for the associated slabs. Then we want to minimize the length of steel with defaults. □

For these different functionals, we obtain the optimal control problems :

$$(1.5)\quad \begin{cases} \text{To find } u \in \mathcal{U} & \text{s.t.} \\ J_i(u) \leq J_i(v) & \forall\, v \in \mathcal{U} \end{cases}$$

II. SOME RESULTS OF EXISTENCE

Firstly, for the variational inequality (1.1), we have the following result of regularity :

Proposition 2.1.

If $f \in L^p(Q)$, $v \in L^p(Q)$ and if the domain Ω is smooth enough, then the solution y of (1.1) belongs to $W_p^{2,1}(Q)$.

This result is obtained by proving the proposition for an associated penalty problem and by passing to the limit □

Now different results of continuous dependance with respect to the control variable are given.

Proposition 2.2 :

If $f \in L^p(Q)$ and if $v_n \to v$ in $L^p(Q)$ weakly, then $y_n = y(v_n) \to y = y(v)$ in $W_p^{2,1}(Q)$ weakly □

Proposition 2.3

If $f \in L^p(Q)$ and if $v_n \to v$ in $L^p(Q)$ strongly with v such that $\mathcal{M}(\{(x,t) \mid f(x,t) + v(x,t) = 0\}) = 0$, then $\chi_{F_n = F(v_n)} \to \chi_{F=F(v)}$ in $L^2(Q)$ strongly.

The demonstration is based on the fact that, with the assumption on v and the regularity of the solution y, the characteristic function χ_F is completely defined by the equation :

$$(2.1)\qquad (f+v)\chi_F = f + v - \frac{\partial y}{\partial t} + Ay$$

Then if we write (2.1) for v_n and if we pass to the limit, we obtain immediatly the proposition 2.1 (C. SAGUEZ [8]) □

Proposition 2.3 :

If $f \in L^p(Q)$ and if $v_n \to v$ in $L^p(Q)$ weakly, then $\lim \mathcal{M}_0(I(v_n)) \geq \mathcal{M}_0(I(v))$

For the proof we refer to F. BONNANS - C. MORENO - C. SAGUEZ [3]. □

With these three last propositions, it is easy to prove the existence of a solution for the optimal control problems (1.5).

III. NECESSARY OPTIMALITY CONDITIONS

To obtain necessary optimality conditions, the difficulty arises from the non differentiability of the state y with respect to the control variables. In this chapter, we consider only the functional J_1. But the difficulties are similar for the others functionals. Two methods are presented. The first one is based on penalty method and has been extensively studied by V. BARBU [1]. The second one is based on a transformation of the initial problem into a linear optimal control problem with state constraints. It has been used in particular by F. MIGNOT - J.P. PUEL [7].

III.1. Penalty method

The main steps of the method are the following :

+ introduction of a regular penalty problem
+ obtention of necessary optimality conditions for the penalty problem
+ obtention of the conditions for the initial problem by passing to the limit.

i) Penalty problem

We introduce the following penalty problem :

$$(2.2)\quad \begin{cases} \dfrac{\partial y_\varepsilon}{\partial t} + Ay_\varepsilon + \dfrac{1}{\varepsilon}\,\phi_\varepsilon(y_\varepsilon) = f + v \\ \\ y_\varepsilon(x,0) = y_0 \end{cases}$$

$$(2.3)\quad J_\varepsilon(v) = \|y_\varepsilon - z_d\|^2_{L^2(Q)} + \nu\,\|v\|^2_{\mathcal{U}}$$

where $\phi_\varepsilon(x)$ is a C^2 regularization of the function $-x^-$

ii) Necessary optimality conditions for the penalty problems

For the penalty problem , the differentiability of the state y_ε with respect to v can be proved. Classically we introduce the following adjoint state equation :

$$(2.4)\quad \begin{cases} -\dfrac{\partial p_\varepsilon}{\partial t} + Ap_\varepsilon + \dfrac{1}{\varepsilon}\,\phi'_\varepsilon(y_\varepsilon)\,p_\varepsilon = y_\varepsilon - z_d \\ p_\varepsilon(x,T) = 0 \end{cases}$$

Then we obtained the necessary optimality conditions :

$$(2.5)\quad p_\varepsilon + \nu\, u_\varepsilon = 0$$

iii) Necessary optimality conditions for the variational inequality

The difficulty is to pass to the limit on the adjoint state equation. For that, we do the following assumptions :

$$(2.6)\quad p > \mathrm{Sup}(n+2/2,\ 2)\ ;\ p \neq n+2$$

$$(2.7)\quad \mathcal{M}(\{(x,t)\mid f(x,t) + u(x,t) = 0\}) = 0$$

$$(2.8)\quad \forall\, \phi\ \ H_0^{1,0}(Q),\ \phi = 0 \ \text{ a.e in } F(u) = \{(x,t)\mid y(u)(x,t) = 0\}$$

$$\Rightarrow \phi \in H_0^{1,0}(\mathcal{O} = Q - F(u))$$

Remark 2.1: The assumptions (2.6) - (2.7) permit to prove the continuous dependance of F(v) with respect to v in the two following senses : L^2 norm for the characteristic function and Hausdorff metric □

With these three assumptions, we deduce the existence of p such that :

$$(2.9)\quad \begin{cases} \int_0^T \{(p, \frac{\partial\phi}{\partial t}) + a(p,\phi)\}\,dt = \int_0^T (y-z_d)\phi\ \ dt \qquad \forall\, \phi \in \mathcal{B}_0(\mathcal{O}) \\ p \quad H_0^{1,0}(\mathcal{O}) \\ p = 0 \quad \text{a.e} \ \text{ in } F(u) \end{cases}$$

with $\mathcal{B}_0(\mathcal{O}) = \{\phi \mid \phi \in \mathcal{F}(\mathcal{O}),\ \frac{\partial\phi}{\partial t} \in \mathcal{F}'(\mathcal{O}),\ \phi(x,0) = 0\}$

$$\mathcal{F}(\mathcal{O}) = \overline{\mathcal{D}(\mathcal{O})}^{H^{1,0}(\mathcal{O})}$$

Then we obtain the necessary optimality conditions :

(2.10) $p + \nu u = 0$

Remark 2.2 : This method is specific to the considered variational inequality and its extension to other problems is extremely difficult. □

Remark 2.3 : The penalty problem may be used to solve numerically the problem □

III.2/ Direct method

The method is based on the transformation of the initial optimal control problem into an equivalent one, with a linear state equation. We suppose that $f \in L^2(Q)$. The principle is the following.

i) Equivalent optimal control problem

If y is solution of the variational inequality (1.1), then there exists $\eta \in L^2(Q)$ such that (y,η) verifies the conditions :

$$(2.11)\qquad \begin{cases} \frac{\partial y}{\partial t} + Ay + \eta = f + v \\ y(0) = y_0 \end{cases}$$

(2.12) $\eta \leq 0 \; ; \; y \geq 0 \; ; \; \langle y,\eta\rangle = 0$

The idea is to consider η as another control variable and (2.12) as new constraints. Then we obtain the equivalent optimal control problem :

+ The state y is solution of (2.11)

+ the functional is given by :

$$(2.13)\qquad G(v,\eta) = \|y(v,\eta) - z_d\|^2_{L^2(Q)} + \nu \|v\|^2_{\mathcal{U}}$$

+ the set of admissible controls is defined by :

$$\{(v,\eta) \in L^2(Q) \times L^2(Q) \mid \eta \leq 0 \; ; \; y \geq 0 \; ; \; \langle y,\eta\rangle = 0\}$$

ii) Necessary optimality conditions

To take into account the constraints, we introduce the functional ψ defined by :

$$\psi(z,\eta) = I_C(z) + I_{C^o}(\eta) + I_{\{o\}}(<z,\eta>)$$

where I_E denotes the indicator function of the set E

$C = \{\phi \in L^2(0,T;V) \mid \phi \geqslant 0 \quad \text{a.e in } Q\}$

C^o is the dual cone of C

Then the optimal control problem is equivalent to the following one :

$$(2.14) \quad \begin{cases} \text{Min } G(v,\eta) + \psi(y(v,\eta),\eta) = F(v,\eta) \\ (v,\eta) \in [L^2(Q)]^2 \end{cases}$$

We verify that the functional $F(v,\eta)$ is convex with respect to v and that $F(u + w,\xi + w)$ is convex with respect to w. Then by writing that $0 \in \partial_v F(u,\xi)$ and $0 \in \partial_w F(u + w,\xi + w)_{|w = o}$, we obtain the necessary optimality conditions :

$$(2.15) \quad \begin{cases} \frac{\partial y}{\partial t} + Ay + \xi = f + u \\ y(0) = y_0 \end{cases}$$

$$(2.16) \quad y \geqslant 0 \ ; \ \xi \leqslant 0 \ ; \ <y,\xi> = 0$$

$$(2.17) \quad \begin{cases} -\frac{\partial p}{\partial t} + Ap = (y - z_d) + \theta \\ p(T) = 0 \end{cases}$$

$$(2.18) \quad <p,\eta> = 0 \quad \forall \eta \in \tilde{\mathscr{C}}_\xi \quad \text{with } <\eta,y> = 0$$

$$(2.19) \quad <\theta,\phi> = 0 \quad \forall \phi \in \mathscr{C}_y \quad \text{with } <\phi,\xi> = 0$$

where $\mathscr{C}_y = \{ \phi \mid \phi \in L^2(0,T \ ; \ V) \ ; \ \exists t, \ y + t\phi \geqslant 0\}$

$\tilde{\mathscr{C}}_y = \{ \eta \mid \eta \in L^2(0,T \ ; \ H^{-1}(\Omega))\} \ ; \ \exists t, \ \xi + t\eta \leqslant 0\}$

(2.20) $p + \nu u = 0$

Remark 2.4 : These conditions can be interpretated as follows. We introduce the three subsets of Q :

$$Q_1 = \{(x,t) \mid y > 0\}$$

$$Q_2 = \{(x,t) \mid y = \xi = 0\}$$

$$Q_3 = \{(x,t) \mid \xi < 0\}$$

then p verifies $p \equiv 0$ in Q_3 and $- \frac{\partial p}{\partial t} + Ap = y - z_d$ in Q_1 □

Remark 2.5 : With this method, we obtain the same conditions than these obtained by F. MIGNOT - J.P. PUEL [6] with the conical derivative. □

Remark 2.6 : This method is relatively general and permits to obtain necessary optimality conditions for a large number of variational inequalities (A. BERMUDEZ - C. SAGUEZ [2]) □

III.3/ Extension to the functional J_2 and J_3

i) When we consider the case of the characteristic function of F(v), we introduce an approximation of the characteristic function, for example $\frac{\varepsilon}{\varepsilon + y}$ with $\varepsilon > 0$. Then we have an optimal control problem of the same type than the previous one.

ii) If we consider the functional $\mathcal{M}_0(I(v))$, firstly we verify that :

$$\mathcal{M}_0(I(v)) = \int_0^T g(\| y - P_{Y_{ad}} y \|^2) dt$$

where g is the characteristic function of strictly positive real numbers and $P_{Y_{ad}}$ the projection on Y_{ad}.
Secondly we use a regularization of g, for example :

$$g_\varepsilon(x) = \frac{x^2}{\varepsilon + x^2} \quad \text{if} \quad x \geqslant 0 \quad \text{and } g_\varepsilon(x) = 0 \text{ if } x < 0.$$

Remark 2.7 : To solve numerically these optimal control problem, we have two main approaches. The first one consists in solving the associated penalty problem by using a classical descent method, for example, a gradient (or a conjugate gradient method). The second one uses the formal interpretation by a Lagrangian of the optimality conditions obtained by the direct method and a duality algorithm (for example an Uzawa algorithm) □

IV/ THE CONTINUOUS CASTING PROCESS

We present an application to the optimization of the secondary cooling system of a continuous casting process. This application has been done in collaboration with IRSID (J. HENRY - M. LARRECQ - J. PETEGNIEFF - C. SAGUEZ [4]). The principle of the continuous casting process is described figure 1. We cast continuously the liquid steel in a mold, the bottom of which is constitued by the solid steel. The system is constitued by two parts :

- the mold in which a thin shell of steel is solidified

- the secondary cooling system in which the steel is cooled by several water-spray systems.

The physical problem is to determine the best water-spray regulation system in order to maximize the productivity taking into account technical and metallurgical constraints. From a mathematical point of view we have to compute the optimal exchange coefficients between steel and water, at each time, for each zones of the water-spray system.

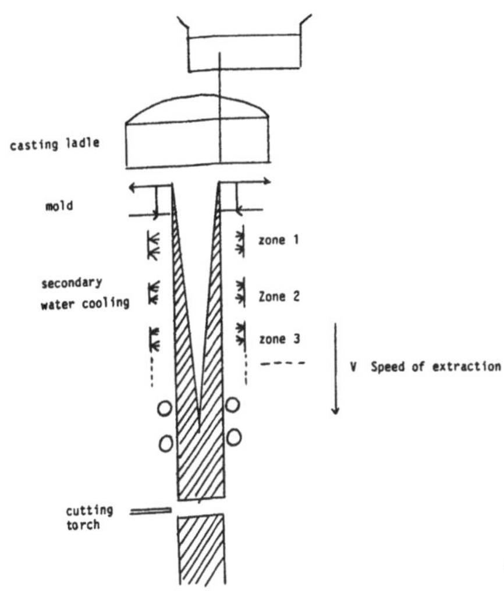

Figure 1 : Continuous casting process

IV.1/ Open-loop control problem

We consider the following mathematical formulation. The state of the system (the temperature of the steel) is solution of a Stefan problem and the control variables are the exhange coefficients for each secondary cooling zone. The purpose is to find the best exhange coefficients to improve the quality of the steel. For that two types of constraints are taken into account :

- the metallurgical constraints, which correspond to constraints on the temperature and on the variation of the temperature
- the technical constraints on the control variables.

The functional, to be minimized, corresponds to the sum of penalty functions associated with the metallurgical constraints. The optimal control is computed by a gradient algorithm with projection at each iteration on the set of the technical constraints. The state system is approximated by a finite element method and solved by a Newton algorithm.

We present figure 2, the results obtained for the USINOR continuous caster at Dunkerque (J. HENRY - M. LARRECQ - J. PETEGNIEFF - C. SAGUEZ [4])

FIGURE 2 - Optimal Solution for Usinor Caster at Dunkerque

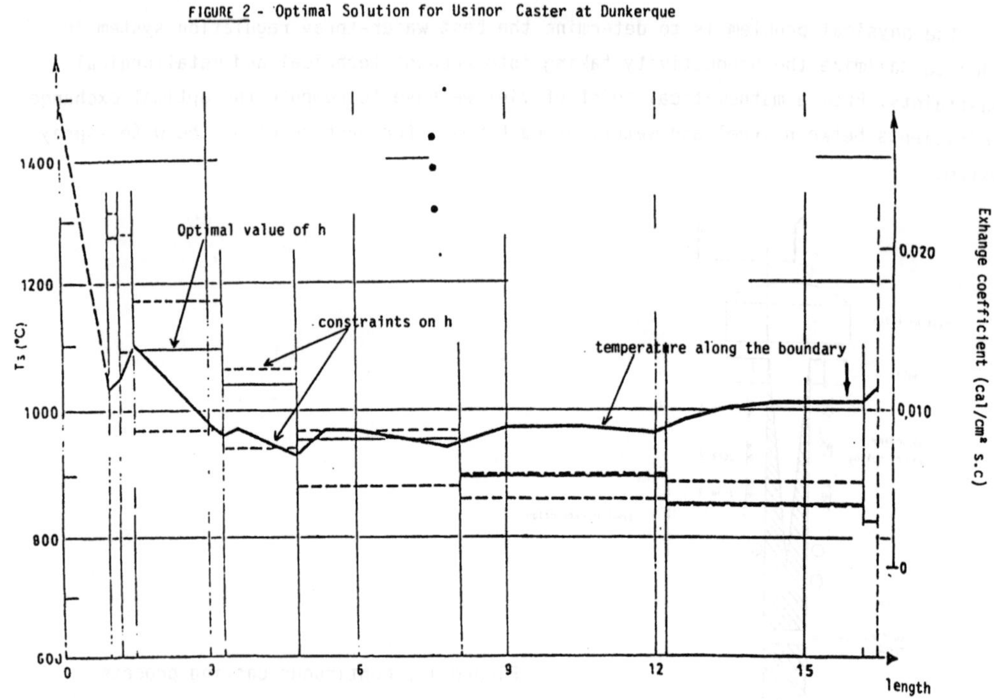

This first experiment proved the efficiency of the method and that significant improvement can be obtained for the control of the continuous casting process.

IV.2/ New model of automatization

Using the results of the open-loop optimal control, IRSID and FCB (M. JOLIVET - C. SAGUEZ - B. LACOSTE - P. BARDET - D. SENANEUCH - A. RUELLE [5]) have developped a new model of automatization. The model is based on the concept of a residence time of the slab in the caster and used the called "spraying curves". The residence time takes into account the past, the present and the future history of slab slives and for each residence time, the spraying curves are, for each zone of the caster, the results of the open-loop optimal control problem, the results of which are calculated off line and stored in a data-base.

This model has been installed on several french slab casters. The experiments prove that, with this new model, the automatization effects have been improved significantly.

REFERENCES

[1] V. BARBU : Optimal Control of Variational inequalities (Research notes in mathematics, Pitman 1984)

[2] A. BERMUDEZ - C. SAGUEZ : Optimal control of a Signorini Problem (to appear)

[3] F. BONNANS - C. MORENO - C. SAGUEZ : Contrôle de domaines temporels (Rapport INRIA N° 308, 1984)

[4] J. HENRY - M. LARRECQ - J. PETEGNIEFF - C. SAGUEZ : Optimisation du refroidissement secondaire en coulée continue de brames (Séminaire Métallurgie, INRIA, 1980)

[5] M. JOLIVET - C. SAGUEZ - B. LACOSTE - P. BARDET - D. SENANEUCH - A. RUELLE : IRSID-FCB Model for Automatization of secondary cooling of slabs, Industrial operation results (Metals Society, London, 1985)

[6] F. MIGNOT - J.P. PUEL : Optimal Control in some variational inequalities (SIAM J on contr.and Optim. 1974)

[7] F. MIGNOT - J.P. PUEL : Contrôle optimal d'un système gouverné par une inéquation variationnelle parabolique (CRAS, t 298, N° 12, 1984)

[8] C. SAGUEZ : Contrôle Optimal de problème à frontière libre (Thèse d'Etat U.T.C., 1981)

[9] J. SOKOLOVSKY - J.P. ZOLESIO : Dérivée par rapport au domaine de la solution d'un problème unilatéral (CRAS t.301 Série 1, 1985)

OPTIMAL CONTROL METHODS
FOR LARGE POWER SYSTEMS
PLANNING AND OPERATION

P. SANDRIN

Electricité de France - Direction des Etudes et Recherches
1, Avenue du Général de Gaulle 92140 CLAMART - FRANCE

SLIDE 1

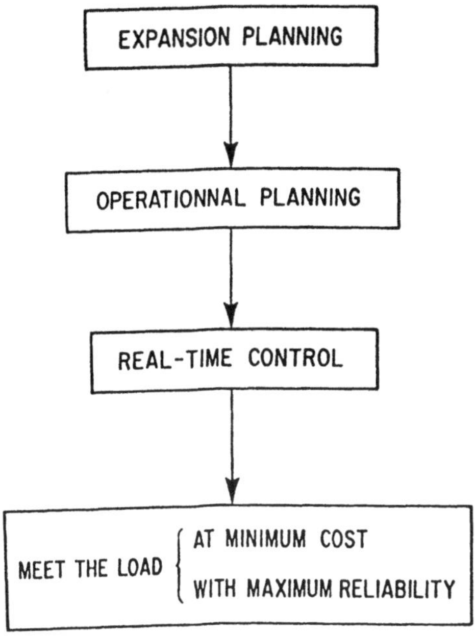

Electricity has become quite essential to the very existence of our developed countries, as much for public security, économic activity and domestic comfort.

The object of power systems is to meet the demand for electricity at lowest cost and highest reliability. To achieve this, a great number of actions are necessary, ranging from expansion planning to operational planning and real-time control.

Expansion planning consists in designing the future system, 5 to 10 or 20 years in advance, deciding which generation, transmission or distribution facilities should be added, where and when.

Operational planning consists in scheduling operation of those existing facilities, one day to one year in advance, optimizing maintenance and management of reserves such as hydro, and preparing real-time control.

Real-time control consists in continuously balancing generation and load, while keeping closed to optimized scheduled operation.

In order to help the planners and the operators to make optimal decisions for each of these tasks, an effort is made to give to these problems a mathematical formulation and to solve them using optimization methods.

There are several difficulties in these optimization problems :

First of all, problems generally have a large size, in time and space. In time, since time step must be small enough to correctly model the system, and optimization period must be long enough to encompass system time constants, (one week, one year), or lifespan of facilities (several decades). Large size in space, due to the number of facilities in the system : several hundred generation units, and transmission lines...
Another difficulty lies in the randomness of the future. Availability of generation units and transmission lines, hydro inflows, and load cannot be exactly predicted, and should be regarded as random variables.
Again, an other difficulty is the discontinuous nature of some decision variables : a lot of them are integer variables : to build a new facility or not, to put it into service or not... and so on.

SLIDE 3

OPTIMAL CONTROL PROBLEM

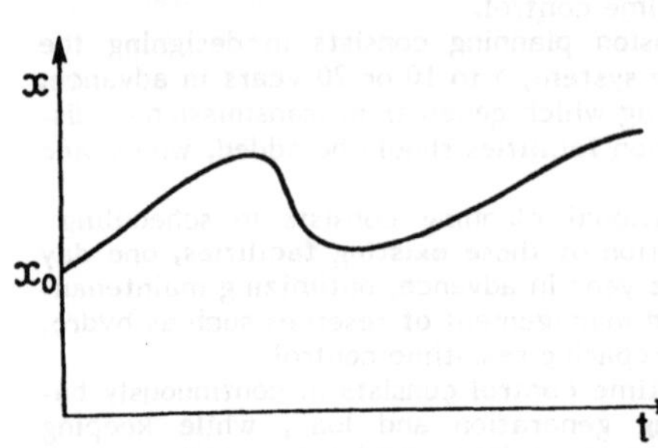

CONTINUOUS CASE		DISCRETE CASE
$\min_{u(t)} \int L(x,u,t)\,dt$	OBJECTIVE	$\min_{u_t} \sum_t L(x_t, u_t, t)$
$\frac{dx}{dt} = f(x,u,t)$	DYNAMICS	$x_{t+1} = f_t(x_t, u_t)$

SLIDE 2

DIFFICULTIES

OF POWER SYSTEM OPTIMIZATION

- LARGE SIZE
- INCERTAINTIES AND RANDOMNESS OF THE FUTURE
- DISCONTINUOUS DECISION VARIABLES

Optimal control methods deal with optimization of time trajectories of dynamic systems. Any optimal control problem can thus be formulated in terms of state variables, control variables, and dynamics.

State variables, x, which summarize past evolution of the system,
- Control variables, u, representing control actions on the system, which have to be optimized,
- Dynamics, or transition equation, modelling system evolution following control actions.
- Dynamics can be deterministic, or stochastic if the system is subject to random factors. In the stochastic case, the control variables should be optimized as functions, at all time of the information known at that time, which is summarized in the state variables of the system : this is what is called a closed loop control, or strategy control, or feedback control.

Time can be modelled as a continuous variable, or as a discrete variable, which is more widely used for power system problems.

Among optimal control methods, we mainly use, for our applications, dynamic programming and maximum's principle.
Two preferential applications of dynamic programming are non-convex problems, for which it can find global optimum solution, and stochastic problems, for which it can find closed-loop optimum solution (or feedback).
But these accomplishements are computationnally expensive, both in time and memory. And the computionnal burden increases exponentially with the number of state variables of the problem. Next slide will show how dynamic programming can be applied to our large size problems when they have some separability properties.
Maximum's principle is not, by itself, an optimization method, but it gives a local optimality condition, from which a gradient can be computed on a given trajectory. This allows for an iterative determination of optimal trajectory. The main advantage of this method is its little computing requirements. In addition, it is still applicable to large size problems, since computing time increases only linearly with system size.
But applications of this method are limited for several reasons :
- First of all, since the optimality condition is only local, this method can give global optimum solution only for convex problems. Otherwise it gives a local optimum solution.
- Another limitation concerns constraints involving state variables, which can not be taken in account, but through penalty techniques, with resulting numerical difficulties.
For stochastic problems, this method cannot give closed-loop solutions, except in the particular case where dynamics is linear, and objective function quadratic with respect to state and control variables. In this latter case, sometimes called tracking problem, or state-regulator problem, this method leads to Riccati's equation, from which closed loop solutions can be derived, even for large size systems.

SLIDE 4

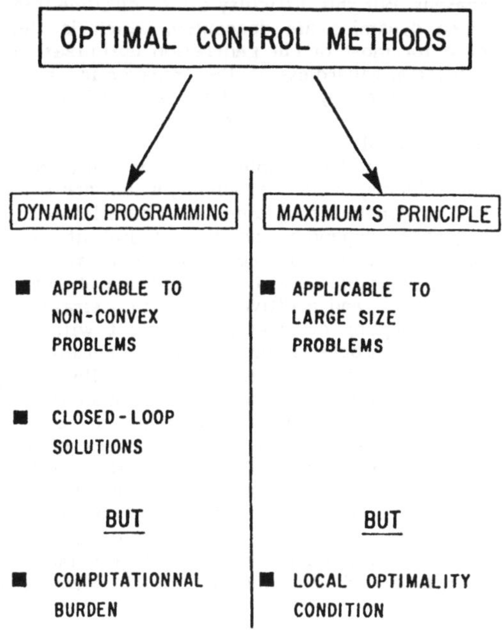

For large size problems, dynamic programming is, in general, inapplicable, for computationnal reasons. But this difficulty can be circumented when dynamics are decoupled, which is the case for most of our generation optimization problems, regardless of the transmission network. In fact, dynamics of one generation unit depends only on local control, and not on other units controls.

Then the global problem can be split into subproblems, or local problems, which are solved by dynamic programming in parallel or in sequence, depending on the separability of objective function and coupling constraints.

In the separable case, that is additive objective function and additive coupling constraints, these coupling constraints are dealt with in the local problems, by means of Lagrangian multipliers. This parallel method is called decomposition coordination, or lagrangian relaxation. In the convex case, it yields the optimum solution, while in the non-convex case, the solution is generally non feasible with regards to coupling constraints due to the duality gap. Therefore, the method has to be complemented by an heuristic procedure to derive a feasible solution. As the lagrangian is a lower bound of unknown optimal objective function, the quality of the feasible solution can be evaluated by comparison with the lagrangian. It is prooved that, the larger problem size, the better solution by this method.

In the non-separable case, subproblems must be solved in sequence, each subproblem requiring the knowledge of other subproblems solutions. This method, using unidirectionnal optimization, is called decomposition-relaxation.

It has a wider field of application than decomposition-coordination, but only yields Nash optimum solutions.

SLIDE 5

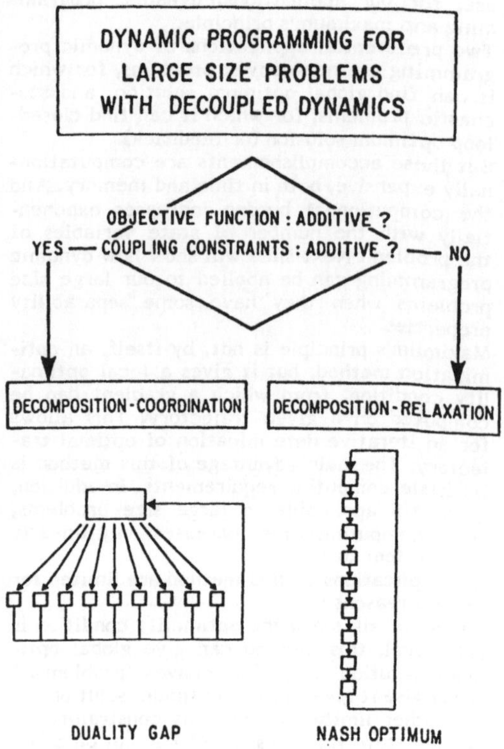

I have presented neither a comprehensive survey of power system optimization problems, nor a comprehensive survey of optimal control methods, since either of them would have taken much more time than was available.

I shall merely present here five applications of optimal control methods to power system optimization problems, both deterministic and stochastic, which can give you an idea of this field of work.

- Unit commitment by decomposition-coordination.
- Scheduling of a tidal power station by the maximum's principle.
- Scheduling strategy of an hydro reservoir by stochastic dynamic programming.
- Scheduling strategy of nuclear reactors by decomposition-relaxation.
- Load-frequency control by Ricatti's equation.

SLIDE 6

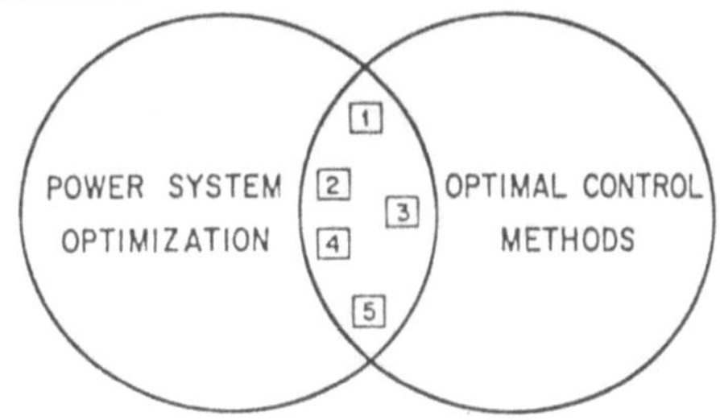

UNIT COMMITMENT	1	DECOMPOSITION COORDINATION
SCHEDULING OF A TIDAL POWER STATION	2	MAXIMUM'S PRINCIPLE
SCHEDULING STRATEGY OF HYDRO RESERVOIR	3	STOCHASTIC DYNAMIC PROGRAMMING
SCHEDULING STRATEGY OF NUCLEAR REACTORS	4	STOCHASTIC DECOMPOSITION RELAXATION
LOAD FREQUENCY CONTROL	5	RICATTI'S EQUATION

SLIDE 7

1 DAILY UNIT COMMITMENT

FIND HOURLY UNIT CONTROL

- TO MEET SYSTEM CONSTRAINTS
  - DEMAND
  - SPINNING RESERVE
- TO MEET UNIT CONSTRAINTS
- TO MINIMIZE OPERATION COSTS

FOR EACH UNIT

- STATE VARIABLE: STATUS OF UNIT (UP, DOWN, ...)
- CONTROL VARIABLE: START-UP/SHUT DOWN GENERATION

First daily unit commitment by decomposition-ccordination. Within one day, the demand for electricity undergoes large variations : low demand during the night, high demand in mid-morning, etc. Therefore, on account of a load forecast, the operator must plan the generation schedule of each unit for the next day, with start-up and shut-down times. In addition, there should be enough ready power reserve, called spinning reserve, in order to balance a possible sudden loss of generation. Each generation unit is subject to local constraints, such as maximum power, or minimum running time.

Of course the objective is to minimize costs, including generation costs, which is taken as a linear function of generation, and start-up costs, an exponential function of down time.

State variables depend, for each unit, on the local constraints to be taken in account : for instance, run-time in the case of a min-run time constraint.

Control variable is, for each unit, status of unit (up or down), and generation schedule if the unit is up.

Unit commitment is a large-scale, deterministic, non-convex problem, with decoupled dynamics.

In addition, objective function and coupling constraints are separable, since total cost is the sum of local costs, and since coupling constraints are additive.

Therefore, a decomposition-coordination method can be applied. Coupling constraints are processed by means of lagrangian multipliers.

The dual problem, which is finding a saddle point of the lagrangian, can be split into n subproblems, or local problems, where n is the number of generation units.

SLIDE 8

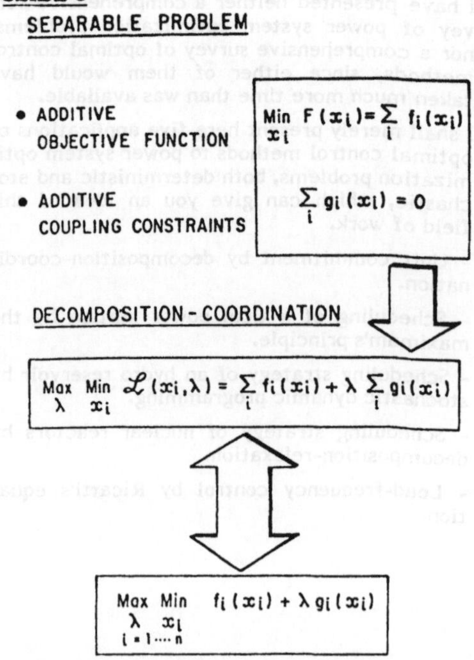

SLIDE 9

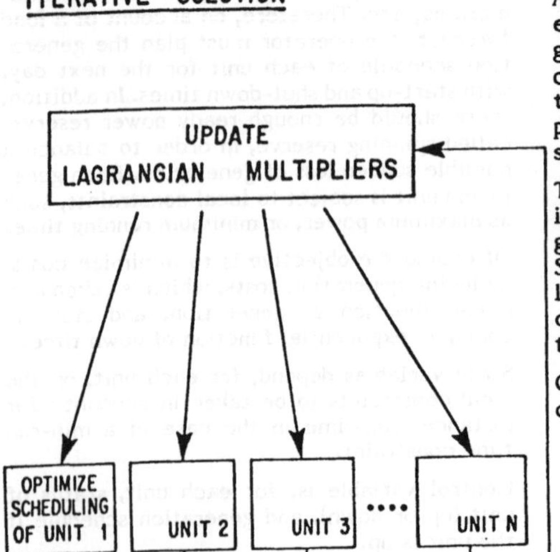

This problem is solved iteratively, by fixing alternatively lagrangian multipliers and unit control variables.

At the lower level, or decomposition level, each local problem consists in optimizing the generation schedule of one unit, on the basis of its costs and of fixed lagrangian multipliers, taking in account local constraints. Each subproblem, requiring mixed-integer solutions, is solved through dynamic programming.

The upper level, or coordination level, consists in improving the lagrangian, by updating lagrangian multipliers, with fixed unit schedule. Since the lagrangian is a convex, piecewise linear function of lagrangian multipliers, it can be improved by a very simple gradient technique.

Convergence is generally obtained after about one hundred iterations.

Since this unit commitment problem is not convex, the solution of the dual problem is generally not feasible with respect to coupling constraints.

An heuristic procedure is then used to derive a feasible solution from the solution of the dual problem.

Since the maximum of the lagrangian is a lower bound of objective function, the gap between it and the value of objective function at the feasible solution is an upper bound of the optimization error. For our problem, with more than one hundred units, this gap is well under. 1 per cent.

SLIDE 10

NON - CONVEX PROBLEM

$\hat{x}_i , \hat{\lambda}$, SOLUTION OF $\underset{\lambda}{\text{Max}} \ \underset{x_i}{\text{Min}} \ \mathcal{L}(x_i, \lambda)$

IS NOT FEASIBLE

HEURISTICS

$\tilde{x}_i$ FEASIBLE

DUALITY GAP = $F(\tilde{x}_i) - \mathcal{L}(\hat{x}_i, \hat{\lambda})$

SLIDE 11

2 SCHEDULING OF A TIDAL POWER STATION

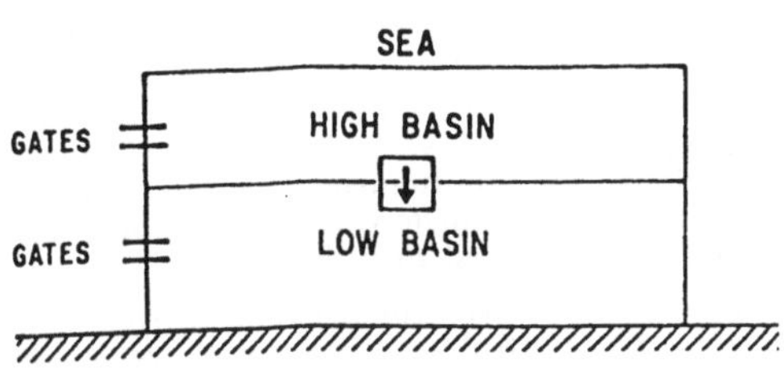

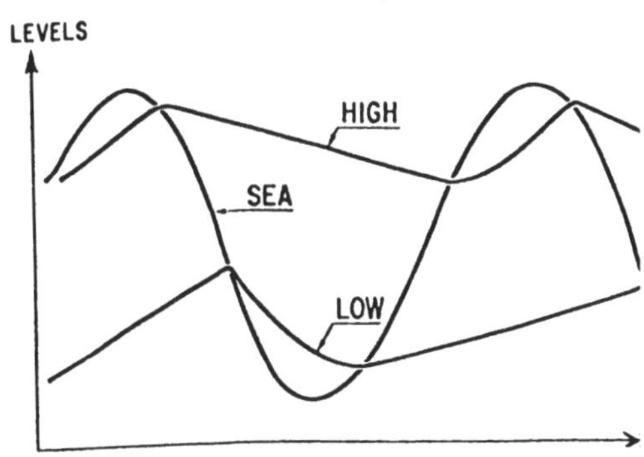

Our second example deals with tidal power optimization.

In some parts of the world, tides ranges can be up to 10, or even 15 meters. The object of a tidal power station is to change this potential energy into electricity.

In former tidal schemes, with only one basin, like la Rance scheme in France, energy could be generated only at some moments, depending on the sea level. Now this scheme with two basins, one high, one low, allows for continuous generation. The bottom figure shows sea level versus time, and high and low basin levels.

Water can be continuously turbined from high basin to low basin. When high basin level is lower than sea level, gates are opened to let water in, until sea level is regained.

In the same way, when low basin level is higher than sea level, gates are opened to let water out, until sea level is regained.

This problem is given a mathematical formulation, with y_s, y_1, y_2 denoting levels of sea and basins, q the water flow in the turbine between high and low basins, and u_1, u_2 integer variables for gates position (open or closed).

At each time, electricity generation is a known function N of the flow q and of the head between the two basins. In fact function N can be approximated by a polynomial of degree 2 with respect to flow and head variables. The problem is to maximize the integral thereof.

Dynamics of the basins are shown at the bottom part of the slide, as a function of the flow through the turbines, and the flow through the gates, which varies linearly with the square root of the level difference.

SLIDE 12

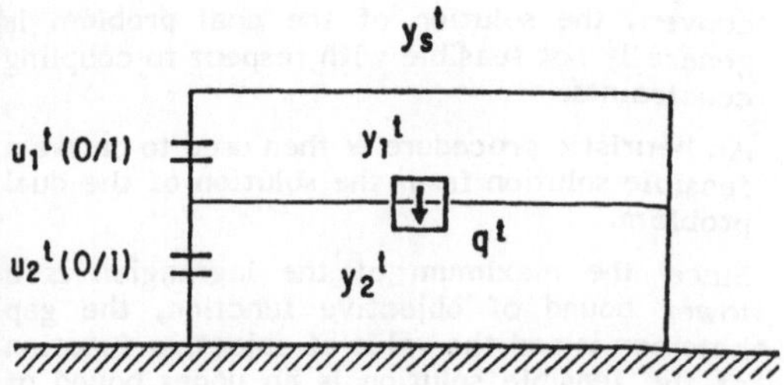

$$\max_{q^t, u_1^t, u_2^t} \int_0^T N(q^t, y_1^t - y_2^t)\, dt$$

$$\frac{dy_1^t}{dt} = -q^t + \alpha u_1^t \sqrt{y_s^t - y_1^t}$$

$$\frac{dy_2^t}{dt} = q^t - \alpha u_2^t \sqrt{y_2^t - y_s^t}$$

SLIDE 13

HAMILTONIAN

$$H^t = N(q^t, y_1^t - y_2^t) + \psi_1^t(-q^t + \alpha u_1^t \sqrt{y_s^t - y_1^t}) + \psi_2^t(q^t - \alpha u_2^t \sqrt{y_2^t - y_s^t})$$

WITH $\quad \psi_1^T = \psi_2^T = 0$

$$\frac{d\psi_1^t}{dt} = -\frac{\delta H}{\delta y_1^t}$$

$$\frac{d\psi_2^t}{dt} = -\frac{\delta H}{\delta y_2^t}$$

Introducing co-state variables, Ψ_1 for high basin, Ψ_2 for low basin, we can form the hamiltonian.

Co-state variable must satisfy differential equations shown at the bottom part of the slide, with final values equal to zero. This is why this problem is known as two points boundary problem.

The maximum's principle states that at each time, the control variables q and u should maximize the hamiltonian. Since function N is a polynomial, optimum values of control variables can be derived analytically.

In addition the hamiltonian being convex with respect to control variables, for all values of state variables, the local optimality condition given by the maximum's principle is at the same time condition for global optimality.

This problem is solved iteratively, starting from an initial guess of state variables, and control variables.

The first phase consists in computing co-state variables Ψ_1 and Ψ_2, through an integration of differential equations defining Ψ_1 and Ψ_2, starting from final time T, going backwards to initial time.

In the second phase, there are simultaneously a maximization of hamiltonian yielding optimum values of control variables q and u, and an integration of basin dynamics, to update state variables y_1 and y_2. This second phase starts at initial time, and goes forwards to final time.

Convergence is obtained, in this problem, after less than ten iterations.

SLIDE 14

ITERATIVE SOLUTION

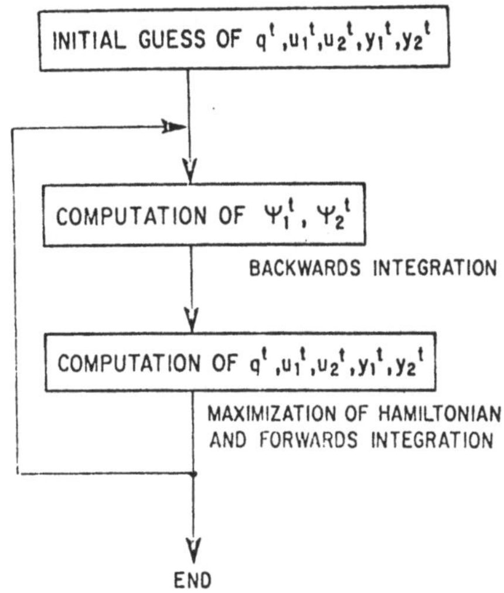

With this third example, we enter the field of stochastic problems.

In a large hydro reservoir, water inflows can be stored, and turbined at the most profitable time. For instance, when demand for electricity is high, hydro generation can reduce utilization of expensive thermal units.

In winter, when there are little inflows and high demand, the reservoir level goes down ; in summer, when there are a lot of inflows and low demand, the reservoir level goes up.

The problem is to optimize reservoir level trajectory to minimize costs, taking in account the randomness of the future, which affects hydro inflows, demand for electricity, and availability of thermal units. Hydro inflows can be represented as the sum of a deterministic term and of a white noise, with standard deviation .

State and control variables are as mentioned on this slide, and reservoir dynamics is the difference between inflows and output, or water release, generating electricity.

SLIDE 15

3 SCHEDULING STRATEGY FOR A HYDRO RESERVOIR

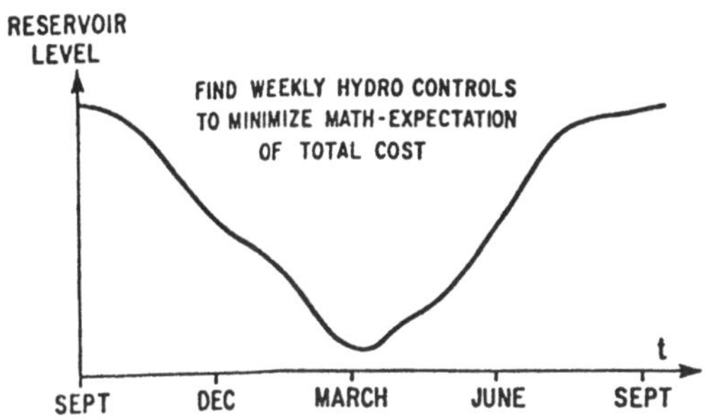

RANDOM VARIABLES

- INFLOWS $I = m(t)\,dt + \sigma(t)\,dB(t)$
- LOAD AND AVAILABILITY OF THERMAL UNITS y

RESERVOIR DYNAMICS

- STATE $x(t)$ RESERVOIR LEVEL
- CONTROL $u(t)$ OUTPUT
- DYNAMICS $dx = m(t)dt + \sigma(t)dB(t) - u(t)dt$

For this stochastic control problem, we are looking for a closed-loop control, or feedback, as a function of reservoir level x and of actual values of randoms affecting the system, y, observed at time t.

This closed-loop control should minimize mathematical expectation of total cost, computed from the probability laws of I, random inflows, and y, random load and availability of thermal units.

This problem is solved by stochastic dynamic programming by defining Bellmann's functions.

In the first phase, we compute Bellmann's functions for all x, with the determination of optimal control, by a backwards integration of Hamilton Jacobi equation.

In the second phase, we compute the probability of state x at time t. To do this a forwards simulation of system operation in a large number of cases of inflows, load and availability of thermal units is required.

SLIDE 17

SIMPLIFICATION

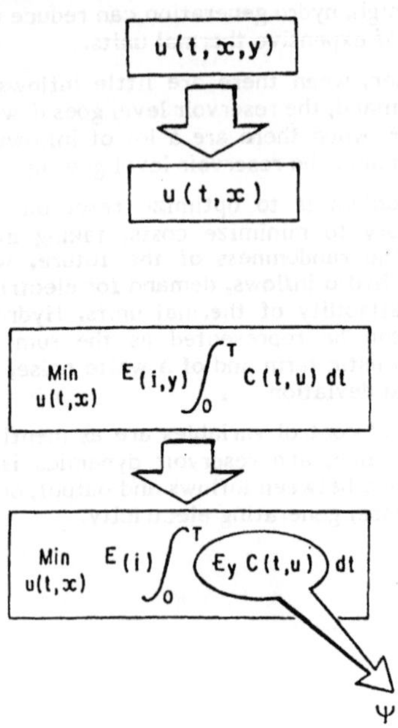

SLIDE 16

OBJECTIVE FUNCTION

$$\min_{u(t,x,y)} E_{(i,y)} \int_0^T C(t,u)\,dt$$

SOLVED BY STOCHASTIC DYNAMIC PROGRAMMING

DEFINE $V(t,x) = \min_{u(s,x,y)} E_{(i,y)} \int_t^T C(s,u)\,ds$

① COMPUTATION OF $V(t,x)$; $u(t,x)$

BACKWARS INTEGRATION OF HAMILTON - JACOBI EQUATION

② COMPUTATION OF $p(t,x)$

PROBABILITY OF STATE x AT TIME t

FORWARDS SIMULATION (MONTE - CARLO)

This problem can be simplified if we are satisfied with a simpler closed-loop control, that is function of reservoir level x only, and not of load and availability of thermal units.

In that case, mathematical expectation of total costs with respect to y, that is load and availability of thermal units, can be pre-computed as a function Ψ of control u, and time t. Then, our objective function is mathematical expectation of that function Ψ, with respect to I, random inflows, only.

This closed-loop control can be termed local, since it is of the closed-loop type as far as local variables are concerned (hydro inflows), and of the open-loop type for other random variables such as load and availability of thermal units.

This simpler problem is solved again in two phases, very similar to these of the more general problem.

The first phase is the computation of Bellmann's functions for all x, by a backwards integration of Kolmogorov equation, differing from Hamilton-Jacobi solely by the last term.

In the second phase, the Monte-Carlo simulation, wich was very computationnally expensive in the previous problem, can be replaced by a direct computation of probability of state x, by a forwards integration of Fokker-Planck equation.

SLIDE 18

SOLUTION

(1) COMPUTATION OF $V(t,x)$; $u(t,x)$

KOLMOGOROV EQUATION

$$\frac{\delta V}{\delta t} + \underset{u}{\mathrm{Min}}\left\{(m(t)-u)\frac{\delta V}{\delta x} + E_{(y)}\,C(t,u)\right\} + \frac{1}{2}\sigma^2\frac{\delta^2 V}{\delta x^2} = 0$$

BACKWARDS INTEGRATION

(2) COMPUTATION OF $p(t,x)$ –
PROBABILITY OF STATE x AT TIME t

FOKKER-PLANCK EQUATION

$$-\frac{\delta p}{\delta t} - \frac{\delta}{\delta x}\left((m-u)p\right) + \frac{1}{2}\frac{\delta^2}{\delta x^2}(\sigma^2.p) = 0$$

FORWARDS INTEGRATION

SLIDE 19

[4] SCHEDULING STRATEGY FOR A GROUP OF NUCLEAR REACTORS

FIND WEEKLY REACTOR CONTROL TO MINIMIZE MATHEMATICAL EXPECTATION OF TOTAL COST

RANDOM VARIABLES

- ω_i AVAILABILITY OF NUCLEAR UNIT i
- y LOAD AND AVAILABILITY OF NON-NUCLEAR UNITS

NUCLEAR UNIT

- STATE x_i (ENERGY IN CORE)
- CONTROL u_i (GENERATION OR REFUELLING)
- DYNAMICS $\dfrac{dx_i}{dt} = f_i\,(t, x_i, u_i)$

Now that is our fourth example. When we go from one reserve, hydro reservoir, to several tens, nuclear reactors, the stochastic optimal control problem becomes much more difficult.

Let us assume that there is no hydro ressource in the system. Then, the random variables are limited to ω, availability of nuclear units, and y, including load and availability of non-nuclear units.

The state variable of a nuclear unit is energy in core, and its control variable is generation, or refuelling, the latter requiring several weeks shut down time.

The problem is to find closed-loop reactor controls, minimizing mathematical expectation of total cost.

Closed-loop reactor controls should be functions of x, nuclear reactor state variable, ω, nuclear reactor availability, and y, load and availability of non-nuclear units.
Mathematical expectation of total cost should be computed from the probability laws of ω and y.
The only optimal method to solve this stochastic optimal control problem is stochastic dynamic programming, but the computational burden would be too heavy for this large-scale problem, with several tens of state variables.
Therefore, the problem has to be simplified.
The simplification, very similar to the one presented for the hydro reservoir, consists in restricting to local closed-loop controls, or local feedbacks. We look for nuclear reactor controls, functions of local information only, that is energy in core and availability of nuclear reactor itself.

This local feedback control is open-loop with regard to other nuclear reactors, load and non-nuclear units.

SLIDE 21

SOLUTION BY DECOMPOSITION-RELAXATION

WITH A GUESS OF :

$$\left[\begin{array}{ll} p(t,x_j) \text{ PROBABILITY OF } x_j, \text{ TIME } t & \\ u_j(t,x_j,\omega_j) & \text{FOR ALL } j \neq i \end{array}\right.$$

$$\underset{(\overline{\omega_i},y)}{E} \; C(t,u_1 \ldots u_n,y)\,dt = \psi_i(t,u_i)$$

SUCCESSIVE SOLUTION OF LOCAL PROBLEMS

$$\text{Min} \int_0^T \underset{(\omega_i)}{E} \; \psi_i(t,u_i)\,dt$$

$$\frac{dx_i}{dt} = f_i(t,x_i,u_i)$$

SLIDE 20

OBJECTIVE FUNCTION

$$\underset{\substack{u_i(t,x_1 \ldots x_n,\\ \omega_1 \ldots \omega_n,y)}}{\text{Min}} \int_0^T \underset{(\omega_1 \ldots \omega_n,y)}{E} \; C(t,u_1 \ldots u_n,y)\,dt$$

RESTRICTION TO LOCAL FEEDBACKS

$$\underset{u_i(t,x_i,\omega_i)}{\text{Min}} \int_0^T \underset{(\omega_1 \ldots \omega_n,y)}{E} \; C(t,u_1 \ldots u_n,y)\,dt$$

$$\underset{u_i(t,x_i,\omega_i)}{\text{Min}} \int_0^T \underset{(\omega_i)}{E}\;\underset{(\overline{\omega_i},y)}{E} \; C(t,u_1 \ldots u_n,y)\,dt$$

$$\overline{\omega_i} = \{\omega_j,\ j \neq i\}$$

But again, this simpler problem is not easy to solve, since its objective function is not separable. Therefore, we have to use a sub-optimal technique, called decomposition-relaxation, consisting in optimizing local problems in sequence.

To solve a local problem, dealing with nuclear unit i, we need a guess of likely operation of other nuclear units, that is, for each of them, probability of state x at time t, and control function u. Then, the objective function of the local problem is a function ψ of local control only.

This local problem can be solved by stochastic dynamic programming, very similar to that used previously for an hydro reservoir.

The solution procedure is shown here, with succesíve solution of local problems, starting from a guess of likely operation of all nuclear units.

Each local problem includes three phases : the first one is the computation of local objective function, taking in account probability laws of operation of other nuclear units, or a guess thereof.

The second phase is the computation of Bellmann's functions and optimum closed-loop control, by backwards integration of Kolmogorov's equation.

The third phase is the computation of the probability law of state x, and therefore of control u, by forwards integration of Fokker-Planck equation.

Convergence is obtained after a little number of iterations, that is 5 to 10.

Although there is no guarantee that this method yields optimum control solution, even local, but merely Nash optimum solution, we observed very good performances on our large-scale problems.

SLIDE 22

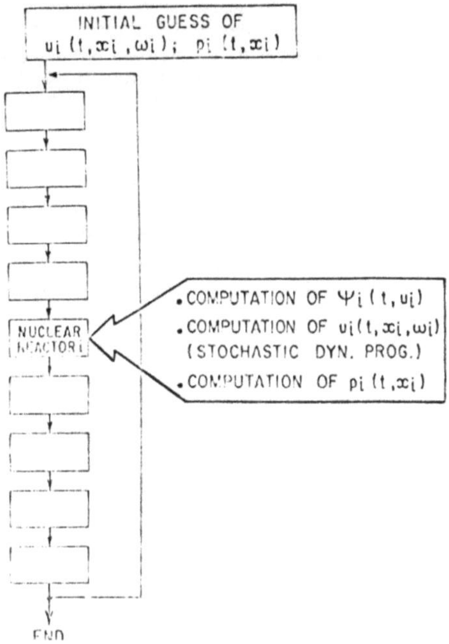

SLIDE 23

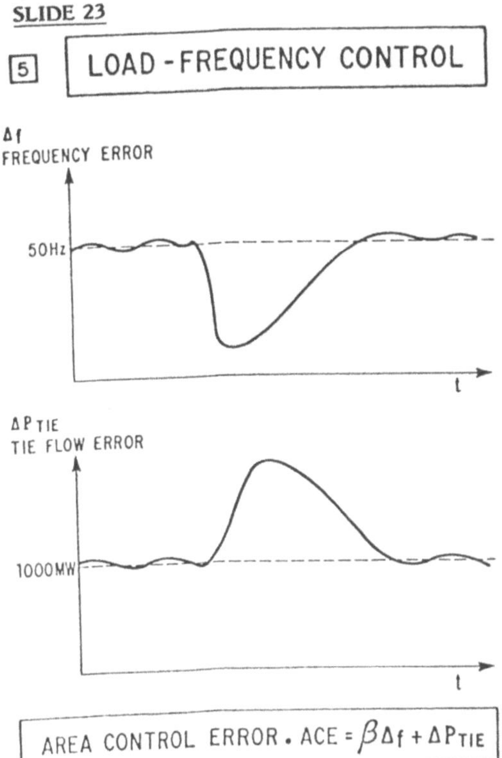

Our last example deals with load-frequency control.

Second after second, minute after minute, the demand for electricity is varying, depending on customers'activity. Electricity generation must be continuously adjusted to it, as accurately as possible. It is the role of the load frequency control to ensure the fine tuning between generation and load. As soon as generation is a little lower than load, frequency automatically goes under its reference value (50 Hz), and tie flows with neighbouring countries automatically increase over their reference values (that resulted from an agreement). In that case, as soon as possible, generation should be increased, to regain reference values of frequency and tie flows. This is represented on this slide with the sudden loss of one generation unit.

The sum of frequency error (with a dimension factor) and of tie flow error is called area control error.

Load-frequency control should minimize area control error.

But if the objective of load-frequency control were only to keep area control error near zero, it could demand fast and frequent changes to generating units, thus resulting in undue stresses on them. In addition, it could lead the system far from scheduled program, supposed to be optimal.

Therefore, the objective of load-frequency control should be changed to minimizing area control error and unit scheduling error, without unnecessary unit variations.

The next slide will show that this tracking problem is of the state-regulator type.

SLIDE 24

OBJECTIVE

KEEP • THE AREA CONTROL ERROR

• THE UNIT SCHEDULING ERROR

NEAR ZERO

WITHOUT UNNECCESSARY UNIT VARIATION

SLIDE 25

STATE

• FREQUENCY ERROR	Δf
• TIE FLOW ERROR	ΔP_{TIE}
• UNIT SCHEDULING ERROR	$x_i (i=1 \dots n)$

CONTROL

• UNIT SCHEDULING CHANGE	$u_i (i=1 \dots n)$
• EXTERNAL UNIT SCHEDULING CHANGE	u_e

DYNAMICS

$$\frac{d(\Delta f)}{dt} = \frac{1}{k+k_e} \left(\sum_i u_i + u_e \right)$$

$$\frac{d(\Delta P_{TIE})}{dt} = \frac{1}{k+k_e} \left(k_e \sum_i u_i - k u_e \right)$$

$$\frac{dx_i}{dt} = u_i$$

OBJECTIVE

$$\underset{u_i}{\mathrm{Min}} \int a \left\{ (\beta \Delta f + \Delta P_{TIE})^2 + \sum_i b_i x_i^2 + \sum_i c_i u_i^2 \right\} dt$$

STATE - REGULATOR PROBLEM

State variables of the system include frequency error, tie flow error, and, for each unit, scheduling error, which is the difference between scheduled generation and actual generation.

Control variables are, mainly, for each unit, schedule change, which is the variation demanded to the unit, over scheduled variation. An other control variable is added, to model the possibility of schedule changes in neighbouring countries.

A nice feature of this problem is that dynamics are linear functions of control variables, and there are no constraints on state or control variables.

In order to fit in a state regulator problem, we write the objective function as a quadratic function, to minimize in successive order, the area control error, the unit scheduling error, and the unit scheduling changes.

This problem has a very nice formulation since dynamics are linear and objective function quadratic, and there are no constraints on state or control variables.

Optimality conditions resulting from the maximum's principle lead to an equation of the Ricatti type, which yields a solution in feedback form : Unit control is a predetermined linear function of state variables, namely frequency error, tie flow error and unit scheduling errors.

This problem could be further sophisticated by introducing gaussian random variables, and by distinguishing between observed state variables and non-observed state variables.

I am coming to the end of my presentation, and I hope that although is was all but a comprehensive survey, it could give you an idea of the interest of optimal control methods for power system optimization.

SLIDE 26

FORMALLY

$$\dot{X} = AX + BU$$

$$\text{Min } \dot{J} = \frac{1}{2}\int ({}^{t}XQX + {}^{t}URU)\,dt$$

SOLUTION : RICATTI EQUATION

$$U = -R^{-1}\,{}^{t}BkX$$

k GIVEN BY

$$\dot{k} = -kA - {}^{t}Ak + kBR^{-1}\,{}^{t}Bk - Q$$

MATHEMATICAL PROGRAMMING SOFTWARE

K. Schittkowski
Mathematisches Institut
Universität Bayreuth
D-8580 Bayreuth, Germany F.R.

Abstract.

The purpose of this paper is to summarize the present state-of-the-art in computational mathematical programming. An attempt is made to characterize the actual trends in developing algorithms and software. The author is aware that all statements represent only his subjective point of view and that for many subtopics a more detailed analysis may be necessary. Specific emphasis is given on model building and intelligent software systems since they play a more and more important role when fitting a practical problem to an abstract mathematical notation, and when transforming a mathematical problem into a computationally solvable form, i.e. a computer program.

1. Introduction.

The present situation in mathematical programming could be summarized as follows:

- We do not observe major activities in some research areas that were very important in the last 15 years, e.g. in smooth, small scale nonlinear programming. It seems that the theory behind the algorithms is well understood now, that only minor subtopics are investigated and that sufficiently qualified software is available, but probably not yet in an optimal form.

- In the most traditional mathematical programming research area, that is in linear programming, some unusual ideas are proposed, e.g. the ellipsoid method a few years ago or Karmarkar's algorithm at present inducing many people to investigate again linear programming both from the theoretical and practical point of view.

- Some other mathematical programming areas become more and more important specifically in practical application models. Among them are large scale, stochastic, global and non-smooth optimization. Although some progress was made in special situations, algorithms solving a larger class of problems are seldom. It is certainly too early to state which type of algorithm is <u>the</u> standard method to solve a problem. Consequently robust and efficient software is available only for certain subclasses.

- Since a few years intelligent software systems are developed that facilitate the transition from a practical problem to a numerical solution. Besides modeling systems that extend techniques known from linear programming, there are decision support systems, programming neighbourhoods, interactive system languages and expert systems.

In the subsequent sections we specify the situation in linear and nonlinear programming a bit more detailed. It is out of the scope of this report to present a list of available mathematical programming software. A collection of more than 60 code descriptions is published in Schittkowski (1985a) and additional information can be found in Einarsson (1979), Waren and Lasdon (1979), Nazareth (1978),Dembo (1978), Land and Powell (1979), Schittkowski (1980), Hiebert (1979), and Lasdon (1985).

2. Some general remarks on algorithms and software.

In this section some basic features of available mathematical programming software are outlined. The first observation is trivial: A general purpose method solving all types of mathematical programming methods, does not exist nor can one predict which code solves a specific problem in the most robust and efficient way. Thus the following remarks are stated to illustrate the specific situation and to get an impression on those mathematical models for which some software is available.

a) Linear and integer programming

Available algorithms are based on various variants of the simplex method, where additional integer variables lead to branch and bound or cutting plane methods, see e.g. Land and Powell (1979) for a review. Most programs require a fixed input format (MPS) and are designed in form of complete systems. Often they contain matrix generators or modeling languages to facilitate the model building phase, see Williams (1985). Reduction techniques are available to reduce the size of linear programming problems in form of preprocessors as described by Karwan, Lotfi, Telgen and Zionts (1983) in form of a survey. It is reported that problems with 250,000 veriables and 50,000 constraints were solved.

b) Linear least squares and quadratic programming

Although the basic mathematical ideas are well known since many years, only very few codes are available that solve the problems under consideration. A linear least squares program was published in Lawson and Hanson (1974) and a few other linear least squares codes were developed since.

For quadratic programming a primal or active set method was implemented by Gill, Murray, Saunders and Wright (1984a) with the name QPSOL. The dual method of Goldfarb and Idnani (1983) got considerable attention since their algorithm seems to be more suitable to use it in the frame of a sequential quadratic programming method for nonlinear programming. A code named ZQPCVX was published by Powell (1983).

c) Nonlinear least squares programming

Nonlinear least squares models arise very frequently in a wide variety of practical applications, e.g. maximum likelihood estimation, parameter identification, data fitting. Thus many well tested and documented computer codes are available and software libraries, e.g. NAG or IMSL, contain a sufficiently large collection of algorithms. They are all based on a Gauß-Newton method, if the residual is expected to be small and contain safeguards in various ways if the residual is large. However most of them are unable to solve problems with linear or nonlinear constraints. A suitable program NSNIP is distributed by Lindström (1983) and another one by Schittkowski (1985b).

d) Unconstrained nonlinear programming

Since most practical mathematical programming problems possess constraints at least in form of simple upper and lower bounds on the variables, methods for constrained optimization are much more important and can be applied to solve unconstrained problems whenever available. But unconstrained nonlinear programming is often a first step when developing methods for the constrained case. This was true in the past for smooth, small scale optimization, and we observe a similar progress in global and large scale optimization, for instance.

e) Constrained nonlinear programming

Many different mathematical algorithms were developed in the past to solve the smooth, small scale nonlinear programming problem, e.g. feasible direction, penalty, multiplier, generalized reduced gradient, projected Lagrangian and sequential quadratic programming methods, to mention only the most important classes. Each of these methods can be specified and implemented in various ways. But there is now a common agreement that the sequential quadratic programming algorithms are the most efficient and reliable ones, although in specific situations which cannot be predicted in general. One of the other methods may be preferred. A few codes are available which differ in algorithmic details, but retain the basic idea to formulate and solve a quadratic programming subproblem in each iteration step. Among them are the programs VFO2AD and VMCWD of Powell (1978,1982), NLPQL of Schittkowski (1984) and QPSOL of Gill, Murray, Saunders and Wright (1984b).

f) Large scale nonlinear programming

Practical large optimization problems of often structured and in many cases, it is necessary to exploit a given structure to be able to solve a problem numerically. Typical structures are network constraints, see Dembo (1983) or partially separable functions, see Griewank and Toint (1985). For a more general application, very few codes are available to solve nonlinearly constrained problems, e.g. MINOS of Murtagh and Saunders (1983) or CONOPT of Drud (1985). The algorithms exploit techniques known from linear programming, e.g. in form of a generalized reduced gradient approach. Large scale nonlinear programming codes solve practical optimization problems with thousands of variables and constraints.

3. Model building and interactive software systems.

An increasing number of software tools is available to support the numerical solution of a practical optimization problem. They can be characterized as modeling systems, programming neighbourhoods, design systems, decision support systems or expert systems depending on the principal design behind the system. They allow a user to omit the organisational "ballast" around the modeling and solution phase, e.g. the work to store and document problem functions, data or results. Since any generated problem is solved automatically by a suitable code, it is not necessary to read and understand a program documentation and the mathematical background. Thus even complex optimization problems can be solved by unexperienced or occasional users.

Well known are model building systems in linear programming which consist of matrix generators or data manipulation languages often combined with report writers, see a.g. Williams (1985). There exist systems that exploit related techniques to the nonlinear programming case, e.g. GINO developed by Liebman, Schrage, Lasdon and Waren (1985) and GAMS written by Brooke, Drud and Meeraus (1984). The latter one contains a couple of different mathematical programming codes, solves also large problems, and is mainly used to formulate and solve operations research models at the World Bank. The basic idea is that the nonlinear programming problem to be solved can be expressed in form of algebraic terms that are then evaluated symbolically.

The system DELIGHT developed by Nye, Polak, Sangiovanni, Vincentelli and Tits (1981) is based on an interactive programming language called RATTLE, which can be used also for developing or modifying an algorithm. Options such as graphics and simulation programs are available. Mathematical programming problems with inequality constraints are solved by a feasible direction method allowing also an interactive multicriteria analysis.

The software system EMP written by the author (1985), supports the solution of linear programming, quadratic programming, linear least squares, L_1, L_2 or L_∞-data fitting, parameter estimation, min-max or general nonlinear programming problems subject ot linear or nonlinear constraints. Nonlinear problem functions are defined interactively in form of sequences of FORTRAN statements each addressing a numerical value to a user provided function name. The problems are stored in form of a data base system, so that they are easily retrieved, modified or deleted on request. EMP selects a suitable algorithm and writes a complete FORTRAN program, which is executed automatically. Numerical results are stored in the problem file and are available for further processing, e.g. to get a problem dependable printed output. Since the generated code may be used outside of the system, one could consider EMP also as a software development tool.

All actions of EMP are controlled by commands which are displayed in form of menues. Step by step a user will be informed how to supply new data which are saved on a user provided file. Besides commands to generate, solve or edit a problem, there are others to transfer data from one problem to another, to delete a problem, to sort problems, to update known "experience-fields", to get a report on problem or solution data and to retrieve some information on the system, the mathematical models and the available algorithms. A user will be asked whether he wants to link the generated FORTRAN-code with some of his own files or whether he wants to insert additional declaration and executable statements. It is possible to provide keywords and to define "experience-fields", so that the system is capable to "learn" from previous solution attempts. A failure analysis explains termination reasons and proposes remedies to overcome failures or numerical difficulties.

Conclusions.

The main conclusions of this paper can be summarized as follows:

a) Reliable and efficient software is available for many mathematical programming areas, e.g. in:
   - linear and integer programming
   - quadratic and linear least squares programming
   - small scale and smooth constrained nonlinear programming
   - large scale nonlinear programming.

b) There is a lack of mathematical programming software that is well tested, applicable on a wide class of practical problems and distributed on request, in some other areas, e.g. in
   - non-smooth optimization
   - stochastic optimization
   - global optimization
   - nonlinear integer programming.

   Nevertheless a lot of experimental software was developed so that for many specific subclasses, special purpose software is available.

c) A couple of differently designed software systems are available to support model building and numerical solution of practical optimization problems.

<u>References:</u>

A.BROOKE, A.DRUD, A.MEERAUS (1984): High level modeling systems and nonlinear programming, Report No. DRD 113, Development Research Dept., World Bank, Washington.

R.S.DEMBO (1978): Current state-of-the-art of algorithms and computer software for geometric programming, J. of Optimization Theory and Applications, Vol. 26, 149 - 184.

R.S.DEMBO (1983): NLPNET - A nonlinear network optimizer - User's guide, SOM Working Paper, Yale University, Yale.

A.DRUD (1985): CONOPT-A GRG code for large sparse dynamic nonlinear optimization problems, to appear: Mathematical Programming.

B.EINARSSON (1979): Bibliography on numerical software, J. on Computational and Applied Mathematics, Vol 5, 145 - 159.

P.E. GILL, W.MURRAY, M.A.SAUNDERS, M.H.WRIGHT (1984a): User's guide for QPSOL: a Fortran package for quadratic programming, Report SOL 84-6, Dept. of Operations Research, Stanford University, Stanford.

P.E. GILL, W.MURRAY, M.A.SAUNDERS, M.H.WRIGHT (1984b): User's guide for NPSOL: a Fortran package to nonlinear programming, Report SOL 84-7, Dept. of Operations Research, Stanford University, Stanford.

D.GOLDFARB, A.IDNANI (1983): A numerically stable dual method for solving strictly convex quadratic programs, Mathematical Programming, Vol. 27, 1 - 33.

A.GRIEWANK, P.L.TOINT (1985): Numerical experiments with partially separable optimization problems, to appear: Numerical Analysis, Dundee 1983. D. Griffiths ed., Springer.

K.HIEBERT (1979): A comparison of nonlinear least squares software, Sandia Technical Report SAND 79-0483, Sandia National Lab., Albuquerque.

M.M.KARWAN, V.LOTFI, J.TELGEN, S.ZIONTS (1983): Redundancy in mathematical programming, Lecture Notes in Economics and Mathematical Systems, Berlin, Heidelberg, New York.

A.LAND, S.POWELL (1979): Computer codes for problems of integer programming, Annals of Discrete Mathematics, Vol. 5, 221 - 269.

L.S.LASDON (1985): Nonlinear programming algorithms - Applications, software and comparisons, Report, Dept. of General Business, University of Texas, Austin.

C.L.LAWSON, R.J.HANSON (1974): Solving least squares problems, Prentice Hall, Englewood Cliffs, New Jersey.

J. LIEBMAN, L.SCHRAGE, L.LASDON, A.WAREN (1985): Applications of modeling and optimization with GINO, Report, Graduate School of Business, University of Chicago, Chicago.

P.LINDSTRÖM (1983): Algorithms for nonlinear least squares-particularly problems with constraints, Report UMINF - 106.83, University of Umea, Umea, Sweden.

B.A.MURTAGH, M.A.SAUNDERS (1983): MINOS 5.0 User's guide, Report SOL 83-20, Dept. of Operations Research, Stanford University, Stanford.

L.NAZARETH (1978): Software for optimization. Report SOL 78-32, Dept. of Operations Research, Stanford University, Stanford.

W.T.NYE, E.POLAK, A.SANGIOVANNI-VINCENTELLIS, A.L.TITS (1981): DELIGHT: An optimization-based computer aided design system, Proceedings of the 1981 IEEE International Symposium on Circuits and Systems, 851 - 855.

M.J.D.POWELL (1978): A fast algorithm for nonlinearly constrained optimization calculations, in: Numerical Analysis. G.A.Watson ed., Lecture Notes in Mathematics, Vol. 630. Springer, Berlin, Heidelberg, New York.

M.J.D.POWELL (1982): VMCWD: A Fortran Subroutine for constrained optimization, Report DAMTP 1982/NA4, DAMTP, University of Cambridge, Cambridge, England.

M.J.D.POWELL (1983): ZQPCVX: A Fortran subroutine for convex quadratic programming. Report DAMTP 83/NA 17, DAMTP, University of Cambridge, England.

K.SCHITTKOWSKI (1980): Nonlinear programming codes, Lecture Notes in Economics and Mathematical Systems, Vol. 183, Springer, Berlin, Heidelberg, New York.

K.SCHITTKOWSKI (1984): NLPQL: A Fortran subroutine solving constrained nonlinear programming problems Report, Institut für Informatik, Universität Stuttgart, Stuttgart.

K.SCHITTKOWSKI (1985a): Software for mathematical programming, in: Computational Mathematical Programming. K. Schittkowski ed., NATO ASI Series F: Computer and System Sciences, Vol. 15, Springer, Berlin, Heidelberg, New York, Tokyo.

K.SCHITTKOWSKI (1985b): Solving nonlinear least squares problems by a general purpose SQP-algorithm, Report,Institut für Informatik, Universität Stuttgart, Stuttgart.

K.SCHITTKOWSKI (1985c): EMP: A software system supporting the numerical solution of mathematical programming problems, Report, Mathematisches Institut, Universität Bayreuth, Bayreuth.

A.D.WARREN, L.S.LASDON (1979): The status of nonlinear programming software, Operations Research, Vol. 27, 431 - 456.

Design, analysis and implementation of thermodynamically motivated simulation for optimization of subgraphs

Christina Schiemangk

Academy of Sciences of the GDR
Central Institute of Cybernetics
and Information Processes

DDR-1086 Berlin, Kurstraße 33

Abstract

A general principle was elaborated to apply thermodynamically motivated strategies on NP-complete subgraph optimization problems with given placing requirements for subsets of vertices. Such problems arise for example in the area of automation equipment, integrated circuit and computer network layout where wiring problems are important. The thermodynamically motivated strategy is a stochastic optimization algorithm by simulated annealing of ideal gases. For fixed temperature T it is a finite homogeneous Markov chain, which converges to the Boltzmann distribution under certain assumption. It is proved, that the sequence of Boltzmann distributions for $T \longrightarrow 0$ converges to all optimal solutions equally distributed and all other feasible solutions are reached with probability 0. Beside, some remarks on the speed of convergence are given.
An interactiv FORTRAN programe - package for solving the Steiner Tree Problem and the optimization of interacting paths systems with help of thermodynamically motivated simulation was developed. Computing results for graphs up to 600 vertices are presented including a comparison with a Greedy algorithm.

1. Convergence of thermodynamically motivated simulation

For a discrete optimization problem of the size n with g_n feasible solutions Z_i the minimum has to be found without loss of generality. $K_i = K(Z_i)$ may denote the costs or the values of the goal function of the feasible solution Z_i.
A transformation set TR over the totality of feasible solutions is called feasible, if the following conditions are fulfilled:

A For each feasible solution Z_i there exists a nonempty partial set $TR_i \subset TR$ so, that for each $T \in TR_i$ holds: $T(Z_i)$ is a feasible solution.

B For two feasible solutions Z_i and Z_j each, there exists a finite sequence of transformations out of TR which transforms one into the other.

Starting from any feasible initial solution z(0) (initial state) the state of the system is changed by a randomly successive choice of exactly one transformation T out of TR_i in the respective N-th step with $z(N) = Z_i$ (N = 0,1,2,...).

$P_{ij} = P\{z(N+1) = Z_j/z(N) = Z_i\}$ may denote the probability of the transformation from the state Z_i into the state Z_j after one step.
From that follows:

$$P_{ij}\begin{cases} = 0, \text{ if } T \notin TR_i \\ > 0, \text{ if } T \in TR_i \end{cases}$$

and

- $$\sum_{j=1}^{g_n} P_{ij} = 1 \qquad (1)$$

- For all index pairs (i,j) with indices i,j = 1,...,g_n, which do not neccessarily differ from each other, it holds: j can be reached from i in M(i,j) > 0 steps. That means:

$$P_{im_1} \cdot P_{m_1m_2} \cdot \ldots \cdot P_{m_{M-1}j} > 0 \qquad (2)$$

- There exists at least one i and one j with

$$K_j \geq K_i \text{ and } P_{ij} > 0. \qquad (3)$$

With the help of the transition probabilities $P_{ij}(T)$ transition probabilities $p_{ij}(T)$ $(0 < T < \infty)$ can now be defined for the thermodynamically motivated simulation of a discrete optimization problem.

$$p_{ij}(T) = \begin{cases} P_{ij}(T), \text{ if } \Delta_{ji} \leq 0 \qquad i \neq j \\ P_{ij}(T) \exp\left(\frac{-\Delta_{ji}}{T}\right), \text{if } \Delta_{ji} > 0 \end{cases}$$

$$p_{ij}(T) = \sum_{K>i}\left[1 - \exp\left(\frac{-\Delta_{ki}}{T}\right)\right] P_{ik}(T) + P_{ii}(T).$$

In the following let be (without loss of generality):
$K_1 \leq K_2 \leq \ldots \leq K_{g_n}$. Then the matrix $p_{ij}(T)$ of the transition probabilities looks as follows:

$$\begin{pmatrix} p_{11} & p_{12}e^{\frac{-\Delta_{21}}{T}} & p_{13}e^{\frac{-\Delta_{31}}{T}} & \cdots & \\ p_{21} & p_{22} & p_{23}e^{\frac{-\Delta_{32}}{T}} & p_{24}e^{\frac{-\Delta_{42}}{T}} & \cdots \\ p_{31} & p_{32} & p_{33} & p_{34}e^{\frac{-\Delta_{43}}{T}} & \cdots \\ p_{41} & p_{42} & p_{43} & p_{44} & \cdots \\ \vdots & \vdots & \vdots & \vdots & \vdots \end{pmatrix}$$

Now it is easy to prove the following

Theorem 1 The thermodynamically motivated simulation is for each fixed T a finite homogeneous Markov chain with transition probabilities $p_{ij}(T)$ and the following properties:

1. $(p_{ij}(T))$ is a stochastic matrix, that means:

$$\sum_{j=1}^{g_n} p_{ij}(T) = 1, \; i = 1,\ldots,g_n$$

2. It is irreducible.
3. There exists at least one i with $p_{ii}(T) > 0$.

That means, that for each $T (0 < T < \infty)$ the assumptions of an ergodic theorem are fulfilled and an unequivocal limit distribution

$$p_j^* = \lim_{N\to\infty} p_j(N) \quad \text{exists,}$$

which is independent of the initial distribution /2/.
In particular it holds:

Theorem 2 For each fixed T the thermodynamically motivated simulation converges to the Boltzmann distribution, if $P_{ij} = P_{ji}$ $i,j = 1,\ldots, g_n$

$$p_j^* = \lim_{N\to\infty} p_j(N) = \frac{\exp(\frac{-K_j}{T})}{S} \quad \text{with } S = \sum_{i=1}^{g_n} \exp(\frac{-K_i}{T}) \; .$$

To prove this theorem, it is sufficient to show the validity of the equation:

$$p_j^* = \sum_{i=1}^{g_n} p_i^* p_{ij} \; .$$

For $T \to \infty$ the sequence of the Boltzmann distribution converges to the uniform distribution:

$$\lim_{T \to \infty} p_j^*(T) = \lim_{T \to \infty} \frac{\exp\left(\frac{-K_j}{T}\right)}{\sum_{i=1}^{g_n} \exp\left(\frac{-K_i}{T}\right)} = \frac{1}{g_n}.$$

Now it is possible to prove the convergence of the thermodynamically motivated simulation:

Theorem 4 For $T \to 0$ the sequence of Boltzmann distributions converges to the optimal distribution:

$$\lim_{T \to 0} p_i^*(T) = \begin{cases} \frac{1}{ANZ}, & \text{if } Z_i \in Opt \\ 0, & \text{otherwise} \end{cases}$$

ANZ - the number of the optimal solutions

Opt - optimum set of the considered discrete optimization problem

$\overline{K} - \max_i \left\{K_i\right\}_{i=1}^{g_n}$ - maximum goal function value

$\underline{K} - \min_i \left\{K_i\right\}_{i=1}^{g_n}$ - minimum goal function value

Proof:

It holds: $\lim_{T \to 0} p_i^*(T) = \frac{1}{\lim_{T \to 0} \frac{1}{p_i^*(T)}}$ and

$$\lim_{T \to 0} \frac{1}{p_i^*(T)} = \sum_{K=1}^{g_n} \lim_{T \to 0} \exp\left(\frac{K_i - K_K}{T}\right).$$

Different cases must be considered,

Case 1: $K_i \neq \overline{K}$ and $K_i \neq \underline{K}$

Case 2: $K_i = \overline{K}$

Case 3: $K_i = \underline{K}$

q.e.d.

The following special case is considered. Every partial set TR_i of transformations may consist of m, $1 \leq m \leq g_n$, transformations, which can be realized with the same probability. Moreover, the difference Δ_{ji} of the goal function values may be constant for all worsenings which are possible after one step

$$X(T) = \exp\left(\frac{-\Delta}{T}\right) = X(T). \text{ Let } X(T) < \frac{1}{2}.$$

Under these assumptions the hypothesis, that with decreasing temperature, with increasing difference Δ and with increasing number m the speed of convergence for reaching the Boltzmann distribution decreases can be confirmed by means of calculating the upper bounds of eigenvalues a ($|a| < 1$) of the transition probabilities $p_{ij}(T)$.

2. Experiments for solving the Steiner-Tree-Problem and the optimization of interacting paths systems

2.1. Problem definition and feasible transformation sets

A connected and weighted graph with vertex set X, arc set U, which has the property: $u = [A,E] \in U \Rightarrow \bar{u} = [E,A] \in U$ and a real valued function $w: U \longrightarrow R$ will be denoted as basic graph G=(X,U). Let $FIX \in X$ be a fixed vertex subset of the basic graph G and let T be a subtree of G.

A path P_{AE} from an initial vertex A to a terminal vertex E is an arc sequence $[u_1,\ldots,u_n]$ with the property, that the terminal vertex of u_k is the initial vertex of u_{k+1} for k=1,...,n-1. Let $\{(A_i,E_i)\}_{1=1}^{n}$ be a fixed set of vertex pairs in G.

We are interested in the following two problems

. Steiner - Tree- Problem

Find a cost minimal subtree T of G, which contains all points of the fixed vertex set FIX.

Let $T = (\hat{X},\hat{U})$, $\hat{X} \subset X$, $\hat{U} \subset U$ be a feasible solution of the Steiner-Tree-Problem in G = (X,U) and $\hat{u} = [A,E]$ an arc of $\hat{U}$. For technical reasons the tree T will be regarded as directed from an initial vertex $I \in FIX$ to the terminal vertices $FIX \setminus \{I\}$. By deleting $\hat{u}$ from U we get two partial trees. As remainder tree $TR^A = (\hat{X}_R, \hat{U}_R)$ will be denoted the partial tree with A as one terminal

vertex and as subtree will be denoted the partial tree $UT_E = (\hat{X}_u, \hat{U}_u)$ with E as initial vertex.
A transformation of T consists in the following operations.

Random selection of an arc $\hat{u} = [A,E] \in \hat{U}$;
Delete the arc $\hat{u}$ from T;
Compute the set W of all paths from $R \in \hat{X}_R$ to E in $G \setminus \{\hat{u}\}$ with a minimal number of arcs;
Random selection of a path $P_{RE} \in W$;
For $u \in P_{RE}$ DO
IF $\bar{u} \in \hat{U}$ THEN $\hat{U} := \hat{U} \setminus \{\bar{u}\} \cup \{u\}$ ELSE $\hat{U} \cup \{u\}$.
In consequence of substituting the arc $\hat{u} = [A,E]$ by the path P_{RE} it is possible, that

- the subtree UT_E contains cycles
- the tree has terminal vertices which are not fixed vertices.

For that reason we perform reductions by deleting arcs:

1. Reduction of the subtree UT_E.
2. Reductions of the path P_{RE} and the path P_{IA} .

Fig. 1 presents a schematic picture of the replacement process.

. Optimization of an interacting path system

Find a cost minimal interacting path system

$P = \{P_i\}_{i=1}^{n}$ denotes a system of n paths $P_i = P_{A_i E_i}$ in G whith initial vertices A_i and terminal vertices E_i .
Special case:
- Star: All paths P_i have the same initial vertex $A_i = I$, $i = 1,2,\ldots,n$

Let S be the set of all arcs of an interacting path system P.
The total cost GK of $P = \{P_i\}_{i=1}^{n}$ can be given by the following formula:

$$GK = \sum_{u \in S} w(u) + \sum_{i=1}^{n} K(P_i)$$

where w is the weight function of the basic graph and $K(P_i)$ means cost of path P_i. (In many applicable cases the connection between the cost computation of u, $u \in S$ and of P is more complicated.)
For calculating GK one needs an algorithm to get the load B of the arcs of the basic graph G

$$B : U \longrightarrow \mathcal{P}(\mathbb{N}) \quad (\mathbb{N} \text{ - natural numbers}).$$

See the following PASCAL-like program fragment:

```
FOR u ∈ U DO B(u) : = Ø ,
FOR i : = 1 TO n DO
            FOR u ∈ P_i DO B(u) : = B(u)∪ i
```

A transformation of P consists in the following operations:
Random selection of an arc a = [A,E] ∈ P;
Random selection of an i ∈ B(a);
B(a) : = B(a)∖{i} ;
Compute the set Q of all paths from A to E in G∖{a} with a minimal number of arcs;
Random selection of a path P_{AE} ∈ Q;
FOR u ∈ P_{AE} DO B(u) : = B(u) ∪ {i}
This part of the replacement process may produce elementary cycles, which must be eleminated.

2.2. Experiments and computing results for solving the Steiner-Tree-Problem

The time-complexity of the thermodynamically motivated simulation for solving the Steiner-Tree-Problem is $O(n^2)$ and the space complexity O(m+n). The following serieses of examples were computed:

- Examples with interactively generated basic graphs.
  Basic graphs were generated by interactively deleting selected arcs out of a basic raster.

Number of vertices/ number of fixed vertices	initial value	terminal value	CPU (sec)
28/4	33	19 *	3.43
29/4	51	25 *	12.87
50/5	52	33	36.35
91/8	93	49 *	21.04
91/6	87	53	11.60

* optimal solution

- Examples with randomly generated basic graphs including a comparison with a Greedy algorithm /8/.

Graph 1/61 Vertices		
	Thermodyn.Mot. Simul.	Greedy-Algorithm
CPU	171 sec	9 sec
COSTS	425	428
Graph 2/360 Vertices		
CPU	44 sec	1,5 sec
COSTS	5769	5772

- Examples with a practically relevant basic graph. The basic graph corresponds to a cable projeȼt which is exactly described in /7/. It contains about 650 vertices. The arc costs are minimal costs of cable channels. Different fixed vertex sets correspond to sets of customers which may be connected with given distributors in the cable projeȼt.
  For each of these fixed vertex sets a solution of the Steiner-Tree-Problem was computed.

- Examples with "Worst Case" Graphs /5/.
  "Worst Case" graphs are such for which the class of Greedy algorithm for solving the Steiner-Tree-Problem described in /8/ takes its lower quality boundary . In all computed examples the Thermodynamically Motivated Simulation reaches nearly the optimal solution.
  Fig. 2 present "Worst case" graph for a class of Greedy algorithms.

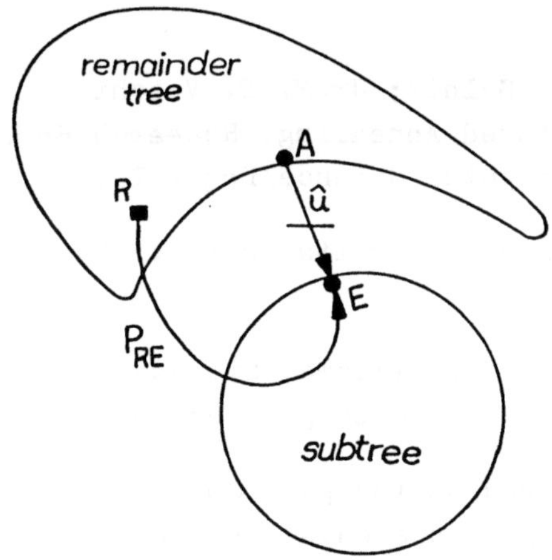

Fig.1: Schematic picture of replacing an arc $\hat{u}$ by a shortest path P_{RE}

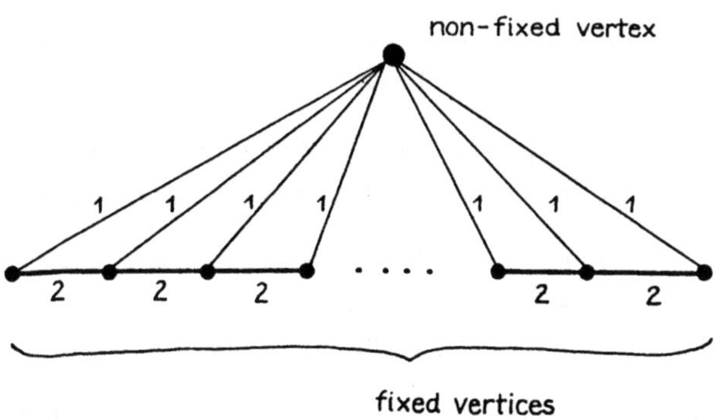

Fig. 2: "Worst case" graph for a class of Greedy algorithms

References:

/1/ Kirkpatrick, S.; C.D. Gelatt; Jr.M. P. Vecchi: Optimization by simulated Annealing; Research Report, RC 9355, 4/2/ 1982 Computer Science/Engm. Techn.

/2/ Rosanow, J.A.: Wahrscheinlichkeitstheorie; Akademie-Verlag GmbH Berlin, 1970

/3/ Fritz, F.-J.; B.Huppert; W.Willems: Stochastische Matrizen; Springer-Verlag Berlin- Heidelberg - New York, 1979

/4/ Schiemangk, Ch.: Remarks on the performance analysis of thermodynamically motivated simulation; ZKI, 1985

/5/ Takahashi,H.; Matsuyama, A.: An Approximate Solution for the Steiner Problem in Graphs, Math. Japonica 24, No. 6 (1980), 573 - 577.

/6/ Schiemangk, Ch.: Thermodynamically motivated simulation for solving the Steiner-Tree-Problem and the optimization of interacting path systems; in: Optimization of connection structures in graphs, ZKI, 1985

/7/ Richter, P.; Taraszow, O.: Random Graph Generation. in: Optimization of connection structures in graphs, ZKI, 1985

/8/ Taraszow, O.; Richter, P.: New Heuristic Algorithms for Solving the Steiner-Tree-Problem in Graphs, in: Optimization of connection structures in graphs, ZKI, 1985

/9/ Tamme, C.: Ein Thermodynamisch motivierter Näherungsalgorithmus für diskrete Optimierungsaufgaben, Diplomarbeit, FSU Jena, 1985

A QUASI-SUBGRADIENT SCHEME FOR CALCULATING SURROGATE CONSTRAINTS

J. Sikorski
Systems Research Institute
Polish Academy of Sciences
Newelska 6, 01-447 Warszawa, Poland

0. Introduction

Calculating an effective bound on the primal optimal value of the candidate problem is of great importance to any branch-and-bound method for solving integer programming problems. In many successful approaches the linear-programming relaxation was applied for this purpose. The Lagrangean dual program as a source of closer bounds has attracted attention since subgradient search methods of solving it were introduced. This led to some effective applications of the Lagrangean duality in solving specially structured problems. Even a more efficient bound is offered by the surrogate dual [1,3÷6,8] because it yields usually a tighter duality gap than the Lagrangean dual. However, it is a rather difficult problem to solve the corresponding mathematical program for it requires maximizing a quasiconcave, often discontinuous, function.

In this paper a certain method for calculating surrogate constraints is discussed theoretically and described in the form of implementable algorithm. The algorithmic scheme of the method is based on the concept of quasi-subgradient [7] which generalizes the subgradient for quasiconcave functions. Hence it is similar to the schemes of subgradient search methods of solving concave maximization problems e.g. [10].

The theoretical framework is presented hereafter for a general formulation of the primal problem so all results are valid in a class of nonconvex mathematical programming problems which includes integer programming problems. Moreover, and this should perhaps be stressed, we use a cone representation of primal constraints and this allows us to discuss uniformly both types of them: equality and inequality constraints. Till now, only the latter type was considered when algorithms for calculating surrogate constraints were constructed [2,9,11].

1. Basic theoretical results

We state the primal problem as

$$\hat{f} = \min \{f(x): x \in X\}, \quad X = \{x \in X_o: -g(x) \in D\}, \qquad \text{(P)}$$

where $f: R^n \to R$, $g: R^n \to R^m$, $X_o \subset R^n$, $D \subset R^m$

We assume that: $X \neq \emptyset$, f is lower semicontinuous and g is continuous on X_o, X_o is compact, D is a convex closed cone with its vertex at $0 \in R^m$. In particular X_o can be a discret set so (P) can represent integer programming problems too. If $D = R^m_+$ (R^m_+ denotes the nonnegative orthant in R^m) then (P) contains inequality constraints but if $D = \{0\}$ then only equality constraints appear in it.

The approach presented here consists in replacing the primal problem by a relaxed one which can be solved more easily. With this purpose we extend the feasible set X by defining a constraint condition weaker than $-g(x) \in D$. Let $D^* = \{w \in R^m: \langle u,w\rangle \geq 0 \text{ for all } u \in D\}$ be a dual cone for D. For any $w \in D^*$ we define the set $S(w) = \{u \in R^m: \langle u,w\rangle \geq 0\}$ that is a closed halfspace for $w \neq 0 \in R^m$ ($S(0) = R^m$). After having defined S(w) we can introduce the following surrogate problem

$$h(w) = \min\{f(x): x \in X(w)\}, \quad X(w) = \{x \in X_o: -g(x) \in S(w)\} \qquad \text{(S)}$$

Thus the cone D has been replaced by S(w) which contains it. The function $h: D^* \to R$ is well defined because all sets X(w) are compact and nonempty. We shall call it the surrogate dual function. An equivalent definition of $X(w) = \{x \in X_o: \langle g(x),w\rangle \leq 0\}$ shows a single inequality called the surrogate constraint. Notice that the surrogate constraint has the form of inequality regardless of the type of primal constraints. This is in accord with Lagrangean duality in which constraints of both types are included in a modified objective function in the same way and by means of multipliers also from D^*. For that reason elements of D^* which parametrize (S) are called the surrogate dual multipliers.

The inclusion $X \subset X(w)$ for all $w \in D^*$ implies that $h(w) \leq \hat{f}$. Thus the surrogate dual problem aimed at finding the best surrogate constraint (if it exists) is stated as

$$\hat{h} = \sup\{h(w): w \in D^*\} \qquad \text{(SD)}$$

All results presented below strictly correspond to the results reported in [5,6] and [2] in the case of inequality constraints ($D=R^m_+$). In most cases they can be proved easily for the present formulation of the primal problem by a straightforward modification of the existing proofs. Hence the proofs are omitted in this section.

Proposition 1.1 $\hat{h} \leq \hat{f}$

If the equality $\hat{h}=\hat{f}$ does not hold the quantity $(\hat{f}-\hat{h})$ is called the surrogate duality gap. The smaller the duality gap, the more valuable is the problem (SD). Even if there is no duality gap $\hat{h}$ may not be attained on D^* by the function h (see [2]).

Proposition 1.2 h is lower semicontinuous and quasiconcave.

The above result states two basic properties of the surrogate dual function. Quasiconcavity of h appears as a mere consequence of the formulation of (S) and is independent of any additional assumptions which can be imposed on the primal problem. Even if (P) is convex i.e. X_o is convex, f is convex on X_o and g is D-convex, h may still remain quasiconcave but not concave.

The properties of h given in Proposition 1.2 imply

Corollary 1.1. For any $q<\hat{h}$ the set $W(q) = \{w \in D^*: h(w) > q\}$ is a nonempty convex cone which is relatively open (i.e. it can be represented as $D^* \cap U$, where U is an open cone).

Proposition 1.3. If the set $F = \{f(x): x \in X_0\}$ has no cluster point then $\hat{h}$ is attained on a relatively open convex cone contained in D^*.

Pure integer-programming problems with finite set F satisfy assumptions of the above proposition so there always exists $w \in D^*$ such that $h(w) = \hat{h}$ in their duals. Thus in that case "sup" in the statement of (SD) can be replaced by "max". Moreover, the equality $h(w) = \hat{h}$ holds in a neighbourhood of any optimal multiplier. This suggests that the surrogate dual can be especially useful for problems of the mentioned class.

2. A scheme of maximization

Let us consider the following mathematical program which can represent several versions of (SD) discussed in the papers adduced in the introduction.

$$\hat{h} = \sup \{h(w): w \in G, \quad p(w) \geq 0\} \tag{MP}$$

Clearly it must be assumed that $h: R^m \to R$ is quasiconcave and lower semicontinuous. We assume additionally that $G \subset R^m$ is convex and closed while $p: R^m \to R^k$ is R^k_+ - concave. Moreover the interior condition must be satisfied for p: there exists $w \in G$ such that $p(w) > 0$.

A recursive scheme which can be useful for the approximative solution of (MP) ought to generate a sequence $\{w^i\} \subset G$ which has the following property: for any $\varepsilon > 0$ there exists i_o such that $h(w^{i_o}) > \hat{h} - \varepsilon$. It is of use to have two search directions at each point of the sequence: one to look for bigger values of h and another one to seek feasible points subject to the condition $p(w) \geq 0$.

In order to construct the first of the two mentioned directions a quasi-subgradient of h at w^i [7] can be used. Transforming to some extent the idea presented in [7] the z-quasiconjugate h^*_z can be defined for h.

Definition 2.1. $h^*_z(y) = z - \sup\{h(u): \langle u,y \rangle \leq z\}$.

The function $h^*_z: R^m \to R$ is quasiconcave for any $z \in R$.

It is known that subgradient is an element satisfying as a equality the Fenchel's inequality which ties together a convex function with its conjugate. A similar statement containing z-quasiconjugate can serve as a definition of the quasi-subgradient.

Definition 2.2. $y^* \in R^m$ is said to be a quasi-subgradient of h at w iff $h(w) + h^*_{\langle y^*,w \rangle}(y^*) = \langle y^*,w \rangle$.

We assume for further presentation of the scheme that it is possible to calculate at least one quasi-subgradient of the function h at each point $w \in G$.

In order to compare a property of quasi-subgradient with an analogous property of subgradient we rewrite the equality in the above definition into an equivalent form using the definition of z-quasiconjugate. This leads to the equality $h(w) = \sup\{h(u): \langle y^*,u\rangle \leq \langle y^*,w\rangle\}$ which can be expressed as an implication:

$$\langle y^*,u\rangle \leq \langle y^*,w\rangle \Rightarrow h(u) \leq h(w) \qquad \text{(for all } u \in R^m) \tag{I}$$

The implication shows that the quasi-subgradient y^* generates a strongly supporting functional for the interior of the upper level set: $\{u: h(u) > h(w)\}$. This statement is not entirely precise but shows the main difference between quasi-subgradient and subgradient. If a subgradient of h exists at w then it generates a supporting functional for the set $\{u: h(u) \geq h(w)\}$. In the case of quasiconcave function a point "w" may not belong to the boundary of the set $\{u: h(u) \geq h(w)\}$ and a subgradient may not exist at this point.

If some point of the sequence generated by a scheme of maximization is infeasible subject to p (i.e. $p(w)<0$) then we need another direction to find the feasible set. The following element is proposed to serve in that case

Definition 2.3. $p^o \in R^m$ is said to be a supporting element of the function p at $w \in R^m$ with respect to R^k_+ iff the following implication holds for all $u \in R^m$:

$$p(u) \geq P_-(p(w)) \Rightarrow \langle p^o,u\rangle \geq \langle p^o,w\rangle$$

The first of above inequalities represents the relation generated by R^k_+. P_- denotes the orthogonal projection on $-R^k_+$. We define p^o to obtain a direction which can lead to the set $\{u: p(u) \geq 0\}$ if we start from a point $w \in R^m$ lying outside of this set. Therefore $p^o = 0 \in R^m$ if $p(w) \in \text{int } R^k_+$ since in this case all points in a neighbourhood of w are feasible.

Usually a subgradient of the function $r(u) = \min\{p_j(u): j=1,\ldots,k\}$ ($r: R^m \to R$) is used for seeking points from the feasible set. The function r is concave and any subgradient of it at w generates a supporting functional of the set $\{u: r(u) \geq r(w)\}$. If $w \in R^m$ is such that $p(w) \notin \text{int } R^k_+$ then $\{u: p(u) \geq P_-(p(w))\} \subset \{u: r(u) \geq r(w)\}$ and any subgradient of r at w is a supporting element of p with respect to R^k_+ but there can exist also such elements p^o which are not subgradients of r.

Having defined the two search directions we can formulate the recursive scheme designed to approximate the optimal value of (MP). We follow the idea from [10].

$$w^{i+1} = P_G(w^i + t_i d^i / \|d^i\|), \quad w^1 \in G, \tag{RS}$$

where $d^i = \begin{cases} \text{a quasi-subgradient of } h \text{ at } w^i, \text{ if } p(w^i) \geq 0 \\ \text{a supporting element of } p \text{ at } w^i \text{ with respect to } R^k_+, \text{ elsewhere.} \end{cases}$

$\{t_i\}$ is a step size sequence which satisfies: $0<t_i \to 0$ for $i\to\infty$ and $\Sigma_{i=1}^{\infty} t_i = \infty$. P_G is the orthogonal projection on G, $\|\cdot\|$ denotes the Euclidean norm in R^m.

Proposition 2.1. If all assumptions imposed on (MP) are satisfied then any sequence generated by the scheme (RS) contains a subsequence $\{w^{i_s}\}$ such that $\lim_{s\to\infty} h(w^{i_s}) = \hat{h}$

Proof: (RS) is defined for $w^i \in G$ such that $h(w^i) \neq \hat{h}$. Choose any $q<\hat{h}$. There exists $\tilde{w} \in G \cap \text{int}\, W_p(q)$, where $W_p(q) = \{w: p(w) \geq 0, \ h(w) \geq q\}$. Set $\delta>0$ such that the closed ball $B(\tilde{w};\delta) \subset W_p(q)$.

Let us suppose that $w^i \notin W_p(q)$ for all i. If $p(w^i) \geq 0$ then $W_p(q) \subset \{w: h(w)>h(w^i)\}$. In this case d^i is a quasi-subgradient of h at w^i and (I) implies $<d^i,w>><d^i,w^i>$. On the other hand, if $p(w^i) \notin R^k_+$ then $W_p(q) \subset \{w: p(w) \geq P_-(p(w^i))\}$. The inequality $<d^i,w>\geq<d^i,w^i>$ holds for all $w \in W_p(q)$ since d^i is a supporting element of p at w^i. This validates the inequality $<d^i,w>\geq<d^i,w^i>$ for all $w \in W_p(q)$ which leads to the contradiction with divergence of Σt_i (see [10]). Hence there exists $w^i \in W_p(q)$ for any $q<\hat{h}$. If we choose the sequence $q_s \to \hat{h}$ for $s\to\infty$ then the corresponding sequence $w^{i_s} \in W_p(q_s)$ is such that $h(w^{i_s}) \to \hat{h}$ for $s\to\infty$. □

The above theorem states the main result concerning the convergence of algorithms designed to solve (MP) and based on the concept of quasi-subgradient.

3. Algorithms for calculating surrogate constraints

The surrogate dual problem (SD) can be directly represented by (MP) if we substitute D^* for G and if we neglect p. We need a quasi-subgradient of h at each point $w \in D^*$ to apply the scheme (RS).

Proposition 3.1. If $\bar{x}$ is an optimal solution for (S) then $g(\bar{x})$ is a quasi-subgradient of h at $w \in D^*$

Proof: $h(w) = f(\bar{x})$ and $<g(x),w>\leq 0$ from the statement of (S). Consider $u \in D^*$ such that $<g(\bar{x})u>\leq<g(\bar{x}),w>$. This means that $\bar{x} \in X(u)$. Hence $h(u) = \min\{f(x): x \in X(u)\} \leq f(\bar{x}) = h(w)$. Thus, the implication (I) is shown for $g(\bar{x})$. □

The above result guarantees the existence of a quasi-subgradient at each point in D^* and moreover shows how to construct it by solving the surrogate problem. The scheme (RS) can be used for the surrogate dual problem but it leads to choosing the sequence $\{w^i\}$ from an unbounded set D^*. This is the main disadvantage of the above approach from the operative point of view. However a simple property of the function h offers the opportunity of restricting the feasible set in (SD) to a compact one: $X(\alpha\cdot w) = X(w)$ for $\alpha>0$ implies that $h(\alpha\cdot w) = h(w)$. This makes an arbitrary normalisation of multipliers possible. Additionally, $X(0) = X_o$ implies that $h(0)\leqslant h(w)$ for all $w\in D^*$ and zero may be excluded from the feasible set.

For example, consider the case where only inequality constraints are present in the primal problem (P). Then $D = R^m_+$ and $D^* = R^m_+$. Without restricting the range of h we can assume that the maximization of h proceeds on the set $K_+ = \{w\in R^m_+ : \|w\| \leqslant 1\}$. In order to apply the scheme (RS) to this program we represent it in terms of (MP). We substitute the unit ball $B(0;1)$ for G and we set $p(w)=w$ $(k=m)$.

Suppose now that in some iteration of (RS) a point w^i has been generated and it is feasible subject to p (i.e. $w^i\in R^m_+$). According to the scheme the step t_i is made along the direction $\bar{d}^i = d^i/\|d^i\|$, where d^i is a quasi-subgradient of h at w^i. If the point $\bar{w}^i = w^i+t_i\bar{d}^i$ is infeasible then the set of indices $J^i = \{j: w^i_j<0\}$ is nonempty and in the next iteration of the scheme we use a supporting element of the identity function. It can be easily checked that $\tilde{d}^i = \sum_{j\in J^i}\bar{w}^i_j e_j$, where $e_j=[0\ldots-1\ldots0]^T$ (the j-th coordinate is nonzero), is a supporting element of p with respect to R^m_+ at $\bar{w}^i$. Thus we can make the next iteration. If we set the step size $\tilde{t}_i = 2\sqrt{\sum_{j\in J^i}(\bar{w}^i_j)^2}$ then the point $\tilde{w}^i = \bar{w}^i + \tilde{t}_i\tilde{d}^i/\|\tilde{d}^i\|$ is feasible because $\tilde{w}^i_j = |\bar{w}^i_j|\geqslant 0$ for $j=1,\ldots,m$. Moreover $\tilde{t}_i\leqslant 2t_i\sqrt{m}$ so it tends to zero if $i\to\infty$. Using the formula proposed in [2] we can choose a quasi-subgradient d^i which is tangent to the unit sphere at w^i if only $\|w^i\|=1$. Hence $\|\tilde{w}^i\|\geqslant 1$. Thus the projection on $G = B(0;1)$ can be calculated by normalisation of $\tilde{w}^i$. Then the next point $w^{i+1} = \tilde{w}^i/\|\tilde{w}^i\|$ belongs to K_+ and $\|w^{i+1}\| = 1$. If we choose the starting point $w^1\in K_+$ such that $\|w^1\| = 1$ then all points generated by the recursive scheme (RS) remain in K_+ and for all $\|w^i\| = 1$.

The above construction can be formulated as animplementable algorithm which strictly corresponds to the algorithm presented in [2]. Some elements of this algorithm can be modified. Any other quasi-subgradient of h at w^i can be used and the step size $\tilde{t}_i$ can be halved that gives $\tilde{w}^i_j = \max(0,\bar{w}^i_j)$ for $j=1,\ldots,m$. The algorithm obtained as a result of such modifications was presented in [11] and it proved to be more efficient numerically. Nevertheless both algorithms are based on the same scheme (RS) and Proposition 2.1 states that the value $\hat{h}$ can be approximated by them with an arbitrary accuracy.

The presented scheme can be also applied for calculating surrogate constraints in the case $D = \{0\}$, where only equality constraints are present in the primal problem (P). The process of maximization can proceed on the set $K = \{w \in R^m: \|w\| \leq 1\}$ and the same property of the sequence of approximations will hold.

References

[1] Bricker P.L.: Bounding a class of nonconvex linearly-constrained resource allocation problems via the surrogate dual, Mathematical Programming 18(1980), 68-83.

[2] Dyer M.E.: Calculating surrogate constraints, Mathematical Programming 19(1980), 255-278.

[3] Fisher M.L., Lageweg B.J., Lenstra J.K., Rinooy kan A.H.G.: Surrogate duality relaxation for job shop scheduling. Discrete Applied Mathematics 5(1983) 1, 65-75.

[4] Glover F.: A multiphase-dual algorithm for the zero-one integer programming problem, Operations Research 13(1965), 879-919.

[5] Glover F.: Surrogate constraint duality in mathematical programming, Operations Research 23(1975), 434-451.

[6] Greenberg H.J., Pierskalla W.P.: Surrogate mathematical programs, Operations Research 18(1970), 924-939.

[7] Greenberg H.J., Pierskalla W.P.: Quasi-conjugate functions and surrogate duality, Cahiers du Centre d'Etudes de Recherche Operationnelle 15(1973), 437-448.

[8] Karwan M.H., Rardin R.L.: Surrogate duality in a branch-and-bound procedure, Naval Research Logistics Quarterly 28(1981)1, 93-101.

[9] Karwan M.H., Rardin R.L.: Surrogate dual multiplier search procedures in integer programming, Operations Research 32(1983), 52-69.

[10] Polyak B.T.: A general method of solving extremal problems, Doklady Akademii Nauk SSSR 174(1967), 33-36.

[11] Sikorski J., Mizukami K.: Three algorithms for calculating surrogate constraints in integer programming problems, Control and Cybernetics 13(1984), 375-397.

Linear programming for electric power distribution system expansion planning: A dynamic expansion model for loop structured networks

by Alfons H. SILLABER, Austria *)

1. Introduction:

Loop structured or ring networks play an essential role for electric power distribution in urban supply areas by reason of their relative low investment costs and the clear and simple operation conditions. The enormous capital requirements of energy supply systems enforce sophisticated planning methods for the expansion process [1].

The optimization problem under consideration is originally a discrete dynamic stochastic problem resulting from discrete investment decisions over a long-term planning period and from uncertainty of load and cost forecasts. Because of the enormous size of practical problems a deterministic model is formulated and in addition a special solution strategy must be developed [2,3].

First the problem is described in detail considering practical requirements of power utilities. Then a mathematical solution method based on linear programming is selected which guarantees a relative low solution effort and a simple application in practice. It follows the presentation of the optimization model, that allows also the solution of symmetric travelling-salesman or transportation routing problems. Some fundamental relations clarify the models size and thus the limits for the practical use. A real-world example demonstrates the abilities of the presented optimization method for power distribution system expansion planning.

2. The problem:

In urban supply areas electric power is distributed by cables from a few transformer stations to a great number of local transformer substations. Electric power distribution networks often have a special topology: Each substation is connected by two branches with the other nodes. One example with such a structure is the ring network.

---

*) A. H. Sillaber, A-6020 INNSBRUCK, Gumppstraße 7; Austria

The investment costs for the erection of a cable line consist of costs for the cable trench, for the cable itself, for the layout of the cable, for refilling the trench and for the restoration of the street surface. Since it is much cheaper to lay two or more cables in the same trench than each cable in its own, the resulting cost function is very similar to that of a fixed-cost transportation problem. Thus also the continuously expanded cost function is a nonconvex one.

The transportation capacity of a cable in an urban supply area with high load density is generally determined by its thermal rating and not by the highest allowable voltage drop. In the normal operation state, all cables are at most half loaded. Only after necessary switching actions in certain outage states some cables are fully loaded. The ring structure guarantees, that after single outages of cables each substation can be resupplied after a short switching time.

In practice most or all substations of an utility are equipped with transformers of the same ratings, so they all have approximately the same load. The future load development and the simultaneousness of peak loads are not known exactly in advance. Therefore it is not a great disadvantage to formulate the transport capacity constraints based on substation numbers per cable ring and not on exact substation loads.

For the distribution system expansion problem, a dynamic investment calculation allows the consideration of the load growth in existing substations and the erection of new substations. Such new substations are often supplied by cutting off an existing cable and connecting the open ends with the new additional node. Thus the whole street network in the planning area must be modelled. By reason of the fixed cost term it is sometimes more economic to lay a cable earlier as needed but together with others into the same trench. Such effects of course require a dynamic investment model.

3. The solution strategy:

After time discretization the problem could be formulated as a large-scale mixed-integer optimization model and solved by one of the standard tree-search strategies implemented in the branch-and-bound algorithms of a commercial linear programming package. With this method a tremendous solution effort is expected to be necessary and the best solution found is probably far away from the continuous optimum, so you get no strong criterion for its quality.

Instead the following idea shall be realized: As far as possible only continuous variables are incorporated into the model. By this reason the model size is increased considerably but this effect must be limited by a sophisticated modelling technique. A continuous model for a symmetric travelling-salesman problem requires special constraints to prevent degeneracy and non-integrity of the optimal solution. Well known is DANTZIGSs formulation of subtour elimination constraints which despite of later improvements causes an exponential model growth. In the model to be presented a dual approach shall be used: Lower limits for flows within selected groups of nodes are set up.

The transport capacity of the cable ring must be modelled by separate flows from each load node to the infeeding points and adequate coupling constraints with the investment decisions. This requires a great number of additional rows but otherwise the optimal continuous solution would be a non-integer one.

To limit the model size reasonably the transport network with ring structure and the citizen street network are modelled in separate parts joined by coupling variables. The first part, the so called transport model, is based on the shortest paths between load nodes and consists of constraints which guarantee the ring structure observing the transport capacity. The coupling variables determine the sequence of load nodes as a base for the second model part, the so called street model. This part models the investment cost function as described before and contains the potential ways for cables, the street crossings form the set of nodes in this network.

This combined model is very large for problems of practical size and only convex cost functions can be included. Thus a two-stage strategy has been developed: First for each time step separately the transport model is solved under the assumption of only one cable per trench. In the second stage the street model including all time steps is solved with the predetermined sequences of load nodes. The non-convex cost function is approximated by a linear one and in most practical applications no or only a few variables must be declared integer for a reiterated solution attempt by a branch-and-bound method.

Another important provision to restrict the model size is the exclusion of long cable lines between distant nodes from the model.

4. The optimization model:

4.1 Part A: The transport model:

The following equations define the ring structure of the transport model. The transport capacity constraints of model part A are not presented here in detail. All variables are assumed to be continuous and nonnegative in the first step.

In each source and load node exactly one cable must begin and one must end:

$$\forall\ t \in T\ ,\ i \in I(t):$$

$$\sum_{\substack{j \in I(t) \\ j \neq i}} U(t,i,j) = 1; \quad \sum_{\substack{j \in I(t) \\ j \neq i}} U(t,j,i) = 1 \qquad \text{(A1)}$$

The above variables U model the investment decisions for a cable line between nodes i and j at the begin of time step t.

The set of nodes I is subdivided into p selected subsets IP and within each subset the variables W define likewise a ring structure. Analogous to A1 we state:

$$\forall\ t \in T;\ p \in P(t);\ i \in IP(t,p):$$

$$\bigcup_{p} IP(t,p) = I(t)$$

$$\sum_{\substack{j \in IP(t,p) \\ j \neq i}} W(t,p,i,j) = 1; \quad \sum_{\substack{j \in IP(t,p) \\ j \neq i}} W(t,p,j,i) = 1 \qquad \text{(A2)}$$

For each subset IP with four or more nodes the following subtour elimination constraints are formulated:

$$\forall\ t \in T;\ p \in P(t);\ l \in L(t,p):$$

$$\sum_{i \in IPL(t,p,l)} \ \sum_{\substack{j \in IPL(t,p,l) \\ j \neq i}} W(t,p,i,j) <= ZIPL(t,p,l) - 1 \qquad \text{(A3)}$$

ZIPL is the number of nodes in the node subset IPL.

The variables V define flows between nodes in the same subset IP with the source node k and the demand node as defined by the variables W. Thus the node incidence relations for this flows are formulated as follows:

$$\forall\ t \in T;\ k \in I(t);\ i \in I(t):$$

$$\underset{\substack{j \in I(t)\\ j \neq i}}{\text{SUM}} \left(V(t,k,i,j) - V(t,k,j,i)\right) = \begin{cases} \text{a)} = 1; & i = k \\ \text{b)} = -W(t,p,k,i); & i \neq k;\ i,k \in IP(t,p) \\ \text{c)} = 0; & i \notin IP(t,p) \end{cases} \qquad \text{(A4)}$$

The flows V must be coupled with the investment decisions U:

$$\forall\ t \in T;\ p \in P(t);\ i,j \in I(t);\ j > i:$$

$$\underset{k \in IP(t,p)}{\text{SUM}} \left(V(t,k,i,j) + V(t,k,j,i)\right) <= U(t,i,j) + U(t,j,i) \qquad \text{(A5)}$$

The objective function of model part A includes the costs of the erection of the shortest possible cable line between two load nodes with at most one cable per trench:

$$\underset{t \in T}{\text{SUM}}\ \underset{i \in I(t)}{\text{SUM}}\ \underset{\substack{j \in I(t)\\ j > i}}{\text{SUM}} \left((KK(t,i,j) + KG(t,j,i)) * (U(t,i,j) + U(t,j,i))\right) \qquad \text{(A6)}$$

KK are the capitalized costs of a cable, KG the capitalized costs of a cable trench between the corresponding load nodes. In general the model part A is used as a static model.

4.2 Part B: The street model:

Each cable line is determined by the solutions of model part A for each time step as a flow from the source node i to the specified demand node j. The following node incidence relations B1 define this flow in the street network. Because of the dynamic nature of the model, each branch flow is modelled by two variables G and H, the first represents the occupation of an existing, earlier erected cable, the second the necessity of a new cable.

$$\forall\ t \in T;\ i \in I(t);\ m \in K(t,i):$$

$$\sum_{n \varepsilon K(t,i,m)} (G(t,i,m,n)+H(t,i,m,n)-G(t,i,n,m)-H(t,i,n,m)) = \begin{cases} \text{a) } = 1;\ m \hat{=} i \\ \text{b) } = -U(t,i,j);\ m \hat{=} j \text{ in the combined model} \\ \text{c) } = -1;\ m \hat{=} j \text{ in the decoupled model} \\ \text{d) } = 0 \text{ elsewhere} \end{cases} \qquad (B1)$$

K(t,i) is the subset of nodes in the street network surrounding substation i in time step t. K(t,i,m) is a subset of K(t,i) with nodes m adjacent to node i.

The above defined flows G and H must be coupled with the variables E, which model the cable layout in the street network:

$$\forall\ t \ \varepsilon\ T,\ m \ \varepsilon\ K(t);\ n \ \varepsilon\ K(t,m);\ n > m:$$

$$\sum_{i \varepsilon I(t,m)} (G(t,i,m,n)+G(t,i,n,m)) - \sum_{\substack{t' \varepsilon T \\ t'<t}} E(t',m,n) <= A(m,n) \qquad (B2)$$

$$\sum_{i \varepsilon I(t,m)} (H(t,i,m,n)+H(t,i,m,n))-E(t,m,n) = 0$$

K(t) is the set of all nodes in the street network, K(t,m) the subset of nodes adjacent to node m.

The necessity of a new cable trench is modelled by the variables F, which are coupled with the variables H by the following restraints:

$$\forall\ t \ \varepsilon\ T;\ i \ \varepsilon\ I(t);\ m \ \varepsilon\ K(t,i);\ n \ \varepsilon\ K(t,i,m);\ n > m:$$

$$H(t,i,m,n)+H(t,i,n,m)-F(t,m,n) <= 0 \qquad (B3)$$

Completed is the model part B by the following objective function:

$$\sum_{t \varepsilon T} \sum_{m \varepsilon K(t)} \sum_{\substack{n \varepsilon K(t,m) \\ n>m}} (KK(t,m,n)*E(t,m,n)+ +KG(t,m,n)*F(t,m,n)) \qquad (B4)$$

KK and KG are capitalized costs for cable layout and cable trench respectively as defined before.

5. Model size and computational requirements:

The complete model part A without further simplifications for one time step and 50 substations would have about 12000 rows and 130000 continuous variables. The a priori exclusion of potential cable lines between distant nodes reduces the model size by a factor of 5 to 10 and the model growth is only quadratic with the substation number.

The complete model part B for 3 time steps and 50 substations would consist of about 63000 rows and 92000 columns. The above mentioned reduction method yields a diminution factor of about 10 to 15 in practice.

The resulting continuous linear models with some thousands of rows and over ten thousand variables can be solved to optimality by standard methods without further difficulties. A commercial LP - code such as MPSX/370 under VM/SP on an IBM 370/158 computer needs only up to about 10 minutes of CPU - time for problems of this size. The provision of a feasible starting solution further reduces the CPU - time up to about 50 %.

If the optimal continuous solution is not all-integer, in model part A a few groups of variables must be declared as special ordered sets; in model part B some variables must be declared integer for a reiterated solution attempt by a branch-and-bound method. If the number of such integer elements is lower than about 10, a reasonable CPU - time is still expectable.

6. Practical example:

The system expansion planning problem to be presented is based on a geografical cut of an urban supply area with a medium load density but a relative fast load growth. The planning horizon covers 6 years and is subdivided into 3 time steps. The problem characteristics are summarized in table 1.

The problem shall be solved by the application of the transport model to each time step separately. After determination of the load node sequences, the street model shall be applied to all 3 time steps simultaneously.

The 4 continuous linear models and their solution processes with MPSX/370 under VM/SP on an IBM 370/158 are characterized in table 2.

Number of	Existing system	T = 1	T = 2	T = 3
Cable rings	1	2	2	3
Central stations	1	1	1	2
Transformer substations	10	12	14	15
Street crossings	41	41	42	42
Street branches	57	57	63	64
Max. stations per ring	10	6	7	5

TABLE 1: Problem characteristics

	Model part A			Model part B
	T = 1	T = 2	T = 3	T = 1,2,3
Number of rows	744	806	1159	1615
Number of columns	1616	1833	2591	8953
Element density	0,48%	0,47%	0,29%	0,21%
CPU - time (min)	1,71	1,82	5,15	4,82
Number of non-integers	0	0	0	10 (2 groups)

TABLE 2: Model and solution characteristcs

All solutions of model part A were full integer already in the first run. The non-integers in the continuous solution of model part B, caused by the approximation of the non-convex cost function could be removed quickly by defining two integer variables and subsequent re-iteration. But this procedure is superfluous, since these degenerations are removable simply by visual inspection of the solution.

The solution of the above problem is presented in FIGURE 1. In comparison to the best solutions found by heuristic methods a reduction of the capitalized costs of about 3 % or 9000 US $ has been reached.

7. Results and conclusion:

A main outcome is, that the selected two stage solution strategy with model decomposition into two parts in general does not cause remarkable deteriorations of the calculated results. This allows a considerable reduction of the model size without need for a mathematical decomposition or partitioning procedure. The two stage optimization strategy first minimizes the sum of cable lengths over all time steps and in the second stage it minimizes the capitalized costs for cables and cable trenches based on the predetermined ring structure.

The transport model is applied sequentially to one time step after another, taking into account the result of the preceding time step. Since now it yielded very good results as long as capacity constraints are not too restrictive. All examples with relaxed capacity constraints had a pure integer solution in the first run. After inserting strong capacity constraints the optimal continuous solution often contains only a few definite investment decisions, but many of them are close to 1. In many cases, these difficulties arose in the surroundings of a central transformer station, the source of several cable rings. Thus it seems profitable to declare the set of investment decisions for cables starting in such a central station as special ordered sets. Another very successful method is a reiteration of the continous problem after fixing the near-to-one variables to one.

The street model shows good results especially for problems with most investment requirements in the first time step as expected for many practical applications. Occasionally a local degeneration appears in the solutions caused by the modelling of a non-convex objective function. These are removable in a very simple way either by visual inspection or by declaring a single variable as integer and subsequent reiteration.

Today the model generator is a computer program with many switches and options for model variations. It is suited both for further model development and for practical use by experienced engineers. A wide spectrum of model elements has been tested and in the present version problems up to about 50 substations and 3 time steps are soluble with reasonable effort. Some problems are treated efficiently, others require some test runs and experiments for their solution. In the next future some model improvements will be realized and further problem oriented reiteration strategies shall be developed.

8. Acknowledgement:

I am very grateful to Professor Harhammer of IBM Austria, who enabled me to perform numerical tests with practical examples and to Mr.

Haschka of IBM Austria for his valuable aid in the practical use of standard LP - software.

9. References:

[1]... SULLIVAN R. L.
Power System Planning
Mc Graw Hill New York 1977

[2]... STEINBAUER E. et al
Menu driven program package "Operation planning of energy systems"
Proceedings of the 8th PSCC Helsinki 1984

[3]... WILLIAMS H. P.
Model building in mathematical programming
Wiley New York 1978

FIGURE 1

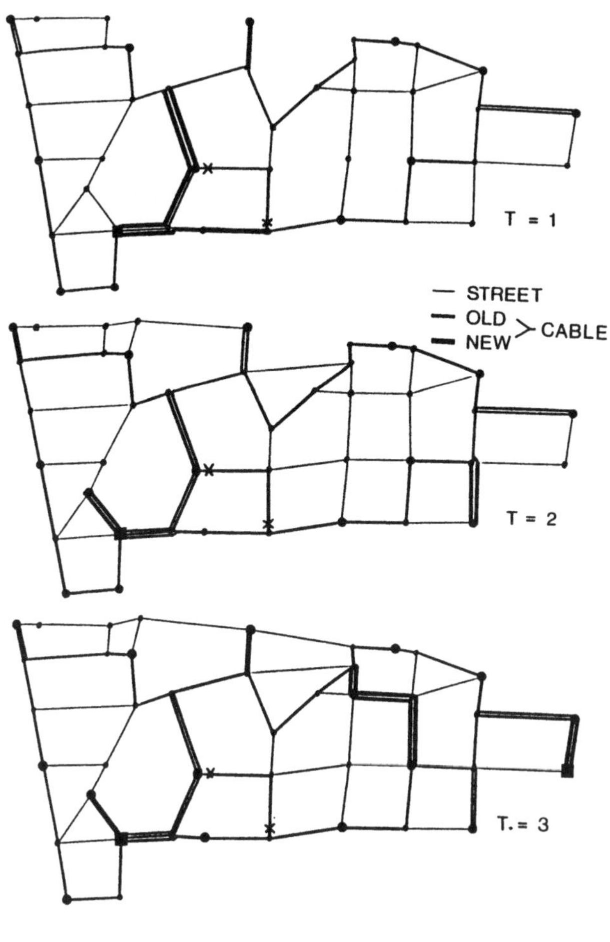

ON EXISTENCE OF ASSIGNMENTS IN ZERO - ONE MATRICES

L. Słomiński
The Systems Research Institute,
Polish Academy of Sciences,
WARSAW, POLAND

1. INTRODUCTION

We prove sufficient conditions for the existence of an assignment in a square, zero-one matrix A_{nxn}. The question of an assignment existence arises for example, in the case of solving the bottleneck assignment problem by the threshold method. The zero-one matrix A_{nxn} represents then the data of some auxiliary problem obtained by the threshold transformation of the initial cost matrix [4]. The answer to the existence question can be easily obtained by solving the maximum cardinality matching problem or the maximum flow problem in the related bipartite graphs, that correspond to the matrix A_{nxn}. The known methods require $0(n^{5/2})$ operations to solve these problems.

However, the sufficient conditions we are going to prove are cheaper, and the number of operations required is proportional to the number of 1's in the matrix A_{nxn}. The computational complexity factor becomes crucial in the case of the great dimensions, say for $n \geqslant 500$. The sufficient conditions we have proposed, can be used as fast feasibility tests in approximate, threshold-type procedures for solving the bottleneck assignment problem.

2. THE PROBLEM

Let us consider a square matrix A_{nxn}, with entries $a_{ij}=0$ or 1 for all i and j, and let us assume that: $\sum_{i=1}^{n} a_{ij} \geqslant 1$, for all j; $\sum_{j=1}^{n} a_{ij} \geqslant 1$, for all i; we also denote by $p=\sum\sum_{i,j=1}^{n} a_{ij}$ - the number of 1's in the matrix and by: $a_i = \sum_{j=1}^{n} a_{ij}$, $b_j = \sum_{i=1}^{n} a_{ij}$ - the number of 1's in the i-th row and in the j-th column respectively.

The assignment (n-assignment) and the partial assignment (k-assignment, $1 \leqslant k \leqslant n-1$) are defined in the usual way.

The problem we would like to solve is in proving sufficient conditions, of complexity $0(p)$, for the existence or nonexistence of an assignment in the matrix A_{nxn}. The proofs of our results are based on the well known Hall's theorem, which gives the necessary and sufficient conditions for the existence of a system of distinct representatives (an SDR - abbreviated) of a family $S = \{S_j | j=1,\ldots,n\}$ of subsets of elements from a set $E = \{e_1,\ldots,e_m\}$.

A list R of distinct elements of E, $R = \{e_{i_1},\ldots,e_{i_n}\}$ is a system of distinct representatives for S if $e_{i_j} \in S_j$, $i=1,\ldots,n$, and e_{i_j} is said to represent S_j.

Necessary and sufficient conditions for the existence of an SDR are contained in the following theorem due to P. Hall, [2].

THEOREM 1. An SDR exists for S if and only if every union of k sets of S contains at least k distinct elements, $k=1,\ldots,n$.

There is close relation between the notion of SDR for family S and the notion of assignment in a matrix A_{nxn}.

To make this relation more apparent we introduce the following correspondences. The set E now consists of the row indices, $i=1,\ldots,n$, the column index j corresponds to the subset S_j index, whereas subset S_j by itself, corresponds to the j-th column of matrix A_{nxn}. We also assume that: $(a_{ij}=1) \Longleftrightarrow (i \in S_j)$, for all i and j.

The following theorem gives necessary and sufficient conditions for the existance of an assignment (an SDR) in a zero-one matrix A_{nxn}.

THEOREM 2. A zero-one matrix A_{nxn} possesses an assignment if and only if there is no any zero submatrix A_{rxs}, such that $r+s \geqslant n+1$, for: $2 \leqslant r,s \leqslant n-1$.

Proof of the THEOREM 2, typical for reasonings we have used in deriving our results, is given in the Appendix.

3. SUFFICIENT CONDITIONS FOR THE EXISTENCE OF AN ASSIGNMENT IN AN A_{nxn} MATRIX.

Let us recall two sufficient conditions for the existence of an assignment, which have already been given in the literature. We denote: $M = \max\{a_1,\ldots,a_n,b_1,\ldots,b_n\}$.

LEMMA 1.([3]). If $(n-1)\,M < p$, then the matrix A_{nxn} has an assignment.

LEMMA 2.([1]). If $a_i=b_j=\text{const.}$, for all i and j, then the matrix A_{nxn} has an assignment.

Now we shall give the proof for two further sufficient conditions that admit an assignment in a matrix A_{nxn}.

LEMMA 3. If the following holds: $b_{j_1} \neq b_{j_2}$, $(a_{i_1} \neq a_{i_2})$ for all $j_1 \neq j_2$, (all $i_1 \neq i_2$), then a matrix A_{nxn} has an assignment.

PROOF. The set union of any k columns (rows), $2 \leqslant k \leqslant n$, contains at least k 1's, thus Hall's condition holds, (see Appendix).

LEMMA 4. If the following inequalities are fulfilled:

$$a_i \geqslant \lceil n/2 \rceil\ ,\ \text{for all } i; \qquad b_j \geqslant \lceil n/2 \rceil,\ \text{for all } j,$$

where: $\lceil x \rceil$ means the greatest integer less or equal to the x, then a matrix A_{nxn} has an assignment.

PROOF. We shall show that under above inequalities Hall's condition must hold. Due to the inequalities, the maximal number of 0's in each line (row or column) of the matrix is not greater than:

$$(n-k) = \begin{cases} \frac{n}{2}\ , & \text{for even } n, \\ \frac{n-1}{2}, & \text{for odd } n. \end{cases}$$

It is evident that the maximal rank of any square zero submatrix A_{rxr} is not greater than n/2 or (n-1)/2 , and:

$$2 \cdot r_{max} = \begin{cases} n, & \text{for even } n, \\ n-1, & \text{for odd } n, \end{cases}$$

thus Hall's condition can not be violated.

The similar arguments can be used for any rectangular-zero submatrix A_{rxs}. Of course: $\max\{s\} = \max\{r\} \leqslant (n-k)$. Let us assume without loss of generality that $r < s$. Hence:

$$\max\{s\} \leqslant (n - \lceil \tfrac{n}{2} \rceil) = \begin{cases} \frac{n}{2}, & \text{for even } n, \\ \frac{n-1}{2}, & \text{for odd } n. \end{cases}$$

The integrity of r and s ensures that:

$$r_{max} = s_{max} - 1 = \begin{cases} \frac{n-2}{2}, & \text{for even } n, \\ \frac{n-3}{2}, & \text{for odd } n. \end{cases}$$

Thus:

$$\max\{r+s\} = \begin{cases} n-1, & \text{for even } n, \\ n-2, & \text{for odd } n, \end{cases}$$

and Hall's condition can not be violated. In the same way the case where $r > s$ can be considered.

We would like to close the present paragraph with a result which addresses the question of nonexistence of an assignment rather, than the question of the existence. For this purpose let us consider the simplest greedy procedure for constructing a partial assignment (possibly an assignment).

GREEDY PROCEDURE

1. Start with all the rows and the columns being admissible in an A_{nxn}.

2. The first positive entry from the first nonzero row is placed into the partial assignment. The row and the column to which used entry has belonged become unadmissible. Repeat the step.

3. Stop when there are no more admissible rows and columns in the A_{nxn}.

The following result is true:

LEMMA 5. If the GREEDY PROCEDURE has stopped with an k-assignment, and $k \leq \lceil n/2 \rceil - 1$, then the matrix A_{nxn} has not an assignment.

PROOF. The matrix A_{nxn} contains a zero submatrix $A_{(n-k)x(n-k)}$. We may assume it to be composed of rows and columns numbered from (k+1) till n. As $k \leq \lceil n/2 \rceil - 1$, so:

$$2(n-k) \geq \begin{cases} n+2, & \text{for even } n, \\ n+1, & \text{for odd } n. \end{cases}$$

And it is evident that Hall's condition is violated.

Whenever the GREEDY PROCEDURE stops and $k \geq \lceil n/2 \rceil$, the question of the existence of an assignment in an A_{nxn} remains opened.

4. THE COMPUTATIONAL COMPLEXITY OF THE SUFFICIENT CONDITIONS

We shall prowe the following:

LEMMA 6. The computational complexities of the sufficient conditions we have considered are as follows:

- Lemma 1 - $O(n)$,
- Lemmas 2÷5 - $O(p)$.

PROOF. The most costly operation for checking the condition of Lemma 1 is that of finding the maximum of 2n integers. The conditions of Lemmas 2÷4 require counting the numbers of 1's in the rows and columns and it takes $O(p)$ calculations.

The GREEDY PROCEDURE is of the complexity $O(p)$. To show this we may assume that the matrix A_{nxn} is stored in a list structure.

Picking one entry costs single operation, whereas readdressing after erasing row and column costs $2(p_i+p_j)$ operations, where p_i is the number of 1's in the i-th row, and p_j - the number of 1's in the j-th column. Each row and column are used at most once.

APPENDIX

We shall prove the THEOREM 2 in slightly modified form, which is equivalent to the original one, and is more convinient for us.

THEOREM. There is no any assignment in a zero-one matrix A_{nxn}, if and only if it contains any zero submatrix A_{rxs}, such that $(r+s)\geqslant(n+1)$, for: $2\leqslant r,s\leqslant n-1$.

PROOF. Necessity. Let us assume that there is no any assignment in a matrix A_{nxn}. From Hall's Theorem we conclude that there must be a subset of at least k columns, such that they contain all their 1's in no more than (k-1) rows. Thus, matrix A_{nxn} contains a zero submatrix A_{rxs}, where: $r\geqslant n-(k+1)$, $s\geqslant k$, and $(r+s)\geqslant(n+1)$.

2. Sufficiency. Let us assume that a zero submatrix A_{rxs}, $(r+s)\geqslant(n+1)$, exists. We have to consider two cases: a) $r=s$, b) $r\neq s$.

Case a. If $r=s$, then $r\geqslant(n+1)/2$, and from the integrity assumption of r and s, the following inequality results;

$$r\geqslant\left\lceil\frac{n+1}{2}\right\rceil = \begin{cases}\frac{n+2}{2}, & \text{for even } n,\\ \frac{n+1}{2}, & \text{for odd } n.\end{cases}$$

Denoting $r_{min}=\left\lceil\frac{n+1}{2}\right\rceil$, and reordering the columns and the rows of A_{nxn} to set the A_{rxr} in the "east-south" corner we easily observe that:

$$\left|\bigcup_{j=n-r_{min}+1} S_j\right| \leqslant r_{min}-1, \quad \text{(see: Fig. 1a, 1b)},$$

and it is clear that Hall's condition is violated.

Case b. We have two subcases: $r>s$ and $r<s$, and it is enough to consider only one of them (see: Fig. 2a, 2b, and Fig. 3a, 3b). In order to avoid a degenerate cases we suppose that $n\geqslant 4$. We would like to show that an assignment can not exist for $r_{min}+s_{min}=n+1$, hence it can not exist for $(r+s)>(n+1)$ and $r>r_{min}$, as well. The following equality holds:

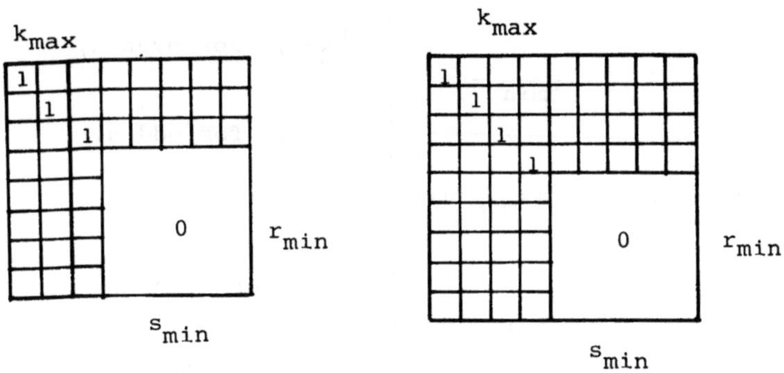

FIGURE 1. CASE $r = s$

1.a. EVEN n 1.b. ODD n.

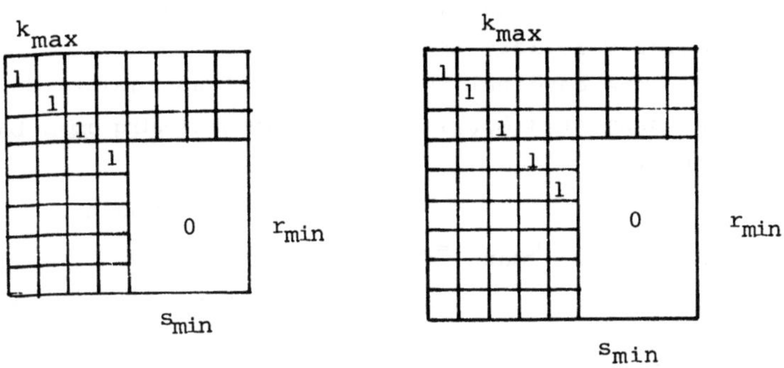

FIGURE 2. CASE r s

2.a. EVEN n 2.b. ODD n

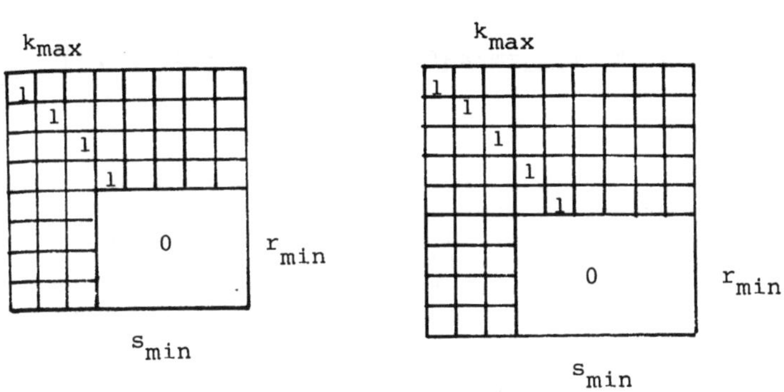

FIGURE 3. CASE r s

3.a. EVEN n 3.b. ODD n

$$r_{min} = \begin{cases} \frac{n+2}{2}, & \text{for even } n, \\ \frac{n+3}{2}, & \text{for odd } n, \end{cases}$$

and assuming that $r_{min}+s_{min}=n+1$, we have:

$$s_{min} = \begin{cases} \frac{n}{2}, & \text{for even } n, \\ \frac{n-1}{2}, & \text{for odd } n. \end{cases}$$

We have to conclude that there are at least $(n+2)/2$ rows in the matrix A_{nxn} such that each of them contains no more than n/2 ones, thus Hall's condition does not hold.

The subcase $r<s$ leads to the same conclusion.

The following Corollary is immediate:

COROLLARY. The cardinality of a partial assignment in a matrix A_{nxn} with zero submatrix A_{rxs}, $(r+s)\geq(n+1)$, has following upper bound:

a) for $r=s$:

$$k_{max} = \begin{cases} \frac{n-2}{2}, & \text{for even } n, \\ \frac{n-1}{2}, & \text{for odd } n, \text{ (see: Fig. 1a,1b)}, \end{cases}$$

b) for $r\neq s$:

$$k_{max} = \begin{cases} \frac{n}{2}, & \text{for even } n, \\ \frac{n+1}{2}, & \text{for odd } n, \text{ (see: Fig. 2a,2b)}, \end{cases}$$

and finally: $1 \leq k \leq k_{max}=\max\{(n-r),(n-s)\}= n-\min\{r,s\}$.

REFERENCES

1. L.R.Ford, Jr., and D.R. Fulkerson: Flows in networks. Princeton University Press. Princeton, New Jersey 1962.
2. P.Hall: On representatives of subsets. Journal of London Mathematical Society, v. 10, 1935, pp. 26-30.
3. N.S.Mendelsohn, and A.L.Dulmage: Some generalizations of the problem of distinct representatives. Canadian Journal of Mathematics. v. 10, 1958, pp. 230-241.
4. L.Słomiński: The bottleneck assignment problem: An efficient algorithm. Archiwum Automatyki i Telemechaniki. t. XXIV, z. 4, 1979.

SENSITIVITY ANALYSIS OF ELASTIC-PLASTIC TORSION PROBLEM

J. SOKOŁOWSKI

Systems Research Institute
Polish Academy of Sciences
ul. Newelska 6, 01-447 Warszawa
P O L A N D

J.P. ZOLESIO

Université de Nice
Département de Mathématiques
Parc Valrose
06034 - NICE-CEDEX, FRANCE

1. Introduction

The paper is concerned with the shape sensitivity analysis of solutions to a variational inequality arising in the modelling of an elastic-plastic torsion problem [5, 6, 7, 22, 23]. The form of the so-called Euler and Lagrange derivatives of the solution of the variational inequality in the direction of a vector field is derived. To this end the material derivative method [25] of the shape sensitivity analysis combined with the results on differential stability of solutions to unilateral problems [15-20] are used. The results presented in this paper are announced in [21]. The related results on the differential stability of solutions to variational inequalities can be found in [1, 8, 13, 14, 26].

Shape optimization problems of plastic design of a cross section under torsion with small bending has been studied in [3] by using a perturbation method. For the related results on optimal plastic design we refer the reader to [10]. The problem of shape optimization of elastic bars in torsion has been investigated e.g. in [2, 11].

The results presented in the paper can be used in order to solve the related optimal design problems.

In the paper standard notation is used. We denote by $H^1(\Omega)$ the Sobolev space given by

$$H^1(\Omega) = \{\phi \in L^2(\Omega) \mid \frac{\partial \phi}{\partial x_i} \in L^2(\Omega),\ i=1,2\}$$

$H^1_o(\Omega)$ is a linear subspace of $H^1(\Omega)$ of the form:

$$H^1_o(\Omega) = \{\phi \in H^1(\Omega) \mid \phi(x) = 0 \quad \text{for a.e. } x \in \partial\Omega\}$$

$H^1_o(\Omega)$ is Hilbert space with the scalar product:

$$(y,z)_{H^1_o(\Omega)} = \int_\Omega \nabla y(x) . \nabla z(x) dx, \quad \forall y,z \in H^1_o(\Omega)$$

where

$$\nabla y = \mathrm{col}\,(\frac{\partial y}{\partial x_1}, \frac{\partial y}{\partial x_2})$$

For an element $y(.) \in H^1(\Omega)$ we denote $y(x) \geq 0$ q.e. on Ω i.e., $y(x) \geq 0$ on Ω possibly except of a set of capacity zero [8].

$$\{S-S\} \overset{def}{=} \{\phi \mid \phi = \phi_1 - \phi_2,\ \phi_i \in S,\ i=1,2\}.$$

2. Elastic-Plastic Torsion Problem

Let $\Omega \subset R^2$ be a given bounded domain with smooth boundary $\Gamma = \partial\Omega$. Let us consider the following variational inequality

$$\text{find } u \in K(\Omega) \text{ such that}$$

$$\int_\Omega \nabla u(x) . \nabla(\phi(x) - u(x))dx \geq \mu \int_\Omega (\phi(x) - u(x))dx \quad \forall \phi \in K(\Omega) \tag{2.1}$$

where $\mu > 0$ is a given constant and $K(\Omega)$ is a closed and convex subset of Sobolev space $H_o^1(\Omega)$ of the form

$$K(\Omega) = \{\phi \in H_o^1(\Omega) \mid |\nabla\phi(x)| \leq k \quad \text{for a.e. } x \in \Omega\} \tag{2.2}$$

here $k > 0$ is a given constant.

It can be shown [5] that there exists a unique solution to (2.1). Denote by $P \subset \Omega$ the so-called plastic region:

$$P = \{x \in \Omega \mid |\nabla u(x)| = k\} \tag{2.3}$$

then $E = \Omega \setminus P$ is the so-called elastic region and we have

$$-\Delta u(x) = \mu \quad \text{in} \quad E \tag{2.4}$$

The elastic region E and the plastic region P are not known a priori and should be determined, therefore (2.1) is a free boundary problem. Let us recall that the solution of the variational inequality (2.1) verifies the following regularity condition [4, 5]

$$u \in H^2(\Omega) \cap H_o^1(\Omega) \tag{2.5}$$

On the other hand it can be shown that the solution of variational inequality (2.1) is simultanously a unique solution of the following variational inequality

$$u \in \tilde{K}(\Omega)$$

$$\int_\Omega \nabla u(x) . \nabla(\phi(x) - u(x))dx \geq \mu \int_\Omega (\phi(x) - u(x))dx , \quad \forall \phi \in \tilde{K}(\Omega) \tag{2.6}$$

here

$$\tilde{K}(\Omega) = \{\phi \in H_o^1(\Omega) \mid \phi(x) \leq \rho(x) \quad \text{for a.e.} \quad x \in \Omega\} \tag{2.7}$$

$$\rho(x) = \min_{\xi \in \partial\Omega} \|x-\xi\|_{R^2} , \quad x \in \bar{\Omega} \tag{2.8}$$

and we assume that $k=1$ in (2.2).

3. Euler (material) derivative $\dot{u}(\Omega)$

In this paper we investigate the local dependence of solutions of variational inequality (2.1) defined in the domain $\Omega \subset R^2$, with respect to the perturbations of the boundary $\partial\Omega$ of the domain of integration Ω. To this end, we define a family of domains $\{\Omega_t\}$ which depends on a real parameter $t \in [0,\delta)$, $\delta > 0$ and on a vector field:

$$V(.,.) : [0,\delta) \times R^2 \to R^2 \tag{3.1}$$

Domains Ω_t are constructed in the following way. Denote by $T_t = T_t(V)$, $t \in [0,\delta)$ the family of mappings

$$T_t \,;\, R^2 \ni X \longrightarrow x(t) \in R^2 \tag{3.2}$$

where the vector function $x(.) : [0,\delta) \to R^2$ satisfies the following ordinary differential equation:

$$\begin{aligned} &\frac{dx}{dt}(s) = V(s, x(s)), \quad s \in [0,\delta) \\ &x(0) = X \end{aligned} \tag{3.3}$$

Define

$$\Omega_t = T_t(V)(\Omega) = \{x \in R^2 \mid \exists X \in \Omega \text{ such that } x = x(t),\ x(0) = X\} \tag{3.4}$$

for $t \in [0,\delta)$, note that $\Omega_o = \Omega$.

We denote by $\rho_t(.)$ the distance function:

$$\rho_t(x) = \inf_{\xi \in \partial\Omega_t} \|x-\xi\|_{R^2} , \quad x \in \Omega_t \tag{3.5}$$

We assume that the following condition is satisfied:

(H1) $\rho_t(.) \in H_o^1(\Omega_t)$ for $t > 0$, t small enough and there exists an element $\dot{\rho}(.) \in H_o^1(\Omega)$ such that

$$\lim_{t \downarrow 0} \|(\rho_t \circ T_t - \rho_o)/t - \dot{\rho}\|_{H_o^1(\Omega)} = 0 \tag{3.6}$$

Here $\dot{\rho}$ denotes the material derivative of the distance function ρ in the direction of the vector field $V(.,.)$.

Let us consider the following variational inequality defined in the domain Ω_t

$$u_t \in K(\Omega_t)$$

$$\int_{\Omega_t} \nabla u_t(x).\nabla(\phi(x)-u_t(x))dx \geq \mu \int_{\Omega_t} (\phi(x)-u_t(x))dx, \quad \forall\phi \in K(\Omega_t) \tag{3.7}$$

Theorem 3.1

For $t>0$, t small enough

$$u_t \circ T_t = u_o + t\dot{u} + o(t) \tag{3.7}$$

where $\|o(t)\|_{H_o^1(\Omega)}/t \to 0$ with $t\downarrow 0$.

The Euler (material) derivative $\dot{u} \in H_o^1(\Omega)$ is given by a unique solution of the following variational inequality

$$\dot{u} \in S(\Omega)$$

$$\int_\Omega \nabla\dot{u}(x).\nabla(\phi(x)-\dot{u}(x))dx \geq \int_\Omega \{F'(x)(\phi(x)-\dot{u}(x))$$

$$-\langle A'(x).\nabla u_o(x), \nabla(\phi(x)-\dot{u}(x))\rangle_{R^2} dx, \quad \forall\phi \in S(\Omega) \tag{3.8}$$

here we denote

$$S(\Omega)=\{\phi \in H_o^1(\Omega) \mid \phi(x) \leq \dot{\rho}(x) \text{ q.e. on } P, \int_P \nabla u_o(x).\nabla\phi(x)dx=\mu\int_P \phi(x)dx\} \tag{3.9}$$

$$F'(x) = \mu \operatorname{div} V(0,x), \quad x \in \Omega \tag{3.10}$$

$$A'(x) = -DV(0,x)-{}^*DV(0,x) + I \operatorname{div} V(0,x) \tag{3.11}$$

Proof:
Let us denote $u^t=u_t \circ T_t \in H_o^1(\Omega)$, $t \in [0,\delta)$. The element u^t is a unique solution of the following variational inequality

$$u^t \in K^t(\Omega)$$

$$\int_\Omega \langle A_t(x)\nabla u^t(x), \nabla(\phi(x)-u^t(x))\rangle_{R^2} dx \geq \int_\Omega F_t(x)(\phi(x)-u^t(x))dx \tag{3.12}$$
$$\forall\phi \in K^t(\Omega)$$

here we denote

$$K^t(\Omega) = \{\phi \in H_o^1(\Omega) \mid \phi(x) \leq \rho^t(x) \quad \text{for a.e.} \quad x \in \Omega\} \tag{3.13}$$

$$\rho^t(x) = (\rho_t \circ T_t)(x), \quad x \in \Omega \tag{3.14}$$

$$A_t(x) = \det(DT_t(x))DT_t^{-1}(x).{}^*DT_t^{-1}(x) \tag{3.15}$$

$$F_t(x) = \mu \det(DT_t(x)) \tag{3.16}$$

We prove that there exists an element $\dot{u} \in H_0^1(\Omega)$ such that

$$\lim_{t \downarrow 0} \| (u^t - u^0)/t - \dot{u} \|_{H_0^1(\Omega)} = 0 \tag{3.17}$$

In order to apply the results on differential stability of solutions to unilateral problems [13-21] we introduce the following variational inequality

$$w^t \in K_0(\Omega)$$

$$\int_\Omega \langle A_t(x) \nabla w^t(x), \nabla(\phi(x) - w^t(x)) \rangle_{R^2} dx \geq \int_\Omega F_t(x)(\phi(x) - w^t(x)) dx$$

$$- \int_\Omega \langle A_t(x) \nabla \rho^t(x), \nabla(\phi(x) - w^t(x)) \rangle_{R^2} dx , \qquad \forall \phi \in K_0(\Omega) \tag{3.18}$$

where

$$K_0(\Omega) = \{\phi \in H_0^1(\Omega) \mid \phi(x) \leq 0 \quad \text{a.e. in} \quad \Omega\} \tag{3.19}$$

It can be verified [19, 20] that for $t > 0$, t small enough

$$w^t = w^0 + t\dot{w} + o(t) \quad \text{in} \quad H_0^1(\Omega) \tag{3.20}$$

where the element $\dot{w}$ is given by a unique solution of the following variational inequality

$$\dot{w} \in S_0(\Omega)$$

$$\int_\Omega \nabla \dot{w}(x) . \nabla(\phi(x) - \dot{w}(x)) dx \geq \int_\Omega F'(x)(\phi(x) - \dot{w}(x)) dx - \int_\Omega \langle A'(x) \nabla \rho(x) + \dot{\rho}(x),$$

$$\nabla(\phi(x) - \dot{w}(x)) \rangle_{R^2} dx , \qquad \forall \phi \in S_0(\Omega) \tag{3.21}$$

here we denote

$$S_0(\Omega) = \{\phi \in H_0^1(\Omega) \mid \phi(x) \leq 0 \quad \text{q.e. on} \quad \Xi , \int_\Xi \nabla w^0(x) . \nabla \phi(x) dx = \mu \int_\Xi \phi(x) dx$$

$$- \int_\Xi \nabla \rho(x) . \nabla \phi(x) dx \} \tag{3.22}$$

$$\Xi = \{x \in \Omega \mid w^0(x) = 0\} \tag{3.23}$$

we have

$$u^t = w^t + \rho^t \tag{3.24}$$

and therefore simple calculations show that (3.7) follows from (3.6), (3.20). ■

Remark 3.2

In order to use the Euler derivative $\dot{u}$ in shape optimization the form of the material derivative $\dot{\rho}(x)$, $x \in P$ should be derived. It can be verified that

$$\dot{\rho}(x) = -\langle n(z(x)), V(0,x)-V(0,z(x))\rangle_{R^2} \qquad x \in P \tag{3.25}$$

where

$$z(x) = \arg\min\{\|\xi-x\|_{R^2} \mid \xi \in \partial\Omega\} \qquad x \in P \tag{3.26}$$

and $n(\xi)$, $\xi \in \partial\Omega$ denotes the outward unit normal vector on $\partial\Omega$.

4. Lagrange Derivative u'

Let us consider a domain functional

$$J(\Omega) = \int_\Omega F(u(x))dx \tag{4.1}$$

where $F(.) \in C^1(R^2)$ and $u \in H_o^1(\Omega)$ is a unique solution of variational inequality (2.1) defined in the domain Ω.
We denote

$$dJ(\Omega;V) \overset{def}{=} \lim_{t\downarrow 0}(J(\Omega_t) - J(\Omega))/t \tag{4.2}$$

It can be verified [24, 25] that

$$dJ(\Omega;V) = \int_\Omega DF(u(x))u'(x)dx + \int_{\partial\Omega} F(u(x)) \langle V(0,x)n(x)\rangle_{R^2} d\Gamma \tag{4.3}$$

where

$$u'(x) \overset{def}{=} \dot{u}(x) - \nabla u(x).V(0,x), \quad x \in \Omega \tag{4.4}$$

here we denote $u(x)=u_o(x)=u^o(x)$, $x \in \Omega_o=\Omega$.

The element $u' \in H^1(\Omega)$ is the so-called Lagrange derivative of the solution u of variational inequality (2.1) in the direction of a vector field $V(.,.)$.

Let $\tilde{u}_t(x)$, $x \in \Omega_t$, $t \in [0,\delta)$ denotes an extension of the element $u_t \in H_o^1(\Omega_t)$ to R^2 defined by

$$\tilde{u}_t(x) = \begin{cases} u_t(x), & x \in \Omega_t,\ t \in [0,\delta) \\ 0 & x \in R^2 \setminus \Omega_t,\ t \in [0,\delta) \end{cases} \tag{4.5}$$

If follows that for $t > 0$, t small enough

$$\tilde{u}_t|_\Omega = u_o + tu' + o(t) \quad \text{in} \quad H^1(\Omega) \tag{4.6}$$

where $\|o(t)\|_{H^1(\Omega)}/t \to 0$ with $t \downarrow 0$.

We derive the form of the Lagrange derivative u'.

Theorem 4.1

The Lagrange derivative u' of the solution u of variational inequality (2.1) in the direction of a vector field $V(.,.)$ is given by a unique solution of the following variational inequality

$$u' \in S_V(\Omega)$$

$$\int_\Omega \nabla u'(x) . \nabla(\phi(x) - u'(x))dx \geq 0 \qquad \forall \phi \in S_V(\Omega) \tag{4.7}$$

where

$$S_V(\Omega) = \{\phi \in H^1(\Omega) \mid \phi(x) = v(x) \text{ on } \partial\Omega,\ \phi(x) \geq \rho'(x) \text{ q.e. on } P$$

$$\int_P (\Delta\rho(x) + \mu)(\phi(x) - \rho'(x))dx = 0\} \tag{4.8}$$

$$v(x) = \langle V(0,x),\ n(x)\rangle_{R^2},\quad x \in \partial\Omega \tag{4.9}$$

$$\rho'(x) = v(z(x)),\quad x \in P \tag{4.10}$$

$$\Delta\rho(x) = 1/(\rho(x) - R(z(x))),\quad x \in P \tag{4.11}$$

$R(.)$ is the radius of curvature of $\partial\Omega$

provided the set

$$\{S_V(\Omega) - S_V(\Omega)\} \cap H^2(\Omega)$$

is dense in the set $\{S_V(\Omega) - S_V(\Omega)\} \subset H^1(\Omega)$. ■

The proof of Theorem 4.1 is omitted here. For the related results on shape sensitivity analysis of variational inequalities we refer the reader to |15-21|.

References

[1] M.P. Bendsøe, N. Olhoff and J. Sokołowski: "Sensitivity analysis of problem of elasticity with unilateral constraints", J. Struct. Mech., 13 (2), 201-222, (1985).

[2] N.V. Banichuk: "Optimization of elastic bars in torsion", Int. J. of Solids and Structures 12 (1976), 275-286.

[3] B. Bochenek, Z. Kordas and M. Życzkowski: "Optimal plastic design of a cross section under with small bending", J. Struct. Mech., 11 (3), 383-400 (1983).

[4] H. Brezis and G. Stampacchia: "Sur la regularité de la solutions d'inequations elliptiques", Bull. Soc. Math. Fr.(96), 153-180, (1968).

|5| A. Friedman: Variational principles and free-boundary problems, J. Wiley & Sons, New York (1982).

|6| L.M. Kachnov: Osnovy teorii plastichnosti, Nauka, Moskva, 1969, (in Russian).

|7| H. Lanchon: "Torsion elastoplastique d'un arbre cylindrique de section simplement ou multiplement connexe", J. de Mecanique, 2, 13 (1974), 267-320.

|8| F. Mignot: "Contrôle dans les inequations variationelles elliptiques", J. Functional Analysis 22, 130-185 (1976).

|9| A. Mioduchowski: "Plastic torsion of rods", Rozpr. Inż., 1, 19, (1971) 129-171, in Polish.

|10| Z. Mroz: "Limit analysis of plastic structures subject to boundary variations", Arch. Mech. Stos. (15), 63-76, (1963).

|11| Moon-Soo Na, Naboru Kikuchi and J.E. Taylor: "Shape optimization for elastic torsion bars", in: Optimization Methods in Structural Design, H. Eschenauer and N. Olhoff eds., Wissenschaftsverlag, (1983), 216-223.

|12| C. Saguez, in the present proceedings.

|13| J. Sokołowski: "Sensitivity analysis for a class of variational inequalities", in: Optimization of Distributed Parameter Structures, Haug E.J. and Cea J. eds., Sijthoff & Noordhoff, Alphen aan den Rijn, Netherlands (1981), 1600-1609.

|14| J. Sokołowski: "Conical differentiability of projection on convex sets - an application to sensitivity analysis of Signorini V.I.", Technical Report, Institute of Mathematics, University of Genoa, (1981).

|15| J. Sokołowski and J.-P. Zolesio: "Shape sensitivity analysis for variational inequalities", in: System Modelling and Optimization, Drenick R.F. and Kozin F. eds., LNCIS, Vol.38, Springer Verlag, New York (1982), 401-407.

|16| J. Sokołowski and J.-P. Zolesio: "Dérivation par rapport au domaine dans les problèmes unilateraux", Rapport de Recherche No.132, INRIA, Rocquencourt (1982).

|17| J. Sokołowski and J.-P. Zolesio: "Shape sensitivity analysis of elastic structures", DCAMM Report No.289, Lyngby (1984).

|18| J. Sokołowski and J.-P. Zolesio: "Differential stability of solutions to unilateral problems", in: Proceedings of Conference on Free Boundary Problems, Maubuisson, France, (1984), Pitman (to appear).

|19| J. Sokołowski and J.-P. Zolesio: "Shape sensitivity analysis of unilateral problems", Publication Mathématiques, No.67, Université de Nice (1985).

|20| J. Sokołowski and J.-P. Zolesio: "Dérivée par rapport au domaine de la solution d'un problème unilateral", C.R. Acad. Sc., Paris, t.301, Serie I, no.4, (1985).

|21| J. Sokołowski and J.-P. Zolesio: "Shape sensitivity analysis of an elastic-plastic torsion problem", Bulletin of the Polish Academy of Sciences, Technical Sciences (in press).

|22| T.W. Ting: "Elastic-plastic torsion of convex cylindrical bars", J. Math. Mech. 19 (1969), 531-551.

|23| T.W. Ting: "Elastic-plastic torsion problem", Arch. Rational Mech. Anal. 34 (1969), 228-244.

|24| J.-P. Zolesio: "Identification de domaines par deformations", Thèse d'Etat, Université de Nice (1979).

|25| J.-P. Zolesio: "The material derivative (or speed) method for shape optimization", in |13|, 1089-1151.

|26| J.-P. Zolesio: "Shape controlability for free bounfaries", in: System Modelling and Optimization, Proceedings of the 11th IFIP Conference, Copenhagen , ed. Thoft-Christensen, Springer Verlag, LNCIS, Vol.59 (1984), 354-361.

DIFFERENTIAL STABILITY OF SOLUTIONS TO BOUNDARY OPTIMAL CONTROL PROBLEMS FOR PARABOLIC SYSTEMS

Jan Sokołowski

Systems Research Institute
Polish Academy of Sciences
ul. Newelska 6
01 - 447 Warszawa, Poland

1. Introduction

The paper is concerned with the differential sensitivity analysis of control constrained optimal control problems with respect to the deformations of the domain of integration [15, 16, 18].

The differential stability of solutions to convex, constrained optimal control problems for systems described by ordinary differential equations is investigated in [10, 11]. The method of sensitivity analysis is exploited in [12] for the sensitivity analysis of control problems for systems described by partial differential equations of elliptic type.

In this paper we use the method of the sensitivity analysis proposed by the author [14-18] combined with the material derivative method [19-21].

The method of sensitivity analysis is based on the concept of the conical differentiability of metric projection in Hilbert space onto a convex and closed set [1, 2, 3, 13, 14]. On the other hand the material derivative method is used to handle the sensitivity analysis with respect to the perturbations of the domain of integration of the state equation.

In this paper some results obtained for convex, boundary optimal control problems with constraints on control are presented. It is shown that the right-derivatives of an optimal control in the direction of a vector field is given by a unique optimal solution of an auxiliary, convex, control constrained optimal control problem.

The outline of the paper is following.

Section 2 is devoted to the analysis of conical differentiability of metric projection in Hilbert space onto convex, closed set. Section 3 describes results obtained for an example of the boundary optimal control problem. In this example a part of the boundary of the domain of integration depends on the parameter. The material derivative method is used for the shape sensitivity analysis of the control problem. The form of the right-derivative of an optimal control in the direction of a vector field is derived.

Throughout the paper standard notation in used [6]. For the related results on optimal control for distributed parameter systems we refer the reader to [5, 7-9].

2. Differential Stability of Projection onto Convex Set in Hilbert Space

In this section the differential stability of solutions to variational inequalities with respect to the perturbations of the right-hand side and the bilinear form is considered.

We need the following notation. Let H be a separable Hilbert space, $K \subset H$ a convex and closed subset. Let there be given a bilinear form

$$a(.,.) : H \times H \to R \tag{2.1}$$

which is coercive and continuous i.e.,

$$a(v,v) \geq \alpha \|v\|_H^2 , \alpha > 0, \forall v \in H \tag{2.2}$$

$$|a(v,z)| \leq M \|v\|_H \|z\|_{H'} \forall v,z \in H \tag{2.3}$$

Let H' denotes the dual space of H and let $f \in H'$ be a given element. We denote by $y=P(f)$ a unique solution of the variational inequality

$$\begin{aligned} & y = P(f) \in K \\ & a(y,v-y) \geq \langle f,v-y\rangle, \quad \forall v \in K \end{aligned} \tag{2.4}$$

where $\langle .,\rangle$ is the duality pairing between H' and H.

Remark 2.1

If the bilinear form $a(.,.)$ is symmetric i.e., $a(v,z)=a(z,v)$, $\forall v,z \in H$ then

$$y = P(f) = \arg\min \{\tfrac{1}{2} a(v,v) - \langle f,v\rangle \mid v \in K\} \tag{2.5}$$

■

It can be verified that the mapping

$$H' \ni f \longrightarrow P(f) \in H \tag{2.6}$$

is Lipschitz continuous

$$\|P(f_1)-P(f_2)\|_H \leq \frac{M}{\alpha}\|f_1-f_2\|_{H'}, \quad \forall f_1,f_2 \in H' \tag{2.7}$$

therefore by a generalization of the Rademacher theorem [13] it follows that there exists a dense subset $A \subset H'$ such that for $f \in A$ we have

$$\forall h \in H' : P(f+\varepsilon h) = P(f) + \varepsilon P'(h) + r(\varepsilon) \quad \text{in} \quad H \tag{2.8}$$

where $r(\varepsilon)/\varepsilon \to 0$ strongly in H with $\varepsilon \downarrow 0$.
The mapping $P'(\cdot)=P'(f;.) : H' \to H$ is linear and continuous.
In the sequel we will use the concept of the so-called conical differentiability of the projection operator.

Definition 2.1 [13]

The mapping (2.6) is conically differentiable at $f \in H'$ if there exists a continuous mappings

$$Q(.) : H' \to H \tag{2.9}$$

such that for $\varepsilon > 0$, ε small enough

$$\forall h \in H' : P(f+\varepsilon h) = P(f)+\varepsilon Q(h)+o(\varepsilon) \quad \text{in} \quad H \tag{2.10}$$

where $\|o(\varepsilon)\|_H/\varepsilon \to 0$ with $\varepsilon \downarrow 0$ uniformly on compact subset of H'. ∎

In order to derive the form of the mapping (2.9) we need the following notation.

For a given element $y \in K$ we denote by $C_K(y)$ the tangent cone

$$C_K(y) = \{\phi \in H \mid \exists\, \varepsilon > 0 \text{ such that } y + \varepsilon\phi \in K\} \tag{2.11}$$

In general the cone (2.11) is not closed, we denote by $\overline{C_K(y)}$ its closure in H. For a given element $f \in H'$ we denote by $T_K(f) \subset H$ a hyperplane of the form:

$$T_K(f) = \{\phi \in H \mid a(P_K(f),\phi) = \langle f,\phi\rangle\} \tag{2.12}$$

Finally we denote by $S_K(f) \subset H$ a convex, closed cone of the form:

$$S_K(f) = \overline{C_K(P_K(f))} \cap T_K(f), \quad \forall f \in H' \tag{2.13}$$

Definition 2.2

The set $K \subset H$ is called polyhedric if for any $f \in H'$ it follows that

$$S_K(f) = \overline{C_K(P_K(f)) \cap T_K(f)} \tag{2.14}$$

Example 2.1

Let there be given a domain $\Omega \subset R^n$ with the boundary $\partial\Omega$. Denote by $K \subset L^2(\partial\Omega)$ the set

$$K = \{u(.) \in L^2(\partial\Omega) \mid 0 \leq u(s) \leq M \quad \text{for a.e. } s \in \partial\Omega, \quad \int_{\partial\Omega} u(s)d\Gamma \leq C\} \tag{2.15}$$

where $C, M > 0$ are given constants.

Assume that

$$a(u,v) = \int_{\partial\Omega} u(s)v(s)d\Gamma, \quad \forall u,v \in L^2(\partial\Omega) \tag{2.16}$$

It can be verified that the set (2.15) is polyhedric, furthermore the cone (2.14) is given by

$$\begin{aligned} S_K(f) = \{v(.) \in L^2(\partial\Omega) \mid\ & v(s) \geq 0 \text{ a.e. on } \Xi_O, \\ & v(s) \leq 0 \text{ a.e. on } \Xi_M, \\ & \int_{\partial\Omega} v(s)d\Gamma \leq 0 \quad \text{whenever} \quad \int_{\partial\Omega} u(s)d\Gamma = C, \\ & \int_{\partial\Omega} u(s)v(s)d\Gamma = \int_{\partial\Omega} f(s)v(s)d\Gamma \} \end{aligned} \tag{2.17}$$

where

$$u = P_K(f), \quad f \in L^2(\partial\Omega)$$

$$\Xi_O = \{s \in \partial\Omega \mid u(s) = 0\}$$

$$\Xi_M = \{s \in \partial\Omega \mid u(s) = M\}$$

■

It can be shown [2, 13] that the projection onto polyhedric convex set is conically differentiable.

Theorem 2.1 [2, 13]

Assume that the set $K \subset H$ is polyhedric. Then for any $f \in H'$ and for $\varepsilon > 0$, ε small enough

$$\forall h \in H' : P_K(f+\varepsilon h) = P_K(f) + \varepsilon P_S(h) + o(\varepsilon) \tag{2.18}$$

where $\|o(\varepsilon)\|_H/\varepsilon \to 0$ with $\varepsilon \downarrow 0$ uniformly with respect to h on compact subsets of H'. Here we denote

$$S = S_K(f), \quad \forall f \in H' \tag{2.19}$$

■

For the related results on the differential stability of projection in Hilbert space onto convex set we refer the reader to [1-3, 13, 17]. Let us consider the following family of variational inequalities depending on a parameter $\varepsilon \in [0,\delta)$, $\delta > 0$:

$$\begin{aligned} & z^{\varepsilon} \in K \\ & a^{\varepsilon}(z^{\varepsilon}, v - z^{\varepsilon}) \geq \langle f^{\varepsilon}, v - z^{\varepsilon} \rangle \qquad \forall v \in K \end{aligned} \tag{2.20}$$

where for each $\varepsilon \in [0,\delta)$, $a^{\varepsilon}(.,.): H \times H \to R$ is a bilinear form and $f^{\varepsilon} \in H'$ is given element. For fixed $\varepsilon \in [0,\delta)$ denote $z^{\varepsilon} = P^{\varepsilon}(f^{\varepsilon})$, $f^{\varepsilon} \in H'$.

Theorem 2.2

Assume that

(i) bilinear forms $a^{\varepsilon}(.,.)$ satisfy (2.2), (2.3) uniformly for $\varepsilon \in [0,\delta)$ and there exists a linear operator $A' \in \mathcal{L}(H,H')$ such that

$$\lim_{\varepsilon \downarrow 0} \sup_{\substack{\|\phi\|_H \leq 1 \\ \|w\|_H \leq 1}} |(a^{\varepsilon}(w,\phi) - a^{o}(w,\phi))/\varepsilon - \langle A'w, \phi \rangle| = 0 \tag{2.21}$$

(ii) there exists an element $f' \in H'$ such that

$$\lim_{\varepsilon \downarrow 0} \sup_{\|\phi\|_H \leq 1} |\langle f^{\varepsilon} - f^{o}, \phi \rangle / \varepsilon - \langle f', \phi \rangle| = 0 \tag{2.22}$$

(iii) the mapping $H' \ni f \to P^{o}(f) \in H$ is conically differentiable at $f^{o} \in H'$ i.e., for $\varepsilon > 0$, ε small enough

$$\forall h \in H' : P^{o}(f^{o} + \varepsilon h) = P^{o}(f^{o}) + \varepsilon Q(h) + o(\varepsilon) \quad \text{in } H \tag{2.23}$$

then the solution z^{ε} to (2.20) is right-differentiable at $\varepsilon = 0$ and it follows that for $\varepsilon > 0$, ε small enough

$$z^{\varepsilon} = z^{o} + \varepsilon Q(f' - A'z^{o}) + o(\varepsilon)$$

where $\|o(\varepsilon)\|_H / \varepsilon \to 0$ with $\varepsilon \downarrow 0$.

The proof of Theorem 2.2 is given in [16, 18, 19]. ∎

3. Shape Sensitivity Analysis of Boundary Optimal Control Problem

In this section the differential sensitivity of an optimal control for the control constrained boundary control problem with respect to the deformations of the domain of integration of the state equation is considered. We restrict ourselves for simplicity to a model problem. For the general results on the shape sensitivity analysis of optimal control problems we refer the reader to [16].

We define a family $\{\Omega_\varepsilon\} \subset R^2$ of domains depending on parameter $\varepsilon \in [0,\delta)$.

3.1. Family of domains $\{\Omega_\varepsilon\}$

Let $\Omega \subset R^2$ be a given domain with smooth boundary $\partial\Omega = \Gamma \cup \Sigma$ (Fig.1)

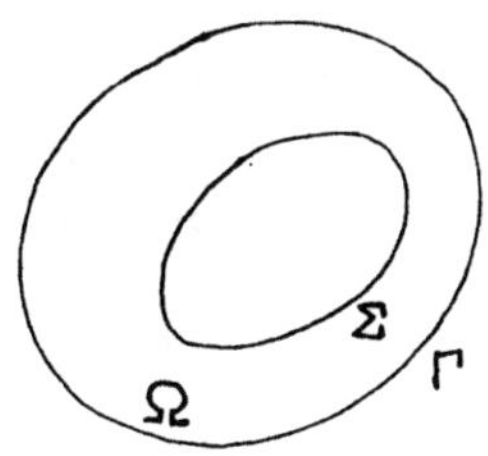

Fig.1 Domain Ω

Assume that there is given a vector field

$$V(.,.) : [0,\delta) \times R^2 \rightarrow R^2 \tag{3.1}$$

The domain $\Omega_\varepsilon \subset R^2$, $\varepsilon \in [0,\delta)$ is constructed in the following way [20, 21]. Denote by $T_\varepsilon = T_\varepsilon(V)$, $\varepsilon \in [0,\delta)$ the family of mappings

$$T_\varepsilon : R^2 \ni X \longrightarrow x(\varepsilon) \in R^2 \tag{3.2}$$

where the vector function $x(.) : [0,\delta) \longrightarrow R^2$ satisfies the following ordinary differential equation

$$\begin{aligned} &\frac{dx}{ds}(\varepsilon) = V(\varepsilon, x(\varepsilon)), \qquad \varepsilon \in (0,\delta) \\ &x(0) = X, \qquad X \in R^2 \end{aligned} \tag{3.3}$$

We define

$$\begin{aligned} \Omega_\varepsilon &= T_\varepsilon(V)(\Omega) \\ &= \{x \in R^2 \mid \text{there exists } X \in \Omega \text{ such that} \\ &\qquad x = x(\varepsilon),\ x(0) = X\} \end{aligned} \tag{3.4}$$

Let us note that $\Omega_o = \Omega$.

We denote by $DT_\varepsilon(X)$ the Jacobian matrix of mapping (3.2) evaluated at $X \in R^2$. It is assumed for simplicity that the vector field $V(.,.)$ has a compact support in a small open neighbourhood of Σ therefore $\partial\Omega_\varepsilon = \Sigma_\varepsilon \cup \Gamma$, $\varepsilon \in [0,\delta)$ where $\Sigma_\varepsilon = T_\varepsilon(V)(\Sigma)$, $\Gamma = T_\varepsilon(V)(\Gamma)$.

3.2. Optimal control problem

In order to define an optimal control problem we introduce state equation, cost functional and set of sdmissible controls of the form:

State equation:

$$\frac{\partial y_\varepsilon}{\partial t}(u;x,t) - \Delta y_\varepsilon(u;x,t) = 0 \quad \text{in} \quad \Omega_\varepsilon \times (0,\tau) \tag{3.5}$$

$$\frac{\partial y_\varepsilon}{\partial n}(u;x,t) = u(x,t) \quad \text{on} \quad \Gamma \times (0,\tau) \tag{3.6}$$

$$y_\varepsilon(u;x,t) = 0 \quad \text{on} \quad \Sigma_\varepsilon \times (0,\ ) \tag{3.7}$$

$$y_\varepsilon(u;x,t) = 0 \quad \text{in} \quad \Omega \tag{3.8}$$

where $\tau > 0$ is given constant, $u \in L^2(\Gamma \times (0,\tau))$ denotes control, $\varepsilon \in [0,\delta)$ is parameter.

It is well known that for any $u \in L^2(\Gamma \times (0,\tau))$ and for a fixed $\varepsilon \in [0,\delta)$ there exists [9] a unique weak solution $y_\varepsilon(u;.,.) \in L^2(\Omega \times (0,\tau))$ to (3.5)-(3.8).

Cost functional:

$$J_\varepsilon(u) = \frac{1}{2}\int_0^\tau \int_{\Omega_\varepsilon} (y_\varepsilon(u;x,t) - z_d(x,t))^2 dx dt + \frac{\alpha}{2}\int_0^\tau \int_\Gamma (u(x,t))^2 d\Gamma dt, \alpha > 0 \tag{3.9}$$

where $z_d \in H^1(R^3)$ is a given element.

Set of admissible controls:

$$U_{ad} = \{u \in L^2(\Gamma \times (0,\tau)) \mid 0 \le u(x,t) \le M \text{ for a.e. } (x,t) \in \Gamma \times (0,\tau), \int_0^\tau \int_\Gamma u(x,t) d\Gamma dt \le C\} \tag{3.10}$$

where $M, C > 0$ are given constants. ■

It can be verified that for any fixed $\varepsilon \in [0,\delta)$ there exists a unique optimal control $u^\varepsilon \in U_{ad}$ such that

$$J_\varepsilon(u^\varepsilon) \le J_\varepsilon(u), \quad \forall u \in U_{ad} \tag{3.11}$$

The optimal control u^ε is given by a unique solution of the following optimality system:

Find $(u^\varepsilon, y^\varepsilon, p^\varepsilon) \in U_{ad} \times L^2(\Omega_\varepsilon \times (0,\tau)) \times L^2(\Omega_\varepsilon \times (0,\tau))$ which satisfy the following system

State equation:

$$\frac{\partial y^\varepsilon}{\partial t}(x,t) - \Delta y^\varepsilon(x,t) = 0 \quad \text{in} \quad \Omega_\varepsilon \times (0,\tau) \tag{3.12}$$

$$\frac{\partial y^\varepsilon}{\partial n}(x,t) = u^\varepsilon(x,t) \quad \text{on} \quad \Gamma \times (0,\tau) \tag{3.13}$$

$$y^\varepsilon(x,t) = 0 \quad \text{on} \quad \Sigma_\varepsilon \times (0,\tau) \tag{3.14}$$

$$y^\varepsilon(x,0) = 0 \quad \text{in} \quad \Omega_\varepsilon \tag{3.15}$$

Adjoint state equation:

$$\frac{\partial p^\varepsilon}{\partial t}(x,t) - \Delta p^\varepsilon(x,t) = y^\varepsilon(x,t) - z_d(x,t) \quad \text{in} \quad \Omega_\varepsilon \times (0,\tau) \tag{3.16}$$

$$\frac{\partial p^\varepsilon}{\partial t}(x,t) = 0 \quad \text{on} \quad \Gamma \times (0,\tau) \tag{3.17}$$

$$p^\varepsilon(x,t) = 0 \quad \text{on} \quad \Sigma_\varepsilon \times (0,\tau) \tag{3.18}$$

$$p^\varepsilon(x,\tau) = 0 \quad \text{in} \quad \Omega_\varepsilon \qquad (3,19)$$

Optimality condition:

$$\int_0^\tau \int_\Gamma (\alpha u^\varepsilon(x,t) - p^\varepsilon(x,t))(\phi(x,t) - u^\varepsilon(x,t))\, d\Gamma dt \geq 0 \quad \forall \phi \in U_{ad} \tag{3.20}$$

■

3.3. Shape sensitivity analysis

In order to derive the form of the right-derivative of an optimal control u^ε with respect to ε at $\varepsilon=0$ we transport [15, 16, 18] the optimality system (3.12)-(3.29) to the fixed domain $\Omega \times (0,\tau)$ using the mapping (3.2).

Denote

$$\tilde{p}_\varepsilon(x,t) = (p^\varepsilon \circ T_\varepsilon(V))(x,t), \quad (x,t) \in \Omega \times (0,\tau)$$

$$\tilde{y}_\varepsilon(x,t) = (y^\varepsilon \circ T_\varepsilon(V))(x,t), \quad (x,t) \in \Omega \times (0,\tau)$$

$$\tilde{u}_\varepsilon(x,t) = (u^\varepsilon \circ T_\varepsilon(V))(x,t), \quad (x,t) \in \Gamma \times (0,\tau)$$

let us recall that we select the vector field $V(.,.) \in C^1([0,\delta);$

$C^2(R^2;R^2))$ in such a way that the following conditions are verified

$$\Gamma = T_\varepsilon(V)(\Gamma), \quad \Sigma_\varepsilon = T_\varepsilon(V)(\Sigma), \quad \varepsilon \in [0,\delta)$$

therefore $\quad \tilde{u}_\varepsilon(x,t) = u^\varepsilon(x,t), \quad (x,t) \in \Gamma \times (0,\tau)$

It can be shown [16] that the elements $(u^\varepsilon, \tilde{y}_\varepsilon, \tilde{p}_\varepsilon)$ are given by a unique solution of the following optimality system which consists of state equation, adjoint state equation and optimality condition of the form:

Find $(u^\varepsilon, \tilde{y}_\varepsilon, \tilde{p}_\varepsilon) \in U_{ad} \times L^2(\Omega \times (0,\tau)) \times L^2(\Omega \times (0,\tau))$ which satisfy the following system

State equation:

$$\gamma_\varepsilon(x)\,\frac{\partial \tilde{y}_\varepsilon}{\partial t}(x,t) - \operatorname{div}(A_\varepsilon(x).\nabla \tilde{y}_\varepsilon(x,t)) = 0 \qquad \text{in} \quad \Omega \times (0,\tau) \tag{3.21}$$

$$\frac{\partial \tilde{y}_\varepsilon}{\partial n}(x,t) = u^\varepsilon(x,t) \qquad \text{on} \quad \Gamma \times (0,\tau) \tag{3.22}$$

$$\tilde{y}_\varepsilon(x,t) = 0 \qquad \text{on} \quad \Sigma \times (0,\tau) \tag{3.23}$$

$$\tilde{y}_\varepsilon(x,0) = 0 \qquad \text{in} \quad \Omega \tag{3.24}$$

Adjoint state equation:

$$-\gamma_\varepsilon(x)\,\frac{\partial \tilde{p}_\varepsilon}{\partial t}(x,t) - \operatorname{div}(A_\varepsilon(x).\nabla \tilde{p}_\varepsilon(x,t)) = \gamma_\varepsilon(x)(\tilde{y}_\varepsilon(x,t) - z_d^\varepsilon(x,t))$$

$$\text{in} \quad \Omega \times (0,\tau) \tag{3.25}$$

$$\frac{\partial \tilde{p}_\varepsilon}{\partial n}(x,t) = 0 \qquad \text{on} \quad \Gamma \times (0,\tau) \tag{3.26}$$

$$\tilde{p}_\varepsilon(x,t) = 0 \qquad \text{on} \quad \Sigma \times (0,\tau) \tag{3.27}$$

$$\tilde{p}_\varepsilon(x,\tau) = 0 \qquad \text{in} \quad \Omega \tag{3.28}$$

Optimality condition:

$$\int_0^\tau \int_\Gamma (\alpha u^\varepsilon(x,t) - \tilde{p}_\varepsilon(x,t))(\phi(x,t) - u^\varepsilon(x,t))\,d\Gamma dt \geq 0, \quad \forall \phi \in U_{ad} \tag{3.29}$$

■

where

$$\gamma_\varepsilon(x) = \det(DT_\varepsilon(x)), \quad x \in \Omega$$

$$A_\varepsilon(x) = \gamma_\varepsilon(x) DT_\varepsilon^{-1}(x).{}^*DT_\varepsilon^{-1}(x), \quad x \in \Omega$$

$$z_d^\varepsilon(x,t) = (z_d \circ T_\varepsilon)(x,t), \quad (x,t) \in \Omega \times (0,\tau)$$

The elements $\gamma_\varepsilon(\cdot)$, $A_\varepsilon(\cdot)$, $z_d^\varepsilon(.,.)$ are sufficiently smooth [16] with respect to space variable $x \in R^2$ and to parameter $\varepsilon \in [0,\delta)$.
It can be shown [16] using the necessary and sufficient optimality condition in the form of variational inequality (2.20) for the optimal control problem corresponding to the system (3.21)-(3.29), that for $\varepsilon > 0$, ε small enough

$$\|u - u^o\|_{L^2(\Gamma \times (0,\tau))} \leq C\varepsilon \tag{3.30}$$

Using (3.30) we can differentiate [16] the system (3.21)-(3.29) with respect to the parameter ε at $\varepsilon=0^+$ and after some simple calculations we obtain the following result.

Theorem 3.1

For $\varepsilon > 0$, ε small enough

$$u^\varepsilon = u^o + \varepsilon q + o(\varepsilon) \quad \text{in} \quad L^2(\Gamma \times (0,\tau))$$

where $\|o(\varepsilon)\|_{L^2(\Gamma \times (0,\tau))} / \varepsilon \to 0$ with $\varepsilon \downarrow 0$.

The element q is given by a unique solution of the following optimal control problem

Problem (Q):

find an element $q \in L^2(\Gamma \times (0,\tau))$ which minimizes the cost functional

$$I(\eta) = \frac{1}{2} \int_o^\tau \int_\Omega (z(\eta;x,t))^2 dxdt + \int_o^\tau \langle g_2(z(\eta;.,t), v(.)\rangle_{\mathcal{D}_1'(\Sigma) \times \mathcal{D}_1(\Sigma)} dt + \frac{\alpha}{2} \int_o^\tau \int_\Gamma (\eta(x,t))^2 d\Gamma dt$$

over the set of admissible controls

$$S(\Gamma) = \{\eta \in L^2(\Gamma \times (0,\tau)) \mid \eta(x,t) \geq 0 \text{ a.e. on } \Xi_o,$$
$$\eta(x,t) \leq 0 \text{ a.e. on } \Xi_M,$$
$$\int_o^\tau \int_\Gamma \eta(x,t) d\Gamma dt \leq 0 \text{ whenever } \int_o^\tau \int_\Gamma u_o(x,t) d\Gamma dt = C,$$
$$\int_o^\tau \int_\Gamma (\alpha u_o(x,t) - \tilde{p}_o(x,t)) \phi(x,t) d\Gamma dt = 0\}.$$

where the element $z(\eta;.,.)$ is given by a unique solution of the following state equation:

$$z(\eta;.,t) \in H^1_\Sigma(\Omega)$$

$$\int_\Omega \{ \frac{\partial z}{\partial t}(\eta;x,t)\phi(x) + \nabla z(\eta;x,t).\nabla\phi(x)\}dx = \int_\Gamma \eta(x,t)\phi(x)d\Gamma +$$

$$+ \langle g_1(\phi),v\rangle_{\mathcal{D}_1'(\Sigma)\times\mathcal{D}_1(\Sigma)} \quad \text{for a.e.} \quad t \in (0,\tau), \quad \forall\phi \in H^1_\Sigma(\Omega)$$

$$z(\eta;x,0) = 0 \quad \text{in} \quad \Omega$$

■

here we denote

$$v(x) = \langle V(0,x),n(x)\rangle_{R^2}, \quad x \in \partial\Omega$$

$$\Xi_o = \{(x,t) \in \Gamma \times (0,\tau) \mid u_o(x,t) = 0\}$$

$$\Xi_M = \{(x,t) \in \Gamma \times (0,\tau) \mid u_o(x,t) = M\}$$

$$H^1_\Sigma(\Omega) = \{\phi \in H^1(\Omega) \mid \phi(x) = 0 \quad \text{on} \quad \Sigma\}$$

It can be shown that for $\phi \in H^2(\Omega) \cap H^1_\Sigma(\Omega)$ the distributions $g_1(\phi)$, $g_2(\phi)$ can be represented in the following form

$$\langle g_1(\phi),v\rangle_{\mathcal{D}_1'(\Sigma)\times\mathcal{D}_1(\Sigma)} = \int_\Sigma v(x)\frac{\partial\tilde{y}_o}{\partial n}(x,t)\frac{\partial\phi}{\partial n}(x,t)d\Sigma$$

$$\langle g_2(\phi),v\rangle_{\mathcal{D}_1'(\Sigma)\times\mathcal{D}_1(\Sigma)} = \int_\Sigma v(x)\frac{\partial\tilde{p}_o}{\partial n}(x,t)\frac{\partial\phi}{\partial n}(x,t)d\Sigma$$

The proof of Theorem 3.1 is omitted here.

Remark :

The method of the sensitivity analysis can be applied in the case when the part Γ of the boundary $\partial\Omega$ depends on the parameter. In this case the form of the so-called Euler and Lagrange derivatives of an optimal control in the direction of a vector field can be derived [16].

References

[1] S. Fitzpatrick and R.R. Phelps: "Differentiability of the metric projection in Hilbert space", Trans. Amer. Math. Soc. 270 (1982), 483-501.

[2] A. Haraux: "How to differentiate the projection on a convex set in Hilbert space. Some applications to variational inequalities", J. Math. Soc. Japan, 29 (4), (1977), 625-631.

[3] R.B. Holmes: "Smoothness of certain metric projections on Hilbert space", Trans. Amer. Math. Soc. 184 (1973), 87-100.

[4] P. Holnicki, J. Sokołowski and A. Żochowski: in the present procedings.

[5] J.L. Lions: Contrôle optimal de systemes gouvernes par des equa-

tions aux dérivées partielles, Dunod, Paris, 1968.

[6] J.L. Lions, Magenes E.: Problèmes aux limites non homogènes et applications, Vol.1, Dunod, Paris, 1968.

[7] J.L. Lions: Perturbations singulieres dans les problèmes aux limites et en contrôle optimal, Lectures Notes in Mathematics, Vol. 323, Springer Verlag, Berlin, 1973.

[8] J.L. Lions: Some Methods in the Mathematical Analysis of Systems and Their Control, Gordon and Breach, New York, 1981.

[9] J.L. Lions: Contrôle des systemes distribues singuliers, Dunod, Paris, 1983.

[10] K. Malanowski: "Differential stability of solutions to convex, control constrained optimal control problems", Applied Mathematics and Optimization, 12 (1984), 1-14.

[11] K. Malanowski: "On differentiability with respect to parameter of solutions to convex optimal control problems subject to state space constraints", Applied Mathematics and Optimization, 12 (1984), 231-245.

[12] K. Malanowski and J. Sokołowski: "Sensitivity of solutions to convex control constrained optimal control problems for distributed parameter systems", to appear in: Journal of Mathematical Analysis and Applications.

[13] F. Mignot: "Controle dans les inequations variationelles elliptiques", J. Functional Analysis, 22, 1976, 130-185.

[14] J. Sokołowski: "Differential stability of solutions to constrained optimization problems", Applied Mathematics and Optimization, 13 (2), 1985, 97-115.

[15] J. Sokołowski: "Differential stability of control constrained optimal control problems for distributed parameter systems", to be published in: Proceedings of 2nd International Conference on Control Theory for Distributed Parameter Systems and Applications, Vorau, Austria, July 9-14, 1984, Springer Verlag.

[16] J. Sokołowski: Sensitivity Analysis and Parametric Optimization of Optimal Control Problems for Distributed Parameter Systems, Zeszyty Naukowe Politechniki Warszawskiej (in Polish), 170 pages (in press).

[17] J. Sokołowski: "Differential stability of projection in Hilbert space onto convex set. Applications to sensitivity analysis of optimal control problems", Publication Mathématiques, Université de Nice (to appear).

[18] J. Sokołowski: "Sensitivity analysis of control constrained optimal control problems for distributed parameter systems", (to appear)

[19] J. Sokołowski and J.-P. Zolesio: "Shape sensitivity analysis of unilateral problems", Publication Mathématiques No.67, Université de Nice, 1985.

[20] J.-P. Zolesio: "Identification de domaines par deformations", These d'Etat, Universite de Nice, 1979.

[21] J.-P. Zolesio: "The material derivative (or speed) method for shape optimization", in: J.E. Haug, J. Cea (eds.): Optimization of Distributed Parameter Structures, Sijthoff-Noordhoff, Alphen aan den Rijn, Netherlands, (1981), 1089-1151.

AN "ANALYTICAL CENTRE" FOR POLYHEDRONS AND NEW CLASSES OF GLOBAL ALGORITHMS FOR LINEAR (SMOOTH, CONVEX) PROGRAMMING

Gy. SONNEVEND

Dept. of Numerical Analysis, Inst. of Mathematics

Eötvös University, 1088. Budapest, Muzeum krt. 6-8.

Abstract

For an arbitrary, finite intersection of halfspaces $b_i \geq \langle a_i, x\rangle$ $i=1,\ldots,m$, $x \in R^n$, i.e. a bounded, convex polyhedron $P(a^m,b^m)$ we define a "central" point $x(a^m,b^m) \in P$, which has the following properties: x depends on (a^m,b^m) analytically (i.e. rather smoothly); x is affinely invariant; there exist ellipsoids containing P and contained in P, centered at x with similarity ratio (m-1); $x(a^m,b^m)$ can be computed effectively by maximizing a strongly concave, analytic function over P. New methods are presented for globalizing (globally accelerating) the convergence of Newton'-s method for finding saddle points of convex concave, smooth functions based on homotopy (i.e. continuation) ideas. The above results are applied to outline new approaches to linear (smooth, convex) programming by constructing affine invariant relatives of Karmarkar'-s interior point, projective invariant method.

1. Introduction

The problem of constructing algorithms for the solution of linear or convex programming problems (where f_i, $i=1,\ldots,m$ are convex)

$$\inf \{f_0(x) \mid f_i(x) \leq 0,\ i=1,\ldots,m,\ x \in R^n \tag{1.1}$$

with good global behaviour is much less understood at present than the problems of local convergence. The popular method of (restricted or "asymptotically exact") line search - to achieve global convergence, say for Newton's method applied to unconstrained minimization on R^n - can be criticized from the point of view of informational efficiency: during line search the information obtained is used only for restricted, "local" prediction (along the line) and not globally over R^n (as e.g. in the ellipsoid method, see [6]). While trust region methods, see e.g. the paper of More in [1] are much more flexible we think that they will be effective only if we are able to provide global approximations for (parts of) the feasible region $\{x \mid f_i(x) \leq 0,\ i=1,\ldots,m\}$ by trust regions and (or) global approximation of the function to be minimized, i.e. f_0 by quadratic functions (together with good error bounds). In fact, any progress (i.e. a new method or tool) in the theory of globalization methods should be based on some fact belonging to the realm of "global convex analysis", in contrast to the case of tools used in the theory of local or asymptotic convergence of minimization methods, where of main interest are the local notions, facts like those concerning first and higher order differentials (their local approximation) or local, piecewise linear models, used in "bundle" methods (see e.g. the papers of Lemarechal and others in [3]). The ellipsoid method is a prototype for such global tools. Homotopy (deformation or continuation) methods give very powerful tools for globalization if they are used with care and, of course, accompanied with global (a priori or a posteriori, i.e. adaptive) estimates for the extrapolations used. They will play an important role in the methods proposed below. These methods should be appreciated (among other reasons) since by their use we can even get methods of "higher order accuracy" (therefore presumably better suited for rather smooth functions) then Newton'-s method (of course - since here we mean "order" intuitively - not by performing two Newton steps to "get" a "forth order" method). Note that the idea to extrapolate function values by values of model functions of higher than second order (i.e. belonging to a class richer than the class of quadratic functions) so that their minimum be easily computable does not seem to be realizable. We hope that the results presented here contribute to a better understanding of the distinction, "numerical" characterization of smoothness versus nondifferentiability in convex programming. Roughly speaking we can say that the solution of a problem (1.1) with smooth, say analytic functions calls for a nonsmooth method (like the ellipsoid method) only if the number of constraints m is large with respect to n, the dimension of x, or if (near x*) the functions f_i, $i=0,\dots,m$, more precisely the problem (1.1) is rather ill conditioned (a notion, surely dependent on the order of smoothness of the problem). In general it is very difficult to construct provably optimal methods for important classes of problems with respect to criterions reflecting the many sided (stability, complexity, ...) restrictions or desires motivated by the applications, thus the solution of a concrete (hard) problem (1.1) remains an art. However the necessity of writing general purpose software (codes) can be met by combining at each "step" several tools (i.e. trial methods), in one and the same general "vehicle" method; this was done - for combining Newton'-s method with a cutting plane method using the ellipsoid method as a vehicle in our paper in [3]. The method see [5], more precisely his potential function could also be used as a vehicle (to accept at each step that trial method which yields the best reduction of the value of a "potential function") of course in each such methods nontrivial problems of "compatibility" for the specific trial updates arise.

The main result of section 2, definition and properties of an "analytic centre" for polyhedrons - by which we provide a global and flexible ellipsoidal approximation for polyhedrons - will be used in section 3 for constructing new classes of methods for the solution of linear (or smooth, convex) programming problems. We propose them only when m is not too large compared to n: a simple operation count shows that for $m > \text{const}\, n^{4/3}$ the arithmetical (worst case) complexity (i.e. guaranted speed in the reduction of the value of the objective function per unit operation) of the ellipsoid method (implemented e.g. for the exact max. penalty function) is smaller than that of Karmarkar'-s method.

2. An analytical centre for polyhedron and its properties

Suppose that we have to find a point x belonging to the set

$$P(a^m,b^m):\{z \mid b_i \geq \langle a_i,z\rangle,\ i=1,\dots,m,\ z \in R^n\}. \tag{2.1}$$

Moreover we wish to obtain - at the same time - an "approximation" for the set $P(a^m,b^m)$. First note that - using the standard imbedding of P into R^m

$$P=\{\xi \mid F\xi = g,\ \xi \geq 0,\ \xi \in R^m\},\ F:R^m \to R^{m-n}, \tag{2.2}$$

where the rows of F, F_j, $j=1,\dots,m-n$, provide a maximal set of linearly independent solutions $v=(v_1,\dots,v_m)$ of

$$\Sigma v_i a_i = 0,\ g_j = \langle F_j, b\rangle,\ j=1,\dots,m-n,$$

assuming, for simplicity, that $a_1, \dots, a_m$ are in general position. We shall select x, i.e. ξ as the solution of

$$\max\{\xi_1\ \dots\ \xi_m \mid F\xi = g,\ \xi \geq 0,\ \xi \in R^m\}. \tag{2.3}$$

A natural generalization of the previous problem to infinite dimensional ξ corresponds to a special solution of moment problems: a measure is sought, for which the values of M=m-n linear functionals are fixed, i.e. prescribed or known. It is rather surprising that the analogue of the solution (2.3) is known to be explicitly (exactly) computable in $O(M^2)$ operations for the moment problem corresponding the "maximum entropy" spectral estimation (for stationary, Gaussion, discrete or continuous time stochastic processes) from generalized correlation data, see [7] for details, where rather strong properties of a "centre" (for obviously not central-symmetrical sets P) are valid for x(P). It is important to note that sometimes in the literature the word "maximum entropy estimation" for a set (2.2), where one of the constraint is $\xi_1+\dots+\xi_m=1$, is defined as the solution of

$$\max\{-\sum_{i=1}^{m} \xi_i \log \xi_i \mid F\xi = g,\ \xi \geq 0\} \quad ;$$

the point $\xi(P) \equiv x(P)$ thus defined is not affinely invariant, i.e.

$x(\tilde{a}^m,\tilde{b}^m)=Tx(a^m,b^m)$, for all linear (nonsingular) transformations defined by matrixes $T \in R^{n\times n}$, and vectors $u \in R^n$ by $\tilde{a}_i:=(T^*)^{-1} a_i$, $\tilde{b}_i:=b_i+\langle a_i,T^{-1}u\rangle$, $i=1,\dots,m$, may not hold, thus (for moment problems) (2.3) yields the "right" solution.

<u>Theorem</u>. For an arbitrary set of data (a^m,b^m) defining a bounded polyhedron $P(a^m,b^m)$ with nonvoid interior in R^n there exists a unique (in P) solution $x(a^m,b^m)$ of the equation

$$E(x):= a_1/(b_1-\langle a_1,x\rangle)+\dots+a_m/(b_m-\langle a_m, x\rangle)=0, \tag{2.4}$$

which is the unique (in P) solution of the problem

$$\max_{x \in P} \Psi(x), \text{ where } \Psi(x):= \prod_{i=1}^{m} (b_i-\langle a_i, x\rangle)^{1/m}. \tag{2.5}$$

The functions $\Psi(x)^{1/m}$ and $\log \Psi(x)$ are strictly concave and analytic inside P. For an ellipsoid $E_{out}(a^m,b^m)$

$$E_{out}:=\{z \mid \langle A^{-1}z,z\rangle\le 1\}, \text{ where } A:=m(m-1)(L^+)^*L^+, \tag{2.6}$$

$Le_i:=a_i/(b_i-\langle a_i,x(P)\rangle$, $e_i=(0,\dots,1^{(i)},\dots,0)$, $i=1,\dots,m$

$$x(P) + \frac{1}{m-1} E_{out} \subsetneq P \subsetneq x(P) + E_{out} \tag{2.7}$$

holds, here-as usual-the notation L^+ is used for the pseudo inverse of the full rank $(n\times m)$ matrix L

$$z=L^+u \leftarrow \min\{\|z\|^2 \mid Lz = u\}.$$

$x(a^m,b^m)$ and $E_{out}(a^m,b^m)$ are invariant under affine transformations of (a^m,b^m).

<u>Proof</u>. The concavity of Ψ and $\log\Psi$ follows from

$$D^2\log\Psi(x)= \frac{1}{m} \sum_{i=1}^{m} -a_ia_i^*/(b_i-\langle a_i,x\rangle)^2=\Psi^{-1}D^2\Psi-\Psi^{-2}(\text{grad}\Psi)(\text{grad}\Psi^*)$$

using the following inequality (for arbitrary vectors $d_1,\dots,d_m$)

$$m \sum_i d_id_i^* \ge (\sum_i d_i)(\sum_i d_i)^* ; \tag{2.8}$$

the latter is obtained (by "multiplying" both sides - from the left and from the right with an arbitrary vector x) from the inequality $m\Sigma\alpha_i^2 \ge (\Sigma\alpha_i)^2$ for arbitrary numbers $\alpha^m \in R^m$. Notice that $E(x) = m \text{ grad} \log \Psi(x)$ and that the equation $E(x) = 0$, (in fact the value of E) are invariant under the transformations $(a^m,b^m)\rightarrow(\tilde{a}^m,\tilde{b}^m)$. The characterization of $x=x(a^m,b^m)$-as solution of (2.4) in P-shows that if x is taken as the origine in R^n, then the Hilbert space dual of P is a polyhedron $\hat{P}$ having $r\le m$ vertices and (m-r) further points belonging to the convex hull of the previous ones, whose arithmetical mean is just zero. From this follows that we can map the standard simplex S^m formed by the endpoints of the ortogonal unit vectors $(e_1, \dots, e_m)$ in R^m affinely into $\hat{P}$ so that the vertices of S^m go into the distinguished points $\hat{P}$ (corresponding to the hyperplanes $b_i \equiv \langle a_i, x\rangle$ by duality). The image of the largest (in volume) ball contained in S^m under this map L is an ellipsoid $\hat{E}$ contained in $\hat{P}$ such that when multiplied with (m-1) it already contains $\hat{P}$. Since duality reverses inclusions and maps an ellipsoid $\langle Bx,x\rangle \le 1$ into the ellipsoid $\langle B^{-1}x,x\rangle \le 1$, we obtain that for the dual of $\hat{E}$, defined as E_{out} the relations (2.6), (2.7) hold true. One can prove easily that the tangential hyperplane to the convex surface $\xi_1, \dots, \xi_m$=const, through the point $\xi(a^m,b^m)$ intersects the positive ortant of R^m in a simplex, which contains the set P, (2.2) and whose centre (of gravity) is $\xi(a^m,b^m)$; (this observation is connected with a special case of Dupin'-s theorem, see [2], §28, and provides an affine "realization" of Karmarkar'-s projective imbedding).

It is important to emphasize that the point x (as well as E_{out}) depends on (a^m,b^m) and not only on P. For m=n+1, $x(a^m,b^m)$, E_{out} can be computed exactly in at most $O(n^4)$ arithmetical operations. This observation can be used for computing $x(a^m,b^m)$ by a continuation method.

Suppose that we have computed a good (i.e. close) approximation $\tilde{x}(a^k,b^k)$ to $x(a^k,b^k)$ for some k≥n+1, let $\hat{a}_j(\lambda):=b_j+\lambda q_j$, j=1,...,m, $q_j=0$, for j=1,...,k and $q_j \ge 0$ is large enough so that for $x(a^m(1), \hat{b}^m(1))$ a good approximation is provided by $x(a^k,b^k)$. Now in order to approximate $x(a^m,b^m)=x(a^m,b^m(0))$ we have to "follow the curve $x(\lambda):=x(a^m,b^m(\lambda))$ from $\lambda=1$, to $\lambda=0$, more precizely: to approximate its "other" end x(0).

Since a continuation technique exploiting the ellipsoidal approximations (2.7) will be described later on, we mention here only that the Jacobi matrix of the parametrized equations for $x(\lambda)$, $E(x,\lambda)=0$ - whose continuously updated inversion constitutes the main complexity of the elementary steps of continuation - is globally regular (well conditioned) and has a simple structure $[\partial E/\partial x](x,\lambda)=AD(x,\lambda)A^*$, where A is a constant matrix and D is a diagonal matrix. Thus we can directly use the update method proposed by Karmankar specially for such matrixes. If we would prefer an other globalization for Newton'-s method to solve the problem (2.4) - (2.5) matrixes of the same type appear (as the ones whose inverses are to be updated). Note that - in order to find the "analytical centre" we work in R^n (and not in R^m as Karmarkar'-s imbedding). It should be noted that the right choice of a continuation method e.g. the "degree" of the extrapolation method, the stepsize selection strategy (as well as exploitation of sparsity) are nontrivial problems of implementation.

3. Applications to linear programming

Suppose that we have to solve the linear programming problem

$$f_o^* = \min\{\langle c,x\rangle \mid x \quad P(a^m,b^m)\}. \tag{3.1}$$

There are many ways to show (or exploit) that linear programming problems are strongly connected (even equivalent - via simultaneous consideration of primal - and dual problems) to the solution of a set of linear inequalities.

For example we can consider the sets $P(\lambda):=P(a^m,b^m)\cap\{x\mid \lambda\geq\langle c,x\rangle\}$ and obtain - by continuation in λ - the set of optimal points as $P(f_o^*)$ and a solution point x^* as the limit of $x(\lambda):=x(P(\lambda))$, for $\lambda\to f_o^*$. The equation for $x(\lambda)$

$$E^c(\lambda,x):=c/(\lambda-\langle c,x\rangle) + \sum_i a_i/(b_i-\langle a_i,x\rangle) = 0 \tag{3.2}$$

is equivalent to a system - for the unknowns $(x,\mu)\in R^{n+m}$

$$c + \sum_{i=1}^{m} \mu_i a_i=0,\ \mu_i(b_i-\langle a_i,x\rangle) = \lambda-\langle c,x\rangle\ . \tag{3.3}$$

At this point we note that - for the convex problem (1.1)-one can consider - as analogons - the concave functions on the intersection of the feasible set with $\{z\mid\lambda>f_o(z)\}$

$$\Phi_\lambda(x)=(\lambda-f_o(x))\prod_{i=1}^{m}(-f_i(x)) \quad \text{and } \log\Phi_\lambda(x) \text{ to obtain}$$

$$f_o'(x)/(\lambda-f_o(x))\sum_{i=1}^{m} f_i'(x)/(-f_i(x)) = 0$$

$$f_o'(x) + \sum_{i=1}^{m}\mu_i\, f_i'(x) = 0\ ,\quad \mu_i(-f_i(x))=\lambda-f_o(x),\ i=1,\dots,m$$

as equations for the continuation curve.

Proposition. The path of (3.3) - for arbitrary $\lambda_o>f_o^*$ - is regular (i.e. $\frac{\partial E}{\partial x}(\lambda,x)$ is nonsingular at $x=x(\lambda)$) on an interval $[-\varepsilon+f_o^*,\ \lambda_o]$ iff there are exactly $(n+1)$ nonzero Lagrange multipliers for the problem (3.1).

Proof. The local unique solvability, regularity - for all values $\lambda>f_o^*$ - follows from theorem 1 (for both systems (3.2) and (3.3)) and $\lambda=f_o^*$ it easy to write down the explicit form of the Jacobi matrix of the arising system and analyse it using the well known results for the extremality (K-T) system, see e.g. [4]. The path $x(\lambda)$, for $\lambda>f_o^*$ is identical with the path of the solutions of the logarithmic penalty function $\hat{x}(r)$, i.e. it is constituted by the solutions of $\min(\langle c,x\rangle + r\sum_i\log(b_i-\langle a_i,x\rangle))$, $r>0$. The new element in our and Karmarkar'-s approach (compared to the previous results of Frisch, Bittner, Fiacco, Mc. Cormick and others) is the availability of the estimation (2.7) which can be used for the selection of the "right" sequence of penalty parameters, and other algorithmic "details". In order that the convergence speed had a lower bound independent of the conditioning of the problem(measured here by that of the Jacobian of (3.3) at the solutions x* and of the "shape" of P), it is necessary that the algorithm (i.e. all its steps) be affine invariant.

A consequence of the estimation (2.7) is the following inequality

$$\frac{1}{m}\lambda \le \langle c, x(P_\lambda^c)\rangle \le (1-1/m)\lambda, \text{ for all } \lambda \ge 0. \qquad (3.4)$$

We describe now a simple (lowest order) "continuation" algorithm, in which, of course, we are not interested to follow $x(\lambda)$ very closely. Suppose - for simplicity - that f_o^* is known to be zero, noting that the general case: f_o^* unknown could be dealt with similarly simple, adaptively updated predictions for f_o^*, i.e. for λ_{k+1} so that one works on R^n - instead of working on R^{n+m} (when the primal-dual system is considered). Let $0<\delta<h<1$ and $0<\beta<\gamma<1$be universal constants. Suppose that x_k and λ_k have been already computed, such that $\|E^c(\lambda_k,x_k)\|=r_k$. Set $\lambda_{k+1}^o:=(1-\delta/m)\lambda_k$ and compute $x_{k+1}^o:=N(x_k,\lambda_{k+1}^o)$, where $N(x,\lambda)$ is the operator giving the result of a Newton iteration for the equation $0=E^c(\lambda,\cdot)$ started from x. Let $d(x):=\min\{\|c\mu + \Sigma_i a_i/(b_i - \langle a_i,x\rangle)\| \mid \mu\in R_+^1\}$ realized for $\mu=\mu(x)$. Set $\lambda_{k+1}^1:=\langle c,x_{k+1}^o\rangle + \mu^{-1}(x_{k+1}^o)$. If $d(x_{k+1}^o) \le$ $\le r_k(1-h/m)$ and $\lambda_{k+1}^1 \le (1-h/m)\lambda_k$, then set $\lambda_{k+1}:=\lambda_{k+1}^o$, otherwise update otherwise update $\lambda_{k+1}^o:=(1-\gamma\delta/m)\lambda_k$ and do the same, stopping the last type of update if $\lambda_{k+1}^o \ge (1-h/m)\lambda_k$. If $\lambda_{k+1}^1 \le (1-h/m)\lambda_k$ never accured update the value of x_k (and so, by definition r_k) - by some globalized Newton method for $E^c(\lambda_k,\cdot)$- so that $r_k^{new} \le \beta r_k$. The performance of this method is easily analyzed for the problem $\min\{\sum_i x_i \mid x_i\ge 0,\ i=1,\dots,m\}$ (at least for $r_k=0$, when the iteration is just "onedimensional") indicating that the same, global convergence rate is obtained as in Karmarkar'-s method. The main point is - besides (3.4) - the stability of the continuation system $\dot{x}(t) = -[\partial E/\partial x]^{-1}(\partial E(x(t))/\partial\lambda)$ and the local quadratic convergence of Newton's method. Thus by appropriate choice of the parameters arbitrarily close and stable following is possible (note that in the simplex method instabilities may occur already far away from the solution point). For small values of λ we "switch" the continuation from the system (3.2) to the system (3.3). The gain in flexibility over the method of Karmarkar is not only that we could use adaptive (i.e. stabilized) higher order extrapolation methods (preferably by rational functions since the curve $x(\lambda)$ may have a singularity at $\lambda=0$, see below) for predicting the path, but also a significant possibility is offered for neglecting (i.e. dropping for ever) those constraints - at specified steps $k=k_r$, $r=1,2,\dots$ - which do not intersect the ellipsoid $E_{out}(P^c(\lambda_k))$.

An other class of methods can be based on the observation that Karmarkar'-s method is just the "steepest descent" method (with a rather clever stepsize selection) for minimizing the potential function $\pi(x):=\log(\langle c,x\rangle^m / \prod_{i=1}^{m}(b_i-\langle a_i,x\rangle))$ over P. This function is not convex over P, however one can show that

$$f(x):=\langle c,x\rangle^{m+1} / \prod_{i=1}^{m}(b_i - \langle a_i,x\rangle) \quad , \quad (\text{for } f^* = f^*_o = 0)$$

is globally strictly convex on the interior of P and its gradient is zero at the optimal points of the LP problem (3.1), so that higher order (e.g. Newton-type) method and simple (spezialised) line search techniques can be used for its "almost" unconstrained minimization. To this idea arrived - independently of the present author-prof. M. Iri (Tokyo University) who mentioned the above choice of f in a personal conversation with the author at the 12 IFIP Conf.). The convexity of f is a consequence of our inequality (2.8) since we have the explcit formulae

$$D^2 f(x)=f(x)(k(x)k^*(x) + \sum_i d_i d_i^* - \frac{1}{m}(\sum_i d_i)(\sum_i d_i)^*)$$

$$k(x)=\sqrt{(m+1)m}\; c/\langle c,x\rangle+\sqrt{\frac{m+1}{m}} \sum_{i=1}^{m} d_i, \quad d_i=a_i/(b_i-\langle a_i,x\rangle) \quad i=1,\dots,m$$

It is easy to check that this function can be used as the "potential" function $\pi(x)$, e.g. we could add a $(m+1)\underline{^{th}}$ constraint $\lambda_0 \geq \langle c,x\rangle$, for some λ_0 large (with respect to $\langle c,x_0\rangle$, $x_0 \in P$). Besides the steepest descent and Newton directions (for onedimensional search) the direction $p=p(x)$

$$p=-\left(\frac{\partial E^c(\lambda(x),x)}{\partial x}\right)^{-1} \frac{\partial E^c(\lambda(x),x)}{\partial \mu}, \quad \min_\mu ||c\mu + \sum_{i=1}^{m} \frac{a_i}{b_i-\langle a_i,x\rangle}||, \quad \lambda=\mu^{-1}+\langle c,x\rangle$$

seems to be of interest.

4. Globally convergent accelerations for Newton'-s method

We assume that $f:R^n \to R^1$ is a sufficiently smooth (at least C^3) function which is strictly convex in u and strictly concave in v, ($x=(u,v)$) being a direct sum decomposition of R^n) such that $g(x):= \text{grad } f(x) = 0$ has a unique solution x^*. The equations $H_0(\lambda,x)=(1-\lambda)g_0(x)+\lambda g(x)$, where where g_0 is the linearization of g at x_0 have for each $0\leq \ \leq 1$ a unique solution $x(\lambda)$, which fill up a (correspondigly) smooth curve (by the implicit function theorem and the convexity of the arising class of

Jacobi matrixes) connecting x_1 and x^*, where x_1 is the solution of $g_0(x) = 0$.

The following "modell algorithm" is interesting, since its "order" can be made arbitrary high, while its global convergence is guaranted. Let $\lambda_0=0$, $\lambda_k=\delta+\ldots+\delta\gamma^{k-1}$, $k\geq 1$, where $\gamma\geq 1$ (near to 1), $k=0,\ldots,d$. It is easy (for small and sequentially updated δ) to compute the solutions x_{k+1} of $H(\lambda_k,.)$ within high accuracy (assuming e.g. that the Hesse matrixes of f are uniformly well conditioned). Now we can build a d-th order extrapolation for the values of $x(\beta^j)$, $j=0,1,\ldots,m$, $\beta^m\geq\lambda_d$ based on the (approximate) values of $x(\lambda_k) = x_{k+1}$ $k=0,\ldots,d$, e.g. by using recursive in k rational interpolation for each of the coordinates, see e.g. the algorithms in [7] (note that in general the analytic continuation of x contain poles beyond $\lambda=1$). Let j be the smallest value for which $||g(x(\beta^j))||\leq\Psi||g(x_0)||$, where Ψ as well as β are fixed positive numbers less than 1 (if this does not happen we let $\delta=\delta/2$ or set $d=2d$). Accept $x(\beta^j)$ as new approximation and continue giving to it the role of the previous x_0. An other method of globalization is obtained if we change the linearization point x_k^ℓ and homotopy paramter λ_k^ℓ simultaneously: supposing that x_k is an approximate solution of $H_{k-1}(\lambda_k,.)=0$, we define $x_k^\ell=\tilde{x}_{k-1}+2^{-\ell}(x_k-\tilde{x}_{k-1})$ $\lambda_k^\ell:=\lambda_k+2^{-\ell}\delta$ and set the latter equal to λ_{k+1} (and x_k^ℓ equal to the next linearization point) only if the Newton step $N(H_k^\ell,\cdot)$ started from x_k gives a small residual (yielding the value of x_{k+1}). More precizely, we assume - at the beginning of the (k+1)th iteration - that for some, fixed $\varepsilon>0$ -

$$||H_{k-1}(\lambda_k,\tilde{x}_{k-1})||\leq\varepsilon,\text{ where } H_{k-1}(\lambda,k)=(1-\lambda)(g(\tilde{x}_{k-1})+\frac{\partial g}{\partial x}(\tilde{x}_{k-1})(x-\tilde{x}_{k-1}))+\lambda g(x)$$

and

$$H_k^\ell(x) := (1-\lambda_k^\ell)(\frac{\partial g}{\partial x}(x_k^\ell)(x-x_k^\ell)) + \lambda_k^\ell g(x)$$

is the equation (for the Newton step) which we try to solve - within residual error ε, for $\ell=0,1,\ldots,\ell_k$, so that

$$\tilde{x}_k := x_k^{\ell_k},\ x_{k+1} = N\ H_k^{\ell_k},\ x_k\ ,\ \lambda_{k+1} = \lambda_k^{\ell_k}.$$

Instead of a fixed δ we could choose - at step k - $\delta=1-\lambda_k$. The starting equation is $g_0(x)=0$, i.e. an easily (explicitly) solvable one.

<u>Proposition</u>. Suppose that for the eigenvalues of the Hesse matrix H of f we have the universal bounds

$$0<m\leq|\lambda_i(H(x))|\leq M<\infty,\ i=\ ,\ldots,n \tag{4.1}$$

$$||H(x) - H(y)|| \leq L||x-y||,\text{ for all } x,y\in R^n, \tag{4.2}$$

then - if $\varepsilon>0$ is sufficiently small (depending only on the values (m,M,L) and $\delta>0$ is arbitrary - the above (second) algorithm reaches, from arbitrary initial vector x_0 in a finite number of steps (thus "ℓ-cycles") a value $k \le N(||x_0 - X||, m, M, L)$ such that $||g(x_k)|| \le \varepsilon$; the length of each of the k "ℓ-cycles" is bounded by a number depending only on (ε, M, m, L). The proof follows from the convexity of the class of symmetric matrixes satisfying (4.1) and (4.2) and from the well known facts about the local convergence of Newton'-s method. In an implementation, especially if we have rather smooth functions, for the update of the inverses of $(\partial H_k(x)/\partial x)^{-1} := B_k^{\ell}(x)$ one can use special methods e.g. interpolations

$$B_k^{\ell}(x_k^{\ell}) = (1-\lambda_k^{\ell})\, B(\tilde{x}_{k-1}) + \lambda_k^{\ell}\, B(x_k).$$

Note that the first step of each cycle is a Newton step for g=0.

Similarly, for the first method we should specify - among other details, the upper bound d^o for d and a lower bound for δ; the use of the information $x(\lambda_k)$, $k=1,\dots,m>d^o$, for a d^o-th order prediction. The main point is that the first method allows for a higher order prediction for (thus convergence to x^*) as the first method. Rational extrapolation is recommended since $x(\lambda)$ may have singularities beyond [0,1]. These methods can be easily generalized to the case when we have a restriction on x, e.g. $x \in P(a^m, b^m)$ thus for the minimization of the functions Ψ or f, at the first step we solve $H(\lambda, x)$ for some $\lambda=\delta$ small so that $x_1 \in P$ is guaranteed.

References

[1] A. Bachem, M. Grötschel, B. Korte (eds.), Mathematical Programming, The State of Art, Springer V., Berlin, 1983.

[2] T. Bonnesen, W. Fenchel, Theorie der konvexen Körper, Springer V., Berlin, 1934.

[3] V. F. Demyanov, D. Pallaschke (eds.), Proc. IIASA Workshop on Nondifferentiable Optimization. Lect. Notes in Economics and Mathematical Systems, Vol. 255, Springer, Berlin, 1985.

[4] A. V. Fiacco, Introduction to Sensitivity and Stability Analysis in Nonlinear Programming, Academic Press, New York, 1983.

[5] N. Karmarkar, A new polynomial-time algorithm for linear programming, Combinatica 3. (4) (1984) pp. 373-395.

[6] Gy. Sonnevend, A modified ellipsoid method for the minimization of convex functions with superlinear convergence (resp. finite termination) for well conditioned C^3 smooth (resp. piecewise linear) functions, see in [3], pp. 264-278.

[7] Gy. Sonnevend, Sequential method for spectral approximation with rational and with Haar functions, to appear in Proc. of the A. Haar Memorial Conference (ed. by. J. Szabados), North Holland Publ. Comp. 1986.

STRUCTURAL OPTIMIZATION WITH RELIABILITY CONSTRAINTS

J. D. Sørensen & P. Thoft-Christensen
University of Aalborg
Sohngaardsholmsvej 57
DK-9000 Aalborg
Denmark

1. INTRODUCTION

During the last 25 years considerable progress has been made in the fields of structural optimization and structural reliability theory. In classical deterministic structural optimization all variables are assumed to be deterministic. Due to the unpredictability of loads and strengths of actual structures it is now widely accepted that structural problems are non-deterministic. Therefore, some of the variables have to be modelled as random variables/processes and a reliability-based design philosophy should be used, Cornell [1], Moses [2], Ditlevsen [3] and Thoft-Christensen & Baker [4].

In this paper we consider only structures which can be modelled as systems of elasto-plastic elements, e.g. frame and truss structures. In section 2 a method to evaluate the reliability of such structural systems is presented. Based on a probabilistic point of view a modern structural optimization problem is formulated in section 3. The formulation is a natural extension of the commonly used formulations in determinstic structural optimization. The mathematical form of the optimization problem is briefly discussed.

In section 4 two new optimization procedures especially designed for the reliability-based optimization problem are presented. In some examples in section 5 the optimization procedures are compared.

2. RELIABILITY OF STRUCTURAL SYSTEMS

The loads on the considered structures and the strengths of the structural elements are modelled as time-invariant stochastic variables. All other variables such as geometrical quantities are assumed to be deterministic.

Failure of the structural system can be defined in a number of ways. For a detailed description, see e.g. Thoft-Christensen [5]. The computationally simplest definition which in statically indeterminate structures can be taken as a serviceability limit state is to define failure of the system as failure of one of the structural elements. This is called a level 1 definition of failure. As a measure of the reliability of a structural element the reliability index β can be used, see Thoft-Christensen & Baker [4]. Let the N basic random variables be collected in the vector

$\underset{\sim}{Y} = (Y_1, Y_2, \ldots, Y_N)$ with given density function $f_{\underset{\sim}{Y}}(\underset{\sim}{y})$ and the failure surface which defines the separation between the safe and failure areas in the sample space be given by the equation $g(\underset{\sim}{y}) = 0$. When $\underset{\sim}{Y}$ is non-normally distributed a transformation from $\underset{\sim}{Y}$ to the normally distributed standardized vector $\underset{\sim}{U} = \underset{\sim}{T}(\underset{\sim}{Y})$ is established (e.g. the Rosenblatt transformation can be used, [6]). The reliability index β is now defined as the shortest distance from the origin in the $\underset{\sim}{u}$-space to the failure surface:

$$\beta = \min_{g(\underset{\sim}{T}(\underset{\sim}{z}))=0} \left(\sum_{i=1}^{N} z_i^2\right)^{\frac{1}{2}} \tag{1}$$

The reliability of the structural system can now be estimated by modelling each structural element as an element in a series system.

The above definition of failure can be generalized to a level m definition of failure. Failure of the structural system is then defined as the event that m structural elements have failed. Such a failure mode can be modelled as a parallel system with m elements. To estimate the reliability of the structural system each parallel system is modelled as an element in a series system.

Usually in elasto-plastic systems the ultimate limit state is defined by the formation of a mechanism (i.e. collapse). The number of possible mechanisms in a structural system is usually very large. For the types of structure considered here the so-called safety margins for the mechanisms can be written

$$M_i = \sum_{j=1}^{N_R} a_{ij} R_j - \sum_{j=1}^{N_P} b_{ij} P_j \quad , \quad i = 1, 2, \ldots, h \tag{2}$$

where $\underset{\sim}{R}$ and $\underset{\sim}{P}$ model the yield strength and load variables. $\underset{\sim}{a}$ and $\underset{\sim}{b}$ are matrices which contain coefficients of influence. $N = N_R + N_P$ and h is the number of mechanisms. Here $\underset{\sim}{R}$ and $\underset{\sim}{P}$ are assumed to be normally distributed with expected values $\underset{\sim}{\mu}_R$ and $\underset{\sim}{\mu}_P$ and covariance matrices $\underset{\sim}{C}_R$ and $\underset{\sim}{C}_P$. $\underset{\sim}{R}$ and $\underset{\sim}{P}$ are assumed independent.

The reliability index β_i for the ith safety margin is then as follows

$$\beta_i = \frac{\mu_i}{\sigma_i} = \frac{\sum_{j=1}^{N_R} a_{ij}\mu_{R_j} - \sum_{j=1}^{N_P} b_{ij}\mu_{P_j}}{\left(\sum_{j=1}^{N_R}\sum_{k=1}^{N_R} a_{ij}a_{ik}C_{R_{jk}} + \sum_{j=1}^{N_P}\sum_{k=1}^{N_P} b_{ij}b_{ik}C_{P_{jk}}\right)^{\frac{1}{2}}} \tag{3}$$

where μ_i and σ_i are the expected value and the standard deviation of M_i. The coefficient of correlation between the ith and j^{th} safety margin is

$$\rho_{ij} = \frac{\sum_{k=1}^{N_R}\sum_{\ell=1}^{N_R} a_{ik}a_{j\ell}C_{R_{k\ell}} + \sum_{k=1}^{N_P}\sum_{\ell=1}^{N_P} b_{ik}b_{j\ell}C_{P_{k\ell}}}{\sigma_i\sigma_j} \tag{4}$$

These failure modes are modelled as elements in a series system, and an upper-bound estimate of the reliability of the elasto-plastic structural system is given by

$$\beta_S = -\Phi^{-1}(1-\Phi_h(\underset{\sim}{\beta};\underset{\sim}{\rho})) \quad (5)$$

where Φ^{-1} is the inverse standard normal distribution function and $\Phi_h(\cdot\,;\underset{\sim}{\rho})$ is the standard distribution function for h normal variables with correlation coefficient matrix $\underset{\sim}{\rho}$.

In real structures (e.g. offshore steel jacket structures) the number of possible failure modes is generally very large. It is therefore important to be able to identify the most significant failure modes. For that purpose the so-called β-unzipping technique has been developed by the authors, Thoft-Christensen & Sørensen [7].

The basic idea in the β-unzipping technique is that a failure tree is successively formed. Each node signifies a modified structure where a number of elements have failed and each branch is an element. The critical elements are selected on the basis of the safety indices of the elements in the modified structures.

3. RELIABILITY-BASED STRUCTURAL OPTIMIZATION

In classical deterministic structural optimization for truss and frame structures the design variables are usually the cross-sectional areas x_i, i = 1, 2, . . . , n, where n is the number of sets of different structural members. Each structural element is characterized by one number. This is fully satisfactory for truss structures where only tensile/compressive forces exist. However, when bending occurs in a structural member, the plastic section moduli w_i, i = 1, 2, ... , n and the second moments of area I_i, i = 1, 2, . . . , n are significant. To maintain the great computational advantage of having only one design variable for each structural member it is often assumed that

$$w_i = k_1 x_i^{2/3} \quad (6)$$

$$I_i = k_2 x_i^2 \quad (7)$$

where k_1 and k_2 are constants.

As objective function a natural choice would be the total cost of the structure. But due to the difficulties in assigning monetary values to failure consequences and to the initial cost we have in this paper used the structural weight as objective function. If the structure is made of only one type of material the weight is proportional to

$$W(\underset{\sim}{x}) = \sum_{i=1}^{n} \ell_i x_i \quad (8)$$

where ℓ_i is the total length of the elements having the area x_i.

In classical structural optimization the constraints usually signify that the stresses and/or displacements should be smaller than some prescribed values. In reliability-based structural optimization a choice for the constraints could be that the reliability index in all elements should be greater than some target value. However, based on the discussion in section 2 a more natural choice would be to use the system reliability index

$$\beta_S(\underset{\sim}{x}) - \beta_S^0 \geq 0 \quad (9)$$

where β_S^0 is some target system reliability index and $\beta_S(\underset{\sim}{x})$ is given by (5). Because the areas of the structural elements have to be non-negative we also have the constraints

$$x_i \geqslant 0 \quad , \quad i = 1, 2, \ldots, n \tag{10}$$

The optimization problem is seen to have a linear, objective function. Since an optimum point is a global optimal point if the optimization problem is convex it is important to investigate if the constraint (9) is concave. A precondition is that the Hessian matrix is negative semi-definite. The elements in the gradient vector and the Hessian matrix are

$$\frac{\partial \beta_S(\underset{\sim}{x})}{\partial x_i} = \frac{1}{\varphi(-\beta_S)}\left[\sum_{k=1}^{n} \frac{\partial \Phi_h}{\partial \beta_k}\frac{\partial \beta_k}{\partial x_i} + \sum_{k=1}^{n-1}\sum_{\ell=k+1}^{n}\frac{\partial \Phi_h}{\partial \rho_{k\ell}}\frac{\partial \rho_{k\ell}}{\partial x_i}\right] \tag{11}$$

where $\underset{\sim}{\beta}$ and $\underset{\sim}{\rho}$ are given by (3) and (4). φ is the standard normal density function.

$$\frac{\partial^2 \beta_S(\underset{\sim}{x})}{\partial x_i \partial x_j} = \frac{1}{\varphi(-\beta_S)}\left[\beta_S \varphi(-\beta_S)\frac{\partial \beta_S}{\partial x_i}\frac{\partial \beta_S}{\partial x_j}\right.$$

$$+ \sum_{k=1}^{n}\sum_{\ell=1}^{n}\frac{\partial^2 \Phi_h}{\partial \beta_k \partial \beta_\ell}\frac{\partial \beta_k}{\partial x_i}\frac{\partial \beta_\ell}{\partial x_j} + \sum_{k=1}^{n}\frac{\partial \Phi_h}{\partial \beta_k}\frac{\partial^2 \beta_k}{\partial x_i \partial x_j}$$

$$\left. + \sum_{k=1}^{n-1}\sum_{K=k+1}^{n}\sum_{\ell=1}^{n-1}\sum_{L=\ell+1}^{n}\frac{\partial^2 \Phi_h}{\partial \rho_{kK}\partial \rho_{\ell L}}\frac{\partial \rho_{kK}}{\partial x_i}\frac{\partial \rho_{\ell L}}{\partial x_j} + \sum_{k=1}^{n-1}\sum_{K=k+1}^{n}\frac{\partial \Phi_h}{\partial \rho_{kK}}\frac{\partial^2 \rho_{kK}}{\partial x_i \partial x_j}\right] \tag{12}$$

The derivatives in (11) and (12) can be derived from the definition of Φ_h, (3) and (4). As seen from (12) it is very difficult to establish whether the constraint in a given problem is concave. To show that the fulfilment of the concavity condition depends on the parameters in the given problem consider the following simple example.

Example

In (2) we assume that $h = 1$, $N_R = 2$, $a_{11} = a_{12} = 1$, $\sum_{j=1}^{N_P} b_{ij}\mu_{P_j} = 2$, $\sum_{j=1}^{N_P}\sum_{k=1}^{N_P} b_{ij}b_{ik}C_{P_{jk}} = 0.05$,

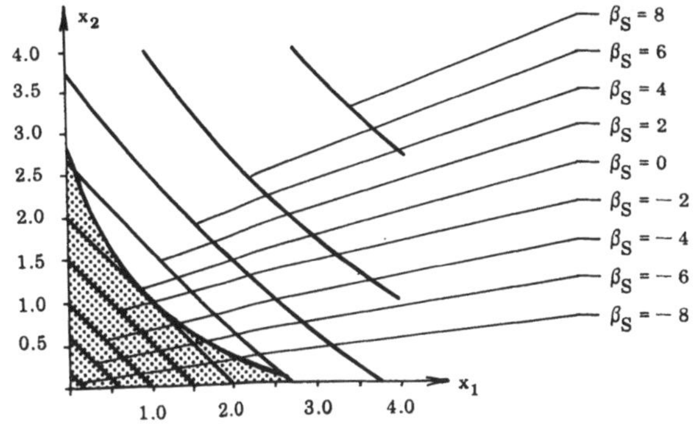

Figure 1.

$E[R_1] = x_1$, $E[R_2] = x_2$, $C_{R_{11}} = (0.1\ x_1)^2$, $C_{R_{22}} = (0.1\ x_2)^2$, $C_{R_{12}} = (0.1)^2 \rho x_1 x_2$, $\rho = 0.3$.
In figure 1 contours of $\beta_S(x_1, x_2)$ are shown. The Hessian matrix is determined according to (12). The hatched area in figure 1 shows the area where the Hessian is not negative semi-definite. It is seen that only for combinations of x_1 and x_2 where the reliability index is relatively small, the Hessian is not negative semi-definite.

The above example indicates that for a structure with a reasonable reliability index ($\beta_S \geq 3$) the reliability-based optimization problem will in most cases be convex.

4. OPTIMIZATION PROCEDURES

In this section we describe two optimization procedures. The computational work involved in solving the optimization problem described above can be divided in three parts:

I Identification of critical failure modes by the β-unzipping method.

II Evaluation of the systems reliability index for a given set of critical failure modes.

III Optimization calculations.

Due to the great complexity of the constraint (9) the derivatives of (9), if needed, will be calculated by using finite differences. Each time calculation of the constraint (9) is requested by the optimization algorithm both part I and II have to be performed. But because part I is very time-consuming compared to part II and because the set of critical failure modes cannot be expected to change significantly due to small changes in the design vector $\underset{\sim}{x}$, the critical failure modes are only identified when one of the following conditions is fulfilled (the latest identification is performed at iteration step i with the design vector $\underset{\sim}{x}^i$):

1. The actual iteration step is equal to $i + I_c$.

2. $\sqrt{\sum_{j=1}^{n} (x_j - x_j^i)^2} \geq x_{max}$, where $\underset{\sim}{x}$ is the actual design vector.

Evaluation of the systems reliability index given a set of significant failure modes (part II) has to be performed many times. Generally, the evaluation has to be made approximately. In this paper we have used partly the so-called PNET method, Ang & Ma [8] and partly the average correlation coefficient method, Thoft-Christensen & Sørensen [9].

The special formulation of the optimization problem where the constraint (9) is very important has caused the following test for optimality to be used:

$$\sum_{j=1}^{n} c_1 \left| \frac{x_j^i - x_j^{i-1}}{x_j^i} \right| + c_2 |\beta_S^0(\underset{\sim}{x}^i) - \beta_S^0| + c_3 \left| \frac{W(\underset{\sim}{x}^i) - W(\underset{\sim}{x}^{i-1})}{W(\underset{\sim}{x}^i)} \right| < \epsilon \qquad (13)$$

where $\underset{\sim}{x}^i$ is the value of $\underset{\sim}{x}$ at iteration level i. This stopping criterion can only be used when the set of significant failure modes is updated. c_1, c_2, and c_3 are prescribed constants.

The two different optimization algorithms which have been used in the above optimization procedure are

a. The non-linear programming code NLPQL developed by Schittkowski [10]. This mathematical method is based on the successive solution of quadratic programming sub-problems and a subsequent one-dimensional line search with an augmented Lagrange function as merit function. The optimality test (13) has been added to the tests in NLPQL.

b. In structural optimization a sequential linear programming technique based on the inverse areas of the structural elements is often used, e.g. Fleury [11]. At the ith iteration level the constraint (9) is linearized

$$\beta_S(\underset{\sim}{x}) - \beta_S^0 \cong \beta_S(\underset{\sim}{x}^i) + \sum_{j=1}^{n} \left(\frac{\partial \beta_S(\underset{\sim}{x})}{\partial(\frac{1}{x_j})} \right|_{\underset{\sim}{x}=\underset{\sim}{x}^i} \left(\frac{1}{x_j} - \frac{1}{x_j^i}\right)\Bigg) - \beta_S^0 = \sum_{j=1}^{n} \frac{d_j^i}{x_j} - \beta^i \qquad (14)$$

At iteration level i the new design vector $\underset{\sim}{x}^{i+1}$ is then found by solving the linearized Problem ((8) + (14) + (10)):

$$x_j^{i+1} = \max\{0, \sqrt{\frac{D_j^i}{\ell_i}} \; \frac{\sum_{k=1}^{n} \sqrt{D_k^i \ell_k}}{\beta^i}\} \qquad (15)$$

where $\underset{\sim}{\ell}$ is defined in (8) and

$$D_j^i = \frac{1}{i} \sum_{k=1}^{i} d_j^k \qquad (16)$$

(16) is added to stabilize the iteration.

5. EXAMPLES

Consider the frame shown in figure 2. It has 4 different structural elements with areas x_1, x_2, x_3, and x_4. k_1 and k_2 in (1) - (2) are chosen as (Gorman [12]) $k_1 = 1.84$ and $k_2 = 3.20$.

The loading (5 concentrated loads) and the 19 failure elements (potential yield hinges) are modelled by 4 + 19 normal stochastic variables with constant coefficients of variation. The expected values of the failure elements are determined by

$$\mu_i = w_i \cdot 270 \cdot 10^3 \text{ kNm}^{-2} \quad , \quad i = 1, 2, 3, 4$$

where w_i is given by (6). Further details concerning the loading, the correlation structure, and the β-unzipping can be found in Thoft-Christensen & Sørensen [13]. The constants in (13) are chosen as $c_1 = c_2 = c_3 = 1$ and $\epsilon = 0.01$.

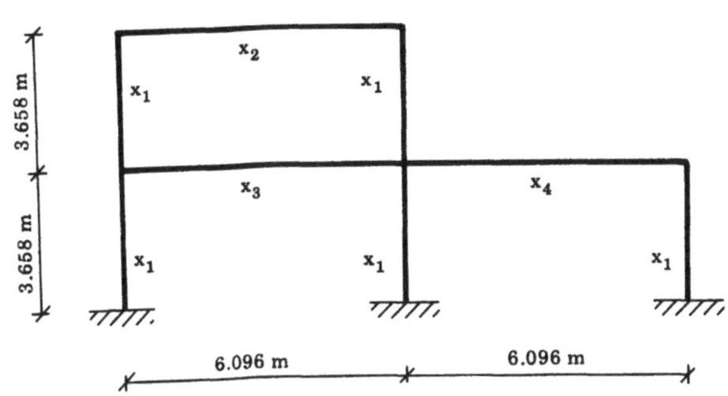

Figure 2. Geometry and optimization variables.

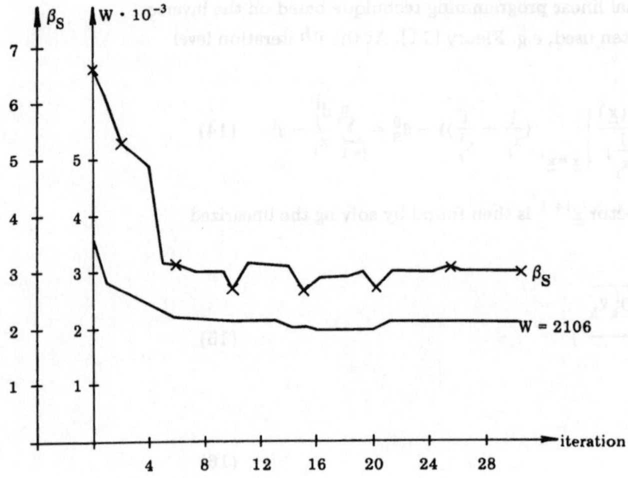

Figure 3. Iteration history for failure defined at level 2. × indicates identification of significant failure modes.

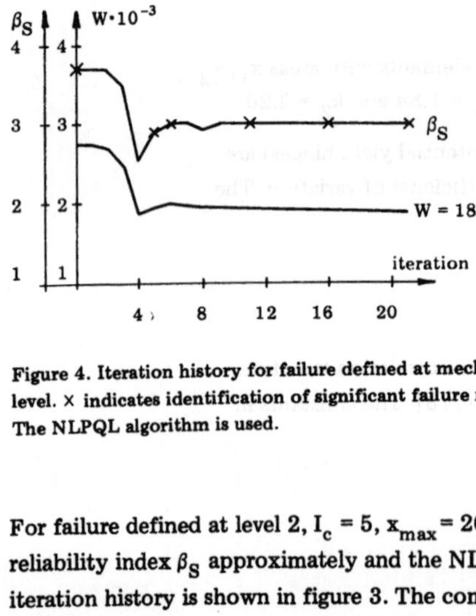

Figure 4. Iteration history for failure defined at mechanism level. × indicates identification of significant failure modes. The NLPQL algorithm is used.

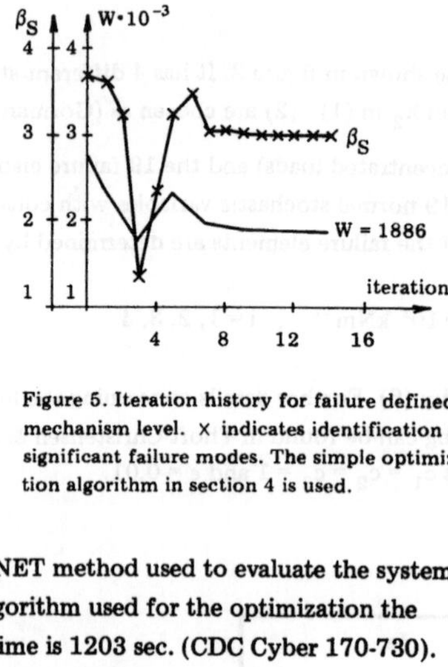

Figure 5. Iteration history for failure defined at mechanism level. × indicates identification of significant failure modes. The simple optimization algorithm in section 4 is used.

For failure defined at level 2, $I_c = 5$, $x_{max} = 20$, the PNET method used to evaluate the systems reliability index β_S approximately and the NLPQL algorithm used for the optimization the iteration history is shown in figure 3. The computer time is 1203 sec. (CDC Cyber 170-730). Some fluctuations are observed. These are mainly due to the stepwise updating of the significant failure modes. The optimal areas are $\underset{\sim}{x} = (52.0, 51.9, 82.6, 55.0)$.

With the same parameters the iteration history for failure defined at mechanism level is shown in figure 4. The computer time is 150 sec. Again it is seen that the process converges, although there is a great fluctuation at iteration no. 4. The optimal areas are $\underset{\sim}{x} = (42.9, 50.7, 70.3, 59.8)$.

In figure 5 the iteration history is shown for the same example as in figure 4. The only differences are that the simple optimization algorithm b in section 4 is used instead of the advanced NLPQL algorithm and $I_c = 1$. Also with this algorithm the process converges. The optimum

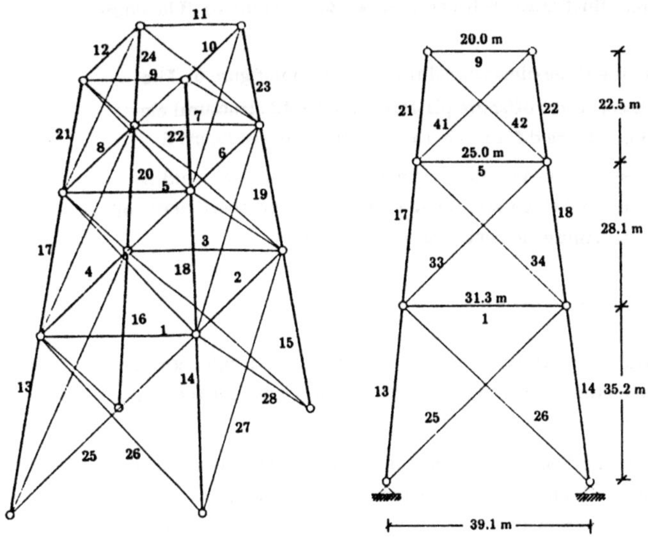

area	elements
x_1	1, . . . , 4
x_2	5, . . . , 8
x_3	9, . . . , 12
x_4	13, . . . , 24
x_5	25, . . . , 32
x_6	33, . . . , 40
x_7	41, . . . , 48

Figure 6. Spatial truss tower.

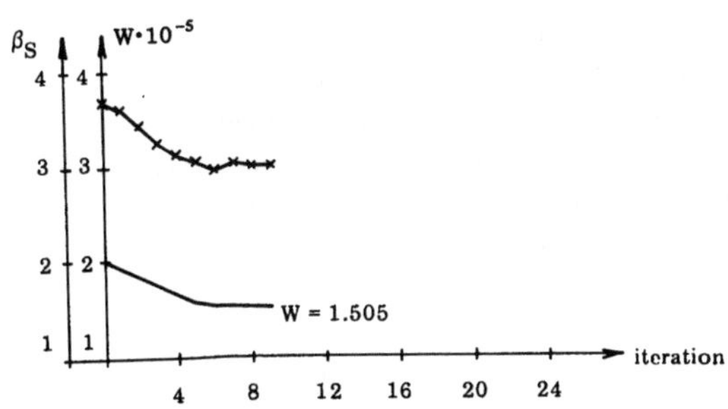

Figure 7. Iteration history for failure defined at mechanism level. × indicates identification of significant failure modes. The simple optimization algorithm in section 4 is used.

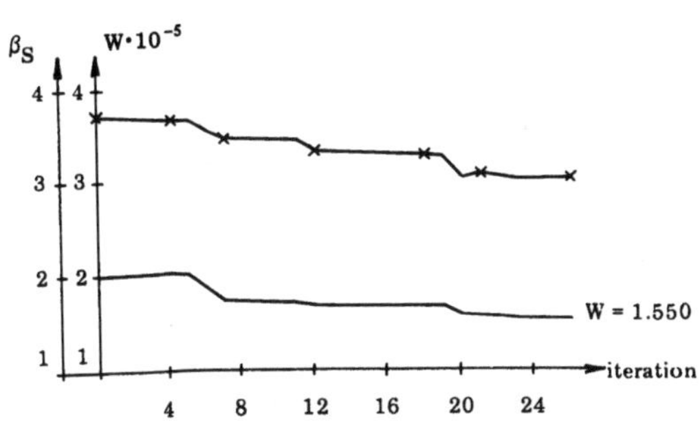

Figure 8. Iteration history for failure defined at mechanism level. × indicates identification of significant failure modes. The NLPQL algorithm is used.

point is the same and the same great fluctuation is found (here at iteration no. 3). The computer time is 124 sec.

In the following example consider the three-dimensional truss structure in figure 6. The structural system is a model of a steel jacket offshore platform and has 48 structural elements. But only 7 of them are chosen to be different in the optimization, see figure 6.

The loading (16 concentrated loads) and the 48 failure elements (potential axial yielding elements) are modelled by 2 + 48 normal stochastic variables with constant coefficients of variation. The expected values of the failure elements are determined by

$$\mu_i = x_i \cdot 270 \cdot 10^3 \ \mathrm{kNm}^{-2} \quad , \quad i = 1, 2, \ldots, 7$$

Further details concerning the loading, the correlation structure, and the β-unzipping can be found in Sørensen et al. [14]. The method of average correlation coefficients, see section 4, is used to evaluate β_S approximately.

With the same parameters as used in the first example and failure defined at mechanism level iteration histories are shown in figures 7 and 8. In figure 7 the result from using the simple optimization algorithm b in section 4 is shown. Convergence is obtained after 9 iterations. The total computer time is 8191 sec. (2805 sec. for identification of failure modes, 4625 sec. for evaluation of β_S, and 761 sec. for optimization calculations). The optimal areas are (0, 105, 127, 172, 0, 2.7, 271).

In figure 8 the result of a run with the NLPQL algorithm is shown. After 26 iterations convergence is not obtained. The run stopped because the algorithm could not find a better point. The reason is probably that the reliability constraint is very flat in the area about the minimum point. The object function value is 3% greater than the value which was found using the simple optimization algorithm and the areas at the point where the algorithm stopped are (0, 60, 88, 248, 0, 5.5, 212). The computer time is 11266 sec. It is seen that three of the areas are almost 0.

The results of the latter example therefore indicate that the simple optimization algorithm (see section 4) is better than the advanced NLPQL algorithm to find the optimal areas in a structural system with reliability constraint. Further investigations of the effect of the choice of the parameters in the algorithms are being performed.

6. CONCLUSION

The optimization problem which is considered in this paper is to find the minimum weight of a structural system subject to the constraint that the reliability of the structure exceeds a critical value.

In this paper it is shown that it is generally not possible to establish that the optimization problem is convex. A simple example demonstrates this.

To solve the optimization problem a new optimization procedure is developed. The procedure is composed of three main parts, namely identification of significant failure modes, evaluation of the systems reliability index, and calculation of the optimal point. Since the first two parts are the most computer time consuming special considerations are given to these parts in designing the procedure. Two different optimization algorithms are investigated, namely a simple procedure based on linearization of the constraint and the NLPQL algorithm which is an advanced procedure based on solving sequential quadratic subproblems.

In two examples the procedure is investigated. In the first example runs with both algorithms converge. In the second example only the run with the simple optimization algorithm converges. The run with the NLPQL algorithm stops at a point with a value of the weight which is 3% greater. This result indicates that for complex structural systems with reliability constraint it is better to use a simple optimization algorithm.

7. REFERENCES

[1] Cornell, C. A.: *A Probability-Based Structural Code.* ACI Journal, American Concrete Institute, Vol. 66, 1969, pp. 974-985.

[2] Moses, F.: *Design for Reliability-Concepts and Applications.* Optimum Structural Design (eds. Gallagher & Zienkiewics). John Wiley, N. Y., 1973, pp. 241-265.

[3] Ditlevsen, O.: *Basic Reliability Concepts.* Proceedings, NATO Advanced Study Institute (ed. P. Thoft-Christensen), Martinus Nijhoff Publishers, The Netherlands, 1983, pp. 1 - 56.

[4] Thoft-Christensen, P. & Baker, M. J.: *Structural Reliability Theory and Its Applications.* Springer-Verlag, Berlin, 1982.

[5] Thoft-Christensen, P.: *Reliability Analysis of Structural Systems by the β-Unzipping Method.* Structural Reliability Theory, Paper no. 3, Institute of Building Technology and Structural Engineering, The University of Aalborg, 1984.

[6] Lind, N. C. & Krenk, S. & Madsen, H. O.: *Safety of Structures.* Prentice Hall, 1985.

[7] Thoft-Christensen, P. & Sørensen, J. D.: *Calculation of Failure Probabilities of Ductile Structures by the β-Unzipping Method.* Institute of Building Technology and Structural Engineering. The University of Aalborg, 1982.

[8] Ang, A. H.-S. & Ma, H. F.: *On the Reliability Analysis of Framed Structures.* Proc. ASCE Specialty Conference on Probabilistic Mechanics and Structural Reliability, Tucson, 1979, pp. 106-111.

[9] Thoft-Christensen, P. & Sørensen, J. D.: *Reliability of Structural Systems with Correlated Elements.* Applied Mathematical Modelling, Vol. 6, 1982, pp. 171-178.

[10] Schittkowski, K.: *Theory, Implementation, and Test of a Nonlinear Programming Algorithm.* Proc. Euromech-Colloquium 164 on »Optimization Methods in Structural Design», 1982, Bibliographisches Institut, Mannheim, 1983, pp. 122-132.

[11] Fleury C.: *Structural Weight Optimization by Dual Methods of Convex Programming.* International Journal for Numerical Methods in Engineering, Vol. 14, 1979, pp. 1761-1783.

[12] Gorman, M. R.: *Reliability of Structural Systems.* Report No. 79-2, Case Western Reserve University, Ohio, Ph.D. Thesis, 1979.

[13] Thoft-Christensen, P. & Sørensen, J. D.: *Optimization of Structural Systems.* Structural Reliability Theory, Paper no. 6, Institute of Building Technology and Structural Engineering, The University of Aalborg, 1984

[14] Sørensen, J. D. & Thoft-Christensen, P. & Sigurdsson, G.: *Development of Applicable Methods for Evaluating the Safety of Offshore Structures, Part 2.* Structural Reliability Theory, Paper no. 11, Institute of Building Technology and Structural Engineering, The University of Aalborg, 1985.

IAEA'S ACTIVITIES IN ELECTRIC SYSTEM EXPANSION PLANNING

J.A. MARQUES DE SOUZA J.P. CHARPENTIER
P. MOLINA K. SCHENK

ECONOMIC STUDIES SECTION
DIVISION OF NUCLEAR POWER
INTERNATIONAL ATOMIC ENERGY AGENCY

ABSTRACT

In meeting its objective to assist its developing Member States in the peaceful uses of nuclear energy, the IAEA conducts an extensive programme of work in nuclear power planning and implementation, including economic assessments to determine the appropriate role of nuclear energy within the national energy plan of developing Member States. Within this framework three major types of interdependent and closely related activities are carried out: (i) development of methodologies appropriate for energy, electricity and nuclear power planning in developing countries; (ii) the conduct of training courses on energy and nuclear power planning techniques, including use of methodologies developed by the Agency; and (iii) the carrying out of nuclear power planning studies in co-operation with requesting Member States.

Development of methodologies

One of the most important determinants of the need for nuclear power is the projected future demand for electrical energy. Experience showed that the electricity demand information supplied by developing countries often was not developed in a systematic procedure which would ensure internal consistency with their overall economic and industrial development objectives and possibilities. Thus, the electricity demand projections often proved to be a weak point in the resulting estimates of the role of nuclear power in the country's energy supply.

The MAED model

To improve the estimates of future electrical energy needs, development was begun in 1980 on a model called MAED (Model for Analysis of Energy Demand), in collaboration with the Institute for Economic and Legal Aspects of Energy (IEJE, Grenoble, France) and the International Institute for Applied Systems Analysis (IIASA, Laxenburg, Austria), based on experience with an existing model called MEDEE (Modele d'Evolution de la Demande d'Energie). Development was completed during 1981 and MAED is now used by the Agency to develop coherent projections of future energy and electricity needs.

The MAED model provides a flexible simulation framework for exploring the influence of social, economic, technological and policy changes on the long-term evolution of energy demand of a country or region. It is somewhat simpler than the

MEDEE model, to facilitate its application with the more limited data base which is typical of developing countries. Also, it has some specific features related to the details of the electrical system, including an estimate of the future annual electrical load duration curves, that is, the power needed hour-by-hour over the whole year.

The MAED model is outlined in Figure 1. In order to analyse the energy demand of a given country, the economy is sub-divided into the major economic sectors (industry, transport, household, service), and the energy needs of each sector are sub-divided into various elementary needs of final and useful energy (needs for space heating, cooking, furnaces, inter-city transport, and so forth).

The useful and final energy requirements are described by two types of parameters: one linked to the technical considerations (such as the efficiency of different appliances) and the other linked to life-style considerations (e.g., average distance travelled by car during a year, size of dwelling, etc.).

Special attention is given to the calculation of the demand for electricity which is not only performed annually as for other energy forms, but also on an hourly basis in terms of the electric power demand of the various consumers in each hour of the year. This in turn, permits the determination of the electrical load coefficients for the generating system, required as input data for the WASP model.

The MAED approach involves the following steps:

- A systematic analysis of the social, economic and technological systems in order to identify the major factors determining the long-term energy demand evolution;

- Disaggregation of the total energy demand into a multiplicity of end-use categories; each one corresponding to a given service or to the production of a certain good. The selection of the categories to be considered depends upon the objectives of the analyst and on the data availability;

- Organization of all determining factors (determinants) into a multi-level structure, from the macro to the micro level, showing how the "macro-determinants" affect each end-use category;

- Construction of a simulation model by simplifying the system structure and grouping the determinants. The determinants chosen as scenario elements are those for which the evolution cannot be extrapolated from past trends because of possible structural changes in the energy demand growth pattern; for example, policy factors. The exogenous determinants encompass those factors for which the evolution is difficult to model (e.g., population growth, number of persons per household), but for which their long-term evolution can be adjusted suitably from past trends or from other studies (e.g., demographic studies);

- Construction of the different scenarios of socio-economic and technological development for the country, and evaluation of the energy demand resulting from each scenario;

- Selection among all possible scenarios proposed, of the "most" probable patterns of development for the country and evaluation of the electric power demand arising from each one;

The WASP model

The WASP [Wien (Vienna) Automatic System Planning] model is a system of computer programs using dynamic programming techniques for economic optimization in electric system expansion planning (ESEP), and may be taken as an example of a supply model, whereas MAED is a demand model. The WASP model was developed for the Agency by the USA Tennessee Valley Authority (TVA) and first used during the Agency's Market Survey for Nuclear Power in Developing Countries (1972-1973). With further assistance from the TVA and the USA Oak Ridge National Laboratory, it was improved to the WASP-II version in 1976, which has been widely used by the Agency and Member States. Finally in a joint effort between the United Nations Economic Commission for Latin America (ECLA) and the IAEA, the most recent version of the program, WASP-III, was completed in 1980. Many improvements of the methodology are being studied by the Agency, including the introduction of improved computational techniques.

The WASP model is structured in a flexible modular system which can treat the following inderdependent parameters in an evaluation:

- load forecast characteristics (electric energy forecast)
- power generation system development
- power plant capital costs
- power plant operating and fuel costs
- power plant technical parameters

- power supply reliability criteria
- power generation system operation practices.

In addition to the data describing the potential power plant system and the related economic information, the major input into the model is the seasonal electrical load duration curve, supplied by the MAED model.

The WASP model is composed of six principal programs (Fig 2):

- A program to describe the annual and seasonal peak load and the seasonal characteristics of the load duration curves.

- A program to describe the existing power system and all plants firmly scheduled for commissioning and decommissioning;

- A program to describe the alternative plants which could be used to expand the power system (plant "candidates");

- A program to generate alternative expansion configurations;

- A program to determine whether a particular configuration has been simulated and, if not, to simulate the system operation with that configuration. Using a probabilistic simulation model, energy generation by each plant and the corresponding operation cost are calculated. The reliability of the generating system and the expected amount of unsatisfied demand are evaluated;

- A program to choose the lowest-cost expansion schedule for adding new units to the system over the period of interest, using the information provided by the other modules together with economic inputs and reliability criteria. The objective function of this dynamic programming optimization is the present-worth discounted value of operating (including fuel) plus capital investment costs, plus a penalty cost for energy not served, minus a salvage-value credit for plant economic life remaining at the planning horizon.

By mid-1985, the Agency had transferred the WASP package to 57 requesting countries and to six international organizations. These countries report having used WASP in over 300 ESEP studies.

Training Courses

The Agency conducts three courses to train specialists from developing Member States in the techniques for energy demand analysis and electric system expansion planning.

Training course on "Energy Planning in Developing Countries with Special Attention to Nuclear Energy"

The major objective of this course is to familiarize energy specialists in developing countries with the fundamental elements of comprehensive national energy planning, with emphasis on developing a better understanding of the appropriate role for nuclear energy. However, the course is not restricted to only those countries already committed to using nuclear energy, but is open to all developing Member States of the Agency, and to participants interested in non-nuclear as well as nuclear energy technologies. The aim is to improve the country's ability to make a careful and objective choice among the various available energy options, including nuclear.

Even among energy planners it is often thought that energy planning is only a question of economic analysis involving sophisticated computer models. This training course is designed to correct this simplistic point of view and to show that good energy planning involves many aspects of technical as well as economic information. Particular attention is given to the link (too often disregarded) between the choice of the primary energy source and the end-use energy needs of the consumer.

Training Course on "Electric System Expansion Planning (ESEP)" (WASP) Course)

In the period 1975 - 1985, some 220 senior engineers and power system planners from 54 countries and three international organizations were trained by the IAEA in the use of the various versions of WASP. During 1975 - 1977, the training was carried out by the Agency at its Headquarters in Vienna. Since 1978 the government of the United States of America has sponsored an IAEA interregional training course on Electric System Expansion Planning which up to 1985 was given seven times at the USA Argonne National Laboratory (ANL), with participation by some 170 engineers and electric system planners from 48 countries. An IAEA Guidebook (TRS-241) on Expansion Planning for Electrical Generating Systems has been used in the course since 1984.

This course has the objective to train specialists in electrical system expansion planning, with emphasis on use of the WASP model. Upon completion, the trainee should be able to carry out studies to determine economically optimal programmes for electric power system expansion, including, in particular, the economically optimal share of nuclear power.

The major subjects focus on technical and economic characteristics of electric power plants, principles of generation expansion planning, electric grid considerations, characteristics of the WASP model and its auxiliary programs, evaluation and presentation of input data for WASP, analysis of optimum solutions and preparation of a study report.

Every year, the training course is open to some 30 candidates from developing countries in all geographical regions. Applications are requested to be presented in national teams, each consisting of at least two persons with experience in power system planning, because the course centres around an ESEP study by each participating national team, based on their own national data.

Interregional Training Course on Energy Demand Forecasting for Nuclear Power Planning (MAED Course)

The purpose of this course will be to train engineers and economists of developing countries in the analysis and forecasting of energy and electricity demands.

This 5-week course, given for the first time in the Fall of 1985, is designed to deal with one of the most important aspects of planning studies: the development of an electricity demand forecast which is consistent with the general economic and energy development of a country.

Like the WASP course it will centre around case studies for the participants' countries using their national data as input to the MAED model.

Energy and Nuclear Power Planning Study

An energy and nuclear power planning (ENPP) study is initiated only upon official request by an Agency Member State and is carried out as a joint project of the Agency and the Member State. The objective is to assist the Member State in detailed economic analyses and planning studies to determine the need and appropriate role for nuclear energy within its national energy plan. This requires assessment in terms of economic plans, and economic comparison with alternative energy sources. The methodologies described previously (MAED and WASP) with improvements or changes as necessary are used during the studies, and released to the country at the end of the study.

Thus, the ENPP study has two specific objectives:

1. To carry out, in co-operation with the requesting country, a study which:
   a. quantifies the future energy requirements consistent with national economic development plans, and the expected share of electrical energy within the overall energy needs;
   b. derives an economically optimum electrical system expansion plan (ESEP), including an assessment of the need for and role of nuclear power.

2. To train a local team of engineers and economists in planning techniques and to supply them with appropriate tools, i.e., the two computer models MAED and WASP, in order that further energy planning studies can be carried out by the country experts.

As such studies are carried out in close co-operation with the requesting country, a joint team is established, composed of:

1. from the IAEA side, two or three staff members familiar with all questions related to energy planning and different models which could be used;

2. from the country side, a team of specialists which could include, at least:

   - 5 to 6 engineers and economists well acquainted with the electricity and energy situation in their country (it is recommended that most of them should have attended the three training courses previously described);
   - a local co-ordinator at a senior level such that he can contribute effectively to the work required and make contacts with different organizations within the country, at the appropriate level, in order to obtain the information and data needed for the study.

An ENPP study requires about two years, during which all members of the joint team work at approximately fifty percent of full time. It is difficult to reduce this time period, due to the time needed for data and information gathering.

Although the exact content, scope and schedule for an ENPP study will vary depending on the country studied, an ENPP study requires about two years of teamwork. Although members of the joint team need not dedicate their full time to this activity, the time period normally cannot be shortened, due to the time needed for data gathering. A typical schedule for an ENPP is outlined in Figure 4.

Conclusions

The Agency has a demonstrated capability to assist its developing Member States in the economic aspects of planning their future electric power system with special emphasis in nuclear power planning. However, the development of an energy planning activity is a long-range undertaking requiring constant review, additions and improvement.

The evaluation of the economic benefits from nuclear energy in a developing country needs a broad-based and in-depth analysis of the total effects of a nuclear power programme on the overall economic development of the country.

Presently emphasis has been placed on three major points:

a. Nuclear energy development in a given country cannot be planned and implemented in an isolated way. Nuclear technology is only one among many means to supply secondary energy (electricity and/or heat), and nuclear power planning should be carried out within the context of all supply options and the various types and forms of energy requirements, as well as within the general energy and economic development planning of a country.

b. Energy, electricity or nuclear power planning are a national problem which can be reasonably and rationally studied only by local energy specialists. The Agency can provide advice and some methodologies but cannot be a substitute for national energy experts, who must take the final responsibility for planning the development of energy supplies in their country. If needed, training can be obtained through the Agency training courses.

   The Agency strongly emphasizes that the joint ENPP study should be carried out mainly by the national team with assistance from Agency-supplied experts. Through this approach, the trained national team will be in a position to better understand the situation in their own country and be able to follow-up on the studies initiated in co-operation with the Agency.

c. Finally, it must be emphasized that economic studies, such as those mentioned in this paper, are only a first step in the long process of energy and nuclear power planning. Many additional studies and analyses should follow, to determine whether nuclear power is a practical option and the broader implications to the country of a decision to undertake a nuclear power programme. However, this deals with questions beyond the scope of this paper (such as impact on the balance of payment, financing constraints, manpower requirements, local industry participation), but which should be kept in mind when a country is evaluating the possibility of embarking in a nuclear power programme.

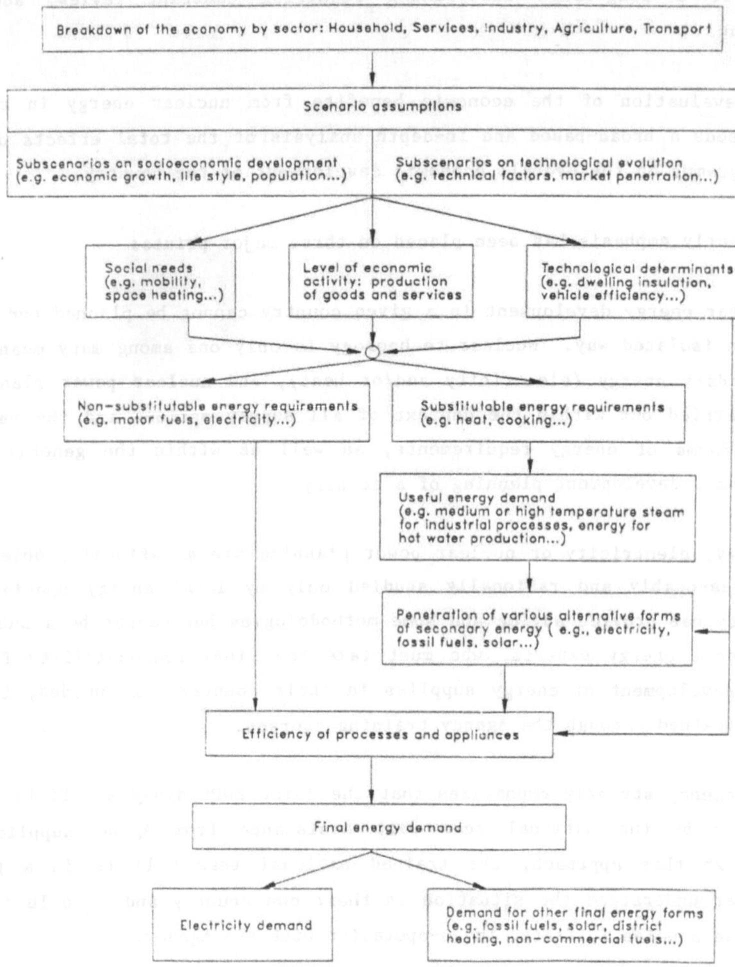

Figure 1. Scheme used in MAED to project useful and final energy demand

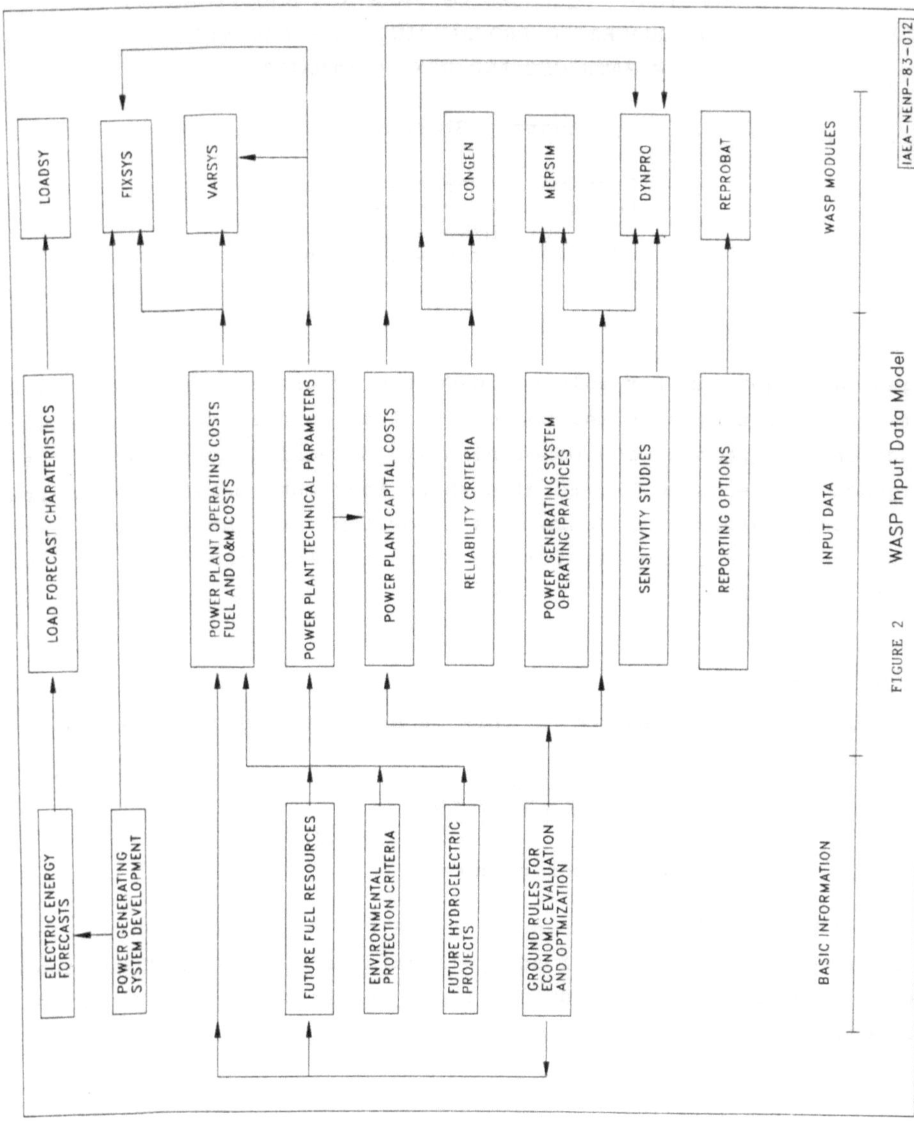

FIGURE 2 WASP Input Data Model

MONOTONE SEQUENCES OF FEASIBLE SOLUTIONS FOR QUADRATIC PROGRAMMING PROBLEMS WITH M-MATRICES AND BOX CONSTRAINTS

Andrzej Stachurski

Systems Research Institute
Polish Academy of Sciences
ul. Newelska 6
01 - 447 Warszawa, Poland

Abstract

A general algorithm for solving quadratic programming problems with an M-matrix as the Hessian and box constraints is presented. It constructs a sequence of points with monotonically increasing components. Its convergence is discussed. It is also shown that many existing algorithms (e.g. Pang's algorithm, Scarpini's algorithm) are the realisations of this general algorithm. An example of a new algorithm belonging to this class is presented.

1. Introduction

The problem considered in this paper is a quadratic program of the following form:

$$\min_x f(x) = \frac{1}{2} x^T Ax - b^T x \tag{1.1}$$

$$\text{subject to: } c \leq x \leq d, \tag{1.2}$$

where $x,b,c,d \in R^n$, A is an $n \times n$ symmetric M-matrix. Such problems arise after approximation of some elliptic problems with free boundary. They include for instance various types of Dirichlet problems with obstacles.

Definition 1 [1]

An $n \times n$ real matrix A is called an M-matrix iff. $a_{ij} < 0$ (for $i=1,\ldots,n$, $j=1,\ldots,n$, $i \neq j$), A is nonsingular and $A^{-1} > 0$. □

For symmetric matrices the notion of M-matrix coincides with the so-called Stieltjes matrix. Throughout the rest of the paper the following notation will be used: i) $N=\{1,\ldots,n\}$, $\overline{J}=N-J$ for any J N, iii) $x_J=(x_{j_1},\ldots,x_{j_p})$, where $J= j_1,\ldots,j_p$, with $j_1 < \ldots < j_p$,iv) $A_{JK}=(c_{rs})$, where $c_{rs}=a_{i_r k_s}$, $j_r \in J=\{j_1,\dots,j_p\} \subseteq N$, $k_s \in K=\{k_1,\dots,k_q\} \subseteq N$.

2. Some properties of the problem

The optimality conditions for problem (1.1-2) have very simple form. $\hat{x}$ is the optimal solution of (1.1-2) if for each $i \in N$ holds:

$$\text{if} \quad \hat{x}_i = c_i \quad \text{then} \quad g_i(\hat{x}) \geq 0, \tag{2.1}$$

$$\text{if} \quad c_i < \hat{x}_i < d_i \quad \text{then} \quad g_i(\hat{x}) = 0, \tag{2.2}$$

$$\text{if} \quad \hat{x}_i = d_i \quad \text{then} \quad g_i(\hat{x}) \leq 0, \tag{2.3}$$

where $g(x) = Ax - b$.

Our further considerations are based on the following main property.

Lemma 2.1 [6]

Every feasible point of (1.2) such that: i) $x_I=d_I$ and $g_I(x) \leq 0$, ii) $c_K < x_K < d_K$ and $g_K(x) \leq 0$, iii) $x_Q=c_Q$, where $I \cup K \cup Q = N$ satisfies componentwise the inequality $x < \hat{x}$, where $\hat{x}$ is the optimal solution of (1.1-2). □

It follows from Lemma 2.1 that one can construct many algorithms such that for every $k > 0$ $x^{k-1} \leq x^k \leq \hat{x}$ if $x^o \leq \hat{x}$. $x^o=c$ is the simplest choice satisfying the requirement $x^o \leq \hat{x}$. One should only ensure that $g_i(x^k) \leq 0$ for each $k > 0$ and $i \in N$ such that $x_i^k > c_i$.

The following results suggest how the requirements can be satisfied.

Theorem 2.1 [3] (Luk and Pagano, 1980)

Let $d=+\infty$ in problem (1.1-2). Let $I \subset N$, $I \neq \emptyset$, $x_I=A_{II}^{-1}(b_I-A_{I\bar{I}}x_{\bar{I}}) > c_I$, $x_{\bar{I}} = c_{\bar{I}}$. Let furthermore $\bar{I}$ be divided into two parts: $\bar{I}=K \cup L$, $K \cap L=\emptyset$ such that $g_K(x) < 0$, $g_L(x) \geq 0$. Let $J=I \cup K$ and y be defined as follows: $y_{\bar{J}} = c_{\bar{J}}$, $y_J = A_{JJ}^{-1}(b_J-A_{J\bar{J}}y_{\bar{J}})$. Then: if $K=\emptyset$ then x solves the problem (1.1-2) else the following inequalities hold: i) $y_I \geq x_I > c_I$, ii) $y_K > c_K$, iii) $g_L(y) \leq g_L(x)$, iv) $f(y) < f(x)$. □

Lemma 2.2 [3]

Let $x,y \in R^n$ be such that $x_I=y_I$, $x_{\bar{I}} \leq y_{\bar{I}}$ for some $I \subset N$ $(I \neq N)$. Then $g_I(x) \geq g_I(y)$, where $g_I(x)=A_{II}x_I + A_{I\bar{I}}x_{\bar{I}}-b_I$. □

Theorem 2.1 and Lemma 2.2 imply that under some conditions the changes in the vector of variables x and the gradient vector will be monotone.

3. A general algorithm

Step 0. Set $x^o := c$, $k := 0$, $Q := N$, $K := \emptyset$.

Step 1. Compute $g_Q(x^k)$. If the optimality conditions are satisfied STOP with $x = x^k$ as the optimal solution; otherwise, set $K := K \cup \{i \in Q;\ g_i(x^k) < 0\}$, $Q := N-K$.

Step 2. Set $z_Q := c_Q$. Find z_K such that

$$x_K^k \le z_K \le d_K \quad \text{and} \quad g_K(z) \le 0. \tag{3.1}$$

Step 3. Set $x^{k+1} := z$, $k := k+1$. Return to Step 1.

In view of Lemma 2.1 it is easily seen that for every $k \ge 0$ the inequality $x^k \le \hat{x}$ holds. Obviously, every limit point of the sequence $\{x^k\}$ will be the optimal solution of problem (1.1-2) if for every k z_K calculated in Step 2 is not equal x_K^k.

Furthermore condition (3.1) ensures that if for some k^* and $i \in N$ $x_i^{k^*} = d_i$ then $x_i^k = d_i$ for every $k \ge k^*$. From Lemma 2.2 it follows that $g_i(x^{k+1}) \le g_i(x^k)$ for each $i \in Q$ corresponding to a variable being on its upper bound. The speed of convergence of the general algorithm depends on the way of generation of the point z_K in Step 2.

In the next sections it will be shown that some of the existing algorithms are the realisations of the general algorithm.

4. The Pang algorithm

Step 0. Set $x^o := c$, $k := 0$, $Q := N$, $K := \emptyset$.

Step 1. Compute $g_Q(x^k)$. If $g_Q(x^k) \ge 0$ then STOP with $x = x^k$ as the optimal solution; otherwise, set $K := K \cup \{i \in Q; g_i(x^k) < 0\}$, $Q := N-K$.

Step 2. Solve the following subproblem in z_K variables:

$$\min \frac{1}{2} z_K^T A_{KK} z_K - (b_K - A_{KQ} x_Q^k)^T z_K \tag{4.1}$$

$$\text{s.t.} \quad z_K \le d_K.$$

Step 3. Set $k := k+1$, $x_Q^k := c_Q$, $x_K^k := z_K$. Return to Step 1.

In Pang's algorithm [3, 4] z_K in Step 2 is calculated as the solution of the quadratic programming subproblem (4.1-2) with upper bounds only (usually the Chandrasekaran method is used for this purpose). The optimality conditions for (4.1-2) imply the conditions of Step 2 of the general slgorithm. In Step 1 of Pang's algorithm it is sufficient to check only the restricted optimality condition $g_Q(x^k) \geq 0$ since x_K^k is the optimal solution of the corresponding subproblem with respect to $x_{\overline{Q}}$.

Hence one can state that Pang's algorithm is an example of the general algorithm.

5. The Scarpini algorithm

Step 0. Set $Q:=N$, $x^o:=c$, $k:=0$. Check the condition: if $g(d) \leq 0$ then STOP with $x=d$ as the optimal solution.

Step 1. If the optimality conditions are satisfied by x^k then STOP with $x=x^k$ as the optimal solution.

Step 2. Repeat Steps 2.1-3 for $i=1,...,n$ $(i \in Q)$.

2.1 $y_i := \frac{1}{A_{ii}}(b_i - \sum_{j \neq i} A_{ij}x_j^k)$,

2.2 if $y_i > d_i$ then $x_i^k := d_i$, $Q:=Q-\{i\}$,

2.3 if $c_i \leq y_i \leq d_i$ then $x_i^k := y_i$.

Step 3. Set $x^{k+1} := x^k$, $k:=k+1$. Return to Step 1.

The original way of presentation of Scarpini's algorithm is a little different. For the sake of clarity we restricted ourselves to the main idea. In this case an iteration of the Gauss-Seidel iterative method for solving systems of linear inequalities is applied in Step 2. Making use of the properties of M-matrices one can prove that in cycle i $(i \in N)$ either x^k remains untouched or x_i^k is increased. In the latter case Lamma 2.2 implies that $g_j(x^k)$ $(j \in N,\ j \neq i)$ decreases.

Hence we can state again that the considered algorithm is the example of the general algorithm.

6. The Chandrasekaran algorithm

The Chandrasekaran method [2, 3] was developed for problem (4.1-2) without upper bounds $(d=+\infty)$. We present it here because its structure resembles the structure of the general algorithm. Furthermore it will be useful for the presentation of the next example of the general algorithm.

Step 0. Set $x^o:=c$, $k:=0$, $Q:=N$, $K:=\emptyset$.

Step 1. Compute $g_Q(x^k)$. If $g_Q(x^k) \geq 0$ then STOP with $x=x^k$ as the optimal solution; otherwise, set $K:=K \cup \{i \in Q;\ g_i(x^k) < 0\}$, $Q:=N-K$.

Step 2. Set $z_Q:=c_Q$. Calculate

$$z_K:=A_{KK}^{-1}(b_K - A_{KQ}c_Q).$$

Step 3. Set $k:=k+1$, $x^{k+1}:=z$. Return to Step 1.

The only difference in comparison with the Pang algorithm occurs in Step 2. Here z_K is the optimal solution of an unconstrained QP subproblem.

7. Other examples of the general algorithm

The Pang algorithm requires in the worst case n iterations to find the optimal solution of problem (1.1-2). The Scarpini algorithm on some problems, at least theoretically, may require an infinite number of iterations to solve the problem. The situation is similar as with the Newton method and the steepest descent method for general unconstrained optimization problems. It seems reasonable to look for something similar to the quasi-Newton methods for solving problem (1.1-2).

A natural way is to make use of the point generated at the k-th iteration of the Chandrasekaran algorithm. The following lemma indicates the possibility of taking a convex combination of x^k and the trial point y generated as in the Chandrasekaran method.

Lemma 7.1 [6]

Let $N=I \cup J$, $I \cap J=\emptyset$. Let $x,y \in R^n$ be such that $g_J(x) \leq 0$, $g_J(y) = 0$ and $x_t=tx + (1-t)y$, where $t \in [0,1]$. Then

$$g_J(x_t) \leq 0.$$

□

Algorithm I

Step 0. Set $x^o:=c$, $k:=0$, $Q:=N$, $K:=\emptyset$, $I:=\emptyset$. Go to Step 3.2.

Step 1. Check the conditions:

1.1 If $S=P=\emptyset$ STOP with $x=x^k$ as the optimal solution. Otherwise go to Step 2.

1.2 If $P=\emptyset$ then STOP with $x=x^k$ as the optimal solution.

Step 2. 2.1 Calculate: $y_K := A_{KK}^{-1}[b_K - A_{KQ}c_Q - A_{KJ}d_J]$.

2.2 Find $S=\{i \in K;\ y_i > d_i\}$.

2.3 If $S=\emptyset$ then set $x_K^k := y_K$ and go to Step 3.2.

2.4 Find $t_i = \min_{j \in S} t_j$, where $t_j = \dfrac{d_j - x_j^k}{y_j - x_j^k}$. Set $S:=S-\{i\}$,

$K:=K-\{i\}$, $J:=J \cup \{i\}$, $x_i^k := d_i$. Calculate: $x_K^k := (1-t_i)x_K^k + t_i y_K$.

Step 3. 3.1 Calculate $g_Q(x^k)$; set $k:=k+1$, $x^k := x^{k-1}$; go to Step 1.1.

3.2 Calculate $g_Q(x^k)$; set $k:=k+1$, $x^k := x^{k-1}$; go to Step 1.2.

Theorem 2.1 and Lemma 7.1 imply that Step 2 of Algorithm I also satisfies the conditions of Step 2 of the general algorithm. Hence Algorithm I is also an example of the general algorithm.

However the numerical experiments described briefly in Section 8 have shown that Algorithm I is in general not competitive with Pang's algorithm. In Step 2.4 only one index is moved from set K to set J. Hence it may require many outer iterations (although the total number of iterations cannot exceed 2n).

We propose to append to Step 2.4 the following:

2.5 If $S=\emptyset$ then go to Step 3.1. Otherwise set $y_S := d_S$, $L:=K-S$, $y_{\bar{K}} := x_{\bar{K}}^k$. Find $y_L := A_{LL}^{-1}|b_L - A_{LS}d_S - A_{L\bar{K}}x_{\bar{K}}^k|$. Calculate $g_S(y)$. Set $L:=\{i \in S;\ g_i(y) \le 0\}$, $x_L^k := d_L$, $K:=K-L$.

□

Algorithm I with Step 2.5 will be referred to as Algorithm II. We expect that it slightly improves on the behaviour of Algorithm I.

8. Numerical experiments

Algorithm I and Pang's algorithm were programmed in FORTRAN IV. They were rum on an SM-4 minicomputer on some test examples. Due to the lack of space we restrict ourselves to the presentation of one example, in which:

$$A = \begin{bmatrix} 2 & -1 & & \\ -1 & 2 & -1 & 0 \\ & & & \\ 0 & -1 & 2 & -1 \\ & & -1 & 2 \end{bmatrix}; \quad c=0; \quad \begin{matrix} b_i = 11-20\, r_i \\ d_i = 7\, s_i \\ n=100. \end{matrix} \quad \text{for } i=1,\dots,n,$$

Pang's algorithm required the solution of 5 subsystems of linear equations (with size = 16, 19, 23, 19, 23 accordingly). Algorithm I required the solution of 14 subsystems (with size = 29, 28, 28, 28, 30, 29, 28, 27, 26, 27, 26, 25, 24, 23). The results on other examples were similar. Only on one tested example Algorithm I appeared to be faster.

9. Concluding remarks

The general algorithm introduced in this paper covers many existing algorithms. The direction of search for the most efficient algorithm belonging to this class is outlined. Such algorithm should be less expensive per iteration than the Pang algorithm. However, to be competitive, it should find the solution after a comparable number of iterations. We hope that Algorithm II may be the desired one.

References

|1| A. Berman, R.J. Plemmons: Nonnegative Matrices in the Mathematical Sciences, Academic Press, New York, 1979.

|2| R. Chandrasekaran: A special case of the complementary pivot problem, Opsearch 7: 263-268, (1970).

|3| F.T. Luk and M. Pagano: Quadratic programming with M-matrices, Linear Algebra and its Applications 33 : 15-40, (1980).

|4| J.S. Pang: On a class of least element complementarity problems, Report SOL-76-10, Systems Optimization Laboratory, Stanford University, 1976.

|5| F. Scarpini: Some algorithms solving the unilateral Dirichlet problems with two constraints, Calcolo 12 : 113-149, (1975).

|6| A. Stachurski: Quadratic programming problems with M-matrices and box constraints, (in Polish), Technical Report, Institute of Automatic Control, Warsaw Technical University, 1984.

Clustering Theory in the Design of Information Systems

by

Larry E. Stanfel, QBA Department
Louisiana State University
Baton Rouge, Louisiana 70803 USA

Abstract

A typical first step in the system analysis or design process is to decompose, hierarchically, the system into subsystems until each subsystem admits no useful decomposition, or is seen to exist, or is recognized as designable. For the actual design work, groups of these pensile subsystems are assigned to design groups and their efforts interconnected to give the overall system. The present paper focuses attention upon the problem of grouping the subsystems--an optimal partition is sought. For a reasonable metric on the set of subsystems it is found they may be treated as objects on a line, a case where exact optimal solutions may be determined. A dynamic programming/LaGrangian algorithm is given and the computer solution to a 100 subsystem example presented.

I. Introduction and Problem Formulation

In the process of analyzing or designing an information system a principle of hierarchial decomposition has been advocated by many authors. The idea is that the complex system or the desired function is partitioned into subsystems (or sub-functions--hereafter we abstain from distinguishing between analysis and design and use the term "system" to indicate both possibilities). These, in turn, are partitioned, and so on, iteratively with a particular subsystem being partitioned only if it is too complex to be treated. A few references are [4], [3], [5], [7].

As Figure 1 shows, these successive operations give rise to a tree where nodes represent subsystems and edges point to new subsystems born in the breaking-down process. Partitioning ends on different levels as a function of the relative complexity of the subsystems.

A diagram such as Fig. 1 does not purport to show interfaces between subsystems; rather the edges are interpreted (in a downward direction) as meaning "is composed of." How the constituents are interconnected is another matter.

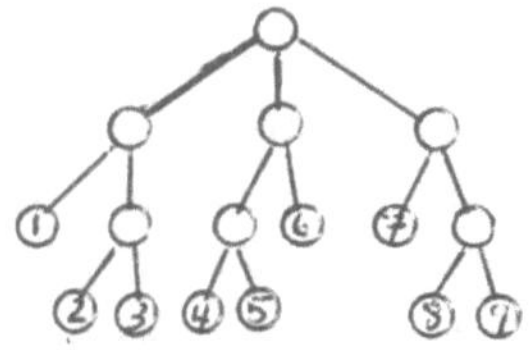

Figure 1. Hierarchical decomposition system.

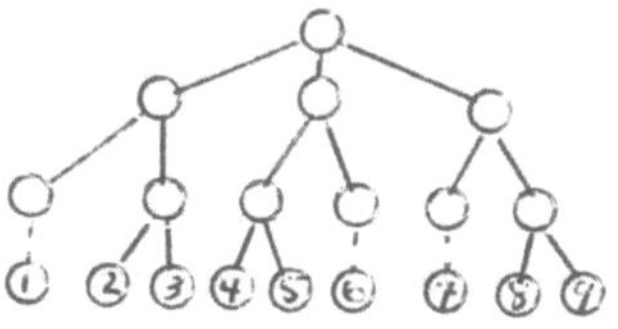

Figure 2. Normalized Fig. 1

In the design problem when all the pensile, non-decomposed subsystems exist or are designable these are obtained or designed, then composed "upward" to give the overall system.

The design of the pensile subsystems is somehow apportioned among the available design manpower and the work proceeds. It is reasonable to expect one design group to undertake more than a single design, and the subject of this investigation is how best to group together the pensile subsystems.

In an earlier paper [9] this topic was addressed relative to systems that have the appearance of Fig. 1; that is, which exist already in hierarchical form. Certainly much greater applicability is achieved by realizing the ubiquity of hierarchical decomposition in systems work. Whatever the configuration of the system the tree gives a scheme for accomplishing the work and therefore a way of deciding upon the grouping of those tasks.

It is desirable to examine the target subsystems in a linear sequence, so those belonging to upper tree levels are extended as shown in Fig. 2.

Any technique for grouping together objects in a natural or useful way must be concerned with the distances between pairs of the objects. Consequently it is mandatory to make a reasonable definition of distance between level 3 subsystems, those that are to be defined.

A reasonable one, and one with metric properties is given by

d_{ij} = highest numbered tree level minus highest numbered level where i and j belong to the same subsystem.

This definition appeared in [6] and, to the extent that the tree gives an accurate picture of the logical relationship of subsystems, gives a consistent measure of the logical proximity of subsystems. Other notions of difference or similarity between subsystems may also be appropriate. Clustering may still be a desirable analytic step to take, but trees such as in Fig. 1 may not reflect the prevailing concept of similarity.

The next, and more difficult, question is how to incorporate the distances into a function giving a reasonable measure of the partition of the subsystems into groups or clusters. First of all it seems incontestable that a subset of uniformly close subsystems would comprise an easier design task than a subset more scattered. For example (2,3,4) should be a logically more compact job than, say, (2,5,9). It is generally arguable in the clustering context that clusters should be pairwise relatively distant; that internal homogeneity is just one important aspect. If we find two dense sets quite close to one another, for example, the defining notion of "cluster" is lost. But have we that concern in system design work? Is there any advantage to different design efforts' being relatively far removed one from the other? This is less obvious to the author, and the benefits must be less tangible than for homogeneous clusters. Two objective functions will be shown, as a result, and the problem-solving method developed will be virtually identical in the two cases.

In either case we shall be interested in averages. If only within cluster homogeneity is considered important, the objective function will be

$$f = \frac{\text{total within group distances}}{\text{number of within group distances}} \qquad (1a)$$

and will measure the average within group distance (wgd) or within group homogeneity. The approach will be to create subsets of design effort so as to minimize f.

If cluster separation is desirable then we define

$$f = \frac{\text{total wgd's}}{\text{number wgd's}} - \frac{\text{total between group distances}}{\text{number bgd's}} \qquad (1b)$$

The reasoning behind the f of (1b) is that a good partition has homogeneous subsets which are also separated.

In the event the value of a unit of average cluster separation is not considered equal to one unit of average cluster homogeneity the expression (1b) lends itself readily to a user - influenced objective function. One may write

$$f = \frac{\text{total wgd's}}{\text{number wgd's}} - \alpha\,\frac{\text{total bgd's}}{\text{number bgd's}} \qquad (1c)$$

By varying the parameter α, $0 \leq \alpha \leq 1$, both (1a), (1b) and an infinite number of objective functions are generatable. This seems desirable for a problem area where that weight is deemed important but where its specification is vague. By solving the clustering problem with various values of α, a spectrum of solutions could be examined in light of whatever subjective criteria are available. If that seems a good approach in an area where defining an objective function may be less than straightforward, then it is valuable that our algorithm will afford that capability, though indirectly. Consequently imagine that (1a) or (1b) is selected, but observe that (1c) will be available for a range of α.

It is unnecessary to display the entire array of distances but it is useful to compare, say, $\underline{d}_1 = (d_{11}, d_{12}, \ldots, d_{19})$ and $\underline{d}_2 = (d_{21}, \ldots, d_{29})$

We get $\underline{d}_1 = (0,2,2,3,3,3,3,3,3)$

$\underline{d}_2 = (2,0,1,3,3,3,3,3,3)$

It happens that collinear sets of objects may be clustered optimally relative to the function f in (1), and it would be convenient to identify the subsystems 1 to 9 with sequential points on a line. It is, however, impossible to attribute real-valued coordinates to them. Nevertheless d_{ij} is a nondecreasing function as j increases for $j \geq i$ and as j decreases for $j \leq i$. As a result, partitions optimal for either function in (1) will have only consecutively numbered subsets, a result of the nature of f. Therefore, while the set of distances corresponding to a particular hierarchical decomposition will never admit realization by a set of points on a line, we may still limit attention to contiguous subsets, as if the objects were points on a line. Any simplifications in clustering problems arising from collinearity (and they are considerable) are amenable to exploitation in this system-partition problem.

We must mention how realistic restrictions would be manifest in a mathematically formulated problem. Generally, it must be supposed that each of the pensile subsystems (henceforth "subsystem" will denote the lowest level structures) has associated a quantity of work, and a reasonable constraint is that the sum of the work associated with any cluster must not exceed some "a priori" bound, say W. That is the only explicit constraint to be formulated, but if the reader has other constraints in mind he can assess their compatibility with this approach once the formulation is apparent.

Imagine that the candidate subsets (each must consist of subsystems that are consecutive on the bottom tree level) are numbered in some fashion.

Define x_j = 1 if the jth subset is selected as a cluster
= 0, otherwise

c_j = total wgd's of the jth subset

A = matrix whose columns $\underline{A}_j$ correspond to the candidate clusters; that is
A_{ij} = 1 if subsystem i belongs to cluster j
= 0, otherwise

$\underline{1}$ = a vector of 1's, D = total distances of all the subsystems

n = number of subsystems, w_j = work associated with cluster j

n_j = number of subsystems in cluster j

d_j = number of distances in cluster j = $\binom{n_j}{2}$

Then with (1b) as objective the problem becomes

$$\min \frac{\Sigma c_j x_j}{\Sigma d_j x_j} - \frac{D - \Sigma c_j x_j}{\binom{n}{2} - \Sigma d_j x_j}$$

$$\text{ST} \quad \Sigma \underline{A}_j x_j = \underline{1} \quad \text{or} \quad A\,\underline{x} = \underline{1}$$

$$\Sigma w_j x_j \le W; \quad x_j = 0 \text{ or } 1, \text{ all } j \tag{2}$$

The nonlinear objective in (2) defies easy accommodation by any integer programming algorithm, so the formulation must be further processed. Note, too, that we either prohibit zero denominators or define f differently for such an event.

If we assumed $\Sigma d_j x_j = K$, a constant, or in words that the number of wgd's is fixed at K, then our problem becomes

$$\begin{aligned} \min \quad & \Sigma c_j x_j \\ \text{ST} \quad & A\underline{x} = \underline{1} \\ & w_j x_j \le W, \text{ all } j \\ & \Sigma d_j x_j = K; \quad x_j = 0 \text{ or } 1 \end{aligned} \tag{3}$$

If one accepts that the inequality constraints in (3) may be handled externally and thereby ignored (for example, by the simple expedient of ignoring any subset that is too heavy - creating no column $\underline{A}_j$ and no variable x_j for those subsets) problem (3) is just a set partitioning problem [2] with an additional constraint.

Incorporating the additional constraint, LaGrangian fashion, into the objective function, problem (3) assumes the form

$$\min \quad \Sigma(c_j + \lambda d_j)x_j$$
$$\text{ST} \quad A\,\underline{x} = \underline{1}; \qquad x_j = 0 \text{ or } 1 \tag{4}$$

The LaGrangian form in problem (4) has a dual usage. It eliminates the only constraint lacking all 0-1 coefficients and thus preserves the set partitioning structure, which will be seen to be even more specialized in our problem. The typical trade-off, of course, is the necessity to vary the multiplier λ and solve a sequence of problems until we realize

$$\Sigma d_j x_j^*(\lambda) = K$$

where $x_j^*(\lambda)$ is a notation to exhibit the dependence of optimal solutions upon the value of the multiplier.

The second advantage to the formulation is that there is no single value of K that must be achieved --we need optimal solutions for every K that represents a number of wgd's realizable by some partition of the subsystems into clusters. (For even moderate n the number of such possibilities is more than sufficient to appear to deal a death blow to the algorithm, but it will be discovered that relatively few subproblems actually require solving. But continue to imagine that we need all these). As a result we must vary λ and generate all these K, anyway. It is the price we are paying for linearity, in truth, not for preserving a set partitioning problem.

Now A has a special structure. Because each candidate subset has a consecutive set of subsystems every column of A has its 1's in consecutive components. This implies that A is unimodular [2], and that is the property that guarantees that the simplex method will give integer solutions to (4). Therefore each subproblem may be solved by ordinary linear programming methods, efficient in practice, we know, even for large problems.

II. Previous Work

In [9] the problem (4) was obtained beginning with the assumption of Fig. 2's representing a system existing already in hierarchical form. (Now it may be seen that, in representing the work involved in system design, which has application to any system, the tree is still appropriate). The solution technique contrived there was rudimentary but deserves some mention. It proposed simply the generation of columns of A and the subsequent solution of a collection of LP's. For 1000 subsystems we would have fifty thousand columns, which is a large problem by any standards. Further, the problem associated with the range of K has not been confronted and the potential computation is huge. Therefore a viable technique for system design problems of reasonable size is needed, and the remainder of the paper deals with that.

III. A New Algorithm

There are two computational aspects demanding redress. First is the large amount of storage and the generation required by a large LP. Secondly, a great concern is the amount of work involved in searching for an optimizing λ^*. An intelligent seeking method is imperative. For the first objection a new algorithm based solely upon dynamic programming has been developed. Because clusters consist only of contiguous objects (subsystems, in this case) DP is efficient and there is no dimension difficulty. Both state and decision variables have unit dimension. The second concern centered upon the number of λ subproblems and that will be discussed later.

Imagine that the stages are labeled in <u>ascending</u> order, left to right. The decision, d_i, at state i is particularly simple. It tells essentially whether point (object) i will be assigned to a cluster of contiguous points being assembled or whether it will become the first point in a new cluster.

$d_i = 1$ means that object i is the next object in the cluster being built.

$d_i = 0$ means that object i is the first object in a new cluster.

The state variable at stage i, say s_i, is equally simple--it is one dimensional and is just an integer. $s_i = k$ means that the cluster being assembled has the previous k objects in it; that is $\{i - k, i - k + 1, \ldots, i - 1\}$. The possible values of k depend upon the stage number, i.

The transformation T_i at stage i is defined by

$$s_{i+1} = T_i(s_i,1) - s_i + 1$$

$$s_{i+1} = T_i(s_i,0) = 1$$

These state that if $d_i = 0$ (the subset is terminated) then the size of the current subset at stage i + 1 is 1 and consists only of the object i; and if $d_i = 1$ (the subset is enlarged by object i) the size of the current subset increases by 1.

Finally, the stage return functions are defined by

$$R_i(s_i,1) = 0$$

$$\text{and } R_i(s_i,0) = C(i,s_i) + \lambda\binom{s_i}{2}$$

$C(i,s_i)$ = total wgd's of the subset terminated

Defining, then, the total return to be the sum of the stage return functions, we have separability, monotonicity, and the objective function in (4). Only two arrays are required for the recursive computation, and these are used in flip-flop fashion, so that one contains the optimization results for all odd numbered stages, and the other, for all the even stages. Each array is (n - 1) by (n + 1) words, so about $2n^2$ words of storage are required by an n-subsystem problem.

Optimization results are maintained cumulatively--that is, the array at stage i maintains the cumulative optimal decisions per state, not just the best decision at stage i for each state. A compact notation carries this information. For example, suppose that for s_{10} we find $d^*_{10} = 0$ corresponding to $s_{11} = 1$, at the previous stage, suppose we find 15, 24, and 30 stored. The meaning is that new clusters <u>begin</u> with objects 15, 24, and 30. For $s_{10} = 5$ we store the optimal partial solution 10, 15, 24, 30 which indicates the clusters (10,...,14), (15,...,23), (24,...,29), (30,...,n). Had d^*_{10} been 1, we would refer to $s_{11} = 6$ at the previous stage. Suppose we found 14, 20, 25 there. Then corresponding to $s_{10} = 5$ we store 14, 20, 25.

As to the ease of locating an optimizing $\lambda \leq 0$ ($\lambda > 0$ always produces the solution corresponding to each object in a different subset. Since the C_j are positive and the object is to minimize, all subset wgd's would be made 0 for positive λ) the algorithm seeks to exploit an empirical property of the function f*(K), the set of optimal f values corresponding to the possible values of K. For every example problem the function f* is nearly unimodal. The paper [8] rationalized that appearance in a grossly quantitative way. Generally it is <u>not</u> unimodal, however, for arbitrary points in Euclidean one-space and has small irregularities that could mislead a search technique which assumed unimodality.

If $f^*(\lambda)$ were unimodal we could select some initial negative value λ_0 near 0 and a value Δ and proceed to find $f^*(\lambda_0)$, $f^*(\lambda_0 - \Delta)$, $f^*(\lambda_0 - 2\Delta)$, etc. so long as f* decreases. When at last $f^*(\lambda_0 - p\Delta)$ increases, then the desired λ^* satisfies

$$\lambda - p\Delta < \lambda^* < \lambda_0 - (p - 2)\Delta.$$

A smaller step size Δ' would be applied to the new interval, and the procedure continued according to whatever tolerances and termination criteria established.

The difference with the actual problem is that when $f^*(\lambda_0 - p\Delta)$ increases, it may be a result of a bump in f^* with λ^* still lying beyond. Consequently, two consecutive increases over the best value for that pass are demanded before a pass is halted. (The bumps tend to be of smaller magnitude). The new Δ' is $\Delta/1.97$, to avoid repetition of trial λ values, and two consecutive passes with no improvement in f^* values are interpreted to mean that an optimum has been found. So also is the creation of a step size Δ' at some stage smaller than an "a prior" tolerance value.

On the topic of tolerances another observed property of f^* is pertinent. Typically f^* is very flat in a substantial neighborhood of λ^*.

The empirical evidence cited above for arbitrary point sets on a line has been contradicted to some degree by the subsystem examples solved to date. First of all, no bumps have been observed. In the few examples, the sequence $f^*(K)$ has been unimodal except for ranges of constant value. Secondly, $f^*(K)$ has <u>not</u> been flat in the neighborhood of the optimum. The drop has been relatively declivitous. The explanation for both these phenomena may lie in the discrete set of distances. The implications for solution are two, assuming these properties to persist. First, we should feel less vulnerable to being mislead by a nonunimodality, but at the same time steepness near the minimum warns that the tolerance for terminating the search must be relatively small.

IV. Data Input and Examples

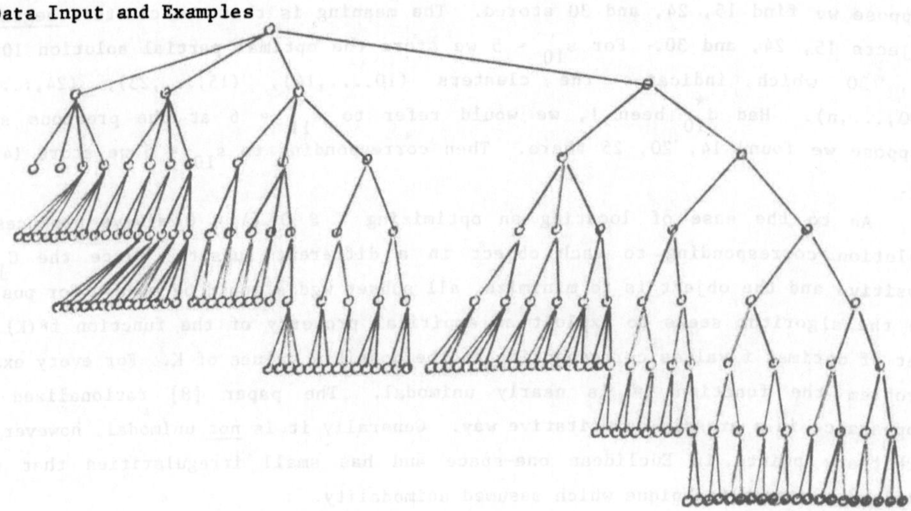

Figure 3. Test problem for computer solution

Data input includes three categories:

1. a numerical description of the tree structure to allow the computation of distances 2. subsystem weights and the value W (A large W and small sybsystem weights were selected so as not to interfere with any natural clusters in the data)
3. λ values and tolerances to initiate, regulate, and terminate the search.

Results

The optimal solution, subject to the cluster weight constraint, has the clusters 1-71--each item in a separate cluster

(72,73), (74,75), (76,77), (78,79), (80), (81,82), (83-86), (87-90), (91-95), (96-100).

The optimal objective function value is -4.8226 and 9 iterations were used by the procedure.

For purposes of comparison, two other partitions are

(1,2), (3-11), (12-15), (16-18), (19-23), (24-30), (31-36), (37-51), (52-60), (61-64), (65-72), (73-80), (81-100) with OFV = -2.2334

and

(1,2), (3-10), (11-14), (15-17), (18-22), (23-28), (29-32), (33-41), (42-52), (53-60), (61), (62-65), (66-73), (74-80), (81-100) with OFV = -2.6332

A first impression is that the optimal partition has done less of a clustering job than the other two solutions portray. Two facts are pertinent in this case. In the first place Fig. 3 is deceptive in a way as regards the metric. Subsystems that appear quite neighborly on an upper tree level such as 1 and 2 or 4 and 5 are actually quite distant relative to many other pairs. More on the distance measure a little later. The second item pertains to the objective function chosen vis-a-vis one's intuitive perception regarding what constitutes a good partition. Perhaps it is that in this problem subset separation is less important than subset homogeneity, so that the objective function should not accord them equal emphasis. The one used in the example does. An advantage of the technique employed is that it displays solutions for a range of λ (The cluster weight restrictions reduce that range, for the effect of decreasing λ is to create larger numbers of wgd's--equivalently, larger clusters--which the weight bound will prohibit). Thus, if subjective valuation contributes to the choice of OF, one sees other solutions, and his predilections may lead to weighting the two OF terms differently. In the present instance, for example, the inference may be that the average bgd term should have less weight than the wgd term.

A difficulty with the distance definition is that it amplifies distances between subsystems lying on different tree levels.

It would be intuitively more satisfying if we let

$$d_{ij} = \max(\text{level of } i, \text{level of } j) - \text{lowest numbered level where } i \text{ and } j \text{ belong to the same subsystem.} \qquad (5)$$

Now this measure is also a metric. Dealing, however, with unnormalized trees creates a new difficulty: the sequence of distances from a particular point may lose the monotonicity property noted earlier and it is not true that optimal clusters must necessarily consist of consecutive subsystems. The algorithm could overlook optimal clusters in that case.

Some trees may be transformed so that only monotonic sequences result, and the subsystem clustering problem is equivalent. We shall denote this process as one of regularization. Here we mean that if subsystems are labeled left-to-right and $i < j$ then $\text{lev}(i) \leq \text{lev}(j)$.

Our example, Fig. 3, is a tree in regular form. Unfortunately, not every tree admits regularization, and if a tree cannot be regularized then the algorithm may overlook optimal clusters if the metric of (5) is employed.

Thus arise questions relative to the uniqueness of decompositions. Roughly, it seems that if pensile sub-systems are equally recognizable at all stages of decomposition, then regular trees could always be given as the result. We simply apply the metric (5) to the former example and resolve the problem.

<u>Results</u>

The optimal solution is now

(1,2), (3-10), (11-14), (15-17), (18,19), (20-22), (23-28), (29-32), (33-36), (37-41), (42-46), (47-52), (53-59), (60-62), (63-65), (66-71), (72,73), (74-77), (78,79), (80), (81,82), (83-86), (87-90), (91-95), (96-100). The OFV is -3.7429, and 9 iterations were required.

Two other solutions are

(1,2), (3-11), (12-15), (16-18), (19-23), (24-30), (31-36), (37-51), (52-60), (61-64), (65-72), (73-80), (81-100) with OFV = -2.3720

and

(1-7), (8-14), (15-17), (18-22), (23-28), (29-31), (32-37), (38-52), (53-60), (61), (62-65), (66-73), (74-82), (83-100) with OFV = -2.6495

V. Conclusions

The effect of the new metric is most dramatically illustrated by the difference between the two optimal solutions. The lower-numbered subsystems became closer immediately and single element clusters disappear. An enlarging of clusters is also noted corresponding to the middle-numbered elements' grouping characteristics.

These problems have been solved with an amount of effort about equal to 9 shortest route problems with n = 100. Therefore, the technique is feasible, computationally, but to be useful, the distance measure and the objective function must reflect realistic properties and concerns. If the definitional aspects of the problem are settled, then since efficient solution for each λ value is guaranteed, the only dubious point is the ability to carry out an effective (hopefully efficient, also) search. With only empirical results available, and few of those, to be sure, conclusions are tenuous. f^* seems free of bumps, but the stretches of constant function value present a potential dilemma. If such a range occurred before an optimum were approached, the search pass could terminate prematurely. If a pass were designed to be more patient, on the other hand, it might futilely peruse unprofitable λ zones.

Empirically, again, the spans of constant function value have occurred beyond the minimum and have consequently proven no jeopardy.

References

1. L. Bodin, A Graph Theoretic Approach to the Grouping of Ordered Data, Networks, 2, (1972), 307-310.
2. R. Garfinkel and G. Nemhauser, Integer Programming, J. Wiley and Sons, New York, 1972.
3. B. Langefors, Theoretical Analysis of Information Systems, Vol. 2, Studentlitteratur, Lund, Sweden, 1970.
4. J. Nordbotten, The Analysis and Design of Computer-Based Information Systems, Houghton Mifflin Co., Boston, 1985.
5. M. Powers, et al, Computer Information Systems Development, South-Western Publishing Company, Cincinnati, Ohio, 1984.
6. S. Romahnowa, Classification as a Kind of Distance Function--Natural Classification, Studia Logica, XII, 1961.
7. J. Senn, Analysis and Design of Information Systems, McGraw-Hill, New York, 1984.
8. L. Stanfel, A LaGrangian Treatment of Certain Nonlinear Clustering Problems, E.J.O.R., 7, (1981), 121-132.
9. L. Stanfel, Applications of Clustering to Information System Design, Information Processing Management, 19, 1, (1983), 37-50.

On Continuous Time Adaptive Impulsive Control

Ł. STETTNER
Institute of Mathematics Polish Academy of Sciences

Suppose we are given a family $\mathcal{T}=\{X^{\alpha},\alpha\in A\}$, A-finite set, of Feller Markov or diffusion processes on a locally compact, separable state space $(E,\mathcal{E})$. The real process, which we observe is X^{α^0}, where $\alpha^0\in A$ is unknown. Assume we have a possibility to control X^{α^0} with the use of impulses, that is at chosen Markov time σ the process X^{α^0} can be shifted to new random state ξ and the evolution starts afresh from ξ. The impulsive strategy is a sequence $V=(\sigma_n, \xi_n)$ of Markov times and adapted to our information random variables ξ_n. Before we consider a functional, or the cost incurred under V, we have to construct a probability space (Ω, P^V) associated with control V. Since the above construction can be found in [7], we shall assume it for granted.
We want to minimize the following long run average functional

$$(1) \qquad J_x^{\alpha^0}(V)=\liminf_{t\to\infty} t^{-1} E_x^{V,\alpha^0}\{\int_0^t f(y_s)\,ds + \sum_{i=1}^{\infty} \chi_{\sigma_i\le t}\, c(x^{i-1}_{\sigma_i}, \xi_i)\}$$

where y_s is the trajectory of controlled process, $y_s=x_s^i$ if $s\in[\sigma_i,\sigma_{i+1}[$, $y_{\sigma_i}=\xi_i$ and x_s^i denotes a copy of X^{α^0} starting from ξ_i at time σ_i.
The function f stands for a "holding" cost, while $c(x,y)$ is the cost we have to pay for impulses.
The impulsive control models have applications in inventory control, stock exchange, cash management and many other problems (for details see [1]). Below we restrict ourselves to mathematical consideration of so called adaptive impulse models.
Assume that for each process X^{α}, $\alpha\in A$ we know the strategy which is optimal for the functional J_x^{α}. Therefore if α^0 is unknown, to minimize $J_x^{\alpha^0}$ we have to estimate the unknown parameter relatively frequently and then to adapt the control to obtained estimation. This means we apply the control which is the best for our estimation.
We will consider four impulsive models, for which we can characterize the optimal strategies of the functional (1), given α^0, and we are able to suggest a reasonable estimation and adaptive control strategies, that lead to any kind of consistency. In every case we will estimate at moments of impulses. Moreover we impose some restrictions on times of impulses. The cost incurred for impulses will be either in separated form $c(x,y)=c(x)+d(y)$, or general, if we restrict shifts to

a finite set $U=\{\zeta_1,\dots,\zeta_r\}$ only.

I. Diffusion models

Suppose the processes from the family $\mathcal{T}$ evolve in a bounded set $E\subset R^n$ and satisfy the following stochastic differential equation

(2) $\quad dX_t^{\alpha}=b(\alpha,X_t^{\alpha})dt+dw_t$, where (w_t) is a Brownian motion and b is a continuous function for $\alpha\in A$. Assume, that if process X^{α^0} enters the boundary ∂E, we have to decide for an impulse.

Using the Girsanov formula on any random interval $[0,\tau]$ we can compare the measures $P_x^{\alpha,\tau}$ that generate processes X^{α}. Namely the likelihood ratio is ([5])

$$(3)\qquad \Lambda^{\alpha}(\omega)=\frac{dP_x^{\alpha,\tau}}{dP_x^{\alpha^0,\tau}}=\exp\Big\{\int_0^{\tau}(b(\alpha,X_s^{\alpha^0})-b(\alpha^0,X_s^{\alpha^0}))dX_s-\frac{1}{2}\int_0^{\tau}b^2(\alpha,X_s^{\alpha^0})-b^2(\alpha^0,X_s^{\alpha^0})ds\Big\}$$

The first integral should be interpreted as an integral with respect to semimartingale (see section 4.2.11 in [5]).

As we suggested earlier, the form of the cost for impulses $c(x,y)$ will determine the model we consider.

1. Separated cost case

1a. Formulation of optimal impulse strategies

Let us consider an impulse control for process X^{α} , and look for a strategy that minimizes the functional (1) with α^0 replaced by α . Using the technics of papers [8],[9] one can prove the following

Theorem 1. Let

$$(4)\qquad \lambda_{\alpha}=\inf_x\inf_{\tau}\,(E_x^{\alpha}(\tau\wedge\tau_E))^{-1}\,E_x^{\alpha}\Big\{\int_0^{\tau\wedge\tau_E}f(x_s)\,ds+c(x_{\tau\wedge\tau_E})+d(x)\Big\}$$

$$(5)\qquad w_{\alpha}(x)=\inf_{\tau}E_x^{\alpha}\Big\{\int_0^{\tau\wedge\tau_E}(f(x_s)-\lambda_{\alpha})ds+c(x_{\tau\wedge\tau_E})\Big\}$$

where $\tau_E=\inf\{t\geqslant 0:\ x_t\in\partial E\}$, and

$$(6)\qquad \tau_{\alpha}=\inf\{t\geqslant 0:\ w_{\alpha}(x_t)=c(x_t)\}$$

$$(7)\qquad x^{\alpha}=\arg\inf\,[w_{\alpha}(x)+d(x)]$$

Then $w_{\alpha}(x^{\alpha})+d(x^{\alpha})=0$,

$$(8)\qquad \inf_V J_x^{\alpha}(V)=\lambda_{\alpha}$$

and the optimal strategy $V^{\alpha}=V(\tau_{\alpha},x^{\alpha})$ consists of shifts at times $\sigma_1=\tau_{\alpha}$, $\sigma_{n+1}=\sigma_n+\tau_{\alpha}\circ\theta_{\sigma_n}$ to the deterministic point x^{α} .

1b. Value biased MLE and the consistency of adaptive control

To apply an adaptive control suggested in our introduction it remains to define an estimation procedure. We can use the maximum likelihood estimation (MLE) i.e. maximize at σ_1 the value $\gamma(\alpha, 0, \sigma_1)$, at σ_n the value $\prod_{i=1}^{n} \gamma(\alpha, \sigma_{i-1}, \sigma_i)$, (we put $\sigma_0 = 0$), where for $\sigma \geq \tau$

(9) $$\gamma(\alpha,\tau,\sigma)=\exp\left\{\int_\tau^\sigma (b(\alpha,X_s^{\alpha^\circ})-b(\alpha^\circ,X_s^{\alpha^\circ}))dX_s-\frac{1}{2}\int_\tau^\sigma (b^2(\alpha,X_s^{\alpha^\circ})-b^2(\alpha^\circ,X_s^{\alpha^\circ}))ds\right\}$$

Nevertheless it will be clear later that an application of the value biased MLE introduced first by P. R. Kumar and A. Becker [4] is more adequate for the adaptive control situation.

Let m be a strictly increasing, positive, real function, and the function $\theta(t)$ satisfies: 1°) $\theta(t) \to \infty$, 2°) $t^{-1}\theta(t) \to 0$ as $t \to \infty$ (for example $\theta(t) = t^q$, $q \in (0,1)$).

We choose arbitrary $\hat{\alpha}_0 \in \Lambda$, wait $\sigma_1 = \tau_{\hat{\alpha}_0}$, and then estimate

(10) $$\hat{\alpha}_{\sigma_1} = \arg\max \left(\frac{m(\lambda_\alpha)}{m(\lambda_{\alpha^\circ})}\right)^{-\theta(\sigma_1)} \gamma(\alpha, 0, \sigma_1)$$

and immediately shift the process to $x^{\hat{\alpha}_{\sigma_1}}$. We wait $\tau_{\hat{\alpha}_{\sigma_1}}$ units of time i.e. up to $\sigma_2 = \sigma_1 + \tau_{\hat{\alpha}_{\sigma_1}} \circ \theta_{\sigma_1}$, then estimate

$$\hat{\alpha}_{\sigma_2} = \arg\max \left(\frac{m(\lambda_\alpha)}{m(\lambda_{\alpha^\circ})}\right)^{-\theta(\sigma_2)} \gamma(\alpha, 0, \sigma_1)\, \gamma(\alpha, \sigma_1, \sigma_2)$$

and shift to $x^{\hat{\alpha}_{\sigma_2}}$. We continue the above procedure estimating at $\sigma_n = \sigma_{n-1} + \tau_{\hat{\alpha}_{\sigma_{n-1}}} \circ \theta_{\sigma_{n-1}}$, the value $\hat{\alpha}_{\sigma_n}$

(11) $$\hat{\alpha}_{\sigma_n} = \arg\max \left(\frac{m(\lambda_\alpha)}{m(\lambda_{\alpha^\circ})}\right)^{-\theta(\sigma_n)} \prod_{i=1}^{n} \gamma(\alpha, \sigma_{i-1}, \sigma_i)$$

We will denote the above adaptive control as $\hat{V}$.

<u>Proposition 1</u>. There exists $N \subset \Omega$, $P_x^{\hat{V}}(N) = 0$ such that if for $\omega \in \Omega \setminus N$

(12) $$\limsup_{n\to\infty} n^{-1} \sum_{i=1}^{n} \chi_{\hat{\alpha}_{\sigma_i} = \alpha}(\omega) > 0, \text{ then}$$

(13) $$E_{x^\alpha}\left\{\int_0^{\tau_\alpha} (b(\alpha, X_s) - b(\alpha^\circ, X_s))^2\, ds\right\} = 0$$

the processes X^α and X^{α°, starting from x^α on the interval $[0, \tau_\alpha]$ are indistinguishable.

Proof. Let us notice first that

(14) $$\gamma(\alpha,\sigma_{i-1},\sigma_i)=\exp\left\{\int_{\sigma_{i-1}}^{\sigma_i} (b(\alpha,y_s)-b(\alpha^\circ,y_s))d\bar{w}_s-\frac{1}{2}\int_{\sigma_{i-1}}^{\sigma_i} (b(\alpha,y_s)-b(\alpha^\circ,y_s))^2 ds\right\}$$

where $(\bar{w}_s)$ is $P_x^{\hat{V},\alpha^\circ}$ Brownian motion.

In the sequel we will need the strong law of large numbers for stochastic integrals.

<u>Lemma 1</u>. If $t^{-1}\int_0^t \eta^2(s,\omega)\, ds$ is bounded P a.s. as $t \to \infty$, then

(15) $$\lim_{t\to\infty} t^{-1} \int_0^t \eta(s,\omega)\, dw_s = 0 \quad P \text{ a.s.}$$

Proof. Using the suitable change of time we reduce the problem to Brownian motion (proof of Lemma 4.1 [2]) . The law of large numbers for a Brownian motion follows from the law of iterated logarithm (section 1.4.2 of [5]).

Because of the definition of the estimator (11)

(16) $\left(\dfrac{m(\lambda_{\hat{\alpha}_{\sigma_n}})}{m(\lambda_{\alpha^0})}\right)^{-\Theta(\sigma_n)} \Lambda_{\sigma_n}(\hat{\alpha}_{\sigma_n}) \geq 1$, where $\Lambda_{\sigma_n}(\alpha) = \prod_{i=1}^{n} \gamma(\alpha, \sigma_{i-1}, \sigma_i)$

Then for $l_{\sigma_n}(\alpha) \overset{\text{def}}{=\!=\!=} \ln \Lambda_{\sigma_n}(\alpha)$ we obtain

(17) $-\Theta(\sigma_n) \ln \dfrac{m(\lambda_{\hat{\alpha}_{\sigma_n}})}{m(\lambda_{\alpha^0})} + l_{\sigma_n}(\hat{\alpha}_{\sigma_n}) \geq 0$, and using 2^0

(18) $\liminf_{n\to\infty} \sigma_n^{-1} l_{\sigma_n}(\hat{\alpha}_{\sigma_n}) \geq 0$

Taking into account (14) , since by (15)

(19) $\lim_{t\to\infty} t^{-1} \int_0^t (b(\alpha, y_s) - b(\alpha^0, y_s))\, d\bar{w}_s = 0$ for every $\alpha \in A$, $\omega \in \Omega \setminus N_1$,

$P_x^{\hat{V}}(N_1) = 0$, we can write (18) in a form

(20) $\liminf_{n\to\infty} \sigma_n^{-1} \left[- \int_0^{\sigma_n} (b(\hat{\alpha}_{\sigma_n}, y_s) - b(\alpha^0, y_s))^2\, ds \right] \geq 0$

But this means, for $\omega \in \Omega \setminus N_1$

(21) $\lim_{n\to\infty} \sigma_n^{-1} \int_0^{\sigma_n} (b(\hat{\alpha}_{\sigma_n}, y_s) - b(\alpha^0, y_s))^2\, ds = 0$

To finish the proof we apply a stability lemma

Lemma 2 ([6] p. 53). Suppose (F_n) is an increasing sequence of σ-fields and F_n-adapted random variables Z_n satisfy

(22) $\sum_{i=1}^{\infty} i^{-2} E\{Z_i^2\} < \infty$

Then

(23) $\lim_{n\to\infty} n^{-1} \sum_{i=1}^{n} (Z_i - E(Z_i | F_{i-1})) = 0$ P a.s.

Namely from (23) since $\sup_{\alpha} E_{x^\alpha}(\tau_\alpha^2) < \infty$ (the conclusion of Theorem 3.7.1 [3])

(24) $n^{-1} \sum_{i=1}^{n} \chi(\hat{\alpha}_{\sigma_{i-1}} = \alpha) \Big[\int_{\sigma_{i-1}}^{\sigma_i} (b(\alpha, y_s) - b(\alpha^0, y_s))^2\, ds +$

$- E_{x^\alpha} \Big\{ \int_0^{\tau_\alpha} (b(\alpha, x_s) - b(\alpha^0, x_s))^2\, ds \Big\} \Big] \longrightarrow 0$ for $\alpha \in A$, $\omega \in \Omega \setminus N_2$, $P_x^{\hat{V}}(N_2) = 0$

and

(25) $n^{-1} \Big(\sigma_n - \sum_{i=1}^{n} E_{x^{\hat{\alpha}_{\sigma_{i-1}}}}(\tau_{\hat{\alpha}_{\sigma_{i-1}}}) \Big) \longrightarrow 0$ for $\omega \in \Omega \setminus N_3$, $P_x^{\hat{V}}(N_3) = 0$, as $n \to \infty$.

Because $w_\alpha(x^\alpha)+d(x^\alpha)=0$, and $c(x)>a>0$, therefore $E^\alpha_{x^\alpha}\tau_\alpha>0$, and using the absolute continuity of measures P^{α° and P^α, also $E^{\alpha^\circ}_{x^\alpha}\tau_\alpha>0$. Thus there exist $0<\varepsilon<K_1\leq K_2$ such that $-\varepsilon+K_1\leq\inf_{\alpha\in A}E^{\alpha^\circ}_{x^\alpha}\tau^\alpha\leq$ $\leq\sup_{\alpha\in A}E^{\alpha^\circ}_x\tau_\alpha\leq K_2+\varepsilon$, and the convergence (25) can be read as

(26) $\qquad K_1\leq\sigma_n n^{-1}\leq K_2$ for sufficently large n, $\omega\in\Omega\setminus N_3$.

Applying (26) to (21) we obtain for $\omega\in\Omega\setminus(N_1\cup N_3)$

(27) $$\lim_{n\to\infty}n^{-1}\int_0^{\sigma_n}\left(b(\hat\alpha_{\sigma_n},y_s)-b(\alpha^\circ,y_s)\right)^2ds=0$$

Comparing (27) and (24) we see that if for $\omega\in\Omega\setminus N$, $N=N_1\cup N_2\cup N_3$, and $\alpha\in A$, (12) is satisfied, then the identity (13) holds. Next proposition explains the sense of bias of the estimation (11).

Proposition 2. There exists a set $\bar N\subset\Omega$, $P^{\hat V}_x(\bar N)=0$, such that if for $\omega\in\Omega\setminus\bar N$, the parameter α is a limit point of the estimation, then $\lambda_\alpha\leq\lambda_{\alpha^\circ}$.

Proof. See [10] Theorem 2.

Now we can summarize the obtained results. From Proposition 1,

$J^{\alpha^\circ}_x(\hat V)=\liminf_{n\to\infty}n^{-1}E^{\alpha^\circ}_x\{\sum_{i=1}^n\lambda_{\hat\alpha_{\sigma_i}}\}\geq\lambda_{\alpha^\circ}$, but taking into account Proposition 2 we obtain the identity, and this way we have proved

Theorem 2. The adaptive strategy $\hat V$ is value consistent, i.e. $J^{\alpha^\circ}_x(\hat V)=\lambda_{\alpha^\circ}$.

2. Nonseparated cost

Assume now, we can shift the process X only to points of the set $U=\{\xi_1,\xi_2,\dots,\xi_r\}$, and consider the general cost for impulses $c(x,y)$. The following theorem characterizes the value and optimal strategies in this case.

Theorem 3. For $\alpha\in A$, there exists a constant λ_α and a continuous, bounded function $w_\alpha(x)$ such that

(28) $$w_\alpha(x)=\inf_\tau E_x\{\int_0^{\tau\wedge\tau_E}(f(x_s)-\lambda_\alpha)ds+Mw_\alpha(x_{\tau\wedge\tau_E})\}$$

where $Mw_\alpha(x)=\inf_{\xi\in U}[c(x,\xi)+w_\alpha(\xi)]$

Moreover

(29) $$\inf_V J^\alpha_x(V)=\lambda_\alpha$$

Let

(30) $$\tau_\alpha=\inf\{s\geq 0:\ w_\alpha(x_s)=Mw_\alpha(x_s)\}$$

(31) $$\xi^\alpha(x)=\arg\inf_{\xi\in U}[c(x,\xi)+w_\alpha(\xi)]$$

Then the impulsive strategy $\sigma_1=\tau_\alpha$, $\xi^\alpha(x^1_{\sigma_1}),\dots,\sigma_{n+1}=\sigma_n+\tau_\alpha\circ\theta_{\sigma_n}$, $\xi^\alpha(x^{n+1}_{\sigma_{n+1}})$ is optimal.

Proof. We adapt the technics of [9] . Namely, let $w_\alpha^\beta(x)$ denotes the value function of β-discounted problem. Put $\lambda_\alpha^\beta = \min_{i=1,\dots,r} w_\alpha^\beta(\xi_i)$. Since $0 \le \bar{w}_\alpha^\beta(\xi_i) = w_\alpha(\xi_i) - \lambda_\alpha^\beta \le \|c\|$, there exists a subsequence $\beta_n \to 0$ such that $\bar{w}_\alpha^{\beta_n}(\xi_i) \to w_\alpha(\xi_i)$, $i=1,\dots,r$, $\beta_n \lambda_\alpha^{\beta_n} \to \lambda_\alpha$. Because $E_x\{\tau_E\} < \infty$, we can let $\beta \to 0$ in the β-discounted Bellman equation

$$(32)\qquad \bar{w}_\alpha^\beta(x) = \inf_\tau E_x^\alpha \left\{ \int_0^{\tau\wedge\tau_E} e^{-\beta t}(f(x_t) - \beta\lambda_\alpha^\beta)dt + e^{-\beta\tau\wedge\tau_E} M\bar{w}_\alpha^\beta(x_{\tau\wedge\tau_E}) \right.$$

and obtain (28). The optimality of (30) and (31) follows in a standard way from (28) (for details see [8],[9]).

We shall apply the same estimation as in the separated cost case (10), (11). Nevertheless we change the adaptive control $\hat{V}$. Namely we choose first arbitrary $\alpha_0 \in A$, wait $\sigma_1 = \tau^{\alpha_0}$ units of time and shift to $\xi^{\alpha_0}(x^1_{\sigma_1})$, and estimate $\hat{\alpha}_{\sigma_1}$. The value of the estimation determines next impulse at $\sigma_2 = \sigma_1 + \tau_{\hat{\alpha}_{\sigma_1}} \circ \theta_{\sigma_1}$ to $\xi^{\hat{\alpha}_{\sigma_1}}(x^2_{\sigma_2})$, and so on. That is at the moment of impulse we estimate the unknown parameter, but shift according to the value of previous estimation.

The following analogs of Proposition 1 and Theorem 2 can be proved in a similar way

Proposition 3. There exists $N \subset \Omega$, $P_x^{\hat{V}}(N) = 0$, such that if $\omega \in \Omega \setminus N$ and

$$(33)\qquad \limsup_{n\to\infty} n^{-1} \sum_{i=1}^n \chi_{y_{\sigma_{i-1}} = \xi_j}\, \chi_{\hat{\alpha}_{\sigma_{i-1}} = \alpha}(\omega) > 0$$

then

$$(34)\qquad E_{\xi_j}\left\{ \int_0^{\tau_\alpha} (b(\alpha, x_s) - b(\alpha^0, x_s))^2\, ds \right\} = 0$$

Theorem 4. We have

$$(35)\qquad J_x^{\alpha^0}(\hat{V}) = \liminf_{t\to\infty} t^{-1} E_x^{\alpha^0}\left\{ \int_0^t \lambda_{\hat{\alpha}_s}\, ds + \sum_{i=1}^\infty \chi_{\sigma_i \le t} \left[w_{\hat{\alpha}_{\sigma_{i-1}}}(x^{i-1}_{\sigma_{i-1}}) - w_{\hat{\alpha}_{\sigma_{i-1}}}(x^{i-1}_{\sigma_i}) \right] \right\}$$

where we consider $(\hat{\alpha}_t)$ as a continuous time process $\hat{\alpha}_t = \hat{\alpha}_{\sigma_i}$, $t \in [\sigma_i, \sigma_{i+1})$.

From Proposition 3, of course $\lambda_{\alpha^0} \le J_x^{\alpha^0}(\hat{V})$, but we don't obtain the value consistency. Under an additional, restrictive assumption

Identifiability assumption:

(I) for each $i=1,\dots,r$, $\alpha \ne \alpha'$, we have $b(\alpha, \xi_i) \ne b(\alpha', \xi_i)$

from Proposition 3 we can get even more, the consistency of parameter i.e. the convergence $\hat{\alpha}_t \to \alpha^0$ as $t \to \infty$.

II. Feller Markov models

Assume now X^α, $\alpha \in A$ are homogeneous, right continuous Feller Markov processes on a compact state space E. Let us impose two restrictions on shifts. Every decision of shift should be taken in a finite deterministic time T, and is realized with a constant delay h. Moreover we put the following assumptions

(36) $P_x^\alpha\{X_s^\alpha \in \Gamma\} = \int_\Gamma p_s(x,y,\alpha)\,\mu(dy),\ s>0$

(37) $\forall\, \mathcal{O}$-open $\mu(\mathcal{O})>0$

(38) if $p_s(x,y,\bar\alpha) > 0$ for some $\bar\alpha \in A$, then $p_s(x,y,\alpha) > 0$ for any $\alpha \in A$

(39) $p_s(x,y,\alpha)$ is x,y -continuous and s-right continuous

(40) $\int_E (\ln(p_s(x,y,\alpha)(p_s(x,y,\alpha^0))^{-1}))^2\, p_s(x,y,\alpha^0)\,\mu(dy)$ is a bounded x-function

One can easy see that (38) together with (36) and (37) imply that the measures that generate on the space of trajectories processes X^α are absolutely continuous, and therefore the processes X^α can be compared. Similarly as in diffusion case we consider two models: with separated cost and shifts to any point of E, and with a general cost and shift to the finite set U only.

1. Separated cost case

1a. Optimal strategies and estimation

Using the technics of [8] we obtain

Theorem 5. Let

(41) $\lambda_\alpha = \inf_x \inf_{\tau \le T} (E_x^\alpha(\tau+h))^{-1} E_x^\alpha\{\int_0^{\tau+h} f(x_s)\,ds + c(x_{\tau+h}) + d(x)\}$

(42) $w_\alpha(x,t) = \inf_{\tau \le T-t} E_x^\alpha\{\int_0^\tau (f(x_s)-\lambda_\alpha)\,ds + \bar c_\alpha(x_\tau) + v_\alpha(x_\tau)\}$

for $\bar c_\alpha(x) = P_h^\alpha c(x)$, $v_\alpha(x) = \int_0^h (P_s^\alpha f(x) - \lambda_\alpha)\,ds$

(43) $\tau_\alpha = \inf\{s \ge 0:\ w_\alpha(x_s,s) = \bar c_\alpha(x_s) + v_\alpha(x_s)\}$

(44) $x^\alpha = \arg\inf [w_\alpha(x,0) + d(x)]$

Then $u_\alpha(x) = \lambda_\alpha$ and the strategy $V(\tau_\alpha, x^\alpha)$ is optimal.

The adaptive strategy $\hat V$ we construct in a similar way as in diffusion case. The only difference is in our estimator. We don't have a formula expressing the densities of measures that processes X^α induce on the space of trajectories. But we can compare the densities $p_s(x,y,\alpha)$ and introduce the following estimation procedure at the moments of impulses

(45) $\hat\alpha_{\sigma_i} = \arg\max \ln m(\lambda_\alpha)^{-\theta(\sigma_i)} + \int_0^{\sigma_i} \chi_{s\in F} \int_0^h \ln p_s(x_t, x_{t+s}, \alpha)\,ds\,dt$

where $F = \bigcup_{i=1}^\infty [\sigma_i - h, \sigma_i[$.

Put

(46) $l_t(\alpha) = 0$ if $t \in F$ and $l_t(\alpha) = \int_0^h \ln(p_s(x_t, x_{t+s}, \alpha)\, p_s(x_t, x_{t+s}, \alpha))^{-1} ds$

otherwise.

Then also

$$(47)\qquad \hat{\alpha}_{\sigma_i} = \arg\max_{\alpha\in\Lambda} \ln\left(\frac{m(\lambda_{\alpha^0})}{m(\lambda_\alpha)}\right)^{\theta(t)} + \int_0^t l_s(\alpha)\,ds$$

and therefore using the definition of $\theta(t)$

$$(48)\qquad \liminf_{i\to\infty} \frac{1}{\sigma_i}\int_0^{\sigma_i} l_s(\hat{\alpha}_{\sigma_i})\,ds \geq 0$$

Let

$$(49)\qquad g(x,\alpha) = \int_0^h \int_E p_s(x,y,\alpha^0) \ln \frac{p_s(x,y,\alpha)}{p_s(x,y,\alpha^0)}\,\mu(dy)\,ds$$

From the Jensen inequality $g(x,\alpha)\leq 0$ and using the stability Lemma 2, for $\omega\in\Omega\setminus N_1$, $P_x^{\hat{V}}(N_1)=0$

$$(50)\qquad \lim_{i\to\infty} \frac{1}{i}\int_0^{\sigma_i} \chi_{s\in F}\left(l_s(\alpha) - g(x_s,\alpha)\right) ds = 0$$

1b. Identification

We will show that the estimation (45) makes possible an identification of the transition densities $p_s(x,y,\alpha)$ and the cost incurred under $\hat{V}$.

<u>Theorem 6.</u> Let $p(x,\alpha,\mathcal{O}) = E_x^{\alpha^0}\{\int_0^{\tau_\alpha} \chi_{x_s\in\mathcal{O}}\,ds\}$ for an open set $\mathcal{O}$,

$$(51)\qquad G_\alpha = \{x:\ \forall_{\mathcal{O}\in B(x)}\ p(x^\alpha,\alpha,\mathcal{O}) > 0\}$$

where $B(x)$ is a basis of open sets at point x.

Then there exists $N\subset\Omega$, $P_x^{\hat{V}}(N)=0$ such that if for $\omega\in\Omega\setminus N$

$$(52)\qquad \limsup_{i\to\infty} \frac{1}{i}\sum_{j=1}^{i} \chi_{\hat{\alpha}_{\sigma_j}=\alpha}(\omega) > 0$$

then

$$(53)\qquad p_s(x,y,\alpha) = p_s(x,y,\alpha^0) \quad \text{for } x\in G_\alpha,\ y\in E,\ s\in[0,h[$$

Proof. Let

$$(54)\qquad z_{j+1}(\alpha,\mathcal{O}) = \int_{\sigma_j}^{\sigma_{j+1}-h} \chi_{\hat{\alpha}_{\sigma_j}=\alpha}\,\chi_{x_t\in\mathcal{O}}\,g(x_t,\alpha)\,dt$$

Then

$$(55)\qquad E_x^{\hat{V}}[z_{j+1}(\alpha,\mathcal{O})\mid F_{\sigma_j}] \leq \chi_{\hat{\alpha}_{\sigma_j}=\alpha} \sup_{x\in\mathcal{O}} g(x,\alpha)\; p(x^\alpha,\alpha,\mathcal{O})$$

From the stability Lemma 2, for each $\alpha\in\Lambda$, $\mathcal{O}$ from the countable basis of the topology on E, outside of $N_2\subset\Omega$, $P_x^{\hat{V}}(N_2)=0$, we have

$$(56)\qquad \lim_{i\to\infty} \frac{1}{i}\sum_{j=1}^{i}\left(z_j(\alpha,\mathcal{O}) - E_x^{\hat{V}}(z_j\mid F_{\sigma_{j-1}})\right) = 0$$

Therefore from (55)

$$(57)\qquad \limsup_{i\to\infty} \frac{1}{i}\sum_{j=1}^{i}\left(z_j(\alpha,\mathcal{O}) - \chi_{\hat{\alpha}_{\sigma_{j-1}}=\alpha} \sup_{x\in\mathcal{O}} g(x,\alpha)\; p(x^\alpha,\alpha,\mathcal{O})\right) \leq 0$$

Taking into account the nonpositivity of $g(x,\alpha)$ and (55), we obtain

$$(58)\qquad \limsup_{i\to\infty} \frac{1}{i}\left(\int_0^{\sigma_i} l_t(\alpha)\,dt - \sum_{j=1}^{i} \chi_{\hat{\alpha}_{\sigma_{j-1}}=\alpha} \sup_{x\in\mathcal{O}} g(x,\alpha)\, p(x^{\alpha},\alpha,\mathcal{O})\right) \leq 0$$

for $\omega \in \Omega \setminus (N_1 \cup N_2)$, $\alpha \in A$.

Suppose α is such that for some $\omega \in \Omega\setminus(N_1 \cup N_2)$, (52) holds. Then one can choose a subsequence $(i_k)_{k\in N}$, $\hat{\alpha}_{\sigma_{i_k}}(\omega) = \alpha$ such that

$$(59)\qquad \lim_{k\to\infty} \frac{1}{i_k} \sum_{j=1}^{i_k} \chi_{\hat{\alpha}_{\sigma_j}=\alpha}(\omega) > 0$$

Since $\sigma_i \leq i\,(T+h)$, then from (48)

$$(60)\qquad \liminf_{k\to\infty} \frac{1}{i_k} \int_0^{\sigma_{i_k}} l_t(\alpha)(\omega)\,dt \geq 0$$

and therefore from (58)

$$(61)\qquad \lim_{k\to\infty} \frac{1}{i_k} \sum_{j=1}^{i_k} \chi_{\hat{\alpha}_{\sigma_{j-1}}=\alpha}(\omega) \sup_{x\in\mathcal{O}} g(x,\alpha)\; p(x^{\alpha},\alpha,\mathcal{O}) = 0$$

Applying (59) we finally obtain

$$(62)\qquad \sup_{x\in\mathcal{O}} g(x,\alpha)\, p(x^{\alpha},\alpha,\mathcal{O}) = 0$$

So if $x \in G_{\alpha}$, $\sup_{y\in\mathcal{O}} g(y,\alpha) = 0$ for all $\mathcal{O} \in B(x)$, and by the continuity $g(x,\alpha)=0$. To finish the proof of Theorem 6 it remains to use the inequality $\exp a > 1+a$ for $a>0$, with the identity for $a=0$, and the assumptions (36) - (39) in a similar way as in the proof of Theorem 2.1 [10]. The next lemma is a continuous version of Proposition 2.1 [10].

Lemma 3. Let $\bar{\alpha}$ be a Cesaro frequent point of the estimation for some $\omega\in\Omega\setminus N$ (N is the same as in Theorem 5)

$$(63)\qquad \limsup_{i\to\infty} \frac{1}{i} \sum_{j=1}^{i} \chi_{\hat{\alpha}_{\sigma_j}=\bar{\alpha}}(\omega) > 0$$

Then for $x \in G_{\bar{\alpha}}$

$$(64)\qquad P_x^{\alpha^0}\{x_t \in G^c_{\bar{\alpha}} \text{ for some } t < \tau_{\bar{\alpha}}\} = P_x^{\bar{\alpha}}\{x_t \in G^c_{\bar{\alpha}} \text{ for some } t<\tau_{\bar{\alpha}}\} = 0$$

The proof is based on the discretization and right continuity of the process X^{α^0}, and similar technics as in [10], is omitted.

Now we can prove

Theorem 7. We have

$$(65)\qquad J_x^{\alpha^0}(\hat{V}) = \liminf_{i\to\infty} \frac{1}{i} \sum_{j=1}^{i} E_x^{\alpha^0}\{\lambda_{\hat{\alpha}_{\sigma_j}}\}$$

Remark. As one can see the result above can be obtained also without the bias $(m(\lambda_{\alpha}))^{-\sigma(t)}$ in the estimation. The identification results, Theorem 6 and Lemma 3 hold also for A countable. If we impose the identifiability assumption

$$(66)\quad \forall_{\alpha\in A}\ \exists_{s\in[0,h[}\ {}_{y\in E}\quad P_s(x^\alpha, y, \alpha) \neq P_s(x^\alpha, y, \alpha^0)$$

we can easy obtain using Theorem 6, the parameter and value consistency.

2. Nonseparated cost case

One can consider also a model with general cost $c(x,v)$, shifts in deterministic time T, realized with delay h to the set $U=\{\xi_1,\dots,\xi_r\}$. The form of optimal strategies can be found as in Theorem 3. The estimation should be applied of the form (45). To the value of the estimation we adapt the strategy as in I.2 . Then using the analogs of of the identification Theorem 6 and Lemma 3 we obtain the results similar to Theorem 4.

References

[1] Bensoussan A. and Lions J.L., Controle impulsionnel et inequations quasi variationnelles, Dunod, 1982,

[2] Borkar V. and Bagchi A., Parameter Estimation in Continuous-Time Stochastic Processes, Stochastics 8, 1982, pp. 193-212,

[3] Hasminskii, R.Z., Stochastic stability of differential equations, Alphen aan den Rijn, Sijthoff and Noordhoff, 1980,

[4] Kumar P.R. and Becker A., A new Family of Optimal Adaptive Controllers for Markov Chains, IEEE Trans. Automat. Contr. 27, 1982, pp. 137-146,

[5] Liptser R.S. and Shiryayev A.N., Statistics of random processes I, Springer-Verlag, New York 1977,

[6] Loeve M., Probability Theory II, Springer, New York 1978,

[7] Robin M., Controle impulsionnelle de processus de Markov, Thesis, University of Paris IX, 1978,

[8] Stettner Ł., On impulsive control with long run average cost criterion, Studia Math. 76, 1983, pp. 279-298,

[9] Stettner Ł., On Ergodic Impulsive Control Problems, submitted to Stochastics,

[10] Stettner Ł., Discrete Time Adaptive Impulsive Control Theory, Preprint IM PAN N° 309, submitted to Stoch Proc. their Appl.

Neuronal Networks Modelled by Petri Type Nets with Controllers

NICOLAE STOICA, Institutul de Medicină şi Farmacie Cluj, Catedra de Fiziologie, str. Clinicilor nr.1, 3400 Cluj-Napoca, R.S.România
GHEORGHE ROTH, Centrul Teritorial de Calcul Electronic, str. Republicii nr.107, 3400 Cluj-Napoca, R.S.România

1. Introduction

The modelling of neuronal networks is a very actual problem in which biologists, electronists and cyberneticians are equally interested. Many things are known about the physiology of nerve cells, but we know very little about the processing, stocking and the manipulation of the informations in the neuronal networks. The various types of nervous systems, even in the inferior animals, are very complex. That is why modelling is one of the principal study methods of networks' functioning. Many types of models were constructed in the past decades: discrete models based on works of Mc.Cullock and Pitts [3], continuous models based on works of Hodgkin and Huxley [7], models based on high abstractions, based on the most general features of the nervous systems. Our model is also based on the basic features of the neurons, as described by the researchers of the field and a series of valid principles of the nervous systems. We mention that we made no experiments; all experimental results we made use of, are taken from the studies listed in the references.

2. The Modelling of Neurons

In a first phase, models were constructed for 6 types of neurons : receptor neurons, somatic motoneurons, automatic neurons, Renshaw interneurons, effectors and ramification nodes.

2.1 Receptor neurons

Receptor neuron's model was constructed based on works of Werner and Mountcastle [10], who found the following relation:

$$R=K*S^n \tag{1}$$

where R is the mean response number, K is a constant of proportionality, S is the stimulus intensity, n is determined by the shape of the fitted straight line in log-log coordinates. The strength of external stimulus is encoded by the firing rate and we took K=9.4,n=0.52 .In our homeostatic model we have 2 receptor types: (a) receptors that can detect the raise of the liquid's level above 0 in a basin (fig.1 a); (b) receptors that can detect the decrease of the liquid's level below 0 in a basin (fig.1 b)

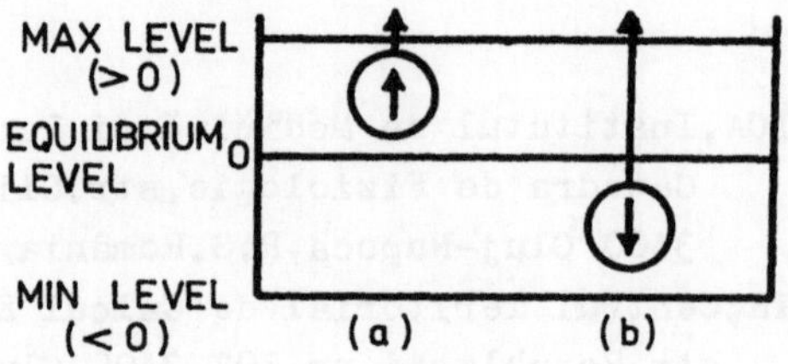

fig.1

Based on the formula (1) we can graphically represent the receptor's functioning .(fig.2)

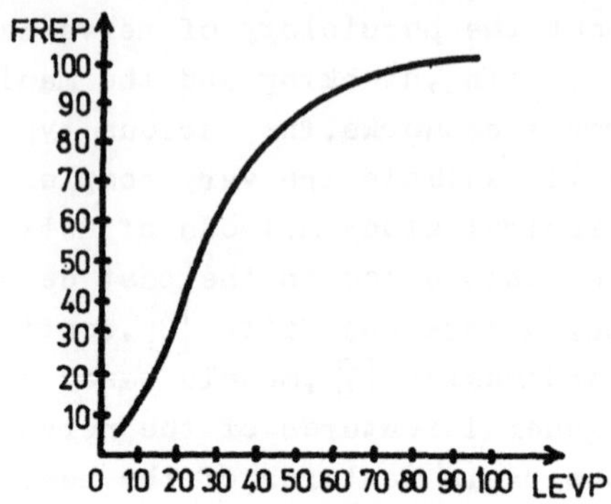

fig.2

FREP is the percentual frequency of spike emitting
FREP=((frequency/max frequency)*100)
LEVP is the percentual level of the liquid
LEVP=((abs(level)/max level)*100) (2)

The physiological proof of modelling is : at Paccini's corpuscles the first Ranvier node is functioning like a summator of the influence of a generator represented by the dendritic extremity (Guyton [6]). After receicing the information about the liquid's level in the basin,the receptor computes the percentual frequency (FREP) based on the formula (2),next from FREP the interspike and finally the adding rate (AR) to the summator,that leads to the firing frequency corresponding to the particular liquid level from the basin.

$$\text{Basin's level} \longrightarrow \text{FREP} = 9.4*\left(\frac{\text{Basin's level}}{\text{max level}}\right)^{0.52} \longrightarrow \text{interspike} \longrightarrow \text{AR}$$

SUMMATOR := SUMMATOR + AR

2.2 The Somatic Motoneuron

In our model the somatic motoneuron has a summator inspired from the real motoneuron's initial segment.The variations of the neurons' membrane potentials (MP) are noted on 2 scales.In the model we use a normalized scale where 0 is the resting potential(-70 mV),1 is the potential corresponding to -60 mV (firing threshold) and -1 is the potential corresponding to -80mV the IPSP's equilibruum potential.The neurons have the possibility to receive signals from other neurons and to send signals to others.(fig.3)

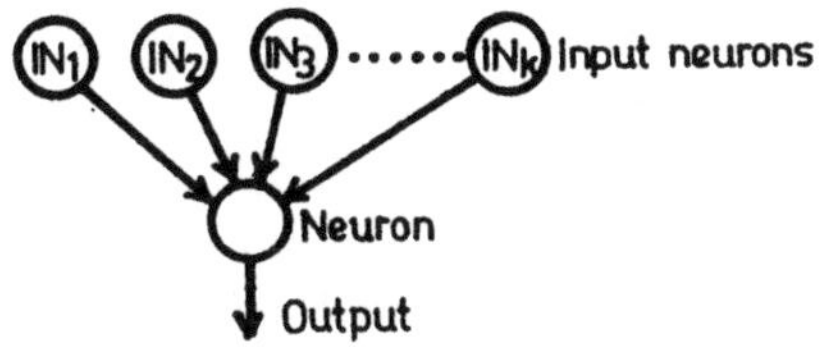

fig.3

Under the excitatory influence from other neurons,the value of membrane potential(noted MP) grows from 0 to 1. When the value 1 is reached a spike is generated toward the output neurons. If the excitatory influence does not reach the threshold (value 1) an EPSP is produced (Excitatory Post Sinaptic Potential), which decreases respecting a time constant law.(fig.4)

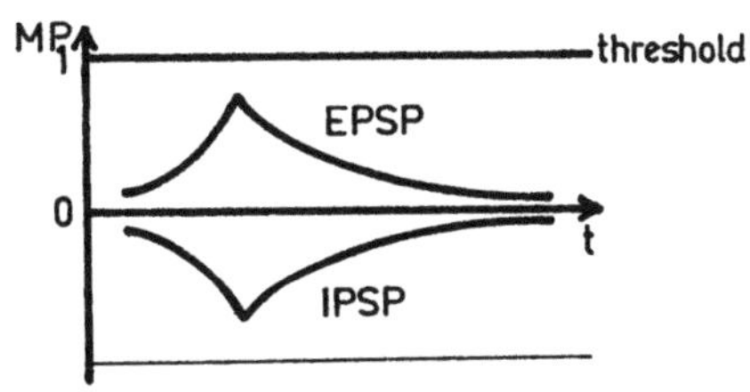

fig.4

The inhibitory influences IPSP coming from the inhibitory neurons(the neurons transmitting negative values) have a similar behaviour with the EPSP,but symmetrical with the time axis.(fig.4) The IPSP is corected by multiplying with a correction factor :

$$cf_m(t)=a*t+b$$

determined by processing of Eccles' experimental results [4].In our model the EPSP-s and IPSP-s are added applying all the principles listed above. If the summator reaches the value 1 a spike is produced. The neuron sending this spike cannot receive any signal from other neurons for 1 ms, the spike's time. The spike is followed by an afterpotential when the neuron returns progresively to the resting excitabillity. (fig.5)

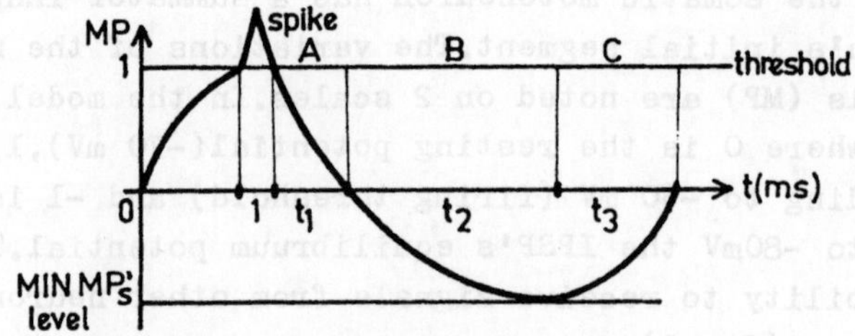

fig.5

This afterpotential was described by means of 3 time constants τ_1, τ_2, and τ_3 determined with the formula (3):

$$\tau_i = t_i / \ln(MAXLEV/MINLEV) \quad i=1,2,3 \tag{3}$$

where t_i are the time periods on fig.5, MAXLEV is the max absolute value of MP on the corresponding interval, MINLEV is the min absolute value of MP on the corresponding interval. In every moment the new value of MP (LEVN) is computed by the following method (fig.6):

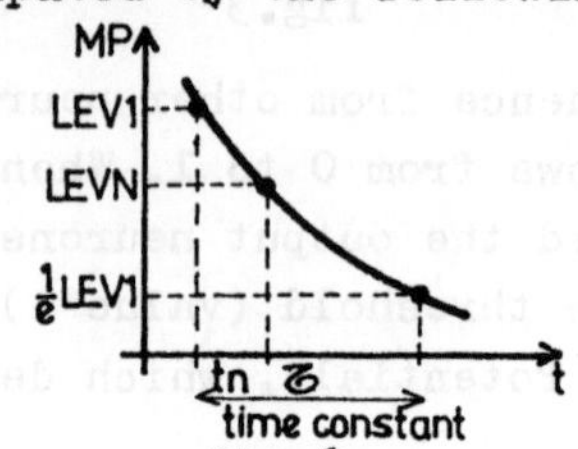

fig.6

$$\Delta LEV = LEV1 - LEV1/e$$
$$LEVN = LEV1 - (\Delta LEV * t_n)/\tau \tag{4}$$

When the curve is ascending the method is almost similar.

2.3 The Automatic Neurons

The automatic neurons (also called "peace makers") were modelled starting from the somato-motoneurons.(fig.7)

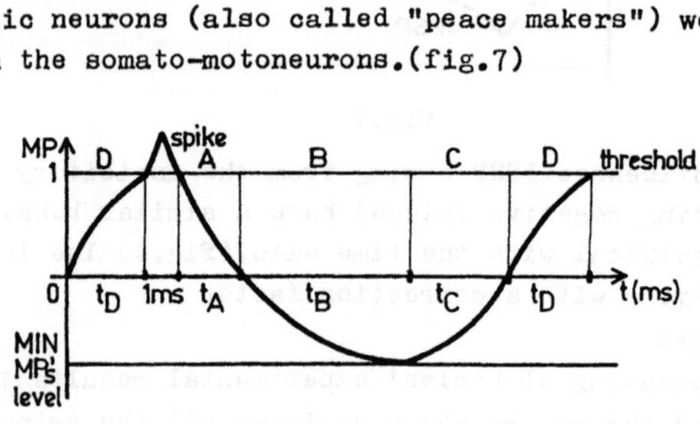

fig.7

The A,B,C intervals were completed by the D interval which assure the automatic charge of the membrane potential so that the spike is produced without any influence from other neurons. The D intercal was realised by a time constant mechanism. Under the excitatory or inhibitory influences of other neurons, the automatic neurons can emite slower or faster so they alter their discharge phase. (fig.8)

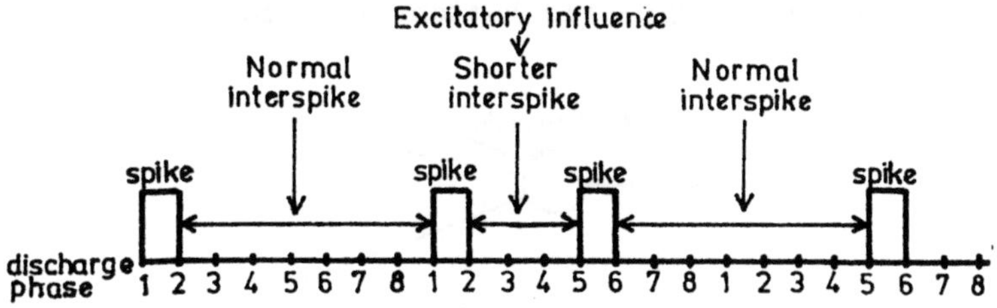

fig.8

By means of this mechanism a phased modulation of signals is realised; this process represents a form of plasticity (memory) (Frazier [5]).

2.4 The Renshaw Interneuron

The Renshaw interneurons are also called "neurons with repetitive discharge". Eccles showed experimentally that if such a neuron receives excitatory impulses, an EPSP is produced that during the decreasing period produces a series of spikes [4]. The Renshaw interneurons were modelled based on works of R.D.Traub [8], so they function according to the following graphics (fig.9):

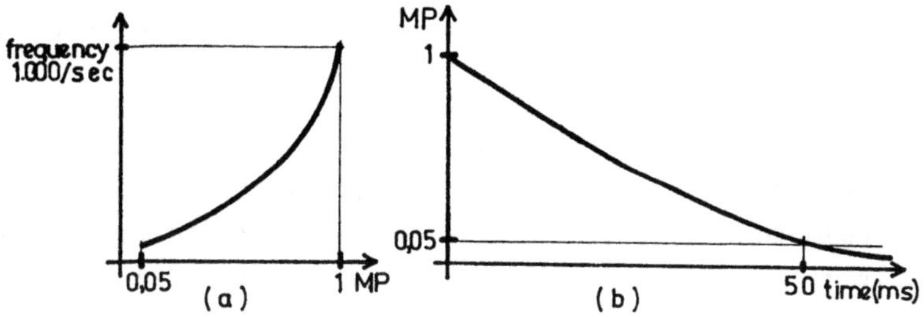

fig.9

In our model this interneuron has 2 summators: S1 is the summator for the membrane's potential ; S2 is the summator for repetitivity. In each moment t, if the membrane's potential is greater than 0.05, is determined from the frequency graphic (fig.9 a) the corresponding discharge frequency which gives also the adding rate to the summator S2. When the summator S2 reaches the threshold a spike is emitted. The summator S1 decreases his value respecting a time constant law, like in fig.9 b.

2.5 The Effectors

We have 2 types of effectors in our model : (a) effectors that pump liquid in the basin ; (b) effectors that absorb liquid from the basin. When these effectors receive an impulse from a neuron,pump(or absorb) a liquid quantity,after a delay.

2.6 The Ramification Nodes

Ramification nodes are elements that permits a distribution of the impulses to some neurons' synaptic buttons (fig.10).

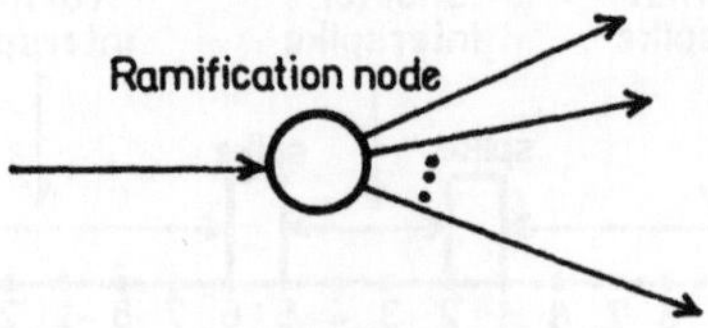

fig.10

For each type of neurons the authors conceived an algorithm which computes the changes of membrane's potential in accordance with each neurons' mechanism of functioning.

3. Petri Type Nets with Controllers

The network model is conceived to discretly simulate the analogue functioning of neurons.That permits determination of every neuron's and the whole network's state at times t,2t,... where t is the basic time unit considered (t=1μs).The neuronal network is modelated by a Petri type net noted NN; NN=(N,S,F) where N is the set of locations that represents neurons; S is the set of transitions that represents synapses; $F \subseteq (N \times S) \cup (S \times N)$ represents the flow relations between N and S. To set N mark function $M: N \rightarrow [a,b]$ is attached,where $[a,b]$ is the interval where the main parameter,characteristic of a neuron's state (the membrane potential),can varry. We also considered the function $W: F \rightarrow [a',b']$ that defines each neuron's firing threshold,so that a neuron emites a spike if and only if $M(n) \geq W(n,t)$. There are 6 types of locations : (a) automatic neurons (b)motoneurons (c)Renshaw interneurons (d)receptors (e) effectors (f)ramification nodes.Each type has a specific algorithm of functioning. These algorithms are called C O N T R O L L E R S. In each moment t acts upon the neurons and the network in 2 phases : Phase 1. Every neuron changes its membrane potential based on the specific algorithm,by means of the corresponding controller : $MP_i(t) := C\ (MP_i(t-1))$. Phase 2. The impulses are transmited through the network according to the functioning of the Petri networks,the membrane potential of every neuron being altered by the inputs. The inputs are either positive if the source neurons are excitatory or negative if the source neurons are inhibitory. For the purpose of simulations,a set of FORTRAN programs called N E T C O N was created. This set of programs makes possible the definition of a network and the simulation of the network's functioning for a given period of time.

4.Experimental Results

We took as external phenomenon for our model the oscillations of a basin's liquid level. The basin is like in fig.11.

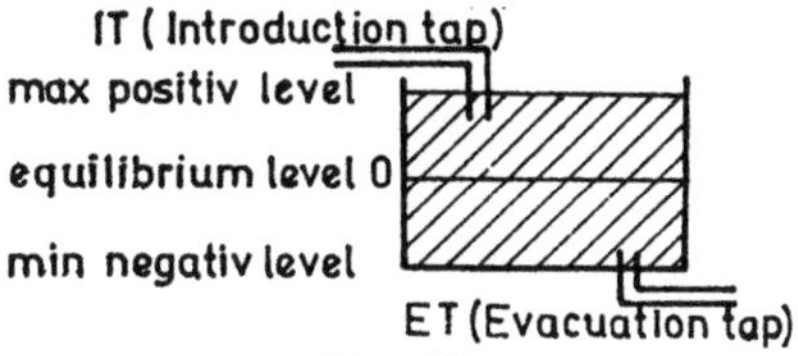

fig.11

If a variation is produced in the basin the phenomenon is coded by means of the receptors by modulation in frequency.The network we considered is a homeostat,that has the function to keep the liquid's level in the basin at level 0. The network is in fact a feed-back loop constructed as based on the neurophysiological principles. The network is represented in fig. 12.

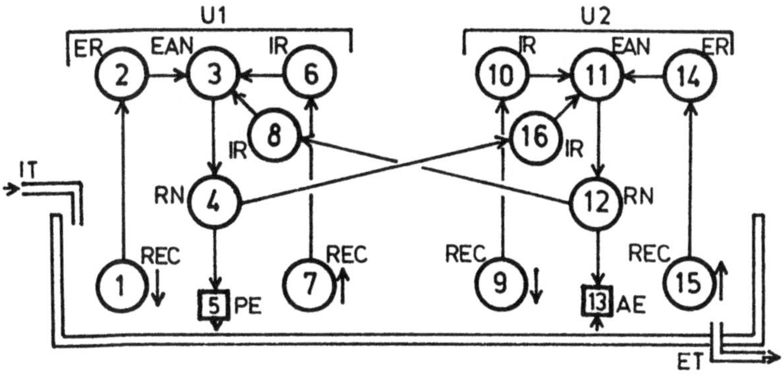

fig.12

REC is receptor; ER is Excitatory Renshaw interneuron; IR is Inhibitory Renshaw interneuron ; EAN is Excitatory Automatic Neuron ; PE is Pumping Effector; AE is Absorbant Effector

When we produced a decrease in the liquid's level,the network presented a tendency to pump liquid by means of PE15,while the effector AE13 left inactive. The result represents a damped oscillation to 0(fig.13). When the liquid's level increased,the network tended to absorb liquid by means of AE13,the effector PE5 being inactive.

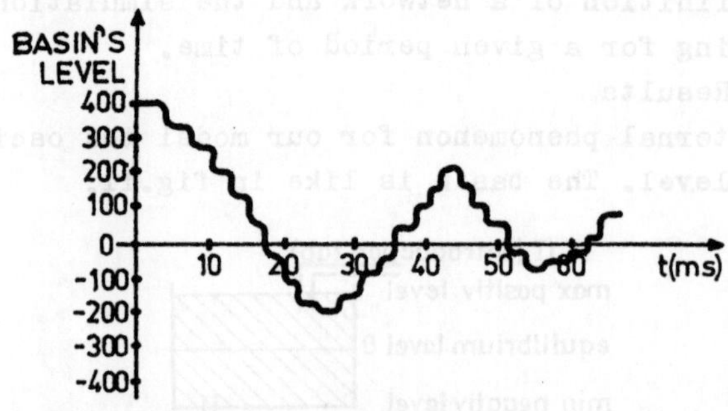

fig.13

When the basin's level decreases, REC1 and REC9 are activated, emitting with a frequency corresponding to that level. By means of ER2, EAN3 is excited, commanding the effector PE5 which pumps liquid in the basin (the REC7 and REC15 were introduced to detect and regulate the level when a liquid increase is produced). The function of REC9 is that of inhibiting by means of IR10 the automatic neuron EAN11 which commands the absorbant effector AE13. The double representation of the information in the presence of the automatic neurons increases the precision and the domains of regulation of the liquid's level. The 2 units U1 and U2 (fig.12) of the network have reverse effects. The 2 automatic neurons EAN3 and EAN11 are mutually inhibated by means of IR8 and IR16 this being the consequence of the reciprocal inhibition principle. The basin's model was inspired by the presence of reciprocal inhibition to the flexor and extensor muscles at the limbs. The pumping effector is similar with the extensor, the absorbant with the flexor. The authors conceived a model with 4 basins (fig.14). By means of a superior level, composed of 2 automatic neurons coupled by reciprocal inhibition, a regulation of the basins pairs(1-2;3-4) is realised. Every neuron from the second level is coupled with all the 4 automatic neurons from the corresponding basins, in accordance with the law of reciprocal inhibition. The decreasing of the first basin's level will produce an increase of the liquid in the second basin, the sum of differences being constant. The global regulation of the 4 basins is produced on the third level by means of 2 automatic neurons A31 and A32 coupled in accordance to reciprocal inhibition principle. That model is suggested by the movement of animals' legs, that's why in fig.14 we presented the association. This model respects the "hyerarchy principle" (Albus [1]) or the principle of re-and-rerepresentation of functions (Jackson [9]). We mention that the authors simulated only 1 basin becouse the time complexity of the programs is prohibitive for networks with a great number of components.

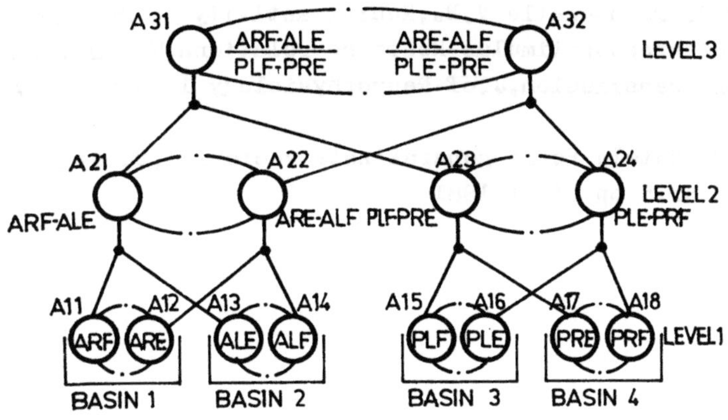

fig.14

Abreviations : A-anterior ; P-posterior ; R-right ; L-left ; E-extensor F-flexor;

-reciprocal inhibition

The programs were runned on a CORAL 4030 computer (compatible PDP 11 /35),for the simulation of a larger network being necesary a faster one.

REFERENCES:

[1] Albus J.S. Brains,Behaviour and Robotics,Byte Books 1981

[2] Bălăceanu C.,Nicolau E. Elemente de Neurocibernetică,Ed.Stiintifică, 1967

[3] Mc Cullock W.S. and W.H.Pitts, A logical calcullus of the ideas immanent in nervous activity,Bull.Math.Biophys.5 pp 115-133

[4] Eccles J.C.,The Physiology of Nerve Cells,Oxford Univ.Press 1957

[5] Frazier W.T. ,Morphological and functional properties of identified neurons in the abdominal ganglion of Aplysia Californica,J.of Neurophysiology 1967,XXX 16,pp 1288-1351

[6] Guyton A.C. ,Organ Physiology,Structure and Function of the Nervous Systems,2-ed,Philadelphia W.B Saunders 1976

[7] Hodgkin A.L and F.Huxley,A quantitive Description of Membrane Current and its application to conduction and excitation in nerve,J.Physiology 117 pp 500-544 (1952)

[8] Traub R.D., Repetitive Firing of Renshaw Spines Interneurons,Biol. Cybernetics 27,71-76(1977)

[9] Jackson,J. Selected Writtings of John Hughlings Jackson.Edited by J.Taylor London: Hodder and Stoughton,1931

[10] Werner G.,Mountcastle V.B.,Neural activity in Mechanoreceptive Cutaneous afferents: Stimulus-Response relations,Weber functions and information transmission,J.of Neurophysiology 1965 vol. XXVIII No.2 pp 359-397

[11] Sandu D.,Reţele Petri generalizate vectorial,Buletinul Român de Informatică nr.5 pp 81-91 1981

THE USE OF FLYING HORIZON METHOD TO TRAFFIC OPTIMIZATION IN PROCESS CONTROL COMMUNICATION SYSTEMS

EMIL I. STOILOV and MARIANA S. STRUGAROVA

Institute of Industrial Cybernetics and Robotics
Bulgarian Academy of Sciences, Sofia, Bulgaria

INTRODUCTION

Considerable development of distributed industrial process control systems has been taking place in recent years. The core of such systems is the communication network, which connects the remote control devices and provides the required quality of the information exchange among them. One of the major problems is the increase of the number of stations which participate in the communication and the amount of the information transmitted by them. This can be solved in some situations by increasing the speed of the data transmission. The difference between the local area networks used in industrial process control and those used in office automation is that the environment of the industrial control is much more hostile than that of the office. Increasing the data transmission speed in industrial process control communication systems while retaining the required reliability of the information exchange, considerably enhances the price of the communication system.

Adaptive and optimal algorithms for information exchange control can be regarded as an alternative. In this case, the capabilities of the communication system are enlarged mainly through software.

The flying horizon method, used for the dynamic optimization of complex systems, has been proposed in [1]. In the present paper we discuss the way by which this method can be utilized as the basis for creating information exchange on-line control algorithms in industrial process control communication systems.

BRIEF DESCRIPTION OF THE COMMUNICATION SYSTEM

The communication system considered here is assumed to have N main stations and K slave stations and a highway architecture. Each of the stations is provided with a private microprocessor performing the exchange protocol over the highway and allowing some on-line simple computations to be done. The stations are divided into main and slave which is essential for the industrial process control systems, where the different types of stations are different sources of traffic. The main stations are usually linked to the operator monitoring stations, while the slaves are directly linked to the industrial process. The message traffic arriving at the main stations is comparatively heavy and varies to a great extent. The incoming messages are mainly commands for observation and control of the industrial process, and are addressed to the slave stations. There are two types of commands [2]: synchronous, where the slave station immediately responds to the command with the needed information, and asynchronous, where the slave station confirms the received command while the information response is transmitted afterwards. Messages addressed to other stations enter the slave stations too. These are messages about the alarms emerging from the industrial process, as well as commands addressed towards other slave stations, which allow the control of complex interconnected industrial processes. This message flow is quite low. The major part of the messages stored in the slaves are responses to the asynchronous commands, sent by the main stations.

The data exchange over the highway is message-oriented and is performed in "command - response" sequences. The communication system has multimaster potential using the token passing for the control of the highway access.

There is one more station in the system, which we shall call "coordinator". No information messages enter this station. Its task is to keep up with the communication system state, to transfer the highway control to the other stations and to optimize the traffic among them.

Receiving a MTF command from the coordinator, each station gets the right to initiate a transmission over the highway. Terminating its transmission, the station returns the control over the highway to the coordinator through a MTF command. The MTF commands contain both the token and the information about the state of the communication process. The coordinator performs its functions in cycles. During one cycle time it transfers the control to all the stations, taking part in the communication. The cycle starts with a successive transfer of the control to the slave stations. These stations have established priorities. In their queues there are either extremely important messages about the alarms and the industrial process control, or responses to the asynchronous commands that have already entered the communication system. Before returning the control to the coordinator, the slave station transmits all the messages from its own queue. As these messages are usually few and the station follows a clear and definite algorithm, we shall not discuss the dynamics of these stations.

The rest of the cycle time is distributed by the coordinator among the main stations. Each main station has access to the highway only once during one cycle. The coordinator determines the order, in which the control is transferred to the main stations, the moment of transfer and the intervals of time, during which the stations are allowed to control the highway.

The dynamics of main station i, where i = 1,...,N, is described by the following state equation:

$$\frac{dx_i(t)}{dt} = v_i(t) - u_i(t), \qquad x_i(t_o) = x_{io} \tag{1}$$

Here $x_i(t)$ denotes the length of the information message queue of main station i at time t, expressed in units of time, necessary for transmitting over the highway. $v_i(t)$ denotes the information messages incoming flow into the station i at time t. It can be seen, that for every t $x_i(t) \geqslant 0$ and $v_i(t) \geqslant 0$. The control of station i at time t is defined as $u_i(t)$.

$$u_i(t) = \begin{cases} 1, & \text{when station } i \text{ controls the highway and transmits an information message at time } t \\ 0, & \text{otherwise} \end{cases} \quad (2)$$

The task of the coordinator is to fulfill the above constraints of the system operation and to choose the transfer moments and access time intervals for the highway control in order to minimize the following cost functional of the communication system.

$$Q = \sum_{i=1}^{N} Q_i = \sum_{i=1}^{N} \int_{t_0}^{t_f} q_i(x_i(t),\ u_i(t),\ v_i(t))dt, \quad (3)$$

where q_i are scalar valued functions. Using equation (1), we can say, that the cost functional depends on the initial state, the incoming flows and the control during the period $[t_0, t_f]$. The interval between t_0 and t_f is generally large.

FLYING HORIZON METHOD

The task of the coordinator is a difficult optimization problem and as noted in [3] it cannot be solved off-line. We usually do not have in advance sufficiently precise information (even statistical) about the incoming message traffic into the communication system for a longer period of time. Because of that, we shall apply the flying horizon method,using frequently enough a feedback from the communication process state. It is necessary to specify the notations, accepted in the description of the method. The superscript of the variable indicates the cycle's number, while the subscript indicates the number of the main station. The use of an interval as a second subscript indicates, that it refers to the trajectory of the variable in that interval. The dash above the variable means, that values of this variable at separate points (or over the entire trajectory) are calculated or predicted on the basis of a model. The maximum length of one coordinator cycle is denoted by η , which is a system parameter.

CONTROL STRATEGY OF MAIN STATION i

Suppose that, at time t_i^k, a main station i receives from the coordinator the control over the highway for the $[t_i^k, \bar{\tau}_i^k]$ time interval. The station transmits information messages from its own queue till the moment τ_i^k, when $x_i(\tau_i^k) = 0$ or $\tau_i^k = \bar{\tau}_i^k$ ($u_{i[t_i^k, \tau_i^k]} = 1$). After that, the station returns the control to the coordinator, transmitting information about its current state $x_i(\tau_i^k)$, the predicted value of the incoming flow $v_i(t)$ during the $[\tau_i^k, \tau_i^k + \sigma]$ interval, i.e. the value of $\bar{v}_{i[\tau_i^k, \tau_i^k + \sigma]}$ ($\sigma \geqslant \eta + \Delta$) as well as the flow's mean value for this interval p_i^{k+1} = $= M\bar{v}_{i[\tau_i^k, \tau_i^k + \sigma]}$. The station does not control the highway until receiving the MTF command during the next cycle, i.e. $u_{i[\tau_i^k, t_i^{k+1}]} = 0$.

STRATEGY OF THE COORDINATOR

Fig.1 shows, that the coordinator minimizes the cost functional within a shortened horizon η. Suppose, that cycle k starts at time t_o^k, when the first slave station receives the control over the highway. Let τ_o^k be the time, when the last slave station ends with the transmission of the information messages. The remaining excess of the cycle time $\gamma^k = t_o^k + \eta - \tau_o^k + (N+1)\Delta$ has to be distributed among the main stations by the coordinator. The Δ interval includes the time for the coordinator to receive the MTF command, to decide to which station the control should be transferred and for how long, as well as to transfer the control to this station through a MTF command.

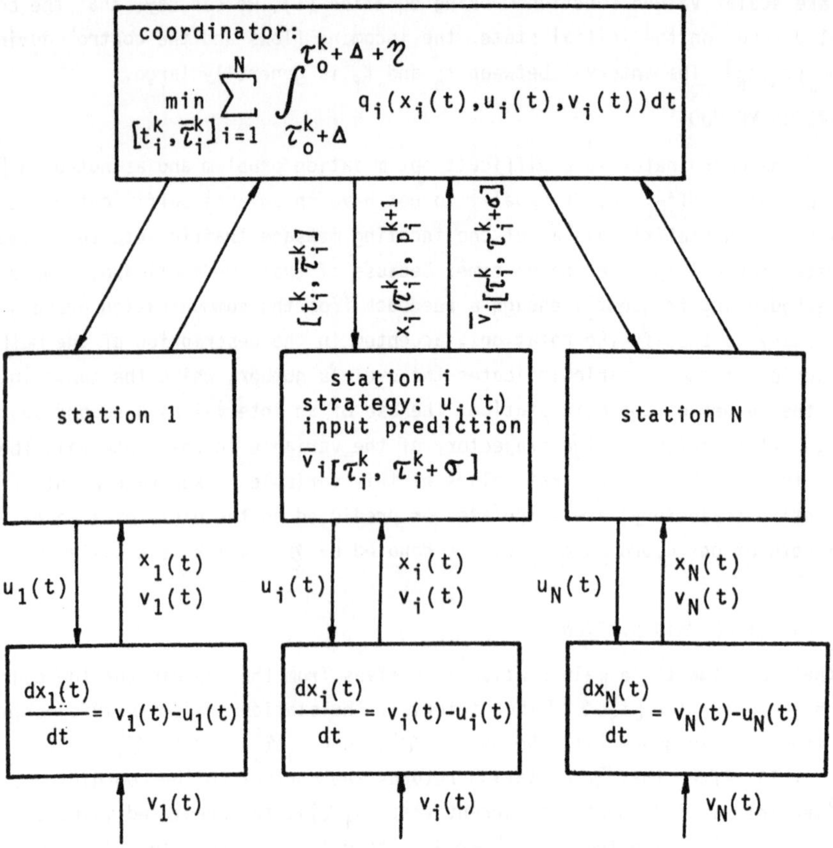

Fig.1 The proposed structure of dynamic on-line coordination

Describing further the algorithm of the coordinator, it is necessary to specify the cost functional. We define it as

$$Q = \sum_{i=1}^{N} \int_{t_0}^{t_f} (x_i(t))^2 dt \tag{4}$$

This cost functional equilizes the lengths of main stations queues by placing high penalty on the stations with long queues. The sudden change of the message flow towards one or several stations is evenly absorbed by all main stations in the communication system.

The coordinator strategy has three basic stages:

1) choosing the station, to which the control to be transferred;
2) determination of the static optimal interval of control of this station;
3) dynamic correction of the determined interval.

CHOOSING THE STATION

The fact is, that the coordinator always chooses two stations: the station to which the control will be transferred (let it be denoted by i) and the station, which will receive the control after station i (denoted by j). At the initial moment we have

$$i = \left\{ i:\ \bar{x}_i(\tau_0^k + \Delta) = \max_{r \in \overline{1,N}} \bar{x}_r(\tau_0^k + \Delta) \right\} , \tag{5}$$

where $\bar{x}_r(t)$ is determined from the equation

$$\bar{x}_r(t) = x_r(\tau_r^{k-1}) - \int_{\tau_r^{k-1}}^{t} \bar{v}_r(\varepsilon) d\varepsilon \tag{6}$$

Choosing i, we have the first value of the interval, i.e. $t_i^k = \tau_0^k + \Delta$. The second station is determined as

$$j = \left\{ j:\ \bar{x}_j(t_i^k) = \max_{r \in S} \bar{x}_r(t_i^k) \right\} \tag{7}$$

where S denotes the set of stations numbers, not having taken part in the current cycle yet. During the following steps of the cycle, station i is already determined (i.e: i = j) and (7) is just used to choose the second station j.

DETERMINATION OF THE STATIC OPTIMAL INTERVAL

Suppose, that within the whole interval $[t_0, t_f]$, the incoming flows $v_i(t)$ are evenly distributed with constant mean values $p_i \neq 0$. Then, arranging the stations in a definite order, the optimum access time δ_i of main station i is defined as

$$\delta_i = \frac{p_i}{\sum_{r=1}^{N} p_r} \gamma ,$$

where γ is the time for highway control shared by the main stations within the $[t_0, t_f]$ interval.

Using the above relation the coordinator determines the value of

$$\beta_{ij}^{k} = \frac{p_i^k + p_j^k}{\sum_{r=1}^{N} p_r^k} \gamma^k + \Delta \tag{8}$$

i.e. the interval, which should be given both to stations i and j (as shown in Fig.2). Determining β_{ij}^k, it is not considered the order in which the control is transferred to the stations.

DYNAMIC CORRECTION

At this stage the coordinator divides the time β_{ij}^k between stations i and j, minimizing the cost functional

$$\int_{t_i^k}^{\tau_0^k + \Delta + \zeta} \left[(\bar{x}_i(t))^2 + (\bar{x}_j(t))^2 \right] dt \tag{9}$$

subject to

$$\bar{x}_i(t) \geqslant 0$$
$$\bar{x}_j(t) \geqslant 0$$

where

$$\bar{x}_i(t) = \begin{cases} \bar{x}_i(t_i^k) + \int_{t_i^k}^{t} \bar{v}_i(\varepsilon)d\varepsilon - t + t_i^k & t \in [t_i^k, t_i^k + \alpha_i^k] \\ \bar{x}_i(t_i^k) + \int_{t_i^k}^{t} \bar{v}_i(\varepsilon)d\varepsilon - \alpha_i^k & t \in [t_i^k + \alpha_i^k, \tau_0^k + \Delta + \zeta] \end{cases} \tag{10}$$

$$\alpha_i^k = \bar{\tau}_i^k - t_i^k$$

$$\bar{x}_j(t) = \begin{cases} \bar{x}_j(t_i^k) + \int_{t_i^k}^{t} \bar{v}_j(\varepsilon)\, d\varepsilon & t \in [t_i^k, t_i^k + \alpha_i^k + \Delta] \\ \bar{x}_j(t_i^k) + \int_{t_i^k}^{t} \bar{v}_j(\varepsilon)d\varepsilon - t + t_i^k + \alpha_i^k + \Delta & t \in [t_i^k + \alpha_i^k + \Delta, t_i^k + \beta_{ij}^k] \\ \bar{x}_j(t_i^k) + \int_{t_i^k}^{t} \bar{v}_j(\varepsilon)d\varepsilon - \beta_{ij}^k + \alpha_i^k + \Delta & t \in [t_i^k + \beta_{ij}^k, \tau_0^k + \Delta + \zeta] \end{cases} \tag{11}$$

The values of $\bar{x}_i(t_i^k)$ and $\bar{x}_j(t_i^k)$ are calculated according to (6), while the minimization of (9) is done for the sake of finding out $\bar{\tau}_i^k$. The meaning of the different time intervals is illustrated in Fig.2.

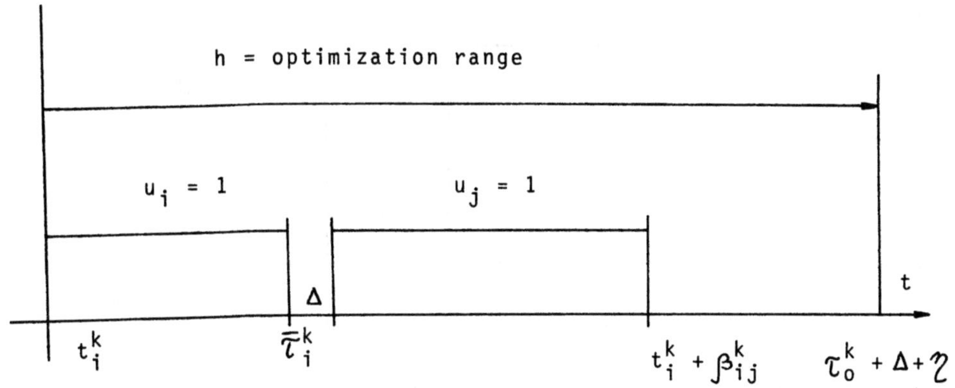

Fig.2 Dynamic correction range

Calculating $\bar{\tau}_i^k$, the coordinator transfers the control to station i for the $[t_i^k, \bar{\tau}_i^k]$ time interval. After getting back the control from station i, the coordinator performs the procedure of new station determination. In case station i is the last one in the cycle, no new station is determined and

$$[t_i^k, \bar{\tau}_i^k] = \gamma^k - \sum_{\substack{r=1 \\ r \neq i}}^{N} (\tau_r^k - t_r^k)$$

APPLICATION OF THE METHOD IN SLINK4 COMMUNICATION SYSTEM

A distributed industrial process control system MIC 2000C has been implemented in the Institute of Industrial Cybernetics and Robotics of the Bulgarian Academy of Sciences. It is intended to control continuous, digital and batch processes mainly in chemical industry. The SLINK system [2] is used as a communication system, comprising two main and eight slave stations. MIC 2000C has already been adopted in several plants in Bulgaria and the results have shown, that in some cases it is necessary to increase the number of the stations taking part in the communication. The communication system SLINK4 has been created for that purpose. The existing hardware and the speed of data transmission (250 Kbps) remained the same. SLINK4 comprises up to 16 main stations and up to 48 slaves, preserving the same information quality. This is achieved mainly by using the described flying horizon method and by giving up the polling. Here on, we shall discuss some details of the implementation of the flying horizon algorithm in the SLINK4 system.

The identification of the incoming flows has shown, that the random functions $v_i(t)$ are evenly distributed and their mean value can be approximated by the function illustrated in Fig.3.

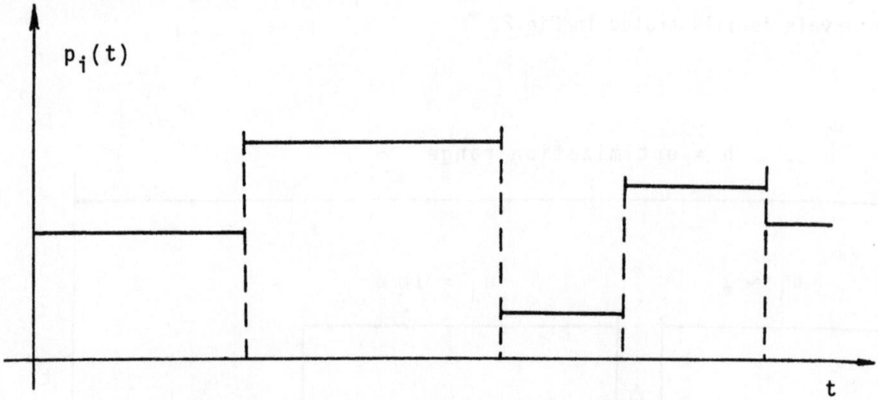

Fig.3 Change of the mean value depending on time

When the operator is watching the information monitor, the mean value of the incoming message flow is constant ($p_i(t)$ = const). Sometimes it lasts as long as several hours. The break points of $p_i(t)$ correspond to the operator's request for new, different kinds of information about the process. Of course, neither the values of $p_i(t)$ nor the break points are known in advance and the task of the main station is to identify on-line the incoming flow. However, the form of $p_i(t)$ suggests that within the interval of $[\tau_i^k, \tau_i^k + \sigma]$ there will be

$$\bar{v}_{i[\tau_i^k, \tau_i^k + \sigma]} = p_i(\tau_i^k) \quad = \text{const}$$

$$p_i^{k+1} = M\bar{v}_{i[\tau_i^k, \tau_i^k + \sigma]} = p_i(\tau_i^k)$$

The main station identifies, at any moment, the mean value of its incoming message flow and transmits to the coordinator only the values of $x_i(\tau_i^k)$ and $p_i^{k+1} = p_i(\tau_i^k)$. Equation (6) is of the form

$$\bar{x}_i(t) = \begin{cases} \bar{x}_i(t_i^k) - (1 - p_i^k)(t - t_i^k) & t \in [t_i^k, t_i^k + \alpha_i^k] \\ \bar{x}_i(t_i^k) + p_i^k(t - t_i^k) - \alpha_i^k & t \in [t_i^k + \alpha_i^k, \tau_o^k + \Delta + \zeta] \end{cases} \quad (13)$$

$$\bar{x}_j(t) = \begin{cases} \bar{x}_j(t_i^k) + p_j^k(t - t_i^k) & t \in [t_i^k, t_i^k + \alpha_i^k + \Delta] \\ \bar{x}_j(t_i^k) - (1 - p_j^k)(t - t_i^k) + \alpha_i^k + \Delta & t \in [t_i^k + \alpha_i^k + \Delta, t_i^k + \beta_{ij}^k] \\ \bar{x}_j(t_i^k) + p_j^k(t - t_i^k) - \beta_{ij}^k + \alpha_i^k + \Delta & t \in [t_i^k + \beta_{ij}^k, \tau_o^k + \Delta + \zeta] \end{cases} \quad (14)$$

The cost functional in (9) can be expressed as

$$F(\alpha_i^k) = A(\alpha_i^k)^3 + B(\alpha_i^k)^2 + C(\alpha_i^k) + \text{const} \quad (15)$$

where

$$h = \tau_o^k + \Delta + \eta - t_i^k$$

$$A = -1 + (p_i^k - p_j^k)/3$$

$$B = \bar{x}_i(t_i^k) - \bar{x}_j(t_i^k) - \Delta p_j^k + 2h - \Delta$$

$$C = 2h(\bar{x}_j(t_i^k) - \bar{x}_i(t_i^k) + \Delta - \beta_{ij}^k) + h^2(p_j^k - p_i^k) - 2\Delta\bar{x}_j(t_i^k) - \Delta^2(1 + p_j^k) + (\beta_{ij}^k)^2$$

and no iterative algorithms are needed for its minimization. The following algorithm, based on the analytical solution of (9) is used in SLINK4:

We compute the local minimum of (15)

$$\hat{\alpha}_i^k = \frac{-B + \sqrt{B^2 - 3AC}}{3A}$$

If $B^2 - 3AC < 0$ or $\hat{\alpha}_i^k \geqslant \beta_{ij}^k - \Delta$, then the global minimum is at the point $\bar{\tau}_i^k = t_i^k + \beta_{ij}^k - \Delta$.

If $\hat{\alpha}_i^k < 0$ then $\bar{\tau}_i^k = t_i^k$ or $\bar{\tau}_i^k = t_i^k + \beta_{ij}^k - \Delta$ depending on, which of the two points provides the minimum of (15).

If $0 \leqslant \hat{\alpha}_i^k < \beta_{ij}^k - \Delta$ then $\bar{\tau}_i^k = t_i^k + \hat{\alpha}_i^k$ or $\bar{\tau}_i^k = t_i^k + \beta_{ij}^k - \Delta$ depending on, which of the two points provides the minimum of (15).

Using the above algorithm we do not take into account the constraint $\bar{x}_i(t_i^k + \bar{\tau}_i^k) \geqslant 0$. But this is not important, as we determine only $\bar{\tau}_i^k$ and transfer the control just to station i. If its message queue is empty, the station returns the control to the coordinator before the settled time is expired. Another peculiarity of SLINK4 system is that the coordinator corrects the mean values of the incoming flows received from the main stations, following the algorithm

$$\text{if } p_i^{k+1} < p_{min} \text{ then } p_i^{k+1} = p_{min}$$

So, the statical optimization problem is always well defined and each station is granted a minimum of time for transmitting single messages, that have entered the station by chance.

CONCLUSIONS

Besides the described operations, the coordinator performs all the additional functions necessary when using the token passing method. It determines the line, over which the information exchange should be done and the station, when it drops out of the communication; on-line attaches new stations to the communication, transferring broadcast commands about the communication system state. The coordinator is not necessarily a separate hardware unit. In the SLINK4 system the coordinator program is stored in each main station. In case the coordinator drops out of the communication its functions are performed by another main station, which at the same time executes its own habitual functions. Thus the system retains the decentralized control ensuring the needed reliability.

When the traffic intensity is low the system behaves as an ordinary one, using the token passing access method: the stations receive the control in logical succession and each station transmits all the messages that have come till that moment. The advantages of the algorithm are obvious when a comparatively intensive traffic exists. Then a station is refused the right to transmit messages from its own queue and its time is given to another station for transmission. Thus the probability some messages to be rejected by from the communication system, at brief dynamic loading of the latter, is decreased.

REFERENCES:

[1] Findeisen W. and K. Malinowski: Two-level control and coordination for dynamical system, Archiwum Automatyki i Telemechaniki, v.24, Nr.1

[2] Stoilov E. and M. Strugarova: SLINK - A communication system for distributed process control, Proceedings of the 13-th International Conference on Fault - Tolerant Computing, Milano, Italy, 1983, IEEE Computer Society

[3] Stoilov E. and M. Strugarova: A theoretical approach to the master transfer technique in data communication local area networks, Proceedings of the TC6 International In-Depth Symposium on Networks in Office Automation, Sofia, Bulgaria, 1984, North-Holland.

ON THE NUMERICAL SOLUTION OF AN OPTIMAL CONTROL PROBLEM

J. SZELEZSÁN

Department of Computer Sciences,
L. Eötvös University, Muzeum krt. 6-8,
Budapest, H u n g a r y

ABSTRACT. In this paper the convergence of an approximate solution for an optimal control problem related to a special distributed parameter system is investigated.

1. PROBLEM STATEMENT

Let us assume that the process is described by the state equation

$$u(x,t)=\int_0^t K(x,t,s)f(s)ds \tag{1}$$

where f is the control function. Let F be a closed, bounded, convex set of control functions.

The cost of a control is given by

$$J(f)=\int_0^T [u(x_o,t)-q(t)]^2 dt \tag{2}$$

where x_o, and q are given.

Problem 1: Determine a control function $f\in F$ such that the performance index $J(f)$ is minimized.

Using the notation

$$u(x_o,t)=\int_0^t K(x_o,t,s)f(s)ds=Kf$$

the functional (2) may be written as

$$J(f)=(Kf-q,Kf-q)=\| Kf-q\|^2_{L_2} \tag{3}$$

Let F_n be a finite ε-net in F, and let $\{K_n\}$ be a sequence of operators such that $\lim_{n\to\infty}\|K_n-K\|_{L_2}=0$

Let us take instead of inner product $(Kf-q,Kf-q)$ an approximation $(K_mf_n-q,K_mf_n-q)_k$ of it such that

$$\lim_{m,n,k\to\infty}(K_mf_n-q,K_mf_n-q)_k=(Kf-q,Kf-q)$$

For brevity let be k=n, m=n.

Introducing the notation

$$J_n(f_n)=(K_nf_n-q,K_nf_n-q)_n$$

instead of *Problem 1* consider

Problem 2: Determine a control f_n such that $f_n \in F_n$ and $J_n(f_n)$ is minimized.

Let us denote the solution of *Problem 1* and *Problem 2* by f^* and f_n^*, respectively.

2. THE CONVERGENCE OF THE APPROXIMATE SOLUTION

Theorem 1.

a) $\lim_{n\to\infty} J_n(f_n^*)=J(f^*)$ and

b) if $|K(t,s)-K(\tau,s)| \leq W(|t-\tau|)\, v(s)$ where W is continuous, $W(0)=0$ and $v(s) \in L_2$, then $Kf_n^* \to Kf^*$ uniformly.

Proof. Instead of functionals $J(f)$ and $J_n(f_n)$ let us take the equivalent functionals

$$J(f)=\|Kf-q\| \text{ and } J_n(f_n)=\|K_nf_n-q\|_n \text{ respectively.}$$

We shall show that for every $\varepsilon>0$ there is an integer N_0 such that $n>N_0$ implies

$$\left|\|K_nf_n^*-q\|_n-\|Kf^*-q\|\right|<\varepsilon. \tag{4}$$

By assumption, for every $\varepsilon_1>0$ there exists an integer N_1 such that $n>N_1$ implies

$$\left|\|K_nf_n^*-q\|_n-\|K_nf_n^*-q\|\right|<\varepsilon_1$$

hence

$$\left|\|K_nf_n^*-q\|_n-\|Kf^*-q\|\right| \leq \left|\|K_nf_n^*-q\|_n-\|K_nf_n^*-q\|\right|+$$

$$+\left|\|K_nf_n^*-q\|-\|Kf^*-q\|\right| \leq \varepsilon_1+\left|\|K_nf_n^*-q\|-\|Kf^*-q\|\right|. \tag{5}$$

For estimation of the difference $\left|\|K_nf_n^*-q\|-\|Kf^*-q\|\right|$ we shall apply the following inequalities.

Since by assumption $\{K_n\}$ is convergent, we can choose N_2 for a given $\varepsilon_2>0$ such that $n>N_2$ implies

$$\|K_nf_n-Kf_n\| <\varepsilon_2.$$

The functional $\|Kf-q\|$ is continuous and F_n is an ε-net, therefore, for a given $\varepsilon_3>0$, there exists an integer N_3 such that

$$\left|\|Kf_n-q\|-\|Kf^*-q\|\right| < \varepsilon_3 \text{ if } n>N_3.$$

Using the foregoing inequalities and the triangle inequality we obtain that

$$\left|\|K_nf_n^*-q\|-\|Kf^*-q\|\right| < \varepsilon_4,\ \varepsilon_4>0,\ n>\max(N_1,N_2,N_3) \tag{6}$$

This together with (5) proves (4).

Now we show that $J(f_n^*) \to J(f^*)$. Since

$$\|Kf_n^*-q\|=\|(Kf_n^*-K_nf_n^*)+(K_nf_n^*-q)\| \leq \varepsilon_1+\|K_nf_n^* -q\|$$

therefore

$$\|Kf_n^*-q\|-\|K_nf_n^*-q\|<\varepsilon \tag{7}$$

and from

$$\| K_n f_n^* - q \| = \| (Kf_n^* - q) + (K_n f_n^* - Kf_n^*) \| \le \| Kf_n^* - q \| + \varepsilon_2$$

we obtain

$$\| Kf_n^* - q \| - \| K_n f_n^* - q \| > -\varepsilon_2$$

which together with (7) gives

$$| \, \|Kf_n^* - q\| - \|K_n f_n^* - q\| \, | < \varepsilon \qquad \text{if } n>N.$$

From this inequality and from (6) we obtain

$$| \, \|Kf_n^* - q\| - \|Kf^* - q\| \, | \le \varepsilon^* \qquad \text{if } n>N^o$$

that is $J(f_n^*) \to J(f^*)$.

It is easy to show that

$$J(f_n^*) - J(f^*) \ge (K(f_n^* - f^*),\ K(f_n^* - f^*))$$

which yields $\lim_{n\to\infty} \|Kf_n^* - Kf^*\| = 0$ that is $\|Kf_n^*\| \to \|Kf^*\|$.

We will now show that the system of functions Kf_n $n=1,2,\dots$ and Kf are uniformly continuous.

Indeed

$$\left| Kf_n^* \big|_{t=t'} - Kf_n^* \big|_{t=t''} \right| = \left| \int_\alpha^\beta K(t',s) f_n^*(s)ds - \right.$$

$$\left. - \int_\alpha^\beta K(t'',s) f_n^*(s)ds \right| \le \sqrt{\int_\alpha^\beta (K(t',s) - K(t'',s))^2 v(s)} \|f_n^*\|$$

$$\le (\beta-\alpha) C_1 C_2 W(|t-\tau|) < \varepsilon$$

if $|t-\tau| < \delta$ where $\|f_n\|_{L_2} < C_1$, $\|v(s)\| < C_2$.

It is known that $g_n \xrightarrow{L_2} g$ implies $g_n \xrightarrow{C} g$ if g_n and g are uniformly continuous, which proves $Kf_n^* \to Kf^*$.

3. A SPECIAL SET OF ADMISSIBLE CONTROLS

Let the set F of admissible controls be

$$F = \{ f(t) \mid f(t) \in C^{(2)},\ a(t) \le f(t) \le b(t) \text{ where } a(t) < b(t),$$

$$a(t),\ b(t) \in C^{(2)} \text{ are given functions} \}.$$

Choosing a spacing $h = \frac{T}{n}$, divide the interval $[0,T]$ by means of equally spaced points $t_i = ih$ $(i=1,2,\dots,n)$ into n equal parts, and consider the set

$$F_n = \{ (f_1, f_2, \dots, f_n) \mid f_i = f(t_i),\ a(t_i) \le f_i \le b(t_i) \}.$$

Replace the functional (2) by the functional

$$J_n(f^n) = \sum_{i=1}^{n} c_i^{(1)} \Big(\sum_{j=1}^{i} c_j^{(2)} [K(t_i, s_j) f_j - q_i]^2 \Big)$$

where $q_i = q(t_i)$, $f^n = (f_1, f_2, \dots, f_n)$ and $c_i^{(1)}$, $c_j^{(2)}$ are certain constants from an approximate quadrature formula. We solve instead of *Problem 1* the *Problem 2'*:

$$\min_{(f_1, f_2, \dots, f_n)} J_n(f^n)$$

Denote $p_n(t)$ an interpolation polynomial on the values $f_1, f_2, \ldots, f_n$. Denote f^* the solution of *Problem 1*, f_n^* the solution of *Problem 2* and p_n^* the interpolation polinom belonging to f_n^*.

Theorem 2. If $Kp_n^* - Kf^* = o(h^\alpha)$ with $\alpha > 2$ then $p_n^*(t_i) - f^*(t_i) = o(h)$.

Proof. The proof is based on the method described in [3]

Applying the trapezoidal formula to approximate the integral $Kp_n^* - Kf^*$ we obtain

$$\frac{h}{2}K(t_i,s_o)[p_n^*(s_o)-f^*(s_o)]+\frac{h}{2}K(t_i,s_i)[p_n^*(s_i)-f^*(s_i)]+$$
$$+h\sum_{j=1}^{i-1}K(t_i,s_j)[p_n^*(s_j)-f^*(s_j)]+o(h^2)=o(h^\alpha)$$

and using the denotation $p_n^*(s_j)-f^*(s_j)=\varepsilon_j$

$$\frac{h}{2}K(t_i,t_i)\varepsilon_i=-[\frac{h}{2}K(t_i,t_o)\varepsilon_o+h\sum_{j=1}^{i-1}K(t_i,s_j)\varepsilon_j]+o(h^\alpha)+o(h^2) \quad (8)$$

Replacing i by i+1 in this equation, and subtracting the equation (8) from the equation obtained, we get

$$\frac{h}{2}[K(t_{i+1},t_{i+1})\varepsilon_{i+1}+K(t_i,t_i)\varepsilon_i]=-\{\frac{h}{2}K(t_{i+1},t_o)-K(t_i,t_o)]\varepsilon_o$$
$$+h\sum_{j=1}^{i-1}[K(t_{i+1},s_j)-K(t_i,s_j)]\varepsilon_j+h[K(t_{i+1},t_i)-K(t_i,t_i)]\ \varepsilon_i\}+$$
$$+o(h^\alpha)+o(h^2) \quad (9)$$

Replacing here i by i+1 once again and subtracting the equation (9) from the equation obtained, introducing the notation

$$\Delta K_j=[K(t_{i+2},t_j)-K(t_{i+1},t_j)]$$

$$\Delta^2K_j=[K(t_{i+2},t_j)-K(t_{i+1},t_j)]-[K(t_{i+1},t_j)-K(t_i,t_j)]$$

and using the triangle equation we obtain

$$\left|\frac{K(t_{i+2},t_{i+2})}{2}\right||\varepsilon_{i+2}|-\left|\frac{K(t_i,t_i)}{2}\right||\varepsilon_i|\le\frac{\varepsilon_o}{2}|\Delta^2K_o|+$$
$$+\sum_{j=1}^{i-1}|\Delta^2K_j||\varepsilon_j|+|\Delta^2K_i||\varepsilon_i|+|\Delta K_{i+1}||\varepsilon_{i+1}|+o(h)+$$
$$+o(h^{\alpha-1}) \quad (10)$$

It is obvious that $\Delta^2K_j=A_jh^2$, $\Delta K_j=B_jh$ where A_j, B_j are certain constans. Choosing a constant A such that

$$A>\max(|A_o|,|A_j|,|B_j|) \text{ and } A=Q\left|\frac{K(t_i,t_i)}{2}\right|$$

where Q>1, the inequality (10) may be written in the form

$$\left|\frac{K(t_{i+2},t_{i+2})}{2}\right||\varepsilon_{i+2}|-\left|\frac{K(t_i,t_i)}{2}\right||\varepsilon_i|\le A\,h^2\sum_{j=0}^{i}|\varepsilon_j|+$$
$$+Ah\,|\varepsilon_{i+1}|+o(h^{\alpha-1})+o(h). \quad (11)$$

Consider the equation

$$z_{i+2}-z_i=Q\ h^2 \sum_{j=0}^{i} z_i+Qh\ z_i+o(h^{\alpha-1})+o(h).$$

Replacing i by i+1 in this equation and subtracting it from the obtained equation we get the difference equation

$$z_{i+3}-(1+Qh)z_{i+2}-(1+Qh^2-Qh)z_{i+1}+z_i=o(h)+o(h^{\alpha-1}). \qquad (12)$$

The solution of equation (12) we can get from the solution of the equation

$$P(x)=x^3-(1+Qh)x^2-(1+Qh^2-Qh)x+1=0$$

whose roots, as it is easy to show, are of the form

$$x_1=-1+o(h),\quad x_2=1-o(h),\quad x_3=1+o(h).$$

It is known, that any solution of the difference equation (12) is of the form

$$z_i=z_i^o+c_1x_1^i+c_2x_2^i+c_3x_3^i$$

where $z_i^o=o(h)$ is the general solution of the equation (12) . The constans c_1,c_2,c_3 can be found from the following system of linear equations

$$o(h)=z_i^o+c_1+c_2+c_3$$

$$o(h)=z_i^o+c_1x_1+c_2x_2+c_3x_3$$

$$o(h)=z_i^o+c_1x_1^2+c_2x_2^2+c_3x_3^2\ .$$

Solving this system of equation we obtain $c_1=o(h)$, $c_2=o(h)$, $c_3=o(h)$ and hence $z_i=o(h)$.

Now we shall show that $|\varepsilon_i|<z_i$. Let be $z=z_o=z_1=z_2=\max(|\varepsilon_o|,|\varepsilon_1|,|\varepsilon_2|)$ and suppose $|\varepsilon_j|\leq z_j$, $j=3,4,\dots,i$ then by virtue of inequality (11)

$$\left|\frac{K(t_{i+2},t_{i+2})}{2}\right||\varepsilon_{i+2}|-\left|\frac{K(t_i,t_i)}{2}\right||\varepsilon_i|\leq Ah^2+\sum_{j=0}^{i} z_i+$$
$$+Ah\ z_{i+1}+o(h)+o(h^{\alpha-1}) \qquad (13)$$

Choosing z_{i+2} so that

$$Ah^2\sum_{j=0}^{i} z_i+Ah\ z_{i+1}+o(h)+o(h^{\alpha-1})=\left|\frac{K(t_i,t_i)}{2}\right|z_{i+2}-z_i$$

from (13) we obtain

$$\left|\frac{K(t_{i+2},t_{i+2})}{K(t_i,t_i)}\right||\varepsilon_{i+2}|\leq z_{i+2}$$

Under our assumption

$$\frac{K(t_{i+2},t_{i+2})}{K(t_i,t_i)}\geq 1$$

therefore $|\varepsilon_{i+2}|\leq z_{i+2}$

EXAMPLE. Consider the system governed by the parabolic partial differential equation

$$\frac{\partial u}{\partial t}=\frac{\partial^2 u}{\partial x^2}\ ,\quad u(x,o)=\left\{\begin{matrix} f(x) & \text{if } o\leq x\leq \beta \\ 0 & \text{if } x<0,\ x>\beta \end{matrix}\right\} \tag{14}$$

Let f be the control function.

As it is known, the solution of (14) can be written in the form

$$u(x,t)=\frac{1}{\sqrt{4\pi t}}\int_0^{\beta} e^{-(x-s)/4t} f(s)ds$$

therefore the solution of *Problem 2* can ben obtained by method described above.

REFERENCES

[1] Butkowski, A.G.: Theory of Optimal Control for Distributed Parameter Systems /in Russian/, Nauka, Moscow, 1965

[2] Ehnes, J.G.: On the Numerical Solution of Convolution Integral Equations, Math. of Comp., Vol 15.

[3] Moiseev, N.N.: Numerical Methods for Optimal Systems /in Russian/, Nauka, Moscow, 1971

[4] Polak, E.: Computational methods in optimization, Academic Press New York, London 1971

LINEAR AND NONLINEAR OLIGOPOLY MODELS

Ferenc Szidarovszky and Koji Okuguchi

Abstract

A new general version of the Cournot equilibrium model is introduced, where each firm simultaneously produces several kinds of items, and it is assumed that the price vectors and cost functions depend on the total and own production levels of the producers, respectively. A new existence theorem is first stated and the relation of the equilibrium problem and fixed-point problems of low dimensional point-to-set mappings is outlined.

Linear complementarity problems, and quadratic programming problems are constructed for the determination of the equilibrium points. On the basis of these methods a new uniqueness theorem is verified.

Dynamic models are examined with discrete time scales. Conditions for stability of the dynamic models are given.

The results presented in the paper are generalizations and extensions of earlier results of the oligopoly theory.

1. Introduction

The purpose of this work is to present some new results on the generalized Cournot oligopoly theory. In this paper the multiproduct oligopoly game is investigated, which is a mathematical game with players 1,2,...,N. The number of products equals M and the strategy of each player is characterized by the production vector

$$\underline{x}_k = (x_k^{(1)}, \dots, x_k^{(M)})^T, \tag{1}$$

where x_k^m is the production level of player k from item m. The total production level vector is defined as

$$\underline{s} = \sum_{k=1}^{N} \underline{x}_k . \tag{2}$$

Dept. of Math. and Comp. Sci., University of Horticulture,
Budapest, Villányi ut 29/35, H-1118, Hungary
Faculty of Economics, Tokyo Metropolitan University,
1-1-1 Yakumo, Meguro-Ku, Tokyo, Japan

The set of strategies of player $k(1 \leq k \leq N)$ is given by the subset S_k of the M dimensional Euclidean space. It is assumed that the payoff function of player $k(1 \leq k \leq N)$ is given as

$$\varphi_k(\underline{x}_1,\ldots,\underline{x}_N) = \underline{x}_k^T \underline{f}(\underline{s}) - K_k(\underline{x}_k), \qquad (3)$$

where the components of function $\underline{f}$ give the unit prices, and K_k is the cost function of player k. Thus φ_k is the profit of this player.

The Nash or Cournot equilibrium point of the N person game with sets S_k of strategies and payoff functions φ_k is any vector $(\underline{x}_1^*,\ldots,\underline{x}_N^*)$ such that for all k, $\underline{x}_k^* \in S_k$ and for arbitrary $\underline{x}_k \in S_k$,

$$\varphi_k(\underline{x}_1^*,\ldots,\underline{x}_{k-1}^*, \underline{x}_k, \underline{x}_{k+1}^*,\ldots,\underline{x}_N^*) \leq \varphi_k(\underline{x}_1^*,\ldots,\underline{x}_N^*) \qquad (4)$$

The classical Cournot oligopoly game is a special case of the above defined game, if M=1. This classical game has been investigated by many authors. In the symmetric case Burger (1959) gave an existence and uniqueness theorem, which has been generalized for the non-symmetric case by Opitz (1970), Szidarovszky (1970). Okuguchi (1976) presented a detailed background on the oligopoly game and also discussed the non-symmetric case. Szidarovszky and Yakowitz (1977) presented a new proof for the existence and uniqueness for the single-product nonsymmetric case. In the linear case Manas 1972 gave a quadratic programming algorithm for finding the only equilibrium, which can also be presented in a closed form as it has been verified by Szidarovszky (1978). The results of Manas were generalized for the multiproduct oligopoly game by Szidarovszky (1978), and Szidarovszky and Okuguchi (1985).

There is a strong relation between equilibrium points of N-person games and point-to-set mappings (Rosen, 1965). Introduce the following notations:

$$\phi(\underline{x},\underline{y}) = \sum_{k=1}^{N} \varphi_k(\underline{x}_1,\ldots,\underline{x}_{k-1}, \underline{y}_k, \underline{x}_{k+1},\ldots,\underline{x}_N), \qquad (5)$$

where for all k, x_k and $y_k \in S_k$. Define the point-to-set mapping

$$H(\underline{x}) = \left\{\underline{u} \mid \underline{u} \in \mathop{X}_{k=1}^{N} S_k,\ \phi(\underline{x},\underline{u}) = \max_{y \in \mathop{X}_{k=1}^{N} S_k} \phi(\underline{x},\underline{y})\right\} \qquad (6)$$

Then the following result is true.

Theorem 1. A vector $\underline{x}^* = \underline{x}_1^*,\ldots,\underline{x}_N^*$ is an equilibrium point of the N-person game if and only if $\underline{x}^* \in H(\underline{x}^*)$, that is, $\underline{x}^*$ is a fixed point of the mapping $H\cdot(\underline{x}^*)$

Proof. See Rosen (1965).

In the next part of this section a similar result will be proven, where the dimension of the point-to-set mapping equals M. Then an existence theorem for the equilibrium points will be introduced. As a special case, the linear game will be solved.

Introduce the following notation. Let

$$\varphi_k(\underline{s},\underline{x}_k,\underline{t}_k) = \underline{t}_k^T \underline{f}(\underline{s}-\underline{x}_k+\underline{t}_k) - K_k(\underline{t}_k) \qquad (7)$$

denote the pay-off of player k, if the total production levels are given by vector $\underline{s}$ and this player replaces his own strategy vector $\underline{x}_k$ by $\underline{t}_k$. For each player k and total production vector $\underline{s}$ define the M-dimensional point-to-set mapping

$$H_k(\underline{s}) = \left\{\underline{x}_k \mid \underline{x}_k \in S_k,\ \varphi_k(\underline{s},\underline{x}_k,\underline{t}_k) = \max_{\underline{t}_k \in S_k} \varphi_k(\underline{s},\underline{x}_k,\underline{t}_k)\right\}. \qquad (8)$$

And finally let

$$H(\underline{s}) = \left\{\underline{t} \mid \text{there exists } \underline{x}_k \subset H_k\ \underline{s}\quad \forall k\text{, such that } \underline{t} = \sum_{k=1}^{N} \underline{x}_k\right\} \qquad (9)$$

From the construction of the point-to-set mappings H_k and H one may easily verify the following result.

Theorem 2. A strategy vector $\underline{x}^* = (\underline{x}_1^*,\ldots,\underline{x}_k^*)$ is an equilibrium point of the game if and only if vector $\underline{s}^* = \sum_{k=1}^{N} \underline{x}_k^*$ is a fixed point of mapping H, and for k=1,...,N, $\underline{x}_k^* \in H_k(\underline{s}^*)$

In comparing the fixed point-problems given in Theorems 1. and 2. observe that the dimensions of $H(\underline{x})$ and $H(\underline{s})$ are MN and M, respectively.

2. An Existence Theorem

Let's make the following assumptions:

A) There exists a convex, closed set $D \subset R^M$ such that for $\underline{s} \in D$, necessarily $\underline{f}(\underline{s}) = \underline{0}$;

B) Each component of $\underline{f}$ is continuously differentiable, concave and for arbitrary $\underline{s} \in S$, matrix

$$\underline{\underline{J}}(\underline{s}) + \underline{\underline{J}}^T(\underline{s})$$

is negative semidefinite, where $\underline{\underline{J}}$ denotes the Jacobian matrix of function $\underline{f}$;

C) For arbitrary $\underline{s} \in D$ and $\underline{0} \leq \underline{\tilde{s}} \leq \underline{s}$, necessarily $\underline{\tilde{s}} \in D$;

D) Functions $K_k (1 \leq k \leq N$ are continuous, convex and strictly increasing in each variable;

E) The set $S = \overset{N}{\underset{k=1}{X}} S_k$ of simultaneous strategies is convex, closed and for all $\underline{x} \in S$ and $\underline{0} \leq \underline{\tilde{x}} \leq \underline{x}$, necessarily $\underline{\tilde{x}} \in S$.

We shall now prove the main result of this section.

Theorem 3. Under the above assumptions the multiproduct oligopoly game has at least one equilibrium point.

Proof. See Szidarovszky and Okuguchi (1985).

3. The linear case

Assume first that the assumptions of the previous section hold and the cost functions K_k are differentiable.

Assume furthermore that $S_k = \overset{M}{\underset{m=1}{X}} [0, L_k^{(m)}]$ and $\overset{N}{\underset{k=1}{X}} S_k \subseteq D$.

Then introduce the following notations:

$$\underline{x}_k = (x_k^{(1)}, \dots, x_k^{(M)}), \qquad \underline{s} = (s^{(1)}, \dots, s^{(M)}).$$

Observe first that the first order optimality conditions for 8 are as follows:

$$\frac{\partial \varphi_k(\underline{x})}{\partial x_k^{(m)}} \begin{cases} \leq 0 & \text{if} \quad x_k^{(m)} = 0 \\ \geq 0 & \text{if} \quad x_k^{(m)} = L_k^{(m)} \\ = 0 & \text{if} \quad 0 < x_k^{(m)} < L_k^{(m)} \end{cases} \tag{10}$$

Thus, $\underline{x}$ is an equilibrium point of the multiproduct oligopoly game if and only if (10) is satisfied for each component $x_k^{(m)}$ of $\underline{x}$.

In the linear case we assume that $\underline{f} = (f_1, \dots, f_M)$,

$$f_\mu(\underline{s}) = \sum_{m=1}^{M} a_\mu^{(m)} s^{(m)} + b_\mu, \quad (1 \leq \mu \leq M) \tag{11}$$

$$K_k(\underline{x}_k) = \sum_{m=1}^{M} A_k^{(m)} x_k^{(m)} + B_k \qquad (1 \leqq k \leqq N) . \tag{12}$$

In this special case relations (10) are equivalent to the conditions

$$\sum_{m=1}^{M} s_k^{(m)} a_m^{(\mu)} + \sum_{m=1}^{M} a_\mu^{(m)} s^{(m)} + b_\mu - A_k^{(\mu)} \begin{cases} \leqq 0 \text{ if } x_k^{(\mu)} = 0 \\ \geqq 0 \text{ if } x_k^{(\mu)} = L_k^{(\mu)} \\ = 0 \text{ if } 0 < x_k^{(\mu)} < L_k^{(\mu)} \end{cases} \tag{13}$$

where $s^{(m)} = \sum_{k=1}^{N} x_k^{(m)}$.

By introducing the slack variables

$$w_k^{(\mu)} = L_k^{(\mu)} - x_k^{(\mu)} \geqq 0, \qquad z_k^{(\mu)} \begin{cases} = 0 \text{ if } x_k^{(\mu)} > 0 \\ \geqq 0 \text{ otherwise} \end{cases}$$

$$v_k^{(\mu)} \begin{cases} = 0 \quad \text{if} \quad x_k^{(\mu)} < L_k^{(\mu)} \\ \geqq 0 \quad \text{otherwise} \end{cases}$$

relations (13) can be rewritten as

$$\sum_{m=1}^{M} s_k^{(m)} a_m^{(\mu)} + \sum_{m=1}^{M} a_\mu^{(m)} s^{(m)} + b_\mu - A_k^{(\mu)} - v_k^{(\mu)} + z_k^{(\mu)} = 0 \tag{14}$$

Introduce the following block-matrix:

$$\underline{\underline{P}} = \begin{bmatrix} \underline{\underline{A}} + \underline{\underline{A}}^T & \underline{\underline{A}} & \cdots & \underline{\underline{A}} \\ \underline{\underline{A}} & \underline{\underline{A}} + \underline{\underline{A}}^T & \cdots & \underline{\underline{A}} \\ \vdots & \vdots & & \vdots \\ \underline{\underline{A}} & \underline{\underline{A}} & \cdots & \underline{\underline{A}} + \underline{\underline{A}}^T \end{bmatrix}, \tag{15}$$

where $\underline{\underline{A}} = (a_\mu^{(m)})$ is the Jacobian of function $\underline{f}$.
Then simple calculation shows that relations (14) and the definition of the slack variables are equivalent to the relations

$$\underline{\underline{P}}\,\underline{x} + \underline{b} - \underline{a} - \underline{v} + \underline{z} = \underline{0}$$
$$\underline{x} + \underline{w} = \underline{L}$$
$$\underline{x}^T\underline{z} = \underline{v}^T\underline{w} = \underline{v}^T\underline{z} = 0 \qquad (16)$$
$$\underline{x}, \underline{v}, \underline{z}, \underline{w} \geq \underline{0},$$

where the components of vectors $\underline{x}$, $\underline{a}$, $\underline{v}$, $\underline{z}$, $\underline{w}$, $\underline{L}$ are $x_k^{(m)}$, $A_k^{(m)}$, $v_k^{(m)}$, z_k^{m}, w_k^{m}, L_k^{m}, respectively, and

$$\underline{b} = (b\ ,\ldots,b_M,\ldots,b\ ,\ldots,b_M).$$

We can now state our next result.

Theorem 4. Under the above assumptions a vector $\underline{x}$ is an equilibrium point of the linear, multiproduct oligopoly game if and only if relations (16) hold.

Proof. See Szidarovszky and Okuguchi (1985).

Assume now that besides the above conditions, matrix $\underline{\underline{A}}$ is symmetric. Then it is also negative semidefinite. In this case the equilibrium point of the game is equivalent to a certain quadratic programming problem.

Theorem 5. Under the previous and this additional assumptions, each equilibrium point of the game is an optimal solution of the quadratic programming problem

$$\text{maximize } \frac{1}{2}\underline{x}^T \underline{\underline{P}}\,\underline{x} + (\underline{b} - \underline{a})^T \underline{x}$$
$$\text{subject to} \qquad (17)$$
$$0 \leq \underline{x} \leq \underline{L}$$

Proof. See Szidarovszky and Okuguchi (1985).

In a further special case the uniqueness of the equilibrium point can also be guaranteed.

Theorem 6. If in addition to the conditions of the previous theorem matrix $\underline{\underline{A}}$ is negative definite, then the equilibrium point of the game is unique.

Proof. It is known (see Szidarovszky, 1978) that matrix $\underline{\underline{P}}$ is negative definite, Consequently problem (17) is strictly concave. Thus the optimal solution is unique.

Consider next the following linear complementarity problem. Find nonnegative vectors $\underline{x}$, $\underline{v}$, $\underline{z}$, $\underline{w}$ such that.

$$\begin{pmatrix} \underline{\underline{P}} & \underline{\underline{I}} \\ -\underline{\underline{I}} & \underline{\underline{O}} \end{pmatrix} \begin{pmatrix} \underline{x} \\ \underline{v} \end{pmatrix} + \begin{pmatrix} \underline{b} - \underline{a} \\ \underline{1} \end{pmatrix} = \begin{pmatrix} \underline{z} \\ \underline{w} \end{pmatrix} \qquad (18)$$

$$\underline{x}'\underline{z} = \underline{v}'\underline{w} = 0 \ .$$

Then the above derivations imply the following result.

Theorem 7. Under the above conditions the Cournot equilibrium point is a solution of (18). If the solution of (18) satisfies the additional constraint $\underline{v}'\underline{z} = 0$, then it is a Cournot equilibrium point.

4. Stability Conditions

Assume that at time t $(t \geq 0)$ a strategy vector $\underline{x}^{(t)} = (\underline{x}_1^{(t)}, \dots, \underline{x}_N^{(t)})$ is given. Then at time t+1 each player maximizes his own pay-off value with fixed strategies $\underline{x}_i^{(t)}$ for all of the other players. This condition is equivalent to the difference equation

$$\underline{x}^{(t+1)} = \begin{pmatrix} \underline{\underline{O}} & \underline{\underline{B}} & \dots & \underline{\underline{B}} \\ \underline{\underline{B}} & \underline{\underline{O}} & \dots & \underline{\underline{B}} \\ & \dots\dots & & \\ \underline{\underline{B}} & \underline{\underline{B}} & \dots & \underline{\underline{O}} \end{pmatrix} \underline{x}^{(t)} + \underline{\alpha} \ , \qquad (19)$$

where $\underline{\underline{B}} = (\underline{\underline{I}} + \underline{\underline{A}}^{-1} \underline{\underline{A}}^T)^{-1}$ and $\underline{\alpha}$ is a constant vector. Starting from an arbitrary strategy vector $\underline{x}^{(o)}$, relation (19) defines a dynamic process. This dynamic game is stable if and only if the eigenvalues of the coefficient matrix of (19) are inside the unit circle. In order to investigate the eigenvalues introduce the matrix

$$\underline{\underline{C}} = \begin{pmatrix} 0 & 1 & \dots & 1 \\ 1 & 0 & \dots & 1 \\ & \dots\dots & & \\ 1 & 1 & \dots & 0 \end{pmatrix},$$

then the coefficient matrix can be rewritten as $\underline{\underline{M}} = \underline{\underline{C}} \times \underline{\underline{B}}$. If β_i and γ_j are the eigenvalues of $\underline{\underline{B}}$ and $\underline{\underline{C}}$, respectively, then the eigenvalues of $\underline{\underline{M}}$ are $\beta_i \gamma_j \ (\forall_{i,j})$. One may verify that $\gamma_j = -1$ or $n-1$, Consequently stability occurs if and only if

$$(\beta_i) < \frac{1}{n-1} \qquad (20)$$

for all i, which is equivalent to the condition that the eigenvalues of

$\underline{\underline{A}}^{-1}\,\underline{\underline{A}}^T$ are greater than N-2. In the special case of symmetric $\underline{\underline{A}}$, $\underline{\underline{A}}^{-1}\,\underline{\underline{A}}^T = \underline{\underline{I}}$. That is, for symmetric $\underline{\underline{A}}$ the dynamic game is stable if and only if $N < 3$. This result is the generalization of the stability theorem of Theocharis (1959).

References

Burger, E. (1959) Einführung in die Theorie der Spiele, de Gruyter, Berlin.

Opitz, O. (1970) Spieltheoretische Aussagen im Oligopol-problem, Zeitschrift für Nationalökonomie, N°-3, pp. 475-482.

Szidarovszky, F. (1970) On the Oligopoly Game, 1970-1, Dept. of Math., K. Marx Univ. of Economics, Budapest.

Okuguchi, K. (1976) Expectations and Stability in Oligopoly Models, Springer Verlag, Berlin/Heidelberg/New York.

Szidarovszky, F. and S. Yakowitz (1977) A New Proof of the Existence and Uniqueness of the Cournot Equilibrium, Internat. Econ. Rev. 18, pp. 787-789.

Manas, M. (1972) A Linear Oligopoly Game, Econometrica, 40, pp. 917-922.

Szidarovszky, F. (1978) Game Theory in Hungarian , textbook for Eötvös L. University, Budapest.

Szidarovszky, F. and Okuguchi, K. (1985) On the Existence and Computation of Equilibrium Points for an Oligopoly Game with Multi-Product Firms, to appear in Annales Univertitatis Scientiarum Budapestinensis De Rolando Eötvös Nominatae, Sectio Computatorica.

Rosen, J.B. (1965) Existence and Uniqueness of Equilibrium Points for Concave n-Person Games, Econometrica, 33, pp. 520-534.

Szidarovszky, F. and S. Yakowitz (1982) Contributions to Cournot Oligopoly Theory, J. of Econ. Theory, 28, pp. 51-70.

Theocharis R.D. (1959) On the Stability of the Cournot Solution on the Oligopoly Problem, Review of Economic Studies, Vol. 27, pp. 133-134.

Solving Scheduling Problems by Computer

JUDIT SZILÁGYI /Budapest, KMF/

Review of the problem

During the last years we have elaborated some production control systems, so we had to solve scheduling problems as underlying problems, too. Although these problems came from different spheres of the industry /textile industry, paper industry, precision engeneering/, there are common features in them; these are:

1/ All the systems are intermittent production systems, that is a large number of products must be made so that processing centers must frequently change from one product to another.

2/ In every case during the routing /routing is the sequence of production centers through which the material flows to produce the products/ there are work centers which are considered buttlenecks.

3/ Setup times are sequence dependent in the buttleneck work centers.

4/ There are several possible routings for the production of one item and one must be selected.

The mathematical model

To formulate the mathematical model we introduce some notations:

$I_N = \{1,2,\ldots,n\}$ the set of the products,

p_{ik}= the production time of product i on machine k,

D_i= the due date of product i,

E_i= the beginning time of the production of product i,

U_i= the completion time of product i,

a_{ijk}= the time required to set up the facility k to produceproduct j after product i,

G= the set of machines used during the routings,

G_o= the set the machines in the seeded work center,

T_k= the number of working hours of the machine k during the scheduling period,

L(k)= a production plan on the machine k,

L={L(k)} = a production plan on G.

The most important goals are:

1/ To minimize the avrage lateness

$$\frac{\sum_{i\in I_N}(U_i-D_i)}{N}$$

2/ To minimize on the set of production plans /on G/ the maximal flow time

$$\max_k F_{L(k)},$$

where

$$F_{L(k)}=\sum_{i\in L(k)} p_{ik} + V_{L(k)}$$

/$V_{L(k)}$ is the summarized slack time according to L(k)/.

3/ To minimize the setup time in the buttleneck work center that is to minimize

$$\sum_{k=1}^{K}\sum_{j=1}^{N}\sum_{i=1}^{N} x(L(k))a_{ijk}$$

where

$$x(L(k))=\begin{cases}1 & \text{if producti is followed by product j on the machine k,}\\ 0 & \text{elsewhere.}\end{cases}$$

4/ To load the machines uniformly that is to minimize

$$\max_{k\in G_o} d_k,$$

as the function of L, where

$$d_k=T_k-\sum_{i\in L(k)} p_{ik}.$$

The simplified problem

To solve the problem we made some simplifications. These are:

1/ The capacity in the work centers except the buttleneckcan be regarded to be infinite, the slack time can be regarded to be zero.

2/ We have recognized that if a setup time between the production of two items is less than the setup time from the idle state to the conditions necessary to process the second one as the first job in the sequence, there are some common features in their production. By the help of the technology we can divide the set of the products into subsets in such a way that the setup time between the elements of one subset can be regarded to be 0 :

$$I_N = \bigcup_{m \in M} S_m \, .$$

The sceleton of the algorithm

Step 1. Formulation of the subsets S_m, calculation of their production times q_m.

Step 2. If $\sum_{m=1}^{M} q_m > \sum_{k=1}^{K} T_k$ then stop.

The numbers q_m must be ordered decreasingly: $q_1' \geq q_2' \geq \ldots \geq q_M'$.

Let $L'(k)$ be the set of the products which have already assigned to T_k,

$$w(k) = T_k - \sum_{m \in L'(k)} q_m \, .$$

The next q_m must be assigned to the T_k for which $w(k)$ has the greatest value. /Of course any q_m can be assigned to T_k only if S_m can be produced on k./ Let us continue this assignement until all the S_m will be assigned to one T_k. If

$$\max_k w(k) > \left(\sum_{k=1}^{K} T_k - \sum_{m=1}^{M} q_m \right) / K + 2$$

then step 3 else step 4.

Step 3. Improve the assignement.

Step 4. The sequence of the dummy products S_m is defined according to the other goals by the help of classical algorithms.

Step 5. The sequence of the jobs inside the sets S_m is defined by the method described in step 4 too.

The computer system

The computer system based on the algorithm described above is written for COMMODORE 64 mostly in BASIC. A Gantt chart, as an output of the scheduling algorithm is produced first. The manager can easily survey the Gantt chart. When he is not satisfied with the result he can transpose the jobs directly or he can change the parameters of the algorithm and let the program run again. When he is satisfied whith the chart he can get a detailed work table for every work day and every job.

References:

[1] L. A. Johnson, D.C.Montgomery: Operations Research in Production Planning, Scheduling and Inventory Control, John Wiley and Sons, New York, 1974.

[2] T.Fiala: Scheduling Problems, Doctoral Dissertation, ELTE, Budapest, 1982.

[3] J.Szilágyi: A Multigoal Scheduling Problem, Sigma 16/1983/,85-91.

MULTIPLE CRITERIA ANALYSIS AND EVALUATION IN DECISION MAKING OF TRANSPORT PLANNING

Dr. Catherine TÁNCZOS
Technical University of Budapest, Hungary

Abstract

This study gives a sketch of the application of multicriteria evaluation in transport.
The author introduces a technique of matrix-analysis developed from a cross-impact method. She makes an investigation of efficiency with this technique on freight transport on Hungarian State Railways.
The paper evaluates the technique of multidimensional data-analysis and introduces the possibilities of their applications in transport planning.
Finally the multicriterial method is illustrated with an example from transport planning.

1. Introduction

Transportation - one of the most important parts of infrastructure - plays a significant role in the development of production, consumption and in the formation of foreign trade balance and so in the development of the whole national economy as well. The decisions connected with planning, organising or developing the different transport systems effect in almost every territory of the social-economical life. The fast development in the fields of decision making methods and computer technique has made favourable conditions to the investigation of complicated systems, like the transport systems. The paper deals with the methods of multiple-criteria analysis and evaluation in decision making of transport planning, organisation and development.

2. A weighted multi-dimensional evaluation of the efficiency in transportation

The method of measuring the efficiency - worked out previously by our department - enabled the complex qualification of efficiency level of differently aggregated transport systems. The multi-dimensional efficiency evaluation method consists of the next steps:

- choosing the criteria of evaluation,
- determining the indicators which express the suitability to the expectation,

- counting the value of the indicators,
- weighing the criteria ,
- analysis: determination of the points according to each criteria,
- synthesis: determination of the total score taking into consideration all the points according to the different criteria.

The complex qualification of the efficiency of the railway freight transport on macroeconomic level takes place by the next criteria. The weights of the criteria are determined on the basis of experts' opinions by the Guilford method.

name	indicator	dimension	weight	value /in 1984/	evaluation	points
1	2	3	4	5	6	7
	of the criteria					
meeting quantitative demand	transported volume / demanded volume	%	6,0	99,1	good	+60
meeting demand in time	complete transport cycle time / average distance of transportation	minute / km	5,2	31,5	bad	-104
protection of goods	conveyance loss / output	Ft / 10^3ton km	1,8	3,20	acceptable	-18
unit cost	total expenditure of transportation on macroeconomic level / output	Ft / ton km	3,0	0,80	good	+30
labour requirement	average number of employees / output	person / 10^6ton km	1,0	2,83	acceptable	-10
capital requirement	capital stock / output	Ft / ton km	2,0	2,48	acceptable	-20

1	2	3	4	5	6	7
profitableness	incomes-expenditures / capital + wages	%	2,9	10,96	medium	0
environmental damage	extent of the damage in the environment / output	Ft / 10^3tonkm	1,3	5,6	good	+13
oil used up	the volume of used petroleum products / output	Ft / 10^6ton km	1,3	5,76	good	+13
contribution to the income in foreign currencies	netto balance of freight rail transport in foreign currencies / the value of export	%	3,7	1,8	good	+37
The komplex qualification of efficiency is almost absolute "medium" because the sum of points is +1.						+153
						-152
						+ 1

The qualification would be "good" /G/ if the sum of points were +282.
The qualification would be "very g."/VG/ if the sum of points were +564.
The qualification would be "accept" /A/ if the sum of points were -282.
The qualification would be "bad" /B/ if the sum of points were -564.

Developing this method it became possible to determin and analyse the most important facts influencing the efficiency. The matrix method - developed from the cross-impact technique - takes into consideration all the criteria influencing the efficiency and so it is possible to deal with different kinds of information.

The application of the method consists of the next steps:

a/ exploration of the determinant factors of the criteria /with the application of brainstorning/;

b/ determination of the closeness of the connections /a_{ij}/ between the criteria and their determinant factors /by experts' estimates/;

c/ counting the vectors characterising the influential role of determinant factors /$\underline{\alpha}$ and $\underline{\beta}$/:

$$\alpha_j = \frac{\sum_{i=1}^{m} i \quad /i=0,\text{if } a_{ij} \neq 3 \text{ and } i=1,\text{if } a_{ij} = 3/}{\sum_{i=1}^{m} i \quad /i=0,\text{if } a_{ij} = 0 \text{ and } i=1,\text{if } a_{ij} \neq 0/} \quad ; \quad j = 1,\dots,n;$$

$$\beta_j = \sum_{i=1}^{m} a_{ij} \quad ; \quad j = 1,\dots,n$$

where:

$j=1,\dots,n$: the number of determinant factors;

$i=1,\dots,m$: the number of criteria ;

a_{ij}: the evaluated closeness between the i-th criteria and the j-th determinant factor;

$a_{ij}=3$: means close connection, when evaluated correlation coefficient between i and j is above 0,7;

$a_{ij}=2$: means moderate connection, when the evaluated correlation coefficient between i and j is between 0,5-0,7;

$a_{ij}=1$: means loose connection, when the evaluated correlation coefficient between i and j is below 0,5;

$a_{ij}=0$: means that there is no connection between i and j;

d/ counting the vectors characterising the impressibility of the criterias / $\underline{\gamma}$, $\underline{\delta}$/:

$$\gamma_i = \frac{\sum_{j=1}^{n} j \quad /j=0,\text{if } a_{ij} \neq 3 \text{ and } j=1,\text{if } a_{ij} = 3/}{\sum_{j=1}^{n} j \quad /j=0,\text{if } a_{ij} = 0 \text{ and } j=1,\text{if } a_{ij} \neq 0/} \quad ; \quad i = 1,\dots,m$$

$$\delta_i = \sum_{j=1}^{n} a_{ij} \ ; \quad i = 1,\dots,m$$

e/ choosing the most important determinant factors of the criteria on the basis of evaluation of $\beta_j \geq \beta_{critical}$;

f/ decision about the operation on the basis of the multi-dimensional evaluation method of the efficiency.

The structure of the matrix is:

		determinant factors of the criteria						vectors characterising the impressibility of the criteria	
		1	2	...	j	...	n	$\underline{\gamma}$	$\underline{\delta}$
criteria of the efficiency	1	a_{11}	a_{12}	...	a_{1j}	...	a_{1n}	γ_1	δ_1
	2	a_{21}	a_{22}	...	a_{2j}	...	a_{2n}	γ_2	δ_2
	⋮	⋮	⋮	...	⋮	...	⋮	⋮	⋮
	i	a_{i1}	a_{i2}	...	a_{ij}	...	a_{in}	γ_i	δ_i
	⋮	⋮	⋮	...	⋮	...	⋮	⋮	⋮
	m	a_{m1}	a_{m2}	...	a_{mj}	...	a_{mn}	γ_m	δ_m
vectors characterising the influential role of determinant factors	$\underline{\alpha}$	α_1	α_2	...	α_j	...	α_n		
	$\underline{\beta}$	β_1	β_2	...	β_j	...	β_n		

3. Evaluation of goods transport at a microeconomic level

The object of the examination is the transport of 20 tons of goods, sensitive to cost, by six different ways, for three different transportation distances. The criteria of evaluation are:

E_1 : cost /Ft/

E_2 : time /hour/

E_3 : safety of goods /conveyance loss/ /Ft/

E_4 : energy /oil/ use /kg/

The weights of the criteria are:

$W_1 = 6,0$

$W_2 = 4,2$

$W_3 = 2,8$

$W_4 = 1,0$

The six alternative solutions:

T_1 : railway + road; loading with mechanization

T_2 : railway + road; loading without mechanization

T_3 : railway; loading with mechanization

T_4 : railway; loading without mechanization

T_5 : road; loading with mechanization

T_6 : road; loading without mechanization

The three different transportation distances are:

$d_1 = 40$ km

$d_2 = 100$ km

$d_3 = 180$ km

The table of points:

Criteria	Weight	Evaluation				
		VG	G	M	A	B
E_1	6,0	+120	+60	0	-60	-120
E_2	4,2	+ 84	+42	0	-42	- 84
E_3	2,8	+ 56	+28	0	-28	- 56
E_4	1,0	+ 20	+10	0	-10	- 20

Intervals of qualification at the d_1 version:

Very good VG	good G	medium M	acceptable A	bad B
1719-2169	2170-2620	2621-3070	3071-3521	3522-3971
10,3-17,0	17,1-23,7	23,8-30,4	30,5-37,1	37,2-43,8
4,2- 5,7	5,8- 7,2	7,3- 8,6	8,7-10,1	10,2-11,6
4,9-11,1	11,2-17,4	17,5-23,6	23,7-29,9	30,0-36,1

Basic data at $d_1 = 40$ km

Criteria	T_1	T_2	T_3	T_4	T_5	T_6
E_1	2683	3971	1719	2189	2643	3287
E_2	38,2	43,8	34,6	37,4	10,3	13,1
E_3	11,6	11,6	8,6	8,6	4,2	4,2
E_4	12,7	12,7	4,9	4,9	36,1	36,1

Results

T_1	T_2	T_3	T_4	T_5	T_6
-	-120	+120	+60	-	-60
-84	- 84	- 42	- 84	+84	+84
-56	- 56	-	-	+56	+56
+10	+ 10	+ 20	+20	-20	-20
-130	-250	+ 98	- 4	+120	+60
V.	VI.	II.	IV.	I.	III.

Basic data d_2 = 100 km

Criteria	alternative solutions					
	T_1	T_2	T_3	T_4	T_5	T_6
E_1	3331	4619	2387	2837	4561	5205
E_2	45,2	50,8	41,6	44,4	6,2	9,0
E_3	23,6	23,6	20,0	20,0	6,6	6,6
E_4	20,1	20,1	12,3	12,3	67,7	67,7

Results

	T_1	T_2	T_3	T_4	T_5	T_6
E_1	+60	-60	+120	+120	-60	-120
E_2	-84	-84	- 42	- 84	+84	+ 84
E_3	-56	-56	- 28	- 28	+56	+ 56
E_4	+20	+20	+ 20	+ 20	-20	- 20
$\sum$	-60	-180	+ 70	+ 28	+60	-
place	V.	VI.	I.	III.	II.	IV.

Basic data d_3 = 180 km

Criteria	alternative solutions					
	T_1	T_2	T_3	T_4	T_5	T_6
E_1	4195	5483	3231	3701	7934	8578
E_2	56,9	62,5	53,3	56,1	9,5	12,3
E_3	39,6	39,6	36,0	36,0	9,8	9,8
E_4	29,9	29,9	22,1	22,1	121,8	121,8

Results

	T_1	T_2	T_3	T_4	T_5	T_6
E_1	+120	-	+120	+120	-120	-120
E_2	- 84	- 84	- 84	- 84	+ 84	+ 84
E_3	- 56	- 56	- 56	- 56	+ 56	+ 56
E_4	+ 20	+ 20	+ 20	+ 20	- 20	- 20
$\sum$	-	-120	-	-	-	-
place	I-V.	VI.	I-V.	I-V.	I-V.	I-V.

In another case the volume of goods is 1000 tons and the alternative solutions are: T_1: railway and road; T_2: railway; T_3: road and the loading happens with mechanization. The criteria of the evaluation are:
E_1: cost /Ft/; $w_1 = 6,0$ E_2: energy /oil/ use /kg/; $w_2 = 4,2$
E_3 : time /hour /; $w_3 = 2,8$ E_4: safety of goods /conveyance loss/ /Ft/; $w_4 = 1,0$

Intervals of qualification at the d_1 version:

Very good VG	good G	medium M	acceptable A	bad B
56796 - 61589	61590 - 66381	66382 - 71174	71175 - 75966	75967-80759
366,9 - 614,1	614,2 - 681,3	681,4 -1109,5	1108,6 -1355,7	1355,8-1602,9
9,7 - 17,3	17,4- 24,8	24,9 - 32,4	32,5 - 39,9	40,0 -47,5
16.3 - 22,3	22,4 - 28,3	28,4- 34,3	34,4 - 40,3	40,4- 46,3

Basic data d_1= 40 km

	T_1	T_2	T_3
E_1	80759	56796	78912
E_2	808,5	366,9	1062,9
E_3	47,5	46,6	9,7
E_4	46,3	16,3	18,3

Results

	T_1	T_2	T_3
E_1	- 120	+ 120	- 120
E_2	+ 42	+ 84	- 84
E_3	- 56	- 56	+ 56
E_4	- 20	+ 20	+ 20
$\sum$	- 154	+ 168	- 128
place	III.	I.	II.

Basic data d_2 = 100 km

	T_1	T_2	T_3
E_1	103586	79622	185067
E_2	1096,2	654,7	3906,0
E_3	51,8	50,9	30,5
E_4	70,8	40,9	29,7

Results

	T_1	T_2	T_3
E_1	+ 60	+ 120	- 120
E_2	+ 84	+ 84	- 84
E_3	- 56	- 56	+56
E_4	- 20	+ 10	+ 20
$\sum$	+ 68	+ 158	- 128
place	II.	I.	III.

Basic data d_3 = 180 km

	T_1	T_2	T_3
E_1	134021	110058	320196
E_2	1479,9	1038,3	6815,3
E_3	58,2	57,3	36,2
E_4	103,5	73,6	44,9

Results

	T_1	T_2	T_3
E_1	+ 120	+ 120	- 120
E_2	+ 84	+ 84	- 84
E_3	- 56	- 56	+ 56
E_4	- 20	-	+ 20
$\sum$	+ 128	+ 148	- 128
place	II.	I.	III.

Literature:

1. K. TÁNCZOS: Többkritériumu döntések a közlekedésben. Közlekedéstudományi Szemle 1984. 1.
2. A közlekedés hatékonyságának mérése. Kutatási jelentés. KTI. Bp. 1983-84.

RELIABILITY MODELLING OF STRUCTURAL SYSTEMS

P. THOFT-CHRISTENSEN
University of Aalborg
DK-9000 Aalborg
Denmark

1. INTRODUCTION

In this paper it is shown how the reliability of a structural system can be evaluated. The most fundamental concepts in modern structural systems reliability theory are introduced and briefly discussed. A much more satisfactory presentation is given by Thoft-Christensen & Murotsu [1] where a large number of examples are used to illustrate the theory. The theory presented belongs to the so-called first order - second moment theories, where the joint probability distribution of the relevant random variables is simplified, and where the failure domain is idealized in such a way that the reliability calculations can be performed with a reasonable amount of work. By using the so-called β-unzipping method it is shown how a measure of the reliability of a structural system can be obtained in a rational way.

2. RELIABILITY THEORY OF STRUCTURAL ELEMENTS

In this chapter the most important assumptions and definitions in the theory of structural elements are presented and discussed.

> Assumption 1: The uncertainties related to the reliability of a structural system can be modelled by a finite number of random variables $\bar{X} = (X_1, \ldots, X_n)$ defined by the joint probability function $f_{\bar{X}}$.

It has for many years been recognized that all quantities that enter into engineering calculations are associated with some uncertainty. It is necessary to distinguish between three types of uncertainty, namely physical uncertainty, statistical uncertainty, and model uncertainty. Physical quanties, such as loads, material properties and dimension always show some variability. This variability can be described in terms of probability distributions or, in the case of time-varying loads, in terms of stochastic processes. However, in reliability analysis it is the extreme value of the load in the reference period that is of interest. When two or more time-varying loads are acting on a structure together, then the distribution of the extreme combined load or load effect is of interest. Therefore the physical variability can be modelled by random variables.

The physical variability is quantified by examining test date; but, since sample sizes are limited, some uncertainty - the statistical uncertainty - must remain. This statistical uncertainty can be taken into account by assuming the distribution parameters themselves to be random variables.

In structural analysis and design output quantities, for example, load effects are related to input quantities by mathematical models. These models are generally deterministic (for example, elastic analysis) and are either based on a deep understanding of the mechanics of the problem or they can be highly empirical. These models will usually give a simplified description. Therefore, the predictions by using these models will contain some additional uncertainty. The model uncertainty associated with a mathematical model may be expressed in terms of a probability distribution of a random variable X_m, defined as the actual response divided by the predicted response using the model.

Definition 1: The random variables $\bar{X} = (X_1, \ldots, X_n)$ used in modelling the uncertainties of a structural system are called basic variables. The outcome of $\bar{X}$ is denoted $\bar{x} = (x_1, \ldots, x_n)$ and $\bar{x}$ is a point in an n-dimensional basic variable space ω.

Assumption 2: The random vector $\bar{X}$ is assumed to be normally distributed.

This assumption 2 is clearly a very severe assumption since, for example, loads usually cannot be satisfactorily modelled by a normally distributed random variable. For such non-normally distributed random variables it is therefore assumed that they are transformed into normally distributed random variables by, for example, the so-called Rosenblatt transformation method (see Thoft-Christensen & Murotsu [1]).

Definition 2: Each failure mode for a structural element is called a failure element.

As an example of structural elements consider a beam. A beam will usually have several potential failure modes, e.g.

- failure in bending at its left end
- failure in bending at its right end
- failure in bending at its midpoint
- failure in tension/compression due to axial forces
- too large deflection at the midpoint

Such failure modes are called failure elements. In the example shown above the beam has 5 failure elements. In each failure element two different forms of failure can usually occur, namely in "tension" or in "compression". The notation "tension" is here used to indicate a tensile force or a positive bending moment, and the notation "compression" indicates a compressive force or a negative moment. To reduce the number of failure elements usually only the most important (highest probability of occurence) form of failure is included. A failure element can also be, for example, a joint between a number of beams. In this way fatique failure can be included.

<u>Assumption 3</u>: The structural system is assumed to contain a finite number of failure elements, say m.

<u>Assumption 4</u>: For each failure element i, i = 1, ..., m there is a function $f_i: \omega \rightarrow R$ such that failure element i is in a state of failure if $f_i(\bar{x}) \leq 0$ and in a safe state if $f_i(\bar{x}) > 0$.

The purpose of introducing the function f_i for failure element i is to be able to decide whether a failure element is in a state of failure or in a safe state. This division can be performed by several different functions.

<u>Definition 3</u>: The function f_i mentioned in assumption 4 is called a <u>failure function</u> of failure element i. The random variable $M_i = f_i(\bar{X})$ is called a <u>safety margin</u> of failure element i.The subspace $\omega_s = \{\bar{x} \in \omega | f(\bar{x}) > 0\}$ is called the <u>safe region</u> and the subspace $\omega_f = \{\bar{x} \in \omega | f(\bar{x}) \leq 0\}$ is called the <u>failure region</u> of failure element i. The safe region and the failure region are divided by a <u>failure surface</u>.

It is important to note that the same failure surface can be described by a number of equivalent failure functions.

<u>Definition 5</u>: The <u>probability of failure</u> $P_{f,i}$ of failure element i

is

$$P_{f,i} = P[M \leq 0] = \int_{\omega_f} f_{\bar{X}}(\bar{x})d\bar{x} \qquad (1)$$

and the <u>reliability</u> is $R_i = 1 - P_{f,i}$

In the special case where the safety margin M_i of failure element i is linear in the normally distributed basic variables $\bar{X} = (X_1, \ldots, X_n)$ it is easy to show that

$$P_{f,i} = \Phi(-\frac{\mu_M}{\sigma_M}) \qquad (2)$$

where Φ is the standard normal distribution function, μ_M the mean value of M, and σ_M the standard deviation of M.

In general, the basic variables $\bar{X} = (X_1, \ldots, X_n)$ are correlated. However, the set of basic variables can by linear transformation be transformed into a set of <u>uncorrelated variables</u> $\bar{Y} = (Y_1, \ldots, Y_n)$. It turns out to be convenient to make a further linear transformation into a set of <u>uncorrelated and normalized variables</u> $\bar{Z} = (Z_1, \ldots, Z_n)$ by the transformation

$$Z_i = \frac{Y_i - \mu_{Y_i}}{\sigma_{Y_i}} \quad , \quad i = 1, \ldots, n$$

where μ_{Y_i} is the mean value of Y_i, and σ_{Y_i} the standard deviation of Y_i. It is easy to show that when the safety margin M_i is linear in $\bar{X}$ and therefore also in $\bar{Z}$ then μ_M/σ_M is equal to the shortest distance β_i from the origin to the failure surface in the z-coordinate system. Therefore in this case

$$P_{f,i} = \Phi(-\beta_i) \qquad (3)$$

<u>Definition 6</u>: The <u>reliability index</u> β_i of failure element i is equal to the shortest distance from the origin to the failure surface for failure element i in the z-coordinate system. The point on the failure surface closest to the origin is called the <u>design point</u>.

As mentioned above, (when the failure surface is linear (a hyperplane)) the simple one-to-one relation (3) between $P_{f,i}$ and β_i exists. For non-linear failure surfaces (3) are not valid so calculation of $P_{f,i}$ on the basis of (3) will only give an approximate value of the correct $P_{f,i}$. The value obtained by using (3) corresponds to approximate the correct failure surface with its tangent plane in the design point. Calculation of the reliability index β_i can be performed by a number of simple iterative methods (see Thoft-Christensen & Baker [2]).

3. STRUCTURAL SYSTEMS RELIABILITY THEORY

In chapter 2 it is shown how the probability of failure $P_{f,i}$ of a failure element can be estimated. The first step is to calculate the reliability index β_i and then an approximate value of $P_{f,i}$ by equation (3). In this chapter a brief presentation of the most important assumptions and definitions in structural systems reliability theory will be given. A more detailed presentation is given by Thoft-Christensen & Murotsu [1].

Consider a statically determinate (non-redundant) structure. Total structural systems failure takes place as soon as any structural element fails. However, failure in a single structural element will for a statically inderterminate (redundant) structure not always result in failure of the total system. The reason for this is that the remaining structural elements may by redistribution of the load effects be able to sustain the external loads. For such structures total failure will usually require that failure takes place in more than one structural element. Formation of a failure mode for the structural system will therefore require simultaneous failure in a number of failure elements. This can be symbolized by a parallel system where the elements are failure elements. In this way the different potential failure modes are all modelled by parallel systems and then combined in a series system where the single elements are the parallel systems (failure modes).

> Definition 7: Let the calculation of the reliability of a structural system be based on an estimate of the probability of failure of a series system, where the elements are failure elements. This analysis is called systems reliability analysis at level 1. At level 2 the systems reliability is estimated on the basis of a series system, where the elements are parallel systems each with two failure elements - so-called critical pairs of failure elements. At level 3 the systems reliability is estimated on the basis of a series system, where the elements are parallel systems each with three failure elements - so-called critical triples of failure elements.

Systems reliability analysis is illustrated in figure 1. Reliability estimate at level 4, 5, etc. can be defined in the same way but in ge-

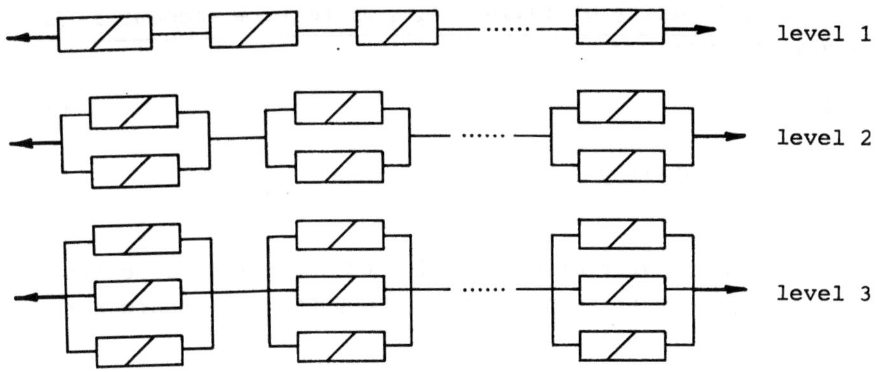

Figure 1

neral reliability analysis beyond level 3 is of minor interest.

Definition 8: Let a failure mode of a structural system be defined as formation of an elasto-plastic mechanism. The reliability of the structural system is then estimated on the basis of a series system with the mechanisms as elements. This analysis is called system reliability analysis at mechanism level.

The number of potential mechanisms for a real structural system will usually be very large. Each mechanism will consist of failure in a number of failure elements but this number of failure elements will vary from mechanism to mechanism. The mechanisms will therefore all be modelled by parallel systems but not all of them will have the same number of elements (failure elements). This is illustrated in figure 2 where r mechanisms are shown.

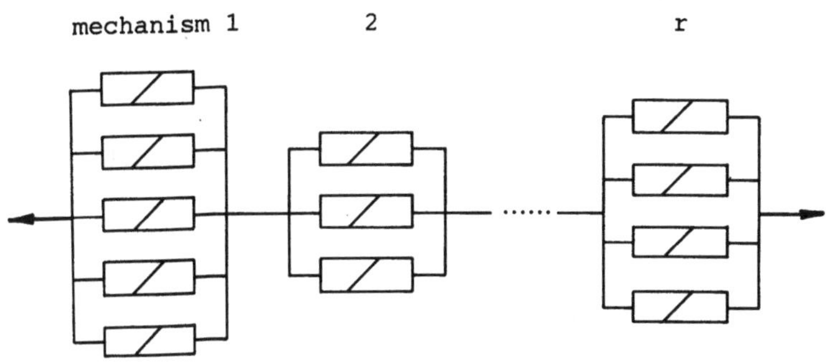

Figure 2

<u>Theorem 1 The probability of failure $P_{f,i}$ of failure element i is approximately given by</u>

$$P_{f,i} \sim \Phi(-\beta_i) \tag{4}$$

where β_i is the reliability index of failure element i.

<u>Proof</u>:

Assume that there is a transformation $\bar{Z} = \bar{T}(\bar{X})$ by which the basic variables $\bar{X} = (X_1, \ldots, X_n)$ are transformed into independent standard normal variables $\bar{Z} = (Z_1, \ldots, Z_n)$ so that

$$P_{f,i} = P(M_i \leq 0) = P(f_i(\bar{X}) \leq 0) = P(f_i(\bar{T}^{-1}(\bar{Z})) \leq 0) =$$
$$P(h_i(\bar{Z}) \leq 0) \tag{5}$$

where h_i is defined by (5). An approximation of $P_{f,i}$ can then be obtained by linearization of h_i in the design point.

$$P_{f,i} = P(h_i(\bar{Z}) \leq 0) \sim P(\bar{\alpha}_i^T\bar{Z} + \beta_i \leq 0) \tag{6}$$

where $\bar{\alpha}_i$ is the unit normal vector in the design point and β_i the reliability index.

The approximation (5) can be written

$$P_{f_i} \approx P(\bar{\alpha}_i^T\bar{Z} + \beta_i \leq 0) = P(\bar{\alpha}_i^T\bar{Z} \leq -\beta_i) = \Phi(-\beta_i) \tag{7}$$

where Φ is the standard normal distribution function.

<u>Theorem 2 The probability of failure P_{f_s} of a series system</u> with n elements is approximately given by

$$P_{f_s} \approx 1 - \Phi_n(\bar{\beta};\bar{\bar{\rho}}) \tag{8}$$

where Φ_n is the n-dimensional standardized distribution function, $\bar{\beta} = (\beta_1, \ldots, \beta_n)$ and $\bar{\bar{\rho}}$ the correlation matrix.

<u>Proof</u>:

$$P_{f_s} = P(\bigcup_{i=1}^{n} \{M_i \leq 0\}) = P(\bigcup_{i=1}^{n} \{f_i(\bar{X}) \leq 0\})$$

$$= P(\bigcup_{i=1}^{n} \{f_i(\bar{T}^{-1}(\bar{Z})) \leq 0\}) \doteq P(\bigcup_{i=1}^{n} \{h_i(\bar{Z}) \leq 0\})$$

$$\approx P\left(\bigcup_{i=1}^{n} \{\bar{\alpha}_i^T \bar{Z} + \beta_i \leq 0\}\right) = P\left(\bigcup_{i=1}^{n} \{\bar{\alpha}_i^T \bar{Z} \leq -\beta_i\}\right)$$

$$= 1 - P\left(\bigcap_{i=1}^{n} \{\bar{\alpha}_i^T \bar{Z} > -\beta_i\}\right) = 1 - P\left(\bigcap_{i=1}^{n} \{-\bar{\alpha}_i^T \bar{Z} < \beta_i\}\right) = 1-\Phi_n(\bar{\beta};\bar{\bar{\rho}})$$

where the same transformation $\bar{T}$ as in the proof of theorem 1 is used and where $\rho_{ij} = \bar{\alpha}_i^T \bar{\alpha}_j$.

Theorem 3 The propability of failure P_{f_p} of a parallel system with n elements is approximately given by

$$P_{f_p} = \Phi_n(-\bar{\beta};\bar{\bar{\rho}}) \tag{9}$$

where Φ_n is the n-dimensional standardized normal distribution function, $\bar{\beta} = (\beta_1, \ldots, \beta_n)$ and $\bar{\bar{\rho}}$ the correlation matrix.

Proof:

$$P_{f_p} = P\left(\bigcap_{i=1}^{n} \{M_i \leq 0\}\right) = P\left(\bigcap_{i=1}^{n} \{g_i(X) \leq 0\}\right)$$

$$= P\left(\bigcap_{i=1}^{n} \{g_i(\bar{T}^{-1}(\bar{Z})) \leq 0\}\right) = P\left(\bigcap_{i=1}^{n} \{h_i(\bar{Z})\}\right)$$

$$\approx P\left(\bigcap_{i=1}^{n} \{\bar{\alpha}_i^T \bar{Z} + \beta_i \leq 0\}\right) = P\left(\bigcap_{i=1}^{n} \{\bar{\alpha}_i^T \bar{Z} \leq -\beta_i\}\right) = \Phi_n(-\bar{\beta};\bar{\bar{\rho}})$$

where the same transformation $\bar{T}$ as in the proof of theorem 1 is used and where $\rho_{ij} = \bar{\alpha}_i^T \bar{\alpha}_j$.

Unfortunately, calculation of values of Φ_n for $n \geq 3$ can generally only be performed in an approximate way. Otherwise upper and lower bounds must be used.

To summarize: The estimate of the reliability of a structural system consists of the following steps

- evaluate the probability of failure of each parallel system by equation (9)
- evaluate the correlation between the parallel systems two and two

• evaluate the probability of failure of the series system by equation (8)

Evaluation of the correlation between a pair of parallel systems can easily be performed if safety margins for the parallel systems are linear. However, in general this will clearly not be the case. It is therefore natural to investigate the possibility of introducing an <u>equivalent linear safety margin</u> for each parallel system. Equivalent linear safety margin have been suggested by several authors. It is beyond the scope of this paper to give a presentation of such approximation. A detailed presentation is given by Thoft-Christensen & Murotsu [1].

4. THE β-UNZIPPING METHOD

In chapter 3 it is shown how the reliability of a structural system can be estimated when the reliability analysis is made at level 1, 2, 3,... or at mechanism level. However, one problem remains, namely <u>identification of the significant failure modes</u>. Several different methods have been suggested in the literature. In this paper the main features of the so-called β-unzipping method are presented. More details are given in Thoft-Christensen & Murotsu [1].

<u>Definition 9</u>: Let a structural system contain m failure elements and let the corresponding reliability indices be β_i, i = 1,2,...,m. Further, let

$$\beta_{min} = \min_{i=1,m} \beta_i$$

Then the <u>critical failure elements</u> are those with $\beta_i \in [\beta_{min}; \beta_{min} + \Delta\beta_1]$, where $\Delta\beta_1$ is a prescribed positive number.

Calculation of β_i, i = 1, ..., m usually requires a finite element analysis. Note that anly static loading is included . A typical value of $\Delta\beta_i$ is 2.00.

<u>Assumption 5</u>: In the β-unzipping method only the critical failure elements are included in the series system used in the reliability analysis at level 1.

Clearly, by including only the critical failure elements in the series system a lower value of the probability of failure will be obtained. However, if $\Delta\beta_i$ is not chosen too small the difference between the correct value and the obtained value will be negligible.

Definition 10: Assume that failure takes place in failure element ℓ and modify the structure by removing failure element ℓ and adding a pair of loads corresponding to the remaining strength of the failure element. These loads are called fictitious loads. For brittle failure elements no loads are added, but if the removed failure element is ductile the fictitious loads F_ℓ are stochastic loads given by $F_\ell = \gamma_\ell R_\ell$, where R_ℓ is the load-carrying capacity of failure element ℓ and where $0 < \gamma_\ell \leq 1$.

Definition 11: Let the smallest reliability indices of all remaining failure elements in a modified structure with failure in failure element ℓ be β_{min}. Critical pairs of failure elements are then obtained by combining ℓ with the remaining failure elements with reliability indices in the interval $[\beta_{min}; \beta_{min} + \Delta\beta_2]$, where $\Delta\beta_2$ is a prescribed positive number.

By assuming in turn failure in all critical failure elements and analysing the modified structures a number of critical pairs of failure elements are identified.

Assumption 6: In the β-unzipping method only the pairs of critical failure elements are included in the series system used in the reliability analysis at level 2.

For each critical pair of failure elements (parallel system with 2 failure elements) the probability of failure is calculated by (9) and an equivalent safety margin is determined. The probability of failure for the series system can then be calculated by (8).

It is now straightforward to generalize the method to higher levels $N > 2$. At level 3 the estimate of the system reliability is based on so-called critical triples of failure elements, i.e. a set of three failure elements. The critical triples of failure elements are identified by the β-unzipping method by assuming that failure takes place in turn in all critical pairs of failure elements and then modify the structure by adding two pairs of fictitious loads and reanalyse the modified structure. The failure elements with β-values in the interval $[\beta_{min}, \beta_{min} + \Delta\beta_3]$, where $\Delta\beta_3$ is a prescribed positive number, are then in turn combined with failure elements ℓ and m to form a number of parallel systems. The system used in the reliability analysis at level 3 is identified in this way and the probability of failure at level 3 can be calculated by the equations derived in chapter 3.

Now turn to reliability analysis at mechanism level where the β-unzipping method is applied in a somewhat different way. Consider an elasto-plastic structure and let the number of potential failure elements (e.g. yield hinges) be n. It is then known from the theory of plasticity that the number of fundamental mechanisms is m = n-r, where r is the degree of redundancy. All other mechanisms can then be formed by linear combinations of the fundamental mechanisms. However, only the most critical or most significant mechanisms need to be included to get a sufficiently accurate estimate of the probability of failure at mechanism level. Note that since some mechanisms are excluded the estimate of the probability of failure by the β-unzipping method is a lower bound for the probability of failure.

The first step is to identify all fundamental mechanisms and calculate the corresponding reliability indices. The next step is then to select a number of fundamental mechanisms as starting points for the unzipping. By the β-unzipping method this is done on the basis of the reliability index β_{min} for the real fundamental mechanism that has the smallest reliability index and on the basis of a preselected constant ε_1 (e.g. $\varepsilon_1 = 0.50$). Only real fundamental mechanisms with β-indices in the interval $[\beta_{min}; \beta_{min} + \varepsilon_1]$ are used as starting mechanisms in the β-unzipping method. Let $\beta_1 \leq \beta_2 \leq \ldots \leq \beta_f$ be an ordered set of reliability indices for f real fundamental mechanisms 1, 2, ..., f, selected by this simple procedure.

The f fundamental mechanisms selected as described above are now in turn combined linearly with all m (real and joint) mechanisms to form new mechanisms. First the fundamental mechanism 1 is combined with the fundamental mechanism 2, 3, ..., m and reliability indices $\beta_{1,2}, \ldots$ $\beta_{1,m}$ for the new mechanisms are calculated. The smallest reliability index is determined, and the new mechanisms with reliability indices within a distance ε_2 from the smallest reliability index are selected for further investigation. The same procedure is then used on the basis of the fundamental mechanisms 2, ..., f and a failure tree is constructed.

More mechanisms can be identified on the basis of the combined mechanisms in the second row of the failure tree in figure 6 by adding or subtracting fundamental mechanisms. By repeating this simple procedure the failure tree for the structure in question can be constructed. The maximum number of rows in the failure tree must be chosen and can typically be m + 2, where m is the number of fundamental mechanisms. A satisfactory estimate of the system reliability can usually be obtained by using the same ε_2-value for all rows in the failure tree.

The final step in the application of the β-unzipping method in evaluating the reliability of an elasto-plastic structure at mechanism level is to select the significant mechanisms from the mechanisms identified in the failure tree. This selection can, in accordance with the selection-criteria used in making the failure tree, e.g. be made by first identifying the smallest β-value, β_{min} of all mechanisms in the failure tree and then selecting a constant ε_3. The significant mechanisms are then by definition those with β-values in the interval $[\beta_{min}; \beta_{min} + \varepsilon_3]$. The probability of failure of the structure is then estimated by modelling the structural system as a series system with the significant mechanisms as elements.

5. REFERENCES

[1] Thoft-Christensen, P. & Murotsu, Y., Application of Structural Systems Reliability Theory. Springer Verlag, 1986.

[2] Thoft-Christensen, P. & Baker, M.J., Structural Reliability Theory and Its Applications. Springer Verlag, 1982.

COMPARISON OF DIFFERENT METHODS FOR SOLVING
A LARGE SCALE, STOCHASTIC DYNAMIC PROBLEM

Ph. TORRION, J. LEVEUGLE, C. HERY
E.D.F. (FRANCE)

INTRODUCTION

The problem of optimizing the yearly operation of a multi-storage power system has received considerable attention in recent research. In this paper, we present various numerical methods which can be considered to solve this problem, and discuss their respective performances, when applied to the French power system, as it should appear in the future.

In the first part, we characterize this system ; its originality results, on the one hand, from the increasing sensitivity of the load to temperature changes and, on the other, from the multiplicity of regulation facilities on the supply side as well as on the demand side. The formulation that follows this description allows us to classify our problem as a large scale, stochastic, dynamic problem.

In the second part, we review the classical available methods for solving such problems. This points out their limitations, that could become really cumbersome when applied to the yearly operation of the French supply-demand system.We were thus led to develop a particular aggregation-decomposition method, that is briefly described and has proved to yield very good results when applied to our case. This manipulation method is the basis of the new National Regulation Model (MNR), that is now currently used at Electricité de France (EDF), more particularly in order to improve the long term expansion strategy.

In order to investigate more thoroughly the advantages of this method, we compare, in the third part, the results obtained with the MNR to those obtained with a local feedback optimization method, relaxation by dynamic programming.

It clearly shows that, implementing a local feedback i.e. without reacting upon the main random fluctuations of demand and thermal availability, is definitely unsuitable to the French system ; furthermore, if only considered as an optimization step in order to determine kilowatthours (kWh) values in each reservoir, this relaxation method yields, through a simulation of the whole system operation, a global feedback far from the optimum.

I - THE PROBLEM

I.1 A brief description

The evolution of the demand for electricity and of the supply system in France ((1), (2), (3)), which began several years ago, leads in the future to a large "nuclear summer" on the one hand, and random and marked alternations of cost in winter, mainly on account of the sensitivity of the load to climatic changes, on the other hand ; in such a system, the optimal operation and development of a number of regulation facilities -conventional and pumped hydro on the supply side, flexible peak tariffs as peak day option (EJP) (introduced in EDF tariff system a couple of years ago) on the demand side- may appear desirable in order to achieve a reliable and economically efficient supply-demand system.

The problem, settled in that manner, seems to be original, since the whole set of "reservoirs", the types and time-scales of which are very different (Table 1), will have to be operated as well as possible, provided that the main random phenomena -climatic conditions and thermal availability- are unforeseeable several days in advance, and will indeed couple the operation of the different storages.

TABLE 1 : Types and Time-Scales of the French Regulation Facilities

Type	Time-scale* average (hours)
Conventional hydro	1 100
Seasonal pumped hydro	200
Weekly pumped hydro	35
Daily pumped hydro	10
EJP tariff option	400

* Time-scale : ratio between the energy in full storage and the maximum capacity of the power-plant

We clearly address a large scale, stochastic dynamic optimization problem.

I.2 Notations and formulation

The problem is to operate thermal units, hydro facilities and tariff mechanisms in order to minimize the yearly mathematical expectation of generation and unsupplied energy cost.

Given N, the number of regulation facilities (N = 4 or 5...) and Tf, the number of time-steps in the year), the regulation facilities can be defined by :

- their maximum energy content : X_i^{max} , $1 \leqslant i \leqslant N$
- their maximum generating capacity : U_i^{max} , $1 \leqslant i \leqslant N$
- their maximum pumping capacity, if applicable :
  $U_i^{min} \leqslant 0, 1 \leqslant i \leqslant N$
- their natural (stochastic) inflows during period t :
  a_i^t , $1 \leqslant i \leqslant N, 1 \leqslant t \leqslant Tf$

Let us denote :

X_i^t : the energy in reservoir i at the beginning of period t ;
$0 \leqslant X^t_i \leqslant X_i^{max}$;

U^t_i : the generation (or demand for pumping, if negative) of reservoir i at period t,

$U_i^{min} \leqslant U^t_i \leqslant U_i^{max}$
$(U^t_i)^+ = \text{Max}\ (0, U^t_i)$; $(U^t_i)^- = \text{Min}\ (0, U^t_i)$;

d^t_i : the energy spillage from reservoir i, at period t ;

D^t : the demand for electrical energy at period t, (in fact,less the generation of the runs of the river) ;

$C^t\ (D^t)$: the fuel and shortage cost function (which depends on thermal availability) ;

ρ : the pumping efficiency.

The state variables, X_i, are evolving as follows :

$$X_i^{t+1} = X^t_i + a^t_i + \rho(U^t_i)^- - (U^t_i)^+ - d^t_i$$

where U^t_i are the control variables of the system. The criterion to be minimized is :

$$J = \underset{(U_1,\dots,U_N)}{\mathrm{Min}} \quad \underset{(D,T,a)}{E} \quad \sum_{t=1}^{Tf} C^t \left(D^t - \sum_{i=1}^{N} U_i^t\right) + F\left(X_1^{Tf},\dots,X_N^{Tf}\right)$$

where the mathematical expectation E is taken with respect to demand D, thermal availability T, and stochastic inflows a, and subject to the following set of constraints :

$$(C) \quad \begin{cases} X_i^{t+1} = X_i^t + a_i^t + \rho\,(U_i^t)^- - (U_i^t)^+ - d_i^t \\ 0 \leqslant X_i^{t+1} \leqslant X_i^{max} \\ U_i^{min} \leqslant U_i^t \leqslant U_i^{max} \\ d_i^t \geqslant 0 \end{cases} \qquad \begin{cases} i = 1,\dots,N \\ t = 1,\dots,Tf \end{cases}$$

F is a salvage value (that can be easily obtained by the operation optimization of the system in the future ; in practice, one year ahead is enough).

II - SELECTING AN APPROPRIATE METHOD

The methods for solving such a large stochastic system are indeed limited to dynamic programming (4), (5). Unfortunately, it is untractable for more than three state variables due to the exponential growth of the computing time and storage requirements.

Since there is no alternative, several compromises have been made in the past. Among them, we should avoid derministic techniques, on account of the marked random fluctuations, as well as total aggregation since we have to operate reservoirs of very different types and time-scales. Therefore, the choice has to be made among other classical techniques that are briefly described hereafter.

II.1 The classical methods

Relaxation techniques can be used to determine the optimal local feedback operating policy for a system of parallel reservoirs, taking into account stochastic inflows (6), (7). They are rather easy to implement and are likely to address very large scale systems since the results can be achieved through the solution of a succession of one dimensional optimization problem. However, it is not possible to take into account, within the optimization step, the capacity of each regulation facility for reacting upon random phenomena -such as temperature and thermal availability- that are coupling indeed the operation of those different means.

The techniques of optimization in a parametrized class of feedback (by a Monte-Carlo method, for instance) are quite interesting, when one has an idea of what a "good" feedback could be. The problem is then much easier to solve since the parameters have to be optimized through an open-loop solution. Practically, one can parametrize either the control itself (8), or the Bellman's function (9), or also the derivative of this function (10), (11). Unfortunately, in this problem, we are dealing with several time-scales, several types of reserves and the available a priori information is very poor.

The method of regular perturbations allows, in case of weak random phenomena, let it be ε, to set up an affin feedback leading to an error, deviating from the optimum in $\varepsilon 4$, (12) ; furthermore, this method is very reasonable in computation time, since it comes to the resolution of a deterministic problem, followed by a linear quadratic problem.

Unfortunately, on the one hand, it is not well adapted to the existence of constraints on control and state variables (constraints which are an inherent characteristic of our problem) and on the other hand, the importance of the random perturbations in the French electric system would probably lead to a solution that would quite clearly deviate from the optimum.

II.2 The National Regulation Model (MNR)

The drawbacks arising from these various methods, regarding the specific problem to be solved by EDF, led us to work out a particular aggregration-decomposition method from an original idea of TURGEON, (13). This method provides a global feedback on all the random perturbations affecting the system and on every means. It constitutes the foundation of the MNR.

To begin with, the principle of the method consists in breaking up the original N dimensional problem into N two-dimensional problems. This enables to solve each of these problems by stochastic dynamic programming.

The problem (Pi) is obtained by aggregating all the reservoirs except reservoir i.

If denoting this partial aggregation by overlining X, U,... variables, the problem (Pi) can be written as follows :

$$\text{(Pi)} \quad \underset{U_i,\ \bar{U}_i}{\text{Min}} \quad \underset{(D,T,a)}{E} \quad \sum_{t=1}^{Tf} C^t \left(D^t - U_i^t - \bar{U}_i^t\right) + F\left(X_i^{Tf}, \bar{X}_i^{Tf}\right)$$

with the set of related constraints, and solved by means of the Bellman's equation :

$$V_i^t \left(X_i^t, \bar{X}_i^t\right) = \underset{(D,T,a)}{E} \quad \underset{U_i^t,\ \bar{U}_i^t}{\text{Min}} \left[C\left(D^t - U_i^t - \bar{U}_i^t\right) + V_i^{-t}\left(X_i^{t+1}, \bar{X}_i^{t+1}\right)\right]$$

Let us remark U_i^t, $\bar{U}_i^t$ do constitute a feedback on the demand and thermal avaibility random perturbations, a key feature as we shall see later.

Then, assuming the optimal kWh values in the reservoirs, the real dynamic programming (in dimension N) would provide, are accurately approximated by the derivatives of V^t_i with respect to X_i, we proceed to a simulation phase for a large amount of yearly realizations of random perturbations. In the simulation we use a mixed scale reserves-thermal facilities, based on the kWh values and the fuel costs.

Thus, we calculate the mathematical expectation of the fuel and outage cost, coming out of the implementation of this admissible strategy.

The justifications for selecting this method and its qualification for the yearly operation optimization of the French system may be found in (14) ; let us just recall this method by itself provides an estimation of the gap to the optimum, which turned out to be lower than 0.8 % : the sub-optimality of this global feedback is particularly low when applied to the French system case.

The purpose of part III is to improve our knowledge of the MNR qualities. Therefore we compare the results provided by the aggregation-decomposition method, that determines a sub-optimal and global feedback to the results issuing from a relaxation, that leads, by means of dynamic programming, to an optimum local feedback.

III - AGGREGATION-DECOMPOSITON AND RELAXATION

Before we start with the comparison and interpretation of the results, let us first recall the relaxation principle.

III.1 Relaxation

The basic idea is to optimize the operation of one reserve i assuming the control laws of the other reserves are known. Therefore we proceed in two steps :

- Firstly, we calculate the Bellman values $V_i^t\ (X_i^t)$ corresponding to reserve i operation, by solving the following Bellman's equation :

$$V_i^t\ (X_i^t) = \underset{U_i^t}{\mathrm{Min}} \quad \underset{(D^t,\ \sum_{j\neq i} U_j^t, T^t, a^t)}{E} \left[C^t\ (D^t - \sum_{j\neq i} U_j^t - U_i^t + V_i^{t+1}\ (X_i^{t+1})\right]$$

with $X_i^{t+1} = X_i^t + a_i^t + \rho\ (U_i^t)^- - (U_i^t)^+ - d_i^t$

and where $\sum_{j\neq i} U_j^t$ is the sum of the independent generation laws of the reserves $j\neq i$.

- Secondly, we simulate the operation of reserve i, using the kWh values $- \partial V_i{}^t / \partial X_i\ (X_i{}^t)$. This produces a local feedback which is the best with regard to the mathematical expectation of the demand and the thermal availability. Thus, we get the corresponding control law $U^t{}_i$ (it turns out to be a deterministic variable if there are no random inflows).

We then move on to the following reserve, using this new control law in order to determine the remaining demand to be faced by reserve i and the thermal facilities ; at each step of iteration, the total operating cost decreases, converging to the optimal cost.

III.2 The results and interpretation

We considered the electrical system characteristics of the year 2000. The yearly consumption amounts to 515 TWh and the peak demand rises to 90 GW.

By operating the French system, firstly with the MNR and secondly with the relaxation which works out a local feedback (with no reaction upon the coupling random phenomena), we obtain the following results :

		MNR (global feedback)	Relaxation (local feedback)
Mathematical expectation of the operating cost (MFF)		30 905	59 295
CPU time on IBM 3081	optimization	94 s	53 s
	simulation	138 s	-

This spectacular result shows a cost obtained by the relaxation almost twice greater than the cost issuing from the MNR model. It obviously illustrates for the French electric system operation the need to avoid a local, even optimum, feedback. One will easily agree that, within the relaxation, each reserve cannot react upon the coupling random phenomena ; this restriction widely contributes to the mediocrity of the results.

It is the reason why we used the kWh values provided by the relaxation after convergence, in order to work out a global simulation of the system, in strategy upon all the random phenomena. Thus we deviate from the classical framework of the relaxation, due to the fact, that as in the case of the MNR, we assume these values are an accurate approximation of the optimum values.

We get the following results :

	Global feedback		
	MNR (1)	relaxation (2)	Gap (2)/(1)
Mathematical expectation of the operating cost (MFF) divided into	30 905	34 792	+ 12.6 %
Fuel cost (MFF)	28 878	32 815	+ 13.6 %
Outage cost (MFF)	2 027	1 977	- 2.5 %

The results are appreciably closer to the criterion value produced by the MNR, owing to the only fact the reserves are this time enabled to react altogether to the random perturbations of demand and thermal availability, and on the only ground of the kWh values given by an optimization, which uses a local feedback.

A brief analysis of kWh values shows clearly the values obtained by means of the relaxation using a local feedback, are greater than the values issuing from the MNR. They are close to, or exceed, the highest fuel costs. The important sub-optimality, characterizing the global simulation from the relaxation, shows these values, highly and quite always, overestimate the mathematical expectation of the future savings on fuel and outage, related to a marginal kWh available in the reserve.

In fact, this overestimation comes from two differents items, whose effects add to each other :

- on the one hand, the operation optimization of one reserve using a local feedback, i.e with no reaction ot the demand and thermal availability, leads systematically, for the time-scales we are interested in, to kWh values slightly larger than the values we would obtain by means of an optimization based on a feedback upon the coupling random perturbations (15),
- on the other hand, because of the relaxation process itself, each reserve has to face a highly random system ; this leads to a more careful operation, which can only be provided by kWh values greater than the optimal values.

The gap thus revealed between the MNR and the global simulation after relaxation is too high (not far from 13 %) to allow the use of such a method, notwithstanding its obvious advantages in regard of implementation.

CONCLUSION

Let us underline the principal result of these experiences : the method used in the MNR clearly proves to be the most suitable method to the optimization of the French system operation.

Indeed, the slight gap to the optimum we had measured (0.8 %), ensured the quality of the global feedback implemented in the aggrepation-decomposition method ; however, the different results we obtained and the interpretation we made, do justify the "a priori" analysis which led us to choose that particular method.

REFERENCES

(1) P. LEDERER, Ph. TORRION, J.P. BOUTTES — Un Feedback Global pour la planification du parc de Production Electrique Français, 6th International Conference on Analysis and Optimization of Systems, JUIN 1984.

(2) B. LESCOEUR, Ph. PENZ — Optimum Control of an Electrical Energy Generation and Consumption System CIGRE, SEPTEMBRE 1982.

(3) Ph. TORRION — Régulation du Système Offre-Demande d'Electricité, un Nouvel Outil d'Analyse : "La Maquette" Rapport d'activité EEG (EDF), 1984.

(4) F. MESLIER — Simulation of the Optimum Control of a Power System : Latest Developments in the Greta Model PSCC, 1978.

(5) P. COLLETER — Optimal Management of an almost purely Hydro System : The Ivory Coast Case PSCC, JUILLET 1983.

(6) P. COLLETER, H. MERCIER — Modèle de Gestion Annuelle des Réservoirs Hydrauliques et des Réacteurs Pressurisés Note Interne EDF, JUIN 1984.

(7) P. SANDRIN, M. JEROSOLIMSKI — A propos des Modèles de Gestion du Système Production-Consommation dans le SGEP Annuel Note Interne EDF, MARS 1985.

(8) P. COLLETER, P. LEDERER — Optimal Operation Feedbacks for the French Hydro-Power System, CORS-TIMS, ORSA, 1981.

(9) S. GAL — Optimal Management of a Multi Reservoir Water supply System
Water Resources Research, 1979.

(10) JM. BOURGUEUIL — Méthode de Valorisaiton d'un Equipement Hydroélectrique par Simulation et utilisant la Notion de Valeur de l'Eau Note Interne EDF, SEPTEMBRE 1982.

(11) C. HERY — Paramétrisation de la Valeur d'Usage Marginale d'un Stock : Application à la Gestion des Capacités d'Effacement de la Pointe Note Interne EDF, JUILLET 1985.

(12) THEOSYS — Commande Optimale de Systèmes Stochastiques RAIRO Automatique/Systems Analysis and Control, 1984.

(13) A. TURGEON — Optimal Operation of Multi reservoir Power Systems with Stochastic Inflows Water Resources Research, AVRIL 1980.

(14) P. LEDERER, Ph. TORRION, JP. BOUTTES — Advances in Multi-Type, Multi Time-Scale, Multi-Reservoir Management Problems IFORS, AOUT 1984.

(15) J. LEVEUGLE, Ph. TORRION — Valeurs du kWh en Stock et Monotones de Coûts Marginaux Note Interne EDF, à paraître.

SEMI-MARKOV ANALYSIS OF THE INHOMOGENEOUS MACHINE INTERFERENCE MODEL

József Tomkó
Budapest/Hungary

ABSTRACT. The aim of the present paper is to treat the inhomogeneous machine interference model by the Markov Renewal Theory. This approach enables one to find the steady-state characteristics of the model without getting involved in analytical questions like smoothness of stationary distribution (cf. theorem 2.1. in [1]). The formulae (2.2) for expected sojourn times in a subset of state space allows an elegant treatment of expected waiting times and busy period length.

1. INTRODUCTION.

The machine interference model that we are going to analyse consists of n machines and a single repairman. The machines may break down in which case they are repaired and put back into operation. If the repairman is busy a broken machine has to wait for service causing a queue to bild up in front of the repairman. In the inhomogeneous case each machine has own operation and repair time distribution. Assume that the operation times are exponentially distributed with parameter λ_i, for the i-th machine, while the repair times have general distribution function F_i, for the i-th machine.

This queueing system plays, recently, an important role in the mathematical modeling of computer systems (networks). For example when n programs (jobs) are executed simultaneously on a multiprogrammed computer, the expected program response time, the average CPU busy period length can be investigated by considering a finite source queueing model in which the jobs play the role of machines, their I/O times are the machine operation times, the CPU times are repair times and the CPU is the repairman.

In this paper we base our analysis on the Markov Renewal Processes with arbitrary state spaces. Thus firstly, in sec.2, we shall list some concepts and results (without proofs) of Semi-Markov Theory. In sec.3. the embedded chain of the model is introduced and an algorithm is given for finding its stationary distribution. Sec.4 is devoted to the main characteristics of the machine interference model. A numerical example is broat up for illustration.

2. MARKOV RENEWAL THEORY. Let $(X_n, T_n;\ n\geq 0)$ be a Markov Renewal Process on a general measurable state space $(E, \mathcal{E})$. The sequence $X=(X_n,\ n\geq 0)$ forms a Markov chain (called the embedded Markov chain). Let $P: E\times\mathcal{E} \to [0,1]$ stands for its transition probability measure. The sequence $(T_n;\ n\geq 0)$ is often referred as sequence of r.v. defined on the chain X, i.e. $T_0, T_1, \ldots, T_k, \ldots$ r.v. independent if the realization of X, $x_0, x_1, \ldots$ is given. Let us denote by $F_{xy}, (x,y\in E)$ the distribution function of T_n provided $X_{n-1}=x$, $X_n=y$. The semi-Markov kernel of $(X_n, T_n;\ n\geq 0)$ is now

$$Q(x, F\times B) = P(X_n\in F,\ T_n\in B \mid X_{n-1}=x)$$

$$= \int_F P(x,dy) \int_B dF_{xy}(t) \quad ,\ (F\in\mathcal{E}, B\in\mathcal{B}^+).$$

Insted of $Q(.,.x.)$ it is more convenient to work with its Laplace-transform operator denoted by $Q[s]$: $\forall f\in b\mathcal{E}$

$$Q[s]f(x) = \int_E P(x,dy) f(y) \int_0^\infty e^{-st} dF_{xy}(t)$$

$$= E(f(X_1) e^{-sT_1} \mid X_0 = x).$$

The n-times interation of $Q[s]$ is

$$Q^n[s]f(x) = E(f(X_n) e^{-s(T_1+\ldots+T_n)} \mid X_0 = x).$$

Associated with the M.R.P. $(X_n, T_n;\ n\geq 0)$ is the semi-Markov process $(\xi_t,\ t\geq 0)$ defined as $\xi_t = X_{N_t}$ for $t\geq T_0$ and some ξ_0 for $0\leq t<T_0$, where

for $t\geq T_0$
$$N_t = \max\{n\geq 0;\ \sum_1^n T_k = S_n \leq t\}.$$

It seems quite frequently natural to assume

Condition (*).

(i) the embedded chain X is Harris and possessed of a stationary measure π;

(ii) there exists $\forall x\in E$ $m_x = E(T_1 \mid X_0 = x)$ and $0 < d = \int_E \pi(dx) m_x < \infty$.

The distribution of ξ_t, $P(\xi_t\in A), (A\in\mathcal{E})$ depends on the initial values (ξ_0, X_0, T_0).

Proposition 1. (Stationarity). Suppose the condition (*) holds and $(I-Q[s])^{-1}$ exists $\forall s>0$. Then if ξ_0, X_0, T_0 are chosen accordingly to the measure

$$P(\xi_0\in A, X_0\in B, T_0\geq t) = \frac{1}{d} \int_A \pi(dx) \int_B P(x,dy) \int_t^\infty [1-F_{xy}(u)]du$$

the probability $P(\xi_t \theta A)$ independent of t and equals to

$$\mu(A)=\frac{1}{d}\int_A \pi(dx)m_x \quad , \quad (A\theta E). \tag{2.1}$$

What concerns the existence of $(I-Q[s])^{-1}$ one can find more about it in [3].

In the application fields we face frequently sojourn times the process ξ_t spents in a subset $H\subset E$. In stationary case the consecutive sojourn times in H form a stationary sequence. Denote the corresponding expected sojourn time in H by W_H. A long-run (ergodic) argument leads to

Proposition 2. Suppose $(X_n, T_n;\ n\geq 0)$ satisfies condition (*). Let $H\theta E$ be such that $\pi(H)\pi(\overline{H})>0$. Then we have

$$W_H/(W_H+W_{\overline{H}})=\mu(H), \tag{2.2}$$

$$W_H+W_{\overline{H}} = d/P(\overline{H}\to H), \tag{2.3}$$

where $P(\overline{H}\to H)=\int_{\overline{H}}\pi(dx)P(x,H)$.

3. THE EMBEDDED CHAIN.

Let the random times τ_k, $(k\geq 0,\ \tau_0=0)$ denote the consecutive transition epochs in our machine interference model. These instants are break down or repair completition times. For $k\geq 0$ define

κ_k - number of broken machines at τ_k+0,

ξ_k - repair time elapsed up to τ_k,

$\nu_1^{(k)},\dots,\nu_{\kappa_k}^{(k)}$ - indicies of broken machines in order of their failure. Assume the repair principle is FIFO. The embedded Markov chain of our machine interference model is the sequence

$$(X_k, k\geq 0) = \{\ (\kappa_k,\ \xi_k,\quad \nu_1^{(k)},\dots,\ \nu_{\kappa_k}^{(k)}\ );\ k\geq 0\ \}. \tag{3.1}$$

The state space E is the collection of half lines

$\{\ x,(i_1,\dots,i_k);\ x\geq 0\}$, $0\leq k\leq n$ and $(i_1,\dots,i_k)$ runs over V_k^n - the all variation of order k without repetition of indicies $\{1,2,\dots,n\}$. For k=0 the half line is just the point $\{0\}$. The transition probabilities are as follows:

$$P(\xi_{i+1}>y,\ \nu_1^{(i+1)}=j_1,\dots,\ \nu_{k+1}^{(i+1)}=j_{k+1}|X_i=(x,j_1,\dots,j_k)) \tag{3.2}$$

$$= \frac{\lambda}{1-F_{j_1}(x)} \int_y^\infty e^{-\lambda(n-k)(z-x)}[1-F_{j_1}(z)]dz, \quad (y \geq x \geq 0);$$

$$P(\xi_{i+1}=0, \nu_1^{(i+1)}=j_2, \ldots, \nu_{k-1}^{(i+1)}=j_k \mid X_i=(x, j_1, \ldots, j_k)) \tag{3.3}$$

$$= \int_x^\infty \frac{dF_{j_1}(z)}{1-F_{j_1}(x)} e^{-\lambda(n-k)(z-x)}, \quad (k \geq 1,\ x \geq 0);$$

$$P(\xi_{i+1}=0, \nu_1^{(i+1)}=j_1 \mid X_i=0) = \frac{1}{n} . \tag{3.4}$$

By some deeper inspection of the chain (3.1) one can see that the point {0} is a positive recurrent state. Hence no any doubt as to the existence of the unique stationary measure for the embedded chain. Let us denote the stationary probabilities of the initial points of lines by

$$\pi(0),\ \pi(0, i_1, \ldots, i_k),\ ((i_1, \ldots, i_k) \in V_k^n,\ 1 \leq k \leq n).$$

The stationary measure arisen on the line $\{x, (i_1, \ldots, i_k);\ x \geq 0\}$ denote by $\pi(B;\ i_1, \ldots, i_k)$ and suppose it has density

$$\pi(x; i_1, \ldots, i_k)dx = \pi(dx; i_1, \ldots, i_k).$$

A rigorous reader can easily justify this assumption. The stationarity equations become simpler in terms of functions

$$\pi^*(x; i_1, \ldots, i_k) = \frac{\pi(x; i_1, \ldots, i_k)}{1-F_{i_1}(x)} .$$

For further simplification assume $\lambda_i = \lambda\ \forall\ 1 \leq i \leq n$ and put

$$m_i = \int_0^\infty (1-F_i(x))dx,\quad \phi_i(s) = \int_0^\infty e^{-su} dF_i(u),\quad (1 \leq i \leq n) .$$

Now the stationary equations are:

$\forall\ x>0,\ k>2$

$$\pi^*(x; i_1, \ldots, i_k) = \int_0^\infty \pi^*(x-u; i_1, \ldots, i_{k-1}) \lambda e^{-\lambda(n-k+1)u} du \tag{s.1}$$

$$+\pi(0,i_1,\ldots,i_{k-1})\lambda e^{-\lambda(n-k+1)x} ;$$

$\forall\ x>0,\ 1\leq i_1,i_2\leq n$

$$\pi^*(x;i_1,i_2)=\pi(0,i_1)\lambda e^{-\lambda(n-1)x} \tag{s.2}$$

$\forall\ 1\leq k\leq n$

$$\pi(0,i_1,\ldots,i_k)=\frac{1}{n}\pi(0)I_{\{k=1\}} \tag{s.3}$$

$$+\sum_{i\neq i_1,\ldots,i_k}\{\int_0^\infty \pi^*(x;i,i_1,\ldots,i_k)\int_x^\infty e^{-\lambda(n-k+1)(z-x)}dF_i(z)$$

$$+\pi(0,i,i_1,\ldots,i_k)\phi_i(\lambda(n-k+1))\}.$$

For the Laplace-transform

$$q(s,i_1,\ldots,i_k)=\int_0^\infty e^{-su}\pi^*(u,i_1,\ldots,i_k)du$$

equations (s.1),(s.2) yields

$$q(s,i_1,\ldots,i_k)=\frac{\lambda}{\lambda(n-k+1)+s}(I_{\{k>2\}}q(s,i_1,\ldots,i_{k-1})+\pi(0,i_1,\ldots,i_{k-1})).$$

By repeated substitutions we obtain

$$q(s,i,\ldots,i_k)=\sum_{j=1}^{k-1}\pi(0,i_1,\ldots,i_j)\prod_{r=1}^{k-j}\frac{\lambda}{\lambda(n-k+r)+s}.$$

For integers $\ell\geq 1$ and $m\geq 0$

$$\prod_{r=0}^{m}\frac{\lambda}{\lambda(\ell+r)+s}=\sum_{r=0}^{m}\frac{\lambda}{m!}(-1)^r\binom{m}{r}\frac{1}{\lambda(\ell+r)+s},$$

thus

$$q(s,i_1,\ldots,i_k)=\sum_{j=1}^{k-1}\pi(0,i_1,\ldots,i_j)\frac{\lambda}{(k-j-1)!}\sum_{r=1}^{k-j}(-1)^{r-1}\frac{1}{\lambda(n-k+r)+s}\binom{k-j-1}{r-1}.$$

Consequently

$$\pi^*(x;i_1,\dots,i_k) \tag{3.6}$$

$$= \sum_{j=1}^{k-1} \pi(0,i_1,\dots,i_j)\frac{\lambda}{(k-j-1)!}\sum_{r=1}^{k-j}(-1)^{r-1}e^{-\lambda(n-k+r)x}\binom{k-i-1}{j-1}.$$

Concerning the probabilities $\pi(0,i_1,\dots,i_k)$ note that they are proportional to those for the restriction of $(X_n, n\geq 0)$ to the subset consisting only the initial points $\{0\}$, $\{0,i,\dots,i_k\}$ of the lines of E. This restriction is a finite (state) chain with transition probabilities:

$$P(\{0\}\to\{0,i\})=\frac{1}{n}$$

$$P(\{0,i_0,i_1,\dots,i_r\}\to\{0,i_1,\dots,i_k\}) = \frac{1}{(k-\ell)!}\int_0^\infty e^{-\lambda(n-k-1)}(1-e^{-x})^{k-\ell}dF_{j_0}(x),$$

$$P(\{0,j\}\to\{0\})=\int_0^\infty e^{-\lambda(n-1)x}dF_j(x)=\phi_j(\lambda(n-1)).$$

For the stationary probabilities $\tilde{\pi}(0,i_1,\dots,i_k)$ we have

for $2\leq k\leq n$

$$\tilde{\pi}(0,j_1,\dots,j_k) \tag{3.7}$$

$$= \sum_{j_0\neq j_1,\dots,j_k}\sum_{\ell=0}^{k}\tilde{\pi}(0,j_0,\dots,j_\ell)\frac{1}{(k-\ell)!}\int_0^\infty e^{-\lambda(n-k-1)x}(1-e^{-x})^{k-\ell}dF_{j_0}(x),$$

$$\tilde{\pi}(0,j)=\frac{1}{n}\tilde{\pi}(0)+\sum_{k\neq j}\tilde{\pi}(0,k,j)\phi_k(\lambda(n-2)), \tag{3.8}$$

$$\tilde{\pi}(0)=\sum_{j=1}\tilde{\pi}(0,j)\phi_j(\lambda(n-1)). \tag{3.9}$$

It is not likely that these equations can be solved explicitly. We describe now an algorithm on the base of which the solution in particular cases can be obtained on computers.

Let Y_k be the $\overline{V}_k^n=\binom{n}{k}k!$ dimensional vector with components

$\tilde{\pi}(0,i_1,\ldots,i_k)$ listed in lexicografical order. Equations under (3.7) turn now into equations

$$Y_k = \sum_{j=1}^{k+1} A_j^k Y_j \;, \quad (2\leq k\leq n-1)$$

$$Y_{n-1} = \sum_{j=1}^{n-1} A_j^{n-1} Y_j, \tag{3.10}$$

with $\bar{V}_k^n \times \bar{V}_j^n$ matrices A_j^k which entries one can pick out from (3.7). The matrix forms of equations (3.8) and (3.9) are

$$Y_1 = \frac{1}{n}\,\tilde{\pi}(0)\begin{pmatrix}1\\ \vdots\\ 1\end{pmatrix} + A_2^1 Y_2 \;,$$

$$\tilde{\pi}(0) = A_1^0 Y_1 \;.$$

Introduce now the $\bar{V}_k^n \times \bar{V}_j^n$ matrices F_j^k by putting $F_j^n = 0 \;\forall\; 1\leq j\leq n$ and

$$F_j^k = (I-(A_k^k + A_{k+1}^k F_k^{k+1}))^{-1}(A_j^k + A_{k+1}^k F_j^{k+1}) \;\forall\; 1\leq j<k<n \;.$$

Then we have

$$Y_k = \sum_{1}^{k-1} F_j^k Y_j \;, \quad (k\geq 2)$$

and

$$Y_1 = (I-A_2^1 F_1^2)^{-1}\begin{pmatrix}1\\ \vdots\\ 1\end{pmatrix}\frac{1}{n}\tilde{\pi}(0)\,.$$

Starting with a fixed value for $\tilde{\pi}(0)$ (e.g. $\tilde{\pi}(0)=1$) one determines Y_k, $(1\leq k\leq n)$ successively. Note that it is not necessary to fulfil normalization.

4. THE MAIN PERFORMANCE MEASURES.

Consider now the M.R.P. $(X_n, T_n; n\geq 0)$ with embedded chain X introduced under (3.1) and with $T_0=0$, $T_n=\tau_n-\tau_{n-1}$, $(n>1)$. Let $(\xi_t, t\geq 0)$ be the associated semi-Markov process. For the stationary distribution of ξ_t we have to know the expected sitting times

$$m_{(x;\, i_1,\ldots,i_k)} = E\,(T_k \mid X_{k-1}=(x,i_1,\ldots,i_k))$$

They are:

$$m_{\{0\}} = \frac{1}{n\lambda} \ , \quad m_{\{0,i_1,\dots,i_k\}} = \frac{1-\phi_{i_1}(\lambda(n-k))}{\lambda(n-k)} \ ,$$

$$m_{\{x,i_1,\dots,i_k\}} = \frac{e^{\lambda(n-k)x}}{1-F_{i_1}(x)} \int_x^\infty e^{-\lambda(n-k)z}(1-F_{i_1}(z))dz \ .$$

Consider now the integral

$$\int_0^\infty \pi(x,i_1,\dots,i_k)\ m_{\{x,i_1,\dots,i_k\}}dx. \tag{4.1}$$

By the virtue of (3.6) this integral is a linear combination of expressions like

$$\int_0^\infty e^{-\lambda ix}\int_x^\infty e^{-\lambda(n-k)z}\ (1-F_{i_1}(z))dzdx$$

$$= \frac{1}{\lambda i}\ (\frac{1-\phi_{i_1}(\lambda(n-k))}{\lambda(n-k)} - \frac{1-\phi_{i_1}(\lambda(n-k+i))}{\lambda(n-k+i)}) \ .$$

Let us now denote by $P_{i_1,\dots,i_k}$ the stationary probability of the line $\{x;(i_1,\dots,i_k);\ x\geq 0\}$ i.e. the probability, that machine i_1 is under repair and machines $i_2,\dots,i_k$ are broken down in this order and waiting for repair.

We have with some $d>0$, that

$$d\cdot P_{i_1,\dots,i_k} = \int_0^\infty \hat{\pi}(x,i_1,\dots,i_k)m_{\{x,i_1,\dots,i_k\}}dx$$

$$+\hat{\pi}(0,i_1,\dots,i_k)\ \frac{1-\phi_{i_1}(\lambda(n-k))}{\lambda(n-k)} \ , \qquad (k\geq 1)\ ;$$

$$d\cdot P_0 = \hat{\pi}(0)\ \frac{1}{n\lambda} \ .$$

Here the $\hat{\pi}(\dots)$ quantities are calculated by using the components of the vectors Y_k $(k\geq 1)$. After normalization we get the proper probabilities $P_{i_1,\dots,i_k}$.

Utilization can now be considered for individual machines and for the repairman. The time during which the i-th machine is out of operation is a sojourn time spent by our semi-Markov process $(\xi_t, t\geq 0)$ in the subset H_i-the collection of all half lines with variation $(i_1,\ldots,i_k)$ such that $i_r=i$ for some $1\leq r\leq k$, $(1\leq k\leq n)$. Let $P^{(i)}$ denote the stationary probability that machine numbered by i is out of operation, i.e.

$$P^{(i)}=\mu(H_i)=\sum_{k=1}^{n}\sum_{(i_1,\ldots,i_k)\ni i} P_{i_1,\ldots,i_k}.$$

The formulae (2.2) enables one to determine the expected time W_i that machine numbered by i is out of operation. As the expected operating time is $1/\lambda$ we have

$$\frac{W_i}{W_i + 1/\lambda} = P^{(i)}.$$

The expected time that the i-th machine has to wait before receiving repair is now $\hat{W}_i=W_i- m_i$.

The utility for the repariman is $1-P_o$ and one determines the corresponding expected busy period length from the relation

$$W_o/(W_o+ 1/\lambda)= 1-P_o.$$

For illustration consider the case when $n=5$, $\lambda=0.08$ and the individual repair time distribution functions are of mixtures of Erlang's distributions $E(n,\mu;x)$ with density

$$\mu\frac{(\mu x)^{n-1}}{(n-1)!}e^{-\mu x}.$$

$F_1(x)=0,4E(2;0,3;x)+0,6E(3;0,25;x)$,

$F_2(x)=0,2E(1;0,6;x)+0,5E(2;0,35;x)+0,3E(4;0,171;x)$,

$F_3(x)=E(3;0,25;x)$,

$F_4(x)=0,3E(1;0,4;x)+0,5E(2;0,170;x)+0,2E(3;0,182;x)$,

$F_6(x)=0,2E(1;0,8;x)+0,3E(2;0,26;x)+0,4E(1;0,168;x)+$
$+0,1E(3;0,65;x)$.

The main characteristics are

$$W_o=450.316$$

$\tilde{W}_1=35.97$, $\tilde{W}_2=35.03$, $\tilde{W}_3=37.11$, $\tilde{W}_4=35.38$, $\tilde{W}_5=33.42$.

Remark. For homogeneous case our result leads to Takács's formulas (cf. [2] , ch. 5, theorem 2. and formulae (34)).

REFERENCES

1. Sztrik, J., Tomkó, J. (1982). Multiprogramozás inhomogén programokkal. Alk. Matematikai Lapok, 8, 285-296.
2. Takács, L. (1962). Introduction to the Theory of Queues. Oxford University Press.
3. Tomkó, J. (1986). Renewal method in the Theory of Semi-Markov Processes on arbitrary spaces. To appear in the Proc. of the fourth Vilnius Conf. on Probab. and Math. Stat. VNU Science Press, Netherlands.

Prof.J. Tomkó,
Budapest University of Technology,
Mechanical Eng. Faculty,
Department of Mathematics,
XI. Müegyetem rkp. 3.
Budapest H-1521

LYAPUNOV FUNCTIONS AND TURNPIKE THEORY

Anatoly D. TSVIRKUN, Sergey Yu. YAKOVENKO
Institute of Control Sciences,
Moscow, USSR

1. INTRODUCTION

This paper is devoted to investigation of the asymptotic properties of optimal trajectories in models of economic dynamics. The main result can be stated as follows: it is possible to distinguish a subset of the phase space such that during the most part of the planning period any optimal trajectory remains close to this subset. This property of dynamical models (known as a turnpike property) has been extencively investigated for discrete time technological models by V.L. Makarov, A.M. Rubinov etc. (see survey [1]) and for optimal consumption models by L.McKenzie, D. Cass, K. Shell etc. [2-4].

In this paper we will deal only with continuous time technological models, that is, no criterion of optimization is a priori fixed except the Pareto-optimality (for precise formulation see below). The turnpike property can be formulated in several different forms (weak, strong, early, etc.). Our method based on constructing Lyapunov functions allows us to prove the corresponding theorems universally. We deal only with finite horizon optimization problems; some results on the convergence of infinite optimal consumption trajectories were obtained by using Lyapunov functions in [2-4]. The used function does not fit to technological models, so one is to look for another function to play the required role. Such a function is constructed below for stationary and variable technology cases.

We are grateful to S.A. Ashmanov, A.M. Rubinov and A.V. Golubkin for helpful discussions.

2. STATEMENT OF THE PROBLEM

Let $\mathbb{R}^n_+ = \{x : x = (x^1, \dots, x^n),\ x^i \geqslant 0\}$ be the phase space of the model endowed with the natural affine structure, usual Pareto order ($x \geqslant y \Leftrightarrow \forall i\ x^i \geqslant y^i$, $x > y \Leftrightarrow \forall i\ x^i > y^i$) and weak partial order ($x \succ y \Leftrightarrow \exists \lambda \in R_+ : \lambda x \geqslant y$).

Define a family of closed convex cones $K_t \subseteq \mathbb{R}^n_+ \times \mathbb{R}^n$, $t \in \mathbb{R}$, such that for every $t \in \mathbb{R}$ and every $x \in \mathbb{R}^n_+$ the set $K_t(x) = \{v \in \mathbb{R}^n : (x, v) \in K_t\}$ is non-empty, $K_t(0) = \{0\}$, $\forall x\ \exists \lambda_t > -\infty$:

$\lambda_t x \in K_t(x)$, $\forall\, v_1, v_2 \in K_t(x)$ $\forall\, v:\ v_1 \geqslant v \geqslant v_2 \Rightarrow v \in K_t(x)$.

An absolutely continuous trajectory $x(t)$, $t \in [t_0, t_1]$, is called technologicaly admissible, if for almost all t $\dot{x}(t) \in K_t(x(t))$. When the family K_t is independent of t, then it is said that the technology is constant (otherwise variable). From now on up to the last section we will deal only with the constant case, $K_t \equiv K$.

There can be formulated different kinds of optimization problems. Given an initial point $x_0 \geqslant 0$ and concave positive homogeneous function $V(x)$ we can look for the admissible trajectories satisfying

$$\dot{x} \in K(x), \quad x(0) = x_0, \tag{1}$$

which are optimal with respect to the terminal criterion:

$$V(x(T)) \longrightarrow \max, \quad t \in [0, T]. \tag{2}$$

Here T stands for the horizon of planning. On the other hand, we can consider Pareto-optimal trajectories in $\mathbb{R}^n_+$ with respect to terminal state: the trajectory $x(t)$ is called Pareto-optimal if for every admissible trajectory $y(t)$

$$y(0) = x(0), \quad y(T) \geqslant x(T) \Rightarrow y(T) = x(T). \tag{3}$$

Let $d(x,y) = |x \cdot |x|^{-1} - y \cdot |y|^{-1}|$ be an angular distance in $\mathbb{R}^n_+$. Denote the set of all solutions of the problem (1,2) (resp., (1,3)) by $X_T(V)$ (resp., X_T(Par)).

D e f i n i t i o n. Let X_T be any of the sets $X_T(V)$, X_T(Par). The closed cone $N \subseteq \mathbb{R}^n_+$ is a weak attractor for X_T if

$$\forall \varepsilon > 0\ \exists \tau = \tau(\varepsilon, x_0): \forall x(\cdot) \in X_T \quad \text{mes}\{t: d(x(t), N) \geqslant \varepsilon\} \leqslant \tau, \tag{4}$$

and a strong attractor if

$$\forall \varepsilon > 0\ \exists \tau: \forall x(\cdot) \in X_T \quad \forall t \in [\tau, T-\tau] \quad d(x(t), N) \leqslant \varepsilon. \tag{5}$$

Note, that an estimate $\tau(\varepsilon)$ is independent of T, but depends generally on x_0, V. In the case of the problem (1,3) it is said that N is a Pareto-attractor.

N o t e. The set X_T(Par) can be represented as the union of the sets $X_T(p^0)$ corresponding to the problem (1,2) with $V(x) = \langle p^0, x \rangle$ (scalar product), with p^0 ranging over the set $\mathbb{R}^{n*}_+$ of non-negative linear functionals. The Pareto-attraction can be deduced from that of $X_T(p^0)$ if the estimate in (4,5) can be chosen uniformly in $p^0 \in \mathbb{R}^{n*}_+$, $p^0 \neq 0$

Naturally, the whole phase space $\mathbb{R}^n_+$ is an attractor; we are looking for a narrower one.

3. NECESSARY CONDITIONS FOR OPTIMALITY

Theorem [5]. If $x(\cdot)\in X_T(V)$, then there is an adjoint trajectory $p(\cdot)$ such that $\forall t\in[0,T]$ $p(t)\geqslant 0$,

$$\dot{x}\in \partial_p H(x,p), \qquad \dot{p}\in \partial_x(-H(x,p)), \tag{6}$$

and

$$x(0)=x_0, \quad -p(T)\in \partial_x(-V(x(T))). \tag{7}$$

Here $H(x,p)$ is a Hamiltonian function defined for $x\in R^n_+$, $p\in R^{n*}_+$ by the formula

$$H(x,p)=\sup_v\{\langle p,v\rangle : v\in K(x)\},$$

and ∂_x, ∂_p stands for subdifferentials. Note (see [6]), that H is convex in p , concave in x , and positive homogeneous in both variables independently.

We prove the turnpike property for all solutions to the boundary value problem (6,7); the turnpike for the initial optimization problem is obtained by projecting onto x -space. Note, that right hand parts of the inclusions (6) are multivalued.

4. LYAPUNOV FUNCTION AND DIFFERENTIAL INCLUSIONS

Let M be any smooth manifold, and consider the following boundary value problem for the differential inclusion:

$$\dot{x}\in F(x), \quad t\in[0,T], \tag{8}$$

$$x(0)\in M_0, \quad x(T)\in M_1, \quad M_i\subseteq M. \tag{9}$$

The set of solutions is again denoted as X_T .

Definition. Smooth function $f: M\to R$ is called Lyapunov function, if $\forall x\in M$ $\forall v\in F(x)$ $\langle df(x),v\rangle\leqslant 0$.

The set $N\subseteq M$ is called neutral (or more precisely f-neutral) if

$$\forall \varepsilon>0 \ \exists\delta>0: \ d(x,N)\geqslant\varepsilon, \ v\in F(x) \Rightarrow \langle df(x),v\rangle\leqslant -\delta$$

(d stands for distance in M).

Theorem 1. If the differential inclusion (8) possesses the Lyapunov function, f , such that $\sup f|_{M_0}<+\infty$, $\inf f|_{M_1}>-\infty$, then the f -neutral set N is a weak attractor to the set X_T of solutions to the problem (8,9) in terms of (4).

If, in addition, there is a limit $c=\lim f(x)$ when $x\to N$, i.e. $\forall\varepsilon>0 \quad \exists\delta>0: \ d(x,N)<\delta\Rightarrow |f(x)-c|<\varepsilon$,

then N is a strong attractor to the problem (8,9).

The general course of reasoning is the following: for the inclusion (6) one has to find the Lyapunov function, f , describe the

ρ- neutral set, N , and verify the boundedness conditions and, if needed, the limit condition. Projecting N to the x -space, one can obtain an attractor for the initial optimization problem.

5. THE TURNPIKE IN AUTONOMOUS CASE

L e m m a 1 . There exist $x_* \in \mathbb{R}^n_+$, $p^* \in \mathbb{R}^{n*}_+$ and $\alpha \in \mathbb{R}$ such that $(x_*, \alpha x_*) \in K$, and $\langle p^*, v\rangle \leq \alpha \langle p^*, x\rangle$ for all $(x,v) \in K$.

Denote

$$N = \{ x \in \mathbb{R}^n_+ : H(x,p^*) = \alpha \langle p^*, x\rangle \},$$

$$N^* = \{ p \in \mathbb{R}^{n*}_+ : H(x_*, p) = \alpha \langle p, x_*\rangle \}.$$

The set N is the projection of the von Neumann facet onto x-space, N^* is a dual object (co-facet). Note, that N, N^* are non-empty closed cones.

L e m m a 2. Let $\frac{\partial H}{\partial x}(x,p)$, $\frac{\partial H}{\partial p}(x,p)$ be subgradients, i.e. elements of the corresponding subdifferentials, and d be an angular distance. Then the following inequalities hold:

$$H(x_*,p) \geq \alpha \langle p, x_*\rangle , \quad H(x,p^*) \leq \alpha \langle p^*, x\rangle,$$

$$\forall \varepsilon > 0 \ \exists \delta > 0: \quad d(p,N^*) \geq \varepsilon \Rightarrow H(x_*,p) \geq \alpha \langle p, x_*\rangle + \delta |p|,$$

$$d(x,N) \geq \varepsilon \Rightarrow H(x,p^*) \leq \alpha \langle p^*, x\rangle - \delta |x|,$$

$$\langle p^*, \frac{\partial H}{\partial p}(x,p)\rangle \leq H(x,p^*), \quad \langle \frac{\partial H}{\partial x}(x,p), x_*\rangle \geq H(x_*,p),$$

$$\langle p, \frac{\partial H}{\partial p}(x,p)\rangle = \langle \frac{\partial H}{\partial x}(x,p), x\rangle = H(x,p) .$$

C o r o l l a r y. The set $\hat{M} = \{ (x,p) : \langle p,x\rangle > 0\}$ is invariant under the admissible trajectories of (6); and $\langle p,x\rangle$ is constant along these trajectories. The function $D(x,p) = \langle p^*, x\rangle \cdot \langle p, x_*\rangle \cdot \langle p,x\rangle^{-1}$ is the Lyapunov function for the inclusion (6).

The lemma is an easy consequence of the convexity/concavity of H and definition of x_*, p^* . The corollary is obtained by differentiating D along the multivalued vector field (6).

L e m m a 3. If $x_0 \succ x_*$, then $D|_{M_0}$ is bounded from above; $D|_{M_1} \geq 0$ and if $p^0 \prec p^*$, $\langle p^0, x_*\rangle > 0$ then $\log D|_{M_1}$ is bounded from below. If $x_* \succ p^*$, then the lower bound is uniform over $p^0 \in \mathbb{R}^{n*}_+$.

L e m m a 4. If $x_* > 0$ then the set $N \times \mathbb{R}^{n*}_+$ is D-neutral; the set $N \times N^*$ is $\log D$ -neutral for the system (6) restricted on dom $\log D$; if $N = \mathbb{R} \cdot x_*$, $x_* > 0$, or $N^* = \mathbb{R} \cdot p^*$, $p^* > 0$, then there is a limit $\lim D(x,p)$, $(x,p) \to N \times N^*$.

These two propositions are easily verified using the explicit expression for D and the inequalities given by Lemma 2.

Applying theorem 1, one obtains the following results describing asymptotical behavior of solutions to problems of dynamic optimization with the constant technology.

Theorem 2. If $x_* > 0$, $x_0 > 0$ then the von Neumann facet N is a weak Pareto-attractor.

Theorem 3. If $x_0 \succ x_*$, $V(x) = \langle p^0, x \rangle$, $p^0 \prec p^*$ and $\langle p^0, x_* \rangle > 0$, then N is a weak attractor for $X_T(V)$.

Theorem 4 A. If $N = \mathbb{R}x_*$, $x_* > 0$, $x_0 > 0$, $p^0 \prec p^*$, then N is a strong $X_T(p^0)$ -attractor. Moreover, if $p^* > 0$, then N is a strong Pareto-attractor.

Theorem 4 B. If $N^* = \mathbb{R}p^*$, $p^* > 0$, $x_0 \succ x_*$, $\langle p^0, x_* \rangle > 0$, then N is a strong $X_T(p^0)$ -attractor. If $x_* > 0$, then N is a strong Pareto-attractor.

Theorems 3,4 are generated by the Lyapunov function $\log D$. Theorem 4 A is a generalization of Nikaido's turnpike theorem for the continuous time case. Theorem 4 B has a dual form. Note, that the condition $N^* = \mathbb{R}p^*$ implies the uniqueness of the supporting hyperplane for the set $K(x_*)$ in the point αx_* . This can be interpreted as a sensitivity: for any $p \notin \mathbb{R}p^*$ the maximum in the definition of H is obtained somewhere but not on αx_* . Theorem 3 is somewhat analogous to the results due to V.L. Makarov (see [7]).

The above results connect the geometry of the set K with the asymptotical behavior of the extremals, and no assumption is made aboout the general set of all admissible trajectories.

6. DISCUSSION

The method used in the analysis of the autonomous case can be extended in many directions.

When a technology is variable, there is no natural way to define the von Neumann facet. Instead, we can consider two different trajectories of (now non-autonomous) Hamiltonian system, $(x_1(t), p_1(t))$ and $(x_2(t), p_2(t))$. The function $\widetilde{D} = \log\{\langle p_1, x_2 \rangle \cdot \langle p_2, x_1 \rangle \cdot \langle p_1, x_1 \rangle^{-1} \cdot \langle p_2, x_2 \rangle^{-1}\}$ is defined (under some assumptions) and monotone (decreasing). Some conditions of uniform strict convexity/concavity met, one can verify that the diagonal $\{x_1 \in \mathbb{R}x_2\}$ is the $\widetilde{D}$--neutral.

The corresponding turnpike theorem is now stating that any two optimal trajectories of a variable technological model are sufficiently close to each other. If the two trajectories have the same initial point, then there is the so-called early turnpike: these two remain close almost up to the end of the planning period.

Instead of the linear functional $\langle p^*, \cdot \rangle$ in the autonomous case, one can consider an effective (nonlinear) functional $s(x)$ introduced by A.M. Rubinov [8], such that $s(x) > 0$, s is positive homogeneous and for every admissible trajectory $\log s(x(t)) - \log s(x(0)) \leq \alpha t$. There is an analogue of the von Neumann facet, i.e. the set of $x \in \mathbb{R}^n_+$ such that exact equality holds for some admissible path with $x(0) = x$. The set can be narrower than N, and the corresponding Lyapunov function $D_s(x, p) = s(x) \cdot \langle p, x_* \rangle \cdot \langle p, x \rangle^{-1}$ generates a turnpike theorem.

The optimal consumption models can be studied using the above technique with the Lyapunov function $\langle p - p^*, x - x_* \rangle$ (see [3,4]). The method allows to investigate not only infinite, but also finite horizon optimal trajectories. Some other Lyapunov functions can also be constructed when utility function is zero or first degree homogene.

REFERENCES

[1.] Rubinov A.M. Economic dynamics.- In: Contemporary problems in mathematics, vol.19, p.59-110. Moscow, VINITI, 1982. (in Russian)

[2.] McKenzie L. Turnpike Theory.- Econometrica, 1976, vol.44, No.5, p.841-866

[3.] Cass D., Shell K. The Structure and Stability of Competitive Dynamical Systems.- Journal of Economic Theory, 1976, v.12, No.1, p.31-70

[4.] Brock W.A., Scheinkman J.A. Global Asymptotic Stability of Optimal Control Systems with Applications to the Theory of Economic Growth.- Journal of Economic Theory, 1976, v.12, No.1, p.164-190

[5.] Rockafellar R.T. Conjugate Convex Functions in Optimal Control and the Calculus of Variations.- Journ. of Math. Anal. Appl., 1970, v.32, No.1, p.174-190

[6.] Rockafellar R.T. Convex Analysis.- Princeton Univ. Press, 1970

[7.] Makarov V.L. Asymptotic behavior of optimal trajectories in linear economic models.- Siberian Math. J., 1966, v.7, No.4, p.832-853 (in Russian)

[8.] Rubinov A.M. Turnpike Sets in Discrete Disperse Dynamical Systems.- Siberian Math. J., 1980, v.21, No.4, p.136-145 (in Russian)

Elasto-Plastic Analysis of Continuum Structures by Nonlinear Programming

A. VÁSÁRHELYI[x] - M. GRÓSZ[xx]

Technical University

Institute for Building Economy

Abstract

In this paper it is presented how the results of classical elastic finite element method can be used if the problem is solved in the stress space. Both, the elastic and elasto-plastic analysis are formulated as mathematical programming problems. Such a kind of feasible direction method is presented the steps of which have a mechanical meaning as well. A thick curvated shell element was computed by this method.

1. Introduction

To solve this problem there are more methods in literature which state the different kinds of plastic stiffness matrices solving nonlinear equations and determine the rates of stresses and displacements with usual iterative methods. The elaboration of a new method is necessary to get the plastic stiffness matrices in every case, for it is not possible to give a general solution (e.g. [2]).

A new approach is presented here using the finite element method; starting by the basic theorems of plasticity the problem is solved by nonlinear programming without using the plastic stiffness matrices.

It is supposed that the material of the structures is linear elastic and ideal plastic. The plastic yield conditions have a quadratic form according to the Mises-Huber-Henky's theory. The loading history is followed by the step by step method. Such approach rests on the general idea of a priori subdividing of the loading history into a sequence of individual proportional loading stages [5].
It is supposed that the small displacements theory is valid.
Notations used in the paper:

ε - vector of strains

B - transmitted matrix

e - vector of displacements

k - stress-freedom at the Gaussian points

l - displacement freedom of the nodes

σ - vector of stresses

x/ H-1126 Budapest, Kiss János altábornagy u. 34. HUNGARY

xx/ H-1025 Budapest, Csalit u. 9. HUNGARY

D - Hook's matrix
F - domain of the element
b - vector of external loads
ϱ_α - Gaussian weight factor at α-th Gaussian point
i - number of Gaussian points on an element
C - $[\ldots \varrho_\alpha B_\alpha \ldots]$
G - $[\ldots B_\alpha \ldots]$
n - number of nodes
m - number of elements
p - k.i.m
q - l.n
j - index of yield conditions
z - i.m: number of yield conditions
s - number of activ yield conditions
$\varrho(G)$ - range of matrix G

2. Elasticity of continuum structures by nonlinear programming

The continuum structure is divided into finite elements by the usual way. The shape functions are given in a curvilinear coordinate system. Executing the necessary deriviations the connections between the displacements and strains are known in the curvilinear system. Using Jacobi transformation and if it is necessary linear transformations the strains can be expressed in the global Cartesian system of the structure (x,y,z). $\underset{(k)}{\underline{\underline{\varepsilon}}} = \underset{(k\cdot l)}{\underline{\underline{B}}} \cdot \underset{(l)}{\underline{\underline{e}}} + \underset{(k)}{\underline{\underline{\varepsilon}}_o}$ (1)

Its usual deduction can be found in many works e.g. [2,6].
Linear elastic material rule is supposed: $\underline{\underline{\sigma}} = \underline{\underline{D}} \cdot \underline{\underline{\varepsilon}} = \underline{\underline{D}}(\underline{\underline{B}}\underline{\underline{e}} + \underline{\underline{\varepsilon}}_o)$ (2)
The eauilibrium equations of an element are: $-\int_F \underline{\underline{B}}^* \underline{\underline{\sigma}} dF + \underline{\underline{b}} = 0$ (3)
According to the Gaussian numerical integral the equilibrium equations get the following form where the stresses are defined at the Gaussian points:

$$\sum_1^i \underline{\underline{C}}\,\underline{\underline{\sigma}} + \underline{\underline{b}} = 0 \qquad \underset{(l,k\cdot i)}{C} = [\varrho_1 \underset{(l,k)}{B_1^*} \ldots \varrho_\alpha \underset{(l,k)}{B_\alpha^*} \cdots \varrho_i \underset{(l,k)}{B_i^*}] \tag{4}$$

The compatibility equations follow from (1) and (2) defining the stresses at the Gaussian points:

$$\underline{\underline{B}}\,\underline{\underline{e}} - \underline{\underline{D}}^{-1}\underline{\underline{\sigma}} + \underline{\underline{\varepsilon}}_o = 0 \tag{5}$$

Assembling the hypermatrixes C into a hypermatrix G and matrices B into hypermatrix H the following equations express the equilibrium and compatibility of the structure in an elastic case:

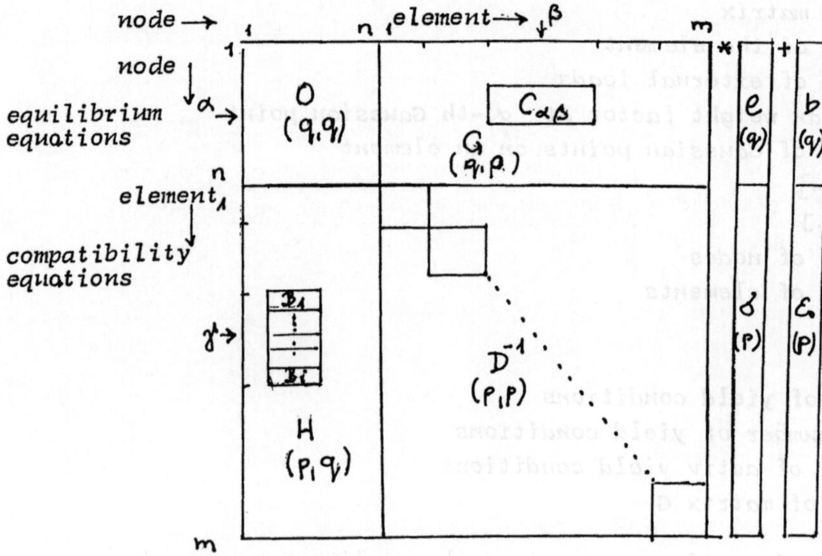

(6)

The boundary conditions are taken into consideration in the usual way.

The symmetricity of the equation system is lost, but solving the problem by mathematical programming doesn't mean any difficulties.

According to the theorem of virtual work the structure gets those stresses which satisfy the equilibrium equations and the virtual energy function has a minimum value.
This means that the elastic analysis of structures can be written as the following mathematical programming problem:

$$\min\{\underline{\underline{\sigma}}^{*}\underline{\underline{D}}^{-1}\underline{\underline{\sigma}} \mid \underline{\underline{G}}\cdot\underline{\underline{\sigma}}+\underline{\underline{b}}=0\} \tag{7}$$

Let's note the solution of this problem (7) by σ_o.

3. Elasto-plastic analysis of continuum structures

In elasto-plastic case the connections between the rate of stresses and external loads are looked for. According to the usual step by step method the equilibrium equations have an unchanged form as in (7) at each stage of external loads. The stresses have to satisfy the yield conditions as well. Using the Mises-Huber-Henky's yield function the following inequalities have to be satisfied at each Gaussian point.

$$f_\alpha(\sigma_\alpha)=\frac{1}{6}\left[(\sigma_{x\alpha}-\sigma_{y\alpha})^2+(\sigma_{y\alpha}-\sigma_{z\alpha})^2+(\sigma_{z\alpha}-\sigma_{x\alpha})^2\right]+\sigma_{xy\alpha}^2+\sigma_{yz\alpha}^2+\sigma_{xz\alpha}^2-\sigma_{L\alpha}^2\leq 0 \tag{8}$$

The structure will get those stresses for which the rate of inner energy function has a minimum value. Using the step by step method the object function is the same as in (7).

Summarizing the elasto-plastic analysis corresponds to the following mathematical programming problem:

$$\min\left\{ g(\sigma)=\underline{\underline{\sigma}}^{*}\underline{\underline{D}}^{-1}\underline{\underline{\sigma}} \;\middle|\; \underline{\underline{G}}\,\underline{\underline{\sigma}}=-\underline{\underline{b}},\; f_j(\sigma_j)\leq 0 \quad j=1\ldots z \quad z=1\ldots m \right\} \tag{9}$$

Plastic yield conditions have positive definite quadratic forms consequently they are convex and so the set that can be described by the intersection of inequalities is convex. The yield conditions are continuous functions in the p-dimensional, space and they contain p stresses. The connections among the stresses are guaranted by the linear equilibrium equations.

This means that the feasible solutions set is convex and bounded if at least one feasible solution exists. The object function is quadratic, namely convex function.

4. Method of solution

To solve the problem (9) such a kind of feasible direction method, [1] is used which takes into consideration the mechanical meaning too.

Starting from the solution of elastic problem (σ_0) a feasible direction (r^0) and the measure of the step(λ^0) are looking for which determines a new point (σ_1) nearer to the feasible solution set: $\sigma^1=\sigma^0+\lambda^0 r^0$ (10)
From this point a new direction and measure is searched. The iteration is continued till the point is in the feasible set. This point will not be the solution of the problem (9) but the iteration can be continued from this inner point.

4.1 Determination of feasible direction

Replacing the solution of the elastic problem (σ_0) into the yield conditions they are either satisfied, or some part of them aren't satisfied. If all of them are satisfied the structure is in elastic state and the next step follows at the external loads. If some part of inequalities are unsatisfied we have to modify vector σ_0, which is an external point from the point of view of mathematical programming.

From all yield conditions let's choose those inequalities which are unsatisfied or become equalities at σ_0, namely the active conditions. For the sake of simplicity of noting we suppose that the first s-th inequalities are active.

$$f_1(\sigma_1^o)>0,\dots,f_t(\sigma_t^o)>0 \quad f_{t+1}(\sigma_{t+1}^o)=0,\dots,f_s(\sigma_s^o)=0 \tag{11}$$

That means the first t-th inequalities are unsatisfied and the inequalities from t+l to s become equalities and the yield conditions from s+l to z are satisfied.

Taking into consideration that the function values of inequalities (11) have to decrease in this direction r^o. $-\nabla f_j(\sigma_j^o)r^o \geq 0 \quad j=1..s$ (12)

The gradient ($\nabla f_j(\sigma_j^o)$) *can't be equal zero in each j-th case, because the feasible set is convex and in the case of the plastic yield conditions from l to t aren't satisfied, so* σ_o *is not a stationary point.*

To form the equality conditions let's replace the expression (10) into the equilibrium equations:

$$\underline{\underline{G}}\underline{\sigma}^1 = \underline{\underline{G}}(\underline{\sigma}^o + \lambda^o\underline{r}^o) = \underline{\underline{G}}\underline{\sigma}^o + \lambda\underline{\underline{G}}\underline{r}^o = \underline{b} \tag{13}$$

At the point σ^o *the equilibrium equations are fulfilled, because this point is the solution of the elastic problem. We can write the following equations instead of (13), because* $\lambda > 0$: $\underline{\underline{G}}\underline{r}^o = 0$ (14)

The objective function was selected with the consideration, that the optimal solution of the starting problem (9) will be the cutting point of the objective function's gradient and of the boundary line of feasible solution's set. So the feasible direction is being searched for to approach the direction of the objective function's gradient as much as possible: $-\underline{\nabla g}(\sigma^o)\underline{r}^o \rightarrow \max$ (15)

On the basis of the above for the selection of the feasible direction the next linear programming problem must be solved:

$$\max\{-\underline{\nabla g}(\sigma^o)\underline{r}^o \,|\, \underline{\underline{G}}\underline{r}^o = 0,\ -\nabla f_j(\sigma_j^o)r^o \geq 0 \quad j=1\dots s\} \tag{16}$$

Further on this problem (16) is transformed into a simpler form. It results from the mechanical meaning of matrix G, that it is a full-row range matrix, which can be divided into two submatrices, a regular and a singular part.

$$\underset{(q,p)}{\underline{\underline{G}}} = [\underset{(q,q)}{\underline{\underline{G}}^1}, \underset{(q,p-q)}{\underline{\underline{G}}^2}],\ \varrho(G^1)=q,\ \underline{\underline{G}}^1\underline{r}^{o1} + \underline{\underline{G}}^2\underline{r}^{o2} = 0,\ \underline{r}^{o1} = -(\underline{\underline{G}}^1)^{-1}\underline{\underline{G}}^2\underline{r}^{o2} \tag{17}$$

In accordance with this problem (16) can be transformed

$$\max\{[-\underline{\nabla g}^1(\sigma^o)\underline{\underline{A}} - \underline{\nabla g}^2(\sigma^o)]\underline{r}^{o2} \,|\, [-\underline{\nabla f}_j^1(\sigma_j^o)\underline{\underline{A}} - \underline{\nabla f}_j^2(\sigma_j^o)]\underline{r}^{o2} \geq 0 \quad j=1..s\} \tag{18}$$

$$\underline{\underline{A}} = -(\underline{\underline{G}}^1)^{-1}\cdot\underline{\underline{G}}^2$$

Let's introduce the following notations:

$$-\underline{\nabla g}^1(\sigma^o)\underline{\underline{A}} - \underline{\nabla g}^2(\sigma^o) = \underline{c}^o$$

$$-\underline{\nabla f}_j^1(\sigma_j^o)\underline{\underline{A}} - \underline{\nabla f}_j^2(\sigma_j^o) = \underline{d}_j^o$$

Accordingly, problem (18) can be written: $\max\{c^o r^o \,|\, d_j^o r^{o2} \geq 0 \quad j=1..s\}$ (19)

Let's transform the problem (19) into a traditional linear programming problem. For this reason the following conditions are introduced:
$|r_j^{o2}| \leq 1/2 \qquad j=1\ldots(p-q)$
On the one hand these guarantee that the set of feasible solutions is bounded, namely that a finite solution will ever exist, on the other hand, by the help of this the satisfaction of the non-negativity conditions can be reached. For this a new unknown vector is introduced: $y=r^{o2}+1/2\ ,\ r^{o2}=y-\frac{1}{2}\ ,\ 0\leq y\leq 1\ ,\ -1/2\leq r^{o2}\leq 1/2$ (20)
After this problem (19) has a traditional linear programming problem: $\max\{c^o y \mid d_i y \geq a_i^o \ i=1,\ldots,s,\ y_j\geq 0,\ y_j\leq 1\ \ j=1\ldots p-q\}$ (21)
where $a_i^o=(d_i^o,\frac{1}{2})$
In so far as the former problem (21) has no solution, this means that the original problem (9) is without a feasible solution either. At the same time this means that a feasible direction, by the help of which the value of the adequate functions $(f_1(\delta_1)\ldots f_t(\delta_t))$ could be reduced, doesn't exist.

4.2 Definition of the measure of step

Let's enter upon the definition of value λ. The least positive λ value is being looked for, for which all plastic yield conditions are satisfied if this is possible.

For this reason relationship (10) is replaced into the yield conditions (11) and the first t inequations are solved as quadratic equations with one unknown.

$$f_i^o(\delta_i^o+\lambda^o r^o)=0 \quad i=1\ldots t \quad f_j^o(\delta_j^o+\lambda^o r^o)\leq 0 \quad j=t+1\ldots z \qquad (22)$$

There are three cases according to the value of the discriminant.

i. If the discriminant is negative this means, that in the feasible direction r^O this plastic yield condition can't be satisfied, this direction avoids the set of feasible solutions. The positive λ is chosen which serves the minimal distance between the $f_i(\delta^o+\lambda r^o)$ function and the feasible direction r^O. Deriving the yield condition according to λ the value is gotten from the following equation:

$$\frac{d f_i(\delta^o+\lambda^o r^o)}{d\lambda}=0 \qquad (23)$$

A positive λ must exist which reduces the value of the function f_j from the point δ^o towards the direction r^O because according to (11) $f_j(\delta_j^o)>0$ and r^O is a feasible direction for function f_j, and $f_j(\delta^o+\lambda r^o)\neq 0$.

ii. If the discriminant is positive, there are two possibilities: both values are positive, or one value is positive, the other is negative. In the first case the less λ value is chosen, which indi-

cates the first point of intersection of the vector taken from point δ_0 towards the feasible direction r^o and of boundary line of the solution-set for the original problem.

In the second case the positive λ is chosen, naturally.

iii. If the discriminant is zero the feasible direction is tangential to the boundary line.

In this way a set of λ -s is determined from the first t inequalities. The least one is chosen and the fulfilment of the yield conditions from t + l to z is checked for the selected λ. If an inequation isn't satisfied the selected λ is reduced accordingly.

Now it must be admitted that the value of least λ isn't equal to zero. The problem (2l) - choosing the feasible direction - contains the first s yield conditions as inequalities and problem (2l) has a solution, so positive values exist in the case of first s yield conditions. The yield conditions s+l to z are satisfied at the δ_0 point in the form of strict inequations. As they are continuous functions, there must be a radius ε where they keep their sign. This means that there exists at least one value of λ in the point δ^o in any direction, so in the feasible direction too.

After defining the λ value the coordinates of point are: $\delta^1 = \delta^0 + \lambda r^o$

In point δ^1 the yield conditions, which have not yet been satified or have become equalities, are searched and the above procedure is repeated, that is problem (2l) is solved by the help of new d_i^1 and c^1 parameters.
This procedure is being continued until each plastic condition is satisfied, that is a δ^t point is got which is an inner point of the feasible solution's set.

Further on, the optimal solution of the original problem (9) is looked for from the inner point δ^t by a similar procedure. The following linear programming problem is solved to determine the feasible solution:

$$\max\{[-\nabla g^1(\delta^t)A - \nabla g^2(\delta^t)]y \mid d_j^\ell y \geq a_j^\ell \quad j = 1 \ldots s\} \qquad (24)$$

$$d_j^\ell = \nabla f_j^1(\delta^\ell)A + \nabla f_j^2(\delta^\ell) \qquad a_j^\ell = (d_j^\ell, 1/2)$$

In this case the inequation must be written to the active part of the yield conditions.

If the problem (24) has no solution, δ^t is already the optimal solution of problem (9). Otherwise, similarly to the former, the least-

positive λ^l value is looked for, for which each plastic yield condition is satisfied.
If we fail to find such a λ^l, δ^l is an optimal solution.
The procedure is carried on until the difference of objective function values in two successive steps is less than a prescribed value.

5. Advantages and disadvantages

Advantages of the recommended method are:

- The yield conditions are used in original form, so it doesn't cause any difficulties if the limit stresses were changed at several steps of external loads and/or in some points of the structure. The quadratic form of the yield conditions is used at the determination of value λ only. In the case of an other kind of yield condition the presented method can be adapted without changes, except the determination of value λ.
- Several kind of strucutre elements can be manipulated together it doesn't cause more difficulties than there are in the case of a usual elastic finite element method.
- Although the knowledge of solution for elastic problems in needed, by omitting equilibrium equotions the size of the linear programming problem (21) to be solved iteratively can be significantly decreased.
- In respect of computing the method can be easily realized and provides fast convergency.

Disadventages of the presented method are:

- It is needed to know the solution of the elastic problems, but there are several effective methods for this purpose.
- The size of matrix G can be very large, but it is a space matrix. Determining the regular part of G and matrix A needs much computation.

6. Numerical example

The geometrical and load data of a curvated thick shell is shown on Figure 1. The transmitted matrix B was computed according to [6].

The results of the computations are presented on the case of first and second load steps on Figure 2. and 3.

The procedure was very efficient. Beyond the solution of elastic problem it needed 400 CPU sec Siemens 4004 to get the result for each load step.

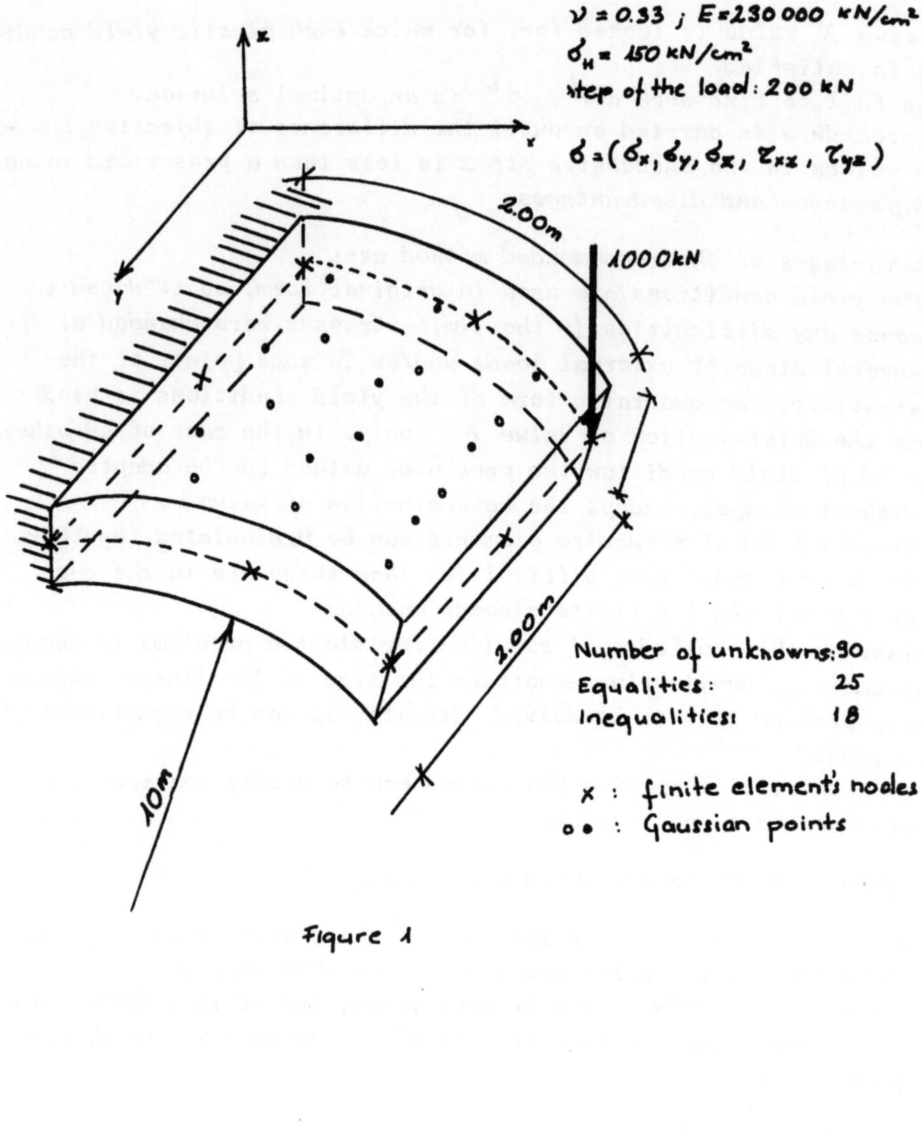
ν = 0.33 ; E=230 000 kN/cm²
σH = 150 kN/cm²
step of the load: 200 kN
σ = (σx, σy, σz, τxz, τyz)
2.00m
1000kN
2.00m
10m
Number of unknowns: 90
Equalities: 25
Inequalities: 18
x : finite element's nodes
o • : Gaussian points

Figure 1

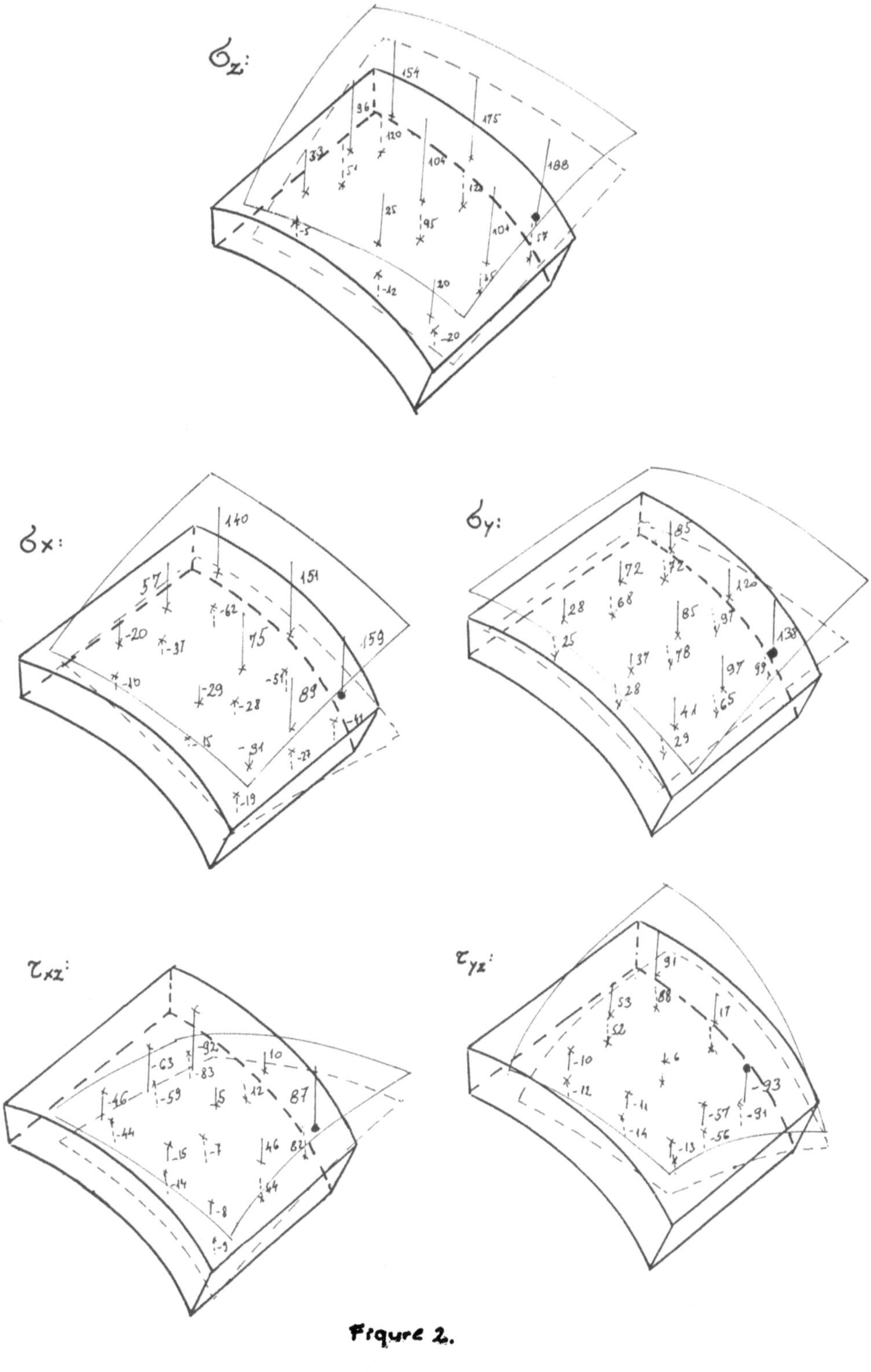

σ_z:
154
96
120
175
33
51
104
188
120
25
95
-5
104
57
-12
20
35
-20
σ_x:
140
57
151
-62
-20
-37
75
159
-10
-29
-51
-28
89
-41
-15
-31
-27
-19
σ_y:
85
72
72
120
28
68
85
97
25
138
37
78
97
99
28
41
65
29
τ_{xz}:
-92
10
-63
-83
-46
-59
15
12
87
-44
-15
-7
46
82
-14
44
-8
-9
τ_{yz}:
91
53
88
17
52
-10
6
-12
-93
-11
-57
-91
-14
-56
-13

Figure 2.

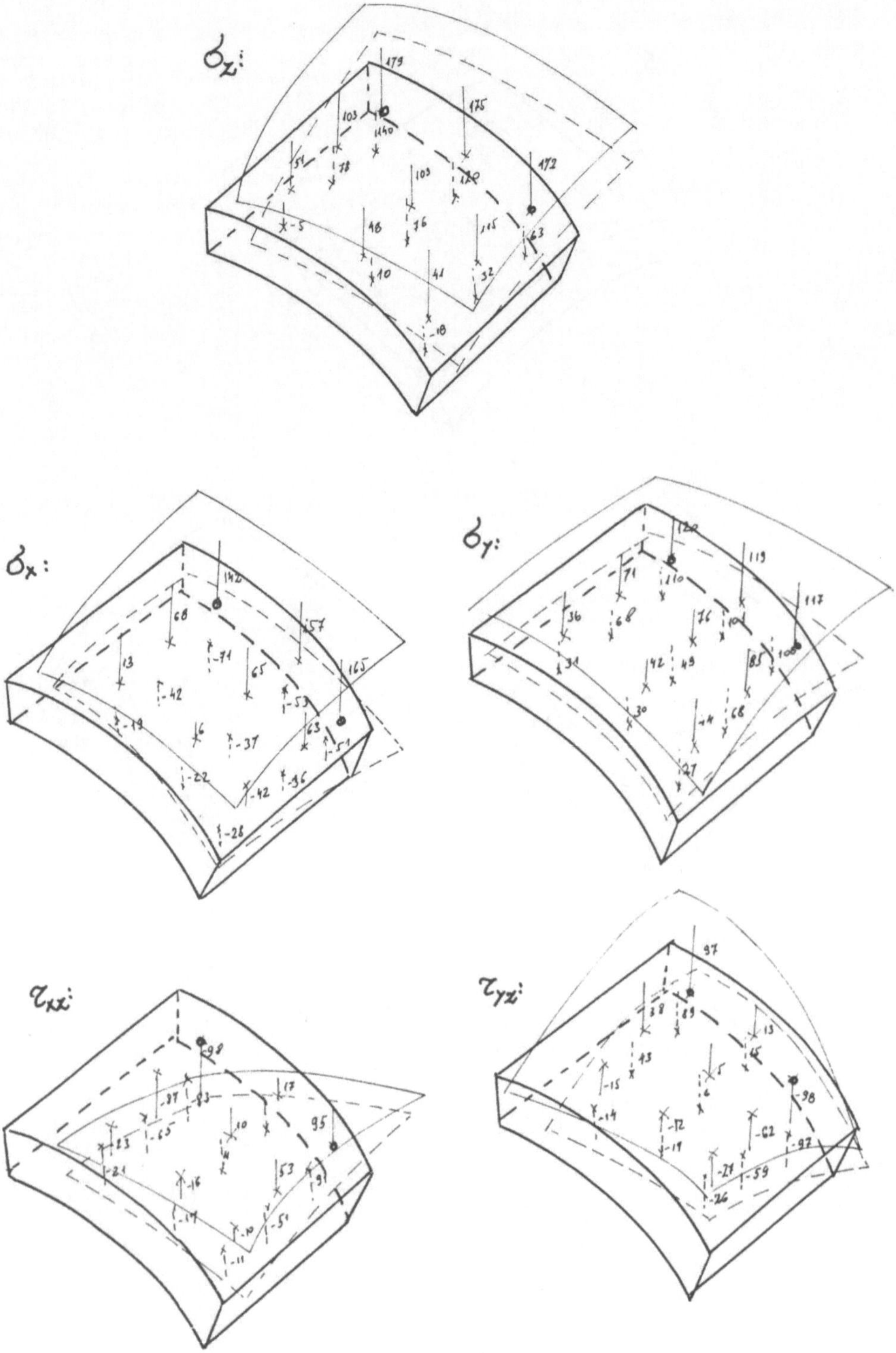

Figure 3.

7. References

1. Zoutendijk, G.: Methods of feasible directions. A study in linear and non linear programming. Amsterdam - London - New York - Princeton, 1960.

2. Zienkiewich, O.C.: The Finite Element Method in Structural and Continuum Mechanics. London - New York - Toronto - McGraw-Hill Publishing Company, 1967.

3. Cohn, M.Z. - Maier, G.: Engineering Plasticity by Mathematical Programming. New York - Toronto - Oxford, Pergamon Press, 1977.

4. Sayegh, A.F. - Rubinstein, M.F.: Elasto-Plastic Analysis by Quadratic Programming. ASCE Journal of Engineering Mechanices Division, Vol.98. No.EM6, 1972 (1547-1572)

5. De Donato, O. - Maier, G.: Historical Deformation Analysis of Elasto-plastic Structures as a Parametric Linear Complementority Problem. Meccanice No.3. Vol.II. 1976 (p.166-171)

6. Karan, S. Surana: Geometrically nonlinear formulation for the tree dimensional solid-shell translation Computer and Structures, Vol.15. No.5. (1982.)

7. Axelsson, K. - Samuelsson, A.: Finite Element Analysis of Elastic-Plastic Matherials Displaying Mixed Hardening. Int.Jou. for Num. Meth. in Eng. Vol.17. No.2. 211-227 (1979)

MODELLING OF METROLOGICAL SYSTEMS

Gábor E. Veress
Technical University of Budapest
Budapest, Hungary

Introduction

An essential problem in natural sciences is the mathematical description of investigated materials using extensive and intensive properties, quantity and homogeneity or distribution. Some set theoretical models have been introduced in physics and systems theory [1]-[5] though the problem remained unsolved till these days.

The purpose of this paper is to introduce a measure theoretical model to describe the characteristics of material in natural sciences [6]-[7]. The proposed measure theoretical model involves four basic concepts: element, property of element, component and quantity of component. Using these concepts the distribution of material and of material quantity can be interpreted.

This measure theoretical model can be used for describing the operation of metrological systems like analytical chemical measuring systems, diagnostical systems and quality control systems. The observed distribution of material or of material quantity is a mixture distribution, so the identification, qualification, diagnosis or measurement is actually the resolution of this observed mixture distribution of material or of material quantity.

1. Model for characterization of materials

1.1. Basic concepts of the model

For the description of material investigation, a measure theoretical model based on the ideas such as element, property of element, component and quantity of component can be used.

The material to be investigated is modelled by the set M, called *material*. The smallest investigated unit of the material is modelled as *element* m, so the material M is the set of elements, that is

$$m \in M.$$

The property of element as the point of view of investigation is modelled by the function $\varphi(m)$, which is called p r o p e r t y o f e l e m e n t , that is

$$\varphi : M \rightarrow \Phi,$$

where Φ is the domain of the element property function. The property of element can be regarded as intensive property of material.

Modelling the idea of component, the c o m p o n e n t set C is defined as a set of elements respecting a value of a property defining component, that is

$$C_i = \{m : \varphi(m) = \Phi_i\}.$$

Introducing the idea of the set of components, $\mathcal{M}$, i.e. the power-set of the material, it is valid that

$$C \in \mathcal{M},$$
$$C \subseteq M.$$

The amount of material and of component is modelled as a measure $q(.)$ on the component set, and it is called q u a n t i t y of component $q(C)$, that is

$$q : \mathcal{M} \rightarrow R^1,$$

where R^1 is the domain of the real numbers.

It follows from these definitions, that the triplet $[M, \mathcal{M}, q]$ is a measure space.

The quantity of component can be regarded as extensive property of material.

1.2. Distribution of material and material quantity

Based on the idea of property defining component, the distribution of material and of material quantity can be defined.

In the case of discrete property, the d i s c r e t e d i s t r i b u t i o n o f m a t e r i a l is the set of components $C_1,\dots,C_J$, where

$$C_j = \left\{ m : \varphi(m) = \phi_j \right\} \qquad j = 1,\dots, J$$

and

$$M = \bigcup_{j=1}^{J} C_j .$$

The d i s c r e t e d i s t r i b u t i o n o f m a t e r i a l q u a n t i t y is the set of component quantities $q_1,\dots,q_J$, where

$$q_j = q\left(C_j\right) \qquad j = 1,\dots, J$$

and

$$q(M) = \sum_{j=1}^{J} q\left(C_j\right) .$$

In the case of continuous property, the c o n t i n u o u s d i s t r i b u t i o n o f m a t e r i a l is the set of components C_r, that is the set of level sets of material, where

$$C_r = \left\{ m : \varphi(m) < r \right\} \qquad r \in R^n$$

and

$$M = C_{\infty} .$$

The c o n t i n u o u s d i s t r i b u t i o n o f m a t e r i a l q u a n t i t y, i.e. t h e d i s t r i b u t i o n f u n c t i o n o f m a t e r i a l q u a n t i t y is the set of component quantities $G(r)$, where

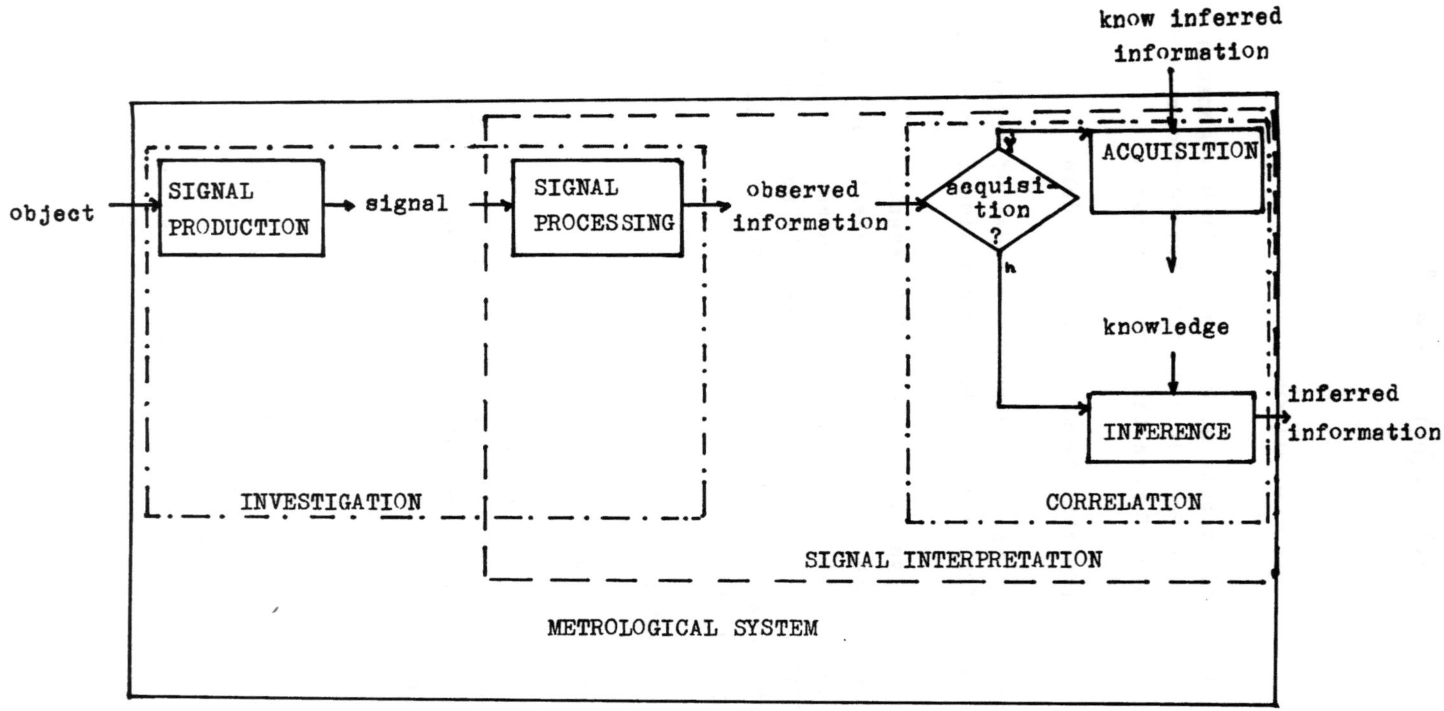

Structure of metrological systems

Figure 1.

$$G(r) = q(C_r) \qquad r \in R^n$$

and

$$q(M) = G(\infty).$$

Using the idea of distribution, a material is homogeneous if it consists of only one component and it is heterogenous or it is a mixture if it consists of more than one component.

1.3. Analogy between probability and material models

There exists a close analogy between the model of probability theory and the model for material characterization introduced in this paper.

The material is equivalent to the certain event, the element to the elementary event and the component to the event. The property of element is equivalent to the random variable and the quantity to the probability. The material distribution is equivalent to the total event system, and the material quantity distribution to the probability distribution function.

2. Model of metrological systems

2.1. Metrological systems

Diagnostic, qualifying, measuring systems can be regarded as special types of metrological systems. M e t r o l o g i c a l s y s t e m s produce inferred information on the investigated object in an indirect way through observed information. The metrological correlation consists of metrological acquisition producing metrological knowledge from observed information and from known inferred information and of metrological inference producing inferred information from observed information using metrological knowledge. The structure of the metrological systems can be seen on Figure 1.

The metrological knowledge is the input-output model of the investigation, so the metrological acquisition is the identification of the investigation model and the metrological inference is the application of the investigation model.

of material quantity.

3. Summary and conclusions

Investigation of materials should be described by the measure theoretical model based on the ideas element, property of element, component and quantity of component. Metrological /diagnostical, measuring, qualifying/ systems consist of investigation and correlation modelling investigation. In the case of linear investigation, the model of investigation can be regarded as mixture distribution.

The advantage of the proposed measure theoretical model is that property, quantity, homogeneity, etc. can unambiguously be defined using this model.

The possibilities of applications are based on the fact, that this model is a unified tool for modelling material investigations, correlations, qualification, or diagnostic systems.

The disadvantage of the proposed model, and so the limit of its applicability is determined by the fact that the model at present is only deterministic and static. It can, however, be extended to becau e a stochastic and dynamic one.

We hope that the ideas stimulated by this measure theoretical model will enhelp the mathematical description of chemometrical, biometrical, psychometrical etc., and qualification problems, and thus, enable a wider application of computing techniques in these fields.

L i t e r a t u r e

1 Gurtin, M.E., Williams, W.O.: An axiomatic foundation for continuum thermodynamics
Arch. Rational Mech. Anal., 26, 83-117 /1967/.

2 Fischer, G.M.C., Leitman, M.J.: On continuum thermodynamics with surfaces
Arch. Rational Mech. Anal., 3o, 225-262 /1968/.

3 Allen, A.D.: Measuring the Empirical Properties of Sets
IEEE Trans. on Systems, Man and Cybernetics, SMC-4,66-73/1974/.

4 Bunge, M., Sangalli, A.A.: Theory of Properties and Kinds
Int. J. General Systems, 3, 183-19o /1977/.

Some examples for metrological systems are given in Table 1. In analytical chemistry, for example, the observed property is an analytical one and the inferred property is a chemical one. Analytical property is, for example, the retention time of molecule and a chemical one is the chemical identity of molecule.

The measure theoretical model introduced in this paper can be used for describing the operation of metrological systems. In metrological systems two types of properties are assigned to the investigated elements: the observed property and the inferred property. Based on these properties, observed components and inferred components of material can be defined. Similarly, two types of quantities, the observed and the inferred quantities are assigned to material or to components.

2.2. Model for linear investigation

Because in metrological systems material is characterized by observed and inferred properties, in modelling investigation, observed and inferred distributions of material or material quantity must be interpeted, based on the measure theoretical model introduced in this paper.

If the investigation is a linear, time-invariant system, then its model as the conclusion of the measure theoretical model is that the observed distribution of material /quantity/ is a mixture distribution of the observed distributions of the inferred components /quantities/, weighed by the inferred distribution of material /quantity/.

The investigation means the determination of the observed distribution of material /quantity/, the acquisition determines the observed distributions of the inferred components and the inference is the determination of the inferred distribution of material /quantity/ using this mixture distribution model, that is the inference is the resolution of the mixture distribution of material /quantity/.

In analytical chemistry, for example, in the case of qualitative analysis the inference is the determination of the chemical distribution of material by the resolution of the analytical distribution of the material, and in the case of quantitative analysis the inference means the determination of the chemical distribution of material quantity by the resolution of the analytical distribution

5 Leitman, M.J.: On the decomposition of measures with applications to continuum physics
Math. Systems Theory, 4/2/, 112-118 /197o/.

6 Almásy, G.A., Veress, G.E., Vadnay, Sz., Ser, V.: True property functions and multicomponent fractionation balances
Hung. J. of Ind. Chem. /Veszprém/, 2/1/, 117-147 /1974/.

7 Veress, G.E.: Mathematical and Systems Theoretical Description of Chemical Measurements /in Hungarian/
VEAB Monográfia, 19, Veszprém, 1984.

OPTIMAL IDENTIFICATION PROBLEMS
OF REGIONALIZED VARIABLES SYSTEMS

Małgorzata Więcek, Antoni Ligęza
Institute of Control and Systems Engineering
University of Mining and Metallurgy, Cracow, Poland

1. INTRODUCTION

In numerous applied sciences such as geology, hydrology, agriculture and environmental researches, a need for adequate description of natural phenomena arises. In order to construct an appropriate model of a phenomenon under consideration a great deal of data has to be collected and analysed. Advanced mathematical methods and computational means are in use.

In geology, for example, one searches for a model of a deposit /as precise as possible, of course/ [2,6]. One of the main problems is to estimate the grade or thickness in a certain point or area on the basis of data obtained in given measurement points /usually drillholes/ [2,6,9,10,11]. In hydrology the water level and in agriculture soil parameters may be the matter of interest [1,9,10]. Finally, in environmental researches the degree of pollution in a certain area may be looked for [5]. The common feature of the subject matter above is that being given a number of values of an unknown function /called regionalized variable/ one is to estimate the value of the function in a desired point or its mean value in some specified area. Since obtaining the data is usually very expensive the application of optimization techniques at the planning and identification stage becomes indispensable.

The presented paper deals with the optimization problem of additional data points choice strategy. A discrete problem is considered. It is presumed that there are N possible locations for observations points and among them there are n, $n<N$, already existing measurement points /e.g. drillholes, pollution measurement points/. On the basis of the data available certain characteristics of a considered phenomenon can be estimated and the mean value of some functions describing certain measurable features of it in a point or area of interest can be calculated. The accuracy of performed estimation strongly depends on the number of available data points and so does the cost of samples. Now, the optimization problem consists in proper selection of certain additional data points to be explored /from the number $N-n$ points left/ such that a given objective function /e.g. the estimation accu-

racy, cost/ should be extremized.

The mathematical formulation of the problem is based on the theory of regionalized variables and kriging [2,6]. Our paper can be viewed as some development of the ideas presented in [10,11]. The problem statement is modified and a new adaptive multistage strategy for solving the problem is introduced. Moreover, a method for dealing with a multicriteria optimization problem is proposed. The paper is completed with some remarks related to the application as well as to possible extension of presented methods.

2. RECAPITULATION OF THE KRIGING METHOD

Let us introduce in brief the formal method we shall base on as well as the way of notation. Let $z(x)$ be the value of the function we are interested in, i.e. z is a function describing the phenomenon under consideration, and x is a point in a one-, two- or three-dimensional space. It may be the thickness of a deposit, the grade of equal-size samples of a certain mineral, the water level or the degree of pollution. Such a function z will be called a regionalized variable.

The fundamental idea of the theory is to consider the function above as one realization of a random function $Z(x)$. The very obvious advantage of such an approach is that now we can make use of the statistical theory in order to analyse the character of the regionalized variable which is usually a very "unpleasant" function from mathematical point of view. As in [2,6,10,11] we shall assume the following properties of the random function $Z(x)$:

$$E\left[Z(x)\right] = \sum_{l=0}^{k} b_l f_l(x) \tag{2.1}$$

$$VAR\left[Z(x+h) - Z(x)\right] = 2\gamma(h) \tag{2.2}$$

where $E[\cdot]$ denotes the expected value, $VAR[\cdot]$ - the variance, f_l are certain arbitrarily chosen functions /usually simple polynomials/, b_l - certain unknown coefficients, $l=0,1,\ldots,k$. $\gamma(h)$ - so-called variogram - is a function of a vector h expressing how the values of a function under consideration differ on average at points x and $x+h$ according to the mutual location of the points - i.e. to the distance and direction [2,6]. Let V be the domain of interest and let $z(V)$ denote the average value of the regionalized variable centred at point x in V, i.e.

$$z(V) = \frac{1}{|V|}\int_V z(x)dx \qquad (2.3)$$

Now, assume that the data from points $x_1, x_2, \ldots, x_n$ are known, where $n < N$. Let $\gamma(W,h)$ denote the variogram obtained in the usual manner on the basis of the experimental data, where $W = \{x_1, x_2, \ldots, x_n\}$. Techniques of variogram identification are presented in [1,2,6] in a comprehensive way.

Our primary goal is to estimate the value $z(V)$ as precisely as possible. The estimated value of $z(V)$ is denoted as Z^*, and it is presumed that it can be calculated as a weighted average of n available data as

$$Z^* = \sum_{i=1}^{n} a_i z(x_i) \qquad (2.4)$$

with coefficients a_i, $i=1,2,\ldots,n$, such that the estimation variance $VAR[Z(V) - Z^*]$ is minimal and the estimator obtained is unbiased.

Formulated as above the optimization problem leads to the so-called kriging system [2,6] which can be expressed in the following form

$$\sum_{i=1}^{n} a_i f_l(x_i) = f_l(V) \qquad l=0,1,\ldots,k \qquad (2.5)$$

$$\sum_{j=1}^{n} a_j \gamma(x_i - x_j) + \sum_{l=0}^{k} \mu_l f_l(x_i) = \gamma_{V,i} \qquad i=1,2,\ldots,n \qquad (2.6)$$

where $\gamma(h) = \gamma(W,h)$, $W = \{x_1, x_2, \ldots, x_n\}$, μ_l are the Lagrangian multipliers and

$$\gamma_{V,i} = \frac{1}{|V|}\int_V \gamma(x_i - x)dx \qquad (2.7)$$

$$f_l(V) = \frac{1}{|V|}\int_V f_l(x)dx \qquad (2.8)$$

The equation defining the minimal value of the estimation variance $VAR[Z(V) - Z^*]$ denoted as $var[Z(V) - Z^*]$, obtained from equations (2.5) and (2.6) can be written in the form [10] :

$$var[Z(V) - Z^*] = \sum_{i=1}^{n} a_i \gamma_{V,i} + \sum_{l=0}^{k} \mu_l f_l(V) - \gamma_{V,V} \qquad (2.9)$$

where

$$\gamma_{V,V} = \frac{1}{|V|^2} \int_V \int_V (x-y)\,dx\,dy \qquad (2.10)$$

In the case we are interested in obtaining the value of the regionalized variable in a single point x_0 instead of the average value for a domain V, we have just to replace $f_l(V)$ by $f_l(x_0)$, $\gamma_{V,i}$ by $\gamma(x_i - x_0)$ and $\gamma_{V,V}$ by $\gamma(x_0 - x_0) = 0$ in equations (2.5), (2.6) and (2.9) respectively.

Now we can move on to the formulation of the optimization problems of the choice of additional data points.

3. THE OPTIMIZATION PROBLEM STATEMENT. A STRATEGY OF ADDITIONAL DATA POINTS SELECTION

Let us recall the problem. There are N possible observation /measurement/ point locations and among them n points already explored. The optimization problem consists in selecting from the $m=N-n$ points left some additional data points in order to minimize a certain objective function.

We shall need some further notation. Let d_i denote a decision variable such that $d_i=1$ if and only if the i-th point has been or is to be explored, else $d_i=0$, for $i=1,2,\dots,N$. Therefore, having given the values of d_i for $i=1,2,\dots,N$, one has determined a variant of identification for the regionalized variable system under consideration - one knows at which points the data should be /have been/ collected. Let c_i denote the measurement cost at point i /e.g. drilling, air pollution measurement etc./, $i=1,2,\dots,N$. Let L denote the maximum number of points that can be explored at a time. The number of drilling rigs available in geology or the number of measurement outfits in environmental researches can serve as typical examples. Let C denote the upper bound of the additional data obtaining cost. We presume that our goal is to identify the phenomenon under consideration as precisely as possible; therefore the estimation variance $VAR[Z(V) - Z^*]$ should be minimized. Now, the optimization problem can be written as follows:

minimize

$$\sum_{i=1}^{N} a_i d_i \gamma_{V,i} + \sum_{l=0}^{k} \mu_l f_l(V) - \gamma_{V,V} \qquad (3.1)$$

subject to

$$\sum_{i=1}^{N} a_i d_i f_l(x_i) = f_l(V) \qquad l=0,1,\dots,k \qquad (3.2)$$

$$\sum_{j=1}^{N} a_j d_j \gamma(x_i - x_j) + \sum_{l=0}^{k} \mu_l f_l(x_i) = \gamma_{V,i} \qquad i=1,2,\dots,N \qquad (3.3)$$

$$\sum_{i=n+1}^{N} c_i d_i \leqslant C \qquad (3.4)$$

where $d_i=1$ for $i=1,2,\dots,n$ and $d_{n+1}, d_{n+2}, \dots, d_N$ are the variables to be determined. It can be seen that objective function (3.1) gives minimal estimation variance (2.9) for a chosen set of d_i parameters, $i=n+1,n+2,\dots,N$, i.e. a given identification variant. Constraints (3.2) and (3.3) are immediate consequence of (2.5) and (2.6) and constraint (3.4) assures keeping within the bounded cost.

The above problem can be solved by use of the branch and bound method [10,11] /for some details about the method see also [4]/. However, taking into account the limitation on the number of simultaneously explored points a hierarchical adaptive multistage strategy for dealing with the problem can be proposed. The strategy is as follows:

Step 1. Set $W=\{x_1,x_2,\dots,x_n\}$, $M=\{n+1,n+2,\dots,N\}$, $\gamma(h)=\gamma(W,h)$, $d_i=1$, $i=1,2,\dots,n$, $d_i=0$, $i=n+1,n+2,\dots,N$.

Step 2. If $M=\emptyset$ or $\sum_{i=n+1}^{N} c_i d_i + \min c_i > C$, then the procedure terminates.
The final estimate Z^*, based on the latest function $\gamma(h)$, should be determined from equations $Z^* = \sum_{i=1}^{N} a_i d_i z(x_i)$ and (3.2), (3.3) for appropriate values of d_i, $i=n+1,n+2,\dots,N$, determined as a result of the procedure application.
Else, go to step 3.

Step 3. Solve optimization problem (3.1) subject to (3.2),(3.3) and (3.4) with the following constraint:

$$\sum_{i\in M} d_i \leqslant L \qquad (3.5)$$

As a result of this one obtains a set $M'=\{j_1,j_2,\dots,j_K\}$, $K \leqslant L$, of the numbers of the points to be explored next. Let $W'=\{x_{j_1},x_{j_2},\dots,x_{j_K}\}$. Set $d_{j_1}=d_{j_2}=\dots=d_{j_K}=1$.
Now, perform sampling at the points belonging to W'. As a result of this one obtains a set of values of a regionalized variable $z(x)$ for any point of W'. Set $W=W\cup W'$, $\gamma(h)=\gamma(W,h)$, $M=M-M'$ and go to step 2.

In the above procedure in step 1 the initial conditions are es-

established. In step 2 the termination conditions are checked. Step 3 corresponds to the solution of one-stage optimization problem. The above procedure offers the advantage of maximal utilization of the experimental data being obtained while performing the identification process.

4. A MULTIOBJECTIVE APPROACH

The optimization problem statement (3.1-3.4) given in section 3 may lead to the multicriteria optimization problem in which regionalized variables describing phenomena or features taken into consideration would become objective functions. In [10,11] a simple method based on scalarization by means of weighting coefficients for solving multicriteria optimization problems was presented. Here we would like to propose a somewhat extended method for dealing with the multicriteria problem designed for regionalized variables systems.

Let there be given a set F of objective functions $F=\{F_1,F_2,...F_P\}$ and set $W_0=\{x_1,x_2,...,x_n\}$ of points which have already been explored. By $F_j(W_0)$, $j=1,2,...,P$, we denote the value of F_j criterion /e.g. estimation variance of the vector objective j-th component/ obtained for the points of set W_0 - this means that $F_j(W_0)$ is the value of F_j objective function calculated on the basis of data collected at the existing points $x_1,x_2,...,x_n$ in the initial situation. Let $Y=\{x_{n+1}, x_{n+2},...,x_N\}$ denote the set of additional admissible data sites.

Then the multicriteria optimization problem /MOP/ consists in the choice of such an additional data points selection in set Y that minimizes objective functions /e.g. of the type (3.1)/

$$F_1,F_2,...,F_P \tag{4.1}$$

subject to given constraints /e.g. (3.2),(3.3),(3.4) and/or some other limitations imposed on the problem in arbitrary way/. Note that in the above problem statement the estimation variance as well as the cost can be the objective functions.

Let assign indices i, $i=1,2,...,Q$, to the elements of set $2^Y - \emptyset$, /2^Y is the set of all subsets of set Y/, where $Q=2^{m-1}$, $m=N-n$. In this way, for any selection Y_i, $i=1,2,...,Q$, of points to be explored /of a solution variant/ we can determine a set W_i as:

$$W_i = W_0 \cup Y_i \qquad i=1,2,...,Q \tag{4.2}$$

We shall refer to W_i as to the i-th realization /feasible selection/ chosen while performing the data points choice.

Let $F=[F_{ij}]$ be a matrix QxP, where $F_{ij}= F_j(W_i)$, $i=1,2,...,Q$,

$j=1,2,\ldots,P$, is the value of F_j criterion for W_i selection.

The method proposed for solving MOP (4.1) is based on the MAPPAC method [8] and consists in a certain ranking of the set of realizations $W=\{W_1,W_2,\ldots,W_Q\}$ with reference to the set of considered criteria $F=\{F_1,F_2,\ldots,F_P\}$. In order to rank the feasible realizations one has to introduce an appropriate order relation with regard to the values of objective functions for the set of feasible solutions.

Assume, that elements F_{ij} of matrix F are calculated for every feasible solution W_j resulting from an appropriate combination of values d_i, $d_i \in \{0,1\}$, $i=n+1,n+2,\ldots,N$. Thus, the considered MOP resolves itself into the discrete choice problem with a finite set of feasible decisions characterized by multiple evaluative criteria.

$$\text{Let } F_j^* = \max\{F_{1j},F_{2j},\ldots,F_{Qj}\} \qquad j=1,2,\ldots,P.$$
$$F_{j*} = \min\{F_{1j},F_{2j},\ldots,F_{Qj}\}$$

F_j^* and F_{j*} are respectively the best and the worst evaluations of Q possible realizations with respect to the j-th criterion.

We shall normalize matrix F transforming it into matrix $T=[T_{ij}]$ as follows:

$$T_{ij} = \begin{cases} \dfrac{F_{ij}-F_{j*}}{F_j^* - F_{j*}} & \text{if } F_{j*} < F_j^* \\ 0 & \text{if } F_{j*} = F_j^* \end{cases} \tag{4.3}$$

where $0 \leqslant T_{ij} \leqslant 1$, $i=1,2,\ldots,Q$, $j=1,2,\ldots,P$.

Let us consider a pair of realizations W_i and W_l, $1 \leqslant i \leqslant Q$, $1 \leqslant l \leqslant Q$, $i \neq l$, and a pair of criteria F_j and F_k, $1 \leqslant j \leqslant P, 1 \leqslant k \leqslant P, j \neq k$.

We shall say that realization W_i dominates realization W_l with reference to criteria F_j and F_k if and only if

$$T_{ij} + T_{ik} < T_{lj} + T_{lk} \tag{4.4}$$

and we shall say that realizations W_i and W_l are indifferent /or equivalent/ with reference to criteria F_j and F_k if and only if

$$T_{ij} + T_{ik} = T_{lj} + T_{lk} \tag{4.5}$$

In order to rank a pair of realizations with reference to all criteria $F_1,F_2,\ldots,F_P$ we have to aggregate data contained in matrix T by means of genaralized preference ranking. We shall say that realization W_i dominates realization W_l with reference to the set of criteria F if and only if

$$\sum_{j=1}^{P} T_{ij} < \sum_{j=1}^{P} T_{lj} \quad (4.6)$$

and we shall say that realizations W_i and W_l are indifferent with reference to the set of criteria F if and only if

$$\sum_{j=1}^{P} T_{ij} = \sum_{j=1}^{P} T_{lj} \quad (4.7)$$

for any $1 \leqslant i \leqslant Q$, $1 \leqslant l \leqslant Q$, $i \neq l$, $1 \leqslant j \leqslant P$, $1 \leqslant k \leqslant P$, $j \neq k$.

The order relation resulting from conditions (4.6) and (4.7) introduces a final ranking in the set of feasible realizations. Moreover, it is to be noted that not only the best selection of data points choice can be chosen but all the selections can be ranked according to decreasing preferences.

5. CONCLUDING REMARKS

In the paper the problems of optimal identification of regionalized variable systems have been discussed. A multistep hierarchical adaptive strategy for dealing with the optimal identification problem has been proposed. Moreover, an approach to the multicriteria optimization problem has been introduced.

A few concluding remarks seem to be in order:

5.1. The presented strategy /section 2/ is not necessarily optimal. It could be optimal in the case it were one stage optimization procedure, after relaxation of condition (3.5) provided that the variogram at our disposal is accurate. Such a case is presented and discussed in [10, 11]. However, as far as physical applications are concerned condition (3.5) expresses an essential constraint /resource limitation/. In addition, once having the new data it seems reasonable to use them not only for the estimation calculation but for updating /recomputing/ the variogram as well. Therefore, although the global optimum would not probably be achieved, the recently obtained data /during the performance of the proposed procedure/ shall be used in a more rational way.

5.2. Condition (3.5) is essential not only from practical point of view /because of the real, physical limitations/ but plays a crucial role for limiting the computational effort decreasing the dimension while solving the identification problem.

5.3. It would be desirable to make use of a recursive procedure for recomputing the variogram.

5.4. The proposed method for dealing with the multicriteria optimization problem requires the full inspection of the set of possible optimization variants - therefore it can be applied directly to a problem of rather small dimension. However, computational effort can be considerably reduced by obvious modification consisting in elimination such optimization variants which are dominated with reference to all criteria considered and in ranking only such variants which are non--comparable. The proposed order relation guarantees the simplicity of computations and results from the natural preference posed on the decision process. Of course, such a relation is defined on the basis of of a certain subjective factor which in the multicriteria decision process appears indispensable [7].

5.5. The application of the proposed multicriteria optimization approach seems to be especially reasonable in a case of somewhat comparable objective functions /e.g. estimation variances of different blocks of the same size - see examples in [10,11]/.

REFERENCES

1. Bastin,G.,Gevers,M. (1985): Identification and Optimal Estimation of Random Fields from Scattered Point-wise Data, Automatica, Vol. 21, pp. 139-155.
2. David,M. (1977): Geostatistical Ore Reserve Estimation, Elseviere Scientific Publishing Company.
3. Dubrule,O. (1983): Two Methods with Different Objectives: Splines and Kriging, Mathematical Geology, Vol.15, No.2, pp. 245-257.
4. Garfinkel,R.S.,Nemhauser,G.L. (1972): Integer Programming, John Wiley and Sons.
5. Gonzalez,P.L.,Chami,H. (1984): Amélioration d'un Réseau de Surveillance de la Pollution Atmospherique, R.A.I.R.O. Automatique/Systems Analysis and Control, Vol.18, No.4, pp. 369-384.
6. Journel,A.G.,Huijbregts,Ch.J. (1978): Mining Geostatistics, Academic Press.
7. Keeney,R.L.,Raiffa,H. (1976): Decision with Multiple Objectives; Preferences and Value Trade-Offs, John Wiley and Sons.
8. Matarazzo,B. (1984): Multicriteria Analysis - The MAPPAC Method, 20th Meeting of the EURO Working Group on Multicriteria Decision Aid, Paris, France.
9. Szczepankiewicz,E. (1985): Stochastic Field Application, Państwowe Wydawnictwo Naukowe, /in Polish/.
10. Szidarovszky,F. (1983): Optimal Observation Network in Geostatistics and Underground Hydrology, Applied Mathematical Modelling, Vol.7, pp. 25-32.
11. Szidarovszky,F. (1983): Multiobjective Observation Network Design for Regionalized Variables, International Journal of Mining Engineering, Vol.1, pp. 331-342.

The Analysis of I/O Configurations: On the Robustness of the Central Server Model *

Rudo J.Wijbrands

Eindhoven University of Technology
Department of Mathematics and Computing Science
P.O. Box 513
5600 MB Eindhoven
The Netherlands

ABSTRACT

The central server model plays an important role in the analysis of computer systems. With the use of product form assumptions this model can be solved efficiently. In more detailed computer models, however, the product form assumptions are often violated. Still, both recursive and iterative approximation methods based on the central server model, are successfully applied in these cases. In this paper it is discussed why these methods are so successful, by studying the approximations made to cope with non exponential and population dependent disk workloads.

1. Introduction

The central server model is a closed queueing network model which gives a simple representation of a computer system (see e.g. Buzen [3]). In its primitive form it consists of a CPU and a number of disk units. Clients in this system alternately visit the CPU and one of the disk units. The disk units are assumed to process jobs according to a first come first served discipline. These jobs are assumed to be independent, identically and exponentially distributed. The CPU is assumed to handle its clients according to a processor sharing service discipline.

The conditions which are listed above, roughly satisfy the product form assumptions (see Baskett, Chandy, Muntz & Palacios [1]), needed for the solution of the model. For the class of product form networks there exists a number of efficient solution algorithms. However, it will be clear that the conditions, which are to be fulfilled in order to obtain a product form network, in practice are violated more often than not.

The central server model cannot be seen independent from its more general surroundings, the computer-terminal model. On a detailed level of the latter model, the computer system may be represented by a central server model. A decomposition/aggregation approach (see e.g. Chandy, Herzog & Woo [4]) may be used to first study the central server model as a stand alone closed system. In a second step this system can be aggregated to obtain a single load dependent station, representing the computer in the computer-terminal model. The service rates of this station are computed using the throughputs of the former central server model under various populations. As the throughput for each possible population is needed, we can exploit the recursiveness of certain solution algorithms, like the mean value algorithm used in this paper, as these recursive algorithms produce all throughputs automatically, while computing the

*) The investigations were supported in part by the Foundation for Computer Science in the Netherlands, SION, with financial aid from the Netherlands Organization for the Advancement of Pure Research (ZWO)

throughput for the maximum population.

If we take a more realistic look at computer systems, we see that the central server model lacks detail. Serious problems, like channel blocking, in which a single channel is to be used for communication purposes by several disk units, are neglected in the model. In the literature much attention is given to this lack of detail problem (e.g. [2],[5], [9]). Several solutions are suggested which are all based on the idea to translate a more detailed computer model in terms of the central server model. As to the computer-terminal model this means that the modeling stretch is extended with an extra step (see Figure 1). The extra step, however, prohibits most algorithms to exploit the recursiveness of the solution technique for the central server model.

From the modeling stretch as depicted in Figure 1, in this paper we focus on the part 1b-1c in which the central server model is involved. It appears from numerical results, such as presented in Section 3, that the central server model and its solution method are quite robust under this translation step. It is this robustness that we want to study in this paper. To solve the central server model, 1c, we use the mean value algorithm. The principles of the mean value algorithm are briefly summarized in Section 2. In Section 3 we indicate three successful ways of translating model 1b into model 1c. Sections 4 and 5 discuss two of the approximations made in this translation; in Section 4 the influence of non exponential disk workloads on the accuracy of the central server model is considered, while in Section 5 the disk workload depending on the system population is studied. In Section 6 we make some concluding remarks about the robustness of the model.

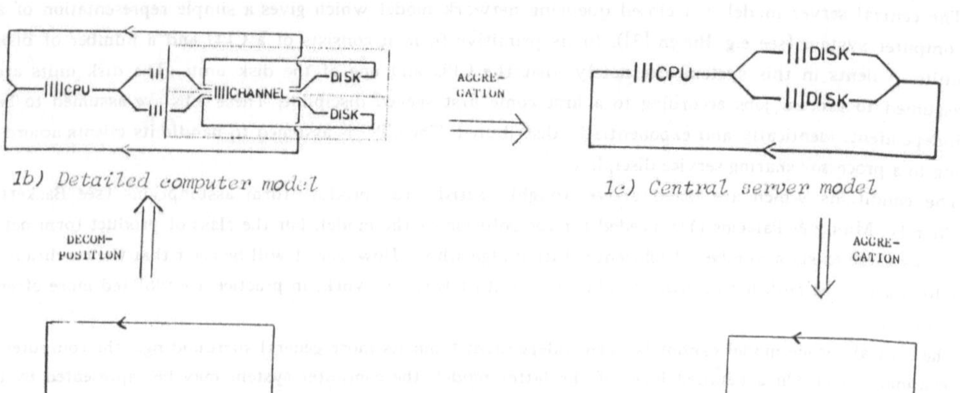

Figure 1: Modeling computer systems

2. The mean value algorithm

Using the product form property of the central server model, the solution of this model can be evaluated e.g. using the mean value algorithm (see Reiser & Lavenberg [8]). As we will use the mean value analysis in the further discussion on the robustness of the central server model, we will briefly summarize the principle relations used in the mean value scheme for single chain closed queueing networks.
The mean value algorithm can be seen as to be based on two theorems. Firstly, we have Little's result (Little [7]), stating that the mean number of clients at a station equals the product of the arrival intensity of clients to that station and the mean sojourn time for clients at that station. And secondly, we have what we shall call the arrival theorem for closed product form queueing networks, which states that a client arriving at a station will observe this station in equilibrium position as if this arriving client was looking at the system from the outside. This last theorem defines the recursion of the mean value algorithm. Now let us define:

k	the number of clients currently in the system, k=1,2,...K
$S_j[k]$	mean sojourn time at station j, given population k
$N_j[k]$	mean number of clients present at station j, given population k
$\Lambda_j[k]$	mean throughput of station j, given population k
$\rho_j[k]$	mean utilization of station j, given population k
w_j	mean workload at station j
σ_j	standard deviation of the workload at station j
f_j	visit frequency for station j

For the station indices we use a 0 to indicate the CPU and the numbers in the range 1 to D to indicate a disk unit. As more general station indices we use j and h. The mean value scheme is then given by the following set of recursive relations:

$$S_j[k] = (1 + N_j[k-1])w_j \tag{1}$$

$$\Lambda_j[k] = \frac{f_j k}{\sum_h f_h S_h[k]} \tag{2}$$

$$N_j[k] = \Lambda_j[k]S_j[k] \tag{3}$$

$$\rho_j[k] = \Lambda_j[k]w_j \tag{4}$$

The first of these relations is based on the arrival theorem, the relation in which the recursion appears. The latter relations are explained by putting Little's result in the appropriate form.
The recursion is initiated by putting k to one and noting that in this situation a client arriving at a station will observe the empty system.

3. A detailed computer model

In Figure 1b a detailed computer model is depicted. The organization of the system underlying the model is as follows. After a client has finished CPU service, he has to visit one of the disk units. The disk units operate the clients in first come first served order, so some queueing might occur. On his turn the client enters the so called I/O subsystem. To get his message to the disk unit, the client needs access to the channel. As there are usually less channels then there are disk units, again some queueing might occur. Next the disk unit seeks the desired information. But to transmit this information the channel is needed; for a third time some queueing might be involved. Finally the data transmission takes place and the client returns to the CPU queue.

The problem is that this organization violates the product form assumptions. This makes it hard to solve the model in a exact way. As indicated in the introduction of this paper, approximate solutions can be obtained by translating the detailed model, Figure 1b, in terms of the central server model, Figure 1c. Some of the major problems which have to be overcome there are the non exponentiality of the disk access time and the dependency between system (channel) utilization and this disk access time.

A well known solution to these problems is the following iterative approach (see e.g. Brandwajn [2]). First a rough estimate is made for the average disk access times, w_d, its variance, σ_d^2, and the throughput, Λ_d, for the situation with a given number of clients in the system. Given these estimates, the disk response times are estimated by considering the individual disk units as M | G | 1 queues. Next a central server like model is constructed, in which the disk units are modeled as infinite server stations with workloads equal to the sojourn times just evaluated. This model can be solved with the mean value scheme, (1)-(4), in which the disk sojourn times equations are replaced by:

$$S_d^{(i)}[k] = w_d^{(i-1)} + \frac{\rho_d^{(i-1)}[K]}{1-\rho_d^{(i-1)}[K]} \frac{(\sigma_d^{(i-1)})^2 + (w_d^{(i-1)})^2}{2\, w_d^{(i-1)}} \tag{5}$$

Here the index (i) is added to denote the iteration. The throughput thus obtained can be used to base new estimates for w_d and σ_d^2 on it, etc.

A second iterative solution technique, discussed in Gürich [5], neglects the non exponentiality of the disk workload, such that the ordinary mean value scheme can be used to solve the central server model. The mean disk access time is estimated using iteration again.

A recursive variant of these approximation methods was suggested in Wijbrands [9]. This variant is based on the arrival theorem which is used as an approximation in a mean value scheme for non product form networks. Both the non exponentiality and the dependency from the disk access time on the system throughput are handled by this approximate arrival theorem, by redefining the disk sojourn time relations as:

$$S_d[k] = (\,1 + N_d[k-1] - \rho_d[k-1]\,)w_d[k-1] + \rho_d[k-1]\frac{\sigma_d^2[k-1] + w_d^2[k-1]}{2\, w_d[k-1]} \tag{6}$$

In the light of the more general computer-terminal problem, the advantage of a recursive algorithm over an iterative algorithm will be clear. At least if both types of algorithms attain a reasonable level of accuracy. A simple example to show this accuracy. In table 1 we gathered some numerical results on an example consisting of a CPU connected via a single channel to 6 identical disk units. We assumed a CPU workload of 20 msec, the positioning of the read/write head to take 32 msec with variance 112 msec^2, a data transmission time exponentially distributed with mean 8 msec, and a disk revolution time taking 16.7 msec. We assumed the disk units to operate with the RPS (Rotational Position Sensing) feature: if the disk unit is ready to transmit its information, it looks whether the channel is busy or not. In case it is busy, the disk unit waits for a full disk revolution before it tries to establish contact again. The

K	Simulation	Wijbrands	Brandwajn	Gürich
1	16.7	16.8	15.3	16.0
2	28.1	27.8	26.3	26.4
3	35.9	35.3	34.2	33.6
4	40.8	40.3	39.8	38.6
5	44.4	43.7	43.8	42.1
6	46.8	46.0	46.5	44.7
7	47.9	47.5	48.2	46.5
8	48.6	48.5	49.2	47.7
9	49.1	49.1	49.7	48.6

Table 1: CPU throughput (per second) for a detailed computer model using different approximation techniques

number of reconnection attempts thus needed to gain access to the channel was approximated by the geometric distribution, and using as parameter the probability that the channel is busy. The time lost in making contact with the disk unit is approximated by the product of the probability that the channel is busy and the average time a data transmission takes.
Both the iterative and the recursive methods attain a reasonable level of accuracy for this example, which suggests a certain robustness of the central server model and the mean value algorithm with respect to the assumptions. In the next two sections we try to answer the question whether the central server model is indeed that robust, or the accuracy of the approximation methods is restricted to special cases.

4. Non exponentially distributed disk workload

Now let us suppose that the real system is represented by a central server like model, where the disk workload is described by a non exponential distribution. Then the product form property of the network is lost and there is hardly any chance that we can give the exact solution. Just in the case that the disk workload has a phase type distribution, this solution can be obtained, at least in principle, by solving the general set of equilibrium equations. In general we shall have to do with approximations. In this section we study three of such approximations methods, directly derived from the approximation methods mentioned in the preceding section.
The most simple approximation we can think of, is just to ignore the non exponentiality of the disk workload. This will give us an idea of the robustness of the exponentiality assumption to be made in the mean value analysis.
A second approximation method is based on the (mis)interpretation of the arrival theorem. If it would hold in the more general case, the sojourn time relation (1) could be replaced for the disk units by:

$$S_d[k] = (1 + N_d[k-1] - \rho_d[k-1])w_d + \rho_d[k-1]\frac{\sigma_d^2 + w_d^2}{2w_d} \qquad (7)$$

Note that this extension is similar to the extension made if an M | M | 1 model is generalized to an M | G | 1

K	I^a	II	III	IV
1	4.00	4.00	4.00	4.39
2	5.30	5.33	5.50	5.58
3	6.81	6.87	7.09	6.95
4	8.47	8.56	8.77	8.47
5	10.27	10.35	10.52	10.19
6	12.14	12.22	12.35	12.07
7	14.07	14.13	14.22	14.02
8	16.03	16.08	16.15	16.00
9	18.02	18.04	18.09	18.00

2a) D=2, $w_0 = 2$, $w_d = 2$
$CV_d = 0.58$, $f_d = 0.5$

K	I^b	II	III	IV
1	10.0	10.0	10.0	10.6
2	10.9	10.9	11.2	11.5
3	11.9	11.9	12.4	12.5
4	13.0	13.1	13.7	13.5
5	14.2	14.2	15.0	14.6
6	15.3	15.5	16.3	15.8
7	16.7	16.9	17.7	17.0
8	18.2	18.3	19.1	18.3
9	19.6	19.8	20.6	19.7

2b) D=8, $w_0 = 2$, $w_d = 8$
$CV_d = 0.58$, $f_d = 0.125$

w_0	I^a	II	III	IV
4.00	12.29	12.35	12.45	12.25
3.50	10.86	10.93	11.03	10.83
3.00	9.46	9.53	9.67	9.44
2.50	8.10	8.17	8.84	8.13
2.00	6.81	6.87	7.09	6.95
1.80	6.33	6.38	6.62	6.52
1.60	5.87	5.91	6.17	6.14
1.40	5.42	5.46	5.74	5.80
1.20	5.04	5.04	5.36	5.51
1.00	4.68	4.65	5.00	5.25
0.80	4.38	4.30	4.69	
0.60	4.13	4.01	4.43	

2c) $D = 2$, $w_d = 2$, $CV_d = 0.58$
$f_d = 0.5$, $K = 3$

w_0	I^b	II	III	IV
4.00	28.4	28.4	28.6	28.4
3.50	25.0	25.2	25.5	25.0
3.00	21.7	22.1	22.5	21.9
2.50	19.1	19.3	19.9	19.2
2.00	16.8	16.9	17.7	17.0
1.80	15.9	16.1	17.0	16.3
1.60	15.3	15.3	16.4	15.8
1.40	14.6	14.7	15.8	15.3
1.20	14.2	14.2	15.4	14.9
1.00	13.8	13.8	15.0	14.6
0.80	13.5	13.4	14.7	14.3
0.60	13.2	13.2	14.5	14.1

2d) $D = 8$, $w_d = 8$, $CV_d = 0.58$
$f_d = 0.125$, $K = 7$

CV_d	I^a	II	III	IV
0.58	6.81	6.87	7.09	6.95
0.71	6.88	6.93	7.09	7.00
1.00	7.09	7.09	7.09	7.21
1.29	7.30	7.30	7.09	7.46
1.76	7.63	7.68	7.09	7.95
2.00	7.79	7.92	7.09	8.25

2e) $D = 2$, $w_0 = 2$, $w_d = 2$
$K = 3$, $f_d = 0.5$

CV_d	I^b	II	III	IV
0.00	16.2	16.4	17.7	16.5
0.58	16.7	16.9	17.7	17.0
0.71	17.0	17.0	17.7	17.3
1.00	17.7	17.7	17.7	18.0
1.29	18.5	18.5	17.7	18.9
1.76	19.9	20.1	17.7	20.6
2.00	21.0	21.0	17.7	21.6

2f) $D = 8$, $w_0 = 2$, $w_d = 8$
$K = 7$, $f_d = 0.125$

I^a Markov chain solution
I^b Simulation (+/- 1%)
II Arrival theorem extension (equation (5))
III Exponentiality assumption
IV Iterative decomposition

Table 2: Cycle times and estimates for central server models with non exponential disk workload model (see e.g. Kleinrock [5]).

The third approximation method is the iterative decomposition technique, in which the disk units are treated as M | G | 1 queues in the decomposition step, and as infinite server stations in the aggregated model. The method can be summarized by the mean value scheme (1)-(4) again, but with disk sojourn times equal to:

$$S_d^{(i)}[k] = w_d + \frac{\rho_d^{(i-1)}[K]}{1-\rho_d^{(i-1)}[K]} \frac{\sigma_d^2 + w_d^2}{2w_d} \tag{8}$$

where (i) is used to indicate the iteration again.

In Table 2 we summarized some numerical results we obtained for this problem. We studied two problems. The first problem consisted of a CPU connected to two disk units. Using phase type distributions this problem could be solved as a Markov chain. A second problem consisted of a CPU connected to eight disk units. Here a comparison was made with simulation results.

For the iterative decomposition approximation, the results were from one point of view disappointing: in most of the cases where the disk unit performed as the bottleneck the method did not converge! This despite of the fact that the iteration was initiated with parameter values rather close to the exact solution of the problem. For the first set of problems, considering models with two disk units, the system cycle time is overestimated by this method if there are few clients in the system, and underestimated if there are many clients. This breakpoint is reached the sooner if the CV (coefficient of variation) of the disk workload is low. This conclusion seems less evident for the second set of examples.

The first two approximation methods performed reasonable. In most cases the arrival theorem based approximation method performed slightly better. The model seems by no means as sensitive for non exponentiality as an M | G | 1 model would suggest. The approximation made through the exponentiality assumption behaves up to a certain extent as expected; if the disk workload has a CV below one, it overestimates the system cycle time and conversely.

The behavior of the approximation method based on the arrival theorem can be explained roughly by the observation that the effect of the CV of the disk workload is overestimated. If a client returns to the disk unit immediately after he finished disk service there, he will observe a client (or not) at the beginning of its service; equation (1) is approximately right. On the other hand, if it takes a long time, e.g. because of the CPU performing as a bottleneck, it is more likely that if he finds a client in service, that this client has been processed for a while; equation (5) is approximately right. On the average the truth will be somewhere in the middle, thus the effect of the CV being unequal to one is overestimated through (5). This observation gives raise to the following argumentation. If the CV of the disk workload is below one, this will yield in first instance an underestimation of the disk sojourn time, and therewith the average number of clients at those stations. This results in an overestimation of the number of clients at the CPU and the CPU sojourn time. If the CPU is the system bottleneck then this implies an overestimation of the system cycle time. All of these arguments may be reversed, but near to the breakpoints other arguments may spoil this way of reasoning.

5. Population dependent disk service times

In Section 3 it was stated that, due to several dependencies, the disk access time increases with the number of clients in the system. The recursive approximation method mentioned there, assumed that this dependency could be dealt with within the recursion of the mean value scheme, thus suggesting a certain

K	Exact $w_d = 3.0$	Approximate m=1	m=3	m=5
1	5.00			
2	6.70	6.75		
3	8.42	8.44		
4	10.15	10.17	10.38	
5	11.91	11.92	12.07	
6	13.68	13.69	13.80	14.18
7	15.48	15.48	15.56	15.84
8	17.29	17.29	17.36	17.57
9	19.12	19.12	19.17	19.34

3a) D=2, $w_0 = 2$, $w_d[0] = 2$, $f_d = 0.5$

K	Exact $w_d = 12.0$	Approximate m=1	m=3	m=5
1	14.00			
2	15.57	15.60		
3	17.15	17.15		
4	18.73	18.73	18.84	
5	20.31	20.31	20.36	
6	21.90	21.90	21.92	22.16
7	23.49	23.49	23.50	23.63
8	25.09	25.09	25.10	25.17
9	26.70	26.70	26.70	26.74

3b) $D = 8$, $w_0 = 2$, $w_d[0] = 8$, $f_d = 0.125$

K	a = 0.2 Exact	Appr.	a = 0.4 Exact	Appr.
1	4.00	4.00	4.00	4.00
2	5.73	5.75	5.96	6.00
3	7.43	7.46	7.76	7.80
4	9.15	9.19	9.56	9.61
5	10.91	10.95	11.36	11.41
6	12.71	12.75	13.17	13.23
7	14.54	14.58	15.00	15.06
8	16.40	16.44	16.85	16.91
9	18.30	18.33	18.71	18.77

3c) $D = 2$, $w_d[0] = 2$, $f_d = 0.5$, $w_0 = 2$

K	a = 0.2 Exact	Appr.	a = 0.4 Exact	Appr.
1	10.00	10.00	10.00	10.00
2	11.55	11.55	11.89	11.90
3	13.07	13.08	13.68	13.69
4	14.58	14.60	15.41	15.44
5	16.10	16.12	17.12	17.15
6	17.62	17.65	18.82	18.86
7	19.15	19.19	20.51	20.55
8	20.70	20.74	22.19	22.24
9	22.26	22.31	23.88	23.93

3d) $D = 8$, $w_d[0] = 8$, $f_d = 0.125$, $w_0 = 2$

w_0	a = 0.2 Exact	Appr.	a = 0.4 Exact	Appr.
4.0	12.66	12.74	12.90	13.04
2.0	7.43	7.46	7.76	7.80
1.8	6.96	6.99	7.30	7.33
1.6	6.52	6.53	6.86	6.88
1.4	6.09	6.10	6.43	6.44
1.2	5.70	5.70	6.12	6.12
1.0	5.39	5.38	5.86	5.85
0.8	5.12	5.12	5.64	5.62
0.6	4.91	4.90	5.46	5.44

3e) $D = 2$, $w_d[0] = 2$, $f_d = 0.5$, $K = 3$

w_0	a = 0.2 Exact	Appr.	a = 0.4 Exact	Appr.
4.0	29.27	29.42	30.15	30.42
2.0	19.15	19.19	20.51	20.55
1.8	18.43	18.45	19.75	19.78
1.6	17.76	17.78	19.04	19.05
1.4	17.15	17.16	18.35	18.35
1.2	16.59	16.59	18.03	18.03
1.0	16.28	16.28	17.77	17.77
0.8	16.03	16.03	17.55	17.55
0.6	15.82	15.82	17.38	17.38

3f) $D = 8$, $w_d[0] = 8$, $f_d = 0.125$, $K = 7$

Table 3: Cycle times and estimates for central server models with population dependent disk workload robustness of this scheme with respect to a population dependent workload. The robustness with respect to this aspect is the topic of this section.

In this section we assume that the mean of the disk workload depends on the system population. However it is assumed that this mean workload is known for each possible system population. Furthermore, the disk workloads are assumed to be exponentially distributed, such that we can compute the performance characteristics related with each possible population using the mean value scheme. If we want to study the approximation method based on equation (6), we find for this problem the equivalent relation :

$$S_d[k] = (1 + N_d[k-1])w_d^*[k-1] \tag{9}$$

In this equation the * is added to indicate that the workload is obtained by looking at the exact throughput of the system (as evaluated with one client removed from the population).

In Table 3a)-3b) some results are summarized for the case where the population dependent disk workload shows the following pattern:

$$w_d^*[k] = \begin{cases} w_d[0] & ,k < m \\ w_d[0](1 + a) & ,k \geqslant m \end{cases} \tag{10}$$

The increment factor a was chosen 0.5. Perhaps to our surprise, the approximation (9) performs very well under this rather extreme workload pattern. Moreover, as the population increases, the error vanishes rapidly; the mean value scheme seems somehow self recovering under this approximation.

In Table 3c)-3f) we summarized some results in which the disk workload shows a pattern which is more like the pattern imposed by the system description of Section 3:

$$w_d^*[k] = w_d[0](1 + b\Lambda_0^*[k]) \tag{11}$$

Again the * is used to indicate exact values. The b is the increment factor such that $w_d^*[k]$ finds an upper bound in $w_d[0](1 + a)$. Under this workload pattern, the mean value approximation seems even more robust. The errors which we find in the summary of results never exceed the 1%.

The pattern of errors is comparable with that of the arrival theorem based approximation method discussed in the preceding section. The explanation is the same, but even more accurate; this time the problem under study does fulfill the product form conditions. If we look at a system with an arbitrary population, we will observe that the approximation method underestimates the number of clients at the disk units. This is due to the fact that the recursion was initiated with a disk workload lower than the workload of the situation of current interest. This automatically causes a shift of clients towards the CPU. If the CPU is the bottleneck then this will produce an overestimation of the cycle time and conversely.

6. Conclusions

In the preceding two sections we have studied some aspects of robustness for the central server model and its method of analysis, the mean value algorithm. If we look at the results in the light of the approximation methods for the more detailed models, as discussed in Section 3, we can say something about the robustness of these approximation methods as well.

The recursive method profits the most from the robustness of the central server model. Two of the approximations made in this scheme, the updating of the workload and the treatment of the non exponentiality produce only modest errors in the cycle time estimation. Using this observation it may be concluded cautiously that the results obtained by this method are potentially accurate if the disk CV is not to far from one, let say between 0.5 and 1.5, and if the actual disk workload can be described in terms of the system throughput satisfactory.

The errors which appear in the solution method of Gürich cannot be explained by the exponentiality assumption. Possibly this method can be improved by estimating the disk workload, like in the recursive algorithm, from the situation with one client less in the system, and thus replacing the iteration by a second recursion. It is clear that this idea is correct in at least one aspect: the disk service time for the situation with a single client in the system is not influenced by the system throughput! A weakness in such an approach is of course that, like in the recursive algorithm, an error implied by an iteration step influences the performance of the next iteration step.

The third method, which we named after Brandwajn, seems to suffer mostly from the fact that the open network approach is in many situations insufficient; the use of this method is restricted to those situations where the disk utilization is relatively low. As was already suggested by Brandwajn in [2], the M | G | 1 approach could be replaced by a decomposition in which the M | G | 1 || K queue is considered, in which case the arrival process is based on a limited population. The computational complexity of the problem will, however, suffer severely from this suggestion. A further improvement of the results is possibly found in the suggestion we made to improve the Gürich method. In the latter case we have the extra advantage that we bypass the problem of convergence.

References

[1] Baskett, F., Chandy, C.M., Muntz, R.R., Palacios, F.G., "Open, Closed and Mixed Networks of Queues with Different Classes of Customers", J.ACM 22 (1975) 248-260.

[2] Brandwajn, A., "Models of DASD Subsystems with Multiple Access Paths: A Throughput-Driven Approach", IEEE Transactions on Computers C-32 (1983) 451-463.

[3] Buzen, J.P., "Queueing Network Models of Multiprogramming" (thesis), Cambridge, Massachusetts (1971).

[4] Chandy, K.M., Herzog, U., Woo, L.S., "Parametric Analysis of Queueing Networks", IBM J. Res. Dev. 19 (1975) 43-49.

[5] Gürich, W., "Eine iterative Methode zur Modellierung gemeinsam benutzter Plattenperipherie bei einem lose gekoppelten Rechnersystem", in Beilner, H., (ed.), "Messung, Modellierung und Bewertung von Rechensystemen", Springer-Verlag, Berlin (1985).

[6] Kleinrock, L., "Queueing Systems Vol. 1", J. Wiley & Sons, New York, (1975).

[7] Little, J.D., "A Proof of the Queueing Formula $L = \lambda W$", Operations Research 9 (1961) 383-387.

[8] Reiser, M., Lavenberg, S.S., "Mean Value Analysis of Closed Multichain Queueing networks", J.ACM 27 (1980) 313-322.

[9] Wijbrands, R.J., "The Analysis of I/O Configurations", in Beilner, H., (ed.), "Messung, Modellierung und Bewertung von Rechensystemen", Springer Verlag, Berlin (1985).

[10] Wijbrands, R.J., "On the Development of a Modeling Tool for the Performance Analysis of Computer Systems", Memorandum COSOR 85-12, Dept. of Math. and Comp.Sc., Eindhoven University of Technology (1985).

Lecture Notes in Control and Information Sciences

Edited by M. Thoma

Vol. 43: Stochastic Differential Systems
Proceedings of the 2nd Bad Honnef Conference
of the SFB 72 of the DFG at the University of Bonn
June 28 – July 2, 1982
Edited by M. Kohlmann and N. Christopeit
XII, 377 pages. 1982.

Vol. 44: Analysis and Optimization of Systems
Proceedings of the Fifth International
Conference on Analysis and Optimization of Systems
Versailles. December 14–17, 1982
Edited by A. Bensoussan and J. L. Lions
XV, 987 pages, 1982

Vol. 45: M. Arató
Linear Stochastic Systems
with Constant Coefficients
A Statistical Approach
IX, 309 pages. 1982

Vol. 46: Time-Scale Modeling of Dynamic Networks
with Applications to Power Systems
Edited by J. H. Chow
X, 218 pages. 1982

Vol. 47: P. A. Ioannou, P. V. Kokotovic
Adaptive Systems with Reduced Models
V, 162 pages. 1983

Vol. 48: Yaakov Yavin
Feedback Strategies for Partially
Observable Stochastic Systems
VI, 233 pages, 1983

Vol. 49: Theory and Application of Random Fields
Proceedings of the IFIP-WG 7/1
Working Conference
held under the joint auspices of the
Indian Statistical Institute
Bangalore, India, January 1982
Edited by G. Kallianpur
VI. 290 pages. 1983

Vol. 50: M. Papageorgiou
Applications of Automatic Control Concepts
to Traffic Flow Modeling and Control
IX, 186 pages. 1983

Vol. 51: Z. Nahorski, H.F. Ravn, R.V.V. Vidal
Optimization of Discrete Time Systems
The Upper Boundary Approach
V, 137 pages 1983

Vol. 52: A. L. Dontchev
Perturbations, Approximations and Sensitivity Analysis
of Optimal Control Systems
IV, 158 pages. 1983

Vol. 53: Liu Chen Hui
General Decoupling Theory of Multivariable
Process Control Systems
XI, 474 pages. 1983

Vol. 54: Control Theory for Distributed
Parameter Systems and Applications
Edited by F. Kappel, K. Kunisch,
W. Schappacher
VII, 245 pages. 1983.

Vol. 55: Ganti Prasada Rao
Piecewise Constant Orthogonal Functions
and Their Application to Systems and Control
VII, 254 pages. 1983.

Vol. 56: Dines Chandra Saha, Ganti Prasada Rao
Identification of Continuous
Dynamical Systems
The Poisson Moment Functional
(PMF) Approach
IX, 158 pages. 1983.

Vol. 57: T. Söderström, P. G. Stoica
Instrumental Variable Methods
for System Identification
VII, 243 pages. 1983.

Vol. 58: Mathematical Theory of
Networks and Systems
Proceedings of the MTNS-83 International
Symposium
Beer Sheva, Israel, June 20–24, 1983
Edited by P. A. Fuhrmann
X, 906 pages. 1984

Vol. 59: System Modelling and Optimization
Proceedings of the 11th IFIP Conference
Copenhagen, Denmark, July 25-29, 1983
Edited by P. Thoft-Christensen
IX, 892 pages. 1984

Vol. 60: Modelling and Performance
Evaluation Methodology
Proceedings of the International Seminar
Paris, France, January 24–26, 1983
Edited by F. Bacelli and G. Fayolle
VII, 655 pages. 1984

Vol. 61: Filtering and Control of Random
Processes
Proceedings of the E.N.S.T.-C.N.E.T. Colloquium
Paris, France, February 23–24, 1983
Edited by H. Korezlioglu, G. Mazziotto, and
J. Szpirglas
V, 325 pages. 1984

Lecture Notes in Control and Information Sciences

Edited by M. Thoma and A. Wyner

Vol. 62: Analysis and Optimization of Systems
Proceedings of the Sixth International Conference on Analysis and Optimization of Systems
Nice, June 19–22, 1984
Edited by A. Bensoussan, J. L. Lions
XIX, 591 pages. 1984.

Vol. 63: Analysis and Optimization of Systems
Proceedings of the Sixth International Conference on Analysis and Optimization of Systems
Nice, June 19–22, 1984
Edited by A. Bensoussan, J. L. Lions
XIX, 700 pages. 1984.

Vol. 64: Arunabha Bagchi
Stackelberg Differential Games in Economic Models
VIII, 203 pages, 1984

Vol. 65: Yaakov Yavin
Numerical Studies in Nonlinear Filtering
VIII, 273 pages, 1985.

Vol. 66: Systems and Optimization
Proceedings of the Twente Workshop
Enschede, The Netherlands, April 16–18, 1984
Edited by A. Bagchi, H. Th. Jongen
X, 206 pages, 1985.

Vol. 67: Real Time Control of Large Scale Systems
Proceedings of the First European Workshop
University of Patras, Greece, Juli 9–12, 1984
Edited by G. Schmidt, M. Singh, A. Titli, S. Tzafestas
XI, 650 pages, 1985.

Vol. 68: T. Kaczorek
Two-Dimensional Linear Systems
IX, 397 pages, 1985.

Vol. 69: Stochastic Differential Systems – Filtering and Control
Proceedings of the IFIP-WG 7/1 Working Conference
Marseille-Luminy, France, March 12-17, 1984
Edited by M. Metivier, E. Pardoux
X, 310 pages, 1985.

Vol. 70: Uncertainty and Control
Proceedings of a DFVLR International Colloquium
Bonn, Germany, March, 1985
Edited by J. Ackermann
IV, 236 pages, 1985.

Vol. 71: N. Baba
New Topics in Learning Automata Theory and Applications
VII, 231 pages, 1985.

Vol. 72: A. Isidori
Nonlinear Control Systems: An Introduction
VI, 297 pages, 1985.

Vol. 73: J. Zarzycki
Nonlinear Prediction Ladder-Filters for Higher-Order Stochastic Sequences
V, 132 pages, 1985.

Vol. 74: K. Ichikawa
Control System Design based on Exact Model Matching Techniques
VII, 129 pages, 1985.

Vol. 75: Distributed Parameter Systems
Proceedings of the 2nd International Conference, Vorau, Austria 1984
Edited by F. Kappel, K. Kunisch, W. Schappacher
VIII, 460 pages, 1985.

Vol. 76: Stochastic Programming
Edited by F. Archetti, G. Di Pillo, M. Lucertini
V, 285 pages, 1986.

Vol. 77: Detection of Abrupt Changes in Signals and Dynamical Systems
Edited by M. Basseville, A. Benveniste
X, 373 pages, 1986.

Vol. 78: Stochastic Differential Systems
Proceedings of the 3rd Bad Honnef Conference, June 3–7, 1985
Edited by N. Christopeit, K. Helmes, M. Kohlmann
V, 372 pages, 1986.

Vol. 79: Signal Processing for Control
Edited by K. Godfrey, P. Jones
XVIII, 413 pages, 1986.

Vol. 80: Artificial Intelligence and Man-Machine Systems
Edited by H. Winter
IV, 211 pages, 1986.

Lecture Notes in Control and Information Sciences

Edited by M. Thoma and A. Wyner

Vol. 81: Stochastic Optimization
Proceedings of the International Conference,
Kiew, 1984
Edited by I. Arkin, A. Shiraev, R. Wets
X, 754 pages. 1986.

Vol. 82: Analysis and Algorithms
of Optimization Problems
Edited by K. Malanowski, K. Mizukami
VIII, 240 pages. 1986.

Vol. 83: Analysis and Optimization
of Systems
Proceedings of the Seventh International
Conference of Analysis and Optimization
of Systems
Antibes, June 26-27, 1986
Edited by A. Benoussan, J. L. Lions
XVI, 901 pages, 1986.

Vol. 84: System Modelling
and Optimization
Proceedings of the 12th IFIP Conference
Budapest, Hungary, September 2-6, 1985
Edited by A. Prékopa, J. Szelezsán, B. Strazicky
XII, 1046 pages, 1986.

MIX
Papier aus verantwortungsvollen Quellen
Paper from responsible sources
FSC® C105338

If you have any concerns about our products,
you can contact us on
ProductSafety@springernature.com

In case Publisher is established outside the EU,
the EU authorized representative is:
Springer Nature Customer Service Center GmbH
Europaplatz 3, 69115 Heidelberg, Germany

Printed by Libri Plureos GmbH
in Hamburg, Germany